Theory of Applied Robotics

Reza N. Jazar

Theory of Applied Robotics

Kinematics, Dynamics, and Control

Third edition

 Springer

Reza N. Jazar (iD)
School of Engineering
RMIT University
Melbourne, VIC, Australia

ISBN 978-3-030-93222-0   ISBN 978-3-030-93220-6   (eBook)
https://doi.org/10.1007/978-3-030-93220-6

This Springer imprint is published by the registered company Springer Nature Switzerland AG
The registered company address is: Gewerbestrasse 11, 6330 Cham, Switzerland

I am Cyrus, king of the world, great king, mighty king,
king of Babylon, king of Sumer and Akkad, king of the four quarters.
I ordered to write books, many books, books to teach my people,
I ordered to make schools, many schools, to educate my people.
Marduk, the lord of the gods, said burning books is the greatest sin.
I, Cyrus, and my people, and my army will protect books and schools.
They will fight whoever burns books and burns schools, the great sin.
Cyrus the great

to:
Vazan
Kavosh
Mojgan

Preface to the Third Edition

Ten years have passed since the second edition of *Theory of Applied Robotics* has been published. The second edition of the book has been strongly accepted in academia for research and teaching. Many universities and colleges adopted this book as a teaching text or as a reference. The book also had great success in Chinese universities, especially after its Chinese translation was officially published by Springer in 2018. During the past 10 years, I have received many constructive comments from all around the world, from colleagues, friends, instructors, researchers, and students. In this third edition, I used those comments to improve the text as well as remove typos and numerical errors. Further, I tried to make the third edition more educational and clearer to make the book more suitable for self-training.

The key concept of robotics' kinematics and dynamics is the understanding of the coordinate frame transformation, a topic that *Theory of Applied Robotics* is covering in the best possible way. We begin this topic by analyzing rotation about a principal axis of global and body coordinate frames. Rotation about an arbitrary axis is the next step. Understanding of rotation about an arbitrary axis is the best way to study kinematics of robots. We will show how to break such a complicated rotation into a series of principal rotations to develop the required mathematical relations by a series of simple steps. To make it possible, we will show how we may introduce extra dummy coordinate frames to simplify moving from initial to final coordinate frame by a modular and structural process. Auxiliary coordinate frames is a unique concept that has been added to the kinematic knowledge body of robot kinematics by this book. Readers of this book will become masters in using this scientific trick in kinematic analysis.

The robot kinematics will be continued by covering the first and second time derivatives of rotation transformation mathematics. The first derivative introduces the complicated and strange concept of angular velocity. Angular velocity is a complicated concept because we do not have any unique mathematical quantity whose time derivative will be angular velocity. Angular velocity is a vectorial quantity, but it is not time derivative of another scalar or vectorial physical quantity. This fact makes angular velocity appear as a new individual concept. It, however, will be connected to a combination of time derivative of rotation transformation matrices. The appearance of rotation transformation matrices in the definition of angular velocity gives us the ability to define time derivative operations in different coordinate frames. Therefore, we will see that there are several different velocities all mathematically correct but not all sensible. The second derivative and introduction of acceleration kinematics make this concept more complicated such that we will be able to define and calculate greater accelerations than those we work with. The freedom of taking a vectorial physical quantity from one coordinate frame and taking derivative in another coordinate and still expressing in a third coordinate frame helped me discover a new acceleration called Razi acceleration (Jazar 2011; Harithuddin et al. 2015).

My intention in this book is to explain robotics in a manner I would have liked to be explained to me as a student. This book can now help students by being a great reference that covers all aspects of robotics and that provides students with detailed explanations and information.

Organization of the Book

The text is organized in such a manner that it can be used for teaching or self-study. Chapter 1 "Introduction," contains general preliminaries with a brief review of the historical development and classification of robots.

Part I, Kinematics, presents the forward and inverse kinematics of robots. Kinematics analysis refers to position, velocity, and acceleration analysis of robots in both joint and base coordinate spaces. It establishes kinematic relations among the end-effecter and the joint variables. The method of Denavit-Hartenberg for representing body coordinate frames is introduced and utilized for forward kinematics analysis. The concept of modular treatment of robots is well covered to show how we may combine simple links to make the forward kinematics of a complex robot. For inverse kinematics analysis, the idea of decoupling, the inverse matrix method, and the iterative technique are introduced. It is shown that the presence of a spherical wrist is what we need to apply analytic methods in inverse kinematics.

Part II, Derivative Kinematics, explains how the derivatives of vectors are calculated and how they are related to each other. It covers angular velocity, velocity, and acceleration kinematics. Definitions of derivatives and coordinate frames are covered in this part. It is fascinating to understand that derivative is a frame-dependent operation.

Part III, Dynamics, presents a detailed discussion of robot dynamics. An attempt is made to review the basic approaches and demonstrate how these can be adapted for the active displacement framework utilized for robot kinematics in the earlier chapters. The concepts of recursive Newton-Euler dynamics, Lagrangian function, manipulator inertia matrix, and generalized forces are introduced and applied for derivation of dynamic equations of motion.

Part IV, Control, presents the floating time technique for time-optimal control of robots. The outcome of the technique is applied to an open-loop control algorithm. Then, a computed-torque method is introduced, in which a combination of feedforward and feedback signals are utilized to render the system error dynamics.

Method of Presentation

The structure of presentation is in a "*fact-reason-application*" fashion. The "fact" is the main subject we introduce in each section. Then the reason is given as a "proof." Finally, the application of the fact is examined in some "examples." The "examples" are a very important part of the book because they show how to implement the knowledge introduced in "facts." They also cover some other facts that are needed to expand the subject.

Level of the Book

This book has evolved from nearly a decade of research in nonlinear dynamic systems and teaching undergraduate- and graduate-level courses in robotics. It is addressed primarily to the last year of undergraduate study and the first-year graduate student in engineering. Hence, it is an intermediate textbook. This book can even be the first exposure to topics in spatial kinematics and dynamics of mechanical systems. Therefore, it provides both fundamental and advanced topics on the kinematics and dynamics of robots. The whole book can be covered in two successive courses; however, it is possible to jump over some sections and cover the book in one course. The students are required to know the fundamentals of kinematics and dynamics, as well as a basic knowledge of numerical methods.

The contents of the book have been kept at a fairly theoretical-practical level. Many concepts are deeply explained and their use emphasized, and most of the related theory and formal proofs have been explained. Throughout the book, a strong emphasis is put on the physical meaning of the concepts introduced. Topics that have been selected are of high interest in the field. An attempt has been made to expose the students to a broad range of topics and approaches.

Prerequisites

Since the book is written for senior undergraduate and first-year graduate level students of engineering, the assumption is that users are familiar with matrix algebra as well as basic feedback control. Prerequisites for readers of this book consist of the fundamentals of kinematics, dynamics, vector analysis, and matrix theory. These basics are usually taught in the first three undergraduate years.

Unit System

The system of units adopted in this book is, unless otherwise stated, the international system of units (SI). The units of degree (deg) or radian (rad) are utilized for variables representing angular quantities.

Symbols

- Lowercase bold letters indicate a vector. Vectors may be expressed in an n dimensional Euclidian space. Example:

$$\mathbf{r} , \mathbf{s} , \mathbf{d} , \mathbf{a} , \mathbf{b} , \mathbf{c}$$
$$\mathbf{p} , \mathbf{q} , \mathbf{v} , \mathbf{w} , \mathbf{y} , \mathbf{z}$$
$$\boldsymbol{\omega} , \boldsymbol{\alpha} , \boldsymbol{\epsilon} , \boldsymbol{\theta} , \boldsymbol{\delta} , \boldsymbol{\phi}$$

- Uppercase bold letters indicate a dynamic vector or a dynamic matrix. Example:

$$\mathbf{F} , \mathbf{M} , \mathbf{I} , \mathbf{L}$$

- Lowercase letters with a hat indicate a unit vector. Unit vectors are not bolded. Example:

$$\hat{i} , \hat{j} , \hat{k} , \hat{e} , \hat{u} , \hat{n}$$
$$\hat{I} , \hat{J} , \hat{K} , \hat{e}_\theta , \hat{e}_\varphi , \hat{e}_\psi$$

- Lowercase letters with a tilde indicate a 3×3 skew symmetric matrix associated to a vector. Example:

$$\tilde{a} = \begin{bmatrix} 0 & -a_3 & a_2 \\ a_3 & 0 & -a_1 \\ -a_2 & a_1 & 0 \end{bmatrix} , \quad \mathbf{a} = \begin{bmatrix} a_1 \\ a_2 \\ a_3 \end{bmatrix}$$

- An arrow above two uppercase letters indicates the start and end points of a position vector. Example:

$$\overrightarrow{ON} = \text{a position vector from point } O \text{ to point } N$$

- A double arrow above a lowercase letter indicates a 4×4 matrix associated to a quaternion. Example:

$$\overleftrightarrow{q} = \begin{bmatrix} q_0 & -q_1 & -q_2 & -q_3 \\ q_1 & q_0 & -q_3 & q_2 \\ q_2 & q_3 & q_0 & -q_1 \\ q_3 & -q_2 & q_1 & q_0 \end{bmatrix}$$

$$q = q_0 + q_1 i + q_2 j + q_3 k$$

- The length of a vector is indicated by a non-bold lowercase letter. Example:

$$r = |\mathbf{r}| \quad , \quad a = |\mathbf{a}| \quad , \quad b = |\mathbf{b}| \quad , \quad s = |\mathbf{s}|$$

- Capital letters A, Q, R, and T indicate rotation or transformation matrices. Example:

$$Q_{Z,\alpha} = \begin{bmatrix} \cos\alpha & -\sin\alpha & 0 \\ \sin\alpha & \cos\alpha & 0 \\ 0 & 0 & 1 \end{bmatrix} \qquad {}^{G}T_B = \begin{bmatrix} c\alpha & 0 & -s\alpha & -1 \\ 0 & 1 & 0 & 0.5 \\ s\alpha & 0 & c\alpha & 0.2 \\ 0 & 0 & 0 & 1 \end{bmatrix}$$

- Capital letter B is utilized to denote a body coordinate frame. Example:

$$B(oxyz) \qquad B(Oxyz) \qquad B_1(o_1x_1y_1z_1)$$

- Capital letter G is utilized to denote a global, inertial, or fixed coordinate frame. Example:

$$G \qquad G(XYZ) \qquad G(OXYZ)$$

- Right subscript on a transformation matrix indicates the *departure* frames. Example:

$$T_B = \text{transformation matrix from frame } B(oxyz)$$

- Left superscript on a transformation matrix indicates the *destination* frame. Example:

$${}^{G}T_B = \text{transformation matrix from frame } B(oxyz)$$
$$\text{to frame } G(OXYZ)$$

- Whenever there is no sub or superscript, the matrices are shown in a bracket. Example:

$$[T] = \begin{bmatrix} \cos\alpha & 0 & -\sin\alpha & -1 \\ 0 & 1 & 0 & 0.5 \\ \sin\alpha & 0 & \cos\alpha & 0.2 \\ 0 & 0 & 0 & 1 \end{bmatrix}$$

- Left superscript on a vector denotes the frame in which the vector is expressed. That superscript indicates the frame that the vector belongs to, and so the vector is expressed using the unit vectors of that frame. Example:

$${}^{G}\mathbf{r} = \text{position vector expressed in frame } G(OXYZ)$$

- Right subscript on a vector denotes the tip point that the vector is referred to. Example:

$${}^{G}\mathbf{r}_P = \text{position vector of point } P$$
$$\text{expressed in coordinate frame } G(OXYZ)$$

- Left subscript on a vector indicates the frame that the angular vector is measured with respect to. Example:

$${}^{G}_{B}\mathbf{v}_P = \text{velocity vector of point } P \text{ in coordinate frame } B(oxyz)$$
$$\text{expressed in the global coordinate frame } G(OXYZ)$$

We drop the left subscript if it is the same as the left superscript. Example:

$${}^{B}_{B}\mathbf{v}_P \equiv {}^{B}\mathbf{v}_P$$

- Unit vectors must always be expressed in their own coordinate frame to be meaningful. Example: ${}^0\hat{\imath}_1$ and ${}^0\hat{\imath}_2$ must be transformed into 0-frame and be expressed by unit vectors of the 0-frame.

$$ {}^0\hat{\imath}_1 = {}^0R_1\,{}^1\hat{\imath}_1 $$
$$ {}^0\hat{\imath}_2 = {}^0R_2\,{}^2\hat{\imath}_2 $$

- Right subscript on an angular velocity vector indicates the frame that the angular vector is referred to. Example:

$$ \boldsymbol{\omega}_B = \text{angular velocity of the body coordinate frame } B(oxyz) $$

- Left subscript on an angular velocity vector indicates the frame that the angular vector is measured with respect to. Example:

$$ {}_G\boldsymbol{\omega}_B = \text{angular velocity of the body coordinate frame } B(oxyz) $$
$$ \text{with respect to the global coordinate frame } G(OXYZ) $$

- Left superscript on an angular velocity vector denotes the frame in which the angular velocity is expressed. Example:

$$ {}_G^{B_2}\boldsymbol{\omega}_{B_1} = \text{angular velocity of the body coordinate frame } B_1 $$
$$ \text{with respect to the global coordinate frame } G $$
$$ \text{and expressed in body coordinate frame } B_2 $$

Whenever the left subscript and superscript of an angular velocity are the same, we usually drop the left superscript. Example:

$$ {}_G\boldsymbol{\omega}_B \equiv {}_G^G\boldsymbol{\omega}_B $$

- If the right subscript on a force vector is a number, it indicates the number of coordinate frame in a serial robot. Coordinate frame B_i is set up at joint $i + 1$. Example:

$$ \mathbf{F}_i = \text{force vector at joint } i + 1 $$
$$ \text{measured at the origin of } B_i(oxyz) $$

At joint i there is always an action force $\mathbf{F}_i$, that link (i) applies on link $(i + 1)$, and a reaction force $-\mathbf{F}_i$, that link $(i + 1)$ applies on link (i). On link (i) there is always an action force $\mathbf{F}_{i-1}$ coming from link $(i - 1)$, and a reaction force $-\mathbf{F}_i$ coming from link $(i + 1)$. Action force is called *driving force*, and reaction force is called *driven force*.

- If the right subscript on a moment vector is a number, it indicates the number of coordinate frames in a serial robot. Coordinate frame B_i is set up at joint $i+1$. Example:

$$ \mathbf{M}_i = \text{moment vector at joint } i + 1 $$
$$ \text{measured at the origin of } B_i(oxyz) $$

At joint i there is always an action moment $\mathbf{M}_i$, that link (i) applies on link $(i + 1)$, and a reaction moment $-\mathbf{M}_i$, that link $(i + 1)$ applies on link (i). On link (i) there is always an action moment $\mathbf{M}_{i-1}$ coming from link $(i - 1)$, and a reaction moment $-\mathbf{M}_i$ coming from link $(i + 1)$. Action moment is called *driving moment*, and reaction moment is called *driven moment*.

- Left superscript on derivative operators indicates the frame in which the derivative of a variable is taken. Example:

$$\frac{{}^G d}{dt} x \qquad \frac{{}^G d}{dt} {}^B \mathbf{r}_P \qquad \frac{{}^B d}{dt} {}^G_B \mathbf{r}_P$$

If the variable is a vector function, and also the frame in which the vector is defined is the same as the frame in which a time derivative is taken, we use the following short notation,

$$\frac{{}^G d}{dt} {}^G \mathbf{r}_P = {}^G \dot{\mathbf{r}}_P \qquad \frac{{}^B d}{dt} {}^B_o \mathbf{r}_P = {}^B_o \dot{\mathbf{r}}_P$$

and write equations simpler. Example:

$${}^G \mathbf{v} = \frac{{}^G d}{dt} {}^G \mathbf{r}(t) = {}^G \dot{\mathbf{r}}$$

If the vector is in different frame than the derivative frame, we have a mixed derivative. The result of a mixed derivative would be a vector indicated by two left indices. Its left subscript indicates the frame in which the derivative is taken, and its left superscript indicates the frame it is expressed in. Derivative operation will not change the expression frame. Example:

$$\frac{{}^G d}{dt} {}^B \mathbf{r} = {}^B_G \dot{\mathbf{r}} = {}^B_G \mathbf{v}$$

$$\frac{{}^B d}{dt} {}^G \mathbf{r} = {}^G_B \dot{\mathbf{r}} = {}^G_B \mathbf{v}$$

- If followed by angles, lowercase c and s denote *cos* and *sin* functions in mathematical equations. Example:

$$c\alpha = \cos\alpha \qquad s\varphi = \sin\varphi$$

- Capital bold letter $\mathbf{I}$ indicates a unit matrix, which, depending on the dimension of the matrix equation, could be a 3×3 or a 4×4 unit matrix. $\mathbf{I}_3$ or $\mathbf{I}_4$ are also being used to clarify the dimension of $\mathbf{I}$. Example:

$$\mathbf{I} = \mathbf{I}_3 = \begin{bmatrix} 1 & 0 & 0 \\ 0 & 1 & 0 \\ 0 & 0 & 1 \end{bmatrix}$$

- An asterisk ★ indicates a more advanced subject or example that is not designed for undergraduate teaching and can be dropped in the first reading.
- Two parallel joint axes are indicated by a parallel sign, (∥).
- Two orthogonal joint axes are indicated by an orthogonal sign, (⊢). Two orthogonal joint axes are intersecting at a right angle.
- Two perpendicular joint axes are indicated by a perpendicular sign, (⊥). Two perpendicular joint axes are at a right angle with respect to their common normal.

Preface to the Second Edition

Second edition of this novel would not be possible without comments and contribution of all my students, especially those at Columbia University in New York. New topics introduced in this edition result from students' feedback which assisted me in clarifying and better presenting certain aspects of this book.

The intent of this book is to explain robotics in a manner I would have liked explained to me as a student. This book can now help students by being a great reference that covers all aspects of robotics and providing students with detailed explanations and information.

The first edition of this book was published in 2006 by Springer. Soon after its publication, the book become very popular in the field of robotics. It was appreciated by many students and instructors in addition to my own students and colleagues. Their questions, comments, and suggestion have helped me in creating the second edition.

Preface to the First Edition

This book is designed to serve as a text for engineering students. It introduces the fundamental knowledge used in robotics. This knowledge can be utilized to develop computer programs for analyzing the kinematics, dynamics, and control of robotic systems.

The subject of robotics may appear overdosed by the number of available texts because the field has been growing rapidly since 1970. However, the topic remains alive with modern developments, which are closely related to the classical material. It is evident that no single text can cover the vast scope of classical and modern materials in robotics. Thus the demand for new books arises because the field continues to progress. Another factor is the trend toward analytical unification of kinematics, dynamics, and control.

Classical kinematics and dynamics of robots has its roots in the work of great scientists of the past four centuries who established the methodology and understanding of the behavior of dynamic systems. The development of dynamic science, since the beginning of the twentieth century, has moved toward analysis of controllable man-made systems. Therefore, merging the kinematics and dynamics with control theory is the expected development for robotic analysis.

The other important development is the fast growing capability of accurate and rapid numerical calculations, along with intelligent computer programming.

Level of the Book

This book has evolved from nearly a decade of research in nonlinear dynamic systems, and teaching undergraduate-graduate level courses in robotics. It is addressed primarily to the last year of undergraduate study and the first year graduate student in engineering. Hence, it is an intermediate textbook. This book can even be the first exposure to topics in spatial kinematics and dynamics of mechanical systems. Therefore, it provides both fundamental and advanced topics on the kinematics and dynamics of robots. The whole book can be covered in two successive courses however, it is possible to jump over some sections and cover the book in one course. The students are required to know the fundamentals of kinematics and dynamics, as well as a basic knowledge of numerical methods.

The contents of the book have been kept at a fairly theoretical-practical level. Many concepts are deeply explained and their use emphasized, and most of the related theory and formal proofs have been explained. Throughout the book, a strong emphasis is put on the physical meaning of the concepts introduced. Topics that have been selected are of high interest in the field. An attempt has been made to expose the students to a broad range of topics and approaches.

Organization of the Book

The text is organized so it can be used for teaching or for self-study. Chapter 1 "Introduction," contains general preliminaries with a brief review of the historical development and classification of robots.

Part I "Kinematics," presents the forward and inverse kinematics of robots. Kinematics analysis refers to position, velocity, and acceleration analysis of robots in both joint and base coordinate spaces. It establishes kinematic relations among the end-effecter and the joint variables. The method of Denavit-Hartenberg for representing body coordinate frames is introduced and utilized for forward kinematics analysis. The concept of modular treatment of robots is well covered to show how we may combine simple links to make the forward kinematics of a complex robot. For inverse kinematics analysis, the idea of decoupling, the inverse matrix method, and the iterative technique are introduced. It is shown that the presence of a spherical wrist is what we need to apply analytic methods in inverse kinematics.

Part II "Dynamics," presents a detailed discussion of robot dynamics. An attempt is made to review the basic approaches and demonstrate how these can be adapted for the active displacement framework utilized for robot kinematics in the earlier chapters. The concepts of the recursive Newton-Euler dynamics, Lagrangian function, manipulator inertia matrix, and generalized forces are introduced and applied for derivation of dynamic equations of motion.

Part III "Control," presents the floating time technique for time-optimal control of robots. The outcome of the technique is applied for an open-loop control algorithm. Then, a computed-torque method is introduced, in which a combination of feedforward and feedback signals are utilized to render the system error dynamics.

Method of Presentation

The structure of presentation is in a "*fact-reason-application*" fashion. The "fact" is the main subject we introduce in each section. Then the reason is given as a "proof." Finally the application of the fact is examined in some "examples." The "examples" are a very important part of the book because they show how to implement the knowledge introduced in "facts." They also cover some other facts that are needed to expand the subject.

Melbourne, VIC, Australia   Reza N. Jazar

How to Use This Book

This book is suitable for the first course in robotics and is written for a full semester of 16 weeks at graduate level. If the level of the course is undergraduate or mixed, and if the length of semester is shorter than 16 weeks, then the following pattern can be suggested to fit the contents to different classes. Teaching Chapter 10, is optional in all cases depending on the background of students. The main topic of this chapter is to review dynamics and remind students the Lagrangean method of deriving equations of motion. Chapter 10 may be left for students to read by themselves.

Undergraduate level, 10 or 12 weeks semester: Skip all asterisk sections and teach the following chapters:

1 Introduction
2 Rotation Kinematics
3 Orientation Kinematics
4 Motion Kinematics
5 Forward Kinematics
6 Inverse Kinematics
8 Velocity Kinematics
11 Robot Dynamics
12 Path Planning

Undergraduate level, 16 weeks semester: Skip all asterisk sections and teach the following chapters:

1 Introduction
2 Rotation Kinematics
3 Orientation Kinematics
4 Motion Kinematics
5 Forward Kinematics
6 Inverse Kinematics
7 Angular Velocity
8 Velocity Kinematics
9 Acceleration Kinematics
11 Robot Dynamics
12 Path Planning

Graduate level, 10 or 12 weeks semester: Skip asterisk sections of Chaps. 3, 4, 5, 7, 9, and 11 and teach the following chapters:

1 Introduction
2 Rotation Kinematics
3 Orientation Kinematics
4 Motion Kinematics

5 Forward Kinematics
6 Inverse Kinematics
7 Angular Velocity
8 Velocity Kinematics
9 Acceleration Kinematics
11 Robot Dynamics
12 Path Planning

Undergraduate level, 16 weeks semester: Skip asterisk sections based on your preference and teach the following chapters:

1 Introduction
2 Rotation Kinematics
3 Orientation Kinematics
4 Motion Kinematics
5 Forward Kinematics
6 Inverse Kinematics
7 Angular Velocity
8 Velocity Kinematics
9 Acceleration Kinematics
11 Robot Dynamics
12 Path Planning
14 Control Techniques

Contents

1 Introduction 1
  1.1   Historical Development 2
  1.2   Robot Components 2
      1.2.1   Link 2
      1.2.2   Joint 3
      1.2.3   Manipulator 5
      1.2.4   Wrist 6
      1.2.5   End-Effector 7
      1.2.6   Actuators 7
      1.2.7   Sensors 8
      1.2.8   Controller 8
  1.3   Robot Classifications 8
      1.3.1   Geometry 9
      1.3.2   Workspace 11
      1.3.3   Actuation 12
      1.3.4   Control 12
      1.3.5   Application 13
  1.4   Robot's Kinematics, Dynamics, and Control 13
  1.5   Principle of Kinematics 14
      1.5.1   ★ Triad 14
      1.5.2   Unit Vectors 15
      1.5.3   Orthogonality Condition 20
      1.5.4   Coordinate Frame and Transformation 20
      1.5.5   ★ Vector Definition 22
      1.5.6   Vector Function 25
  1.6   Summary 28
  1.7   Key Symbols 29
  Exercises 31

Part I   Kinematics

2 Rotation Kinematics 37
  2.1   Rotation About Global Cartesian Axes 37
  2.2   Successive Rotation About Global Cartesian Axes 43
  2.3   Rotation About Local Cartesian Axes 52
  2.4   Successive Rotation About Local Cartesian Axes 55
  2.5   Euler Angles 59
  2.6   Local Axes Versus Global Axes Rotation 69
  2.7   General Transformation 71
  2.8   ★ Active and Passive Transformation 79

2.9   Summary 80
2.10   Key Symbols 82
        Exercises 83

3   Orientation Kinematics 91
3.1   Axis–Angle Rotation 91
3.2   ★ Order-Free Rotation 102
3.3   ★ Euler Parameters 109
3.4   ★ Quaternions.. 117
3.5   ★ Spinors and Rotators 124
3.6   ★ Problems in Representing Rotations 126
        3.6.1   ★ Rotation Matrix 127
        3.6.2   ★ Axis–Angle 127
        3.6.3   ★ Euler Angles 128
        3.6.4   ★ Quaternion 129
        3.6.5   ★ Euler Parameters 130
3.7   ★ Composition and Decomposition of Rotations 132
        3.7.1   ★ Composition of Rotations 132
        3.7.2   ★ Decomposition of Rotations 134
3.8   Summary 137
3.9   Key Symbols 139
        Exercises 140

4   Motion Kinematics 149
4.1   Rigid Body Motion 149
4.2   Homogenous Transformation................................... 153
4.3   Inverse and Reverse Homogenous Transformation 162
4.4   Combined Homogenous Transformation 167
4.5   ★ Order-Free Transformation 176
4.6   ★ Screw Coordinates 182
4.7   ★ Inverse Screw 197
4.8   ★ Combined Screw Transformation 200
4.9   ★ The Plücker Line Coordinate 202
4.10   ★ The Geometry of Plane and Line 207
        4.10.1   ★ Moment 208
        4.10.2   ★ Angle and Distance 208
        4.10.3   ★ Plane and Line 209
4.11   ★ Screw and Plücker Coordinate 212
4.12   Summary 213
4.13   Key Symbols 215
        Exercises 216

5   Forward Kinematics 225
5.1   Denavit–Hartenberg Notation 225
5.2   Transformation Between Adjacent Coordinate Frames............ 232
5.3   Forward Position Kinematics of Robots 247
5.4   Spherical Wrist 269
5.5   Assembling Kinematics 276
5.6   ★ Coordinate Transformation Using Screws.................... 289
5.7   ★ Non-Denavit–Hartenberg Methods 293
5.8   Summary 299
5.9   Key Symbols 300
        Exercises 302

6   Inverse Kinematics 313
   6.1   Decoupling Technique 313
   6.2   Inverse Transformation Technique 329
   6.3   ★ Iterative Technique 343
   6.4   ★ Comparison of the Inverse Kinematics Techniques 347
          6.4.1   ★ Existence and Uniqueness of Solution 347
          6.4.2   ★ Inverse Kinematics Techniques 347
   6.5   ★ Singular Configuration 347
   6.6   Summary 349
   6.7   Key Symbols 350
   Exercises 352

Part II   Derivative Kinematics

7   Angular Velocity 361
   7.1   Angular Velocity Vector and Matrix 361
   7.2   ★ Time Derivative and Coordinate Frames 376
   7.3   Rigid Body Velocity 388
   7.4   ★ Velocity Transformation Matrix 392
   7.5   Derivative of a Homogenous Transformation Matrix 400
   7.6   Summary 405
   7.7   Key Symbols 407
   Exercises 409

8   Velocity Kinematics 415
   8.1   ★ Rigid Link Velocity 415
   8.2   Forward Velocity Kinematics 419
   8.3   Jacobian Generating Vectors 429
   8.4   Inverse Velocity Kinematics 442
   8.5   ★ Linear Algebraic Equations 448
   8.6   Matrix Inversion 458
   8.7   Nonlinear Algebraic Equations 463
   8.8   ★ Jacobian Matrix From Link Transformation Matrices 469
   8.9   Summary 476
   8.10   Key Symbols 477
   Exercises 479

9   Acceleration Kinematics 489
   9.1   Angular Acceleration Vector and Matrix 489
   9.2   Rigid Body Acceleration 508
   9.3   ★ Acceleration Transformation Matrix 510
   9.4   Forward Acceleration Kinematics 518
   9.5   Inverse Acceleration Kinematics 520
   9.6   ★ Rigid Link Recursive Acceleration 526
   9.7   ★ Second Derivative and Coordinate Frames 534
   9.8   Summary 543
   9.9   Key Symbols 545
   Exercises 547

Part III   Dynamics

10   Applied Dynamics 557
    10.1   Force and Moment 557
        10.1.1   Force and Moment 557
        10.1.2   Momentum 558
        10.1.3   Equation of Motion 558
        10.1.4   Work and Energy 559
    10.2   Rigid Body Translational Kinetics 565
    10.3   Rigid Body Rotational Kinetics 567
    10.4   Mass Moment Matrix 576
    10.5   Lagrange's Form of Newton's Equations 584
    10.6   Lagrangian Mechanics 591
    10.7   Summary 597
    10.8   Key Symbols 600
    Exercises 602

11   Robot Dynamics 609
    11.1   Rigid Link Newton–Euler Dynamics 609
    11.2   ★ Recursive Newton–Euler Dynamics 626
    11.3   Robot Lagrange Dynamics 632
    11.4   ★ Lagrange Equations and Link Transformation Matrices 659
    11.5   Robot Statics 667
    11.6   Summary 673
    11.7   Key Symbols 676
    Exercises 678

Part IV   Control

12   Path Planning 687
    12.1   Cubic Path 687
    12.2   Polynomial Path 692
    12.3   ★ Non-polynomial Path Planning 702
    12.4   ★ Spatial Path Design 705
    12.5   Forward Path Robot Motion 708
    12.6   Inverse Path Robot Motion 712
    12.7   ★ Rotational Path 720
    12.8   Summary 724
    12.9   Key Symbols 725
    Exercises 726

13   ★ Time Optimal Control 731
    13.1   ★ Minimum Time and Bang-Bang Control 731
    13.2   ★ Floating Time Method 738
    13.3   ★ Time Optimal Control for Robots 746
    13.4   Summary 752
    13.5   Key Symbols 753
    Exercises 754

14   Control Techniques 759
    14.1   Open- and Closed-Loop Control 759
    14.2   Computed Torque Control 764
    14.3   Linear Control Technique 767
        14.3.1   Proportional Control 768
        14.3.2   Integral Control 768
        14.3.3   Derivative Control 768
    14.4   Sensing and Control 770
        14.4.1   Position Sensors 771
        14.4.2   Speed Sensors 771
        14.4.3   Acceleration Sensors 771
    14.5   Summary 772
    14.6   Key Symbols 773
    Exercises 774

A   Global Frame Triple Rotation 777

B   Local Frame Triple Rotation 779

C   Principal Central Screws Triple Combination 781

D   Industrial Link DH Matrices 783

E   Matrix Calculus 791

F   Trigonometric Formula 797

G   Algebraic Formula 805

H   Unit Conversions 807

Bibliography 809

Index 815

About the Author

Reza N. Jazar is Professor of Mechanical Engineering. Reza received his PhD degree from Sharif University of Technology, and MSc and BSc from Tehran Polytechnic. His areas of expertise include nonlinear dynamic systems and applied mathematics. He obtained original results in non-smooth dynamic systems, applied nonlinear vibrating problems, time optimal control, and mathematical modeling of vehicle dynamics and stability. He has authored several monographs in vehicle dynamics, robotics, dynamics, vibrations, and mathematics and published numerous professional articles, as well as book chapters, in research volumes. Most of his textbooks have been adopted by many universities for teaching and research, and by many research agencies as standard model for research results.

Dr. Jazar has the pleasure to work in several Canadian, American, Asian, Middle Eastern, and Australian universities, as well as several years in automotive industries all around the world. Working in different engineering firms and educational systems provided him a vast experience and knowledge to publish his researches on important topics in engineering and science. His unique style of writing helps readers to learn the topics deeply in an easy way.

Introduction

Isaac Asimov's Laws of Robotics:

Law Zero: A robot may not harm humanity, or, by inaction, allow humanity to come to harm.

Law One: A robot may not injure a human being, or, through inaction, allow a human being to come to harm, unless this would violate a higher order law.

Law Two: A robot must obey orders given by human beings, except where such orders would conflict with a higher order law.

Law Three: A robot must protect its own existence as long as such protection does not conflict with a higher order law.

Fig. 1.1 A high performance robot hand

The great Czech writer Karel Čapek (1890–1938) is best known for his 1920 play Rossumovi univerzální roboti, *RUR*, (*RUR*: Rossum's Universal Robots), in which he introduced the word "Robot" and "Robotess" to the world for the first time (Čapek, 1994; Husbands et al., 2008). The word "robot" is of Slavic origin close to Russian, in which the word "pa6oTa" (rabota) means labor. Later, Isaac Asimov (1920–1992) adopted the term "Robot" and proposed Laws 1–3 in his short story "Runaround," which was first published in the March 1942 issue of Astounding Science Fiction magazine (Asimov, 1942, 1950a). In 1985 Asimov added a Zeroth Law later, in "Robots and Empire," (Asimov, 1950b). Isaac Asimov proposed these four refined laws of "Robotics" to protect us from intelligent generations of robots. These laws, although look logical, have been challenged by others in late 20th and early 21st centuries.

In science and engineering, the term *Robotics* refers to the study and use of *Robots*. Based on definition of the Robotics Institute of America (*RIA*): "A robot is a reprogrammable multifunctional manipulator designed to move material, parts, tools, or specialized devices through variable programmed motions for the performance of a variety of tasks." It means, Robots are similar to mechanisms but also capable to change the task they will be doing. Hence, Robot is a mechanism that we can change its operation task, usually by only a computer program.

From an engineering point of view, Robots are complex, versatile devices that contain a mechanical structure, a sensory system, and an automatic control system. Theoretical fundamentals of Robotics rely on the results of research in mechanics, electric, electronics, automatic control, mathematics, and computer sciences.

© The Author(s), under exclusive license to Springer Nature Switzerland AG 2022
R. N. Jazar, *Theory of Applied Robotics*, https://doi.org/10.1007/978-3-030-93220-6_1

1.1   Historical Development

The first position controlling apparatus was invented around 1938 for spray painting. However, the first modern industrial Robot were the Unimates, made by Joseph Engelberger (1925–2015) in the early 1960s. Unimation was the first to market robots. Therefore, Engelberger has been called the father of Robotics. In the 1980s the robot industry grew very fast primarily because of the huge investments by the automotive industry (Rosheim, 1994).

In research communities the first automata were probably William Grey Walter's (1910–1977) machina in 1940s and the Johns Hopkins beast built in the 1960s. The first programmable Robot was designed by George Devol (1912–2011) in 1954. Devol funded Unimation. In 1959 the first commercially available Robot appeared on the market. Robotic manipulators were used in industries after 1960 and faced sky rocketing growth in the 1980s.

Robots appeared as a result of combination of two technologies: teleoperators and computer numerical control (CNC) of milling machines. Teleoperators were developed during World War II to handle radioactive materials, and CNC was developed to increase the precision required in machining of new technologic parts. Therefore, the first Robots were numerical controlled mechanical linkages that were designed to transfer material from point A to B (Niku, 2020).

Today, more complicated applications, such as welding, painting, and assembling, require much more motion capability and sensing. Hence, a Robot is a multi-disciplinary engineering device. Mechanical engineering deals with the design of mechanical components, arms, end-effectors, and also is responsible for kinematics, dynamics, and control analyses of Robots. Electrical engineering works on Robot actuators, sensors, power, and control systems. System design engineering deals with perception, sensing, and control methods of Robots. Programming, or software engineering, is responsible for logic, intelligence, communication, and networking.

Today we have more than 1000 robotics-related organizations, associations, and clubs; more than 500 robotics-related magazines, journals, and newsletters; more than 100 robotics-related conferences and competitions every year; and more than 50 robotics-related courses in colleges. Robots find a vast amount of industrial applications and are used for various technological operations. Robots enhance labor productivity in industry and deliver relief from tiresome, monotonous, or hazardous works. Moreover, Robots perform many operations better than human do, and they provide higher accuracy and repeatability. In many engineering fields, high technological standards are hardly attainable without Robots. Apart from industry, Robots are used in extreme environments. They can work at low and high temperatures; they usually do not need lights, rest, fresh air, and more importantly they do not need salary, or promotions, and do not get sad. Robots are prospective machines whose application area is widening and their structures getting more complex. Figure 1.1 illustrates a high performance robot hand.

It is claimed that robots appeared to perform in 4A for 4D, or 3D3H environments. 4A performances are Automation, Augmentation, Assistance, Autonomous; and 4D environments are Dangerous, Dirty, Dull, Difficult. 3D3H means Dull, Dirty, Dangerous, Hot, Heavy, Hazardous (Tsai, 1999; Veit, 1992).

1.2   Robot Components

We kinematically model a Robotic manipulator as a multibody system made of connected rigid bodies by revolute or prismatic joints that allow the bodies to relatively move. We apply rigid body kinematics to the connected bodies and show how to determine their relative motions. However, a Robot as a system also consists of a *manipulator* or *rover*, *wrist*, *end-effector*, *actuators*, *sensors*, *controllers*, *processors*, and *software*. Figure 1.2 illustrates the Shuttle arm, which is a robotic manipulator as a multibody system.

1.2.1   Link

Every individual rigid member of a robot that can move relative to all other members is called a *link*. In robotics we may use *bar* and *arm* and *object* equal to link. A robot arm or a robot link is a rigid member that is able to have relative motion with respect to all other links. Any two or more connected links, such that no relative motion can occur among them, are considered a single compound link.

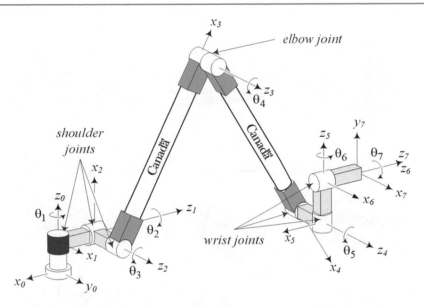

Fig. 1.2 Shuttle arm is a robotic manipulator as a multibody system

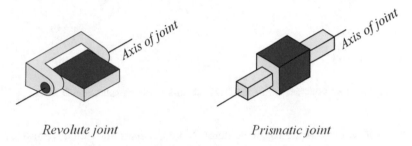

Fig. 1.3 Illustration of revolute and prismatic joints

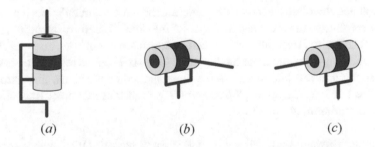

Fig. 1.4 Symbolic illustration of revolute joints in robotic models

1.2.2 Joint

Two rigid bodies that are permanently in contact with a possible relative motion is called a *kinematic pair*. Two links are connected by contact at a *joint* where their relative motion can be expressed by a single joint coordinate. Joints are typically *revolute* (rotary) or *prismatic* (translatory). Figure 1.3 depicts the geometric form of a revolute and a prismatic joint. A *revolute joint* (R) is like a hinge and allows relative rotation between two links. A *prismatic joint* (P) allows a translation of relative motion between two links.

Relative rotation of connected links by a revolute joint occurs about a line called *axis of joint*. Also, translation of two connected links by a prismatic joint occurs along a line also called *axis of joint*. The value of the coordinate describing the relative position of two connected links at a joint is called *joint coordinate* or *joint variable*. It is an *angle* for a revolute joint, and a *distance* for a prismatic joint. A symbolic illustration of revolute and prismatic joints is shown in Figs. 1.4a–c and 1.5a–c, respectively.

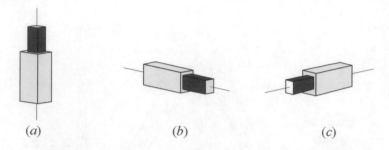

Fig. 1.5 Symbolic illustration of prismatic joints in robotic models

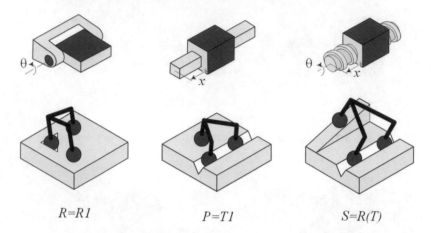

Fig. 1.6 Joints of class 1 with one DOF: Revolute, R, Prismatic, P, and Screw, S

The coordinate of an *active joint* is controlled by an actuator. A *passive joint* does not have any actuator. The coordinate of a passive joint is a function of the coordinates of active joints and the geometry of the robot arms. Passive joints are also called *inactive* or *free joints*. Active joints are usually prismatic or revolute; however, passive joints may be any of the *lower pair joints* that provide surface contact. Revolute and prismatic joints are the most common joints that are utilized in serial robotic manipulators. Prismatic and revolute joints provide one degree of freedom. Therefore, the number of joints of a manipulator is the *degrees-of-freedom* (DOF) of the manipulator. Typically a spacial manipulator possesses at least six DOF: three for positioning and three for orientation. A manipulator having more than six DOF is referred to as a kinematically *redundant* manipulator. Although prismatic and revolute joints are the most applied connections, there are other types of joints that are classified according to the number of *degrees of freedom* (DOF) they eliminate or they allow. A joint can provide a maximum of five DOF and a minimum of one DOF.

Proof When there is no contact between two bodies A and B, then B has six DOF with respect to A. Every permanent contact eliminates some rotational or translational DOF. All possible kinematic pairs can be classified by the number of rotational, R, or translational, T, degrees of freedom the contact provides.

Class 1 are joints that provide one DOF between links A and B. To get one DOF we kinematically need to have five contact points between the two links. The class 1 joints and their kinematic models are shown in Fig. 1.6. There are three types of joints in class 1: Revolute, R, with a rotational freedom $R1$; Prismatic, P, with a translational freedom $T1$; and Helical or Screw, S, with a proportional rotational-translational freedom $R(T)$.

Class 2 are joints that provide two DOF between links A and B. To get two DOF we kinematically need to have four contact points between the two links. The class 2 joints and their kinematic models are shown in Fig. 1.7. There are three types of joints in class 2: Sphere in slot, with two rotational freedom $R2$; Cylindrical, C, with a coaxial rotational and a translational freedom $R1T1$; and disc in slot, with a perpendicular rotational and translational freedom $R1T1$.

Class 3 are joints that provide three DOF between links A and B. To get three DOF we kinematically need to have three contact points between the two links. The class 3 joints and their kinematic models are shown in Fig. 1.8. There are three types of joints in class 3: Spherical, S, with three rotational freedom $R3$; Sphere in slot, with two rotational and a translational freedom $R2T1$; and Disc in slot, with a rotational and two translational freedom $R1T2$.

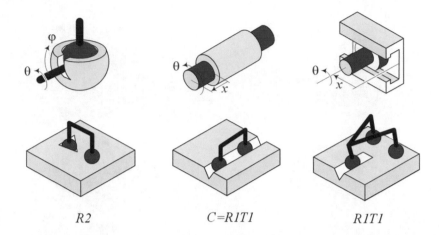

Fig. 1.7 Joints of class 2 with two DOF: Sphere in slot $R2$; Cylindrical, C, and Disc in slot $R1T$

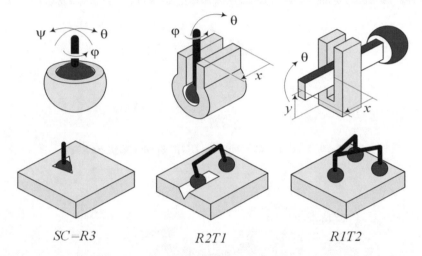

Fig. 1.8 Joints of class 3 with three DOF: Spherical, S, Sphere in slot, $R2T1$; and Disc in slot, $R1T2$

Class 4 are joints that provide four DOF between links A and B. To get two DOF we kinematically need to have two contact points between the two links. The class 4 joints and their kinematic models are shown in Fig. 1.9. There are two types of joints in class 4: Sphere in slot, with three rotational and one translational freedom $R3T1$, and Cylinder in slot, with two rotational and two translational freedom $R2T2$. These joints provide four DOF between two links A and B.

Class 5 are joints that provide five DOF between links A and B. To get one DOF we kinematically need to have one contact point between the two links. The kinematic model of class 5 joint is shown in Fig. 1.10. There is only one type of joints in class 5: Sphere on plane, with three rotational and two translational freedom $R3T2$.

The DOF of a mechanism f is

$$f = 6n - j_5 - 2j_4 - 3j_3 - 4j_2 - 5j_1 \tag{1.1}$$

where, n is the number of links and j_k, $k = 1, 2, 3, 4, 5$, indicates the number of joints of class j_k. ∎

1.2.3 Manipulator

The main body of a robot consisting of the links, joints, and other structural elements is called the *manipulator*. A manipulator becomes a robot when we attach wrist and gripper, and install its control system. However, in literature robots and manipulators are utilized equivalently and both refer to robots. Figure 1.11 shows kinematic model of a $3R$ manipulator, and Fig. 1.16 illustrates a $3R$ manipulator.

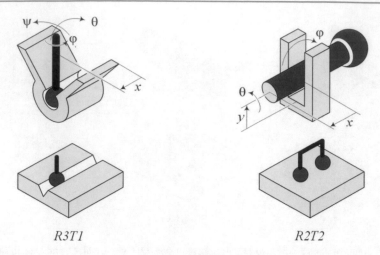

Fig. 1.9 Joints of class 4 with four DOF: sphere in slot $R3T1$ and cylinder in slot $R2T2$

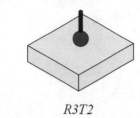

Fig. 1.10 Joint of class 5 with five DOF: Sphere on plane $R3T2$

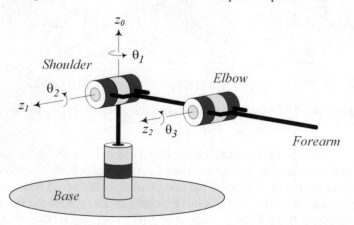

Fig. 1.11 Model of a $3R$ manipulator

1.2.4   Wrist

The joints in the kinematic chain of a robot between the forearm and end-effector are referred to as the *wrist*. It is common to design manipulators with spherical wrists by which it means three revolute joint axes intersecting at a common point called the *wrist point*. Figure 1.12 shows kinematic model of a spherical wrist, and Fig. 1.13 illustrates a spherical wrist at rest position. It is made of three revolute joints with orthogonal axes of rotations and shown by R⊢R⊢R mechanism.

The spherical wrist greatly simplifies the kinematic analysis by allowing us to decouple the positioning and orienting of the end-effector. Therefore, the manipulator will possess three degrees-of-freedom for position of the wrist point. Positioning is set by controlling three joints of three arms. The number of DOF for orientation will then depend on the wrist. We may design a wrist having one, two, or three DOF depending on the application.

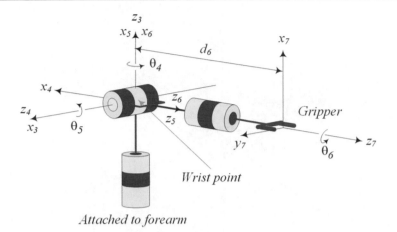

Fig. 1.12 Model of a spherical wrist kinematics

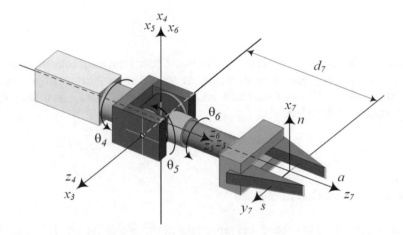

Fig. 1.13 A spherical wrist kinematics

1.2.5 End-Effector

The *end-effector* is the part mounted on the last link to do the required job of the robot. The simplest end-effector is a *gripper*, which is usually capable of two actions: opening and closing. The arm and wrist assemblies of a robot are used primarily for positioning the end-effector and any tool it may carry. It is the end-effector that performs the work. A great deal of research is devoted to the design of special purpose end-effectors and tools. There is also extensive research on the development of anthropomorphic hands. Such hands have been developed for prosthetic use in manufacturing. Hence, a robot is composed of a manipulator or *mainframe* and a wrist plus an end-effector. The wrist and end-effector assembly is also called a *hand*. Figure 1.14 illustrates an end-effector.

1.2.6 Actuators

Actuators are drivers that act as muscles of robots to change their configuration. The actuators provide power to act on the mechanical structure against gravity, inertia, and other external forces to modify the geometric location and orientation of the robot's hand. The actuators can be of electric, hydraulic, or pneumatic, and have to be controllable.

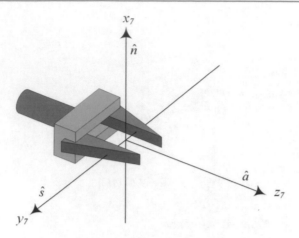

Fig. 1.14 An end-effector

1.2.7   Sensors

The elements that are utilized to detect and collect information about internal and environmental states are *sensors*. According to the scope of this book, joints' positions, velocities, accelerations, and forces are the most important information to be sensed and measured. Sensors, integrated into the robot, send information about each link and joint to the control unit, and the control unit determines the configuration of the robot.

1.2.8   Controller

The *controller* or *control unit* of a robot has three roles:

1. 1-*Information role*, which consists of collecting and processing the information provided by the robot's sensors.
2. 2-*Decision role*, which consists of planning the geometric motion of the robot structure.
3. 3-*Communication role*, which consists of organizing the information between the robot and its environment. The control unit includes the processor and software.

1.3   Robot Classifications

The Robotics Institute of America (RIA) considers classes 3-6 of the following classification to be Robots, and the Association Francaise de Robotique (AFR) combines classes 2-4, as the same type and divides robots in 4 classes. However, the Japanese Industrial Robot Association divides robots in 6 different classes:

Class 1:   *Manual handling devices*: A device with multi degrees of freedom that is actuated by an operator.
Class 2:   *Fixed sequence robot*: A device that performs successive stages of a task according to a predetermined and fixed program.
Class 3:   *Variable sequence robot*: A device that performs successive stages of a task according to a predetermined but programmable method.
Class 4:   *Playback robot*: A human operator performs the task manually by leading the robot, which records the motions for later playback. The robot repeats the same motions according to the recorded information.
Class 5:   *Numerical control robot*: The operator supplies the robot with a motion program rather than teaching it the task manually.
Class 6:   *Intelligent robot*: A robot with the ability to understand its environment and the ability to successfully complete a task despite changes in the surrounding conditions under which it is to be performed.

Other than these official classifications, robots can be classified by other criteria such as geometry, workspace, actuation, control, and application.

1.3.1 Geometry

A robot is called a *serial* or *open-loop* manipulator if its kinematic structure does not make a loop chain. It is called a *parallel* or *closed-loop* manipulator if its structure makes a loop chain. A robot is a *hybrid* manipulator if its structure consists of both open and closed-loop chains.

As a mechanical system, we may think of a robot as a set of rigid bodies connected together at some joints. The joints can be either *revolute* (R) or *prismatic* (P), because all other kinds of joint can be modeled as a combination of these two simple joints.

Most industrial manipulators have six DOF. The open-loop manipulators can be classified based on their first three joints starting from the grounded joint. Using the two types of joints, there are mathematically 72 different industrial manipulator configurations, because each joint can be P or R, and the axes of two adjacent joints can be *parallel* ($\parallel$), *orthogonal* ($\vdash$), or *perpendicular* ($\perp$). Two orthogonal joint axes intersect at a right angle; however, two perpendicular joint axes are in right-angle with respect to their common normal. Two perpendicular joint axes become parallel if one axis turns 90 deg about the common normal. Two perpendicular joint axes become orthogonal if the length of their common normal tends to zero.

Out of the 72 possible manipulators, the important ones are $R\parallel R\parallel P$ (*SCARA*), $R\vdash R\parallel R$ (*articulated*), $R\vdash R\perp P$ (*spherical*), $R\parallel P\vdash P$ (*cylindrical*), and $P\vdash P\vdash P$ (*Cartesian*).

Proof Most of the industrial robots are made of a wrist with tree rotational degrees-of-freedom to be attached to the tip point of a manipulator with three degrees-of-freedom. The duty of the manipulator is to move the tip point to a given coordinate in the $3D$ space. Starting from the ground, the first link of the manipulator is connected to the ground by the first joint, which may be a P or an R joint. The second link of the manipulator will be connected to the first link by the second joint P or R whose axis may be parallel ($\parallel$), orthogonal ($\vdash$), or perpendicular ($\perp$) to the axis of the first joint. The third link will be attached to the second link by a joint R or P whose axis may also be parallel ($\parallel$), orthogonal ($\vdash$), or perpendicular ($\perp$) to the axis of the second joint.

Hence, a $3D$ industrial manipulator will be made of three links and two joints. Considering every joint with two options, P or R, every two axes with three options, parallel ($\parallel$), orthogonal ($\vdash$), or perpendicular ($\perp$), we will have a total of 72 configurations.

$$2 \times 3 \times 2 \times 3 \times 2 = 72 \tag{1.2}$$

Therefore, we show a manipulator by the type of the first joint, axis type of first and second joint, the type of second joint, axis type of second and third joint, followed by the type of the third joint. As an example, an articulated manipulator will be shown by $R\vdash R\parallel R$, because it is made by three links. The first link will be connected to the ground by a revolute joint R. The second link is connected to the first one whose axis is orthogonal to the axis of the first joint, $R \vdash R$. The third link will be connected to the second link via another revolute joint whose axis is parallel to the axis of the second joint, $R \vdash R \parallel R$. ∎

1. 2RP manipulator, $R\parallel R\parallel P$
   The *SCARA* arm (Selective Compliant Articulated Robot for Assembly) shown in Fig. 1.15 is a popular manipulator, which, as its name suggests, is made for assembly operations.
2. Articulated manipulator, $R\vdash R\parallel R$
   The $R\vdash R\parallel R$ configuration, illustrated in Figs. 1.11 and 1.16, is called *elbow, revolute, articulated,* or *anthropomorphic*. It is a suitable configuration for industrial robots. Almost 25% of industrial robots, *PUMA* for instance, are made of this kind.
3. Spherical manipulator, $R\vdash R\perp P$
   The spherical manipulator is a suitable configuration for small robots. Almost 15% of industrial robots, Stanford arm for instance, are made of this configuration. The $R\vdash R\perp P$ configuration is illustrated in Fig. 1.17.
   By replacing the third joint of an articulate manipulator with a prismatic joint, we obtain a spherical manipulator. The term spherical manipulator derives from the fact that the spherical coordinates define the position of the end-effector with respect to the base frame. Figure 1.18 schematically illustrates the Stanford arm, one of the most well-known spherical robots.
4. Cylindrical manipulator, $R\parallel P\vdash P$
   The cylindrical manipulator is a suitable configuration for medium load capacity robots. Almost 45% of industrial robots are made of this kind. The $R\parallel P\vdash P$ configuration is illustrated in Fig. 1.19. The first joint of a cylindrical manipulator is revolute and produces a rotation about the base, while the second and third joints are prismatic. As the name suggests, the joint variables are the cylindrical coordinates of the end-effector with respect to the base.

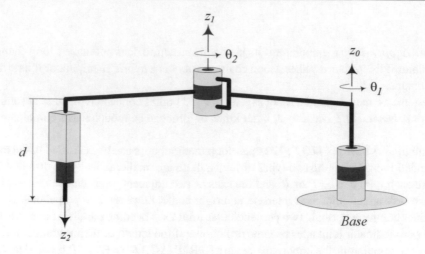

Fig. 1.15 An R‖R‖P manipulator

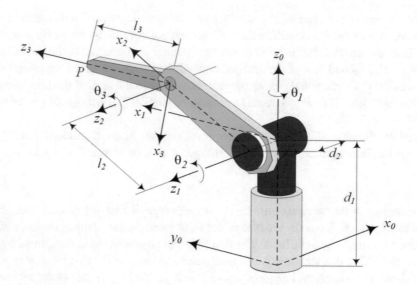

Fig. 1.16 Structure and terminology of an R⊢R‖R elbow manipulator

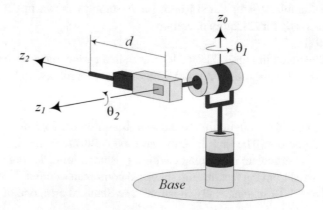

Fig. 1.17 The R⊢R⊥P spherical configuration of robotic manipulators

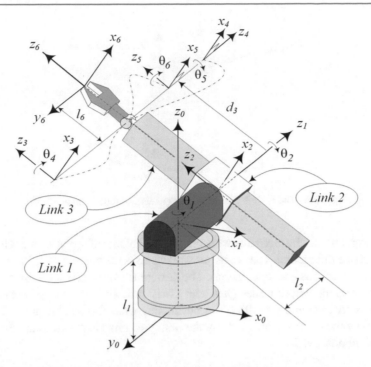

Fig. 1.18 Illustration of Stanford arm; an R⊢R⊥P spherical manipulator

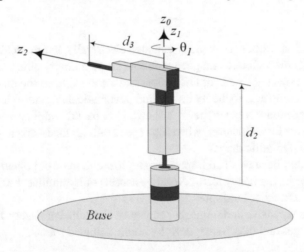

Fig. 1.19 The R∥P⊥P configuration of robotic manipulators

5. Cartesian manipulator, P⊢P⊢P

The Cartesian configuration is a suitable configuration for heavy load capacity and large robots. Almost 15% of industrial robots are made of this configuration. The P⊢P⊢P configuration is illustrated in Fig. 1.20.

For a Cartesian manipulator, the joint variables are the Cartesian coordinates of the end-effector with respect to the base. The kinematic description of this manipulator is the simplest of all manipulators. Cartesian manipulators are useful for table-top assembly applications and, as gantry robots, for transfer of cargo.

1.3.2 Workspace

The *workspace* of a manipulator is the total volume of space the end-effector can reach. The workspace is constrained by the geometry of the manipulator as well as the mechanical constraints on the joints. The workspace is broken into a *reachable*

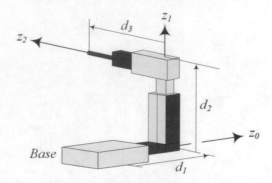

Fig. 1.20 The P⊢P⊢P Cartesian configuration of robotic manipulators

workspace and a *dexterous* workspace. The reachable workspace is the volume of space within which every point is reachable by the end-effector in at least one orientation. The dexterous workspace is the volume of space within which every point can be reached by the end-effector in all possible orientations. The dexterous workspace is a subset of the reachable workspace.

Most of the open-loop chain manipulators are designed with a wrist subassembly attached to the main three links assembly. Therefore, the first three links are long and are utilized for positioning while the wrist is utilized for control and orientation of the end-effector. This is why the subassembly made by the first three links is the arm, and the subassembly made by the other links is the wrist (Shahinpoor, 1987).

1.3.3   Actuation

Actuators translate power into motion. Robots are usually actuated electrically, hydraulically, or pneumatically. Other types of actuation might be considered as piezoelectric, magnetostriction, shape memory alloy, and polymeric.

Electrically actuated robots are powered by AC or DC motors and are considered the most acceptable actuators. They are cleaner, quieter, and more precise compared to the hydraulic and pneumatic actuators. Electric motors are efficient at high speeds so a high ratio gearbox is needed to reduce the high speed. Non-backdrivability and self-braking is an advantage of high ratio gearboxes in case of power loss. However, when high speed or high load-carrying capabilities are needed, electric drivers are unable to compete with hydraulic drivers.

Hydraulic actuators are satisfactory because of high speed and high torque/mass or power/mass ratios. Therefore, hydraulic driven robots are used primarily for lifting heavy loads. Negative aspects of hydraulics, besides their noisiness and tendency to leak, include a necessary pump and other hardware.

Pneumatic actuated robots are inexpensive and simple but cannot be controlled precisely. Besides the lower precise motion, they have almost the same advantages and disadvantages as hydraulic actuated robots.

1.3.4   Control

Robots can be classified by control method into *servo* (closed-loop control) and *non-servo* (open-loop control) robots. Servo robots use closed-loop computer control to determine their motion, and hence, they are capable of being truly multifunctional reprogrammable devices. Servo controlled robots are further classified according to the method that the controller uses to guide the end-effector.

The simplest type of a servo robots is the *point-to-point* robot. A point-to-point robot can be taught a discrete set of points, called *control points*, but there is no control on the path of the end-effector in between the points. On the other hand, in *continuous path* robots, the entire path of the end-effector can be controlled. For example, the robot end-effector can be taught to follow a straight line between two points or even to follow a contour such as a welding seam. In addition, the velocity and/or acceleration of the end-effector can often be controlled. These are the most advanced robots and require the most sophisticated computer controllers and software development.

Non-servo robots are essentially open-loop devices whose movement is limited to predetermined mechanical stops, and they are primarily used for material transfer.

1.3.5   Application

Regardless of size, robots can mainly be classified according to their application into *assembly* and *non-assembly* robots. However, in the industry they are classified by the category of application such as *machine loading, pick and place, welding, painting, assembling, inspecting, sampling, manufacturing, biomedical, assisting, remote controlled mobile,* and *telerobot.*

According to design characteristics, most industrial robot arms are anthropomorphic, in the sense that they have a "shoulder," (first two joints) an "elbow," (third joint), and a "wrist" (last three joints). Therefore, in total, they usually have six degrees of freedom needed to put an object in any position and orientation.

Most manufacturers of commercial serial manipulators try to use revolute joints more than prismatic. Compared to prismatic joints, revolute joints cost less and provide a larger dextrous workspace for the same robot volume. Serial robots are very heavy; compared to the maximum load they can move without losing their accuracy. Their useful load-to-weight ratio is less than $1/10$. The robots are so heavy because the links must be stiff in order to work rigidly. Simplicity of the forward and inverse position and velocity kinematics has always been one of the major design criteria for industrial manipulators. Hence, almost all of them have a special kinematic structure.

1.4   Robot's Kinematics, Dynamics, and Control

The forward kinematics problem is when the kinematical data are known for the joint coordinates and are utilized to find the data in the base Cartesian coordinate frame. The inverse kinematics problem is when the kinematics data are known for the end-effecter in Cartesian space and the kinematic data are needed in joint space. Inverse kinematics is highly nonlinear and usually a much more difficult problem than the forward kinematics problem. The inverse velocity and acceleration problems are linear, and much simpler, once the inverse position problem has been solved (Chernousko et al., 1994).

Kinematics, which is the English version of the French word *cinématique* from the Greek $\kappa\acute{\iota}\upsilon\eta\mu\alpha$ (movement), is a branch of science that analyzes motion with no attention to what causes the motion. By *motion* we mean any type of displacement, which includes changes in position and orientation. Therefore, *displacement*, and its successive derivatives with respect to time, velocity, acceleration, and jerk, all are included in kinematics (Erdman, 1993; Dugas, 1995).

Positioning is to bring the end-effector to an arbitrary point within dextrose, while *orientation* is to move the end-effector to the required orientation at the position. The positioning is the job of the arm, and orientation is the job of the wrist. To simplify the kinematic analysis, we may decouple the positioning and orientation of the end-effector.

In terms of the kinematic formation, a 6 DOF robot comprises six sequential moveable links and six joints. Generally speaking, almost all problems of kinematics can be interpreted as a vector addition. However, every vector in a vectorial equation must be transformed and expressed in a common reference frame. Therefore, reference and coordinate frames transformation is the main part of kinematic analysis of robots.

Dynamics is the study of systems that undergo changes of state as time evolves. In mechanical systems such as robots, the change of states involves motion. Derivation of the equations of motion for the system is the main step in dynamic analysis of the system, as the equations of motion are essential in the design, analysis, and control of the system. The dynamic equations of motion describe dynamic behavior of a robot. They can be used for computer simulation of the robot's motion, design of suitable control equations, and evaluation of the dynamic performance of the design.

Similar to kinematics, the problem of robot dynamics may be considered as *direct* and *inverse dynamics* problems. In direct dynamics, we should predict the motion of the robot for a given set of initial conditions and torques at active joints. In the inverse dynamics problem, we should compute the forces and torques necessary to generate the prescribed trajectory for a given set of positions, velocities, and accelerations.

The robot control problem may be characterized as the desired motion of the end-effector. Such a desired motion is specified as a trajectory in Cartesian coordinates while the control system requires input in joint coordinates (Fahimi, 2009).

Sensors generate data to find the actual state of the robot at joint space. This implies a requirement for expressing the kinematic variables in Cartesian space to be transformed into their equivalent joint coordinate space. These transformations are highly dependent on the kinematic geometry of the manipulator. Hence, the robot control comprises three computational problems (Aspragathos and Dimitros, 1998):

1. Determination of the trajectory in Cartesian coordinate space.
2. Transformation of the Cartesian trajectory into equivalent joint coordinate space.
3. Generation of the motor torque commands to realize the trajectory.

1.5   Principle of Kinematics

Kinematics is the study of positions, velocities, and accelerations, regardless of the forces that cause these motions. Vectors and reference frames are essential tools for analyzing motions of complex systems, especially when the motion is three dimensional and involves many parts. Kinematics is the language with which scalars, vectors, and tensors are expressed, transformed, and communicated. To make the science of kinematics, we need to define space, axes, scales, coordinate frames, and unit vectors. The communication between coordinate frames will be covered in next part of the book.

1.5.1   ★ Triad

To indicate the position of a point P relative to another point O in a three dimensional ($3D$) space, we need to establish a coordinate frame and provide three relative coordinates. The three coordinates are scalar functions and can be used to define a position vector and derive other kinematic characteristics.

Take any four non-coplanar points O, A, B, C. The *triad* $OABC$ is defined by three lines OA, OB, OC forming a rigid body. Rotate OB about O in the plane OAB such that the angle AOB becomes 90 deg, the direction of rotation of OB being such that OB moves through an angle less than 90 deg. Next, rotate OC about a perpendicular line in AOB until OC becomes perpendicular to the plane AOB, in such a way that OC moves through an angle less than 90 deg. Calling now the new position of $OABC$ an *orthogonal triad*. Any other orthogonal triad can be superposed on the $OABC$. Given an orthogonal triad $OABC$, another triad $OA'BC$ may be derived by moving A to the other side of O to make the *opposite triad* $OA'BC$.

All orthogonal triads can be superposed either on a given orthogonal triad $OABC$ or on its opposite $OA'BC$. One of the two triads $OABC$ and $OA'BC$ is defined as the *positive triad* and used as standard triad. The other is then defined as *negative triad*. It is immaterial which one is chosen as positive; however, usually the *right-handed convention* is chosen as positive, the one for which the direction of rotation from OA to OB propels a *right-handed screw* in the direction OC. A right-handed (positive) orthogonal triad cannot be superposed to a left-handed (negative) triad. Thus there are just two essentially distinct types of triad. This is an essential property of three dimensional space.

We use an orthogonal triad $OABC$ with scaled lines OA, OB, OC to locate a point in a $3D$ space. When the three lines OA, OB, OC have scales, then such a triad is called a *coordinate frame*.

Every moving link of a robot caries a *moving frame* that is rigidly attached to the link and moves with the link. A moving frame may also be called a *body frame* or *local frame*. *The body frame* accepts every motion of the link. The position and orientation of a link with respect to other frames are expressed by the position and orientation of its local coordinate frame.

When there are several relatively moving coordinate frames, we choose one of them as a *reference frame* in which we express motions of other local frames and measure kinematic information. The motion of a body may be observed and measured in different reference frames; however, we usually compare the motion of different links in the *global reference frame*. A global reference frame is assumed to be motionless and attached to the ground (Jazar, 2011).

Example 1   ★ Cyclic interchange of letters. Symmetric equations.

Cyclic interchanging of the letters ABC in any orthogonal triad $OABC$ produces another orthogonal triad superposable on the original triad. Cyclic interchanging means relabeling A as B, B as C, and C as A or selecting any three consecutive letters from $ABCABCABC\cdots$. Instead of letters, A, B, C, we may use numbers $1, 2, 3$ to indicated axes of triads. Then, assigning the axes of an orthogonal trial by any three consecutive numbers from $123123123\cdots$, makes a positive orthogonal trial. When an equation is invariant by cyclic interchanging of the letters or numbers, it is called a symmetric equation. Cyclic interchanging simplifies expression of equations.

Example 2   Right-hand rule. Examples and origin.

The right-hand rule states: If we indicate the OC axis of an orthogonal triad by the thumb of the right hand, the other fingers should turn from OA to OB to close our fist. A right-handed triad is identified by the right-hand rule.

The right-hand rule also shows the rotation of Earth when the thumb of the right hand indicates the north pole.

Point your right thumb to the center of a clock, then the other fingers simulate the rotation of the clock's hands.

Point your index finger of the right hand in the direction of an electric current. Then point your middle finger in the direction of the magnetic field. Your thumb now points the direction of the magnetic force.

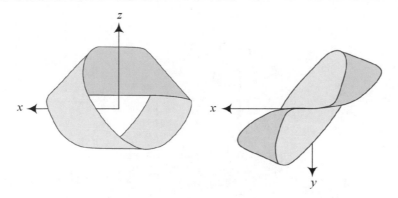

Fig. 1.21 A mobius strip

If the thumb, index finger, and middle finger of the right hand are held so that they form three right angles, then the thumb indicates the Z-axis when the index finger indicates the X-axis and the middle finger the Y-axis.

The right-hand rule was introduced by English electrical engineer and physicist Sir John Ambrose Fleming (1849–1945) in the late nineteenth century for use in electromagnetism.

Example 3 ★ Independent orthogonal coordinate frames.

Having only two types of orthogonal triads in $3D$ space is associated with the fact that a plane has only two sides. In other words, there are two opposite normal directions to a plane. This may also be interpreted as: We can arrange the letters A, B, and C in only two orders when cyclic interchange is allowed:

$$ABC \qquad ACB \tag{1.3}$$

In a $4D$ space, there are six cyclic orders for four letters A, B, C, and D:

$$ABCD, \ ABDC, \ ACBD, \ ACDB, \ ADBC, \ ADCB \tag{1.4}$$

So, there are six different *tetrads* in a $4D$ space.

In an nD space there are $(n-1)!$ cyclic orders for n letters, so there are $(n-1)!$ different coordinate frames in an nD space.

These theories are only valid in Cartesian n dimensional space, which is considered as linear Euclidean space. Non-Newtonian spaces are nonlinear and the number of possible independent coordinate frames are different (Jazar, 2011). As an example, there is only one possible coordinate frame on a Mobius strip or Mobius surface, which has only one side. Figure 1.21 illustrates a Mobius strip plotted by the following parametric equations (Fosdick and Fried, 2016):

$$x = \left(1 + \frac{\alpha}{2}\cos\frac{\beta}{2}\right)\cos\beta \tag{1.5}$$

$$y = \left(1 + \frac{\alpha}{2}\cos\frac{\beta}{2}\right)\sin\beta \tag{1.6}$$

$$z = \frac{\alpha}{2}\sin\frac{\beta}{2} \tag{1.7}$$

1.5.2 Unit Vectors

Figure 1.22 illustrates a positive orthogonal triad $OABC$. We select a *unit length* and define a *directed line* $\hat{\imath}$ on OA with the unit length. A point P_1 on OA is at a distance x from O such that the directed line $\overrightarrow{OP_1}$ from O to P_1 is x times $\hat{\imath}$, indicated by $\overrightarrow{OP_1} = x\hat{\imath}$. The directed line $\hat{\imath}$ is called a *unit vector* on OA, the unit length is called the *scale*, point O is called the *origin*, and the real number x is called the $\hat{\imath}$-*coordinate* of P_1. The distance x may also be called the $\hat{\imath}$ *measure number* of $\overrightarrow{OP_1}$.

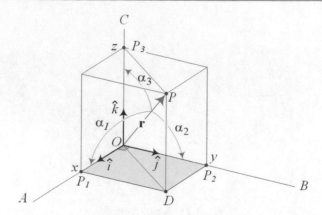

Fig. 1.22 A positive orthogonal triad $OABC$, unit vectors $\hat{\imath}, \hat{\jmath}, \hat{k}$, and a position vector $\mathbf{r}$ with components x, y, z

Similarly, we define the unit vectors $\hat{\jmath}$ on OB and $\hat{k}$ on OC and use y and z as their coordinates, respectively. Although it is not necessary, we usually use the same scale for $\hat{\imath}, \hat{\jmath}, \hat{k}$ and refer to OA, OB, OC by $\hat{\imath}, \hat{\jmath}, \hat{k}$ as well as x, y, z. The scalar coordinates x, y, z are respectively the length of projections of a point P on OA, OB, OC and may be called the *components* of $\overrightarrow{OP}$. The components x, y, z are independent and we may vary any of them while keeping the others unchanged.

Hence, a scaled positive orthogonal triad with unit vectors $\hat{\imath}, \hat{\jmath}, \hat{k}$ is called an *orthogonal coordinate frame*. The position of a point P with respect to O is defined by three coordinates x, y, z and is shown by a *position vector* $\mathbf{r} = \mathbf{r}_P$:

$$\mathbf{r} = \mathbf{r}_P = x\hat{\imath} + y\hat{\jmath} + z\hat{k} \tag{1.8}$$

We indicate coordinate frames by a capital letter, such as G and B, to clarify the coordinate frame in which the vector $\mathbf{r}$ is expressed. We show the name of the frame as a left superscript to the vector name.

$$^{B}\mathbf{r} = x\hat{\imath} + y\hat{\jmath} + z\hat{k} \tag{1.9}$$

A vector $\mathbf{r}$ is expressed in a coordinate frame B only if its unit vectors $\hat{\imath}, \hat{\jmath}, \hat{k}$ belong to the axes of B-frame. If necessary, we also use a left superscript B and show the unit vectors as $^{B}\hat{\imath}, {}^{B}\hat{\jmath}, {}^{B}\hat{k}$ to indicate that $\hat{\imath}, \hat{\jmath}, \hat{k}$ belong to B-frame:

$$^{B}\mathbf{r} = x\ {}^{B}\hat{\imath} + y\ {}^{B}\hat{\jmath} + z\ {}^{B}\hat{k} \tag{1.10}$$

However, we may drop the frame indicator superscript as long as we have only one coordinate frame.

The distance between O and P is a scalar number r that is called the *length, magnitude, modulus, norm,* or *absolute value* of the vector $\mathbf{r}$:

$$r = |\mathbf{r}| = \sqrt{x^2 + y^2 + z^2} \tag{1.11}$$

We may also define a new unit vector $\hat{u}_r$ on $\mathbf{r}$ and show $\mathbf{r}$ by

$$\mathbf{r} = r\hat{u}_r \tag{1.12}$$

The equation $\mathbf{r} = r\hat{u}_r$ is called the *natural expression* of vector $\mathbf{r}$, while the equation $\mathbf{r} = x\hat{\imath} + y\hat{\jmath} + z\hat{k}$ is called the *decomposition* or *decomposed expression* of vector $\mathbf{r}$ over the axes $\hat{\imath}, \hat{\jmath}, \hat{k}$. From Eqs. (1.11) and (1.12) we have

$$\hat{u}_r = \frac{x\hat{\imath} + y\hat{\jmath} + z\hat{k}}{r} = \frac{x\hat{\imath} + y\hat{\jmath} + z\hat{k}}{\sqrt{x^2 + y^2 + z^2}}$$

$$= \frac{x}{\sqrt{x^2 + y^2 + z^2}}\hat{\imath} + \frac{y}{\sqrt{x^2 + y^2 + z^2}}\hat{\jmath} + \frac{z}{\sqrt{x^2 + y^2 + z^2}}\hat{k} \tag{1.13}$$

Because the length of $\hat{u}_r$ is unity, the components of $\hat{u}_r$ are the cosines of the angles $\alpha_1, \alpha_2, \alpha_3$ between $\hat{u}_r$ and $\hat{\imath}, \hat{\jmath}, \hat{k}$, respectively:

$$\cos \alpha_1 = \frac{x}{r} = \frac{x}{\sqrt{x^2 + y^2 + z^2}} \tag{1.14}$$

$$\cos \alpha_2 = \frac{y}{r} = \frac{y}{\sqrt{x^2 + y^2 + z^2}} \tag{1.15}$$

$$\cos \alpha_3 = \frac{z}{r} = \frac{z}{\sqrt{x^2 + y^2 + z^2}} \tag{1.16}$$

The cosines of the angles $\alpha_1, \alpha_2, \alpha_3$ are called the *directional cosines* of $\hat{u}_r$, which, as is shown in Fig. 1.22, are the same as the directional cosines of any other vector on the same axis as $\hat{u}_r$, including **r**.

Equations (1.14)–(1.16) indicate that the three directional cosines are related by the equation

$$\cos^2 \alpha_1 + \cos^2 \alpha_2 + \cos^3 \alpha_3 = 1 \tag{1.17}$$

From definition of the unit vectors $\hat{\imath}, \hat{\jmath}, \hat{k}$, we have

$$\hat{\imath}^2 = 1 \qquad \hat{\jmath}^2 = 1 \qquad \hat{k}^2 = 1 \tag{1.18}$$

Moreover, based on the right-hand rule and definition of vector inner and outer product we have

$$\hat{\jmath} \times \hat{k} = \hat{\imath} \qquad \hat{k} \times \hat{\imath} = \hat{\jmath} \qquad \hat{\imath} \times \hat{\jmath} = \hat{k} \tag{1.19}$$

$$\hat{\imath} \cdot \hat{\jmath} = 0 \qquad \hat{\jmath} \cdot \hat{k} = 0 \qquad \hat{k} \cdot \hat{\imath} = 0 \tag{1.20}$$

Employing the inner product relationship,

$$\mathbf{r} \cdot \hat{\imath} = \left(x\hat{\imath} + y\hat{\jmath} + z\hat{k} \right) \cdot \hat{\imath} = x \tag{1.21}$$

$$\mathbf{r} \cdot \hat{\jmath} = \left(x\hat{\imath} + y\hat{\jmath} + z\hat{k} \right) \cdot \hat{\jmath} = y \tag{1.22}$$

$$\mathbf{r} \cdot \hat{k} = \left(x\hat{\imath} + y\hat{\jmath} + z\hat{k} \right) \cdot \hat{k} = z \tag{1.23}$$

we may show any vector **r** in the following form:

$$\mathbf{r} = (\mathbf{r} \cdot \hat{\imath})\hat{\imath} + (\mathbf{r} \cdot \hat{\jmath})\hat{\jmath} + (\mathbf{r} \cdot \hat{k})\hat{k} \tag{1.24}$$

Vector addition is the key operation in kinematics. Vectors can be added only when they are expressed in the same frame. Thus, a vector equation such as

$$\mathbf{a} = \mathbf{b} + \mathbf{c} \tag{1.25}$$

is meaningless without indicating the frame they are expressed in.

$$^B\mathbf{a} = {}^B\mathbf{b} + {}^B\mathbf{c} \tag{1.26}$$

Example 4 Position vector of a point P. Length, directional cosines.

Assume a point P with coordinates $x = 1, y = 2, z = 3$. The position vector of P is

$$\mathbf{r} = \hat{\imath} + 2\hat{\jmath} + 3\hat{k} \tag{1.27}$$

The distance between O and P is

$$r = |\mathbf{r}| = \sqrt{1^2 + 2^2 + 3^2} = \sqrt{14} \simeq 3.7416 \tag{1.28}$$

and the unit vector $\hat{u}_r$ on $\mathbf{r}$ is

$$\hat{u}_r = \frac{x}{r}\hat{i} + \frac{y}{r}\hat{j} + \frac{z}{r}\hat{k} = \frac{1}{3.7416}\hat{i} + \frac{2}{3.7416}\hat{j} + \frac{2}{3.7416}\hat{k}$$

$$= 0.26726\hat{i} + 0.53452\hat{j} + 0.80178\hat{k} \tag{1.29}$$

The directional cosines of $\hat{u}_r$ are

$$\cos \alpha_1 = \frac{x}{r} = 0.26726$$

$$\cos \alpha_2 = \frac{y}{r} = 0.53452 \tag{1.30}$$

$$\cos \alpha_3 = \frac{z}{r} = 0.80178$$

and therefore, the angles between $\mathbf{r}$ and the x, y, z axes are:

$$\alpha_1 = \cos^{-1}\frac{x}{r} = \cos^{-1} 0.26726 = 1.30025\,\text{rad} \approx 74.498\,\text{deg}$$

$$\alpha_2 = \cos^{-1}\frac{y}{r} = \cos^{-1} 0.53452 = 1.00685\,\text{rad} \approx 57.688\,\text{deg}$$

$$\alpha_3 = \cos^{-1}\frac{z}{r} = \cos^{-1} 0.80178 = 0.64052\,\text{rad} \approx 36.699\,\text{deg} \tag{1.31}$$

Example 5 ★ Inner and outer products of two vectors. Scalar product, vector product, right-hand rule.

We may prove the result of the inner and outer products of two vectors by using decomposed expression and expansion:

$$\mathbf{r}_1 \cdot \mathbf{r}_2 = \left(x_1\hat{i} + y_1\hat{j} + z_1\hat{k}\right) \cdot \left(x_2\hat{i} + y_2\hat{j} + z_2\hat{k}\right)$$

$$= x_1x_2\hat{i} \cdot \hat{i} + x_1y_2\hat{i} \cdot \hat{j} + x_1z_2\hat{i} \cdot \hat{k}$$

$$+ y_1x_2\hat{j} \cdot \hat{i} + y_1y_2\hat{j} \cdot \hat{j} + y_1z_2\hat{j} \cdot \hat{k}$$

$$+ z_1x_2\hat{k} \cdot \hat{i} + z_1y_2\hat{k} \cdot \hat{j} + z_1z_2\hat{k} \cdot \hat{k}$$

$$= x_1x_2 + y_1y_2 + z_1z_2 \tag{1.32}$$

$$\mathbf{r}_1 \cdot \mathbf{r}_2 = |\mathbf{r}_1|\ |\mathbf{r}_2|\cos\theta \tag{1.33}$$

$$\mathbf{r}_1 \times \mathbf{r}_2 = \left(x_1\hat{i} + y_1\hat{j} + z_1\hat{k}\right) \times \left(x_2\hat{i} + y_2\hat{j} + z_2\hat{k}\right)$$

$$= x_1x_2\hat{i} \times \hat{i} + x_1y_2\hat{i} \times \hat{j} + x_1z_2\hat{i} \times \hat{k}$$

$$+ y_1x_2\hat{j} \times \hat{i} + y_1y_2\hat{j} \times \hat{j} + y_1z_2\hat{j} \times \hat{k}$$

$$+ z_1x_2\hat{k} \times \hat{i} + z_1y_2\hat{k} \times \hat{j} + z_1z_2\hat{k} \times \hat{k}$$

$$= (y_1z_2 - y_2z_1)\,\hat{i} + (x_2z_1 - x_1z_2)\,\hat{j}$$

$$+ (x_1y_2 - x_2y_1)\,\hat{k} \tag{1.34}$$

$$\mathbf{r}_1 \times \mathbf{r}_2 = \hat{\mathbf{n}}\ |\mathbf{r}_1|\ |\mathbf{r}_2|\sin\theta \tag{1.35}$$

where, θ is the angle between $\mathbf{r}_1$ and $\mathbf{r}_2$, and $\hat{\mathbf{n}}$ is a unit vector, perpendicular to the plane of $(\mathbf{r}_1, \mathbf{r}_2)$. We may also find the outer product of two vectors by expanding a determinant and derive the same result as Eq. (1.34):

$$\mathbf{r}_1 \times \mathbf{r}_2 = \begin{vmatrix} \hat{i} & \hat{j} & \hat{k} \\ x_1 & y_1 & z_1 \\ x_2 & y_2 & z_2 \end{vmatrix} \tag{1.36}$$

The outer vector product has the following three laws:

1. Antisymmetric law:

$$\mathbf{r}_1 \times \mathbf{r}_2 = -\mathbf{r}_2 \times \mathbf{r}_1 \tag{1.37}$$

2. Associative law for scalar multiplication:

$$(c\mathbf{r}_1 \times \mathbf{r}_2) = c (\mathbf{r}_1 \times \mathbf{r}_2) \tag{1.38}$$

3. Distributive law:

$$\mathbf{r}_1 \times (\mathbf{r}_2 + \mathbf{r}_3) = \mathbf{r}_1 \times \mathbf{r}_2 + \mathbf{r}_1 \times \mathbf{r}_3 \tag{1.39}$$

Example 6 ★ bac-cab rule. Expansion of vector triple product, $\mathbf{a} \times (\mathbf{b} \times \mathbf{c})$. Lagrange formula.
    If $\mathbf{a}, \mathbf{b}, \mathbf{c}$ are three vectors, we may expand their triple cross product and show that

$$\mathbf{a} \times (\mathbf{b} \times \mathbf{c}) = \mathbf{b} (\mathbf{a} \cdot \mathbf{c}) - \mathbf{c} (\mathbf{a} \cdot \mathbf{b}) \tag{1.40}$$

$$\mathbf{a} \times (\mathbf{b} \times \mathbf{c}) = \begin{vmatrix} \hat{i} & \hat{j} & \hat{k} \\ a_1 & a_2 & a_3 \\ \begin{vmatrix} b_2 & b_3 \\ c_2 & c_3 \end{vmatrix} & \begin{vmatrix} b_3 & b_1 \\ c_3 & c_1 \end{vmatrix} & \begin{vmatrix} b_1 & b_2 \\ c_1 & c_2 \end{vmatrix} \end{vmatrix} \tag{1.41}$$

because

$$\begin{bmatrix} a_1 \\ a_2 \\ a_3 \end{bmatrix} \times \left(\begin{bmatrix} b_1 \\ b_2 \\ b_3 \end{bmatrix} \times \begin{bmatrix} c_1 \\ c_2 \\ c_3 \end{bmatrix} \right)$$

$$= \begin{bmatrix} a_2 (b_1 c_2 - b_2 c_1) + a_3 (b_1 c_3 - b_3 c_1) \\ a_3 (b_2 c_3 - b_3 c_2) - a_1 (b_1 c_2 - b_2 c_1) \\ -a_1 (b_1 c_3 - b_3 c_1) - a_2 (b_2 c_3 - b_3 c_2) \end{bmatrix}$$

$$= \begin{bmatrix} b_1 (a_1 c_1 + a_2 c_2 + a_3 c_3) - c_1 (a_1 b_1 + a_2 b_2 + a_3 b_3) \\ b_2 (a_1 c_1 + a_2 c_2 + a_3 c_3) - c_2 (a_1 b_1 + a_2 b_2 + a_3 b_3) \\ b_3 (a_1 c_1 + a_2 c_2 + a_3 c_3) - c_3 (a_1 b_1 + a_2 b_2 + a_3 b_3) \end{bmatrix}$$

$$\tag{1.42}$$

Equation (1.40) may be referred to as the *bac-cab rule*, which makes it easy to remember. It is also called Lagrange formula for vector triple product. The bac-cab rule is the most important equation in $3D$ vector algebra. It is the key to prove a great number of other theorems.
    This identity can be generalized to n dimensions space.

$$\mathbf{a}_2 \times \cdots \times \mathbf{a}_{n-1} \times (\mathbf{b}_1 \times \cdots \times \mathbf{b}_{n-1})$$

$$= (-1)^{n-1} \begin{vmatrix} \mathbf{b}_1 & \cdots\cdots & \mathbf{b}_{n-1} \\ \mathbf{a}_2 \cdot \mathbf{b}_1 & \cdots\cdots & \mathbf{a}_2 \cdot \mathbf{b}_{n-1} \\ \cdots & \cdots\cdots & \cdots \\ \mathbf{a}_{n-1} \cdot \mathbf{b}_1 & \cdots\cdots & \mathbf{a}_{n-1} \cdot \mathbf{b}_{n-1} \end{vmatrix} \tag{1.43}$$

The vector triple product also makes Jacobi symmetric identity by cyclic interchanging.

$$\mathbf{a} \times (\mathbf{b} \times \mathbf{c}) = \mathbf{b} \times (\mathbf{c} \times \mathbf{a}) = \mathbf{c} \times (\mathbf{a} \times \mathbf{b}) \tag{1.44}$$

1.5.3   Orthogonality Condition

Orthogonal coordinate frames are the most important type of coordinates and usually the only type we use in Robotics. It is compatible to our everyday life and our sense of dimensions. There is an orthogonality condition that is the principal equation to express any vector in an orthogonal coordinate frame. Consider a coordinate system $(Ouvw)$ with unit vectors $\hat{u}_u, \hat{u}_v, \hat{u}_w$. The condition for the coordinate system $(Ouvw)$ to be orthogonal is that $\hat{u}_u, \hat{u}_v, \hat{u}_w$ are mutually perpendicular and hence,

$$
\begin{aligned}
\hat{u}_u \cdot \hat{u}_v &= 0 \\
\hat{u}_v \cdot \hat{u}_w &= 0 \\
\hat{u}_w \cdot \hat{u}_u &= 0
\end{aligned}
\tag{1.45}
$$

In an orthogonal coordinate system, every vector $\mathbf{r}$ can be shown in its decomposed expression as

$$
\mathbf{r} = (\mathbf{r} \cdot \hat{u}_u)\hat{u}_u + (\mathbf{r} \cdot \hat{u}_v)\hat{u}_v + (\mathbf{r} \cdot \hat{u}_w)\hat{u}_w
\tag{1.46}
$$

We call Eq. (1.46) the *orthogonality condition* of the coordinate system $(Ouvw)$. The orthogonality condition for a Cartesian coordinate system reduces to

$$
\mathbf{r} = (\mathbf{r} \cdot \hat{\imath})\hat{\imath} + (\mathbf{r} \cdot \hat{\jmath})\hat{\jmath} + (\mathbf{r} \cdot \hat{k})\hat{k}
\tag{1.47}
$$

Proof Assume that the coordinate system $(Ouvw)$ is an orthogonal frame. Using the unit vectors $\hat{u}_u, \hat{u}_v, \hat{u}_w$ and the components u, v, w, we can show any vector $\mathbf{r}$ in the coordinate system $(Ouvw)$ as

$$
\mathbf{r} = u\,\hat{u}_u + v\,\hat{u}_v + w\,\hat{u}_w
\tag{1.48}
$$

Because of orthogonality, we have

$$
\hat{u}_u \cdot \hat{u}_v = 0 \qquad \hat{u}_v \cdot \hat{u}_w = 0 \qquad \hat{u}_w \cdot \hat{u}_u = 0
\tag{1.49}
$$

Therefore, the inner product of $\mathbf{r}$ by $\hat{u}_u, \hat{u}_v, \hat{u}_w$ would be equal to

$$
\begin{aligned}
\mathbf{r} \cdot \hat{u}_u &= \left(u\,\hat{u}_u + v\,\hat{u}_v + w\,\hat{u}_w\right) \cdot \left(1\hat{u}_u + 0\hat{u}_v + 0\hat{u}_w\right) = u \\
\mathbf{r} \cdot \hat{u}_v &= \left(u\,\hat{u}_u + v\,\hat{u}_v + w\,\hat{u}_w\right) \cdot \left(0\hat{u}_u + 1\hat{u}_v + 0\hat{u}_w\right) = v \\
\mathbf{r} \cdot \hat{u}_v &= \left(u\,\hat{u}_u + v\,\hat{u}_v + w\,\hat{u}_w\right) \cdot \left(0\hat{u}_u + 0\hat{u}_v + 1\hat{u}_w\right) = w
\end{aligned}
\tag{1.50}
$$

Substituting for the components u, v, and w in Eq. (1.48), we may show the vector $\mathbf{r}$ as

$$
\mathbf{r} = (\mathbf{r} \cdot \hat{u}_u)\hat{u}_u + (\mathbf{r} \cdot \hat{u}_v)\hat{u}_v + (\mathbf{r} \cdot \hat{u}_w)\hat{u}_w
\tag{1.51}
$$

If vector $\mathbf{r}$ is expressed in a Cartesian coordinate system, then $\hat{u}_u = \hat{\imath}$, $\hat{u}_v = \hat{\jmath}$, $\hat{u}_w = \hat{k}$, and therefore,

$$
\mathbf{r} = (\mathbf{r} \cdot \hat{\imath})\hat{\imath} + (\mathbf{r} \cdot \hat{\jmath})\hat{\jmath} + (\mathbf{r} \cdot \hat{k})\hat{k}
\tag{1.52}
$$

The orthogonality condition is the most important reason for employing orthogonal coordinate systems.   ■

1.5.4   Coordinate Frame and Transformation

In robotics, we attach at least one coordinate frame to every link of a robot and a coordinate frame to every object of the robot's environment. The relative positions and motions of the links are solely calculated by position and orientation of

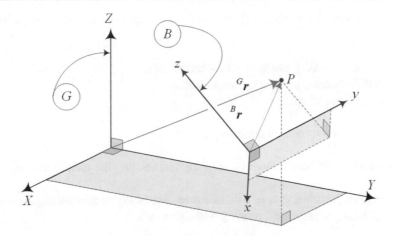

Fig. 1.23 Position vector of a point P may be decomposed in body frame B, or global frame G

their coordinate frames. Communication among the coordinate frames, which is called *transformation of frames*, is the fundamental concept in the analysis of kinematics, dynamics, modeling, and programming of a robot.

A *fixed frame* is a motionless reference frame that is attached to the ground. The motion of a robot takes place in the fixed frame called the *global reference frame*. A *moving frame* is a coordinate frame that moves with a link. Every moving link has an attached coordinate frame that sticks to the link and accepts every motion of the link. The moving coordinate frame is called the *local coordinate frame*. The position and orientation of a link with respect to the ground is expressed by the position and orientation of its local coordinate frame in the global coordinate frame. We use "reference frame," "coordinate frame," and "coordinate system" equivalently, because a Cartesian system is the only system we will use.

A *coordinate system* is different from coordinate frames. A coordinate system determines the way we describe the motion in each coordinate frame. A *Cartesian system* is the most popular coordinate system used in robotics, but cylindrical, spherical, and other systems may also be used.

The position of a point P of a rigid body B is indicated by a vector $\mathbf{r}$. As shown in Fig. 1.23, the position vector of P can be decomposed either in global coordinate frame

$$^{G}\mathbf{r} = X\hat{I} + Y\hat{J} + Z\hat{K} \tag{1.53}$$

or in body coordinate frame.

$$^{B}\mathbf{r} = x\hat{i} + y\hat{j} + z\hat{k} \tag{1.54}$$

The coefficients (X, Y, Z) and (x, y, z) are called *coordinates* or *components* of the point P in global and local coordinate frames, respectively. It is efficient for mathematical calculations to show vectors $^{G}\mathbf{r}$ and $^{B}\mathbf{r}$ by vertical arrays made by its components.

$$^{G}\mathbf{r} = \begin{bmatrix} X \\ Y \\ Z \end{bmatrix} \qquad ^{B}\mathbf{r} = \begin{bmatrix} x \\ y \\ z \end{bmatrix} \tag{1.55}$$

A coordinate frame is defined by a set of basis vectors, such as unit vectors along the three coordinate axes. Coordinate transformation is mathematically equivalent to multiplication of the coordinates of a vector in a coordinate frame, say B, by a transformation matrix R, to derive the coordinates of the same vector in another coordinate frame, say G.

$$^{G}\mathbf{r} = [R] \, ^{B}\mathbf{r} \tag{1.56}$$

So, a transformation matrix can also be viewed as defining a change of basis from one coordinate frame to another (Denavit and Hartenberg, 1955; Hunt, 1978).

The angular motion of a rigid link with respect to another link can be expressed in several ways. The most popular methods are:

1. A set of rotations about a right-handed globally fixed Cartesian axes.
2. A set of rotations about a right-handed moving Cartesian axes.
3. Angular rotation about a fixed axis in space.

A transformation matrix can be interpreted in three ways:

1. *Mapping*. It represents a coordinate transformation, mapping and relating the coordinates of a point P in two different frames.
2. *Description of a frame*. It gives the orientation of a transformed coordinate frame with respect to a fixed coordinate frame.
3. *Operator*. It is an operator taking a vector and rotating it to a new vector.

Rotation of a rigid link can be expressed by several methods such as *rotation matrix*, *Euler angles*, *angle-axis convention*, *Euler parameters*, *quaternion*, each with advantages and disadvantages.

The advantage of rotation matrix method is direct interpretation in change of basis while its disadvantage is that nine dependent parameters must be stored for only three independent parameters. The physical role of individual parameters is lost, and only the matrix as a whole has meaning.

Euler angles are defined by three successive rotations about three axes of local coordinate frames. The advantage of using Euler angles is that the rotation is described by three independent parameters with clear physical interpretations. Their disadvantage is that their representation is not unique and leads to a problem with singularities. There is also no simple way to compute multiple rotations unless they are converted into matrix representation.

Angle-axis convention is the most intuitive representation of rotations. However, it requires four parameters to store for a single rotation, computation of combined rotations is not simple, and it is ill-conditioned for small rotations.

Euler parameters and quaternions are good in preserving most of the intuition of the angle-axis representation while overcoming the ill conditioning for small rotations and admitting a group structure that allows computation of combined rotations. The disadvantage of Euler parameters and quaternion is that four parameters are needed to express a rotation. The parameterization is more complicated than angle-axis and sometimes loses physical meaning. Euler parameters and Quaternion multiplication is not as clear as matrix multiplication.

1.5.5   ★ Vector Definition

By a vector we mean any physical quantity that can be represented by a directed section of a line with a start point, such as O, and an end point, such as P. We may show a vector by an ordered pair of points with an arrow, such as $\overrightarrow{OP}$. Hence, the sign $\overrightarrow{PP}$ indicates a zero vector at point P (Jazar, 2011; Hunt, 1978).

Length and direction are necessary to have a vector; however, a vector may have up to five characteristics:

1. *Length*. The length of section OP corresponds to the magnitude of the physical quantity that the vector is representing.
2. *Axis*. A straight line that indicates the line on which the vector is sitting. The vector axis is also called the *line of action*.
3. *End point*. A start or an end point indicates the point at which the vector is applied. Such a point is called the *affecting point*.
4. *Direction*. The direction indicates at what direction on the axis the vector is pointing.
5. *Physical quantity*. Any vector represents a physical quantity. If a physical quantity can be represented by a vector, it is called a *vectorial physical quantity*. The value of the quantity is proportional to the length of the vector. Having a vector that represents no physical quantity is meaningless, although a vector may be dimensionless.

Vectors serve as the basis of our study of kinematics and dynamics. Positions, velocities, accelerations, momenta, forces, and moments all are vectors. Position vectors locate a point according to a given reference. Only vectors representing same physical quantity can be added with each other. Depending on the physical quantity and application, there are seven types of vectors:

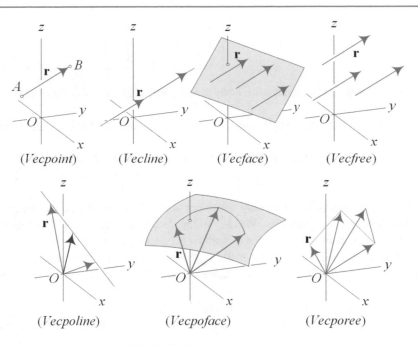

Fig. 1.24 Seven types of vectors

1. *Vecpoint.* When all of the five characteristics of length, axis, end point, direction, physical quantity, are specified, the vector is called a *bounded vector*, *point vector*, or *vecpoint*. Such a vector is fixed at a point with no movability.
2. *Vecline.* If the start and end points of a vector are not fixed on the vector axis, the vector is called a *sliding vector*, *line vector,* or *vecline*. A sliding vector is free to slide on its axis.
3. *Vecface.* When the affecting point of a vector can move on a surface while the vector displaces parallel to itself, the vector is called a *surface vector* or *vecface*. If the surface is a plane, then the vector is a *plane vector* or *veclane*.
4. *Vecfree.* If the axis of a vector is not fixed, the vector is called a *free vector*, *direction vector,* or *vecfree*. Such a vector can move to any point of a specified space while it remains parallel to itself and keeps its direction.
5. *Vecpoline.* If the start point of a vector is fixed while the end point can slide on a line, the vector is a *point-line vector* or *vecpoline*. Such a vector has a constraint variable length and orientation. However, if the start and end points of a vecpoline are on the sliding line, its orientation is fixed.
6. *Vecpoface.* If the start point of a vector is fixed, while the end point can slide on a surface, the vector is a *point-surface vector* or *vecpoface*. Such a vector has a constraint variable length and orientation. The start and end points of a vecpoface may both be on the sliding surface. If the surface is a plane, the vector is called a *point-plane vector* or *vecpolane*.
7. *Vecporee.* When the start point of a vector is fixed and the end point can move anywhere in a specified space, the vector is called a *point-free vector* or *vecporee*. Such a vector has a variable length and orientation.

Figure 1.24 illustrates a vecpoint, a vecline, a vecface, a vecfree, a vecpoline, a vecpoface, and a vecporee.

We may compare two vectors only if they represent the same physical quantity and are expressed in the same coordinate frame. Two vectors are equal if they are comparable and are the same type and have the same characteristics. Two vectors are equivalent if they are comparable and the same type and can be substituted with each other.

Example 7 Physical examples of vector types.

Moving from a point A to a point B is called the displacement. Displacement is equal to the difference of two position vectors. Displacement is a vecpoint. A position vector $\mathbf{r}_P$ of a point P starts from the origin of a coordinate frame G and ends at the point in the frame G. If point A is at $\mathbf{r}_A$ and point B at $\mathbf{r}_B$, then displacement from A to B is

$$\mathbf{r}_{A/B} = {}_B\mathbf{r}_A = \mathbf{r}_A - \mathbf{r}_B \tag{1.57}$$

Assume a point P moves from the origin of a global coordinate frame G to a point at $(2, 2, 0)$ and then moves to $(4, 3, 0)$. If we express the first displacement by a vector $\mathbf{r}_1$ and its final position by $\mathbf{r}_3$, the second displacement is $\mathbf{r}_2$.

$$\mathbf{r}_2 = \mathbf{r}_3 - \mathbf{r}_1 = \begin{bmatrix} 4 \\ 3 \\ 0 \end{bmatrix} - \begin{bmatrix} 2 \\ 2 \\ 0 \end{bmatrix} = \begin{bmatrix} 2 \\ 1 \\ 0 \end{bmatrix} \tag{1.58}$$

In Newtonian mechanics, a force can be applied on a body at any point of its axis of action and provides the same reaction. Force is a vecline.

In Newtonian mechanics, a moment can be applied on a body at any point parallel to itself and provides the same reaction. Torque is an example of vecfree.

A space curve is expressed by a vecpoline, a surface is expressed by a vecpoface, and a field is expressed by a vecporee.

Example 8 Scalars.

Any physical quantity that can be expressed by only a number is called scalar. If a physical quantity can be represented by a scalar, it is called a scalaric physical quantity. We may compare two scalars only if they represent the same physical quantity. Temperature, density, and work are examples of scalaric physical quantities.

Two scalars are equal if they represent the same scalaric physical quantity and they have the same number in the same system of units. Two scalars are equivalent if we can substitute one with the other. Scalars must be equal to be equivalent.

Example 9 ★ Vector addition, inner multiplication, and linear space.

Vectors and adding operation make a *linear space*. For any two vectors $\mathbf{r}_1$, $\mathbf{r}_2$ we have the following properties:
1. Commutative:

$$\mathbf{r}_1 + \mathbf{r}_2 = \mathbf{r}_2 + \mathbf{r}_1 \tag{1.59}$$

2. Associative:

$$\mathbf{r}_1 + (\mathbf{r}_2 + \mathbf{r}_3) = (\mathbf{r}_1 + \mathbf{r}_2) + \mathbf{r}_3 \tag{1.60}$$

3. Null element:

$$\mathbf{0} + \mathbf{r} = \mathbf{r} \tag{1.61}$$

4. Inverse element:

$$\mathbf{r} + (-\mathbf{r}) = \mathbf{0} \tag{1.62}$$

Vector addition and scalar multiplication make a linear space.

$$k_1 (k_2 \mathbf{r}) = (k_1 k_2) \mathbf{r} \tag{1.63}$$

$$(k_1 + k_2) \mathbf{r} = k_1 \mathbf{r} + k_2 \mathbf{r} \tag{1.64}$$

$$k (\mathbf{r}_1 + \mathbf{r}_2) = k \mathbf{r}_1 + k \mathbf{r}_2 \tag{1.65}$$

$$1 \cdot \mathbf{r} = \mathbf{r} \tag{1.66}$$

$$(-1) \cdot \mathbf{r} = -\mathbf{r} \tag{1.67}$$

$$0 \cdot \mathbf{r} = \mathbf{0} \tag{1.68}$$

$$k \cdot \mathbf{0} = \mathbf{0} \tag{1.69}$$

Example 10 ★ Norm of a vector and vector space.

Assume $\mathbf{r}$, $\mathbf{r}_1$, $\mathbf{r}_2$, $\mathbf{r}_3$ are four arbitrary vectors and c, c_1, c_3 are three arbitrary scalars. The *norm* of a vector $\|\mathbf{r}\|$ is defined as a real-valued function on a vector space V such that for all $\{\mathbf{r}_1, \mathbf{r}_2\} \in V$ and all $c \in \mathbb{R}$ we have

1. Positive definition: $\|\mathbf{r}\| > 0$ if $\mathbf{r} \neq 0$ and $\|\mathbf{r}\| = 0$ if $\mathbf{r} = 0$.
2. Homogeneity: $\|c\mathbf{r}\| = \|c\| \|\mathbf{r}\|$.
3. Triangle inequality: $\|\mathbf{r}_1 + \mathbf{r}_2\| \leq \|\mathbf{r}_1\| + \|\mathbf{r}_2\|$.

The most common definition of the norm of a vector is the length. The definition of norm is up to the investigator and may vary depending on the application.

$$\|\mathbf{r}\| = |\mathbf{r}| = \sqrt{r_1^2 + r_2^2 + r_3^2} \tag{1.70}$$

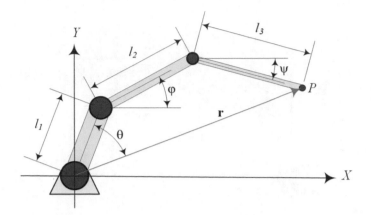

Fig. 1.25 A planar $3R$ manipulator and position vector of the tip point P in global coordinate $G(X, Y)$

The set V with vector elements is called a *vector space* if the following conditions are fulfilled:

1. Addition: If $\{\mathbf{r}_1, \mathbf{r}_2\} \in V$ and $\mathbf{r}_1 + \mathbf{r}_2 = \mathbf{r}$, then $\mathbf{r} \in V$.
2. Commutativity: $\mathbf{r}_1 + \mathbf{r}_2 = \mathbf{r}_2 + \mathbf{r}_1$.
3. Associativity: $\mathbf{r}_1 + (\mathbf{r}_2 + \mathbf{r}_3) = (\mathbf{r}_1 + \mathbf{r}_2) + \mathbf{r}_3$ and $c_1(c_2\mathbf{r}) = (c_1 c_2)\mathbf{r}$.
4. Distributivity: $c(\mathbf{r}_1 + \mathbf{r}_2) = c\mathbf{r}_1 + c\mathbf{r}_2$ and $(c_1 + c_2)\mathbf{r} = c_1\mathbf{r} + c_2\mathbf{r}$.
5. Identity element: $\mathbf{r} + \mathbf{0} = \mathbf{r}$, $1\mathbf{r} = \mathbf{r}$, and $\mathbf{r} - \mathbf{r} = \mathbf{r} + (-1)\mathbf{r} = \mathbf{0}$.

1.5.6   Vector Function

If either the magnitude of a vector $\mathbf{r}$ and/or the direction of $\mathbf{r}$ in a coordinate frame B depends on a scalar variable, say θ, then $\mathbf{r}$ is called a *vector function* of θ in B. A vector $\mathbf{r}$ may be a function of a variable in one coordinate frame, but be independent of the variable in another frame.

Example 11 Coordinate frame and vector functions.

In Fig. 1.25, P represents the tip point of a planar $3R$ robot made by three links, joined by revolute joint. The point P is ideally free to move on and in a circle with radius $l_1 + l_2 + l_3$. The variables θ, φ, and ψ are the angles of links with respect to horizon. Then the position vector $\mathbf{r}$ of the point P is a function of θ, φ, and ψ in the coordinate frame $G(X, Y)$. The length and direction of $\mathbf{r}$ depend on θ, φ, and ψ. If $G(X, Y)$, and $B_2(x, y)$ designate coordinate frames attached to the ground and link (2), and P the tip point of link (3) as shown in Fig. 1.26, then the position vector $\mathbf{r}$ of point P in coordinate frame B_2 is only a function of φ and ψ, and is independent of θ.

Example 12 Variable vectors.

There are two ways that a vector can vary: 1–length and 2–direction. A variable-length vector is a vector in the natural expression where its magnitude is variable, say of time t.

$$\mathbf{r} = r(t)\,\hat{u}_r \tag{1.71}$$

The axis of a variable-length vector is fixed.

A variable-direction vector is a vector in its natural expression where the axis of its unit vector varies. Let us use the decomposed expression of a vector $\mathbf{r}$ and show that its directional cosines are variable.

$$\mathbf{r} = r\,\hat{u}_r(t) = r\left(u_1(t)\hat{i} + u_2(t)\hat{j} + u_3(t)\hat{k}\right) \tag{1.72}$$

$$\sqrt{u_1^2 + u_2^2 + u_3^2} = 1 \tag{1.73}$$

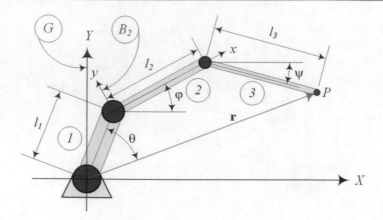

Fig. 1.26 A planar $3R$ manipulator and position vector of the tip point P in second link local coordinate $B_2(x, y)$

The axis and direction characteristics are not fixed for a variable-direction vector, while its magnitude remains constant. The end point of a variable-direction vector slides on a sphere with a center at the starting point.

A variable vector may have both the length and direction variable. Such a vector is shown in its decomposed expression with variable components:

$$\mathbf{r} = x(t)\hat{\imath} + y(t)\hat{\jmath} + z(t)\hat{k} \tag{1.74}$$

It can also be shown in its natural expression with variable length and direction:

$$\mathbf{r} = r(t)\,\hat{u}_r(t) \tag{1.75}$$

Example 13 Vector function and derivative.

If a vector is expressed in a Cartesian coordinate frame, its derivative can be found by taking derivative of its components. The Cartesian unit vectors are invariant and have zero derivative with respect to any parameter. A vector $\mathbf{r} = \mathbf{r}(t)$ is a vector function of the scalar variable t if there is a definite vector for every value of t. In a Cartesian coordinate frame G, the specification of the vector function $\mathbf{r}(t)$ is equivalent to the specification of three scalar functions $x(t)$, $y(t)$, $z(t)$ (Milne, 1948; Jazar, 2012):

$$^G\mathbf{r}(t) = x(t)\,\hat{\imath} + y(t)\,\hat{\jmath} + z(t)\,\hat{k} \tag{1.76}$$

If the vector $\mathbf{r}$ is expressed in Cartesian decomposition form, then the derivative $d\mathbf{r}/dt$ in the coordinate frame G is

$$\frac{^Gd}{dt}\,^G\mathbf{r} = \frac{dx(t)}{dt}\hat{\imath} + \frac{dy(t)}{dt}\hat{\jmath} + \frac{dz(t)}{dt}\hat{k} \tag{1.77}$$

and if $\mathbf{r}$ is expressed in its natural form,

$$^G\mathbf{r} = r\hat{u}_r = r(t)\left(u_1(t)\,\hat{\imath} + u_2(t)\,\hat{\jmath} + u_3(t)\,\hat{k}\right) \tag{1.78}$$

then, using the chain rule, the derivative $d\mathbf{r}/dt$ is

$$\begin{aligned}
\frac{^Gd}{dt}\,^G\mathbf{r} &= \frac{dr}{dt}\hat{u}_r + r\frac{d}{dt}\hat{u}_r \\
&= \frac{dr}{dt}\left(u_1\hat{\imath} + u_2\hat{\jmath} + u_3\hat{k}\right) + r\left(\frac{du_1}{dt}\hat{\imath} + \frac{du_2}{dt}\hat{\jmath} + \frac{du_3}{dt}\hat{k}\right) \\
&= \left(\frac{dr}{dt}u_1 + r\frac{du_1}{dt}\right)\hat{\imath} + \left(\frac{dr}{dt}u_2 + r\frac{du_2}{dt}\right)\hat{\jmath} \\
&\quad + \left(\frac{dr}{dt}u_3 + r\frac{du_3}{dt}\right)\hat{k}
\end{aligned} \tag{1.79}$$

If the independent variable t is time, an overdot $\dot{\mathbf{r}}(t)$ is used as a shorthand notation to indicate the time derivative (Harithuddin et al., 2015).

$$^G\mathbf{v} = \frac{^Gd}{dt}\,^G\mathbf{r}(t) = \dot{x}(t)\,\hat{\imath} + \dot{y}(t)\,\hat{\jmath} + \dot{z}(t)\,\hat{k} \tag{1.80}$$

If a variable vector $^G\mathbf{r}$ is expressed in a natural form

$$^G\mathbf{r} = r(t)\,\hat{u}_r(t) \tag{1.81}$$

we express the unit vector $\hat{u}_r(t)$ in its decomposed form

$$\begin{aligned}
^G\mathbf{r} &= r(t)\,\hat{u}_r(t) \\
&= r(t)\left(u_1(t)\,\hat{\imath} + u_2(t)\,\hat{\jmath} + u_3(t)\,\hat{k}\right)
\end{aligned} \tag{1.82}$$

and take the derivative using the chain rule and variable-length vector derivative.

$$\begin{aligned}
^G\mathbf{v} &= \frac{^Gd}{dt}\,^G\mathbf{r} = \dot{r}\hat{u}_r + r\frac{^Gd}{dt}\hat{u}_r \\
&= \dot{r}\left(u_1\hat{\imath} + u_2\hat{\jmath} + u_3\hat{k}\right) + r\left(\dot{u}_1\hat{\imath} + \dot{u}_2\hat{\jmath} + \dot{u}_3\hat{k}\right) \\
&= (\dot{r}u_1 + r\dot{u}_1)\,\hat{\imath} + (\dot{r}u_2 + r\dot{u}_2)\,\hat{\jmath} + (\dot{r}u_3 + r\dot{u}_3)\,\hat{k}
\end{aligned} \tag{1.83}$$

Consider a moving point P with a continuously varying position vector $\mathbf{r} = \mathbf{r}(t)$. When the starting point of $\mathbf{r}$ is fixed at the origin of G, its end point traces a continuous curve C. The curve C is called a *configuration path* that describes the motion of P, and the vector function $\mathbf{r}(t)$ is its vector representation. At each point of the continuously smooth curve $C = \{\mathbf{r}(t), t \in [\tau_1, \tau_2]\}$ there exists a tangent line and a derivative vector $d\mathbf{r}(t)/dt$ that is directed along the tangent line and directed toward increasing the parameter t. If the parameter is the arc length s of the curve that is measured from a convenient point on the curve, the derivative of $^G\mathbf{r}$ with respect to s is the tangential unit vector $\hat{u}_t$ to the curve at $^G\mathbf{r}$:

$$\frac{^Gd}{ds}\,^G\mathbf{r} = \hat{u}_t \tag{1.84}$$

Example 14 Velocity, acceleration, jerk, and other derivatives.

If the vector $\mathbf{r} = \,^G\mathbf{r}(t)$ is a position vector in a coordinate frame G, then its time derivative is a *velocity* vector $^G\mathbf{v}$. It shows the speed and the direction of motion of the tip point of $^G\mathbf{r}$:

$$^G\mathbf{v} = \frac{^Gd}{dt}\,^G\mathbf{r}(t) = \dot{x}(t)\,\hat{\imath} + \dot{y}(t)\,\hat{\jmath} + \dot{z}(t)\,\hat{k} \tag{1.85}$$

A time derivative of a velocity vector $^G\mathbf{v}$ is called the *acceleration* $^G\mathbf{a}$,

$$^G\mathbf{a} = \frac{^Gd}{dt}\,^G\mathbf{v}(t) = \ddot{x}(t)\,\hat{\imath} + \ddot{y}(t)\,\hat{\jmath} + \ddot{z}(t)\,\hat{k} \tag{1.86}$$

and the time derivative of an acceleration vector $^G\mathbf{a}$ is called the *jerk* $^G\mathbf{j}$,

$$^G\mathbf{j} = \frac{^Gd}{dt}\,^G\mathbf{a}(t) = \dddot{x}(t)\,\hat{\imath} + \dddot{y}(t)\,\hat{\jmath} + \dddot{z}(t)\,\hat{k} \tag{1.87}$$

As an example, consider a moving point P on a helix with position vector $\mathbf{r}$ in a coordinate frame G.

$$^G\mathbf{r}(t) = \cos(\omega t)\,\hat{\imath} + \sin(\omega t)\,\hat{\jmath} + 2t\,\hat{k} \tag{1.88}$$

The helix is uniformly turning on a circle in the (x, y)-plane while the circle is moving with a constant speed in the z-direction. Taking time derivatives show that the velocity, acceleration, and jerk of the point P are:

$$^G\mathbf{v}(t) = -\omega \sin(\omega t)\,\hat{\imath} + \omega \cos(\omega t)\,\hat{\jmath} + 2\,\hat{k} \tag{1.89}$$

$$^G\mathbf{a}(t) = -\omega^2 \cos(\omega t)\,\hat{\imath} - \omega^2 \sin(\omega t)\,\hat{\jmath} \tag{1.90}$$

$$^G\mathbf{j}(t) = \omega^3 \sin(\omega t)\,\hat{\imath} - \omega^3 \cos(\omega t)\,\hat{\jmath} \tag{1.91}$$

The derivative of acceleration or the third time derivative of the position vector $\mathbf{r}$ is called the *jerk* $\mathbf{j}$; in England the word *jolt* is used instead of jerk. The third derivative may also wrongly be called pulse, impulse, bounce, surge, shock, or super-acceleration.

In engineering, acceleration is important because of equations of motion, and jerk is important for evaluating the destructive effects of motion of a moving object. High jerk is a reason for discomfort of passengers in a vehicle. Jerk is the reason for liquid splashing from an open container. The movement of fragile objects, such as eggs, needs to be kept within specified limits of jerk to avoid damage. It is required that engineers keep the jerk of public transportation vehicles less than $2\,\mathrm{m/s^3}$ for passenger comfort (Jazar, 2017).

Although there are no universally accepted names for the fourth and higher derivatives of a position vector $\mathbf{r}$, the terms *snap* $\mathbf{s}$ and *jounce* $\mathbf{s}$ have been used for derivatives of jerk. The fifth derivative of $\mathbf{r}$ is *crackle* $\mathbf{c}$, the sixth derivative is *pop* $\mathbf{p}$, the seventh derivative is *larz* $\mathbf{z}$, the eight derivative is *bong* $\mathbf{b}$, the ninth derivative is *jeeq* $\mathbf{q}$, and the tenth derivative is *sooz* $\mathbf{u}$ (Jazar, 2011). Not much application for fourth and higher derivatives of position vector have been reported.

1.6 Summary

There are two kinds of robots: serial and parallel. A serial robot is made from a series of rigid links, where each pair of links is connected by a revolute (R) or prismatic (P) joint. An R or P joint provides only one degree of freedom, which is rotational or translational, respectively. The final link of a robot, also called the end-effector, is the operating member of a robot that interacts with the environment.

To reach any point in a desired orientation, within a robot's workspace, a robot needs at least $6\ DOF$. Hence, it must have at least 6 links and 6 joints. Most robots use $3\ DOF$ to position the wrist point, and use the other $3\ DOF$ to orient the end-effector about the wrist point.

We attach a Cartesian coordinate frame to each link of a robot and determine the position and orientation of each frame with respect to the others. Therefore, to determine the position and orientation of the end-effector, we need to find the end-effector frame in the base frame.

Any physical quantity that can be represented by a directed section of a line with a start and an end point is a vector quantity. A vector may have five characteristics: length, axis, end point, direction, and physical quantity. The length and direction are necessary. There are seven types of vectors: vecpoint, vecline, vecface, vecfree, vecpoline, vecpoface, and pecporee. Vectors can be added when they are coaxial. In case the vectors are not coaxial, the decomposed expression of vectors must be used to add the vectors.

1.7   Key Symbols

$\mathbf{0}$	Zero vector
$a, \ddot{x}, \mathbf{a}, \dot{\mathbf{v}}$	Acceleration
a_{ijk}	Inner product constant of $\mathbf{x}_i$
$\mathbf{a}, \mathbf{b}, \mathbf{c}, \mathbf{p}, \mathbf{q}$	Vectors, constant vectors
A, B	Points
A, B, C	Axes of triad, constant parameters
A, B, C, D	Axes of tetrad, constant parameters
$\mathbf{b}$	Bong
$B(oxyz), B_1, B_2$	Body coordinate frames
$\mathbf{c}$	Crackle
C	Space curve
$d\mathbf{r}$	Infinitesimal displacement
ds	Arc length element
$^G\mathbf{d}_o$	Position vector of the origin of B in G
D	Dimension
g	Gravitational acceleration
G	Gravitational constant
$G(OXYZ)$	Global coordinate frame
$\hat{\imath}, \hat{\jmath}, \hat{k}$	Unit vectors of a Cartesian coordinate frame
$\hat{I}, \hat{J}, \hat{K}$	Unit vectors of a global Cartesian system G
$\mathbf{j}, \dot{a}, \ddot{v}, \dddot{r}$	Jerk
k	Scalar coefficient
l	Length, a line
n	Number of dimensions of an nD space
$\hat{n}$	Perpendicular unit vector
O	Origin of a triad, origin of a coordinate frame
$OABC$	A triad with axes A, B, C
$(Ouvw)$	An orthogonal coordinate frame
$(Oq_1q_2q_3)$	An orthogonal coordinate system
P	Prismatic joint, point, particle
q, p	Parameters, variables
$\mathbf{q} = \dot{\mathbf{b}}$	Jeeq
$r = |\mathbf{r}|$	Length of $\mathbf{r}$
$\mathbf{r}$	Position vector
$_B\mathbf{r}_A$	Position vector of point A relative to B
$\mathbf{r}_p, \mathbf{r}_q$	Partial derivatives of $^G\mathbf{r}$
R	Revolute joint, transformation matrix
s	Arc length parameter
$\mathbf{s} = d\mathbf{j}/dt$	Snap, jounce
S	Screw joint, surface
t	Time
T	Translatory
u, v, w	Components of a vector $\mathbf{r}$ in $(Ouvw)$
u_1, u_2, u_3	Components of $\hat{u}_r$
$\hat{u}^T$	Transpose of $\hat{u}$
$\hat{u}_l$	Unit vector on a line l
$\hat{u}_r$	A unit vector on $\mathbf{r}$
$\hat{u}_1, \hat{u}_2, \hat{u}_3$	Unit vectors along the axes q_1, q_2, q_3
$\hat{u}_u, \hat{u}_v, \hat{u}_w$	Unit vectors of $(Ouvw)$
$\mathbf{u}$	Sooz
v	Speed

$v, \dot{x}, \mathbf{v}$	Velocity
V	Vector space
x, y, z	Axes of orthogonal Cartesian coordinate frames
$\mathbf{x}$	Vector functions, unknown vector
X, Y, Z	Global coordinate axes
$\mathbf{z} = \dot{\mathbf{þ}}$	Larz
Z	Short notation symbol

Greek

α	Angle between two vectors
α, β, γ	Directional cosines of a line, angles
$\alpha_1, \alpha_2, \alpha_3$	Directional cosines of $\mathbf{r}$ and $\hat{u}_r$
θ, φ, ψ	Angle, angular coordinate, angular parameter, joint variable

Symbol

$\cdot$	Inner product of two vectors
$\times$	Outer product of two vectors
$\vdash$	Orthogonal
$\parallel$	Parallel sign
$\perp$	Perpendicular
$\rightarrow$	Vector sign
$\mathbf{þ}$	Pop

Exercises

1. Meaning of indexes for a position vector.
   Explain the meaning of ${}^G\mathbf{r}_P$, ${}^G_B\mathbf{r}_P$, and ${}^G_G\mathbf{r}_P$, if $\mathbf{r}$ is the position vector of a point P.
2. Meaning of indexes for a velocity vector.
   Explain the meaning of ${}^G\mathbf{v}_P$, ${}^G_B\mathbf{v}_P$, and ${}^B_B\mathbf{v}_P$, if $\mathbf{v}$ is a velocity vector.
3. Meaning of indexes for an angular velocity vector.
   Explain the meaning of ${}^{B_2}\boldsymbol{\omega}_{B_1}$, ${}^G_{B_2}\boldsymbol{\omega}_{B_1}$, ${}^B_G\boldsymbol{\omega}_B$, and ${}^{B_2}_{B_3}\boldsymbol{\omega}_{B_1}$, if $\boldsymbol{\omega}$ is an angular velocity vector.
4. Meaning of indexes for a transformation matrix.
   Explain the meaning of ${}^{B_2}T_{B_1}$, and ${}^G T_B$, if T is a transformation matrix.
5. Laws of robotics.
   What is the difference between law zero and law one of robotics?
6. New law of robotics.
   What do you think about adding a fourth law to robotics, such as: A robot must protect the other robots as long as such protection does not conflict with a higher order law.
7. Robot classification.
   Explain why a crane is not considered as a robot.
8. Robot market.
   Most small robot manufacturers went out of the market around 1990, and only a few large companies remained. Why do you think this happened?
9. Humanoid robots.
   The mobile robot industry is trying to make robots as similar to humans as possible. What do you think is the reason for this. Does humans have the best structural design? Does humans have the simplest design? Do these kinds of robots can be sold in the market better?
10. Robotic person.
    Why do you think we call somebody who works or behaves mechanically, showing no emotion and often responding to orders without question, a robotic person?
11. Robotic journals.
    Find the name of 10 technical journals related to robotics.
12. Number of robots in industrial poles.
    Search the robotic literatures and find out, approximately, how many industrial robots are currently in operation in the USA, Europe, and Japan.
13. Robotic countries.
    Search the robotic literatures and find out what countries are ranked first, second, and third according to the number of industrial robots in use?
14. Advantages and disadvantages of robots.
    Search the robotic literatures and name 10 advantages and 2 disadvantages of robots compared to mechanisms.
15. Mechanisms and robots.
    Why do we not replace every mechanism, in an assembly line for example, with robots? Are robots substituting mechanisms?
16. ★ Higher pairs and lower pairs.
    Joints can be classified as *lower pairs* and *higher pairs*. Find the meaning of "lower pairs" and "higher pairs" in mechanics of machinery.
17. Human wrist DOF.
    Examine and count the number of DOF of your wrist.
18. Human hand is a redundant manipulator.
    An arm (including shoulder, elbow, and wrist) has 7 DOF. What is the advantage of having one extra DOF (with respect to 6 DOF) in our hands?
19. Multiple DOF robot.
    Sometimes we do not need many DOF robots to do a specific job that can be done by a low DOF robot. What do you think about having a robot with variable DOF?
20. Usefulness of redundant manipulators.
    Discuss possible applications in which the redundant manipulators would be useful.

21. ★ Disadvantages of a non-orthogonal triad.

    Why do we use an orthogonal triad to define a Cartesian space? Can we define a $3D$ space with non-orthogonal triads?

22. ★ Usefulness of an orthogonal triad.

    Orthogonality is the common property of all useful coordinate systems such as Cartesian, cylindrical, spherical, parabolical, and ellipsoidal coordinate systems. Why do we only define and use orthogonal coordinate systems? Do you think ability to define a vector, based on inner product and unit vectors of the coordinate system, such as

    $$\mathbf{r} = (\mathbf{r} \cdot \hat{\imath})\hat{\imath} + (\mathbf{r} \cdot \hat{\jmath})\hat{\jmath} + (\mathbf{r} \cdot \hat{k})\hat{k} \tag{1.92}$$

    is the main reason for defining the orthogonal coordinate systems?

23. ★ Three coplanar vectors.

    Show that if $\mathbf{a} \times \mathbf{b} \cdot \mathbf{c} = 0$, then $\mathbf{a}$, $\mathbf{b}$, $\mathbf{c}$ are coplanar.

24. ★ Vector function, vector variable.

    A vector function is defined as a dependent vectorial variable that relates to a scalar independent variable.

    $$\mathbf{r} = \mathbf{r}(t) \tag{1.93}$$

    Describe the meaning and define an example for a vector function of a vector variable,

    $$\mathbf{a} = \mathbf{a}(\mathbf{b}) \tag{1.94}$$

    and a scalar function of a vector variable.

    $$f = f(\mathbf{b}) \tag{1.95}$$

25. ★ Frame dependent and frame independent.

    A vector function of scalar variables is a frame-dependent quantity. Is a vector function of vector variables frame dependent? What about a scalar function of vector variables?

26. ★ Coordinate frame and vector function.

    Explain the meaning of ${}^{B}\mathbf{v}_P({}^{G}\mathbf{r}_P)$, if $\mathbf{r}$ is a position vector, $\mathbf{v}$ is a velocity vector, and $\mathbf{v}(\mathbf{r})$ means $\mathbf{v}$ is a function of $\mathbf{r}$.

27. Vector multiplication.

    Prove that:

    $$\mathbf{r}^2 = \mathbf{r} \cdot \mathbf{r} = r^2 \tag{1.96}$$

    $$(\mathbf{r}_1 + \mathbf{r}_2)^2 = \mathbf{r}_1^2 + 2\mathbf{r}_1 \cdot \mathbf{r}_2 + \mathbf{r}_2^2 \tag{1.97}$$

    $$(\mathbf{r}_1 - \mathbf{r}_2) \cdot (\mathbf{r}_1 + \mathbf{r}_2) = \mathbf{r}_1^2 - \mathbf{r}_2^2 \tag{1.98}$$

28. Vector equation.

    Assume $\mathbf{x}$ is an unknown vector, k is a scalar, and $\mathbf{a}$, $\mathbf{b}$, and $\mathbf{c}$ are three constant vectors in the following vector equation:

    $$k\mathbf{x} + (\mathbf{b} \cdot \mathbf{x})\,\mathbf{a} = \mathbf{c} \tag{1.99}$$

    (a) Solve the equation for $\mathbf{x}$.
    (b) Solve for $\mathbf{x}$.

    $$a\mathbf{x} + \mathbf{x} \times \mathbf{b} = \mathbf{c} \tag{1.100}$$

29. Vector outer product.

    Consider a point P at $x = 10$, $y = 15$, $z = 5$.

    (a) Write the position vector of P in natural expression,

    $$\mathbf{r}_P = r\,\hat{u}_r \tag{1.101}$$

    and determine $\hat{u}_r$ and r.

(b) Determine the directional cosines of $\mathbf{r}_P$.

(c) Determine a vector $\mathbf{r}_Q$, perpendicular to $\mathbf{r}_P$ such that their outer product is a vector in z-direction with a length of 10π.

30. Vectors and geometry.

(a) Assume $\mathbf{r}_Q$ and $\mathbf{r}_P$ to be position vectors of points Q and P. Employing vector arithmetic, determine an equation to calculate the area A of the triangle OPD.

(b) Determine a vector from O and perpendicular to PQ. The length of that vector would be the height of the triangle OPD.

(c) Assume $\mathbf{r}_M = \mathbf{r}_P \times \mathbf{r}_Q$. Determine area of triangle MPQ.

(d) Assume $\mathbf{r}_N = \mathbf{r}_Q \times \mathbf{r}_P$. Determine area of triangle NPQ.

31. Cosine law.

Consider a triangle ABC where its sides are expressed by vectors as

$$\overrightarrow{AB} = \mathbf{c} \qquad \overrightarrow{AC} = \mathbf{b} \qquad \overrightarrow{CB} = \mathbf{a} \qquad \mathbf{c} = \mathbf{a} + \mathbf{b} \tag{1.102}$$

Use vector algebra and prove the cosine law,

$$c^2 = a^2 + b^2 - 2ab\cos\alpha \tag{1.103}$$

where

$$\alpha = \angle ACB \tag{1.104}$$

32. A trigonometric equation.

Assume two planar vectors $\mathbf{a}$ and $\mathbf{b}$ that respectively make angles α and β with the x-axis. Prove the following trigonometric identity:

$$\cos(\alpha - \beta) = \cos\alpha \cos\beta + \sin\alpha \sin\beta \tag{1.105}$$

33. ★ Spherical trigonometric equations.

Prove the following spherical trigonometric equations in a spherical triangle $\triangle ABC$ with sides a, b, c and angle α, β, γ:

$$\cos a = \cos b \cos c + \sin b \sin c \cos\alpha \tag{1.106}$$

$$\cos b = \cos c \cos a + \sin c \sin a \cos\beta \tag{1.107}$$

$$\cos c = \cos a \cos b + \sin a \sin b \cos\gamma \tag{1.108}$$

34. Three colinear points.

Consider three points A, B, and C indicated by vectors $\mathbf{a}$, $\mathbf{b}$, and $\mathbf{c}$. If the points are colinear, then

$$\frac{c_x - a_x}{b_x - a_x} = \frac{c_y - a_y}{b_y - a_y} = \frac{c_z - a_z}{b_z - a_z} \tag{1.109}$$

Show that this condition can be expressed by

$$(\mathbf{a} \times \mathbf{b}) + (\mathbf{b} \times \mathbf{c}) + (\mathbf{c} \times \mathbf{a}) = 0 \tag{1.110}$$

35. A derivative identity.

Assume a variable vector $\mathbf{a} = \mathbf{a}(t)$ and a constant vector $\mathbf{b}$ to prove

$$\frac{d}{dt}(\mathbf{a} \cdot (\dot{\mathbf{a}} \times \mathbf{b})) = \mathbf{a} \cdot (\ddot{\mathbf{a}} \times \mathbf{b}) \tag{1.111}$$

36. ★ Lagrange and Jacobi identities.

(a) Show that for any four vectors $\mathbf{a}, \mathbf{b}, \mathbf{c}, \mathbf{d}$, the Lagrange identity is correct:

$$(\mathbf{a} \times \mathbf{b}) \cdot (\mathbf{c} \times \mathbf{d}) = (\mathbf{a} \cdot \mathbf{c})(\mathbf{b} \cdot \mathbf{d}) - (\mathbf{a} \cdot \mathbf{d})(\mathbf{b} \cdot \mathbf{c}) \tag{1.112}$$

(b) Show that for any four vectors **a**, **b**, **c**, **d** the following identities are correct:

$$(\mathbf{a} \times \mathbf{b}) \times (\mathbf{c} \times \mathbf{d}) = [\mathbf{abd}]\,\mathbf{c} - [\mathbf{abc}]\,\mathbf{d} \tag{1.113}$$

$$[\mathbf{abc}]\,\mathbf{d} = [\mathbf{dbc}]\,\mathbf{a} + [\mathbf{dca}]\,\mathbf{b} + [\mathbf{dab}]\,\mathbf{c} \tag{1.114}$$

where, [**abc**] is the scalar triple product. This product for the vectors $\mathbf{r}_1$, $\mathbf{r}_2$, $\mathbf{r}_3$ is shown by the $[\mathbf{r}_1\mathbf{r}_2\mathbf{r}_3]$ and is equal to

$$[\mathbf{r}_1\mathbf{r}_2\mathbf{r}_3] = \mathbf{r}_1 \cdot \mathbf{r}_2 \times \mathbf{r}_3 \tag{1.115}$$

(c) Show that for any four vectors **a**, **b**, **c**, **d**, the Jacobi identity is correct:

$$\mathbf{a} \times (\mathbf{b} \times \mathbf{c}) + \mathbf{c} \times (\mathbf{a} \times \mathbf{b}) + \mathbf{b} \times (\mathbf{c} \times \mathbf{a}) = 0 \tag{1.116}$$

37. ★ Moving a point on a given curve.

Assume a particle is moving on a planar curve $y = f(x)$ such that the x-component of its velocity vector **v**, remains constant. Determine the acceleration and jerk of the particle if:

(a) $y = x^2$

(b) $y = x^3$

(c) $y = e^x$

(d) Determine the angle between velocity vectors of curves (a) and (b) at their intersection.

Kinematics is the science of *geometry in motion*. It is restricted to a pure geometrical description of motion by means of position, orientation, and their time derivatives. In robotics, the kinematic descriptions of manipulators and their assigned tasks are utilized to set up the fundamental equations for dynamics and control.

Because the links and arms in a robotic system are modeled as rigid bodies, the properties of rigid body displacement take a central place in robotics. Vector and matrix algebra are utilized to develop a systematic and generalized approach to describe and present the location of the arms of a robot with respect to a global fixed reference frame G. Because the arms of a robot may rotate or translate with respect to each other, body-attached coordinate frames A, B, C, $\cdots$ or B_1, B_2, B_3, $\cdots$ will be established for each link to find their relative configurations, and in the reference frame G. The position of a link B relative to another link A is defined kinematically by a coordinate transformation ^AT_B between coordinate frames attached to the links.

The direct kinematics problem is reduced to finding a transformation matrix ^GT_B that relates the body local coordinate frame B to the global reference coordinate frame G. A 3×3 rotation matrix is utilized to describe the rotational operations of the local frame with respect to the global frame. The homogenous coordinates are then introduced to represent position vectors and directional vectors in a three dimensional space. The rotation matrices are expanded to 4×4 homogenous transformation matrices to include both, rotational and translational motions. Homogenous matrices that express the relative rigid links of a robot are made by a special set of rules and are called *Denavit–Hartenberg matrices* after Denavit and Hartenberg (1955). The advantage of using the Denavit–Hartenberg method is its algorithmic and universality in deriving the kinematic equation of robot links.

The analytical description of displacement of a rigid body is based on the notion that all points in a rigid body must retain their original relative positions regardless of the new position and orientation of the body. The total rigid body displacement can always be reduced to the sum of its two basic components: the translation displacement of an arbitrary reference point fixed in the rigid body plus the unique rotation of the body about a line through that point.

Study of motion of rigid bodies leads to the relation between the time rate of change of a vector in a global frame and the time rate of change of the same vector in a local frame. *Transformation* from a local coordinate frame B to a global coordinate frame G is expressed by

$$^G\mathbf{r} = {}^GR_B\,{}^B\mathbf{r} + {}^G\mathbf{d}_B$$

where $^B\mathbf{r}$ is the position vector of a point in B, $^G\mathbf{r}$ is the position vector of the same point expressed in G, and $\mathbf{d}$ is the position vector of the origin o of the body coordinate frame $B(oxyz)$ with respect to the origin O of the global coordinate frame $G(OXYZ)$. Therefore, a transformation has two parts: a *translation* $\mathbf{d}$ that brings the origin o on the origin O, plus a *rotation*, ^GR_B that brings the axes of $oxyz$ on the corresponding axes of $OXYZ$.

The transformation formula $^G\mathbf{r} = {}^GR_B\,{}^B\mathbf{r} + {}^G\mathbf{d}_B$ can be expanded to connect more than two coordinate frames. The combination formula for a transformation from a local coordinate frame B_1 to another coordinate frame B_2 followed by a transformation from B_2 to the global coordinate G is

$$^G\mathbf{r} = \left({}^GR_{B_2}\,{}^{B_2}\mathbf{r} + {}^G\mathbf{d}_{B_2}\right) + \left({}^{B_2}R_{B_1}\,{}^{B_1}\mathbf{r} + {}^{B_2}\mathbf{d}_{B_1}\right)$$

$$= {}^GR_{B_2}\,{}^{B_2}R_{B_1}\,{}^{B_1}\mathbf{r} + {}^G\mathbf{d}_{B_2} + {}^{B_2}\mathbf{d}_{B_1}$$

$$= {}^GR_{B_1}\,{}^{B_1}\mathbf{r} + {}^G\mathbf{d}_{B_1}$$

A robot consists of n rigid links with relative motions. The link attached to the ground is link (0) and the link attached to the final moving link, the end-effector, is link (n). There are two important problems in kinematic analysis of robots: the *forward kinematics problem* and the *inverse kinematics problem*.

In forward kinematics, the problem is that the position vector of a point P is in the coordinate frame B_n attached to the end-effector at $^n\mathbf{r}_p$, and we are looking for the position of P in the base frame B_0 shown by $^0\mathbf{r}_p$. The forward kinematics problem is equivalent to having the values of joint variables and asking for the position of the end-effector.

In inverse kinematics, the problem is that we have the position vector of a point P in the base coordinate frame B_0 as $^0\mathbf{r}_p$, and we are looking for $^n\mathbf{r}_p$, the position of P in the end-effector frame B_n. This problem is equivalent to having the position of the end-effector and asking for a set of joint variables that make the robot reach the point P.

In this part, we develop the transformation formula to move the kinematic information back and forth from a coordinate frame to another coordinate frame.

Rotation Kinematics

Consider a rigid body B with a fixed point O. Rotation about the fixed point O is the only possible motion of the body. We represent the rigid body by a body coordinate frame $B\,(O, x, y, z)$, which rotates in another coordinate frame $G\,(O, X, Y, Z)$, as is shown in Fig. 2.1. In this chapter, we develop the rotation calculus based on transformation matrices to determine the orientation of B in G and relate the coordinates of a body point P in both frames.

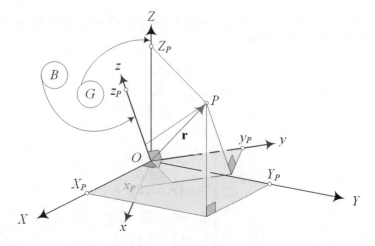

Fig. 2.1 A rotated body frame B in a fixed global frame G, about a fixed point at O

2.1 Rotation About Global Cartesian Axes

Consider a rigid body B with a local coordinate frame $Oxyz$ that is originally coincident with a global coordinate frame $OXYZ$. Point O is the origin of both coordinate frames and is the only point of the body B to be fixed in the global frame G. If the rigid body B rotates α radians about the Z-axis of the global coordinate frame, then coordinates of any point P of the rigid body in the local and global coordinate frames are related by

$$^{G}\mathbf{r} = Q_{Z,\alpha}\,{}^{B}\mathbf{r} \tag{2.1}$$

where $^{B}\mathbf{r}$ is the B-expression of the position vector of the point P in B, and $^{G}\mathbf{r}$ is the G-expression of the same vector,

$$^{G}\mathbf{r} = \begin{bmatrix} X \\ Y \\ Z \end{bmatrix} \qquad ^{B}\mathbf{r} = \begin{bmatrix} x \\ y \\ z \end{bmatrix} \tag{2.2}$$

© The Author(s), under exclusive license to Springer Nature Switzerland AG 2022
R. N. Jazar, *Theory of Applied Robotics*, https://doi.org/10.1007/978-3-030-93220-6_2

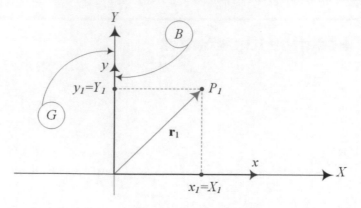

Fig. 2.2 Position vector of point P when local and global frames are coincident

and $Q_{Z,\alpha}$ is the Z-rotation transformation matrix.

$$Q_{Z,\alpha} = \begin{bmatrix} \cos\alpha & -\sin\alpha & 0 \\ \sin\alpha & \cos\alpha & 0 \\ 0 & 0 & 1 \end{bmatrix} \tag{2.3}$$

Similarly, rotation β about the Y-axis and rotation γ about the X-axis of the global frame relate the local and global coordinates of point P by the following equations:

$$^{G}\mathbf{r} = Q_{Y,\beta}\,^{B}\mathbf{r} \tag{2.4}$$

$$^{G}\mathbf{r} = Q_{X,\gamma}\,^{B}\mathbf{r} \tag{2.5}$$

where

$$Q_{Y,\beta} = \begin{bmatrix} \cos\beta & 0 & \sin\beta \\ 0 & 1 & 0 \\ -\sin\beta & 0 & \cos\beta \end{bmatrix} \tag{2.6}$$

$$Q_{X,\gamma} = \begin{bmatrix} 1 & 0 & 0 \\ 0 & \cos\gamma & -\sin\gamma \\ 0 & \sin\gamma & \cos\gamma \end{bmatrix} \tag{2.7}$$

Equations (2.1), (2.4), and (2.5) show how to go from B coordinate frame to the G coordinate frame. In other words, we have coordinates of a point of the B-frame as $^{B}\mathbf{r}$ and we are to determine its coordinates in the G-frame as $^{G}\mathbf{r}$.

Proof Let $(\hat{\imath}, \hat{\jmath}, \hat{k})$ and $(\hat{I}, \hat{J}, \hat{K})$ be the unit vectors along the coordinate axes of $Oxyz$ and $OXYZ$, respectively. The rigid body has a globally fixed point O, which is the common origin of $Oxyz$ and $OXYZ$. Figure 2.2 illustrates the top view of the system when the coordinate frames B and G are initially coincident. The initial position of a point P is indicated by P_1. The position vector $\mathbf{r}_1$ of P_1 can be expressed in body and global coordinate frames by

$$^{B}\mathbf{r}_1 = x_1\hat{\imath} + y_1\hat{\jmath} + z_1\hat{k} \tag{2.8}$$

$$^{G}\mathbf{r}_1 = X_1\hat{I} + Y_1\hat{J} + Z_1\hat{K} \tag{2.9}$$

where

$$x_1 = X_1 \qquad y_1 = Y_1 \qquad z_1 = Z_1 \tag{2.10}$$

The vector $^{B}\mathbf{r}_1$ refers to the position vector $\mathbf{r}_1$ expressed in the body coordinate frame B, and $^{G}\mathbf{r}_1$ refers to the position vector $\mathbf{r}_1$ expressed in the global coordinate frame G.

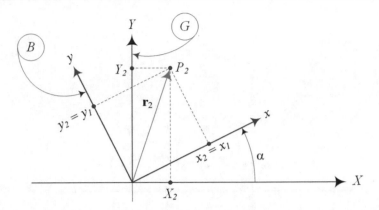

Fig. 2.3 Position vector of point P when the body and its frame B are rotated α about the Z-axis

If the rigid body undergoes a rotation of α radians about the Z-axis, then the local frame $Oxyz$, point P, and the position vector $\mathbf{r}$ will be seen in a second position, as shown in Fig. 2.3. Now the position vector $\mathbf{r}_2$ of P_2 is expressed in both coordinate frames by

$$^{B}\mathbf{r}_2 = x_2\hat{\imath} + y_2\hat{\jmath} + z_2\hat{k} \tag{2.11}$$

$$^{G}\mathbf{r}_2 = X_2\hat{I} + Y_2\hat{J} + Z_2\hat{K} \tag{2.12}$$

Using the definition of the inner product and Eq. (2.11), we may write

$$X_2 = \hat{I} \cdot \mathbf{r}_2 = \hat{I} \cdot x_2\hat{\imath} + \hat{I} \cdot y_2\hat{\jmath} + \hat{I} \cdot z_2\hat{k} \tag{2.13}$$

$$Y_2 = \hat{J} \cdot \mathbf{r}_2 = \hat{J} \cdot x_2\hat{\imath} + \hat{J} \cdot y_2\hat{\jmath} + \hat{J} \cdot z_2\hat{k} \tag{2.14}$$

$$Z_2 = \hat{K} \cdot \mathbf{r}_2 = \hat{K} \cdot x_2\hat{\imath} + \hat{K} \cdot y_2\hat{\jmath} + \hat{K} \cdot z_2\hat{k} \tag{2.15}$$

or equivalently,

$$\begin{bmatrix} X_2 \\ Y_2 \\ Z_2 \end{bmatrix} = \begin{bmatrix} \hat{I}\cdot\hat{\imath} & \hat{I}\cdot\hat{\jmath} & \hat{I}\cdot\hat{k} \\ \hat{J}\cdot\hat{\imath} & \hat{J}\cdot\hat{\jmath} & \hat{J}\cdot\hat{k} \\ \hat{K}\cdot\hat{\imath} & \hat{K}\cdot\hat{\jmath} & \hat{K}\cdot\hat{k} \end{bmatrix} \begin{bmatrix} x_2 \\ y_2 \\ z_2 \end{bmatrix} \tag{2.16}$$

The elements of the Z-*rotation matrix*, $Q_{Z,\alpha}$, are *direction cosines* of $^{B}\mathbf{r}_2$ with respect to $OXYZ$. Figure 2.4 shows the top view of the initial and final configurations of $\mathbf{r}$ in both coordinate systems $B\,(Oxyz)$ and $G\,(OXYZ)$. Figure 2.4 indicates that

$$\begin{array}{lll} \hat{I}\cdot\hat{\imath} = \cos\alpha & \hat{I}\cdot\hat{\jmath} = \cos(\pi+\alpha) & \hat{I}\cdot\hat{k} = \cos\pi = 0 \\ & \qquad = -\sin\alpha & \\ \hat{J}\cdot\hat{\imath} = \cos(\pi-\alpha) & \hat{J}\cdot\hat{\jmath} = \cos\alpha & \hat{J}\cdot\hat{k} = \cos\pi = 0 \\ \qquad = \sin\alpha & & \\ \hat{K}\cdot\hat{\imath} = \cos\pi = 0 & \hat{K}\cdot\hat{\jmath} = \cos\pi = 0 & \hat{K}\cdot\hat{k} = \cos 0 = 1 \end{array} \tag{2.17}$$

Combining Eqs. (2.16) and (2.17) shows that we can find the components of $^{G}\mathbf{r}_2$ by multiplying the Z-rotation matrix, $Q_{Z,\alpha}$, and the vector $^{B}\mathbf{r}_2$.

$$\begin{bmatrix} X_2 \\ Y_2 \\ Z_2 \end{bmatrix} = \begin{bmatrix} \cos\alpha & -\sin\alpha & 0 \\ \sin\alpha & \cos\alpha & 0 \\ 0 & 0 & 1 \end{bmatrix} \begin{bmatrix} x_2 \\ y_2 \\ z_2 \end{bmatrix} \tag{2.18}$$

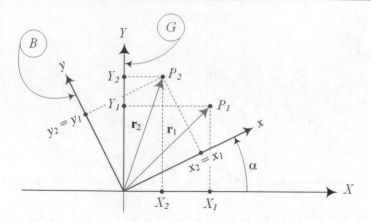

Fig. 2.4 Position vectors of point P before and after the rotation α of the local frame B about the Z-axis of the global frame G

It can also be shown in a short notation.

$$^G\mathbf{r}_2 = Q_{Z,\alpha} \, ^B\mathbf{r}_2 \tag{2.19}$$

Equation (2.19) says that the vector $\mathbf{r}$ at the second position in the global coordinate frame $^G\mathbf{r}$ is equal to Q_Z times the position vector in the local coordinate frame $^B\mathbf{r}$. Hence, we are able to find the global coordinates of a point of a rigid body after rotation about the Z-axis, if we have its local coordinates.

Similarly, rotation β about the Y-axis and rotation γ about the X-axis are expressed by the *Y-rotation matrix $Q_{Y,\beta}$* and the *X-rotation matrix $Q_{X,\gamma}$*. The rotation matrices $Q_{Z,\alpha}$, $Q_{Y,\beta}$, and $Q_{X,\gamma}$

$$Q_{Z,\alpha} = \begin{bmatrix} \cos\alpha & -\sin\alpha & 0 \\ \sin\alpha & \cos\alpha & 0 \\ 0 & 0 & 1 \end{bmatrix} \tag{2.20}$$

$$Q_{Y,\beta} = \begin{bmatrix} \cos\beta & 0 & \sin\beta \\ 0 & 1 & 0 \\ -\sin\beta & 0 & \cos\beta \end{bmatrix} \tag{2.21}$$

$$Q_{X,\gamma} = \begin{bmatrix} 1 & 0 & 0 \\ 0 & \cos\gamma & -\sin\gamma \\ 0 & \sin\gamma & \cos\gamma \end{bmatrix} \tag{2.22}$$

are called *basic global rotation matrices*. We usually refer to the first, second, and third rotations about the axes of the global coordinate frame by α, β, and γ, respectively. All of the basic global rotation matrices $Q_{Z,\alpha}$, $Q_{Y,\beta}$, and $Q_{X,\gamma}$ transform a B-expression vector to its G-expression. ∎

Example 15 Successive rotation about global axes. Intermediate and final position of the corner of a brick.

The final position of the corner $P(5, 30, 10)$ of the slab shown in Fig. 2.5 after 30 deg rotation about the Z-axis, followed by 30 deg about the X-axis, and then 90 deg about the Y-axis can be found by first multiplying $Q_{Z,\pi/6}$ by $[5, 30, 10]^T$ to get the new global position after the first rotation

$$\begin{bmatrix} X_2 \\ Y_2 \\ Z_2 \end{bmatrix} = \begin{bmatrix} \cos\dfrac{\pi}{6} & -\sin\dfrac{\pi}{6} & 0 \\ \sin\dfrac{\pi}{6} & \cos\dfrac{\pi}{6} & 0 \\ 0 & 0 & 1 \end{bmatrix} \begin{bmatrix} 5 \\ 30 \\ 10 \end{bmatrix} = \begin{bmatrix} -10.67 \\ 28.48 \\ 10.0 \end{bmatrix} \tag{2.23}$$

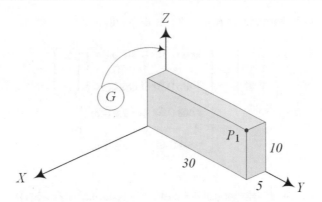

Fig. 2.5 Corner P of the slab at first position

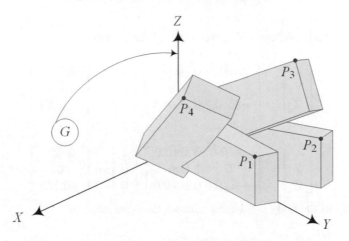

Fig. 2.6 Corner P and the slab at first, second, third, and final positions

and then multiplying $Q_{X,\pi/6}$ and $[-10.67, 28.48, 10.0]^T$ to get the position of P after the second rotation

$$
\begin{bmatrix} X_3 \\ Y_3 \\ Z_3 \end{bmatrix} = \begin{bmatrix} 1 & 0 & 0 \\ 0 & \cos\dfrac{\pi}{6} & -\sin\dfrac{\pi}{6} \\ 0 & \sin\dfrac{\pi}{6} & \cos\dfrac{\pi}{6} \end{bmatrix} \begin{bmatrix} -10.67 \\ 28.48 \\ 10.0 \end{bmatrix} = \begin{bmatrix} -10.67 \\ 19.66 \\ 22.9 \end{bmatrix} \tag{2.24}
$$

and finally multiplying $Q_{Y,\pi/2}$ and $[-10.67, 19.66, 22.9]^T$ to get the final position of P after the third rotation. The slab and the point P in first, second, third, and fourth positions are shown in Fig. 2.6.

$$
\begin{bmatrix} X_4 \\ Y_4 \\ Z_4 \end{bmatrix} = \begin{bmatrix} \cos\dfrac{\pi}{2} & 0 & \sin\dfrac{\pi}{2} \\ 0 & 1 & 0 \\ -\sin\dfrac{\pi}{2} & 0 & \cos\dfrac{\pi}{2} \end{bmatrix} \begin{bmatrix} -10.68 \\ 19.66 \\ 22.9 \end{bmatrix} = \begin{bmatrix} 22.90 \\ 19.66 \\ 10.67 \end{bmatrix} \tag{2.25}
$$

Example 16 Time dependent global rotation.

Consider a rigid body B that is continuously turning about the Y-axis of the G-frame at a rate of 0.3 rad/s. The rotation transformation matrix of the body is

$$
{}^G Q_B = \begin{bmatrix} \cos 0.3t & 0 & \sin 0.3t \\ 0 & 1 & 0 \\ -\sin 0.3t & 0 & \cos 0.3t \end{bmatrix} \tag{2.26}
$$

Any point of B will move on a circle with radius $R = \sqrt{X^2 + Z^2}$ parallel to (X, Z)-plane.

$$
\begin{bmatrix} X \\ Y \\ Z \end{bmatrix} = \begin{bmatrix} \cos 0.3t & 0 & \sin 0.3t \\ 0 & 1 & 0 \\ -\sin 0.3t & 0 & \cos 0.3t \end{bmatrix} \begin{bmatrix} x \\ y \\ z \end{bmatrix}
$$

$$
= \begin{bmatrix} x \cos 0.3t + z \sin 0.3t \\ y \\ z \cos 0.3t - x \sin 0.3t \end{bmatrix} \tag{2.27}
$$

$$
X^2 + Z^2 = (x \cos 0.3t + z \sin 0.3t)^2 + (z \cos 0.3t - x \sin 0.3t)^2
$$

$$
= x^2 + z^2 = R^2 \tag{2.28}
$$

Consider a point P at $^B\mathbf{r} = \begin{bmatrix} 1 & 0 & 0 \end{bmatrix}^T$. After $t = 1$ s, the point will be seen at

$$
\begin{bmatrix} X \\ Y \\ Z \end{bmatrix} = \begin{bmatrix} \cos 0.3 & 0 & \sin 0.3 \\ 0 & 1 & 0 \\ -\sin 0.3 & 0 & \cos 0.3 \end{bmatrix} \begin{bmatrix} 1 \\ 0 \\ 0 \end{bmatrix} = \begin{bmatrix} 0.955 \\ 0 \\ -0.295 \end{bmatrix} \tag{2.29}
$$

and after $t = 2$ s, at

$$
\begin{bmatrix} X \\ Y \\ Z \end{bmatrix} = \begin{bmatrix} \cos 0.6 & 0 & \sin 0.6 \\ 0 & 1 & 0 \\ -\sin 0.6 & 0 & \cos 0.6 \end{bmatrix} \begin{bmatrix} 1 \\ 0 \\ 0 \end{bmatrix} = \begin{bmatrix} 0.825 \\ 0 \\ -0.565 \end{bmatrix} \tag{2.30}
$$

We can find the global velocity of the body point P by taking a time derivative of

$$
^G\mathbf{r}_P = Q_{Y,\beta}\,^B\mathbf{r}_P \tag{2.31}
$$

$$
Q_{Y,\beta} = \begin{bmatrix} \cos 0.3t & 0 & \sin 0.3t \\ 0 & 1 & 0 \\ -\sin 0.3t & 0 & \cos 0.3t \end{bmatrix} \tag{2.32}
$$

Therefore, the global expression of its velocity vector is

$$
^G\mathbf{v}_P = \frac{d}{dt}\,^G\mathbf{r}_P = \dot{Q}_{Y,\beta}\,^B\mathbf{r}_P = 0.3 \begin{bmatrix} -0.3 \sin 0.3t \\ 0 \\ -0.3 \cos 0.3t \end{bmatrix} \tag{2.33}
$$

where

$$
\dot{Q}_{Y,\beta} = 0.3 \begin{bmatrix} -\sin 0.3t & 0 & \cos 0.3t \\ 0 & 0 & 0 \\ -\cos 0.3t & 0 & -\sin 0.3t \end{bmatrix} \tag{2.34}
$$

The velocity of a general point at $^B\mathbf{r} = \begin{bmatrix} x & y & z \end{bmatrix}^T$ will be

$$
^G\mathbf{v} = \frac{d}{dt}\,^G\mathbf{r} = \dot{Q}_{Y,\beta}\,^B\mathbf{r} = 0.3 \begin{bmatrix} z \cos 0.3t - x \sin 0.3t \\ 0 \\ -x \cos 0.3t - z \sin 0.3t \end{bmatrix} \tag{2.35}
$$

Example 17 Global rotation is given, and local position is requested.

If a point P is moved to $^G\mathbf{r}_2 = [4, 3, 2]^T$ after a 60 deg rotation about the Z-axis, its position in the local coordinate is

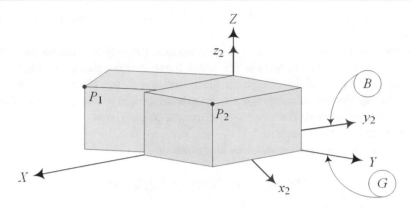

Fig. 2.7 Positions of point P in Example 17 before and after rotation

$$^B\mathbf{r}_2 = Q_{Z,\pi/3}^{-1}\,{}^G\mathbf{r}_2 \tag{2.36}$$

$$\begin{bmatrix} x_2 \\ y_2 \\ z_2 \end{bmatrix} = \begin{bmatrix} \cos\dfrac{\pi}{3} & -\sin\dfrac{\pi}{3} & 0 \\ \sin\dfrac{\pi}{3} & \cos\dfrac{\pi}{3} & 0 \\ 0 & 0 & 1 \end{bmatrix}^{-1} \begin{bmatrix} 4 \\ 3 \\ 2 \end{bmatrix} = \begin{bmatrix} 4.60 \\ -1.95 \\ 2.0 \end{bmatrix}$$

The local coordinate frame was coincident with the global coordinate frame before rotation, and thus the global coordinates of P before rotation was also $^G\mathbf{r}_1 = [4.60, -1.95, 2.0]^T$. Positions of P before and after rotation are shown in Fig. 2.7.

2.2 Successive Rotation About Global Cartesian Axes

The final global position of a point P of a rigid body B with position vector $\mathbf{r}$, after a series of rotations $Q_1, Q_2, Q_3, ..., Q_n$ about the global axes can be found by one transformation:

$$^G\mathbf{r} = {}^GQ_B\,{}^B\mathbf{r} \tag{2.37}$$

where

$$^GQ_B = Q_n \cdots Q_3 Q_2 Q_1 \tag{2.38}$$

The vectors $^G\mathbf{r}$ and $^B\mathbf{r}$ indicate the position vector $\mathbf{r}$ in the global and local coordinate frames. The matrix ^GQ_B, which transforms the local coordinates to their corresponding global coordinates, is called the *global rotation matrix*. Because matrix multiplications do not commute, the order of performing rotations is important.

Proof Assume a body frame B that undergoes two sequential rotations Q_1 and Q_2 about the global axes. The body coordinate frame B is initially coincident with the global coordinate frame G. The rigid body rotates about a global axis, and the global rotation matrix Q_1 gives us the new global coordinate $^G\mathbf{r}_1$ of the body point at $^B\mathbf{r}$.

$$^G\mathbf{r}_1 = Q_1\,{}^B\mathbf{r} \tag{2.39}$$

Before the second rotation, we put the B-frame aside and assume a new body coordinate frame B_1 to be coincident with the global frame. Therefore, the second rotation is treated similar to the first rotation. The new body coordinate would be $^{B_1}\mathbf{r} \equiv {}^G\mathbf{r}_1$. The second global rotation matrix Q_2 provides the new global position $^G\mathbf{r}_2$ of the body points $^{B_1}\mathbf{r}$:

$$^{B_1}\mathbf{r} = Q_2\,{}^{B_1}\mathbf{r} \tag{2.40}$$

Substituting (2.39) into (2.40) shows that

$$^G\mathbf{r} = Q_2 Q_1\,{}^B\mathbf{r} \tag{2.41}$$

Following the same procedure, we can determine the final global position of a body point after a series of sequential rotations $Q_1, Q_2, Q_3, ..., Q_n$ as (2.38).

Rotation about global coordinate axes is conceptually simple because the axes of rotations are fixed in space. Assume we have the coordinates of every point of a rigid body in the global frame that is equal to the local coordinates initially. The rigid body rotates about a global axis, and then the proper global rotation matrix gives us the new global coordinate of the points. Then, our situation before the second rotation is similar to what we had before the first rotation. This process may be continued until all rotations are performed. The resultant transformation matrix would be equal to matrix multiplication of all individual transformation matrices in order.

The multiplication of rotation transformation matrices is associative.

$$Q_3 Q_2 Q_1 = Q_3 [Q_2 Q_1] = [Q_3 Q_2] Q_1 \tag{2.42}$$

■

Example 18 Successive global rotation matrix. Several rotations about the global principal axes and an equivalent transformation matrix.

The global rotation matrix $^G Q_B$ after a rotation $Q_{Z,\alpha}$ followed by $Q_{Y,\beta}$ and then $Q_{X,\gamma}$ is

$$^G Q_B = Q_{X,\gamma} \, Q_{Y,\beta} \, Q_{Z,\alpha} \tag{2.43}$$

$$= \begin{bmatrix} 1 & 0 & 0 \\ 0 & \cos\gamma & -\sin\gamma \\ 0 & \sin\gamma & \cos\gamma \end{bmatrix} \begin{bmatrix} \cos\beta & 0 & \sin\beta \\ 0 & 1 & 0 \\ -\sin\beta & 0 & \cos\beta \end{bmatrix} \begin{bmatrix} \cos\alpha & -\sin\alpha & 0 \\ \sin\alpha & \cos\alpha & 0 \\ 0 & 0 & 1 \end{bmatrix}$$

$$= \begin{bmatrix} c\alpha\,c\beta & -c\beta\,s\alpha & s\beta \\ c\gamma\,s\alpha + c\alpha\,s\beta\,s\gamma & c\alpha c\gamma - s\alpha\,s\beta\,s\gamma & -c\beta\,s\gamma \\ s\alpha\,s\gamma - c\alpha\,c\gamma\,s\beta & c\alpha s\gamma + c\gamma\,s\alpha s\beta & c\beta\,c\gamma \end{bmatrix}$$

where c stands for cos and s stands for sin.

$$c \equiv \cos \qquad s \equiv \sin \tag{2.44}$$

If we change the order of rotation, then the global rotation matrix $^G Q_B$ after a rotation $Q_{X,\gamma}$ followed by $Q_{Y,\beta}$ and *then* $Q_{Z,\alpha}$ is

$$^G Q_B = Q_{Z,\alpha} \, Q_{Y,\beta} \, Q_{X,\gamma} \tag{2.45}$$

$$= \begin{bmatrix} c\alpha\,c\beta & c\alpha\,s\beta\,s\gamma - c\gamma\,s\alpha & s\alpha\,s\gamma + c\alpha\,c\gamma\,s\beta \\ c\beta\,s\alpha & c\alpha\,c\gamma + s\alpha\,s\beta\,s\gamma & c\gamma\,s\alpha\,s\beta - c\alpha\,s\gamma \\ -s\beta & c\beta\,s\gamma & c\beta\,c\gamma \end{bmatrix}$$

A negative rotation cancels a positive rotation and the resultant rotation matrix will be a unity matrix. The global rotation matrix $^G Q_B$ after a rotation $Q_{X,\gamma}$ followed by $Q_{Y,\beta}$ and $Q_{Z,\alpha}$ and then a rotation $Q_{Z,-\alpha}$ followed by $Q_{Y,-\beta}$ and then $Q_{X,-\gamma}$ is

$$^G Q_B = Q_{X,-\gamma} \, Q_{Y,-\beta} \, Q_{Z,-\alpha} Q_{Z,\alpha} \, Q_{Y,\beta} \, Q_{X,\gamma} = \begin{bmatrix} 1 & 0 & 0 \\ 0 & 1 & 0 \\ 0 & 0 & 1 \end{bmatrix} \tag{2.46}$$

Example 19 Successive global rotations, global position. Numerical example of position of the tip point of a bar after several rotations about the global principal axes.

The end point P of the arm shown in Fig. 2.8 is initially located at

$$\begin{bmatrix} X_1 \\ Y_1 \\ Z_1 \end{bmatrix} = \begin{bmatrix} 0 \\ l\cos\theta \\ l\sin\theta \end{bmatrix} = \begin{bmatrix} 0 \\ 1\cos 75 \\ 1\sin 75 \end{bmatrix} = \begin{bmatrix} 0.0 \\ 0.26 \\ 0.97 \end{bmatrix} \tag{2.47}$$

The rotation matrix to find the new position of the end point after -29 deg rotation about the X-axis, followed by 30 deg about the Z-axis, and again 132 deg about the X-axis is

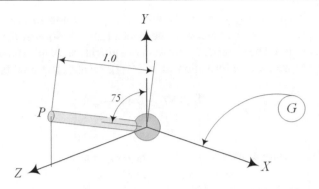

Fig. 2.8 The arm of Example 19

$$^G Q_B = Q_{X,132} \, Q_{Z,30} \, Q_{X,-29} \tag{2.48}$$

$$= \begin{bmatrix} 1 & 0 & 0 \\ 0 & \cos\frac{132\pi}{180} & -\sin\frac{132\pi}{180} \\ 0 & \sin\frac{132\pi}{180} & \cos\frac{132\pi}{180} \end{bmatrix} \begin{bmatrix} \cos\frac{\pi}{6} & -\sin\frac{\pi}{6} & 0 \\ \sin\frac{\pi}{6} & \cos\frac{\pi}{6} & 0 \\ 0 & 0 & 1 \end{bmatrix}$$

$$\times \begin{bmatrix} 1 & 0 & 0 \\ 0 & \cos\frac{-29\pi}{180} & -\sin\frac{-29\pi}{180} \\ 0 & \sin\frac{-29\pi}{180} & \cos\frac{-29\pi}{180} \end{bmatrix}$$

$$= \begin{bmatrix} 0.87 & -0.44 & -0.24 \\ -0.33 & -0.15 & -0.93 \\ 0.37 & 0.89 & -0.27 \end{bmatrix}$$

and its new position is at

$$\begin{bmatrix} X_2 \\ Y_2 \\ Z_2 \end{bmatrix} = \begin{bmatrix} 0.87 & -0.44 & -0.24 \\ -0.33 & -0.15 & -0.93 \\ 0.37 & 0.89 & -0.27 \end{bmatrix} \begin{bmatrix} 0.0 \\ 0.26 \\ 0.97 \end{bmatrix} = \begin{bmatrix} -0.35 \\ -0.94 \\ -0.031 \end{bmatrix} \tag{2.49}$$

Example 20 Twelve independent triple global rotations. Triple global rotations are just enough to move a rigid body to any desired orientation. Inverse trigonometric functions are multi-valued, and this fact makes inverse kinematics problem complicated.

Consider a rigid body in a final orientation after a series of rotations about global axes. We may transform its body coordinate frame B from the coincident position with a global frame G to any final orientation by only three rotations about the global axes provided that no two consequence rotations are about the same axis. In general, there are 12 different independent combinations of triple rotations about the global axes. They are

$$\begin{aligned}
1 &- Q_{X,\gamma} \, Q_{Y,\beta} \, Q_{Z,\alpha} \\
2 &- Q_{Y,\gamma} \, Q_{Z,\beta} \, Q_{X,\alpha} \\
3 &- Q_{Z,\gamma} \, Q_{X,\beta} \, Q_{Y,\alpha} \\
4 &- Q_{Z,\gamma} \, Q_{Y,\beta} \, Q_{X,\alpha} \\
5 &- Q_{Y,\gamma} \, Q_{X,\beta} \, Q_{Z,\alpha} \\
6 &- Q_{X,\gamma} \, Q_{Z,\beta} \, Q_{Y,\alpha} \\
7 &- Q_{X,\gamma} \, Q_{Y,\beta} \, Q_{X,\alpha} \\
8 &- Q_{Y,\gamma} \, Q_{Z,\beta} \, Q_{Y,\alpha} \\
9 &- Q_{Z,\gamma} \, Q_{X,\beta} \, Q_{Z,\alpha} \\
10 &- Q_{X,\gamma} \, Q_{Z,\beta} \, Q_{X,\alpha} \\
11 &- Q_{Y,\gamma} \, Q_{X,\beta} \, Q_{Y,\alpha} \\
12 &- Q_{Z,\gamma} \, Q_{Y,\beta} \, Q_{Z,\alpha}
\end{aligned} \tag{2.50}$$

In other words, we may move a rigid body to any given final orientation by minimum three rotations about the global principal axes. There are 12 independent options for such maneuver. The value of the rotations α, β, and γ is different in the given 12 maneuvers for the same final orientation. The expanded form of the 12 global axes triple rotations is presented in Appendix A.

As an example, assume we are given a transformation matrix ${}^GQ_B$ relating the final orientation of a rigid body to the global coordinate frame.

$$
{}^GQ_B = \begin{bmatrix} 0.87 & -0.44 & -0.24 \\ -0.33 & -0.15 & -0.93 \\ 0.37 & 0.89 & -0.27 \end{bmatrix}
\tag{2.51}
$$

Selecting the first sequence of three rotations of case $1 - Q_{X,\gamma} Q_{Y,\beta} Q_{Z,\alpha}$ yields

$$
{}^GQ_B = Q_{X,\gamma} \, Q_{Y,\beta} \, Q_{Z,\alpha}
\tag{2.52}
$$

$$
= \begin{bmatrix} c\alpha\,c\beta & -c\beta\,s\alpha & s\beta \\ c\gamma\,s\alpha + c\alpha\,s\beta\,s\gamma & c\alpha\,c\gamma - s\alpha\,s\beta\,s\gamma & -c\beta\,s\gamma \\ s\alpha\,s\gamma - c\alpha\,c\gamma\,s\beta & c\alpha\,s\gamma + c\gamma\,s\alpha\,s\beta & c\beta\,c\gamma \end{bmatrix}
$$

The element Q_{13} gives β, the ratio Q_{12}/Q_{11} gives α, and the ratio Q_{23}/Q_{33} gives γ.

$$
\sin\beta = -0.24 \qquad \beta = -(-1)^k \, 0.242 + k\pi \,\text{rad}
\tag{2.53}
$$

$$
\tan\alpha = \frac{0.44}{0.87} \qquad \alpha = 0.468 + k\pi \,\text{rad}
\tag{2.54}
$$

$$
\tan\gamma = \frac{0.93}{-0.27} \qquad \gamma = -1.288 + k\pi \,\text{rad}
\tag{2.55}
$$

It is a problem of trigonometric functions that their inverse functions are multi-valued. There is no straightforward method to determine the right solutions matching the problem in hand. Trial and error is usually the easiest way to find the right solution, as the matching one is associated with $k = 0$ or $k = \pm 1$. In this problem, we may check to find the matching solutions as follows:

$$
\sin\beta = -0.24 \qquad \beta = -(-1)\, 0.242 + \pi \,\text{rad} = 3.383 \,\text{rad}
\tag{2.56}
$$

$$
\tan\alpha = \frac{0.44}{0.87} \qquad \alpha = 0.468 + \pi \,\text{rad} = 3.6096 \,\text{rad}
\tag{2.57}
$$

$$
\tan\gamma = \frac{0.93}{-0.27} \qquad \gamma = -1.288 \,\text{rad}
\tag{2.58}
$$

Therefore, we will make the same overall transformation matrix (2.51) by turning the rigid body 3.6096 rad about the Z-axis, then 3.383 rad about the Y-axis, and then -1.288 rad about the X-axis.

$$
{}^GQ_B = Q_{X,\gamma} \, Q_{Y,\beta} \, Q_{Z,\alpha} = Q_{X,-1.29} \, Q_{Y,3.38} \, Q_{Z,3.61}
\tag{2.59}
$$

$$
= \begin{bmatrix} 0.86659 & -0.43803 & -0.23907 \\ -0.33077 & -0.14547 & -0.93243 \\ 0.37365 & 0.88711 & -0.27095 \end{bmatrix}
$$

Example 21 ★ Problem of inverse of trigonometric functions. It is always needed to examine all answers of inverse trigonometric functions to determine the correct ones.

The general solutions of $\sin x = y$, $\cos x = y$, and $\tan x = y$ are

$$
\arcsin y = (-1)^k \, x + k\pi
\tag{2.60}
$$

$$
\arccos y = \pm x + 2k\pi
\tag{2.61}
$$

$$
\arctan y = x + k\pi
\tag{2.62}
$$

$$
k = 0, 1, 2, 3, \cdots
$$

A big problem in mathematics that specifically affects kinematics of mechanisms and robots is the multi-valued inverse trigonometric functions. There must always be a checking process to realize which solution is the one that matches with the problem in hand.

As a standard guidance, the inverse trigonometric functions usually denote the principal value. The principal values of the inverse trigonometric functions are defined as

$$
\begin{aligned}
-1 \le y \le 1 \quad & -\frac{\pi}{2} \le \arcsin y \le \frac{\pi}{2} \\
-1 \le y \le 1 \quad & 0 \le \arccos y \le \pi \\
-\infty \le y \le \infty \quad & -\frac{\pi}{2} \le \arctan y \le \frac{\pi}{2} \\
-\infty \le y \le \infty \quad & 0 \le \operatorname{arccot} y \le \pi
\end{aligned}
\tag{2.63}
$$

$$
\begin{aligned}
1 \le y \quad & 0 \le \operatorname{arccsc} y \le \frac{\pi}{2} \\
y \le -1 \quad & -\frac{\pi}{2} \le \operatorname{arccsc} y \le 0 \\
1 \le y \quad & 0 \le \operatorname{arcsec} y \le \frac{\pi}{2} \\
y \le -1 \quad & \frac{\pi}{2} \le \operatorname{arcsec} y \le \pi
\end{aligned}
\tag{2.64}
$$

Example 22 ★ Problem of inverse function symbol. Exponent -1 is not a correct sign to indicate inverse of a function. Exponent -1 stands for radiomimetic inverse.

$$
x^{-1} = \frac{1}{x}
\tag{2.65}
$$

Imposing more than one operation to a symbol is a mistake making students and computers confused.

$$
f^{-1}(y) = \frac{1}{f(y)}
\tag{2.66}
$$

$$
y = f(x) \qquad x \ne f^{-1}(y)
\tag{2.67}
$$

Therefore, using exponent -1 to indicate inverse functions is not wise. It was wrongly introduced and accepted to show inverse functions by exponent -1. However, inverse of a function is an operator such as derivative and integral and needs to have a proper symbol. Let us for the moment use In as the symbol of inverse function operator. The inverse trigonometric functions also have special names. If

$$
\sin x = y \qquad \cos x = y \qquad \tan x = y
\tag{2.68}
$$

then

$$
x = \arcsin y = \operatorname{Inv} \sin y
\tag{2.69}
$$

$$
x = \arccos y = \operatorname{Inv} \cos y
\tag{2.70}
$$

$$
x = \arctan y = \operatorname{Inv} \tan y
\tag{2.71}
$$

As an example, let us find the inverse of a function Inv y. If

$$
y = f(x) = 2x + 3 \qquad x = \frac{y - 3}{2}
\tag{2.72}
$$

then

$$
\operatorname{Inv} y = \frac{x - 3}{2}
\tag{2.73}
$$

because

$$
x = \frac{y - 3}{2}
\tag{2.74}
$$

and we switch x and y to have

$$y = \frac{x - 3}{2} \tag{2.75}$$

If

$$y = 2^x \tag{2.76}$$

then

$$\text{Inv } y = \log_2 x \tag{2.77}$$

If

$$y = x^2 \qquad x = \text{Inv } y = \sqrt{y} \tag{2.78}$$

then

$$\frac{dy}{dx} = 2x \qquad \frac{dx}{dy} = \frac{1}{2\sqrt{y}} = \frac{1}{2x} \tag{2.79}$$

$$\frac{dy}{dx}\frac{dx}{dy} = 2x\frac{1}{2x} = 1 \tag{2.80}$$

Example 23 Order of rotation and order of matrix multiplication. Order of rotation must match with order of rotation transformation matrix multiplication.

Changing the order of global rotation matrices is equivalent to changing the order of rotations. Assume the position of a point P of a rigid body B to be at $^B\mathbf{r}_P = \begin{bmatrix} 1 & 2 & 3 \end{bmatrix}^T$. Its global position after rotation 30 deg about X-axis and then 60 deg about Y-axis is at

$$\left(^G\mathbf{r}_P\right)_1 = Q_{Y,60}\, Q_{X,30}\, ^B\mathbf{r}_P \tag{2.81}$$

$$= \begin{bmatrix} \cos\frac{\pi}{3} & 0 & \sin\frac{\pi}{3} \\ 0 & 1 & 0 \\ -\sin\frac{\pi}{3} & 0 & \cos\frac{\pi}{3} \end{bmatrix} \begin{bmatrix} 1 & 0 & 0 \\ 0 & \cos\frac{\pi}{6} & -\sin\frac{\pi}{6} \\ 0 & \sin\frac{\pi}{6} & \cos\frac{\pi}{6} \end{bmatrix} \begin{bmatrix} 1 \\ 2 \\ 3 \end{bmatrix}$$

$$= \begin{bmatrix} 0.5 & 0.433 & 0.75 \\ 0 & 0.866 & -0.5 \\ -0.866 & 0.25 & 0.433 \end{bmatrix} \begin{bmatrix} 1 \\ 2 \\ 3 \end{bmatrix} = \begin{bmatrix} 3.616 \\ 0.232 \\ 0.933 \end{bmatrix}$$

If we change the order of rotations, then its position would be at

$$\left(^G\mathbf{r}_P\right)_2 = Q_{X,30}\, Q_{Y,60}\, ^B\mathbf{r}_P \tag{2.82}$$

$$= \begin{bmatrix} 0.5 & 0 & 0.866 \\ 0.433 & 0.866 & -0.25 \\ -0.75 & 0.5 & 0.433 \end{bmatrix} \begin{bmatrix} 1 \\ 2 \\ 3 \end{bmatrix} = \begin{bmatrix} 3.0981 \\ 1.4151 \\ 1.549 \end{bmatrix}$$

These two final positions of P are $d = \left|\left(^G\mathbf{r}_P\right)_1 - \left(^G\mathbf{r}_P\right)_2\right| = 1.4309$ apart.

Example 24 ★ An unsolved problem, repeated rotation about global axes. The general form of this problem has never been solved in mathematics.

If we turn a body frame B about X-axis γ rad, where

$$\gamma = \frac{2\pi}{n} \qquad n \in \mathbb{N} \tag{2.83}$$

then we need to repeat the rotation n times to turn the body back to its original configuration. We can check it by multiplying $Q_{X,\gamma}$ by itself until we achieve an identity matrix. So, any body point of B will be mapped to the same point in the global frame. To show this, we may find $Q_{X,\gamma}$ to the power m as

$$Q_{X,\gamma}^m = \begin{bmatrix} 1 & 0 & 0 \\ 0 & \cos\gamma & -\sin\gamma \\ 0 & \sin\gamma & \cos\gamma \end{bmatrix}^m = \begin{bmatrix} 1 & 0 & 0 \\ 0 & \cos\dfrac{2\pi}{n} & -\sin\dfrac{2\pi}{n} \\ 0 & \sin\dfrac{2\pi}{n} & \cos\dfrac{2\pi}{n} \end{bmatrix}^m$$

$$= \begin{bmatrix} 1 & 0 & 0 \\ 0 & \cos m\dfrac{2\pi}{n} & -\sin m\dfrac{2\pi}{n} \\ 0 & \sin m\dfrac{2\pi}{n} & \cos m\dfrac{2\pi}{n} \end{bmatrix} \tag{2.84}$$

If $m = n$, then we have an identity matrix.

$$Q_{X,\gamma}^n = \begin{bmatrix} 1 & 0 & 0 \\ 0 & \cos n\dfrac{2\pi}{n} & -\sin n\dfrac{2\pi}{n} \\ 0 & \sin n\dfrac{2\pi}{n} & \cos n\dfrac{2\pi}{n} \end{bmatrix} = \begin{bmatrix} 1 & 0 & 0 \\ 0 & 1 & 0 \\ 0 & 0 & 1 \end{bmatrix} \tag{2.85}$$

Repeated rotation about any other global axis provides the same result.

Let us now rotate B about two global axes repeatedly, such as turning α about Z-axis followed by a rotation γ about X-axis, such that

$$\alpha = \frac{2\pi}{n_1} \qquad \gamma = \frac{2\pi}{n_2} \qquad \{n_1, n_2\} \in \mathbb{N} \tag{2.86}$$

We may guess that repeating the rotations $n = n_1 \times n_2$ times will turn B back to its original configuration, but it is not always a right solution. As an example, consider $\alpha = \frac{2\pi}{3}$ and $\gamma = \frac{2\pi}{4}$. We need 13 times combined rotations to achieve the original configuration.

$$^G Q_B = Q_{X,2\pi/4}\, Q_{Z,2\pi/3} = \begin{bmatrix} -0.5 & -0.86603 & 0 \\ 0 & 0 & -1 \\ 0.86603 & -0.5 & 0 \end{bmatrix} \tag{2.87}$$

$$^G Q_B^{13} = \begin{bmatrix} 0.9997 & -0.01922 & -0.01902 \\ 0.01902 & 0.99979 & -0.0112 \\ 0.01922 & 0.01086 & 0.9998 \end{bmatrix} \approx \mathbf{I} \tag{2.88}$$

We may turn B to its original configuration by lower number of combined rotations if n_1 and n_2 have a common divisor. For example, if $n_1 = n_2 = 4$, we only need to apply the combined rotations three times.

$$^G Q_B = Q_{X,2\pi/4}\, Q_{Z,2\pi/4} = \begin{bmatrix} 0 & -1 & 0 \\ 0 & 0 & -1 \\ 1 & 0 & 0 \end{bmatrix} \tag{2.89}$$

$$^G Q_B^3 = \begin{bmatrix} 1.0 & 0 & 0 \\ 0 & 1.0 & 0 \\ 0 & 0 & 1.0 \end{bmatrix} = \mathbf{I} \tag{2.90}$$

There is an unsolved problem. In a general case, determination of the minimum required number n to repeat a general combined rotation $^G Q_B$ to turn back to the original orientation is an unsolved question.

$$^G Q_B = \prod_{j=1}^{m} Q_{X_i,\alpha_j} \qquad i = 1, 2, 3 \tag{2.91}$$

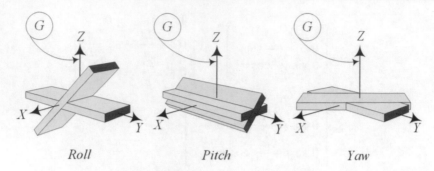

Fig. 2.9 Global roll, pitch, and yaw rotations

$$\alpha_j = \frac{2\pi}{n_j} \qquad\qquad m, n_j \in \mathbb{N} \tag{2.92}$$

$${}^G Q_B^n = [I] \qquad\qquad n = ? \tag{2.93}$$

Example 25 Global roll–pitch–yaw angles. Rotation about the X-axis, then Y-axis, and then Z-axis has special name as the same rotations about the body axes, but not as important.

The rotation about the X-axis of the global coordinate frame is called a *roll*, the rotation about the Y-axis is called a *pitch*, and the rotation about the Z-axis is called a *yaw*. The global *roll–pitch–yaw rotation matrix* is

$$
\begin{aligned}
{}^G Q_B &= Q_{Z,\gamma}\, Q_{Y,\beta}\, Q_{X,\alpha} \\[4pt]
&= \begin{bmatrix} c\beta\, c\gamma & -c\alpha\, s\gamma + c\gamma\, s\alpha\, s\beta & s\alpha\, s\gamma + c\alpha\, c\gamma\, s\beta \\ c\beta\, s\gamma & c\alpha\, c\gamma + s\alpha\, s\beta\, s\gamma & -c\gamma\, s\alpha + c\alpha\, s\beta\, s\gamma \\ -s\beta & c\beta\, s\alpha & c\alpha\, c\beta \end{bmatrix}
\end{aligned}
\tag{2.94}
$$

Figure 2.9 illustrates 45 deg roll, pitch, and yaw rotations about the axes of a global coordinate frame. Given the roll, pitch, and yaw angles, we can calculate the overall rotation matrix using Eq. (2.94). Also we are able to compute the equivalent roll, pitch, and yaw angles when a rotation matrix is given. Suppose that r_{ij} indicates the element of row i and column j of the roll–pitch–yaw rotation matrix (2.94), then the roll angle α is

$$\alpha = \arctan \tan \frac{r_{32}}{r_{33}} \tag{2.95}$$

and the pitch angle β is

$$\beta = -\arcsin r_{31} \tag{2.96}$$

and the yaw angle γ is

$$\gamma = \arctan \frac{r_{21}}{r_{11}} \tag{2.97}$$

provided that $\cos \beta \neq 0$.

Example 26 Determination of roll–pitch–yaw angles. A numerical example for determination of global roll–pitch–yaw angles for a given rotated rigid body.

Let us determine the required roll–pitch–yaw angles that move the B-frame to an orientation at which the x-axis of the body frame B is parallel to **u**, while y-axis lays in (X, Y)-plane.

$$\mathbf{u} = \hat{I} + 2\hat{J} + 3\hat{K} \tag{2.98}$$

Because x-axis must be along $\mathbf{u}$, we have

$$^{G}\hat{\imath} = \frac{\mathbf{u}}{|\mathbf{u}|} = \frac{1}{\sqrt{14}}\hat{I} + \frac{2}{\sqrt{14}}\hat{J} + \frac{3}{\sqrt{14}}\hat{K} \tag{2.99}$$

$$= 0.267\hat{I} + 0.534\hat{J} + 0.802\hat{K} \tag{2.100}$$

and because y-axis is in (X, Y)-plane, we have

$$^{G}\hat{\jmath} = \left(\hat{I}\cdot\hat{\jmath}\right)\hat{I} + \left(\hat{J}\cdot\hat{\jmath}\right)\hat{J} = \cos\theta\,\hat{I} + \sin\theta\,\hat{J} \tag{2.101}$$

The axes $^{G}\hat{\imath}$ and $^{G}\hat{\jmath}$ must be orthogonal, therefore,

$$\begin{bmatrix} 0.267 \\ 0.534 \\ 0.802 \end{bmatrix} \cdot \begin{bmatrix} \cos\theta \\ \sin\theta \\ 0 \end{bmatrix} = 0 \tag{2.102}$$

$$0.267\cos\theta + 0.534\sin\theta = 0 \tag{2.103}$$

$$\theta = -0.464\,\text{rad} = -26.56\,\text{deg} \tag{2.104}$$

We may find $^{G}\hat{k}$ by a cross product.

$$^{G}\hat{k} = {}^{G}\hat{\imath} \times {}^{G}\hat{\jmath} = \begin{bmatrix} 0.267 \\ 0.534 \\ 0.802 \end{bmatrix} \times \begin{bmatrix} 0.894 \\ -0.447 \\ 0 \end{bmatrix} = \begin{bmatrix} 0.358 \\ 0.717 \\ -0.597 \end{bmatrix} \tag{2.105}$$

Hence, the transformation matrix $^{G}Q_{B}$ is

$$^{G}Q_{B} = \begin{bmatrix} \hat{I}\cdot\hat{\imath} & \hat{I}\cdot\hat{\jmath} & \hat{I}\cdot\hat{k} \\ \hat{J}\cdot\hat{\imath} & \hat{J}\cdot\hat{\jmath} & \hat{J}\cdot\hat{k} \\ \hat{K}\cdot\hat{\imath} & \hat{K}\cdot\hat{\jmath} & \hat{K}\cdot\hat{k} \end{bmatrix} = \begin{bmatrix} 0.267 & 0.894 & 0.358 \\ 0.534 & -0.447 & 0.717 \\ 0.802 & 0 & -0.597 \end{bmatrix} \tag{2.106}$$

Now it is possible to determine the required roll–pitch–yaw angles to move the body coordinate frame B from the coincidence orientation with G to the final orientation.

$$\alpha = \arctan\frac{r_{32}}{r_{33}} = \arctan\frac{0}{-0.597} = 0 \tag{2.107}$$

$$\beta = -\arcsin r_{31} = -\arcsin 0.802 \approx -0.93\,\text{rad} \tag{2.108}$$

$$\gamma = \arctan\frac{r_{21}}{r_{11}} = \arctan\frac{0.534}{0.267} \approx 1.1071\,\text{rad} \tag{2.109}$$

$$^{G}Q_{B} = Q_{Z,\gamma}\,Q_{Y,\beta}\,Q_{X,\alpha} \tag{2.110}$$

$$= \begin{bmatrix} 0.267\,39 & -0.894\,41 & -0.358\,53 \\ 0.534\,71 & 0.447\,26 & -0.716\,97 \\ 0.801\,62 & 0 & 0.597\,83 \end{bmatrix}$$

2.3   Rotation About Local Cartesian Axes

Consider a rigid body B with a local coordinate frame $B(Oxyz)$ that is originally coincident with a global coordinate frame $G(OXYZ)$. Point O is the origin of both coordinate frames and is the only point of the body B to be fixed in the global frame G. If the body undergoes a rotation φ about the z-axis of its local coordinate frame, as can be seen in the top view shown in Fig. 2.10, then coordinates of any point of the rigid body in local and global coordinate frames are related by the following equation:

$$^{B}\mathbf{r} = A_{z,\varphi}\,^{G}\mathbf{r} \tag{2.111}$$

where $^{B}\mathbf{r}$ is the B-expression of the position vector of the point P in B, and $^{G}\mathbf{r}$ is the G-expression of the same vector,

$$^{G}\mathbf{r} = \begin{bmatrix} X \\ Y \\ Z \end{bmatrix} \qquad ^{B}\mathbf{r} = \begin{bmatrix} x \\ y \\ z \end{bmatrix} \tag{2.112}$$

and $A_{z,\varphi}$ is the *z-rotation matrix.*

$$A_{z,\varphi} = \begin{bmatrix} \cos\varphi & \sin\varphi & 0 \\ -\sin\varphi & \cos\varphi & 0 \\ 0 & 0 & 1 \end{bmatrix} \tag{2.113}$$

Similarly, rotation θ about the y-axis and rotation ψ about the x-axis of the local frame relate the local and global coordinates of point P by the following equations:

$$^{B}\mathbf{r} = A_{y,\theta}\,^{G}\mathbf{r} \tag{2.114}$$

$$^{B}\mathbf{r} = A_{x,\psi}\,^{G}\mathbf{r} \tag{2.115}$$

where $A_{y,\theta}$ is the *y-rotation matrix* and $A_{x,\psi}$ is the *x-rotation matrix.*

$$A_{y,\theta} = \begin{bmatrix} \cos\theta & 0 & -\sin\theta \\ 0 & 1 & 0 \\ \sin\theta & 0 & \cos\theta \end{bmatrix} \tag{2.116}$$

$$A_{x,\psi} = \begin{bmatrix} 1 & 0 & 0 \\ 0 & \cos\psi & \sin\psi \\ 0 & -\sin\psi & \cos\psi \end{bmatrix} \tag{2.117}$$

Proof Assume that vector $\mathbf{r}$ indicates the position of a point P of the rigid body B where it is initially at P_1. The rigid body turns α about the z-axis of the body coordinate frame B, and hence, point P goes from P_1 to P_2 indicated by vector $\mathbf{r}_2$. Using the unit vectors $(\hat{\imath}, \hat{\jmath}, \hat{k})$ along the axes of local coordinate frame $B(Oxyz)$ and $(\hat{I}, \hat{J}, \hat{K})$ along the axes of global coordinate frame $G(OXYZ)$, the initial and final position vectors $\mathbf{r}_1$ and $\mathbf{r}_2$ in both coordinate frames can be expressed by

$$^{B}\mathbf{r}_1 = x_1\hat{\imath} + y_1\hat{\jmath} + z_1\hat{k} \tag{2.118}$$

$$^{G}\mathbf{r}_1 = X_1\hat{I} + Y_1\hat{J} + Z_1\hat{K} \tag{2.119}$$

$$^{B}\mathbf{r}_2 = x_2\hat{\imath} + y_2\hat{\jmath} + z_2\hat{k} \tag{2.120}$$

$$^{G}\mathbf{r}_2 = X_2\hat{I} + Y_2\hat{J} + Z_2\hat{K}. \tag{2.121}$$

The vectors $^{B}\mathbf{r}_1$ and $^{B}\mathbf{r}_2$ are the initial and final positions of the vector $\mathbf{r}$ expressed in body coordinate frame $B(Oxyz)$, and $^{G}\mathbf{r}_1$ and $^{G}\mathbf{r}_2$ are the initial and final positions of the vector $\mathbf{r}$ expressed in the global coordinate frame $G(OXYZ)$.

The components of $^{B}\mathbf{r}_2$ can be found if we have the components of $^{G}\mathbf{r}_2$. Using Eqs. (2.120) and (2.121) and the definition of the inner product, we may write

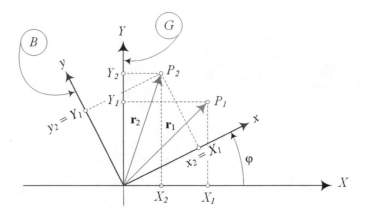

Fig. 2.10 Position vectors of point P before and after rotation of the local frame about the z-axis of the local frame

$$x_2 = \hat{\imath} \cdot \mathbf{r}_2 = \hat{\imath} \cdot X_2 \hat{I} + \hat{\imath} \cdot Y_2 \hat{J} + \hat{\imath} \cdot Z_2 \hat{K} \tag{2.122}$$

$$y_2 = \hat{\jmath} \cdot \mathbf{r}_2 = \hat{\jmath} \cdot X_2 \hat{I} + \hat{\jmath} \cdot Y_2 \hat{J} + \hat{\jmath} \cdot Z_2 \hat{K} \tag{2.123}$$

$$z_2 = \hat{k} \cdot \mathbf{r}_2 = \hat{k} \cdot X_2 \hat{I} + \hat{k} \cdot Y_2 \hat{J} + \hat{k} \cdot Z_2 \hat{K} \tag{2.124}$$

or equivalently

$$\begin{bmatrix} x_2 \\ y_2 \\ z_2 \end{bmatrix} = \begin{bmatrix} \hat{\imath} \cdot \hat{I} & \hat{\imath} \cdot \hat{J} & \hat{\imath} \cdot \hat{K} \\ \hat{\jmath} \cdot \hat{I} & \hat{\jmath} \cdot \hat{J} & \hat{\jmath} \cdot \hat{K} \\ \hat{k} \cdot \hat{I} & \hat{k} \cdot \hat{J} & \hat{k} \cdot \hat{K} \end{bmatrix} \begin{bmatrix} X_2 \\ Y_2 \\ Z_2 \end{bmatrix}. \tag{2.125}$$

The elements of the z-rotation matrix $A_{z,\varphi}$ are the *direction cosines* of $^G\mathbf{r}_2$ with respect to axes of $B(Oxyz)$. So, the elements of the matrix in Eq. (2.125) are

$$\begin{array}{lll} \hat{\imath} \cdot \hat{I} = \cos\varphi & \hat{\imath} \cdot \hat{J} = \sin\varphi & \hat{\imath} \cdot \hat{K} = 0 \\ \hat{\jmath} \cdot \hat{I} = -\sin\varphi & \hat{\jmath} \cdot \hat{J} = \cos\varphi & \hat{\jmath} \cdot \hat{K} = 0 \\ \hat{k} \cdot \hat{I} = 0 & \hat{k} \cdot \hat{J} = 0 & \hat{k} \cdot \hat{K} = 1 \end{array} \tag{2.126}$$

Combining Eqs. (2.125) and (2.126), we find the components of $^B\mathbf{r}_2$ by multiplying z-rotation matrix $A_{z,\varphi}$ and vector $^G\mathbf{r}_2$.

$$^B\mathbf{r}_2 = A_{z,\varphi} \, ^G\mathbf{r}_2 \tag{2.127}$$

$$\begin{bmatrix} x_2 \\ y_2 \\ z_2 \end{bmatrix} = \begin{bmatrix} \cos\varphi & \sin\varphi & 0 \\ -\sin\varphi & \cos\varphi & 0 \\ 0 & 0 & 1 \end{bmatrix} \begin{bmatrix} X_2 \\ Y_2 \\ Z_2 \end{bmatrix} \tag{2.128}$$

Equation (2.127) indicates that after rotation about the z-axis of the local coordinate frame, the position vector in the local frame is equal to $A_{z,\varphi}$ times the position vector in the global frame. Hence, after rotation about the z-axis, we are able to find the coordinates of any point of a rigid body in local coordinate frame, if we have its coordinates in the global frame.

Similarly, rotation θ about the y-axis and rotation ψ about the x-axis are described by the y-rotation matrix $A_{y,\theta}$ and the x-rotation matrix $A_{x,\psi}$, respectively. The rotation matrices $A_{z,\varphi}$, $A_{y,\theta}$, and $A_{x,\psi}$,

$$A_{z,\varphi} = \begin{bmatrix} \cos\varphi & \sin\varphi & 0 \\ -\sin\varphi & \cos\varphi & 0 \\ 0 & 0 & 1 \end{bmatrix} \tag{2.129}$$

$$A_{y,\theta} = \begin{bmatrix} \cos\theta & 0 & -\sin\theta \\ 0 & 1 & 0 \\ \sin\theta & 0 & \cos\theta \end{bmatrix} \tag{2.130}$$

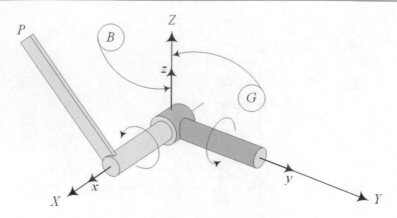

Fig. 2.11 Arm of Example 28

$$
A_{x,\psi} = \begin{bmatrix} 1 & 0 & 0 \\ 0 & \cos\psi & \sin\psi \\ 0 & -\sin\psi & \cos\psi \end{bmatrix}
\tag{2.131}
$$

are called *basic body rotation matrices*. We usually refer to the first, second, and third rotations about the axes of the body coordinate frame by φ, θ, and ψ, respectively. All of the basic global rotation matrices $A_{z,\varphi}$, $A_{y,\theta}$, and $A_{x,\psi}$ transform a G-expression vector to its B-expression. We use local and body coordinate frames equally. ∎

Example 27 Local rotation, local position. Local rotation, global position. A numerical example to determine position of a given point after rotation.

If a body coordinate frame $Oxyz$ has been rotated 60 deg about the z-axis and a point P in the global coordinate frame $OXYZ$ is at $(4, 3, 2)$, then its coordinates in the body coordinate frame $Oxyz$ are

$$
\begin{bmatrix} x \\ y \\ z \end{bmatrix} = \begin{bmatrix} \cos\dfrac{\pi}{3} & \sin\dfrac{\pi}{3} & 0 \\ -\sin\dfrac{\pi}{3} & \cos\dfrac{\pi}{3} & 0 \\ 0 & 0 & 1 \end{bmatrix} \begin{bmatrix} 4 \\ 3 \\ 2 \end{bmatrix} = \begin{bmatrix} 4.60 \\ -1.96 \\ 2.0 \end{bmatrix}
\tag{2.132}
$$

If a local coordinate frame $Oxyz$ has been rotated 60 deg about the z-axis and a point P in the local coordinate frame $Oxyz$ is at $(4, 3, 2)$, its position in the global coordinate frame $OXYZ$ is at

$$
\begin{bmatrix} X \\ Y \\ Z \end{bmatrix} = \begin{bmatrix} \cos\dfrac{\pi}{3} & \sin\dfrac{\pi}{3} & 0 \\ -\sin\dfrac{\pi}{3} & \cos\dfrac{\pi}{3} & 0 \\ 0 & 0 & 1 \end{bmatrix}^{T} \begin{bmatrix} 4 \\ 3 \\ 2 \end{bmatrix} = \begin{bmatrix} -0.60 \\ 4.96 \\ 2.0 \end{bmatrix}
\tag{2.133}
$$

Example 28 Successive local rotation, global position. Global position of the tip point of a robotic arm after few rotations.

The arm shown in Fig. 2.11 has two actuators. The first actuator rotates the arm -90 deg about y-axis, and then the second actuator rotates the arm 90 deg about x-axis. If, before the rotations, the end point P is at

$$
{}^{B}\mathbf{r}_P = \begin{bmatrix} 9.5 & -10.1 & 10.1 \end{bmatrix}^{T}
\tag{2.134}
$$

then its position in the global coordinate frame is at

$$
{}^{G}\mathbf{r}_P = \left[A_{x,\pi/2}\, A_{y,-\pi/2} \right]^{-1}\, {}^{B}\mathbf{r}_P = A_{y,-\pi/2}^{-1}\, A_{x,\pi/2}^{-1}\, {}^{B}\mathbf{r}_P
$$

$$
= \begin{bmatrix} 0 & -1 & 0 \\ 0 & 0 & -1 \\ 1 & 0 & 0 \end{bmatrix} \begin{bmatrix} 9.5 \\ -10.1 \\ 10.1 \end{bmatrix} = \begin{bmatrix} 10.1 \\ -10.1 \\ 9.5 \end{bmatrix}
\tag{2.135}
$$

2.4   Successive Rotation About Local Cartesian Axes

Consider a point P in a rigid body $B(Oxyz)$ at position vector $\mathbf{r}$. Having the final global position vector $^G\mathbf{r}$ of P, we can determine its local position vector $^B\mathbf{r}$ after a series of sequential rotations $A_1, A_2, A_3, \ldots, A_n$ about the local axes, by

$$^B\mathbf{r} = {}^B A_G \, {}^G\mathbf{r} \tag{2.136}$$

where

$$^B A_G = A_n \cdots A_3 A_2 A_1 \tag{2.137}$$

The matrix $^B A_G$ is the *local rotation matrix* that maps the global coordinates to their corresponding local coordinates.

Proof Assume the body coordinate frame $B(Oxyz)$ is initially coincident with the global coordinate frame $G(OXYZ)$. The rigid body rotates about a local axis such that a local rotation matrix A_1 relates the global coordinates of a body point P at $^G\mathbf{r}$ to the associated local coordinates.

$$^B\mathbf{r} = R_1 \, {}^G\mathbf{r} \tag{2.138}$$

Let us introduce an intermediate space-fixed frame G_1 coincident with the new position of the body coordinate frame,

$$^{G_1}\mathbf{r} \equiv {}^B\mathbf{r} \tag{2.139}$$

and give the rigid body a second rotation about a local coordinate axis. Then another local rotation matrix A_2 relates the coordinates in the intermediate fixed frame to the corresponding local coordinates:

$$^B\mathbf{r} = A_2 \, {}^{G_1}\mathbf{r} \tag{2.140}$$

To relate the final coordinates of the point, we must first transform its global coordinates to the intermediate fixed frame and then transform to the original body frame. Substituting (2.138) in (2.140) shows that

$$^B\mathbf{r} = A_2 \, A_1 \, {}^G\mathbf{r} \tag{2.141}$$

Following the same procedure, we can determine the final global position of a body point after a series of sequential rotations $A_1, A_2, A_3, \ldots, A_n$ as (2.137).

Rotation about the local coordinate axes is conceptually more interesting than rotation about global coordinate axes. It is because in a sequence of rotations every rotation is about an axis, which has been moved to its new global position during the previous rotation.

The multiplication of rotation transformation matrices is associative.

$$A_3 A_2 A_1 = A_3 \left[A_2 A_1 \right] = \left[A_3 A_2 \right] A_1 \tag{2.142}$$

∎

Example 29 Successive local rotation, local position. A numerical example to determine coordinates of a body point of a given global point after several rotations.

A local coordinate frame $B(Oxyz)$ that initially is coincident with a global coordinate frame $G(OXYZ)$ undergoes a rotation $\varphi = 30$ deg about the z-axis, then $\theta = 30$ deg about the x-axis, and then $\psi = 30$ deg about the y-axis. The local coordinates of a point P located at $X = 5$, $Y = 30$, and $Z = 10$ can be found by

$$\begin{bmatrix} x & y & z \end{bmatrix}^T = A_{y,\psi} \, A_{x,\theta} \, A_{z,\varphi} \begin{bmatrix} 5 & 30 & 10 \end{bmatrix}^T \tag{2.143}$$

The local rotation matrix is

$$^{B}A_G = A_{y,30}\, A_{x,30}\, A_{z,30} = \begin{bmatrix} 0.63 & 0.65 & -0.43 \\ -0.43 & 0.75 & 0.50 \\ 0.65 & -0.125 & 0.75 \end{bmatrix} \tag{2.144}$$

and coordinates of P in the local frame are

$$\begin{bmatrix} x \\ y \\ z \end{bmatrix} = \begin{bmatrix} 0.63 & 0.65 & -0.43 \\ -0.43 & 0.75 & 0.50 \\ 0.65 & -0.125 & 0.75 \end{bmatrix} \begin{bmatrix} 5 \\ 30 \\ 10 \end{bmatrix} = \begin{bmatrix} 18.28 \\ 25.33 \\ 7.0 \end{bmatrix} \tag{2.145}$$

Example 30 Twelve independent triple local rotations. Any triple local rotations is just enough to move a rigid body to any desired orientation. Inverse trigonometric functions are multi-valued, and this fact makes inverse kinematics problem complicated.

Leonhard Euler (1707–1783) proved that any two independent orthogonal coordinate frames with a common origin can be related by a sequence of three rotations about the local coordinate axes, where no two successive rotations may be about the same axis. In other words, we may transform a body coordinate frame B from the coincident position with a global frame G to any final orientation by minimum three rotations about the local axes provided that no two consequence rotations are about the same axis. In general, there are 12 different independent combinations of triple rotation about local axes. They are

$$
\begin{aligned}
&1 - A_{x,\psi}\, A_{y,\theta}\, A_{z,\varphi} \\
&2 - A_{y,\psi}\, A_{z,\theta}\, A_{x,\varphi} \\
&3 - A_{z,\psi}\, A_{x,\theta}\, A_{y,\varphi} \\
&4 - A_{z,\psi}\, A_{y,\theta}\, A_{x,\varphi} \\
&5 - A_{y,\psi}\, A_{x,\theta}\, A_{z,\varphi} \\
&6 - A_{x,\psi}\, A_{z,\theta}\, A_{y,\varphi} \\
&7 - A_{x,\psi}\, A_{y,\theta}\, A_{x,\varphi} \\
&8 - A_{y,\psi}\, A_{z,\theta}\, A_{y,\varphi} \\
&9 - A_{z,\psi}\, A_{x,\theta}\, A_{z,\varphi} \\
&10 - A_{x,\psi}\, A_{z,\theta}\, A_{x,\varphi} \\
&11 - A_{y,\psi}\, A_{x,\theta}\, A_{y,\varphi} \\
&12 - A_{z,\psi}\, A_{y,\theta}\, A_{z,\varphi}
\end{aligned}
\tag{2.146}
$$

The expanded form of the 12 local axes' triple rotation is presented in Appendix B.

As an example, assume we are given a transformation matrix $^{G}A_B$ relating the final orientation of a rigid body to the global coordinate frame.

$$^{G}A_B = \begin{bmatrix} 0.63 & 0.65 & -0.43 \\ -0.43 & 0.75 & 0.50 \\ 0.65 & -0.125 & 0.75 \end{bmatrix} \tag{2.147}$$

Selecting the first sequence of three rotations of case $1 - A_{x,\psi}\, A_{y,\theta}\, A_{z,\varphi}$ yields

$$^{G}A_B = A_{x,\psi}\, A_{y,\theta}\, A_{z,\varphi} \tag{2.148}$$

$$= \begin{bmatrix} c\theta\, c\varphi & c\theta\, s\varphi & -s\theta \\ -c\psi\, s\varphi + c\varphi\, s\theta\, s\psi & c\varphi\, c\psi + s\theta\, s\varphi\, s\psi & c\theta\, s\psi \\ s\varphi\, s\psi + c\varphi\, s\theta\, c\psi & -c\varphi\, s\psi + s\theta\, c\psi\, s\varphi & c\theta\, c\psi \end{bmatrix}$$

The element A_{13} gives θ, the ratio A_{12}/A_{11} gives φ, and the ratio A_{23}/A_{33} gives ψ.

$$\sin\theta = 0.43 \qquad \theta = (-1)^k\, 0.444 + k\pi\,\text{rad} \tag{2.149}$$

$$\tan\varphi = \frac{0.65}{0.63} \qquad \varphi = 0.801 + k\pi\,\text{rad} \tag{2.150}$$

$$\tan\psi = \frac{0.50}{0.75} \qquad \psi = 0.588 + k\pi\,\text{rad} \tag{2.151}$$

In this problem, we may check to find the matching solutions as follows:

$$\theta = (-1)\,0.444 + \pi\,\text{rad} = 2.697\,\text{rad} \tag{2.152}$$

$$\varphi = 0.801 + \pi\,\text{rad} = 3.943\,\text{rad} \tag{2.153}$$

$$\psi = 0.588 + \pi\,\text{rad} = 3.7296\,\text{rad} \tag{2.154}$$

Therefore, we will make the same overall transformation matrix (2.147) by turning the rigid body 3.943 rad about the z-axis, then 2.697 rad about the y-axis, and then 3.7296 rad about the x-axis.

$$^G A_B = A_{x,\psi}\, A_{y,\theta}\, A_{z,\varphi} = A_{x,3.73}\, A_{y,2.69}\, A_{z,3.94} \tag{2.155}$$

$$= \begin{bmatrix} 0.62806 & 0.6485 & -0.43009 \\ -0.43172 & 0.75023 & 0.50078 \\ 0.64742 & -0.12885 & 0.75116 \end{bmatrix}$$

Example 31 ★ Repeated rotation about body axes. A numerical example.

Consider a body frame B that turns φ about the z-axis. If n times fraction of 2π

$$\varphi = \frac{2\pi}{n} \qquad n \in \mathbb{N} \tag{2.156}$$

then we need to repeat the rotation n times to get back the body B to the original configuration. We can check it by multiplying $R_{z,\varphi}$ by itself until an identity matrix is achieved. To show this, we may find that

$$R_{z,\varphi}^m = \begin{bmatrix} \cos\varphi & \sin\varphi & 0 \\ -\sin\varphi & \cos\varphi & 0 \\ 0 & 0 & 1 \end{bmatrix}^m = \begin{bmatrix} \cos\dfrac{2\pi}{n} & \sin\dfrac{2\pi}{n} & 0 \\ -\sin\dfrac{2\pi}{n} & \cos\dfrac{2\pi}{n} & 0 \\ 0 & 0 & 1 \end{bmatrix}^m$$

$$= \begin{bmatrix} \cos\dfrac{2\pi m}{n} & \sin\dfrac{2\pi m}{n} & 0 \\ -\sin\dfrac{2\pi m}{n} & \cos\dfrac{2\pi m}{n} & 0 \\ 0 & 0 & 1 \end{bmatrix} \tag{2.157}$$

and when $m = n$, we have

$$R_{z,\varphi}^n = \begin{bmatrix} \cos n\dfrac{2\pi}{n} & \sin n\dfrac{2\pi}{n} & 0 \\ -\sin n\dfrac{2\pi}{n} & \cos n\dfrac{2\pi}{n} & 0 \\ 0 & 0 & 1 \end{bmatrix} = \begin{bmatrix} 1 & 0 & 0 \\ 0 & 1 & 0 \\ 0 & 0 & 1 \end{bmatrix} \tag{2.158}$$

Repeated rotation about any other body axis provides the same result.

Example 32 ★ Open problem. An unsolved problem, repeated rotation about local axes. The general form of this problem has never been solved in mathematics.

Consider a body frame B that turns φ about the x_i-axis followed by a rotation θ about the x_j-axis and ψ about the x_k-axis such that

$$\varphi = \frac{2\pi}{n_1} \qquad \theta = \frac{2\pi}{n_2} \qquad \psi = \frac{2\pi}{n_3} \qquad \{n_1, n_2, n_3\} \in \mathbb{N} \tag{2.159}$$

Although it seems that we need to repeat the rotations $n = n_1 \times n_2 \times n_3$ times to get back to the original configuration, it is not true in general. The determination of the required number n to repeat a general combined rotation to get back to the original orientation has not solved yet and is still an open question:

$$^B R_G = \prod_{j=1}^m R_{x_i,\varphi_j} \qquad i = 1, 2, 3 \tag{2.160}$$

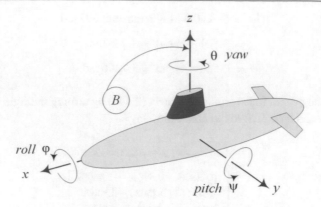

Fig. 2.12 Local roll–pitch–yaw angles

$$\varphi_j = \frac{2\pi}{n_j} \qquad\qquad m, n_j \in \mathbb{N} \tag{2.161}$$

$$^B R_G^n = [I] \qquad\qquad n = ? \tag{2.162}$$

Example 33 Local roll–pitch–yaw angles. Orientation expression of rigid bodies about local principal axes is very applied method in study of all types of vehicle dynamics.

Rotation about the x-axis of the local frame is called *roll* or *bank*, rotation about y-axis of the local frame is called *pitch* or *attitude*, and rotation about the z-axis of the local frame is called *yaw, spin,* or *heading*. The local roll–pitch–yaw angles are shown in Fig. 2.12.

The local roll–pitch–yaw rotation matrix is

$$^B A_G = A_{z,\psi}\, A_{y,\theta}\, A_{x,\varphi}$$

$$= \begin{bmatrix} c\theta\,c\psi & c\varphi\,s\psi + s\theta\,c\psi vs\varphi & s\varphi\,s\psi - c\varphi\,s\theta\,c\psi \\ -c\theta\,s\psi & c\varphi\,c\psi - s\theta\,s\varphi\,s\psi & c\psi\,s\varphi + c\varphi\,s\theta vs\psi \\ s\theta & -c\theta s\varphi & c\theta\,c\varphi \end{bmatrix} \tag{2.163}$$

$$^G Q_B = {}^B A_G^T = \left[A_{z,\psi}\, A_{y,\theta}\, A_{x,\varphi}\right]^T = A_{x,\varphi}^T\, A_{y,\theta}^T\, A_{z,\psi}^T$$

$$= Q_{X,\varphi}\, Q_{Y,\theta}\, Q_{Z,\psi}$$

$$= \begin{bmatrix} c\theta\,c\psi & -c\theta\,s\psi & s\theta \\ c\varphi\,s\psi + s\theta\,c\psi\,s\varphi & c\varphi\,c\psi - s\theta\,s\varphi\,s\psi & -c\theta\,s\varphi \\ s\varphi\,s\psi - c\varphi\,s\theta\,c\psi & c\psi\,s\varphi + c\varphi\,s\theta\,s\psi & c\theta\,c\varphi \end{bmatrix} \tag{2.164}$$

Note the difference between roll–pitch–yaw and Euler angles, although we show both utilizing $\varphi, \theta,$ and ψ. Having a rotation matrix $^B Q_G = [r_{ij}]$, provided that $\cos\theta \neq 0$, we are able to determine the equivalent local roll, pitch, and yaw angles by

$$\theta = \arcsin r_{31} \qquad \varphi = -\arctan\frac{r_{32}}{r_{33}} \qquad \psi = -\arctan\frac{r_{21}}{r_{11}} \tag{2.165}$$

Example 34 ★ Angular velocity and local roll–pitch–yaw rate. This is the first time introducing angular velocity without definition. Angular velocity is expressed as time rate of roll–pitch–yaw angles.

Using the roll–pitch–yaw frequencies, the angular velocity $_G\boldsymbol{\omega}_B$ of a body B with respect to the global reference frame G is

$$_G\boldsymbol{\omega}_B = \omega_x \hat{\imath} + \omega_y \hat{\jmath} + \omega_z \hat{k} = \dot{\varphi}\hat{e}_\varphi + \dot{\theta}\hat{e}_\theta + \dot{\psi}\hat{e}_\psi \tag{2.166}$$

Relationships between the components of $_G\boldsymbol{\omega}_B$ in body frame and roll–pitch–yaw components are found when the local roll unit vector $\hat{e}_\varphi$ and pitch unit vector $\hat{e}_\theta$ are transformed to the body frame. The roll unit vector $\hat{e}_\varphi = \begin{bmatrix} 1 & 0 & 0 \end{bmatrix}^T$ transforms to the body frame after rotation θ and then rotation ψ.

$$^B\hat{e}_\varphi = A_{z,\psi}\, A_{y,\theta}\begin{bmatrix}1\\0\\0\end{bmatrix} = \begin{bmatrix}\cos\theta\cos\psi\\-\cos\theta\sin\psi\\\sin\theta\end{bmatrix} \tag{2.167}$$

The pitch unit vector $\hat{e}_\theta = \begin{bmatrix}0 & 1 & 0\end{bmatrix}^T$ transforms to the body frame after rotation ψ.

$$^B\hat{e}_\theta = A_{z,\psi}\begin{bmatrix}0\\1\\0\end{bmatrix} = \begin{bmatrix}\sin\psi\\\cos\psi\\0\end{bmatrix} \tag{2.168}$$

The yaw unit vector $\hat{e}_\psi = \begin{bmatrix}0 & 0 & 1\end{bmatrix}^T$ is already along the local z-axis. Hence, $_G\omega_B$ can be expressed in body frame $B\,(Oxyz)$ as

$$
\begin{aligned}
^B_G\boldsymbol{\omega}_B &= \begin{bmatrix}\omega_x\\\omega_y\\\omega_z\end{bmatrix} = \dot\varphi\begin{bmatrix}\cos\theta\cos\psi\\-\cos\theta\sin\psi\\\sin\theta\end{bmatrix} + \dot\theta\begin{bmatrix}\sin\psi\\\cos\psi\\0\end{bmatrix} + \dot\psi\begin{bmatrix}0\\0\\1\end{bmatrix} \\
&= \begin{bmatrix}\cos\theta\cos\psi & \sin\psi & 0\\-\cos\theta\sin\psi & \cos\psi & 0\\\sin\theta & 0 & 1\end{bmatrix}\begin{bmatrix}\dot\varphi\\\dot\theta\\\dot\psi\end{bmatrix}
\end{aligned} \tag{2.169}
$$

and therefore, $_G\omega_B$ in global frame $G\,(OXYZ)$ in terms of local roll–pitch–yaw frequencies is

$$
\begin{aligned}
^G_G\boldsymbol{\omega}_B &= \begin{bmatrix}\omega_X\\\omega_Y\\\omega_Z\end{bmatrix} = {}^B A_G^{-1}\begin{bmatrix}\omega_x\\\omega_y\\\omega_z\end{bmatrix} = {}^B A_G^{-1}\begin{bmatrix}\dot\theta\sin\psi + \dot\varphi\cos\theta\cos\psi\\\dot\theta\cos\psi - \dot\varphi\cos\theta\sin\psi\\\dot\psi + \dot\varphi\sin\theta\end{bmatrix} \\
&= \begin{bmatrix}\dot\varphi + \dot\psi\sin\theta\\\dot\theta\cos\varphi - \dot\psi\cos\theta\sin\varphi\\\dot\theta\sin\varphi + \dot\psi\cos\theta\cos\varphi\end{bmatrix} \\
&= \begin{bmatrix}1 & 0 & \sin\theta\\0 & \cos\varphi & -\cos\theta\sin\varphi\\0 & \sin\varphi & \cos\theta\cos\varphi\end{bmatrix}\begin{bmatrix}\dot\varphi\\\dot\theta\\\dot\psi\end{bmatrix}
\end{aligned} \tag{2.170}
$$

2.5   Euler Angles

The rotation about the Z-axis of the global coordinate is called *precession*, the rotation about the x-axis of the local coordinate is called *nutation*, and the rotation about the z-axis of the local coordinate is called *spin*. The *precession–nutation–spin rotation* angles are also called *Euler angles*. Euler angles rotation matrix has many application in rigid body kinematics. The kinematics and dynamics of axisymmetric rigid bodies have simpler and more understandable expression based on Euler angles.

The Euler angle rotation matrix $^B A_G$ to transform a position vector from $G(OXYZ)$ to $B(Oxyz)$

$$^B\mathbf{r} = {}^B A_G\, {}^G\mathbf{r} \tag{2.171}$$

is

$$
\begin{aligned}
^B A_G &= A_{z,\psi}\, A_{x,\theta}\, A_{z,\varphi} \\
&= \begin{bmatrix}c\varphi\, c\psi - c\theta\, s\varphi\, s\psi & c\psi\, s\varphi + c\theta\, c\varphi\, s\psi & s\theta\, s\psi\\-c\varphi\, s\psi - c\theta\, c\psi\, s\varphi & -s\varphi\, s\psi + c\theta\, c\varphi\, c\psi & s\theta\, c\psi\\s\theta\, s\varphi & -c\varphi\, s\theta & c\theta\end{bmatrix}
\end{aligned} \tag{2.172}
$$

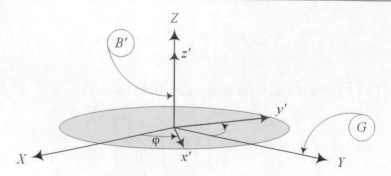

Fig. 2.13 First Euler angle

and hence, the Euler angle rotation matrix $^{G}Q_{B}$ to transform a position vector from $B(Oxyz)$ to $G(OXYZ)$

$$^{G}\mathbf{r} = \,^{G}Q_{B}\,^{B}\mathbf{r} \tag{2.173}$$

is

$$^{G}Q_{B} = \,^{B}A_{G}^{-1} = \,^{B}A_{G}^{T} = \left[A_{z,\psi}\,A_{x,\theta}\,A_{z,\varphi}\right]^{T}$$

$$= \begin{bmatrix} c\varphi\,c\psi - c\theta\,s\varphi\,s\psi & -c\varphi\,s\psi - c\theta\,c\psi\,s\varphi & s\theta\,s\varphi \\ c\psi\,s\varphi + c\theta\,c\varphi\,s\psi & -s\varphi\,s\psi + c\theta\,c\varphi\,c\psi & -c\varphi\,s\theta \\ s\theta\,s\psi & s\theta\,c\psi & c\theta \end{bmatrix} \tag{2.174}$$

Proof To find the Euler angle rotation matrix $^{B}R_{G}$ to go from the global frame $G(OXYZ)$ to the final body frame $B(Oxyz)$, we employ a temporary body frame $B'(Ox'y'z')$ as shown in Fig. 2.13 that before the first rotation coincides with the global frame. Let there be at first a rotation φ about the z'-axis. The first rotation φ about the Z-axis relates the coordinate systems by

$$^{G}\mathbf{r} = \,^{G}Q_{B'}\,^{B'}\mathbf{r} \tag{2.175}$$

Using matrix inversion, we find the body position for a given global position vector:

$$^{B'}\mathbf{r} = \,^{G}Q_{B'}^{-1}\,^{G}\mathbf{r} = \,^{B'}A_{G}\,^{G}\mathbf{r} \tag{2.176}$$

Therefore, the first rotation is equivalent to a rotation about the local z-axis by looking for $^{B'}\mathbf{r}$:

$$^{B'}\mathbf{r} = \,^{B'}A_{G}\,^{G}\mathbf{r} \tag{2.177}$$

$$^{B'}A_{G} = A_{z,\varphi} = \begin{bmatrix} \cos\varphi & \sin\varphi & 0 \\ -\sin\varphi & \cos\varphi & 0 \\ 0 & 0 & 1 \end{bmatrix} \tag{2.178}$$

Next we assume the frame $B'(Ox'y'z')$ to be a new fixed global frame and introduce a new body frame $B''(Ox''y''z'')$. Before the second rotation, the two frames coincide. Then, we apply a rotation θ about x''-axis as shown in Fig. 2.14. The transformation between $B'(Ox'y'z')$ and $B''(Ox''y''z'')$ is

$$^{B''}\mathbf{r} = \,^{B''}A_{B'}\,^{B'}\mathbf{r} \tag{2.179}$$

$$^{B''}A_{B'} = A_{x,\theta} = \begin{bmatrix} 1 & 0 & 0 \\ 0 & \cos\theta & \sin\theta \\ 0 & -\sin\theta & \cos\theta \end{bmatrix} \tag{2.180}$$

Finally we consider the frame $B''(Ox''y''z'')$ as a new fixed global frame and consider the final body frame $B(Oxyz)$ to coincide with B'' before the third rotation. We now execute a ψ rotation about the z''-axis as shown in Fig. 2.15. The transformation between $B''(Ox''y''z'')$ and $B(Oxyz)$ is

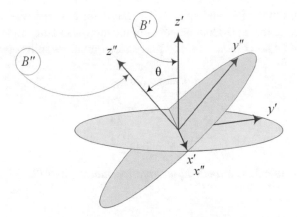

Fig. 2.14 Second Euler angle

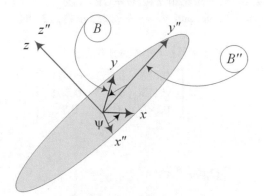

Fig. 2.15 Third Euler angle

$$^B\mathbf{r} = {}^B A_{B''} {}^{B''}\mathbf{r} \tag{2.181}$$

$$^B A_{B''} = A_{z,\psi} = \begin{bmatrix} \cos\psi & \sin\psi & 0 \\ -\sin\psi & \cos\psi & 0 \\ 0 & 0 & 1 \end{bmatrix} \tag{2.182}$$

By the rule of composition of rotations, the transformation from $G(OXYZ)$ to $B(Oxyz)$ is

$$^B\mathbf{r} = {}^B A_G {}^G\mathbf{r} \tag{2.183}$$

where

$$\begin{aligned}
^B A_G &= A_{z,\psi}\, A_{x,\theta}\, A_{z,\varphi} \\
&= \begin{bmatrix} c\varphi\,c\psi - c\theta\,s\varphi\,s\psi & c\psi\,s\varphi + c\theta\,c\varphi\,s\psi & s\theta\,s\psi \\ -c\varphi\,s\psi - c\theta\,c\psi\,s\varphi & -s\varphi\,s\psi + c\theta\,c\varphi\,c\psi & s\theta\,c\psi \\ s\theta\,s\varphi & -c\varphi\,s\theta & c\theta \end{bmatrix}
\end{aligned} \tag{2.184}$$

and therefore,

$$\begin{aligned}
^G Q_B &= {}^B A_G^{-1} = {}^B A_G^T = \left[A_{z,\psi}\, A_{x,\theta}\, A_{z,\varphi} \right]^T \\
&= \begin{bmatrix} c\varphi\,c\psi - c\theta\,s\varphi\,s\psi & -c\varphi\,s\psi - c\theta\,c\psi\,s\varphi & s\theta\,s\varphi \\ c\psi\,s\varphi + c\theta\,c\varphi\,s\psi & -s\varphi\,s\psi + c\theta\,c\varphi\,c\psi & -c\varphi\,s\theta \\ s\theta\,s\psi & s\theta\,c\psi & c\theta \end{bmatrix}
\end{aligned} \tag{2.185}$$

Given the angles of precession φ, nutation θ, and spin ψ, we can determine the overall rotation matrix using Equation (2.184). Also we are able to calculate the equivalent precession, nutation, and spin angles when a rotation matrix is given. If r_{ij} indicates the element of row i and column j of the precession–nutation–spin rotation matrix (2.185), provided that $\sin\theta \neq 0$, we have

$$\theta = \arccos r_{33} \tag{2.186}$$

$$\varphi = -\arctan\frac{r_{31}}{r_{32}} \tag{2.187}$$

$$\psi = \arctan\frac{r_{13}}{r_{23}} \tag{2.188}$$

The transformation matrix (2.184) is called a *local Euler rotation matrix*, and (2.185) is called a *global Euler rotation matrix*.   ∎

Example 35 Euler angle rotation matrix. A numerical example for local and global Euler angles transformation.

The Euler or precession–nutation–spin rotation matrix for $\varphi = 79.15$ deg, $\theta = 41.41$ deg, and $\psi = -40.7$ deg would be found by substituting $\varphi = 1.38$ rad, $\theta = 0.72$ rad, and $\psi = -0.71$ rad in Eq. (2.184)

$$^B A_G = A_{z,-0.71}\, A_{x,0.72}\, A_{z,1.38} \tag{2.189}$$

$$= \begin{bmatrix} 0.623 & 0.652 & -0.431 \\ -0.436 & 0.747 & 0.501 \\ 0.649 & -0.124 & 0.75 \end{bmatrix}$$

and the inverse of the transformation will be

$$^G Q_B = {}^B A_G^{-1} \tag{2.190}$$

$$= \begin{bmatrix} 0.623 & -0.436 & 0.65 \\ 0.653 & 0.748 & -0.124 \\ -0.431 & 0.501 & 0.751 \end{bmatrix}$$

Example 36 Euler angles of a local transformation matrix. A numerical example to determine Euler angles to provide the same rotation matrix made of given rotations.

The local rotation matrix after rotation of 30 deg about the z-axis, then 30 deg about the x-axis, and then 30 deg about the y-axis is

$$^B A_G = A_{y,\frac{\pi}{6}}\, A_{x,\frac{\pi}{6}}\, A_{z,\frac{\pi}{6}} \tag{2.191}$$

$$= \begin{bmatrix} 0.63 & 0.65 & -0.43 \\ -0.43 & 0.75 & 0.50 \\ 0.65 & -0.125 & 0.75 \end{bmatrix}$$

and therefore, the local coordinates of a sample point at $X = 5$, $Y = 30$, and $Z = 10$ are

$$\begin{bmatrix} x \\ y \\ z \end{bmatrix} = \begin{bmatrix} 0.63 & 0.65 & -0.43 \\ -0.43 & 0.75 & 0.50 \\ 0.65 & -0.125 & 0.75 \end{bmatrix} \begin{bmatrix} 5 \\ 30 \\ 10 \end{bmatrix} = \begin{bmatrix} 18.35 \\ 25.35 \\ 7.0 \end{bmatrix} \tag{2.192}$$

The Euler angles of the corresponding precession–nutation–spin rotation matrix are

$$\theta = \arccos(r_{33}) = \arccos 0.75 = 0.72\,\text{rad} = 41.4\,\text{deg} \tag{2.193}$$

$$\varphi = -\arctan\frac{r_{31}}{r_{32}} = -\arctan\frac{0.65}{-0.125} = 1.38\,\text{rad} = 79.1\,\text{deg} \tag{2.194}$$

$$\psi = \arctan\frac{r_{13}}{r_{23}} = \arctan\frac{-0.43}{0.50} = -0.71\,\text{rad} = -40.7\,\text{deg} \tag{2.195}$$

Hence, $A_{y,\frac{\pi}{6}} A_{x,\frac{\pi}{6}} A_{z,\frac{\pi}{6}} = A_{z,\psi} A_{x,\theta} A_{z,\varphi}$ when $\varphi = 1.38$ rad, $\theta = 0.72$ rad, and $\psi = -0.71$ rad.

$$^B A_G = A_{z,-0.71} A_{x,0.72} A_{z,1.38} \tag{2.196}$$

$$= \begin{bmatrix} 0.62306 & 0.65253 & -0.43128 \\ -0.43545 & 0.74741 & 0.50176 \\ 0.64976 & -0.12482 & 0.74982 \end{bmatrix}$$

In other words, the rigid body attached to the local frame moves to the final configuration by undergoing either three consecutive rotations $\varphi = 79.1$ deg, $\theta = 41.4$ deg, and $\psi = -40.7$ deg about z, x, and z axes, respectively, or three consecutive rotations 30 deg, 30 deg, and 30 deg about z, x, and y axes.

Example 37 Relative rotation matrix of two bodies. A numerical example to relate two sets of Euler angles rotations.

Consider a rigid body B_1 with an orientation matrix $^{B_1} A_G$ made by Euler angles $\varphi = 30$ deg, $\theta = -45$ deg, $\psi = 60$ deg, and another rigid body B_2 having $\varphi = 10$ deg, $\theta = 25$ deg, $\psi = -15$ deg, with respect to the global frame. To find the relative rotation matrix $^{B_1} A_{B_2}$ to map the coordinates of second body frame B_2 to the first body frame B_1, we need to find the individual rotation matrices first.

$$^{B_1} A_G = A_{z,60} A_{x,-45} A_{z,30} \tag{2.197}$$

$$= \begin{bmatrix} 0.127 & 0.78 & -0.612 \\ -0.927 & -0.127 & -0.354 \\ -0.354 & 0.612 & 0.707 \end{bmatrix}$$

$$^{B_2} A_G = A_{z,10} A_{x,25} A_{z,-15} \tag{2.198}$$

$$= \begin{bmatrix} 0.992 & -0.0633 & -0.109 \\ 0.103 & 0.907 & 0.408 \\ 0.0734 & -0.416 & 0.906 \end{bmatrix}$$

The desired rotation matrix $^{B_1} A_{B_2}$ may be found by

$$^{B_1} A_{B_2} = {}^{B_1} A_G \, {}^G A_{B_2} \tag{2.199}$$

which is equal to

$$^{B_1} A_{B_2} = {}^{B_1} A_G \, {}^{B_2} A_G^{-1} = {}^{B_1} A_G \, {}^{B_2} A_G^T \tag{2.200}$$

$$= \begin{bmatrix} 0.14353 & 0.47084 & -0.87026 \\ -0.87277 & -0.35498 & -0.33587 \\ -0.46737 & 0.80748 & 0.36049 \end{bmatrix}$$

Example 38 ★ Euler angle rotation matrix for small angles. It shows what happens when Euler angles are very small.

The Euler rotation matrix $^B A_G = A_{z,\psi} A_{x,\theta} A_{z,\varphi}$ for very small Euler angles φ, θ, and ψ is approximated by

$$^B A_G = \begin{bmatrix} 1 & \gamma & 0 \\ -\gamma & 1 & \theta \\ 0 & -\theta & 1 \end{bmatrix} \tag{2.201}$$

where

$$\gamma = \varphi + \psi \tag{2.202}$$

Therefore, in case of small angles of rotation, the angles φ and ψ are indistinguishable.

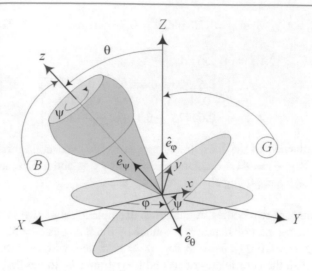

Fig. 2.16 Application of Euler angles in describing the configuration of a top

If only $\theta \to 0$, then the Euler rotation matrix $^B A_G = A_{z,\psi} A_{x,\theta} A_{z,\varphi}$ approaches to

$$
^B A_G = \begin{bmatrix} c\varphi\,c\psi - s\varphi\,s\psi & c\psi\,s\varphi + c\varphi\,s\psi & 0 \\ -c\varphi\,s\psi - c\psi\,s\varphi & -s\varphi\,s\psi + c\varphi\,c\psi & 0 \\ 0 & 0 & 1 \end{bmatrix}
$$

$$
= \begin{bmatrix} \cos(\varphi+\psi) & \sin(\varphi+\psi) & 0 \\ -\sin(\varphi+\psi) & \cos(\varphi+\psi) & 0 \\ 0 & 0 & 1 \end{bmatrix} \tag{2.203}
$$

and therefore, the angles φ and ψ are indistinguishable even if the value of φ and ψ is finite. Hence, the Euler set of angles in rotation matrix (2.172) is not unique when $\theta = 0$.

Example 39 Euler angles application in motion of rigid bodies. Euler angles are the best way to express orientation of a top.

The Zxz Euler angles are good parameters to describe the configuration of a rigid body with a fixed point. The Euler angles to show the configuration of a top are shown in Fig. 2.16 as an example.

The Zxz Euler angles seem to be natural parameters to express the configuration of rotating axisymmetric rigid bodies. Euler angles that show the expression of a top in Fig. 2.16 are a good example. The rotation of the top about its axis of symmetry is called spin ψ. The angle between the axis of symmetry and the Z-axis is called nutation θ, and the rotation of the axis of symmetry about the Z-axis is called precession φ. The motion of the Earth will also be expressed by Euler angles better.

Example 40 ★ Angular velocity vector in terms of Euler frequencies. Angular velocity is expressed as time rate of Euler angles.

An Eulerian local frame $E\left(O, \hat{e}_\varphi, \hat{e}_\theta, \hat{e}_\psi\right)$ can be introduced by defining unit vectors $\hat{e}_\varphi$, $\hat{e}_\theta$, and $\hat{e}_\psi$ as shown in Fig. 2.17. Although the Eulerian frame is not necessarily orthogonal, it is very useful in rigid body kinematic analysis. The angular velocity vector $_G\boldsymbol{\omega}_B$ of the body frame $B(Oxyz)$ with respect to the global frame $G(OXYZ)$ can be expressed in the non-orthogonal Euler angles frame E as the sum of three Euler angle rate vectors.

$$
^E_G\boldsymbol{\omega}_B = \dot{\varphi}\,\hat{e}_\varphi + \dot{\theta}\,\hat{e}_\theta + \dot{\psi}\,\hat{e}_\psi \tag{2.204}
$$

The rate of Euler angles, $\dot{\varphi}$, $\dot{\theta}$, and $\dot{\psi}$, are called *Euler frequencies*.

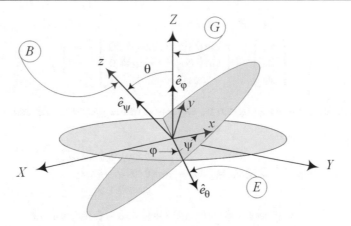

Fig. 2.17 Euler angles frame $\hat{e}_\varphi, \hat{e}_\theta, \hat{e}_\psi$

To find $_G\boldsymbol{\omega}_B$ in body frame, we must express the unit vectors $\hat{e}_\varphi$, $\hat{e}_\theta$, and $\hat{e}_\psi$ shown in Fig. 2.17, in the body frame B. The unit vector $\hat{e}_\varphi = \hat{K}$ is in the global frame G.

$$\hat{e}_\varphi = \begin{bmatrix} 0 & 0 & 1 \end{bmatrix}^T = \hat{K} \tag{2.205}$$

It can be transformed to the body frame after three rotations.

$$^B\hat{e}_\varphi = {}^B A_G \, \hat{K} = A_{z,\psi} \, A_{x,\theta} \, A_{z,\varphi} \, \hat{K} = \begin{bmatrix} \sin\theta \sin\psi \\ \sin\theta \cos\psi \\ \cos\theta \end{bmatrix} \tag{2.206}$$

The unit vector $\hat{e}_\theta = \hat{\imath}'$ is in the intermediate frame $Ox'y'z'$,

$$\hat{e}_\theta = \begin{bmatrix} 1 & 0 & 0 \end{bmatrix}^T = \hat{\imath}' \tag{2.207}$$

and it needs to get two rotations $A_{x,\theta}$ and $A_{z,\psi}$ to be transformed to the body frame.

$$^B\hat{e}_\theta = {}^B A_{Ox'y'z'} \, \hat{\imath}' = A_{z,\psi} \, A_{x,\theta} \, \hat{\imath}' = \begin{bmatrix} \cos\psi \\ -\sin\psi \\ 0 \end{bmatrix} \tag{2.208}$$

The unit vector $\hat{e}_\psi$ is already in the body frame.

$$\hat{e}_\psi = \begin{bmatrix} 0 & 0 & 1 \end{bmatrix}^T = \hat{k} \tag{2.209}$$

Therefore, $_G\boldsymbol{\omega}_B$ is expressed in body coordinate frame.

$$\begin{aligned}
{}^B_G\boldsymbol{\omega}_B &= \dot{\varphi} \begin{bmatrix} \sin\theta \sin\psi \\ \sin\theta \cos\psi \\ \cos\theta \end{bmatrix} + \dot{\theta} \begin{bmatrix} \cos\psi \\ -\sin\psi \\ 0 \end{bmatrix} + \dot{\psi} \begin{bmatrix} 0 \\ 0 \\ 1 \end{bmatrix} \\
&= \left(\dot{\varphi} \sin\theta \sin\psi + \dot{\theta} \cos\psi \right) \hat{\imath} + \left(\dot{\varphi} \sin\theta \cos\psi - \dot{\theta} \sin\psi \right) \hat{\jmath} \\
&\quad + \left(\dot{\varphi} \cos\theta + \dot{\psi} \right) \hat{k}
\end{aligned} \tag{2.210}$$

Therefore, components of $_G\boldsymbol{\omega}_B$ in body frame $B\,(Oxyz)$ are related to the Euler angle frame $E\,(O\varphi\theta\psi)$ by the following equation, in which, the transformation matrix $^B A_E$ is not orthogonal and hence is not a rotation matrix.

$$_G^B\boldsymbol{\omega}_B = {}^BA_E \, _G^E\boldsymbol{\omega}_B \tag{2.211}$$

$$\begin{bmatrix} \omega_x \\ \omega_y \\ \omega_z \end{bmatrix} = \begin{bmatrix} \sin\theta\sin\psi & \cos\psi & 0 \\ \sin\theta\cos\psi & -\sin\psi & 0 \\ \cos\theta & 0 & 1 \end{bmatrix} \begin{bmatrix} \dot\varphi \\ \dot\theta \\ \dot\psi \end{bmatrix} \tag{2.212}$$

Then, G-expression of the angular velocity $_G\boldsymbol{\omega}_B$ can be expressed in the global frame using an inverse transformation of Euler rotation matrix.

$$_G^G\boldsymbol{\omega}_B = {}^BA_G^{-1} \, _G^B\boldsymbol{\omega}_B = {}^BA_G^{-1} \begin{bmatrix} \dot\varphi\sin\theta\sin\psi + \dot\theta\cos\psi \\ \dot\varphi\sin\theta\cos\psi - \dot\theta\sin\psi \\ \dot\varphi\cos\theta + \dot\psi \end{bmatrix}$$

$$= \left(\dot\theta\cos\varphi + \dot\psi\sin\theta\sin\varphi\right)\hat{I} + \left(\dot\theta\sin\varphi - \dot\psi\cos\varphi\sin\theta\right)\hat{J}$$

$$+ \left(\dot\varphi + \dot\psi\cos\theta\right)\hat{K} \tag{2.213}$$

The components of $_G\boldsymbol{\omega}_B$ in the global coordinate frame $G\,(OXYZ)$ are related to the Euler angle coordinate frame $E\,(O\varphi\theta\psi)$ by the following relationship, in which, the transformation matrix ^GQ_E is again a non-orthogonal matrix.

$$_G^G\boldsymbol{\omega}_B = {}^GQ_E \, _G^E\boldsymbol{\omega}_B \tag{2.214}$$

$$\begin{bmatrix} \omega_X \\ \omega_Y \\ \omega_Z \end{bmatrix} = \begin{bmatrix} 0 & \cos\varphi & \sin\theta\sin\varphi \\ 0 & \sin\varphi & -\cos\varphi\sin\theta \\ 1 & 0 & \cos\theta \end{bmatrix} \begin{bmatrix} \dot\varphi \\ \dot\theta \\ \dot\psi \end{bmatrix} \tag{2.215}$$

Example 41 ★ Euler frequencies based on a Cartesian frequencies. Relationship between angular velocity expression by Euler frequencies and Cartesian frequencies.

The vector $_G^B\boldsymbol{\omega}_B$ that indicates the angular velocity of a rigid body B with respect to the global frame G expressed in frame B is related to the Euler frequencies by

$$_G^B\boldsymbol{\omega}_B = {}^BA_E \, _G^E\boldsymbol{\omega}_B \tag{2.216}$$

$$_G^B\boldsymbol{\omega}_B = \begin{bmatrix} \omega_x \\ \omega_y \\ \omega_z \end{bmatrix} = \begin{bmatrix} \sin\theta\sin\psi & \cos\psi & 0 \\ \sin\theta\cos\psi & -\sin\psi & 0 \\ \cos\theta & 0 & 1 \end{bmatrix} \begin{bmatrix} \dot\varphi \\ \dot\theta \\ \dot\psi \end{bmatrix} \tag{2.217}$$

The coefficient matrix ^BA_E is not an orthogonal matrix because its transpose is not equal to its inverse.

$$^BA_E^T \neq {}^BA_E^{-1} \tag{2.218}$$

$$^BA_E^T = \begin{bmatrix} \sin\theta\sin\psi & \sin\theta\cos\psi & \cos\theta \\ \cos\psi & -\sin\psi & 0 \\ 0 & 0 & 1 \end{bmatrix} \tag{2.219}$$

$$^BA_E^{-1} = \frac{1}{\sin\theta} \begin{bmatrix} \sin\psi & \cos\psi & 0 \\ \sin\theta\cos\psi & -\sin\theta\sin\psi & 0 \\ -\cos\theta\sin\psi & -\cos\theta\cos\psi & 1 \end{bmatrix} \tag{2.220}$$

It is because the Euler angles coordinate frame $E\,(O\varphi\theta\psi)$ is not an orthogonal frame. For the same reason, the matrix of coefficients that relates the Euler frequencies and the components of $_G^G\boldsymbol{\omega}_B$ is also not an orthogonal matrix.

$$_G^G\boldsymbol{\omega}_B = {}^GQ_E \, _G^E\boldsymbol{\omega}_B \tag{2.221}$$

$$_G^G\boldsymbol{\omega}_B = \begin{bmatrix} \omega_X \\ \omega_Y \\ \omega_Z \end{bmatrix} = \begin{bmatrix} 0 & \cos\varphi & \sin\theta\sin\varphi \\ 0 & \sin\varphi & -\cos\varphi\sin\theta \\ 1 & 0 & \cos\theta \end{bmatrix} \begin{bmatrix} \dot\varphi \\ \dot\theta \\ \dot\psi \end{bmatrix} \tag{2.222}$$

Therefore, the Euler frequencies based on local and global decomposition of the angular velocity vector $_G\boldsymbol{\omega}_B$ must solely be found by the inverse of coefficient matrices

$$_G^E\boldsymbol{\omega}_B = {}^BA_E^{-1}\,{}_G^B\boldsymbol{\omega}_B \tag{2.223}$$

$$\begin{bmatrix} \dot{\varphi} \\ \dot{\theta} \\ \dot{\psi} \end{bmatrix} = \frac{1}{\sin\theta} \begin{bmatrix} \sin\psi & \cos\psi & 0 \\ \sin\theta\cos\psi & -\sin\theta\sin\psi & 0 \\ -\cos\theta\sin\psi & -\cos\theta\cos\psi & 1 \end{bmatrix} \begin{bmatrix} \omega_x \\ \omega_y \\ \omega_z \end{bmatrix} \tag{2.224}$$

$$_G^E\boldsymbol{\omega}_B = {}^GQ_E^{-1}\,{}_G^G\boldsymbol{\omega}_B \tag{2.225}$$

$$\begin{bmatrix} \dot{\varphi} \\ \dot{\theta} \\ \dot{\psi} \end{bmatrix} = \frac{1}{\sin\theta} \begin{bmatrix} -\cos\theta\sin\varphi & \cos\theta\cos\varphi & 1 \\ \sin\theta\cos\varphi & \sin\theta\sin\varphi & 0 \\ \sin\varphi & -\cos\varphi & 0 \end{bmatrix} \begin{bmatrix} \omega_X \\ \omega_Y \\ \omega_Z \end{bmatrix} \tag{2.226}$$

Using (2.223) and (2.225), it can be verified that the transformation matrix $^BA_G = {}^BA_E\,{}^GQ_E^{-1}$ would be the same as Euler transformation matrix (2.184). The angular velocity vector can thus be expressed as

$$_G^B\boldsymbol{\omega}_B = \begin{bmatrix} \hat{\imath} & \hat{\jmath} & \hat{k} \end{bmatrix} \begin{bmatrix} \omega_x \\ \omega_y \\ \omega_z \end{bmatrix} \tag{2.227}$$

$$_G\boldsymbol{\omega}_B = \begin{bmatrix} \hat{I} & \hat{J} & \hat{K} \end{bmatrix} \begin{bmatrix} \omega_X \\ \omega_Y \\ \omega_Z \end{bmatrix} \tag{2.228}$$

$$_G^E\boldsymbol{\omega}_B = \begin{bmatrix} \hat{K} & \hat{e}_\theta & \hat{k} \end{bmatrix} \begin{bmatrix} \dot{\varphi} \\ \dot{\theta} \\ \dot{\psi} \end{bmatrix} \tag{2.229}$$

Example 42 ★ Integrability of the angular velocity components. The definition of integrability and proving angular velocity is integrable if it is based on Euler frequencies.

The integrability condition for an arbitrary total differential of $f = f(x, y)$

$$df = f_1 dx + f_2 dy = \frac{\partial f}{\partial x} dx + \frac{\partial f}{\partial y} dy \tag{2.230}$$

is

$$\frac{\partial f_1}{\partial y} = \frac{\partial f_2}{\partial x} \tag{2.231}$$

The angular velocity components ω_x, ω_y, and ω_z along the body coordinate axes x, y, and z cannot be integrated to obtain the associated angles because they are not total differentials. As an example,

$$\omega_x dt = \sin\theta\sin\psi\, d\varphi + \cos\psi\, d\theta \tag{2.232}$$

$$\frac{\partial(\sin\theta\sin\psi)}{\partial\theta} \neq \frac{\partial\cos\psi}{\partial\varphi} \tag{2.233}$$

However, the integrability condition (2.231) is satisfied by the Euler frequencies. From (2.224), we have

$$d\varphi = \frac{\sin\psi}{\sin\theta}(\omega_x dt) + \frac{\cos\psi}{\sin\theta}(\omega_y dt) \tag{2.234}$$

$$d\theta = \cos\psi\,(\omega_x dt) - \sin\psi\,(\omega_y dt) \tag{2.235}$$

$$d\psi = \frac{-\cos\theta\sin\psi}{\sin\theta}(\omega_x dt) + \frac{-\cos\theta\cos\psi}{\sin\theta}(\omega_y dt) + \frac{(\omega_z dt)}{\sin\theta} \tag{2.236}$$

These equations indicate that

$$\frac{\sin \psi}{\sin \theta} = \frac{\partial \varphi}{\partial (\omega_x \, dt)} \qquad \frac{\cos \psi}{\sin \theta} = \frac{\partial \varphi}{\partial (\omega_y \, dt)} \tag{2.237}$$

$$\cos \psi = \frac{\partial \theta}{\partial (\omega_x \, dt)} \qquad - \sin \psi = \frac{\partial \theta}{\partial (\omega_y \, dt)} \tag{2.238}$$

$$\frac{- \cos \theta \sin \psi}{\sin \theta} = \frac{\partial \psi}{\partial (\omega_x \, dt)} \qquad \frac{- \cos \theta \cos \psi}{\sin \theta} = \frac{\partial \psi}{\partial (\omega_y \, dt)} \tag{2.239}$$

$$\frac{1}{\sin \theta} = \frac{\partial \psi}{\partial (\omega_z \, dt)} \tag{2.240}$$

and therefore,

$$
\begin{aligned}
\frac{\partial}{\partial (\omega_y \, dt)} \left(\frac{\sin \psi}{\sin \theta} \right) &= \frac{1}{\sin^2 \theta} \left(\cos \psi \sin \theta \frac{\partial \psi}{\partial (\omega_y \, dt)} - \cos \theta \sin \psi \frac{\partial \theta}{\partial (\omega_y \, dt)} \right) \\
&= \frac{1}{\sin^2 \theta} (- \cos \psi \cos \theta \cos \psi + \cos \theta \sin \psi \sin \psi) \\
&= \frac{- \cos \theta \cos 2\psi}{\sin^2 \theta}
\end{aligned} \tag{2.241}
$$

$$
\begin{aligned}
\frac{\partial}{\partial (\omega_x \, dt)} \left(\frac{\cos \psi}{\sin \theta} \right) &= \frac{-1}{\sin^2 \theta} \left(\sin \psi \sin \theta \frac{\partial \psi}{\partial (\omega_x \, dt)} + \cos \theta \cos \psi \frac{\partial \theta}{\partial (\omega_x \, dt)} \right) \\
&= \frac{1}{\sin^2 \theta} (\sin \psi \cos \theta \sin \psi - \cos \theta \cos \psi \cos \psi) \\
&= \frac{- \cos \theta \cos 2\psi}{\sin^2 \theta}
\end{aligned} \tag{2.242}
$$

It can be checked that $d\theta$ and $d\psi$ are also integrable (Rimrott, 1989).

Example 43 ★ Cardan angles and frequencies. The definition of Cardan angles and relationship between angular velocity expression by Euler frequencies and Cardan frequencies.

The system of Euler angles is singular at $\theta = 0$, and as a consequence, φ and ψ become coplanar and indistinguishable. From the 12 angle systems of Appendix B, each with certain names, characteristics, advantages, and disadvantages, the rotations about three different axes such as ${}^B A_G = A_{z,\psi} A_{y,\theta} A_{x,\varphi}$ are called *Cardan angles* after the Italian mathematician Geronimo Cardano (1501–1576) or *Bryant angles*. The Cardan angle system is not singular at $\theta = 0$ and has some application in mechatronics and attitude analysis of satellites in a central force field.

$$
\begin{aligned}
{}^B A_G &= A_{z,\psi} A_{y,\theta} A_{x,\varphi} \\
&= \begin{bmatrix} c\theta \, c\psi & c\varphi \, s\psi + s\theta \, c\psi \, s\varphi & s\varphi \, s\psi - c\varphi \, s\theta \, c\psi \\ -c\theta \, s\psi & c\varphi \, c\psi - s\theta \, s\varphi \, s\psi & c\psi \, s\varphi + c\varphi \, s\theta \, s\psi \\ s\theta & -c\theta \, s\varphi & c\theta \, c\varphi \end{bmatrix}
\end{aligned} \tag{2.243}
$$

The angular velocity ${}_G \boldsymbol{\omega}_B$ of a rigid body with respect to the global frame can either be expressed in terms of the components along the axes of $B(Oxyz)$ or in terms of the *Cardan frequencies* along the axes of the non-orthogonal Cardan frame. The angular velocity in terms of Cardan frequencies is

$$
{}_G \boldsymbol{\omega}_B = \dot{\varphi} \, A_{z,\psi} A_{y,\theta} \begin{bmatrix} 1 \\ 0 \\ 0 \end{bmatrix} + \dot{\theta} \, A_{z,\psi} \begin{bmatrix} 0 \\ 1 \\ 0 \end{bmatrix} + \dot{\psi} \begin{bmatrix} 0 \\ 0 \\ 1 \end{bmatrix} \tag{2.244}
$$

and therefore,

$$\begin{bmatrix} \omega_x \\ \omega_y \\ \omega_z \end{bmatrix} = \begin{bmatrix} \cos\theta\cos\psi & \sin\psi & 0 \\ -\cos\theta\sin\psi & \cos\psi & 0 \\ \sin\theta & 0 & 1 \end{bmatrix} \begin{bmatrix} \dot\varphi \\ \dot\theta \\ \dot\psi \end{bmatrix} \tag{2.245}$$

$$\begin{bmatrix} \dot\varphi \\ \dot\theta \\ \dot\psi \end{bmatrix} = \begin{bmatrix} \dfrac{\cos\psi}{\cos\theta} & -\dfrac{\sin\psi}{\cos\theta} & 0 \\ \sin\psi & \cos\psi & 0 \\ -\tan\theta\cos\psi & \tan\theta\sin\psi & 1 \end{bmatrix} \begin{bmatrix} \omega_x \\ \omega_y \\ \omega_z \end{bmatrix} \tag{2.246}$$

In case of small Cardan angles, we have

$$^{B}A_G = \begin{bmatrix} 1 & \psi & -\theta \\ -\psi & 1 & \varphi \\ \theta & -\varphi & 1 \end{bmatrix} \tag{2.247}$$

and

$$\begin{bmatrix} \omega_x \\ \omega_y \\ \omega_z \end{bmatrix} = \begin{bmatrix} 1 & \psi & 0 \\ -\psi & 1 & 0 \\ \theta & 0 & 1 \end{bmatrix} \begin{bmatrix} \dot\varphi \\ \dot\theta \\ \dot\psi \end{bmatrix} \tag{2.248}$$

$$\begin{bmatrix} \dot\varphi \\ \dot\theta \\ \dot\psi \end{bmatrix} = \begin{bmatrix} \psi & -\psi & 0 \\ \psi & 1 & 0 \\ -\theta & 0 & 1 \end{bmatrix} \begin{bmatrix} \omega_x \\ \omega_y \\ \omega_z \end{bmatrix} \tag{2.249}$$

2.6 Local Axes Versus Global Axes Rotation

The global rotation matrix $^{G}Q_B$ is equal to the inverse of the local rotation matrix $^{B}A_G$ and vice versa,

$$^{G}Q_B = {}^{B}A_G^{-1} \qquad {}^{B}A_G = {}^{G}Q_B^{-1} \tag{2.250}$$

where

$$^{G}Q_B = A_1^{-1} A_2^{-1} A_3^{-1} \cdots A_n^{-1} \tag{2.251}$$

$$^{B}A_G = Q_1^{-1} Q_2^{-1} Q_3^{-1} \cdots Q_n^{-1} \tag{2.252}$$

Also, premultiplication of the global rotation matrix is equal to postmultiplication of the local rotation matrix.

Proof Consider a sequence of global rotations and their resultant rotation matrix $^{G}Q_B$ to transform a position vector $^{B}\mathbf{r}$ to $^{G}\mathbf{r}$.

$$^{G}\mathbf{r} = {}^{G}Q_B\,{}^{B}\mathbf{r} \tag{2.253}$$

The global position vector $^{G}\mathbf{r}$ can also be transformed to $^{B}\mathbf{r}$ using a local rotation matrix $^{B}A_G$.

$$^{B}\mathbf{r} = {}^{B}A_G\,{}^{G}\mathbf{r} \tag{2.254}$$

Combining Eqs. (2.253) and (2.254) yields

$$^{G}\mathbf{r} = {}^{G}Q_B\,{}^{B}A_G\,{}^{G}\mathbf{r} \tag{2.255}$$

$$^{B}\mathbf{r} = {}^{B}A_G\,{}^{G}Q_B\,{}^{B}\mathbf{r} \tag{2.256}$$

and hence,

$$^{G}Q_B\,{}^{B}A_G = {}^{B}A_G\,{}^{G}Q_B = \mathbf{I} \tag{2.257}$$

Therefore, the global and local rotation matrices are inverse of each other.

$$^GQ_B = {}^BA_G^{-1}$$
$$^GQ_B^{-1} = {}^BA_G \tag{2.258}$$

Assume $^GQ_B = Q_n \cdots Q_3 Q_2 Q_1$ and $^BA_G = A_n \cdots A_3 A_2 A_1$, and we have

$$^GQ_B = {}^BA_G^{-1} = A_1^{-1} A_2^{-1} A_3^{-1} \cdots A_n^{-1} \tag{2.259}$$
$$^BA_G = {}^GQ_B^{-1} = Q_1^{-1} Q_2^{-1} Q_3^{-1} \cdots Q_n^{-1} \tag{2.260}$$

and Eq. (2.257) becomes

$$Q_n \cdots Q_2 Q_1 A_n \cdots A_2 A_1 = A_n \cdots A_2 A_1 Q_n \cdots Q_2 Q_1 = \mathbf{I} \tag{2.261}$$

and therefore,

$$Q_n \cdots Q_3 Q_2 Q_1 = A_1^{-1} A_2^{-1} A_3^{-1} \cdots A_n^{-1}$$
$$A_n \cdots A_3 A_2 A_1 = Q_1^{-1} Q_2^{-1} Q_3^{-1} \cdots Q_n^{-1} \tag{2.262}$$

or

$$Q_1^{-1} Q_2^{-1} Q_3^{-1} \cdots Q_n^{-1} Q_n \cdots Q_3 Q_2 Q_1 = \mathbf{I} \tag{2.263}$$
$$A_1^{-1} A_2^{-1} A_3^{-1} \cdots A_n^{-1} A_n \cdots A_3 A_2 A_1 = \mathbf{I} \tag{2.264}$$

Hence, the effect of in order rotations about the global coordinate axes is equivalent to the effect of the same rotations about the local coordinate axes performed in the reverse order.

$$^GQ_B = A_1^{-1} A_2^{-1} A_3^{-1} \cdots A_n^{-1} \tag{2.265}$$
$$^BA_G = Q_1^{-1} Q_2^{-1} Q_3^{-1} \cdots Q_n^{-1} \tag{2.266}$$

$\blacksquare$

Example 44 Global position and postmultiplication of rotation matrix. A numerical example to show how post and premultiplication by rotation matrix.

The local position of a point P after rotation is at $^B\mathbf{r} = \begin{bmatrix} 1 & 2 & 3 \end{bmatrix}^T$. If the local rotation matrix to transform $^G\mathbf{r}$ to $^B\mathbf{r}$ is given as

$$^BA_{z,\varphi} = \begin{bmatrix} \cos\varphi & \sin\varphi & 0 \\ -\sin\varphi & \cos\varphi & 0 \\ 0 & 0 & 1 \end{bmatrix} = \begin{bmatrix} \cos 30 & \sin 30 & 0 \\ -\sin 30 & \cos 30 & 0 \\ 0 & 0 & 1 \end{bmatrix} \tag{2.267}$$

then we may find the global position vector $^G\mathbf{r}$ by postmultiplication $^BA_{z,\varphi}$ and the local position vector $^B\mathbf{r}^T$,

$$^G\mathbf{r}^T = {}^B\mathbf{r}^T \, {}^BA_{z,\varphi} = \begin{bmatrix} 1 & 2 & 3 \end{bmatrix} \begin{bmatrix} \cos 30 & \sin 30 & 0 \\ -\sin 30 & \cos 30 & 0 \\ 0 & 0 & 1 \end{bmatrix}$$

$$= \begin{bmatrix} -0.13 & 2.23 & 3.0 \end{bmatrix} \tag{2.268}$$

instead of premultiplication of $^BA_{z,\varphi}^{-1}$ by $^B\mathbf{r}$.

$$^G\mathbf{r} = {}^BA_{z,\varphi}^{-1} \, {}^B\mathbf{r}$$

$$= \begin{bmatrix} \cos 30 & -\sin 30 & 0 \\ \sin 30 & \cos 30 & 0 \\ 0 & 0 & 1 \end{bmatrix} \begin{bmatrix} 1 \\ 2 \\ 3 \end{bmatrix} = \begin{bmatrix} -0.13 \\ 2.23 \\ 3.0 \end{bmatrix} \tag{2.269}$$

2.7   General Transformation

Consider a general situation in which two coordinate frames, $G(OXYZ)$ and $B(Oxyz)$ with a common origin O, are employed to express the components of a given vector $\mathbf{r}$. There is always a *transformation matrix* $^G R_B$ to map the components of $\mathbf{r}$ from the reference frame $B(Oxyz)$ to the other reference frame $G(OXYZ)$.

$$^G \mathbf{r} = {}^G R_B \, {}^B \mathbf{r} \tag{2.270}$$

We may also reverse the equation, $^B \mathbf{r} = {}^G R_B^{-1} \, {}^G \mathbf{r}$, to map the components of $\mathbf{r}$ from the reference frame $G(OXYZ)$ to the other reference frame $B(Oxyz)$ by introducing $^B R_G$

$$^B \mathbf{r} = {}^B R_G \, {}^G \mathbf{r} \tag{2.271}$$

where

$$^B R_G = {}^G R_B^{-1} = {}^G R_B^T \tag{2.272}$$

$$\left| {}^G R_B \right| = \left| {}^B R_G \right| = 1 \tag{2.273}$$

When the coordinate frames B and G are orthogonal, the rotation matrix $^G R_B$ is called an *orthogonal matrix*. The transpose R^T and inverse R^{-1} of an orthogonal matrix $[R]$ are equal:

$$R^T = R^{-1} \tag{2.274}$$

Because of the matrix orthogonality condition, only three of the nine elements of a transformation matrix $^G R_B$ are independent.

Proof Decomposition of the unit vectors of $G(OXYZ)$ along the axes of $B(Oxyz)$

$$\hat{I} = (\hat{I} \cdot \hat{\imath})\hat{\imath} + (\hat{I} \cdot \hat{\jmath})\hat{\jmath} + (\hat{I} \cdot \hat{k})\hat{k} \tag{2.275}$$

$$\hat{J} = (\hat{J} \cdot \hat{\imath})\hat{\imath} + (\hat{J} \cdot \hat{\jmath})\hat{\jmath} + (\hat{J} \cdot \hat{k})\hat{k} \tag{2.276}$$

$$\hat{K} = (\hat{K} \cdot \hat{\imath})\hat{\imath} + (\hat{K} \cdot \hat{\jmath})\hat{\jmath} + (\hat{K} \cdot \hat{k})\hat{k} \tag{2.277}$$

introduces the transformation matrix $^G R_B$ to map the local frame to the global frame

$$\begin{bmatrix} \hat{I} \\ \hat{J} \\ \hat{K} \end{bmatrix} = \begin{bmatrix} \hat{I} \cdot \hat{\imath} & \hat{I} \cdot \hat{\jmath} & \hat{I} \cdot \hat{k} \\ \hat{J} \cdot \hat{\imath} & \hat{J} \cdot \hat{\jmath} & \hat{J} \cdot \hat{k} \\ \hat{K} \cdot \hat{\imath} & \hat{K} \cdot \hat{\jmath} & \hat{K} \cdot \hat{k} \end{bmatrix} \begin{bmatrix} \hat{\imath} \\ \hat{\jmath} \\ \hat{k} \end{bmatrix} = {}^G R_B \begin{bmatrix} \hat{\imath} \\ \hat{\jmath} \\ \hat{k} \end{bmatrix} \tag{2.278}$$

where

$$^G R_B = \begin{bmatrix} \hat{I} \cdot \hat{\imath} & \hat{I} \cdot \hat{\jmath} & \hat{I} \cdot \hat{k} \\ \hat{J} \cdot \hat{\imath} & \hat{J} \cdot \hat{\jmath} & \hat{J} \cdot \hat{k} \\ \hat{K} \cdot \hat{\imath} & \hat{K} \cdot \hat{\jmath} & \hat{K} \cdot \hat{k} \end{bmatrix}$$

$$= \begin{bmatrix} \cos(\hat{I}, \hat{\imath}) & \cos(\hat{I}, \hat{\jmath}) & \cos(\hat{I}, \hat{k}) \\ \cos(\hat{J}, \hat{\imath}) & \cos(\hat{J}, \hat{\jmath}) & \cos(\hat{J}, \hat{k}) \\ \cos(\hat{K}, \hat{\imath}) & \cos(\hat{K}, \hat{\jmath}) & \cos(\hat{K}, \hat{k}) \end{bmatrix} \tag{2.279}$$

Each column of $^G R_B$ is decomposition of a unit vector of the local frame $B(Oxyz)$ in the global frame $G(OXYZ)$.

$$^{G}R_B = \begin{bmatrix} ^{G}\hat{\imath} & ^{G}\hat{\jmath} & ^{G}\hat{k} \end{bmatrix} \tag{2.280}$$

Similarly, each row of $^{G}R_B$ is decomposition of a unit vector of the global frame $G(OXYZ)$ in the local frame $B(Oxyz)$.

$$^{G}R_B = \begin{bmatrix} ^{B}\hat{I}^{T} \\ ^{B}\hat{J}^{T} \\ ^{B}\hat{K}^{T} \end{bmatrix} \tag{2.281}$$

The elements of $^{G}R_B$ are directional cosines of the axes of $G(OXYZ)$ in frame $B(Oxyz)$ or B in G. This set of nine directional cosines completely defines the orientation of $B(Oxyz)$ in $G(OXYZ)$ and can be used to map the coordinates of any point (x, y, z) to its corresponding coordinates (X, Y, Z).

Alternatively, using the method of unit vector decomposition to develop the matrix $^{B}R_G$ yields

$$^{B}\mathbf{r} = {^{B}R_G}\,{^{G}\mathbf{r}} = {^{G}R_B^{-1}}\,{^{G}\mathbf{r}} \tag{2.282}$$

$$^{B}R_G = \begin{bmatrix} \hat{\imath} \cdot \hat{I} & \hat{\imath} \cdot \hat{J} & \hat{\imath} \cdot \hat{K} \\ \hat{\jmath} \cdot \hat{I} & \hat{\jmath} \cdot \hat{J} & \hat{\jmath} \cdot \hat{K} \\ \hat{k} \cdot \hat{I} & \hat{k} \cdot \hat{J} & \hat{k} \cdot \hat{K} \end{bmatrix}$$

$$= \begin{bmatrix} \cos(\hat{\imath}, \hat{I}) & \cos(\hat{\imath}, \hat{J}) & \cos(\hat{\imath}, \hat{K}) \\ \cos(\hat{\jmath}, \hat{I}) & \cos(\hat{\jmath}, \hat{J}) & \cos(\hat{\jmath}, \hat{K}) \\ \cos(\hat{k}, \hat{I}) & \cos(\hat{k}, \hat{J}) & \cos(\hat{k}, \hat{K}) \end{bmatrix} \tag{2.283}$$

It shows that the inverse of a transformation matrix is equal to the transpose of the transformation matrix,

$$^{G}R_B^{-1} = {^{G}R_B^{T}} \tag{2.284}$$

or

$$^{G}R_B \cdot {^{G}R_B^{T}} = \mathbf{I} \tag{2.285}$$

A matrix with condition (2.284) is called an *orthogonal matrix*. Orthogonality of $^{G}R_B$ comes from the fact that it maps an orthogonal coordinate frame to another orthogonal coordinate frame.

Applying Eq. (2.285) yields

$$\begin{bmatrix} r_{11} & r_{12} & r_{13} \\ r_{21} & r_{22} & r_{23} \\ r_{31} & r_{32} & r_{33} \end{bmatrix} \begin{bmatrix} r_{11} & r_{21} & r_{31} \\ r_{12} & r_{22} & r_{32} \\ r_{13} & r_{23} & r_{33} \end{bmatrix} = \begin{bmatrix} 1 & 0 & 0 \\ 0 & 1 & 0 \\ 0 & 0 & 1 \end{bmatrix} \tag{2.286}$$

$$r_{11}^2 + r_{12}^2 + r_{13}^2 = 1 \tag{2.287}$$

$$r_{21}^2 + r_{22}^2 + r_{23}^2 = 1 \tag{2.288}$$

$$r_{31}^2 + r_{32}^2 + r_{33}^2 = 1 \tag{2.289}$$

$$r_{11}r_{21} + r_{12}r_{22} + r_{13}r_{23} = 0 \tag{2.290}$$

$$r_{21}r_{31} + r_{22}r_{32} + r_{23}r_{33} = 0 \tag{2.291}$$

$$r_{31}r_{11} + r_{32}r_{12} + r_{33}r_{13} = 0 \tag{2.292}$$

Therefore, the inner product of any two different rows of $^{G}R_B$ is zero, and the inner product of any row of $^{G}R_B$ by itself is unity. These constraint relations are also true for columns of $^{G}R_B$ and evidently for rows and columns of $^{B}R_G$. Because of six independent constraints among the nine elements of $^{G}R_B$, an orthogonal transformation matrix $^{G}R_B$ has only three *independent* elements. The orthogonality condition can be summarized by the equation

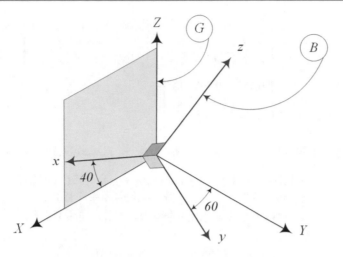

Fig. 2.18 Body and global coordinate frames of Example 45

$$\sum_{i=1}^{3} r_{ij} r_{ik} = \delta_{jk} \qquad j, k = 1, 2, 3 \tag{2.293}$$

where r_{ij} is the element of row i and column j of the transformation matrix R, and δ_{jk} is *Kronecker's delta*.

$$\delta_{ij} = \delta_{ji} = \begin{cases} 1 & i = j \\ 0 & i \neq j \end{cases} \tag{2.294}$$

It states that $\delta_{jk} = 1$ if $j = k$ and $\delta_{jk} = 0$ if $j \neq k$.

Equation (2.293) provides six independent relations that must be satisfied by the nine directional cosines. Therefore, there are only three independent directional cosines. Because of the constraint equations, the independent elements of the matrix R cannot all be in the same row or column or any diagonal.

Because of Eq. (2.285), the determinant of a transformation matrix is equal to one,

$$\left| {}^{G}R_{B} \right| = \pm 1 \tag{2.295}$$

because

$$\left| {}^{G}R_{B} \cdot {}^{G}R_{B}^{T} \right| = \left| {}^{G}R_{B} \right| \cdot \left| {}^{G}R_{B}^{T} \right| = \left| {}^{G}R_{B} \right| \cdot \left| {}^{G}R_{B} \right| = \left| {}^{G}R_{B} \right|^{2} = 1 \tag{2.296}$$

Using linear algebra and column vectors ${}^{G}\hat{\imath}$, ${}^{G}\hat{\jmath}$, and ${}^{G}\hat{k}$ of ${}^{G}R_{B}$, we know that

$$\left| {}^{G}R_{B} \right| = {}^{G}\hat{\imath} \cdot \left({}^{G}\hat{\jmath} \times {}^{G}\hat{k} \right) \tag{2.297}$$

and because the coordinate system is right-handed, we have ${}^{G}\hat{\jmath} \times {}^{G}\hat{k} = {}^{G}\hat{\imath}$, and therefore,

$$\left| {}^{G}R_{B} \right| = {}^{G}\hat{\imath}^{T} \cdot {}^{G}\hat{\imath} = +1 \tag{2.298}$$

∎

Example 45 Elements of transformation matrix. A numerical example to determine rotation matrix by comprehending indirect information.

The position vector **r** of a point P may be expressed in terms of its components with respect to either $G\,(OXYZ)$ or $B\,(Oxyz)$ frame. Body and global coordinate frames are shown in Fig. 2.18. If ${}^{G}\mathbf{r} = 100\hat{I} - 50\hat{J} + 150\hat{K}$, and we are looking for components of **r** in the $B\,(Oxyz)$ frame, then we need to find the proper rotation matrix ${}^{B}R_{G}$ first.

The row elements of ${}^{B}R_{G}$ are the direction cosines of the axes of $Oxyz$ in the coordinate frame $G\,(OXYZ)$. The x-axis lies in the XZ-plane at 40 deg from the X-axis, and the angle between y and Y is 60 deg. Therefore,

$$
{}^B R_G = \begin{bmatrix} \hat{\imath} \cdot \hat{I} & \hat{\imath} \cdot \hat{J} & \hat{\imath} \cdot \hat{K} \\ \hat{\jmath} \cdot \hat{I} & \hat{\jmath} \cdot \hat{J} & \hat{\jmath} \cdot \hat{K} \\ \hat{k} \cdot \hat{I} & \hat{k} \cdot \hat{J} & \hat{k} \cdot \hat{K} \end{bmatrix} = \begin{bmatrix} \cos 40 & 0 & \sin 40 \\ \hat{\jmath} \cdot \hat{I} & \cos 60 & \hat{\jmath} \cdot \hat{K} \\ \hat{k} \cdot \hat{I} & \hat{k} \cdot \hat{J} & \hat{k} \cdot \hat{K} \end{bmatrix}
$$

$$
= \begin{bmatrix} 0.766 & 0 & 0.643 \\ \hat{\jmath} \cdot \hat{I} & 0.5 & \hat{\jmath} \cdot \hat{K} \\ \hat{k} \cdot \hat{I} & \hat{k} \cdot \hat{J} & \hat{k} \cdot \hat{K} \end{bmatrix} \tag{2.299}
$$

and by using the orthogonality condition, ${}^B R_G \, {}^G R_B = {}^B R_G \, {}^B R_G^T = I$, we obtain a set of equations to find the missing elements.

$$
\begin{bmatrix} 0.766 & 0 & 0.643 \\ r_{21} & 0.5 & r_{23} \\ r_{31} & r_{32} & r_{33} \end{bmatrix} \begin{bmatrix} 0.766 & r_{21} & r_{31} \\ 0 & 0.5 & r_{32} \\ 0.643 & r_{23} & r_{33} \end{bmatrix} = \begin{bmatrix} 1 & 0 & 0 \\ 0 & 1 & 0 \\ 0 & 0 & 1 \end{bmatrix} \tag{2.300}
$$

$$
0.766\, r_{21} + 0.643\, r_{23} = 0
$$
$$
0.766\, r_{31} + 0.643\, r_{33} = 0
$$
$$
r_{21}^2 + r_{23}^2 + 0.25 = 1
$$
$$
r_{21} r_{31} + 0.5 r_{32} + r_{23} r_{33} = 0
$$
$$
r_{31}^2 + r_{32}^2 + r_{33}^2 = 1 \tag{2.301}
$$

These set of five coupled algebraic equations are nonlinear, and hence, they provide us with multiple solutions.

$$
\left({}^B R_G \right)_1 = \begin{bmatrix} 0.766 & 0 & 0.643 \\ 0.5568 & 0.5 & -0.6633 \\ -0.32146 & 0.866 & 0.38296 \end{bmatrix} \tag{2.302}
$$

$$
\left({}^B R_G \right)_2 = \begin{bmatrix} 0.766 & 0 & 0.643 \\ 0.5568 & 0.5 & 0.6633 \\ 0.32146 & 0.866 & -0.38296 \end{bmatrix} \tag{2.303}
$$

$$
\left({}^B R_G \right)_3 = \begin{bmatrix} 0.766 & 0 & 0.643 \\ 0.5568 & 0.5 & -0.6633 \\ 0.32146 & -0.866 & -0.38296 \end{bmatrix} \tag{2.304}
$$

$$
\left({}^B R_G \right)_4 = \begin{bmatrix} 0.766 & 0 & 0.643 \\ -0.5568 & 0.5 & 0.6633 \\ -0.32146 & -0.866 & 0.38296 \end{bmatrix} \tag{2.305}
$$

The second solution $\left({}^B R_G \right)_2$ will be eliminated because it does not fit the orthogonality condition (2.285). Considering the acute angle between the z and Z axes, we need a positive r_{33} and that eliminates $\left({}^B R_G \right)_3$. Also, considering the obtuse angle between the z and X axes, we need a negative r_{31} and that eliminates $\left({}^B R_G \right)_1$. Therefore, the right transformation matrix between $G\,(OXYZ)$ and $B\,(Oxyz)$ will be

$$
{}^B R_G = \begin{bmatrix} 0.766 & 0 & 0.643 \\ 0.5568 & 0.5 & -0.6633 \\ -0.32146 & 0.866 & 0.38296 \end{bmatrix} \tag{2.306}
$$

and then we can find the components of ${}^B \mathbf{r}$.

$$^B\mathbf{r} = {}^BR_G\,{}^G\mathbf{r} = \begin{bmatrix} 0.766 & 0 & 0.643 \\ 0.5568 & 0.5 & -0.6633 \\ -0.32146 & 0.866 & 0.38296 \end{bmatrix} \begin{bmatrix} 100 \\ -50 \\ 150 \end{bmatrix}$$

$$= \begin{bmatrix} 173.05 \\ -68.82 \\ -18.0 \end{bmatrix} \tag{2.307}$$

Example 46 Two points transformation matrix. A numerical example of determination of rotation matrix by indirect information.

The global position vector of two points, P_1 and P_2, of a rigid body B are

$$^G\mathbf{r}_{P_1} = \begin{bmatrix} 1.077 \\ 1.365 \\ 2.666 \end{bmatrix} \qquad ^G\mathbf{r}_{P_2} = \begin{bmatrix} -0.473 \\ 2.239 \\ -0.959 \end{bmatrix} \tag{2.308}$$

The origin of the body B ($Oxyz$) is fixed on the origin of G ($OXYZ$), and the points P_1 and P_2 are lying on the local x-axis and y-axis, respectively.

To find ^GR_B, we use the local unit vectors $^G\hat{\imath}$ and $^G\hat{\jmath}$ to obtain $^G\hat{k}$.

$$^G\hat{\imath} = \frac{^G\mathbf{r}_{P_1}}{|^G\mathbf{r}_{P_1}|} = \begin{bmatrix} 0.338 \\ 0.429 \\ 0.838 \end{bmatrix} \qquad ^G\hat{\jmath} = \frac{^G\mathbf{r}_{P_2}}{|^G\mathbf{r}_{P_2}|} = \begin{bmatrix} -0.191 \\ 0.902 \\ -0.387 \end{bmatrix} \tag{2.309}$$

$$^G\hat{k} = \hat{\imath} \times \hat{\jmath} = \begin{bmatrix} -0.922 \\ -0.029 \\ 0.387 \end{bmatrix} \tag{2.310}$$

Hence, the transformation matrix using the coordinates of two points $^G\mathbf{r}_{P_1}$ and $^G\mathbf{r}_{P_2}$ would be

$$^GR_B = \begin{bmatrix} ^G\hat{\imath} & ^G\hat{\jmath} & ^G\hat{k} \end{bmatrix}$$

$$= \begin{bmatrix} 0.338 & -0.191 & -0.922 \\ 0.429 & 0.902 & -0.029 \\ 0.838 & -0.387 & 0.387 \end{bmatrix} \tag{2.311}$$

Example 47 Length invariant of a position vector. The proof of an invariant property of vectors under rotation transformation.

Expressing a vector in different frames utilizing rotation matrices does not affect the length and direction properties of the vector. Therefore, the length of a vector is an invariant.

$$|\mathbf{r}| = |^G\mathbf{r}| = |^B\mathbf{r}| \tag{2.312}$$

The length invariant property can be shown by

$$|\mathbf{r}|^2 = {}^G\mathbf{r}^T\,{}^G\mathbf{r} = \left[{}^GR_B\,{}^B\mathbf{r}\right]^T\,{}^GR_B\,{}^B\mathbf{r} = {}^B\mathbf{r}^T\,{}^GR_B^T\,{}^GR_B\,{}^B\mathbf{r}$$

$$= {}^B\mathbf{r}^T\,{}^B\mathbf{r} \tag{2.313}$$

Example 48 ★ Group property of transformations. The definition of group and showing rotation transformations make a group.

A set S together with a binary operation $\otimes$ defined on elements of S is called a group $(S, \otimes)$ if it satisfies the following four axioms:

1. **Closure**: If $s_1, s_2 \in S$, then $s_1 \otimes s_2 \in S$.
2. **Identity**: There exists an identity element s_0 such that $s_0 \otimes s = s \otimes s_0 = s$ for $\forall s \in S$.

3. **Inverse**: For each $s \in S$, there exists a unique inverse $s^{-1} \in S$ such that $s^{-1} \otimes s = s \otimes s^{-1} = s_0$.
4. **Associativity**: If $s_1, s_2, s_3 \in S$, then $(s_1 \otimes s_2) \otimes s_3 = s_1 \otimes (s_2 \otimes s_3)$.

Three dimensional coordinate transformations make a group if we define the set of rotation matrices by

$$S = \left\{ R \in \mathbb{R}^{3 \times 3} : RR^T = R^T R = \mathbf{I}, |R| = 1 \right\}. \tag{2.314}$$

Therefore, the elements of the set S are transformation matrices R_i, the binary operator $\otimes$ is matrix multiplication, the identity matrix is $\mathbf{I}$, and the inverse of element R is $R^{-1} = R^T$.

S is also a continuous group because

5. The binary matrix multiplication is a continuous operation and
6. The inverse of any element in S is a continuous function of that element.

Therefore, S is a differentiable manifold. A group that is a differentiable manifold is called a Lie group.

Example 49 ★ Transformation with determinant -1. All physical rotation matrices have $+1$ as an eigenvalue. If it is -1, then the matrix would be a reflection.

An orthogonal matrix with determinant $+1$ corresponds to a rotation as described in Eq. (2.295). In contrast, an orthogonal matrix with determinant -1 describes a *reflection*. It transforms a right-handed coordinate system into a left-handed and vice versa. This transformation does not correspond to any possible physical action on rigid bodies.

Example 50 Alternative proof for transformation matrix. Re-definition of rotation matrices based on unit vectors of the two coordinate frames.

Starting with an identity

$$\begin{bmatrix} \hat{\imath} & \hat{\jmath} & \hat{k} \end{bmatrix} \begin{bmatrix} \hat{\imath} \\ \hat{\jmath} \\ \hat{k} \end{bmatrix} = 1 \tag{2.315}$$

we may write

$$\begin{bmatrix} \hat{I} \\ \hat{J} \\ \hat{K} \end{bmatrix} = \begin{bmatrix} \hat{I} \\ \hat{J} \\ \hat{K} \end{bmatrix} \begin{bmatrix} \hat{\imath} & \hat{\jmath} & \hat{k} \end{bmatrix} \begin{bmatrix} \hat{\imath} \\ \hat{\jmath} \\ \hat{k} \end{bmatrix} \tag{2.316}$$

Triple matrix multiplication can be performed in different order.

$$A (BC) = (AB) C \tag{2.317}$$

Therefore,

$$\begin{bmatrix} \hat{I} \\ \hat{J} \\ \hat{K} \end{bmatrix} = \begin{bmatrix} \hat{I} \cdot \hat{\imath} & \hat{I} \cdot \hat{\jmath} & \hat{I} \cdot \hat{k} \\ \hat{J} \cdot \hat{\imath} & \hat{J} \cdot \hat{\jmath} & \hat{J} \cdot \hat{k} \\ \hat{K} \cdot \hat{\imath} & \hat{K} \cdot \hat{\jmath} & \hat{K} \cdot \hat{k} \end{bmatrix} \begin{bmatrix} \hat{\imath} \\ \hat{\jmath} \\ \hat{k} \end{bmatrix} = {}^G R_B \begin{bmatrix} \hat{\imath} \\ \hat{\jmath} \\ \hat{k} \end{bmatrix} \tag{2.318}$$

where

$$ {}^G R_B = \begin{bmatrix} \hat{I} \\ \hat{J} \\ \hat{K} \end{bmatrix} \begin{bmatrix} \hat{\imath} & \hat{\jmath} & \hat{k} \end{bmatrix} \tag{2.319}$$

Following the same method, we can show that

$$ {}^B R_G = \begin{bmatrix} \hat{\imath} \\ \hat{\jmath} \\ \hat{k} \end{bmatrix} \begin{bmatrix} \hat{I} & \hat{J} & \hat{K} \end{bmatrix} \tag{2.320}$$

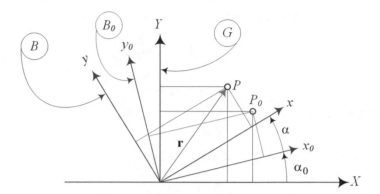

Fig. 2.19 Rotation of a rotated rigid body about the Z-axis

Example 51 ★ Rotation of a rotated body. How to express rotation of a B-frame that is not coincident by the G-frame.

Consider a rigid body with a fixed point at a general position B_0 that is not necessarily coincident with the global frame G. Assume the rotation matrix $^G R_0$ between B_0 and G is known. If the body rotates from this initial position, then the transformation matrix $^G R_B$ between the final position of the body and the global frame is

$$^G R_B = {^G R_0} \, {^0 R_B} \tag{2.321}$$

The matrix $^0 R_B$ is the transformation between initial and final positions of the body with the assumption that B_0 is a fixed position. The principal assumption in the theory of rotation kinematics is that the body coordinate frame B always begins to rotate from a coincident configuration with the global frame G. If the body frame is not initially on the global frame, we show it by B_0 and assume that the body is already rotated from the coincident configuration with G. The rotated body B_0 is then going to turn a second rotation.

Figure 2.19 illustrates a body frame B at its final position after a rotation of α about the Z-axis from an initial position B_0. The initial position of the body is after a first rotation of α_0 about the Z-axis. In this example, both the first and second rotations are about the global Z-axis. To prevent confusion and keep the previous definitions, we define three transformation matrices $^G R_0$, $^0 R_B$, and $^G R_B$. The matrix $^G R_0$ indicates the transformation between G and the initial position of the body at B_0. The matrix $^0 R_B$ is the transformation between the initial position of the body at B_0 and the final position of the body B, and the matrix $^G R_B$ is the transformation between G and the final position of the body B.

Let us assume $^G R_0$ is given as

$$^G R_0 = \begin{bmatrix} \cos\alpha_0 & -\sin\alpha_0 & 0 \\ \sin\alpha_0 & \cos\alpha_0 & 0 \\ 0 & 0 & 1 \end{bmatrix} \tag{2.322}$$

The matrix $^G R_B$ can be found using the regular method. Consider a body frame starts from G and goes to the final position by two rotations α_0 and α:

$$^G R_B = R_{Z,\alpha_0} \, R_{Z,\alpha} = {^G R_0} \, {^0 R_B}$$

$$= \begin{bmatrix} \cos(\alpha+\alpha_0) & -\sin(\alpha+\alpha_0) & 0 \\ \sin(\alpha+\alpha_0) & \cos(\alpha+\alpha_0) & 0 \\ 0 & 0 & 1 \end{bmatrix} \tag{2.323}$$

Therefore, the rotation matrix $^B R_0$ will be

$$^B R_0 = {^G R_B^T} \, {^G R_0} = \begin{bmatrix} \cos\alpha & \sin\alpha & 0 \\ -\sin\alpha & \cos\alpha & 0 \\ 0 & 0 & 1 \end{bmatrix} \tag{2.324}$$

This is correct for a general case in which the first and second rotations are about the same axis. To see the application, let us assume that a point P to be at

$$^B\mathbf{r} = x\hat{i} + y\hat{j} + z\hat{k} \tag{2.325}$$

Its global position is

$$^G\mathbf{r} = {}^GR_B \,{}^B\mathbf{r} \tag{2.326}$$

and its position in the initial frame B_0 is

$$^0\mathbf{r} = {}^BR_0^T \,{}^B\mathbf{r} \tag{2.327}$$

When the body is at its initial position, the global coordinates of P are

$$^{G_0}\mathbf{r} = {}^GR_0 \,{}^0\mathbf{r} \tag{2.328}$$

Example 52 Rotation about X of a rotated body about Z. A numerical example for rotation of a rotated body about global axes.

A rigid body is already rotated 30 deg about the Z-axis and is at B_0:

$$^GR_0 = \begin{bmatrix} \cos\dfrac{\pi}{6} & -\sin\dfrac{\pi}{6} & 0 \\ \sin\dfrac{\pi}{6} & \cos\dfrac{\pi}{6} & 0 \\ 0 & 0 & 1 \end{bmatrix} \tag{2.329}$$

The body rotates 60 deg about the X-axis from the position B_0. The rotation matrix ^GR_B is given as

$$^GR_B = R_{X,\frac{\pi}{6}} \, R_{Z,\frac{\pi}{3}} = \begin{bmatrix} 0.86603 & -0.5 & 0 \\ 0.35355 & 0.61237 & -0.70711 \\ 0.35355 & 0.61237 & 0.70711 \end{bmatrix} \tag{2.330}$$

Therefore,

$$^0R_B = {}^GR_0^T \,{}^GR_B = \begin{bmatrix} 0.92678 & -0.12683 & -0.35356 \\ -0.12683 & 0.78033 & -0.61238 \\ 0.35355 & 0.61237 & 0.70711 \end{bmatrix} \tag{2.331}$$

Example 53 Rotation about x of a rotated body about Z. A numerical example for rotation about body axes of a rotated body about global axes.

Consider a rigid body that is already rotated 45 deg about the Z-axis and is at B_0:

$$^GR_0 = \begin{bmatrix} \cos\dfrac{\pi}{4} & -\sin\dfrac{\pi}{4} & 0 \\ \sin\dfrac{\pi}{4} & \cos\dfrac{\pi}{4} & 0 \\ 0 & 0 & 1 \end{bmatrix} \approx \begin{bmatrix} 0.707 & -0.707 & 0 \\ 0.707 & 0.707 & 0 \\ 0 & 0 & 1 \end{bmatrix} \tag{2.332}$$

The body rotates 45 deg about the x-axis from the position B_0. Considering B_0 as a fixed frame, we can determine ^0R_B:

$$^0R_B = {}^BR_0^T = R_{x,\frac{\pi}{4}}^T = \begin{bmatrix} 1 & 0 & 0 \\ 0 & \cos\dfrac{\pi}{4} & \sin\dfrac{\pi}{4} \\ 0 & -\sin\dfrac{\pi}{4} & \cos\dfrac{\pi}{4} \end{bmatrix}^T \tag{2.333}$$

$$^GR_B = {}^GR_0 \,{}^0R_B = \begin{bmatrix} 0.70711 & -0.5 & 0.5 \\ 0.70711 & 0.5 & -0.5 \\ 0 & 0.70711 & 0.70711 \end{bmatrix} \tag{2.334}$$

Assume a triangle in the body frame with corners at O, P_1, and P_2 where

$$^B\mathbf{r}_1 = 1\hat{\imath} \tag{2.335}$$

$$^B\mathbf{r}_2 = 1\hat{\imath} + 1\hat{\jmath} + 1\hat{k} \tag{2.336}$$

The global coordinates of P_1 and P_2 are

$$^G\mathbf{r}_1 = {}^GR_B\,{}^B\mathbf{r}_1 = {}^GR_B \begin{bmatrix} 1 \\ 0 \\ 0 \end{bmatrix} = \begin{bmatrix} 0.707\,11 \\ 0.707\,11 \\ 0 \end{bmatrix} \tag{2.337}$$

$$^G\mathbf{r}_2 = {}^GR_B\,{}^B\mathbf{r}_2 = {}^GR_B \begin{bmatrix} 1 \\ 1 \\ 1 \end{bmatrix} = \begin{bmatrix} 0.707\,11 \\ 0.707\,11 \\ 1.414\,2 \end{bmatrix} \tag{2.338}$$

The coordinates of the current positions of P_1 and P_2 in B_0 are

$$^0\mathbf{r}_1 = {}^0R_B\,{}^B\mathbf{r}_1 = {}^0R_B \begin{bmatrix} 1 \\ 0 \\ 0 \end{bmatrix} = \begin{bmatrix} 1 \\ 0 \\ 0 \end{bmatrix} \tag{2.339}$$

$$^0\mathbf{r}_2 = {}^0R_B\,{}^B\mathbf{r}_2 = {}^0R_B \begin{bmatrix} 1 \\ 1 \\ 1 \end{bmatrix} = \begin{bmatrix} 1 \\ 0 \\ 1.414\,2 \end{bmatrix} \tag{2.340}$$

2.8 ★ Active and Passive Transformation

Rotation of a local frame B in a global frame G is an *active rotation* if the position vector $^B\mathbf{r}$ of a point P is fixed in the local frame B and rotates with B. Rotation of a local frame B in a global frame G is called a *passive rotation* if the position vector $^G\mathbf{r}$ of a point P is fixed in the global frame and does not rotate with the local frame.

The passive and active transformations are mathematically equivalent. The coordinates of a point P can be transformed from one coordinate to the other by a proper rotation matrix. In an active rotation, the observer is standing in a global frame G and calculating the position of body points in G. However, in a passive rotation, the observer is standing on the body frame B and calculating the position of global points in B. If $R_1 = {}^GR_B$ is an active rotation matrix about an axis, such as the Z-axis, then

$$^G\mathbf{r} = R_1\,{}^B\mathbf{r} = R_{Z,\alpha}\,{}^B\mathbf{r} \tag{2.341}$$

and if $R_2 = {}^BR_G$ is the passive rotation matrix of the same rotation, then

$$^G\mathbf{r} = R_2^T\,{}^B\mathbf{r} = R_{Z,-\alpha}\,{}^B\mathbf{r} \tag{2.342}$$

In robotics, we usually work with active rotations and examine the rotation of a robot's lines in a global frame.

Proof Assume a local frame $B(Oxyz)$ that is initially coincident with a global frame $G(OXYZ)$ performs a rotation. In an active rotation, a body point P will move with B. The body coordinates of P will remain unchanged, while its global coordinate will be found by the proper rotation matrix $R_1 = {}^GR_B$. Let us assume that the axis of rotation is the Z-axis.

$$^G\mathbf{r} = R_1\,{}^B\mathbf{r} = {}^GR_B\,{}^B\mathbf{r} = R_{Z,\alpha}\,{}^B\mathbf{r} \tag{2.343}$$

In a passive rotation, the global coordinates of point P will remain unchanged. We may switch the roll of frames B and G to consider the passive rotation as an active rotation of $-\alpha$ for G in B about the z-axis. The coordinate of P in B will be found by the proper rotation matrix $R_2 = {}^BR_G$:

$$^B\mathbf{r} = R_2\,{}^G\mathbf{r} = {}^BR_G\,{}^G\mathbf{r} = R_{z,-\alpha}\,{}^G\mathbf{r} \tag{2.344}$$

Therefore, the global coordinates of P in the passive rotation may be found from (2.344) as

$$^G\mathbf{r} = R_2^T \, {}^B\mathbf{r} = R_{z,-\alpha}^T \, {}^B\mathbf{r} \tag{2.345}$$

However, because of $R_{Z,\alpha} = R_{z,\alpha}^T$, we have

$$^G\mathbf{r} = R_2^T \, {}^B\mathbf{r} = R_{Z,-\alpha} \, {}^B\mathbf{r} \tag{2.346}$$

∎

Example 54 ★ Active and passive rotation about X-axis. A numerical example for active and passive rotations.
    Consider local and global frames B and G that are coincident. A body point P is at $^B\mathbf{r}$.

$$^B\mathbf{r} = \begin{bmatrix} 1 \\ 2 \\ 1 \end{bmatrix} \tag{2.347}$$

A rotation of 90 deg about the X-axis will move the point to $^G\mathbf{r}$.

$$^G\mathbf{r} = R_{X,90} \, {}^B\mathbf{r} \tag{2.348}$$

$$= \begin{bmatrix} 1 & 0 & 0 \\ 0 & \cos\dfrac{\pi}{2} & -\sin\dfrac{\pi}{2} \\ 0 & \sin\dfrac{\pi}{2} & \cos\dfrac{\pi}{2} \end{bmatrix} \begin{bmatrix} 1 \\ 2 \\ 1 \end{bmatrix} = \begin{bmatrix} 1 \\ -1 \\ 2 \end{bmatrix}$$

Now assume that P is fixed in G. When B rotates 90 deg about X-axis, the coordinates of P in the local frame will change to

$$^B\mathbf{r} = R_{X,-90} \, {}^G\mathbf{r} \tag{2.349}$$

$$= \begin{bmatrix} 1 & 0 & 0 \\ 0 & \cos\dfrac{-\pi}{2} & -\sin\dfrac{-\pi}{2} \\ 0 & \sin\dfrac{-\pi}{2} & \cos\dfrac{-\pi}{2} \end{bmatrix} \begin{bmatrix} 1 \\ 2 \\ 1 \end{bmatrix} = \begin{bmatrix} 1 \\ 1 \\ -2 \end{bmatrix}$$

2.9   Summary

This chapter is about Rotation Kinematics in which we learn the following two objectives:

1. To learn how to determine the transformation matrix between two Cartesian coordinate frames B and G with a common origin by applying rotations about principal axes of either frames
2. To decompose a given transformation matrix to a series of principal rotations

Two Cartesian coordinate frames B and G with a common origin are related by nine directional cosines of a frame in the other. The conversion of coordinates in the two frames can be casted in a matrix transformation

$$^G\mathbf{r} = {}^GR_B \, {}^B\mathbf{r} \tag{2.350}$$

$$\begin{bmatrix} X_2 \\ Y_2 \\ Z_2 \end{bmatrix} = \begin{bmatrix} \hat{I}\cdot\hat{\imath} & \hat{I}\cdot\hat{\jmath} & \hat{I}\cdot\hat{k} \\ \hat{J}\cdot\hat{\imath} & \hat{J}\cdot\hat{\jmath} & \hat{J}\cdot\hat{k} \\ \hat{K}\cdot\hat{\imath} & \hat{K}\cdot\hat{\jmath} & \hat{K}\cdot\hat{k} \end{bmatrix} \begin{bmatrix} x_2 \\ y_2 \\ z_2 \end{bmatrix} \tag{2.351}$$

where

$$
{}^G R_B = \begin{bmatrix} \cos(\hat{I}, \hat{\imath}) & \cos(\hat{I}, \hat{\jmath}) & \cos(\hat{I}, \hat{k}) \\ \cos(\hat{J}, \hat{\imath}) & \cos(\hat{J}, \hat{\jmath}) & \cos(\hat{J}, \hat{k}) \\ \cos(\hat{K}, \hat{\imath}) & \cos(\hat{K}, \hat{\jmath}) & \cos(\hat{K}, \hat{k}) \end{bmatrix}
\tag{2.352}
$$

The transformation matrix ${}^G R_B$ is orthogonal. That means the determinant of ${}^G R_B$ is one, and its inverse is equal to its transpose.

$$
\left| {}^G R_B \right| = 1
\tag{2.353}
$$

$$
{}^G R_B^{-1} = {}^G R_B^T
\tag{2.354}
$$

The orthogonality condition generates six equations between the nine elements of ${}^G R_B$ that shows only three elements of ${}^G R_B$ are independent.

Any relative orientation of B in G can be achieved by three consecutive principal rotations about the coordinate axes in either frames, B or G. If B is the body coordinate frame, and G is the globally fixed frame, the global principal rotation transformation matrices are

$$
R_{X,\gamma} = {}^G R_B = \begin{bmatrix} 1 & 0 & 0 \\ 0 & \cos\gamma & -\sin\gamma \\ 0 & \sin\gamma & \cos\gamma \end{bmatrix}
\tag{2.355}
$$

$$
R_{Y,\beta} = {}^G R_B = \begin{bmatrix} \cos\beta & 0 & \sin\beta \\ 0 & 1 & 0 \\ -\sin\beta & 0 & \cos\beta \end{bmatrix}
\tag{2.356}
$$

$$
R_{Z,\alpha} = {}^G R_B = \begin{bmatrix} \cos\alpha & -\sin\alpha & 0 \\ \sin\alpha & \cos\alpha & 0 \\ 0 & 0 & 1 \end{bmatrix}
\tag{2.357}
$$

and the body principal rotation transformation matrices are

$$
R_{x,\psi} = {}^B R_G = \begin{bmatrix} 1 & 0 & 0 \\ 0 & \cos\psi & \sin\psi \\ 0 & -\sin\psi & \cos\psi \end{bmatrix}
\tag{2.358}
$$

$$
R_{y,\theta} = {}^B R_G = \begin{bmatrix} \cos\theta & 0 & -\sin\theta \\ 0 & 1 & 0 \\ \sin\theta & 0 & \cos\theta \end{bmatrix}
\tag{2.359}
$$

$$
R_{z,\varphi} = {}^B R_G = \begin{bmatrix} \cos\varphi & \sin\varphi & 0 \\ -\sin\varphi & \cos\varphi & 0 \\ 0 & 0 & 1 \end{bmatrix}
\tag{2.360}
$$

The global and local rotation transformations are inverse of each other.

$$
R_{X,\gamma} = R_{x,\gamma}^T
\tag{2.361}
$$

$$
R_{Y,\beta} = R_{y,\beta}^T
\tag{2.362}
$$

$$
R_{Z,\alpha} = R_{z,\alpha}^T
\tag{2.363}
$$

2.10   Key Symbols

a	A general vector
$\tilde{a}$	Skew symmetric matrix of the vector **a**
A	Transformation matrix of rotation about a local axis
B	Body coordinate frame, local coordinate frame
c	cos
d	Distance between two points
$\hat{e}_\varphi, \hat{e}_\theta, \hat{e}_\psi$	Coordinate axes of E, local roll–pitch–yaw coordinate axes
E	Eulerian local frame
f, f_1, f_2	A function of x and y
G	Global coordinate frame, fixed coordinate frame
$\mathbf{I} = [I]$	Identity matrix
$\hat{\imath}, \hat{\jmath}, \hat{k}$	Local coordinate axes unit vectors
$\tilde{\imath}, \tilde{\jmath}, \tilde{k}$	Skew symmetric matrices of the unit vector $\hat{\imath}, \hat{\jmath}, \hat{k}$
$\hat{I}, \hat{J}, \hat{K}$	Global coordinate axes unit vectors
l	Length
m	Number of repeating rotation
n	Fraction of 2π, number of repeating rotation
$\mathbb{N}$	The set of natural numbers
O	Common origin of B and G
$O\varphi\theta\psi$	Euler angle frame
P	A body point, a fixed point in B, a partial derivative
Q	Transformation matrix of rotation about a global axis, a partial derivative
r	Position vector
r_{ij}	The element of row i and column j of a matrix
R	Rotation transformation matrix, radius of a circle
$\mathbb{R}$	The set of real numbers
s	sin, a member of S
S	A set
t	Time
u	A general axis
v	Velocity vector
x, y, z	Local coordinate axes
X, Y, Z	Global coordinate axes

Greek

α, β, γ	Rotation angles about global axes
δ_{ij}	Kronecker's delta
φ, θ, ψ	Rotation angles about local axes, Euler angles
$\dot{\varphi}, \dot{\theta}, \dot{\psi}$	Euler frequencies
$\omega_x, \omega_y, \omega_z$	Angular velocity components
$\boldsymbol{\omega}$	Angular velocity vector

Symbol

$[\ ]^{-1}$	Inverse of the matrix $[\ ]$
$[\ ]^{T}$	Transpose of the matrix $[\ ]$
$\otimes$	A binary operation
$(S, \otimes)$	A group

Exercises

1. Notation and symbols.
   Describe the meaning of these notations.

   a- $^G\mathbf{r}$ b- $^G\mathbf{r}_P$ c- $^B\mathbf{r}_P$ d- $^G R_B$ e- $^G R_B^T$ f- $^B R_G$
   g- $^B R_G^{-1}$ h- $^G\mathbf{d}_B$ i- $^2\mathbf{d}_1$ j- Q_X k- $Q_{Y,\beta}$ l- $Q_{Y,45}^{-1}$
   m- $\hat{k}$ n- $\hat{J}$ o- $A_{z,\varphi}^T$ p- $\hat{e}_\psi$ q- $\tilde{\imath}$ r- $\mathbf{I}$

2. Body point and global rotations.
   The point P is at $^B\mathbf{r}_P = [1, 2, 1]^T$ in a body coordinate $B(Oxyz)$. Find the final global position of P after,
   (a) A rotation of 30 deg about the X-axis, followed by a 45 deg rotation about the Z-axis to move P from P_1 to P_2.
   (b) A rotation of 30 deg about the Z-axis, followed by a 45 deg rotation about the X-axis to move P from P_1 to P_3.
   (c) ★ Point P will move on a sphere. Let us name the initial global position of P by P_1, the second position by P_2, and the third position by P_3. Determine the angles of $\angle P_1 O P_2$, $\angle P_2 O P_3$, and $\angle P_3 O P_1$.
   (d) ★ Determine the area of the triangle made by points P_1, P_2, and P_3.

3. Body point after global rotation.
   Find the position of a point P in the local coordinate, if it is moved to $^G\mathbf{r}_P = [1, 3, 2]^T$ after
   (a) A rotation of 60 deg about Z-axis
   (b) A rotation of 60 deg about X-axis
   (c) A rotation of 60 deg about Y-axis
   (d) Rotations of 60 deg about Z-axis, 60 deg about X-axis, and 60 deg about Y-axis

4. Invariant of a vector.
   A point was at $^B\mathbf{r}_P = [1, 2, z]^T$. After a rotation of 60 deg about X-axis, followed by a 30 deg rotation about Z-axis, it is at

   $$^G\mathbf{r}_P = \begin{bmatrix} X \\ Y \\ 2.933 \end{bmatrix} \tag{2.364}$$

   Find z, X, and Y.

5. Global rotation of a cube.
   Figure 2.20 illustrates the original position of a cube with a fixed point at D and edges of length $l = 1$.
   Determine:
   (a) Coordinates of the corners after rotation of 30 deg about the X-axis.
   (b) Coordinates of the corners after rotation of 30 deg about the Y-axis.
   (c) Coordinates of the corners after rotation of 30 deg about the Z-axis.
   (d) Coordinates of the corners after rotation of 30 deg about the X-axis, then 30 deg about the Y-axis, and then 30 deg about the Z-axis.

6. Constant length vector.
   (a) Show that the length of a vector will not change by rotation.

   $$\left| ^G\mathbf{r} \right| = \left| ^G R_B \, ^B\mathbf{r} \right| \tag{2.365}$$

   (b) Show that the distance between two body points will not change by rotation.

   $$\left| ^B\mathbf{p}_1 - ^B\mathbf{p}_2 \right| = \left| ^G R_B \, ^B\mathbf{p}_1 - ^G R_B \, ^B\mathbf{p}_2 \right| \tag{2.366}$$

7. Coordinate invariant of inner product.
   Knowing

   $$\mathbf{r}_1 \cdot \mathbf{r}_2 = \mathbf{r}_1^T \, \mathbf{r}_2 \tag{2.367}$$

   prove that the inner product of two vectors does not depend on the choice of coordinate frames in which they are expressed.

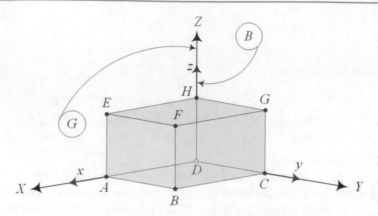

Fig. 2.20 A cube at its initial position

8. Repeated global rotations.

   Rotate $^B\mathbf{r}_P = [2, 2, 3]^T$, 60 deg about the X-axis, followed by 30 deg about the Z-axis. Then, repeat the sequence of rotations for 60 deg about the X-axis, followed by 30 deg about the Z-axis. After how many rotations will point P be back to its initial global position?

9. ★ Alternative motions to reach an orientation.

   The coordinates of a body point P in B and G frames are

$$
^B\mathbf{r}_P = \begin{bmatrix} 1.23 \\ 4.56 \\ 7.89 \end{bmatrix} \qquad ^G\mathbf{r}_P = \begin{bmatrix} 4.56 \\ 7.89 \\ 1.23 \end{bmatrix} \tag{2.368}
$$

   Determine:

   (a) If it is possible to transform $^B\mathbf{r}_P$ to $^G\mathbf{r}_P$?
   (b) A transformation matrix $^G R_B$ between $^B\mathbf{r}_P$ and $^G\mathbf{r}_P$.
   (c) Euler angles to transform $^B\mathbf{r}_P$ to $^G\mathbf{r}_P$.
   (d) Global roll–pitch–yaw to transform $^B\mathbf{r}_P$ to $^G\mathbf{r}_P$.
   (e) Body roll–pitch–yaw to transform $^B\mathbf{r}_P$ to $^G\mathbf{r}_P$.

10. ★ Repeated global rotations.

    How many rotations of $\alpha = \pi/m$ deg about X-axis, followed by $\beta = \pi/n$ deg about Z-axis, are needed to bring a body point to its initial global position, if $m, n \in \mathbb{N}$?

11. Triple global rotations.

    Verify the equations in Appendix A.

12. ★ Special triple rotation.

    Assume that the first triple rotation in Appendix A brings a body point back to its initial global position. What are the possible values of the angles $\alpha \neq 0$, $\beta \neq 0$, and $\gamma \neq 0$?

13. ★ Combination of triple rotations.

    Any triple rotation in Appendix A can move a body point to its new global position. Assume α_1, β_1, and γ_1 for the case $1 - Q_{X,\gamma_1} Q_{Y,\beta_1} Q_{Z,\alpha_1}$ are given. What can α_2, β_2, and γ_2 be (in terms of α_1, β_1, and γ_1) to get the same global position using the case $2 - Q_{Y,\gamma_2} Q_{Z,\beta_2} Q_{X,\alpha_2}$?

14. Global roll–pitch–yaw rotation angles.

    Calculate the role, pitch, and yaw angles for the following rotation matrix:

$$
^B R_G = \begin{bmatrix} 0.53 & -0.84 & 0.13 \\ 0.0 & 0.15 & 0.99 \\ -0.85 & -0.52 & 0.081 \end{bmatrix} \tag{2.369}
$$

15. ★ Rotation matrix and spherical trigonometry.

A plane cutting a sphere and passing through the center of the sphere is called a great circle. Any other plane intersecting the sphere but not passing through the center cuts the surface in a small circle. A spherical triangle is made up when three great circles intersect. If we are given any three points on the surface of a sphere, we can join them by great circle arcs to form a spherical triangle.

Consider a vector $^B\mathbf{r}$ in a body coordinate frame B. The tip point of $^B\mathbf{r}$ will move on a sphere when the body frame performs multiple rotations. The vector $^B\mathbf{r}$ goes to three positions $\mathbf{r}_1$, $\mathbf{r}_2$, and $\mathbf{r}_3$. The vector $\mathbf{r}_1$ indicates the original position of a body point, $\mathbf{r}_2$ indicates the position vector of the body point after a rotation about the Z-axis, and $\mathbf{r}_3$ indicates the position vector of the point after another rotation about the X-axis. However, these rotations may be about any arbitrary axes. The tip point of the vectors $\mathbf{r}_1$, $\mathbf{r}_2$, and $\mathbf{r}_3$ makes a spherical triangle that lies on a sphere with radius $R = |\mathbf{r}_1| = |\mathbf{r}_2| = |\mathbf{r}_3|$. Assume $\triangle ABC$ is a triangle on a sphere of unit radius $R = 1$. Let us show the angles of the triangle by α, β, and γ and the length of its sides by a, b, and c. The arc lengths a, b, and c are, respectively, equal to the plane angles $\angle BOC$, $\angle AOC$, and $\angle AOB$ for the unit sphere. Show that

$$\cos a = \cos b \cos c + \sin b \sin c \cos \alpha \tag{2.370}$$

$$\cos b = \cos c \cos a + \sin c \sin a \cos \beta \tag{2.371}$$

$$\cos c = \cos a \cos b + \sin a \sin b \cos \gamma \tag{2.372}$$

$$\frac{\sin \alpha}{\sin a} = \frac{\sin \beta}{\sin b} = \frac{\sin \gamma}{\sin c} \tag{2.373}$$

16. ★ Back to the initial orientation and Appendix A.

Assume we turn a rigid body B using the first set of Appendix A. How can we turn it back to its initial orientation by applying

   (a) The first set of Appendix A

   (b) The second set of Appendix A

   (c) The third set of Appendix A

   (d) ★ Assume that we have turned a rigid body B by $\alpha_1 = 30$ deg, $\beta_1 = 30$ deg, and $\gamma_1 = 30$ deg using the first set of Appendix A. We want to turn B back to its original orientation. Which one of the second or third set of Appendix A does it faster? Let us assume that the fastest set is the one with minimum sum of $s = \alpha_2 + \beta_2 + \gamma_2$.

17. ★ Back to the original orientation and Appendix B.

Assume we turn a rigid body B using the first set of Appendix A. How can we turn it back to its initial orientation by applying

   (a) The first set of Appendix B

   (b) The second set of Appendix B

   (c) The third set of Appendix B

   (d) ★ Assume that we have turned a rigid body B by $\alpha = 30$ deg, $\beta = 30$ deg, and $\gamma = 30$ deg using the first set of Appendix A. We want to turn B back to its original orientation. Which one of the first, second, or third set of Appendix B does it faster? Let us assume that the fastest set is the one with minimum sum of $s = \varphi + \theta + \psi$.

18. Two local rotations.

Find the global coordinates of a body point at $^B\mathbf{r}_P = [2, 2, 3]^T$ after

   (a) A rotation of 60 deg about x-axis followed by 60 deg about z-axis

   (b) A rotation of 60 deg about z-axis followed by 60 deg about x-axis

   (c) A rotation of 60 deg about z-axis followed by 60 deg about x-axis, and a rotation of 60 deg about z-axis

19. Local rotation of a cube.

Figure 2.20 illustrates the initial position of a cube with a fixed point at D and edges of length $l = 1$. Determine:

   (a) Coordinates of the corners after rotation of 30 deg about x-axis.

   (b) Coordinates of the corners after rotation of 30 deg about y-axis.

   (c) Coordinates of the corners after rotation of 30 deg about z-axis.

   (d) Coordinates of the corners after rotation of 30 deg about x-axis, then 30 deg about y-axis, and then 30 deg about z-axis.

20. Global and local rotation of a cube.

   Figure 2.20 illustrates the initial position of a cube with a fixed point at D and edges of length $l = 1$.
   Determine:

   (a) Coordinates of the corners after rotation of 30 deg about x-axis followed by rotation of 30 deg about X-axis.
   (b) Coordinates of the corners after rotation of 30 deg about y-axis followed by rotation of 30 deg about X-axis.
   (c) Coordinates of the corners after rotation of 30 deg about z-axis followed by rotation of 30 deg about X-axis.
   (d) Coordinates of the corners after rotation of 30 deg about x-axis, then 30 deg about X-axis, and then 30 deg about x-axis.
   (e) Coordinates of the corners after rotation of 30 deg about x-axis, then 30 deg about Y-axis, and then 30 deg about z-axis.

21. Body point, local rotation.

   What is the global coordinates of a body point at $^B\mathbf{r}_P = [2, 2, 3]^T$, after

   (a) A rotation of 60 deg about the x-axis
   (b) A rotation of 60 deg about the y-axis
   (c) A rotation of 60 deg about the z-axis

22. Unknown rotation angle 1.

   Transform $^B\mathbf{r}_P = [2, 2, 3]^T$ to $^G\mathbf{r}_P = [2, Y_P, 0]^T$ by a rotation about x-axis and determine Y_P and the angle of
   rotation.

23. Unknown rotation angle 2.

   Consider a point P at $^B\mathbf{r}_P = [2, \sqrt{3}, \sqrt{2}]^T$. Determine:

   (a) The required principal global rotations in order X, Y, Z, to move P to $^G\mathbf{r}_P = [\sqrt{2}, 2, \sqrt{3}]^T$.
   (b) The required principal global rotations in order Z, Y, Z, to move P to $^G\mathbf{r}_P = [\sqrt{2}, 2, \sqrt{3}]^T$.
   (c) The required principal global rotations in order Z, X, Z, to move P to $^G\mathbf{r}_P = [\sqrt{2}, 2, \sqrt{3}]^T$.

24. Triple local rotations.

   Verify the equations in Appendix B.

25. Combination of local and global rotations.

   Find the final global position of a body point at $^B\mathbf{r}_P = [10, 10, -10]^T$ after

   (a) A rotation of 45 deg about the x-axis followed by 60 deg about the Z-axis.
   (b) A rotation of 45 deg about the z-axis followed by 60 deg about the Z-axis.
   (c) A rotation of 45 deg about the x-axis followed by 45 deg about the Z-axis and 60 deg about the X-axis.

26. Combination of global and local rotations.

   Find the final global position of a body point at $^B\mathbf{r}_P = [10, 10, -10]^T$ after

   (a) A rotation of 45 deg about the X-axis followed by 60 deg about the z-axis.
   (b) A rotation of 45 deg about the Z-axis followed by 60 deg about the z-axis.
   (c) A rotation of 45 deg about the X-axis followed by 45 deg about the x-axis and 60 deg about the z-axis.

27. Repeated local rotations.

   Rotate $^B\mathbf{r}_P = [2, 2, 3]^T$, 60 deg about the x-axis, followed by 30 deg about the z-axis. Then repeat the sequence of
   rotations for 60 deg about the x-axis, followed by 30 deg about the z-axis. After how many rotations will point P move
   back to its initial global position?

28. ★ Repeated local rotations.

   How many rotations of $\alpha = \pi/m$ deg about the x-axis, followed by $\beta = \pi/n$ deg about the z-axis, are needed to
   bring a body point to its initial global position if $m, n \in \mathbb{N}$?

29. ★ Remaining rotation.

   Find the result of the following sequence of rotations:

$$^G R_B = A_{y,\theta}^T A_{z,\psi}^T A_{y,-\theta}^T \tag{2.374}$$

30. Angles from rotation matrix.

   Find the angles φ, θ, and ψ if the rotation transformation matrices of Appendix B are given.

31. Euler angles from rotation matrix.

   (a) Check if the following matrix $^G R_B$ is a rotation transformation:

$$^G R_B = \begin{bmatrix} 0.53 & -0.84 & 0.13 \\ 0.0 & 0.15 & 0.99 \\ -0.85 & -0.52 & 0.081 \end{bmatrix} \tag{2.375}$$

(b) Find the Euler angles for $^G R_B$.

(c) Find the local roll–pitch–yaw angles for $^G R_B$.

32. Equivalent Euler angles to two rotations.

Find the Euler angles corresponding to the following rotation matrix:

(a) $^B R_G = A_{y,45} A_{x,30}$

(b) $^B R_G = A_{x,45} A_{y,30}$

(c) $^B R_G = A_{y,45} A_{z,30}$

33. Equivalent Euler angles to three rotations.

Find the Euler angles corresponding to the following rotation matrix:

(a) $^B R_G = A_{z,60} A_{y,45} A_{x,30}$

(b) $^B R_G = A_{z,60} A_{y,45} A_{z,30}$

(c) $^B R_G = A_{x,60} A_{y,45} A_{x,30}$

34. ★ A cube rotation and forbidden space of $z < 0$.

Figure 2.20 illustrates the initial position of a cube with a fixed point at D and edges of length $l = 1$.

Assume that none of the corners is allowed to have a negative z-components at any time.

(a) Present a series of global principal rotations to make the line FH parallel to z-axis.

(b) Present a series of global principal rotations to make the line DB on the z-axis and point A in (Z, Y)-plane.

(c) Present a series of local principal rotations to make the line FH parallel to z-axis.

(d) Present a series of local principal rotations to make the line DB on the z-axis and point A in (Z, Y)-plane.

35. ★ Local and global positions, Euler angles.

Find the conditions between the Euler angles,

(a) To transform $^G \mathbf{r}_P = [1, 1, 0]^T$ to $^B \mathbf{r}_P = [0, 1, 1]^T$

(b) To transform $^G \mathbf{r}_P = [1, 1, 0]^T$ to $^B \mathbf{r}_P = [1, 0, 1]^T$

36. ★ Equivalent Euler angles to a triple rotations.

Find the Euler angles for the rotation matrix of the fourth case in Appendix B.

$$4 - A_{z,\psi'} A_{y,\theta'} A_{x,\varphi'} \tag{2.376}$$

37. ★ Integrability of Euler frequencies.

Show that $d\varphi$ and $d\psi$ are integrable, if φ and ψ are first and third Euler angles.

38. ★ Cardan angles for Euler angles.

(a) Find the Cardan angles for a given set of Euler angles.

(b) Find the Euler angles for a given set of Cardan angles.

39. ★ Cardan frequencies for Euler frequencies.

(a) Find the Euler frequencies in terms of Cardan frequencies.

(b) Find the Cardan frequencies in terms of Euler frequencies.

40. ★ Transformation matrix and three rotations.

Figure 2.20 illustrates the original position of a cube with a fixed point at D and edges of length $l = 1$.

Assume a new orientation in which points D and F are on Z-axis and point A is in (X, Z)-plane. Determine:

(a) Transformation matrix between initial and new orientations.

(b) Euler angles to move the cube to its new orientation.

(c) Global roll–pitch–yaw angles to move the cube to its new orientation.

(d) Local roll–pitch–yaw angles to move the cube to its new orientation.

41. ★ Alternative maneuvers.

Figure 2.20 illustrates the initial position of a cube with a fixed point at D and edges of length $l = 1$.

Assume a new orientation in which points D and F are on Z-axis and point A is in (X, Z)-plane. Determine:

(a) Angles for maneuver $Y - X - Z$ as first–second–third rotations.

(b) Angles for maneuver $Y - Z - X$ as first–second–third rotations.

(c) Angles for maneuver $y - x - z$ as first–second–third rotations.

(d) Angles for maneuver $y - z - x$ as first–second–third rotations.
(e) Angles for maneuver $y - Z - x$ as first–second–third rotations.
(f) Angles for maneuver $Y - z - X$ as first–second–third rotations.
(g) Angles for maneuver $x - X - x$ as first–second–third rotations.
42. Elements of rotation matrix.

   The elements of rotation matrix $^G R_B$ are

$$^G R_B = \begin{bmatrix} \cos(\hat{I}, \hat{\imath}) & \cos(\hat{I}, \hat{\jmath}) & \cos(\hat{I}, \hat{k}) \\ \cos(\hat{J}, \hat{\imath}) & \cos(\hat{J}, \hat{\jmath}) & \cos(\hat{J}, \hat{k}) \\ \cos(\hat{K}, \hat{\imath}) & \cos(\hat{K}, \hat{\jmath}) & \cos(\hat{K}, \hat{k}) \end{bmatrix} \tag{2.377}$$

   Find $^G R_B$ if $^G \mathbf{r}_{P_1} = [0.7071, -1.2247, 1.4142]^T$ is a point on the x-axis, and $^G \mathbf{r}_{P_2} = [2.7803, 0.38049, -1.0607]^T$ is a point on the y-axis.
43. Linearly independent vectors.

   A set of vectors $\mathbf{a}_1, \mathbf{a}_2, \cdots, \mathbf{a}_n$ is considered linearly independent if the equation

$$k_1 \mathbf{a}_1 + k_2 \mathbf{a}_2 + \cdots + k_n \mathbf{a}_n = 0 \tag{2.378}$$

   in which $k_1, k_2, \cdots, k_n$ are unknown coefficients, has only one solution.

$$k_1 = k_2 = \cdots = k_n = 0 \tag{2.379}$$

   Verify that the unit vectors of a body frame $B(Oxyz)$, expressed in the global frame $G(OXYZ)$, are linearly independent.
44. Product of orthogonal matrices.

   A matrix R is called orthogonal if $R^{-1} = R^T$ where $\left(R^T\right)_{ij} = R_{ji}$. Prove that the product of two orthogonal matrices is also orthogonal.
45. Vector identity.

   The formula $(a + b)^2 = a^2 + b^2 + 2ab$ for scalars is equivalent to

$$(\mathbf{a} + \mathbf{b})^2 = \mathbf{a} \cdot \mathbf{a} + \mathbf{b} \cdot \mathbf{b} + 2\mathbf{a} \cdot \mathbf{b} \tag{2.380}$$

   for vectors. Show that this formula is equal to

$$(\mathbf{a} + \mathbf{b})^2 = \mathbf{a} \cdot \mathbf{a} + \mathbf{b} \cdot \mathbf{b} + 2\,^G R_B \mathbf{a} \cdot \mathbf{b} \tag{2.381}$$

   if $\mathbf{a}$ is a vector in local frame and $\mathbf{b}$ is a vector in global frame.
46. Rotation as a linear operation.

   Show that
$$R\left(\mathbf{a} \times \mathbf{b}\right) = R\mathbf{a} \times R\mathbf{b} \tag{2.382}$$

   where R is a rotation matrix and $\mathbf{a}$ and $\mathbf{b}$ are two vectors defined in a coordinate frame.
47. Scalar triple product.

   Show that for three arbitrary vectors $\mathbf{a}$, $\mathbf{b}$, and $\mathbf{c}$, we have

$$\mathbf{a} \cdot (\mathbf{b} \times \mathbf{c}) = (\mathbf{a} \times \mathbf{b}) \cdot \mathbf{c} \tag{2.383}$$

48. ★ Euler angles and minimization distances.

   Figure 2.20 illustrates the initial position of a cube with a fixed point at D and edges of length $l = 1$.
   Assume a new orientation in which points D and F are on Z-axis and point A is in (X, Y)-plane. Determine:
   (a) Transformation matrix between initial and new orientations.
   (b) Euler angles to move the cube to its new orientation.

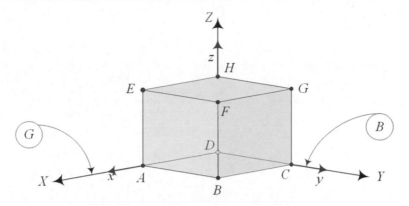

Fig. 2.21 A cube with sides of length $l = 1$

(c) Choose three non-coplanar corners and determine their position using Euler transformation matrix with unknown Euler angles. Define the distance between the initial and final positions of the points as d_1, d_2 and d_3. Is it possible to determine the Euler angles by minimizing a sum of distances objective function $J = d_1^2 + d_2^2 + d_3^2$?

49. ★ Continues rotation.

Figure 2.20 illustrates the initial position of a cube with a fixed point at D and edges of length $l = 1$.

Assume that the cube is turning about x-axis with angular speed of ω_1 and at the same time it is turning about Z-axis with angular speed of ω_2. Determine the path of motion of point F. What is the path for $\omega_1 = \omega_2$, $\omega_1 = 2\omega_2$, $\omega_1 = 3\omega_2$, and $\omega_1 = 4\omega_2$?

50. ★ Project: Quickest triple global rotations.

Consider a body frame B coincident with a global frame G. The frame B undergoes a rotation of 45 deg about the X-axis followed by a rotation of 30 deg about the Z-axis, then 30 deg about the Y-axis, and 60 deg about the X-axis. If the body was supposed to be moved to the final orientation by only three principal global rotations of Appendix A, which one is the most efficient triple rotation? Assume efficiency means minimization of the sum of total rotation angles s.

$$s = |\alpha| + |\beta| + |\gamma| \tag{2.384}$$

51. ★ Project: Quickest triple body rotations.

Consider a body frame B coincident with a global frame G. The frame B undergoes a rotation of 45 deg about the X-axis followed by a rotation of 30 deg about the Z-axis, then 30 deg about the Y-axis, and 60 deg about the X-axis. If the body was supposed to be moved to the final orientation by only three principal body rotations of Appendix B, which one is the most efficient triple rotation? Assume efficiency means minimization of the sum of total rotation angles s.

$$s = |\varphi| + |\theta| + |\psi| \tag{2.385}$$

52. ★ Project: Rotations of a cube.

Consider a cube with sides of length $l = 1$ as is shown in Fig. 2.21. The corners of the cube are at $A\,(1, 0, 0)$, $B\,(1, 1, 0)$, $C\,(0, 1, 0)$, $D\,(0, 0, 0)$, $E\,(1, 0, 1)$, $F\,(1, 1, 1)$, $G\,(0, 1, 1)$, $H\,(0, 0, 1)$. Let us call this original configuration the rest position.

(a) Move the longest diagonal DF to lay on the Z-axis only by two rotations about global principal axes. There are multiple solutions. Either one is acceptable. Determine and report the coordinates of all corners.

(b) If the x-axis is not in (XZ)-plane, turn the cube at the end of case a about the Z-axis until the x-axis is in (XZ)-plane. Determine and report the coordinates of all corners.

(c) Return the cube at the end of case b to its rest position only by two rotations about body principal axes. Determine the axes and angles for this maneuver with the minimum sum of total rotation angles.

Any finial orientation of a rigid body with a fixed point O, after a finite number of rotations is equivalent to a unique ϕ about a fixed axis $\hat{u}$. Also any rotation ϕ of a rigid body with a fixed point O about a fixed axis $\hat{u}$ can be decomposed into three rotations about three given non-coplanar axes including the global or body principal exes. Determination of the angle and axis is called the *orientation kinematics* of rigid bodies.

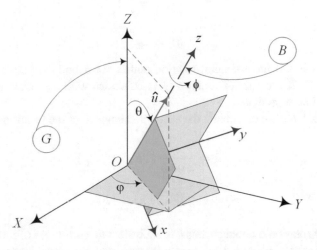

Fig. 3.1 Axis of rotation $\hat{u}$ when it is coincident with the local z-axis

3.1 Axis–Angle Rotation

Assume a body frame $B(Oxyz)$ rotates ϕ about a line indicated by a unit vector $\hat{u}$ with direction cosines u_1, u_2, u_3.

$$\hat{u} = u_1 \hat{I} + u_2 \hat{J} + u_3 \hat{K} = \begin{bmatrix} u_1 \\ u_2 \\ u_3 \end{bmatrix} \tag{3.1}$$

$$\sqrt{u_1^2 + u_2^2 + u_3^2} = 1 \tag{3.2}$$

This is called *axis–angle* representation of a rotation. Two parameters are necessary to define the unit vector $\hat{u}$ through O, and one is necessary to define the amount of rotation ϕ of the rigid body about $\hat{u}$.

The *axis–angle* transformation matrix $^G R_B$ that maps the coordinates in the local frame $B(Oxyz)$ to the corresponding coordinates in the global frame $G(OXYZ)$,

$$^G\mathbf{r} = {}^G R_B \, {}^B\mathbf{r} \tag{3.3}$$

is

$$^G R_B = \begin{bmatrix} u_1^2 \operatorname{vers}\phi + c\phi & u_1 u_2 \operatorname{vers}\phi - u_3 s\phi & u_1 u_3 \operatorname{vers}\phi + u_2 s\phi \\ u_1 u_2 \operatorname{vers}\phi + u_3 s\phi & u_2^2 \operatorname{vers}\phi + c\phi & u_2 u_3 \operatorname{vers}\phi - u_1 s\phi \\ u_1 u_3 \operatorname{vers}\phi - u_2 s\phi & u_2 u_3 \operatorname{vers}\phi + u_1 s\phi & u_3^2 \operatorname{vers}\phi + c\phi \end{bmatrix} \tag{3.4}$$

$$^G R_B = R_{\hat{u},\phi} = \mathbf{I}\cos\phi + \hat{u}\hat{u}^T \operatorname{vers}\phi + \tilde{u}\sin\phi \tag{3.5}$$

with inverse

$$^B R_G = {}^G R_B^T = R_{\hat{u},-\phi} = \mathbf{I}\cos\phi + \hat{u}\hat{u}^T \operatorname{vers}\phi - \tilde{u}\sin\phi \tag{3.6}$$

where,

$$\operatorname{vers}\phi = versine\,\phi = 1 - \cos\phi = 2\sin^2\frac{\phi}{2} \tag{3.7}$$

and $\tilde{u}$ is the skew symmetric matrix corresponding to the vector $\hat{u}$.

$$\hat{u} = \begin{bmatrix} u_1 \\ u_2 \\ u_3 \end{bmatrix} \qquad \tilde{u} = \begin{bmatrix} 0 & -u_3 & u_2 \\ u_3 & 0 & -u_1 \\ -u_2 & u_1 & 0 \end{bmatrix} \tag{3.8}$$

A matrix $\tilde{u}$ is *skew symmetric* if:

$$\tilde{u}^T = -\tilde{u} \tag{3.9}$$

The transformation matrix (3.4) is the most general rotation matrix for a body frame B with a fixed point O, rotating with respect to a global frame G. If the axis of rotation (3.1) coincides with a global coordinate axis Z, Y, or X, then Eqs. (2.20), (2.21), or (2.22) will be reproduced.

Given a transformation matrix $^G R_B$ we can obtain the axis $\hat{u}$ and angle ϕ of the rotation by

$$\tilde{u} = \frac{1}{2\sin\phi}\left({}^G R_B - {}^G R_B^T\right) \tag{3.10}$$

$$\cos\phi = \frac{1}{2}\left(\operatorname{tr}\left({}^G R_B\right) - 1\right) \tag{3.11}$$

Proof Interestingly, the effect of rotation ϕ about an axis $\hat{u}$ is equivalent to a sequence of rotations about the axes of the body frame such that the body frame is first rotated to bring one of its axes, say the z-axis, into coincidence with the rotation axis $\hat{u}$, followed by a rotation ϕ about that local axis, then the reverse of the first sequence of rotations. The remaining rotation after these actions would be a rotation ϕ about $\hat{u}$.

Figure 3.1 illustrates an axis of rotation $\hat{u} = u_1\hat{I} + u_2\hat{J} + u_3\hat{K}$, the global frame $G\,(OXYZ)$, and the rotated local frame $B\,(Oxyz)$ when the local z-axis is coincident with $\hat{u}$. The local frame $B\,(Oxyz)$ undergoes a sequence of rotations φ about the z-axis and θ about the y-axis to bring the local z-axis into coincidence with the rotation axis $\hat{u}$, followed by rotation ϕ about $\hat{u}$, and then perform the sequence backward by turning $-\theta$ about the y-axis and then $-\varphi$ about the z-axis. Therefore, using (2.259), the rotation matrix $^G R_B$ to map coordinates in body frame to their coordinates in global frame after rotation ϕ about $\hat{u}$ is

$$\begin{aligned} ^G R_B = {}^B R_G^{-1} &= {}^B R_G^T = R_{\hat{u},\phi} \\ &= \left[A_{z,-\varphi}\, A_{y,-\theta}\, A_{z,\phi}\, A_{y,\theta}\, A_{z,\varphi}\right]^T \\ &= A_{z,\varphi}^T\, A_{y,\theta}^T\, A_{z,\phi}^T\, A_{y,-\theta}^T\, A_{z,-\varphi}^T \end{aligned} \tag{3.12}$$

but we have

$$\sin \varphi = \frac{u_2}{\sqrt{u_1^2 + u_2^2}} \quad \cos \varphi = \frac{u_1}{\sqrt{u_1^2 + u_2^2}}$$

$$\sin \theta = \sqrt{u_1^2 + u_2^2} \quad \cos \theta = u_3 \tag{3.13}$$

$$\sin \theta \sin \varphi = u_2 \quad \sin \theta \cos \varphi = u_1$$

and hence,

$$^G R_B = R_{\hat{u},\phi} \tag{3.14}$$

$$= \begin{bmatrix} u_1^2 \operatorname{vers} \phi + c\phi & u_1 u_2 \operatorname{vers} \phi - u_3 s\phi & u_1 u_3 \operatorname{vers} \phi + u_2 s\phi \\ u_1 u_2 \operatorname{vers} \phi + u_3 s\phi & u_2^2 \operatorname{vers} \phi + c\phi & u_2 u_3 \operatorname{vers} \phi - u_1 s\phi \\ u_1 u_3 \operatorname{vers} \phi - u_2 s\phi & u_2 u_3 \operatorname{vers} \phi + u_1 s\phi & u_3^2 \operatorname{vers} \phi + c\phi \end{bmatrix}$$

The matrix (3.14) can be decomposed to

$$R_{\hat{u},\phi} = \cos \phi \begin{bmatrix} 1 & 0 & 0 \\ 0 & 1 & 0 \\ 0 & 0 & 1 \end{bmatrix} + (1 - \cos \phi) \begin{bmatrix} u_1 \\ u_2 \\ u_3 \end{bmatrix} \begin{bmatrix} u_1 & u_2 & u_3 \end{bmatrix}$$

$$+ \sin \phi \begin{bmatrix} 0 & -u_3 & u_2 \\ u_3 & 0 & -u_1 \\ -u_2 & u_1 & 0 \end{bmatrix} \tag{3.15}$$

to be equal to

$$^G R_B = R_{\hat{u},\phi} = \mathbf{I} \cos \phi + \hat{u}\, \hat{u}^T \operatorname{vers} \phi + \tilde{u} \sin \phi \tag{3.16}$$

Equation (3.5) is called the *Rodriguez rotation formula* or the *Euler–Lexell–Rodriguez formula*, after French mathematician Olinde Rodriguez (1795–1851), Swiss mathematician Leonhard Euler (1707–1783), Finnish–Swedish astronomer Anders Johan Lexell (1740–1784).

To show the rules (3.10) and (3.11), we expand $^G R_B - {}^G R_B^T$ to determine the axis of rotation $\hat{u}$

$$^G R_B - {}^G R_B^T = \begin{bmatrix} 0 & -2(\sin \phi) u_3 & 2(\sin \phi) u_2 \\ 2(\sin \phi) u_3 & 0 & -2(\sin \phi) u_1 \\ -2(\sin \phi) u_2 & 2(\sin \phi) u_1 & 0 \end{bmatrix}$$

$$= 2 \sin \phi \begin{bmatrix} 0 & -u_3 & u_2 \\ u_3 & 0 & -u_1 \\ -u_2 & u_1 & 0 \end{bmatrix} = 2\tilde{u} \sin \phi \tag{3.17}$$

and expand $\operatorname{tr}\left({}^G R_B\right)$ to provide the angle of rotation ϕ.

$$\operatorname{tr}\left({}^G R_B\right) = r_{11} + r_{22} + r_{33}$$

$$= 3 \cos \phi + u_1^2 (1 - \cos \phi) + u_2^2 (1 - \cos \phi)$$

$$+ u_3^2 (1 - \cos \phi)$$

$$= 3 \cos \phi + u_1^2 + u_2^2 + u_3^2 - \left(u_1^2 + u_2^2 + u_3^2\right) \cos \phi$$

$$= 2 \cos \phi + 1 \tag{3.18}$$

The *Rodriguez rotation formula* may also be reported in literature by the following equivalent forms:

$$R_{\hat{u},\phi} = \mathbf{I} + \tilde{u}\sin\phi + 2\tilde{u}^2\sin^2\frac{\phi}{2} \tag{3.19}$$

$$R_{\hat{u},\phi} = \mathbf{I} + 2\tilde{u}\sin\frac{\phi}{2}\left(\mathbf{I}\cos\frac{\phi}{2} + \tilde{u}\sin\frac{\phi}{2}\right) \tag{3.20}$$

$$R_{\hat{u},\phi} = \mathbf{I} + \tilde{u}\sin\phi + \tilde{u}^2\,\text{vers}\,\phi \tag{3.21}$$

$$R_{\hat{u},\phi} = \left[\mathbf{I} - \hat{u}\,\hat{u}^T\right]\cos\phi + \tilde{u}\sin\phi + \hat{u}\,\hat{u}^T \tag{3.22}$$

$$R_{\hat{u},\phi} = \mathbf{I} + \tilde{u}^2 + \tilde{u}\sin\phi - \tilde{u}^2\cos\phi \tag{3.23}$$

The *inverse of an angle-axis rotation* is

$$^G R_B^T = {}^B R_G = R_{\hat{u},-\phi} = \mathbf{I}\cos\phi + \hat{u}\,\hat{u}^T\,\text{vers}\,\phi - \tilde{u}\sin\phi \tag{3.24}$$

It means orientation of B in G, when B is rotated ϕ about $\hat{u}$, is the same as the orientation of G in B, when B is rotated $-\phi$ about $\hat{u}$.

The 3×3 real orthogonal transformation matrix R is also called a *rotator* and the skew symmetric matrix $\tilde{u}$ is called a *spinor*, or the *Euler* axis or the *eigenaxis* of rotation.

We can verify that

$$\tilde{u}\,\hat{u} = 0 \tag{3.25}$$

$$\hat{u}\,\hat{u}^T - \mathbf{I} = \tilde{u}^2 \tag{3.26}$$

$$\mathbf{r}^T\,\tilde{u}\,\mathbf{r} = 0 \tag{3.27}$$

$$\hat{u} \times \mathbf{r} = \tilde{u}\,\mathbf{r} = -\tilde{r}\,\hat{u} = -\mathbf{r} \times \hat{u} \tag{3.28}$$

The angle-axis representation of rotation is not unique. Rotation $(\theta, \hat{u})$ is equal to rotation $(-\theta, -\hat{u})$, and $(\theta + 2\pi, \hat{u})$.   ∎

Example 55 Axis–angle rotation when $\hat{u} = \hat{K}$. Simplifying the axis–angle rotation matrix for a special known case axis of rotation.

If the local frame B $(Oxyz)$ rotates about the Z-axis, then

$$\hat{u} = \hat{K} = \begin{bmatrix} 0 & 0 & 1 \end{bmatrix}^T \tag{3.29}$$

and the transformation matrix (3.4) reduces to

$$
\begin{aligned}
^G R_B &= \begin{bmatrix} 0\,\text{vers}\,\phi + \cos\phi & 0\,\text{vers}\,\phi - 1\sin\phi & 0\,\text{vers}\,\phi + 0\sin\phi \\ 0\,\text{vers}\,\phi + 1\sin\phi & 0\,\text{vers}\,\phi + \cos\phi & 0\,\text{vers}\,\phi - 0\sin\phi \\ 0\,\text{vers}\,\phi - 0\sin\phi & 0\,\text{vers}\,\phi + 0\sin\phi & 1\,\text{vers}\,\phi + \cos\phi \end{bmatrix} \\
&= \begin{bmatrix} \cos\phi & -\sin\phi & 0 \\ \sin\phi & \cos\phi & 0 \\ 0 & 0 & 1 \end{bmatrix}
\end{aligned} \tag{3.30}
$$

which is equivalent to the rotation matrix about the Z-axis of global frame in (2.20).

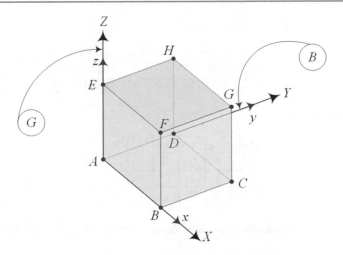

Fig. 3.2 A cube with a fixed point at A

Example 56 Rotation about a rotated local axis. Axis of rotation if it is a body axis after a rotation about global axes.

If the body coordinate frame B $(Oxyz)$ rotates φ about the global Z-axis, then the x-axis would be along $\hat{u}_x$.

$$\hat{u}_x = {}^G R_{Z,\varphi}\, \hat{\imath} = \begin{bmatrix} \cos\varphi & -\sin\varphi & 0 \\ \sin\varphi & \cos\varphi & 0 \\ 0 & 0 & 1 \end{bmatrix} \begin{bmatrix} 1 \\ 0 \\ 0 \end{bmatrix} = \begin{bmatrix} \cos\varphi \\ \sin\varphi \\ 0 \end{bmatrix} \tag{3.31}$$

Rotation θ about $\hat{u}_x$ can be defined by *Rodriguez's formula* (3.4).

$$ {}^G R_{\hat{u}_x,\theta} = \begin{bmatrix} \cos^2\varphi\, \mathrm{vers}\,\theta + \cos\theta & \cos\varphi\sin\varphi\, \mathrm{vers}\,\theta & \sin\varphi\sin\theta \\ \cos\varphi\sin\varphi\, \mathrm{vers}\,\theta & \sin^2\varphi\, \mathrm{vers}\,\theta + \cos\theta & -\cos\varphi\sin\theta \\ -\sin\varphi\sin\theta & \cos\varphi\sin\theta & \cos\theta \end{bmatrix} \tag{3.32}$$

Now, rotation φ about the global Z-axis followed by rotation θ about the local x-axis is transformed by

$$\begin{aligned} {}^G R_B &= {}^G R_{\hat{u}_x,\theta}\, {}^G R_{Z,\varphi} \\ &= \begin{bmatrix} \cos\varphi & -\cos\theta\sin\varphi & \sin\theta\sin\varphi \\ \sin\varphi & \cos\theta\cos\varphi & -\cos\varphi\sin\theta \\ 0 & \sin\theta & \cos\theta \end{bmatrix} \end{aligned} \tag{3.33}$$

that must be equal to $\left[A_{x,\theta} A_{z,\varphi} \right]^{-1} = A_{z,\varphi}^T A_{x,\theta}^T$.

Example 57 Axis and angle of rotation. Determine all important points of a rigid body after a given axis–angle rotation.

Consider a cubic rigid body with a fixed point at A and a unit length of edges as is shown in Fig. 3.2. If we turn the cube 45 deg about $\mathbf{u}$,

$$\mathbf{u} = \begin{bmatrix} 1 & 1 & 1 \end{bmatrix}^T \tag{3.34}$$

then we can find the global coordinates of its corners using Rodriguez transformation matrix.

$$\phi = \frac{\pi}{4} \qquad \hat{u} = \frac{\mathbf{u}}{\sqrt{3}} = \begin{bmatrix} 0.577\,35 \\ 0.577\,35 \\ 0.577\,35 \end{bmatrix} \tag{3.35}$$

$$R_{\hat{u},\phi} = \mathbf{I}\cos\phi + \hat{u}\,\hat{u}^T \operatorname{vers}\phi + \tilde{u}\sin\phi$$

$$= \begin{bmatrix} 0.804\,74 & -0.310\,62 & 0.505\,88 \\ 0.505\,88 & 0.804\,74 & -0.310\,62 \\ -0.310\,62 & 0.505\,88 & 0.804\,74 \end{bmatrix} \tag{3.36}$$

The local coordinates of the corners are

	$^B\mathbf{r}_B$	$^B\mathbf{r}_C$	$^B\mathbf{r}_D$	$^B\mathbf{r}_E$	$^B\mathbf{r}_F$	$^B\mathbf{r}_G$	$^B\mathbf{r}_H$
x	1	1	0	0	1	1	0
y	0	1	1	0	0	1	1
z	0	0	0	1	1	1	1

(3.37)

and therefore, using $^G\mathbf{r} = R_{\hat{u},\phi}\,^B\mathbf{r}$ the global coordinates of the corners after the rotation will be

	$^G\mathbf{r}_B$	$^G\mathbf{r}_C$	$^G\mathbf{r}_D$	$^G\mathbf{r}_E$	$^G\mathbf{r}_F$	$^G\mathbf{r}_G$	$^G\mathbf{r}_H$
X	0.804	0.495	-0.31	0.505	1.310	1	0.196
Y	0.505	1.31	0.804	-0.31	0.196	1	0.495
Z	-0.31	0.196	0.505	0.804	0.495	1	1.31

(3.38)

Point G is on the axis of rotation, so its coordinates will not change. Points B, D, F, and H are in a symmetric plane indicated by $\hat{u}$. Therefore, they will move on a circle. To check this fact, we may find the midpoint of BH, or FD and see if it is on the $\hat{u}$-axis. Let us call the midpoint of the cube by P.

$$^B\mathbf{r}_P = \frac{1}{2}\left(^B\mathbf{r}_B + {}^B\mathbf{r}_H\right) = \frac{1}{2}\left(^B\mathbf{r}_F + {}^B\mathbf{r}_D\right) = \begin{bmatrix} 0.5 \\ 0.5 \\ 0.5 \end{bmatrix} \tag{3.39}$$

$$^G\mathbf{r}_P = \frac{1}{2}\left(^G\mathbf{r}_B + {}^G\mathbf{r}_H\right) = \frac{1}{2}\left(^G\mathbf{r}_F + {}^G\mathbf{r}_D\right) = \begin{bmatrix} 0.5 \\ 0.5 \\ 0.5 \end{bmatrix} \tag{3.40}$$

Example 58 Axis and angle of a rotation matrix. Numerical example for determination of axis–angle rotation for a set of given Euler angles.

A body coordinate frame B undergoes three Euler rotations $(\varphi, \theta, \psi) = (30, 45, 60)$ deg with respect to a global frame G. The rotation matrix to transform coordinates of B to G is

$$^G R_B = {}^B R_G^T = \left[R_{z,\psi}\, R_{x,\theta}\, R_{z,\varphi}\right]^T = R_{z,\varphi}^T\, R_{x,\theta}^T\, R_{z,\psi}^T$$

$$= \begin{bmatrix} 0.126\,83 & -0.926\,78 & 0.353\,55 \\ 0.780\,33 & -0.126\,83 & -0.612\,37 \\ 0.612\,37 & 0.353\,55 & 0.707\,11 \end{bmatrix} \tag{3.41}$$

The unique angle-axis of rotation for this rotation matrix can then be found by Eqs. (3.10) and (3.11).

$$\phi = \arccos\left(\frac{1}{2}\left(\operatorname{tr}\left(^G R_B\right) - 1\right)\right)$$

$$= \arccos\left(-0.146\,45\right) = 1.7178\,\text{rad} = 98\,\text{deg} \tag{3.42}$$

$$\tilde{u} = \frac{1}{2\sin\phi} \left({}^G R_B - {}^G R_B^T \right)$$

$$= \begin{bmatrix} 0.0 & -0.862\,85 & -0.130\,82 \\ 0.862\,85 & 0.0 & -0.488\,22 \\ 0.130\,82 & 0.488\,22 & 0.0 \end{bmatrix} \tag{3.43}$$

$$\hat{u} = \begin{bmatrix} 0.488\,22 \\ -0.130\,82 \\ 0.862\,85 \end{bmatrix} \tag{3.44}$$

As a double check, we may verify the angle-axis rotation formula and derive the same rotation matrix.

$$^G R_B = R_{\hat{u},\phi} = \mathbf{I}\cos\phi + \hat{u}\,\hat{u}^T \text{ vers}\,\phi + \tilde{u}\sin\phi$$

$$= \begin{bmatrix} 0.12682 & -0.92677 & 0.35354 \\ 0.78032 & -0.12683 & -0.61237 \\ 0.61236 & 0.35355 & 0.70709 \end{bmatrix} \tag{3.45}$$

Example 59 ★ Skew symmetric characteristic of rotation matrix. Time derivative of orthogonality condition gives a skew symmetric combination of rotation matrices that defines angular velocity.

Time derivative of the orthogonality condition of rotation matrix (2.285)

$$^G R_B^T \, {}^G R_B = \mathbf{I} \tag{3.46}$$

leads to

$$\frac{d}{dt} \left({}^G R_B^T \, {}^G R_B \right) = {}^G \dot{R}_B^T \, {}^G R_B + {}^G R_B^T \, {}^G \dot{R}_B = 0 \tag{3.47}$$

$$\left[{}^G R_B^T \, {}^G \dot{R}_B \right]^T = - {}^G R_B^T \, {}^G \dot{R}_B \tag{3.48}$$

showing that $\left[{}^G R_B^T \, {}^G \dot{R}_B \right]$ is a skew symmetric matrix. A matrix is skew symmetric if its transpose is equal to its negative. If we show the rotation matrix by its elements, $^G R_B = [r_{ij}]$, then $^G \dot{R}_B = [\dot{r}_{ij}]$. Rewriting the orthogonality condition as

$$^G R_B \, {}^G R_B^T = \mathbf{I} \tag{3.49}$$

provides another form of the identity.

$$^G \dot{R}_B \, {}^G R_B^T + {}^G R_B \, {}^G \dot{R}_B^T = 0 \tag{3.50}$$

$$\left[{}^G \dot{R}_B \, {}^G R_B^T \right]^T = - {}^G R_B \, {}^G \dot{R}_B^T \tag{3.51}$$

Let us show the skew symmetric matrix $\left[{}^G R_B^T \, {}^G \dot{R}_B \right]$ by $\tilde{\omega}$,

$$\tilde{\omega} = {}^G R_B^T \, {}^G \dot{R}_B \tag{3.52}$$

then, we find the following equation for the time derivative of rotation matrix:

$$^G \dot{R}_B = {}^G R_B \, \tilde{\omega} \tag{3.53}$$

where ω is the vector of angular velocity of the frame $B(Oxyz)$ with respect to frame $G(OXYZ)$, and $\tilde{\omega}$ is its skew symmetric matrix expression of ω.

Example 60 ★ Differentiating a rotation matrix with respect to a parameter. The skew symmetric characteristic of derivative of rotation matrix is correct for derivative with respect to any parameter.

Suppose that a rotation matrix R is a function of a variable τ; hence, $R = R(\tau)$. To find the differential of R with respect to τ, we use the orthogonality characteristic

$$R R^T = I \tag{3.54}$$

and take derivative of both sides

$$\frac{dR}{d\tau} R^T + R \frac{dR^T}{d\tau} = 0 \tag{3.55}$$

which can be rewritten in the following form:

$$\frac{dR}{d\tau} R^T + \left[\frac{dR}{d\tau} R^T \right]^T = 0 \tag{3.56}$$

showing that $[\frac{dR}{d\tau} R^T]$ is a skew symmetric matrix.

Example 61 ★ Eigenvalues and eigenvectors of $^G R_B$. Any rotation matrix has a real eigenvalue of $+1$ and its associated real eigenvector $\hat{u}$ which is the axis of rotation ϕ. The other two eigenvalues are complex conjugate and their associated eigenvectors are complex vectors making an orthogonal triad with $\hat{u}$.

Consider a rotation matrix $^G R_B$. Applying a rotation on the axis of rotation $\hat{u}$ cannot change the direction of $\hat{u}$.

$$^G R_B \hat{u} = \lambda \hat{u} \tag{3.57}$$

Therefore, the transformation equation implies that

$$\left| ^G R_B - \lambda \mathbf{I} \right| = 0 \tag{3.58}$$

The characteristic equation of this determinant is

$$- \lambda^3 + \mathrm{tr}(^G R_B)\lambda^2 - \mathrm{tr}(^G R_B)\lambda + 1 = 0 \tag{3.59}$$

Factoring the left-hand side, yields

$$(\lambda - 1) \left[\lambda^2 - \lambda \left(\mathrm{tr}(^G R_B) - 1 \right) + 1 \right] = 0 \tag{3.60}$$

It shows that $\lambda_1 = 1$ is always an eigenvalue of $^G R_B$. Hence, there exist a real vector $\hat{u}$, such that every point on the line indicated by the vector $\mathbf{n}_1 = \hat{u}$ remains fixed and invariant under transformation $^G R_B$. The rotation angle of this rotation ϕ is defined by

$$\cos \phi = \frac{1}{2} \left(\mathrm{tr}(^G R_B) - 1 \right) = \frac{1}{2} \left(r_{11} + r_{22} + r_{33} - 1 \right) \tag{3.61}$$

The remaining eigenvalues are

$$\lambda_2 = e^{i\phi} = \cos \phi + i \sin \phi \tag{3.62}$$

$$\lambda_3 = e^{-i\phi} = \cos \phi - i \sin \phi \tag{3.63}$$

and their associated eigenvectors are $\mathbf{v}$ and $\bar{\mathbf{v}}$, where $\bar{\mathbf{v}}$ is the complex conjugate of $\mathbf{v}$. Because $^G R_B$ is orthogonal, $\mathbf{n}_1$, $\mathbf{v}$, and $\bar{\mathbf{v}}$ are also orthogonal. The eigenvectors $\mathbf{v}$ and $\bar{\mathbf{v}}$ span a plane perpendicular to the axis of rotation $\mathbf{n}_1$. A real basis for this plane can be found by using the following vectors:

$$\mathbf{n}_2 = \frac{1}{2} |\mathbf{v} + \bar{\mathbf{v}}| \tag{3.64}$$

$$\mathbf{n}_3 = \frac{i}{2} |\mathbf{v} - \bar{\mathbf{v}}| \tag{3.65}$$

The basis $\mathbf{n}_2$ and $\mathbf{n}_3$ transform to

$$
\begin{aligned}
{}^{G}R_B\, \mathbf{n}_2 &= \frac{1}{2}\left|\lambda_2 \mathbf{v} + \lambda_3 \bar{\mathbf{v}}\right| = \frac{1}{2}\left|e^{i\phi}\mathbf{v} + \overline{e^{i\phi}\mathbf{v}}\right| \\
&= \mathbf{v}\cos\phi + \bar{\mathbf{v}}\sin\phi
\end{aligned}
\tag{3.66}
$$

$$
\begin{aligned}
{}^{G}R_B\, \mathbf{n}_3 &= \frac{i}{2}\left|\lambda_2 \mathbf{v} - \lambda_3 \bar{\mathbf{v}}\right| = \frac{1}{2}\left|e^{i\phi}\mathbf{v} - \overline{e^{i\phi}\mathbf{v}}\right| \\
&= -\mathbf{v}\cos\phi + \bar{\mathbf{v}}\sin\phi
\end{aligned}
\tag{3.67}
$$

Therefore, the effect of transformation ${}^{G}R_B$ is to rotate vectors in the plane spanned by $\mathbf{n}_2$ and $\mathbf{n}_3$ through angle ϕ about $\mathbf{n}_1$, while vectors along $\mathbf{n}_1$ are invariant.

Example 62 ★ **Final rotation formula.** The rotation kinematics is based on the assumption that before applying a rotation, the body B and global G frames are coincident. Here we learn about breaking this assumption, and changes to axis-angle rotation formula.

The assumption to apply any rotation of a body B in a fixed frame G is that the coordinate frames B and G should be coincident before the rotation. We can imagine a situation in which B and G are not coincident and we want to rotate B about a globally fixed axis ${}^{G}\hat{u}$. Consider a global frame G and a body frame B_0 at a non-coincident configuration. The body frame is supposed to turn ϕ about an axis ${}^{G}\hat{u}$ from its current position at B_0.

$$
{}^{G}\hat{u} = u_1\hat{I} + u_2\hat{J} + u_3\hat{K}
\tag{3.68}
$$

$$
\sqrt{u_1^2 + u_2^2 + u_3^2} = 1
\tag{3.69}
$$

We can always assume that the body B has come to the position B_0, from a coincident configuration with G, by a rotation α about z_0 followed by a rotation β about x_0 and then a rotation γ about z_0.

Consider the body frame B at the coincident position with B_0. When we apply a sequence of rotations φ about the z-axis and θ about the y-axis on the body frame, the local z-axis will coincide with the rotation axis ${}^{G}\hat{u}$. Let us imagine B at this time and indicate it by B_1. Then we apply the rotation ϕ about $z \equiv \hat{u}$ and perform the sequence of rotations $-\theta$ about the y-axis and $-\varphi$ about the z-axis. The resultant of this maneuver would be a rotation ϕ of B about $\hat{u}$, starting from B_0.

The initial relative orientation of the body must be known; therefore, the transformation ${}^{G}R_0$ between B_0 and G is a given matrix.

$$
{}^{G}R_0 = [b_{ij}] = \begin{bmatrix} b_{11} & b_{12} & b_{13} \\ b_{21} & b_{22} & b_{23} \\ b_{31} & b_{32} & b_{33} \end{bmatrix}
\tag{3.70}
$$

Having ${}^{G}R_0$, we can determine the angles α, β, and γ.

$$
\begin{aligned}
{}^{G}R_0 &= R_{z_0,\gamma}\, R_{x_0,\beta}\, R_{z_0,\alpha} \\
&= \begin{bmatrix} c\alpha\, c\gamma - c\beta\, s\alpha\, s\gamma & c\gamma\, s\alpha + c\alpha\, c\beta\, s\gamma & s\beta\, s\gamma \\ -c\alpha\, s\gamma - c\beta\, c\gamma\, s\alpha & c\alpha\, c\beta\, c\gamma - s\alpha\, s\gamma & s\beta\, c\gamma \\ s\alpha\, s\beta & -c\alpha\, s\beta & c\beta \end{bmatrix}
\end{aligned}
\tag{3.71}
$$

$$
\alpha = -\arctan\frac{b_{31}}{b_{32}} \qquad \beta = \arccos b_{33} \qquad \gamma = \arctan\frac{b_{13}}{b_{23}}
\tag{3.72}
$$

The transformation matrix between B_0 and B comes from Rodriguez formula (3.5). However, ${}^{G}\hat{u}$ must be expressed in B_0 to apply the Rodriguez formula.

$$^0\hat{u} = {}^GR_0^T \, {}^G\hat{u} \tag{3.73}$$

$$
\begin{aligned}
^0R_B &= \mathbf{I}\cos\phi + {}^0\hat{u}\,{}^0\hat{u}^T \text{ vers }\phi + {}^0\tilde{u}\sin\phi \\
&= \mathbf{I}\cos\phi + \left({}^GR_0^T \, {}^G\hat{u}\right)\left({}^GR_0^T \, {}^G\hat{u}\right)^T \text{ vers }\phi \\
&\quad + {}^GR_0^T \, \tilde{u} \, {}^GR_0 \sin\phi \\
&= \mathbf{I}\cos\phi + {}^GR_0^T \, {}^G\hat{u}\,{}^G\hat{u}^T \, {}^GR_0 \text{ vers }\phi \\
&\quad + {}^GR_0^T \, \tilde{u} \, {}^GR_0 \sin\phi
\end{aligned}
\tag{3.74}
$$

The transformation matrix ^GR_B between the final position of the body and global frame would be

$$
\begin{aligned}
^GR_B &= {}^GR_0\,{}^0R_B \\
&= {}^GR_0\left[\mathbf{I}\cos\phi + {}^GR_0^T \, {}^G\hat{u}\,{}^G\hat{u}^T \, {}^GR_0 \text{ vers }\phi\right. \\
&\quad \left. + {}^GR_0^T \, \tilde{u} \, {}^GR_0 \sin\phi\right] \\
&= {}^GR_0\cos\phi + \left[{}^G\hat{u}\,{}^G\hat{u}^T\right]{}^GR_0 \text{ vers }\phi + \tilde{u}\,{}^GR_0\sin\phi \\
&= {}^GR_{\hat{u},\phi}\,{}^GR_0
\end{aligned}
\tag{3.75}
$$

We call this equation the **final rotation formula**. It determines the transformation matrix between a body frame B and the global frame G after the rotation ϕ of B about $\hat{u} = {}^G\hat{u}$, starting from a given in coincident position with global frame $B_0 \neq G$, expressed by a given transformation matrix $^GR_0$.

As an example, consider a body that is rotated 30 deg about the Z axis and is at B_0.

$$
^GR_0 = \begin{bmatrix} \cos\dfrac{\pi}{6} & -\sin\dfrac{\pi}{6} & 0 \\ \sin\dfrac{\pi}{6} & \cos\dfrac{\pi}{6} & 0 \\ 0 & 0 & 1 \end{bmatrix} \approx \begin{bmatrix} 0.866 & -0.5 & 0 \\ 0.5 & 0.866 & 0 \\ 0 & 0 & 1 \end{bmatrix}
\tag{3.76}
$$

The body is then supposed to turn 90 deg about $^G\hat{u}$.

$$\phi = \frac{\pi}{2} \qquad {}^G\hat{u} = \hat{I} \tag{3.77}$$

Therefore,

$$
\begin{aligned}
^GR_B &= \left(\begin{bmatrix} 1 & 0 & 0 \\ 0 & 1 & 0 \\ 0 & 0 & 1 \end{bmatrix}\cos\frac{\pi}{2} + \begin{bmatrix} 1 & 0 & 0 \\ 0 & 0 & 0 \\ 0 & 0 & 0 \end{bmatrix}\left(1-\cos\frac{\pi}{2}\right)\right. \\
&\quad \left. + \begin{bmatrix} 0 & 0 & 0 \\ 0 & 0 & -1 \\ 0 & 1 & 0 \end{bmatrix}{}^GR_0\sin\frac{\pi}{2}\right)\begin{bmatrix} 0.866 & -0.5 & 0 \\ 0.5 & 0.866 & 0 \\ 0 & 0 & 1 \end{bmatrix} \\
&= \begin{bmatrix} 0.866 & -0.5 & 0 \\ 0 & 0 & -1 \\ 0.5 & 0.866 & 0 \end{bmatrix}
\end{aligned}
\tag{3.78}
$$

A body point at $^B\mathbf{r} = 2\hat{\imath}$ will be seen at

$$^G\mathbf{r} = {}^GR_B\,{}^B\mathbf{r} \tag{3.79}$$

$$
= \begin{bmatrix} 0.866 & -0.5 & 0 \\ 0 & 0 & -1 \\ 0.5 & 0.866 & 0 \end{bmatrix}\begin{bmatrix} 2 \\ 0 \\ 0 \end{bmatrix} = \begin{bmatrix} 1.732 \\ 0 \\ 1 \end{bmatrix}
$$

Example 63 ★ Rotation of a rotated body. A numerical example for final rotation formula (3.75).

A rigid body B has already turned 30 deg about Y-axis. We need to turn the body 45 deg about $\hat{u}$.

$$\hat{u} = {}^G\hat{u} = \frac{1}{\sqrt{3}}\hat{I} + \frac{1}{\sqrt{3}}\hat{J} + \frac{1}{\sqrt{3}}\hat{K} \tag{3.80}$$

Because of the first rotation, we have

$$
{}^G R_0 = \begin{bmatrix} \cos\dfrac{\pi}{6} & 0 & \sin\dfrac{\pi}{6} \\ 0 & 1 & 0 \\ -\sin\dfrac{\pi}{6} & 0 & \cos\dfrac{\pi}{6} \end{bmatrix} = \begin{bmatrix} 0.866 & 0 & 0.5 \\ 0 & 1 & 0 \\ -0.5 & 0 & 0.866 \end{bmatrix} \tag{3.81}
$$

Using the final rotation formula (3.75), we are able to determine the required rotation transformation matrix, ${}^G R_B$.

$$
{}^G R_{\hat{u},\phi} = \mathbf{I}\cos\phi + \left[\hat{u}\,\hat{u}^T\right]\text{vers}\,\phi + \tilde{u}\sin\phi
$$

$$
= \begin{bmatrix} 0.804\,74 & -0.310\,62 & 0.505\,88 \\ 0.505\,88 & 0.804\,74 & -0.310\,62 \\ -0.310\,62 & 0.505\,88 & 0.804\,74 \end{bmatrix} \tag{3.82}
$$

$$
{}^G R_B = {}^G R_{\hat{u},\phi}\ {}^G R_0
$$

$$
= \begin{bmatrix} 0.443\,99 & -0.310\,62 & 0.840\,47 \\ 0.593\,41 & 0.804\,74 & -1.606\,5\times10^{-2} \\ -0.671\,37 & 0.505\,88 & 0.541\,62 \end{bmatrix} \tag{3.83}
$$

Example 64 Orthogonality of Rodriguez rotation matrix. Here is to examine that Rodriguez rotation matrix is orthogonal and discover a few identities.

The orthogonality characteristic of the rotation matrix must be consistent with the Rodriguez formula as well. We show that we can multiply Eqs. (3.5) and (3.6):

$$
\begin{aligned}
{}^G R_B\,{}^B R_G &= R_{\hat{u},\phi}\,R_{\hat{u},-\phi} \\
&= \left(\mathbf{I}\cos\phi + \hat{u}\hat{u}^T\,\text{vers}\,\phi + \tilde{u}\sin\phi\right) \\
&\quad \times \left(\mathbf{I}\cos\phi + \hat{u}\hat{u}^T\,\text{vers}\,\phi - \tilde{u}\sin\phi\right) \\
&= \mathbf{I}\cos^2\phi + \hat{u}\hat{u}^T\,\text{vers}\,\phi\cos\phi - \tilde{u}\sin\phi\cos\phi \\
&\quad + \hat{u}\hat{u}^T\,\text{vers}\,\phi\cos\phi + \hat{u}\hat{u}^T\hat{u}\hat{u}^T\,\text{vers}\,\phi\,\text{vers}\,\phi \\
&\quad - \tilde{u}\hat{u}\hat{u}^T\,\text{vers}\,\phi\sin\phi + \tilde{u}\sin\phi\cos\phi \\
&\quad + \hat{u}\hat{u}^T\tilde{u}\sin\phi\,\text{vers}\,\phi - \tilde{u}\tilde{u}\sin^2\phi = \mathbf{1}
\end{aligned} \tag{3.84}
$$

That is because of

$$
\hat{u}\,\hat{u}^T = \begin{bmatrix} u_1^2 & u_1u_2 & u_1u_3 \\ u_1u_2 & u_2^2 & u_2u_3 \\ u_1u_3 & u_2u_3 & u_3^2 \end{bmatrix} \tag{3.85}
$$

$$
\hat{u}\,\hat{u}^T\,\hat{u}\,\hat{u}^T = \begin{bmatrix} u_1^2 & u_1u_2 & u_1u_3 \\ u_1u_2 & u_2^2 & u_2u_3 \\ u_1u_3 & u_2u_3 & u_3^2 \end{bmatrix} = \hat{u}\hat{u}^T \tag{3.86}
$$

$$
\tilde{u}\,\hat{u}\,\hat{u}^T = \hat{u}\,\hat{u}^T\,\tilde{u} = \begin{bmatrix} 0 & 0 & 0 \\ 0 & 0 & 0 \\ 0 & 0 & 0 \end{bmatrix} \tag{3.87}
$$

Example 65 ★ Euler rotation theorem 66. Expression and proof of Euler theorems about rigid body rotation.

Motion of a rigid body with a fixed point can be summarized in Euler rigid body rotation theorem proven in 1775 by Swiss mathematician Leonhard Euler (1707–1783).

Theorem 66 *A rigid body rotation about a point is equivalent to a rotation about a line passing through the point.*

This theorem was originally stated by Euler as below:

Theorem 67 *Quomodocunque sphaera circa centrum suum conuertatur, semper assignari potest diameter, cuius directio in situ translato conueniat cum situ initiali.*

When a sphere is moved around its center it is always possible to find a diameter whose direction in the displaced position is the same as in the initial position.

Proof Assume a body has been rotated about the fixed point O. Let us select an arbitrary segment $A_1 B_1$ in the initial position of the body that is not passing through O. Then $A_2 B_2$ would be the corresponding segment at the final position. We draw the plane of symmetry π of segment $A_1 B_1$ and $A_2 B_2$ by points O and midpoints of $A_1 A_2$ and $B_1 B_2$. The planes $O A_1 B_1$ and $O A_2 B_2$ intersect at a line l, which is also in the plane of symmetry π. The line l passes through O because l is the locus of points equidistant from A_1 and A_2 as well as B_1 and B_2. While $O A_1 = O A_2$ and $O B_1 = O B_2$, let us pick a point C on l different than O.

The tetrahedrons $O C A_1 B_1$ and $O C A_2 B_2$ are equal and superposable. This is because the vertices O, C, A_1 and O, C, A_2 are placed symmetrically with respect to π. Hence, if the tetrahedrons $O C A_1 B_1$ and $O C A_2 B_2$ were not superposable, the vertices B_1 and B_2 would have to be placed asymmetrically with respect to π. If we now rotate the rigid body about the axis l so that A_1 falls on A, then the tetrahedron $O C A_1 B_1$ will fall on the tetrahedron $O C A_2 B_2$. After this rotation of the body we will know the positions of three points O, A_2, B_2 of the body and therefore we know the position of every point of the body.

A consequence of the Euler theorem is that: During a rotation of a body about a point, there exists a certain line in the body having the property that its points do not change their position.

Another consequence of the Euler theorem is that: If a rigid body makes two successive rotations about two axes passing through the fixed point O, then the body can be displaced from its initial position to its final position by means of one rotation about an axis passing through O. Therefore, the combination of two rotations about an axis passing through one point is a rotation about an axis passing through the same point. ∎

Example 68 ★ Euler rotation theorem 2. Expression and proof of Euler theorem about rigid body rotation.

Theorem 69 *A rigid body can be displaced from an initial position to a final position by means of two successive rotations.*

Proof Let P_1 be a point of the rigid body B at its initial position B_1, and P_2 the corresponding point in the final position B_2. We rotate the body 180 deg about the axis l, which is the axis of symmetry of the segment $P_1 P_2$. By this rotation, point P_1 will fall on point P_2. The body will assume the position B', which has the point P_2 in common with position B_2. Consequently, we can go from position B' to B_2 by a rotation about an axis l passing through P_2. ∎

3.2   ★ Order-Free Rotation

Most industrial multibodies with a common fixed point are similar to spherical wrists with one, two, three, or more rotary actuators that can work independently. Practically, the order of actuation of the motors is not important if motor number i is carrying all other bodies numbers $i + 1$, $i + 2$, $\cdots$. The orientation of the body would be the same, no matter which motor acts first. We introduce a theorem to generalize and simplify the applied rotation transformation of multibodies. Here is the mathematical explanation.

Theorem Order-free rotations.

Consider n rigid bodies with coordinate frames $B_1, B_2, \cdots, B_n$ with a common fixed point. The body B_1 carries the bodies $B_2, B_3, \cdots, B_n$ and turns α_1 about a fixed axis in the global coordinate frame G. The body B_2 carries $B_3, B_4, \cdots, B_n$ and turns α_2 about a fixed axis in B_1, and so on. The transformation matrix ${}^G R_n = {}^G R_n {}^n R_{n-1} \cdots {}^2 R_1 = {}^n R_G^T$ is independent of the order of rotations $\alpha_1, \alpha_2, \cdots, \alpha_n$.

Proof Consider a globally fixed coordinate frame $G \, (OXYZ)$ and two body frames $B_1 \, (Ox_1y_1z_1)$ and $B_2 \, (Ox_2y_2z_2)$, all with a common origin O. The body B_1 caries B_2 and rotates with respect to G. The body B_2 rotates in B_1. The first rotation α of B_1 in G is about a G-fixed axis ${}^G \hat{u}_1$. The second motion is a rotation β of B_2 about a B_1-fixed axis ${}^1 \hat{u}_2$, or equivalently about ${}^2 \hat{u}_2$ in B_2. The frames B_2, B_1, G are coincident before rotations. A transformation matrix ${}^2 R_1$ for rotation $\beta = \alpha_2$ about a body axis ${}^2 \hat{u}_2$ relates B_2 and B_1. Let us show this rotation matrix by R_β, and call it the β-rotation.

$$ {}^2 R_1 = R_{{}^2\hat{u}_2, \beta} = R_\beta \tag{3.88} $$

Now assume that a γ-rotation of $R_{{}^2\hat{u}_2, \gamma}$ similar to (3.88) has happened and an α-rotation of $R_{{}^2\hat{u}_1, \alpha}$ is going to follow. To determine the overall transformation matrix ${}^1 R_G$ to relate the frame B_1 to the global frame G, we apply a rotation $-\gamma$ on B_2 about the local axis ${}^2 \hat{u}_2$ to bring B_2 on B_1 and G. The rotation $\alpha = \alpha_1$ will turn B_1 and B_2 together, considering B_2 is carried by B_1. The axis ${}^G \hat{u}_1$ would coincide with an axis ${}^2 \hat{u}_1$ that would be about a local axis of B_2. A rotation α of B_2 about ${}^2 \hat{u}_1$ will equivalently rotate B_1 and B_2 in G. Then, we apply the rotation α about ${}^2 \hat{u}_1$ and then apply a reverse rotation γ about ${}^2 \hat{u}_2$ to remove the effect of the rotation $-\gamma$ about ${}^2 \hat{u}_2$. Let us show the transformation matrix ${}^1 R_G$ for rotation α by R_α, and call it the α-rotation.

$$ {}^1 R_G = R_{{}^2\hat{u}_2, \gamma} \, R_{{}^2\hat{u}_1, \alpha} \, R_{{}^2\hat{u}_2, -\gamma} = R_\alpha \tag{3.89} $$

Therefore a rotation R_α of B_1 in G is being considered as a combination of three rotations from B_2 viewpoint. A rotation $-\gamma$ of B_2 with respect to B_1, followed by a rotation α of B_2 and B_1 with respect to G, and then a rotation γ of B_2 with respect to B_1.

The rotations α and β can be interchanged and performed in any order. The final transformation matrix will not be altered by changing the order of rotations α and β. To show this fact, let us assume that the frames B_1 and B_2 are on G. Applying the rotation α and then the rotation β provides the following transformation matrix:

$$ {}^2 R_G = R_\beta \, R_\alpha = R_{{}^2\hat{u}_2, \beta} \left[R_{{}^2\hat{u}_2, 0} \, R_{{}^2\hat{u}_1, \alpha} \, R_{{}^2\hat{u}_2, -0} \right] $$
$$ = R_{{}^2\hat{u}_2, \beta} \left[\mathbf{I} \, R_{{}^2\hat{u}_1, \alpha} \, \mathbf{I} \right] = R_{{}^2\hat{u}_2, \beta} \, R_{{}^2\hat{u}_1, \alpha} \tag{3.90} $$

By changing the order of rotations and applying the rotation β first and then α, we find the same transformation matrix.

$$ {}^2 R_G = R_\alpha \, R_\beta = \left[R_{{}^2\hat{u}_2, \beta} \, R_{{}^2\hat{u}_1, \alpha} \, R_{{}^2\hat{u}_2, -\beta} \right] R_{{}^2\hat{u}_2, \beta} $$
$$ = R_{{}^2\hat{u}_2, \beta} \, R_{{}^2\hat{u}_1, \alpha} \tag{3.91} $$

Now consider ${}^3 R_2$ as a transformation matrix for rotation of B_3 about ${}^2 \hat{u}_3$ in B_2, and ${}^2 R_G$ as the matrix for rotation of B_2 in G. Employing the same procedure, we can find the order-free transformation ${}^3 R_G$.

$$ {}^3 R_G = {}^3 R_2 \, {}^2 R_G = {}^2 R_G \, {}^3 R_2 \tag{3.92} $$

We can similarly expand and apply the proof of the order-free rotations theorem to any number of bodies that are similarly connected. ∎

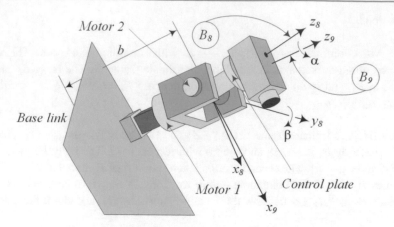

Fig. 3.3 An inspection camera

Example 70 ★ Order-free rotations in directional control systems. A detection device needs only two rotary actuators. The axes of actuators are better to be set orthogonal. Here the order-free rotation theorem is examined on an inspecting camera.

Consider the inspection camera in Fig. 3.3 with a direction control system to direct the camera to a point. To have a directional control system, we need to control the angles of a flat plate and direct an axis on the plate or its normal vector to a desired direction. Figure 3.3 shows that the two motors of the directional control system are independent. Therefore, we may command motors 1 and 2 to turn in different order or together. The final configuration of the camera must be the same if we command motor 1 to turn α and motor 2 to turn β in any order. To show this fact, we must define the transformation matrices such that the matrix multiplication is independent of the order of the matrices.

Consider the camera frame B_8 is not coincident with the fixed Sina frame B_9. We may always bring the local z_8-axis of B_8 on the global z_9-axis of B_9 by a rotation $-\beta$ about the y_8-axis. Now rotation α about the globally fixed z_9-axis becomes a local rotation about the z_8-axis. A reverse rotation β about the y_8-axis eliminates the first and third rotations and a rotation α about the fixed z_9-axis remains:

$$^8R_9 = R_{Z,\alpha} = R_{y,\beta} R_{z,\alpha} R_{y,-\beta} \tag{3.93}$$

$$= \begin{bmatrix} c\beta & 0 & -s\beta \\ 0 & 1 & 0 \\ s\beta & 0 & c\beta \end{bmatrix} \begin{bmatrix} c\alpha & -s\alpha & 0 \\ s\alpha & c\alpha & 0 \\ 0 & 0 & 1 \end{bmatrix} \begin{bmatrix} c(-\beta) & 0 & -s(-\beta) \\ 0 & 1 & 0 \\ s(-\beta) & 0 & c(-\beta) \end{bmatrix}$$

$$= \begin{bmatrix} c\alpha c^2\beta + s^2\beta & -c\beta s\alpha & c\alpha c\beta s\beta - c\beta s\beta \\ c\beta s\alpha & c\alpha & s\alpha s\beta \\ c\alpha c\beta s\beta - c\beta s\beta & -s\alpha s\beta & c^2\beta + c\alpha s^2\beta \end{bmatrix}$$

It reduces to the principal $R_{z,\alpha}$ for $\beta = 0$.

Let us apply a rotation α to the camera frame B_8 about the z_9-axis from a coincident configuration with the Sina frame B_9 followed by a rotation β about the y_8-axis:

$$^8R_9 = R_{y_8,\beta} R_{z_9,\alpha}^T \tag{3.94}$$

$$= \begin{bmatrix} \cos\beta & 0 & -\sin\beta \\ 0 & 1 & 0 \\ \sin\beta & 0 & \cos\beta \end{bmatrix} \begin{bmatrix} \cos\alpha & -\sin\alpha & 0 \\ \sin\alpha & \cos\alpha & 0 \\ 0 & 0 & 1 \end{bmatrix}^T$$

$$= \begin{bmatrix} \cos\alpha \cos\beta & \cos\beta \sin\alpha & -\sin\beta \\ -\sin\alpha & \cos\alpha & 0 \\ \cos\alpha \sin\beta & \sin\alpha \sin\beta & \cos\beta \end{bmatrix}$$

Now let us turn the camera frame B_8 an angle β from a coincident configuration with frame B_9 about the y_8-axis followed by a rotation α about the z_9-axis. However, the rotation about the z_9-axis must be considered as (3.93):

$$^8R_9 = R_{z9,\alpha}^T \, R_{y8,\beta} \tag{3.95}$$

$$= \begin{bmatrix} c\alpha c^2\beta + s^2\beta & -c\beta s\alpha & c\alpha c\beta s\beta - c\beta s\beta \\ c\beta s\alpha & c\alpha & s\alpha s\beta \\ c\alpha c\beta s\beta - c\beta s\beta & -s\alpha s\beta & c^2\beta + c\alpha s^2\beta \end{bmatrix} \begin{bmatrix} c\beta & 0 & -s\beta \\ 0 & 1 & 0 \\ s\beta & 0 & c\beta \end{bmatrix}$$

$$= \begin{bmatrix} \cos\alpha\cos\beta & -\cos\beta\sin\alpha & -\sin\beta \\ \sin\alpha & \cos\alpha & 0 \\ \cos\alpha\sin\beta & -\sin\alpha\sin\beta & \cos\beta \end{bmatrix}$$

The final rotation matrix (3.95) is the same as (3.94). So, it is immaterial which motor turns first provided that the correct matrices are used. Therefore, we may define the transformation matrix between the camera frame B_8 and frame B_9 by matrix multiplication in any order,

$$^8R_9 = R_{z9,\alpha}^T \, R_{y8,\beta} = R_{y8,\beta} \, R_{z9,\alpha}^T \tag{3.96}$$

where,

$$R_{z9,\alpha}^T = \begin{bmatrix} c\alpha c^2\beta + s^2\beta & -c\beta s\alpha & c\alpha c\beta s\beta - c\beta s\beta \\ c\beta s\alpha & c\alpha & s\alpha s\beta \\ c\alpha c\beta s\beta - c\beta s\beta & -s\alpha s\beta & c^2\beta + c\alpha s^2\beta \end{bmatrix} \tag{3.97}$$

and

$$R_{y8,\beta} = \begin{bmatrix} \cos\beta & 0 & -\sin\beta \\ 0 & 1 & 0 \\ \sin\beta & 0 & \cos\beta \end{bmatrix} \tag{3.98}$$

However, to evaluate $^8R_9$, we must substitute the current value of β in $R_{z9,\alpha}^T$.

To aim the camera at from a point to the other, let us assume the camera frame B_8 of the directional control system has already been rotated and is not coincident with the Sina frame B_9. If the rotation α_1 happened before rotation β_1, the rotation matrix between B_8 and B_9 would be

$$R_1 = \,^8R_9 = \begin{bmatrix} \cos\alpha_1\cos\beta_1 & -\cos\beta_1\sin\alpha_1 & -\sin\beta_1 \\ \sin\alpha_1 & \cos\alpha_1 & 0 \\ \cos\alpha_1\sin\beta_1 & -\sin\alpha 1\sin\beta_1 & \cos\beta_1 \end{bmatrix} \tag{3.99}$$

otherwise, we may use Eq. (3.96) to reverse the order of rotations. To aim the camera to a new direction, we may turn the camera β_2 about y_8 and α_2 about z_9. The rotation matrix between the camera and the base would then be

$$^8R_9 = R_1 R_{y8,\beta_2} R_{z9,\alpha_2}^T = R_1 R_{y8,\beta_2} \left[R_{y8,\beta_1} R_{z8,\alpha_2} R_{y8,-\beta_1} \right]$$

$$= R_{y8,\beta_1} R_{z8,\alpha_1} R_{y8,\beta_2} R_{y8,\beta_1} R_{z8,\alpha_2} R_{y8,-\beta_1}$$

$$= R_{y8,\beta_1} R_{z8,\alpha_1} R_{y8,\beta_1+\beta_2} R_{z8,\alpha_2} R_{y8,-\beta_1} \tag{3.100}$$

or if we change the order of second rotations β_2 and α_2 and use Eq. (3.93), then

$$^8R_9 = R_1 R_{z9,\alpha_2}^T R_{y8,\beta_2} = R_1 \left[R_{y8,\beta_1+\beta_2} R_{z8,\alpha_2} R_{y8,-\beta_1-\beta_2} \right] R_{y8,\beta_2}$$

$$= R_{y8,\beta_1} R_{z8,\alpha_1} R_{y8,\beta_1+\beta_2} R_{z8,\alpha_2} R_{y8,-\beta_1-\beta_2} R_{y8,\beta_2}$$

$$= R_{y8,\beta_1} R_{z8,\alpha_1} R_{y8,\beta_1+\beta_2} R_{z8,\alpha_2} R_{y8,-\beta_1} \tag{3.101}$$

Example 71 ★ Spherical wrist and order free rotations. Spherical wrist is a good example for real dynamic system with order-free rotations.

A spherical wrist, as is shown in Fig. 3.4, will act by three independent rotary actuators. The independency of the actuators means we may run the three actuators in any arbitrary order. This freedom requires that the transformation matrix between B_0 and B_2 be order free and independent of the order of rotations. Let us assume none of the angles is zero, and hence, the coordinate frames B_0, B_1, B_2 are not coincident. The rotation ψ of B_2 is always about the local z_2-axis. This rotation will not move any axis of B_2.

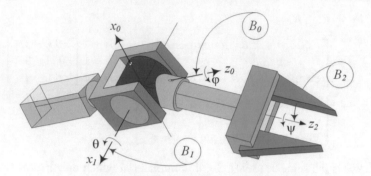

Fig. 3.4 Eulerian spherical wrist or type roll–pitch–roll

$$^2R_1 = R_{z_2,\psi} = R_{z,\psi} = \begin{bmatrix} \cos\psi & \sin\psi & 0 \\ -\sin\psi & \cos\psi & 0 \\ 0 & 0 & 1 \end{bmatrix} \tag{3.102}$$

To find the order-free transformation matrix for rotation θ about x_1, we turn B_2 an angle $-\psi$ about z_2 to bring x_2 on x_1. Now rotation θ about x_1 becomes a local rotation about x_2. A reverse rotation ψ about z_2 eliminates the first and third rotations and a rotation θ about x_1 remains:

$$^2R_1 = R_{x_1,\theta} = R_{z_2,\psi} R_{x_2,\theta} R_{z_2,-\psi} = R_{z,\psi} R_{x,\theta} R_{z,-\psi} \tag{3.103}$$

$$= \begin{bmatrix} c\psi & s\psi & 0 \\ -s\psi & c\psi & 0 \\ 0 & 0 & 1 \end{bmatrix} \begin{bmatrix} 1 & 0 & 0 \\ 0 & c\theta & s\theta \\ 0 & -s\theta & c\theta \end{bmatrix} \begin{bmatrix} c(-\psi) & s(-\psi) & 0 \\ -s(-\psi) & c(-\psi) & 0 \\ 0 & 0 & 1 \end{bmatrix}$$

$$= \begin{bmatrix} c^2\psi + c\theta s^2\psi & c\theta c\psi s\psi - c\psi s\psi & s\theta s\psi \\ c\theta c\psi s\psi - c\psi s\psi & c\theta c^2\psi + s^2\psi & c\psi s\theta \\ -s\theta s\psi & -c\psi s\theta & c\theta \end{bmatrix}$$

The matrix (3.103) is the order-free transformation for turning θ about x_1. We can examine and check that

$$^2R_1 = R_{x_1,\theta} R_{z_2,\psi} = R_{z_2,\psi} R_{x_1,\theta} \tag{3.104}$$

provided that we use (3.103) for $R_{x_1,\theta}$ and substitute the current value of ψ before multiplication. Assuming a rotation ψ about z_2 and then a rotation θ about x_1, we find

$$^2R_1 = R_{x_1,\theta} R_{z_2,\psi}$$

$$= \begin{bmatrix} c^2\psi + c\theta s^2\psi & c\theta c\psi s\psi - c\psi s\psi & s\theta s\psi \\ c\theta c\psi s\psi - c\psi s\psi & c\theta c^2\psi + s^2\psi & c\psi s\theta \\ -s\theta s\psi & -c\psi s\theta & c\theta \end{bmatrix} \begin{bmatrix} c\psi & s\psi & 0 \\ -s\psi & c\psi & 0 \\ 0 & 0 & 1 \end{bmatrix}$$

$$= \begin{bmatrix} \cos\psi & \cos\theta\sin\psi & \sin\theta\sin\psi \\ -\sin\psi & \cos\theta\cos\psi & \cos\psi\sin\theta \\ 0 & -\sin\theta & \cos\theta \end{bmatrix} \tag{3.105}$$

Now assuming a rotation θ about x_1 and then a rotation ψ about z_2, we find

$$^2R_1 = R_{z_2,\psi} R_{x_1,\theta}$$

$$= \begin{bmatrix} c\psi & s\psi & 0 \\ -s\psi & c\psi & 0 \\ 0 & 0 & 1 \end{bmatrix} \begin{bmatrix} c^2 0 + c\theta s^2 0 & c\theta c0s0 - c0s0 & s\theta s0 \\ c\theta c0s0 - c0s0 & c\theta c^2 0 + s^2 0 & c0s\theta \\ -s\theta s0 & -c0s\theta & c\theta \end{bmatrix}$$

$$= \begin{bmatrix} \cos\psi & \cos\theta\sin\psi & \sin\theta\sin\psi \\ -\sin\psi & \cos\theta\cos\psi & \cos\psi\sin\theta \\ 0 & -\sin\theta & \cos\theta \end{bmatrix} \tag{3.106}$$

because we substitute $\psi = 0$ in $R_{x_1,\theta}$ to evaluate $R_{z_2,\psi} R_{x_1,\theta}$.

To find the order-free transformation matrix for rotation φ about x_0, we turn B_2 an angle $-\psi$ about the local z_2 to bring x_2 on x_1. Another rotation $-\theta$ about x_2 brings z_2 on z_0. Now rotation φ about x_0 becomes a local rotation about x_2. A reverse rotation θ about x_2 and then ψ about z_2 eliminates the first, second, fourth, and fifth rotations and a rotation φ about x_0 remains:

$$^1R_0 = R_{x_0,\theta} = R_{z_2,\psi} R_{x_2,\theta} R_{z_2,\varphi} R_{x_2,-\theta} R_{z_2,-\psi} \tag{3.107}$$

$$= R_{z,\psi} R_{x,\theta} R_{z,\varphi} R_{x,-\theta} R_{z,-\psi} = \begin{bmatrix} r_{11} & r_{12} & r_{13} \\ r_{21} & r_{22} & r_{23} \\ r_{31} & r_{32} & r_{33} \end{bmatrix}$$

where

$$r_{11} = -(1 - \cos\varphi)\cos^2\psi \sin^2\theta \cos^2\theta + 1 \tag{3.108}$$

$$r_{21} = (1 - \cos\varphi)\sin^2\theta \cos\psi \sin\psi - \cos\theta \sin\varphi \tag{3.109}$$

$$r_{31} = ((1 - \cos\varphi)\cos\theta \sin\psi + \cos\psi \sin\varphi)\sin\theta \tag{3.110}$$

$$r_{12} = (1 - \cos\varphi)\sin^2\theta \cos\psi \sin\psi + \cos\theta \sin\varphi \tag{3.111}$$

$$r_{22} = \cos\varphi + \sin^2\theta \cos^2\psi (1 - \cos\varphi) \tag{3.112}$$

$$r_{32} = ((1 - \cos\varphi)\cos\theta \cos\psi - \sin\psi \sin\varphi)\sin\theta \tag{3.113}$$

$$r_{13} = ((1 - \cos\varphi)\cos\theta \sin\psi - \sin\varphi \cos\psi)\sin\theta \tag{3.114}$$

$$r_{23} = ((1 - \cos\varphi)\cos\theta \cos\psi + \sin\psi \sin\varphi)\sin\theta \tag{3.115}$$

$$r_{33} = \cos\varphi + (1 - \cos\varphi)\cos^2\theta \tag{3.116}$$

The matrix (3.107) is the order-free transformation for turning φ about z_0. We can check that the order of rotations φ, θ, ψ are not important and we have

$$^2R_0 = R_{x_1,\theta} R_{z_2,\psi} R_{z_0,\varphi} = R_{z_2,\psi} R_{z_0,\varphi} R_{x_1,\theta}$$

$$= R_{z_0,\varphi} R_{x_1,\theta} R_{z_2,\psi} = R_{z_0,\varphi} R_{z_2,\psi} R_{x_1,\theta}$$

$$= R_{z_2,\psi} R_{x_1,\theta} R_{z_0,\varphi} = R_{x_1,\theta} R_{z_0,\varphi} R_{z_2,\psi} \tag{3.117}$$

We need to substitute the current values of $\varphi, \theta,$ and ψ before multiplying any transformation matrices.

Example 72 ★ Order-free Euler angle matrix. Here is to show order-free rotation on Euler angles.

The Euler angle transformation with the order φ, θ, ψ is

$$^B R_G = R_{z,\psi} R_{x,\theta} R_{z,\varphi} \tag{3.118}$$

$$= \begin{bmatrix} c\varphi c\psi - c\theta s\varphi s\psi & c\psi s\varphi + c\theta c\varphi s\psi & s\theta s\psi \\ -c\varphi s\psi - c\theta c\psi s\varphi & -s\varphi s\psi + c\theta c\varphi c\psi & s\theta c\psi \\ s\theta s\varphi & -c\varphi s\theta & c\theta \end{bmatrix}$$

where,

$$R_{z,\varphi} = \begin{bmatrix} \cos\varphi & \sin\varphi & 0 \\ -\sin\varphi & \cos\varphi & 0 \\ 0 & 0 & 1 \end{bmatrix} \tag{3.119}$$

$$R_{x,\theta} = \begin{bmatrix} 1 & 0 & 0 \\ 0 & \cos\theta & \sin\theta \\ 0 & -\sin\theta & \cos\theta \end{bmatrix} \tag{3.120}$$

$$R_{z,\psi} = \begin{bmatrix} \cos\psi & \sin\psi & 0 \\ -\sin\psi & \cos\psi & 0 \\ 0 & 0 & 1 \end{bmatrix} \tag{3.121}$$

These are in-order rotation matrices associated with φ, θ, ψ. The rotation φ transforms the global coordinate frame to B', rotation θ transforms B' to B'', and rotation ψ transforms B'' to B. To make them to be order free, we should define them as a rotation about a local axis of the final coordinate frame. The rotation ψ is already about the local z-axis and is order free. The order-free rotation θ is

$$R_{x,\theta} = R_{z,\psi} R_{x,\theta} R_{z,-\psi} \tag{3.122}$$

$$= \begin{bmatrix} c^2\psi + c\theta s^2\psi & c\theta c\psi s\psi - c\psi s\psi & s\theta s\psi \\ c\theta c\psi s\psi - c\psi s\psi & c\theta c^2\psi + s^2\psi & c\psi s\theta \\ -s\theta s\psi & -c\psi s\theta & c\theta \end{bmatrix}$$

The matrix (3.122) is the order-free transformation for rotation θ about x:

$$^{B}R_{G} = R_{x,\theta} R_{z,\psi} = R_{z,\psi} R_{x,\theta} \tag{3.123}$$

where,

$$^{B}R_{G} = R_{x,\theta} R_{z,\psi}$$

$$= \begin{bmatrix} c^2\psi + c\theta s^2\psi & (c\theta - 1) & s\theta s\psi \\ (c\theta - 1) c\psi s\psi & c\theta c^2\psi + s^2\psi & c\psi s\theta \\ -s\theta s\psi & -c\psi s\theta & c\theta \end{bmatrix} \begin{bmatrix} c\psi & s\psi & 0 \\ -s\psi & c\psi & 0 \\ 0 & 0 & 1 \end{bmatrix}$$

$$= \begin{bmatrix} \cos\psi & \cos\theta\sin\psi & \sin\theta\sin\psi \\ -\sin\psi & \cos\theta\cos\psi & \cos\psi\sin\theta \\ 0 & -\sin\theta & \cos\theta \end{bmatrix} \tag{3.124}$$

$$^{B}R_{G} = R_{z,\psi} R_{x,\theta}$$

$$= \begin{bmatrix} c\psi & s\psi & 0 \\ -s\psi & c\psi & 0 \\ 0 & 0 & 1 \end{bmatrix} \begin{bmatrix} c^2 0 + c\theta s^2 0 & (c\theta - 1) c0s0 & s\theta s0 \\ (c\theta - 1) c0s0 & c\theta c^2 0 + s^2 0 & c0s\theta \\ -s\theta s0 & -c0s\theta & c\theta \end{bmatrix}$$

$$= \begin{bmatrix} \cos\psi & \cos\theta\sin\psi & \sin\theta\sin\psi \\ -\sin\psi & \cos\theta\cos\psi & \cos\psi\sin\theta \\ 0 & -\sin\theta & \cos\theta \end{bmatrix} \tag{3.125}$$

The order-free transformation matrix for rotation φ is

$$R_{z,\varphi} = R_{z,\psi} R_{x,\theta} R_{z,\varphi} R_{x,-\theta} R_{z,-\psi} \tag{3.126}$$

$$= \begin{bmatrix} r_{11} & r_{12} & r_{13} \\ r_{21} & r_{22} & r_{23} \\ r_{31} & r_{32} & r_{33} \end{bmatrix}$$

where,

$$r_{11} = -(1 - \cos\varphi)\cos^2\psi\sin^2\theta\cos^2\theta + 1 \tag{3.127}$$

$$r_{21} = (1 - \cos\varphi)\sin^2\theta\cos\psi\sin\psi - \cos\theta\sin\varphi \tag{3.128}$$

$$r_{31} = ((1 - \cos\varphi)\cos\theta\sin\psi + \cos\psi\sin\varphi)\sin\theta \tag{3.129}$$

$$r_{12} = (1 - \cos\varphi)\sin^2\theta\cos\psi\sin\psi + \cos\theta\sin\varphi \tag{3.130}$$

$$r_{22} = \cos\varphi + \sin^2\theta\cos^2\psi\,(1 - \cos\varphi) \tag{3.131}$$

$$r_{32} = ((1 - \cos\varphi)\cos\theta\cos\psi - \sin\psi\sin\varphi)\sin\theta \tag{3.132}$$

$$r_{13} = ((1 - \cos\varphi)\cos\theta\sin\psi - \sin\varphi\cos\psi)\sin\theta \tag{3.133}$$

$$r_{23} = ((1 - \cos\varphi)\cos\theta\cos\psi + \sin\psi\sin\varphi)\sin\theta \tag{3.134}$$

$$r_{33} = \cos\varphi + (1 - \cos\varphi)\cos^2\theta \tag{3.135}$$

The matrix (3.126) is the order-free transformation for rotation φ about z. Using the order-free matrices (3.121), (3.122), and (3.126) we find the Euler angle transformation matrix (3.122) by rotations φ, θ, ψ in any order:

$$\begin{aligned}
{}^{B}R_{G} &= R_{x,\theta}R_{z,\psi}R_{z,\varphi} = R_{z,\psi}R_{z,\varphi}R_{x,\theta} \\
&= R_{z,\varphi}R_{x,\theta}R_{z,\psi} = R_{z,\varphi}R_{z,\psi}R_{x,\theta} \\
&= R_{z,\psi}R_{x,\theta}R_{z,\varphi} = R_{x,\theta}R_{z,\varphi}R_{z,\psi}
\end{aligned} \tag{3.136}$$

Equations (3.127)–(3.135) are the same as (3.108)–(3.116), and it justifies why we call the spherical wrist of Fig. 3.4 an Eulerian wrist.

3.3 ★ Euler Parameters

Assume ϕ to be the angle of rotation of a body coordinate frame $B(Oxyz)$ about $\hat{u} = u_1\hat{I} + u_2\hat{J} + u_3\hat{K}$ relative to a global frame $G(OXYZ)$. The existence of such an axis of rotation is the analytical representation of the Euler's theorem about rigid body rotation:

Euler Rigid Body Rotation Theorem *The most general displacement of a rigid body with one fixed point is a rotation about an axis.*

To find the axis and angle of rotation we introduce the *Euler parameters* e_0, e_1, e_2, e_3 such that e_0 is a scalar and e_1, e_2, e_3 are components of a vector **e**,

$$e_0 = \cos\frac{\phi}{2} \tag{3.137}$$

$$\mathbf{e} = e_1\hat{I} + e_2\hat{J} + e_3\hat{K} = \hat{u}\sin\frac{\phi}{2} \tag{3.138}$$

The Euler parameters have a constraint among them.

$$e_1^2 + e_2^2 + e_3^2 + e_0^2 = e_0^2 + \mathbf{e}^T\mathbf{e} = 1 \tag{3.139}$$

Then, the transformation matrix ^GR_B to satisfy the equation $^G\mathbf{r} = {}^GR_B\,{}^B\mathbf{r}$ can be derived based on Euler parameters

$$^GR_B = R_{\hat{u},\phi} = \left(e_0^2 - \mathbf{e}^2\right)\mathbf{I} + 2\mathbf{e}\,\mathbf{e}^T + 2e_0\tilde{e} \tag{3.140}$$

$$= \begin{bmatrix} e_0^2 + e_1^2 - e_2^2 - e_3^2 & 2\,(e_1e_2 - e_0e_3) & 2\,(e_0e_2 + e_1e_3) \\ 2\,(e_0e_3 + e_1e_2) & e_0^2 - e_1^2 + e_2^2 - e_3^2 & 2\,(e_2e_3 - e_0e_1) \\ 2\,(e_1e_3 - e_0e_2) & 2\,(e_0e_1 + e_2e_3) & e_0^2 - e_1^2 - e_2^2 + e_3^2 \end{bmatrix}$$

where $\tilde{e}$ is the *skew symmetric* matrix corresponding to vector $\mathbf{e}$.

$$\tilde{e} = \begin{bmatrix} 0 & -e_3 & e_2 \\ e_3 & 0 & -e_1 \\ -e_2 & e_1 & 0 \end{bmatrix} \tag{3.141}$$

Given a transformation matrix ^GR_B we may obtain Euler parameters:

$$e_0^2 = \frac{1}{4}\left(\mathrm{tr}\left({}^GR_B\right) + 1\right) \tag{3.142}$$

$$\tilde{e} = \frac{1}{4e_0}\left({}^GR_B - {}^GR_B^T\right) \tag{3.143}$$

and determine the angle ϕ and axis of rotation $\hat{u}$.

$$\cos\phi = \frac{1}{2}\left(\mathrm{tr}\left({}^GR_B\right) - 1\right) \tag{3.144}$$

$$\tilde{u} = \frac{1}{2\sin\phi}\left({}^GR_B - {}^GR_B^T\right) \tag{3.145}$$

Euler parameters provide a well-suited, redundant, and non-singular rotation description for arbitrary and large rotations.

Assume a rotation transformation matrix ^GR_B is given.

$$^GR_B = \begin{bmatrix} r_{11} & r_{12} & r_{13} \\ r_{21} & r_{22} & r_{23} \\ r_{31} & r_{32} & r_{33} \end{bmatrix} \tag{3.146}$$

It is then possible to calculate the Euler parameters e_0, e_1, e_2, e_3 and indicate the axis and angle of rotation to derive ^GR_B by utilizing any one of the following four sets of equations:

$$e_0 = \pm\frac{1}{2}\sqrt{1 + r_{11} + r_{22} + r_{33}}$$

$$e_1 = \frac{1}{4}\frac{r_{32} - r_{23}}{e_0} \qquad e_2 = \frac{1}{4}\frac{r_{13} - r_{31}}{e_0} \qquad e_3 = \frac{1}{4}\frac{r_{21} - r_{12}}{e_0} \tag{3.147}$$

$$e_1 = \pm\frac{1}{2}\sqrt{1 + r_{11} - r_{22} - r_{33}}$$

$$e_2 = \frac{1}{4}\frac{r_{21} + r_{12}}{e_1} \qquad e_3 = \frac{1}{4}\frac{r_{31} + r_{13}}{e_1} \qquad e_0 = \frac{1}{4}\frac{r_{32} + r_{23}}{e_1} \tag{3.148}$$

$$e_2 = \pm\frac{1}{2}\sqrt{1 - r_{11} + r_{22} - r_{33}}$$

$$e_3 = \frac{1}{4}\frac{r_{32} + r_{23}}{e_2} \qquad e_0 = \frac{1}{4}\frac{r_{13} - r_{31}}{e_2} \qquad e_1 = \frac{1}{4}\frac{r_{21} + r_{12}}{e_2} \tag{3.149}$$

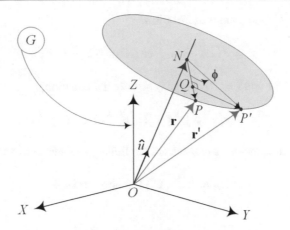

Fig. 3.5 Axis and angle of rotation

$$e_3 = \pm \frac{1}{2}\sqrt{1 - r_{11} - r_{22} + r_{33}}$$

$$e_0 = \frac{1}{4}\frac{r_{21} - r_{12}}{e_3} \quad e_1 = \frac{1}{4}\frac{r_{31} + r_{13}}{e_3} \quad e_2 = \frac{1}{4}\frac{r_{32} + r_{23}}{e_3} \tag{3.150}$$

Although Eqs. (3.147)–(3.150) present four different sets of solutions, their resulting Euler parameters e_0, e_1, e_2, e_3 are identical. Therefore, numerical inaccuracies can be minimized by using the set with maximum divisor. The plus and minus signs indicate that rotation ϕ about $\hat{u}$ is equivalent to rotation $-\phi$ about $-\hat{u}$.

Proof Figure 3.5 depicts a point P of a rigid body with position vector $\mathbf{r}$, and the unit vector $\hat{u}$ along the axis of rotation ON fixed in the global frame. The point moves to P' with position vector $\mathbf{r}'$ after an active rotation ϕ about $\hat{u}$. To obtain the relationship between $\mathbf{r}$ and $\mathbf{r}'$, we express $\mathbf{r}'$ by the following vector equation:

$$\mathbf{r}' = \overrightarrow{ON} + \overrightarrow{NQ} + \overrightarrow{QP'} \tag{3.151}$$

From Fig. 3.5 we may describe Eq. (3.151) utilizing $\mathbf{r}$, $\mathbf{r}'$, $\hat{u}$, and ϕ.

$$\begin{aligned}\mathbf{r}' &= (\mathbf{r} \cdot \hat{u})\,\hat{u} + \hat{u} \times (\mathbf{r} \times \hat{u})\cos\phi - (\mathbf{r} \times \hat{u})\sin\phi \\ &= (\mathbf{r} \cdot \hat{u})\,\hat{u} + [\mathbf{r} - (\mathbf{r} \cdot \hat{u})\,\hat{u}]\cos\phi + (\hat{u} \times \mathbf{r})\sin\phi \end{aligned} \tag{3.152}$$

Rearranging Eq. (3.152) leads to a new form of the Rodriguez rotation formula.

$$\mathbf{r}' = \mathbf{r}\cos\phi + (1 - \cos\phi)(\hat{u} \cdot \mathbf{r})\,\hat{u} + (\hat{u} \times \mathbf{r})\sin\phi \tag{3.153}$$

Using the Euler parameters in (3.137) and (3.138), along with the following trigonometric relations:

$$\cos\phi = 2\cos^2\frac{\phi}{2} - 1 \tag{3.154}$$

$$\sin\phi = 2\sin\frac{\phi}{2}\cos\frac{\phi}{2} \tag{3.155}$$

$$1 - \cos\phi = 2\sin^2\frac{\phi}{2} \tag{3.156}$$

converts the Rodriguez formula (3.153) to a more useful form.

$$\mathbf{r}' = \mathbf{r}\left(2e_0^2 - 1\right) + 2\mathbf{e}\left(\mathbf{e} \cdot \mathbf{r}\right) + 2e_0\left(\mathbf{e} \times \mathbf{r}\right) \tag{3.157}$$

Defining new notations for $\mathbf{r}' = {}^G\mathbf{r}$ and $\mathbf{r} = {}^B\mathbf{r}$, we may rewrite the equation,

$$^G\mathbf{r} = \left(e_0^2 - \mathbf{e}^2\right) {}^B\mathbf{r} + 2\mathbf{e}\left(\mathbf{e}^T {}^B\mathbf{r}\right) + 2e_0\left(\tilde{\mathbf{e}} {}^B\mathbf{r}\right) \tag{3.158}$$

that allows us to factor out the position vector ${}^B\mathbf{r}$ and extract the Euler parameter transformation matrix ${}^G R_B$

$$^G R_B = \left(e_0^2 - \mathbf{e}^2\right) \mathbf{I} + 2\mathbf{e}\,\mathbf{e}^T + 2e_0\tilde{\mathbf{e}} \tag{3.159}$$

where,

$$^G\mathbf{r} = {}^G R_B {}^B\mathbf{r} = R_{\hat{u},\phi} {}^B\mathbf{r} \tag{3.160}$$

To prove the equations of the angle ϕ and axis of rotation $\hat{u}$ for a given transformation matrix ${}^G R_B$, let us calculate the trace of ${}^G R_B$ and find e_0 and ϕ,

$$\text{tr}\left({}^G R_B\right) = 4e_0^2 - 1 = 2\cos\phi + 1 \tag{3.161}$$

and calculate ${}^G R_B - {}^G R_B^T$ to find $\mathbf{e}$ and $\hat{u}$.

$$^G R_B - {}^G R_B^T = \begin{bmatrix} 0 & -4e_0e_3 & 4e_0e_2 \\ 4e_0e_3 & 0 & -4e_0e_1 \\ -4e_0e_2 & 4e_0e_1 & 0 \end{bmatrix}$$

$$= 2\sin\phi \begin{bmatrix} 0 & -u_3 & u_2 \\ u_3 & 0 & -u_1 \\ -u_2 & u_1 & 0 \end{bmatrix} \tag{3.162}$$

$$\tilde{e} = \frac{1}{4e_0} \begin{bmatrix} r_{32} - r_{23} \\ r_{13} - r_{31} \\ r_{21} - r_{12} \end{bmatrix} \tag{3.163}$$

$$\hat{u} = \frac{1}{2\sin\phi} \begin{bmatrix} r_{32} - r_{23} \\ r_{13} - r_{31} \\ r_{21} - r_{12} \end{bmatrix} \tag{3.164}$$

Comparing (3.140) and (3.146) shows that for the first set of Eqs. (3.147), e_0 can be found by summing the diagonal elements r_{11}, r_{22}, and r_{33} to have $\text{tr}\left({}^G R_B\right) = 4e_0^2 - 1$. To find e_1, e_2, and e_3 we need to simplify $r_{32} - r_{23}$, $r_{13} - r_{31}$, and $r_{21} - r_{12}$, respectively. The other sets of solutions (3.148)–(3.150) can also be found by comparison. ∎

Example 73 ★ Axis–angle rotation ${}^G R_B$ for a ϕ and $\hat{u}$. Numerical example for Euler parameters based on a given axis–angle of rotations.

Euler parameters for rotation $\phi = 30$ deg about $\hat{u} = \left(\hat{I} + \hat{J} + \hat{K}\right)/\sqrt{3}$ are

$$e_0 = \cos\frac{\pi}{12} = 0.966 \tag{3.165}$$

$$\mathbf{e} = \hat{u}\sin\frac{\phi}{2} = e_1\hat{I} + e_2\hat{J} + e_3\hat{K} = 0.149\left(\hat{I} + \hat{J} + \hat{K}\right) \tag{3.166}$$

therefore, the corresponding transformation matrix ${}^G R_B$ is

$$^G R_B = \left(e_0^2 - \mathbf{e}^2\right) \mathbf{I} + 2\mathbf{e}\,\mathbf{e}^T + 2e_0\tilde{e}$$

$$= \begin{bmatrix} 0.91 & -0.244 & 0.333 \\ 0.333 & 0.91 & -0.244 \\ -0.244 & 0.333 & 0.91 \end{bmatrix} \tag{3.167}$$

Example 74 ★ Euler parameters and Euler angles relationship.

Comparing the Euler angles rotation matrix (2.172) and the Euler parameter transformation matrix (3.140) we determine the following relationships between Euler angles and Euler parameters:

$$e_0 = \cos\frac{\theta}{2}\cos\frac{\psi + \varphi}{2} \tag{3.168}$$

$$e_1 = \sin\frac{\theta}{2}\cos\frac{\psi - \varphi}{2} \tag{3.169}$$

$$e_2 = \sin\frac{\theta}{2}\sin\frac{\psi - \varphi}{2} \tag{3.170}$$

$$e_3 = \cos\frac{\theta}{2}\sin\frac{\psi + \varphi}{2} \tag{3.171}$$

or

$$\varphi = \cos^{-1}\frac{2\left(e_2 e_3 + e_0 e_1\right)}{\sin\theta} \tag{3.172}$$

$$\theta = \cos^{-1}\left[2\left(e_0^2 + e_3^2\right) - 1\right] \tag{3.173}$$

$$\psi = \cos^{-1}\frac{-2\left(e_2 e_3 - e_0 e_1\right)}{\sin\theta} \tag{3.174}$$

Example 75 ★ Rotation matrix for angle of rotation $\phi = k\pi$. Euler parameter transformation matrix for ± 90 deg rotations.

When the angle of rotation is $\phi = k\pi$, $k = \pm 1, \pm 3, ...$, then $e_0 = 0$. Therefore, the Euler parameter transformation matrix (3.140) becomes

$$^G R_B = 2 \begin{bmatrix} e_1^2 - \frac{1}{2} & e_1 e_2 & e_1 e_3 \\ e_1 e_2 & e_2^2 - \frac{1}{2} & e_2 e_3 \\ e_1 e_3 & e_2 e_3 & e_3^2 - \frac{1}{2} \end{bmatrix} \tag{3.175}$$

which is a symmetric matrix and indicates that rotation $\phi = k\pi$, and $\phi = -k\pi$ are equivalent.

Example 76 ★ Vector of infinitesimal rotation. For infinitesimal rotations, the axis of rotation $\hat{u}$ is axis of angular velocity $\boldsymbol{\omega}$.

Consider the Rodriguez rotation formula (3.153) for a differential rotation $d\phi$

$$\mathbf{r}' = \mathbf{r} + \left(\hat{u} \times \mathbf{r}\right)d\phi \tag{3.176}$$

In this case the difference between $\mathbf{r}'$ and $\mathbf{r}$ is also very small,

$$d\mathbf{r} = \mathbf{r}' - \mathbf{r} = d\phi\,\hat{u} \times \mathbf{r} \tag{3.177}$$

and hence, a differential rotation $d\phi$ about an axis indicated by the unit vector $\hat{u}$ is a vector along $\hat{u}$ with magnitude $d\phi$. Dividing both sides by dt leads to

$$\dot{\mathbf{r}} = \dot{\phi}\,\hat{u} \times \mathbf{r} = \boldsymbol{\omega} \times \mathbf{r} \tag{3.178}$$

$$\boldsymbol{\omega} = \dot{\phi}\,\hat{u} \tag{3.179}$$

which represents the global velocity vector of any point in a rigid body rotating about $\hat{u}$.

Example 77 ★ Exponential form of rotation $e^{\phi\tilde{u}}$. Axis–angle of rotation can be expressed by an exponential function equivalently. It is an interesting expression with not much advantages over traditional methods.

Consider a point P in the body frame B with a position vector $\mathbf{r}$. If the rigid body has an angular velocity $\boldsymbol{\omega}$, then the velocity of P in the global coordinate frame is

$$\mathbf{P} = \boldsymbol{\omega} \times \mathbf{r} = \tilde{\omega}\mathbf{r} \tag{3.180}$$

This is a first-order linear differential equation that may be integrated to have

$$\mathbf{r}(t) = e^{\tilde{\omega}t}\mathbf{r}(0) \tag{3.181}$$

where $\mathbf{r}(0)$ is the initial position vector of P, and $e^{\tilde{\omega}t}$ is a matrix exponential.

$$e^{\tilde{\omega}t} = \mathbf{I} + \tilde{\omega}t + \frac{(\tilde{\omega}t)^2}{2!} + \frac{(\tilde{\omega}t)^3}{3!} + \cdots \tag{3.182}$$

The angular velocity $\boldsymbol{\omega}$ has a magnitude ω and direction indicated by a unit vector $\hat{u}$. Therefore,

$$\boldsymbol{\omega} = \omega\hat{u} \tag{3.183}$$

$$\tilde{\omega} = \omega\tilde{u} \tag{3.184}$$

$$\tilde{\omega}t = \omega t\tilde{u} = \phi\tilde{u} \tag{3.185}$$

and hence

$$e^{\tilde{\omega}t} = e^{\phi\tilde{u}} \tag{3.186}$$

$$= \mathbf{I} + \left(\phi - \frac{\phi^3}{3!} + \frac{\phi^5}{5!} - \cdots\right)\tilde{u} + \left(\frac{\phi^2}{2!} - \frac{\phi^4}{4!} + \frac{\phi^6}{6!}\cdots\right)\tilde{u}^2$$

or equivalently

$$e^{\phi\tilde{u}} = \mathbf{I} + \tilde{u}\sin\phi + \tilde{u}^2(1 - \cos\phi) \tag{3.187}$$

It is an alternative form of the Rodriguez formula showing that $e^{\phi\tilde{u}}$ is the rotation transformation to map $^B\mathbf{r} = \mathbf{r}(0)$ to $^G\mathbf{r} = \mathbf{r}(t)$.

Expanding $e^{\phi\tilde{u}}$ gives

$$e^{\phi\tilde{u}} = \begin{bmatrix} u_1^2\,\mathrm{vers}\,\phi + c\phi & u_1u_2\,\mathrm{vers}\,\phi - u_3s\phi & u_1u_3\,\mathrm{vers}\,\phi + u_2s\phi \\ u_1u_2\,\mathrm{vers}\,\phi + u_3s\phi & u_2^2\,\mathrm{vers}\,\phi + c\phi & u_2u_3\,\mathrm{vers}\,\phi - u_1s\phi \\ u_1u_3\,\mathrm{vers}\,\phi - u_2s\phi & u_2u_3\,\mathrm{vers}\,\phi + u_1s\phi & u_3^2\,\mathrm{vers}\,\phi + c\phi \end{bmatrix} \tag{3.188}$$

which is equal to the axis–angle Eq. (3.4), and therefore, $e^{\phi\tilde{u}}$ is equivalent to the rotation matrix $^G R_B$.

$$e^{\phi\tilde{u}} = R_{\hat{u},\phi} = {}^G R_B = \mathbf{I}\cos\phi + \hat{u}\hat{u}^T\,\mathrm{vers}\,\phi + \tilde{u}\sin\phi \tag{3.189}$$

To show that $e^{\phi\tilde{u}} \in S$, where S is the set of rotation matrices,

$$S = \left\{R \in \mathbb{R}^{3\times3} : RR^T = \mathbf{I}, |R| = 1\right\} \tag{3.190}$$

we have to show that $R = e^{\phi\tilde{u}}$ has the orthogonality property $R^T R = I$ and its determinant is $|R| = 1$. The orthogonality can be verified by the fact that $R^T = R^{-1}$ and consequently $RR^T = \mathbf{I}$.

$$\left[e^{\phi\tilde{u}}\right]^{-1} = e^{-\phi\tilde{u}} = e^{\phi\tilde{u}^T} = \left[e^{\phi\tilde{u}}\right]^T \tag{3.191}$$

From orthogonality, it follows that $|R| = \pm1$, and from continuity of exponential function, it follows that $|e^0| = 1$. Therefore, $|R| = 1$.

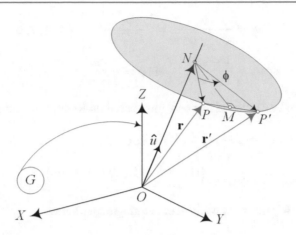

Fig. 3.6 Illustration of a rotation of a rigid body to derive a new form of the *Rodriguez rotation formula* in Example 78

Example 78 ★ New form of Rodriguez rotation formula.Rodriguez rotation formula may also be expressed by vectors.
Considering Fig. 3.6 we may write

$$\cos \frac{\phi}{2} \left| \overrightarrow{MP'} \right| = \sin \frac{\phi}{2} \left| \overrightarrow{NM} \right| \tag{3.192}$$

$$\left| \overrightarrow{NM} \right| \overrightarrow{MP'} = \left| \overrightarrow{MP'} \right| \hat{u} \times \overrightarrow{NM} \tag{3.193}$$

to find

$$\left(\cos \frac{\phi}{2} \right) \overrightarrow{MP'} = \left(\sin \frac{\phi}{2} \right) \hat{u} \times \overrightarrow{NM}. \tag{3.194}$$

Now using the following expressions:

$$2\overrightarrow{MP'} = \overrightarrow{NP'} - \overrightarrow{NP} \tag{3.195}$$

$$2\overrightarrow{NM} = \overrightarrow{NP'} + \overrightarrow{NP} \tag{3.196}$$

$$\overrightarrow{NP'} - \overrightarrow{NP} = \mathbf{r}' - \mathbf{r} \tag{3.197}$$

$$\hat{u} \times \left(\overrightarrow{NP'} + \overrightarrow{NP} \right) = \hat{u} \times \left(\mathbf{r}' + \mathbf{r} \right) \tag{3.198}$$

we can write an alternative form of the *Rodriguez rotation formula*

$$\cos \frac{\phi}{2} \left(\mathbf{r}' - \mathbf{r} \right) = \sin \frac{\phi}{2} \hat{u} \times \left(\mathbf{r}' + \mathbf{r} \right) \tag{3.199}$$

or

$$\left(\mathbf{r}' - \mathbf{r} \right) = \tan \frac{\phi}{2} \hat{u} \times \left(\mathbf{r}' + \mathbf{r} \right) \tag{3.200}$$

Example 79 ★ Rodriguez vector. A combination of axis and angle of rotation may be used to define a vector to represent rotations. Then the rotation vector would be the only operator to apply rotations.
The vector **w** is called the Rodriguez vector associated to a rotation $(\mathbf{u}, \phi)$.

$$\mathbf{w} = \frac{\mathbf{e}}{e_0} = \hat{u} \tan \frac{\phi}{2} \tag{3.201}$$

$$e_0 = \cos \frac{\phi}{2} = \frac{1}{\sqrt{1 + \mathbf{w}^T \mathbf{w}}} = \frac{1}{\sqrt{1 + w^2}} \tag{3.202}$$

$$e_i = \frac{w_i}{\sqrt{1 + \mathbf{w}^T \mathbf{w}}} \doteq \frac{w_i}{\sqrt{1 + w^2}} \qquad i = 1, 2, 3 \tag{3.203}$$

$$\cos\phi = \frac{1 - w^2}{1 + w^2} \qquad \sin\phi = \frac{2w}{1 + w^2} \tag{3.204}$$

The Rodriguez formula (3.140) can be converted to a new form based on Rodriguez vector.

$$
\begin{aligned}
{}^G R_B = R_{\hat{u},\phi} &= \left(e_0^2 - \mathbf{e}^2\right) \mathbf{I} + 2\mathbf{e}\,\mathbf{e}^T + 2e_0\tilde{e} \\
&= \frac{1}{1 + \mathbf{w}^T \mathbf{w}} \left(\left(1 - \mathbf{w}^T \mathbf{w}\right) \mathbf{I} + 2\mathbf{w}\mathbf{w}^T + 2\tilde{w}\right)
\end{aligned}
\tag{3.205}
$$

The combination of two rotations, $\mathbf{w}'$ and $\mathbf{w}''$, is equivalent to a single rotation $\mathbf{w}$.

$$\mathbf{w} = \frac{\mathbf{w}'' + \mathbf{w}' - \mathbf{w}'' \times \mathbf{w}'}{1 - \mathbf{w}'' \cdot \mathbf{w}'} \tag{3.206}$$

Example 80 ★ Elements of ${}^G R_B$. Employing permutation symbol ϵ_{ijk} and Kronecker delta δ_{ij}, to express elements of rotation matrices. The expression is good for tensor calculus and contracting long equations.

Introducing Levi-Civita density or permutation symbol ϵ_{ijk},

$$\epsilon_{ijk} = \begin{cases} 1 & ijk = 123 = 231 = 312 \\ 0 & i = j, j = k, k = i \\ -1 & ijk = 321 = 213 = 132 \end{cases} \tag{3.207}$$

$$= \frac{1}{2}(i - j)(j - k)(k - i) \qquad i, j, k = 1, 2, 3 \tag{3.208}$$

and recalling Kronecker delta (2.294)

$$\delta_{ij} = 1 \text{ if } i = j, \text{ and } \delta_{ij} = 0 \text{ if } i \neq j \tag{3.209}$$

we can redefine the elements of the Rodriguez rotation matrix.

$$r_{ij} = \delta_{ij} \cos\phi + (1 - \cos\phi)\, u_i u_j + \epsilon_{ijk} u_k \sin\phi \tag{3.210}$$

Example 81 ★ Euler parameters from ${}^G R_B$. Numerical example to determine Euler parameters for a given rotation matrix. A transformation matrix ${}^G R_B$ is given.

$$
{}^G R_B = \begin{bmatrix} 0.5449 & -0.5549 & 0.6285 \\ 0.3111 & 0.8299 & 0.4629 \\ -0.7785 & -0.0567 & 0.6249 \end{bmatrix}
\tag{3.211}
$$

To calculate the corresponding Euler parameters, we use Eq. (3.147) and find

$$
\begin{aligned}
\operatorname{tr}\left({}^G R_B\right) &= r_{11} + r_{22} + r_{33} \\
&= 0.5449 + 0.8299 + 0.6249 = 1.9997
\end{aligned}
\tag{3.212}
$$

therefore,

$$e_0 = \sqrt{\left(\operatorname{tr}\left({}^G R_B\right) + 1\right)/4} = 0.866 \tag{3.213}$$

and

$$e_1 = -0.15 \qquad e_2 = 0.406 \qquad e_3 = 0.25 \tag{3.214}$$

Example 82 ★ Euler parameters when we have one of them. Symmetric equations to determine Euler parameters from a rotation matrix.

Consider the Euler parameter rotation matrix (3.140) corresponding to rotation ϕ about an axis indicated by a unit vector $\hat{u}$. The off-diagonal elements of $^{G}R_{B}$

$$e_0 e_1 = \frac{1}{4}(r_{32} - r_{23}) \qquad e_0 e_2 = \frac{1}{4}(r_{13} - r_{31})$$

$$e_0 e_3 = \frac{1}{4}(r_{21} - r_{12}) \qquad e_1 e_2 = \frac{1}{4}(r_{12} + r_{21})$$

$$e_1 e_3 = \frac{1}{4}(r_{13} + r_{31}) \qquad e_2 e_3 = \frac{1}{4}(r_{23} + r_{32}) \tag{3.215}$$

can be utilized to calculate e_i, $i = 0, 1, 2, 3$ if we have one of them.

Example 83 ★ Euler parameters by Stanley method. A numerical efficient method to determine Euler parameters for a given rotation matrix.

Following an effective method developed by Stanley (1978), we may first find the four e_i^2

$$e_0^2 = \frac{1}{2}\left(1 + \mathrm{tr}\left(^{G}R_{B}\right)\right) \tag{3.216}$$

$$e_1^2 = \frac{1}{4}\left(1 + 2r_{11} - \mathrm{tr}\left(^{G}R_{B}\right)\right) \tag{3.217}$$

$$e_2^2 = \frac{1}{4}\left(1 + 2r_{22} - \mathrm{tr}\left(^{G}R_{B}\right)\right) \tag{3.218}$$

$$e_3^2 = \frac{1}{4}\left(1 + 2r_{33} - \mathrm{tr}\left(^{G}R_{B}\right)\right) \tag{3.219}$$

and take the positive square root of the largest e_i^2. Then the other e_i are found by dividing the appropriate three of the six equations (3.215) by the largest e_i.

3.4 ★ Quaternions

A global *quaternion q* is defined as a scalar+vector quantity, where q_0 is a scalar and $\mathbf{q}$ is a vector.

$$q = q_0 + \mathbf{q} = q_0 + q_1 \hat{I} + q_2 \hat{J} + q_3 \hat{K} \tag{3.220}$$

A quaternion can also be shown by a four-element vector

$$q = \begin{bmatrix} q_0 \\ q_1 \\ q_2 \\ q_3 \end{bmatrix} \tag{3.221}$$

in a *flag* form

$$q = q_0 + q_1 i + q_2 j + q_3 k \tag{3.222}$$

where, i, j, k are *flags* and defined as follows:

$$i^2 = j^2 = k^2 = ijk = -1 \tag{3.223}$$

$$ij = -ji = k \tag{3.224}$$

$$jk = -kj = i \tag{3.225}$$

$$ki = -ik = j \tag{3.226}$$

Addition and multiplication of two quaternions q and p are quaternions:

$$
\begin{aligned}
q + p &= (q_0 + \mathbf{q}) + (p_0 + \mathbf{p}) \\
&= q_0 + q_1 i + q_2 j + q_3 k + p_0 + p_1 i + p_2 j + p_3 k \\
&= (q_0 + p_0) + (q_1 + p_1) i + (q_2 + p_2) j + (q_3 + p_3) k
\end{aligned}
\tag{3.227}
$$

$$
\begin{aligned}
qp &= (q_0 + \mathbf{q})(p_0 + \mathbf{p}) = q_0 p_0 + q_0 \mathbf{p} + p_0 \mathbf{q} + \mathbf{q}\mathbf{p} \\
&= q_0 p_0 + q_0 \mathbf{p} + p_0 \mathbf{q} + \mathbf{q} \times \mathbf{p} - \mathbf{q} \cdot \mathbf{p} \\
&= (q_0 p_0 - q_1 p_1 - q_2 p_2 - q_3 p_3) \\
&\quad + (p_0 q_1 + p_1 q_0 - p_2 q_3 + p_3 q_2) i \\
&\quad + (p_0 q_2 + p_1 q_3 + p_2 q_0 - p_3 q_1) j \\
&\quad + (p_0 q_3 - p_1 q_2 + p_2 q_1 + p_3 q_0) k
\end{aligned}
\tag{3.228}
$$

where $\mathbf{qp}$ is the *quaternion vector product* and is equal to the outer vector product minus the inner vector product of $\mathbf{q}$ and $\mathbf{p}$.

$$\mathbf{qp} = \mathbf{q} \times \mathbf{p} - \mathbf{q} \cdot \mathbf{p} \tag{3.229}$$

Quaternion addition is associative and commutative,

$$q + p = p + q \tag{3.230}$$

$$q + (p + r) = (q + p) + r \tag{3.231}$$

but quaternion multiplication is not commutative,

$$qp \neq pq \tag{3.232}$$

However, quaternion multiplication is associative and distributes over addition,

$$(pq)r = p(qr) \tag{3.233}$$

$$(p + q)r = pr + qr \tag{3.234}$$

A quaternion q has a conjugate q^*.

$$q^* = q_0 - \mathbf{q} = q_0 - q_1 \hat{I} - q_2 \hat{J} - q_3 \hat{K} \tag{3.235}$$

Therefore,

$$
\begin{aligned}
qq^* &= (q_0 + \mathbf{q})(q_0 - \mathbf{q}) = q_0 q_0 + q_0 \mathbf{q} - q_0 \mathbf{q} - \mathbf{q}\mathbf{q} \\
&= q_0 q_0 + \mathbf{q} \cdot \mathbf{q} - \mathbf{q} \times \mathbf{q} = q_0^2 + q_1^2 + q_2^2 + q_3^2 \\
&= |q|^2
\end{aligned}
\tag{3.236}
$$

and hence, we may define the quaternion inverse and quaternion division:

$$|q| = \sqrt{qq^*} = \sqrt{q_0^2 + q_1^2 + q_2^2 + q_3^2} \tag{3.237}$$

$$q^{-1} = \frac{1}{q} = \frac{q^*}{|q|^2} \tag{3.238}$$

$$pq^{-1} = \frac{pq^*}{|q|^2} \tag{3.239}$$

$$q^{-1}p = \frac{q^*p}{|q|^2} \tag{3.240}$$

A quaternion q is a unit quaternion if $|q| = 1$. When a quaternion q is a unit, then we have $q^{-1} = q^*$.

Let $e(\phi, \hat{u})$ be a unit quaternion, $|e(\phi, \hat{u})| = 1$, based on Euler parameters, we have

$$e(\phi, \hat{u}) = e_0 + \mathbf{e} = e_0 + e_1\hat{I} + e_2\hat{J} + e_3\hat{K} \tag{3.241}$$

$$= \cos\frac{\phi}{2} + \sin\frac{\phi}{2}\hat{u}$$

The $\mathbf{r} = 0 + \mathbf{r}$ is a quaternion corresponding to a vector $\mathbf{r}$. The vector $\mathbf{r}$ after a rotation ϕ about $\hat{u}$ would be

$$\mathbf{r}' = e(\phi, \hat{u}) \, \mathbf{r} \, e^*(\phi, \hat{u}) \tag{3.242}$$

which is equivalent to

$$^G\mathbf{r} = e(\phi, \hat{u}) \, ^B\mathbf{r} \, e^*(\phi, \hat{u}) \tag{3.243}$$

Therefore, a rotation $R_{\hat{u},\phi}$ can be defined by a quaternion $e(\phi, \hat{u}) = \cos\frac{\phi}{2} + \sin\frac{\phi}{2}\hat{u}$, and consequently, two consecutive rotations $R = R_2R_1$ are defined by

$$e(\phi, \hat{u}) = e_2(\phi_2, \hat{u}_2) \, e_1(\phi_1, \hat{u}_1) \tag{3.244}$$

A rotation notation is shown by its scalar and vector, equal to the angle and unit vector on axis of rotation, $e_1(\phi_1, \hat{u}_1)$, $e_2(\phi_2, \hat{u}_2)$, $\cdots$ while e_0, e_1, e_2, e_3 are Euler parameters.

If we show a quaternion by its equivalent a 4×4 matrix,

$$\overleftrightarrow{q} = \begin{bmatrix} q_0 & -q_1 & -q_2 & -q_3 \\ q_1 & q_0 & -q_3 & q_2 \\ q_2 & q_3 & q_0 & -q_1 \\ q_3 & -q_2 & q_1 & q_0 \end{bmatrix} \tag{3.245}$$

it provides the important orthogonality property.

$$\overleftrightarrow{q}^{-1} = \overleftrightarrow{q}^T \tag{3.246}$$

The matrix quaternion (3.245) can also be represented by

$$\overleftrightarrow{q} = \begin{bmatrix} q_0 & -\mathbf{q}^T \\ \mathbf{q} & q_0\mathbf{I}_3 - \tilde{q} \end{bmatrix} \tag{3.247}$$

where,

$$\tilde{q} = \begin{bmatrix} 0 & -q_3 & q_2 \\ q_3 & 0 & -q_1 \\ -q_2 & q_1 & 0 \end{bmatrix} \tag{3.248}$$

Employing the *matrix quaternion* $\overleftrightarrow{q}$, we can show quaternion multiplication by matrix multiplication.

$$qp = \overleftrightarrow{q}\, p = \begin{bmatrix} q_0 & -q_1 & -q_2 & -q_3 \\ q_1 & q_0 & -q_3 & q_2 \\ q_2 & q_3 & q_0 & -q_1 \\ q_3 & -q_2 & q_1 & q_0 \end{bmatrix} \begin{bmatrix} p_0 \\ p_1 \\ p_2 \\ p_3 \end{bmatrix} \tag{3.249}$$

The matrix expression of quaternions relates quaternion manipulations and matrix manipulations. If p, q, and v are three quaternions such that

$$qp = v \tag{3.250}$$

then,

$$\overleftrightarrow{q}\,\overleftrightarrow{p} = \overleftrightarrow{v} \tag{3.251}$$

Hence, the quaternion representation of transformation between coordinate frames (3.244) can also be defined by matrix multiplication.

$$\overleftrightarrow{{}^G\mathbf{r}} = \overleftarrow{e\left(\phi,\hat{u}\right)}\;\overleftrightarrow{{}^B\mathbf{r}}\;\overrightarrow{e^*\left(\phi,\hat{u}\right)} = \overleftarrow{e\left(\phi,\hat{u}\right)}\;\overleftrightarrow{{}^B\mathbf{r}}\;\overrightarrow{e\left(\phi,\hat{u}\right)}^T \tag{3.252}$$

Proof Sir William Rowan Hamilton (1805–1865), an Irish mathematician, suggested we have three different numbers that are all square roots of -1, labeled i, j, k and called quaternion.

$$i^2 = j^2 = k^2 = ijk = -1 \tag{3.253}$$

Quaternion may be considered as an expansion of complex numbers. Quaternions represent rotations better than Euler angles as they do not have singularity of Euler angles for $\pm\pi/2$.

To prove that a unit quaternion $e\left(\phi,\hat{u}\right)$ can work as a rotation matrix ${}^G R_B$, let us consider a quaternion $e\left(\phi,\hat{u}\right)$ as in Eq. (3.241) and employ the quaternion multiplication (3.228) to calculate ere^*.

$$
\begin{aligned}
\mathbf{r}e^* &= e_0\mathbf{r} + \mathbf{r}\times\mathbf{e}^* - \mathbf{r}\cdot\mathbf{e}^* \\
&= e_0\begin{bmatrix} r_1 \\ r_2 \\ r_3 \end{bmatrix} + \begin{bmatrix} r_1 \\ r_2 \\ r_3 \end{bmatrix} \times \begin{bmatrix} -e_1 \\ -e_2 \\ -e_3 \end{bmatrix} - \begin{bmatrix} r_1 \\ r_2 \\ r_3 \end{bmatrix} \cdot \begin{bmatrix} -e_1 \\ -e_2 \\ -e_3 \end{bmatrix} \\
&= (e_1 r_1 + e_2 r_2 + e_3 r_3) + \begin{bmatrix} e_0 r_1 + e_2 r_3 - e_3 r_2 \\ e_0 r_2 - e_1 r_3 + e_3 r_1 \\ e_0 r_3 + e_1 r_2 - e_2 r_1 \end{bmatrix}
\end{aligned} \tag{3.254}
$$

$$
\begin{aligned}
e\mathbf{r}e^* &= e_0(e_1 r_1 + e_2 r_2 + e_3 r_3) - \begin{bmatrix} e_1 \\ e_2 \\ e_3 \end{bmatrix} \cdot \begin{bmatrix} e_0 r_1 + e_2 r_3 - e_3 r_2 \\ e_0 r_2 - e_1 r_3 + e_3 r_1 \\ e_0 r_3 + e_1 r_2 - e_2 r_1 \end{bmatrix} \\
&+ e_0\begin{bmatrix} e_0 r_1 + e_2 r_3 - e_3 r_2 \\ e_0 r_2 - e_1 r_3 + e_3 r_1 \\ e_0 r_3 + e_1 r_2 - e_2 r_1 \end{bmatrix} + (e_1 r_1 + e_2 r_2 + e_3 r_3)\begin{bmatrix} e_1 \\ e_2 \\ e_3 \end{bmatrix} \\
&+ \begin{bmatrix} e_1 \\ e_2 \\ e_3 \end{bmatrix} \times \begin{bmatrix} e_0 r_1 + e_2 r_3 - e_3 r_2 \\ e_0 r_2 - e_1 r_3 + e_3 r_1 \\ e_0 r_3 + e_1 r_2 - e_2 r_1 \end{bmatrix} = {}^G R_B \begin{bmatrix} r_1 \\ r_2 \\ r_3 \end{bmatrix}
\end{aligned} \tag{3.255}
$$

which ${}^G R_B$ is equivalent to the Euler parameter transformation matrix (3.140).

$$
{}^G R_B = \begin{bmatrix} e_0^2 + e_1^2 - e_2^2 - e_3^2 & 2(e_1 e_2 - e_0 e_3) & 2(e_0 e_2 + e_1 e_3) \\ 2(e_0 e_3 + e_1 e_2) & e_0^2 - e_1^2 + e_2^2 - e_3^2 & 2(e_2 e_3 - e_0 e_1) \\ 2(e_1 e_3 - e_0 e_2) & 2(e_0 e_1 + e_2 e_3) & e_0^2 - e_1^2 - e_2^2 + e_3^2 \end{bmatrix} \tag{3.256}
$$

Using a similar method we can also show that $\mathbf{r} = e^*\left(\phi,\hat{u}\right)\mathbf{r}'e\left(\phi,\hat{u}\right)$ is the inverse transformation of $\mathbf{r}' = e\left(\phi,\hat{u}\right)\mathbf{r}\,e^*\left(\phi,\hat{u}\right)$, which is equivalent to

$$
{}^B\mathbf{r} = e^*\left(\phi,\hat{u}\right)\,{}^G\mathbf{r}\,e\left(\phi,\hat{u}\right) \tag{3.257}
$$

Now assume $e_1(\phi_1, \hat{u}_1)$ and $e_2(\phi_2, \hat{u}_2)$ are the quaternions corresponding to the rotation matrix $R_{\hat{u}_1, \phi_1}$ and $R_{\hat{u}_2, \phi_2}$, respectively. The first rotation maps $^{B_1}\mathbf{r}$ to $^{B_2}\mathbf{r}$, and the second rotation maps $^{B_2}\mathbf{r}$ to $^{B_3}\mathbf{r}$. Therefore,

$$^{B_2}\mathbf{r} = e_1(\phi_1, \hat{u}_1) \ ^{B_1}\mathbf{r} \ e_1^*(\phi_1, \hat{u}_1) \tag{3.258}$$

$$^{B_3}\mathbf{r} = e_2(\phi_2, \hat{u}_2) \ ^{B_2}\mathbf{r} \ e_2^*(\phi_2, \hat{u}_2) \tag{3.259}$$

which implies

$$^{B_3}\mathbf{r} = e_2(\phi_2, \hat{u}_2) \ e_1(\phi_1, \hat{u}_1) \ ^{B_1}\mathbf{r} \ e_1^*(\phi_1, \hat{u}_1) \ e_2^*(\phi_2, \hat{u}_2)$$
$$= e(\phi, \hat{u}) \ ^{B_1}\mathbf{r} \ e^*(\phi, \hat{u}) \tag{3.260}$$

showing that

$$e(\phi, \hat{u}) = e_2(\phi_2, \hat{u}_2) \ e_1(\phi_1, \hat{u}_1) \tag{3.261}$$

It is the quaternion equation corresponding to $R = R_2 R_1$. Simplifying Eq. (3.255) recovers Rodriguez formula.

$$\mathbf{r}' = e(\phi, \hat{u}) \ \mathbf{r} \ e^*(\phi, \hat{u})$$
$$= (e_0^2 - \mathbf{e} \cdot \mathbf{e})\mathbf{r} + 2e_0(\mathbf{e} \times \mathbf{r}) + 2\mathbf{e}(\mathbf{e} \cdot \mathbf{r}) \tag{3.262}$$

Using matrix definition of quaternions, we have

$$\overleftrightarrow{e(\phi, \hat{u})} = \begin{bmatrix} e_0 & -e_1 & -e_2 & -e_3 \\ e_1 & e_0 & -e_3 & e_2 \\ e_2 & e_3 & e_0 & -e_1 \\ e_3 & -e_2 & e_1 & e_0 \end{bmatrix} \tag{3.263}$$

$$\overleftrightarrow{^B\mathbf{r}} = \begin{bmatrix} 0 & -^Br_1 & -^Br_2 & -^Br_3 \\ ^Br_1 & 0 & -^Br_3 & ^Br_2 \\ ^Br_2 & ^Br_3 & 0 & -^Br_1 \\ ^Br_3 & -^Br_2 & ^Br_1 & 0 \end{bmatrix} \tag{3.264}$$

$$\overleftrightarrow{e(\phi, \hat{u})}^T = \begin{bmatrix} e_0 & e_1 & e_2 & e_3 \\ -e_1 & e_0 & e_3 & -e_2 \\ -e_2 & -e_3 & e_0 & e_1 \\ -e_3 & e_2 & -e_1 & e_0 \end{bmatrix} \tag{3.265}$$

Therefore,

$$^G\mathbf{r} = \overleftrightarrow{e(\phi, \hat{u})} \ \overleftrightarrow{^B\mathbf{r}} \ \overleftrightarrow{e(\phi, \hat{u})}^T = \begin{bmatrix} 0 & -^Gr_1 & -^Gr_2 & -^Gr_3 \\ ^Gr_1 & 0 & -^Gr_3 & ^Gr_2 \\ ^Gr_2 & ^Gr_3 & 0 & -^Gr_1 \\ ^Gr_3 & -^Gr_2 & ^Gr_1 & 0 \end{bmatrix} \tag{3.266}$$

where

$$^Gr_1 = \ ^Br_1\left(e_0^2 + e_1^2 - e_2^2 - e_3^2\right) + \ ^Br_2\left(2e_1e_2 - 2e_0e_3\right)$$
$$+ \ ^Br_3\left(2e_0e_2 + 2e_1e_3\right) \tag{3.267}$$

$$^Gr_2 = \ ^Br_1\left(2e_0e_3 + 2e_1e_2\right) + \ ^Br_2\left(e_0^2 - e_1^2 + e_2^2 - e_3^2\right)$$
$$+ \ ^Br_3\left(2e_2e_3 - 2e_0e_1\right) \tag{3.268}$$

$$
\begin{aligned}
{}^{G}r_3 = {}^{B}r_1 \left(2e_1 e_3 - 2e_0 e_2\right) + {}^{B}r_2 \left(2e_0 e_1 + 2e_2 e_3\right) \\
+ {}^{B}r_3 \left(e_0^2 - e_1^2 - e_2^2 + e_3^2\right)
\end{aligned}
\tag{3.269}
$$

Equation (3.266) is compatible with Eqs. (3.252) and (3.244).   ∎

Example 84 ★ Composition of rotations using quaternions. How to combine rotations by rotation quaternion and determine the equivalent rotation quaternion.

Using quaternions to represent rotations makes it easier to calculate the composition of rotations. If the quaternion $e_1\left(\phi_1, \hat{u}_1\right)$ represents the rotation $R_{\hat{u}_1, \phi_1}$ and $e_2\left(\phi_2, \hat{u}_2\right)$ represents $R_{\hat{u}_2, \phi_2}$, then the product

$$
e\left(\phi, \hat{u}\right) = e_2\left(\phi_2, \hat{u}_2\right) e_1\left(\phi_1, \hat{u}_1\right)
\tag{3.270}
$$

represents $R_{\hat{u}_2, \phi_2} R_{\hat{u}_1, \phi_1}$ because:

$$
\begin{aligned}
e\left(\phi, \hat{u}\right) &= R_{\hat{u}_2, \phi_2} R_{\hat{u}_1, \phi_1} \mathbf{r} \\
&= R_{\hat{u}_2, \phi_2} \left(e_1\left(\phi_1, \hat{u}_1\right) \mathbf{r} \, e_1^*\left(\phi_1, \hat{u}_1\right)\right) \\
&= e_2\left(\phi_2, \hat{u}_2\right) \left(e_1\left(\phi_1, \hat{u}_1\right) \mathbf{r} \, e_1^*\left(\phi_1, \hat{u}_1\right)\right) e_2^*\left(\phi_2, \hat{u}_2\right) \\
&= \left(e_2\left(\phi_2, \hat{u}_2\right) e_1\left(\phi_1, \hat{u}_1\right)\right) \mathbf{r} \left(e_1^*\left(\phi_1, \hat{u}_1\right) e_2^*\left(\phi_2, \hat{u}_2\right)\right) \\
&= \left(e_2\left(\phi_2, \hat{u}_2\right) e_1\left(\phi_1, \hat{u}_1\right)\right) \mathbf{r} \left(e_1\left(\phi_1, \hat{u}_1\right) e_2\left(\phi_2, \hat{u}_2\right)\right)^*
\end{aligned}
\tag{3.271}
$$

Example 85 ★ Principal global rotation matrices. Rotation quaternion for rotation about global axes.

The associated quaternion to the principal global rotation matrices $R_{Z,\alpha}$, $R_{Y,\beta}$, and $R_{X,\gamma}$ are

$$
e\left(\alpha, \hat{K}\right) = \left(\cos\frac{\alpha}{2}, \sin\frac{\alpha}{2} \begin{bmatrix} 0 \\ 0 \\ 1 \end{bmatrix}\right)
\tag{3.272}
$$

$$
e\left(\beta, \hat{J}\right) = \left(\cos\frac{\beta}{2}, \sin\frac{\beta}{2} \begin{bmatrix} 0 \\ 1 \\ 0 \end{bmatrix}\right)
\tag{3.273}
$$

$$
e\left(\gamma, \hat{I}\right) = \left(\cos\frac{\gamma}{2}, \sin\frac{\gamma}{2} \begin{bmatrix} 1 \\ 0 \\ 0 \end{bmatrix}\right)
\tag{3.274}
$$

Employing (3.272)–(3.274), we are able to derive the principal transformation matrices. As an example, let us find $R_{Z,\alpha}$:

$$
\begin{aligned}
R_{Z,\alpha} &= {}^{G}R_B = \left(e_0^2 - \mathbf{e}^2\right) \mathbf{I} + 2\mathbf{e}\,\mathbf{e}^T + 2e_0 \tilde{e} \\
&= \left(\cos^2\frac{\alpha}{2} - \sin^2\frac{\alpha}{2}\right) \mathbf{I} + 2\hat{K}\hat{K}^T + 2\cos\frac{\alpha}{2}\sin\frac{\alpha}{2}\tilde{K} \\
&= \begin{bmatrix} \cos\alpha & -\sin\alpha & 0 \\ \sin\alpha & \cos\alpha & 0 \\ 0 & 0 & 1 \end{bmatrix}
\end{aligned}
\tag{3.275}
$$

Example 86 ★ Inner automorphism property of $e\left(\phi, \hat{u}\right)$. Introducing automorphism property and showing rotation quaternion acts based on automorphism property.

Because $e\left(\phi, \hat{u}\right)$ is a unit quaternion,

$$
e^*\left(\phi, \hat{u}\right) = e^{-1}\left(\phi, \hat{u}\right)
\tag{3.276}
$$

we may write

$$^G\mathbf{r} = e\left(\phi, \hat{u}\right) \, ^B\mathbf{r} \, e^{-1}\left(\phi, \hat{u}\right) \tag{3.277}$$

In abstract algebra, a mapping of the form $\mathbf{r} = q\,\mathbf{r}\,q^{-1}$, calculated by multiplying on the left by an element and on the right by its inverse, is called an inner automorphism. Thus, $^G\mathbf{r}$ is the inner automorphism of $^B\mathbf{r}$ based on the rotation quaternion $e\left(\phi, \hat{u}\right)$.

Example 87 Global roll–pitch–yaw quaternions. An example of showing combined rotation matrices by combination of rotation quaternions.

The global *roll–pitch–yaw* rotations about the global coordinate axes X, Y, and Z is

$$
\begin{aligned}
^G R_B &= R_{Z,\gamma} \, R_{Y,\beta} \, R_{X,\alpha} \\
&= \begin{bmatrix}
c\beta\,c\gamma & -c\alpha\,s\gamma + c\gamma\,s\alpha\,s\beta & s\alpha\,s\gamma + c\alpha\,c\gamma\,s\beta \\
c\beta\,s\gamma & c\alpha\,c\gamma + s\alpha\,s\beta\,s\gamma & -c\gamma\,s\alpha + c\alpha\,s\beta\,s\gamma \\
-s\beta & c\beta\,s\alpha & c\alpha\,c\beta
\end{bmatrix}
\end{aligned}
\tag{3.278}
$$

The roll quaternion $e\left(\alpha, \hat{I}\right)$, pitch quaternion $e\left(\beta, \hat{J}\right)$, and yaw quaternion $e\left(\gamma, \hat{K}\right)$ are given in (3.272)–(3.274). Multiplying these quaternions creates the global roll–pitch–yaw quaternion $e\left(\phi, \hat{u}\right)$:

$$
\begin{aligned}
e\left(\phi, \hat{u}\right) &= e\left(\gamma, \hat{K}\right) e\left(\beta, \hat{J}\right) e\left(\alpha, \hat{I}\right) \\
&= \cos\frac{\alpha}{2} \cos\frac{\beta}{2} \cos\frac{\gamma}{2} + \sin\frac{\alpha}{2} \sin\frac{\beta}{2} \sin\frac{\gamma}{2} \\
&\quad + \begin{bmatrix}
\cos\frac{\beta}{2} \cos\frac{\gamma}{2} \sin\frac{\alpha}{2} - \cos\frac{\alpha}{2} \sin\frac{\beta}{2} \sin\frac{\gamma}{2} \\
\cos\frac{\alpha}{2} \cos\frac{\gamma}{2} \sin\frac{\beta}{2} + \cos\frac{\beta}{2} \sin\frac{\alpha}{2} \sin\frac{\gamma}{2} \\
\cos\frac{\alpha}{2} \cos\frac{\beta}{2} \sin\frac{\gamma}{2} - \cos\frac{\gamma}{2} \sin\frac{\alpha}{2} \sin\frac{\beta}{2}
\end{bmatrix}
\end{aligned}
\tag{3.279}
$$

As a double check, we must get the same matrix (3.278) by substituting (3.279) in (3.256).

Example 88 ★ Polar expression of quaternions. Although we are only interested in rotation quaternions, general quaternions can be expressed in general polar forms. Here is how.

Consider a quaternion q in flag and polar forms.

$$q = q_0 + q_1 i + q_2 j + q_3 k \tag{3.280}$$

$$= r\left(\cos\theta + \mathbf{u}\sin\theta\right) \qquad 0 \le \theta \le 2\pi \tag{3.281}$$

$$\mathbf{u} = \pm \frac{q_1 i + q_2 j + q_3 k}{\sqrt{q_1^2 + q_2^2 + q_3^2}} \tag{3.282}$$

$$r = |q| = \sqrt{q_0^2 + q_1^2 + q_2^2 + q_3^2} \tag{3.283}$$

$$\cos\theta = \frac{q_0}{r} \qquad \sin\theta = \frac{\pm\sqrt{q_1^2 + q_2^2 + q_3^2}}{r} \tag{3.284}$$

$$\cot \theta = \pm \frac{q_0}{r} \qquad \tan \theta = \pm \frac{r}{q_0} \tag{3.285}$$

For example,

$$q = 3 + i + j + k \tag{3.286}$$

can be shown in polar form.

$$q = 2\sqrt{3} \left(\cos \frac{\pi}{6} + \frac{i+j+k}{\sqrt{3}} \sin \frac{\pi}{6} \right) \tag{3.287}$$

3.5   ★ Spinors and Rotators

Finite rotations can be expressed in two general ways: first, using 3×3 real orthogonal matrices R, which is called *rotator*. Rotator is an abbreviation for *rotation tensor*; and second, using 3×3 real skew symmetric matrices $\tilde{u}$ called *spinor*. Spinor is an abbreviation for *spin tensor*.

A rotator is a linear operator that maps $^B\mathbf{r}$ to $^G\mathbf{r}$ when a rotation axis $\hat{u}$ and a rotation angle ϕ are given.

$$^G\mathbf{r} = {}^G R_B \, {}^B\mathbf{r} \tag{3.288}$$

The spinor $\tilde{u}$ is corresponding to the unit vector $\hat{u}$, which, along with ϕ, can be utilized to describe a rotation.

$$^G R_B = \left(\mathbf{I} \cos \phi + \hat{u} \, \hat{u}^T \operatorname{vers} \phi + \tilde{u} \sin \phi \right) \tag{3.289}$$

Spinors and rotators are functions of each other so R must be expandable in a Taylor series of $\tilde{u}$.

$$R = \mathbf{I} + c_1 \tilde{u} + c_2 \tilde{u}^2 + c_3 \tilde{u}^3 + \cdots \tag{3.290}$$

However, because of (3.294), all powers of order three or higher may be eliminated. The rotator R is a linear function of $\mathbf{I}$, $\tilde{u}$, and $\tilde{u}^2$:

$$R = \mathbf{I} + a(\lambda \tilde{u}) + b(\lambda \tilde{u})^2 \tag{3.291}$$

$$= \begin{bmatrix} -b\lambda^2(u_2^2 + u_3^2) + 1 & -a\lambda u_3 + b\lambda^2 u_1 u_2 & a\lambda u_2 + b\lambda^2 u_1 u_3 \\ a\lambda u_3 + b\lambda^2 u_1 u_2 & -b\lambda^2(u_1^2 + u_3^2) + 1 & -a\lambda u_1 + b\lambda^2 u_2 u_3 \\ -a\lambda u_2 + b\lambda^2 u_1 u_3 & a\lambda u_1 + b\lambda^2 u_2 u_3 & -b\lambda^2(u_1^2 + u_2^2) + 1 \end{bmatrix}$$

where λ is the spinor normalization factor, and $a = a(\phi)$ and $b = b(\phi)$ are scalar functions of the rotation angle ϕ.

Table 3.1 collects some representations of rotator R as a function of the coefficients a, b, and the spinor $\lambda \tilde{u}$.

Table 3.1 Rotator R as a function of spinors

a	b	λ	R
$\sin \phi$	$\sin^2 \frac{\phi}{2}$	1	$\mathbf{I} + \sin \phi \tilde{u} + 2\sin^2 \frac{\phi}{2} \tilde{u}^2$
$2\cos^2 \frac{\phi}{2}$	$2\cos^2 \frac{\phi}{2}$	$\tan \frac{\phi}{2}$	$\mathbf{I} + 2\cos^2 \frac{\phi}{2}[\tan \frac{\phi}{2}\tilde{u} + \tan^2 \frac{\phi}{2}\tilde{u}^2]$ $= [\mathbf{I} + \tan \frac{\phi}{2}\tilde{u}][\mathbf{I} - \tan \frac{\phi}{2}\tilde{u}]^{-1}$
$2\cos \frac{\phi}{2}$	2	$\sin \frac{\phi}{2}$	$\mathbf{I} + 2\cos \frac{\phi}{2}\sin \frac{\phi}{2}\tilde{u} + 2\sin^2 \frac{\phi}{2}\tilde{u}^2$
$\frac{1}{\phi}\sin \phi$	$\frac{2}{\phi^2}\sin^2 \frac{\phi}{2}$	ϕ	$\mathbf{I} + \sin \phi \tilde{u} + 2\sin^2 \frac{\phi}{2}\tilde{u}^2$

Proof Square of $\tilde{u}$, calculated by direct multiplication, is

$$\tilde{u}^2 = \begin{bmatrix} -u_2^2 - u_3^2 & u_1 u_2 & u_1 u_3 \\ u_1 u_2 & -u_1^2 - u_3^2 & u_2 u_3 \\ u_1 u_3 & u_2 u_3 & -u_1^2 - u_2^2 \end{bmatrix}$$

$$= -\tilde{u}\,\tilde{u}^T = -\tilde{u}^T\tilde{u} = \hat{u}\hat{u}^T - u^2\mathbf{I} \tag{3.292}$$

This is a symmetric matrix whose trace is

$$\text{tr}[\tilde{u}^2] = -2|\hat{u}|^2 = -2u^2 = -2(u_1^2 + u_2^2 + u_3^2) \tag{3.293}$$

and its eigenvalues are $0, -u^2, -u^2$. In other words, $\tilde{u}$ satisfies its own characteristic equation.

$$\tilde{u}^2 = -u^2\mathbf{I}, \tilde{u}^3 = -u^2\tilde{u}, \cdots, \tilde{u}^n = -u^2\tilde{u}^{n-2}, n \geq 3 \tag{3.294}$$

The odd powers of $\tilde{u}$ are skew symmetric with distinct purely imaginary eigenvalues, while even powers of $\tilde{u}$ are symmetric with repeated real eigenvalues.

A rotator is a function of a spinor, so R can be expanded in a Taylor series of $\tilde{u}$.

$$R = \mathbf{I} + c_1\tilde{u} + c_2\tilde{u}^2 + c_3\tilde{u}^3 + \cdots \tag{3.295}$$

However, because of (3.294), all powers of order 3 or higher can be eliminated. Hence, R is a quadratic function of $\tilde{u}$:

$$R = \mathbf{I} + a(\lambda\tilde{u}) + b(\lambda\tilde{u})^2 \tag{3.296}$$

where the parameter λ is the spinor normalization factor, and $a = a(\phi, u)$ and $b = b(\phi, u)$ are scalar functions of rotation angle ϕ and vector $\tilde{u}$. Assuming $\lambda = 1$, we may find $\text{tr}\, R = 1 + 2\cos\phi$, which, because of (3.291), is equal to

$$\text{tr}\, R = 1 + 2\cos\phi = 3 - 2bu^2 \tag{3.297}$$

and therefore,

$$b = \frac{1 - \cos\phi}{u^2} = \frac{2}{u^2}\sin^2\frac{\phi}{2} \tag{3.298}$$

Now the orthogonality condition

$$\mathbf{I} = R^T R = \left(\mathbf{I} - a\tilde{u} + b\tilde{u}^2\right)\left(\mathbf{I} + a\tilde{u} + b\tilde{u}^2\right)$$

$$= \mathbf{I} + (2b - a^2)\tilde{u}^2 + b^2\tilde{u}^4$$

$$= \mathbf{I} + (2b - a^2 - b^2 u^2)\tilde{u}^2 \tag{3.299}$$

yields,

$$a = \sqrt{2b - b^2 u^2} = \frac{1}{u}\sin\phi \tag{3.300}$$

and therefore,

$$R = \mathbf{I} + \frac{1}{u}\sin\phi\tilde{u} + \frac{2}{u^2}\sin^2\frac{\phi}{2}\tilde{u}^2$$

$$= \mathbf{I} + \sin\phi\,\tilde{u} + \text{ver}\,\phi\,\tilde{u}^2 \tag{3.301}$$

From a numerical analysis viewpoint, the sine-squared form is preferred to avoid the cancelation in calculating $1 - \cos\phi$ for small ϕ. Replacing a and b in (3.296) provides the explicit rotator in terms of $\tilde{u}$ and ϕ.

$$R = \begin{bmatrix} u_1^2 + \left(u_2^2 + u_3^2\right) c\phi & 2u_1u_2s^2\frac{\phi}{2} - u_3s\phi & 2u_1u_3s^2\frac{\phi}{2} + u_2s\phi \\ 2u_1u_2s^2\frac{\phi}{2} + u_3s\phi & u_2^2 + \left(u_3^2 + u_1^2\right) c\phi & 2u_2u_3s^2\frac{\phi}{2} - u_1s\phi \\ 2u_1u_3s^2\frac{\phi}{2} - u_2s\phi & 2u_2u_3s^2\frac{\phi}{2} + u_1s\phi & u_3^2 + \left(u_1^2 + u_2^2\right) c\phi \end{bmatrix} \tag{3.302}$$

This is equivalent to Eq. (3.14).

If $\lambda \neq 1$ and $\lambda \neq 0$, then

$$a = \frac{1}{\lambda u} \sin\phi \qquad b = \frac{2}{(\lambda u)^2} \sin^2 \frac{\phi}{2} \tag{3.303}$$

which do not affect R.

∎

Example 89 ★ Eigenvalues of a spinor. Numerical example for eigenvalues and eigenvectors of a spinor.

Consider the axis of rotation indicated by

$$\mathbf{u} = \begin{bmatrix} 6 \\ 2 \\ 3 \end{bmatrix} \qquad u = |\mathbf{u}| = 7 \tag{3.304}$$

The associated spin matrix $\tilde{u}$ and its square $\tilde{u}^2$ are

$$\tilde{u} = \begin{bmatrix} 0 & -3 & 2 \\ 3 & 0 & -6 \\ -2 & 6 & 0 \end{bmatrix} \tag{3.305}$$

$$\tilde{u}^2 = \begin{bmatrix} -13 & 12 & 18 \\ 12 & -45 & 6 \\ 18 & 6 & -40 \end{bmatrix} \tag{3.306}$$

where the eigenvalues of $\tilde{u}$ are $(0, 7i, -7i)$ while those of $\tilde{u}^2$ are $(0, -49, -49)$.

3.6   ★ Problems in Representing Rotations

There are a number of different methods to express rotations, however only a few of them are fundamentally distinct. The coordinates required to completely describe the orientation of a rigid body relative to a reference frames are called *attitude coordinates*. There are two inherent problems in representing rotations, both related to incontrovertible properties of rotations.

1. Rigid body rotations do not commute.
2. Rigid body rotations cannot map smoothly in three dimensional Euclidean space.

The non-commutativity of rotations forces us to obey the order of rotations. The lack of a smooth mapping in three dimensional Euclidean space means we cannot smoothly represent every kind of rotation by using only one set of three numbers. Any set of three rotational coordinates contains at least one geometrical orientation where the coordinates are singular. Kinematic singularity is the orientation at which coordinates are undefined or not unique. The problem is similar to defining a coordinate system to locate a point on the Earth's surface. Using *longitude* and *latitude* becomes problematic at the north and south poles, where a small displacement can produce a radical change in longitude. We cannot find a superior system because it is not possible to smoothly wrap a sphere with a plane. Similarly, it is not possible to smoothly wrap the space of spatial rotations with three dimensional Euclidean space.

Singularity is the reason why we sometimes describe rotations by using four numbers. We may use only three-number systems and expect to see the resulting singularity, or use four numbers, and cope with the redundancy. The choice depends on the application and method of calculation. For computer applications, the redundancy is not an important problem, so most algorithms use representations with extra numbers. However, engineers prefer to work with the minimum set of numbers. Therefore, there is no unique and superior method for representing rotations.

3.6.1 ★ Rotation Matrix

The rotation matrix representation based on directional cosines is the most useful representation method of rigid body rotations for many applications. The two reference frames G and B, having a common origin, are defined through orthogonal right-handed sets of unit vectors $\{G\} = \{\hat{I}, \hat{J}, \hat{K}\}$ and $\{B\} = \{\hat{i}, \hat{j}, \hat{k}\}$. The rotation or transformation matrix between the two frames is found by using the orthogonality condition of B and G and describing the unit vectors of one of them in the other:

$$\hat{I} = (\hat{I} \cdot \hat{i})\hat{i} + (\hat{I} \cdot \hat{j})\hat{j} + (\hat{I} \cdot \hat{k})\hat{k}$$
$$= \cos(\hat{I}, \hat{i})\hat{i} + \cos(\hat{I}, \hat{j})\hat{j} + \cos(\hat{I}, \hat{k})\hat{k} \tag{3.307}$$

$$\hat{J} = (\hat{J} \cdot \hat{i})\hat{i} + (\hat{J} \cdot \hat{j})\hat{j} + (\hat{J} \cdot \hat{k})\hat{k}$$
$$= \cos(\hat{J}, \hat{i})\hat{i} + \cos(\hat{J}, \hat{j})\hat{j} + \cos(\hat{J}, \hat{k})\hat{k} \tag{3.308}$$

$$\hat{K} = (\hat{K} \cdot \hat{i})\hat{i} + (\hat{K} \cdot \hat{j})\hat{j} + (\hat{K} \cdot \hat{k})\hat{k}$$
$$= \cos(\hat{K}, \hat{i})\hat{i} + \cos(\hat{K}, \hat{j})\hat{j} + \cos(\hat{K}, \hat{k})\hat{k} \tag{3.309}$$

Therefore, having the rotation matrix $^G R_B$,

$$
^G R_B = \begin{bmatrix} \hat{I} \cdot \hat{i} & \hat{I} \cdot \hat{j} & \hat{I} \cdot \hat{k} \\ \hat{J} \cdot \hat{i} & \hat{J} \cdot \hat{j} & \hat{J} \cdot \hat{k} \\ \hat{K} \cdot \hat{i} & \hat{K} \cdot \hat{j} & \hat{K} \cdot \hat{k} \end{bmatrix}
$$
$$
= \begin{bmatrix} \cos(\hat{I}, \hat{i}) & \cos(\hat{I}, \hat{j}) & \cos(\hat{I}, \hat{k}) \\ \cos(\hat{J}, \hat{i}) & \cos(\hat{J}, \hat{j}) & \cos(\hat{J}, \hat{k}) \\ \cos(\hat{K}, \hat{i}) & \cos(\hat{K}, \hat{j}) & \cos(\hat{K}, \hat{k}) \end{bmatrix} \tag{3.310}
$$

would be enough to find the coordinates of a point in the reference frame G, when its coordinates are given in the other frame B.

$$^G\mathbf{r} = {}^G R_B \, {}^B\mathbf{r} \tag{3.311}$$

The rotation matrices convert the composition of rotations to matrix multiplication. It is simple and convenient especially when rotations are about the principal axes of global or body coordinate frames.

Orthogonality is the most important and useful property of rotation matrices, which shows that the inverse of a rotation matrix is equivalent to its transpose, $^G R_B^{-1} = {}^G R_B^T$. The null rotation is represented by the identity matrix, $\mathbf{I}$. The primary disadvantage of rotation matrices is that there are so many numbers, which often make rotation matrices hard to interpret. Numerical errors may build up until a normalization is necessary.

3.6.2 ★ Axis–Angle

Axis–angle representation, described by the Rodriguez formula, is a direct result of the Euler rigid body rotation theorem, explained in Example (98). In axis–angle method a rotation is described by the magnitude of rotation, ϕ, with the positive right-hand direction about an axis directed by the unit vector, $\hat{u}$.

$$^G R_B = R_{\hat{u},\phi} = \mathbf{I} \cos\phi + \hat{u}\hat{u}^T \text{ vers}\phi + \tilde{u} \sin\phi \tag{3.312}$$

Converting the axis–angle representation to matrix form is done by expanding.

$$
^G R_B = \begin{bmatrix} u_1^2 \text{ vers}\phi + c\phi & u_1 u_2 \text{ vers}\phi - u_3 s\phi & u_1 u_3 \text{ vers}\phi + u_2 s\phi \\ u_1 u_2 \text{ vers}\phi + u_3 s\phi & u_2^2 \text{ vers}\phi + c\phi & u_2 u_3 \text{ vers}\phi - u_1 s\phi \\ u_1 u_3 \text{ vers}\phi - u_2 s\phi & u_2 u_3 \text{ vers}\phi + u_1 s\phi & u_3^2 \text{ vers}\phi + c\phi \end{bmatrix} \tag{3.313}
$$

Converting the matrix representation to axis–angle form is shown in Example 57 using matrix manipulation. However, it is easier if we convert the matrix to a quaternion and then convert the quaternion to axis–angle form.

Axis–angle representation of rotation has some problems. First, the rotation axis is indeterminate when $\phi = 0$. Second, the axis–angle representation is a two-to-one mapping system because,

$$R_{-\hat{u},-\phi} = R_{\hat{u},\phi} \tag{3.314}$$

and it is redundant because for any integer k, we have

$$R_{\hat{u},\phi+2k\pi} = R_{\hat{u},\phi} \tag{3.315}$$

However, both of these problems can be improved by restricting ϕ to some suitable range such as $[0, \pi]$ or $[-\frac{\pi}{2}, \frac{\pi}{2}]$. Finally, axis–angle representation is not efficient to find the composition of two rotations and determine the equivalent axis–angle of rotations.

3.6.3 ★ Euler Angles

Euler angles are also employed to describe the rotation matrix of rigid bodies utilizing only three numbers.

$$
\begin{aligned}
{}^{G}R_{B} &= \left[R_{z,\psi} R_{x,\theta} R_{z,\varphi} \right]^{T} \\
&= \begin{bmatrix}
c\varphi c\psi - c\theta s\varphi s\psi & -c\varphi s\psi - c\theta c\psi s\varphi & s\theta s\varphi \\
c\psi s\varphi + c\theta c\varphi s\psi & -s\varphi s\psi + c\theta c\varphi c\psi & -c\varphi s\theta \\
s\theta s\psi & s\theta c\psi & c\theta
\end{bmatrix}
\end{aligned} \tag{3.316}
$$

Euler angles and rotation matrices are not generally one-to-one, and also, they are not convenient representations of rotations or good enough for constructing composite rotations. The angles φ and ψ are not distinguishable when $\theta \to 0$.

The equivalent rotation matrix is directly obtained by matrix multiplication, while the inverse conversion, from rotation matrix to a set of Euler angles is not straightforward. It is also not applicable when $\sin \theta = 0$. Employing (3.316) we can find the Euler angles as follows:

$$\varphi = -\arctan \frac{r_{13}}{r_{23}} \tag{3.317}$$

$$\theta = \arccos r_{33} \tag{3.318}$$

$$\psi = \arctan \frac{r_{31}}{r_{32}} \tag{3.319}$$

It is possible to use a more efficient method that handles all cases uniformly. The main idea is to work with sum and difference of φ and ψ

$$\sigma = \varphi + \psi \tag{3.320}$$

$$\upsilon = \varphi - \psi \tag{3.321}$$

and then,

$$\varphi = \frac{\sigma - \upsilon}{2} \tag{3.322}$$

$$\psi = \frac{\sigma + \upsilon}{2} \tag{3.323}$$

Therefore,

$$r_{11} + r_{22} = \cos\sigma (1 + \cos\theta) \tag{3.324}$$

$$r_{11} - r_{22} = \cos\upsilon (1 - \cos\theta) \tag{3.325}$$

$$r_{21} - r_{12} = \sin\sigma (1 + \cos\theta) \tag{3.326}$$

$$r_{21} + r_{12} = \sin\upsilon (1 - \cos\theta), \tag{3.327}$$

which leads to

$$\sigma = \tan^{-1} \frac{r_{21} - r_{12}}{r_{11} + r_{22}} \tag{3.328}$$

$$\upsilon = \tan^{-1} \frac{r_{21} + r_{12}}{r_{11} - r_{22}} \tag{3.329}$$

This approach resolves the problem at $\sin\theta = 0$. At $\theta = 0$, we can find σ, but υ is undetermined, and at $\theta = \pi$, we can find υ, but σ is undetermined. The undetermined values are consequence of $\arctan\frac{0}{0}$. Besides this singularity, both σ and υ are uniquely determined. The middle rotation angle, θ, can also be found using arctan operator.

$$\theta = \arctan\left(\frac{r_{13}\sin\varphi - r_{23}\cos\varphi}{r_{33}}\right) \tag{3.330}$$

The main advantage of Euler angles is that they use only three numbers. They are integrable, and they provide a good visualization of spatial rotation with no redundancy. Euler angles are used in dynamic analysis of spinning bodies. The other combinations of Euler angles as well as roll–pitch–yaw angles have the same kind of problems, and similar advantages.

3.6.4 ★ Quaternion

Quaternion uses four numbers to represent a rotation according to a special rules for addition and multiplication. Rotation quaternion is a unit quaternion that may be described by Euler parameters, or the axis and angle of rotation.

$$e\left(\phi, \hat{u}\right) = e_0 + \mathbf{e} = e_0 + e_1\hat{\imath} + e_2\hat{\jmath} + e_3\hat{k}$$

$$= \cos\frac{\phi}{2} + \sin\frac{\phi}{2}\hat{u} \tag{3.331}$$

We can also define a 4×4 matrix to represent a quaternion

$$\overleftrightarrow{q} = \begin{bmatrix} q_0 & -q_1 & -q_2 & -q_3 \\ q_1 & q_0 & -q_3 & q_2 \\ q_2 & q_3 & q_0 & -q_1 \\ q_3 & -q_2 & q_1 & q_0 \end{bmatrix} \tag{3.332}$$

which provides the important orthogonality property.

$$\overleftrightarrow{q}^{-1} = \overleftrightarrow{q}^{T} \tag{3.333}$$

The matrix quaternion (3.332) can also be represented by 4×4 matrix quaternion, $\overleftrightarrow{q}$,

$$\overleftrightarrow{q} = \begin{bmatrix} q_0 & -\mathbf{q}^T \\ \mathbf{q} & q_0\mathbf{I}_3 - \tilde{q} \end{bmatrix} \tag{3.334}$$

where,

$$\tilde{q} = \begin{bmatrix} 0 & -q_3 & q_2 \\ q_3 & 0 & -q_1 \\ -q_2 & q_1 & 0 \end{bmatrix} \tag{3.335}$$

Employing the 4×4 matrix quaternion, $\overleftrightarrow{q}$, we can show quaternion multiplication by matrix multiplication.

$$qp = \overleftrightarrow{q}\, p = \begin{bmatrix} q_0 & -q_1 & -q_2 & -q_3 \\ q_1 & q_0 & -q_3 & q_2 \\ q_2 & q_3 & q_0 & -q_1 \\ q_3 & -q_2 & q_1 & q_0 \end{bmatrix} \begin{bmatrix} p_0 \\ p_1 \\ p_2 \\ p_3 \end{bmatrix} \tag{3.336}$$

The matrix expression of quaternions ties the quaternion manipulations and matrix manipulations. If p, q, and v are three quaternions and

$$qp = v \tag{3.337}$$

then,

$$\overleftrightarrow{q}\, \overleftrightarrow{p} = \overleftrightarrow{v} \tag{3.338}$$

Hence, quaternion representation of transformation between coordinate frames,

$$^G\mathbf{r} = e\left(\phi, \hat{u}\right)\ ^B\mathbf{r}\ e^*\left(\phi, \hat{u}\right) \tag{3.339}$$

may also be defined by matrix multiplication.

$$\overleftrightarrow{^G\mathbf{r}} = \overleftrightarrow{e\left(\phi, \hat{u}\right)}\ \overleftrightarrow{^B\mathbf{r}}\ \overleftrightarrow{e^*\left(\phi, \hat{u}\right)} = \overleftrightarrow{e\left(\phi, \hat{u}\right)}\ \overleftrightarrow{^B\mathbf{r}}\ \overleftrightarrow{e\left(\phi, \hat{u}\right)}^T \tag{3.340}$$

3.6.5 ★ Euler Parameters

Euler parameters are the elements of rotation quaternions. Therefore, there is a direct conversion between rotation quaternion and Euler parameters, which in turn are related to angle-axis parameters. We can obtain the axis and angle of rotation $\left(\phi, \hat{u}\right)$, from Euler parameters or from rotation quaternion $e\left(\phi, \hat{u}\right)$.

$$\phi = 2 \tan^{-1} \frac{|\mathbf{e}|}{e_0} \tag{3.341}$$

$$\hat{u} = \frac{\mathbf{e}}{|\mathbf{e}|} \tag{3.342}$$

Unit quaternion provides a suitable base for expressing spatial rotations, although it needs normalization due to the error pile-up problem. In general, in most applications quaternions offer superior computational efficiency.

Leonhard Euler (1707–1783) was the first to derive the Rodriguez formula, while Benjamin Rodriguez (1795–1851) was the first to discover the Euler parameters. William Hamilton (1805–1865) introduced quaternions, although Friedrich Gauss (1777–1855) discovered them but never published.

Example 90 ★ Taylor expansion of rotation matrix. Time varying rotation matrices can be expanded by Taylor series. Such expansion provides several interesting identities.

Assume the rotation matrix $R = R(t)$ is a time dependent transformation between coordinate frames B and G. The body frame B is coincident with G at $t = 0$. Therefore, $R(0) = \mathbf{I}$, and we may expand the elements of R in a Taylor series expansion

$$R(t) = \mathbf{I} + R_1 t + \frac{1}{2!} R_2 t^2 + \frac{1}{3!} R_3 t^3 + \cdots \tag{3.343}$$

in which R_i, $(i = 1, 2, 3, \cdots)$ is a constant matrix. The rotation matrix $R(t)$ is orthogonal for all t, hence,

$$RR^T = \mathbf{I} \tag{3.344}$$

$$\left(\mathbf{I} + R_1 t + \frac{1}{2!} R_2 t^2 + \cdots\right)\left(\mathbf{I} + R_1^T t + \frac{1}{2!} R_2^T t^2 + \cdots\right) = \mathbf{I} \tag{3.345}$$

The coefficient of t^i, $(i = 1, 2, 3, \cdots)$ must vanish on the left-hand side. This gives us

$$R_1 + R_1^T = 0 \tag{3.346}$$

$$R_2 + 2R_1 R_1^T + R_2^T = 0 \tag{3.347}$$

$$R_3 + 3R_2 R_1^T + 3R_1 R_2^T + R_3^T = 0 \tag{3.348}$$

or in general

$$\sum_{i=0}^{n} \binom{n}{i} R_{n-i} R_i^T = 0 \tag{3.349}$$

where,

$$R_0 = R_0^T = \mathbf{I} \tag{3.350}$$

Equation (3.346) shows that R_1 is a skew symmetric matrix, and therefore, $R_1 R_1^T = -R_1^2 = C_1$ is symmetric. Now Eq. (3.347)

$$R_2 + R_2^T = -2R_1 R_1^T = -[R_1 R_1^T + [R_1 R_1^T]^T] = 2C_1 \tag{3.351}$$

leads to

$$R_2 = C_1 + [C_1 - R_2^T] \tag{3.352}$$

$$R_2^T = C_1 + [C_1 - R_2] = C_1 + [C_1 - R_2^T]^T \tag{3.353}$$

that shows $[C_1 - R_2^T]$ is skew symmetric because we must have

$$[C_1 - R_2^T] + [C_1 - R_2^T]^T = 0 \tag{3.354}$$

Therefore, we have a symmetric matrix product:

$$[C_1 - R_2^T][C_1 - R_2^T]^T = -[C_1 - R_2^T]^2 \tag{3.355}$$

The next step shows

$$R_3 + R_3^T = -3[R_1 R_2^T + R_2 R_1^T] = -3[R_1 R_2^T + [R_1 R_2^T]^T]$$
$$= 3[R_1[R_1^2 - R_2^T] + [R_1^2 - R_2^T]R_1] = 2C_2 \tag{3.356}$$

that leads to

$$R_3 = C_2 + \left[C_2 - R_3^T\right] \tag{3.357}$$

$$R_3^T = C_2 + [C_2 - R_3] = C_2 + \left[C_2 - R_3^T\right]^T \tag{3.358}$$

that shows $\left[C_2 - R_3^T\right]$ is also skew symmetric because we must have

$$\left[C_2 - R_3^T\right] + \left[C_2 - R_3^T\right]^T = 0. \tag{3.359}$$

Therefore, the matrix product

$$\left[C_2 - R_3^T\right]\left[C_2 - R_3^T\right]^T = -\left[C_2 - R_3^T\right]^2 \tag{3.360}$$

is also symmetric.

Continuing this procedure shows that the expansion of a rotation matrix $R(t)$ around the unit matrix can be written in the form of

$$R(t) = \mathbf{I} + C_1 t + \frac{1}{2!}\left[C_1 + [C_1 - R_2^T]\right]t^2$$
$$+ \frac{1}{3!}\left[C_2 + [C_2 - R_3^T]\right]t^3 + \cdots \tag{3.361}$$

where C_i are symmetric and $[C_i - R_{i+1}^T]$ are skew symmetric matrices and,

$$C_i = \frac{1}{2}\left[R_{i-1} + R_{1-1}^T\right] \tag{3.362}$$

Therefore, the expansion of an inverse rotation matrix can be written as

$$R^T(t) = \mathbf{I} + C_1 t + \frac{1}{2!}\left[C_1 + [C_1 - R_2^T]\right]t^2$$
$$+ \frac{1}{3!}\left[C_2 + [C_2 - R_3^T]\right]t^3 + \cdots \tag{3.363}$$

3.7   ★ Composition and Decomposition of Rotations

Determination of a rotation to be equivalent to some given rotations and determination of some rotations to be equivalent to a given rotation matrix are challenging problems when the axes of rotations are not orthogonal. This problem is called the *composition and decomposition of rotations*.

3.7.1   ★ Composition of Rotations

Rotation ϕ_1 about $\hat{u}_1$ of a rigid body with a fixed point followed by a rotation ϕ_2 about $\hat{u}_2$ can be composed to a unique rotation ϕ_3 about $\hat{u}_3$. In other words, when a rigid body rotates from an initial position to a middle position, $^{B_2}\mathbf{r} = {}^{B_2}R_{B_1}{}^{B_1}\mathbf{r}$, and then rotates to a final position, $^{B_3}\mathbf{r} = {}^{B_3}R_{B_2}{}^{B_2}\mathbf{r}$, the middle position can be skipped to rotate directly to the final position, $^{B_3}\mathbf{r} = {}^{B_3}R_{B_1}{}^{B_1}\mathbf{r}$.

Proof Let us rewrite the Rodriguez rotation formula (3.200) as

$$\left(\mathbf{r}' - \mathbf{r}\right) = \mathbf{w} \times \left(\mathbf{r}' + \mathbf{r}\right) \tag{3.364}$$

using the Rodriguez vector $\mathbf{w}$.

$$\mathbf{w} = \tan\frac{\phi}{2}\hat{u} \tag{3.365}$$

Rotation $\mathbf{w}_1$ followed by rotation $\mathbf{w}_2$ are

$$\left(\mathbf{r}_2 - \mathbf{r}_1\right) = \mathbf{w}_1 \times \left(\mathbf{r}_2 + \mathbf{r}_1\right) \tag{3.366}$$

$$\left(\mathbf{r}_3 - \mathbf{r}_2\right) = \mathbf{w}_2 \times \left(\mathbf{r}_3 + \mathbf{r}_2\right) \tag{3.367}$$

The right-hand side of the first one is perpendicular to $\mathbf{w}_1$ and the second one is perpendicular to $\mathbf{w}_2$. Hence, inner product of the first one with $\mathbf{w}_1$ and the second one with $\mathbf{w}_2$ show that

$$\mathbf{w}_1 \cdot \mathbf{r}_2 = \mathbf{w}_1 \cdot \mathbf{r}_1 \tag{3.368}$$

$$\mathbf{w}_2 \cdot \mathbf{r}_3 = \mathbf{w}_2 \cdot \mathbf{r}_2 \tag{3.369}$$

and outer product of the first one with $\mathbf{w}_2$ and the second one with $\mathbf{w}_1$ show that

$$\mathbf{w}_2 \times (\mathbf{r}_2 - \mathbf{r}_1) - \mathbf{w}_1 \times (\mathbf{r}_3 - \mathbf{r}_2) = \mathbf{w}_1 \left[\mathbf{w}_2 \cdot (\mathbf{r}_2 + \mathbf{r}_1) \right]$$
$$- (\mathbf{w}_1 \cdot \mathbf{w}_2)(\mathbf{r}_2 + \mathbf{r}_1) - \mathbf{w}_2 \left[\mathbf{w}_1 \cdot (\mathbf{r}_3 + \mathbf{r}_2) \right]$$
$$+ (\mathbf{w}_1 \cdot \mathbf{w}_2)(\mathbf{r}_3 + \mathbf{r}_2) \tag{3.370}$$

Rearranging while using Eqs. (3.368) and (3.369) gives us

$$\mathbf{w}_2 \times (\mathbf{r}_2 - \mathbf{r}_1) - \mathbf{w}_1 \times (\mathbf{r}_3 - \mathbf{r}_2)$$
$$= (\mathbf{w}_2 \times \mathbf{w}_1) \times (\mathbf{r}_1 + \mathbf{r}_3) + (\mathbf{w}_1 \cdot \mathbf{w}_2)(\mathbf{r}_3 - \mathbf{r}_1) \tag{3.371}$$

which can be written as

$$(\mathbf{w}_1 + \mathbf{w}_2) \times \mathbf{r}_2 = \mathbf{w}_2 \times \mathbf{r}_1 + \mathbf{w}_1 \times \mathbf{r}_3$$
$$+ (\mathbf{w}_2 \times \mathbf{w}_1) \times (\mathbf{r}_1 + \mathbf{r}_3) + (\mathbf{w}_1 \cdot \mathbf{w}_2)(\mathbf{r}_3 - \mathbf{r}_1) \tag{3.372}$$

Adding Eqs. (3.366) and (3.367) to obtain $(\mathbf{w}_1 + \mathbf{w}_2) \times \mathbf{r}_2$ yields

$$(\mathbf{w}_1 + \mathbf{w}_2) \times \mathbf{r}_2 = \mathbf{r}_3 - \mathbf{r}_1 - \mathbf{w}_1 \times \mathbf{r}_1 - \mathbf{w}_2 \times \mathbf{r}_3 \tag{3.373}$$

Therefore, we obtain the required Rodriguez rotation formula to rotate $\mathbf{r}_1$ to $\mathbf{r}_3$

$$\mathbf{r}_3 - \mathbf{r}_1 = \mathbf{w}_3 \times (\mathbf{r}_3 + \mathbf{r}_1) \tag{3.374}$$

where,

$$\mathbf{w}_3 = \frac{\mathbf{w}_1 + \mathbf{w}_2 + \mathbf{w}_2 \times \mathbf{w}_1}{1 - \mathbf{w}_1 \cdot \mathbf{w}_2} \tag{3.375}$$

∎

Example 91 Equivalent rotation to two individual rotations. Vector expression of rotations and combination of rotations.

A rigid body B that undergoes two rotations. First rotation is $\alpha = 30$ deg about $\hat{u}$,

$$\hat{u} = \frac{1}{\sqrt{3}} \left(\hat{I} + \hat{J} + \hat{K} \right) \tag{3.376}$$

followed by a second rotation $\beta = 45$ deg about $\hat{v}$.

$$\hat{v} = \hat{K} \tag{3.377}$$

To determine the equivalent single rotation, we define the vectors $\mathbf{u}$ and $\mathbf{v}$,

$$\mathbf{u} = \tan\frac{\alpha}{2}\hat{u} = \begin{bmatrix} \sqrt{3}/3 \\ \sqrt{3}/3 \\ \sqrt{3}/3 \end{bmatrix} \tan\frac{\pi/6}{2} = \begin{bmatrix} 0.1547 \\ 0.1547 \\ 0.1547 \end{bmatrix} \tag{3.378}$$

$$\mathbf{v} = \tan\frac{\beta}{2}\hat{u} = \begin{bmatrix} 0 \\ 0 \\ 1 \end{bmatrix} \tan\frac{\pi/4}{2} = \begin{bmatrix} 0 \\ 0 \\ 0.41421 \end{bmatrix} \tag{3.379}$$

and calculate the $\mathbf{w}$ from (3.375).

$$\mathbf{w} = \frac{\mathbf{u} + \mathbf{v} + \mathbf{v} \times \mathbf{u}}{1 - \mathbf{u} \cdot \mathbf{v}} = \begin{bmatrix} 0.23376 \\ 9.6827 \times 10^{-2} \\ 0.60786 \end{bmatrix} \tag{3.380}$$

The equivalent Rodriguez vector **w** may be decomposed to determine the axis and angle of rotations $\hat{w}$ and ϕ.

$$\mathbf{w} = \begin{bmatrix} 0.23376 \\ 9.68 \times 10^{-2} \\ 0.60786 \end{bmatrix} = 0.658 \begin{bmatrix} 0.355 \\ 0.147 \\ 0.923 \end{bmatrix} = \tan\frac{\phi}{2}\hat{w} \tag{3.381}$$

$$\phi = 2\arctan 0.658 = 1.1645\,\text{rad} \approx 66.72\,\text{deg} \tag{3.382}$$

3.7.2 ★ Decomposition of Rotations

A rotation ϕ_1 of a rigid body with a fixed point about $\hat{u}_1$ can be decomposed into three successive rotations about three arbitrary axes $\hat{a}$, $\hat{b}$, and $\hat{c}$ through unique angles α, β, and γ. Let $^G R_{\hat{a},\alpha}$, $^G R_{\hat{b},\beta}$, and $^G R_{\hat{c},\gamma}$ be any three successive rotation matrices about non-coaxis non-coplanar unit vectors $\hat{a}$, $\hat{b}$, and $\hat{c}$ through non-vanishing values α, β, and γ. Then, any other rotation $^G R_{\hat{u},\phi}$ can be expressed in terms of $^G R_{\hat{a},\alpha}$, $^G R_{\hat{b},\beta}$, and $^G R_{\hat{c},\gamma}$, if α, β, and γ are properly chosen numbers.

$$^G R_{\hat{u},\phi} = {}^G R_{\hat{c},\gamma}\,{}^G R_{\hat{b},\beta}\,{}^G R_{\hat{a},\alpha} \tag{3.383}$$

Proof Using the definition of quaternion rotations, we may write

$$^G\mathbf{r} = {}^G R_{\hat{u},\phi}\,{}^B\mathbf{r} = e\left(\phi,\hat{u}\right)\,{}^B\mathbf{r}\,e^*\left(\phi,\hat{u}\right) \tag{3.384}$$

Let us assume $\mathbf{r}_1$ indicate the position vector $\mathbf{r}$ before rotation, and $\mathbf{r}_2$, $\mathbf{r}_3$, and $\mathbf{r}_4$ indicate the position vector $\mathbf{r}$ after rotation $R_{\hat{a},\alpha}$, $R_{\hat{b},\beta}$, and $R_{\hat{c},\gamma}$, respectively. Hence,

$$\mathbf{r}_2 = a\,\mathbf{r}_1\,a^* \tag{3.385}$$

$$\mathbf{r}_3 = b\,\mathbf{r}_2\,b^* \tag{3.386}$$

$$\mathbf{r}_4 = c\,\mathbf{r}_3\,c^* \tag{3.387}$$

$$\mathbf{r}_4 = e\,\mathbf{r}_1\,e^* \tag{3.388}$$

where, $\left(\alpha,\hat{a}\right)$, $(\beta,\hat{b})$, $\left(\gamma,\hat{c}\right)$, and $\left(\phi,\hat{u}\right)$ are associated rotation quaternions.

$$a\left(\alpha,\hat{a}\right) = a_0 + \mathbf{a} = \cos\frac{\alpha}{2} + \sin\frac{\alpha}{2}\hat{a} \tag{3.389}$$

$$b(\beta,\hat{b}) = b_0 + \mathbf{b} = \cos\frac{\beta}{2} + \sin\frac{\beta}{2}\hat{b} \tag{3.390}$$

$$c\left(\gamma,\hat{c}\right) = c_0 + \mathbf{c} = \cos\frac{\gamma}{2} + \sin\frac{\gamma}{2}\hat{c} \tag{3.391}$$

$$e\left(\phi,\hat{u}\right) = e_0 + \mathbf{e} = \cos\frac{\phi}{2} + \sin\frac{\phi}{2}\hat{u} \tag{3.392}$$

We define the following scalars to simplify the equations:

$$\cos\frac{\alpha}{2} = C_1 \quad \cos\frac{\beta}{2} = C_2 \quad \cos\frac{\gamma}{2} = C_3 \quad \cos\frac{\phi}{2} = C \tag{3.393}$$

$$\sin\frac{\alpha}{2} = S_1 \quad \sin\frac{\beta}{2} = S_2 \quad \sin\frac{\gamma}{2} = S_3 \quad \sin\frac{\phi}{2} = S \tag{3.394}$$

$$\frac{b_2 c_3 - b_3 c_2}{S_2 S_3} = f_1 \quad \frac{b_3 c_1 - b_1 c_3}{S_2 S_3} = f_2 \quad \frac{b_1 c_2 - b_2 c_1}{S_2 S_3} = f_3 \tag{3.395}$$

$$\frac{c_2 a_3 - c_3 a_2}{S_3 S_1} = g_1 \quad \frac{a_3 c_1 - a_1 c_3}{S_3 S_1} = g_2 \quad \frac{a_1 c_2 - a_2 c_1}{S_3 S_1} = g_3 \tag{3.396}$$

$$\frac{a_2 b_3 - a_3 b_2}{S_1 S_2} = h_1 \quad \frac{a_3 b_1 - a_1 b_3}{S_1 S_2} = h_2 \quad \frac{a_1 b_2 - a_2 b_1}{S_1 S_2} = h_3 \tag{3.397}$$

$$\mathbf{b} \cdot \mathbf{c} = n_1 S_2 S_3 \tag{3.398}$$

$$\mathbf{c} \cdot \mathbf{a} = n_2 S_3 S_1 \tag{3.399}$$

$$\mathbf{a} \cdot \mathbf{b} = n_3 S_1 S_2 \tag{3.400}$$

$$(\mathbf{a} \times \mathbf{b}) \cdot \mathbf{c} = n_4 S_1 S_2 S_3 \tag{3.401}$$

Direct substitution shows that

$$\mathbf{r}_4 = e \mathbf{r}_1 e^* = cba \mathbf{r}_1 a^* b^* c^* \tag{3.402}$$

and therefore,

$$\begin{aligned}
e = cba &= c \left(b_0 a_0 - \mathbf{b} \cdot \mathbf{a} + b_0 \mathbf{a} + a_0 \mathbf{b} + \mathbf{b} \times \mathbf{a} \right) \\
&= c_0 b_0 a_0 - a_0 \mathbf{b} \cdot \mathbf{c} - b_0 \mathbf{c} \cdot \mathbf{a} - c_0 \mathbf{a} \cdot \mathbf{b} + (\mathbf{a} \times \mathbf{b}) \cdot \mathbf{c} \\
&\quad + a_0 b_0 \mathbf{c} + b_0 c_0 \mathbf{a} + c_0 a_0 \mathbf{b} \\
&\quad + a_0 (\mathbf{b} \times \mathbf{c}) + b_0 (\mathbf{c} \times \mathbf{a}) + c_0 (\mathbf{b} \times \mathbf{a}) \\
&\quad - (\mathbf{a} \cdot \mathbf{b}) \mathbf{c} - (\mathbf{b} \cdot \mathbf{c}) \mathbf{a} + (\mathbf{c} \cdot \mathbf{a}) \mathbf{b}.
\end{aligned} \tag{3.403}$$

Hence,

$$e_0 = c_0 b_0 a_0 - a_0 n_1 S_2 S_3 - b_0 n_2 S_3 S_1 - c_0 n_3 S_1 S_2 + n_4 S_1 S_2 S_3 \tag{3.404}$$

and

$$\begin{aligned}
\mathbf{e} &= a_0 b_0 \mathbf{c} + b_0 c_0 \mathbf{a} + c_0 a_0 \mathbf{b} \\
&\quad + a_0 (\mathbf{b} \times \mathbf{c}) + b_0 (\mathbf{c} \times \mathbf{a}) + c_0 (\mathbf{b} \times \mathbf{a}) \\
&\quad - n_1 S_2 S_3 \mathbf{a} + n_2 S_3 S_1 \mathbf{b} - n_3 S_1 S_2 \mathbf{c}
\end{aligned} \tag{3.405}$$

which generate four equations

$$C_1 C_2 C_3 - n_1 C_1 S_2 S_3 - n_2 S_1 C_2 S_3 + n_3 S_1 S_2 C_3 - n_4 S_1 S_2 S_3 = C \tag{3.406}$$

$$\begin{aligned}
& a_1 S_1 C_2 C_3 + b_1 C_1 S_2 C_3 + c_1 C_1 C_2 S_3 \\
& + f_1 C_1 S_2 S_3 + g_1 S_1 C_2 S_3 + h_1 S_1 S_2 C_3 \\
& - n_1 a_1 S_1 S_2 S_3 + n_2 b_1 S_1 S_2 S_3 - n_3 c_1 S_1 S_2 S_3 = u_1 S
\end{aligned} \tag{3.407}$$

$$\begin{aligned}
& a_2 S_1 C_2 C_3 + b_2 C_1 S_2 C_3 + c_2 C_1 C_2 S_3 \\
& + f_2 C_1 S_2 S_3 + g_2 S_1 C_2 S_3 + h_2 S_1 S_2 C_3 \\
& - n_1 a_2 S_1 S_2 S_3 + n_2 b_2 S_1 S_2 S_3 - n_3 c_2 S_1 S_2 S_3 = u_2 S
\end{aligned} \tag{3.408}$$

$$\begin{aligned}
& a_3 S_1 C_2 C_3 + b_{13} C_1 S_2 C_3 + c_3 C_1 C_2 S_3 \\
& + f_3 C_1 S_2 S_3 + g_3 S_1 C_2 S_3 + h_3 S_1 S_2 C_3 \\
& - n_1 a_3 S_1 S_2 S_3 + n_2 b_3 S_1 S_2 S_3 - n_3 c_3 S_1 S_2 S_3 = u_3 S
\end{aligned} \tag{3.409}$$

Since $e_0^2 + e_1^2 + e_2^2 + e_3^2 = 1$, only the first equation and two out of the other three equations, along with

$$C_1 = \sqrt{1 - S_1^2} \qquad C_2 = \sqrt{1 - S_2^2} \qquad C_3 = \sqrt{1 - S_3^2} \tag{3.410}$$

must be utilized to determine C_1, C_2, C_3, S_1, S_2, and S_3.   ∎

Example 92 ★ Decomposition of a vector in a non-orthogonal coordinate frame. Decomposition of a rotation matrix into rotations about three arbitrary axes is an interesting and complicated rotation theorem. This is an example to show the equation is compatible with rotations about principal axes.

Let **a**, **b**, and **c** be any three non-coplanar, non-vanishing vectors. Any other vector **r** can be expressed in terms of **a**, **b**, and **c**, in the following form, provided u, v, and w are properly chosen numbers:

$$\mathbf{r} = u\mathbf{a} + v\mathbf{b} + w\mathbf{c} \tag{3.411}$$

If $(\mathbf{a}, \mathbf{b}, \mathbf{c}) = (\hat{I}, \hat{J}, \hat{K})$ coordinate system is a Cartesian coordinate system, then decomposition reduces to orthogonality condition.

$$\mathbf{r} = (\mathbf{r} \cdot \hat{I})\hat{I} + (\mathbf{r} \cdot \hat{J})\hat{J} + (\mathbf{r} \cdot \hat{K})\hat{K} \tag{3.412}$$

To show this, we start with finding the vector inner product of Eq. (3.411) by $(\mathbf{b} \times \mathbf{c})$,

$$\mathbf{r} \cdot (\mathbf{b} \times \mathbf{c}) = u\mathbf{a} \cdot (\mathbf{b} \times \mathbf{c}) + v\mathbf{b} \cdot (\mathbf{b} \times \mathbf{c}) + w\mathbf{c} \cdot (\mathbf{b} \times \mathbf{c})$$

$$\tag{3.413}$$

and noting that $(\mathbf{b} \times \mathbf{c})$ is perpendicular to both **b** and **c**, consequently. we have

$$\mathbf{r} \cdot (\mathbf{b} \times \mathbf{c}) = u\mathbf{a} \cdot (\mathbf{b} \times \mathbf{c}) \tag{3.414}$$

Therefore,

$$u = \frac{[\mathbf{rbc}]}{[\mathbf{abc}]} \tag{3.415}$$

where,

$$[\mathbf{abc}] = \mathbf{a} \cdot \mathbf{b} \times \mathbf{c} = \mathbf{a} \cdot (\mathbf{b} \times \mathbf{c}) = \begin{vmatrix} a_1 & b_1 & c_1 \\ a_2 & b_2 & c_2 \\ a_3 & b_3 & c_3 \end{vmatrix} \tag{3.416}$$

Similarly v and w would be

$$v = \frac{[\mathbf{rca}]}{[\mathbf{abc}]} \qquad w = \frac{[\mathbf{rab}]}{[\mathbf{abc}]} \tag{3.417}$$

Hence,

$$\mathbf{r} = \frac{[\mathbf{rbc}]}{[\mathbf{abc}]}\mathbf{a} + \frac{[\mathbf{rca}]}{[\mathbf{abc}]}\mathbf{b} + \frac{[\mathbf{rab}]}{[\mathbf{abc}]}\mathbf{c} \tag{3.418}$$

that can also be written as

$$\mathbf{r} = \left(\mathbf{r} \cdot \frac{\mathbf{b} \times \mathbf{c}}{[\mathbf{abc}]}\right)\mathbf{a} + \left(\mathbf{r} \cdot \frac{\mathbf{c} \times \mathbf{a}}{[\mathbf{abc}]}\right)\mathbf{b} + \left(\mathbf{r} \cdot \frac{\mathbf{a} \times \mathbf{b}}{[\mathbf{abc}]}\right)\mathbf{c} \tag{3.419}$$

Multiplying (3.419) by $[\mathbf{abc}]$ gives a symmetric equation.

$$[\mathbf{abc}]\,\mathbf{r} - [\mathbf{bcr}]\,\mathbf{a} + [\mathbf{cra}]\,\mathbf{b} - [\mathbf{rab}]\,\mathbf{c} = 0 \tag{3.420}$$

If the $(\mathbf{a}, \mathbf{b}, \mathbf{c})$ coordinate system is a Cartesian system $(\hat{I}, \hat{J}, \hat{K})$, which is a mutually orthogonal system of unit vectors, then

$$\left[\hat{I}\hat{J}\hat{K}\right] = 1 \qquad \hat{I} \times \hat{J} = \hat{K} \qquad \hat{J} \times \hat{K} = \hat{I} \qquad \hat{K} \times \hat{I} = \hat{J} \tag{3.421}$$

and Eq. (3.419) becomes equal to (3.412).

$$\mathbf{r} = \left(\mathbf{r} \cdot \hat{I}\right) \hat{I} + \left(\mathbf{r} \cdot \hat{J}\right) \hat{J} + \left(\mathbf{r} \cdot \hat{K}\right) \hat{K} \tag{3.422}$$

3.8 Summary

The objectives of this chapter are

1. To determine the transformation matrix $^G R_B$ between two Cartesian coordinate frames B and G with a common origin when B is turning ϕrad about a given axis indicated by the unit vector $^G \hat{u} = \left[u_1 \ u_2 \ u_3\right]^T$.
2. To determine the axis $^G \hat{u}$ and angle ϕ of rotation for a given transformation matrix $^G R_B$.

There are several methods to express rotation of a body frame B with respect to the global frame G. The most applied method is the Rodriguez rotation formula that provides us with a 3×3 transformation matrix made of three independent elements. One element to indicate the angle ϕrad and two elements to indicate the unit vector on the axis of rotation $\hat{u}$.

$$^G R_B = R_{\hat{u},\phi} = \mathbf{I}\cos\phi + \hat{u}\hat{u}^T \operatorname{vers}\phi + \tilde{u}\sin\phi \tag{3.423}$$

$$= \begin{bmatrix} u_1^2 \operatorname{vers}\phi + c\phi & u_1 u_2 \operatorname{vers}\phi - u_3 s\phi & u_1 u_3 \operatorname{vers}\phi + u_2 s\phi \\ u_1 u_2 \operatorname{vers}\phi + u_3 s\phi & u_2^2 \operatorname{vers}\phi + c\phi & u_2 u_3 \operatorname{vers}\phi - u_1 s\phi \\ u_1 u_3 \operatorname{vers}\phi - u_2 s\phi & u_2 u_3 \operatorname{vers}\phi + u_1 s\phi & u_3^2 \operatorname{vers}\phi + c\phi \end{bmatrix}$$

$$\operatorname{vers}\phi = versine\,\phi = 1 - \cos\phi = 2\sin^2\frac{\phi}{2} \tag{3.424}$$

The skew symmetric matrix $\tilde{u}$ is the matrix expression of the vector $\hat{u}$.

$$\hat{u} = \begin{bmatrix} u_1 \\ u_2 \\ u_3 \end{bmatrix} \qquad \tilde{u} = \begin{bmatrix} 0 & -u_3 & u_2 \\ u_3 & 0 & -u_1 \\ -u_2 & u_1 & 0 \end{bmatrix} \tag{3.425}$$

On the other hand, we can determine the angle ϕ and axis of rotation $\hat{u}$ from a given transformation matrix $^G R_B$.

$$\cos\phi = \frac{1}{2}\left(\operatorname{tr}\left(^G R_B\right) - 1\right) \tag{3.426}$$

$$\tilde{u} = \frac{1}{2\sin\phi}\left(^G R_B - {}^G R_B^T\right) \tag{3.427}$$

The transformation matrix, angle and axis of rotation, as well as Rodriguez rotation formula can also be defined by Euler parameters e_0, e_1, e_2, e_3.

$$e_0 = \cos\frac{\phi}{2} \tag{3.428}$$

$$\mathbf{e} = e_1 \hat{I} + e_2 \hat{J} + e_3 \hat{K} = \hat{u}\sin\frac{\phi}{2} \tag{3.429}$$

$$e_1^2 + e_2^2 + e_3^2 + e_0^2 = e_0^2 + \mathbf{e}^T \mathbf{e} = 1 \tag{3.430}$$

$$^{G}R_{B} = R_{\hat{u},\phi} = \left(e_0^2 - \mathbf{e}^2\right)\mathbf{I} + 2\mathbf{e}\,\mathbf{e}^T + 2e_0\tilde{e}$$

$$(3.431)$$

Another method to present a rotation transformation between B and G is the unit quaternions $e\left(\phi, \hat{u}\right)$.

$$e\left(\phi, \hat{u}\right) = e_0 + \mathbf{e} = e_0 + e_1\hat{I} + e_2\hat{J} + e_3\hat{K}$$

$$= \cos\frac{\phi}{2} + \sin\frac{\phi}{2}\hat{u} \tag{3.432}$$

$$e^*\left(\phi, \hat{u}\right) = e_0 - \mathbf{e} = e_0 - e_1\hat{I} - e_2\hat{J} - e_3\hat{K} \tag{3.433}$$

$$^{G}\mathbf{r} = e\left(\phi, \hat{u}\right)\ ^{B}\mathbf{r}\ e^*\left(\phi, \hat{u}\right) \tag{3.434}$$

3.9   **Key Symbols**

A	Transformation matrix of rotation about a local axis
B	Body coordinate frame, local coordinate frame
C	Constant value, cosine of half angle
c	cos
e	Unit quaternion, rotation quaternion, exponential
e_0, e_1, e_2, e_3	Euler parameters
G	Global coordinate frame, fixed coordinate frame
i, j, k	Flags of a quaternion
$\mathbf{I} = [I]$	Identity matrix
$\hat{\imath}, \hat{\jmath}, \hat{k}$	Local coordinate axes unit vectors
$\tilde{\imath}, \tilde{\jmath}, \tilde{k}$	Skew symmetric matrices of the unit vector $\hat{\imath}, \hat{\jmath}, \hat{k}$
$\hat{I}, \hat{J}, \hat{K}$	Global coordinate axes unit vectors
$\mathbf{n}$	Eigenvectors of R
l	Length
O	Common origin of B and G
P	A body point, a fixed point in B
p, q, r	General quaternions
$\mathbf{r}$	Position vector
r_{ij}	The element of row i and column j of a matrix
R	Rotation transformation matrix
$\mathbb{R}$	The set of real numbers
s	sin
S	Sine of half angle
t	Time
$\hat{u}$	A unit vector on axis of rotation
$\tilde{u}$	Skew symmetric matrix of the vector $\hat{u}$
$\mathbf{v}$	Velocity vector, eigenvectors of R
$\mathbf{w}$	Rodriguez vector
x, y, z	Local coordinate axes
X, Y, Z	Global coordinate axes

Greek

α, β, γ	Rotation angles about global axes
ϵ_{ijk}	Permutation symbol
δ_{ij}	Kronecker's delta
λ	Eigenvalues of R
ϕ	Angle of rotation about $\hat{u}$
φ, θ, ψ	Rotation angles about local axes, Euler angles
$\boldsymbol{\omega}$	Angular velocity vector
$\tilde{\omega}$	Skew symmetric matrix of the vector $\boldsymbol{\omega}$

Symbol

tr	Trace operator
vers	$1 - \cos$
$[\ ]^{-1}$	Inverse of the matrix $[\ ]$
$[\ ]^T$	Transpose of the matrix $[\ ]$
$\overleftrightarrow{q}$	Matrix form of a quaternion q

Exercises

1. Notation and symbols.
   Describe the meaning of the following notations:

$$a - \hat{u} \qquad b - \text{vers}\,\phi \quad c - \tilde{u} \qquad d - R_{\hat{u},\phi} \quad e - e_0 \qquad f - \mathbf{e}$$

$$g - {}^B R_G^{-1} \quad h - d\phi \qquad i - q_0 + \mathbf{q} \quad j - ij \qquad k - q^* \quad l - e\left(\phi, \hat{u}\right)$$

$$m - \hat{J} \qquad n - \overleftrightarrow{q} \qquad o - \tilde{q} \qquad p - |q| \quad q - \mathbb{R}_{3\times3}$$

2. Invariant axis of rotation.
   Show that the axis of rotation $\hat{u}$ is fixed in $B(Oxyz)$ or $G(OXYZ)$.
3. z-axis–angle rotation matrix.
   Expand

$$
\begin{aligned}
{}^G R_B = {}^B R_G^{-1} &= {}^B R_G^T = R_{\hat{u},\phi} \\
&= \left[A_{z,-\varphi} A_{y,-\theta} A_{z,\phi} A_{y,\theta} A_{z,\varphi} \right]^T \\
&= A_{z,\varphi}^T A_{y,\theta}^T A_{z,\phi}^T A_{y,-\theta}^T A_{z,-\varphi}^T
\end{aligned}
\tag{3.435}
$$

and verify the axis–angle rotation matrix.

$$
{}^G R_B = R_{\hat{u},\phi} =
\tag{3.436}
$$

$$
\begin{bmatrix}
u_1^2 \,\text{vers}\,\phi + c\phi & u_1 u_2 \,\text{vers}\,\phi - u_3 s\phi & u_1 u_3 \,\text{vers}\,\phi + u_2 s\phi \\
u_1 u_2 \,\text{vers}\,\phi + u_3 s\phi & u_2^2 \,\text{vers}\,\phi + c\phi & u_2 u_3 \,\text{vers}\,\phi - u_1 s\phi \\
u_1 u_3 \,\text{vers}\,\phi - u_2 s\phi & u_2 u_3 \,\text{vers}\,\phi + u_1 s\phi & u_3^2 \,\text{vers}\,\phi + c\phi
\end{bmatrix}
$$

4. Axis–angle decomposition.
   A body frame B turns 30 deg about X-axis and then 45 deg about Z-axis.
   (a) Determine the rotation transformation matrix ${}^G R_B$.
   (b) Determine the angle and axis of rotation to provide the same ${}^G R_B$.
   (c) Determine the Euler angles for ${}^G R_B$.
   (d) Determine the Euler parameters for ${}^G R_B$.
5. ★ x-axis–angle and y-axis–angle rotation matrix.
   (a) Find the axis–angle rotation matrix by transforming the x-axis on the axis of rotation $\hat{u}$.
   (b) Find the axis–angle rotation matrix by transforming the y-axis on the axis of rotation $\hat{u}$.
6. Singular axis–angle rotation.
   Determine the angle and axis of rotation for the given ${}^G R_B$.

$$
{}^G R_B = R_{\hat{u},\phi} =
\begin{bmatrix}
0 & 0 & 1 \\
0 & -1 & 0 \\
1 & 0 & 0
\end{bmatrix}
\tag{3.437}
$$

7. Axis–angle rotation and Euler angles.
   (a) Find the Euler angles corresponding to the rotation 45 deg about $\mathbf{u} = \begin{bmatrix} 1 & 1 & 1 \end{bmatrix}^T$.
   (b) Determine the axis and angle of rotation for a combined rotation of 45 deg about x-axis, then 45 deg about y-axis, and then 45 deg about z-axis.
   (c) Find the Euler angles corresponding to the rotations in section b.
   (d) Determine the required angles to turn about x, then y, and then z axes to have the same final orientation as section a.

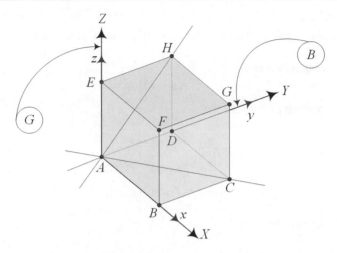

Fig. 3.7 A cube with edge length $l = 1$ at original configuration

8. Euler angles between two local frames.
   The Euler angles between the coordinate frame B_1 and G are 20 deg, 35 deg, and -40 deg. The Euler angles between the coordinate frame B_2 and G are 60 deg, -30 deg, and -10 deg. Find the angle and axis of rotation that transform B_2 to B_1.

9. Global rotation of a cube.
   Figure 3.7 illustrates the original position of a cube with a fixed point at A and edges of length $l = 1$.
   (a) Turn the cube 45 deg about AC and determine the global coordinates of the corners.
   (b) Turn the cube 45 deg about AH and determine the global coordinates of the corners.
   (c) Turn the cube 45 deg about AG and determine the global coordinates of the corners.

10. Axis–angle of a series of rotation.
    The cube in Fig. 3.7 with a fixed point at A has an edge length of $l = 1$.
    (a) Determine the angle and axis of rotation when we turn the cube 45 deg about x-axis followed by a rotation of 45 deg about y-axis.
    (b) Determine the angle and axis of rotation when we turn the cube 45 deg about x-axis followed by a rotation of 45 deg about AH.

11. Global decomposition of the rotation of a cube.
    The cube in Fig. 3.7 with a fixed point at A has an edge length of $l = 1$.
    (a) Decompose a rotation of 45 deg about AG into a rotation about X-axis, then Y-axis, and then Z-axis.
    (b) Decompose a rotation of 45 deg about AG into a rotation about Y-axis, then Z-axis, and then X-axis.
    (c) Decompose a rotation of 45 deg about AG into a rotation about Z-axis, then Y-axis, and then X-axis.
    (d) ★ Decompose a rotation of 45 deg about AG into a rotation about AC and then AH.

12. $-Z$ volume in a rotation of a cube.
    The cube in Fig. 3.7 with a fixed point at A has an edge length of $l = 1$.
    (a) How much should be the angle of rotation ϕ for the axis AG to move AE on X-axis?
    (b) ★ Calculate the volume of the cube that gets a negative Z coordinate during the rotation ϕ.

13. Rotation of a cube about a fixed and a body axis.
    The cube in Fig. 3.7 with a fixed point at A has an edge length of $l = 1$.
    (a) Turn the cube 45 deg about AC and then 45 deg about AH. Determine the global coordinates of the corners after the rotations.
    (b) Turn the cube 45 deg about AC and then 45 deg about AG. Determine the global coordinates of the corners after the rotations.
    (c) ★ Turn the cube 45 deg about AC, then 45 deg about AG, and then 45 deg about AH. Determine the global coordinates of the corners after the rotations.

14. ★ Angular velocity vector.
    Use the definition ${}^G R_B = [r_{ij}]$ and ${}^G \dot{R}_B = [\dot{r}_{ij}]$, and find the angular velocity vector $\boldsymbol{\omega}$, where $\tilde{\omega} = {}^G \dot{R}_B \, {}^G R_B^T$.

15. Sum of two orthogonal matrices.
Show that the sum of two orthogonal matrices is not, in general, an orthogonal matrix, but their product is.

16. Characteristic equation of $^G R_B$.
Expand the following determinant to derive

$$\left| ^G R_B - \lambda \mathbf{I} \right| = 0 \tag{3.438}$$

to derive the characteristic equation of $^G R_B$. Prove

$$r_{11} = r_{22} r_{33} - r_{23} r_{32} \tag{3.439}$$

$$r_{22} = r_{33} r_{11} - r_{31} r_{13} \tag{3.440}$$

$$r_{33} = r_{11} r_{22} - r_{12} r_{21} \tag{3.441}$$

to reduce the characteristic equation to

$$-\lambda^3 + \text{tr}(^G R_B)\lambda^2 - \text{tr}(^G R_B)\lambda + 1 = 0 \tag{3.442}$$

and

$$(\lambda - 1)\left[\lambda^2 - \lambda\left(\text{tr}(^G R_B) - 1\right) + 1\right] = 0 \tag{3.443}$$

17. Equivalent vector cross product.
Show that if $\mathbf{a} = \begin{bmatrix} a_1 & a_2 & a_3 \end{bmatrix}^T$ and $\mathbf{b} = \begin{bmatrix} b_1 & b_2 & b_3 \end{bmatrix}^T$ are two arbitrary vectors, and $\tilde{\mathbf{a}}$ is the skew symmetric matrix corresponding to $\mathbf{a}$,

$$\tilde{\mathbf{a}} = \begin{bmatrix} 0 & -a_3 & a_2 \\ a_3 & 0 & -a_1 \\ -a_2 & a_1 & 0 \end{bmatrix} \tag{3.444}$$

then

$$\tilde{\mathbf{a}}\mathbf{b} = \mathbf{a} \times \mathbf{b} \tag{3.445}$$

18. ★ Combined angle-axis rotations.
The rotation ϕ_1 about $\hat{u}_1$ followed by rotation ϕ_2 about $\hat{u}_2$ is equivalent to a rotation ϕ about $\hat{u}$. Find the angle ϕ and axis $\hat{u}$ in terms of ϕ_1, $\hat{u}_1$, ϕ_2, and $\hat{u}_2$.

19. ★ Rodriguez vector.
Using the Rodriguez rotation formula show that

$$\mathbf{r}' - \mathbf{r} = \tan\frac{\phi}{2}\hat{u} \times (\mathbf{r}' + \mathbf{r}) \tag{3.446}$$

20. ★ Equivalent Rodriguez rotation matrices.
Show that the Rodriguez rotation matrix

$$^G R_B = \mathbf{I}\cos\phi + \hat{u}\hat{u}^T \text{ vers }\phi + \tilde{u}\sin\phi \tag{3.447}$$

can also be written as

$$^G R_B = \mathbf{I} + (\sin\phi)\,\tilde{u} + (\text{vers}\,\phi)\,\tilde{u}^2 \tag{3.448}$$

21. ★ Rotation matrix and Rodriguez formula.
Knowing the alternative definition of the Rodriguez formula

$$^G R_B = \mathbf{I} + (\sin\phi)\,\tilde{u} + (\text{vers}\,\phi)\,\tilde{u}^2 \tag{3.449}$$

and

$$\tilde{u}^{2n-1} = (-1)^{n-1}\,\tilde{u} \tag{3.450}$$

$$\tilde{u}^{2n} = (-1)^{n-1}\,\tilde{u}^2 \tag{3.451}$$

examine the following equation:

$$^G R_B^T \, ^G R_B = \, ^G R_B \, ^G R_B^T \tag{3.452}$$

22. ★ Rodriguez formula application.
Use the alternative definition of the Rodriguez formula

$$^G R_B = \mathbf{I} + (\sin\phi)\, \tilde{u} + (\text{vers}\,\phi)\, \tilde{u}^2 \tag{3.453}$$

and find the global position of a body point at

$$^B \mathbf{r} = \begin{bmatrix} 1 & 3 & 4 \end{bmatrix}^T \tag{3.454}$$

after a rotation of 45 deg about the axis indicated by

$$\hat{u} = \begin{bmatrix} \frac{1}{\sqrt{3}} & \frac{1}{\sqrt{3}} & \frac{1}{\sqrt{3}} \end{bmatrix}^T \tag{3.455}$$

23. Associative property of rotation transformation matrices multiplication.
Using Rodriguez rotation formula, prove that multiplication of rotation transformation matrices is associative.

$$R_3 R_2 R_1 = R_3 \, [R_2 R_1] = [R_3 R_2] \, R_1 \tag{3.456}$$

24. Axis and angle of rotation.
Find the axis and angle of rotation for the following transformation matrix:

$$R = \begin{bmatrix} \frac{3}{4} & \frac{\sqrt{6}}{4} & \frac{1}{4} \\ \frac{-\sqrt{6}}{4} & \frac{2}{4} & \frac{\sqrt{6}}{4} \\ \frac{1}{4} & \frac{-\sqrt{6}}{4} & \frac{3}{4} \end{bmatrix} \tag{3.457}$$

25. ★ Axis of rotation multiplication.
Show that

$$\tilde{u}^{2k+1} = (-1)^k \, \tilde{u} \tag{3.458}$$

and

$$\tilde{u}^{2k} = (-1)^k \left(\mathbf{I} - \hat{u}\hat{u}^T \right) \tag{3.459}$$

26. ★ Stanley method.
Find the Euler parameters of the following rotation matrix based on the Stanley method:

$$^G R_B = \begin{bmatrix} 0.5449 & -0.5549 & 0.6285 \\ 0.3111 & 0.8299 & 0.4629 \\ -0.7785 & -0.0567 & 0.6249 \end{bmatrix} \tag{3.460}$$

27. ★ Stanley method.
The cube in Fig. 3.7 with a fixed point at A has an edge length of $l = 1$.
(a) Determine the matrix $^G R_B$ for a rotation of 45 deg about AG.
(b) Determine the Euler angles for $^G R_B$ of section a.
(c) Determine the Euler parameters for $^G R_B$ of section a.
(d) Determine the Euler parameters from $^G R_B$ of section a using Stanley method.
28. ★ Rotation matrices identity.
Show that if A, B, and C are three rotation matrices then,
(a)

$$(AB)\,C = A\,(BC) = ABC \tag{3.461}$$

(b)

$$(A + B)^T = A^T + B^T \tag{3.462}$$

(c)

$$(AB)^T = B^T A^T \tag{3.463}$$

(d)

$$\left(A^{-1}\right)^T = \left(A^T\right)^{-1} \tag{3.464}$$

29. ★ Angle-axis of rotation and Euler angles.
   Compare the Euler angles rotation matrix with the angle-axis rotation matrix and find the angle and axis of rotation based on Euler angles. What are the axis and angle of rotation in terms of Euler angles, and what are the Euler angles for a given angle and axis of rotation?

30. ★ Repeating global-local rotations.
   Rotate $^B\mathbf{r}_P = [6, 2, -3]^T$, 60 deg about the Z-axis, followed by 30 deg about the x-axis. Then repeat the sequence of rotations for 60 deg about the Z-axis, followed by 30 deg about the x-axis. After how many rotations will point P be back to its initial global position?

31. ★ Repeating global-local rotations.
   How many rotations of $\alpha = \pi/m$ deg about the X-axis, followed by $\beta = \pi/k$ deg about the z axis are needed to bring a body point to its initial global position, if $m, k \in \mathbb{N}$?

32. ★ Small rotation angles.
   Show that for very small angles of rotation φ, θ, and ψ about the axes of the body coordinate frame, the first and third rotations are indistinguishable when they are about the same axis.

33. ★ Inner automorphism property of $\tilde{\mathbf{a}}$.
   If R is a rotation matrix and $\mathbf{a}$ is a vector, show that

$$R\tilde{\mathbf{a}}R^T = \widetilde{R\mathbf{a}} \tag{3.465}$$

34. ★ Angle-derivative of principal rotation matrices.
   Show that

$$\frac{dR_{Z,\alpha}}{d\alpha} = \tilde{K} R_{Z,\alpha} \tag{3.466}$$

$$\frac{dR_{Y,\beta}}{d\beta} = \tilde{J} R_{Y,\beta} \tag{3.467}$$

$$\frac{dR_{X,\gamma}}{d\gamma} = \tilde{I} R_{X,\gamma} \tag{3.468}$$

35. ★ Euler angles, Euler parameters.
   Compare the Euler angles rotation matrix and Euler parameter transformation matrix to verify the following relationships between Euler angles and Euler parameters:

$$e_0 = \cos\frac{\theta}{2}\cos\frac{\psi + \varphi}{2} \tag{3.469}$$

$$e_1 = \sin\frac{\theta}{2}\cos\frac{\psi - \varphi}{2} \tag{3.470}$$

$$e_2 = \sin\frac{\theta}{2}\sin\frac{\psi - \varphi}{2} \tag{3.471}$$

$$e_3 = \cos\frac{\theta}{2}\sin\frac{\psi + \varphi}{2} \tag{3.472}$$

$$\varphi = \cos^{-1}\frac{2\left(e_2 e_3 + e_0 e_1\right)}{\sin\theta} \tag{3.473}$$

$$\theta = \cos^{-1}\left[2\left(e_0^2 + e_3^2\right) - 1\right] \tag{3.474}$$

$$\psi = \cos^{-1}\frac{-2\left(e_2 e_3 - e_0 e_1\right)}{\sin\theta} \tag{3.475}$$

36. ★ Quaternion definition.
    Find the unit quaternion $e\left(\phi, \hat{u}\right)$ associated to

$$\hat{u} = \left[\, 1/\sqrt{3} \;\; 1/\sqrt{3} \;\; 1/\sqrt{3}\,\right]^{T} \qquad \phi = \frac{\pi}{3} \tag{3.476}$$

    and find the result of $e\left(\phi, \hat{u}\right) \hat{\imath}\, e^{*}\left(\phi, \hat{u}\right)$.

37. ★ Quaternion product.
    (a) Calculate $pq, qp, p^{*}q, qp^{*}, p^{*}p, qq^{*}, p^{*}q^{*}$, and $p^{*}rq^{*}$ if:

$$p = 3 + i - 2j + 2k \tag{3.477}$$
$$q = 2 - i + 2j + 4k \tag{3.478}$$
$$r = -1 + i + j - 3k \tag{3.479}$$

    (b) The dot product of two quaternions q and p is defined as

$$p \cdot q = p_0 q_0 + \mathbf{p} \cdot \mathbf{q} \tag{3.480}$$

    If $p \cdot q = 0$, then q and p are orthogonal quaternions. Calculate, $p \cdot q, q \cdot p, r \cdot q$.
    (c) Prove these identities.

$$p \cdot q = \frac{1}{2}\left(p^{*}q + q^{*}p\right) \tag{3.481}$$
$$|q|^{2} = q_0^{2} + \mathbf{q} \cdot \mathbf{q} \tag{3.482}$$

38. ★ Quaternion inverse.
    (a) Calculate $q^{-1}, p^{-1}, p^{-1}q^{-1}, q^{-1}p^{*}, p^{*}p^{-1}, q^{-1}q^{*}$, and $p^{*^{-1}}q^{*^{-1}}$ if:

$$p = 3 + i - 2j + 2k \tag{3.483}$$
$$q = 2 - i + 2j + 4k \tag{3.484}$$

    (b) Compute the right and left quotients for the given quaternions p and q.

$$pq^{-1} = \frac{pq^{*}}{|q|^{2}} \qquad q^{-1}p = \frac{q^{*}p}{|q|^{2}} \tag{3.485}$$

39. ★ Quaternion and angle-axis rotation.
    (a) Find the unit quaternion associated to p and q, and find the angle and axis of rotation for each of the unit quaternions.

$$p = 3 + i - 2j + 2k \tag{3.486}$$
$$q = 2 - i + 2j + 4k \tag{3.487}$$

    (b) Use the unit quaternion p

$$p = \frac{1 + i - j + k}{2} \tag{3.488}$$

    and find the global position of

$$^{B}\mathbf{r} = \left[\, 2 \;\; -2 \;\; 6\,\right]^{T} \tag{3.489}$$

40. ★ Quaternion matrix.
    Use the unit quaternion matrices associated to

$$p = 3 + i - 2j + 2k \tag{3.490}$$

$$q = 2 - i + 2j + 4k \tag{3.491}$$

$$r = -1 + i + j - 3k \tag{3.492}$$

and find $\overleftrightarrow{p}\,\overleftrightarrow{r}\,\overleftrightarrow{q}$, $\overleftrightarrow{q}\,\overleftrightarrow{p}$, $\overleftrightarrow{p^*}\,\overleftrightarrow{q}$, $\overleftrightarrow{q}\,\overleftrightarrow{p^*}$, $\overleftrightarrow{p^*}\,\overleftrightarrow{p}$, $\overleftrightarrow{q}\,\overleftrightarrow{q^*}$, $\overleftrightarrow{p^*}\,\overleftrightarrow{q^*}$, and $\overleftrightarrow{p^*}\,\overleftrightarrow{r}\,\overleftrightarrow{q^*}$.

41. ★ Quaternion properties.
    Show the following properties. q and p are quaternions

$$q = q_0 + q_1 i + q_2 j + q_3 k$$

$$p = p_0 + p_1 i + p_2 j + p_3 k$$

and $a \in \mathbb{R}$ is a real number, and $n \in \mathbb{N}$ is a natural number.

$$(p \pm q)^* = p^* \pm q^* \tag{3.493}$$

$$\left(q^*\right)^* = q \tag{3.494}$$

$$(pq)^* = q^* p^* \tag{3.495}$$

$$(aq)^* = aq^* \tag{3.496}$$

$$|q + p|^2 + |q - p|^2 = 2\left(|q|^2 + |p|^2\right) \tag{3.497}$$

$$(|q| + |p|)^2 = |q|^2 + 2\left|qp^*\right| + |p|^2 \tag{3.498}$$

$$q^{2n} = (-1)^n |q|^{2n} \tag{3.499}$$

$$\sum_{n=0}^{\infty} \frac{q^n}{n!} = \cos |q| + \frac{q}{|q|} \sin |q| \tag{3.500}$$

42. ★ Quaternion equations.
    Find the quaternion $= q_0 + q_1 i + q_2 j + q_3 k$ in the following equations:

$$q^2 = -j \tag{3.501}$$

$$\left(q^*\right)^2 = i - j \tag{3.502}$$

$$q - \frac{1}{4}q^* + k = 0 \tag{3.503}$$

$$\frac{1}{2}qq^* - ijk + q^2 = 0 \tag{3.504}$$

$$q\left(q - q^*\right)^2 = i + j + k \tag{3.505}$$

43. ★ Euler angles and quaternion.
    Find quaternion components in terms of Euler angles, and Euler angles in terms of quaternion components.

44. ★ *bac-cab* rule application.
    (a) Use the Levi-Civita density ϵ_{ijk} to prove the *bac-cab* rule.

$$\mathbf{a} \times (\mathbf{b} \times \mathbf{c}) = \mathbf{b} (\mathbf{a} \cdot \mathbf{c}) - \mathbf{c} (\mathbf{a} \cdot \mathbf{b}) \tag{3.506}$$

    (b) Use the bac-cab rule to show that

$$\mathbf{a} = \hat{n} (\mathbf{a} \cdot \hat{n}) + \hat{n} \times (\mathbf{a} \times \hat{n}) \tag{3.507}$$

   where $\hat{n}$ is any unit vector. What is the geometric significance of this equation?
45. ★ Two rotations are not enough.
    Show that, in general, it is impossible to move a point $P(X, Y, Z)$ from the initial position $P(X_i, Y_i, Z_i)$ to the final position $P(X_f, Y_f, Z_f)$ only by *two* rotations about the global axes.
46. ★ Three rotations are enough.
    Show that, in general, it is possible to move a point $P(X, Y, Z)$ from the initial position $P(X_i, Y_i, Z_i)$ to the final position $P(X_f, Y_f, Z_f)$ by *three* rotations about different global axes.
47. ★ Closure property.
    Show the closure property of transformation matrices.

$$\text{If } q_1, q_2 \in S, \text{ then } q_1 \otimes q_2 \in S \tag{3.508}$$

   A set S together with a binary operation $\otimes$ defined on elements of S is called a group $(S, \otimes)$. Closure property is one of the necessary axioms to have a group.
48. ★ Skew symmetric matrices.
    Use $\mathbf{a} = \begin{bmatrix} a_1 & a_2 & a_3 \end{bmatrix}^T$ and $\mathbf{b} = \begin{bmatrix} b_1 & b_2 & b_3 \end{bmatrix}^T$ to show that
    (a)

$$\tilde{\mathbf{a}}\mathbf{b} = -\tilde{\mathbf{b}}\mathbf{a} \tag{3.509}$$

    (b)

$$\left(\widetilde{\mathbf{a} + \mathbf{b}}\right) = \tilde{\mathbf{a}} + \tilde{\mathbf{b}} \tag{3.510}$$

    (c)

$$\left(\widetilde{\tilde{\mathbf{a}}\mathbf{b}}\right) = \mathbf{b}\mathbf{a}^T - \mathbf{a}\mathbf{b}^T \tag{3.511}$$

49. ★ Skew symmetric matrix multiplication.
    Verify that
    (a)

$$\mathbf{a}^T \tilde{\mathbf{a}}^T = -\mathbf{a}^T \tilde{\mathbf{a}} = 0 \tag{3.512}$$

    (b)

$$\tilde{\mathbf{a}}\tilde{\mathbf{b}} = \mathbf{b}\mathbf{a}^T - \mathbf{a}^T\mathbf{b}\,\mathbf{I} \tag{3.513}$$

50. ★ Skew symmetric matrix derivative.
    Show that

$$\dot{\tilde{\mathbf{a}}} = \widetilde{\dot{\mathbf{a}}} \tag{3.514}$$

51. ★ Time derivative of $A = \begin{bmatrix} \mathbf{a}, \tilde{\mathbf{a}} \end{bmatrix}$.
    Assume that $\mathbf{a}$ is a time dependent vector and $A = \begin{bmatrix} \mathbf{a}, \tilde{\mathbf{a}} \end{bmatrix}$ is a 3×4 matrix. What is the time derivative of $C = AA^T$?
52. ★ Rotation for a rotated position.
    Consider a rigid body B at a position that is not coincident with the global coordinate frame G. The directional cosines of $\hat{i}, \hat{j}$ are

$$\hat{i} = \begin{bmatrix} 0.5 \\ 0.25 \\ \cos(\hat{i}, \hat{K}) \end{bmatrix} \qquad \hat{j} = \begin{bmatrix} 0.25 \\ \cos(\hat{j}, \hat{J}) \\ \cos(\hat{j}, \hat{K}) \end{bmatrix} \tag{3.515}$$

   Calculate the missing terms, and determine the global coordinates of a body point P at $^B\mathbf{r}_P$

$$^{B}\mathbf{r}_P = \begin{bmatrix} 1 & 2 & 3 \end{bmatrix}^T \tag{3.516}$$

after rotation of 45 deg about **u**.

$$\mathbf{u} = \begin{bmatrix} 1 & 1 & 1 \end{bmatrix}^T \tag{3.517}$$

53. ★ Rotation from a rotated position.
   (a) Calculate the transformation matrix for rotation 30 deg about z-axis, followed by a rotation 45 deg about x-axis.
   (b) Assume the body B is given to us after the first rotation of 30 deg about z-axis in section a. Determine the axis $\hat{u}$ that at this time coincides with x-axis and turns the body 45 deg about $\hat{u}$.
   (c) Calculate the transformation matrix for rotation 30 deg about z-axis, followed by a rotation 45 deg about x-axis, and then, 60 deg about z-axis.
   (d) Assume the body B is given to us after the second rotation in section b. Determine the axis $\hat{u}$ that at this time coincides with z-axis and turns the body 60 deg about $\hat{u}$.

54. ★ Global roll–pitch–yaw quaternions.
   Derive matrix (3.278) by substituting (3.279) in (3.256).

55. ★ Project: Axis–angle rotations for body axes rotations.
   Consider a body frame B coincident with a global frame G.
   (a) The frame B undergoes a rotation of 45 deg about the X-axis followed by a rotation of 90 deg about the Z-axis. Determine the axis $\hat{u}_1$ and angle ϕ_1 to move B to the final orientation only by one rotation.
   (b) The frame B undergoes a rotation of 90 deg about the Z-axis followed by a rotation of 45 deg about the X-axis. Determine the axis $\hat{u}_2$ and angle ϕ_2 to move B to the final orientation only by one rotation.
   (c) The frame B undergoes a rotation of ϕ_1 about $\hat{u}_1$ followed by a rotation ϕ_2 about $\hat{u}_2$. Turn B to its coincident position with G by turning it ϕ_3 about $\hat{u}_3$. Determine ϕ_3 about $\hat{u}_3$.

56. ★ Project: Axis–angle and incremental body axes rotations.
   Consider a body frame B coincident with a global frame G.
   (a) The frame B undergoes a rotation of 45 deg about the X-axis followed by a rotation of 90 deg about the Z-axis. Determine the axis $\hat{u}$ and angle ϕ to move B to the final orientation only by one rotation.
   (b) The frame B undergoes a rotation of 45/2 deg about the X-axis followed by a rotation of 90/2 deg about the Z-axis and then a rotation of 45/2 deg about the X-axis followed by a rotation of 90/2 deg about the Z-axis. Determine the axis $\hat{u}$ and angle ϕ to move B to the final orientation only by one rotation.
   (c) The frame B undergoes a rotation of 45/3 deg about the X-axis followed by a rotation of 90/3 deg about the Z-axis, then a rotation of 45/3 deg about the X-axis followed by a rotation of 90/3 deg about the Z-axis, and then a rotation of 45/3 deg about the X-axis followed by a rotation of 90/3 deg about the Z-axis. Determine the axis $\hat{u}$ and angle ϕ to move B to the final orientation only by one rotation.
   (d) Keep doing this process of turning frame B a rotation of 45/n deg about the X-axis followed by a rotation of 90/n deg about the Z-axis, and repeat the rotations n times, $n = 1, 2, 3, \cdots, 45$. Determine the axis $\hat{u}_n$ and angle ϕ_n to move B to the final orientation only by one rotation for every n. How $\hat{u}_n$ and angle ϕ_n are related to each other? Plot ϕ_n versus n. Plot directional cosines of $\hat{u}_n$ versus n.

The most general motion of a rigid body B in a global frame G is made by a rotation ϕ about an axis $\hat{u}$, plus a displacement **d**. Kinematically, the rigid body motion may be expressed by a 3×3 rotation matrix plus a 3×1 displacement vector. It may also be expressed by a 4×4 homogenous transformation matrix. This chapter will review all methods to express rigid body motions. Every rigid body represents a link of a robot.

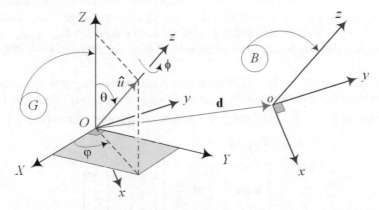

Fig. 4.1 Rotation and translation of a local frame with respect to a global frame

4.1 Rigid Body Motion

Consider a rigid body with an attached body coordinate frame $B\,(oxyz)$ moving in a fixed global coordinate frame $G(OXYZ)$. The rigid body can rotate in the global frame, while the origin point o of the body frame B can translate relative to the origin O of G as shown in Fig. 4.1. If the displacement vector $^G\mathbf{d}$ indicates the position of the moving origin o relative to the fixed origin O, then the coordinates of a body point P in local and global frames are related by the following equation:

$$^G\mathbf{r}_P = {}^G R_B\,{}^B\mathbf{r}_P + {}^G\mathbf{d} \tag{4.1}$$

where

$$^G\mathbf{r}_P = \begin{bmatrix} X_P \\ Y_P \\ Z_P \end{bmatrix} \qquad ^B\mathbf{r}_P = \begin{bmatrix} x_P \\ y_P \\ z_P \end{bmatrix} \qquad ^G\mathbf{d} = \begin{bmatrix} X_o \\ Y_o \\ Z_o \end{bmatrix} \tag{4.2}$$

© The Author(s), under exclusive license to Springer Nature Switzerland AG 2022
R. N. Jazar, *Theory of Applied Robotics*, https://doi.org/10.1007/978-3-030-93220-6_4

The vector $^G\mathbf{d}$ is called the *displacement* or *translation* of B with respect to G, and ^GR_B is the *rotation matrix* to transform $^B\mathbf{r}$ to $^G\mathbf{r}$ when $^G\mathbf{d} = 0$. Such a combination of a translation and a rotation in Eq. (4.1) is called *rigid motion*. Decomposition of a rigid motion into a rotation and a translation is the simplest method for representing spatial motion. We show the translation by a vector, and the rotation by a coordinate transformation matrix.

Proof Figure 4.1 illustrates a rotated and translated body frame B in the global frame G. The most general rotation is represented by the *Rodriguez rotation formula* (3.153), which needs an axis of rotation $\hat{u}$ and an angle of rotation ϕ. The translation $^G\mathbf{d}$ displaces the whole rigid body parallel to $^G\mathbf{d}$ such that all points of the body have the same displacement. Hence, the most general displacement of a rigid body is expressed by the following equation and has two independent parts: a rotation and a translation.

$$^G\mathbf{r} = {}^B\mathbf{r}\cos\phi + (1 - \cos\phi)\left(\hat{u} \cdot {}^B\mathbf{r}\right)\hat{u} + \left(\hat{u} \times {}^B\mathbf{r}\right)\sin\phi + {}^G\mathbf{d}$$

$$= {}^GR_B\,{}^B\mathbf{r} + {}^G\mathbf{d} \tag{4.3}$$

Equation (4.3) shows that the most general displacement of a rigid body is a rotation about an axis and a translation along an axis. The choice of the point of reference o is arbitrary, but when this point is chosen and the body coordinate frame is set up, the rotation and translation are uniquely determined.

Based on translation and rotation, the position of a body can be uniquely determined by six independent parameters: three translation components X_o, Y_o, Z_o; and three rotational components. The three rotational components can be any two coordinates of the unit vector $\hat{u}$ and the angle of rotation ϕ. If a body moves in such a way that its rotational components remain constant, the motion is a *pure translation*; and if it moves in such a way that X_o, Y_o, Z_o remain constant, the motion is a *pure rotation*. Therefore, a rigid body has three translational and three rotational degrees of freedom. ∎

Example 93 Translation and rotation of a body coordinate frame. Numerical example to apply rigid body motion.

A body coordinate frame $B(oxyz)$, that is originally coincident with global coordinate frame $G(OXYZ)$, rotates 45 deg about the X-axis and translates to $\begin{bmatrix} 3 & 5 & 7 \end{bmatrix}^T$. Find, the global position of a point at $^B\mathbf{r} = \begin{bmatrix} x & y & z \end{bmatrix}^T$.

$$^G\mathbf{r} = {}^GR_B\,{}^B\mathbf{r} + {}^G\mathbf{d}$$

$$= \begin{bmatrix} 1 & 0 & 0 \\ 0 & \cos\dfrac{\pi}{4} & -\sin\dfrac{\pi}{4} \\ 0 & \sin\dfrac{\pi}{4} & \cos\dfrac{\pi}{4} \end{bmatrix} \begin{bmatrix} x \\ y \\ z \end{bmatrix} + \begin{bmatrix} 3 \\ 5 \\ 7 \end{bmatrix}$$

$$= \begin{bmatrix} x + 3 \\ \frac{1}{2}\sqrt{2}y - \frac{1}{2}\sqrt{2}z + 5 \\ \frac{1}{2}\sqrt{2}y + \frac{1}{2}\sqrt{2}z + 7 \end{bmatrix}.$$

$$= (x + 3)\,\hat{I} + \left(\frac{\sqrt{2}}{2}y - \frac{\sqrt{2}}{2}z + 5\right)\hat{J}$$

$$+ \left(\frac{\sqrt{2}}{2}y + \frac{\sqrt{2}}{2}z + 7\right)\hat{K} \tag{4.4}$$

Example 94 Moving body coordinate frame. A numerical example of rotation and translation of a rigid body.

Figure 4.2 illustrates a point P at $^B\mathbf{r}_P = 0.1\hat{i} + 0.3\hat{j} + 0.3\hat{k}$ in a body frame B, which is rotated 50 deg about the Z-axis and translated -1 along X, 0.5 along Y, and 0.2 along the Z axes. Find the position $^G\mathbf{r}_P$ of P in global coordinate frame.

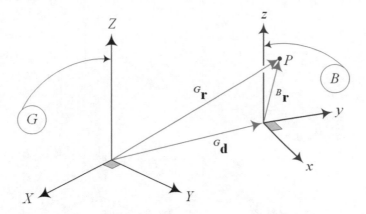

Fig. 4.2 A translating and rotating body B in a global coordinate frame G

$$^G\mathbf{r}_P = {}^G R_B \, {}^B\mathbf{r}_P + {}^G\mathbf{d}$$

$$= \begin{bmatrix} \cos\dfrac{50\pi}{180} & -\sin\dfrac{50\pi}{180} & 0 \\ \sin\dfrac{50\pi}{180} & \cos\dfrac{50\pi}{180} & 0 \\ 0 & 0 & 1 \end{bmatrix} \begin{bmatrix} 0.1 \\ 0.3 \\ 0.3 \end{bmatrix} + \begin{bmatrix} -1 \\ 0.5 \\ 0.2 \end{bmatrix}$$

$$= \begin{bmatrix} -1.166 \\ 0.769 \\ 0.5 \end{bmatrix} \tag{4.5}$$

Example 95 Rotation of a translated rigid body. Numerical example to apply rigid body motion.

Point P of a rigid body B has an initial position vector $^B\mathbf{r}_P$.

$$^B\mathbf{r}_P = \begin{bmatrix} 1 & 2 & 3 \end{bmatrix}^T \tag{4.6}$$

If the body rotates 45 deg about the x-axis and then translates to $^G\mathbf{d} = \begin{bmatrix} 4 & 5 & 6 \end{bmatrix}^T$, find the final global position of P.

$$^G\mathbf{r} = {}^B R^T_{x,\pi/4} \, {}^B\mathbf{r}_P + {}^G\mathbf{d}$$

$$= \begin{bmatrix} 1 & 0 & 0 \\ 0 & \cos\dfrac{\pi}{4} & -\sin\dfrac{\pi}{4} \\ 0 & \sin\dfrac{\pi}{4} & \cos\dfrac{\pi}{4} \end{bmatrix} \begin{bmatrix} 1 \\ 2 \\ 3 \end{bmatrix} + \begin{bmatrix} 4 \\ 5 \\ 6 \end{bmatrix} = \begin{bmatrix} 5.0 \\ 4.29 \\ 9.53 \end{bmatrix} \tag{4.7}$$

The rotation occurs first when $^G\mathbf{d} = \mathbf{0}$ and then translation happens.

Example 96 Arm rotation plus elongation. Numerical example to apply rigid body motion illustrated by a robotic arm.

Position vector of point P_1 at the tip of a PR arm shown in Fig. 4.3a is at:

$$^G\mathbf{r}_{P_1} = {}^B\mathbf{r}_{P_1} = \begin{bmatrix} 1350 & 0 & 900 \end{bmatrix}^T \quad \text{ıılıı} \tag{4.8}$$

The arm rotates 60 deg about the global Z-axis and elongates by $\mathbf{d} = 720.2\hat{\imath}$ mm. The final configuration of the arm is shown in Fig. 4.3b.

The new position vector of P is

$$^G\mathbf{r}_{P_2} = {}^G R_B \, {}^B\mathbf{r}_{P_1} + {}^G\mathbf{d} \tag{4.9}$$

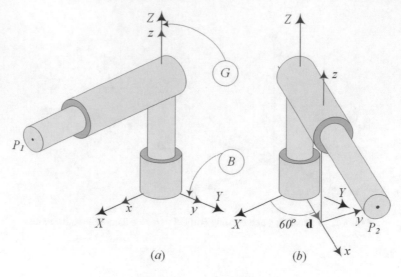

Fig. 4.3 A polar RP arm

where $^{G}R_{B} = R_{Z,60}$ is the rotation matrix to transform $\mathbf{r}_{P_2}$ to $\mathbf{r}_{P_1}$ when $^{G}\mathbf{d} = 0$,

$$
^{G}R_{B} = \begin{bmatrix} \cos\dfrac{\pi}{3} & -\sin\dfrac{\pi}{3} & 0 \\ \sin\dfrac{\pi}{3} & \cos\dfrac{\pi}{3} & 0 \\ 0 & 0 & 1 \end{bmatrix}
\tag{4.10}
$$

and $^{G}\mathbf{d}$ is the translation vector of o with respect to O in the global frame. The translation vector in the body coordinate frame is $^{B}\mathbf{d} = \begin{bmatrix} 720.2 & 0 & 0 \end{bmatrix}^{T}$ so $^{G}\mathbf{d}$ would be found by a transformation.

$$
^{G}\mathbf{d} = {}^{G}R_{B}\,{}^{B}\mathbf{d}
\tag{4.11}
$$

$$
= \begin{bmatrix} \cos\dfrac{\pi}{3} & -\sin\dfrac{\pi}{3} & 0 \\ \sin\dfrac{\pi}{3} & \cos\dfrac{\pi}{3} & 0 \\ 0 & 0 & 1 \end{bmatrix} \begin{bmatrix} 720.2 \\ 0.0 \\ 0.0 \end{bmatrix} = \begin{bmatrix} 360.10 \\ 623.71 \\ 0.0 \end{bmatrix}
$$

Therefore, the final global position of the tip of the arm is at:

$$
^{G}\mathbf{r}_{P_2} = {}^{G}R_{B}\,{}^{B}\mathbf{r}_{P_1} + {}^{G}\mathbf{d}
\tag{4.12}
$$

$$
= \begin{bmatrix} c60 & -s60 & 0 \\ s60 & c60 & 0 \\ 0 & 0 & 1 \end{bmatrix} \begin{bmatrix} 1350 \\ 0 \\ 900 \end{bmatrix} + \begin{bmatrix} 360.1 \\ 623.7 \\ 0.0 \end{bmatrix} = \begin{bmatrix} 1035.1 \\ 1792.8 \\ 900.0 \end{bmatrix}
$$

Example 97 Composition of transformations. How to combine two rigid body motions and determine an equivalent rigid body motion.

Assume $^{2}\mathbf{r}$ indicates the rigid motion of body B_1 with respect to body B_2, and $^{G}\mathbf{r}$ indicates the rigid motion of body B_2 with respect to frame G.

$$
^{2}\mathbf{r} = {}^{2}R_{1}\,{}^{1}\mathbf{r} + {}^{2}\mathbf{d}_{1}
\tag{4.13}
$$

$$
^{G}\mathbf{r} = {}^{G}R_{2}\,{}^{2}\mathbf{r} + {}^{G}\mathbf{d}_{2}
\tag{4.14}
$$

Composition of these two rigid motions can be defined by a third rigid motion, which will be calculated by substituting the expression for $^{2}\mathbf{r}$ into the equation for $^{G}\mathbf{r}$.

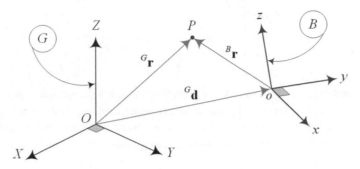

Fig. 4.4 Representation of a point P in coordinate frames B and G

$$
\begin{aligned}
^G\mathbf{r} &= {}^G R_2 \left({}^2 R_1 \, {}^1\mathbf{r} + {}^2\mathbf{d}_1 \right) + {}^G\mathbf{d}_2 \\
&= {}^G R_2 \, {}^2 R_1 \, {}^1\mathbf{r} + {}^G R_2 \, {}^2\mathbf{d}_1 + {}^G\mathbf{d}_2 \\
&= {}^G R_1 \, {}^1\mathbf{r} + {}^G\mathbf{d}_1
\end{aligned}
\tag{4.15}
$$

Therefore,

$$
^G R_1 = {}^G R_2 \, {}^2 R_1
\tag{4.16}
$$

$$
^G\mathbf{d}_1 = {}^G R_2 \, {}^2\mathbf{d}_1 + {}^G\mathbf{d}_2
\tag{4.17}
$$

which shows that the transformation from frame B_1 to frame G can be done by a rotation $^G R_1$ and a translation $^G\mathbf{d}_1$.

Example 98 ★ Euler rigid body theorem. Expression and proof of Euler rigid body motion theorem.

Rigid body motion is well expressed by Leonhard Euler (1707-1783) theorem.

4.2 Homogenous Transformation

A rigid body with coordinate frame B is moving in a globally fixed coordinate frame G. The position vector of an arbitrary point P of the rigid body is denoted by $^B\mathbf{r}_P$ and $^G\mathbf{r}_P$ in the frames, as is shown in Fig. 4.4. The translation vector $^G\mathbf{d}$ indicates the position of origin o of the body frame B in the global frame G. The general motion of a rigid body B $(oxyz)$ in the global frame G $(OXYZ)$ is a combination of rotation $^G R_B$ and translation $^G\mathbf{d}$.

$$
^G\mathbf{r} = {}^G R_B \, {}^B\mathbf{r} + {}^G\mathbf{d}
\tag{4.18}
$$

Combining a rotation matrix plus a vector can be expressed better by homogenous transformation matrix. Introducing a 4×4 *homogenous transformation matrix* $^G T_B$ helps us show a rigid motion by a single matrix transformation:

$$
^G\mathbf{r} = {}^G T_B \, {}^B\mathbf{r}
\tag{4.19}
$$

where

$$
^G T_B = \begin{bmatrix} r_{11} & r_{12} & r_{13} & X_o \\ r_{21} & r_{22} & r_{23} & Y_o \\ r_{31} & r_{32} & r_{33} & Z_o \\ 0 & 0 & 0 & 1 \end{bmatrix}
$$

$$
\equiv \left[\begin{array}{ccc|c} & {}^G R_B & & {}^G\mathbf{d} \\ 0 & 0 & 0 & 1 \end{array} \right] \equiv \begin{bmatrix} {}^G R_B & {}^G\mathbf{d} \\ 0 & 1 \end{bmatrix}
\tag{4.20}
$$

$$
{}^G\mathbf{r} = \begin{bmatrix} X_P \\ Y_P \\ Z_P \\ 1 \end{bmatrix} \qquad {}^B\mathbf{r} = \begin{bmatrix} x_P \\ y_P \\ z_P \\ 1 \end{bmatrix} \qquad {}^G\mathbf{d} = \begin{bmatrix} X_o \\ Y_o \\ Z_o \\ 1 \end{bmatrix} \tag{4.21}
$$

The homogenous transformation matrix ${}^G T_B$ is a 4×4 matrix that maps a homogenous position vector from one frame to another. This extension of matrix representation of rigid motions simplifies numerical calculations significantly.

Representation of an n-component position vector by an $(n + 1)$-component vector is called *homogenous coordinate representation*. The appended element is a *scale factor*, w; hence, in general, homogenous representation of a position vector $\mathbf{r} = \begin{bmatrix} x & y & z \end{bmatrix}^T$ is

$$
\mathbf{r} = \begin{bmatrix} wx \\ wy \\ wz \\ w \end{bmatrix} = \begin{bmatrix} r_1 \\ r_2 \\ r_3 \\ w \end{bmatrix} \tag{4.22}
$$

Using homogenous coordinates shows that the absolute values of the four coordinates are not important. Instead, it is the three ratios, r_1/w, r_2/w, and r_3/w, that are important because, provided $w \neq 0$, and $w \neq \infty$, we have

$$
\begin{bmatrix} wx \\ wy \\ wz \\ w \end{bmatrix} = \begin{bmatrix} x \\ y \\ z \\ 1 \end{bmatrix} \tag{4.23}
$$

Therefore, the homogenous vector $w\mathbf{r}$ refers to the same point as $\mathbf{r}$ does. If $w = 1$, then the homogenous coordinates of a position vector are the same as physical coordinates of the vector and the space is the Euclidean configuration space (Jazar, 2011).

Hereafter, if no confusion exists and $w = 1$, we will use regular vectors, and their homogenous representation equivalently.

Proof We append a 1 to the coordinates of a point and define 4×1 *homogenous position vectors* as follows:

$$
{}^G\mathbf{r}_P = \begin{bmatrix} X_P \\ Y_P \\ Z_P \\ 1 \end{bmatrix} \qquad {}^B\mathbf{r}_P = \begin{bmatrix} x_P \\ y_P \\ z_P \\ 1 \end{bmatrix} \qquad {}^G\mathbf{d} = \begin{bmatrix} X_o \\ Y_o \\ Z_o \\ 1 \end{bmatrix} \tag{4.24}
$$

Employing the homogenous transformation matrix to transform homogenous expressions of position vector of a body point P from B-frame to G-frame yields

$$
{}^G\mathbf{r}_P = {}^G T_B \; {}^B\mathbf{r}_P \tag{4.25}
$$

$$
\begin{bmatrix} X_P \\ Y_P \\ Z_P \\ 1 \end{bmatrix} = \begin{bmatrix} r_{11} & r_{12} & r_{13} & X_o \\ r_{21} & r_{22} & r_{23} & Y_o \\ r_{31} & r_{32} & r_{33} & Z_o \\ 0 & 0 & 0 & 1 \end{bmatrix} \begin{bmatrix} x_P \\ y_P \\ z_P \\ 1 \end{bmatrix}
$$

$$
= \begin{bmatrix} X_o + r_{11}x_P + r_{12}y_P + r_{13}z_P \\ Y_o + r_{21}x_P + r_{22}y_P + r_{23}z_P \\ Z_o + r_{31}x_P + r_{32}y_P + r_{33}z_P \\ 1 \end{bmatrix} \tag{4.26}
$$

However, the standard method of rotation $^G R_B$ plus displacement $^G \mathbf{d}$ reduces to:

$$^G \mathbf{r}_P = {}^G R_B \, {}^B \mathbf{r}_p + {}^G \mathbf{d} \tag{4.27}$$

$$\begin{bmatrix} X_P \\ Y_P \\ Z_P \end{bmatrix} = \begin{bmatrix} r_{11} & r_{12} & r_{13} \\ r_{21} & r_{22} & r_{23} \\ r_{31} & r_{32} & r_{33} \end{bmatrix} \begin{bmatrix} x_P \\ y_P \\ z_P \end{bmatrix} + \begin{bmatrix} X_o \\ Y_o \\ Z_o \end{bmatrix}$$

$$= \begin{bmatrix} X_o + r_{11}x_P + r_{12}y_P + r_{13}z_P \\ Y_o + r_{21}x_P + r_{22}y_P + r_{23}z_P \\ Z_o + r_{31}x_P + r_{32}y_P + r_{33}z_P \end{bmatrix} \tag{4.28}$$

which is compatible with the definition of homogenous vector and homogenous transformation. The advantage of homogenous transformation is combination of rotation and translation and expressing rigid body motions by matrix calculus. The main disadvantage of homogenous transformation matrix is that its inverse is not equal to its transpose. ∎

Example 99 Rotation and translation of a body coordinate frame. Numerical example of a rigid body motion.

A body coordinate frame $B(oxyz)$, that is originally coincident with global coordinate frame $G(OXYZ)$, rotates 45 deg about the X-axis and translates to $\begin{bmatrix} 3 & 5 & 7 & 1 \end{bmatrix}^T$. Find the matrix representation of the global position of a body point at $^B \mathbf{r} = \begin{bmatrix} x & y & z & 1 \end{bmatrix}^T$.

$$^G \mathbf{r} = {}^G T_B \, {}^B \mathbf{r}$$

$$= \begin{bmatrix} X \\ Y \\ Z \\ 1 \end{bmatrix} = \begin{bmatrix} 1 & 0 & 0 & 3 \\ 0 & \cos\frac{\pi}{4} & -\sin\frac{\pi}{4} & 5 \\ 0 & \sin\frac{\pi}{4} & \cos\frac{\pi}{4} & 7 \\ 0 & 0 & 0 & 1 \end{bmatrix} \begin{bmatrix} x \\ y \\ z \\ 1 \end{bmatrix}$$

$$= \begin{bmatrix} x + 3 \\ 0.707y - 0.707z + 5 \\ 0.707y + 0.707z + 7 \\ 1 \end{bmatrix} \tag{4.29}$$

Example 100 An axis–angle rotation and a translation. Numerical example or rigid body motion applied on a cube and determination coordinates of critical points.

A cubic rigid body with a unit length of edges sits at the corner of the first quadrant as is shown in Fig. 4.5. If we turn the cube 45 deg about $\mathbf{u}$ and translate it by $^G \mathbf{d}$, determine the coordinates of the corners of the cube after the rigid body motion.

$$\mathbf{u} = \begin{bmatrix} 1 & 1 & 1 \end{bmatrix}^T \tag{4.30}$$

$$^G \mathbf{d} = \begin{bmatrix} 1 & 1 & 1 \end{bmatrix}^T \tag{4.31}$$

The axis $\hat{u}$ and angle ϕ of rotation are

$$\phi = \frac{\pi}{4} \qquad \hat{u} = \frac{\mathbf{u}}{\sqrt{3}} = \begin{bmatrix} 0.577\,35 \\ 0.577\,35 \\ 0.577\,35 \end{bmatrix} \tag{4.32}$$

The Rodriguez transformation matrix of the rotation will be

$$R_{\hat{u},\phi} = \mathbf{I}\cos\phi + \hat{u}\hat{u}^T \operatorname{vers}\phi + \tilde{u}\sin\phi$$

$$= \begin{bmatrix} 0.804\,74 & -0.310\,62 & 0.505\,88 \\ 0.505\,88 & 0.804\,74 & -0.310\,62 \\ -0.310\,62 & 0.505\,88 & 0.804\,74 \end{bmatrix} \tag{4.33}$$

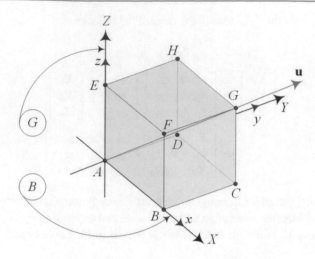

Fig. 4.5 A cubic rigid body with a unit length of edges

Translating the cube by $^G\mathbf{d}$ makes the following homogenous transformation matrix $^G T_B$.

$$
^G T_B =
\begin{bmatrix}
0.804\,74 & -0.310\,62 & 0.505\,88 & 1 \\
0.505\,88 & 0.804\,74 & -0.310\,62 & 1 \\
-0.310\,62 & 0.505\,88 & 0.804\,74 & 1 \\
0 & 0 & 0 & 1
\end{bmatrix}
\tag{4.34}
$$

The local coordinates of the corners of the cube are

	$^B\mathbf{r}_A$	$^B\mathbf{r}_B$	$^B\mathbf{r}_C$	$^B\mathbf{r}_D$	$^B\mathbf{r}_E$	$^B\mathbf{r}_F$	$^B\mathbf{r}_G$	$^B\mathbf{r}_H$
x	0	1	1	0	0	1	1	0
y	0	0	1	1	0	0	1	1
z	0	0	0	0	1	1	1	1

$$\tag{4.35}$$

Using the transformation equation $^G\mathbf{r} = {}^G T_B \, {}^B\mathbf{r}$, the global coordinates of the corners after the motion will be

	$^B\mathbf{r}_A$	$^B\mathbf{r}_B$	$^B\mathbf{r}_C$	$^B\mathbf{r}_D$	$^G\mathbf{r}_E$	$^G\mathbf{r}_F$	$^G\mathbf{r}_G$	$^G\mathbf{r}_H$
X	1	1.805	1.49	0.689	1.505	2.311	2	1.195
Y	1	1.506	2.31	1.805	0.689	1.195	2	1.494
Z	1	0.689	1.195	1.506	1.804	1.494	2	2.31

$$\tag{4.36}$$

Example 101 Decomposition of $^G T_B$ into translation and rotation. How to decompose a rigid body motion into combination of a rotation and a translation matrices.

homogenous transformation matrix $^G T_B$ can be decomposed into a matrix multiplication of a pure rotation matrix $^G R_B$ and a pure translation matrix $^G D_B$. We use R to indicate rotation matrix and D to indicate displacement equivalent to translation.

$$
^G T_B = {}^G D_B \, {}^G R_B
\tag{4.37}
$$

$$
=
\begin{bmatrix}
1 & 0 & 0 & X_o \\
0 & 1 & 0 & Y_o \\
0 & 0 & 1 & Z_o \\
0 & 0 & 0 & 1
\end{bmatrix}
\begin{bmatrix}
r_{11} & r_{12} & r_{13} & 0 \\
r_{21} & r_{22} & r_{23} & 0 \\
r_{31} & r_{32} & r_{33} & 0 \\
0 & 0 & 0 & 1
\end{bmatrix}
$$

$$
=
\begin{bmatrix}
r_{11} & r_{12} & r_{13} & X_o \\
r_{21} & r_{22} & r_{23} & Y_o \\
r_{31} & r_{32} & r_{33} & Z_o \\
0 & 0 & 0 & 1
\end{bmatrix}
$$

Therefore, a homogenous transformation can be achieved by a pure rotation first, followed by a pure translation. The formula for decomposing transformation matrix $^G T_B$ to a rotation $^G R_B$ and a translation $^G D_B$ is

Rotation first, translation second $(RFDS)$, $^G T_B = {}^G D_B {}^G R_B$

Decomposition of a homogenous transformation to translation and rotation is not interchangeable

$$^G T_B = {}^G D_B {}^G R_B \neq {}^G R_B {}^G D_B \tag{4.38}$$

However, according to the definition of $^G R_B$ and $^G D_B$, we have

$$^G T_B = {}^G D_B {}^G R_B = {}^G D_B + {}^G R_B - \mathbf{I} = {}^G R_B + {}^G D_B - \mathbf{I} \tag{4.39}$$

If a rigid body with its coordinate frame $B(oxyz)$, that is originally coincident with global coordinate frame $G(OXYZ)$, translates by $^G\mathbf{d}$,

$$^G\mathbf{d} = \begin{bmatrix} d_X & d_Y & d_Z \end{bmatrix}^T \tag{4.40}$$

the motion of the rigid body is a pure translation and the transformation matrix for a point of a rigid body in the global frame is

$$^G\mathbf{r} = {}^G T_B {}^B\mathbf{r} = {}^G D_B {}^B\mathbf{r} \tag{4.41}$$

$$= \begin{bmatrix} X \\ Y \\ Z \\ 1 \end{bmatrix} = \begin{bmatrix} 1 & 0 & 0 & d_X \\ 0 & 1 & 0 & d_Y \\ 0 & 0 & 1 & d_Z \\ 0 & 0 & 0 & 1 \end{bmatrix} \begin{bmatrix} x \\ y \\ z \\ 1 \end{bmatrix} = \begin{bmatrix} x + d_X \\ y + d_Y \\ z + d_Z \\ 1 \end{bmatrix}$$

Example 102 Rotation about and translation along a global and local axes. Numerical examples for working with decomposition of homogenous transformation matrix of rotation and translation along a local or global axes.

A point P is located at $^B\mathbf{r} = (0, 0, 20)$ in a body coordinate frame. If the rigid body rotates 30 deg about the global X-axis and the origin of the body frame translates to $(X, Y, Z) = (50, 0, 60)$, then the coordinates of the point in the global frame are

$$\begin{bmatrix} 1 & 0 & 0 & 50 \\ 0 & 1 & 0 & 0 \\ 0 & 0 & 1 & 60 \\ 0 & 0 & 0 & 1 \end{bmatrix} \begin{bmatrix} 1 & 0 & 0 & 0 \\ 0 & \cos\frac{\pi}{6} & -\sin\frac{\pi}{6} & 0 \\ 0 & \sin\frac{\pi}{6} & \cos\frac{\pi}{6} & 0 \\ 0 & 0 & 0 & 1 \end{bmatrix} \begin{bmatrix} 0 \\ 0 \\ 20 \\ 1 \end{bmatrix} = \begin{bmatrix} 50 \\ -10 \\ 77.3 \\ 1 \end{bmatrix} \tag{4.42}$$

A point P is located at $(0, 0, 20)$ in a body coordinate frame. If the origin of the body frame translates to $(X, Y, Z) = (50, 0, 60)$ and rotates 30 deg about the local x-axis, then the coordinates of the point in global frame are

$$\begin{bmatrix} 1 & 0 & 0 & 50 \\ 0 & 1 & 0 & 0 \\ 0 & 0 & 1 & 60 \\ 0 & 0 & 0 & 1 \end{bmatrix} \begin{bmatrix} 1 & 0 & 0 & 0 \\ 0 & \cos\frac{\pi}{6} & \sin\frac{\pi}{6} & 0 \\ 0 & -\sin\frac{\pi}{6} & \cos\frac{\pi}{6} & 0 \\ 0 & 0 & 0 & 1 \end{bmatrix}^T \begin{bmatrix} 0 \\ 0 \\ 20 \\ 1 \end{bmatrix} = \begin{bmatrix} 50 \\ -10 \\ 77.3 \\ 1 \end{bmatrix} \tag{4.43}$$

Example 103 Translation. Numerical example for a pure translational homogenous transformation matrix.

If a body point at

$$^B\mathbf{r} = \begin{bmatrix} -1 & 0 & 2 & 1 \end{bmatrix}^T \tag{4.44}$$

is translated to $^G\mathbf{r}$ without rotation,

$$^G\mathbf{r} = \begin{bmatrix} 0 & 10 & -5 & 1 \end{bmatrix}^T \tag{4.45}$$

find the displacement vector $^G\mathbf{d}$.

$$^G\mathbf{d} = \begin{bmatrix} d_X & d_Y & d_Z & 1 \end{bmatrix}^T \tag{4.46}$$

The corresponding transformation is

$$\begin{bmatrix} 2 \\ 10 \\ -5 \\ 1 \end{bmatrix} = \begin{bmatrix} 1 & 0 & 0 & d_X \\ 0 & 1 & 0 & d_Y \\ 0 & 0 & 1 & d_Z \\ 0 & 0 & 0 & 1 \end{bmatrix} \begin{bmatrix} 1 \\ 4 \\ 2 \\ 1 \end{bmatrix} \tag{4.47}$$

Therefore,

$$1 + d_X = 2 \qquad 4 + d_Y = 10 \qquad 2 + d_Z = -5 \tag{4.48}$$

and

$$d_X = 1 \qquad d_Y = 6 \qquad d_Z = -7 \tag{4.49}$$

Example 104 Pure rotation and pure translation. Principal global rotation and translation homogenous matrices are collected here for reference.

A set of basic homogenous transformations for translation along and rotation about X, Y, and Z axes are given below.

$$^{G}T_B = D_{X,a} = \begin{bmatrix} 1 & 0 & 0 & a \\ 0 & 1 & 0 & 0 \\ 0 & 0 & 1 & 0 \\ 0 & 0 & 0 & 1 \end{bmatrix} \tag{4.50}$$

$$^{G}T_B = R_{X,\gamma} = \begin{bmatrix} 1 & 0 & 0 & 0 \\ 0 & \cos\gamma & -\sin\gamma & 0 \\ 0 & \sin\gamma & \cos\gamma & 0 \\ 0 & 0 & 0 & 1 \end{bmatrix} \tag{4.51}$$

$$^{G}T_B = D_{Y,b} = \begin{bmatrix} 1 & 0 & 0 & 0 \\ 0 & 1 & 0 & b \\ 0 & 0 & 1 & 0 \\ 0 & 0 & 0 & 1 \end{bmatrix} \tag{4.52}$$

$$^{G}T_B = R_{Y,\beta} = \begin{bmatrix} \cos\beta & 0 & \sin\beta & 0 \\ 0 & 1 & 0 & 0 \\ -\sin\beta & 0 & \cos\beta & 0 \\ 0 & 0 & 0 & 1 \end{bmatrix} \tag{4.53}$$

$$^{G}T_B = D_{Z,c} = \begin{bmatrix} 1 & 0 & 0 & 0 \\ 0 & 1 & 0 & 0 \\ 0 & 0 & 1 & c \\ 0 & 0 & 0 & 1 \end{bmatrix} \tag{4.54}$$

$$^{G}T_B = R_{Z,\alpha} = \begin{bmatrix} \cos\alpha & -\sin\alpha & 0 & 0 \\ \sin\alpha & \cos\alpha & 0 & 0 \\ 0 & 0 & 1 & 0 \\ 0 & 0 & 0 & 1 \end{bmatrix} \tag{4.55}$$

Example 105 Homogenous transformation as a vector addition. Position of a point P in B-frame, which is at $^{G}\mathbf{d}$ in G-frame makes it possible to show rigid body motion by vectors.

As is shown in Fig. 4.4, the position of point P can be expressed by a vector addition.

$$^{G}\mathbf{r}_P = {}^{G}\mathbf{d} + {}^{B}\mathbf{r}_P \tag{4.56}$$

A vector equation is only meaningful when all the vectors are expressed in the same coordinate frame. Hence, we need to transform either $^B\mathbf{r}_P$ to G or $^G\mathbf{d}$ to B. Therefore, the applied vector equation is

$$^G\mathbf{r}_P = {}^GR_B\,{}^B\mathbf{r}_P + {}^G\mathbf{d} \tag{4.57}$$

or

$$^BR_G\,{}^G\mathbf{r}_P = {}^BR_G\,{}^G\mathbf{d} + {}^B\mathbf{r}_P \tag{4.58}$$

The first one defines a homogenous transformation from B to G,

$$^G\mathbf{r}_P = {}^GT_B\,{}^B\mathbf{r}_P \tag{4.59}$$

$$^GT_B = \begin{bmatrix} {}^GR_B & {}^G\mathbf{d} \\ 0 & 1 \end{bmatrix} \tag{4.60}$$

and the second one defines a transformation from G to B.

$$^B\mathbf{r}_P = {}^BT_G\,{}^G\mathbf{r}_P \tag{4.61}$$

$$^BT_G = \begin{bmatrix} {}^BR_G & -{}^BR_G\,{}^G\mathbf{d} \\ 0 & 1 \end{bmatrix} = \begin{bmatrix} {}^GR_B^T & -{}^GR_B^T\,{}^G\mathbf{d} \\ 0 & 1 \end{bmatrix} \tag{4.62}$$

Example 106 ★ Point at infinity and direction line. The meaning of the fourth element of homogenous expression of vectors to be 0.

Points at infinity have a convenient representation with homogenous coordinates. Consider the scale factor w as the fourth coordinate of a point, then the homogenous representation of the point is given by:

$$\begin{bmatrix} x \\ y \\ z \\ w \end{bmatrix} = \begin{bmatrix} x/w \\ y/w \\ z/w \\ 1 \end{bmatrix} \tag{4.63}$$

When w tends to zero, the point goes to infinity, and we may adapt the convention that the homogenous coordinate

$$\begin{bmatrix} x \\ y \\ z \\ 0 \end{bmatrix} \tag{4.64}$$

represents a point at infinity. More importantly, a point at infinity indicates a direction. In this case, it indicates all lines parallel to the vector $\mathbf{r} = \begin{bmatrix} x & y & z \end{bmatrix}^T$, which intersect at a point at infinity. The homogenous coordinate transformation of points at infinity introduces a neat decomposition of the homogenous transformation matrices.

$$^GT_B = \begin{bmatrix} r_{11} & r_{12} & r_{13} & X_o \\ r_{21} & r_{22} & r_{23} & Y_o \\ r_{31} & r_{32} & r_{33} & Z_o \\ 0 & 0 & 0 & 1 \end{bmatrix} \tag{4.65}$$

The first three columns have zero as the fourth coordinate. Therefore, they represent points at infinity, which are the directions corresponding to the three coordinate axes. The fourth column has one as the fourth coordinate and represents the location of the origin of coordinate frame.

Example 107 ★ The most general homogenous transformation. Inhomogeneous transformation matrices are important tools in computer graphics.

The homogenous transformation matrix (4.20), that represents a rigid body motion, is a special case of the general homogenous transformation. The most general homogenous transformation, which has been extensively used in the field of computer graphics, is

$$^A T_B = \left[\begin{array}{c|c} ^A R_B \,(3 \times 3) & ^A \mathbf{d} \,(3 \times 1) \\ \hline p\,(1 \times 3) & w\,(1 \times 1) \end{array} \right] \tag{4.66}$$

$$= \left[\begin{array}{c|c} \text{rotation} & \text{translation} \\ \hline \text{perspective} & \text{scale factor} \end{array} \right]$$

For the purpose of robotics, we always take the last row vector of $[T]$ to be $(0, 0, 0, 1)$. However, the more general form of (4.66) could be useful, for example, when a graphical simulator or a vision system is added to the overall robotic system.

The upper left 3×3 submatrix $^A R_B$ denotes the orientation of a moving frame B with respect to a reference frame A. The upper right submatrix 3×1 is $^A \mathbf{d}$ and denotes the position of the origin of the moving frame B relative to the reference frame A. The lower left submatrix 1×3 is p and denotes a perspective transformation, and the lower right element w is a scaling factor.

Consider an identity transformation. If we change the main diagonal elements of the homogenous transformation matrix to any other number than 1, the matrix will define scaling transformation. Assume to show the scaling factors in x, y, z, by s_x, s_y, s_z, respectively, the point (x, y, z) will be mapped to $(s_x x, s_y y, s_z z)$. If scale factors are greater than 1, the transformation will be magnification and if the scale factors are less than 1, the transformation will be shrinking. Scale factor 0 is not defined.

$$\left[\begin{array}{c} s_x x \\ s_y y \\ s_z z \\ 0 \end{array} \right] = \left[\begin{array}{cccc} s_x & 0 & 0 & 0 \\ 0 & s_y & 0 & 0 \\ 0 & 0 & s_z & 0 \\ 0 & 0 & 0 & 1 \end{array} \right] \left[\begin{array}{c} x \\ y \\ z \\ 0 \end{array} \right] \tag{4.67}$$

Every negative values on the main diagonal elements will show reflection associated to its coordinate. Adjusting the negative coefficients enables us to make reflection with respect to any principal plane we wish. A reflection with respect to origin will be made by all negative diagonal elements.

$$\left[\begin{array}{c} -x \\ -y \\ -z \\ 0 \end{array} \right] = \left[\begin{array}{cccc} -1 & 0 & 0 & 0 \\ 0 & -1 & 0 & 0 \\ 0 & 0 & -1 & 0 \\ 0 & 0 & 0 & 1 \end{array} \right] \left[\begin{array}{c} x \\ y \\ z \\ 0 \end{array} \right] \tag{4.68}$$

Reflection about the x-axis will be made by

$$\left[\begin{array}{c} x \\ -y \\ -z \\ 0 \end{array} \right] = \left[\begin{array}{cccc} 1 & 0 & 0 & 0 \\ 0 & -1 & 0 & 0 \\ 0 & 0 & -1 & 0 \\ 0 & 0 & 0 & 1 \end{array} \right] \left[\begin{array}{c} x \\ y \\ z \\ 0 \end{array} \right] \tag{4.69}$$

and reflection about the (x, y)-plane will be

$$\left[\begin{array}{c} x \\ y \\ -z \\ 0 \end{array} \right] = \left[\begin{array}{cccc} 1 & 0 & 0 & 0 \\ 0 & 1 & 0 & 0 \\ 0 & 0 & -1 & 0 \\ 0 & 0 & 0 & 1 \end{array} \right] \left[\begin{array}{c} x \\ y \\ z \\ 0 \end{array} \right] \tag{4.70}$$

Off-diagonal elements will be used to make shear transformation in graphic. For example, a shear transformation in y and z coordinates, proportional to x, is made by change in off-diagonal elements of the first column.

$$\left[\begin{array}{c} x \\ y + c_x x \\ z + c_x x \\ 0 \end{array} \right] = \left[\begin{array}{cccc} 1 & 0 & 0 & 0 \\ c_x & 1 & 0 & 0 \\ c_x & 0 & 1 & 0 \\ 0 & 0 & 0 & 1 \end{array} \right] \left[\begin{array}{c} x \\ y \\ z \\ 0 \end{array} \right] \tag{4.71}$$

Example 108 Rigid body based on corner coordinates. Numerical example of calculating position of corners of a box, using homogenous transformation method, after rotation and translation.

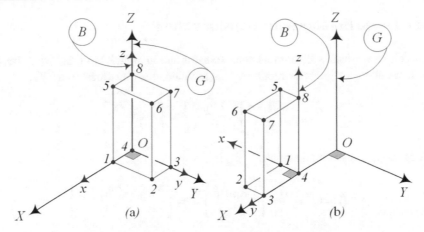

Fig. 4.6 Expressing the motion of a rigid body in terms of some body points

Employing the coordinates of sharp corners of a rigid body, we may express a rigid body by an array of homogenous coordinates of the corners in body coordinate frame B. Figure 4.6a illustrates a box. The coordinates of the corners of the box in its body frame B are collected in the matrix $[P]$:

$$
{}^B P = \begin{bmatrix} \mathbf{r}_1\ \mathbf{r}_2\ \mathbf{r}_3\ \mathbf{r}_4\ \mathbf{r}_5\ \mathbf{r}_6\ \mathbf{r}_7\ \mathbf{r}_8 \end{bmatrix}
$$

$$
= \begin{bmatrix} 1\ 1\ 0\ 0\ 1\ 1\ 0\ 0 \\ 0\ 3\ 3\ 0\ 0\ 3\ 3\ 0 \\ 0\ 0\ 0\ 0\ 5\ 5\ 5\ 5 \\ 1\ 1\ 1\ 1\ 1\ 1\ 1\ 1 \end{bmatrix}
\tag{4.72}
$$

The configuration of the box after a rotation of -90 deg about the Z-axis and a translation of three units along the X-axis is shown in Fig. 4.6b. The new coordinates of the corners in the global frame G can be found by multiplying the corresponding transformation matrix and $[P]$:

$$
{}^G T_B = D_{X,3} R_{Z,90}
$$

$$
= \begin{bmatrix} 1\ 0\ 0\ 3 \\ 0\ 1\ 0\ 0 \\ 0\ 0\ 1\ 0 \\ 0\ 0\ 0\ 1 \end{bmatrix}
\begin{bmatrix} \cos\left(-\frac{\pi}{2}\right)\ -\sin\left(-\frac{\pi}{2}\right)\ 0\ 0 \\ \sin\left(-\frac{\pi}{2}\right)\ \cos\left(-\frac{\pi}{2}\right)\ 0\ 0 \\ 0\ \ \ \ \ 0\ \ \ \ \ 1\ 0 \\ 0\ \ \ \ \ 0\ \ \ \ \ 0\ 1 \end{bmatrix}
$$

$$
= \begin{bmatrix} 0\ \ \ 1\ 0\ 3 \\ -1\ 0\ 0\ 0 \\ 0\ \ \ 0\ 1\ 0 \\ 0\ \ \ 0\ 0\ 1 \end{bmatrix}
\tag{4.73}
$$

Therefore, the global coordinates of corners 1–8 after the motion are

$$
{}^G P = {}^G T_B\, {}^B P
$$

$$
= \begin{bmatrix} 0\ \ \ 1\ 0\ 3 \\ -1\ 0\ 0\ 0 \\ 0\ \ \ 0\ 1\ 0 \\ 0\ \ \ 0\ 0\ 1 \end{bmatrix}
\begin{bmatrix} 1\ 1\ 0\ 0\ 1\ 1\ 0\ 0 \\ 0\ 3\ 3\ 0\ 0\ 3\ 3\ 0 \\ 0\ 0\ 0\ 0\ 5\ 5\ 5\ 5 \\ 1\ 1\ 1\ 1\ 1\ 1\ 1\ 1 \end{bmatrix}
$$

$$
= \begin{bmatrix} 3\ \ \ 6\ \ 6\ 3\ 3\ \ \ 6\ \ 6\ 3 \\ -1\ -1\ 0\ 0\ -1\ -1\ 0\ 0 \\ 0\ \ \ 0\ \ 0\ 0\ 5\ \ \ 5\ \ 5\ 5 \\ 1\ \ \ 1\ \ 1\ 1\ 1\ \ \ 1\ \ 1\ 1 \end{bmatrix}
\tag{4.74}
$$

4.3   Inverse and Reverse Homogenous Transformation

The advantage of simplicity to work with homogenous transformation matrices come with the penalty of losing the orthogonality property. If we show a rigid body motion by the homogenous transformation $^G T_B$,

$$^G T_B = \begin{bmatrix} \mathbf{I} & ^G\mathbf{d} \\ 0 & 1 \end{bmatrix} \begin{bmatrix} ^G R_B & 0 \\ 0 & 1 \end{bmatrix} = \begin{bmatrix} ^G R_B & ^G\mathbf{d} \\ 0 & 1 \end{bmatrix} \tag{4.75}$$

then the *inverse* of homogenous transformation matrix $^G T_B$ is

$$^B T_G = {^G T_B^{-1}} = \begin{bmatrix} ^G R_B & ^G\mathbf{d} \\ 0 & 1 \end{bmatrix}^{-1} = \begin{bmatrix} ^G R_B^T & -{^G R_B^T}\,{^G\mathbf{d}} \\ 0 & 1 \end{bmatrix} \tag{4.76}$$

and the *reverse* motion of $^G T_B$ would be $^G T_{-B}$

$$^G T_{-B} = \begin{bmatrix} ^G R_B^T & 0 \\ 0 & 1 \end{bmatrix} \begin{bmatrix} \mathbf{I} & -{^G\mathbf{d}} \\ 0 & 1 \end{bmatrix} = \begin{bmatrix} ^G R_B^T & -{^G R_B^T}\,{^G\mathbf{d}} \\ 0 & 1 \end{bmatrix} \tag{4.77}$$

showing that

$$^G T_B^{-1}\,{^G T_B} = \mathbf{I}_4 \tag{4.78}$$

$$^G T_{-B}\,{^G T_B} = \mathbf{I}_4 \tag{4.79}$$

where $\mathbf{I}_4$ is the identity matrix of rank 4.

The homogenous transformation matrix $^G T_B$ presents rigid body motions very well; however, a shortcoming is that they lose the orthogonality property and, hence, its inverse is not equal to its transpose:

$$^G T_B^{-1} \neq {^G T_B^T} \tag{4.80}$$

Although it is traditional to use the inverse matrix notation $^G T_B^{-1}$ for $^B T_G$,

$$^G T_B^{-1} = {^B T_G} \tag{4.81}$$

calculating $^G T_B^{-1}$ must be done according to Eq. (4.76), and not by regular matrix inversion of a 4×4 matrix. The matrix inversion notation makes equations consistent with the multiplication of a matrix $[T]$ by its inverse, $[T]^{-1}$, because

$$^G T_B^{-1}\,{^G T_B} = {^B T_G}\,{^G T_B} = \mathbf{I}_4 \tag{4.82}$$

Proof 1. Figure 4.7 depicts a rotated and translated body frame $B(oxyz)$ with respect to a global frame $G(OXYZ)$. Transformation of the coordinates of a point P from the global frame to the body frame will be expressed by homogenous transformation matrix $^B T_G$, which is the inverse of the transformation $^G T_B$.

To find $^B T_G$, let us start with expression of $^G\mathbf{r}$ and definition of $^G T_B$ for mapping $^B\mathbf{r}$ to $^G\mathbf{r}$

$$^G\mathbf{r} = {^G R_B}\,{^B\mathbf{r}} + {^G\mathbf{d}} \tag{4.83}$$

$$\begin{bmatrix} ^G\mathbf{r} \\ 1 \end{bmatrix} = \begin{bmatrix} ^G R_B & ^G\mathbf{d} \\ 0 & 1 \end{bmatrix} \begin{bmatrix} ^B\mathbf{r} \\ 1 \end{bmatrix} \tag{4.84}$$

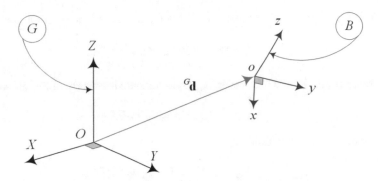

Fig. 4.7 A rotated and translated body frame $B(oxyz)$ with respect to a global frame $G(OXYZ)$

and calculate $^B\mathbf{r}$

$$^B\mathbf{r} = {}^GR_B^{-1}\,({}^G\mathbf{r} - {}^G\mathbf{d}) = {}^GR_B^T\,{}^G\mathbf{r} - {}^GR_B^T\,{}^G\mathbf{d} \tag{4.85}$$

to express the transformation matrix ^BT_G for mapping $^G\mathbf{r}$ to $^B\mathbf{r}$.

$$^BT_G = {}^GT_B^{-1} = \begin{bmatrix} {}^GR_B^T & -\,{}^GR_B^T\,{}^G\mathbf{d} \\ 0 & 1 \end{bmatrix} \tag{4.86}$$

A matrix multiplication indicates that

$$
\begin{aligned}
{}^GT_B^{-1}\,{}^GT_B &= \begin{bmatrix} {}^GR_B^T & -\,{}^GR_B^T\,{}^G\mathbf{d} \\ 0 & 1 \end{bmatrix} \begin{bmatrix} {}^GR_B & {}^G\mathbf{d} \\ 0 & 1 \end{bmatrix} \\
&= \begin{bmatrix} {}^GR_B^T\,{}^GR_B & {}^GR_B^T\,{}^G\mathbf{d} - {}^GR_B^T\,{}^G\mathbf{d} \\ 0 & 1 \end{bmatrix} = \begin{bmatrix} \mathbf{I}_3 & 0 \\ 0 & 1 \end{bmatrix} = \mathbf{I}_4
\end{aligned}
\tag{4.87}
$$

Decomposing a homogenous motion ^GT_B by a rotation ^GR_B plus a translation ^GD_B,

$$^GT_B = {}^GD_B\,{}^GR_B = \begin{bmatrix} \mathbf{I} & {}^G\mathbf{d} \\ 0 & 1 \end{bmatrix} \begin{bmatrix} {}^GR_B & \mathbf{0} \\ 0 & 1 \end{bmatrix} \tag{4.88}$$

we define the reverse motion by a translation $-\,{}^GD_B$ followed by a reverse rotation $^GR_{-B} = {}^GR_B^T$.

$$^GT_{-B} = \begin{bmatrix} {}^GR_{-B} \end{bmatrix} \begin{bmatrix} -\,{}^GD_B \end{bmatrix} = \begin{bmatrix} {}^GR_B^T & \mathbf{0} \\ 0 & 1 \end{bmatrix} \begin{bmatrix} \mathbf{I} & -\,{}^G\mathbf{d} \\ 0 & 1 \end{bmatrix} \tag{4.89}$$

The result of a rigid body motion ^GT_B and then the reverse motion $^GT_{-B}$ would also be an identity matrix, indicating no resultant motion.

$$
\begin{aligned}
{}^GT_{-B}\,{}^GT_B &= {}^GT_B\,{}^GT_{-B} \\
&= \begin{bmatrix} {}^GR_B^T & \mathbf{0} \\ 0 & 1 \end{bmatrix} \begin{bmatrix} \mathbf{I} & -\,{}^G\mathbf{d} \\ 0 & 1 \end{bmatrix} \begin{bmatrix} \mathbf{I} & {}^G\mathbf{d} \\ 0 & 1 \end{bmatrix} \begin{bmatrix} {}^GR_B & \mathbf{0} \\ 0 & 1 \end{bmatrix} \\
&= \begin{bmatrix} {}^GR_B^T & \mathbf{0} \\ 0 & 1 \end{bmatrix} \begin{bmatrix} \mathbf{I}_3 & 0 \\ 0 & 1 \end{bmatrix} \begin{bmatrix} {}^GR_B & \mathbf{0} \\ 0 & 1 \end{bmatrix} = \mathbf{I}_4
\end{aligned}
\tag{4.90}
$$

∎

***Proof* 2.** $^B T_G$ can also be found by a geometric expression. Using the inverse of the rotation matrix $^G R_B$,

$$^G R_B^{-1} = {}^G R_B^T = {}^B R_G \tag{4.91}$$

and transforming $^G \mathbf{d}$ to the body frame B, and then reversing $^B \mathbf{d}$ to indicate the origin of the global frame with respect to the origin of the body frame

$$- {}^B \mathbf{d} = - {}^B R_G {}^G \mathbf{d} = - {}^G R_B^T {}^G \mathbf{d} \tag{4.92}$$

allow us to define the homogenous transformation $^B T_G$.

$$^B T_G = \begin{bmatrix} {}^B R_G & -{}^B \mathbf{d} \\ 0 & 1 \end{bmatrix} = \begin{bmatrix} {}^G R_B^T & -{}^G R_B^T {}^G \mathbf{d} \\ 0 & 1 \end{bmatrix} \tag{4.93}$$

We use the notation

$$^B T_G = {}^G T_B^{-1} \tag{4.94}$$

and remember that the inverse of a homogenous transformation matrix must be calculated according to Eq. (4.76), and not by regular inversion of a 4×4 matrix. This notation is consistent with the multiplication of a T matrix by its inverse T^{-1}.

$$^G T_B^{-1} {}^G T_B = \mathbf{I}_4 \tag{4.95}$$

∎

Example 109 Inverse of a homogenous transformation matrix. It is a numerical example for calculation of inverse of a homogenous transformation matrix.

Assume that

$$^G T_B = \begin{bmatrix} 0.643 & -0.766 & 0 & -1 \\ 0.766 & 0.643 & 0 & 0.5 \\ 0 & 0 & 1 & 0.2 \\ 0 & 0 & 0 & 1 \end{bmatrix} = \begin{bmatrix} {}^G R_B & {}^G \mathbf{d} \\ 0 & 1 \end{bmatrix} \tag{4.96}$$

then

$$^G R_B = \begin{bmatrix} 0.643 & -0.766 & 0 \\ 0.766 & 0.643 & 0 \\ 0 & 0 & 1 \end{bmatrix} \qquad {}^G \mathbf{d} = \begin{bmatrix} -1 \\ 0.5 \\ 0.2 \end{bmatrix} \tag{4.97}$$

and, therefore,

$$^B T_G = {}^G T_B^{-1} = \begin{bmatrix} {}^G R_B^T & -{}^G R_B^T {}^G \mathbf{d} \\ 0 & 1 \end{bmatrix}$$

$$= \begin{bmatrix} 0.643 & 0.766 & 0 & 0.26 \\ -0.766 & 0.643 & 0 & -1.087 \\ 0 & 0 & 1 & -0.2 \\ 0 & 0 & 0 & 1 \end{bmatrix} \tag{4.98}$$

Example 110 Transformation matrix and coordinate of points. It is a numerical example to determine transformation matrix when coordinates of few point are given before and after motion.

It is possible and sometimes convenient to express a rigid body motion in terms of known displacement of specified points of the body. Assume A, B, C, D are four points with the known coordinates at two different positions:

$$A_1(2, 4, 1) \quad B_1(2, 6, 1) \quad C_1(1, 5, 1) \quad D_1(3, 5, 2) \tag{4.99}$$

$$A_2(5, 1, 1) \quad B_2(7, 1, 1) \quad C_2(6, 2, 1) \quad D_2(6, 2, 3) \tag{4.100}$$

There must be a transformation matrix T to map the initial positions to the final positions.

$$[T]\begin{bmatrix} 2&2&1&3 \\ 4&6&5&5 \\ 1&1&2&2 \\ 1&1&1&1 \end{bmatrix} = \begin{bmatrix} 5&7&6&6 \\ 1&1&2&2 \\ 1&1&1&3 \\ 1&1&1&1 \end{bmatrix} \tag{4.101}$$

It is easy to find $[T]$ by matrix multiplication.

$$\begin{aligned}[T] &= \begin{bmatrix} 5&7&6&6 \\ 1&1&2&2 \\ 1&1&1&3 \\ 1&1&1&1 \end{bmatrix}\begin{bmatrix} 2&2&1&3 \\ 4&6&5&5 \\ 1&1&2&2 \\ 1&1&1&1 \end{bmatrix}^{-1} \\ &= \begin{bmatrix} 5&7&6&6 \\ 1&1&2&2 \\ 1&1&1&3 \\ 1&1&1&1 \end{bmatrix}\begin{bmatrix} 0&-1/2&-1/2&7/2 \\ 0&1/2&-1/2&-3/2 \\ -1/2&0&1/2&1/2 \\ 1/2&0&1/2&-3/2 \end{bmatrix} \\ &= \begin{bmatrix} 0&1&0&1 \\ 0&0&1&0 \\ 1&0&1&-2 \\ 0&0&0&1 \end{bmatrix}\end{aligned} \tag{4.102}$$

What if we had more than 4 points? In that case, we only pick four points to determine $[T]$.

Example 111 Quick inverse transformation. An easier method to determine inverse of homogenous matrices by breaking it into displacement and rotation.

For numerical calculation, it is more practical to decompose a transformation matrix into rotation $[R]$ and displacement $[D]$ and take advantage of the inverse of matrix multiplication.

$$T^{-1} = [DR]^{-1} = R^{-1}D^{-1} = R^T D^{-1} \tag{4.103}$$

Consider a homogenous matrix $[T]$

$$[T] = \begin{bmatrix} r_{11}&r_{12}&r_{13}&r_{14} \\ r_{21}&r_{22}&r_{23}&r_{24} \\ r_{31}&r_{32}&r_{33}&r_{34} \\ 0&0&0&1 \end{bmatrix} = \begin{bmatrix} 1&0&0&r_{14} \\ 0&1&0&r_{24} \\ 0&0&1&r_{34} \\ 0&0&0&1 \end{bmatrix}\begin{bmatrix} r_{11}&r_{12}&r_{13}&0 \\ r_{21}&r_{22}&r_{23}&0 \\ r_{31}&r_{32}&r_{33}&0 \\ 0&0&0&1 \end{bmatrix} \tag{4.104}$$

therefore,

$$\begin{aligned}T^{-1} &= [DR]^{-1} = R^{-1}D^{-1} = R^T D^{-1} \\ &= \begin{bmatrix} r_{11}&r_{21}&r_{31}&0 \\ r_{12}&r_{22}&r_{32}&0 \\ r_{13}&r_{23}&r_{33}&0 \\ 0&0&0&1 \end{bmatrix}\begin{bmatrix} 1&0&0&-r_{14} \\ 0&1&0&-r_{24} \\ 0&0&1&-r_{34} \\ 0&0&0&1 \end{bmatrix} \\ &= \begin{bmatrix} r_{11}&r_{21}&r_{31}&-r_{11}r_{14}-r_{21}r_{24}-r_{31}r_{34} \\ r_{12}&r_{22}&r_{32}&-r_{12}r_{14}-r_{22}r_{24}-r_{32}r_{34} \\ r_{13}&r_{23}&r_{33}&-r_{13}r_{14}-r_{23}r_{24}-r_{33}r_{34} \\ 0&0&0&1 \end{bmatrix}\end{aligned} \tag{4.105}$$

Example 112 ★ Inverse of a general matrix. How to determine inverse of large matrices by breaking it into four smaller matrices. Ideal for computerized calculations.

Let $[T]$ be defined as a general matrix combined by four submatrices $[A]$, $[B]$, $[C]$, and $[D]$.

$$[T] = \begin{bmatrix} \mathbf{A} & \mathbf{B} \\ \mathbf{C} & \mathbf{D} \end{bmatrix} \tag{4.106}$$

Then, the inverse of T is given by:

$$T^{-1} = \begin{bmatrix} A^{-1} + A^{-1}BE^{-1}CA^{-1} & -A^{-1}BE^{-1} \\ -E^{-1}CA^{-1} & E^{-1} \end{bmatrix} \tag{4.107}$$

$$[E] = D - CA^{-1}B \tag{4.108}$$

In case of the most general homogenous transformation matrix,

$$^AT_B = \left[\begin{array}{c|c} ^AR_B\,(3 \times 3) & ^A\mathbf{d}\,(3 \times 1) \\ \hline p\,(1 \times 3) & w\,(1 \times 1) \end{array} \right] \tag{4.109}$$

$$= \left[\begin{array}{c|c} \text{rotation} & \text{translation} \\ \hline \text{perspective} & \text{scale factor} \end{array} \right]$$

we have

$$[T] = {}^AT_B \tag{4.110}$$

$$[A] = {}^AR_B \tag{4.111}$$

$$[B] = {}^A\mathbf{d} \tag{4.112}$$

$$[C] = \begin{bmatrix} p_1 & p_2 & p_3 \end{bmatrix} \tag{4.113}$$

$$[D] = [w_{1\times1}] = w \tag{4.114}$$

and, therefore,

$$\begin{aligned} [E] &= [E_{1\times1}] = E \\ &= w - \begin{bmatrix} p_1 & p_2 & p_3 \end{bmatrix} {}^AR_B^{-1}\,{}^A\mathbf{d} \\ &= w - \begin{bmatrix} p_1 & p_2 & p_3 \end{bmatrix} {}^AR_B^{T}\,{}^A\mathbf{d} \\ &= 1 - p_1\,(d_1 r_{11} + d_2 r_{21} + d_3 r_{31}) \\ &\quad + p_2\,(d_1 r_{12} + d_2 r_{22} + d_3 r_{32}) \\ &\quad + p_3\,(d_1 r_{13} + d_2 r_{23} + d_3 r_{33}) \end{aligned} \tag{4.115}$$

$$-E^{-1}CA^{-1} = \frac{1}{E} \begin{bmatrix} g_1 & g_2 & g_3 \end{bmatrix} \tag{4.116}$$

$$g_1 = p_1 r_{11} + p_2 r_{12} + p_3 r_{13}$$

$$g_2 = p_1 r_{21} + p_2 r_{22} + p_3 r_{23}$$

$$g_3 = p_1 r_{31} + p_2 r_{32} + p_3 r_{33}$$

$$-A^{-1}BE^{-1} = -\frac{1}{E} \begin{bmatrix} d_1 r_{11} + d_2 r_{21} + d_3 r_{31} \\ d_1 r_{12} + d_2 r_{22} + d_3 r_{32} \\ d_1 r_{13} + d_2 r_{23} + d_3 r_{33} \end{bmatrix} \tag{4.117}$$

$$A^{-1} + A^{-1}BE^{-1}CA^{-1} = [F] \tag{4.118}$$

$$= \begin{bmatrix} f_{11} & f_{12} & f_{13} \\ f_{21} & f_{22} & f_{23} \\ f_{31} & f_{32} & r_{33} \end{bmatrix}$$

where

$$f_{11} = r_{11} + \frac{1}{E}(p_1 r_{11} + p_2 r_{12} + p_3 r_{13})(d_1 r_{11} + d_2 r_{21} + d_3 r_{31})$$

$$f_{12} = r_{21} + \frac{1}{E}(p_1 r_{21} + p_2 r_{22} + p_3 r_{23})(d_1 r_{11} + d_1 r_{21} + d_1 r_{31})$$

$$f_{13} = r_{31} + \frac{1}{E}(p_1 r_{31} + p_2 r_{32} + p_3 r_{33})(d_1 r_{11} + d_2 r_{21} + d_3 r_{31})$$

$$f_{21} = r_{12} + \frac{1}{E}(p_1 r_{11} + p_2 r_{12} + p_3 r_{13})(d_1 r_{12} + d_2 r_{22} + d_3 r_{32})$$

$$f_{22} = r_{22} + \frac{1}{E}(p_1 r_{21} + p_2 r_{22} + p_3 r_{23})(d_1 r_{12} + d_2 r_{22} + d_3 r_{32})$$

$$f_{23} = r_{32} + \frac{1}{E}(p_1 r_{31} + p_2 r_{32} + p_3 r_{33})(d_1 r_{12} + d_2 r_{22} + d_3 r_{32})$$

$$f_{31} = r_{13} + \frac{1}{E}(p_1 r_{11} + p_2 r_{12} + p_3 r_{13})(d_1 r_{13} + d_2 r_{23} + d_3 r_{33})$$

$$f_{32} = r_{23} + \frac{1}{E}(p_1 r_{21} + p_2 r_{22} + p_3 r_{23})(d_1 r_{13} + d_2 r_{23} + d_3 r_{33})$$

$$f_{33} = r_{33} + \frac{1}{E}(p_1 r_{31} + p_2 r_{32} + p_3 r_{33})(d_1 r_{13} + d_2 r_{23} + d_3 r_{33}) \tag{4.119}$$

In the case of a coordinate homogenous transformation,

$$[T] = {}^A T_B \qquad [A] = {}^A R_B \qquad [B] = {}^A \mathbf{d} \tag{4.120}$$

$$[C] = \begin{bmatrix} 0 & 0 & 0 \end{bmatrix} \qquad [D] = [1] \tag{4.121}$$

we have

$$[E] = [1] \tag{4.122}$$

and, therefore,

$${}^A T_B^{-1} = \begin{bmatrix} {}^A R_B^T & -{}^A R_B^T {}^A \mathbf{d} \\ 0 & 1 \end{bmatrix} \tag{4.123}$$

4.4 Combined Homogenous Transformation

Figure 4.8 illustrates three reference frames: A, B, and C. The transformation matrices to transform coordinates from frame B to A and from frame C to B are

$${}^A T_B = \begin{bmatrix} {}^A R_B & {}^A \mathbf{d}_1 \\ 0 & 1 \end{bmatrix} \tag{4.124}$$

$${}^B T_C = \begin{bmatrix} {}^B R_C & {}^B \mathbf{d}_2 \\ 0 & 1 \end{bmatrix} \tag{4.125}$$

Hence, the transformation matrix from C to A is

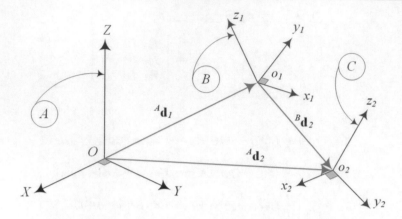

Fig. 4.8 Three coordinate frames to analyze compound transformations

$$
{}^A T_C = {}^A T_B \, {}^B T_C = \begin{bmatrix} {}^A R_B & {}^A \mathbf{d}_1 \\ 0 & 1 \end{bmatrix} \begin{bmatrix} {}^B R_C & {}^B \mathbf{d}_2 \\ 0 & 1 \end{bmatrix}
$$

$$
= \begin{bmatrix} {}^A R_B \, {}^B R_C & {}^A R_B \, {}^B \mathbf{d}_2 + {}^A \mathbf{d}_1 \\ 0 & 1 \end{bmatrix}
$$

$$
= \begin{bmatrix} {}^A R_C & {}^A \mathbf{d}_2 \\ 0 & 1 \end{bmatrix} \tag{4.126}
$$

and, therefore, the inverse transformation is

$$
{}^C T_A = \begin{bmatrix} {}^B R_C^T \, {}^A R_B^T & -{}^B R_C^T \, {}^A R_B^T \left[{}^A R_B \, {}^B \mathbf{d}_2 + {}^A \mathbf{d}_1 \right] \\ 0 & 1 \end{bmatrix}
$$

$$
= \begin{bmatrix} {}^B R_C^T \, {}'^A R_B^T & -{}^B R_C^T \, {}^B \mathbf{d}_2 - {}^A R_C^T \, {}^A \mathbf{d}_1 \\ 0 & 1 \end{bmatrix}
$$

$$
= \begin{bmatrix} {}^A R_C^T & -{}^A R_C^T \, {}^A \mathbf{d}_2 \\ 0 & 1 \end{bmatrix} \tag{4.127}
$$

The value of homogenous coordinates are better appreciated when several displacements occur in succession, which, for instance, can be written as

$$
{}^G T_4 = {}^G T_1 \, {}^1 T_2 \, {}^2 T_3 \, {}^3 T_4 \tag{4.128}
$$

rather than:

$$
{}^G R_4 \, {}^4 \mathbf{r}_P + {}^G \mathbf{d}_4
$$
$$
= {}^G R_1 \left({}^1 R_2 \left({}^2 R_3 \left({}^3 R_4 \, {}^4 \mathbf{r}_P + {}^3 \mathbf{d}_4 \right) + {}^2 \mathbf{d}_3 \right) + {}^1 \mathbf{d}_2 \right) + {}^G \mathbf{d}_1 \tag{4.129}
$$

Example 113 Homogenous transformation for multiple frames. How to combine rigid body motions when more than two coordinate frames are involved.

Figure 4.9 depicts a point P in a local frame $B_2 \, (x_2 y_2 z_2)$. The coordinates of P in the global frame $G(OXYZ)$ can be found by using the homogenous transformation matrices. The position of P in frame $B_2 \, (x_2 y_2 z_2)$ is indicated by ${}^2 \mathbf{r}_P$. Therefore, its position in frame $B_1 \, (x_1 y_1 z_1)$ is

$$
\begin{bmatrix} x_1 \\ y_1 \\ z_1 \\ 1 \end{bmatrix} = \begin{bmatrix} {}^1 R_2 & {}^1 \mathbf{d}_2 \\ 0 & 1 \end{bmatrix} \begin{bmatrix} x_2 \\ y_2 \\ z_2 \\ 1 \end{bmatrix} \tag{4.130}
$$

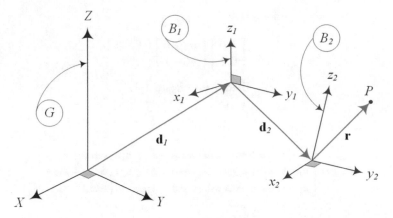

Fig. 4.9 Point P in a local frame B_2 $(x_2 y_2 z_2)$

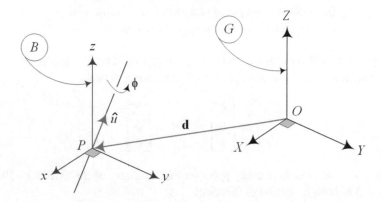

Fig. 4.10 Rotation about an axis not going through origin

and, hence, its position in the global frame $G(OXYZ)$ would be

$$
\begin{bmatrix} X \\ Y \\ Z \\ 1 \end{bmatrix} = \begin{bmatrix} {}^G R_1 & {}^G \mathbf{d}_1 \\ 0 & 1 \end{bmatrix} \begin{bmatrix} x_1 \\ y_1 \\ z_1 \\ 1 \end{bmatrix}
$$

$$
= \begin{bmatrix} {}^G R_1 & {}^G \mathbf{d}_1 \\ 0 & 1 \end{bmatrix} \begin{bmatrix} {}^1 R_2 & {}^1 \mathbf{d}_2 \\ 0 & 1 \end{bmatrix} \begin{bmatrix} x_2 \\ y_2 \\ z_2 \\ 1 \end{bmatrix}
$$

$$
= \begin{bmatrix} {}^G R_1 \, {}^1 R_2 & {}^G R_1 \, {}^1 \mathbf{d}_2 + {}^G \mathbf{d}_1 \\ 0 & 1 \end{bmatrix} \begin{bmatrix} x_2 \\ y_2 \\ z_2 \\ 1 \end{bmatrix} \tag{4.131}
$$

Example 114 ★ Rotation about an axis not going through origin. This is a good application of homogenous transformation matrix to determine rotation of a rigid body about a non-central axis. This example is a reference for rigid body rotational motions about any arbitrary axis.

The homogenous transformation matrix can also be used to present rotation about an axis not going through the origin. Such an axis is called an off-center axis. Figure 4.10 indicates an angle of rotation, ϕ, around an axis $\hat{u}$, passing through a point P at a position ${}^G \mathbf{d}_Q$.

Let us set a local frame B at point P parallel to the global frame G. Then, a rotation around $\hat{u}$ can be expressed as a translation along $-\mathbf{d}$, to bring the body frame B to the global frame G, followed by a rotation ϕ about $\hat{u}$, and a reverse translation along $\mathbf{d}$ to move origin of the body frame B back to where it was.

$$^G T_B = D_{\hat{d},d}\, R_{\hat{u},\phi}\, D_{\hat{d},-d}$$

$$= \begin{bmatrix} \mathbf{I} & \mathbf{d} \\ 0 & 1 \end{bmatrix} \begin{bmatrix} R_{\hat{u},\phi} & 0 \\ 0 & 1 \end{bmatrix} \begin{bmatrix} \mathbf{I} & -\mathbf{d} \\ 0 & 1 \end{bmatrix}$$

$$= \begin{bmatrix} R_{\hat{u},\phi} & \mathbf{d} - R_{\hat{u},\phi}\mathbf{d} \\ 0 & 1 \end{bmatrix} \tag{4.132}$$

$$R_{\hat{u},\phi} = \begin{bmatrix} u_1^2 \operatorname{vers}\phi + c\phi & u_1 u_2 \operatorname{vers}\phi - u_3 s\phi & u_1 u_3 \operatorname{vers}\phi + u_2 s\phi \\ u_1 u_2 \operatorname{vers}\phi + u_3 s\phi & u_2^2 \operatorname{vers}\phi + c\phi & u_2 u_3 \operatorname{vers}\phi - u_1 s\phi \\ u_1 u_3 \operatorname{vers}\phi - u_2 s\phi & u_2 u_3 \operatorname{vers}\phi + u_1 s\phi & u_3^2 \operatorname{vers}\phi + c\phi \end{bmatrix} \tag{4.133}$$

$$\mathbf{d} - R_{\hat{u},\phi}\mathbf{d} =$$

$$\begin{bmatrix} d_1(1 - u_1^2) \operatorname{vers}\phi - u_1 \operatorname{vers}\phi\,(d_2 u_2 + d_3 u_3) + s\phi\,(d_2 u_3 - d_3 u_2) \\ d_2(1 - u_2^2) \operatorname{vers}\phi - u_2 \operatorname{vers}\phi\,(d_3 u_3 + d_1 u_1) + s\phi\,(d_3 u_1 - d_1 u_3) \\ d_3(1 - u_3^2) \operatorname{vers}\phi - u_3 \operatorname{vers}\phi\,(d_1 u_1 + d_2 u_2) + s\phi\,(d_1 u_2 - d_2 u_1) \end{bmatrix} \tag{4.134}$$

Example 115 ★ A rotating cylinder. Numerical example for off-center rigid body rotation.
   Imagine a cylinder with radius $R = 2$ that is going to turn about the axis $\hat{u}$ at $\mathbf{d}$.

$$\hat{u} = \begin{bmatrix} 0 \\ 0 \\ 1 \end{bmatrix} \qquad \mathbf{d} = \begin{bmatrix} 2 \\ 0 \\ 0 \end{bmatrix} \tag{4.135}$$

If the cylinder turns 90 deg about its axis, then every point on the periphery of the cylinder will move 90 deg on circular paths parallel to (x, y)-plane. The transformation of this motion is

$$^G T_B = D_{\hat{d},d}\, R_{\hat{u},\phi}\, D_{\hat{d},-d} = \begin{bmatrix} \mathbf{I} & \mathbf{d} \\ 0 & 1 \end{bmatrix} \begin{bmatrix} R_{\hat{K},\frac{\pi}{2}} & 0 \\ 0 & 1 \end{bmatrix} \begin{bmatrix} \mathbf{I} & -\mathbf{d} \\ 0 & 1 \end{bmatrix}$$

$$= \begin{bmatrix} 1 & 0 & 0 & 2 \\ 0 & 1 & 0 & 0 \\ 0 & 0 & 1 & 0 \\ 0 & 0 & 0 & 1 \end{bmatrix} \begin{bmatrix} c\frac{\pi}{2} & -s\frac{\pi}{2} & 0 & 0 \\ s\frac{\pi}{2} & c\frac{\pi}{2} & 0 & 0 \\ 0 & 0 & 1 & 0 \\ 0 & 0 & 0 & 1 \end{bmatrix} \begin{bmatrix} 1 & 0 & 0 & -2 \\ 0 & 1 & 0 & 0 \\ 0 & 0 & 1 & 0 \\ 0 & 0 & 0 & 1 \end{bmatrix}$$

$$= \begin{bmatrix} 0 & -1 & 0 & 2 \\ 1 & 0 & 0 & -2 \\ 0 & 0 & 1 & 0 \\ 0 & 0 & 0 & 1 \end{bmatrix} \tag{4.136}$$

Consider a point of cylinder that was on the origin. After the rotation, the point would be seen at:

$$^G\mathbf{r} = {}^G T_B\, {}^G\mathbf{r}$$

$$= \begin{bmatrix} 0 & -1 & 0 & 2 \\ 1 & 0 & 0 & -2 \\ 0 & 0 & 1 & 0 \\ 0 & 0 & 0 & 1 \end{bmatrix} \begin{bmatrix} 0 \\ 0 \\ 0 \\ 1 \end{bmatrix} = \begin{bmatrix} 2 \\ -2 \\ 0 \\ 1 \end{bmatrix} \tag{4.137}$$

As another example, let a body frame B sitting on G-frame turn 90 deg about an axis $\hat{u}$ parallel to $\hat{K}$ that is at $^G\mathbf{d}$. Determine the new location of a body point P at $^B\mathbf{r}_P$.

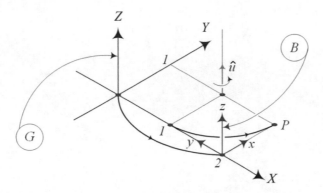

Fig. 4.11 Rotating a body frame B about an off-center axis

$$^B\mathbf{r}_P = \begin{bmatrix} 1 & 0 & 0 \end{bmatrix}^T \tag{4.138}$$

$$^G\mathbf{d} = \begin{bmatrix} 1 & 1 & 0 \end{bmatrix}^T \tag{4.139}$$

The associated homogenous transformation ^GT_B from (4.132) is

$$
\begin{aligned}
^GT_B &= D_{\hat{d},d}\, R_{\hat{u},\phi}\, D_{\hat{d},-d} \\
&= \begin{bmatrix} 1 & 0 & 0 & 1 \\ 0 & 1 & 0 & 1 \\ 0 & 0 & 1 & 0 \\ 0 & 0 & 0 & 1 \end{bmatrix}
\begin{bmatrix} \cos\frac{\pi}{2} & -\sin\frac{\pi}{2} & 0 & 0 \\ \sin\frac{\pi}{2} & \cos\frac{\pi}{2} & 0 & 0 \\ 0 & 0 & 1 & 0 \\ 0 & 0 & 0 & 1 \end{bmatrix}
\begin{bmatrix} 1 & 0 & 0 & -1 \\ 0 & 1 & 0 & -1 \\ 0 & 0 & 1 & 0 \\ 0 & 0 & 0 & 1 \end{bmatrix} \\
&= \begin{bmatrix} 0 & -1 & 0 & 2 \\ 1 & 0 & 0 & 0 \\ 0 & 0 & 1 & 0 \\ 0 & 0 & 0 & 1 \end{bmatrix}
\end{aligned}
\tag{4.140}
$$

After the rotation, the point P will be seen at $^G\mathbf{r}_P$:

$$
^G\mathbf{r}_P = {}^GT_B\, {}^B\mathbf{r}_P = \begin{bmatrix} 0 & -1 & 0 & 2 \\ 1 & 0 & 0 & 0 \\ 0 & 0 & 1 & 0 \\ 0 & 0 & 0 & 1 \end{bmatrix} \begin{bmatrix} 1 \\ 0 \\ 0 \\ 1 \end{bmatrix} = \begin{bmatrix} 2 \\ 1 \\ 0 \\ 1 \end{bmatrix}
\tag{4.141}
$$

and the origin o of the body coordinate frame B would be at

$$
^G\mathbf{r}_o = {}^GT_B\, {}^B\mathbf{r}_o = \begin{bmatrix} 0 & -1 & 0 & 2 \\ 1 & 0 & 0 & 0 \\ 0 & 0 & 1 & 0 \\ 0 & 0 & 0 & 1 \end{bmatrix} \begin{bmatrix} 0 \\ 0 \\ 0 \\ 1 \end{bmatrix} = \begin{bmatrix} 2 \\ 0 \\ 0 \\ 1 \end{bmatrix}
\tag{4.142}
$$

The final positions of point P, o, and frame B are shown in Fig. 4.11.

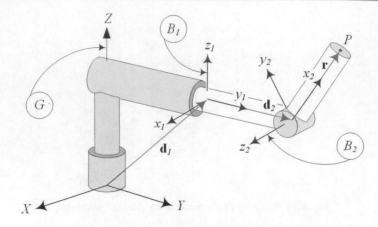

Fig. 4.12 An RPR manipulator robot

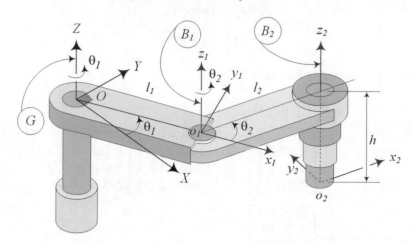

Fig. 4.13 The $SCARA$ robot R‖R‖P of Example 117

Example 116 End-effector of an RPR robot in a global frame. How to analyze a simple robotic arm, using rigid body motion and homogenous transformation matrices

Point P is at tip of the end arm of the robot shown in Fig. 4.12. Position vector of P in frame B_2 $(x_2y_2z_2)$ is $^2\mathbf{r}_P$. Frame B_2 $(x_2y_2z_2)$ at location $^G\mathbf{d}_1$ can rotate about z_2 and slide along y_1. Frame B_1 $(x_1y_1z_1)$ can rotate about the Z-axis of the global frame $G(OXYZ)$. The position of the origin of B_1 is shown by $^1\mathbf{d}_2$ in B_2. To determine the position of P in $G(OXYZ)$, we add $^G\mathbf{d}_1$ and $^G\mathbf{d}_2$ and $^G\mathbf{r}_P$.

$$
^G\mathbf{r} = {}^GR_1\,{}^1R_2\,{}^2\mathbf{r}_P + {}^GR_1\,{}^1\mathbf{d}_2 + {}^G\mathbf{d}_1 = {}^GT_1\,{}^1T_2\,{}^2\mathbf{r}_P
$$
$$
= {}^GT_2\,{}^2\mathbf{r}_P \tag{4.143}
$$

where

$$
^GT_1 = \begin{bmatrix} {}^GR_1 & {}^G\mathbf{d}_1 \\ 0 & 1 \end{bmatrix} \qquad {}^1T_2 = \begin{bmatrix} {}^1R_2 & {}^1\mathbf{d}_2 \\ 0 & 1 \end{bmatrix} \tag{4.144}
$$

and

$$
^GT_2 = \begin{bmatrix} {}^GR_1\,{}^1R_2 & {}^GR_1\,{}^1\mathbf{d}_2 + {}^G\mathbf{d}_1 \\ 0 & 1 \end{bmatrix} \tag{4.145}
$$

Example 117 End-effector of a $SCARA$ robot in a global frame. How to analyze relative DH homogenous matrices for a robotic arm. $SCARA$ robot is an important robotic arm.

Figure 4.13 depicts an R‖R‖P ($SCARA$) robot with a global coordinate frame $G(OXYZ)$ attached to the base link along with the coordinate frames $B_1(o_1x_1y_1z_1)$ and $B_2(o_2x_2y_2z_2)$ attached to link (1) and the tip of link (3).

The transformation matrix $^G T_1$ to map points in frame B_1 to the base frame G is

$$
^G T_1 = \begin{bmatrix} \cos\theta_1 & -\sin\theta_1 & 0 & l_1\cos\theta_1 \\ \sin\theta_1 & \cos\theta_1 & 0 & l_1\sin\theta_1 \\ 0 & 0 & 1 & 0 \\ 0 & 0 & 0 & 1 \end{bmatrix}
\tag{4.146}
$$

and the transformation matrix $^1 T_2$ to map points in frame B_2 to the frame B_1 is

$$
^1 T_2 = \begin{bmatrix} \cos\theta_2 & -\sin\theta_2 & 0 & l_2\cos\theta_2 \\ \sin\theta_2 & \cos\theta_2 & 0 & l_2\sin\theta_2 \\ 0 & 0 & 1 & -h \\ 0 & 0 & 0 & 1 \end{bmatrix}
\tag{4.147}
$$

Therefore, the transformation matrix $^G T_2$ to map points in the end-effector frame B_2 to the base frame G is

$$
^G T_2 = {}^G T_1 \, {}^1 T_2
\tag{4.148}
$$

$$
= \begin{bmatrix} c(\theta_1+\theta_2) & -s(\theta_1+\theta_2) & 0 & l_1 c\theta_1 + l_2 c(\theta_1+\theta_2) \\ s(\theta_1+\theta_2) & c(\theta_1+\theta_2) & 0 & l_1 s\theta_1 + l_2 s(\theta_1+\theta_2) \\ 0 & 0 & 1 & -h \\ 0 & 0 & 0 & 1 \end{bmatrix}
$$

The origin of the last frame is at $^2\mathbf{r}_{o_2} = \begin{bmatrix} 0 & 0 & 0 & 1 \end{bmatrix}^T$ in its local frame B_2. Hence, the position of o_2 in the base coordinate frame is at

$$
^G\mathbf{r}_2 = {}^G T_2 \, {}^2\mathbf{r}_{o_2} = \begin{bmatrix} l_1\cos\theta_1 + l_2\cos(\theta_1+\theta_2) \\ l_1\sin\theta_1 + l_2\sin(\theta_1+\theta_2) \\ -h \\ 1 \end{bmatrix}
\tag{4.149}
$$

Example 118 Object manipulation. Numerical example for motion of a rigid body and determining the coordinates of its corners after motion.

The geometry of a rigid body may be represented by an array of the homogenous coordinates of some specific points of the body expressed in a local coordinate frame. Figure 4.14a illustrates the configuration of a wedge. The coordinates of the corners of the wedge in its body frame B are collected in the matrix P, standing for Points

$$
^B P = \begin{bmatrix} 1 & 1 & 1 & 0 & 0 & 0 \\ 0 & 0 & 3 & 3 & 0 & 0 \\ 4 & 0 & 0 & 0 & 0 & 4 \\ 1 & 1 & 1 & 1 & 1 & 1 \end{bmatrix}
\tag{4.150}
$$

The configuration of the wedge after a rotation of -90 deg about the Z-axis and a translation of three units along the X-axis is shown in Fig. 4.14b. The new coordinates of the corners in the global frame G can be found by multiplying the homogenous transformation matrix $^G T_B$ by the P matrix.

$$
^G T_B = D_{X,3} R_{Z,-\pi}
$$

$$
= \begin{bmatrix} 1 & 0 & 0 & 3 \\ 0 & 1 & 0 & 0 \\ 0 & 0 & 1 & 0 \\ 0 & 0 & 0 & 1 \end{bmatrix} \begin{bmatrix} \cos-\pi & -\sin-\pi & 0 & 0 \\ \sin-\pi & \cos-\pi & 0 & 0 \\ 0 & 0 & 1 & 0 \\ 0 & 0 & 0 & 1 \end{bmatrix}
$$

$$
= \begin{bmatrix} 0 & 1 & 0 & 3 \\ -1 & 0 & 0 & 0 \\ 0 & 0 & 1 & 0 \\ 0 & 0 & 0 & 1 \end{bmatrix}
\tag{4.151}
$$

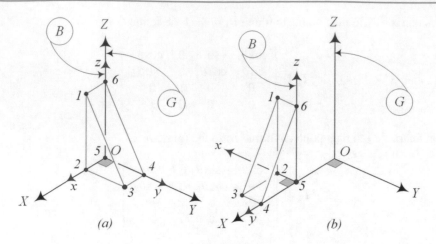

Fig. 4.14 Describing the motion of a rigid body in terms of some body points

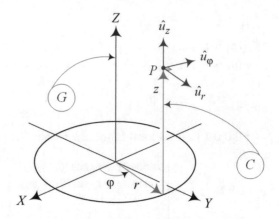

Fig. 4.15 Cylindrical coordinates of a point P

$$
{}^G P = {}^G T_B \, {}^B P = \begin{bmatrix} 0 & 1 & 0 & 3 \\ -1 & 0 & 0 & 0 \\ 0 & 0 & 1 & 0 \\ 0 & 0 & 0 & 1 \end{bmatrix} \begin{bmatrix} 1 & 1 & 1 & 0 & 0 & 0 \\ 0 & 0 & 3 & 3 & 0 & 0 \\ 4 & 0 & 0 & 0 & 0 & 4 \\ 1 & 1 & 1 & 1 & 1 & 1 \end{bmatrix}
$$

$$
= \begin{bmatrix} 3 & 3 & 6 & 6 & 3 & 3 \\ -1 & -1 & -1 & 0 & 0 & 0 \\ 4 & 0 & 0 & 0 & 0 & 4 \\ 1 & 1 & 1 & 1 & 1 & 1 \end{bmatrix} \tag{4.152}
$$

Example 119 Cylindrical coordinates. Cartesian–Cylindrical coordinate system transformation using homogenous transformation method.

A set of cylindrical coordinates, as shown in Fig. 4.15, can be achieved by a translation r along the X-axis, followed by a rotation φ about the Z-axis, and finally a translation z parallel to the Z-axis. Therefore, the homogenous transformation matrix to go from cylindrical coordinates $C(Or\varphi z)$ to Cartesian coordinates $G(OXYZ)$ is

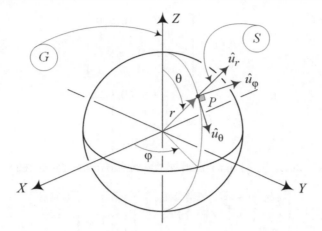

Fig. 4.16 Spherical coordinates of a point P

$${}^{G}T_C = D_{Z,z}\, R_{Z,\varphi}\, D_{X,r}$$

$$= \begin{bmatrix} 1 & 0 & 0 & 0 \\ 0 & 1 & 0 & 0 \\ 0 & 0 & 1 & z \\ 0 & 0 & 0 & 1 \end{bmatrix} \begin{bmatrix} \cos\varphi & -\sin\varphi & 0 & 0 \\ \sin\varphi & \cos\varphi & 0 & 0 \\ 0 & 0 & 1 & 0 \\ 0 & 0 & 0 & 1 \end{bmatrix} \begin{bmatrix} 1 & 0 & 0 & r \\ 0 & 1 & 0 & 0 \\ 0 & 0 & 1 & 0 \\ 0 & 0 & 0 & 1 \end{bmatrix}$$

$$= \begin{bmatrix} \cos\varphi & -\sin\varphi & 0 & r\cos\varphi \\ \sin\varphi & \cos\varphi & 0 & r\sin\varphi \\ 0 & 0 & 1 & z \\ 0 & 0 & 0 & 1 \end{bmatrix} \tag{4.153}$$

As an example, consider a point P at $(1, \frac{\pi}{3}, 2)$ in a cylindrical coordinate frame. Then, the Cartesian coordinates of P would be

$$\begin{bmatrix} \cos\frac{\pi}{3} & -\sin\frac{\pi}{3} & 0 & \cos\frac{\pi}{3} \\ \sin\frac{\pi}{3} & \cos\frac{\pi}{3} & 0 & \sin\frac{\pi}{3} \\ 0 & 0 & 1 & 2 \\ 0 & 0 & 0 & 1 \end{bmatrix} \begin{bmatrix} 0 \\ 0 \\ 0 \\ 1 \end{bmatrix} = \begin{bmatrix} 0.5 \\ 0.866 \\ 2.0 \\ 1.0 \end{bmatrix} \tag{4.154}$$

Example 120 Spherical coordinates. Cartesian–Spherical coordinate system transformation using homogenous transformation method.

A set of spherical coordinates, as shown in Fig. 4.16, can be achieved by a translation r along the Z-axis, followed by a rotation θ about the Y-axis, and finally a rotation φ about the Z-axis. Therefore, the homogenous transformation matrix to go from spherical coordinates $S(Or\theta\varphi)$ to Cartesian coordinates $G(OXYZ)$ is

$$\begin{aligned} {}^{G}T_S &= R_{Z,\varphi}\, R_{Y,\theta}\, D_{Z,r} \\[4pt] &= \begin{bmatrix} \cos\theta\cos\varphi & -\sin\varphi & \cos\varphi\sin\theta & r\cos\varphi\sin\theta \\ \cos\theta\sin\varphi & \cos\varphi & \sin\theta\sin\varphi & r\sin\theta\sin\varphi \\ -\sin\theta & 0 & \cos\theta & r\cos\theta \\ 0 & 0 & 0 & 1 \end{bmatrix} \end{aligned} \tag{4.155}$$

where

$$R_{Z,\varphi} = \begin{bmatrix} \cos\varphi & -\sin\varphi & 0 & 0 \\ \sin\varphi & \cos\varphi & 0 & 0 \\ 0 & 0 & 1 & 0 \\ 0 & 0 & 0 & 1 \end{bmatrix} \tag{4.156}$$

$$R_{Y,\theta} = \begin{bmatrix} \cos\theta & 0 & \sin\theta & 0 \\ 0 & 1 & 0 & 0 \\ -\sin\theta & 0 & \cos\theta & 0 \\ 0 & 0 & 0 & 1 \end{bmatrix} \qquad (4.157)$$

$$D_{Z,r} = \begin{bmatrix} 1 & 0 & 0 & 0 \\ 0 & 1 & 0 & 0 \\ 0 & 0 & 1 & r \\ 0 & 0 & 0 & 1 \end{bmatrix} \qquad (4.158)$$

As an example, consider a point P at $(2, \frac{\pi}{3}, \frac{\pi}{3})$ in a spherical coordinate frame. Then, the Cartesian coordinates of P would be

$$\begin{bmatrix} c\frac{\pi}{3}c\frac{\pi}{3} & -s\frac{\pi}{3} & c\frac{\pi}{3}s\frac{\pi}{3} & 2c\frac{\pi}{3}s\frac{\pi}{3} \\ c\frac{\pi}{3}s\frac{\pi}{3} & c\frac{\pi}{3} & s\frac{\pi}{3}s\frac{\pi}{3} & 2s\frac{\pi}{3}s\frac{\pi}{3} \\ -s\frac{\pi}{3} & 0 & c\frac{\pi}{3} & 2c\frac{\pi}{3} \\ 0 & 0 & 0 & 1 \end{bmatrix} \begin{bmatrix} 0 \\ 0 \\ 0 \\ 1 \end{bmatrix} = \begin{bmatrix} 0.866 \\ 1.5 \\ 1.0 \\ 1.0 \end{bmatrix} \qquad (4.159)$$

4.5 ★ Order-Free Transformation

A serial robotic manipulator with n controllable joints can be moved with any arbitrary order of joint actuations to achieve a desired set of relative joint variables. The final orientation and configuration of the links are independent of the order of activation of the joints. We introduce a theorem and develop order-free transformation matrices to show that under specific conditions, the order of transformations of such robotic multibody is immaterial.

Theorem Order-free transformations

Consider n connected links $B_1, B_2, \cdots, B_n$ such that the link B_1 carries the links $B_2, B_3, \cdots, B_n$. The link B_1 can move with respect to G with one DOF that can be a rotation θ_1 about a fixed axis in G or a translation d_1 along a fixed axis in G. The link B_2 carries the links $B_3, B_4, \cdots, B_n$ and moves with respect to B_1 that can be a rotation θ_2 about a fixed axis in B_1, or a translation d_2 along a fixed axis in B_1, and so on. The homogenous transformation matrix $^G T_n = {}^n T_G^{-1}$ is independent of the order of motions θ_1 or d_1, θ_2 or d_2, $\cdots$, θ_n or d_n.

Proof Let us consider a global frame G $(OXYZ)$ and two body frames B_1 $(Ox_1y_1z_1)$ and B_2 $(Ox_2y_2z_2)$. The body B_1 carries B_2 and moves with respect to G, and the body B_2 moves in B_1. The transformation ^1T_G of G to B_1 is a rotation ^1R_G plus a translation ^1D_G. The only variable of this transformation is either a rotation α or a translation d_1 about or along a globally fixed axis $^G\hat{u}_1$:

$$^1T_G = {}^1D_G \, {}^1R_G = {}^1D_{d_1} \, {}^1R_\alpha \qquad (4.160)$$

The second transformation $^2T_1$ of B_1 to B_2 is a rotation $^2R_1$ plus a translation $^2D_1$. The only variable of $^2T_1$ is either a rotation β or a translation d_2 about or along a fixed axis $^1\hat{u}_2$ in B_1:

$$^2T_1 = {}^2D_1 \, {}^2R_1 = {}^2D_{d_2} \, {}^2R_\beta = T_\beta \qquad (4.161)$$

To make ^1T_G an order-free transformation, we modify the first motion to be about or along a local axis $^2\hat{u}_2$ in B_2. To redefine ^1T_G, we move B_2 to B_1 by a translation $-d_2$ plus a rotation $-\beta$, both along and about their associated local axes in B_2. Now the axes of B_2 are the same as B_1 and ^1T_G can be performed by a rotation α about a local axis in B_2 plus a translation d_1 along a local axis B_2. Then, we apply the inverse rotation β plus an inverse translation d_2 about and along their axes to move B_2 back to its position. The resultant of these rotations and translations would be an order-free ^1T_G:

$$^1T_G = {}^2D_{d_2} \, {}^2R_\beta \, {}^2D_{d_1} \, {}^2R_\alpha \, {}^2R_{-\beta} \, {}^2D_{-d_2} \qquad (4.162)$$

$$= {}^2T_1 \, {}^1T_G \, {}^2T_1^{-1} = T_\alpha \qquad (4.163)$$

The transformation matrices (4.161) and (4.162) are indicating the order-free motions that can be performed in any order to calculate ^1T_G. The final transformation matrix will not be altered by changing the order of motions T_α and T_β. To examine this fact, let us assume that the frames B_1 and B_2 are on G. Applying the motion T_α and then the motion T_β provides

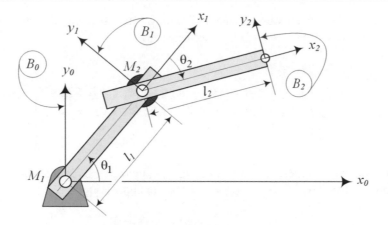

Fig. 4.17 A 2R or R∥R planar manipulator

$$
{}^{2}T_{G} = {}^{2}T_{1}{}^{1}T_{G} = T_{\beta} T_{\alpha}
$$
$$
= {}^{2}D_{d_{2}}{}^{2}R_{\beta} \left[{}^{2}D_{d_{2}}{}^{2}R_{\beta}{}^{1}D_{d_{1}}{}^{1}R_{\alpha}{}^{2}R_{-\beta}{}^{2}D_{-d_{2}} \right]
$$
$$
= {}^{2}D_{d_{2}}{}^{2}R_{\beta} \left[{}^{2}D_{d_{2}}{}^{2}R_{\beta}{}^{1}D_{d_{1}}{}^{1}R_{\alpha} \right]{}^{2}R_{-\beta}{}^{2}D_{-d_{2}}
$$
$$
= {}^{2}D_{d_{2}} \left[{}^{2}D_{d_{2}}{}^{2}R_{\beta}{}^{1}D_{d_{1}}{}^{1}R_{\alpha} \right]{}^{2}D_{-d_{2}}
$$
$$
= {}^{2}D_{d_{2}}{}^{2}R_{\beta}{}^{1}D_{d_{1}}{}^{1}R_{\alpha} \tag{4.164}
$$

By changing the order of motions and applying first T_{β} and then T_{α}, we find the same transformation matrix:

$$
{}^{2}T_{G} = {}^{1}T_{G}{}^{2}T_{1} = T_{\alpha} T_{\beta}
$$
$$
= \left[{}^{2}D_{d_{2}}{}^{2}R_{\beta}{}^{1}D_{d_{1}}{}^{1}R_{\alpha}{}^{2}R_{-\beta}{}^{2}D_{-d_{2}} \right]{}^{2}D_{d_{2}}{}^{2}R_{\beta}
$$
$$
= \left[{}^{2}D_{d_{2}}{}^{2}R_{\beta}{}^{1}D_{d_{1}}{}^{1}R_{\alpha} \right] \mathbf{I}
$$
$$
= {}^{2}D_{d_{2}}{}^{2}R_{\beta}{}^{1}D_{d_{1}}{}^{1}R_{\alpha} \tag{4.165}
$$

Now consider ${}^{3}T_{2}$ a transformation matrix for motion of B_{3} in B_{2} and ${}^{2}T_{G}$ a matrix for motion of B_{2} in G. Employing the same procedure, we can find the order-free transformation matrix ${}^{3}T_{G}$ as:

$$
{}^{3}T_{G} = {}^{3}T_{2}{}^{2}T_{G} = {}^{2}T_{G}{}^{3}T_{2} \tag{4.166}
$$

We can similarly expand and apply the proof of the theory of order-free transformations to any number of bodies that are similarly connected. ∎

Example 121 ★ Order-free transformations of a planar R∥R manipulator.

Figure 4.17 illustrates an R∥R planar manipulator with two parallel revolute joints and variables θ_{1} and θ_{2}. Links (1) and (2) are both R∥R(0) and their transformation matrices ${}^{0}T_{1}$, ${}^{1}T_{2}$ are

$$
{}^{0}T_{1} = \begin{bmatrix} \cos\theta_{1} & -\sin\theta_{1} & 0 & l_{1}\cos\theta_{1} \\ \sin\theta_{1} & \cos\theta_{1} & 0 & l_{1}\sin\theta_{1} \\ 0 & 0 & 1 & 0 \\ 0 & 0 & 0 & 1 \end{bmatrix} \tag{4.167}
$$

$$
{}^{1}T_{2} = \begin{bmatrix} \cos\theta_{2} & -\sin\theta_{2} & 0 & l_{2}\cos\theta_{2} \\ \sin\theta_{2} & \cos\theta_{2} & 0 & l_{2}\sin\theta_{2} \\ 0 & 0 & 1 & 0 \\ 0 & 0 & 0 & 1 \end{bmatrix} \tag{4.168}
$$

Employing $^0T_1$ and $^1T_2$, we can find the forward kinematics $^0T_2$ only with the assumption that first, the grounded actuator M_1 acts to change θ_1 and then, the second motor M_2 acts to change θ_2:

$$^0T_2 = {}^0T_1 \, {}^1T_2 \tag{4.169}$$

$$= \begin{bmatrix} c\,(\theta_1 + \theta_2) & -s\,(\theta_1 + \theta_2) & 0 & l_2 c\,(\theta_1 + \theta_2) + l_1 c\theta_1 \\ s\,(\theta_1 + \theta_2) & c\,(\theta_1 + \theta_2) & 0 & l_2 s\,(\theta_1 + \theta_2) + l_1 s\theta_1 \\ 0 & 0 & 1 & 0 \\ 0 & 0 & 0 & 1 \end{bmatrix}$$

However, the actuators M_1 and M_2 are independent and may act in any order. To find the order-free transformations, we redefine the transformation matrices $^0T_1$, $^1T_2$ to go from lower body to upper body:

$$^1T_0 = {}^0T_1^{-1} = \begin{bmatrix} \cos\theta_1 & \sin\theta_1 & 0 & -l_1 \\ -\sin\theta_1 & \cos\theta_1 & 0 & 0 \\ 0 & 0 & 1 & 0 \\ 0 & 0 & 0 & 1 \end{bmatrix} \tag{4.170}$$

$$^2T_1 = {}^1T_2^{-1} = \begin{bmatrix} \cos\theta_2 & \sin\theta_2 & 0 & -l_2 \\ -\sin\theta_2 & \cos\theta_2 & 0 & 0 \\ 0 & 0 & 1 & 0 \\ 0 & 0 & 0 & 1 \end{bmatrix} \tag{4.171}$$

Using $^0T_1$, $^1T_2$, the forward kinematics of the manipulator can be found as

$$^2T_0 = {}^2T_1 \, {}^1T_0 = {}^0T_2^{-1} \tag{4.172}$$

$$= \begin{bmatrix} \cos\,(\theta_1 + \theta_2) & \sin\,(\theta_1 + \theta_2) & 0 & -l_2 - l_1 \cos\theta_2 \\ -\sin\,(\theta_1 + \theta_2) & \cos\,(\theta_1 + \theta_2) & 0 & l_1 \sin\theta_2 \\ 0 & 0 & 1 & 0 \\ 0 & 0 & 0 & 1 \end{bmatrix}$$

We employ the rule $(RFDS)$ to redefine the homogenous transformation matrices $^0T_1$, and $^1T_2$. The transformation $^2T_1 = {}^2D_1 \, {}^2R_1$ is made by a rotation θ_2 of B_2 about the local axis z_2 plus a translation l_2 of B_2 along the local axis x_2:

$$^2T_1 = {}^2D_1 \, {}^2R_1 = D_{x,l_2} R_{z,\theta_2} \tag{4.173}$$

$$= R_{z,\alpha} = \begin{bmatrix} 1 & 0 & 0 & -l_2 \\ 0 & 1 & 0 & 0 \\ 0 & 0 & 1 & 0 \\ 0 & 0 & 0 & 1 \end{bmatrix} \begin{bmatrix} \cos\theta_2 & \sin\theta_2 & 0 & 0 \\ -\sin\theta_2 & \cos\theta_2 & 0 & 0 \\ 0 & 0 & 1 & 0 \\ 0 & 0 & 0 & 1 \end{bmatrix}$$

The other transformation $^1T_0 = {}^1D_0 \, {}^1R_0$ is made by a rotation θ_1 of B_1 about the local axis z_1 plus a translation l_1 of B_1 along the local axis x_1. To redefine $^1T_0$, we move B_2 to B_1 by a translation l_2 along the local axis x_2 plus a rotation θ_2 about the local axis z_2. Now the axes of B_2 are the same as B_1 and $^1T_0$ can be performed by a rotation θ_1 about the local axis z_2 plus a translation l_1 along the local axis x_2. Then, we apply the inverse rotation θ_2 about z_2 plus an inverse translation l_2 along x_2 to move B_2 back to its position. The resultant of these rotations and translations would be $^1T_0$:

$$^1T_0 = D_{x,l_2} R_{z,\theta_2} D_{x,l_1} R_{z_2,\theta_1} R_{z,-\theta_2} D_{x,-l_2} = {}^2T_1 \, {}^1T_0 \, {}^2T_1^{-1} \tag{4.174}$$

$$= \begin{bmatrix} \cos\theta_1 & \sin\theta_1 & 0 & l_2 \cos\theta_1 - l_1 \cos\theta_2 - l_2 \\ -\sin\theta_1 & \cos\theta_1 & 0 & l_1 \sin\theta_2 - l_2 \sin\theta_1 \\ 0 & 0 & 1 & 0 \\ 0 & 0 & 0 & 1 \end{bmatrix}$$

where

$$D_{x,-l_2} = \begin{bmatrix} 1 & 0 & 0 & l_2 \\ 0 & 1 & 0 & 0 \\ 0 & 0 & 1 & 0 \\ 0 & 0 & 0 & 1 \end{bmatrix}$$

(4.175)

$$R_{z,-\theta_2} = \begin{bmatrix} \cos-\theta_2 & \sin-\theta_2 & 0 & 0 \\ -\sin-\theta_2 & \cos-\theta_2 & 0 & 0 \\ 0 & 0 & 1 & 0 \\ 0 & 0 & 0 & 1 \end{bmatrix}$$

(4.176)

$$R_{z_2,\theta_1} = \begin{bmatrix} \cos\theta_1 & \sin\theta_1 & 0 & 0 \\ -\sin\theta_1 & \cos\theta_1 & 0 & 0 \\ 0 & 0 & 1 & 0 \\ 0 & 0 & 0 & 1 \end{bmatrix}$$

(4.177)

$$D_{x,l_1} = \begin{bmatrix} 1 & 0 & 0 & -l_1 \\ 0 & 1 & 0 & 0 \\ 0 & 0 & 1 & 0 \\ 0 & 0 & 0 & 1 \end{bmatrix}$$

(4.178)

$$R_{z,\theta_2} = \begin{bmatrix} \cos\theta_2 & \sin\theta_2 & 0 & 0 \\ -\sin\theta_2 & \cos\theta_2 & 0 & 0 \\ 0 & 0 & 1 & 0 \\ 0 & 0 & 0 & 1 \end{bmatrix}$$

(4.179)

$$D_{x,l_2} = \begin{bmatrix} 1 & 0 & 0 & -l_2 \\ 0 & 1 & 0 & 0 \\ 0 & 0 & 1 & 0 \\ 0 & 0 & 0 & 1 \end{bmatrix}$$

(4.180)

The matrices (4.171) and (4.174) are order free and we can check the result of their product with different orders:

$${}^2T_0 = {}^1T_0 \, {}^2T_1$$

(4.181)

$$= \begin{bmatrix} c\theta_1 & s\theta_1 & 0 & l_2c\theta_1 - l_1c\theta_2 - l_2 \\ -s\theta_1 & c\theta_1 & 0 & l_1s\theta_2 - l_2s\theta_1 \\ 0 & 0 & 1 & 0 \\ 0 & 0 & 0 & 1 \end{bmatrix} \begin{bmatrix} c\theta_2 & s\theta_2 & 0 & -l_2 \\ -s\theta_2 & c\theta_2 & 0 & 0 \\ 0 & 0 & 1 & 0 \\ 0 & 0 & 0 & 1 \end{bmatrix}$$

$$= \begin{bmatrix} c(\theta_1+\theta_2) & s(\theta_1+\theta_2) & 0 & -l_2 - l_1c\theta_2 \\ -s(\theta_1+\theta_2) & c(\theta_1+\theta_2) & 0 & l_1s\theta_2 \\ 0 & 0 & 1 & 0 \\ 0 & 0 & 0 & 1 \end{bmatrix}$$

$${}^2T_0 = {}^1T_0 \, {}^2T_1$$

(4.182)

$$= \begin{bmatrix} c\theta_2 & s\theta_2 & 0 & -l_2 \\ -s\theta_2 & c\theta_2 & 0 & 0 \\ 0 & 0 & 1 & 0 \\ 0 & 0 & 0 & 1 \end{bmatrix} \begin{bmatrix} c\theta_1 & s\theta_1 & 0 & 0c\theta_1 - l_1c0 - 0 \\ -s\theta_1 & c\theta_1 & 0 & l_1s0 - 0s\theta_1 \\ 0 & 0 & 1 & 0 \\ 0 & 0 & 0 & 1 \end{bmatrix}$$

$$= \begin{bmatrix} c(\theta_1+\theta_2) & s(\theta_1+\theta_2) & 0 & -l_2 - l_1c\theta_2 \\ -s(\theta_1+\theta_2) & c(\theta_1+\theta_2) & 0 & l_1s\theta_2 \\ 0 & 0 & 1 & 0 \\ 0 & 0 & 0 & 1 \end{bmatrix}$$

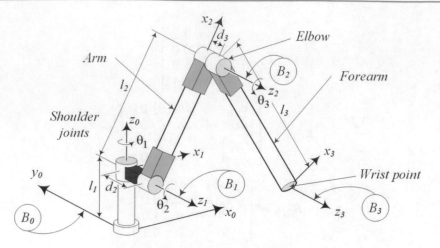

Fig. 4.18 An articulated manipulator

Example 122 ★ Order-free transformations of an articulated manipulator.

An articulated manipulator is any serial multibody with three links and three joints to reach a point in three dimensional space. However, an articulated arm is usually referred to a three-link multibody that is similar to a human hand with a shoulder, arm, and forearm such as the one shown in Fig. 4.18.

The first link of the arm is an R⊢R(90) with a distance l_1 between x_0 and x_1, and, therefore,

$$
{}^0T_1 = \begin{bmatrix} \cos\theta_1 & 0 & \sin\theta_1 & 0 \\ \sin\theta_1 & 0 & -\cos\theta_1 & 0 \\ 0 & 1 & 0 & l_1 \\ 0 & 0 & 0 & 1 \end{bmatrix} \tag{4.183}
$$

$$
{}^1T_0 = {}^0T_1^{-1} = \begin{bmatrix} \cos\theta_1 & \sin\theta_1 & 0 & 0 \\ 0 & 0 & 1 & -l_1 \\ \sin\theta_1 & -\cos\theta_1 & 0 & 0 \\ 0 & 0 & 0 & 1 \end{bmatrix} \tag{4.184}
$$

The second link of the arm is an R∥R(90) with a distance d_2 between x_1 and x_2 and a distance l_2 between z_1 and z_2:

$$
{}^1T_2 = \begin{bmatrix} \cos\theta_2 & -\sin\theta_2 & 0 & l_2\cos\theta_2 \\ \sin\theta_2 & \cos\theta_2 & 0 & l_2\sin\theta_2 \\ 0 & 0 & 1 & d_2 \\ 0 & 0 & 0 & 1 \end{bmatrix} \tag{4.185}
$$

$$
{}^2T_1 = {}^1T_2^{-1} = \begin{bmatrix} \cos\theta_2 & \sin\theta_2 & 0 & -l_2 \\ -\sin\theta_2 & \cos\theta_2 & 0 & 0 \\ 0 & 0 & 1 & -d_2 \\ 0 & 0 & 0 & 1 \end{bmatrix} \tag{4.186}
$$

The third link is not connected to any other link at its distal end, and we may set B_3 at its proximal end and attach a takht frame at its tip point. We may also attach only one coordinate frame at the tip point and consider the link as an R⊢R(90) with a distance d_3 between x_2 and x_3 and a distance l_3 between z_2 and z_3:

$$^2T_3 = \begin{bmatrix} \cos\theta_3 & -\sin\theta_3 & 0 & l_3\cos\theta_3 \\ \sin\theta_3 & \cos\theta_3 & 0 & l_3\sin\theta_3 \\ 0 & 0 & 1 & d_3 \\ 0 & 0 & 0 & 1 \end{bmatrix} \tag{4.187}$$

$$^3T_2 = {}^2T_3^{-1} = \begin{bmatrix} \cos\theta_3 & \sin\theta_3 & 0 & -l_3 \\ -\sin\theta_3 & \cos\theta_3 & 0 & 0 \\ 0 & 0 & 1 & -d_3 \\ 0 & 0 & 0 & 1 \end{bmatrix} \tag{4.188}$$

The final to base coordinate frame of the manipulator is

$$^0T_3 = {}^0T_1\,{}^1T_2\,{}^2T_3 \tag{4.189}$$

$$= \begin{bmatrix} c\,(\theta_2+\theta_3)\,c\theta_1 & -s\,(\theta_2+\theta_3)\,c\theta_1 & s\theta_1 & {}^0d_x \\ c\,(\theta_2+\theta_3)\,s\theta_1 & -s\,(\theta_2+\theta_3)\,s\theta_1 & -c\theta_1 & {}^0d_y \\ s\,(\theta_2+\theta_3) & c\,(\theta_2+\theta_3) & 0 & {}^0d_z \\ 0 & 0 & 0 & 1 \end{bmatrix}$$

$$\begin{bmatrix} {}^0d_x \\ {}^0d_y \\ {}^0d_z \end{bmatrix} = \begin{bmatrix} (l_2 c\theta_2 + l_3 c\,(\theta_2+\theta_3))\,c\theta_1 + (d_2+d_3)\,s\theta_1 \\ (l_2 c\theta_2 + l_3 c\,(\theta_2+\theta_3))\,s\theta_1 - (d_3+d_2)\,c\theta_1 \\ l_1 + l_2 s\theta_2 + l_3 s\,(\theta_2+\theta_3) \end{bmatrix} \tag{4.190}$$

or

$$^3T_0 = {}^3T_2\,{}^2T_1\,{}^1T_0 \tag{4.191}$$

$$= \begin{bmatrix} c\,(\theta_2+\theta_3)\,c\theta_1 & c\,(\theta_2+\theta_3)\,s\theta_1 & s\,(\theta_2+\theta_3) & {}^3d_x \\ -s\,(\theta_2+\theta_3)\,c\theta_1 & -s\,(\theta_2+\theta_3)\,s\theta_1 & c\,(\theta_2+\theta_3) & {}^3d_y \\ s\theta_1 & -c\theta_1 & 0 & {}^3d_z \\ 0 & 0 & 0 & 1 \end{bmatrix}$$

$$\begin{bmatrix} {}^3d_x \\ {}^3d_y \\ {}^3d_z \end{bmatrix} = \begin{bmatrix} -l_1 s\,(\theta_2+\theta_3) - l_2 c\theta_3 - l_3 \\ -l_1 c\,(\theta_2+\theta_3) + l_2 s\theta_3 \\ -d_2 - d_3 \end{bmatrix} \tag{4.192}$$

To determine the order-free transformations $^3T_2$, $^2T_1$, $^1T_0$, we begin with the final coordinate frame B_3. The transformation $^3T_2$ is already a motion about and along the axes of B_3. The order-free transformation of the second matrix is

$$^2T_1 = {}^3T_2\,{}^2T_1\,{}^3T_2^{-1} = {}^3T_2\,{}^2T_1\,{}^2T_3 \tag{4.193}$$

$$= \begin{bmatrix} \cos\theta_2 & \sin\theta_2 & 0 & l_3\cos\theta_2 - l_2\cos\theta_3 - l_3 \\ -\sin\theta_2 & \cos\theta_2 & 0 & l_2\sin\theta_3 - l_3\sin\theta_2 \\ 0 & 0 & 1 & -d_2 \\ 0 & 0 & 0 & 1 \end{bmatrix}$$

We may check to see that:

$$^3T_1 = {}^3T_2\,{}^2T_1 = {}^2T_1\,{}^3T_2 \tag{4.194}$$

$$= \begin{bmatrix} \cos\,(\theta_2+\theta_3) & \sin\,(\theta_2+\theta_3) & 0 & -l_3 - l_2\cos\theta_3 \\ -\sin\,(\theta_2+\theta_3) & \cos\,(\theta_2+\theta_3) & 0 & l_2\sin\theta_3 \\ 0 & 0 & 1 & -d_2 - d_3 \\ 0 & 0 & 0 & 1 \end{bmatrix}$$

However, when $^2T_1$ is the first motion, we must substitute the parameters l_3 and θ_3 of the second motion equal to zero.

The order-free transformation of the third matrix is

$$^1T_0 = \, ^3T_2 \, ^2T_1 \, ^1T_0 \, ^2T_1^{-1} \, ^3T_2^{-1} = \, ^3T_2 \, ^2T_1 \, ^1T_0 \, ^1T_2 \, ^2T_3 \tag{4.195}$$

$$= \frac{1}{2} \begin{bmatrix} c\gamma + c\theta_1 & s\gamma + s\theta_1 & 2s\,(\theta_2 + \theta_3) & ^1d_x \\ s\gamma - s\theta_1 & c\theta_1 - c\gamma & 2c\,(\theta_2 + \theta_3) & ^1d_y \\ 2s\,(\theta_1 - \theta_2 - \theta_3) & -2c\,(\theta_1 - \theta_2 - \theta_3) & 0 & ^1d_z \\ 0 & 0 & 0 & 1 \end{bmatrix}$$

$$\gamma = \theta_1 - 2\theta_2 - 2\theta_3 \tag{4.196}$$

where

$$^1d_x = 2\,(d_2 + d_3 - l_1)\sin(\theta_2 + \theta_3) + l_3\cos\theta_1 - 2l_2\cos\theta_3 - 2l_3$$
$$+ (l_2 + l_3)\cos(\theta_1 - 2\theta_2 - 2\theta_3) + l_2\cos(\theta_1 + \theta_3) \tag{4.197}$$

$$^1d_y = 2\,(d_2 + d_3 - l_1)\cos(\theta_2 + \theta_3) - l_3\sin\theta_1 + 2l_2\sin\theta_3$$
$$+ (l_2 + l_3)\sin(\theta_1 - 2\theta_2 - 2\theta_3) - l_2\sin(\theta_1 + \theta_3) \tag{4.198}$$

$$^1d_z = 2\,(l_3\sin(\theta_1 - \theta_2 - \theta_3) - d_3 - d_2 + l_2\sin(\theta_1 - \theta_2)) \tag{4.199}$$

We may check to see that the transformation $^3T_0 = \, ^3T_2 \, ^2T_1 \, ^1T_0$ can be found by multiplying $^3T_2, \, ^2T_1, \, ^1T_0$ in any order:

$$^3T_0 = \, ^3T_2 \, ^2T_1 \, ^1T_0 = \, ^2T_1 \, ^3T_2 \, ^1T_0$$
$$= \, ^3T_2 \, ^1T_0 \, ^2T_1 = \, ^1T_0 \, ^3T_2 \, ^2T_1$$
$$= \, ^2T_1 \, ^1T_0 \, ^3T_2 = \, ^1T_0 \, ^2T_1 \, ^3T_2 \tag{4.200}$$

It indicates that we can run the three motors of the manipulator in any arbitrary order. As long as the associated angles of rotation of the motors remain the same, the final configuration of the manipulator would be the same.

4.6 ★ Screw Coordinates

Any rigid body motion can be replaced by a single translation along an axis combined with a unique rotation about that axis. Such a motion is called *screw* motion. Consider the *screw* motion illustrated in Fig. 4.19. Screw is made by motion of a point P that rotates about the screw axis $\hat{u}$ and simultaneously translates along the same axis $\hat{u}$. Hence, any point on the *screw axis* moves along the axis, while any point off the axis moves along a *helix*.

The angular rotation of the rigid body about the screw is called *twist*. A screw motion is indicated by its *pitch*, p, that is the ratio of *translation*, h, to *rotation*, ϕ.

$$p = \frac{h}{\phi} \tag{4.201}$$

Hence, the rectilinear distance h through which the rigid body translates parallel to the axis of screw $\hat{u}$ for a unit rotation ϕ is the pitch p. If $p > 0$, then the screw is *right-handed*, and if $p < 0$, it is *left-handed*.

A screw motion $\check{s}$ is shown by $\check{s}(h, \phi, \hat{u}, \mathbf{s})$ and is indicated by a twist axis unit vector $\hat{u}$, a *location* vector $\mathbf{s}$, a twist angle ϕ, and a translation h (or a pitch p). The location vector $\mathbf{s}$ indicates the global position of a point on the screw axis. The twist angle ϕ, the twist axis $\hat{u}$, and the pitch p (or translation h) are called *screw parameters*. The screw motion is another transformation method to present a general motion of a rigid body. Screw motion may be expressed by homogenous matrix screw transformation, which is a combination of rotation and translation about the screw axis.

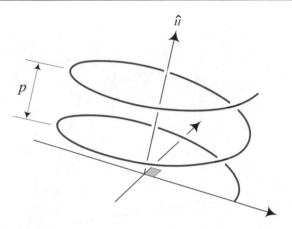

Fig. 4.19 A screw motion is translation along a line combined with a rotation about the line

If $\hat{u}$ passes through the origin of the coordinate frame, then $\mathbf{s} = 0$ and the screw motion is called *central screw* $\check{s}(h, \phi, \hat{u})$. A *reverse central screw* is defined as $\check{s}(-h, -\phi, \hat{u})$. A translation displacement $D_{\hat{u},h}$ along an axis $\hat{u}$, combined with an angular rotation $R_{\hat{u},\phi}$ about the same axis appears repeatedly in robotic application. For a central screw motion, we have

$$^{G}\check{s}_{B}(h, \phi, \hat{u}) = D_{\hat{u},h} \, R_{\hat{u},\phi} \tag{4.202}$$

where

$$D_{\hat{u},h} = \begin{bmatrix} 1 & 0 & 0 & hu_1 \\ 0 & 1 & 0 & hu_2 \\ 0 & 0 & 1 & hu_3 \\ 0 & 0 & 0 & 1 \end{bmatrix} \tag{4.203}$$

$$R_{\hat{u},\phi} = \begin{bmatrix} u_1^2 \operatorname{vers}\phi + c\phi & u_1 u_2 \operatorname{vers}\phi - u_3 s\phi & u_1 u_3 \operatorname{vers}\phi + u_2 s\phi & 0 \\ u_1 u_2 \operatorname{vers}\phi + u_3 s\phi & u_2^2 \operatorname{vers}\phi + c\phi & u_2 u_3 \operatorname{vers}\phi - u_1 s\phi & 0 \\ u_1 u_3 \operatorname{vers}\phi - u_2 s\phi & u_2 u_3 \operatorname{vers}\phi + u_1 s\phi & u_3^2 \operatorname{vers}\phi + c\phi & 0 \\ 0 & 0 & 0 & 1 \end{bmatrix} \tag{4.204}$$

and, hence,

$$^{G}\ddot{s}_{B}(h, \phi, \hat{u}) = \begin{bmatrix} u_1^2 \operatorname{vers}\phi + c\phi & u_1 u_2 \operatorname{vers}\phi - u_3 s\phi & u_1 u_3 \operatorname{vers}\phi + u_2 s\phi & hu_1 \\ u_1 u_2 \operatorname{vers}\phi + u_3 s\phi & u_2^2 \operatorname{vers}\phi + c\phi & u_2 u_3 \operatorname{vers}\phi - u_1 s\phi & hu_2 \\ u_1 u_3 \operatorname{vers}\phi - u_2 s\phi & u_2 u_3 \operatorname{vers}\phi + u_1 s\phi & u_3^2 \operatorname{vers}\phi + c\phi & hu_3 \\ 0 & 0 & 0 & 1 \end{bmatrix} \tag{4.205}$$

As a result, a central screw transformation matrix indicates a pure translation corresponds to $\phi = 0$, and a pure rotation corresponds to $h = 0$ (or $p = \infty$).

A general screw motion is expressed by a homogenous transformation matrix $^{G}T_{B}$,

$$^{G}T_{B} = {}^{G}\check{s}_{B}(h, \phi, \hat{u}, \mathbf{s}) = \begin{bmatrix} {}^{G}R_{B} & {}^{G}\mathbf{s} - {}^{G}R_{B} \, {}^{G}\mathbf{s} + h\hat{u} \\ 0 & 1 \end{bmatrix}$$

$$= \begin{bmatrix} {}^{G}R_{B} & {}^{G}\mathbf{d} \\ 0 & 1 \end{bmatrix} \tag{4.206}$$

where

$$^{G}R_{B} = \mathbf{I} \cos\phi + \hat{u}\hat{u}^{T} (1 - \cos\phi) + \tilde{u} \sin\phi \tag{4.207}$$

$$^{G}\mathbf{d} = \left((\mathbf{I} - \hat{u}\hat{u}^{T}) (1 - \cos\phi) - \tilde{u} \sin\phi \right) {}^{G}\mathbf{s} + h\hat{u} \tag{4.208}$$

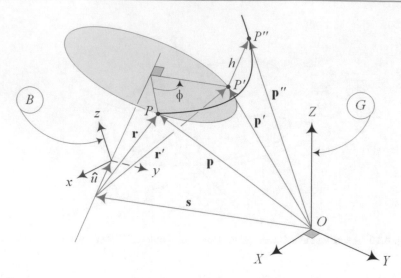

Fig. 4.20 Screw motion of a rigid body

When the screw motion is not central and $\hat{u}$ is not passing through the origin, and a screw motion to move a position vector $\mathbf{p}$ to $\mathbf{p}''$ is denoted by

$$\mathbf{p}'' = (\mathbf{p} - \mathbf{s}) \cos \phi + (1 - \cos \phi) \left(\hat{u} \cdot (\mathbf{p} - \mathbf{s}) \right) \hat{u}$$
$$+ \left(\hat{u} \times (\mathbf{p} - \mathbf{s}) \right) \sin \phi + \mathbf{s} + h\hat{u} \tag{4.209}$$

$$= {}^{G}R_{B} \left(\mathbf{p} - \mathbf{s} \right) + \mathbf{s} + h\hat{u} \tag{4.210}$$

$$= {}^{G}R_{B} \, \mathbf{p} + \mathbf{s} - {}^{G}R_{B} \, \mathbf{s} + h\hat{u} \tag{4.211}$$

and, therefore,

$$\mathbf{p}'' = \check{s}(h, \phi, \hat{u}, \mathbf{s}) \, \mathbf{p} = {}^{G}T_{B} \, \mathbf{p} \tag{4.212}$$

$${}^{G}\mathbf{p} = \check{s}(h, \phi, \hat{u}, \mathbf{s}) \, {}^{B}\mathbf{p} = {}^{G}T_{B} \, \mathbf{p} \tag{4.213}$$

where ${}^{G}T_{B}$ is the screw motion homogenous transformation matrix of Eq. (4.206).

The *location vector* ${}^{G}\mathbf{s}$ is the global position vector of the body frame B before screw motion. The vectors $\mathbf{p}''$ and $\mathbf{p}$ are positions of a point P after and before screw motion, as are shown in Fig. 4.20. The screw axis is indicated by the unit vector $\hat{u}$. Now a body point P moves from its first position to its second position P' by a rotation about $\hat{u}$. Then it moves to P'' by a translation h parallel to $\hat{u}$. The initial position of P is pointed by $\mathbf{p}$ and its final position is pointed by $\mathbf{p}''$.

Proof Let us break the screw motion to a twist ϕ about $\hat{u}$ and a translation h along $\hat{u}$. A body point P moves from its first position at $\mathbf{p}$ to its second position P' at $\mathbf{p}'$ by a rotation about $\hat{u}$. Then it moves to P'' at $\mathbf{p}''$ by a translation h parallel to $\hat{u}$.

The angle–axis rotation formula (3.5) relates $\mathbf{r}'$ and $\mathbf{r}$, which are position vectors of P after and before rotation ϕ about $\hat{u}$ when $\mathbf{s} = 0$, $h = 0$.

$$\mathbf{r}' = \mathbf{r} \cos \phi + (1 - \cos \phi) \left(\hat{u} \cdot \mathbf{r} \right) \hat{u} + \left(\hat{u} \times \mathbf{r} \right) \sin \phi \tag{4.214}$$

However, when the screw axis does not pass through the origin of $G(OXYZ)$, then $\mathbf{r}'$ and $\mathbf{r}$ must accordingly be substituted with the following equations:

$$\mathbf{r} = \mathbf{p} - \mathbf{s} \tag{4.215}$$

$$\mathbf{r}' = \mathbf{p}'' - \mathbf{s} - h\hat{u} \tag{4.216}$$

where $\mathbf{r}'$ is a vector after rotation and hence in G coordinate frame, and $\mathbf{r}$ is a vector before rotation and hence in B coordinate frame.

Therefore, the relationship between the new and old positions of the body point P after a screw motion is

$$\mathbf{p}'' = (\mathbf{p} - \mathbf{s})\cos\phi + (1 - \cos\phi)\left(\hat{u} \cdot (\mathbf{p} - \mathbf{s})\right)\hat{u}$$
$$+ \left(\hat{u} \times (\mathbf{p} - \mathbf{s})\right)\sin\phi + (\mathbf{s} + h\hat{u}) \tag{4.217}$$

Equation (4.217) is the *Rodriguez formula* for the most general rigid body motion. Defining new notations $^{G}\mathbf{p} = \mathbf{p}''$ and $^{B}\mathbf{p} = \mathbf{p}$ and also noting that $\mathbf{s}$ indicates a point on the rotation axis and therefore rotation does not affect $\mathbf{s}$, we may factor out $^{B}\mathbf{p}$ and write the Rodriguez formula in the following form:

$$^{G}\mathbf{p} = \left(\mathbf{I}\cos\phi + \hat{u}\hat{u}^{T}(1 - \cos\phi) + \tilde{u}\sin\phi\right){}^{B}\mathbf{p}$$
$$- \left(\mathbf{I}\cos\phi + \hat{u}\hat{u}^{T}(1 - \cos\phi) + \tilde{u}\sin\phi\right){}^{G}\mathbf{s} + {}^{G}\mathbf{s} + h\hat{u} \tag{4.218}$$

which can be rearranged to show that a screw can be represented by a homogenous transformation

$$^{G}\mathbf{p} = {}^{G}R_{B}\,{}^{B}\mathbf{p} + {}^{G}\mathbf{s} - {}^{G}R_{B}\,{}^{G}\mathbf{s} + h\hat{u} = {}^{G}R_{B}\,{}^{B}\mathbf{p} + {}^{G}\mathbf{d}$$
$$= {}^{G}T_{B}\,{}^{B}\mathbf{p} \tag{4.219}$$

$$^{G}T_{B} = {}^{G}\check{s}_{B}(h, \phi, \hat{u}, \mathbf{s}) \tag{4.220}$$
$$= \begin{bmatrix} {}^{G}R_{B} & {}^{G}\mathbf{s} - {}^{G}R_{B}\,{}^{G}\mathbf{s} + h\hat{u} \\ 0 & 1 \end{bmatrix} = \begin{bmatrix} {}^{G}R_{B} & {}^{G}\mathbf{d} \\ 0 & 1 \end{bmatrix}$$

where

$$^{G}R_{B} = \mathbf{I}\cos\phi + \hat{u}\hat{u}^{T}(1 - \cos\phi) + \tilde{u}\sin\phi \tag{4.221}$$
$$^{G}\mathbf{d} = \left((\mathbf{I} - \hat{u}\hat{u}^{T})(1 - \cos\phi) - \tilde{u}\sin\phi\right){}^{G}\mathbf{s} + h\hat{u} \tag{4.222}$$

Direct substitution shows that

$$^{G}R_{B} = \begin{bmatrix} u_{1}^{2}\text{ vers }\phi + c\phi & u_{1}u_{2}\text{ vers }\phi - u_{3}s\phi & u_{1}u_{3}\text{ vers }\phi + u_{2}s\phi \\ u_{1}u_{2}\text{ vers }\phi + u_{3}s\phi & u_{2}^{2}\text{ vers }\phi + c\phi & u_{2}u_{3}\text{ vers }\phi - u_{1}s\phi \\ u_{1}u_{3}\text{ vers }\phi - u_{2}s\phi & u_{2}u_{3}\text{ vers }\phi + u_{1}s\phi & u_{3}^{2}\text{ vers }\phi + c\phi \end{bmatrix} \tag{4.223}$$

$$^{G}\mathbf{d} =$$
$$\begin{bmatrix} hu_{1} - u_{1}(s_{3}u_{3} + s_{2}u_{2} + s_{1}u_{1})\text{ vers }\phi + (s_{2}u_{3} - s_{3}u_{2})s\phi + s_{1}\text{ vers }\phi \\ hu_{2} - u_{2}(s_{3}u_{3} + s_{2}u_{2} + s_{1}u_{1})\text{ vers }\phi + (s_{3}u_{1} - s_{1}u_{3})s\phi + s_{2}\text{ vers }\phi \\ hu_{3} - u_{3}(s_{3}u_{3} + s_{2}u_{2} + s_{1}u_{1})\text{ vers }\phi + (s_{1}u_{2} - s_{2}u_{1})s\phi + s_{3}\text{ vers }\phi \end{bmatrix} \tag{4.224}$$

This representation of a rigid motion requires six independent parameters, namely one for rotation angle ϕ, one for translation h, two for screw axis $\hat{u}$, and two for location vector $^{G}\mathbf{s}$. It is because three components of $\hat{u}$ are related to each other according to

$$\hat{u}^{T}\hat{u} = 1 \tag{4.225}$$

and the location vector $^{G}\mathbf{s}$ can locate any arbitrary point on the screw axis. It is convenient to choose the point where it has the minimum distance from O to make $^{G}\mathbf{s}$ perpendicular to u. Let us indicate the *shortest location vector* by $^{G}\mathbf{s}_{0}$, then there is a constraint among the components of the location vector.

$$^{G}\mathbf{s}_{0}^{T}\hat{u} = 0 \tag{4.226}$$

If $\mathbf{s} = 0$, then the screw axis passes through the origin of G and (4.220) reduces to (4.205).

The screw parameters ϕ and h, together with the screw axis $\hat{u}$ and location vector $^{G}\mathbf{s}$, completely define a rigid motion of $B(oxyz)$ in $G(OXYZ)$. It means, given the screw parameters and screw axis, we can find the elements of the transformation matrix by Eqs. (4.223) and (4.224). On the other hand, given the transformation matrix $^G T_B$, we can find the screw angle and axis by

$$\cos \phi = \frac{1}{2} \left(\text{tr} \left(^G R_B \right) - 1 \right) = \frac{1}{2} \left(\text{tr} \left(^G T_B \right) - 2 \right)$$

$$= \frac{1}{2} \left(r_{11} + r_{22} + r_{33} - 1 \right) \tag{4.227}$$

$$\tilde{u} = \frac{1}{2 \sin \phi} \left(^G R_B - ^G R_B^T \right) \tag{4.228}$$

hence,

$$\hat{u} = \frac{1}{2 \sin \phi} \begin{bmatrix} r_{32} - r_{23} \\ r_{13} - r_{31} \\ r_{21} - r_{12} \end{bmatrix} \tag{4.229}$$

To find all the required screw parameters, we must also find h and coordinates of one point on the screw axis. Since the points on the screw axis are invariant under the rotation, we must have

$$\begin{bmatrix} r_{11} & r_{12} & r_{13} & r_{14} \\ r_{21} & r_{22} & r_{23} & r_{24} \\ r_{31} & r_{32} & r_{33} & r_{34} \\ 0 & 0 & 0 & 1 \end{bmatrix} \begin{bmatrix} X \\ Y \\ Z \\ 1 \end{bmatrix} = \begin{bmatrix} 1 & 0 & 0 & hu_1 \\ 0 & 1 & 0 & hu_2 \\ 0 & 0 & 1 & hu_3 \\ 0 & 0 & 0 & 1 \end{bmatrix} \begin{bmatrix} X \\ Y \\ Z \\ 1 \end{bmatrix} \tag{4.230}$$

where (X, Y, Z) are coordinates of points on the screw axis.

As a sample point, we may find the intersection point of the screw line with (Y, Z)-plane, by setting $X_s = 0$ and searching for $\mathbf{s} = \begin{bmatrix} 0 & Y_s & Z_s \end{bmatrix}^T$. Therefore,

$$\begin{bmatrix} r_{11} - 1 & r_{12} & r_{13} & r_{14} - hu_1 \\ r_{21} & r_{22} - 1 & r_{23} & r_{24} - hu_2 \\ r_{31} & r_{32} & r_{33} - 1 & r_{34} - hu_3 \\ 0 & 0 & 0 & 0 \end{bmatrix} \begin{bmatrix} 0 \\ Y_s \\ Z_s \\ 1 \end{bmatrix} = \begin{bmatrix} 0 \\ 0 \\ 0 \\ 0 \end{bmatrix} \tag{4.231}$$

which generates three equations to be solved for Y_s, Z_s, and h.

$$\begin{bmatrix} h \\ Y_s \\ Z_s \end{bmatrix} = \begin{bmatrix} u_1 & -r_{12} & -r_{13} \\ u_2 & 1 - r_{22} & -r_{23} \\ u_3 & -r_{32} & 1 - r_{33} \end{bmatrix}^{-1} \begin{bmatrix} r_{14} \\ r_{24} \\ r_{34} \end{bmatrix} \tag{4.232}$$

Now we can find the shortest location vector $^G \mathbf{s}_0$ by

$$^G \mathbf{s}_0 = \mathbf{s} - (\mathbf{s} \cdot \hat{u}) \hat{u} \tag{4.233}$$

A screw motion $\check{s}(h, \phi, \hat{u}, \mathbf{s})$ has a line of action $\hat{u}$ at $^G \mathbf{s}$, a twist ϕ, and a translation h. The screw motion was first used by the Italian mathematician, Giulio Mozzi (1730–1813) in 1763 although the French mathematician Michel Chasles (1793–1880) is credited for this discovery because of his publication in 1830.

There are three principal central screws, namely the X-screw, Y-screw, Z-screw, respectively, having pitches p_X, p_Y, p_Z, and twist angles γ, β, α, which are

$$\check{s}(h_Z, \alpha, \hat{K}) = \begin{bmatrix} \cos \alpha & -\sin \alpha & 0 & 0 \\ \sin \alpha & \cos \alpha & 0 & 0 \\ 0 & 0 & 1 & p_Z \, \alpha \\ 0 & 0 & 0 & 1 \end{bmatrix} \tag{4.234}$$

$$\check{s}(h_Y, \beta, \hat{J}) = \begin{bmatrix} \cos \beta & 0 & \sin \beta & 0 \\ 0 & 1 & 0 & p_Y \beta \\ -\sin \beta & 0 & \cos \beta & 0 \\ 0 & 0 & 0 & 1 \end{bmatrix} \tag{4.235}$$

$$\check{s}(h_X, \gamma, \hat{I}) = \begin{bmatrix} 1 & 0 & 0 & p_X \gamma \\ 0 & \cos \gamma & -\sin \gamma & 0 \\ 0 & \sin \gamma & \cos \gamma & 0 \\ 0 & 0 & 0 & 1 \end{bmatrix} \tag{4.236}$$

∎

Example 123 ★ Euler–Chasles theorem. Definition and proof of Euler-Chasles theorem about rigid body motions.
Rigid body motion is well expressed by Euler and Michel Chasles (1793–1880) in a theorem.

Theorem 124 *A rigid body can be displaced from an initial position to another position by means of one translation and one rotation.*

Proof Let P_1 be a point of the rigid body B in its initial position B_1, and P_2 the corresponding point in the final position B_2. Let us first translate the body to an intermediate position B' so that P_1 falls on P_2. If the position B' is identical with B_2, then we have displaced the body by means of one translation, conformably to the requirements of the theorem. Let us assume that the position B' is different than B_2. Because the positions B' and B_2 of the rigid body have a common point P_2, then we can displace B' to B_2 by a rotation about an axis passing through P_2.
If we choose a different P_1, then in general we will have a different translation and a different rotation about a different axis. ∎

Expression of rigid body motion by screw motion may have any of the following expressions:
Every rigid body motion can be realized by a rotation about an axis combined with a translation parallel to the same axis.
The most general rigid body displacement can be produced by a translation along a line followed by a rotation about that line.
The displacement in the Euclidean space of a rigid body can be represented by a rotation along a unique axis and a translation parallel to the axis.

Example 125 ★ Alternative proof of Euler-Chasles theorem. Another proof for Euler-Chasles theorem on rigid body motions, using homogenous transformation matrix.
Let $[T]$ be an arbitrary spatial displacement, and decompose it into a rotation R about $\hat{u}$ and a translation D.

$$[T] = [D][R] \tag{4.237}$$

We may also decompose the translation $[D]$ into two components $[D_{\parallel}]$ and $[D_{\perp}]$, parallel and perpendicular to $\hat{u}$, respectively.

$$[T] = [D_{\parallel}][D_{\perp}][R] \tag{4.238}$$

Now $[D_{\perp}][R]$ is a planar motion and is therefore equivalent to some rotation $[R'] = [D_{\perp}][R]$ about an axis parallel to the rotation axis $\hat{u}$. This yields the decomposition $[T] = [D_{\parallel}][R']$. This decomposition completes the proof, since the axis of $[D_{\parallel}]$ can be taken equal to $\hat{u}$.

Example 126 ★ Poinsot theorem. Definition of Poinsot theorem about rigid body motion to be equivalent to screw motion.
Similar to the screw motion of kinematics of rigid body motions, there is a theorem about dynamics of rigid bodies, called Poinsot theorem.

Theorem 127 *Rigid body action is equivalent to a wrench on a screw, that is, a force along a unique line and a couple parallel to the line.*

Example 128 ★ Central screw transformation of a base unit vector. Numerical example for a central screw motion of a unit vector.

Consider two initially coincident frames $G(OXYZ)$ and $B(oxyz)$. The body performs a screw motion along the Y-axis for $h = 2$ cm and $\phi = 90$ deg. The position of a body point at $\begin{bmatrix} 1 & 0 & 0 & 1 \end{bmatrix}^T$ can be found by applying the central screw transformation.

$$\check{s}(h, \phi, \hat{u}) = \check{s}(2, \frac{\pi}{2}, \hat{J}) = D(2\hat{J}) R(\hat{J}, \frac{\pi}{2}) \tag{4.239}$$

$$= \begin{bmatrix} 1 & 0 & 0 & 0 \\ 0 & 1 & 0 & 2 \\ 0 & 0 & 1 & 0 \\ 0 & 0 & 0 & 1 \end{bmatrix} \begin{bmatrix} 0 & 0 & 1 & 0 \\ 0 & 1 & 0 & 0 \\ -1 & 0 & 0 & 0 \\ 0 & 0 & 0 & 1 \end{bmatrix} = \begin{bmatrix} 0 & 0 & 1 & 0 \\ 0 & 1 & 0 & 2 \\ -1 & 0 & 0 & 0 \\ 0 & 0 & 0 & 1 \end{bmatrix}$$

Therefore,

$$^G\hat{\imath} = \check{s}(2, \frac{\pi}{2}, \hat{J}) \, ^B\hat{\imath} \tag{4.240}$$

$$= \begin{bmatrix} 0 & 0 & 1 & 0 \\ 0 & 1 & 0 & 2 \\ -1 & 0 & 0 & 0 \\ 0 & 0 & 0 & 1 \end{bmatrix} \begin{bmatrix} 1 \\ 0 \\ 0 \\ 1 \end{bmatrix} = \begin{bmatrix} 0 \\ 2 \\ -1 \\ 1 \end{bmatrix}$$

The pitch of this screw is

$$p = \frac{h}{\phi} = \frac{2}{\pi/2} = \frac{4}{\pi} = 1.2732 \, \text{cm/rad} \tag{4.241}$$

Example 129 ★ Screw transformation of a point. Numerical example for a screw motion of a point of a rigid body.

Consider two initially parallel frames $G(OXYZ)$ and $B(oxyz)$. The body performs a screw motion along $X = 2$ and parallel to the Y-axis for $h = 2$ and $\phi = 90$ deg. Therefore, the body coordinate frame is at location $\mathbf{s} = \begin{bmatrix} 2 & 0 & 0 \end{bmatrix}^T$. The position of a body point at $^B\mathbf{r} = \begin{bmatrix} 3 & 0 & 0 & 1 \end{bmatrix}^T$ can be found by applying the screw transformation, which is

$$^G T_B = \begin{bmatrix} ^G R_B & \mathbf{s} - \,^G R_B \, \mathbf{s} + h\hat{u} \\ 0 & 1 \end{bmatrix} = \begin{bmatrix} 0 & 0 & 1 & 2 \\ 0 & 1 & 0 & 2 \\ -1 & 0 & 0 & 2 \\ 0 & 0 & 0 & 1 \end{bmatrix} \tag{4.242}$$

because

$$^G R_B = \begin{bmatrix} 0 & 0 & 1 \\ 0 & 1 & 0 \\ -1 & 0 & 0 \end{bmatrix} \qquad \mathbf{s} = \begin{bmatrix} 2 \\ 0 \\ 0 \end{bmatrix} \qquad \hat{u} = \begin{bmatrix} 0 \\ 1 \\ 0 \end{bmatrix} \tag{4.243}$$

Therefore, the position vector of $^G\mathbf{r}$ would be

$$^G\mathbf{r} = \,^G T_B \, ^B\mathbf{r} \tag{4.244}$$

$$= \begin{bmatrix} 0 & 0 & 1 & 2 \\ 0 & 1 & 0 & 2 \\ -1 & 0 & 0 & 2 \\ 0 & 0 & 0 & 1 \end{bmatrix} \begin{bmatrix} 3 \\ 0 \\ 0 \\ 1 \end{bmatrix} = \begin{bmatrix} 2 \\ 2 \\ -1 \\ 1 \end{bmatrix}$$

Example 130 ★ The screw motion of a cube. Numerical example for a screw motion of a rigid body and calculation coordinates of important points.

Consider the cubic rigid body of Fig. 4.21 that has a unit length and is at the corner of the first quadrant. If we turn the cube 45 deg about $\mathbf{u}$ and translating it by $^G\mathbf{d} = h\hat{u}$,

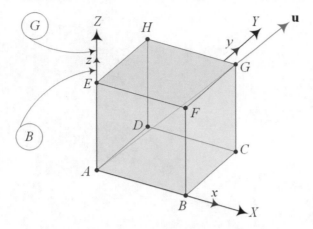

Fig. 4.21 A cubic rigid body at the corner of the first quadrant

$$\mathbf{u} = \begin{bmatrix} 1 & 1 & 1 \end{bmatrix}^T \tag{4.245}$$

$$h\hat{u} = \begin{bmatrix} 1 & 1 & 1 \end{bmatrix}^T \tag{4.246}$$

then the associated central screw motion would be

$${}^{G}T_B = {}^{G}\check{s}_B(h, \phi, \hat{u}) \tag{4.247}$$

$$= \begin{bmatrix} {}^{G}R_B & h\hat{u} \\ 0 & 1 \end{bmatrix} = \begin{bmatrix} {}^{G}R_B & {}^{G}\mathbf{d} \\ 0 & 1 \end{bmatrix}$$

where ${}^{G}R_B$ is

$${}^{G}R_B = \mathbf{I}\cos\phi + \hat{u}\hat{u}^T(1 - \cos\phi) + \tilde{u}\sin\phi$$

$$= \begin{bmatrix} 0.804\,74 & -0.310\,62 & 0.505\,88 \\ 0.505\,88 & 0.804\,74 & -0.310\,62 \\ -0.310\,62 & 0.505\,88 & 0.804\,74 \end{bmatrix} \tag{4.248}$$

$$\phi = \frac{\pi}{4} \qquad \hat{u} = \frac{\mathbf{u}}{\sqrt{3}} = \begin{bmatrix} 0.577\,35 \\ 0.577\,35 \\ 0.577\,35 \end{bmatrix} \tag{4.249}$$

Therefore, the central screw homogenous transformation is

$${}^{G}T_B = {}^{G}\check{s}_B(h, \phi, \hat{u}) =$$

$$= \begin{bmatrix} 0.804\,74 & -0.310\,62 & 0.505\,88 & 1 \\ 0.505\,88 & 0.804\,74 & -0.310\,62 & 1 \\ -0.310\,62 & 0.505\,88 & 0.804\,74 & 1 \\ 0 & 0 & 0 & 1 \end{bmatrix} \tag{4.250}$$

Figure 4.22 depicts the cube after the central screw ${}^{G}\check{s}_B(h, \phi, \hat{u})$.

Now suppose we would like to turn the cube about $\mathbf{u}$ and translate it by $h\hat{u} = \mathbf{u}$ where $\mathbf{u}$ is at:

$${}^{G}\mathbf{s} = \begin{bmatrix} 1 & 0 & 0 \end{bmatrix}^T \tag{4.251}$$

The screw of this motion has the same rotation matrix ${}^{G}R_B$ of (4.248) with a new translation vector ${}^{G}\mathbf{d}$.

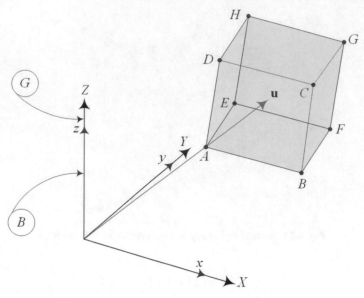

Fig. 4.22 A unit cube after a central screw $^G\breve{s}_B(h, \phi, \hat{u})$, where $h = 1, \phi = \frac{\pi}{4}, \hat{u} = \frac{1}{\sqrt{3}}\begin{bmatrix} 1 & 1 & 1 \end{bmatrix}^T$

$$
\begin{aligned}
{}^G\mathbf{d} &= {}^G\mathbf{s} - {}^GR_B\, {}^G\mathbf{s} + h\hat{u} \\
&= \left(\left(\mathbf{I} - \hat{u}\hat{u}^T\right)(1 - \cos\phi) - \tilde{u}\sin\phi\right){}^G\mathbf{s} + h\hat{u} \\
&= \begin{bmatrix} 1.1953 \\ 0.4941 \\ 1.3106 \end{bmatrix}
\end{aligned}
\tag{4.252}
$$

It provides the following central screw transformation.

$$
\begin{aligned}
{}^GT_B &= {}^G\breve{s}_B(h, \phi, \hat{u}, \mathbf{s}) = \begin{bmatrix} {}^GR_B & {}^G\mathbf{s} - {}^GR_B\, {}^G\mathbf{s} + h\hat{u} \\ 0 & 1 \end{bmatrix} \\[4pt]
&= \begin{bmatrix} 0.8047 & -0.3106 & 0.5059 & 1.1953 \\ 0.5059 & 0.8047 & -0.3106 & 0.4941 \\ -0.3106 & 0.5059 & 0.8047 & 1.3106 \\ 0 & 0 & 0 & 1 \end{bmatrix}
\end{aligned}
\tag{4.253}
$$

The local coordinates of the corners of the upper face are

	$^B\mathbf{r}_A$	$^B\mathbf{r}_B$	$^B\mathbf{r}_C$	$^B\mathbf{r}_D$	$^B\mathbf{r}_E$	$^B\mathbf{r}_F$	$^B\mathbf{r}_G$	$^B\mathbf{r}_H$
x	0	1	1	0	0	1	1	0
y	0	0	1	1	0	0	1	1
z	0	0	0	0	1	1	1	1

$$\tag{4.254}$$

which after the screw motion will be at $^G\mathbf{r} = {}^G\breve{s}_B(h, \phi, \hat{u}, \mathbf{s})\,^B\mathbf{r}$:

	$^B\mathbf{r}_A$	$^B\mathbf{r}_B$	$^B\mathbf{r}_C$	$^B\mathbf{r}_D$	$^G\mathbf{r}_E$	$^G\mathbf{r}_F$	$^G\mathbf{r}_G$	$^G\mathbf{r}_H$
X	1.195	2	1.689	0.885	1.701	2.506	2.195	1.390
Y	0.494	1	1.805	1.299	0.183	0.689	1.494	0.988
Z	1.311	1	1.506	1.816	2.115	1.804	2.311	2.62

$$\tag{4.255}$$

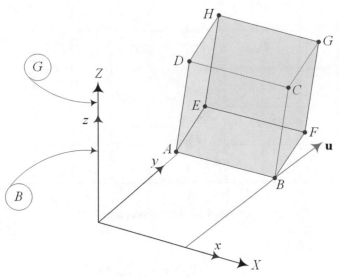

Fig. 4.23 A unit cube after a screw motion $^G\breve{s}_B(h, \phi, \hat{u})$, where $h = 1$, $\phi = \dfrac{\pi}{4}$, $\hat{u} = \dfrac{1}{\sqrt{3}}\begin{bmatrix} 1 & 1 & 1 \end{bmatrix}^T$, $\mathbf{s} = \begin{bmatrix} 1 & 0 & 0 \end{bmatrix}^T$

because:

$$
\begin{bmatrix}
0.804\,74 & -0.310\,62 & 0.505\,88 & 1.1953 \\
0.505\,88 & 0.804\,74 & -0.310\,62 & 0.49412 \\
-0.310\,62 & 0.505\,88 & 0.804\,74 & 1.3106 \\
0 & 0 & 0 & 1
\end{bmatrix}
$$

$$
\times
\begin{bmatrix}
0 & 1 & 1 & 0 & 0 & 1 & 1 & 0 \\
0 & 0 & 1 & 1 & 0 & 0 & 1 & 1 \\
0 & 0 & 0 & 0 & 1 & 1 & 1 & 1 \\
1 & 1 & 1 & 1 & 1 & 1 & 1 & 1
\end{bmatrix}
$$

$$
=
\begin{bmatrix}
1.195 & 2.0 & 1.689 & 0.885 & 1.701 & 2.506 & 2.195 & 1.391 \\
0.494 & 1 & 1.805 & 1.299 & 0.183 & 0.689 & 1.494 & 0.988 \\
1.311 & 1 & 1.506 & 1.816 & 2.115 & 1.805 & 2.312 & 2.621 \\
1 & 1 & 1 & 1 & 1 & 1 & 1 & 1
\end{bmatrix}
\tag{4.256}
$$

Figure 4.23 depicts the cube after the screw motion $^G\breve{s}_B(h, \phi, \hat{u}, \mathbf{s})$.

Example 131 ★ Rotation of a vector. Applying rigid body motion on two pints of a rigid body.

Transformation equation $^G\mathbf{r} = {}^GR_B \, {}^B\mathbf{r}$ and Rodriguez rotation formula (3.5) describe the rotation of any vector fixed in a rigid body. However, the vector can conveniently be described in terms of two points fixed in the body to derive screw equation.

A reference point P_1 with position vector $\mathbf{r}_1$ at the tail and a point P_2 with position vector $\mathbf{r}_2$ at the head define a vector in the rigid body. Then the transformation equation between body and global frames can be written as:

$$
^G(\mathbf{r}_2 - \mathbf{r}_1) = {}^GR_B \, {}^B(\mathbf{r}_2 - \mathbf{r}_1)
\tag{4.257}
$$

Assume the original and final positions of the reference point P_1 are along the rotation axis. Equation (4.257) can then be rearranged in a form suitable for calculating coordinates of the new position of point P_2 in a transformation matrix form

$$
\begin{aligned}
^G\mathbf{r}_2 &= {}^GR_B \, {}^B(\mathbf{r}_2 - \mathbf{r}_1) + {}^G\mathbf{r}_1 \\
&= {}^GR_B \, {}^B\mathbf{r}_2 + {}^G\mathbf{r}_1 - {}^GR_B \, {}^B\mathbf{r}_1 \\
&= {}^GT_B \, {}^B\mathbf{r}_2
\end{aligned}
\tag{4.258}
$$

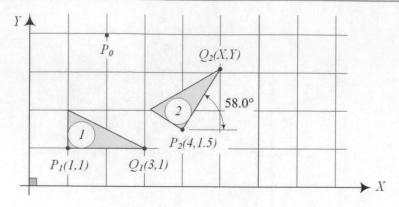

Fig. 4.24 Motion in a plane

where

$$^G T_B = \begin{bmatrix} ^G R_B \ ^G \mathbf{r}_1 - \ ^G R_B \ ^B \mathbf{r}_1 \\ 0 & 1 \end{bmatrix} \tag{4.259}$$

It is compatible with screw motion (4.220) for $h = 0$.

Example 132 ★ Special cases for screw determination. Screw axis for special cases of $\phi = 0$ and $\phi = \pi$.

There are two special cases for screws. The first one occurs when $r_{11} = r_{22} = r_{33} = 1$, then, $\phi = 0$ and the motion is a pure translation h parallel to $\hat{u}$, where

$$\hat{u} = \frac{r_{14} - s_1}{h} \hat{I} + \frac{r_{24} - s_2}{h} \hat{J} + \frac{r_{34} - s_3}{h} \hat{K} \tag{4.260}$$

As there is no unique screw axis in this case, we cannot locate any specific point on the screw axis.

The second special case occurs when $\phi = 180$ deg. In this case

$$\hat{u} = \begin{bmatrix} \sqrt{\frac{1}{2}(r_{11} + 1)} \\ \sqrt{\frac{1}{2}(r_{22} + 1)} \\ \sqrt{\frac{1}{2}(r_{33} + 1)} \end{bmatrix} \tag{4.261}$$

however, h and (X, Y, Z) can again be calculated from (4.232).

Example 133 ★ Pole of planar rotation and translation. Any planar rigid body motion may be interpreted by rotation about a point called pole. How to locate pole of a motion.

Assume a flat planar shape in (X, Y)-plane is displaced from position 1 to position 2 according to Fig. 4.24. New coordinates of point Q_2 are

$$\mathbf{r}_{Q_2} = \ ^2 R_1 \left(\mathbf{r}_{Q_1} - \mathbf{r}_{P_1} \right) + \mathbf{r}_{P_2} \tag{4.262}$$

$$= \begin{bmatrix} \cos 58 & -\sin 58 & 0 \\ \sin 58 & \cos 58 & 0 \\ 0 & 0 & 1 \end{bmatrix} \left(\begin{bmatrix} 3 \\ 1 \\ 0 \end{bmatrix} - \begin{bmatrix} 1 \\ 1 \\ 0 \end{bmatrix} \right) + \begin{bmatrix} 4 \\ 1.5 \\ 0 \end{bmatrix}$$

$$= \begin{bmatrix} 1.06 \\ 1.696 \\ 0 \end{bmatrix} + \begin{bmatrix} 4 \\ 1.5 \\ 0 \end{bmatrix} = \begin{bmatrix} 5.06 \\ 3.196 \\ 0.0 \end{bmatrix}$$

or equivalently

$$\mathbf{r}_{Q_2} = {}^2T_1 \, \mathbf{r}_{Q_1} = \begin{bmatrix} {}^2R_1 \, \mathbf{r}_{P_2} - {}^2R_1 \, \mathbf{r}_{P_1} \\ 0 \qquad\qquad 1 \end{bmatrix} \mathbf{r}_{Q_1} \tag{4.263}$$

$$= \begin{bmatrix} \cos 58 & -\sin 58 & 0 & 4.318 \\ \sin 58 & \cos 58 & 0 & 0.122 \\ 0 & 0 & 1 & 0 \\ 0 & 0 & 0 & 1 \end{bmatrix} \begin{bmatrix} 3 \\ 1 \\ 0 \\ 1 \end{bmatrix} = \begin{bmatrix} 5.06 \\ 3.196 \\ 0 \\ 1 \end{bmatrix}$$

In the planar motion of a rigid body, going from position 1 to position 2, there is always one point in the plane of motion that does not change its location. Hence, the body can be considered as having rotated about that point. The point is known as the *finite rotation pole*. The transformation matrix can be used to locate the pole. Figure 4.24 depicts a planar motion of a triangle. To locate the pole of motion, $P_0(X_0, Y_0)$, we need the transformation matrix of the motion. Using the data given in Fig. 4.24, we have

$$^2T_1 = \begin{bmatrix} {}^2R_1 \, \mathbf{r}_{P_2} - {}^2R_1 \, \mathbf{r}_{P_1} \\ 0 \qquad\qquad 1 \end{bmatrix} \tag{4.264}$$

$$= \begin{bmatrix} \cos\alpha & -\sin\alpha & 0 & -\cos\alpha + \sin\alpha + 4 \\ \sin\alpha & \cos\alpha & 0 & -\cos\alpha - \sin\alpha + 1.5 \\ 0 & 0 & 1 & 0 \\ 0 & 0 & 0 & 1 \end{bmatrix}$$

The location of the pole would be conserved under the transformation. Therefore,

$$\mathbf{r}_{P_0} = {}^2T_1 \, \mathbf{r}_{P_0} \tag{4.265}$$

$$\begin{bmatrix} X_0 \\ Y_0 \\ 0 \\ 1 \end{bmatrix} = \begin{bmatrix} \cos\alpha & -\sin\alpha & 0 & -\cos\alpha + \sin\alpha + 4 \\ \sin\alpha & \cos\alpha & 0 & -\cos\alpha - \sin\alpha + 1.5 \\ 0 & 0 & 1 & 0 \\ 0 & 0 & 0 & 1 \end{bmatrix} \begin{bmatrix} X_0 \\ Y_0 \\ 0 \\ 1 \end{bmatrix}$$

which for $\alpha = 58$ deg provides

$$X_0 = -1.5 \sin\alpha + 1 - 4\cos\alpha = 2.049 \tag{4.266}$$

$$Y_0 = 4 \sin\alpha + 1 - 1.5\cos\alpha = 3.956 \tag{4.267}$$

The location of the pole $P_0(X_0, Y_0)$ is indicated in Fig. 4.24.

Example 134 ★ Determination of screw parameters. How to determine screw parameters when motion of three points are given.

We are able to determine screw parameters when we have the original and final positions of three non-colinear points of a rigid body. Assume $\mathbf{p}_0, \mathbf{q}_0, \mathbf{r}_0$ denote the position of points P, Q, R before the screw motion, and $\mathbf{p}_1, \mathbf{q}_1$, and $\mathbf{r}_1$ denote their positions after the screw motion. To determine screw parameters, $\phi, \hat{u}, h, \mathbf{s}$, we should solve the following three simultaneous Rodriguez equations:

$$\mathbf{p}_1 - \mathbf{p}_0 = \tan\frac{\psi}{2}\hat{u} \times (\mathbf{p}_1 + \mathbf{p}_0 - 2\mathbf{s}) + h\hat{u} \tag{4.268}$$

$$\mathbf{q}_1 - \mathbf{q}_0 = \tan\frac{\phi}{2}\hat{u} \times (\mathbf{q}_1 + \mathbf{q}_0 - 2\mathbf{s}) + h\hat{u} \tag{4.269}$$

$$\mathbf{r}_1 - \mathbf{r}_0 = \tan\frac{\phi}{2}\hat{u} \times (\mathbf{r}_1 + \mathbf{r}_0 - 2\mathbf{s}) + h\hat{u} \tag{4.270}$$

Let begin by subtracting Eq. (4.270) from (4.268) and (4.269).

$$(\mathbf{p}_1 - \mathbf{p}_0) - (\mathbf{r}_1 - \mathbf{r}_0) = \tan\frac{\phi}{2}\hat{u} \times [(\mathbf{p}_1 + \mathbf{p}_0) - (\mathbf{r}_1 - \mathbf{r}_0)] \tag{4.271}$$

$$(\mathbf{q}_1 - \mathbf{q}_0) - (\mathbf{r}_1 - \mathbf{r}_0) = \tan\frac{\phi}{2}\hat{u} \times [(\mathbf{q}_1 + \mathbf{q}_0) - (\mathbf{r}_1 - \mathbf{r}_0)] \tag{4.272}$$

Multiplying both sides of (4.271) by $[(\mathbf{q}_1 - \mathbf{q}_0) - (\mathbf{r}_1 - \mathbf{r}_0)]$, which is perpendicular to $\hat{u}$

$$\begin{aligned}
&[(\mathbf{q}_1 - \mathbf{q}_0) - (\mathbf{r}_1 - \mathbf{r}_0)] \times [(\mathbf{p}_1 - \mathbf{p}_0) - (\mathbf{r}_1 - \mathbf{r}_0)] \\
&= \tan\frac{\phi}{2}[(\mathbf{q}_1 - \mathbf{q}_0) - (\mathbf{r}_1 - \mathbf{r}_0)] \times \{\hat{u} \times [(\mathbf{p}_1 + \mathbf{p}_0) - (\mathbf{r}_1 - \mathbf{r}_0)]\}
\end{aligned} \tag{4.273}$$

gives us

$$\begin{aligned}
&[(\mathbf{q}_1 - \mathbf{q}_0) - (\mathbf{r}_1 - \mathbf{r}_0)] \times [(\mathbf{p}_1 + \mathbf{p}_0) - (\mathbf{r}_1 - \mathbf{r}_0)] \\
&= \tan\frac{\phi}{2}[(\mathbf{q}_1 - \mathbf{q}_0) - (\mathbf{r}_1 - \mathbf{r}_0)] \cdot [(\mathbf{p}_1 + \mathbf{p}_0) - (\mathbf{r}_1 - \mathbf{r}_0)]\hat{u}
\end{aligned} \tag{4.274}$$

and, therefore, the rotation angle can be found by equating $\tan\frac{\phi}{2}$ and the norm of the right-hand side of the following equation:

$$\tan\frac{\phi}{2}\hat{u} = \frac{[(\mathbf{q}_1 - \mathbf{q}_0) - (\mathbf{r}_1 - \mathbf{r}_0)] \times [(\mathbf{p}_1 + \mathbf{p}_0) - (\mathbf{r}_1 - \mathbf{r}_0)]}{[(\mathbf{q}_1 - \mathbf{q}_0) - (\mathbf{r}_1 - \mathbf{r}_0)] \cdot [(\mathbf{p}_1 + \mathbf{p}_0) - (\mathbf{r}_1 - \mathbf{r}_0)]} \tag{4.275}$$

To find $\mathbf{s}$, we may start with the cross product of $\hat{u}$ with Eq. (4.268).

$$\begin{aligned}
\hat{u} \times (\mathbf{p}_1 - \mathbf{p}_0) &= \hat{u} \times \left[\tan\frac{\phi}{2}\hat{u} \times (\mathbf{p}_1 + \mathbf{p}_0 - 2\mathbf{s}) + h\hat{u}\right] \\
&= \tan\frac{\phi}{2}\{[\hat{u} \cdot (\mathbf{p}_1 + \mathbf{p}_0)]\hat{u} - (\mathbf{p}_1 + \mathbf{p}_0) \\
&\quad + 2[\mathbf{s} - (\hat{u} \cdot \mathbf{s})\hat{u}]\}
\end{aligned} \tag{4.276}$$

The term $\mathbf{s} - (\hat{u} \cdot \mathbf{s})\hat{u}$ is the component of $\mathbf{s}$ perpendicular to $\hat{u}$, where $\mathbf{s}$ is a vector from the origin of the global frame $G(OXYZ)$ to an arbitrary point on the screw axis. This perpendicular component indicates a vector with the shortest distance between O and $\hat{u}$. Let us assume $\mathbf{s}_0$ is the name of the shortest $\mathbf{s}$. Therefore,

$$\begin{aligned}
\mathbf{s}_0 &= \mathbf{s} - (\hat{u} \cdot \mathbf{s})\hat{u} \\
&= \frac{1}{2}\left[\frac{\hat{u} \times \mathbf{p}_1 - \mathbf{p}_0}{\tan\frac{\phi}{2}} - [\hat{u} \cdot (\mathbf{p}_1 + \mathbf{p}_0)]\hat{u} + \mathbf{p}_1 + \mathbf{p}_0\right]
\end{aligned} \tag{4.277}$$

The last parameter of the screw is the pitch h, which can be found from any one of the Eqs. (4.268), (4.269), or (4.270).

$$h = \hat{u} \cdot (\mathbf{p}_1 - \mathbf{p}_0) = \hat{u} \cdot (\mathbf{q}_1 - \mathbf{q}_0) = \hat{u} \cdot (\mathbf{r}_1 - \mathbf{r}_0) \tag{4.278}$$

Example 135 ★ Alternative derivation of screw transformation. Another proof of screw motion homogenous transformation expression.

Assume the screw axis does not pass through the origin of G. If $^G\mathbf{s}$ is the position vector of some point on the axis $\hat{u}$, then we can derive the matrix representation of screw $\check{s}(h, \phi, \hat{u}, \mathbf{s})$ by translating the screw axis back to the origin, performing the central screw motion, and translating the line back to its original position. This is how a general and a central screw are related.

$$\check{s}(h, \phi, \hat{u}, \mathbf{s}) = D(^G\mathbf{s})\ \check{s}(h, \phi, \hat{u})\ D(-\,^G\mathbf{s})$$

$$= D(^G\mathbf{s})\ D(h\hat{u})\ R(\hat{u}, \phi)\ D(-\,^G\mathbf{s})$$

$$= \begin{bmatrix} \mathbf{I} & ^G\mathbf{s} \\ 0 & 1 \end{bmatrix} \begin{bmatrix} ^GR_B & h\hat{u} \\ 0 & 1 \end{bmatrix} \begin{bmatrix} \mathbf{I} & -\,^G\mathbf{s} \\ 0 & 1 \end{bmatrix}$$

$$= \begin{bmatrix} ^GR_B & ^G\mathbf{s} - \,^GR_B\,^G\mathbf{s} + h\hat{u} \\ 0 & 1 \end{bmatrix} \tag{4.279}$$

Example 136 ★ Rotation about an off-center axis. Screw transformation for rotation about a non-central axis.

Rotation of a rigid body about an axis indicated by $\hat{u}$ and passing through a point at $^G\mathbf{s}$, where $^G\mathbf{s} \times \hat{u} \neq 0$, is a rotation about an off-center axis. The transformation matrix associated to an off-center rotation can be obtained from the screw transformation by setting $h = 0$. Therefore, an off-center rotation transformation is

$$^GT_B = \begin{bmatrix} ^GR_B & ^G\mathbf{s} - \,^GR_B\,^G\mathbf{s} \\ 0 & 1 \end{bmatrix} \tag{4.280}$$

Example 137 ★ Every rigid motion is a screw. An illustrative proof of rigid body motion to be expressed by screw motion.

To show that any proper rigid motion can be considered as a screw motion, we must show that a homogenous transformation matrix

$$^GT_B = \begin{bmatrix} ^GR_B & ^G\mathbf{d} \\ 0 & 1 \end{bmatrix} \tag{4.281}$$

can be written in the form of an screw motion transformation.

$$^GT_B = \begin{bmatrix} ^GR_B & (\mathbf{I} - \,^GR_B)\,\mathbf{s} + h\hat{u} \\ 0 & 1 \end{bmatrix} \tag{4.282}$$

This problem is then equivalent to the following equation to find h and $\hat{u}$.

$$^G\mathbf{d} = (\mathbf{I} - \,^GR_B)\,\mathbf{s} + h\hat{u} \tag{4.283}$$

The matrix $[\mathbf{I} - \,^GR_B]$ is singular because ^GR_B always has 1 as an eigenvalue. This eigenvalue corresponds to $\hat{u}$ as eigenvector. Therefore,

$$[\mathbf{I} - \,^GR_B]\hat{u} = [\mathbf{I} - \,^GR_B^T]\hat{u} = 0 \tag{4.284}$$

and an inner product shows that

$$\hat{u} \cdot \,^G\mathbf{d} = \hat{u} \cdot \left[\mathbf{I} - \,^GR_B\right]\mathbf{s} + \hat{u} \cdot h\hat{u}$$

$$= \left[\mathbf{I} - \,^GR_B\right]\hat{u} \cdot \mathbf{s} + \hat{u} \cdot h\hat{u} \tag{4.285}$$

which leads to:

$$h = \hat{u} \cdot \,^G\mathbf{d} \tag{4.286}$$

Now we may use h to find $\mathbf{s}$.

$$\mathbf{s} = \left[\mathbf{I} - \,^GR_B\right]^{-1} (^G\mathbf{d} - h\hat{u}) \tag{4.287}$$

Example 138 ★ Classification of motions of a rigid body. This is a summary of all possible type of rigid body motions and how to express them.

Imagine a body coordinate frame $B(oxyz)$ is moving with respect to a global frame $G(OXYZ)$. Point P with position vector $\mathbf{p}_B = \begin{bmatrix} p_1 & p_2 & p_3 \end{bmatrix}^T$ is an arbitrary point in $B(oxyz)$. The possible motions of $B(oxyz)$ and the required transformation between frames can be classified as:

1. Rotation ϕ about an axis passing through the origin and indicating by the unit vector $\hat{u} = \begin{bmatrix} u_1 & u_2 & u_3 \end{bmatrix}^T$

$$
{}^G\mathbf{p} = {}^G R_B \, {}^B\mathbf{p} \tag{4.288}
$$

where ${}^G R_B$ comes from the Rodriguez rotation formula (3.5).

2. Rotation ${}^G R_B$ plus a translation by ${}^G\mathbf{d} = \begin{bmatrix} d_1 & d_2 & d_3 \end{bmatrix}^T$.

$$
{}^G\mathbf{p} = {}^G R_B \, {}^B\mathbf{p} + {}^G\mathbf{d} \tag{4.289}
$$

3. Rotation ϕ about an axis on the unit vector $\hat{u} = \begin{bmatrix} u_1 & u_2 & u_3 \end{bmatrix}^T$ passing through an arbitrary point indicated by ${}^G\mathbf{s} = \begin{bmatrix} s_1 & s_2 & s_3 \end{bmatrix}^T$.

$$
{}^G\mathbf{p} = {}^G R_B \, {}^B\mathbf{p} + {}^G\mathbf{s} - {}^G R_B \, {}^G\mathbf{s} \tag{4.290}
$$

4. Screw motion with angle ϕ and displacement h, about and along an axis directed by $\hat{u} = \begin{bmatrix} u_1 & u_2 & u_3 \end{bmatrix}^T$ passing through an arbitrary point indicated by ${}^G\mathbf{s} = \begin{bmatrix} s_1 & s_2 & s_3 \end{bmatrix}^T$.

$$
{}^G\mathbf{p} = {}^G R_B \, {}^B\mathbf{p} + {}^G\mathbf{s} - {}^G R_B \, {}^G\mathbf{s} + h\hat{u} \tag{4.291}
$$

5. Reflection
   (a) in the (x, y)-plane:

$$
{}^G\mathbf{p} = {}^G R_B \, {}^B\mathbf{p}_{(-z)} \qquad \mathbf{p}_{(-z)} = \begin{bmatrix} p_1 \\ p_2 \\ -p_3 \end{bmatrix} \tag{4.292}
$$

   (b) in the (y, z)-plane:

$$
{}^G\mathbf{p} = {}^G R_B \, {}^B\mathbf{p}_{(-x)} \qquad \mathbf{p}_{(-z)} = \begin{bmatrix} -p_1 \\ p_2 \\ p_3 \end{bmatrix} \tag{4.293}
$$

   (c) in the xz-plane:

$$
{}^G\mathbf{p} = {}^G R_B \, {}^B\mathbf{p}_{(-y)} \qquad \mathbf{p}_{(-z)} = \begin{bmatrix} p_1 \\ -p_2 \\ p_3 \end{bmatrix} \tag{4.294}
$$

   (d) in a plane with equation $u_1 x + u_2 y + u_3 z + h = 0$;

$$
{}^G\mathbf{p} = \frac{1}{u_1^2 + u_2^2 + u_3^2} \left({}^G R_B \, {}^B\mathbf{p} - 2h\hat{u} \right)
$$

$$
= {}^G R_B \, {}^B\mathbf{p} - 2h\hat{u} \tag{4.295}
$$

   where

$$
{}^G R_B = \begin{bmatrix} -u_1^2 + u_2^2 + u_3^2 & -2u_1 u_2 & -2u_3 u_1 \\ -2u_2 u_1 & u_1^2 - u_2^2 + u_3^2 & -2u_2 u_3 \\ -2u_1 u_3 & -2u_3 u_2 & u_1^2 + u_2^2 - u_3^2 \end{bmatrix} \tag{4.296}
$$

   (e) in a plane going through the point $\begin{bmatrix} s_1 & s_2 & s_3 \end{bmatrix}^T$ and normal to $\hat{u} = \begin{bmatrix} u_1 & u_2 & u_3 \end{bmatrix}^T$,

$$
{}^G\mathbf{p} = {}^G R_B \, {}^B\mathbf{p} + {}^G\mathbf{s} - {}^G R_B \, {}^G\mathbf{s} \tag{4.297}
$$

   where ${}^G R_B$ is as in (4.296).

4.7 ★ Inverse Screw

Inverse of the general screw motion $\check{s}(h, \phi, \hat{u}, \mathbf{s})$ is defined by

$$
\begin{aligned}
{}^G\check{s}_B^{-1}(h, \phi, \hat{u}, \mathbf{s}) &= {}^B\check{s}_G(h, \phi, \hat{u}, \mathbf{s}) \\
&= \begin{bmatrix} {}^GR_B^T & {}^G\mathbf{s} - {}^GR_B^T\,{}^G\mathbf{s} - h\hat{u} \\ 0 & 1 \end{bmatrix}
\end{aligned}
\tag{4.298}
$$

where $\hat{u}$ is a unit vector indicating the axis of screw, $\mathbf{s}$ is the location vector of a point on the axis of screw, ϕ is the screw angle, and h is the screw translation. If the screw motion is central, the axis of screw passes through the origin and $\mathbf{s} = 0$. Therefore, the inverse of a central screw motion is

$$
{}^G\check{s}_B^{-1}(h, \phi, \hat{u}) = \begin{bmatrix} {}^GR_B^T & -h\hat{u} \\ 0 & 1 \end{bmatrix}
\tag{4.299}
$$

Proof The homogenous matrix expression of a screw motion $\check{s}(h, \phi, \hat{u}, \mathbf{s})$ is

$$
\begin{aligned}
{}^GT_B &= {}^G\check{s}_B(h, \phi, \hat{u}, \mathbf{s}) \\
&= \begin{bmatrix} {}^GR_B & {}^G\mathbf{s} - {}^GR_B\,{}^G\mathbf{s} + h\hat{u} \\ 0 & 1 \end{bmatrix}
\end{aligned}
\tag{4.300}
$$

A homogenous transformation matrix ${}^GT_B$

$$
{}^GT_B = \begin{bmatrix} {}^GR_B & {}^G\mathbf{d} \\ 0 & 1 \end{bmatrix}
\tag{4.301}
$$

can be inverted according to:

$$
{}^BT_G = {}^GT_B^{-1} = \begin{bmatrix} {}^GR_B^T & -{}^GR_B^T\,{}^G\mathbf{d} \\ 0 & 1 \end{bmatrix}
\tag{4.302}
$$

To show the correctness of Eq. (4.298), we need to calculate $-{}^GR_B^T\,{}^G\mathbf{d}$:

$$
\begin{aligned}
-{}^GR_B^T\,{}^G\mathbf{d} &= -{}^GR_B^T\left({}^G\mathbf{s} - {}^GR_B\,{}^G\mathbf{s} + h\hat{u}\right) \\
&= -{}^GR_B^T\,{}^G\mathbf{s} + {}^GR_B^T\,{}^GR_B\,{}^G\mathbf{s} - {}^GR_B^T h\hat{u} \\
&= -{}^GR_B^T\,{}^G\mathbf{s} + {}^G\mathbf{s} - {}^GR_B^T h\hat{u}
\end{aligned}
\tag{4.303}
$$

Because $\hat{u}$ is an invariant vector in both coordinate frames B and G, we have

$$
\hat{u} = {}^GR_B\,\hat{u} = {}^GR_B^T\,\hat{u}
\tag{4.304}
$$

and, therefore,

$$
-{}^GR_B^T\,{}^G\mathbf{d} = {}^G\mathbf{s} - {}^GR_B^T\,{}^G\mathbf{s} - h\hat{u}
\tag{4.305}
$$

This completes the inversion of a general screw motion:

$$
{}^G\check{s}_B^{-1}(h, \phi, \hat{u}, \mathbf{s}) = \begin{bmatrix} {}^GR_B^T & {}^G\mathbf{s} - {}^GR_B^T\,{}^G\mathbf{s} - h\hat{u} \\ 0 & 1 \end{bmatrix}
\tag{4.306}
$$

If the screw is central, the location vector $\mathbf{s}$ is zero and the inverse of the screw motion will be

$$
{}^G\check{s}_B^{-1}(h, \phi, \hat{u}) = \begin{bmatrix} {}^GR_B^T & -h\hat{u} \\ 0 & 1 \end{bmatrix}
\tag{4.307}
$$

As the inversion of a rotation matrix ${}^G R_B = R_{\hat{u},\phi}$ indicates a rotation $-\phi$ about $\hat{u}$

$$
{}^G R_B^{-1} = {}^G R_B^T = {}^B R_G = R_{\hat{u},-\phi} \tag{4.308}
$$

the inversion of a screw motion can also be interpreted as a rotation $-\phi$ about $\hat{u}$, plus a translation $-h$ along $\hat{u}$.

$$
{}^G \check{s}_B^{-1}(h, \phi, \hat{u}, \mathbf{s}) = \check{s}(-h, -\phi, \hat{u}, \mathbf{s}) \tag{4.309}
$$

To examine the screw inversion formula, we must have

$$
\check{s}(h, \phi, \hat{u}, \mathbf{s}) \, {}^G \check{s}_B^{-1}(h, \phi, \hat{u}, \mathbf{s}) = \mathbf{I}_4 \tag{4.310}
$$

where $\mathbf{I}_4$ is a 4 by 4 identity matrix. It can be examined by matrix multiplication.

$$
\begin{bmatrix} {}^G R_B & {}^G \mathbf{s} - {}^G R_B \, {}^G \mathbf{s} + h\hat{u} \\ 0 & 1 \end{bmatrix} \begin{bmatrix} {}^G R_B^T & {}^G \mathbf{s} - {}^G R_B^T \, {}^G \mathbf{s} - h\hat{u} \\ 0 & 1 \end{bmatrix}
$$
$$
= \begin{bmatrix} \mathbf{I}_3 & {}^G R_B \left({}^G \mathbf{s} - {}^G R_B^T \, {}^G \mathbf{s} - h\hat{u} \right) + \left({}^G \mathbf{s} - {}^G R_B \, {}^G \mathbf{s} + h\hat{u} \right) \\ 0 & 1 \end{bmatrix}
$$
$$
= \begin{bmatrix} \mathbf{I}_3 & {}^G R_B \, {}^G \mathbf{s} - {}^G \mathbf{s} - h \, {}^G R_B \, \hat{u} + {}^G \mathbf{s} - {}^G R_B \, {}^G \mathbf{s} + h\hat{u} \\ 0 & 1 \end{bmatrix}
$$
$$
= \begin{bmatrix} \mathbf{I}_3 & 0 \\ 0 & 1 \end{bmatrix} = \mathbf{I}_4 \tag{4.311}
$$

∎

Example 139 ★ Inversion of a central screw. Numerical example of inversion of a central screw.
Assume the unit cubic of Fig. 4.21 turns $\phi = 45$ deg about $\mathbf{u}$ and translates by ${}^G \mathbf{d} = h\hat{u}$.

$$
\mathbf{u} = \begin{bmatrix} 1 & 1 & 1 \end{bmatrix}^T \qquad h\hat{u} = \begin{bmatrix} 1 & 1 & 1 \end{bmatrix}^T \tag{4.312}
$$

The associated central screw motion is

$$
{}^G T_B = {}^G \check{s}_B(h, \phi, \hat{u}) = \begin{bmatrix} R_{\hat{u},\phi} & h\hat{u} \\ 0 & 1 \end{bmatrix} = \begin{bmatrix} {}^G R_B & {}^G \mathbf{d} \\ 0 & 1 \end{bmatrix}
$$
$$
= \begin{bmatrix} 0.8047 & -0.3106 & 0.5059 & 1 \\ 0.5059 & 0.8047 & -0.3106 & 1 \\ -0.3106 & 0.5059 & 0.8047 & 1 \\ 0 & 0 & 0 & 1 \end{bmatrix} \tag{4.313}
$$

The inverse of this screw motion is

$$
{}^G \check{s}_B^{-1}(h, \phi, \hat{u}) = \begin{bmatrix} {}^G R_B^T & -h\hat{u} \\ 0 & 1 \end{bmatrix}
$$
$$
= \begin{bmatrix} 0.8047 & 0.5059 & -0.3106 & -1 \\ -0.3106 & 0.8047 & 0.5059 & -1 \\ 0.5059 & -0.3106 & 0.8047 & -1 \\ 0 & 0 & 0 & 1 \end{bmatrix} \tag{4.314}
$$

We may check the inverse screw by a matrix multiplication.

$$
{}^G \check{s}_B(h, \phi, \hat{u}) \, {}^G \check{s}_B^{-1}(h, \phi, \hat{u}) = {}^G \check{s}_B^{-1}(h, \phi, \hat{u}) \, {}^G \check{s}_B(h, \phi, \hat{u}) = \mathbf{I}_4 \tag{4.315}
$$

Example 140 ★ Inversion of a general screw. Numerical example of inversion of a general screw.

Figure 4.23 shows the unit cubic of Fig. 4.21 after a rotation $\phi = 45$ deg about $^G\mathbf{u}$ and a translation $h\hat{u} = \mathbf{u}$ where $\mathbf{u}$ is at $^G\mathbf{s}$.

$$
{}^G\mathbf{u} = \begin{bmatrix} 1 \\ 1 \\ 1 \end{bmatrix} \qquad h\hat{u} = \begin{bmatrix} 1 \\ 1 \\ 1 \end{bmatrix} \qquad {}^G\mathbf{s} = \begin{bmatrix} 1 \\ 0 \\ 0 \end{bmatrix} \tag{4.316}
$$

The screw matrix of this motion is

$$
{}^G T_B = {}^G\breve{s}_B(h, \phi, \hat{u}, \mathbf{s})
$$

$$
= \begin{bmatrix} 0.8047 & -0.3106 & 0.5059 & 1.1953 \\ 0.5059 & 0.8047 & -0.3106 & 0.4941 \\ -0.3106 & 0.5059 & 0.8047 & 1.3106 \\ 0 & 0 & 0 & 1 \end{bmatrix} \tag{4.317}
$$

The inverse of this screw is

$$
{}^G\breve{s}_B^{-1}(h, \phi, \hat{u}, \mathbf{s}) = \begin{bmatrix} {}^G R_B^T & {}^G\mathbf{s} - {}^G R_B^T \, {}^G\mathbf{s} - h\hat{u} \\ 0 & 1 \end{bmatrix}
$$

$$
= \begin{bmatrix} 0.8047 & 0.5059 & -0.3106 & -0.8047 \\ -0.3106 & 0.8047 & 0.5059 & -0.6894 \\ 0.5059 & -0.3106 & 0.8047 & -1.5059 \\ 0 & 0 & 0 & 1 \end{bmatrix} \tag{4.318}
$$

We must also be able to turn the cube back to its original position by a rotation $\phi = -45$ deg about $^G\mathbf{u}$ and a translation $h\hat{u} = -\mathbf{u}$ where $\mathbf{u}$ is at $^G\mathbf{s}$. Such a screw motion would be

$$
{}^G T_B = \begin{bmatrix} 0.8047 & 0.5059 & -0.3106 & -0.8047 \\ -0.3106 & 0.8047 & 0.5059 & -0.6894 \\ 0.5059 & -0.3106 & 0.8047 & -1.5059 \\ 0 & 0 & 0 & 1 \end{bmatrix} \tag{4.319}
$$

because:

$$
{}^G R_B = \mathbf{I} \cos\phi + \hat{u}\hat{u}^T (1 - \cos\phi) + \tilde{u} \sin\phi
$$

$$
= \begin{bmatrix} 0.8047 & 0.5059 & -0.3106 \\ -0.3106 & 0.8047 & 0.5059 \\ 0.5059 & -0.3106 & 0.8047 \end{bmatrix} \tag{4.320}
$$

$$
{}^G\mathbf{s} - {}^G R_B \, {}^G\mathbf{s} + h\hat{u} = \begin{bmatrix} 1 \\ 0 \\ 0 \end{bmatrix} - {}^G R_B \begin{bmatrix} 1 \\ 0 \\ 0 \end{bmatrix} + \begin{bmatrix} -1 \\ -1 \\ -1 \end{bmatrix}
$$

$$
= \begin{bmatrix} -0.80474 \\ -0.68938 \\ -1.5059 \end{bmatrix} \tag{4.321}
$$

The screw (4.319) is the inverse of (4.317), so their multiplication is equal to $\mathbf{I}_4$.

4.8 ★ Combined Screw Transformation

Assume ${}^1\check{s}_2(h_1, \phi_1, \hat{u}_1, \mathbf{s}_1)$ is a screw motion to move from coordinate frame B_2 to B_1 and ${}^G\check{s}_1(h_0, \phi_0, \hat{u}_0, \mathbf{s}_0)$ is a screw motion to move from coordinate frame B_1 to G. Then, the combined screw motion ${}^G\check{s}_2(h, \phi, \hat{u}, \mathbf{s})$ to move from B_2 to G is

$$
{}^G\check{s}_2(h, \phi, \hat{u}, \mathbf{s}) = {}^G\check{s}_1(h_0, \phi_0, \hat{u}_0, \mathbf{s}_0) \, {}^1\check{s}_2(h_1, \phi_1, \hat{u}_1, \mathbf{s}_1) \tag{4.322}
$$

$$
= \begin{bmatrix} {}^G R_2 & {}^G R_1(\mathbf{I} - {}^1 R_2)\mathbf{s}_1 + (\mathbf{I} - {}^G R_1)\mathbf{s}_0 + h_1 \, {}^G R_1 \hat{u}_1 + h_0 \hat{u}_0 \\ 0 & 1 \end{bmatrix}
$$

Proof Direct substitution for ${}^1 s_2(h_1, \phi_1, \hat{u}_1)$ and ${}^G s_1(h_0, \phi_0, \hat{u}_0)$

$$
{}^G\check{s}_1(h_0, \phi_0, \hat{u}_0, \mathbf{s}_0) = \begin{bmatrix} {}^G R_1 & \mathbf{s}_0 - {}^G R_1 \mathbf{s}_0 + h_0 \hat{u}_0 \\ 0 & 1 \end{bmatrix} \tag{4.323}
$$

$$
{}^1\check{s}_2(h_1, \phi_1, \hat{u}_1, \mathbf{s}_1) = \begin{bmatrix} {}^1 R_2 & \mathbf{s}_1 - {}^1 R_2 \mathbf{s}_1 + h_1 \hat{u}_1 \\ 0 & 1 \end{bmatrix} \tag{4.324}
$$

shows that

$$
{}^G\check{s}_2(h, \phi, \hat{u}, \mathbf{s}) = {}^G\check{s}_1(h_0, \phi_0, \hat{u}_0, \mathbf{s}_0) \, {}^1\check{s}_2(h_1, \phi_1, \hat{u}_1, \mathbf{s}_1) \tag{4.325}
$$

$$
= \begin{bmatrix} {}^G R_1 & \mathbf{s}_0 - {}^G R_1 \mathbf{s}_0 + h_0 \hat{u}_0 \\ 0 & 1 \end{bmatrix} \begin{bmatrix} {}^1 R_2 & \mathbf{s}_1 - {}^1 R_2 \mathbf{s}_1 + h_1 \hat{u}_1 \\ 0 & 1 \end{bmatrix}
$$

$$
= \begin{bmatrix} {}^G R_2 & {}^G R_1\left(\mathbf{s}_1 - {}^1 R_2 \mathbf{s}_1 + h_1 \hat{u}_1\right) + \mathbf{s}_0 - {}^G R_1 \mathbf{s}_0 + h_0 \hat{u}_0 \\ 0 & 1 \end{bmatrix}
$$

$$
= \begin{bmatrix} {}^G R_2 & {}^G R_1(\mathbf{I} - {}^1 R_2)\mathbf{s}_1 + (\mathbf{I} - {}^G R_1)\mathbf{s}_0 + h_1 \, {}^G R_1 \hat{u}_1 + h_0 \hat{u}_0 \\ 0 & 1 \end{bmatrix}
$$

where

$$
{}^G R_2 = {}^G R_1 \, {}^1 R_2 \tag{4.326}
$$

To find the screw parameters of the combined screw motion ${}^G\check{s}_2(h, \phi, \hat{u}, \mathbf{s})$, we start by obtaining $\hat{u}$ and ϕ from ${}^G R_2$ based on (4.229) and (4.227). Then, utilizing (4.286) and (4.287) we can find h and $\mathbf{s}$

$$
h = \hat{u} \cdot {}^G\mathbf{d} \tag{4.327}
$$

$$
\mathbf{s} = \left[\mathbf{I} - {}^G R_2\right]^{-1}\left({}^G\mathbf{d} - h\hat{u}\right) \tag{4.328}
$$

where

$$
{}^G\mathbf{d} = {}^G R_1(\mathbf{I} - {}^1 R_2)\mathbf{s}_1 + (\mathbf{I} - {}^G R_1)\mathbf{s}_0 + h_1 \, {}^G R_1 \hat{u}_1 + h_0 \hat{u}_0
$$

$$
= \left({}^G R_1 - {}^G R_2\right)\mathbf{s}_1 + {}^G R_1\left(h_1 \hat{u}_1 - \mathbf{s}_0\right) + \mathbf{s}_0 + h_0 \hat{u}_0 \tag{4.329}
$$

∎

Example 141 ★ Exponential representation of a screw. Screw motion can be expressed by exponential function. This shows how.

To compute a rigid body motion associated with a screw, consider the motion of point P in Fig. 4.19. The final position of the point can be given by

$$
\mathbf{p}'' = \mathbf{s} + e^{\phi\tilde{u}}\mathbf{r} + h\hat{u} = \mathbf{s} + e^{\phi\tilde{u}}(\mathbf{p} - \mathbf{s}) + h\hat{u} = [T]\mathbf{p} \tag{4.330}
$$

where $[T]$ is the exponential representation of screw motion.

$$[T] = \begin{bmatrix} e^{\phi\tilde{u}} & (\mathbf{I} - e^{\phi\tilde{u}})\,\mathbf{s} + h\hat{u} \\ 0 & 1 \end{bmatrix} \tag{4.331}$$

The exponential screw transformation matrix (4.331) is based on the exponential form of the Rodriguez formula (3.187).

$$e^{\phi\tilde{u}} = I + \tilde{u}\sin\phi + \tilde{u}^2\,(1 - \cos\phi) \tag{4.332}$$

Therefore, the combination of two screws can also be found by

$$\begin{aligned}
[T] &= T_1 T_2 \\
&= \begin{bmatrix} e^{\phi_1\tilde{u}_1} & (\mathbf{I} - e^{\phi_1\tilde{u}_1})\,\mathbf{s}_1 + h_1\hat{u}_1 \\ 0 & 1 \end{bmatrix} \begin{bmatrix} e^{\phi_2\tilde{u}_2} & (\mathbf{I} - e^{\phi_2\tilde{u}_2})\,\mathbf{s}_2 + h_2\hat{u}_2 \\ 0 & 1 \end{bmatrix} \\
&= \begin{bmatrix} e^{\phi_1\tilde{u}_1 + \phi_2\tilde{u}_2} & {}^G\mathbf{d} \\ 0 & 1 \end{bmatrix}
\end{aligned} \tag{4.333}$$

where

$$^G\mathbf{d} = \left(e^{\phi_1\tilde{u}_1} - e^{\phi_1\tilde{u}_1 + \phi_2\tilde{u}_2} \right) \mathbf{s}_2 + e^{\phi_1\tilde{u}_1}\left(h_2\hat{u}_2 - \mathbf{s}_1 \right) + \mathbf{s}_1 + h_1\hat{u}_1 \tag{4.334}$$

Example 142 ★ Combination of two principal central screws. Combination of two simple principal screws. An illustration.

Combination of every two principal central screws can be found by matrix multiplication. As an example, a screw motion about Y followed by another screw motion about X is

$$\check{s}(h_X, \gamma, \hat{I})\,\check{s}(h_Y, \beta, \hat{J}) \tag{4.335}$$

$$= \begin{bmatrix} 1 & 0 & 0 & \gamma\,p_X \\ 0 & c\gamma & -s\gamma & 0 \\ 0 & s\gamma & c\gamma & 0 \\ 0 & 0 & 0 & 1 \end{bmatrix} \begin{bmatrix} c\beta & 0 & s\beta & 0 \\ 0 & 1 & 0 & \beta\,p_Y \\ -s\beta & 0 & c\beta & 0 \\ 0 & 0 & 0 & 1 \end{bmatrix}$$

$$= \begin{bmatrix} \cos\beta & 0 & \sin\beta & \gamma\,p_X \\ \sin\beta\sin\gamma & \cos\gamma & -\cos\beta\sin\gamma & \beta\,p_Y\cos\gamma \\ -\cos\gamma\sin\beta & \sin\gamma & \cos\beta\cos\gamma & \beta\,p_Y\sin\gamma \\ 0 & 0 & 0 & 1 \end{bmatrix}$$

Screw motion combination is not commutative and, therefore,

$$\check{s}(h_X, \gamma, \hat{I})\,\check{s}(h_Y, \beta, \hat{J}) \neq \check{s}(h_Y, \beta, \hat{J})\,\check{s}(h_X, \gamma, \hat{I}) \tag{4.336}$$

Example 143 ★ Decomposition of a screw. How to determine screw parameters from a given homogenous transformation matrix.

Every general screw can be decomposed into three principal central screws,

$$^G\check{s}_B(h, \phi, \hat{u}, \mathbf{s}) = \begin{bmatrix} {}^G R_B & \mathbf{s} - {}^G R_B\,\mathbf{s} + h\hat{u} \\ 0 & 1 \end{bmatrix}$$

$$= \check{s}(h_X, \gamma, \hat{I})\,\check{s}(h_Y, \beta, \hat{J})\,\check{s}(h_Z, \alpha, \hat{K}) \tag{4.337}$$

where

$$^G R_B = \begin{bmatrix} c\alpha c\beta & -c\beta s\alpha & s\beta \\ c\gamma s\alpha + c\alpha s\beta s\gamma & c\alpha c\gamma - s\alpha s\beta s\gamma & -c\beta s\gamma \\ s\alpha s\gamma - c\alpha c\gamma s\beta & c\alpha s\gamma + c\gamma s\alpha s\beta & c\beta c\gamma \end{bmatrix} \tag{4.338}$$

and

$$\mathbf{s} - {}^G R_B \mathbf{s} + h\hat{u} = \begin{bmatrix} \gamma p_X + \alpha p_Z \sin\beta \\ \beta p_Y \cos\gamma - \alpha p_Z \cos\beta \sin\gamma \\ \beta p_Y \sin\gamma + \alpha p_Z \cos\beta \cos\gamma \end{bmatrix} = \begin{bmatrix} d_X \\ d_Y \\ d_Z \end{bmatrix} \tag{4.339}$$

The twist angles α, β, γ can be found from ${}^G R_B$, then the pitches p_X, p_Y, p_Z can be found as follows.

$$p_Z = \frac{d_Z \cos\gamma - d_Y \sin\gamma}{\alpha \cos\beta} \tag{4.340}$$

$$p_Y = \frac{d_Z \sin\gamma + d_Y \cos\gamma}{\beta} \tag{4.341}$$

$$p_X = \frac{d_X}{\gamma} - \frac{d_Z \cos\gamma - d_Y \sin\gamma}{\gamma \cos\beta} \sin\beta \tag{4.342}$$

Example 144 ★ Decomposition of a screw to principal central screws. Any given screw motion can be reproduced by combining three principal screw motions in 6 ways.

In general, there are six different independent combinations of triple principal central screws and, therefore, there are six different methods to decompose a general screw into a combination of principal central screws. They are

$$
\begin{aligned}
&1 - \check{s}(h_X, \gamma, \hat{I})\,\check{s}(h_Y, \beta, \hat{J})\,\check{s}(h_Z, \alpha, \hat{K}) \\
&2 - \check{s}(h_Y, \beta, \hat{J})\,\check{s}(h_Z, \alpha, \hat{K})\,\check{s}(h_X, \gamma, \hat{I}) \\
&3 - \check{s}(h_Z, \alpha, \hat{K})\,\check{s}(h_X, \gamma, \hat{I})\,\check{s}(h_Y, \beta, \hat{J}) \\
&4 - \check{s}(h_Z, \alpha, \hat{K})\,\check{s}(h_Y, \beta, \hat{J})\,\check{s}(h_X, \gamma, \hat{I}) \\
&5 - \check{s}(h_Y, \beta, \hat{J})\,\check{s}(h_X, \gamma, \hat{I})\,\check{s}(h_Z, \alpha, \hat{K}) \\
&6 - \check{s}(h_X, \gamma, \hat{I})\,\check{s}(h_Z, \alpha, \hat{K})\,\check{s}(h_Y, \beta, \hat{J})
\end{aligned}
\tag{4.343}
$$

The expanded form of the six combinations of principal central screws are presented in Appendix C. It indicates that every screw can be decomposed into three principal central screws.

4.9   ★ The Plücker Line Coordinate

Plücker coordinates are a set of six coordinates used to define and show a directed line in space. Analytical representation of a line in space can be found if we have the position of two different points of that line. Such a line definition is more natural than the traditional methods and simplifies kinematics. Assume $P_1(X_1, Y_1, Z_1)$ and $P_2(X_2, Y_2, Z_2)$ at $\mathbf{r}_1$ and $\mathbf{r}_2$ are two different points on the line l as shown in Fig. 4.25.

Using the position vectors $\mathbf{r}_1$ and $\mathbf{r}_2$, the equation of line l can be defined by six elements of two vectors

$$l = \begin{bmatrix} \hat{u} \\ \boldsymbol{\rho} \end{bmatrix} = \begin{bmatrix} L \\ M \\ N \\ P \\ Q \\ R \end{bmatrix} \tag{4.344}$$

referred to as *Plücker coordinates* of the *directed* line l, where $\hat{u}$ is a unit vector along the line l referred to as *direction vector*,

$$\hat{u} = \frac{\mathbf{r}_2 - \mathbf{r}_1}{|\mathbf{r}_2 - \mathbf{r}_1|} = L\hat{I} + M\hat{J} + N\hat{K} \tag{4.345}$$

and $\boldsymbol{\rho}$ is the *moment vector* of $\hat{u}$ about the origin.

$$\boldsymbol{\rho} = \mathbf{r}_1 \times \hat{u} = P\hat{I} + Q\hat{J} + R\hat{K} \tag{4.346}$$

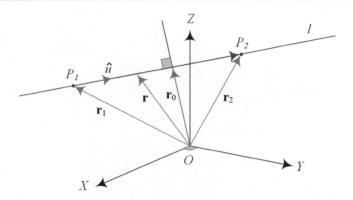

Fig. 4.25 A line indicated by two points

The Plücker method is a canonical representation of line definition and therefore is more efficient than the other methods such as parametric form, $l(t) = \mathbf{r}_1 + t\hat{u}$, point and direction form $(\mathbf{r}_1, \hat{u})$, or two-point representation form $(\mathbf{r}_1, \mathbf{r}_2)$.

Proof The unit vector $\hat{u}$, whose direction is along the line connecting P_1 and P_2, is

$$\hat{u} = \frac{\mathbf{r}_2 - \mathbf{r}_1}{|\mathbf{r}_2 - \mathbf{r}_1|} = L\hat{I} + M\hat{J} + N\hat{K}$$

$$= \frac{X_2 - X_1}{d}\hat{I} + \frac{Y_2 - Y_1}{d}\hat{J} + \frac{Z_2 - Z_1}{d}\hat{K} \tag{4.347}$$

where

$$L^2 + M^2 + N^2 = 1 \tag{4.348}$$

and the distance between P_1 and P_2 is

$$d = \sqrt{(X_2 - X_1)^2 + (Y_2 - Y_1)^2 + (Z_2 - Z_1)^2} \tag{4.349}$$

If $\mathbf{r}$ represents a vector from the origin O to a point on the line l, then the vector $\mathbf{r} - \mathbf{r}_1$ is parallel to $\hat{u}$ and therefore, the equation of the line can be written as

$$(\mathbf{r} - \mathbf{r}_1) \times \hat{u} = 0 \tag{4.350}$$

or equivalently

$$\mathbf{r} \times \hat{u} = \rho \tag{4.351}$$

where

$$\rho = \mathbf{r}_1 \times \hat{u} \tag{4.352}$$

ρ is the moment of the direction vector $\hat{u}$ about O. Because vectors ρ and $\hat{u}$ are perpendicular, there is a constraint among their components.

$$\hat{u} \cdot \rho = 0 \tag{4.353}$$

Expanding (4.346) yields

$$\rho = \begin{vmatrix} \hat{I} & \hat{J} & \hat{K} \\ X_1 & Y_1 & Z_1 \\ L & M & N \end{vmatrix} = P\hat{I} + Q\hat{J} + R\hat{K} \tag{4.354}$$

where

$$P = Y_1 N - Z_1 M$$

$$Q = Z_1 L - X_1 N \tag{4.355}$$

$$R = X_1 M - Y_1 L$$

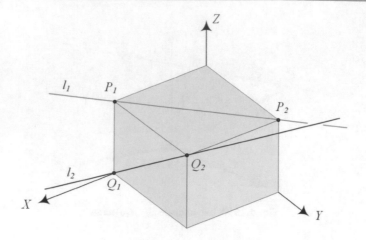

Fig. 4.26 A unit cube

and, therefore, the orthogonality condition (4.353) can be expressed as:

$$LP + MQ + NR = 0 \tag{4.356}$$

The Plücker coordinates of the line (4.344) have four independent coordinates because of the two constraints (4.348) and (4.356).

Our arrangement of Plücker coordinates in the form of (4.344) is the *line arrangement* and is called *ray coordinates*; however, sometimes the reverse order in *axis arrangement* $l = \begin{bmatrix} \boldsymbol{\rho} & \hat{u} \end{bmatrix}^T$ is also used by some other textbooks. In either case, a vertical line $\begin{bmatrix} \hat{u} \,|\, \boldsymbol{\rho} \end{bmatrix}^T$ or a semi-colon $\begin{bmatrix} \hat{u} \,;\, \boldsymbol{\rho} \end{bmatrix}^T$ may be utilized to separate the first three elements from the last three. Both arrangements can be used in kinematics efficiently.

The Plücker line coordinates $\begin{bmatrix} \hat{u} & \boldsymbol{\rho} \end{bmatrix}^T$ are homogenous because Eq. (4.346) shows that the coordinates are $\begin{bmatrix} w\hat{u} & w\boldsymbol{\rho} \end{bmatrix}^T$, where $w \in \mathbb{R}$, determines the same line.

Force–moment, angular velocity–translational velocity, and rigid motion act like a line vector and can be expressed by Plücker coordinates. Julius Plücker (1801–1868), a German mathematician and physicist, invented the line geometry in 1836.   ∎

Example 145 ★ Plücker coordinates of a line connecting two points. How to determine Plücker coordinates when positions of two points are given.

Plücker line coordinates of the line connecting points $P_1(1, 0, 0)$ and $P_2(0, 1, 1)$ are

$$l = \begin{bmatrix} \hat{u} \\ \boldsymbol{\rho} \end{bmatrix} = \begin{bmatrix} -1 & 1 & 1 & 0 & -1 & 1 \end{bmatrix}^T \tag{4.357}$$

because

$$\sqrt{3}\hat{u} = \frac{\mathbf{r}_2 - \mathbf{r}_1}{|\mathbf{r}_2 - \mathbf{r}_1|} = -\hat{I} + \hat{J} + \hat{K} \tag{4.358}$$

and

$$\sqrt{3}\boldsymbol{\rho} = \mathbf{r}_1 \times \sqrt{3}\hat{u} = -\hat{J} + \hat{K} \tag{4.359}$$

Example 146 ★ Plücker coordinates of diagonals of a cube. Numerical example of Plücker coordinates for diagonals of a cube.

Figure 4.26 depicts a unit cube and two lines on diagonals of two adjacent faces. Line l_1 connecting corners $P_1(1, 0, 1)$ and $P_2(0, 1, 1)$ is

$$l_1 = \begin{bmatrix} \hat{u}_1 \\ \boldsymbol{\rho}_1 \end{bmatrix} = \begin{bmatrix} -\frac{\sqrt{2}}{2} & \frac{\sqrt{2}}{2} & 0 & -\frac{\sqrt{2}}{2} & -\frac{\sqrt{2}}{2} & \frac{\sqrt{2}}{2} \end{bmatrix}^T \tag{4.360}$$

because

$$\hat{u}_1 = \frac{\mathbf{p}_2 - \mathbf{p}_1}{|\mathbf{p}_2 - \mathbf{p}_1|} = \frac{-\hat{I} + \hat{J}}{\sqrt{2}} \tag{4.361}$$

$$\boldsymbol{\rho}_1 = \mathbf{p}_1 \times \hat{u}_1 = \frac{-\hat{I} - \hat{J} + \hat{K}}{\sqrt{2}} \tag{4.362}$$

Line l_2 connecting corners $Q_1(1, 0, 0)$ and $Q_2(1, 1, 1)$ is

$$l_2 = \begin{bmatrix} \hat{u}_2 \\ \boldsymbol{\rho}_2 \end{bmatrix} = \begin{bmatrix} 0 & \frac{\sqrt{2}}{2} & \frac{\sqrt{2}}{2} & 0 & -\frac{\sqrt{2}}{2} & \frac{\sqrt{2}}{2} \end{bmatrix}^T \tag{4.363}$$

because

$$\hat{u}_2 = \frac{\mathbf{q}_2 - \mathbf{q}_1}{|\mathbf{q}_2 - \mathbf{q}_1|} = \frac{\hat{J} + \hat{K}}{\sqrt{2}} \tag{4.364}$$

$$\boldsymbol{\rho}_2 = \mathbf{q}_1 \times \hat{u}_2 = \frac{-\hat{J} + \hat{K}}{\sqrt{2}} \tag{4.365}$$

Example 147 ★ Grassmannian matrix to show Plücker coordinates. Grassmannian matrix is another method to express lines. It is connected to Plücker coordinates.

It can be verified that the Grassmannian matrix for coordinates of two points,

$$\begin{bmatrix} w_1 & X_1 & Y_1 & Z_1 \\ w_2 & X_2 & Y_2 & Z_2 \end{bmatrix} \tag{4.366}$$

is a short notation for the Plücker coordinates if we define Plücker coordinates as follows.

$$L = \begin{vmatrix} w_1 & X_1 \\ w_2 & X_2 \end{vmatrix} \qquad P = \begin{vmatrix} Y_1 & Z_1 \\ Y_2 & Z_2 \end{vmatrix} \tag{4.367}$$

$$M = \begin{vmatrix} w_1 & Y_1 \\ w_2 & Y_2 \end{vmatrix} \qquad Q = \begin{vmatrix} Z_1 & X_1 \\ Z_2 & X_2 \end{vmatrix} \tag{4.368}$$

$$N = \begin{vmatrix} w_1 & Z_1 \\ w_2 & Z_2 \end{vmatrix} \qquad R = \begin{vmatrix} X_1 & Y_1 \\ X_2 & Y_2 \end{vmatrix} \tag{4.369}$$

A German mathematician, Hermann Günther Grassmann (1809-1877), is credited for the development of a general calculus of vectors exposed in his masterpiece "The Theory of Linear Extension: A New Branch of Mathematics." In that contribution Grassmann proposed a new foundation for all of mathematics, showing that once geometry is put into the algebraic context he advocated, the number of possible dimensions in space is unbounded. Grassmann developed a ground breaking n-dimensional algebra of space. In his 1862, Grassmann reported a study about the reduction and decomposition of geometric structures into two components with applications to screw theory.

Example 148 ★ Ray–axis arrangement transformation. different expressions of Plücker coordinates.

It can be verified that the ray arrangement of Plücker coordinates,

$$l_{ray} = \begin{bmatrix} \hat{u} \\ \boldsymbol{\rho} \end{bmatrix} \tag{4.370}$$

can be transformed to the axis arrangement,

$$l_{axis} = \begin{bmatrix} \boldsymbol{\rho} \\ \hat{u} \end{bmatrix} \tag{4.371}$$

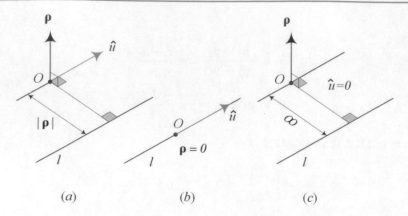

Fig. 4.27 Three cases of Plücker coordinates; (**a**) general case, (**b**) line through origin, (**c**) line at infinity

and vice versa utilizing the following 6×6 transformation matrix Δ.

$$\begin{bmatrix} \boldsymbol{\rho} \\ \hat{u} \end{bmatrix} = \Delta \begin{bmatrix} \hat{u} \\ \boldsymbol{\rho} \end{bmatrix} \tag{4.372}$$

$$\Delta = \begin{bmatrix} 0 & 0 & 0 & 1 & 0 & 0 \\ 0 & 0 & 0 & 0 & 1 & 0 \\ 0 & 0 & 0 & 0 & 0 & 1 \\ 1 & 0 & 0 & 0 & 0 & 0 \\ 0 & 1 & 0 & 0 & 0 & 0 \\ 0 & 0 & 1 & 0 & 0 & 0 \end{bmatrix} = \begin{bmatrix} 0 & \mathbf{I}_3 \\ \mathbf{I}_3 & 0 \end{bmatrix} \tag{4.373}$$

The transformation matrix Δ is symmetric and satisfies the following equations:

$$\Delta^2 = \Delta\Delta = \mathbf{I} \qquad \Delta^T = \Delta \tag{4.374}$$

Example 149 ★ Classification of Plücker coordinates. Graphical expression of special cases of Plücker coordinates.

There are three cases of Plücker coordinates as shown in Fig. 4.27. They are: general case, line through origin, and line at infinity.

The general case of $l = \begin{bmatrix} \hat{u} & \boldsymbol{\rho} \end{bmatrix}^T$, illustrated in Fig. 4.27a, happens when both $\hat{u}$ and $\boldsymbol{\rho}$ are nonzero. The direction vector $\hat{u}$ is parallel to the line, $\boldsymbol{\rho}$ is normal to the plane including the origin and the line, and $|\boldsymbol{\rho}|$ gives the distance from the origin to the line l.

Line through origin $l = \begin{bmatrix} \hat{u} & \mathbf{0} \end{bmatrix}^T$, illustrated in Fig. 4.27b, happens when the line l passes through the origin and $\boldsymbol{\rho}$ is zero.

Line at infinity $l = \begin{bmatrix} 0 & \boldsymbol{\rho} \end{bmatrix}^T$, illustrated in Fig. 4.27c, happens when the distance of the line l from origin tends to infinity. In this case we may assume $\hat{u}$ is zero. When the line l is at infinity, it is better to redefine the Plücker coordinates by normalizing the moment vector.

$$l = \begin{bmatrix} \dfrac{\hat{u}}{|\boldsymbol{\rho}|} & \dfrac{\boldsymbol{\rho}}{|\boldsymbol{\rho}|} \end{bmatrix}^T \tag{4.375}$$

Therefore, the direction components of the line l tend to zero by increasing the distance, while the moment components remain finite.

No line is defined by the zero Plücker coordinates $\begin{bmatrix} 0 & 0 & 0 & 0 & 0 & 0 \end{bmatrix}^T$.

Example 150 ★ Transformation of a line vector. How to employ Plücker coordinates to show motion of a vector.

Consider the line ${}^B l$ in Fig. 4.28 that is defined in a local frame $B(oxyz)$ by

$$ {}^B l = \begin{bmatrix} {}^B \hat{u} \\ {}^B \boldsymbol{\rho} \end{bmatrix} = \begin{bmatrix} {}^B \hat{u} \\ {}^B \mathbf{r}_P \times {}^B \hat{u} \end{bmatrix} \tag{4.376}$$

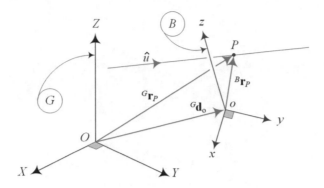

Fig. 4.28 A line vector in B and G frames

where $\hat{u}$ is a unit vector parallel to the line l, and P is any point on the line. The Plücker coordinates of the line in the global frame $G(OXYZ)$ is expressed by

$$^G l = \begin{bmatrix} ^G\hat{u} \\ ^G\boldsymbol{\rho} \end{bmatrix} = \begin{bmatrix} ^G\hat{u} \\ ^G\mathbf{r}_P \times {}^G\hat{u} \end{bmatrix} \tag{4.377}$$

where

$$^G\hat{u} = {}^G R_B \, {}^B\hat{u} \tag{4.378}$$

and $^G\boldsymbol{\rho}$ is the moment of $^G\hat{u}$ about O.

$$\begin{aligned} ^G\boldsymbol{\rho} &= {}^G\mathbf{r}_P \times {}^G\hat{u} = ({}^G\mathbf{d}_o + {}^G R_B \, {}^B\mathbf{r}_P) \times {}^G R_B \, {}^B\hat{u} \\ &= {}^G\mathbf{d}_o \times {}^G R_B \, {}^B\hat{u} + {}^G R_B({}^B\mathbf{r}_P \times {}^B\hat{u}) \\ &= {}^G\mathbf{d}_o \times {}^G R_B \, {}^B\hat{u} + {}^G R_B \, {}^B\boldsymbol{\rho} \\ &= {}^G\tilde{\mathbf{d}}_o \, {}^G R_B \, {}^B\hat{u} + {}^G R_B \, {}^B\boldsymbol{\rho} \end{aligned} \tag{4.379}$$

The 6×1 Plücker coordinates $\begin{bmatrix} \hat{u} & \boldsymbol{\rho} \end{bmatrix}^T$ for a line vector can be transformed from a frame B to another frame G by a 6×6 transformation matrix $^G\Gamma_B$ defined as

$$^G l = {}^G\Gamma_B \, {}^B l \tag{4.380}$$

$$\begin{bmatrix} ^G\hat{u} \\ ^G\boldsymbol{\rho} \end{bmatrix} = {}^G\Gamma_B \begin{bmatrix} ^B\hat{u} \\ ^B\boldsymbol{\rho} \end{bmatrix} \tag{4.381}$$

$$^G\Gamma_B = \begin{bmatrix} ^G R_B & 0 \\ ^G\tilde{\mathbf{d}}_o \, {}^G R_B & {}^G R_B \end{bmatrix} \tag{4.382}$$

where

$$^G R_B = \begin{bmatrix} r_{11} & r_{12} & r_{13} \\ r_{21} & r_{22} & r_{23} \\ r_{31} & r_{32} & r_{33} \end{bmatrix} \qquad ^G\tilde{\mathbf{d}}_o = \begin{bmatrix} 0 & -d_3 & d_2 \\ d_3 & 0 & -d_1 \\ -d_2 & d_1 & 0 \end{bmatrix} \tag{4.383}$$

$$^G\tilde{\mathbf{d}}_o \, {}^G R_B = \begin{bmatrix} d_2 r_{31} - d_3 r_{21} & d_2 r_{32} - d_3 r_{22} & d_2 r_{33} - d_3 r_{23} \\ -d_1 r_{31} + d_3 r_{11} & -d_1 r_{32} + d_3 r_{12} & -d_1 r_{33} + d_3 r_{13} \\ d_1 r_{21} - d_2 r_{11} & d_1 r_{22} - d_2 r_{12} & d_1 r_{23} - d_2 r_{13} \end{bmatrix} \tag{4.384}$$

4.10 ★ The Geometry of Plane and Line

Plücker coordinates introduce a suitable method to define the moment between two lines, shortest distance between two lines, and the angle between two lines.

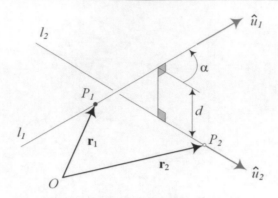

Fig. 4.29 Two skew lines

4.10.1 ★ Moment

Consider two arbitrary lines $l_1 = \begin{bmatrix} \hat{u}_1 & \boldsymbol{\rho}_1 \end{bmatrix}^T$ and $l_2 = \begin{bmatrix} \hat{u}_2 & \boldsymbol{\rho}_2 \end{bmatrix}^T$ as shown in Fig. 4.29. Points P_1 and P_2 on l_1 and l_2 are indicated by vectors $\mathbf{r}_1$ and $\mathbf{r}_2$, respectively. Direction vectors of the lines are $\hat{u}_1$ and $\hat{u}_2$.

The moment of the line l_2 about P_1 is $(\mathbf{r}_2 - \mathbf{r}_1) \times \hat{u}_2$ and we can define the moment of the line l_2 about l_1 by $l_2 \times l_1$,

$$l_2 \times l_1 = \hat{u}_1 \cdot (\mathbf{r}_2 - \mathbf{r}_1) \times \hat{u}_2 \tag{4.385}$$

which, because of

$$\hat{u}_1 \cdot \mathbf{r}_1 \times \hat{u}_2 = \hat{u}_2 \cdot \hat{u}_1 \times \mathbf{r}_1 \tag{4.386}$$

simplifies to:

$$l_2 \times l_1 = \hat{u}_1 \cdot \boldsymbol{\rho}_2 + \hat{u}_2 \cdot \boldsymbol{\rho}_1 \tag{4.387}$$

The *reciprocal product* or *virtual product* of two lines described by Plücker coordinates is defined as:

$$l_2 \times l_1 = \begin{bmatrix} \hat{u}_2 \\ \boldsymbol{\rho}_2 \end{bmatrix} \otimes \begin{bmatrix} \hat{u}_1 \\ \boldsymbol{\rho}_1 \end{bmatrix} = \hat{u}_2 \cdot \boldsymbol{\rho}_1 + \hat{u}_1 \cdot \boldsymbol{\rho}_2 \tag{4.388}$$

The reciprocal product is commutative and gives the moment between two directed lines.

4.10.2 ★ Angle and Distance

If d is the shortest distance between two lines $l_1 = \begin{bmatrix} \hat{u}_1 & \boldsymbol{\rho}_1 \end{bmatrix}^T$ and $l_2 = \begin{bmatrix} \hat{u}_2 & \boldsymbol{\rho}_2 \end{bmatrix}^T$, and $\alpha \in [0, \pi]$ is the angle between l_1 and l_2, then

$$\sin \alpha = \left| \hat{u}_2 \times \hat{u}_1 \right| \tag{4.389}$$

and

$$d = \frac{1}{\sin \alpha} \left| \hat{u}_2 \cdot \boldsymbol{\rho}_1 + \hat{u}_1 \cdot \boldsymbol{\rho}_2 \right| = \frac{1}{\sin \alpha} \left| \begin{bmatrix} \hat{u}_2 \\ \boldsymbol{\rho}_2 \end{bmatrix} \otimes \begin{bmatrix} \hat{u}_1 \\ \boldsymbol{\rho}_1 \end{bmatrix} \right|$$
$$= \frac{1}{\sin \alpha} \left| l_2 \times l_1 \right| \tag{4.390}$$

Therefore, the lines l_1 and l_2 intersect if and only if their reciprocal product is zero. Two parallel lines may be assumed to intersect at infinity. The distance expression does not work for parallel lines.

4.10.3 ★ Plane and Line

The equation of a plane $^G\pi$ having a normal unit vector $\hat{n}$ is

$$n_1 X + n_2 Y + n_3 Z = s \tag{4.391}$$

$$\hat{n} = n_1 \hat{I} + n_2 \hat{J} + n_3 \hat{K} \tag{4.392}$$

where s is the minimum distance of the plane to the origin O. We may indicate a plane using a homogenous representation,

$$\pi = \begin{bmatrix} n_1 & n_2 & n_3 & s \end{bmatrix}^T \tag{4.393}$$

and write the condition $\pi^T \cdot r = 0$ for a point $\mathbf{r} = \begin{bmatrix} X & Y & Z & w \end{bmatrix}^T$ to be in the plane by

$$\pi^T \cdot \mathbf{r} = \begin{bmatrix} n_1 & n_2 & n_3 & s \end{bmatrix} \begin{bmatrix} X \\ Y \\ Z \\ w \end{bmatrix} = 0 \tag{4.394}$$

Moreover, $w = 0$ indicates all points at infinity, and $s = 0$ indicates all planes containing the origin.

The intersection of π-plane with the X-axis, or the X-intercept, is $X = -s/n_1$, the Y-intercept is $Y = -s/n_2$, and the Z-intercept is $Z = -s/n_3$. The plane is perpendicular to (X, Y)-plane if $n_3 = 0$. It is perpendicular to the X-axis if $n_2 = n_3 = 0$. There are similar conditions for the other planes and axes. If (X_0, Y_0, Z_0) is a point in the π-plane (4.393), then

$$n_1(X - X_0) + n_2(Y - Y_0) + n_3(Z - Z_0) = s \tag{4.395}$$

The distance of a point $\begin{bmatrix} X & Y & Z & w \end{bmatrix}^T$ from the origin is

$$d = \sqrt{\frac{X^2 + Y^2 + Z^2}{w^2}} \tag{4.396}$$

while the distance of a plane $\pi = \begin{bmatrix} m_1 & m_2 & m_3 & s \end{bmatrix}^T$ from the origin is

$$s = \sqrt{\frac{s^2}{m_1^2 + m_2^2 + m_3^2}} \tag{4.397}$$

The equation of a line connecting two points $P_1(X_1, Y_1, Z_1)$ and $P_2(X_2, Y_2, Z_2)$ at $\mathbf{r}_1$ and $\mathbf{r}_2$ can also be expressed by

$$l = \mathbf{r}_1 + m(\mathbf{r}_2 - \mathbf{r}_1) \tag{4.398}$$

and the distance of a point $P(X, Y, Z)$ from any point on l is given by

$$\begin{aligned} d^2 = &(X_1 + m(X_2 - X_1) - X)^2 \\ &+ (Y_1 + m(Y_2 - Y_1) - Y)^2 \\ &+ (Z_1 + m(Z_2 - Z_1) - Z)^2 \end{aligned} \tag{4.399}$$

which is minimum for

$$m = -\frac{(X_2 - X_1)(X_1 - X) + (Y_2 - Y_1)(Y_1 - Y) + (Z_2 - Z_1)(Z_1 - Z)}{(X_2 - X_1)^2 + (Y_2 - Y_1)^2 + (Z_2 - Z_1)^2} \tag{4.400}$$

To find the minimum distance of the origin, we set $X = Y = Z = 0$.

Example 151 ★ Angle and distance between two diagonals of a cube. Numerical example of application of Plücker coordinates to determine the distance between two lines.

The Plücker coordinates of the two diagonals of the unit cube shown in Fig. 4.26 are

$$l_1 = \begin{bmatrix} \hat{u}_1 \\ \boldsymbol{\rho}_1 \end{bmatrix} = \begin{bmatrix} -\frac{\sqrt{2}}{2} & \frac{\sqrt{2}}{2} & 0 & -\frac{\sqrt{2}}{2} & -\frac{\sqrt{2}}{2} & \frac{\sqrt{2}}{2} \end{bmatrix}^T \tag{4.401}$$

$$l_2 = \begin{bmatrix} \hat{u}_2 \\ \boldsymbol{\rho}_2 \end{bmatrix} = \begin{bmatrix} 0 & \frac{\sqrt{2}}{2} & \frac{\sqrt{2}}{2} & 0 & -\frac{\sqrt{2}}{2} & \frac{\sqrt{2}}{2} \end{bmatrix}^T \tag{4.402}$$

The angle between l_1 and l_2 is

$$\alpha = \arcsin|\hat{u}_2 \times \hat{u}_1| = \arcsin\frac{\sqrt{3}}{2} = \frac{\pi}{3}\,\text{rad} = 60\,\text{deg} \tag{4.403}$$

and the distance between them is

$$d = \frac{1}{\sin\alpha}\left|\begin{bmatrix} \hat{u}_1 \\ \boldsymbol{\rho}_1 \end{bmatrix} \otimes \begin{bmatrix} \hat{u}_2 \\ \boldsymbol{\rho}_2 \end{bmatrix}\right| = \frac{1}{\sin\alpha}|\hat{u}_1 \cdot \boldsymbol{\rho}_2 + \hat{u}_2 \cdot \boldsymbol{\rho}_1|$$

$$= \frac{2}{\sqrt{3}}|-0.5| = \frac{1}{\sqrt{3}} \tag{4.404}$$

Example 152 Distance of a point from a line. Numerical example of application of Plücker coordinates to determine the distance between a point and a line.

The equation of the line connecting two points $\mathbf{r}_1$ and $\mathbf{r}_2$

$$\mathbf{r}_1 = \begin{bmatrix} -1 & 2 & 1 \end{bmatrix}^T \qquad \mathbf{r}_2 = \begin{bmatrix} 1 & -2 & -1 & 1 \end{bmatrix}^T \tag{4.405}$$

is

$$l = \mathbf{r}_1 + m(\mathbf{r}_2 - \mathbf{r}_1) = \begin{bmatrix} -1 + 2m \\ 2 - 4m \\ 1 - 2m \\ 1 \end{bmatrix} \tag{4.406}$$

Now the distance between point $\mathbf{r}_3 = \begin{bmatrix} 1 & 1 & 0 & 1 \end{bmatrix}^T$ and l is given by

$$s^2 = (X_1 + m(X_2 - X_1) - X_3)^2$$
$$+ (Y_1 + m(Y_2 - Y_1) - Y_3)^2$$
$$+ (Z_1 + m(Z_2 - Z_1) - Z_3)^2$$
$$= 24m^2 - 20m + 6 \tag{4.407}$$

which is minimum for:

$$m = \frac{5}{12} \tag{4.408}$$

Thus, the point on the line at a minimum distance from $\mathbf{r}_3$ is

$$\mathbf{r} = \begin{bmatrix} -1/6 & 1/3 & 1/6 & 1 \end{bmatrix}^T \tag{4.409}$$

Example 153 Distance between two lines. Numerical example of application of Plücker coordinates to determine the distance between two lines.

The line connecting $\mathbf{r}_1$ and $\mathbf{r}_2$

$$\mathbf{r}_1 = \begin{bmatrix} -1 & 2 & 1 \end{bmatrix}^T \qquad \mathbf{r}_2 = \begin{bmatrix} 1 & -2 & -1 & 1 \end{bmatrix}^T \tag{4.410}$$

is

$$l = \mathbf{r}_1 + m(\mathbf{r}_2 - \mathbf{r}_1) = \begin{bmatrix} -1 + 2m \\ 2 - 4m \\ 1 - 2m \\ 1 \end{bmatrix} \tag{4.411}$$

and the line connecting $\mathbf{r}_3$ and $\mathbf{r}_4$

$$\mathbf{r}_3 = \begin{bmatrix} 1 & 1 & 0 & 1 \end{bmatrix}^T \qquad \mathbf{r}_4 = \begin{bmatrix} 0 & -1 & 2 & 1 \end{bmatrix}^T \tag{4.412}$$

is

$$l = \mathbf{r}_1 + m(\mathbf{r}_2 - \mathbf{r}_1) = \begin{bmatrix} 1 - n \\ 1 - 2n \\ 2n \\ 1 \end{bmatrix} \tag{4.413}$$

The distance between two arbitrary points on the lines is

$$s^2 = \left(-1 + 2m - 1 + n^2\right) + (2 - 4m - 1 + 2n)^2 + (1 - 2m - 2n)^2 \tag{4.414}$$

The minimum distance is found by minimizing s^2 with respect to m and n.

$$m = 0.443 \qquad n = 0.321 \tag{4.415}$$

Hence, the two points on the lines at a minimum distance apart are at:

$$\mathbf{r}_m = \begin{bmatrix} -0.114 \\ 0.228 \\ 0.114 \\ 1 \end{bmatrix} \qquad \mathbf{r}_n = \begin{bmatrix} 0.679 \\ 0.358 \\ 0.642 \\ 1 \end{bmatrix} \tag{4.416}$$

Example 154 ★ Intersection condition for two lines. Expression of intersection of two lines by Plücker coordinates.

If two lines $l_1 = \begin{bmatrix} \hat{u}_1 & \boldsymbol{\rho}_1 \end{bmatrix}^T$ and $l_2 = \begin{bmatrix} \hat{u}_2 & \boldsymbol{\rho}_2 \end{bmatrix}^T$ intersect, and the position of their common point is at $\mathbf{r}$, then

$$\boldsymbol{\rho}_1 = \mathbf{r} \times \hat{u}_1 \qquad \boldsymbol{\rho}_2 = \mathbf{r} \times \hat{u}_2 \tag{4.417}$$

and, therefore,

$$\boldsymbol{\rho}_1 \cdot \hat{u}_2 = \left(\mathbf{r} \times \hat{u}_1\right) \cdot \hat{u}_2 = \mathbf{r} \cdot \left(\hat{u}_1 \times \hat{u}_2\right) \tag{4.418}$$

$$\boldsymbol{\rho}_2 \cdot \hat{u}_1 = \left(\mathbf{r} \times \hat{u}_2\right) \cdot \hat{u}_1 = \mathbf{r} \cdot \left(\hat{u}_2 \times \hat{u}_1\right) \tag{4.419}$$

which implies

$$\hat{u}_1 \cdot \boldsymbol{\rho}_2 + \hat{u}_2 \cdot \boldsymbol{\rho}_1 = 0 \tag{4.420}$$

or equivalently:

$$\begin{bmatrix} \hat{u}_1 \\ \boldsymbol{\rho}_1 \end{bmatrix} \otimes \begin{bmatrix} \hat{u}_2 \\ \boldsymbol{\rho}_2 \end{bmatrix} = 0 \tag{4.421}$$

Example 155 ★ Plücker coordinates of the axis of rotation. Expression of rotations by Plücker coordinates.

Consider a homogenous transformation matrix corresponding to a rotation α about Z, along with a translation in (X, Y)-plane

$$^G T_B = \begin{bmatrix} r_{11} & r_{12} & 0 & X_o \\ r_{21} & r_{22} & 0 & Y_0 \\ 0 & 0 & 1 & 0 \\ 0 & 0 & 0 & 1 \end{bmatrix} = \begin{bmatrix} \cos\alpha & -\sin\alpha & 0 & X_o \\ \sin\alpha & \cos\alpha & 0 & Y_o \\ 0 & 0 & 1 & 0 \\ 0 & 0 & 0 & 1 \end{bmatrix} \tag{4.422}$$

The angle of rotation can be obtained by comparison.

$$\alpha = \arctan \frac{r_{21}}{r_{11}} \tag{4.423}$$

The pole of rotation can be found by searching for a point that has the same coordinates in both frames,

$$\begin{bmatrix} r_{11} & r_{12} & 0 & X_o \\ r_{21} & r_{22} & 0 & Y_o \\ 0 & 0 & 1 & 0 \\ 0 & 0 & 0 & 1 \end{bmatrix} \begin{bmatrix} X_p \\ Y_p \\ 0 \\ 1 \end{bmatrix} = \begin{bmatrix} X_p \\ Y_p \\ 0 \\ 1 \end{bmatrix} \tag{4.424}$$

that can be written as:

$$\begin{bmatrix} r_{11} - 1 & r_{12} & 0 & X_o \\ r_{21} & r_{22} - 1 & 0 & Y_o \\ 0 & 0 & 1 & 0 \\ 0 & 0 & 0 & 1 \end{bmatrix} \begin{bmatrix} X_p \\ Y_p \\ 0 \\ 1 \end{bmatrix} = \begin{bmatrix} 0 \\ 0 \\ 0 \\ 1 \end{bmatrix} \tag{4.425}$$

The solutions of these equations are

$$X_p = \frac{1}{2}X_o - \frac{1}{2}\frac{r_{21}}{1 - r_{11}}Y_o = \frac{1}{2}X_o - \frac{1}{2}\frac{\sin\alpha}{\text{vers}\,\alpha}Y_o \tag{4.426}$$

$$Y_p = \frac{1}{2}Y_o + \frac{1}{2}\frac{r_{21}}{1 - r_{11}}X_o = \frac{1}{2}Y_o + \frac{1}{2}\frac{\sin\alpha}{\text{vers}\,\alpha}X_o \tag{4.427}$$

The Plücker line coordinates $l = \begin{bmatrix} \hat{u} & \boldsymbol{\rho} \end{bmatrix}^T$ of the pole axis is then equal to:

$$l = \begin{bmatrix} 0 & 0 & 1 & Y_p & -X_p & 0 \end{bmatrix}^T \tag{4.428}$$

4.11   ★ Screw and Plücker Coordinate

Consider a screw motion $\check{s}(h, \phi, \hat{u}, \mathbf{s})$ whose line of action is given by Plücker coordinates $\begin{bmatrix} \hat{u} & \boldsymbol{\rho} \end{bmatrix}^T$, and its pitch is $p = \frac{h}{\phi}$. The screw motion can be defined by a set of Plücker coordinates.

$$\check{s}(h, \phi, \hat{u}, \mathbf{s}) = \begin{bmatrix} \hat{u} \\ \boldsymbol{\xi} \end{bmatrix} = \begin{bmatrix} \hat{u} \\ \boldsymbol{\rho} + p\hat{u} \end{bmatrix} = \begin{bmatrix} \phi\hat{u} \\ \phi\boldsymbol{\rho} + h\hat{u} \end{bmatrix} \tag{4.429}$$

If the pitch is infinite, $p = \infty$, then the screw motion reduces to a pure translation, or equivalently, a line at infinity.

$$\check{s}(h, 0, \hat{u}, \mathbf{r}) = \begin{bmatrix} 0 \\ h\hat{u} \end{bmatrix} \tag{4.430}$$

A zero pitch screw $p = 0$ corresponds to a pure rotation, then the screw coordinates are identical to the Plücker coordinates of the screw line.

$$\check{s}(0, \phi, \hat{u}, \mathbf{s}) = \begin{bmatrix} \phi\hat{u} \\ \phi\boldsymbol{\rho} \end{bmatrix} = \begin{bmatrix} \hat{u} \\ \boldsymbol{\rho} \end{bmatrix} \tag{4.431}$$

A central screw motion is defined by a line through origin.

$$\check{s}(h, \phi, \hat{u}) = \check{s}(h, \phi, \hat{u}, 0) = \begin{bmatrix} \hat{u} \\ p\hat{u} \end{bmatrix} = \begin{bmatrix} \phi\hat{u} \\ h\hat{u} \end{bmatrix}$$

$$= D(h\hat{u})\,R(\hat{u}, \phi) \tag{4.432}$$

Screw coordinates for differential screw motion are useful in velocity analysis of robots. Consider a screw axis l, an angular velocity $\boldsymbol{\omega} = \omega \hat{u} = \dot{\phi} \hat{u}$ about l, and a velocity $\mathbf{v}$ along l. If the location vector $\mathbf{s}$ is the position of a point on l, then the Plücker coordinates of the line l are

$$l = \begin{bmatrix} \hat{u} \\ \boldsymbol{\rho} \end{bmatrix} = \begin{bmatrix} \hat{u} \\ \mathbf{s} \times \hat{u} \end{bmatrix} \tag{4.433}$$

The pitch of screw is

$$p = \frac{|\mathbf{v}|}{|\boldsymbol{\omega}|} \tag{4.434}$$

and the direction of screw is defined by

$$\hat{u} = \frac{\boldsymbol{\omega}}{|\boldsymbol{\omega}|} \tag{4.435}$$

so the instantaneous screw coordinates $\check{v}(p, \omega, \hat{u}, \mathbf{s})$ are

$$\begin{aligned} \check{v}(p, \omega, \hat{u}, \mathbf{r}) &= \begin{bmatrix} \omega \hat{u} & \dfrac{\mathbf{r} \times \boldsymbol{\omega} + |\mathbf{v}| \, \boldsymbol{\omega}}{|\boldsymbol{\omega}|} \end{bmatrix}^T \\ &= \begin{bmatrix} \omega \hat{u} \\ \mathbf{s} \times \hat{u} + \mathbf{v} \end{bmatrix} = \begin{bmatrix} \boldsymbol{\omega} \\ \mathbf{s} \times \hat{u} + p\boldsymbol{\omega} \end{bmatrix} \\ &= \begin{bmatrix} \boldsymbol{\omega} \\ \boldsymbol{\rho} + p\boldsymbol{\omega} \end{bmatrix} \end{aligned} \tag{4.436}$$

Example 156 ★ Pitch of an instantaneous screw. Screw motion may be expressed by Plücker coordinates. This is showing a special situation.

The pitch of an instantaneous screw motion, defined by Plücker coordinates, can be found by

$$p = \hat{u} \cdot \boldsymbol{\xi} \tag{4.437}$$

because two Plücker vectors are orthogonal, $\hat{u} \cdot \boldsymbol{\rho} = 0$, and therefore:

$$\phi \hat{u} \cdot \phi \boldsymbol{\xi} = \phi \hat{u} \cdot \left(\boldsymbol{\rho} + \phi p \hat{u} \right) = \left(\phi \hat{u} \cdot \boldsymbol{\rho} + \phi^2 \, p \right) = \phi^2 \, p \tag{4.438}$$

Example 157 ★ Nearest point on a screw axis to the origin. An application of Plücker coordinates to screw motions.

The point on the instantaneous screw axis, nearest to the origin, is indicated by the following position vector:

$$\mathbf{s}_0 = \phi \hat{u} \times \phi \boldsymbol{\xi} \tag{4.439}$$

4.12 Summary

Arbitrary motion of a body B with respect to another body G is called rigid body motion and can be expressed by a rotation $^G R_B \, ^B \mathbf{r}$ plus a translation $^G \mathbf{d}$

$$^G \mathbf{r} = {}^G R_B \, ^B \mathbf{r} + {}^G \mathbf{d} \tag{4.440}$$

where

$$^G \mathbf{r} = \begin{bmatrix} X & Y & Z \end{bmatrix}^T \tag{4.441}$$

$$^B \mathbf{r} = \begin{bmatrix} x & y & z \end{bmatrix}^T \tag{4.442}$$

$$^G \mathbf{d} = \begin{bmatrix} X_o & Y_o & Z_o \end{bmatrix}^T \tag{4.443}$$

The vector $^G \mathbf{d}$ is translation of B with respect to G, and $^G R_B$ is the rotation transformation matrix to map $^B \mathbf{r}$ to $^G \mathbf{r}$ when $^G \mathbf{d} = 0$.

By introducing the homogenous coordinate representation for a point

$$\mathbf{r} = \begin{bmatrix} x \\ y \\ z \\ 1 \end{bmatrix} \tag{4.444}$$

we may combine the rotation and translation of a rigid body motion by a 4×4 homogenous transformation matrix $^G T_B$ and show the coordinate transformation by

$$^G\mathbf{r} = {}^G T_B \, {}^B\mathbf{r} \tag{4.445}$$

where

$$^G T_B = \begin{bmatrix} ^G R_B & {}^G\mathbf{d} \\ 0 & 1 \end{bmatrix} \tag{4.446}$$

As the homogenous transformation matrix $^G T_B$ is not orthogonal, its inverse obeys a specific rule:

$$^G T_B^{-1} = {}^B T_G = \begin{bmatrix} ^G R_B^T & -{}^G R_B^T \, {}^G\mathbf{d} \\ 0 & 1 \end{bmatrix} \tag{4.447}$$

to be consistent with

$$^G T_B^{-1} \, {}^G T_B = \mathbf{I}_4 \tag{4.448}$$

The rigid motion can also be expressed with screw motion $\check{s}(h, \phi, \hat{u}, \mathbf{s})$ and screw transformation

$$^G\mathbf{r} = {}^G\check{s}_B(h, \phi, \hat{u}, \mathbf{s}) \, {}^B\mathbf{r} \tag{4.449}$$

$$\check{s}(h, \phi, \hat{u}, \mathbf{s}) = \begin{bmatrix} ^G R_B & {}^G\mathbf{s} - {}^G R_B \, {}^G\mathbf{s} + h\hat{u} \\ 0 & 1 \end{bmatrix} \tag{4.450}$$

The screw motion $\check{s}(h, \phi, \hat{u}, \mathbf{s})$ is indicated by screw parameters: a unit vector on the axis of rotation $\hat{u}$, a location vector $\mathbf{s}$, a twist angle ϕ, and a translation h (or pitch p). The location vector $\mathbf{s}$ indicates the global position of a point on the screw axis. When $\mathbf{s} = 0$, then $\hat{u}$ passes through the origin of the coordinate frame and the screw motion is called a central screw $\check{s}(h, \phi, \hat{u})$. Every screw motion can be decomposed into three principal central screws about the three axes of the coordinate frame G.

A rigid body motion can be expressed more effectively by screw and Plücker coordinates of directed lines.

$$\check{s}(h, \phi, \hat{u}, \mathbf{s}) = \begin{bmatrix} \hat{u} \\ \boldsymbol{\xi} \end{bmatrix} = \begin{bmatrix} \hat{u} \\ \boldsymbol{\rho} + p\hat{u} \end{bmatrix} = \begin{bmatrix} \phi\hat{u} \\ \phi\boldsymbol{\rho} + h\hat{u} \end{bmatrix} \tag{4.451}$$

$$\boldsymbol{\rho} = \mathbf{r} \times \hat{u} \tag{4.452}$$

4.13   Key Symbols

B	Body coordinate frame, local coordinate frame
c	cos
$\mathbf{d}$	Translation vector, displacement vector
D	Displacement matrix
e	Exponential
G	Global coordinate frame, fixed coordinate frame
h	Translation of a screw
$\mathbf{I} = [I]$	Identity matrix
$\hat{\imath}, \hat{\jmath}, \hat{k}$	Local coordinate axes' unit vectors
$\tilde{\imath}, \tilde{\jmath}, \tilde{k}$	Skew symmetric matrices of the unit vector $\hat{\imath}, \hat{\jmath}, \hat{k}$
$\hat{I}, \hat{J}, \hat{K}$	Global coordinate axes' unit vectors
l	Line, Plücker coordinates, Plücker line
L, M, N	Components of $\hat{u}$
$\hat{n}$	Normal unit vector to a plane
p	Pitch of a screw
P	A body point, a fixed point in B, point matrix
P, Q, R	Components of $\boldsymbol{\rho}$
$\mathbf{p}, \mathbf{q}, \mathbf{r}$	Position vectors, homogenous position vector
r_{ij}	The element of row i and column j of a matrix
R	Rotation transformation matrix
s	sin
$\mathbf{s}$	Location vector of a screw
$\check{s}$	Screw
T	homogenous transformation matrix
$\hat{u}$	A unit vector on axis of rotation
$\tilde{u}$	Skew symmetric matrix of the vector $\hat{u}$
$\mathbf{v}$	Velocity vector
w	Weight factor of a homogenous vector
x, y, z	Local coordinate axes
X, Y, Z	Global coordinate axes

Greek

α, β, γ	Rotation angles about global axes
$^{G}\Gamma_{B}$	Transformation matrix for Plücker coordinates
λ	Eigenvalues of R
$\boldsymbol{\xi}$	Moment vector of a Plücker line
$\boldsymbol{\pi}$	Homogenous expression of a plane
$\boldsymbol{\rho}$	Moment vector of $\hat{u}$ about origin
ϕ	Angle of rotation about $\hat{u}$, rotation of a screw
$\boldsymbol{\omega}$	Angular velocity vector
$\tilde{\omega}$	Skew symmetric matrix of the vector $\boldsymbol{\omega}$

Symbol

tr	Trace operator
vers	$1 - \cos$
$[\ ]^{-1}$	Inverse of the matrix $[\ ]$
$[\ ]^{T}$	Transpose of the matrix $[\ ]$
$\triangle$	Transformation matrix of ray to axis arrangement

Exercises

1. Notation and symbols.
   Describe the meaning of:

   $$\text{a} - {}^2\mathbf{d}_1 \quad \text{b} - {}^GT_B^{-1} \quad \text{c} - {}^BT_G \quad \text{d} - {}^GD_B \quad \text{e} - {}^G\check{s}_B(h, \phi, \hat{u})$$

   $$\text{f} - D_{\hat{u},h} \quad \text{g} - \check{s}^{-1} \quad \text{h} - {}^G\Gamma_B \quad \text{i} - \begin{bmatrix} \hat{u} \\ \rho \end{bmatrix} \quad \text{j} - {}^G\check{s}_B(2, \tfrac{\pi}{3}, \hat{K})$$

   $$\text{k} - p = \tfrac{h}{\phi} \quad \text{l} - {}^G\mathbf{s}_0^T \quad \text{m} - \check{s}^{-T} \quad \text{n} - \begin{bmatrix} \hat{u} \\ \xi \end{bmatrix} \quad \text{o} - \check{s}(h, \phi, \hat{u}, \mathbf{s})$$

2. Global position in a rigid body motion.
   We move the body coordinate frame B to

   $$^G\mathbf{d} = \begin{bmatrix} 4 & -3 & 7 \end{bmatrix}^T \tag{4.453}$$

   Find $^G\mathbf{r}_P$ if the local position of a point P is

   $$^B\mathbf{r}_P = \begin{bmatrix} 7 & 3 & 2 \end{bmatrix}^T \tag{4.454}$$

   The orientation of B with respect to the global frame G can be found by a rotation 45 deg about the X-axis, and then 45 deg about the Y-axis.

3. Global rotation and global translation.
   A body frame B turns 90 deg about Z-axis, then 90 deg about X-axis.
   (a) Determine the transformation matrix ^GT_B.
   (b) Determine the global coordinates of $^B\mathbf{r}_P = \begin{bmatrix} 1 & 1 & 1 \end{bmatrix}^T$ after the rotations.
   (c) Determine a unique rotation about a unique axis to turn B to its new position.
   (d) If the body is a cube at the positive corner of the global coordinate frame such that P is the furthest corner of the cube from the origin, then determine the required translation vector $^G\mathbf{d}$ to move the cube back to the corner of the first quadrant.
   (e) What are the global coordinates of P after moving the cube back to the corner of the first quadrant.

4. ★ Rotation matrix compatibility.
   It is not possible to find ^GR_B from the equation of rigid body motion

   $$^G\mathbf{r}_P = {}^GR_B \, {}^B\mathbf{r}_P + {}^G\mathbf{d} \tag{4.455}$$

   if $^G\mathbf{r}_P$, $^B\mathbf{r}_P$, and $^G\mathbf{d}$ are given. Explain why and find the required conditions to be able to find ^GR_B.

5. Global position with local rotation in a rigid body motion.
   Assume a body coordinate frame B is at $^G\mathbf{d}$.

   $$^G\mathbf{d} = \begin{bmatrix} 4 & -3 & 7 \end{bmatrix}^T \tag{4.456}$$

   Find $^G\mathbf{r}_P$ if the local position of a point is

   $$^B\mathbf{r}_P = \begin{bmatrix} 7 & 3 & 2 \end{bmatrix}^T \tag{4.457}$$

   and the orientation of B with respect to the global frame G can be found by a rotation 45 deg about the x-axis, and then 45 deg about the y-axis.

6. Local rotation and global translation.
   A body frame B turns 90 deg about z-axis, then 90 deg about x-axis.
   (a) Determine the transformation matrix ^GT_B.
   (b) Determine the global coordinates of $^B\mathbf{r}_P = \begin{bmatrix} 1 & 1 & 1 \end{bmatrix}^T$ after the rotations.
   (c) Determine a unique rotation about a unique axis to turn B to its new configuration.

7. Repeating global rotation.
   A body frame B turns 90 deg about Z-axis, then 90 deg about Y-axis.
   (a) Determine the transformation matrix ^GT_B.
   (b) Determine the global coordinates of $^B\mathbf{r}_P = \begin{bmatrix} 1 & 1 & 1 \end{bmatrix}^T$ after the rotations.

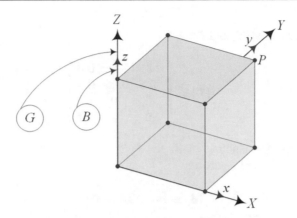

Fig. 4.30 A cube at the positive corner of the global coordinate frame

(c) Determine a unique rotation about a unique axis to turn B to its new position.

(d) ★ Assume that the body is a cube at the positive corner of the global coordinate frame as shown in Fig. 4.30. The point P is the furthest corner of the cube from the origin. Show that if we repeat the combined rotations of section (a) three times, the cube will be back to its initial position and orientation.

(e) ★ After rotations 90 deg about Z-axis, then 90 deg about Y-axis, turn the cube ϕ about $^G\mathbf{u} = \begin{bmatrix} 1 & 1 & 1 \end{bmatrix}^T$. How much should ϕ be to turn the cube back to its initial position and orientation?

8. Global and local rotation in a rigid body motion.
   A body coordinate frame B is translated to $^G\mathbf{d}$.

$$^G\mathbf{d} = \begin{bmatrix} 4 & -3 & 7 \end{bmatrix}^T \tag{4.458}$$

   Find $^G\mathbf{r}_P$ if the local position of a point is

$$^B\mathbf{r}_P = \begin{bmatrix} 7 & 3 & 2 \end{bmatrix}^T \tag{4.459}$$

   and the orientation of B with respect to the global frame G can be found by a rotation 45 deg about the X-axis, then 45 deg about the y-axis, and finally 45 deg about the z-axis.

9. Combination of rigid motions.
   The frame B_1 is rotated 35 deg about the z_1-axis and translated to

$$^2\mathbf{d} = \begin{bmatrix} -40 & 30 & 20 \end{bmatrix}^T \tag{4.460}$$

   with respect to another frame B_2. The orientation of the frame B_2 in the global frame G can be found by a rotation of 55 deg about

$$\mathbf{u} = \begin{bmatrix} 2 & -3 & 4 \end{bmatrix}^T \tag{4.461}$$

   Calculate $^G\mathbf{d}_1$ and $^G R_1$.

10. Global rotation and translation.
    Determine the transformation matrix for the unit cube of Fig. 4.30 after:
    (a) A rotation 90 deg about Z-axis followed by a unit translation along x-axis.
    (b) Determine the global coordinates of point P after the motion in (a).
    (c) ★ Repeat the motion in (a) four times. The cube will be back to its initial position. Determine the coordinates of P after each motion.

11. Rotation submatrix in a homogenous transformation matrix.
    Find the missing elements in this homogenous transformation matrix $[T]$.

$$[T] = \begin{bmatrix} ? & 0 & ? & 4 \\ 0.707 & ? & ? & 3 \\ ? & ? & 0 & 1 \\ 0 & 0 & 0 & 1 \end{bmatrix} \tag{4.462}$$

12. Angle and axis of rotation.
    Find the angle and axis of rotation for $[T]$ and T^{-1}.

$$[T] = \begin{bmatrix} 0.866 & -0.5 & 0 & 4 \\ 0.5 & 0.866 & 0 & 3 \\ 0 & 0 & 1 & 1 \\ 0 & 0 & 0 & 1 \end{bmatrix} \tag{4.463}$$

13. Combination of homogenous transformations.
    Assume the origin of the frames B_1, and B_2 are at:

$$^2\mathbf{d}_1 = \begin{bmatrix} -10 & 20 & -20 \end{bmatrix}^T \tag{4.464}$$

$$^G\mathbf{d}_2 = \begin{bmatrix} -15 & -20 & 30 \end{bmatrix}^T \tag{4.465}$$

The orientation of B_1 in B_2 can be found by a rotation of 60 deg about $^2\mathbf{u}$,

$$^2\mathbf{u} = \begin{bmatrix} 1 & -2 & 4 \end{bmatrix}^T \tag{4.466}$$

and the orientation of B_2 in the global frame G can be found by a rotation of 30 deg about $^G\mathbf{u}$.

$$^G\mathbf{u} = \begin{bmatrix} 4 & 3 & -4 \end{bmatrix}^T \tag{4.467}$$

Calculate the transformation matrix $^GT_2$, $^GT_2^{-1}$, $^GT_1$, and $^GT_1^{-1}$.

14. Rotation about an axis not going through origin.
    Find the global position of a body point at $^B\mathbf{r}_P$

$$^B\mathbf{r}_P = \begin{bmatrix} 7 & 3 & 2 \end{bmatrix}^T \tag{4.468}$$

after a rotation of 30 deg about an axis parallel to $^G\mathbf{u}$

$$^G\mathbf{u} = \begin{bmatrix} 4 & 3 & -4 \end{bmatrix}^T \tag{4.469}$$

and passing through a point at $(3, 3, 3)$.

15. ★ Box arrangement.
    Figure 4.31 illustrates two adjacent unit boxes. The vectors $^G\mathbf{u}_1$, $^G\mathbf{u}_2$, and $^G\mathbf{u}_3$ are globally fixed.
    (a) Put the second box on top of the first box by a rotation about $\mathbf{u}_3$. Determine the transformation matrix and the coordinates of the four top points.
    (b) Put the second box on top of the first box by a rotation about $\mathbf{u}_2$, $\mathbf{u}_3$, and $\mathbf{u}_2$. Determine the transformation matrix and the coordinates of the four top points.
    (c) Put the second box on top of the first box by a rotation about Z, then $\mathbf{u}_3$, and a translation along Z-axis. Determine the transformation matrix and the name of the top points.
    (d) Put the first box on top of the second box by a rotation abut $\mathbf{u}_3$. Determine the transformation matrix and the coordinates of the four top points.
    (e) Put the first box on top of the second box by a rotation about Z, $\mathbf{u}_1$, $\mathbf{u}_3$, and $\mathbf{u}_2$. Determine the transformation matrix and the coordinates of the four top points.
    (f) Determine one rotation and one translation to put the second box on top of the first box.
    (g) Determine one rotation and one translation to put the first box on top of the second box.

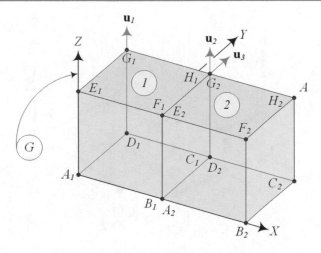

Fig. 4.31 Two adjacent unit boxes

16. Inversion of a square matrix.

    Knowing that the inverse of a 2×2 matrix

    $$[A] = \begin{bmatrix} a & b \\ c & d \end{bmatrix} \tag{4.470}$$

    is

    $$A^{-1} = \begin{bmatrix} -\dfrac{d}{-ad+bc} & \dfrac{b}{-ad+bc} \\ \dfrac{c}{-ad+bc} & -\dfrac{d}{-ad+bc} \end{bmatrix} \tag{4.471}$$

    use the inverse method of splitting a matrix $[T]$ into

    $$[T] = \begin{bmatrix} A & B \\ C & D \end{bmatrix} \tag{4.472}$$

    and applying the inverse technique (4.107), calculate the inverse of $[T]$.

    $$[T] = \begin{bmatrix} 1 & 2 & 3 & 4 \\ 12 & 13 & 14 & 5 \\ 11 & 16 & 15 & 6 \\ 10 & 9 & 8 & 7 \end{bmatrix} \tag{4.473}$$

17. ★ Combination of rotations about non-central axes.

    Consider a rotation 30 deg about an axis at the global point $(3, 0, 0)$ followed by another rotation 30 deg about an axis at the global point $(0, 3, 0)$. Find the possible transformations such that the final global coordinates of

    $$^{B}\mathbf{r}_P = \begin{bmatrix} 1 & 1 & 0 \end{bmatrix}^T \tag{4.474}$$

    to be

    $$^{G}\mathbf{r}_P = \begin{bmatrix} \sqrt{2} & 0 & 3 \end{bmatrix}^T \tag{4.475}$$

18. Transformation matrix from body points.

    Figure 4.32a shows a cube at initial configurations. Label the corners of the cube, at the final configuration shown in Fig. 4.32b, and find the associated homogenous transformation matrix. The length of each side of the cube is 2.

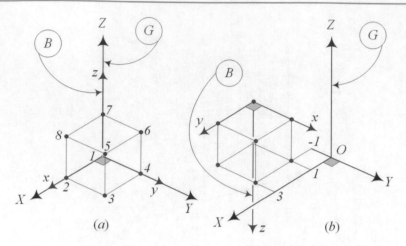

Fig. 4.32 A cube, before and after a rigid motion

19. Eulerian angles and translations.
    (a) Determine the homogenous transformation matrix ^GT_B for a set of Eulerian rotations φ, θ, ψ, plus a unit translations along $\hat{e}_\varphi, \hat{e}_\theta, \hat{e}_\psi$.
    (b) Determine the inverse of the transformation matrix in (a).
    (c) Simplify the ^GT_B and $^GT_B^{-1}$ for $\varphi = 45$ deg, $\theta = 45$ deg, $\psi = 45$ deg.
20. Euler angles and global translations.
    (a) Let turn a rigid body 45 deg about $^G\mathbf{u}$ and translate one unit along $^G\mathbf{u}$.

$$^G\mathbf{u} = \begin{bmatrix} 1 & 2 & 3 \end{bmatrix}^T \tag{4.476}$$

   (b) Determine the homogenous transformation matrix ^GT_B.
   (c) What are the Eulerian rotations φ, θ, ψ, and translations one unit along $^G\mathbf{u}$ to have the same ^GT_B?
   (d) What are the translations one unit along the Eulerian axes $\hat{e}_\varphi, \hat{e}_\theta, \hat{e}_\psi$ to have the same ^GT_B?
   (e) What are the Eulerian rotations φ, θ, ψ, and translations along $\hat{e}_\varphi, \hat{e}_\theta, \hat{e}_\psi$ to have the same ^GT_B?
21. Wrong order of rotation and translation.
    When we split a homogenous transformation to a rotation and a translation, we must follow the rule (4.38) and apply the rotation first.

$$^GT_B = {}^GD_B \, {}^GR_B \neq {}^GR_B \, {}^GD_B \tag{4.477}$$

   (a) Apply a rotation $^GR_{Z,45}$ and translation $^GD_{X,2}$ and determine the global coordinates of a point at $^B\mathbf{r} = \begin{bmatrix} 1 & 1 & 1 \end{bmatrix}^T$.
   (b) Apply the translation $^GD_{X,2}$ on $^B\mathbf{r}$, followed by the rotation $^GR_{Z,45}$ and determine the global coordinates of a point at $^B\mathbf{r}$.
   (c) Determine the required rigid body motion to move $^G\mathbf{r}_b$ to $^G\mathbf{r}_a$.
   (d) Determine the required rigid body motion to move $^G\mathbf{r}_a$ to $^G\mathbf{r}_b$.
22. Determination of homogenous transformation matrix.
    Figure 4.33 illustrates 4 body frames in a G-frame. Determine the homogenous transformation matrices $^GT_1$, $^GT_2$, $^GT_3, {}^GT_4, {}^2T_1, {}^3T_1, {}^4T_1, {}^3T_2, {}^1T_3, {}^3T_G \, {}^4T_2, {}^4T_3, {}^4T_G$.
23. ★ Change the order of rotation and translation.
    When we split a homogenous transformation to a rotation and a translation, we must follow the rule (4.38) and apply the rotation first.

$$^GT_B = {}^GD_B \, {}^GR_B \neq {}^GR_B \, {}^GD_B \tag{4.478}$$

   (a) Apply a general rotation ^GR_B and translation ^GD_B and determine the global coordinates of a point at $^B\mathbf{r} = \begin{bmatrix} x & y & z \end{bmatrix}^T$.
   (b) Apply the translation ^GD_B on $^B\mathbf{r}$, followed by the rotation ^GR_B and determine the global coordinates of a point at $^B\mathbf{r}$.
   (c) Determine the required rotation and translation to move $^G\mathbf{r}_b$ to $^G\mathbf{r}_a$.
   (d) Introduce a new equation for splitting a homogenous transformation matrix to:

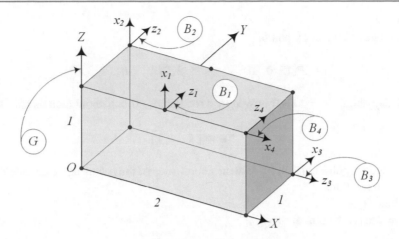

Fig. 4.33 4 body frames in a G-frame

$$^{G}T_{B} \overset{?}{=} \, ^{G}D_{2} \, ^{G}R_{2} \, ^{G}R_{B} \, ^{G}D_{B} \tag{4.479}$$

(e) Introduce a new equation for splitting a homogenous transformation matrix to:

$$^{G}T_{B} \overset{?}{=} \, ^{G}R_{2} \, ^{G}D_{2} \, ^{G}R_{B} \, ^{G}D_{B} \tag{4.480}$$

(f) Prove

$$^{G}T_{B} = \, ^{G}D_{B} + \, ^{G}R_{B} - \mathbf{I} = \, ^{G}R_{B} + \, ^{G}D_{B} - \mathbf{I} \tag{4.481}$$

24. ★ Principal central screw.
Find the principal central screw that moves the point $^{B}\mathbf{r}_{P}$

$$^{B}\mathbf{r}_{P} = \begin{bmatrix} 1 & 0 & 0 \end{bmatrix}^{T} \tag{4.482}$$

to $^{G}\mathbf{r}_{P}$.

$$^{G}\mathbf{r}_{P} = \begin{bmatrix} 0 & 1 & 4 \end{bmatrix}^{T} \tag{4.483}$$

25. ★ Screw motion.
Find the global position of

$$^{B}\mathbf{r}_{P} = \begin{bmatrix} 1 & 0 & 0 & 1 \end{bmatrix}^{T} \tag{4.484}$$

after a screw motion $^{G}\check{s}_{B}(h, \phi, \hat{u}, \mathbf{s}) = \, ^{G}\check{s}_{B}(4, 60\ \deg, \hat{u}, \mathbf{s})$, where

$$^{G}\mathbf{s} = \begin{bmatrix} 3 & 0 & 0 \end{bmatrix}^{T} \tag{4.485}$$

$$^{G}\mathbf{u} = \begin{bmatrix} 1 & 1 & 1 \end{bmatrix}^{T} \tag{4.486}$$

26. ★ Pole of a planar motion.
(a) Find the pole position of a planar motion if we have the coordinates of two body points, before and after the motion, as given below.

$$P_{1}\,(1, 1, 1) \qquad P_{2}\,(5, 2, 1) \tag{4 487}$$

$$Q_{1}\,(4, 1, 1) \qquad Q_{2}\,(7, 2 + \sqrt{5}, 1) \tag{4.488}$$

(b) Show that the pole is at the intersection of lines $P_{1}P_{2}$ and $Q_{1}Q_{2}$.

27. ★ Screw parameters
    Find the global coordinates of the body points

$$P_1\,(5, 0, 0) \qquad Q_1\,(5, 5, 0) \qquad R_1\,(0, 5, 0) \tag{4.489}$$

after a rotation of 30 deg about the x-axis followed by a rotation of 40 deg about an axis parallel to the vector $^G\mathbf{u}$

$$^G\mathbf{u} = \begin{bmatrix} 4 & 3 & -4 \end{bmatrix}^T \tag{4.490}$$

and passing through a global point at $(0, 3, 0)$. Use the coordinates of the points and calculate the screw parameters that are equivalent to the two rotations.

28. ★ Non-central rotation.
    Find the global coordinates of a point at $^B\mathbf{r}_P$

$$^B\mathbf{r}_P = \begin{bmatrix} 10 & 20 & -10 \end{bmatrix}^T \tag{4.491}$$

when the body frame rotates about $^G\mathbf{u}$

$$^G\mathbf{u} = \begin{bmatrix} 1 & 1 & 1 \end{bmatrix}^T \tag{4.492}$$

which passes through a point at $(-1, -4, 2)$.

29. ★ Equivalent screw.
    Calculate the transformation matrix ^GT_B for a rotation of 30 deg about the x-axis followed by a translation parallel to $^G\mathbf{d}$

$$^G\mathbf{d} = \begin{bmatrix} 3 & 2 & -1 \end{bmatrix}^T \tag{4.493}$$

and then a rotation of 40 deg about an axis parallel to the vector $^G\mathbf{u}$.

$$^G\mathbf{u} = \begin{bmatrix} 2 & -1 & 1 \end{bmatrix}^T \tag{4.494}$$

Find the screw parameters of ^GT_B.

30. ★ Central screw decomposition.
    Find a triple central screws for the case 1 in Appendix C,

$$^G\check{s}_B(h, \phi, \hat{u}, \mathbf{s}) = \check{s}(h_X, \gamma, \hat{I})\,\check{s}(h_Y, \beta, \hat{J})\,\check{s}(h_Z, \alpha, \hat{K}) \tag{4.495}$$

to get the same screw as

$$^G\check{s}_B(h, \phi, \hat{u}, \mathbf{s}) = {}^G\check{s}_B(4, 60, \hat{u}, \mathbf{s}) \tag{4.496}$$

where

$$^G\mathbf{s} = \begin{bmatrix} 3 & 0 & 0 \end{bmatrix}^T \qquad {}^G\mathbf{u} = \begin{bmatrix} 1 & 1 & 1 \end{bmatrix}^T \tag{4.497}$$

31. ★ Central screw composition.
    What is the final position of a point at

$$^B\mathbf{r}_P = \begin{bmatrix} 10 & 0 & 0 \end{bmatrix}^T \tag{4.498}$$

after the central screw $\check{s}(4, 30\,\text{deg}, \hat{J})$ followed by $\check{s}(2, 30\,\text{deg}, \hat{I})$ and $\check{s}(-6, 120\,\text{deg}, \hat{K})$?

32. ★ Screw composition.
    Find the final position of a point at

$$^B\mathbf{r}_P = \begin{bmatrix} 10 & 0 & 0 \end{bmatrix}^T \tag{4.499}$$

after a screw

$$^1\check{s}_2(h_1, \phi_1, \hat{u}_1, \mathbf{s}_1) = {}^1\check{s}_2\left(1, 40\,\text{deg}, \begin{bmatrix} 1/\sqrt{3} \\ -1/\sqrt{3} \\ 1/\sqrt{3} \end{bmatrix}, \begin{bmatrix} 2 \\ 3 \\ -1 \end{bmatrix}\right) \tag{4.500}$$

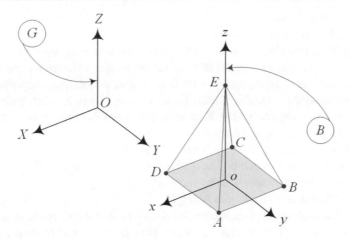

Fig. 4.34 A pyramid

followed by

$$^G\check{s}_1(h_0, \phi_0, \hat{u}_0, s_0) = {}^G\check{s}_1\left(-1, 45\deg, \begin{bmatrix} 1/9 \\ 4/9 \\ 4/9 \end{bmatrix}, \begin{bmatrix} -3 \\ 1 \\ 5 \end{bmatrix}\right) \tag{4.501}$$

33. ★ Zero translation screw motion.
    If the translation h is zero, then the screw motion is a rotation about a non-central axis. Show that a zero translation screw is equivalent to Eq. (4.132) for rotation about a non-central axis.

34. ★ Plücker line coordinate.
    Find the missing numbers in l.

$$l = \begin{bmatrix} 1/3 & 1/5 & ? & ? & 2 & -1 \end{bmatrix}^T \tag{4.502}$$

35. ★ Plücker lines.
    Find the Plücker lines for AE, BE, CE, DE in the local coordinate B, and calculate the angle between AE, and the Z-axis in the pyramid shown in Fig. 4.34. The local coordinate of the edges are

$$\begin{array}{ll} A(1, 1, 0) & \\ B(-1, 1, 0) & D(1, -1, 0) \\ C(-1, -1, 0) & E(0, 0, 3) \end{array} \tag{4.503}$$

Transform the Plücker lines AE, BE, CE, DE to the global coordinate G. The global position of o is at:

$$o(1, 10, 2) \tag{4.504}$$

36. ★ Angle between two lines.
    Find the angle between OE, OD, of the pyramid shown in Fig. 4.34. The coordinates of the points are

$$D(1, -1, 0) \qquad E(0, 0, 3) \tag{4.505}$$

37. ★ Distance from the origin.
    The equation of a plane is given as

$$4X - 5Y - 12Z - 1 = 0 \tag{4.506}$$

Determine the perpendicular distance of the plane from the origin.

38. ★Project: Continuous transformation.

Figure 4.33 illustrates 4 body frames in a G-frame.

(a) Determine the homogenous transformation $^{G}T_{1}$. Then define a continuous function of time that moves the origin of B_{1} from O to o_{1} on a straight line all in a unit of time, say one second. Determine the unique axis–angle of rotation that turns B_{1} from the coincident configuration with G to its final orientation. Applying the functions will move B_{1} from G to its final position and orientation. Divide the unit of time by 10 and determine $^{G}T_{1}$ at every 0.1 s and show B_{1} at those incremental times. Determine the directional cosines of the axes of B_{1} and plot them versus time.

(b) Repeat part (a) for B_{2}-frame moving in G-frame.

(c) Repeat part (a) for B_{2}-frame moving from B_{1}-frame.

(d) Can you find the result of part (c) by combining the results of parts (a) and (b)?

39. ★Project: Rigid body motion and incremental screw motions.

Consider a body frame B coincident with a global frame G.

(a) The frame B undergoes a rotation of 45 deg about the X-axis followed by a rotation of 90 deg about the Z-axis and then frame B moves to point $(1, 1, 1)$ in frame G. Determine the axis $\hat{u}$ and angle ϕ and translation h to move B to the final position only by one central screw motion.

(b) The frame B undergoes a rotation of $45/2$ deg about the X-axis followed by a rotation of $90/2$ deg about the Z-axis and then frame B moves to point $(1/2, 1/2, 1/2)$ in frame G. Determine the axis $\hat{u}_{1}$ and angle ϕ_{1} and translation h_{1} to move B to its position B_{1} by one central screw motion. Then B undergoes a rotation of $45/2$ deg about the X-axis followed by a rotation of $90/2$ deg about the Z-axis and then frame B moves another displacement $(1/2, 1/2, 1/2)$ in addition to the previous displacement in frame G. Determine the axis $\hat{u}_{2}$ and angle ϕ_{2} and translation h_{2} to move B_{1} to the position B_{2} only by one central screw motion. Also determine the axis $\hat{u}$ and angle ϕ and translation h to move B to the final position only by one central screw motion.

(c) The frame B undergoes a rotation of $45/3$ deg about the X-axis followed by a rotation of $90/3$ deg about the Z-axis and then frame B moves to point $(1/3, 1/3, 1/3)$ in frame G. Determine the axis $\hat{u}_{1}$ and angle ϕ_{1} and translation h_{1} to move B to its position B_{1} by one central screw motion. Then B undergoes a rotation of $45/3$ deg about the X-axis followed by a rotation of $90/3$ deg about the Z-axis and then frame B moves another displacement $(1/3, 1/3, 1/3)$ in addition to the previous displacement in frame G. Determine the axis $\hat{u}_{2}$ and angle ϕ_{2} and translation h_{2} to move B_{1} to the position B_{2} only by one central screw motion. Then B undergoes a rotation of $45/3$ deg about the X-axis followed by a rotation of $90/3$ deg about the Z-axis and then frame B moves another displacement $(1/3, 1/3, 1/3)$ in addition to the previous displacement in frame G. Determine the axis $\hat{u}_{3}$ and angle ϕ_{3} and translation h_{3} to move B_{2} to the position B_{3} only by one central screw motion. Also determine the axis $\hat{u}$ and angle ϕ and translation h to move B to the final position only by one central screw motion.

(d) Keep doing this process of turning frame B a rotation of $45/n$ deg about the X-axis followed by a rotation of $90/n$ deg about the Z-axis and then frame B moves another displacement $(1/n, 1/n, 1/n)$ in addition to the previous displacement in frame G. Repeat the rotations and displacement n times, $n = 1, 2, 3, \cdots, 45$. Determine the axis $\hat{u}_{n}$ and angle ϕ_{n} and translation h_{n} to move B to B_{n} by one screw motion for every n. How $\hat{u}_{n}$ and angle ϕ_{n} and translation h_{n} are related to each other? Plot ϕ_{n} versus n. Plot h_{n} versus n. Plot directional cosines of $\hat{u}_{n}$ versus n. Determine the axis $\hat{u}$ and angle ϕ and translation h that moves B to the final position only by one central screw motion for $n = 1, 2, 3, \cdots, 45$.

Forward kinematics means having the joint variables of a robot, we are able to determine the position and orientation of every link of the robot, including the end-effector. We attach a coordinate frame to every link and determine its configuration in the neighbor frames using rigid motion method. The analysis of determination of position and orientation of all links of a robot relative to each other is called *forward kinematics*. However, the first job of forward kinematics is to determine the position and orientation of the end-effector in base coordinate frame.

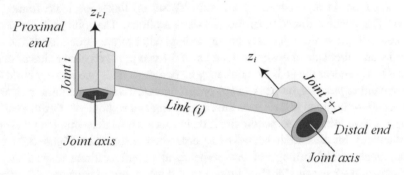

Fig. 5.1 Link (i) and its beginning joint $i - 1$ and its end joint i

5.1 Denavit–Hartenberg Notation

A serial robot with n joints will have $n + 1$ links. Numbering of links starts from link (0) for the immobile grounded *base link*. The moveable link jointed to the base link would be link number (1). The numbering of links increases sequentially up to link (n) for the *end-effector*. Numbering of joints starts from 1, for the joint connecting the first movable link (1) to the base link (0), and increases sequentially up to joint n. Therefore, the link (i) is connected to its *lower link* $(i - 1)$ at its *proximal end* by joint i and is connected to its *upper link* $(i + 1)$ at its *distal end* by joint $i + 1$, as shown in Fig. 5.1. For clarification, we show link numbers in parenthesis, such as link (i), and joint numbers free, such as joint i.

Figure 5.3 illustrates links $(i - 1)$, (i), and $(i + 1)$ of a serial robot, along with joints $i - 1$, i, and $i + 1$. Every link of a robot needs a coordinate frame to be kinematically distinguished from other links. We use the coordinate frame transformation calculus to determine the relative position and orientation of the links. Although we are free to attach a coordinate frame at any point of a link, a point on a joint axis provides some advantages. Every joint is indicated by a joint axis, which will be either translational or rotational. To determine the kinematic information of a robot component, we rigidly attach a local coordinate frame B_i to every link (i) at joint $i + 1$ based on the following *standard method*, known as *Denavit–Hartenberg* (DH) method.

© The Author(s), under exclusive license to Springer Nature Switzerland AG 2022
R. N. Jazar, *Theory of Applied Robotics*, https://doi.org/10.1007/978-3-030-93220-6_5

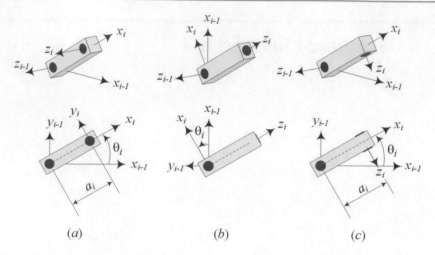

Fig. 5.2 A link with two revolute joints at its ends. The joint axes of link (**a**) are parallel, link (**b**) are orthogonal, and link (**c**) are perpendicular

Denavit–Hartenberg (DH) principles to set up a link coordinate frame:

1. The z_i-axis is aligned with the $i + 1$ joint axis.

   Every joint, without exception, is represented by a z-axis. To set up link coordinate frames, we always begin with identifying the z_i-axes. The positive direction of the z_i-axis is arbitrary. The joint axis for revolute joints is the pin's centerline axis. The axis of a prismatic joint may be any axis parallel to the direction of translation. By assigning the z_i-axes, the pairs of links on either side of every joint, and also the two joints on either side of every link, are identified. Although the z_i-axes of two joints at the ends of a link may be two skew lines, we usually build industrial robots in such a way that the z_i-axes are either *parallel*, *perpendicular*, or *orthogonal*. Two parallel joint axes are indicated by a parallel sign, ($\parallel$). Two joint axes are orthogonal if their axes are intersecting at a right angle. Orthogonal joint axes are indicated by an orthogonal sign, ($\vdash$). Two joints are perpendicular if their axes are not intersecting but their perpendicular planes are at right angle. Perpendicular joint axes are indicated by a perpendicular sign, ($\perp$). Figure 5.2a–c illustrates two revolute joints at two ends of a link at parallel, orthogonal, and perpendicular configurations, respectively.

   Every link of a robot, except the base and the final links, has two joints, one at each end. They are either revolute (R) or prismatic (P). The axes of the joints can be parallel ($\parallel$), perpendicular ($\perp$), or orthogonal ($\vdash$). Therefore, we may indicate a link by the type and configuration of the end joints. A P$\parallel$R indicates a link where its lower joint is prismatic, its upper joint is revolute, and the joint axes are parallel.

2. The x_i-axis is defined along the *common normal* between the z_{i-1} and z_i axes, pointing from the z_{i-1} to the z_i-axis.

   In general, the z axes may be skew lines; however, there is always one line mutually perpendicular to any two skew lines, called the common normal. The common normal has the shortest distance between the two skew lines.

   When the two z axes are parallel, there are an infinite number of common normals. In that case, we pick the common normal that is collinear with the common normal of the previous joint axes.

   When the two z axes are intersecting, there is no common normal between them. In that case, we assign the x_i-axis perpendicular to the plane formed by the two z axes in the direction of $z_{i-1} \times z_i$.

   In case the two z axes are collinear, the only nontrivial arrangement of joints is either P$\parallel$R or R$\parallel$P. We assign the x_i-axis for these arrangements in such a way that we have the joint variable equal to $\theta_i = 0$ in the rest position of the robot.

3. The y_i-axis is determined by the right-hand rule, $y_i = z_i \times x_i$.

   We assign coordinate frames to every link such that one of the three coordinate axes x_i, y_i, or z_i (usually x_i) is along the axis of the distal joint.

By applying the DH method, the origin o_i of the frame $B_i(o_i, x_i, y_i, z_i)$, attached to the link (i), is placed at the intersection of the joint axis $i + 1$ with the common normal between the z_{i-1} and z_i axes.

A DH coordinate frame is identified by four parameters: a_i, α_i, θ_i, and d_i.

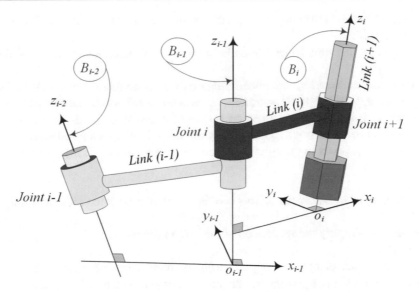

Fig. 5.3 Links $(i-1)$, (i), and $(i+1)$ along with coordinate frames B_i and B_{i+1}

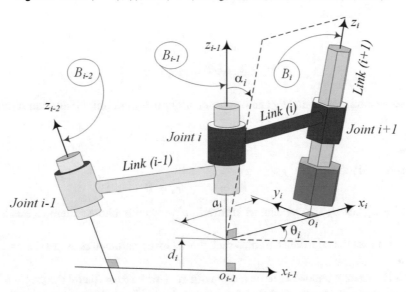

Fig. 5.4 DH parameters $a_i, \alpha_i, d_i, \theta_i$ defined for joint i and link (i)

1. *Link length a_i* is the distance between z_{i-1} and z_i axes along the x_i-axis. a_i is the *kinematic length* of the link (i). Hence, a_i is along the x_i-axis, from z_{i-1} to z_i axes.
2. *Link twist α_i* is the required rotation of the z_{i-1}-axis about the x_i-axis to become parallel to the z_i-axis. Hence, α_i is about the x_i-axis, from z_{i-1} to z_i axes.
3. *Joint distance d_i* is the distance between x_{i-1} and x_i axes along the z_{i-1}-axis. Joint distance is also called *link offset*. Hence, d_i is along the z_{i-1}-axis, from x_{i-1} to x_i axes.
4. *Joint angle θ_i* is the required rotation of x_{i-1}-axis about the z_{i-1}-axis to become parallel to the x_i-axis. Hence, θ_i is about the z_{i-1}-axis, from x_{i-1} to x_i axes.

Figure 5.4 illustrates the DH frame parameters $(a_i, \alpha_i, \theta_i, d_i)$ of the links shown in Fig. 5.3. The parameters θ_i and d_i are called *joint parameters*, because they define the relative position of two adjacent links connected at joint i. In a given design for a robot, every joint is revolute or prismatic. Thus, for every joint, it will always be the case that either θ_i or d_i is fixed and the other is the joint variable. For a revolute joint (R) at joint i, the θ_i is the unique joint variable, and the value of d_i is fixed. For a prismatic joint (P), the d_i is the only joint variable, while the value of θ_i is fixed. The variable parameter, either θ_i or d_i, is called the *joint variable*.

The joint parameters θ_i and d_i define a screw motion because θ_i is a rotation about the z_{i-1}-axis and d_i is a translation along the z_{i-1}-axis.

The parameters α_i and a_i are called *link parameters*, because they define relative positions of joints i and $i+1$ at two ends of link (i). The link twist α_i is the angle of rotation z_{i-1}-axis about x_i to become parallel with the z_i-axis. The other link kinematic length parameter, a_i, is the translation along the x_i-axis to bring the z_{i-1}-axis on the z_i-axis. The link parameters α_i and a_i define a screw motion because α_i is a rotation about the x_i-axis and a_i is a translation along the x_i-axis.

Therefore, we can move the z_{i-1}-axis to the z_i-axis by a central screw $\check{s}(a_i, \alpha_i, \hat{\imath})$ and move the x_{i-1}-axis to the x_i-axis by a central screw $\check{s}(d_i, \theta_i, \hat{k}_{i-1})$.

Example 158 Simplification tricks in DH method. This is reference to remind some tricks and simplified cases in applying DH method.

There are some comments to simplify the application of the DH frame method:

1. Showing only z and x axes is sufficient to identify a coordinate frame. Drawing is made clearer by not showing y axes. The DH coordinate frames are not unique because the direction of z_i axes are arbitrary. Every link, except the base and the last, is a binary link and is connected to two other links.
2. If the first and last joints are R, then

$$a_0 = 0 \qquad a_n = 0 \tag{5.1}$$

$$\alpha_0 = 0 \qquad \alpha_n = 0 \tag{5.2}$$

In these cases, the zero position for θ_1 and θ_n can be chosen arbitrarily, and link offsets can be set to zero.

$$d_1 = 0 \qquad d_n = 0 \tag{5.3}$$

3. If the first and last joints are P, then

$$\theta_1 = 0 \qquad \theta_n = 0 \tag{5.4}$$

and the zero position for d_1 and d_n can be chosen arbitrarily. We usually choose them to make as many parameters as possible to be zero.
4. A general applied trick is to set the coordinate frames such that as many parameters as possible are zero at the rest position of the system.

    If the final joint n is R, we may choose x_n to be along with x_{n-1} when $\theta_n = 0$ and the origin of B_n is such that $d_n = 0$. If the final joint n is P, we may choose x_n such that $\theta_n = 0$ and the origin of B_n is at the intersection of x_{n-1} and the axis of joint n to make $d_n = 0$.
5. The parameters a_i and α_i are determined by the geometric design of a robot and are always constant. The distance d_i is the offset of the frame B_i with respect to B_{i-1} along the z_{i-1}-axis. Because a_i is a length, $a_i \geq 0$.
6. The angles α_i and θ_i are directional. Their positive directions are determined by the right-hand rule according to the directions of x_i and z_{i-1}, respectively. For industrial robots, the link twist angle, α_i, is usually a multiple of $\pi/2$ radians.
7. The base frame $B_0(x_0, y_0, z_0) = G(X, Y, Z)$ is the global frame for an immobile robot. It is convenient to choose the Z-axis along the axis of joint 1 and set the origin O where the axes of the G frame are colinear or parallel with the axes of the frame B_1 at rest position.
8. The configuration of a robot at which all the joint variables are zero is called the *home configuration* or *rest position*, which is the reference configuration for all motions of a robot. The best rest position is where it makes as many axes parallel to each other and coplanar as possible.
9. For convenience, we can relax the strict DH definition for the direction of x_i, so that it points from z_i to z_{i-1} and still obtains a valid DH parameterization. A flip direction of the x_i-axis is to set a more convenient reference frame when most of the joint parameters are zero.

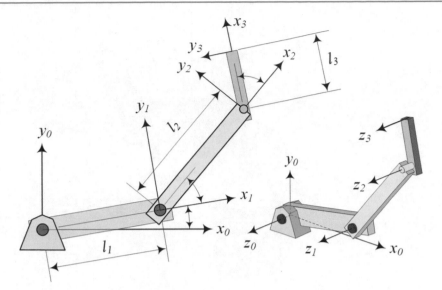

Fig. 5.5 Illustration of a $3R$ planar manipulator robot and DH frames of each link

Table 5.1 DH parameter table for establishment of link frames

Frame no.	a_i	α_i	d_i	θ_i
1	a_1	α_1	d_1	θ_1
2	a_2	α_2	d_2	θ_2
......	......	......	......	......
j	a_j	α_j	d_j	θ_j
......	......	......	......	......
n	a_n	α_n	d_n	θ_n

Table 5.2 DH table for the $3R$ planar manipulator robot of Fig. 5.5

Frame no.	a_i	α_i	d_i	θ_i
1	l_1	0	0	θ_1
2	l_2	0	0	θ_2
3	l_3	0	0	θ_3

10. A DH parameter table helps to establish a systematic link frame. As shown in Table 5.1, a DH table has five columns for frame index and four DH parameters. The rows of the four DH parameters for each frame will be filled by constant parameters and the joint variable. The joint variable can be found by considering what frames and links will move with each varying active joint.

Example 159 DH table and coordinate frames for $3R$ planar manipulator. An application of DH method on a planar multi-DOF robot.

An R∥R∥R manipulator is a planar robot with three parallel revolute joints. Figure 5.5 illustrates a $3R$ planar manipulator robot. The DH table can be filled as shown in Table 5.2, and the link coordinate frames can be set up as shown in Fig. 5.5.

Example 160 Coordinate frames for a $3R$ $PUMA$ robot. Application of DH method on a manipulator robot. This is a very important robot in industry.

A $PUMA$ manipulator is shown in Fig. 5.6. It has R⊢R∥R main structure, without considering the end-effector of the robot. Coordinate frames attached to the links of the robot are indicated in the figure and tabulated in Table 5.3.

The joint axes of an R⊥R∥R manipulator are called *waist* z_0, *shoulder* z_1, and *elbow* z_2. Typically, the joint axes z_1 and z_2 are parallel, and z_0 is perpendicular to z_1.

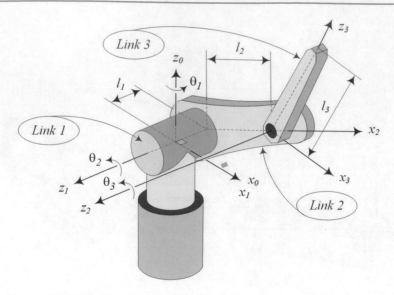

Fig. 5.6 $3R\ PUMA$ manipulator and links coordinate frame

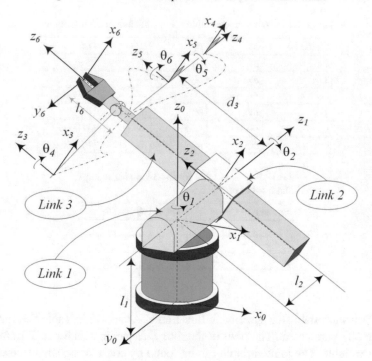

Fig. 5.7 Stanford arm R⊢R⊢P:R⊢R⊢R

Table 5.3 DH table for the $3R\ PUMA$ manipulator shown in Fig. 5.6

Frame no.	a_i	α_i	d_i	θ_i
1	0	90 deg	0	θ_1
2	l_2	0	l_1	θ_2
3	0	−90 deg	0	θ_3

Example 161 Stanford arm. Application of DH method on Stanford robot. This is one of the very first robots in robotic research.

Stanford arm is an R⊢R⊢P robot. A Stanford arm with a spherical wrist R⊢R⊢R is illustrated in Fig. 5.7.

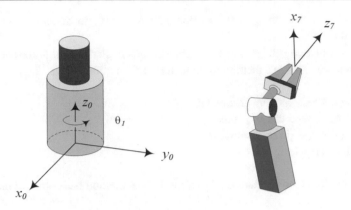

Fig. 5.8 Base and tools frames

Table 5.4 DH table for Stanford arm shown in Fig. 5.7

Frame no.	a_i	α_i	d_i	θ_i
1	0	-90 deg	l_1	θ_1
2	0	90 deg	l_2	θ_2
3	0	0	d_3	0
4	0	-90 deg	0	θ_4
5	0	90 deg	0	θ_5
6	0	0	l_6	θ_6

The DH parameters of the Stanford arm are tabulated in Table 5.4. This robot has 6 DOFs indicated by joint variables $\theta_1, \theta_2, d_3, \theta_4, \theta_5, \theta_6$.

Example 162 Special coordinate frames. The base and end-effector links have special coordinate frame because they connected to another link only from one side.

In a robotic manipulator, some frames have special names. The *base frame* B_0 or G is the grounded link in which the robot is installed. It is in the base frame that every kinematic information must be calculated. The departure point, path of motion, and the arrival point of the end-effector are all defined in this frame. The base frame is illustrated in Fig. 5.8. The *station frame S*, also called *world frame* or *universe frame*, is the frame in which the kinematics of the robot are calculated. Station frame is important when several robots are installed in a workshop, or the robot is mobile.

The *wrist frames W* are installed at the wrist point where the hand is attached to the last arm of a robot. The hands of industrial robots are usually attached to a wrist mechanism with three orthogonal revolute axes. The *tools frame T* is attached to the end of any tools the robot is holding. It may also be called the *end-effector frame* or *final frame*. When the hand is empty, the tools frame is chosen with its origin between the fingertips of the robot. The tools frame is illustrated in Fig. 5.8.

The *goal frame F* is the location where the robot is to move a tool. The goal frame is specified relative to the station frame.

Example 163 Transforming the frame B_{i-1} into B_i. The steps to construct DH transformation matrix between two neighbor coordinate frames.

Two neighbor coordinate frames can be brought into coincidence by several sequences of translations and rotations. However, to make a transformation matrix $^{i-1}T_i$ we should start from the moment that coordinate frame B_i is coincident with B_{i-1}, and then move B_i to come to its present position. Therefore, we must follow the following sequence of motions:

1. Rotate frame B_i through α_i about the x_{i-1}-axis.
2. Translate frame B_i along the x_{i-1}-axis by distance a_i.
3. Rotate frame B_i through θ_i about the z_{i-1}-axis.
4. Translate frame B_i along the z_{i-1}-axis by distance d_i.

During these maneuvers, B_{i-1} acts as the global coordinate frame for the local coordinate frame B_i, and these motions are about and along the global axes.

The following prescribed set of two rotations and two translations is also a straightforward method to move the frame B_{i-1} to coincide with the frame B_i. This is a method to make a transformation matrix $^iT_{i-1}$:

1. Translate frame B_{i-1} along the z_{i-1}-axis by distance d_i.
2. Rotate frame B_{i-1} through θ_i around the z_{i-1}-axis.
3. Translate frame B_{i-1} along the x_i-axis by distance a_i.
4. Rotate frame B_{i-1} through α_i about the x_i-axis.

Example 164 Shortcomings of the Denavit–Hartenberg method. DH method has some problems. Here are two important ones.

The Denavit–Hartenberg method for describing link coordinate frames is neither unique nor the best method. The drawbacks of the DH method are:

1. The successive coordinate axes are defined in such a way that the origin o_i and axis x_i are defined on the common perpendicular to adjacent link axes. This may be a difficult task depending on the geometry of the links and may produce singularity.
2. The DH notation cannot be extended to ternary and compound links.

5.2   Transformation Between Adjacent Coordinate Frames

The coordinate frame B_i is fixed to the link (i) and the coordinate frame B_{i-1} is fixed to the link $(i - 1)$. Based on the Denavit–Hartenberg convention, the transformation matrix $^{i-1}T_i$ to transform coordinate frames B_i into B_{i-1} is represented as a product of four basic transformations using the parameters of link (i) and joint i.

$$^{i-1}T_i = D_{z_{i-1},d_i} \; R_{z_{i-1},\theta_i} \; D_{x_{i-1},a_i} \; R_{x_{i-1},\alpha_i} \tag{5.5}$$

$$= \begin{bmatrix} \cos\theta_i & -\sin\theta_i \cos\alpha_i & \sin\theta_i \sin\alpha_i & a_i \cos\theta_i \\ \sin\theta_i & \cos\theta_i \cos\alpha_i & -\cos\theta_i \sin\alpha_i & a_i \sin\theta_i \\ 0 & \sin\alpha_i & \cos\alpha_i & d_i \\ 0 & 0 & 0 & 1 \end{bmatrix}$$

$$R_{x_{i-1},\alpha_i} = \begin{bmatrix} 1 & 0 & 0 & 0 \\ 0 & \cos\alpha_i & -\sin\alpha_i & 0 \\ 0 & \sin\alpha_i & \cos\alpha_i & 0 \\ 0 & 0 & 0 & 1 \end{bmatrix} \tag{5.6}$$

$$D_{x_{i-1},a_i} = \begin{bmatrix} 1 & 0 & 0 & a_i \\ 0 & 1 & 0 & 0 \\ 0 & 0 & 1 & 0 \\ 0 & 0 & 0 & 1 \end{bmatrix} \tag{5.7}$$

$$R_{z_{i-1},\theta_i} = \begin{bmatrix} \cos\theta_i & -\sin\theta_i & 0 & 0 \\ \sin\theta_i & \cos\theta_i & 0 & 0 \\ 0 & 0 & 1 & 0 \\ 0 & 0 & 0 & 1 \end{bmatrix} \tag{5.8}$$

$$D_{z_{i-1},d_i} = \begin{bmatrix} 1 & 0 & 0 & 0 \\ 0 & 1 & 0 & 0 \\ 0 & 0 & 1 & d_i \\ 0 & 0 & 0 & 1 \end{bmatrix} \tag{5.9}$$

Therefore, the transformation equation from coordinate frame $B_i(x_i, y_i, z_i)$, to its previous coordinate frame $B_{i-1}(x_{i-1}, y_{i-1}, z_{i-1})$, is

$$\begin{bmatrix} x_{i-1} \\ y_{i-1} \\ z_{i-1} \\ 1 \end{bmatrix} = {}^{i-1}T_i \begin{bmatrix} x_i \\ y_i \\ z_i \\ 1 \end{bmatrix} \tag{5.10}$$

where

$$ {}^{i-1}T_i = \begin{bmatrix} \cos\theta_i & -\sin\theta_i\cos\alpha_i & \sin\theta_i\sin\alpha_i & a_i\cos\theta_i \\ \sin\theta_i & \cos\theta_i\cos\alpha_i & -\cos\theta_i\sin\alpha_i & a_i\sin\theta_i \\ 0 & \sin\alpha_i & \cos\alpha_i & d_i \\ 0 & 0 & 0 & 1 \end{bmatrix} \tag{5.11}$$

This 4×4 homogenous transformation matrix may be partitioned into two submatrices, which represent a unique rotation combined with a unique translation to produce the same rigid motion required to move from B_i to B_{i-1}.

$$ {}^{i-1}T_i = \begin{bmatrix} {}^{i-1}R_i & {}^{i-1}\mathbf{d}_i \\ 0 & 1 \end{bmatrix} \tag{5.12}$$

$$ {}^{i-1}R_i = \begin{bmatrix} \cos\theta_i & -\sin\theta_i\cos\alpha_i & \sin\theta_i\sin\alpha_i \\ \sin\theta_i & \cos\theta_i\cos\alpha_i & -\cos\theta_i\sin\alpha_i \\ 0 & \sin\alpha_i & \cos\alpha_i \end{bmatrix} \tag{5.13}$$

$$ {}^{i-1}\mathbf{d}_i = \begin{bmatrix} a_i\cos\theta_i \\ a_i\sin\theta_i \\ d_i \end{bmatrix} \tag{5.14}$$

The inverse of the homogenous transformation matrix (5.11) is

$$ {}^{i}T_{i-1} = {}^{i-1}T_i^{-1} \tag{5.15}$$

$$ = \begin{bmatrix} \cos\theta_i & \sin\theta_i & 0 & -a_i \\ -\sin\theta_i\cos\alpha_i & \cos\theta_i\cos\alpha_i & \sin\alpha_i & -d_i\sin\alpha_i \\ \sin\theta_i\sin\alpha_i & -\cos\theta_i\sin\alpha_i & \cos\alpha_i & -d_i\cos\alpha_i \\ 0 & 0 & 0 & 1 \end{bmatrix} $$

Proof 1

Equation (5.5) indicates the direct method to prove the DH homogenous transformation matrix (5.11). However, we may provide alternative proofs to look at this matrix from different viewpoints.

Assume the coordinate frames $B_2(x_2, y_2, z_2)$ and $B_1(x_1, y_1, z_1)$ in Fig. 5.9 are set up based on Denavit–Hartenberg rules. The position vector of point P can be expressed in frame $B_1(x_1, y_1, z_1)$ using ${}^2\mathbf{r}_P$ and ${}^1\mathbf{s}_2$,

$$ {}^1\mathbf{r}_P = {}^1R_2 \, {}^2\mathbf{r}_P + {}^1\mathbf{s}_2 \tag{5.16}$$

which, in homogenous coordinate representation, is equal to

$$ \begin{bmatrix} x_1 \\ y_1 \\ z_1 \\ 1 \end{bmatrix} = \begin{bmatrix} \cos(\hat{i}_2, \hat{i}_1) & \cos(\hat{j}_2, \hat{i}_1) & \cos(\hat{k}_2, \hat{i}_1) & s_{2_x} \\ \cos(\hat{i}_2, \hat{j}_1) & \cos(\hat{j}_2, \hat{j}_1) & \cos(\hat{k}_2, \hat{j}_1) & s_{2_y} \\ \cos(\hat{i}_2, \hat{k}_1) & \cos(\hat{j}_2, \hat{k}_1) & \cos(\hat{k}_2, \hat{k}_1) & s_{2_z} \\ 0 & 0 & 0 & 1 \end{bmatrix} \begin{bmatrix} x_2 \\ y_2 \\ z_2 \\ 1 \end{bmatrix} \tag{5.17}$$

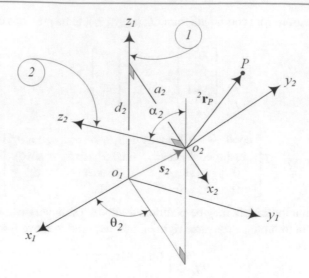

Fig. 5.9 Two coordinate frames based on Denavit–Hartenberg rules

Using the parameters introduced in Fig. 5.9, Eq. (5.17) becomes

$$
\begin{bmatrix} x_1 \\ y_1 \\ z_1 \\ 1 \end{bmatrix} = \begin{bmatrix} c\theta_2 & -s\theta_2 c\alpha_2 & s\theta_2 s\alpha_2 & a_2 c\theta_2 \\ s\theta_2 & c\theta_2 c\alpha_2 & -c\theta_2 s\alpha_2 & a_2 s\theta_2 \\ 0 & s\alpha_2 & c\alpha_2 & d_2 \\ 0 & 0 & 0 & 1 \end{bmatrix} \begin{bmatrix} x_2 \\ y_2 \\ z_2 \\ 1 \end{bmatrix}
\tag{5.18}
$$

If we substitute the coordinate frame B_1 by $B_{i-1}(x_{i-1}, y_{i-1}, z_{i-1})$, and B_2 by $B_i(x_i, y_i, z_i)$, we may rewrite the above equation in the required form:

$$
\begin{bmatrix} x_{i-1} \\ y_{i-1} \\ z_{i-1} \\ 1 \end{bmatrix} = \begin{bmatrix} c\theta_i & -s\theta_i c\alpha_i & s\theta_i s\alpha_i & a_i c\theta_i \\ s\theta_i & c\theta_i c\alpha_i & -c\theta_i s\alpha_i & a_i s\theta_i \\ 0 & s\alpha_i & c\alpha_i & d_i \\ 0 & 0 & 0 & 1 \end{bmatrix} \begin{bmatrix} x_i \\ y_i \\ z_i \\ 1 \end{bmatrix}
\tag{5.19}
$$

Following the inversion rule of homogenous transformation matrix,

$$
{}^{i-1}T_i = \begin{bmatrix} {}^G R_B & {}^G \mathbf{s} \\ 0 & 1 \end{bmatrix}
\tag{5.20}
$$

$$
{}^{i-1}T_i^{-1} = \begin{bmatrix} {}^G R_B^T & -{}^G R_B^T \, {}^G \mathbf{s} \\ 0 & 1 \end{bmatrix}
\tag{5.21}
$$

we also find the required inverse transformation (5.15).

$$
\begin{bmatrix} x_i \\ y_i \\ z_i \\ 1 \end{bmatrix} = \begin{bmatrix} c\theta_i & s\theta_i & 0 & -a_i \\ -s\theta_i c\alpha_i & c\theta_i c\alpha_i & s\alpha_i & -d_i s\alpha_i \\ s\theta_i s\alpha_i & -c\theta_i s\alpha_i & c\alpha_i & -d_i c\alpha_i \\ 0 & 0 & 0 & 1 \end{bmatrix} \begin{bmatrix} x_{i-1} \\ y_{i-1} \\ z_{i-1} \\ 1 \end{bmatrix}
\tag{5.22}
$$

■

Proof 2

An alternative method to find ${}^i T_{i-1}$ is to follow the sequence of translations and rotations that brings the frame B_{i-1} to the present configuration starting from a coincident position with the frame B_i. Working with two frames can also be equivalently described by $B \equiv B_i$ and $G \equiv B_{i-1}$, so all the following rotations and translations are about and along the local coordinate frame axes. Inspecting Fig. 5.9 shows that:

1. Frame B_{i-1} is translated along the local z_i-axis by distance $-d_i$.
2. The displaced frame B_{i-1} is rotated through $-\theta_i$ about the local z_i-axis.
3. The displaced frame B_{i-1} is translated along the local x_i-axis by distance $-a_i$.
4. The displaced frame B_{i-1} is rotated through $-\alpha_i$ about the local x_i-axis.

Following these displacement sequences, the transformation matrix $^iT_{i-1}$ would be

$$^iT_{i-1} = R_{x_i,-\alpha_i}\, D_{x_i,-a_i}\, R_{z_i,-\theta_i}\, D_{z_i,-d_i} \tag{5.23}$$

$$= \begin{bmatrix} \cos\theta_i & \sin\theta_i & 0 & -a_i \\ -\sin\theta_i\cos\alpha_i & \cos\theta_i\cos\alpha_i & \sin\alpha_i & -d_i\sin\alpha_i \\ \sin\theta_i\sin\alpha_i & -\cos\theta_i\sin\alpha_i & \cos\alpha_i & -d_i\cos\alpha_i \\ 0 & 0 & 0 & 1 \end{bmatrix}$$

where

$$D_{z_i,-d_i} = \begin{bmatrix} 1 & 0 & 0 & 0 \\ 0 & 1 & 0 & 0 \\ 0 & 0 & 1 & -d_i \\ 0 & 0 & 0 & 1 \end{bmatrix} \tag{5.24}$$

$$R_{z_i,-\theta_i} = \begin{bmatrix} \cos\theta_i & -\sin\theta_i & 0 & 0 \\ \sin\theta_i & \cos\theta_i & 0 & 0 \\ 0 & 0 & 1 & 0 \\ 0 & 0 & 0 & 1 \end{bmatrix} \tag{5.25}$$

$$D_{x_i,-a_i} = \begin{bmatrix} 1 & 0 & 0 & -a_i \\ 0 & 1 & 0 & 0 \\ 0 & 0 & 1 & 0 \\ 0 & 0 & 0 & 1 \end{bmatrix} \tag{5.26}$$

$$R_{x_i,-\alpha_i} = \begin{bmatrix} 1 & 0 & 0 & 0 \\ 0 & \cos\alpha_i & -\sin\alpha_i & 0 \\ 0 & \sin\alpha_i & \cos\alpha_i & 0 \\ 0 & 0 & 0 & 1 \end{bmatrix} \tag{5.27}$$

Using $^{i-1}T_i = {}^iT_{i-1}^{-1}$ we find:

$$^{i-1}T_i = {}^iT_{i-1}^{-1} \tag{5.28}$$

$$= \begin{bmatrix} \cos\theta_i & -\sin\theta_i\cos\alpha_i & \sin\theta_i\sin\alpha_i & a_i\cos\theta_i \\ \sin\theta_i & \cos\theta_i\cos\alpha_i & -\cos\theta_i\sin\alpha_i & a_i\sin\theta_i \\ 0 & \sin\alpha_i & \cos\alpha_i & d_i \\ 0 & 0 & 0 & 1 \end{bmatrix}$$

∎

Example 165 DH transformation matrices for a $2R$ planar manipulator. Application of DH method and forward kinematics of a planar two DOF, $2R$ planar robot.

Figure 5.10 illustrates an R‖R planar manipulator and its DH link coordinate frames.

Based on DH Table 5.5 we can find the transformation matrices from frame B_i to frame B_{i-1} by direct substitution of DH parameters in Eq. (5.11). Therefore,

$$^1T_2 = \begin{bmatrix} \cos\theta_2 & -\sin\theta_2 & 0 & l_2\cos\theta_2 \\ \sin\theta_2 & \cos\theta_2 & 0 & l_2\sin\theta_2 \\ 0 & 0 & 1 & 0 \\ 0 & 0 & 0 & 1 \end{bmatrix} \tag{5.29}$$

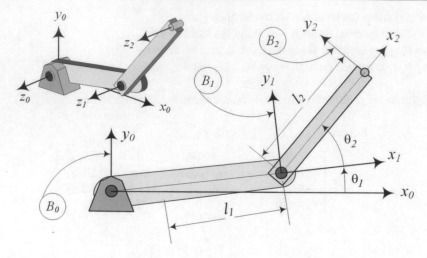

Fig. 5.10 Illustration of a $2R$ planar manipulator robot and DH frames of every link

Table 5.5 DH table for $2R$ planar manipulator shown in Fig. 5.10

Frame no.	a_i	α_i	d_i	θ_i
1	l_1	0	0	θ_1
2	l_2	0	0	θ_2

$$
{}^0T_1 = \begin{bmatrix} \cos\theta_1 & -\sin\theta_1 & 0 & l_1\cos\theta_1 \\ \sin\theta_1 & \cos\theta_1 & 0 & l_1\sin\theta_1 \\ 0 & 0 & 1 & 0 \\ 0 & 0 & 0 & 1 \end{bmatrix} \tag{5.30}
$$

and consequently the forward kinematics of the $2R$ planar robot will be

$$
\begin{aligned}
{}^0T_2 &= {}^0T_1\, {}^1T_2 \\
&= \begin{bmatrix} c\,(\theta_1+\theta_2) & -s\,(\theta_1+\theta_2) & 0 & l_1c\theta_1 + l_2c\,(\theta_1+\theta_2) \\ s\,(\theta_1+\theta_2) & c\,(\theta_1+\theta_2) & 0 & l_1s\theta_1 + l_2s\,(\theta_1+\theta_2) \\ 0 & 0 & 1 & 0 \\ 0 & 0 & 0 & 1 \end{bmatrix}
\end{aligned} \tag{5.31}
$$

Example 166 Classification of industrial robot links. Most of industrial robots are built by assembling the links that are classified here. This example is important to be used as a reference to model all industrial robots easily.

A robot link is identified and classified by its joints at both ends. Such classification determines the DH homogenous transformation matrix to go from the distal joint coordinate frame B_i to the proximal joint coordinate frame B_{i-1}. There are 12 types of links to make an industrial robot. The transformation matrix for each type depends solely on the proximal joint and angle between the z axes. The 12 types of transformation matrices are summarized in Table 5.6.

Their geometric illustration and their associated DH matrices are presented in Appendix D of the book as a reference to be used to solve forward kinematics of industrial robots.

Table 5.6 DH classification of industrial robot links. The angle in parenthesis is α_i, the angle from z_{i-1} to z_i about x_i

1	R∥R(0)	or	R∥P(0)
2	R∥R(180)	or	R∥P(180)
3	R⊥R(90)	or	R⊥P(90)
4	R⊥R(−90)	or	R⊥P(−90)
5	R⊢R(90)	or	R⊢P(90)
6	R⊢R(−90)	or	R⊢P(−90)
7	P∥R(0)	or	P∥P(0)
8	P∥R(180)	or	P∥P(180)
9	P⊥R(90)	or	P⊥P(90)
10	P⊥R(−90)	or	P⊥P(−90)
11	P⊢R(90)	or	P⊢P(90)
12	P⊢R(−90)	or	P⊢P(−90)

1. Link with R∥R(0) or R∥P(0) joints: $a = const$, $\alpha_i = 0$, $d_i = const$, θ_i is the variable.

$$
{}^{i-1}T_i = \begin{bmatrix} \cos\theta_i & -\sin\theta_i & 0 & a_i\cos\theta_i \\ \sin\theta_i & \cos\theta_i & 0 & a_i\sin\theta_i \\ 0 & 0 & 1 & d_i \\ 0 & 0 & 0 & 1 \end{bmatrix} \tag{5.32}
$$

2. Link with R∥R(180) or R∥P(180): $a = const$, $\alpha_i = \pi$, $d_i = const$, θ_i is the variable.

$$
{}^{i-1}T_i = \begin{bmatrix} \cos\theta_i & \sin\theta_i & 0 & a_i\cos\theta_i \\ \sin\theta_i & -\cos\theta_i & 0 & a_i\sin\theta_i \\ 0 & 0 & -1 & d_i \\ 0 & 0 & 0 & 1 \end{bmatrix} \tag{5.33}
$$

3. Link with R⊥R(90) or R⊥P(90) joints: $a = const$, $\alpha_i = \pi/2$, $d_i = const$, θ_i is the variable.

$$
{}^{i-1}T_i = \begin{bmatrix} \cos\theta_i & 0 & \sin\theta_i & a_i\cos\theta_i \\ \sin\theta_i & 0 & -\cos\theta_i & a_i\sin\theta_i \\ 0 & 1 & 0 & d_i \\ 0 & 0 & 0 & 1 \end{bmatrix} \tag{5.34}
$$

4. Link with R⊥R(−90) or R⊥P(−90) joints: $a = const$, $\alpha_i = -\pi/2$, $d_i = const$, θ_i is the variable.

$$
{}^{i-1}T_i = \begin{bmatrix} \cos\theta_i & 0 & -\sin\theta_i & a_i\cos\theta_i \\ \sin\theta_i & 0 & \cos\theta_i & a_i\sin\theta_i \\ 0 & -1 & 0 & d_i \\ 0 & 0 & 0 & 1 \end{bmatrix} \tag{5.35}
$$

5. Link with R⊢R(90) or R⊢P(90) joints: $a = 0$, $\alpha_i = \pi/2$, $d_i = const$, θ_i is the variable.

$$
{}^{i-1}T_i = \begin{bmatrix} \cos\theta_i & 0 & \sin\theta_i & 0 \\ \sin\theta_i & 0 & -\cos\theta_i & 0 \\ 0 & 1 & 0 & d_i \\ 0 & 0 & 0 & 1 \end{bmatrix} \tag{5.36}
$$

6. Link with R⊢R(−90) or R⊢P(−90) joints: $a = 0, \alpha_i = -\pi/2, d_i = const, \theta_i$ is the variable.

$$
{}^{i-1}T_i = \begin{bmatrix} \cos\theta_i & 0 & -\sin\theta_i & 0 \\ \sin\theta_i & 0 & \cos\theta_i & 0 \\ 0 & -1 & 0 & d_i \\ 0 & 0 & 0 & 1 \end{bmatrix}
\tag{5.37}
$$

7. Link with P∥R(0) or P∥P(0) joints. $a = const, \alpha_i = 0, d_i$ is the variable, $\theta_i = 0$.

$$
{}^{i-1}T_i = \begin{bmatrix} 1 & 0 & 0 & a_i \\ 0 & 1 & 0 & 0 \\ 0 & 0 & 1 & d_i \\ 0 & 0 & 0 & 1 \end{bmatrix}
\tag{5.38}
$$

8. Link with P∥R(180) or P∥P(180) joints. $a = const, \alpha_i = \pi, d_i$ is the variable, $\theta_i = 0$.

$$
{}^{i-1}T_i = \begin{bmatrix} 1 & 0 & 0 & a_i \\ 0 & -1 & 0 & 0 \\ 0 & 0 & -1 & d_i \\ 0 & 0 & 0 & 1 \end{bmatrix}
\tag{5.39}
$$

9. Link with P⊥R(90) or P⊥P(90) joints. $a = const, \alpha_i = \pi/2, d_i$ is the variable, $\theta_i = 0$.

$$
{}^{i-1}T_i = \begin{bmatrix} 1 & 0 & 0 & a_i \\ 0 & 0 & -1 & 0 \\ 0 & 1 & 0 & d_i \\ 0 & 0 & 0 & 1 \end{bmatrix}
\tag{5.40}
$$

10. Link with P⊥R(−90) or P⊥P(−90) joints. $a = const, \alpha_i = -\pi/2, d_i$ is the variable, $\theta_i = 0$.

$$
{}^{i-1}T_i = \begin{bmatrix} 1 & 0 & 0 & a_i \\ 0 & 0 & 1 & 0 \\ 0 & -1 & 0 & d_i \\ 0 & 0 & 0 & 1 \end{bmatrix}
\tag{5.41}
$$

11. Link with P⊢R(90) or P⊢P(90) joints. $a = 0, \alpha_i = \pi/2, d_i$ is the variable, $\theta_i = 0$.

$$
{}^{i-1}T_i = \begin{bmatrix} 1 & 0 & 0 & 0 \\ 0 & 0 & -1 & 0 \\ 0 & 1 & 0 & d_i \\ 0 & 0 & 0 & 1 \end{bmatrix}
\tag{5.42}
$$

12. Link with P⊢R(−90) or P⊢P(−90) joints. $a = 0, \alpha_i = -\pi/2, d_i$ is the variable, $\theta_i = 0$.

$$
{}^{i-1}T_i = \begin{bmatrix} 1 & 0 & 0 & 0 \\ 0 & 0 & 1 & 0 \\ 0 & -1 & 0 & d_i \\ 0 & 0 & 0 & 1 \end{bmatrix}
\tag{5.43}
$$

Example 167 Assembling industrial links to make a manipulator. Illustration of making a robot by assembling classified links.

Industrial manipulators are usually made by connecting the introduced industrial links in Example 166. A manipulator is a combination of three links that provide three $DOFs$ to the tip point in a Cartesian space. The articulated and spherical manipulators are two common and practical manipulators. Figure 5.11a and b illustrates how we make these manipulators by connecting the proper industrial links.

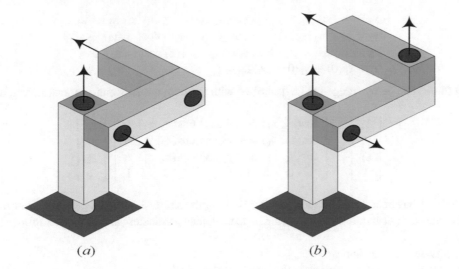

Fig. 5.11 The articulated and spherical manipulators are two common and practical manipulators

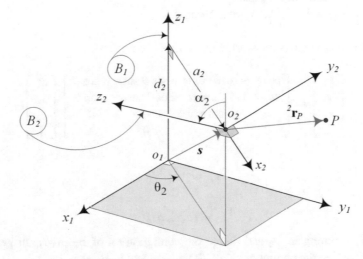

Fig. 5.12 Alternative method to derive the Denavit–Hartenberg coordinate transformation

Example 168 DH coordinate transformation based on vector addition. Alternative method to construct DH transformation matrix using vector addition.

The DH transformation from a coordinate frame to another may be described by a vector addition. The coordinates of a point P in frame B_1, as shown in Fig. 5.12, are given by a vector equation,

$$^{B_1}\overrightarrow{o_1 P} = {}^{B_2}\overrightarrow{o_2 P} + {}^{B_1}\overrightarrow{o_1 o_2} \tag{5.44}$$

where

$$^{B_1}\overrightarrow{o_1 o_2} = \begin{bmatrix} s_1 & s_2 & s_3 \end{bmatrix}^T \tag{5.45}$$

$$^{B_1}\overrightarrow{o_1 P} = \begin{bmatrix} x_1 & y_1 & z_1 \end{bmatrix}^T \tag{5.46}$$

$$^{B_2}\overrightarrow{o_2 P} = \begin{bmatrix} x_2 & y_2 & z_2 \end{bmatrix}^T \tag{5.47}$$

They must be expressed in the same coordinate frame. Employing cosines of the angles between axes of the two coordinate frames, we have

$$
\begin{aligned}
x_1 &= x_2 \cos(x_2, x_1) + y_2 \cos(y_2, x_1) + z_2 \cos(z_2, x_1) + s_1 \\
y_1 &= x_2 \cos(x_2, y_1) + y_2 \cos(y_2, y_1) + z_2 \cos(z_2, y_1) + s_2 \\
z_1 &= x_2 \cos(x_2, z_1) + y_2 \cos(y_2, z_1) + z_2 \cos(z_2, z_1) + s_3 \\
1 &= x_2(0) + y_2(0) + z_2(0) + 1
\end{aligned}
\tag{5.48}
$$

The transformation (5.48) can be rearranged to be described with the homogenous matrix transformation.

$$
\begin{bmatrix} x_1 \\ y_1 \\ z_1 \\ 1 \end{bmatrix} =
\begin{bmatrix}
\cos(x_2, x_1) & \cos(y_2, x_1) & \cos(z_2, x_1) & s_1 \\
\cos(x_2, y_1) & \cos(y_2, y_1) & \cos(z_2, y_1) & s_2 \\
\cos(x_2, z_1) & \cos(y_2, z_1) & \cos(z_2, z_1) & s_3 \\
0 & 0 & 0 & 1
\end{bmatrix}
\begin{bmatrix} x_2 \\ y_2 \\ z_2 \\ 1 \end{bmatrix}
\tag{5.49}
$$

In Fig. 5.12 the axis x_2 has been selected such that it lies along the shortest common perpendicular between axes z_1 and z_2. The axis y_2 completes a right-handed set of coordinate axes. Other parameters are defined as follows:

1. a is the distance between axes z_1 and z_2.
2. α is the twist angle that screws the z_1-axis into the z_2-axis along a.
3. d is the distance from the x_1-axis to the x_2-axis.
4. θ is the angle that screws the x_1-axis into the x_2-axis along d.

Using these definitions, the homogenous transformation matrix becomes

$$
\begin{bmatrix} x_1 \\ y_1 \\ z_1 \\ 1 \end{bmatrix} =
\begin{bmatrix}
\cos\theta & -\sin\theta\cos\alpha & -\sin\theta\sin\alpha & a\cos\theta \\
\sin\theta & \cos\theta\cos\alpha & \cos\theta\sin\alpha & a\sin\theta \\
0 & -\sin\alpha & \cos\alpha & d \\
0 & 0 & 0 & 1
\end{bmatrix}
\begin{bmatrix} x_2 \\ y_2 \\ z_2 \\ 1 \end{bmatrix}
\tag{5.50}
$$

or

$$
{}^1\mathbf{r}_P = {}^1T_2 \, {}^2\mathbf{r}_P
\tag{5.51}
$$

where

$$
{}^1T_2 = (a, \alpha, d, \theta)
\tag{5.52}
$$

The DH parameters a, α, θ, d belong to B_2 and define the configuration of B_2 in B_1. In general, the DH parameters $a_i, \alpha_i, \theta_i, d_i$ belong to B_i and define the configuration of B_i with respect to B_{i-1}:

$$
{}^{i-1}T_i = (a_i, \alpha_i, d_i, \theta_i)
\tag{5.53}
$$

Example 169 Trebuchet kinematics as a multibody. How to apply DH on multibody systems. Introducing Sina coordinate frames.

The trebuchet, shown in Fig. 5.13, is a shooting weapon of war powered by a falling massive counterweight m_1. The beam AB is pivoted to the chassis with two unequal sections a and b. The counterweight m_1 is hinged at the shorter arm of the beam at a distance c from the end A. The mass of the projectile is m_2, and it is at the end of a rope with a length l attached to the end B of the longer arm of the beam.

To analyze the trebuchet as a robotic system, we attach a global coordinate frame G at the fixed pivot M and three body frames B, B_1, B_2 to the three moving bodies as shown in Fig. 5.13. The three independent variables α, θ, and γ define the relative angular positions of the bodies. Let us assume the parameters a, b, c, l, m_1, m_2 to be constant and find the transformation matrices to determine the coordinates in the relative coordinate frames. We also attach two Sina coordinate frames B_3 and B_4 to the beam at joints A and B to simplify the coordinate transformations. A Sina coordinate frame is an extra frame that will be installed to simplify calculations of the main coordinate frames transformations.

We may find ${}^BT_G$ by using the transformation matrix of a zero length link R$\|$R(0) or equivalently by turning B a rotation $-\theta$ about the z-axis:

$$
{}^BT_G = R_{z,-\theta} =
\begin{bmatrix}
\cos-\theta & \sin-\theta & 0 & 0 \\
-\sin-\theta & \cos-\theta & 0 & 0 \\
0 & 0 & 1 & 0 \\
0 & 0 & 0 & 1
\end{bmatrix}
\tag{5.54}
$$

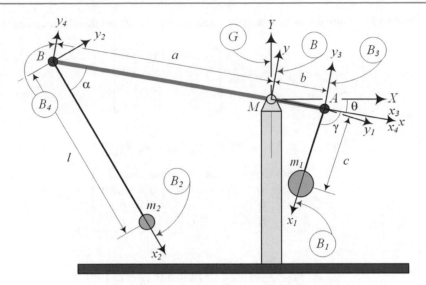

Fig. 5.13 The kinematic model of a trebuchet

The Sina frame B_3 is connected to the main frame B by a translation b along x_3-axis. Similarly, the Sina frame B_4 is connected to B by a translation $-a$ along x_4-axis.

$$^3T_B = D_{x_3,b} = \begin{bmatrix} 1 & 0 & 0 & -b \\ 0 & 1 & 0 & 0 \\ 0 & 0 & 1 & 0 \\ 0 & 0 & 0 & 1 \end{bmatrix} \tag{5.55}$$

$$^4T_B = D_{x_4,-a} = \begin{bmatrix} 1 & 0 & 0 & a \\ 0 & 1 & 0 & 0 \\ 0 & 0 & 1 & 0 \\ 0 & 0 & 0 & 1 \end{bmatrix} \tag{5.56}$$

We find $^1T_3$ by turning B_1 a rotation $-\gamma$ about the z_1-axis:

$$^1T_3 = R_{z_1,-\gamma} = \begin{bmatrix} \cos -\gamma & \sin -\gamma & 0 & 0 \\ -\sin -\gamma & \cos -\gamma & 0 & 0 \\ 0 & 0 & 1 & 0 \\ 0 & 0 & 0 & 1 \end{bmatrix} \tag{5.57}$$

Using the same method, we find $^2T_4$ by turning B_2 a rotation $-\alpha$ about the z_2-axis:

$$^2T_4 = R_{z_2,-\alpha} = \begin{bmatrix} \cos -\alpha & \sin -\alpha & 0 & 0 \\ -\sin -\alpha & \cos -\alpha & 0 & 0 \\ 0 & 0 & 1 & 0 \\ 0 & 0 & 0 & 1 \end{bmatrix} \tag{5.58}$$

Therefore, ^1T_B and ^2T_B are

$$^1T_B = {}^1T_3\,{}^3T_B = \begin{bmatrix} \cos \gamma & -\sin \gamma & 0 & b\cos \gamma \\ \sin \gamma & \cos \gamma & 0 & -b\sin \gamma \\ 0 & 0 & 1 & 0 \\ 0 & 0 & 0 & 1 \end{bmatrix} \tag{5.59}$$

$$^2T_B = {}^2T_4\,{}^4T_B = \begin{bmatrix} \cos \alpha & -\sin \alpha & 0 & a\cos \alpha \\ \sin \alpha & \cos \alpha & 0 & a\sin \alpha \\ 0 & 0 & 1 & 0 \\ 0 & 0 & 0 & 1 \end{bmatrix} \tag{5.60}$$

By matrix multiplication and a homogenous matrix inverse calculation, we determine the required transformation matrices to go to the G frame:

$$^G T_B = {}^B T_G^{-1} = \begin{bmatrix} \cos\theta & \sin\theta & 0 & 0 \\ -\sin\theta & \cos\theta & 0 & 0 \\ 0 & 0 & 1 & 0 \\ 0 & 0 & 0 & 1 \end{bmatrix} \tag{5.61}$$

$$^G T_1 = {}^G T_B \, {}^B T_1 = {}^B T_G^{-1} \, {}^1 T_B^{-1} = \left[{}^1 T_B \, {}^B T_G \right]^{-1}$$

$$= \begin{bmatrix} \cos(\theta+\gamma) & \sin(\theta+\gamma) & 0 & b\cos\theta \\ -\sin(\theta+\gamma) & \cos(\theta+\gamma) & 0 & -b\sin\theta \\ 0 & 0 & 1 & 0 \\ 0 & 0 & 0 & 1 \end{bmatrix} \tag{5.62}$$

$$^G T_2 = {}^G T_B \, {}^B T_2 = {}^B T_G^{-1} \, {}^2 T_B^{-1} = \left[{}^2 T_B \, {}^B T_G \right]^{-1}$$

$$= \begin{bmatrix} \cos(\theta+\alpha) & \sin(\theta+\alpha) & 0 & -a\cos\theta \\ -\sin(\theta+\alpha) & \cos(\theta+\alpha) & 0 & a\sin\theta \\ 0 & 0 & 1 & 0 \\ 0 & 0 & 0 & 1 \end{bmatrix} \tag{5.63}$$

Therefore, the global coordinates of m_1 at ${}^1\mathbf{r}_{m_1} = \begin{bmatrix} c & 0 & 0 \end{bmatrix}^T$ and m_2 at ${}^2\mathbf{r}_{m_2} = \begin{bmatrix} l & 0 & 0 \end{bmatrix}^T$ are

$$^G\mathbf{r}_{m_1} = {}^G T_1 \, {}^1\mathbf{r}_{m_1} = \begin{bmatrix} b\cos\theta + c\cos(\theta+\gamma) \\ -b\sin\theta - c\sin(\theta+\gamma) \\ 0 \\ 1 \end{bmatrix} \tag{5.64}$$

$$^G\mathbf{r}_{m_2} = {}^G T_2 \, {}^2\mathbf{r}_{m_2} = \begin{bmatrix} l\cos(\theta+\alpha) - a\cos\theta \\ a\sin\theta - l\sin(\theta+\alpha) \\ 0 \\ 1 \end{bmatrix} \tag{5.65}$$

Example 170 The same DH transformation matrix. Decomposition of DH matrix in different methods, all equivalent.

In the DH method of setting coordinate frames, because a translation D and a rotation R are along and about the same axis, it is immaterial if we first apply the translation D and then the rotation R or vice versa. Therefore, we can interchange the order of D and R along and about the same axis and obtain the same DH transformation matrix 5.11.

$$^{i-1} T_i = D_{z_i, d_i} \, R_{z_i, \theta_i} \, D_{x_i, a_i} \, R_{x_i, \alpha_i}$$

$$= R_{z_i, \theta_i} \, D_{z_i, d_i} \, D_{x_i, a_i} \, R_{x_i, \alpha_i}$$

$$= D_{z_i, d_i} \, R_{z_i, \theta_i} \, R_{x_i, \alpha_i} \, D_{x_i, a_i}$$

$$= R_{z_i, \theta_i} \, D_{z_i, d_i} \, R_{x_i, \alpha_i} \, D_{x_i, a_i} \tag{5.66}$$

Example 171 Non-DH transformation. DH is not the only way to construct transformation matrices between links of a robot, and it is sometimes not possible as well. Here is an illustration how to deal with such cases.

The robot kinematic problems are always related to a correct expression of the transformation matrix between two coordinate frames. The classical method is to set up the coordinate frames based on DH rules and use the DH transformation matrix. However, when the DH rules cannot be applied, we recommend determining the homogenous transformation matrix by a proper combination of principal rotations and translations. Consider two arbitrary coordinate frames B and G. The

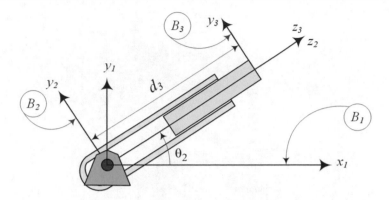

Fig. 5.14 A 2 DOF polar manipulator with non-DH coordinate frames

homogenous transformation matrix $^B T_G$ can be found by moving B to its present position from the coincident configuration with G. The rotations are positive about the local axes of B and the translations are negative along the local axes of B.

Figure 5.14 depicts a polar manipulator with two DOFs and a set of coordinate frames that are not set up exactly according to the DH rules. To determine the link's transformation matrices $^1 T_2$, $^2 T_3$, and $^1 T_3$ in such cases, we may always follow the method of homogenous transformations. The frame B_1 can go to B_2 by a rotation R_{z_2,θ_2} followed by $R_{y_2,90}$. There is no translation between B_1 and B_2. Therefore,

$$^1 R_2 = \left[R_{y_2,90} \, R_{z_2,\theta_2} \right]^T = R_{z_2,\theta_2}^T \, R_{y_2,90}^T$$

$$= \begin{bmatrix} \cos\theta_2 & \sin\theta_2 & 0 \\ -\sin\theta_2 & \cos\theta_2 & 0 \\ 0 & 0 & 1 \end{bmatrix}^T \begin{bmatrix} \cos\dfrac{\pi}{2} & 0 & -\sin\dfrac{\pi}{2} \\ 0 & 1 & 0 \\ \sin\dfrac{\pi}{2} & 0 & \cos\dfrac{\pi}{2} \end{bmatrix}^T$$

$$= \begin{bmatrix} 0 & -\sin\theta_2 & \cos\theta_2 \\ 0 & \cos\theta_2 & \sin\theta_2 \\ -1 & 0 & 0 \end{bmatrix} \tag{5.67}$$

and hence,

$$^1 T_2 = \begin{bmatrix} 0 & -\sin\theta_2 & \cos\theta_2 & 0 \\ 0 & \cos\theta_2 & \sin\theta_2 & 0 \\ -1 & 0 & 0 & 0 \\ 0 & 0 & 0 & 1 \end{bmatrix} \tag{5.68}$$

It is also possible to determine $^1 R_2$ directly from the definition of the directional cosines:

$$^1 R_2 = \begin{bmatrix} \hat{I} \cdot \hat{i} & \hat{I} \cdot \hat{j} & \hat{I} \cdot \hat{k} \\ \hat{J} \cdot \hat{i} & \hat{J} \cdot \hat{j} & \hat{J} \cdot \hat{k} \\ \hat{K} \cdot \hat{i} & \hat{K} \cdot \hat{j} & \hat{K} \cdot \hat{k} \end{bmatrix} = \begin{bmatrix} \hat{i}_1 \cdot \hat{i}_2 & \hat{i}_1 \cdot \hat{j}_2 & \hat{i}_1 \cdot \hat{k}_2 \\ \hat{j}_1 \cdot \hat{i}_2 & \hat{j}_1 \cdot \hat{j}_2 & \hat{j}_1 \cdot \hat{k}_2 \\ \hat{k}_1 \cdot \hat{i}_2 & \hat{k}_1 \cdot \hat{j}_2 & \hat{k}_1 \cdot \hat{k}_2 \end{bmatrix}$$

$$= \begin{bmatrix} \cos\dfrac{\pi}{2} & \cos\left(\dfrac{\pi}{2} + \theta_2\right) & \cos\theta_2 \\ \cos\dfrac{\pi}{2} & \cos\theta_2 & \cos\left(\dfrac{\pi}{2} - \theta_2\right) \\ \cos\pi & \cos\dfrac{\pi}{2} & \cos\dfrac{\pi}{2} \end{bmatrix}$$

$$= \begin{bmatrix} 0 & -\sin\theta_2 & \cos\theta_2 \\ 0 & \cos\theta_2 & \sin\theta_2 \\ -1 & 0 & 0 \end{bmatrix} \tag{5.69}$$

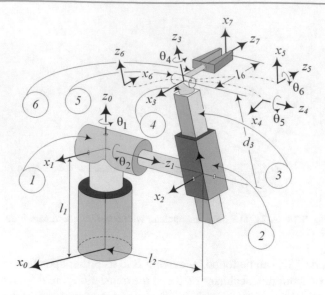

Fig. 5.15 A spherical robot made by a spherical manipulator equipped with a spherical wrist

Table 5.7 DH parameter table for Stanford arm

Frame no.	a_i	α_i	d_i	θ_i
1	0	−90 deg	l_1	θ_1
2	0	90 deg	l_2	θ_2
3	0	0	d_3	0
4	0	−90 deg	0	θ_4
5	0	90 deg	0	θ_5
6	0	0	0	θ_6

The final transformation matrix is only a translation along z_3,

$$^2T_3 = \begin{bmatrix} 1 & 0 & 0 & 0 \\ 0 & 1 & 0 & 0 \\ 0 & 0 & 1 & d_3 \\ 0 & 0 & 0 & 1 \end{bmatrix} \tag{5.70}$$

which provides

$$^1T_3 = {}^1T_2\,{}^2T_3 = \begin{bmatrix} 0 & -\sin\theta_2 & \cos\theta_2 & d_3\cos\theta_2 \\ 0 & \cos\theta_2 & \sin\theta_2 & d_3\sin\theta_2 \\ -1 & 0 & 0 & 0 \\ 0 & 0 & 0 & 1 \end{bmatrix} \tag{5.71}$$

Example 172 ★ *DH* application for spherical robot. Applying *DH* method on a 6 *DOF* spherical robot.

Figure 5.15 illustrates a spherical manipulator equipped with a spherical wrist. A spherical manipulator is an R⊢R⊢P arm. The associated *DH* table of the robot is given in Table 5.7.

The link-joint classifications of the robot are tabulated in Table 5.8.

Using Appendix D, we determine the homogenous transformation matrices for the link-joint classification in Table 5.8 to move from B_i to B_{i-1}:

$$^0T_1 = \begin{bmatrix} \cos\theta_1 & 0 & -\sin\theta_1 & 0 \\ \sin\theta_1 & 0 & \cos\theta_1 & 0 \\ 0 & -1 & 0 & l_1 \\ 0 & 0 & 0 & 1 \end{bmatrix} \tag{5.72}$$

Table 5.8 Link-joint classifications for Stanford arm

Link no.	Type
1	R⊢R(−90)
2	R⊢P(90)
3	P‖R(0)
4	R⊢R(−90)
5	R⊢R(90)
6	R‖R(0)

$$
^1T_2 = \begin{bmatrix} \cos\theta_2 & 0 & \sin\theta_2 & 0 \\ \sin\theta_2 & 0 & -\cos\theta_2 & 0 \\ 0 & 1 & 0 & l_2 \\ 0 & 0 & 0 & 1 \end{bmatrix}
\tag{5.73}
$$

$$
^2T_3 = \begin{bmatrix} 1 & 0 & 0 & 0 \\ 0 & 1 & 0 & 0 \\ 0 & 0 & 1 & d_3 \\ 0 & 0 & 0 & 1 \end{bmatrix}
\tag{5.74}
$$

$$
^3T_4 = \begin{bmatrix} \cos\theta_4 & 0 & -\sin\theta_4 & 0 \\ \sin\theta_4 & 0 & \cos\theta_4 & 0 \\ 0 & -1 & 0 & 0 \\ 0 & 0 & 0 & 1 \end{bmatrix}
\tag{5.75}
$$

$$
^4T_5 = \begin{bmatrix} \cos\theta_5 & 0 & \sin\theta_5 & 0 \\ \sin\theta_5 & 0 & -\cos\theta_5 & 0 \\ 0 & 1 & 0 & 0 \\ 0 & 0 & 0 & 1 \end{bmatrix}
\tag{5.76}
$$

$$
^5T_6 = \begin{bmatrix} \cos\theta_6 & -\sin\theta_6 & 0 & 0 \\ \sin\theta_6 & \cos\theta_6 & 0 & 0 \\ 0 & 0 & 1 & 0 \\ 0 & 0 & 0 & 1 \end{bmatrix}
\tag{5.77}
$$

Example 173 ★ *DH* application for a slider-crank planar linkage. How to apply *DH* method on closed-loop mechanisms.

For a closed-loop robot or a mechanism there would also be a connection between the first and last links. The *DH* convention will not be satisfied by this connection. Figure 5.16 depicts a planar slider-crank linkage R⊥P⊢R‖R‖R and *DH* coordinate frames installed on each link.

Table 5.9 summarizes the *DH* parameters of all links of the mechanism. A closed-loop robot provides a constraint on transformation matrices,

$$
[T] = {}^1T_2 \, {}^2T_3 \, {}^3T_4 \, {}^4T_1 = \mathbf{I}_4
\tag{5.78}
$$

where the transformation matrix $[T]$ contains elements that are functions of $a_2, d, a_3, \theta_3, a_4, \theta_4, \theta_1$. The parameters a_2, a_3, and a_4 are constant, while $d, \theta_3, \theta_4, \theta_1$ are variables. Assuming θ_1 is input and specified, we may solve for other unknown variables θ_3, θ_4, d by equating the corresponding elements of $[T]$ and **I**.

Example 174 ★ Non-standard Denavit–Hartenberg notation. Instead of installing coordinate frame at the end of links, they could be set at the beginning. Such installation was also presented in the beginning of robotic application, but they are non-standard today. This is how *DH* matrix would be if we use the non-standard way to set coordinate frames.

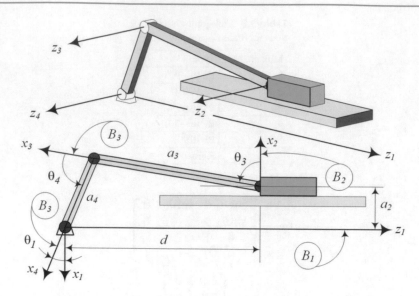

Fig. 5.16 A planar slider-crank linkage, making a closed-loop or parallel mechanism

Table 5.9 DH table for the slider-crank planar linkage shown in Fig. 5.16

Frame no.	a_i	α_i	d_i	θ_i
1	a_2	-90 deg	d	180 deg
2	a_3	0	0	θ_3
3	a_4	0	0	θ_4
4	0	-90 deg	0	θ_1

The Denavit–Hartenberg notation presented in this section is the standard DH method. However, we may adopt a different set of DH parameters, by setting the link coordinate frame B_i at proximal joint i instead of the distal joint $i + 1$ as shown in Fig. 5.17. The z_i-axis is along the axis of joint i and the x_i-axis is along the common normal of the z_i and z_{i+1} axes, directed from z_i to z_{i+1} axes. The y_i-axis makes the B_i frame a right-handed coordinate frame.

The parameterization of this shift of coordinate frames is:

1. a_i is the distance between the z_i and z_{i+1} axes along the x_i-axis.
2. α_i is the angle from z_i to z_{i+1} axes about the x_i-axis.
3. d_i is the distance between the x_{i-1} and x_i axes along the z_i-axis.
4. θ_i is the angle from the x_{i-1} and x_i axes about the z_i-axis.

The transformation matrix from B_{i-1} to B_i utilizing the non-standard DH method is made of two rotations and two translations about and along the local coordinate axes of B_{i-1}. 1—Rotate α_{i-1} about x_{i-1}. 2—Translate a_{i-1} along x_{i-1}. 3—Translate d_i along z_{i-1}. 4—Rotate θ_i about z_{i-1}.

$$^{i}T_{i-1} = R_{z_{i-1},\theta_i}\, D_{z_{i-1},-d_i}\, D_{x_{i-1},a_{i-1}}\, R_{x_{i-1},-\alpha_{i-1}} \tag{5.79}$$

$$= \begin{bmatrix} \cos\theta_i & \sin\theta_i \cos\alpha_{i-1} & \sin\theta_i \sin\alpha_{i-1} & -a_{i-1}\cos\theta_i \\ -\sin\theta_i & \cos\theta_i \cos\alpha_{i-1} & \cos\theta_i \sin\alpha_{i-1} & a_{i-1}\sin\theta_i \\ 0 & -\sin\alpha_{i-1} & \cos\alpha_{i-1} & -d_i \\ 0 & 0 & 0 & 1 \end{bmatrix}$$

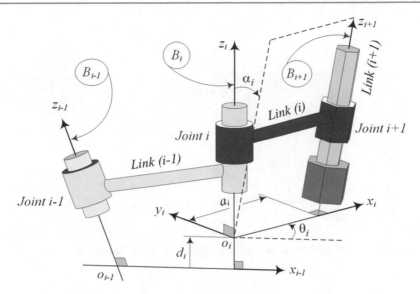

Fig. 5.17 Non-standard definition of DH parameters $a_i, \alpha_i, d_i, \theta_i$ defined for joint i and link (i)

Therefore, the transformation matrix from the B_i to B_{i-1} for the non-standard DH method is

$$
{}^{i-1}T_i = \begin{bmatrix}
\cos\theta_i & -\sin\theta_i & 0 & a_{i-1} \\
\sin\theta_i \cos\alpha_{i-1} & \cos\theta_i \cos\alpha_{i-1} & -\sin\alpha_{i-1} & -d_i \sin\alpha_{i-1} \\
\sin\theta_i \sin\alpha_{i-1} & \cos\theta_i \sin\alpha_{i-1} & \cos\alpha_{i-1} & d_i \cos\alpha_{i-1} \\
0 & 0 & 0 & 1
\end{bmatrix} \tag{5.80}
$$

An advantage of the non-standard DH method is that the rotation θ_i is around the z_i-axis and the joint number is the same as the coordinate number. Actuation force, which is exerted at joint i, is also at the same place as the coordinate frame B_i. Addressing the link's geometrical characteristics, such as center of gravity, is more natural in this non-standard DH system.

A disadvantage of the non-standard DH method is that the transformation matrix is a mix of $i-1$ and i indices. Applying the standard or non-standard notation is a personal preference, as both methods can be applied effectively.

5.3 Forward Position Kinematics of Robots

Determining the end-effector position and orientation for a given set of joint variables is the main problem in *forward kinematics*. The *forward* or *direct kinematics* is the transformation of kinematic information from the robot joint variable space to the Cartesian coordinate space. This problem can be solved by determining transformation matrices ${}^0T_i$ to express the kinematic information of link (i) in the base link coordinate frame. The traditional way of producing forward kinematic equations for robotic manipulators is to proceed link by link using the Denavit–Hartenberg transformation matrices.

For an n-DOF robot, at least n transformation matrices, one for every link, are required to determine the global coordinate of any point in any frame. The last frame attached to the final frame is usually set at the center of the gripper as is shown in Fig. 5.18. For a given set of joint variables and a set of link coordinate frames, the transformation matrices ${}^jT_i$ are uniquely determined. Therefore, the transformation matrices ${}^jT_i = {}^jT_i(q_k)$ are functions of n joint variables q_k, $k = 1, 2, 3, \cdots, n$. Therefore, the position and orientation of the end-effector are also a unique function of the joint variables. The configuration of the multibody when all the joint variables are zero is called the *rest position*. Determination of the transformation matrices at the rest position is an applied checking procedure.

If the links of a robot are arranged such that every link (i) has only one coordinate frame B_i and the frames are arranged sequentially, then:

$$
{}^0T_i = {}^0T_1 \, {}^1T_2 \, {}^2T_3 \, {}^3T_4 \cdots {}^{i-1}T_i \qquad i = 1, 2, 3, \cdots, n \tag{5.81}
$$

Having ${}^0T_i$, we can determine the coordinates of any point P of link (i) in the base frame when its coordinates are given in frame B_i:

$$
{}^0\mathbf{r}_P = {}^0T_i \, {}^i\mathbf{r}_P \qquad i = 1, 2, 3, \cdots, n \tag{5.82}
$$

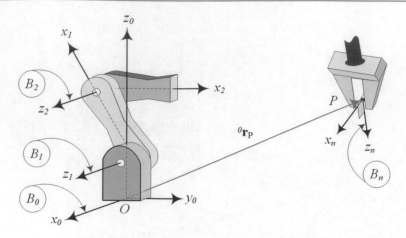

Fig. 5.18 The position of the final frame in the base frame

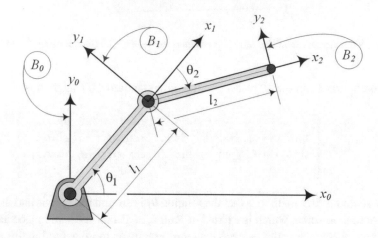

Fig. 5.19 A $2R$ or R∥R planar manipulator

Generally speaking, the number and labels of coordinate frames do not need to be consequential or increasing. The designer is free to number the frames in any order. However, assigning them sequentially provides simpler and more meaningful transformations.

Example 175 A $2R$ planar manipulator. The $2R$ robot is the simplest practical robot that is used to demonstrate classical methods of analysis and control. The number of links and transformation matrices are minimum, so the principle of forward kinematics will be clear. This example will be reference for forward kinematics of $2R$ planar robots.

Figure 5.19 illustrates a $2R$ or R∥R planar manipulator with two parallel revolute joints. Links (1) and (2) are both R∥R(0), and therefore, the transformation matrices ${}^0T_1$, ${}^1T_2$ are

$${}^0T_1 = \begin{bmatrix} \cos\theta_1 & -\sin\theta_1 & 0 & l_1\cos\theta_1 \\ \sin\theta_1 & \cos\theta_1 & 0 & l_1\sin\theta_1 \\ 0 & 0 & 1 & 0 \\ 0 & 0 & 0 & 1 \end{bmatrix} \tag{5.83}$$

$${}^1T_2 = \begin{bmatrix} \cos\theta_2 & -\sin\theta_2 & 0 & l_2\cos\theta_2 \\ \sin\theta_2 & \cos\theta_2 & 0 & l_2\sin\theta_2 \\ 0 & 0 & 1 & 0 \\ 0 & 0 & 0 & 1 \end{bmatrix} \tag{5.84}$$

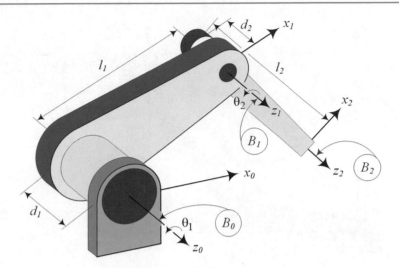

Fig. 5.20 A $3D$ R∥R planar manipulator

The forward kinematics of the manipulator is to determine the transformation matrices ^0T_i, $i = 1, 2$. The matrix $^0T_1$ is given in (5.83) and $^0T_2$ can be found by matrix multiplication:

$$^0T_2 = {}^0T_1\,{}^1T_2 \tag{5.85}$$

$$= \begin{bmatrix} \cos(\theta_1 + \theta_2) & -\sin(\theta_1 + \theta_2) & 0 & l_2\cos(\theta_1 + \theta_2) + l_1\cos\theta_1 \\ \sin(\theta_1 + \theta_2) & \cos(\theta_1 + \theta_2) & 0 & l_2\sin(\theta_1 + \theta_2) + l_1\sin\theta_1 \\ 0 & 0 & 1 & 0 \\ 0 & 0 & 0 & 1 \end{bmatrix}$$

If the robot had depth and nonzero offsets as is shown in Fig. 5.20 , the robot becomes a $3D$ R∥R planar manipulator. Links (1) and (2) are both R∥R(0) and move in parallel planes. Their transformation matrices $^0T_1$, $^1T_2$ are

$$^0T_1 = \begin{bmatrix} \cos\theta_1 & -\sin\theta_1 & 0 & l_1\cos\theta_1 \\ \sin\theta_1 & \cos\theta_1 & 0 & l_1\sin\theta_1 \\ 0 & 0 & 1 & d_1 \\ 0 & 0 & 0 & 1 \end{bmatrix} \tag{5.86}$$

$$^1T_2 = \begin{bmatrix} \cos\theta_2 & -\sin\theta_2 & 0 & l_2\cos\theta_2 \\ \sin\theta_2 & \cos\theta_2 & 0 & l_2\sin\theta_2 \\ 0 & 0 & 1 & d_2 \\ 0 & 0 & 0 & 1 \end{bmatrix} \tag{5.87}$$

The forward kinematics of the manipulator is to determine the transformation matrices ^0T_i, $i = 1, 2$. The matrix $^0T_1$ is given in (5.83) and $^0T_2$ can be found by matrix multiplication:

$$^0T_2 = {}^0T_1\,{}^1T_2 \tag{5.88}$$

$$= \begin{bmatrix} \cos(\theta_1 + \theta_2) & -\sin(\theta_1 + \theta_2) & 0 & l_2\cos(\theta_1 + \theta_2) + l_1\cos\theta_1 \\ \sin(\theta_1 + \theta_2) & \cos(\theta_1 + \theta_2) & 0 & l_2\sin(\theta_1 + \theta_2) + l_1\sin\theta_1 \\ 0 & 0 & 1 & d_1 + d_2 \\ 0 & 0 & 0 & 1 \end{bmatrix}$$

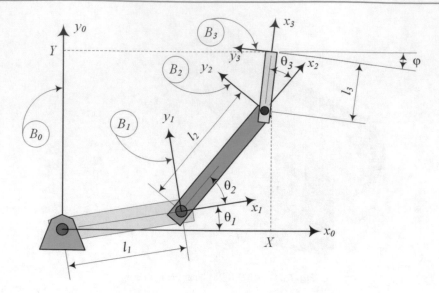

Fig. 5.21 An R∥R∥R planar manipulator

Example 176 3*R*, R∥R∥R, planar manipulator forward kinematics. Application of *DH* matrices in forward kinematic analysis of a planar 3 *DOF* robot.

Figure 5.21 illustrates an R∥R∥R planar manipulator. Utilizing the *DH* parameters indicated in Example 159 and applying Eq. (5.11), we can find the transformation matrices $^{i-1}T_i$ for $i = 3, 2, 1$. It is also possible to use the transformation matrix (5.32) of Example 166, because links (1) and (2) are both R∥R(0).

$$
^2T_3 = \begin{bmatrix} \cos\theta_3 & -\sin\theta_3 & 0 & l_3\cos\theta_3 \\ \sin\theta_3 & \cos\theta_3 & 0 & l_3\sin\theta_3 \\ 0 & 0 & 1 & 0 \\ 0 & 0 & 0 & 1 \end{bmatrix} \tag{5.89}
$$

$$
^1T_2 = \begin{bmatrix} \cos\theta_2 & -\sin\theta_2 & 0 & l_2\cos\theta_2 \\ \sin\theta_2 & \cos\theta_2 & 0 & l_2\sin\theta_2 \\ 0 & 0 & 1 & 0 \\ 0 & 0 & 0 & 1 \end{bmatrix} \tag{5.90}
$$

$$
^0T_1 = \begin{bmatrix} \cos\theta_1 & -\sin\theta_1 & 0 & l_1\cos\theta_1 \\ \sin\theta_1 & \cos\theta_1 & 0 & l_1\sin\theta_1 \\ 0 & 0 & 1 & 0 \\ 0 & 0 & 0 & 1 \end{bmatrix} \tag{5.91}
$$

Therefore, the transformation matrix to relate the end-effector frame to the base frame is

$$
\begin{aligned}
^0T_3 &= {}^0T_1\,{}^1T_2\,{}^2T_3 \\
&= \begin{bmatrix} \cos(\theta_1+\theta_2+\theta_3) & -\sin(\theta_1+\theta_2+\theta_3) & 0 & r_{14} \\ \sin(\theta_1+\theta_2+\theta_3) & \cos(\theta_1+\theta_2+\theta_3) & 0 & r_{24} \\ 0 & 0 & 1 & 0 \\ 0 & 0 & 0 & 1 \end{bmatrix}
\end{aligned} \tag{5.92}
$$

$$
r_{14} = l_1\cos\theta_1 + l_2\cos(\theta_1+\theta_2) + l_3\cos(\theta_1+\theta_2+\theta_3)
$$

$$
r_{24} = l_1\sin\theta_1 + l_2\sin(\theta_1+\theta_2) + l_3\sin(\theta_1+\theta_2+\theta_3)
$$

The origin of the frame B_3 is the tip point of the robot. Its position is at

$$^0T_3 \begin{bmatrix} 0 \\ 0 \\ 0 \\ 1 \end{bmatrix} = \begin{bmatrix} l_1 c\theta_1 + l_2 c\,(\theta_1 + \theta_2) + l_3 c\,(\theta_1 + \theta_2 + \theta_3) \\ l_1 s\theta_1 + l_2 s\,(\theta_1 + \theta_2) + l_3 s\,(\theta_1 + \theta_2 + \theta_3) \\ 0 \\ 1 \end{bmatrix} \tag{5.93}$$

It means we can find the coordinate of the tip point in the base Cartesian coordinate frame if we have the geometry of the robot and all joint variables.

$$X = l_1 \cos\theta_1 + l_2 \cos(\theta_1 + \theta_2) + l_3 \cos(\theta_1 + \theta_2 + \theta_3) \tag{5.94}$$

$$Y = l_1 \sin\theta_1 + l_2 \sin(\theta_1 + \theta_2) + l_3 \sin(\theta_1 + \theta_2 + \theta_3) \tag{5.95}$$

The rest position of the manipulator is lying on the x_0-axis where $\theta_1 = 0$, $\theta_2 = 0$, $\theta_3 = 0$ because $^0T_3$ becomes

$$^0T_3 = \begin{bmatrix} 1 & 0 & 0 & l_1 + l_2 + l_3 \\ 0 & 1 & 0 & 0 \\ 0 & 0 & 1 & 0 \\ 0 & 0 & 0 & 1 \end{bmatrix} \tag{5.96}$$

Having the transformation matrices $^{i-1}T_i$ are enough to determine the configuration of every link in other links' frames. The configuration of the link (2) in (0) is

$$^0T_2 = {}^0T_1\,{}^1T_2$$

$$= \begin{bmatrix} \cos(\theta_1 + \theta_2) & -\sin(\theta_1 + \theta_2) & 0 & f_{14} \\ \sin(\theta_1 + \theta_2) & \cos(\theta_1 + \theta_2) & 0 & f_{24} \\ 0 & 0 & 1 & 0 \\ 0 & 0 & 0 & 1 \end{bmatrix} \tag{5.97}$$

$$f_{14} = l_1 \cos\theta_1 + l_2 \cos(\theta_1 + \theta_2)$$

$$f_{24} = l_1 \sin\theta_1 + l_2 \sin(\theta_1 + \theta_2)$$

and the configuration of the link (3) in (1) is

$$^1T_3 = {}^1T_2\,{}^2T_3$$

$$= \begin{bmatrix} \cos(\theta_2 + \theta_3) & -\sin(\theta_2 + \theta_3) & 0 & g_{14} \\ \sin(\theta_2 + \theta_3) & \cos(\theta_2 + \theta_3) & 0 & g_{24} \\ 0 & 0 & 1 & 0 \\ 0 & 0 & 0 & 1 \end{bmatrix} \tag{5.98}$$

$$g_{14} = l_3 \cos(\theta_2 + \theta_3) + l_2 \cos\theta_2$$

$$g_{24} = l_3 \sin(\theta_2 + \theta_3) + l_2 \sin\theta_2$$

Example 177 Redundancy. If a robot has degrees of freedom more than required, it has redundant freedom.

A manipulator with n joints has n DOFs. It is redundant if it can perform a task that requires less than the n available degrees of freedom. For example, ignoring motions of fingers, a human hand has 7 DOFs, namely, 3 rotational DOFs at shoulder joint, 1 rotational DOF at elbow joint, 3 rotational DOFs at wrist joint. Therefore, a frozen palm may be moved to any location and any orientation with 7 DOFs which is one DOF more than need. Including fingers, a hand has 27 DOFs.

In robotics, any extra DOF provides us with ability to add a constraint or a condition to share the motions between the joints such that a particular path or a particular condition such as minimum energy to be satisfied.

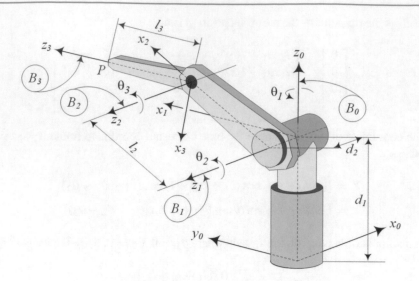

Fig. 5.22 An R⊢R∥R articulated arm

Table 5.10 DH parameter table for setting up the link frames

Frame no.	a_i	α_i	d_i	θ_i
1	0	−90 deg	d_1	θ_1
2	l_2	0	d_2	θ_2
3	0	90 deg	l_3	θ_3

Table 5.11 Link classification for setup of the link frames

Link no.	Type
1	R⊢R(−90)
2	R∥R(0)
3	R⊢R(90)

Example 178 $3R$, R⊢R∥R, articulated arm forward kinematics. Articulated robot is very applied in industry. This is how to determine forward kinematics of the robot.

Consider an R⊢R∥R arm as shown schematically in Fig. 5.22. To develop forward kinematics of the robot, the DH parameter table of the robot at rest position is set up as indicated in Table 5.10. The rest position of the robot could be set at any configuration where the joint axes z_1, z_2, z_3 are coplanar; however, two rest positions are mostly being used. The first rest position is where x_1 is coplanar with z_1, z_2, z_3, and the second rest position is where x_0 is coplanar with z_1, z_2, z_3.

We apply the link-joints classification in Examples 166. Therefore, we must be able to determine the type of link-joints combination as shown in Table 5.11.

Therefore, the successive transformation matrices have the following expressions:

$$
{}^{0}T_1 = \begin{bmatrix} \cos\theta_1 & 0 & -\sin\theta_1 & 0 \\ \sin\theta_1 & 0 & \cos\theta_1 & 0 \\ 0 & -1 & 0 & d_1 \\ 0 & 0 & 0 & 1 \end{bmatrix}
\tag{5.99}
$$

$$
{}^{1}T_2 = \begin{bmatrix} \cos\theta_2 & -\sin\theta_2 & 0 & l_2\cos\theta_2 \\ \sin\theta_2 & \cos\theta_2 & 0 & l_2\sin\theta_2 \\ 0 & 0 & 1 & d_2 \\ 0 & 0 & 0 & 1 \end{bmatrix}
\tag{5.100}
$$

$$^2T_3 = \begin{bmatrix} \cos\theta_3 & 0 & \sin\theta_3 & 0 \\ \sin\theta_3 & 0 & -\cos\theta_3 & 0 \\ 0 & 1 & 0 & 0 \\ 0 & 0 & 0 & 1 \end{bmatrix} \tag{5.101}$$

To express the complete forward kinematics transformation

$$^0T_3 = {}^0T_1 \, {}^1T_2 \, {}^2T_3 \tag{5.102}$$

we need to determine the result of a matrix multiplication.

$$^0T_3 = {}^0T_1 \, {}^1T_2 \, {}^2T_3$$

$$= \begin{bmatrix} r_{11} & r_{12} & r_{13} & r_{14} \\ r_{21} & r_{22} & r_{23} & r_{24} \\ r_{31} & r_{32} & r_{33} & r_{34} \\ 0 & 0 & 0 & 1 \end{bmatrix} \tag{5.103}$$

$$r_{11} = \cos\theta_1 \cos(\theta_2 + \theta_3) \tag{5.104}$$

$$r_{21} = \sin\theta_1 \cos(\theta_2 + \theta_3) \tag{5.105}$$

$$r_{31} = -\sin(\theta_2 + \theta_3) \tag{5.106}$$

$$r_{12} = -\sin\theta_1 \qquad r_{22} = \cos\theta_1 \qquad r_{32} = 0 \tag{5.107}$$

$$r_{13} = \cos\theta_1 \sin(\theta_2 + \theta_3) \tag{5.108}$$

$$r_{23} = \sin\theta_1 \sin(\theta_2 + \theta_3) \tag{5.109}$$

$$r_{33} = \cos(\theta_2 + \theta_3) \tag{5.110}$$

$$r_{14} = l_2 \cos\theta_1 \cos\theta_2 - d_2 \sin\theta_1 \tag{5.111}$$

$$r_{24} = l_2 \cos\theta_2 \sin\theta_1 + d_2 \cos\theta_1 \tag{5.112}$$

$$r_{34} = d_1 - l_2 \sin\theta_2 \tag{5.113}$$

The tip point P of the third arm is at $^3\mathbf{r}_P = \begin{bmatrix} 0 & 0 & l_3 \end{bmatrix}^T$ in B_3. So, its position in the base frame would be at

$$^0\mathbf{r}_P = {}^0T_3 \, {}^3\mathbf{r}_P$$

$$= {}^0T_3 \begin{bmatrix} 0 \\ 0 \\ l_3 \\ 1 \end{bmatrix} = \begin{bmatrix} -d_2 s\theta_1 + l_2 c\theta_1 c\theta_2 + l_3 c\theta_1 s\,(\theta_2 + \theta_3) \\ d_2 c\theta_1 + l_2 c\theta_2 s\theta_1 + l_3 s\theta_1 s\,(\theta_2 + \theta_3) \\ d_1 - l_2 s\theta_2 + l_3 c\,(\theta_2 + \theta_3) \\ 1 \end{bmatrix} \tag{5.114}$$

The transformation matrix at rest position, where $\theta_1 = 0$, $\theta_2 = 0$, $\theta_3 = 0$, is

$$^0T_3 = \begin{bmatrix} 1 & 0 & 0 & l_2 \\ 0 & 1 & 0 & d_2 \\ 0 & 0 & 1 & d_1 \\ 0 & 0 & 0 & 1 \end{bmatrix} \tag{5.115}$$

This setup of coordinate frames shows that at the rest position, x_1, x_2, x_3 are colinear and parallel to x_0; furthermore, z_1, z_3 are colinear, and z_1, z_2 are parallel.

Example 179 An articulated arm kinematics. Numerical example for forward kinematics of an articulated arm.

Consider the R⊢R‖R arm of Fig. 5.22 with the following dimensions:

$$l_2 = 0.75\,\mathrm{m} \qquad l_3 = 0.65\,\mathrm{m}$$
$$d_1 = 0.48\,\mathrm{m} \qquad d_2 = 0.174\,\mathrm{m} \tag{5.116}$$

Using the link-joints combination of Table 5.11, we have

$$
{}^0T_1 = \begin{bmatrix} \cos\theta_1 & 0 & -\sin\theta_1 & 0 \\ \sin\theta_1 & 0 & \cos\theta_1 & 0 \\ 0 & -1 & 0 & 0.48 \\ 0 & 0 & 0 & 1 \end{bmatrix} \tag{5.117}
$$

$$
{}^1T_2 = \begin{bmatrix} \cos\theta_2 & -\sin\theta_2 & 0 & 0.75\cos\theta_2 \\ \sin\theta_2 & \cos\theta_2 & 0 & 0.75\sin\theta_2 \\ 0 & 0 & 1 & 0.174 \\ 0 & 0 & 0 & 1 \end{bmatrix} \tag{5.118}
$$

$$
{}^2T_3 = \begin{bmatrix} \cos\theta_3 & 0 & \sin\theta_3 & 0 \\ \sin\theta_3 & 0 & -\cos\theta_3 & 0 \\ 0 & 1 & 0 & 0 \\ 0 & 0 & 0 & 1 \end{bmatrix} \tag{5.119}
$$

Therefore, the transformation matrix of B_3 to B_0 is

$$
\begin{aligned}
{}^0T_3 &= {}^0T_1\,{}^1T_2\,{}^2T_3 \\
&= \begin{bmatrix} c\theta_1 c(\theta_2+\theta_3) & -s\theta_1 & c\theta_1 s(\theta_2+\theta_3) & r_{11} \\ s\theta_1 c(\theta_2+\theta_3) & c\theta_1 & s\theta_1 s(\theta_2+\theta_3) & r_{12} \\ -s(\theta_2+\theta_3) & 0 & c(\theta_2+\theta_3) & r_{13} \\ 0 & 0 & 0 & 1 \end{bmatrix}
\end{aligned} \tag{5.120}
$$

$$r_{11} = 0.75\cos\theta_1\cos\theta_2 - 0.174\sin\theta_1$$
$$r_{12} = 0.174\cos\theta_1 + 0.75\cos\theta_2\sin\theta_1$$
$$r_{13} = 0.48 - 0.75\sin\theta_2$$

The tip point P of the third link is at

$$
{}^0\mathbf{r}_P = {}^0T_3\,{}^3\mathbf{r}_P = {}^0T_3 \begin{bmatrix} 0 \\ 0 \\ 0.65 \\ 1 \end{bmatrix} = \begin{bmatrix} r_1 \\ r_1 \\ r_1 \\ 1 \end{bmatrix} \tag{5.121}
$$

$$r_1 = 0.75\cos\theta_1\cos\theta_2 - 0.174\sin\theta_1 + 0.65\cos\theta_1\sin(\theta_2+\theta_3)$$
$$r_2 = 0.174\cos\theta_1 + 0.75\cos\theta_2\sin\theta_1 + 0.65\sin\theta_1\sin(\theta_2+\theta_3)$$
$$r_3 = 0.65\cos(\theta_2+\theta_3) - 0.75\sin\theta_2 + 0.48$$

At the rest position, where $\theta_1 = 0, \theta_2 = 0, \theta_3 = 0$, we have

$$^0T_3 = \begin{bmatrix} 1 & 0 & 0 & 0.75 \\ 0 & 1 & 0 & 0.174 \\ 0 & 0 & 1 & 0.48 \\ 0 & 0 & 0 & 1 \end{bmatrix} \tag{5.122}$$

that shows the tip point is at

$$^0\mathbf{r}_P = {}^0T_3\,{}^3\mathbf{r}_P = {}^0T_3 \begin{bmatrix} 0 \\ 0 \\ 0.65 \\ 1 \end{bmatrix} = \begin{bmatrix} 0.75 \\ 0.174 \\ 1.13 \\ 1 \end{bmatrix} \tag{5.123}$$

Example 180 Working space of an articulated arm. How to determine the space volume that a robot can reach. It is the result of a kinematic analysis.

Consider the R⊢R∥R arm of Fig. 5.22 with the following dimensions:

$$l_2 = 0.75\,\text{m} \qquad l_3 = 0.65\,\text{m}$$

$$d_1 = 0.48\,\text{m} \qquad d_2 = 0.174\,\text{m} \tag{5.124}$$

Link transformation matrices are given in (5.117)–(5.119). The manipulator's transformation matrix $^0T_3$ at the rest position is (5.122) when point P is at (5.123).

Assume that every joint can turn 360 deg. Theoretically, point P must be able to reach any point in the sphere S_1,

$$\left(\mathbf{r} - {}^0\mathbf{d}_1 - d_2\,{}^0\hat{k}_1\right)^2 = (l_2 + l_3)^2 \tag{5.125}$$

$$x^2 + (y - 0.174)^2 + (z - 0.48)^2 = 1.96 \tag{5.126}$$

$$^0\mathbf{d}_1 = \begin{bmatrix} 0 \\ 0 \\ d_1 \end{bmatrix} = \begin{bmatrix} 0 \\ 0 \\ 0.48 \end{bmatrix} \tag{5.127}$$

$$d_2\,{}^0\hat{k}_1 = {}^0R_1\,d_2\,{}^1\hat{k}_1 \tag{5.128}$$

$$= \begin{bmatrix} 1 & 0 & 0 \\ 0 & 0 & 1 \\ 0 & -1 & 0 \end{bmatrix} \begin{bmatrix} 0 \\ 0 \\ 0.174 \end{bmatrix} = \begin{bmatrix} 0 \\ 0.174 \\ 0 \end{bmatrix}$$

and out of the sphere S_2.

$$\left(\mathbf{r} - {}^0\mathbf{d}_1 - d_2\,{}^0\hat{k}_1\right)^2 = (l_2 - l_3)^2 \tag{5.129}$$

$$x^2 + (y - 0.174)^2 + (z - 0.48)^2 = 0.01 \tag{5.130}$$

The reachable space between S_1 and S_2 is called working space of the manipulator.

Example 181 SCARA robot (R∥R∥R∥P). Forward kinematics.

Consider the R∥R∥R∥P robot shown in Fig. 5.23. The forward kinematics of the robot can be solved by obtaining individual transformation matrices $^{i-1}T_i$. The first link is an R∥R(0), which has the following transformation matrix:

$$^0T_1 = \begin{bmatrix} \cos\theta_1 & -\sin\theta_1 & 0 & l_1\cos\theta_1 \\ \sin\theta_1 & \cos\theta_1 & 0 & l_1\sin\theta_1 \\ 0 & 0 & 1 & 0 \\ 0 & 0 & 0 & 1 \end{bmatrix} \tag{5.131}$$

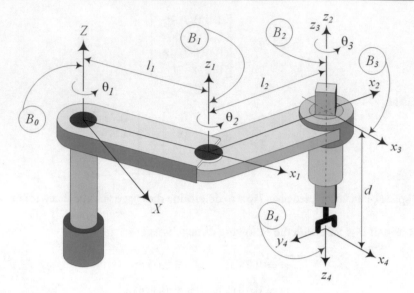

Fig. 5.23 An R‖R‖R‖P *SCARA* manipulator robot

The second link is also an R‖R(0).

$$
^1T_2 = \begin{bmatrix} \cos\theta_2 & -\sin\theta_2 & 0 & l_2\cos\theta_2 \\ \sin\theta_2 & \cos\theta_2 & 0 & l_2\sin\theta_2 \\ 0 & 0 & 1 & 0 \\ 0 & 0 & 0 & 1 \end{bmatrix}
\tag{5.132}
$$

The third link is an R‖R(0) with zero length,

$$
^2T_3 = \begin{bmatrix} \cos\theta_3 & -\sin\theta_3 & 0 & 0 \\ \sin\theta_3 & \cos\theta_3 & 0 & 0 \\ 0 & 0 & 1 & 0 \\ 0 & 0 & 0 & 1 \end{bmatrix}
\tag{5.133}
$$

and the fourth link is an R‖P(180). From a coincident position with B_3, the frame B_4 must turn π about the x_3-axis and translate $-d$ along the z_3-axis to move to the current position.

$$
^3T_4 = \begin{bmatrix} 1 & 0 & 0 & 0 \\ 0 & \cos\pi & -\sin\pi & 0 \\ 0 & \sin\pi & \cos\pi & -d \\ 0 & 0 & 0 & 1 \end{bmatrix} = \begin{bmatrix} 1 & 0 & 0 & 0 \\ 0 & -1 & 0 & 0 \\ 0 & 0 & -1 & -d \\ 0 & 0 & 0 & 1 \end{bmatrix}
\tag{5.134}
$$

Therefore, the configuration of the end-effector in the base coordinate frame is

$$
^0T_4 = {}^0T_1\,{}^1T_2\,{}^2T_3\,{}^3T_4
\tag{5.135}
$$

$$
= \begin{bmatrix} c(\theta_1+\theta_2+\theta_3) & s(\theta_1+\theta_2+\theta_3) & 0 & l_1c\theta_1 + l_2c(\theta_1+\theta_2) \\ s(\theta_1+\theta_2+\theta_3) & -c(\theta_1+\theta_2+\theta_3) & 0 & l_1s\theta_1 + l_2s(\theta_1+\theta_2) \\ 0 & 0 & -1 & -d \\ 0 & 0 & 0 & 1 \end{bmatrix}
$$

It shows the rest position of the robot $\theta_1 = \theta_2 = \theta_3 = d = 0$ is at

$$
^0T_4 = \begin{bmatrix} 1 & 0 & 0 & l_1+l_2 \\ 0 & -1 & 0 & 0 \\ 0 & 0 & -1 & 0 \\ 0 & 0 & 0 & 1 \end{bmatrix}
\tag{5.136}
$$

Example 182 Space station remote manipulator system (*SSRMS*). Forward kinematics of the 7 *DOF* Space station arm.

The Space Shuttle Remote Manipulator system (*SSRMS*), also known as the Shuttle Remote Manipulator System (*SRMS*), is a robotic arm to act as the hand of the Shuttle or a space station. It is utilized for several purposes such as: satellite deployment, construction of a space station, transporting a crew member at the end of the arm, and surveying and inspecting the outside of the station. A simplified model of the *SSRMS*, schematically shown in Fig. 5.24, and the approximate values of its kinematic characteristics are given in Table 5.12.

Table 5.13 indicates the *DH* parameters of *SSRMS*, and Table 5.14 names the link-joints classifications of *SSRMS*. Utilizing these values and indicating the type of each link enable us to determine the required transformation matrices to move from B_i to B_{i-1} and solve the forward kinematics problem.

Links (1) and (2) are R⊢R(−90).

$$
{}^0T_1 = \begin{bmatrix} \cos\theta_1 & 0 & -\sin\theta_1 & 0 \\ \sin\theta_1 & 0 & \cos\theta_1 & 0 \\ 0 & -1 & 0 & d_1 \\ 0 & 0 & 0 & 1 \end{bmatrix} \tag{5.137}
$$

$$
{}^1T_2 = \begin{bmatrix} \cos\theta_2 & 0 & -\sin\theta_2 & 0 \\ \sin\theta_2 & 0 & \cos\theta_2 & 0 \\ 0 & -1 & 0 & d_2 \\ 0 & 0 & 0 & 1 \end{bmatrix} \tag{5.138}
$$

Links (3) and (4) are R‖R(0).

$$
{}^2T_3 = \begin{bmatrix} \cos\theta_3 & -\sin\theta_3 & 0 & a_3\cos\theta_3 \\ \sin\theta_3 & \cos\theta_3 & 0 & a_3\sin\theta_3 \\ 0 & 0 & 1 & d_3 \\ 0 & 0 & 0 & 1 \end{bmatrix} \tag{5.139}
$$

$$
{}^3T_4 = \begin{bmatrix} \cos\theta_4 & -\sin\theta_4 & 0 & a_4\cos\theta_4 \\ \sin\theta_4 & \cos\theta_4 & 0 & a_4\sin\theta_4 \\ 0 & 0 & 1 & d_4 \\ 0 & 0 & 0 & 1 \end{bmatrix} \tag{5.140}
$$

Link (5) is R⊢R(90), and link (6) is R⊢R(−90).

$$
{}^4T_5 = \begin{bmatrix} \cos\theta_5 & 0 & \sin\theta_5 & 0 \\ \sin\theta_5 & 0 & -\cos\theta_5 & 0 \\ 0 & 1 & 0 & d_5 \\ 0 & 0 & 0 & 1 \end{bmatrix} \tag{5.141}
$$

$$
{}^5T_6 = \begin{bmatrix} \cos\theta_6 & 0 & -\sin\theta_6 & 0 \\ \sin\theta_6 & 0 & \cos\theta_6 & 0 \\ 0 & -1 & 0 & d_6 \\ 0 & 0 & 0 & 1 \end{bmatrix} \tag{5.142}
$$

Finally, link (7) is R⊢R(0) and the coordinate frame attached to the end-effector has a translation d_7 with respect to the coordinate frame B_6.

$$
{}^6T_7 = \begin{bmatrix} \cos\theta_7 & -\sin\theta_7 & 0 & 0 \\ \sin\theta_7 & \cos\theta_7 & 0 & 0 \\ 0 & 0 & 1 & d_7 \\ 0 & 0 & 0 & 1 \end{bmatrix} \tag{5.143}
$$

The forward kinematics of *SSRMS* can be found by direct multiplication of ${}^{i-1}T_i$ ($i = 1, 2, \cdots, 7$).

$$
{}^0T_7 = {}^0T_1\,{}^1T_2\,{}^2T_3\,{}^3T_4\,{}^4T_5\,{}^5T_6\,{}^6T_7 \tag{5.144}
$$

The forward kinematics of the first three frames are given as

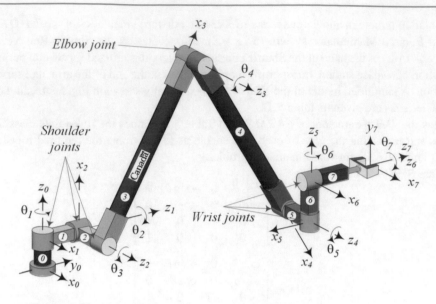

Fig. 5.24 Illustration of the space station remote manipulator system (not in scale)

Table 5.12 Space station's robot arm characteristics

Length	14.22 m
Diameter	38.1 cm
Weight	1336 kg
Number of joints	Seven
Handling capacity	116000 kg (in space)
Maximum velocity of end of arm	Carrying nothing: 37 cm/s
	Full capacity: 1.2 cm/s
Maximum rotational speed	Approximately: 4 deg /s

Table 5.13 DH parameters for $SSRMS$

Frame no.	a_i	α_i	d_i	θ_i
1	0	−90 deg	380 mm	θ_1
2	0	−90 deg	1360 mm	θ_2
3	7110 mm	0	570 mm	θ_3
4	7110 mm	0	475 mm	θ_4
5	0	90 deg	570 mm	θ_5
6	0	−90 deg	635 mm	θ_6
7	0	0	d_7	0

Table 5.14 Link-joints classifications for $SSRMS$

Link no.	Type
1	R⊢R(−90)
2	R⊢R(−90)
3	P∥R(0)
4	P∥R(0)
5	R⊢R(90)
6	R⊢R(−90)
7	R⊢R(0)

$$
{}^{0}T_1 = \begin{bmatrix} \cos\theta_1 & 0 & -\sin\theta_1 & 0 \\ \sin\theta_1 & 0 & \cos\theta_1 & 0 \\ 0 & -1 & 0 & d_1 \\ 0 & 0 & 0 & 1 \end{bmatrix} \tag{5.145}
$$

$$
{}^{0}T_2 = {}^{0}T_1\,{}^{1}T_2 = \begin{bmatrix} c\theta_1 c\theta_2 & s\theta_1 & -c\theta_1 s\theta_2 & -d_2 s\theta_1 \\ c\theta_2 s\theta_1 & -c\theta_1 & -s\theta_1 s\theta_2 & d_2 c\theta_1 \\ -s\theta_2 & 0 & -c\theta_2 & d_1 \\ 0 & 0 & 0 & 1 \end{bmatrix} \tag{5.146}
$$

$$
{}^{0}T_3 = {}^{0}T_1\,{}^{1}T_2\,{}^{2}T_3 \tag{5.147}
$$

$$
= \begin{bmatrix} s\theta_1 s\theta_3 + c\theta_1 c\theta_2 c\theta_3 & c\theta_3 s\theta_1 - c\theta_1 c\theta_2 s\theta_3 & -c\theta_1 s\theta_2 & {}^{0}d_{3x} \\ c\theta_2 c\theta_3 s\theta_1 - c\theta_1 s\theta_3 & -c\theta_1 c\theta_3 - c\theta_2 s\theta_1 s\theta_3 & -s\theta_1 s\theta_2 & {}^{0}d_{3y} \\ -c\theta_3 s\theta_2 & s\theta_2 s\theta_3 & -c\theta_2 & {}^{0}d_{3z} \\ 0 & 0 & 0 & 1 \end{bmatrix}
$$

where

$$
{}^{0}\mathbf{d}_3 = \begin{bmatrix} a_3 s\theta_1 s\theta_3 - d_3 c\theta_1 s\theta_2 - d_2 s\theta_1 + a_3 c\theta_1 c\theta_2 c\theta_3 \\ d_2 c\theta_1 - a_3 c\theta_1 s\theta_3 - d_3 s\theta_1 s\theta_2 + a_3 c\theta_2 c\theta_3 s\theta_1 \\ d_1 - d_3 c\theta_2 - a_3 c\theta_3 s\theta_2 \\ 1 \end{bmatrix} \tag{5.148}
$$

and the forward kinematics of the last three frames as

$$
{}^{6}T_7 = \begin{bmatrix} \cos\theta_7 & -\sin\theta_7 & 0 & 0 \\ \sin\theta_7 & \cos\theta_7 & 0 & 0 \\ 0 & 0 & 1 & d_7 \\ 0 & 0 & 0 & 1 \end{bmatrix} \tag{5.149}
$$

$$
{}^{5}T_7 = {}^{5}T_6\,{}^{6}T_7 = \begin{bmatrix} c\theta_6 c\theta_7 & -c\theta_6 s\theta_7 & -s\theta_6 & -d_7 s\theta_6 \\ c\theta_7 s\theta_6 & -s\theta_6 s\theta_7 & c\theta_6 & d_7 c\theta_6 \\ -s\theta_7 & -c\theta_7 & 0 & d_6 \\ 0 & 0 & 0 & 1 \end{bmatrix} \tag{5.150}
$$

$$
{}^{4}T_7 = {}^{4}T_5\,{}^{5}T_6\,{}^{6}T_7 \tag{5.151}
$$

$$
= \begin{bmatrix} c\theta_5 c\theta_6 c\theta_7 - s\theta_5 s\theta_7 & -c\theta_7 s\theta_5 - c\theta_5 c\theta_6 s\theta_7 & -c\theta_5 s\theta_6 & {}^{4}d_{7x} \\ c\theta_5 s\theta_7 + c\theta_6 c\theta_7 s\theta_5 & c\theta_5 c\theta_7 - c\theta_6 s\theta_5 s\theta_7 & -s\theta_5 s\theta_6 & {}^{4}d_{7y} \\ c\theta_7 s\theta_6 & -s\theta_6 s\theta_7 & c\theta_6 & {}^{4}d_{7z} \\ 0 & 0 & 0 & 1 \end{bmatrix}
$$

where

$$
{}^{4}\mathbf{d}_7 = \begin{bmatrix} d_6 s\theta_5 - d_7 c\theta_5 s\theta_6 \\ -d_6 c\theta_5 - d_7 s\theta_5 s\theta_6 \\ d_5 + d_7 c\theta_6 \\ 1 \end{bmatrix} \tag{5.152}
$$

The transformation matrices ${}^{0}T_4$, ${}^{0}T_5$, ${}^{0}T_6$ are determined by matrix multiplication,

$$
{}^{0}T_4 = {}^{0}T_3\,{}^{3}T_4 \tag{5.153}
$$

$$
{}^{0}T_5 = {}^{0}T_4\,{}^{4}T_5 \tag{5.154}
$$

$$
{}^{0}T_6 = {}^{0}T_5\,{}^{5}T_6 \tag{5.155}
$$

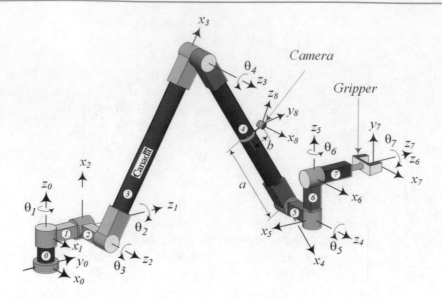

Fig. 5.25 The space shuttle remote manipulator system ($SSRMS$) with a camera attached to link (4)

and finally, we can find the coordinates of a point in the gripper frame B_7 in the base frame by using ${}^0T_3\,{}^3T_4\,{}^4T_7$:

$$
{}^0T_7 = {}^0T_3
\begin{bmatrix}
\cos\theta_4 & -\sin\theta_4 & 0 & a_4\cos\theta_4 \\
\sin\theta_4 & \cos\theta_4 & 0 & a_4\sin\theta_4 \\
0 & 0 & 1 & d_4 \\
0 & 0 & 0 & 1
\end{bmatrix}
{}^4T_7
\tag{5.156}
$$

Example 183 Camera on an arm of $SSRMS$. We may attach several accessories as different points of links of a robot. Every accessory needs a coordinate frame to be located kinematically. A good example is an inspecting camera. This example shows how an accessory will be defined kinematically on a robot.

The $SSRMS$ is shown in Fig. 5.25, with the link transformation matrices given in Eqs. (5.137)–(5.150). Using the numerical values of Table 5.12 and assuming the following values for the joint variables and gripper length,

$$
\theta_1 = 0 \qquad \theta_2 = -\frac{\pi}{2}\ \text{rad}\ \ \theta_3 = -\frac{\pi}{4}\ \text{rad}\ \ \theta_4 = -\frac{\pi}{2}\ \text{rad}
$$
$$
\theta_5 = -\frac{3\pi}{4}\ \ \theta_6 = \frac{\pi}{2}\ \text{rad}\ \ \ \theta_7 = 0 \qquad\ \ d_7 = 500\ \text{mm}
\tag{5.157}
$$

we find the following link transformation matrices:

$$
{}^0T_1 =
\begin{bmatrix}
1 & 0 & 0 & 0 \\
0 & 0 & 1 & 0 \\
0 & -1 & 0 & 380 \\
0 & 0 & 0 & 1
\end{bmatrix}
\tag{5.158}
$$

$$
{}^1T_2 =
\begin{bmatrix}
0 & 0 & 1 & 0 \\
-1 & 0 & 0 & 0 \\
0 & -1 & 0 & 1360 \\
0 & 0 & 0 & 1
\end{bmatrix}
\tag{5.159}
$$

$$
{}^2T_3 =
\begin{bmatrix}
0.707\,11 & 0.707\,11 & 0 & 5027.5 \\
-0.707\,11 & 0.707\,11 & 0 & -5027.5 \\
0 & 0 & 1 & 570 \\
0 & 0 & 0 & 1
\end{bmatrix}
\tag{5.160}
$$

$$^3T_4 = \begin{bmatrix} 0 & 1 & 0 & 0 \\ -1 & 0 & 0 & -7110 \\ 0 & 0 & 1 & 475 \\ 0 & 0 & 0 & 1 \end{bmatrix}$$ (5.161)

$$^4T_5 = \begin{bmatrix} -0.707\,11 & 0 & -0.707\,11 & 0 \\ -0.707\,11 & 0 & 0.707\,11 & 0 \\ 0 & 1 & 0 & 570 \\ 0 & 0 & 0 & 1 \end{bmatrix}$$ (5.162)

$$^5T_6 = \begin{bmatrix} 0 & 0 & -1 & 0 \\ 1 & 0 & 0 & 0 \\ 0 & -1 & 0 & 635 \\ 0 & 0 & 0 & 1 \end{bmatrix}$$ (5.163)

$$^6T_7 = \begin{bmatrix} 1 & 0 & 0 & 0 \\ 0 & 1 & 0 & 0 \\ 0 & 0 & 1 & 500 \\ 0 & 0 & 0 & 1 \end{bmatrix}$$ (5.164)

Assume there is a camera attached to link (4) to inspect the gripper's operation. The camera is at a point P in B_4 with position vector $^4\mathbf{r}_P$ and is aiming the origin of the gripper's frame B_7:

$$^4\mathbf{r}_P = \begin{bmatrix} -a & b & 0 & 1 \end{bmatrix}^T$$ (5.165)

$$a = 696\,\text{mm} \qquad b = 154\,\text{mm}$$ (5.166)

To determine the position of the camera in the base frame, $^0\mathbf{r}_P$, we need to calculate $^0T_4$,

$$^0T_4 = {}^0T_1\,{}^1T_2\,{}^2T_3\,{}^3T_4$$

$$= \begin{bmatrix} 0 & 0 & 1 & 1045 \\ 0.707\,11 & 0.707\,11 & 0 & 11415 \\ -0.707\,11 & 0.707\,11 & 0 & 379.95 \\ 0 & 0 & 0 & 1 \end{bmatrix}$$ (5.167)

and find $^0\mathbf{r}_P$,

$$^0\mathbf{r}_P = {}^0T_4\,{}^4\mathbf{r}_P = {}^0T_4 \begin{bmatrix} -696 \\ 154 \\ 0 \\ 1 \end{bmatrix} = \begin{bmatrix} 1045 \\ 11032 \\ 980.99 \\ 1 \end{bmatrix}$$ (5.168)

At the configuration (5.157) the transformation matrix $^0T_7$ is given as

$$^0T_7 = {}^0T_1\,{}^1T_2\,{}^2T_3\,{}^3T_4\,{}^4T_5\,{}^5T_6\,{}^6T_7$$

$$= \begin{bmatrix} 1 & 0 & 0 & 1615 \\ 0 & 0 & 1 & 11915 \\ 0 & -1 & 0 & 1015 \\ 0 & 0 & 0 & 1 \end{bmatrix}$$ (5.169)

where the position of the gripper point in the base frame is at

$$^0\mathbf{d}_7 = \begin{bmatrix} 1615 \\ 11915 \\ 1015 \\ 1 \end{bmatrix}$$ (5.170)

Therefore, the position of the camera in the gripper frame B_7 would be at

$$^7\mathbf{r}_P = {}^7T_4 \, {}^4\mathbf{r}_P \tag{5.171}$$

$$= {}^7T_6 \, {}^6T_5 \, {}^5T_4 \, {}^4\mathbf{r}_P = {}^6T_7^{-1} \, {}^5T_6^{-1} \, {}^4T_5^{-1} \, {}^4\mathbf{r}_P = \left[{}^4T_5 \, {}^5T_6 \, {}^6T_7\right]^{-1} \, {}^4\mathbf{r}_P$$

$$= \begin{bmatrix} 0 & 0 & 1 & -570 \\ 0.707 & -0.707 & 0 & 635 \\ 0.707 & 0.707 & 0 & -500 \\ 0 & 0 & 0 & 1 \end{bmatrix} \begin{bmatrix} -696 \\ 154 \\ 0 \\ 1 \end{bmatrix} = \begin{bmatrix} -570 \\ 33.965 \\ -883.25 \\ 1 \end{bmatrix}$$

To use the camera kinematically, we attach a coordinate frame B_8 to the camera at P and determine the configuration of the gripper frame B_7 in the camera frame. The position vectors of the camera in the base and gripper frames are calculated in (5.168) and (5.171) as

$$^0\mathbf{r}_P = \begin{bmatrix} 1045 \\ 11032 \\ 980.99 \\ 1 \end{bmatrix} \qquad {}^7\mathbf{r}_P = \begin{bmatrix} -570 \\ 33.965 \\ -883.25 \\ 1 \end{bmatrix} \tag{5.172}$$

Assume the camera is equipped with two electric motors that give it two rotational degrees of freedom. To determine the coordinate frame B_8 at the present configuration, we begin with calculating the direction of the x_8-axis in the gripper frame B_7,

$$^7\hat{\imath}_8 = \frac{-^7\mathbf{r}_P}{\left|^7\mathbf{r}_P\right|} = \begin{bmatrix} 0.54193 \\ -3.2292 \times 10^{-2} \\ 0.83975 \end{bmatrix} \tag{5.173}$$

and determine the x_8-axis in frame B_4.

$$^4\hat{\imath}_8 = {}^4R_7 \, {}^7\hat{\imath}_8 = \begin{bmatrix} 0.57096 \\ 0.61663 \\ 0.54193 \\ 0 \end{bmatrix} \tag{5.174}$$

Let us assume the first motor of the camera turns the camera about z_8 and the second motor turns it about the displaced y_8. The B_4 frame may act as a global frame to the camera. So, we can find the matrix $^8R_4$ by a rotation α about z_8 followed by a rotation β about y_8:

$$^8R_4 = R_{y,\beta} \, R_{z,\alpha}$$

$$= \begin{bmatrix} \cos\alpha\cos\beta & \cos\beta\sin\alpha & -\sin\beta \\ -\sin\alpha & \cos\alpha & 0 \\ \cos\alpha\sin\beta & \sin\alpha\sin\beta & \cos\beta \end{bmatrix} \tag{5.175}$$

Using $^8\hat{\imath}_8 = \begin{bmatrix} 1 & 0 & 0 \end{bmatrix}$, we have

$$^4\hat{\imath}_8 = {}^4R_8 \, {}^8\hat{\imath}_8 = {}^8R_4^T \, {}^8\hat{\imath}_8 \tag{5.176}$$

$$\begin{bmatrix} 0.57096 \\ 0.61663 \\ 0.54193 \end{bmatrix} = {}^8R_4^T \begin{bmatrix} 1 \\ 0 \\ 0 \end{bmatrix} = \begin{bmatrix} \cos\alpha\cos\beta \\ \cos\beta\sin\alpha \\ -\sin\beta \end{bmatrix} \tag{5.177}$$

where

$$^4R_8 = {}^8R_4^T = \begin{bmatrix} \cos\alpha\cos\beta & -\sin\alpha & \cos\alpha\sin\beta \\ \cos\beta\sin\alpha & \cos\alpha & \sin\alpha\sin\beta \\ -\sin\beta & 0 & \cos\beta \end{bmatrix} \tag{5.178}$$

The third component of (5.177) provides β,

$$\beta = -\arcsin 0.54193 = -0.5727 \, \text{rad} \approx -32.8 \, \text{deg} \tag{5.179}$$

and the first and second components provide α.

$$\alpha = \arctan \frac{0.61663}{0.57096} = 0.8238 \, \text{rad} \approx 47.2 \, \text{deg} \tag{5.180}$$

The angles α and β determine the orientation of B_8 at ${}^4\mathbf{d}_8 = \begin{bmatrix} -a & b & 0 & 1 \end{bmatrix}^T$. Therefore,

$$
{}^4T_8 = \begin{bmatrix}
0.57103 & -0.73372 & -0.36821 & -696 \\
0.61664 & 0.67945 & -0.39762 & 154 \\
0.54193 & 0 & 0.84042 & 0 \\
0 & 0 & 0 & 1
\end{bmatrix} \tag{5.181}
$$

$$
{}^8T_4 = \begin{bmatrix}
0.57103 & 0.61664 & 0.54192 & 302.47 \\
-0.73372 & 0.67945 & -6.58 \times 10^{-8} & -615.31 \\
-0.36822 & -0.39763 & 0.84043 & -195.04 \\
0 & 0 & 0 & 1
\end{bmatrix} \tag{5.182}
$$

Now, we can calculate the position and orientation of the gripper frame in the eye of camera as well as the camera in the base coordinate frame:

$$
{}^8T_7 = {}^8T_4 \, {}^4T_5 \, {}^5T_6 \, {}^6T_7 \tag{5.183}
$$

$$
\approx \begin{bmatrix}
0.54192 & -3.225 \times 10^{-2} & 0.83981 & 1051.8 \\
0 & -0.99927 & -3.837 \times 10^{-2} & 0 \\
0.84043 & 2.08 \times 10^{-2} & -0.54154 & 0 \\
0 & 0 & 0 & 1
\end{bmatrix}
$$

$$
{}^0T_8 = {}^0T_7 \, {}^7T_8 = {}^0T_7 \, {}^8T_7^{-1} \tag{5.184}
$$

$$
= \begin{bmatrix}
0.54192 & -0.03225 & 0.83981 & 2666.8 \\
0.84043 & 0.0208 & -0.54154 & 11915 \\
0 & 0.99927 & 0.03837 & 1015 \\
0 & 0 & 0 & 1
\end{bmatrix}
$$

Example 184 ★ Directional control system. A detailed robotic analysis of kinematic and control of a motorized camera. Such cameras are always applied in robotics for vision and inspection.

A radar, satellite antenna, or inspection camera are samples of direction control and detection systems for which we need to control the angles of a disc and direct an axis normal vector to a desired direction. Such a directional control system is usually equipped with two motors that provide two rotational motions, as shown in Fig. 5.26. The axes of the rotations intersect at a *wrist point*. Let attach a coordinate frame B_8 to the camera at the wrist point and a Sina coordinate frame B_9 to the base link at the same point. The fixed frame B_9 $(x_9, y_9, z_9) \equiv B_9$ (X, Y, Z) acts as the global frame for the local frame B_8 (x_8, y_8, z_8). Sina frames are auxiliary coordinate frames we attach to any point of a robot to simplify transformations. Sina frames usually have no variable parameter. The Sina frame B_9 is a fixed frame on the base link on which the camera is installed.

The first motor turns the control plate by α about the fixed axis z_9 that is initially coincident with z_8. The second motor turns the plate by β about the y_8-axis. The transformation matrix ${}^8R_9$ between B_8 and B_9 from a coincident configuration may be found by a rotation α about z_9 followed by a rotation β about y_8:

$$
{}^8R_9 = R_{y,\beta} \, R_{z,\alpha} = \begin{bmatrix}
\cos\alpha\cos\beta & \cos\beta\sin\alpha & -\sin\beta \\
-\sin\alpha & \cos\alpha & 0 \\
\cos\alpha\sin\beta & \sin\alpha\sin\beta & \cos\beta
\end{bmatrix} \tag{5.185}
$$

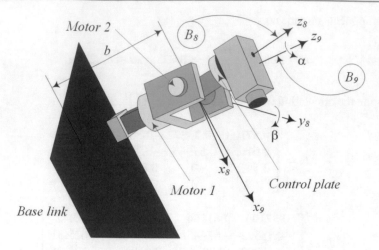

Fig. 5.26 A directional control system that is equipped with two motors

To clarify the transformation matrix $^8R_9$ let us determine $^8R_9$ by rotations about the actual rotation axes z_9 and y_8. The first rotation $R_{Z,\alpha}$ is about the global Z-axis,

$$^9R_8 = R_{Z,\alpha} = \begin{bmatrix} \cos\alpha & -\sin\alpha & 0 \\ \sin\alpha & \cos\alpha & 0 \\ 0 & 0 & 1 \end{bmatrix} \tag{5.186}$$

and the second rotation $R_{y,\beta}$ is about the local y-axis,

$$^8R_9 = R_{y,\beta} = \begin{bmatrix} \cos\beta & 0 & -\sin\beta \\ 0 & 1 & 0 \\ \sin\beta & 0 & \cos\beta \end{bmatrix} \tag{5.187}$$

To determine the combined rotation matrix, we may find $^9R_8 = R_{y,\beta}^T R_{Z,\alpha}$ or $^8R_9 = R_{y,\beta} R_{Z,\alpha}^T$.

$$^8R_9 = R_{y,\beta} R_{Z,\alpha}^T \tag{5.188}$$

$$= \begin{bmatrix} \cos\beta & 0 & -\sin\beta \\ 0 & 1 & 0 \\ \sin\beta & 0 & \cos\beta \end{bmatrix} \begin{bmatrix} \cos\alpha & -\sin\alpha & 0 \\ \sin\alpha & \cos\alpha & 0 \\ 0 & 0 & 1 \end{bmatrix}^T$$

$$= \begin{bmatrix} \cos\alpha\cos\beta & \cos\beta\sin\alpha & -\sin\beta \\ -\sin\alpha & \cos\alpha & 0 \\ \cos\alpha\sin\beta & \sin\alpha\sin\beta & \cos\beta \end{bmatrix}$$

The transformation matrix (5.188) is the same as (5.185) and works only when the initial configuration of the camera frame B_8 coincides with the Sina frame B_9. Furthermore, the rotations must be in order.

Assume the camera aims the desired direction when motor 1 turns α and motor 2 turns β. Because the two motors can act independently, the order of action of motors is immaterial in determination of the final aiming direction as well as the transformation matrix between B_8 and B_9. The two motors may also act together and turn their associated angles together. In this case, the motor will perform an axis–angle of rotation to the camera. Having the rotation matrix as (5.188), we are able to determine the equivalent angle-axis of rotation by Eqs. (3.10) and (3.11).

$$\cos\phi = \frac{1}{2}\left(\mathrm{tr}\left({}^{G}R_{B}\right) - 1\right)$$

$$= \frac{1}{2}\left(\cos\alpha + \cos\beta + \cos\alpha\cos\beta - 1\right) \tag{5.189}$$

$$\tilde{u} = \frac{1}{2\sin\phi}\left({}^{G}R_{B} - {}^{G}R_{B}^{T}\right) = \frac{1}{2\sin\phi}\left({}^{8}R_{9}^{T} - {}^{8}R_{9}\right)$$

$$= \frac{\begin{bmatrix} 0 & -s\alpha - c\beta s\alpha & s\beta + c\alpha s\beta \\ s\alpha + c\beta s\alpha & 0 & s\alpha s\beta \\ -s\beta - c\alpha s\beta & -s\alpha s\beta & 0 \end{bmatrix}}{2\sqrt{1 - \left(\frac{1}{2}\left(\cos\alpha + \cos\beta + \cos\alpha\cos\beta - 1\right)\right)^{2}}} \tag{5.190}$$

Let us assume that

$$\alpha = 0.823\,78\,\mathrm{rad} \approx 47.199\,\mathrm{deg} \tag{5.191}$$

$$\beta = -0.57273\,\mathrm{rad} \approx -32.815\,\mathrm{deg} \tag{5.192}$$

therefore, the rotation matrix (5.188) is

$$^{8}R_{9} = R_{y,\beta}\,R_{z,\alpha} = \begin{bmatrix} 0.57103 & 0.61664 & 0.54193 \\ -0.73372 & 0.67945 & 0 \\ -0.36821 & -0.39762 & 0.84042 \end{bmatrix} \tag{5.193}$$

and the associated axis and angle are

$$\cos\phi = \frac{1}{2}\left(\mathrm{tr}\left({}^{8}R_{9}^{T}\right) - 1\right) = 0.54545\,\mathrm{rad} \approx 31.252\,\mathrm{deg} \tag{5.194}$$

$$\tilde{u} = \frac{1}{2\sin\phi}\left({}^{8}R_{9}^{T} - {}^{8}R_{9}\right)$$

$$= \begin{bmatrix} 0 & -1.301\,5 & -0.877\,15 \\ 1.301\,5 & 0 & -0.383\,21 \\ 0.87715 & 0.383\,21 & 0 \end{bmatrix} \tag{5.195}$$

To activate both motors together, we may command motors 1 and 2 to act according to the following suggested functions:

$$\alpha(t) = 2.4713\frac{t^{2}}{t_{f}^{2}} - 1.6476\frac{t^{3}}{t_{f}^{3}} \tag{5.196}$$

$$\beta(t) = -1.7182\frac{t^{2}}{t_{f}^{2}} + 1.1455\frac{t^{3}}{t_{f}^{3}} \tag{5.197}$$

where t is the time and t_{f} is the total time it takes for the motors to finish their rotations. These equations guarantee that both motors start from zero angular velocity at $t = 0$ and reach their final angular rotations at $t = t_{f}$ and stop there. Let us assume

$$t_{f} = 10\,\mathrm{s} \tag{5.198}$$

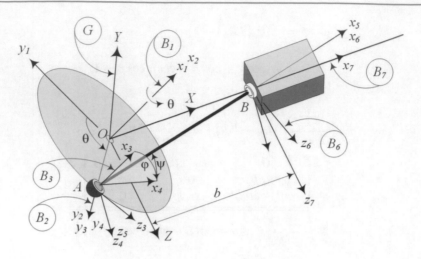

Fig. 5.27 A 3D slider-crank mechanism

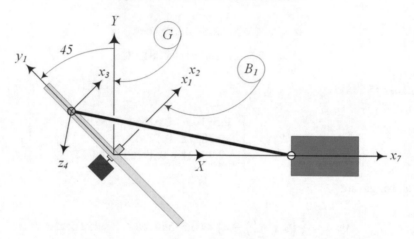

Fig. 5.28 Side view of the 3D slider-crank mechanism of Fig. 5.27

to have

$$\alpha(t) = 2.471\,3 \times 10^{-2}t^2 - 1.6476 \times 10^{-3}t^3 \tag{5.199}$$

$$\beta(t) = -1.718\,2 \times 10^{-2}t^2 + 1.1455 \times 10^{-3}t^3 \tag{5.200}$$

Example 185 ★ A three dimensional slider-crank mechanism. We may treat a mechanism as a robot to determine its kinematics similar to a robot forward kinematic analysis. This is a good sample of three dimensional motion of a floating bar with one end sliding on a fixed axis, while the other end is turning on a spatial circle. This example shows how auxiliary Sina coordinate frames simplify kinematic analysis of a complicated robotic system.

Figure 5.27 illustrates a slider that is moving back and forth on the global X-axis by the connecting rod AB. The rod is attached to a spinning disc that its angular rotation is measured by an input angle θ. The plane of the disc is at an angle 45 deg about the Z-axis. A side view of the mechanism along the Z-axis is shown in Fig. 5.28. The global coordinate frame G is set at the fixed center of the disc. The displacement of the slider is measured by the distance b.

We have four relatively moving bodies, including the ground. So, we technically need four coordinate frames. However, it might be clearer to set up some auxiliary Sina coordinate frames to simplify the kinematic of the system with the penalty of extra matrix multiplications. The frame B_1 is another globally fixed frame at an angle 45 deg about the Z-axis such that the x_1-axis is perpendicular to the disc and indicates its spin axis. The body frame B_2 is attached to the disc such that x_1 and x_2 are coincident and y_2 points the joint A. The angle between y_1 and y_2 is θ. The connecting rod AB has a constant length l and is attached to the disc at point A at a distance R from the disc center:

$$^2\mathbf{r}_A = \begin{bmatrix} 0 & R & 0 \end{bmatrix}^T \tag{5.201}$$

A Sina frame B_3 parallel to B_2 is attached to the disc at point A. The frame B_4 is also at A such that y_4 is coincident with y_3 and x_4 is seen coincident with AB from the Z viewpoint. The angle between x_3 and x_4 is φ. There is another frame B_5 at A such that z_4 and z_5 are coincident and x_5 is along AB. The angle between x_4 and x_5 is ψ. At point B of the bar AB, we attach a coordinate frame B_6 parallel to B_5. The slider frame B_7, which is parallel to G, is attached to the box at point B:

$$^G\mathbf{r}_B = \begin{bmatrix} b\ 0\ 0 \end{bmatrix}^T \tag{5.202}$$

The transformation between G and B_1 is a constant 45 deg rotation about the Z-axis:

$$^G T_1 = \begin{bmatrix} R_{Z,\pi/4} & 0 \\ 0 & 1 \end{bmatrix} = \begin{bmatrix} \cos\frac{\pi}{4} & -\sin\frac{\pi}{4} & 0 & 0 \\ \sin\frac{\pi}{4} & \cos\frac{\pi}{4} & 0 & 0 \\ 0 & 0 & 1 & 0 \\ 0 & 0 & 0 & 1 \end{bmatrix} \tag{5.203}$$

The transformation between B_1 and B_2 is a variable rotation θ about the x_1-axis:

$$^1 T_2 = \begin{bmatrix} R_{x_1,\theta} & 0 \\ 0 & 1 \end{bmatrix} = \begin{bmatrix} 1 & 0 & 0 & 0 \\ 0 & \cos\theta & -\sin\theta & 0 \\ 0 & \sin\theta & \cos\theta & 0 \\ 0 & 0 & 0 & 1 \end{bmatrix} \tag{5.204}$$

The transformation between B_2 and B_3 is a constant translation R along the x_2-axis.

$$^2 T_3 = \begin{bmatrix} 1 & 0 & 0 & R \\ 0 & 1 & 1 & 0 \\ 0 & 0 & 1 & 0 \\ 0 & 0 & 0 & 1 \end{bmatrix} \tag{5.205}$$

The transformation between B_3 and B_4 is a variable rotation φ about the y_3-axis:

$$^3 T_4 = \begin{bmatrix} R_{y_3,\varphi} & 0 \\ 0 & 1 \end{bmatrix} = \begin{bmatrix} \cos\varphi & 0 & \sin\varphi & 0 \\ 0 & 1 & 0 & 0 \\ -\sin\varphi & 0 & \cos\varphi & 0 \\ 0 & 0 & 0 & 1 \end{bmatrix} \tag{5.206}$$

The transformation between B_4 and B_5 is a variable rotation ψ about the z_4-axis:

$$^4 T_5 = \begin{bmatrix} R_{z_4,\psi} & 0 \\ 0 & 1 \end{bmatrix} = \begin{bmatrix} \cos\psi & -\sin\psi & 0 & 0 \\ \sin\psi & \cos\psi & 0 & 0 \\ 0 & 0 & 1 & 0 \\ 0 & 0 & 0 & 1 \end{bmatrix} \tag{5.207}$$

The transformation between B_5 and B_6 is a constant translation l along the x_5-axis:

$$^5 T_6 = \begin{bmatrix} 1 & 0 & 0 & l \\ 0 & 1 & 1 & 0 \\ 0 & 0 & 1 & 0 \\ 0 & 0 & 0 & 1 \end{bmatrix} \tag{5.208}$$

The transformation between B_7 and G is a variable translation b along the X-axis:

$$^G T_7 = \begin{bmatrix} 1 & 0 & 0 & b \\ 0 & 1 & 1 & 0 \\ 0 & 0 & 1 & 0 \\ 0 & 0 & 0 & 1 \end{bmatrix} \tag{5.209}$$

To find the global position of joint B, we go through B_1, B_2, $\cdots$, B_6 and determine $^G T_6$:

$$^G T_6 = {}^G T_1 \, {}^1 T_2 \, {}^2 T_3 \, {}^3 T_4 \, {}^4 T_5 \, {}^5 T_6$$

$$= \begin{bmatrix} r_{11} & r_{12} & r_{13} & d_{6X} \\ r_{21} & r_{22} & r_{23} & d_{6Y} \\ r_{31} & r_{32} & r_{33} & d_{6Z} \\ 0 & 0 & 0 & 1 \end{bmatrix} \tag{5.210}$$

where

$$r_{11} = \frac{\sqrt{2}}{2} \left(\cos\theta \cos\psi - \cos\psi \sin\theta \right) \sin\varphi$$

$$+ \frac{\sqrt{2}}{2} \left(\cos\psi \cos\varphi - \cos\theta \sin\psi \right) \tag{5.211}$$

$$r_{21} = \frac{\sqrt{2}}{2} \left(-\cos\theta \cos\psi + \cos\psi \sin\theta \right) \sin\varphi$$

$$+ \frac{\sqrt{2}}{2} \left(\cos\psi \cos\varphi + \cos\theta \sin\psi \right) \tag{5.212}$$

$$r_{31} = \frac{1}{2} \left(\cos(\theta - \psi) - \cos(\theta + \psi) \right)$$

$$- \left(\cos\theta \cos\psi + \cos\psi \sin\theta \right) \sin\varphi \tag{5.213}$$

$$r_{12} = \frac{\sqrt{2}}{2} \left(\sin\theta \sin\psi - \cos\theta \sin\psi \right) \sin\varphi$$

$$- \frac{\sqrt{2}}{2} \left(\cos\varphi \sin\psi + \cos\theta \cos\psi \right) \tag{5.214}$$

$$r_{22} = \frac{\sqrt{2}}{2} \left(\cos\theta \sin\psi - \sin\theta \sin\psi \right) \sin\varphi$$

$$+ \frac{\sqrt{2}}{2} \left(\cos\theta \cos\psi - \cos\varphi \sin\psi \right) \tag{5.215}$$

$$r_{32} = \frac{1}{4} \sin(\theta + \psi - \varphi) + \frac{1}{4} \sin(\theta - \psi + \varphi) - \frac{1}{4} \sin(\theta - \psi - \varphi)$$

$$- \frac{1}{4} \sin(\theta + \psi + \varphi) + \cos\psi \sin\theta + \cos\theta \sin\psi \sin\varphi \tag{5.216}$$

$$r_{13} = -\frac{\sqrt{2}}{2} \cos\theta \cos\psi + \frac{\sqrt{2}}{2} \left(\sin\theta \sin\psi - \cos\theta \sin\psi \right) \sin\varphi$$

$$+ \frac{\sqrt{2}}{2} \left(\sin\theta - \cos\theta - \sin\psi \right) \cos\varphi + \frac{\sqrt{2}}{2} \sin\varphi \tag{5.217}$$

$$r_{23} = \frac{\sqrt{2}}{2} \sin\varphi + \frac{\sqrt{2}}{2} \left(\cos\theta \sin\psi - \sin\theta \sin\psi \right) \sin\varphi$$

$$+ \frac{\sqrt{2}}{2} \left(\cos\psi + \cos\varphi \right) \cos\theta - \frac{\sqrt{2}}{2} \left(\sin\theta + \sin\psi \right) \cos\varphi \tag{5.218}$$

$$r_{33} = \frac{1}{4} \left(\sin(\theta + \psi - \varphi) + \sin(\theta - \psi + \varphi) - \sin(\theta - \psi - \varphi) \right)$$

$$- \frac{1}{4} \sin(\theta + \psi + \varphi) + \cos\theta \left(\cos\varphi + \sin\psi \sin\varphi \right)$$

$$+ \left(\cos\psi + \cos\varphi \right) \sin\theta \tag{5.219}$$

$$d_{6X} = \frac{\sqrt{2}}{2}R + \frac{\sqrt{2}}{2}l\cos\psi\cos\varphi - \frac{\sqrt{2}}{2}l\cos\theta\sin\psi$$

$$+ \frac{\sqrt{2}}{2}l\left(\cos\theta\cos\psi - \cos\psi\sin\theta\right)\sin\varphi \tag{5.220}$$

$$d_{6Y} = \frac{\sqrt{2}}{2}R + \frac{\sqrt{2}}{2}l\cos\psi\cos\varphi + \frac{\sqrt{2}}{2}l\cos\theta\sin\psi$$

$$- \frac{\sqrt{2}}{2}l\left(\cos\theta\cos\psi + \cos\psi\sin\theta\right)\sin\varphi \tag{5.221}$$

$$d_{6Z} = l\sin\theta\sin\psi - l\left(\cos\theta\cos\psi + \cos\psi\sin\theta\right)\sin\varphi \tag{5.222}$$

The X-component of $^G\mathbf{d}_6$ must be equal to b and the Y- and Z-components of $^G\mathbf{d}_6$ must be zero:

$$d_{6X} = b \qquad d_{6Y} = 0 \qquad d_{6Z} = 0 \tag{5.223}$$

We should be able to solve these three equations to calculate b, φ, ψ for a given θ.

In a different design, we may assume that the angle between G and B_1 is a controllable angle α. In this case the transformation between G and B_1 is based on a given angle α rotation about the Z-axis:

$$^G T_1 = \begin{bmatrix} R_{Z,\alpha} & 0 \\ 0 & 1 \end{bmatrix} = \begin{bmatrix} \cos\alpha & -\sin\alpha & 0 & 0 \\ \sin\alpha & \cos\alpha & 0 & 0 \\ 0 & 0 & 1 & 0 \\ 0 & 0 & 0 & 1 \end{bmatrix} \tag{5.224}$$

Re-evaluating both sides of Eq. (5.210) provides us with three new equations instead of (5.220)–(5.222) that must be solved for b, φ, ψ, for given θ and α. When $\alpha = 90$ deg, the mechanism reduces to the planar slider-crank mechanism, and when $\alpha = 0$, the slider will not move and we will have $b = const$.

5.4 Spherical Wrist

Figure 5.29 illustrates a *spherical wrist* with a *spherical joint*. A *spherical wrist* is a combination of links and joints that provides three rotations about three orthogonal axes intersecting at the wrist point. The wrist is a practical device for an industrial robot to hold, adjust, and control the orientation of a tool. The spherical joint connects two links: the *forearm* link and *hand* link. The axis of the forearm and hand are colinear at the rest position of the hand. The axis of the hand is called the *gripper axis*. An industrial *spherical wrist* is to simulate a spherical joint and provide three rotational DOF for the gripper link. At the wrist point, we define two coordinate frames. The first one is the *wrist dead frame*, attached to the forearm link, and the second frame is the *wrist live frame*, attached to the hand link. We also introduce a *tool* or *gripper frame*. The gripper frame of the wrist is denoted by three vectors, $\mathbf{a} \equiv \hat{k}$, $\mathbf{s} \equiv \hat{\imath}$, $\mathbf{n} \equiv \hat{\jmath}$. It is set at a symmetric point between the fingers of an empty hand or at the tip of the tools held by the fingers. The vector $\mathbf{n}$ is called *tilt* and is the normal vector perpendicular to the fingers or jaws. The vector $\mathbf{s}$ is called *twist* and the vector $\mathbf{a}$ is called *turn*.

Figure 5.29 illustrates the ideal case of a spherical joint; however, a spherical joint is not mechanically controllable. We need to replace the ideal spherical joint with three joints, every one with only one rotational DOF. It is made by three R⊢R links with zero lengths and zero offset where their joint axes are mutually orthogonal and intersecting at the wrist point. The wrist point is invariant in a robot structure and will not move by wrist angular motions.

Figure 5.30 illustrates a schematic of a practical spherical wrist configuration. It is made of an R⊢R(−90) link, attached to another R⊢R(90) link, that finally is attached to a spinning gripper link R∥R(0). The gripper coordinate frame B_7 is always parallel to B_6 and is attached at a distance d_7 from the wrist point.

To classify spherical wrists, let us decompose the rotations of the spherical wrist into three rotations about three orthogonal axes. We call the rotations, Roll, Pitch, and Yaw as are shown in Fig. 5.29. The *Roll* is any rotation that turns the gripper about its axis when the wrist is at the rest position. The gripper axis $\mathbf{a}$ defines a perpendicular plane $\mathbf{s} \times \mathbf{n}$ to the gripper axis that is

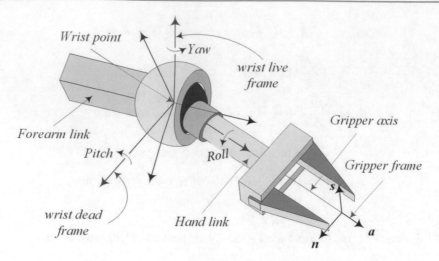

Fig. 5.29 A spherical joint provides roll, pitch, and yaw rotations

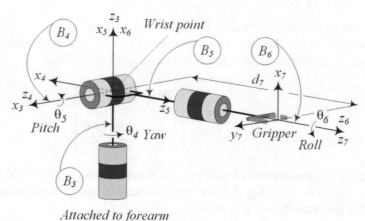

Fig. 5.30 Spherical wrist kinematics

Table 5.15 Spherical wrists classification

Type	Rotation order
1	Roll–Pitch–Roll
2	Roll–Pitch–Yaw
3	Pitch–Yaw–Roll

called the *gripper wall*. The *Pitch* and *Yaw* are rotations about two perpendicular axes in the gripper wall at the wrist point. We can consider the first rotation about an axis in the gripper wall as pitch and the rotation about the axis perpendicular to the first, as yaw. The Roll, Pitch, Yaw rotations are defined at the rest position of the wrist.

There are three types of practical spherical wrists that are classified in Table 5.15.

Figure 5.30 illustrates a Roll–Pitch–Roll spherical wrist with the following transformation matrix.

$$
{}^3T_6 = {}^3T_4\,{}^4T_5\,{}^5T_6 \tag{5.225}
$$

$$
= \begin{bmatrix}
c\theta_4 c\theta_5 c\theta_6 - s\theta_4 s\theta_6 & -c\theta_6 s\theta_4 - c\theta_4 c\theta_5 s\theta_6 & c\theta_4 s\theta_5 & 0 \\
c\theta_4 s\theta_6 + c\theta_5 c\theta_6 s\theta_4 & c\theta_4 c\theta_6 - c\theta_5 s\theta_4 s\theta_6 & s\theta_4 s\theta_5 & 0 \\
-c\theta_6 s\theta_5 & s\theta_5 s\theta_6 & c\theta_5 & 0 \\
0 & 0 & 0 & 1
\end{bmatrix}
$$

Proof We provide the roll, pitch, and yaw rotations by introducing two links and three frames between the dead and live frames. The links will be connected by three revolute joints. The joint axes must always intersect at the wrist point and be orthogonal when the wrist is at the rest position. The kinematic analysis of such wrists should be conducted by attaching a proper DH coordinate frame to every link. Then the associated DH transformation matrices can be combined to develop the overall wrist transformation matrix.

Utilizing the transformation matrix (5.37) for link (4), (5.36) for link (5), R_{Z,θ_6} for link (6), and a D_{Z,d_6} for frame B_7, we find the following transformation matrices:

$$
{}^3T_4 = \begin{bmatrix} \cos\theta_4 & 0 & -\sin\theta_4 & 0 \\ \sin\theta_4 & 0 & \cos\theta_4 & 0 \\ 0 & -1 & 0 & 0 \\ 0 & 0 & 0 & 1 \end{bmatrix}
\tag{5.226}
$$

$$
{}^4T_5 = \begin{bmatrix} \cos\theta_5 & 0 & \sin\theta_5 & 0 \\ \sin\theta_5 & 0 & -\cos\theta_5 & 0 \\ 0 & 1 & 0 & 0 \\ 0 & 0 & 0 & 1 \end{bmatrix}
\tag{5.227}
$$

$$
{}^5T_6 = \begin{bmatrix} \cos\theta_6 & -\sin\theta_6 & 0 & 0 \\ \sin\theta_6 & \cos\theta_6 & 0 & 0 \\ 0 & 0 & 1 & 0 \\ 0 & 0 & 0 & 1 \end{bmatrix}
\tag{5.228}
$$

$$
{}^6T_7 = \begin{bmatrix} 1 & 0 & 0 & 0 \\ 0 & 1 & 0 & 0 \\ 0 & 0 & 1 & d_7 \\ 0 & 0 & 0 & 1 \end{bmatrix}
\tag{5.229}
$$

The matrix ${}^3T_6 = {}^3T_4\,{}^4T_5\,{}^5T_6$ provides the wrist's orientation in the forearm coordinate frame B_3,

$$
{}^3T_6 = {}^3T_4\,{}^4T_5\,{}^5T_6
\tag{5.230}
$$

$$
= \begin{bmatrix} c\theta_4 c\theta_5 c\theta_6 - s\theta_4 s\theta_6 & -c\theta_6 s\theta_4 - c\theta_4 c\theta_5 s\theta_6 & c\theta_4 s\theta_5 & 0 \\ c\theta_4 s\theta_6 + c\theta_5 c\theta_6 s\theta_4 & c\theta_4 c\theta_6 - c\theta_5 s\theta_4 s\theta_6 & s\theta_4 s\theta_5 & 0 \\ -c\theta_6 s\theta_5 & s\theta_5 s\theta_6 & c\theta_5 & 0 \\ 0 & 0 & 0 & 1 \end{bmatrix}
$$

and the following transformation matrix provides the configuration of the tool frame B_7 in the forearm coordinate frame B_3.

$$
{}^3T_7 = {}^3T_4\,{}^4T_5\,{}^5T_6\,{}^6T_7
\tag{5.231}
$$

$$
= \begin{bmatrix} c\theta_4 c\theta_5 c\theta_6 - s\theta_4 s\theta_6 & -c\theta_6 s\theta_4 - c\theta_4 c\theta_5 s\theta_6 & c\theta_4 s\theta_5 & d_7 c\theta_4 s\theta_5 \\ c\theta_4 s\theta_6 + c\theta_5 c\theta_6 s\theta_4 & c\theta_4 c\theta_6 - c\theta_5 s\theta_4 s\theta_6 & s\theta_4 s\theta_5 & d_7 s\theta_4 s\theta_5 \\ -c\theta_6 s\theta_5 & s\theta_5 s\theta_6 & c\theta_5 & d_7 c\theta_5 \\ 0 & 0 & 0 & 1 \end{bmatrix}
$$

It is also possible to define a compact ${}^5T_6$ to include rotation θ_6 and translation d_7.

$$
{}^5T_6 = \begin{bmatrix} \cos\theta_6 & -\sin\theta_6 & 0 & 0 \\ \sin\theta_6 & \cos\theta_6 & 0 & 0 \\ 0 & 0 & 1 & d_7 \\ 0 & 0 & 0 & 1 \end{bmatrix}
\tag{5.232}
$$

Employing the compact matrix ${}^5T_6$ reduces the number of matrices, and therefore the number of numerical calculations.

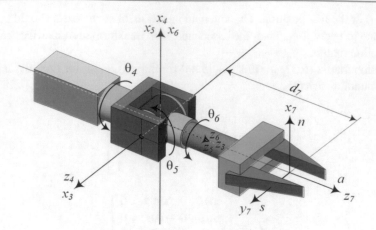

Fig. 5.31 Hand of a robot at rest position

The transformation matrix at rest position, where $\theta_4 = 0, \theta_5 = 0, \theta_6 = 0$, is

$$
{}^3T_7 = \begin{bmatrix} 1 & 0 & 0 & 0 \\ 0 & 1 & 0 & 0 \\ 0 & 0 & 1 & d_7 \\ 0 & 0 & 0 & 1 \end{bmatrix} \tag{5.233}
$$

To show there are only three types of spherical wrists, we start with the first rotation of the wrist that is always about a fixed axis on the forearm link. If the first joint axis is along the forearm axis, then the first motion is a 1−Roll. This axis would also be along the gripper axis at the rest position. If the first axis of rotation is perpendicular to the forearm axis, then we consider the first rotation as 1−Pitch. If the first rotation is a 1−Roll, then the second rotation is perpendicular to the forearm axis and will be a 2−Pitch. There are two possible situations for the third rotation. It is a 3−Roll, if it is about the gripper axis, or a 3−Yaw, if it is orthogonal to the axis of the first two rotations. In case the first rotation is 1−Pitch, then the second one practically can only be a 2−Yaw, and the third one a 3−Roll. Other combinations are only possible mathematically, not practically. If the first rotation is a Pitch, the second rotation can be a Roll or a Yaw. If it is a Yaw, then the third rotation must be a Roll to have independent rotations. If it is a Roll, then the third rotation must be a Yaw. The Pitch−Yaw−Roll and Pitch−Roll−Yaw are not distinguishable, and we may pick Pitch−Yaw−Roll as the only possible spherical wrist with the first rotation as a Pitch. Therefore, we may have three types of wrists: (1−Roll, 2−Pitch, 3−Roll), (1−Roll, 2−Pitch, 3−Yaw), and (1−Pitch, 2−Yaw, 3−Roll), as are shown in (5.234).

$$
wrist\ types = \begin{cases} Roll \quad Pitch \begin{cases} Roll \\ Yaw \end{cases} \\ Pitch \quad Yaw \quad Roll \end{cases} \tag{5.234}
$$

Figure 5.31 illustrates a Roll−Pitch−Roll wrist at the rest position, and Fig. 5.32 illustrates it in motion. This type of wrist is also called Eulerian wrist just because Roll−Pitch−Roll reminds $Z − x − z$ rotation axes.

We model the Roll, Pitch, and Yaw rotations by introducing two links and three frames between the dead and live frames of a wrist. The links will be connected by three revolute joints. The joints' axes intersect at the wrist point and are orthogonal when the wrist is at the rest position.

It is simpler if we kinematically analyze a spherical wrist by defining three non-DH coordinate frames at the wrist point and determine their relative transformations. Figure 5.33 shows a Roll−Pitch−Roll wrist with three coordinate frames. The first orthogonal frame $B_0 (x_0, y_0, z_0)$ is fixed to the forearm and acts as the wrist dead frame such that z_0 is the joint axis of the forearm and a rotating link. The rotating link is the first wrist link and the joint is the first wrist joint. The direction of the axes x_0 and y_0 is arbitrary. The second frame $B_1 (x_1, y_1, z_1)$ is defined such that z_1 is along the gripper axis and x_1 is the axis of the second joint. B_1 always turns φ about z_0 and θ about x_1 relative to B_0. The third frame $B_2 (x_2, y_2, z_2)$ is the wrist live frame and is defined such that z_2 is always along the gripper axis. If the third joint provides a Roll, then z_2 is the joint axis; otherwise, the third joint provides Yaw and x_2 is the joint axis. Therefore, B_2 always turns ψ about z_2 or x_2, relative to B_1. Introducing the coordinate frames B_1 and B_2 simplifies the spherical wrist kinematics by not seeing the interior links

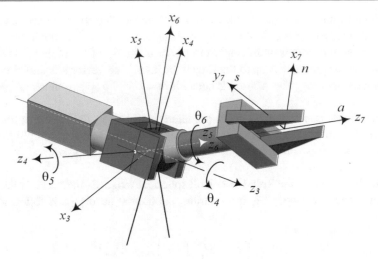

Fig. 5.32 Hand of a robot in motion

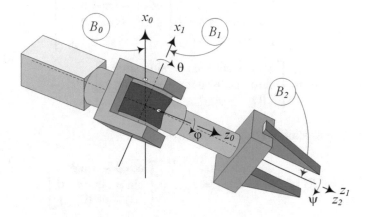

Fig. 5.33 Spherical wrist of the Roll–Pitch–Roll or Eulerian type

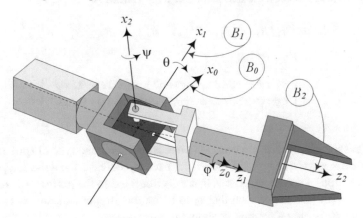

Fig. 5.34 Spherical wrist of the Roll–Pitch–Yaw type

of the wrist. Considering the definition and rotations of B_2 relative to B_1, and B_1 relative to B_0, there are only three types of practical spherical wrists as are classified in Table 5.15. These three wrists are shown in Figs. 5.33, 5.34, and 5.35. ∎

Example 186 Roll–Pitch–Roll or Eulerian wrist and *DH* frames of spherical wrist. This example shows the forward kinematics of the Roll–Pitch–Roll wrist between live and dead frames in detail and step by step. The transformation matrices between frames need special attention to be used as classical method for other types of wrists.

Figures 5.31 and 5.32 depict illustrations of a spherical wrist of type 1. The common origin of frames B_4, B_5, and B_6 is at the wrist point. The final frame, which is called the *tool* or *end-effector frame*, is denoted by three vectors, **a**, **s**, **n**, and is set at a symmetric point between the fingers of an empty gripper or at the tip of the tools hold by the hand. The vector **n** is called tilt and is the normal vector perpendicular to the fingers or jaws. The vector **s** is called twist and is the slide vector showing the direction of fingers opening. The vector **a** is called turn and is the approach vector perpendicular to the palm of the hand.

The placement of internal links' coordinate frames is predetermined by the DH method; however, for the end link the placement of the tool's frame B_n is arbitrary. This arbitrariness may be resolved through simplifying choices or by placement at a distinguished location in the gripper. It is easier to work with the coordinate system B_n if z_n is made coincident with z_{n-1}. This choice sets $a_n = 0$ and $\alpha_n = 0$.

Figure 5.33 illustrates another view of the Roll–Pitch–Roll spherical wrist. B_0 indicates its dead and B_2 indicates its live coordinate frames. The transformation matrix ${}^0R_1$ is a rotation φ about the dead z_0-axis followed by a rotation θ about the x_1-axis.

$$
{}^0R_1 = {}^1R_0^T = \left[R_{x_1,\theta}\, R_{z_0,\varphi}^T \right]^T = \left[R_{x,\theta}\, R_{Z,\varphi}^T \right]^T \tag{5.235}
$$

$$
= \left[\begin{bmatrix} 1 & 0 & 0 \\ 0 & \cos\theta & \sin\theta \\ 0 & -\sin\theta & \cos\theta \end{bmatrix} \begin{bmatrix} \cos\varphi & -\sin\varphi & 0 \\ \sin\varphi & \cos\varphi & 0 \\ 0 & 0 & 1 \end{bmatrix}^T \right]^T
$$

$$
= \begin{bmatrix} \cos\varphi & -\cos\theta\sin\varphi & \sin\theta\sin\varphi \\ \sin\varphi & \cos\theta\cos\varphi & -\cos\varphi\sin\theta \\ 0 & \sin\theta & \cos\theta \end{bmatrix}
$$

The transformation matrix ${}^1R_2$ is a rotation ψ about the local axis z_2.

$$
{}^1R_2 = {}^2R_1^T = R_{z_2,\psi}^T = R_{z,\psi}^T = \begin{bmatrix} \cos\psi & -\sin\psi & 0 \\ \sin\psi & \cos\psi & 0 \\ 0 & 0 & 1 \end{bmatrix} \tag{5.236}
$$

Therefore, the transformation matrix between the live and dead wrist frames is

$$
{}^0R_2 = {}^0R_1\,{}^1R_2 = \left[R_{x_1,\theta}\, R_{z_0,\varphi}^T \right]^T R_{z_2,\psi}^T = R_{z_0,\varphi}\, R_{x_1,\theta}^T\, R_{z_2,\psi}^T
$$

$$
= R_{Z,\varphi}\, R_{x,\theta}^T\, R_{z,\psi}^T \tag{5.237}
$$

$$
= \begin{bmatrix} c\psi c\varphi - c\theta s\psi s\varphi & -c\varphi s\psi - c\theta c\psi s\varphi & s\theta s\varphi \\ c\psi s\varphi + c\theta c\varphi s\psi & c\theta c\psi c\varphi - s\psi s\varphi & -c\varphi s\theta \\ s\theta s\psi & c\psi s\theta & c\theta \end{bmatrix}
$$

Example 187 Roll–Pitch–Yaw spherical wrist. Forward kinematics of the second type of spherical wrist.

Figure 5.34 illustrates a spherical wrist of type 2, Roll–Pitch–Yaw. B_0 indicates the wrist dead coordinate frame. The main kinematic disadvantage of this type of spherical wrist is that z_1 is not fixed to the gripper. However, we attach a coordinate frame B_2 to the gripper as the wrist live frame such that z_2 to be on the gripper axis and x_2 to be the third joint axis. The transformation between B_2 and B_1 is only a rotation ψ about the x_2-axis.

$$
{}^1R_2 = R_{x_2,\psi}^T = R_{x,\psi}^T = \begin{bmatrix} 1 & 0 & 0 \\ 0 & \cos\psi & \sin\psi \\ 0 & -\sin\psi & \cos\psi \end{bmatrix}^T \tag{5.238}
$$

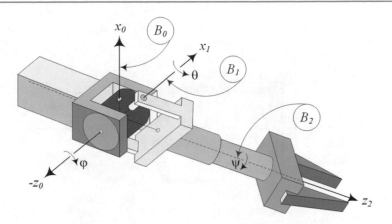

Fig. 5.35 Spherical wrist of the Pitch–Yaw–Roll type

To determine the transformation matrix $^{0}R_{1}$, we turn B_{1} first φ about the z_{0}-axis and then θ about the x_{1}-axis.

$$
^{0}R_{1} = {}^{1}R_{0}^{T} = \left[R_{x_{1},\theta}\, R_{z_{0},\varphi}^{T}\right]^{T} = \left[R_{x,\theta}\, R_{Z,\varphi}^{T}\right]^{T} \tag{5.239}
$$

$$
= \left[\begin{bmatrix} 1 & 0 & 0 \\ 0 & c\theta & s\theta \\ 0 & -s\theta & c\theta \end{bmatrix} \begin{bmatrix} c\varphi & -s\varphi & 0 \\ s\varphi & c\varphi & 0 \\ 0 & 0 & 1 \end{bmatrix}^{T}\right]^{T}
$$

$$
= \begin{bmatrix} \cos\varphi & -\cos\theta\sin\varphi & \sin\theta\sin\varphi \\ \sin\varphi & \cos\theta\cos\varphi & -\cos\varphi\sin\theta \\ 0 & \sin\theta & \cos\theta \end{bmatrix}
$$

Therefore, the transformation matrix between the live and dead wrist's frames is

$$
^{0}R_{2} = {}^{0}R_{1}\,{}^{1}R_{2} = \left[R_{x_{1},\theta}\, R_{z_{0},\varphi}^{T}\right]^{T} R_{x_{2},\psi}^{T} = R_{z_{0},\varphi}\, R_{x_{1},\theta}^{T}\, R_{x_{2},\psi}^{T}
$$

$$
= R_{Z,\varphi}\, R_{x,\theta}^{T}\, R_{x,\psi}^{T} \tag{5.240}
$$

$$
= \begin{bmatrix} c\varphi & s\theta s\psi s\varphi - c\theta c\psi s\varphi & c\theta s\psi s\varphi + c\psi s\theta s\varphi \\ s\varphi & c\theta c\psi c\varphi - c\varphi s\theta s\psi & -c\theta c\varphi s\psi - c\psi c\varphi s\theta \\ 0 & c\theta s\psi + c\psi s\theta & c\theta c\psi - s\theta s\psi \end{bmatrix}
$$

Example 188 Pitch–Yaw–Roll spherical wrist. Forward kinematics of the third type of spherical wrist.

Figure 5.35 illustrates a spherical wrist of the type 3, Pitch–Yaw–Roll. B_{0} indicates its dead and B_{2} indicates its live coordinate frames. The transformation matrix $^{1}R_{2}$ is a rotation ψ about the local z_{2}-axis.

$$
^{1}R_{2} = {}^{2}R_{1}^{T} = R_{z_{2},\psi}^{T} = R_{z,\psi}^{T} = \begin{bmatrix} \cos\psi & -\sin\psi & 0 \\ \sin\psi & \cos\psi & 0 \\ 0 & 0 & 1 \end{bmatrix} \tag{5.241}
$$

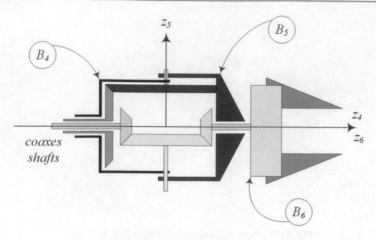

Fig. 5.36 A practical spherical wrist

To determine the transformation matrix $^0R_1$, we turn B_1 first φ about the z_0-axis and then θ about the x_1-axis.

$$^0R_1 = {}^1R_0^T = \left[R_{x_1,\theta}\, R_{z_0,\varphi}^T\right]^T = \left[R_{x,\theta}\, R_{Z,\varphi}^T\right]^T \tag{5.242}$$

$$= \left[\begin{bmatrix} 1 & 0 & 0 \\ 0 & c\theta & s\theta \\ 0 & -s\theta & c\theta \end{bmatrix} \begin{bmatrix} c\varphi & -s\varphi & 0 \\ s\varphi & c\varphi & 0 \\ 0 & 0 & 1 \end{bmatrix}^T\right]^T$$

$$= \begin{bmatrix} \cos\varphi & -\cos\theta\sin\varphi & \sin\theta\sin\varphi \\ \sin\varphi & \cos\theta\cos\varphi & -\cos\varphi\sin\theta \\ 0 & \sin\theta & \cos\theta \end{bmatrix}$$

Therefore, the transformation matrix between the live and dead wrist's frames is

$$^0R_2 = {}^0R_1\,{}^1R_2 = \left[R_{x_1,\theta}\, R_{z_0,\varphi}^T\right]^T R_{z_2,\psi}^T = R_{z_0,\varphi}\, R_{x_1,\theta}^T\, R_{z_2,\psi}^T$$

$$= R_{Z,\varphi}\, R_{x,\theta}^T\, R_{z,\psi}^T \tag{5.243}$$

$$= \begin{bmatrix} c\psi c\varphi - c\theta s\psi s\varphi & -c\varphi s\psi - c\theta c\psi s\varphi & s\theta s\varphi \\ c\psi s\varphi + c\theta c\varphi s\psi & c\theta c\psi c\varphi - s\psi s\varphi & -c\varphi s\theta \\ s\theta s\psi & c\psi s\theta & c\theta \end{bmatrix}$$

Example 189 Practical design of a spherical wrist. An illustration of how a spherical wrist is designed and works. The three rotational motions are all controlled by three coaxes shafts through gear arrangements.

Figure 5.36 illustrates a practical Eulerian spherical wrist. The three rotations of Roll–Pitch–Roll are controlled by three coaxes shafts. The first rotation is a Roll of B_4 about z_4. The second rotation is a Pitch of B_5 about z_5. The third rotation is a roll of B_6 about z_6. The figure shows the wrist at rest position.

5.5　Assembling Kinematics

Most modern industrial robots have a main manipulator and a series of changeable *wrists*. The *manipulator* is multibody so that holds the main power units and provides a powerful motion for the wrist point. Figure 5.37 illustrates an example of an articulated manipulator with three DOFs. This manipulator can rotate θ_1 about the z_0-axis relative to the global frame by the base motor M_1. It carries the other motors M_2 and M_3. Motor M_2 is the base motor of a $2R$ manipulator with arms l_1 and l_8. The motor M_2 turns the arms about the z_1-axis by angle θ_2 and carries motor M_3. The motor M_3 controls the motion of the last arm by turning θ_3 about the z_2-axis. The tip point of the last arm of the manipulator will be the point at which a wrist will be attached.

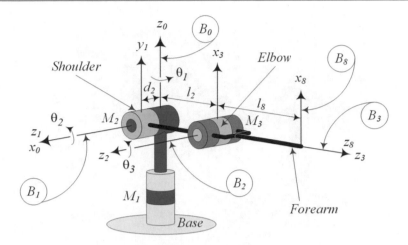

Fig. 5.37 An articulator manipulator with three DOFs

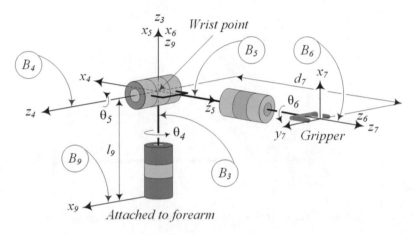

Fig. 5.38 A spherical wrist and its kinematics

Changeable wrists are complex multibodies that are made to provide three rotational DOFs about the wrist point. The base of the wrist will be attached to the tip point of the manipulator. The wrist is the actual operator of the robot that is also called the *end-effector, gripper, hand,* or *tool*. Figure 5.38 illustrates a sample of a spherical wrist that is supposed to be attached to the manipulator in Fig. 5.37.

To solve the kinematics of a modular robot, we consider the manipulator and the wrist as individual multibodies. We attach a temporary coordinate frame at the tip point of the manipulator and another temporary frame at the base point of the wrist. The temporary coordinate frame at the tip point is called the *takht frame*, and the coordinate frame at the base of the wrist is called the *neshin frame*. Mating the neshin and takht frames assembles the robot kinematically. The kinematic mating of the wrist and arm is called *assembling*.

The coordinate frame B_8 in Fig. 5.37 is the takht frame of the manipulator, and the coordinate frame B_9 in Fig. 5.38 is the neshin frame of the wrist. In the assembling process, the neshin frame B_9 sits on the takht frame B_8 such that z_8 be coincident with z_9, and x_8 be coincident with x_9. The articulated robot that is made by assembling the spherical wrist and articulated manipulator is shown in Fig. 5.39.

The assembled multibody will always have some additional coordinate frames. The extra frames require extra transformation matrices that may increase the number of mathematical calculations. It is possible to make a recommendation to eliminate the neshin coordinate frame and keep the takht frame at the connection point. However, as long as the transformation matrices between the frames are known, having extra coordinate frames is not a significant disadvantage. In Fig. 5.39, we may ignore B_8 and directly go from B_3 to B_4 and substitute l_8 and l_9 with $l_3 = l_8 + l_9$.

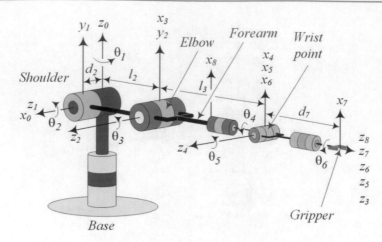

Fig. 5.39 An articulated robot that is made by assembling a spherical wrist to an articulated manipulator

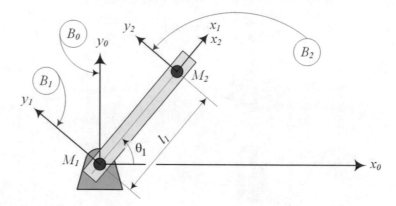

Fig. 5.40 A single DOF arm as the base for a R∥R planar manipulator

Every link needs at least one coordinate frame, called the *main frame* of the link. However, we may also attach some other frames to the link for better expression of coordinates or a simpler transformation. An extra coordinate frame on a link is called the *Sina frame*. A Sina coordinate frame is a spare frame that may be used instead of the main frame. Usually Sina frames have no variables and are related to the main frame of the same link by a constant translation and constant rotation. In fact, whenever we need to install a sensor, a camera, or any equipment to an arm of a robot, we must attach a Sina coordinate frame to determine the position and orientation of the equipment. Such auxiliary coordinate frame is a Sina coordinate frame. The Sina frame may also be called the spare, extra, dummy, temporary, intermediate, or jump frame. The frames B_3 and B_4 are the Sina frames for the link AB, while B is its main frame.

The word "*takht*" means "*chair,*" and the word "*neshin*" means "*sit,*" both from Persian. Sina (Avicenna or Abu Ali Sina) $(980 - 1037)$, a Persian polymath, also known as the "Prince of physicians," and "Father of early modern medicine," is known as one of the foremost philosophers and scientists of the world in the tenth century.

Example 190 A planar $2R$ manipulator assembling. A first example to show the principle of assembling process.

Figure 5.40 illustrates an example of a one DOF arm as the base for an R∥R planar manipulator. This arm can rotate relative to the global frame by a motor at M_1 and carries another motor at M_2. Figure 5.41 illustrates a sample of a planar wrist that is supposed to be attached to the arm in Fig. 5.40. The coordinate frame B_2 in Fig. 5.40 is the takht frame of the arm, and the coordinate frame B_3 in Fig. 5.41 is the neshin frame of the wrist.

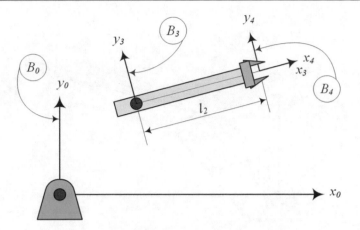

Fig. 5.41 A planar wrist

In Fig. 5.40, transformation between B_1 and B_0 is a rotation θ_1 about z_0.

$$
{}^0T_1 = \begin{bmatrix} \cos\theta_1 & -\sin\theta_1 & 0 & 0 \\ \sin\theta_1 & \cos\theta_1 & 0 & 0 \\ 0 & 0 & 1 & 0 \\ 0 & 0 & 0 & 1 \end{bmatrix}
\tag{5.244}
$$

Transformation between the takht frame B_2 and B_1 is a displacement l_1 along x_1.

$$
{}^1T_2 = \begin{bmatrix} 1 & 0 & 0 & l_1 \\ 0 & 1 & 0 & 0 \\ 0 & 0 & 1 & 0 \\ 0 & 0 & 0 & 1 \end{bmatrix}
\tag{5.245}
$$

In Fig. 5.41, transformation between B_4 and the neshin frame B_3 is a displacement l_1 along x_1.

$$
{}^3T_4 = \begin{bmatrix} 1 & 0 & 0 & l_2 \\ 0 & 1 & 0 & 0 \\ 0 & 0 & 1 & 0 \\ 0 & 0 & 0 & 1 \end{bmatrix}
\tag{5.246}
$$

The R∥R planar manipulator that is made by assembling the wrist and arm is shown in Fig. 5.42.

In Fig. 5.42, transformation between the takht frame B_2 and the neshin frame B_3 is a rotation θ_2 about x_2.

$$
{}^2T_3 = \begin{bmatrix} \cos\theta_2 & -\sin\theta_2 & 0 & 0 \\ \sin\theta_2 & \cos\theta_2 & 0 & 0 \\ 0 & 0 & 1 & 0 \\ 0 & 0 & 0 & 1 \end{bmatrix}
\tag{5.247}
$$

The forward kinematics of the end-effector of the assembled $2R$ robot will be

$$
{}^0T_4 = {}^0T_1 \, {}^1T_2 \, {}^2T_3 \, {}^3T_4
\tag{5.248}
$$

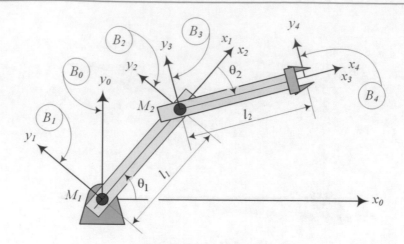

Fig. 5.42 The R‖R planar manipulator that is made by assembling a wrist and an arm

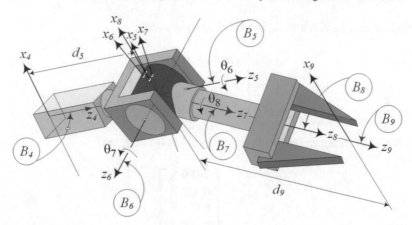

Fig. 5.43 A spherical wrist

We may eliminate B_1 and connect the takht frame B_2 to the ground frame B_0, directly.

$$
{}^0T_2 = {}^0T_1 \, {}^1T_2 = \begin{bmatrix} \cos\theta_1 & -\sin\theta_1 & 0 & l_1\cos\theta_1 \\ \sin\theta_1 & \cos\theta_1 & 0 & l_1\sin\theta_1 \\ 0 & 0 & 1 & 0 \\ 0 & 0 & 0 & 1 \end{bmatrix} \tag{5.249}
$$

We may also eliminate the neshin frame B_3 and connect the end-effector frame B_4 to B_2.

$$
{}^2T_4 = {}^2T_3 \, {}^3T_4 = \begin{bmatrix} \cos\theta_2 & -\sin\theta_2 & 0 & l_2\cos\theta_2 \\ \sin\theta_2 & \cos\theta_2 & 0 & l_2\sin\theta_2 \\ 0 & 0 & 1 & 0 \\ 0 & 0 & 0 & 1 \end{bmatrix} \tag{5.250}
$$

These matrices match with Eqs. (5.83) and (5.84) after a rename of the remaining coordinate frames.

Example 191 Spherical wrist. In this example, a spherical wrist will be decomposed into its component to see how kinematically, we assemble the components to make the wrist working.

A spherical wrist is a combined multibody that simulates a spherical joint. Such a combination gives three rotational DOFs to a final link. The link may be a gripper if the wrist is being used in an assembling and production line. Figure 5.43 illustrates a sample of spherical wrists. The axes z_5, z_6, z_7 are the rotation axes of the wrists.

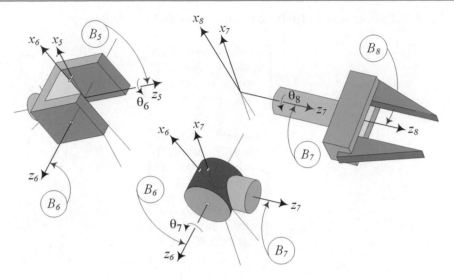

Fig. 5.44 The rotating bodies of a spherical wrist

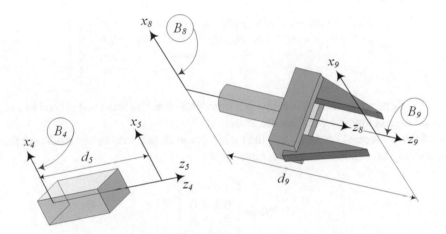

Fig. 5.45 The neshin, dead, living, and gripper frames of a spherical wrist

The coordinate frame B_4 is the neshin frame of the wrist, and B_9 is the tool frame of the wrist. The coordinate frame B_5 is fixed with respect to B_4 and may be considered as the wrist *dead* frame. B_6 is the frame of the first link in shape $\sqsubset$ that rotates about z_5. B_7 is the frame of the middle cylindrical link $\odot$ that rotates about z_6. B_8 is the *wrist live frame* and is the fixed frame to the shaft that supports the gripper. B_8 rotates about z_7. Therefore, B_6, B_7, and B_8 are the frames of the three rotating links. The joint axes of the wrist are intersecting at the *wrist point*. These three links and their associated frames are illustrated in Fig. 5.44.

The spherical wrist of Fig. 5.43 is made of a link R⊢R(-90) attached to another link R⊢R(90) that finally is attached to a spinning gripper link R∥R(0). The final frame B_9 is always parallel to B_8 and is attached to the gripper at a distance d_9 from the wrist point. The final frame B_9 is the *gripper* or *tool frame*. It is set at a symmetric point between the fingers of an empty gripper. The wrist will be attached to the final link of a manipulator arm, which usually provides three DOFs for positioning the wrist point at a desired coordinate in space.

The kinematic analysis of the spherical wrist begins by deriving the required transformation matrices $^5T_6$, $^6T_7$, and $^7T_8$. The matrix $^4T_5$ determines how the wrist dead frame B_5 relates to the neshin frame B_4, and the matrix $^8T_9$ determines how the tool frame B_9 connects to the wrist live frame B_8, as are shown in Fig. 5.45.

Link (6) is an R⊢R(-90), link (7) is an R⊢R(90), and link (8) is an R∥R(0); therefore,

$$
{}^5T_6 = \begin{bmatrix} \cos\theta_6 & 0 & -\sin\theta_6 & 0 \\ \sin\theta_6 & 0 & \cos\theta_6 & 0 \\ 0 & -1 & 0 & 0 \\ 0 & 0 & 0 & 1 \end{bmatrix}
$$
(5.251)

$$
{}^6T_7 = \begin{bmatrix} \cos\theta_7 & 0 & \sin\theta_7 & 0 \\ \sin\theta_7 & 0 & -\cos\theta_7 & 0 \\ 0 & 1 & 0 & 0 \\ 0 & 0 & 0 & 1 \end{bmatrix}
$$
(5.252)

$$
{}^7T_8 = \begin{bmatrix} \cos\theta_8 & -\sin\theta_8 & 0 & 0 \\ \sin\theta_8 & \cos\theta_8 & 0 & 0 \\ 0 & 0 & 1 & 0 \\ 0 & 0 & 0 & 1 \end{bmatrix}
$$
(5.253)

$$
{}^5T_8 = {}^5T_6 \, {}^6T_7 \, {}^7T_8
$$
(5.254)

$$
= \begin{bmatrix} c\theta_6 c\theta_7 c\theta_8 - s\theta_6 s\theta_8 & -c\theta_8 s\theta_6 - c\theta_6 c\theta_7 s\theta_8 & c\theta_6 s\theta_7 & 0 \\ c\theta_6 s\theta_8 + c\theta_7 c\theta_8 s\theta_6 & c\theta_6 c\theta_8 - c\theta_7 s\theta_6 s\theta_8 & s\theta_6 s\theta_7 & 0 \\ -c\theta_8 s\theta_7 & s\theta_7 s\theta_8 & c\theta_7 & 0 \\ 0 & 0 & 0 & 1 \end{bmatrix}
$$

The matrix ${}^5T_8 = {}^5T_6 \, {}^6T_7 \, {}^7T_8$ provides the spherical wrist's transformation. ${}^5T_8$ must be reduced to an identity matrix when the wrist is at rest position and all angular variables are zero.

The transformation of the wrist dead frame B_5 into the neshin frame B_4 and the transformation of the wrist live frame B_8 into the tool frame B_9 are only translations d_5 and d_9, respectively.

$$
{}^4T_5 = \begin{bmatrix} 1 & 0 & 0 & 0 \\ 0 & 1 & 0 & 0 \\ 0 & 0 & 1 & d_5 \\ 0 & 0 & 0 & 1 \end{bmatrix}
$$
(5.255)

$$
{}^8T_9 = \begin{bmatrix} 1 & 0 & 0 & 0 \\ 0 & 1 & 0 & 0 \\ 0 & 0 & 1 & d_9 \\ 0 & 0 & 0 & 1 \end{bmatrix}
$$
(5.256)

Example 192 Assembling a wrist mechanism to a manipulator. Forward kinematics of a wrist and a manipulator have been calculated independently and in this example they are attached to each other to make a full robot.

Consider a robot made by mounting the hand shown in Fig. 5.31, to the tip point of the articulated arm shown in Fig. 5.22. The resulting robot would have six $DOFs$ to reach any point within the working space in a desired orientation. The robot's forward kinematics can be found by combining the wrist transformation matrix (5.225) and the manipulator transformation matrix (5.103).

$$
{}^0T_7 = T_{arm} \, T_{wrist} \, {}^6T_7 = {}^0T_3 \, {}^3T_6 \, {}^6T_7
$$
(5.257)

The wrist transformation matrix ${}^3T_7$ has been given in Eq. (5.225), and the arm transformation matrix ${}^0T_3$ has been found in Example 178. However, because we are attaching the wrist at point P of the frame B_3, the transformation matrix ${}^3T_4$ in (5.226) must include this joint distance. So, we substitute matrix (5.226) with

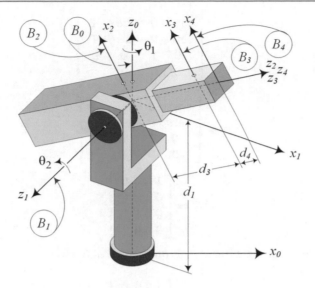

Fig. 5.46 A spherical manipulator

$$
{}^3T_4 = \begin{bmatrix} \cos\theta_4 & 0 & -\sin\theta_4 & 0 \\ \sin\theta_4 & 0 & \cos\theta_4 & 0 \\ 0 & -1 & 0 & l_3 \\ 0 & 0 & 0 & 1 \end{bmatrix}
\tag{5.258}
$$

to find

$$
T_{wrist} = {}^3T_6
\tag{5.259}
$$

$$
= \begin{bmatrix} -s\theta_4 s\theta_6 + c\theta_4 c\theta_5 c\theta_6 & -c\theta_6 s\theta_4 - c\theta_4 c\theta_5 s\theta_6 & c\theta_4 s\theta_5 & 0 \\ c\theta_4 s\theta_6 + c\theta_5 c\theta_6 s\theta_4 & c\theta_4 c\theta_6 - c\theta_5 s\theta_4 s\theta_6 & s\theta_4 s\theta_5 & 0 \\ -c\theta_6 s\theta_5 & s\theta_5 s\theta_6 & c\theta_5 & l_3 \\ 0 & 0 & 0 & 1 \end{bmatrix}
$$

The rest position of the robot can be checked to be at

$$
{}^0T_7 = \begin{bmatrix} 1 & 0 & 0 & l_2 \\ 0 & 1 & 0 & d_2 + l_3 \\ 0 & 0 & 1 & d_1 + d_7 \\ 0 & 0 & 0 & 1 \end{bmatrix}
\tag{5.260}
$$

because

$$
{}^6T_7 = \begin{bmatrix} 1 & 0 & 0 & 0 \\ 0 & 1 & 0 & 0 \\ 0 & 0 & 1 & d_7 \\ 0 & 0 & 0 & 1 \end{bmatrix}
\tag{5.261}
$$

Example 193 A spherical manipulator. A spherical manipulator simulates the spherical coordinate for positioning a point in a $3D$ space to move the tip point of the final link to a desired point in space by two rotations and a translation.

Figure 5.46 illustrates a spherical manipulator. The coordinate frame B_0 is the global or base frame of the manipulator. The link (1) can turn about z_0 and the link (2) can turn about z_1 that is perpendicular to z_0. These two rotations simulate the two angular motions of spherical coordinates. The radial coordinate is simulated by link (3) that has a prismatic joint with link (2). There is also a takht coordinate frame at the tip point of link (3) at which a wrist can be attached.

The link (1) in Fig. 5.46 is an R⊢R(90), link (2) is also an R⊢P(90), and link (3) is a P∥R(0); therefore,

$$
{}^0T_1 = \begin{bmatrix} \cos\theta_1 & 0 & \sin\theta_1 & 0 \\ \sin\theta_1 & 0 & -\cos\theta_1 & 0 \\ 0 & 1 & 0 & d_1 \\ 0 & 0 & 0 & 1 \end{bmatrix}
\tag{5.262}
$$

$$
{}^1T_2 = \begin{bmatrix} \cos\theta_2 & 0 & \sin\theta_2 & 0 \\ \sin\theta_2 & 0 & -\cos\theta_2 & 0 \\ 0 & 1 & 0 & 0 \\ 0 & 0 & 0 & 1 \end{bmatrix}
\tag{5.263}
$$

$$
{}^2T_3 = \begin{bmatrix} 1 & 0 & 0 & 0 \\ 0 & 1 & 0 & 0 \\ 0 & 0 & 1 & d_3 \\ 0 & 0 & 0 & 1 \end{bmatrix}
\tag{5.264}
$$

The transformation matrix from B_3 to the takht frame B_4 is only a translation d_4.

$$
{}^3T_4 = \begin{bmatrix} 1 & 0 & 0 & 0 \\ 0 & 1 & 0 & 0 \\ 0 & 0 & 1 & d_4 \\ 0 & 0 & 0 & 1 \end{bmatrix}
\tag{5.265}
$$

The transformation matrix of the takht frame B_4 to the base frame B_0 is

$$
{}^0T_4 = {}^0T_1 \; {}^1T_2 \; {}^2T_3 \; {}^3T_4
\tag{5.266}
$$

$$
= \begin{bmatrix} c\theta_1 c\theta_2 & s\theta_1 & c\theta_1 s\theta_2 & (d_3+d_4)\,(c\theta_1 s\theta_2) \\ c\theta_2 s\theta_1 & -c\theta_1 & s\theta_1 s\theta_2 & (d_3+d_4)\,(s\theta_2 s\theta_1) \\ s\theta_2 & 0 & -c\theta_2 & d_1 - d_3 c\theta_2 - d_4 c\theta_2 \\ 0 & 0 & 0 & 1 \end{bmatrix}
$$

${}^0T_4$ at the rest position reduces to

$$
{}^0T_4 = \begin{bmatrix} 1 & 0 & 0 & 0 \\ 0 & -1 & 0 & 0 \\ 0 & 0 & -1 & d_1 - d_4 \\ 0 & 0 & 0 & 1 \end{bmatrix}
\tag{5.267}
$$

As a general recommendation, the setup of the DH coordinate frames is such that the overall transformation matrix at the rest position becomes an identity matrix. If we rearrange the coordinate frame of the link (1) to make it an R⊢R(-90), then ${}^0T_1$ becomes

$$
{}^0T_1 = \begin{bmatrix} \cos\theta_1 & 0 & -\sin\theta_1 & 0 \\ \sin\theta_1 & 0 & \cos\theta_1 & 0 \\ 0 & -1 & 0 & d_1 \\ 0 & 0 & 0 & 1 \end{bmatrix}
\tag{5.268}
$$

and the overall transformation matrix at the rest position becomes an identity matrix.

To make link (1) to be R⊢R(-90), we may reverse the direction of z_1- or x_1-axis. Figure 5.47 illustrates the new arrangement of the coordinate frames. Therefore, the transformation matrix of the takht frame B_4 to the base frame B_0 at the rest position reduces to

$$
{}^0T_4 = \begin{bmatrix} 1 & 0 & 0 & 0 \\ 0 & 1 & 0 & 0 \\ 0 & 0 & 1 & d_1 + d_4 \\ 0 & 0 & 0 & 1 \end{bmatrix}
\tag{5.269}
$$

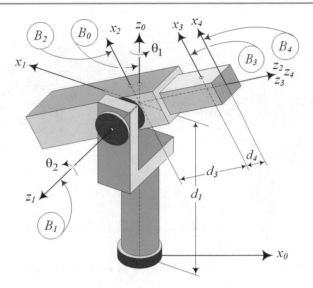

Fig. 5.47 A spherical arm with the arrangement of coordinate frames such that the overall transformation matrix reduces to an identity at rest position

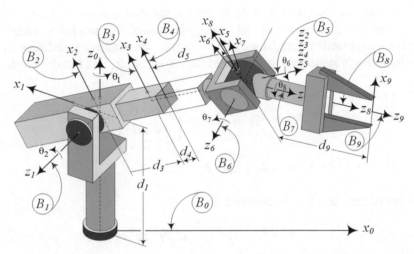

Fig. 5.48 Assembling of a spherical hand and arm

Example 194 ★ Assembling of a spherical wrist to a spherical manipulator. The spherical wrist will be attached to the spherical manipulator by a kinematic surgery to make a complete robot. This example indicates how easier will be such attachment by introducing auxiliary Sina frame whenever needed. This example will be a reference for similar operations.

To transform the manipulator of Fig. 5.47 into a robot, we need to attach a hand to it. Let kinematically assemble the Eulerian spherical wrist of Example 191 or 186 to the spherical manipulator. The wrist, manipulator, and their associated DH coordinate frames are shown in Figs. 5.43 and 5.47, respectively. Assembling a hand to a manipulator is a kinematic surgery in which during a kinematic operation we attach a multibody to the other. In this example we attach a spherical hand to a spherical manipulator to make a spherical arm-hand robot.

The takht coordinate frame B_4 of the manipulator and the neshin coordinate frame B_4 of the wrist are set up to be exactly the same. Therefore, we may assemble the manipulator and wrist by matching these two frames and make a combined manipulator-wrist robot as is shown in Fig. 5.48. However, in general case the takht and neshin coordinate frames may have different labels and there be a constant transformation matrix between them.

The forward kinematics of the robot for tool frame B_9 can be found by a matrix multiplication.

$$^0T_9 = {}^0T_1 \, {}^1T_2 \, {}^2T_3 \, {}^3T_4 \, {}^4T_5 \, {}^5T_6 \, {}^6T_7 \, {}^7T_8 \, {}^8T_9 \tag{5.270}$$

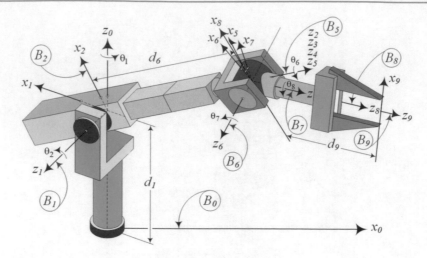

Fig. 5.49 Simplification of the coordinate frames for an assembled of a spherical hand and arm

The matrices $^{i-1}T_i$ are given in Examples 191 and 193.

We can eliminate the coordinate frames B_3 and B_4 to reduce the total number of frames, and simplify the matrix calculations. However, we may prefer to keep them and simplify the assembling process of changing the wrist with a new one. If this assembled robot is supposed to work for a while, we may do the elimination and simplify the robot to the one in Fig. 5.49. We should mathematically substitute the eliminated frames B_3 and B_4 by a transformation matrix $^2T_5$.

$$^2T_5 = {^2T_3}\ {^3T_4}\ {^4T_5} = \begin{bmatrix} 1 & 0 & 0 & 0 \\ 0 & 1 & 0 & 0 \\ 0 & 0 & 1 & d_6 \\ 0 & 0 & 0 & 1 \end{bmatrix} \tag{5.271}$$

$$d_6 = d_3 + d_4 + d_5 \tag{5.272}$$

Now the forward kinematics of the tool frame B_9 becomes

$$^0T_9 = {^0T_1}\ {^1T_2}\ {^2T_5}\ {^5T_6}\ {^6T_7}\ {^7T_8}\ {^8T_9} \tag{5.273}$$

Example 195 Spherical robot forward kinematics. Technically there is nothing new in this example that readers cannot extract from previous example. However, it is assumed here that a new spherical robot with different designs is given and we are supposed to determine its forward kinematics. Here is a detailed analysis.

Figure 5.50 illustrates a spherical manipulator equipped with a spherical wrist to make an R⊢R⊢P robot. The associated DH parameters are shown in Table 5.16.

However, we recommend applying the link-joints classification of Example 166. The link-joint combinations are shown in Table 5.17.

Employing the associated transformation matrices for moving from B_i to B_{i-1} provides the following matrices:

$$^0T_1 = \begin{bmatrix} \cos\theta_1 & 0 & -\sin\theta_1 & 0 \\ \sin\theta_1 & 0 & \cos\theta_1 & 0 \\ 0 & -1 & 0 & 0 \\ 0 & 0 & 0 & 1 \end{bmatrix} \tag{5.274}$$

$$^1T_2 = \begin{bmatrix} \cos\theta_2 & 0 & \sin\theta_2 & 0 \\ \sin\theta_2 & 0 & -\cos\theta_2 & 0 \\ 0 & 1 & 0 & l_2 \\ 0 & 0 & 0 & 1 \end{bmatrix} \tag{5.275}$$

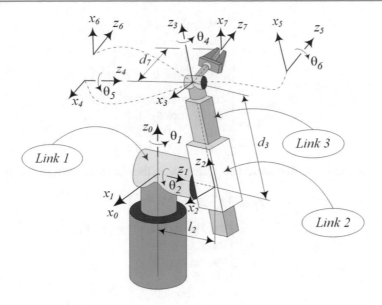

Fig. 5.50 A spherical robot made by a spherical manipulator equipped with a spherical wrist

Table 5.16 DH parameter table for Stanford arm

Frame no.	a_i	α_i	d_i	θ_i
1	0	-90 deg	0	θ_1
2	0	90 deg	l_2	θ_2
3	0	0	d_3	0
4	0	-90 deg	0	θ_4
5	0	90 deg	0	θ_5
6	0	0	0	θ_6

Table 5.17 DH parameter table for setting up the link frames

Link no.	Type
1	R⊢R(-90)
2	R⊢P(90)
3	P∥R(0)
4	R⊢R(-90)
5	R⊢R(90)
6	R∥R(0)

$$^{2}T_3 = \begin{bmatrix} 1 & 0 & 0 & 0 \\ 0 & 1 & 0 & 0 \\ 0 & 0 & 1 & d_3 \\ 0 & 0 & 0 & 1 \end{bmatrix} \tag{5.276}$$

$$^{3}T_4 = \begin{bmatrix} \cos\theta_4 & 0 & -\sin\theta_4 & 0 \\ \sin\theta_4 & 0 & \cos\theta_4 & 0 \\ 0 & -1 & 0 & 0 \\ 0 & 0 & 0 & 1 \end{bmatrix} \tag{5.277}$$

$$
{}^4T_5 = \begin{bmatrix} \cos\theta_5 & 0 & \sin\theta_5 & 0 \\ \sin\theta_5 & 0 & -\cos\theta_5 & 0 \\ 0 & 1 & 0 & 0 \\ 0 & 0 & 0 & 1 \end{bmatrix} \tag{5.278}
$$

$$
{}^5T_6 = \begin{bmatrix} \cos\theta_6 & -\sin\theta_6 & 0 & 0 \\ \sin\theta_6 & \cos\theta_6 & 0 & 0 \\ 0 & 0 & 1 & 0 \\ 0 & 0 & 0 & 1 \end{bmatrix} \tag{5.279}
$$

Therefore, the configuration of the wrist final coordinate frame B_6 in the global coordinate frame is

$$
{}^0T_6 = {}^0T_1\,{}^1T_2\,{}^2T_3\,{}^3T_4\,{}^4T_5\,{}^5T_6
$$

$$
= \begin{bmatrix} r_{11} & r_{12} & r_{13} & r_{14} \\ r_{21} & r_{22} & r_{23} & r_{24} \\ r_{31} & r_{32} & r_{33} & r_{34} \\ 0 & 0 & 0 & 1 \end{bmatrix} \tag{5.280}
$$

where

$$
r_{11} = s\theta_6\left(-c\theta_4 s\theta_1 - c\theta_1 c\theta_2 s\theta_4\right)
$$
$$
+ c\theta_6\left(-c\theta_1 s\theta_2 s\theta_5 + c\theta_5\left(-s\theta_1 s\theta_4 + c\theta_1 c\theta_2 c\theta_4\right)\right) \tag{5.281}
$$
$$
r_{21} = s\theta_6\left(c\theta_1 c\theta_4 - c\theta_2 s\theta_1 s\theta_4\right)
$$
$$
+ c\theta_6\left(-s\theta_1 s\theta_2 s\theta_5 + c\theta_5\left(c\theta_1 s\theta_4 + c\theta_2 c\theta_4 s\theta_1\right)\right) \tag{5.282}
$$
$$
r_{31} = s\theta_2 s\theta_4 s\theta_6 + c\theta_6\left(-c\theta_2 s\theta_5 - c\theta_4 c\theta_5 s\theta_2\right) \tag{5.283}
$$

$$
r_{12} = c\theta_6\left(-c\theta_4 s\theta_1 - c\theta_1 c\theta_2 s\theta_4\right)
$$
$$
- s\theta_6\left(-c\theta_1 s\theta_2 s\theta_5 + c\theta_5\left(-s\theta_1 s\theta_4 + c\theta_1 c\theta_2 c\theta_4\right)\right) \tag{5.284}
$$
$$
r_{22} = c\theta_6\left(c\theta_1 c\theta_4 - c\theta_2 s\theta_1 s\theta_4\right)
$$
$$
- s\theta_6\left(-s\theta_1 s\theta_2 s\theta_5 + c\theta_5\left(c\theta_1 s\theta_4 + c\theta_2 c\theta_4 s\theta_1\right)\right) \tag{5.285}
$$
$$
r_{32} = c\theta_6 s\theta_2 s\theta_4 - s\theta_6\left(-c\theta_2 s\theta_5 - c\theta_4 c\theta_5 s\theta_2\right) \tag{5.286}
$$

$$
r_{13} = c\theta_1 c\theta_5 s\theta_2 + s\theta_5\left(-s\theta_1 s\theta_4 + c\theta_1 c\theta_2 c\theta_4\right) \tag{5.287}
$$
$$
r_{23} = c\theta_5 s\theta_1 s\theta_2 + s\theta_5\left(c\theta_1 s\theta_4 + c\theta_2 c\theta_4 s\theta_1\right) \tag{5.288}
$$
$$
r_{33} = c\theta_2 c\theta_5 - c\theta_4 s\theta_2 s\theta_5 \tag{5.289}
$$

$$
r_{14} = -l_2 s\theta_1 + d_3 c\theta_1 s\theta_2 \tag{5.290}
$$
$$
r_{24} = l_2 c\theta_1 + d_3 s\theta_1 s\theta_2 \tag{5.291}
$$
$$
r_{34} = d_3 c\theta_2 \tag{5.292}
$$

The end-effector kinematics can be solved by multiplying the position of the tool frame B_7 with respect to the wrist point, by ${}^0T_6$,

$$
{}^0T_7 = {}^0T_6\,{}^6T_7 \tag{5.293}
$$

where

$$^6T_7 = \begin{bmatrix} 1 & 0 & 0 & 0 \\ 0 & 1 & 0 & 0 \\ 0 & 0 & 1 & d_7 \\ 0 & 0 & 0 & 1 \end{bmatrix} \tag{5.294}$$

To check the correctness of the final transformation matrix to map the coordinates in tool frame into the base frame, we may set the joint variables at a specific rest position. Let us substitute the joint rotational angles of the spherical robot equal to zero.

$$\theta_1 = 0, \ \theta_2 = 0, \ \theta_3 = 0, \ \theta_4 = 0, \ \theta_5 = 0 \tag{5.295}$$

Therefore, the transformation matrix (5.280) would be

$$^0T_6 = {}^0T_1 \, {}^1T_2 \, {}^2T_3 \, {}^3T_4 \, {}^4T_5 \, {}^5T_6$$

$$= \begin{bmatrix} 1 & 0 & 0 & 0 \\ 0 & 1 & 0 & l_2 \\ 0 & 0 & 1 & d_3 \\ 0 & 0 & 0 & 1 \end{bmatrix} \tag{5.296}$$

that correctly indicates the origin of the tool frame in robot's stretched-up configuration, at

$$^G\mathbf{r}_{o_6} = \begin{bmatrix} 0 \\ l_2 \\ d_3 \end{bmatrix} \tag{5.297}$$

5.6   ★ Coordinate Transformation Using Screws

It is possible to use screws to describe a transformation matrix between two adjacent coordinate frames B_i and B_{i-1}. We can move B_i to B_{i-1} by a central screw $\breve{s}(a_i, \alpha_i, \hat{\imath}_{i-1})$ followed by another central screw $\breve{s}(d_i, \theta_i, \hat{k}_{i-1})$.

$$^{i-1}T_i = \breve{s}(d_i, \theta_i, \hat{k}_{i-1}) \, \breve{s}(a_i, \alpha_i, \hat{\imath}_{i-1}) \tag{5.298}$$

$$= \begin{bmatrix} \cos\theta_i & -\sin\theta_i \cos\alpha_i & \sin\theta_i \sin\alpha_i & a_i \cos\theta_i \\ \sin\theta_i & \cos\theta_i \cos\alpha_i & -\cos\theta_i \sin\alpha_i & a_i \sin\theta_i \\ 0 & \sin\alpha_i & \cos\alpha_i & d_i \\ 0 & 0 & 0 & 1 \end{bmatrix}$$

Proof A Denavit–Hartenberg (DH) frame is based on four parameters $(a_i, \alpha_i, \theta_i, d_i)$ of the link (i). The parameters θ_i and d_i are joint parameters, because they define the relative position of two adjacent links connected at joint i. Assuming every joint to be revolute or prismatic, it will always be the case that either θ_i or d_i is fixed and the other is the joint variable. For a revolute joint (R) at joint i, the θ_i is the unique joint variable, and the value of d_i is fixed. For a prismatic joint (P), the d_i is the only joint variable, while the value of θ_i is fixed. The variable parameter, either θ_i or d_i, is the joint variable. The parameters α_i and a_i are link parameters, because they define relative positions of joints i and $i + 1$ at two ends of link (i). The link twist α_i is the angle of rotation z_{i-1}-axis about x_i to become parallel with the z_i-axis. The kinematic length parameter, a_i, is the translation along the x_i-axis to bring the z_{i-1}-axis on the z_i-axis.

The joint parameters θ_i and d_i define a screw motion because θ_i is a rotation about the z_{i-1}-axis, and d_i is a translation along the z_{i-1}-axis. The link parameters α_i and a_i also define a screw motion because α_i is a rotation about the x_i-axis, and a_i is a translation along the x_i-axis. Therefore, we can move the z_{i-1}-axis to the z_i-axis by a central screw $\breve{s}(a_i, \alpha_i, \hat{\imath}_i)$, and move the x_{i-1}-axis to the x_i-axis by a central screw $\breve{s}(d_i, \theta_i, \hat{k}_{i-1})$.

The central screw $\check{s}(a_i, \alpha_i, \hat{\imath}_i)$ is

$$\check{s}(a_i, \alpha_i, \hat{\imath}_i) = D(a_i, \hat{\imath}_i) R(\hat{\imath}_i, \alpha_i) = D_{x_i, a_i} R_{x_i, \alpha_i} \tag{5.299}$$

$$= \begin{bmatrix} 1 & 0 & 0 & a_i \\ 0 & 1 & 0 & 0 \\ 0 & 0 & 1 & 0 \\ 0 & 0 & 0 & 1 \end{bmatrix} \begin{bmatrix} 1 & 0 & 0 & 0 \\ 0 & \cos\alpha_i & -\sin\alpha_i & 0 \\ 0 & \sin\alpha_i & \cos\alpha_i & 0 \\ 0 & 0 & 0 & 1 \end{bmatrix}$$

$$= \begin{bmatrix} 1 & 0 & 0 & a_i \\ 0 & \cos\alpha_i & -\sin\alpha_i & 0 \\ 0 & \sin\alpha_i & \cos\alpha_i & 0 \\ 0 & 0 & 0 & 1 \end{bmatrix}$$

and the central screw $\check{s}(d_i, \theta_i, \hat{k}_{i-1})$ is

$$\check{s}(d_i, \theta_i, \hat{k}_{i-1}) = D(d_i, \hat{k}_{i-1}) R(\hat{k}_{i-1}, \theta_i) = D_{z_{i-1}, d_i} R_{z_{i-1}, \theta_i} \tag{5.300}$$

$$= \begin{bmatrix} 1 & 0 & 0 & 0 \\ 0 & 1 & 0 & 0 \\ 0 & 0 & 1 & d_i \\ 0 & 0 & 0 & 1 \end{bmatrix} \begin{bmatrix} \cos\theta_i & -\sin\theta_i & 0 & 0 \\ \sin\theta_i & \cos\theta_i & 0 & 0 \\ 0 & 0 & 1 & 0 \\ 0 & 0 & 0 & 1 \end{bmatrix}$$

$$= \begin{bmatrix} \cos\theta_i & -\sin\theta_i & 0 & 0 \\ \sin\theta_i & \cos\theta_i & 0 & 0 \\ 0 & 0 & 1 & d_i \\ 0 & 0 & 0 & 1 \end{bmatrix}$$

Therefore, the transformation matrix $^{i-1}T_i$ made by two screw motions would be

$$^{i-1}T_i = \check{s}(d_i, \theta_i, \hat{k}_{i-1})\, \check{s}(a_i, \alpha_i, \hat{\imath}_i) \tag{5.301}$$

$$= \begin{bmatrix} \cos\theta_i & -\sin\theta_i & 0 & 0 \\ \sin\theta_i & \cos\theta_i & 0 & 0 \\ 0 & 0 & 1 & d_i \\ 0 & 0 & 0 & 1 \end{bmatrix} \begin{bmatrix} 1 & 0 & 0 & a_i \\ 0 & \cos\alpha_i & -\sin\alpha_i & 0 \\ 0 & \sin\alpha_i & \cos\alpha_i & 0 \\ 0 & 0 & 0 & 1 \end{bmatrix}$$

$$= \begin{bmatrix} \cos\theta_i & -\cos\alpha_i \sin\theta_i & \sin\theta_i \sin\alpha_i & a_i \cos\theta_i \\ \sin\theta_i & \cos\theta_i \cos\alpha_i & -\cos\theta_i \sin\alpha_i & a_i \sin\theta_i \\ 0 & \sin\alpha_i & \cos\alpha_i & d_i \\ 0 & 0 & 0 & 1 \end{bmatrix}$$

The resultant transformation matrix $^{i-1}T_i$ is equivalent to a general screw whose parameters can be found based on Eqs. (4.223) and (4.224). The twist of screw, ϕ, can be computed based on Eq. (4.227),

$$\cos\phi = \frac{1}{2}\left(\text{tr}\left(^G R_B\right) - 1\right)$$

$$= \frac{1}{2}(\cos\theta_i + \cos\theta_i \cos\alpha_i + \cos\alpha_i - 1) \tag{5.302}$$

and the axis of screw, $\hat{u}$, can be found by using Eq. (4.229)

$$
\tilde{u} = \frac{1}{2\sin\phi} \left({}^G R_B - {}^G R_B^T \right)
$$

$$
= \frac{1}{2s\phi} \begin{bmatrix} c\theta_i & -c\alpha_i s\theta_i & s\theta_i s\alpha_i \\ s\theta_i & c\theta_i c\alpha_i & -c\theta_i s\alpha_i \\ 0 & s\alpha_i & c\alpha_i \end{bmatrix} - \begin{bmatrix} c\theta_i & s\theta_i & 0 \\ -c\alpha_i s\theta_i & c\theta_i c\alpha_i & s\alpha_i \\ s\theta_i s\alpha_i & -c\theta_i s\alpha_i & c\alpha_i \end{bmatrix}
$$

$$
= \frac{1}{2s\phi} \begin{bmatrix} 0 & -s\theta_i - c\alpha_i s\theta_i & s\theta_i s\alpha_i \\ s\theta_i + c\alpha_i s\theta_i & 0 & -s\alpha_i - c\theta_i s\alpha_i \\ -s\theta_i s\alpha_i & s\alpha_i + c\theta_i s\alpha_i & 0 \end{bmatrix} \tag{5.303}
$$

and therefore,

$$
\hat{u} = \frac{1}{2s\phi} \begin{bmatrix} \sin\alpha_i + \cos\theta_i \sin\alpha_i \\ \sin\theta_i \sin\alpha_i \\ \sin\theta_i + \cos\alpha_i \sin\theta_i \end{bmatrix} \tag{5.304}
$$

The translation parameter, h, and the position vector of a point on the screw axis, for instance $\begin{bmatrix} 0 & y_{i-1} & z_{i-1} \end{bmatrix}$, can be found based on Eq. (4.232).

$$
\begin{bmatrix} h \\ y_{i-1} \\ y_{i-1} \end{bmatrix} = \begin{bmatrix} u_1 & -r_{12} & -r_{13} \\ u_2 & 1-r_{22} & -r_{23} \\ u_3 & -r_{32} & 1-r_{33} \end{bmatrix}^{-1} \begin{bmatrix} r_{14} \\ r_{24} \\ r_{34} \end{bmatrix} \tag{5.305}
$$

$$
= \frac{1}{2s\phi} \begin{bmatrix} s\alpha_i + c\theta_i s\alpha_i & -s\theta_i c\alpha_i & s\theta_i s\alpha_i \\ s\theta_i s\alpha_i & 1-c\theta_i c\alpha_i & -c\theta_i s\alpha_i \\ s\theta_i + c\alpha_i s\theta_i & s\alpha_i & c\alpha_i \end{bmatrix}^{-1} \begin{bmatrix} a_i c\theta_i \\ a_i s\theta_i \\ d_i \end{bmatrix}
$$

∎

Example 196 ★ Classification of industrial robot links by screws. This example indicates how we may determine DH transformation matrices of the 12 types of industrial links, using screw method.

There are 12 different link configurations that are mostly used for industrial robots. Every type has its own class of geometrical configuration and transformation. Each class is identified by its joints at both ends and has its own transformation matrix to go from the distal joint coordinate frame B_i to the proximal joint coordinate frame B_{i-1}. The transformation matrix of each class depends solely on the proximal joint and the angle between z axes. The screw expression for two arbitrary coordinate frames is

$$
{}^{i-1}T_i = \check{s}(d_i, \theta_i, \hat{k}_{i-1})\, \check{s}(a_i, \alpha_i, \hat{\imath}_i) \tag{5.306}
$$

where

$$
\check{s}(d_i, \theta_i, \hat{k}_{i-1}) = D(d_i, \hat{k}_{i-1}) R(\hat{k}_{i-1}, \theta_i) \tag{5.307}
$$

$$
\check{s}(a_i, \alpha_i, \hat{\imath}_i) = D(a_i, \hat{\imath}_i) R(\hat{\imath}_i, \alpha_i) \tag{5.308}
$$

The screw expression of the frame transformation can be simplified for each class according to Table 5.18.

Table 5.18 Classification of industrial robot link by screws

No.	Type	of	Link	$^{i-1}T_i$
1	R∥R(0)	or	R∥P(0)	$\check{s}(0, \theta_i, \hat{k}_{i-1})\,\check{s}(a_i, 0, \hat{\imath}_i)$
2	R∥R(180)	or	R∥P(180)	$\check{s}(0, \theta_i, \hat{k}_{i-1})\,\check{s}(a_i, 2\pi, \hat{\imath}_i)$
3	R⊥R(90)	or	R⊥P(90)	$\check{s}(0, \theta_i, \hat{k}_{i-1})\,\check{s}(a_i, \pi, \hat{\imath}_i)$
4	R⊥R(−90)	or	R⊥P(−90)	$\check{s}(0, \theta_i, \hat{k}_{i-1})\,\check{s}(a_i, -\pi, \hat{\imath}_i)$
5	R⊢R(90)	or	R⊢P(90)	$\check{s}(0, \theta_i, \hat{k}_{i-1})\,\check{s}(0, \pi, \hat{\imath}_i)$
6	R⊢R(−90)	or	R⊢P(−90)	$\check{s}(0, \theta_i, \hat{k}_{i-1})\,\check{s}(0, -\pi, \hat{\imath}_i)$
7	P∥R(0)	or	P∥P(0)	$\check{s}(d_i, 0, \hat{k}_{i-1})\,\check{s}(a_i, 0, \hat{\imath}_i)$
8	P∥R(180)	or	P∥P(180)	$\check{s}(d_i, 0, \hat{k}_{i-1})\,\check{s}(a_i, 2\pi, \hat{\imath}_i)$
9	P⊥R(90)	or	P⊥P(90)	$\check{s}(d_i, 0, \hat{k}_{i-1})\,\check{s}(a_i, \pi, \hat{\imath}_i)$
10	P⊥R(−90)	or	P⊥P(−90)	$\check{s}(d_i, 0, \hat{k}_{i-1})\,\check{s}(a_i, -\pi, \hat{\imath}_i)$
11	P⊢R(90)	or	P⊢P(90)	$\check{s}(d_i, 0, \hat{k}_{i-1})\,\check{s}(0, \pi, \hat{\imath}_i)$
12	P⊢R(−90)	or	P⊢P(−90)	$\check{s}(d_i, 0, \hat{k}_{i-1})\,\check{s}(0, -\pi, \hat{\imath}_i)$

Table 5.19 Screw transformation for the spherical robot shown in Fig. 5.50

Link no.	Class	Screw transformation
1	R⊢R(−90)	$^0T_1 = \check{s}(0, \theta_i, \hat{k}_{i-1})\,\check{s}(0, -\pi, \hat{\imath}_i)$
2	R⊢P(90)	$^1T_2 = \check{s}(0, \theta_i, \hat{k}_{i-1})\,\check{s}(0, \pi, \hat{\imath}_i)$
3	P∥R(0)	$^2T_3 = \check{s}(d_i, 0, \hat{k}_{i-1})\,\check{s}(a_i, 0, \hat{\imath}_i)$
4	R⊢R(−90)	$^3T_4 = \check{s}(0, \theta_i, \hat{k}_{i-1})\,\check{s}(0, -\pi, \hat{\imath}_i)$
5	R⊢R(90)	$^4T_5 = \check{s}(0, \theta_i, \hat{k}_{i-1})\,\check{s}(0, \pi, \hat{\imath}_i)$
6	R∥R(0)	$^5T_6 = \check{s}(0, \theta_i, \hat{k}_{i-1})\,\check{s}(a_i, 0, \hat{\imath}_i)$

As an example, we may examine the first class,

$$
\begin{aligned}
^{i-1}T_i &= \check{s}(0, \theta_i, \hat{k}_{i-1})\,\check{s}(a_i, 0, \hat{\imath}_i) \\
&= D(0, \hat{k}_{i-1})\,R(\hat{k}_{i-1}, \theta_i)\,D(a_i, \hat{\imath}_i)\,R(\hat{\imath}_i, 0) \\
&= \begin{bmatrix} \cos\theta_i & -\sin\theta_i & 0 & 0 \\ \sin\theta_i & \cos\theta_i & 0 & 0 \\ 0 & 0 & 1 & 0 \\ 0 & 0 & 0 & 1 \end{bmatrix} \begin{bmatrix} 1 & 0 & 0 & a_i \\ 0 & 1 & 0 & 0 \\ 0 & 0 & 1 & 0 \\ 0 & 0 & 0 & 1 \end{bmatrix} \\
&= \begin{bmatrix} \cos\theta_i & -\sin\theta_i & 0 & a_i\cos\theta_i \\ \sin\theta_i & \cos\theta_i & 0 & a_i\sin\theta_i \\ 0 & 0 & 1 & 0 \\ 0 & 0 & 0 & 1 \end{bmatrix}
\end{aligned}
\tag{5.309}
$$

and find the same result as Eq. (5.32).

Example 197 ★ Spherical robot forward kinematics based on screws. An example to show how we are able to determine forward kinematics of robots by screw method easier.

Application of screws in forward kinematics can be done by determining the class of each link of a robot and applying the associated screws. The class of links for the spherical robot shown in Fig. 5.50 is indicated in Table 5.19. Therefore, the configuration of the end-effector frame of the spherical robot in the base frame is

$$
\begin{aligned}
^0T_6 &= {}^0T_1\,{}^1T_2\,{}^2T_3\,{}^3T_4\,{}^4T_5\,{}^5T_6 \\
&= \check{s}(0, \theta_i, \hat{k}_{i-1})\,\check{s}(0, -\pi, \hat{\imath}_i)\,\check{s}(0, \theta_i, \hat{k}_{i-1})\,\check{s}(0, \pi, \hat{\imath}_i) \\
&\quad \times \check{s}(d_i, 0, \hat{k}_{i-1})\,\check{s}(a_i, 0, \hat{\imath}_i)\,\check{s}(0, \theta_i, \hat{k}_{i-1})\,\check{s}(0, -\pi, \hat{\imath}_i) \\
&\quad \times \check{s}(0, \theta_i, \hat{k}_{i-1})\,\check{s}(0, \pi, \hat{\imath}_i)\,\check{s}(0, \theta_i, \hat{k}_{i-1})\,\check{s}(a_i, 0, \hat{\imath}_i)
\end{aligned}
\tag{5.310}
$$

Example 198 ★ Plücker coordinate of a central screw. Plücker coordinate is an alternative to work with screws. In general, whenever in science we are working with lines and rotations, Plücker coordinate is considered a smart alternative method of expression of operations.

Utilizing Plücker coordinates we can define a central screw as

$$\check{s}(h, \phi, \hat{u}) = \begin{bmatrix} \phi\hat{u} \\ h\hat{u} \end{bmatrix} \tag{5.311}$$

which is equal to

$$\begin{bmatrix} \phi\hat{u} \\ h\hat{u} \end{bmatrix} = D(h\hat{u})\, R(\hat{u}, \phi) \tag{5.312}$$

The central screw $\check{s}(a_i, \alpha_i, \hat{i}_i)$ can also be expressed by a proper Plücker coordinate.

$$\check{s}(a_i, \alpha_i, \hat{i}_i) = \begin{bmatrix} \alpha_i\,\hat{i}_i \\ a_i\,\hat{i}_i \end{bmatrix} = D(a_i, \hat{i}_i)R(\hat{i}_i, \alpha_i) \tag{5.313}$$

Similarly, the central screw $\check{s}(d_i, \theta_i, \hat{k}_{i-1})$ can also be expressed by a proper Plücker coordinate.

$$\check{s}(d_i, \theta_i, \hat{k}_{i-1}) = \begin{bmatrix} \theta_i\,\hat{k}_{i-1} \\ d_i\,\hat{k}_{i-1} \end{bmatrix} = D(d_i, \hat{k}_{i-1})R(\hat{k}_{i-1}, \theta_i) \tag{5.314}$$

Example 199 ★ Intersecting two central screws. The condition of intersecting screw axes by Plücker coordinate is shown here.

Two lines (and therefore two screws) are intersecting if their reciprocal product is zero. We can check that the reciprocal product of the screws $\check{s}(a_i, \alpha_i, \hat{i}_i)$ and $\check{s}(d_i, \theta_i, \hat{k}_{i-1})$ is zero.

$$\check{s}(d_i, \theta_i, \hat{k}_{i-1}) \times \check{s}(a_i, \alpha_i, \hat{i}_i) = \begin{bmatrix} \theta_i\,\hat{k}_{i-1} \\ d_i\,\hat{k}_{i-1} \end{bmatrix} \otimes \begin{bmatrix} \alpha_i\,\hat{i}_i \\ a_i\,\hat{i}_i \end{bmatrix} \tag{5.315}$$

$$= \theta_i\,\hat{k}_{i-1} \cdot a_i\,\hat{i}_i + \alpha_i\,\hat{i}_i \cdot \theta_i\,\hat{k}_{i-1} = 0$$

5.7 ★ Non-Denavit–Hartenberg Methods

The Denavit–Hartenberg (DH) method of assigning relative coordinate frames of links of a robot is the most common method. However, the DH method is not the only applied method, nor necessarily the best. There are other methods with advantages and disadvantages when compared to the DH method.

The *Sheth method* is an alternative method that can overcome the limitations of the DH method for higher order links, by introducing a number of frames equal to the number of joints on the link. It also provides more flexibility to specify the link geometry. In the Sheth method, we define a coordinate frame at every joint of a link. Hence, an n joint robot would have $2n$ frames. Figure 5.51 shows the case of a binary link (i) where a first frame (x_i, y_i, z_i) is attached at the *origin* of the link and a second frame (u_i, v_i, w_i) to the *end* of the link. The assignment of the origin joint and the end joint is arbitrary; however, it is easier if they are in the direction of base-to-tool frames.

To describe the geometry, first we locate the joint axes by z_i and w_i and then determine the common perpendicular to both joint axes z_i and w_i. The common normal is indicated by a unit vector $\hat{n}_i$. Specifying the link geometry requires six parameters that are determined as follows:

1. a_i is the distance from z_i to w_i, measured along $\hat{n}_i$. It is the kinematic distance between z_i and w_i.
2. b_i is the distance from $\hat{n}_i$ to u_i, measured along w_i. It is the elevation of the w_i-axis.
3. c_i is the distance from x_i to $\hat{n}_i$, measured along z_i. It is the elevation of the z_i-axis.
4. α_i is the angle made by axes z_i and w_i, measured positively from z_i to w_i about $\hat{n}_i$.
5. β_i is the angle made by axes $\hat{n}_i$ and u_i, measured positively from $\hat{n}_i$ to u_i about w_i.
6. γ_i is the angle made by axes x_i and $\hat{n}_i$, measured positively from x_i to $\hat{n}_i$ about z_i.

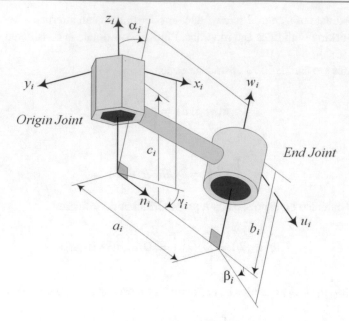

Fig. 5.51 Sheth method for defining the origin and end coordinate frames on a binary link

The Sheth parameters generate a homogenous transformation matrix to map the end coordinate frame B_e to the origin coordinate frame B_o,

$$^oT_e = {^oT_e}\left(a_i, b_i, c_i, \alpha_i, \beta_i, \gamma_i\right) \tag{5.316}$$

$$\begin{bmatrix} x_i \\ y_i \\ z_i \\ 1 \end{bmatrix} = {^oT_e} = \begin{bmatrix} u_i \\ v_i \\ w_i \\ 1 \end{bmatrix} \tag{5.317}$$

where ^oT_e denotes the Sheth transformation matrix

$$^oT_e = \begin{bmatrix} r_{11} & r_{12} & r_{13} & r_{14} \\ r_{21} & r_{22} & r_{23} & r_{24} \\ r_{31} & r_{32} & r_{33} & r_{34} \\ 0 & 0 & 0 & 1 \end{bmatrix} \tag{5.318}$$

$$r_{11} = \cos\beta_i \cos\gamma_i - \cos\alpha_i \sin\beta_i \sin\gamma_i \tag{5.319}$$

$$r_{21} = \cos\beta_i \sin\gamma_i + \cos\alpha_i \cos\gamma_i \sin\beta_i \tag{5.320}$$

$$r_{31} = \sin\alpha_i \sin\beta_i \tag{5.321}$$

$$r_{12} = -\cos\gamma_i \sin\beta_i - \cos\alpha_i \cos\beta_i \sin\gamma_i \tag{5.322}$$

$$r_{22} = -\sin\beta_i \sin\gamma_i + \cos\alpha_i \cos\beta_i \cos\gamma_i \tag{5.323}$$

$$r_{32} = \cos\beta_i \sin\alpha_i \tag{5.324}$$

$$r_{13} = \sin\alpha_i \sin\gamma_i \tag{5.325}$$

$$r_{23} = -\cos\gamma_i \sin\alpha_i \tag{5.326}$$

$$r_{33} = \cos\alpha_i \tag{5.327}$$

$$r_{14} = a_i \cos\gamma_i + b_i \sin\alpha_i \sin\gamma_i \tag{5.328}$$

$$r_{24} = a_i \sin \gamma_i - b_i \cos \gamma_i \sin \alpha_i \tag{5.329}$$

$$r_{34} = c_i + b_i \cos \alpha_i \tag{5.330}$$

The Sheth transformation matrix for two coordinate frames at a joint is simplified to a translation for a prismatic joint and a rotation about the Z-axis for a revolute joint.

Proof The homogenous transformation matrix to provide the coordinates in B_i, when the coordinates in B_j are given, is

$$^oT_e = D_{z_i,c_i} \, R_{z_i,\gamma_i} \, D_{x_i,a_i} \, R_{x_i,\alpha_i} \, D_{z_i,b_i} \, R_{z_i,\beta_i} \tag{5.331}$$

Employing the associated transformation matrices

$$R_{z_i,\beta_i} = \begin{bmatrix} \cos \beta_i & -\sin \beta_i & 0 & 0 \\ \sin \beta_i & \cos \beta_i & 0 & 0 \\ 0 & 0 & 1 & 0 \\ 0 & 0 & 0 & 1 \end{bmatrix} \tag{5.332}$$

$$D_{z_i,b_i} = \begin{bmatrix} 1 & 0 & 0 & 0 \\ 0 & 1 & 0 & 0 \\ 0 & 0 & 1 & b_i \\ 0 & 0 & 0 & 1 \end{bmatrix} \tag{5.333}$$

$$R_{x_i,\alpha_i} = \begin{bmatrix} 1 & 0 & 0 & 0 \\ 0 & \cos \alpha_i & -\sin \alpha_i & 0 \\ 0 & \sin \alpha_i & \cos \alpha_i & 0 \\ 0 & 0 & 0 & 1 \end{bmatrix} \tag{5.334}$$

$$D_{x_i,a_i} = \begin{bmatrix} 1 & 0 & 0 & a_i \\ 0 & 1 & 0 & 0 \\ 0 & 0 & 1 & 0 \\ 0 & 0 & 0 & 1 \end{bmatrix} \tag{5.335}$$

$$R_{z_i,\gamma_i} = \begin{bmatrix} \cos \gamma_i & -\sin \gamma_i & 0 & 0 \\ \sin \gamma_i & \cos \gamma_i & 0 & 0 \\ 0 & 0 & 1 & 0 \\ 0 & 0 & 0 & 1 \end{bmatrix} \tag{5.336}$$

$$D_{z_i,c_i} = \begin{bmatrix} 1 & 0 & 0 & 0 \\ 0 & 1 & 0 & 0 \\ 0 & 0 & 1 & c_i \\ 0 & 0 & 0 & 1 \end{bmatrix} \tag{5.337}$$

we can verify that the Sheth transformation matrix is

$$^oT_e = \begin{bmatrix} c\beta_i \, c\gamma_i & -c\gamma_i \, s\beta_i & & a_i \, c\gamma_i \\ -c\alpha_i \, s\beta_i \, s\gamma_i & -c\alpha_i \, c\beta_i \, s\gamma_i & s\alpha_i \, s\gamma_i & +b_i \, s\alpha_i \, s\gamma_i \\ & & & \\ c\beta_i \, s\gamma_i & -s\beta_i \, s\gamma_i & & a_i \, s\gamma_i \\ +c\alpha_i \, c\gamma_i \, s\beta_i & +c\alpha_i \, c\beta_i \, c\gamma_i & -c\gamma_i \, s\alpha_i & b_i \, c\gamma_i \, s\alpha_i \\ & & & \\ s\alpha_i \, s\beta_i & c\beta_i \, s\alpha_i & c\alpha_i & c_i + b_i \, c\alpha_i \\ 0 & 0 & 0 & 1 \end{bmatrix} \tag{5.338}$$

■

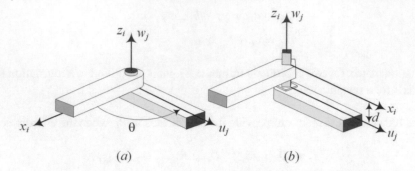

Fig. 5.52 Illustration of (**a**) a revolute joint and (**b**) a prismatic joint to define Sheth coordinate transformation

Example 200 ★ Sheth transformation matrix at revolute and prismatic joints.

Two links connected by a revolute joint are shown in Fig. 5.52a. The coordinate frames of the two links at the common joint are set such that the axes z_i and w_j coincide with the rotation axis, and both frames have the same origin. The Sheth parameters are $a = 0$, $b = 0$, $c = 0$, $\alpha = 0$, $\beta = 0$, $\gamma = \theta$, and therefore, the transformation matrix will be

$$
{}^iT_j = \begin{bmatrix} \cos\theta & -\sin\theta & 0 & 0 \\ \sin\theta & \cos\theta & 0 & 0 \\ 0 & 0 & 1 & 0 \\ 0 & 0 & 0 & 1 \end{bmatrix} \tag{5.339}
$$

Two links connected by a prismatic joint are illustrated in Fig. 5.52b. The Sheth variable at this joint is d along the joint axis. The coordinate frames of the two links at the common joint are set such that the axes z_i and w_j coincide with the translational axis, and axes x_i and u_j are chosen parallel in the same direction. The Sheth parameters are $a = 0$, $b = 0$, $c = d$, $\alpha = 0$, $\beta = 0$, $\gamma = 0$, and therefore, the transformation matrix will be

$$
{}^iT_j = \begin{bmatrix} 1 & 0 & 0 & 0 \\ 0 & 1 & 0 & 0 \\ 0 & 0 & 1 & d \\ 0 & 0 & 0 & 1 \end{bmatrix} \tag{5.340}
$$

Example 201 ★ Sheth transformation matrix at a cylindrical joint.

A cylindrical joint provides two DOFs, a rotational and a translational about the same axis. Two links connected by a cylindrical joint are shown in Fig. 5.53. The transformation matrix for a cylindrical joint can be described by combining a revolute and a prismatic joint. Therefore, the Sheth parameters are $a = 0$, $b = 0$, $c = d$, $\alpha = 0$, $\beta = 0$, $\gamma = \theta$.

$$
{}^iT_j = \begin{bmatrix} \cos\theta & -\sin\theta & 0 & 0 \\ \sin\theta & \cos\theta & 0 & 0 \\ 0 & 0 & 1 & d \\ 0 & 0 & 0 & 1 \end{bmatrix} \tag{5.341}
$$

Example 202 ★ Sheth transformation matrix at a screw joint.

A screw joint, as is shown in Fig. 5.54, provides us with a proportional rotation and translation motion, which has one DOF. The relationship between translation h and rotation θ is called pitch of screw and is defined by

$$
p = \frac{h}{\theta} \tag{5.342}
$$

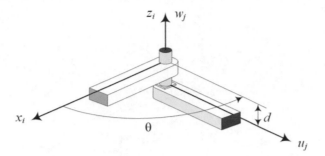

Fig. 5.53 Illustration of a cylindrical joint to define Sheth coordinate transformation

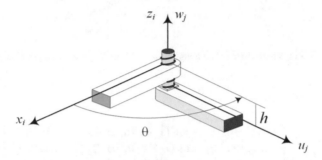

Fig. 5.54 Illustration of a screw joint to define Sheth coordinate transformation

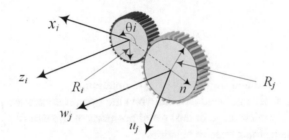

Fig. 5.55 Illustration of a gear joint to define Sheth coordinate transformation

The transformation for a screw joint may be expressed in terms of the relative rotation θ,

$$
{}^{i}T_{j} = \begin{bmatrix} \cos\theta & -\sin\theta & 0 & 0 \\ \sin\theta & \cos\theta & 0 & 0 \\ 0 & 0 & 1 & p\theta \\ 0 & 0 & 0 & 1 \end{bmatrix}
\tag{5.343}
$$

or displacement h.

$$
{}^{i}T_{j} = \begin{bmatrix} \cos\frac{h}{p} & -\sin\frac{h}{p} & 0 & 0 \\ \sin\frac{h}{p} & \cos\frac{h}{p} & 0 & 0 \\ 0 & 0 & 1 & h \\ 0 & 0 & 0 & 1 \end{bmatrix}
\tag{5.344}
$$

The coordinate frames are installed on the two connected links at the screw joint such that the axes w_j and z_i are aligned along the screw axis, and the axes u_j and x_i coincide at rest position.

Example 203 ★ Sheth transformation matrix at a gear joint.

The Sheth method can also be utilized to describe the relative motion of two links connected by a gear joint. A gear joint, as is shown in Fig. 5.55, provides us with a proportional rotation, which has $1\ DOF$. The axes of rotations indicate the axes w_j and z_i, and their common perpendicular shown by the $\hat{n}$ vector. Then, the Sheth parameters are defined as $a = R_i + R_j$,

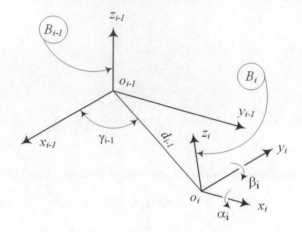

Fig. 5.56 Hayati–Roberts (HR) notation to avoid the singularity in the DH method

$b = 0, c = 0, \alpha = 0, \beta = \theta_j, \gamma = \theta_i$, to have

$$
{}^iT_j = \begin{bmatrix} \cos(1+\varepsilon)\theta_i & -\sin(1+\varepsilon)\theta_i & 0 & R_j(1+\varepsilon)\cos\theta_i \\ \sin(1+\varepsilon)\theta_i & \cos(1+\varepsilon)\theta_i & 0 & R_j(1+\varepsilon)\sin\theta_i \\ 0 & 0 & 1 & 0 \\ 0 & 0 & 0 & 1 \end{bmatrix} \tag{5.345}
$$

where

$$
\varepsilon = \frac{R_i}{R_j} \tag{5.346}
$$

Example 204 ★ Hayati–Roberts method and singularity of DH notation.

In DH notation, the common normal is not well defined when the two joint axes are parallel. In this condition, the DH notation has a singularity, because a small change in the spatial positions of the parallel joint axes can cause a large change in the DH coordinate representation of their relative position.

The Hayati–Roberts (HR) notation is another convention to represent subsequent links. HR avoids the coordinate singularity in the DH method for the case of parallel lines. In the HR method, the direction of the z_i-axis is defined in the B_{i-1} frame using roll and pitch angles α_i and β_i as are shown in Fig. 5.56. The origin of the B_i frame is chosen to lie in the (x_{i-1}, y_{i-1})-plane where the distance d_i is measured between o_{i-1} and o_i.

Similar to DH convention, there is no unique HR convention concerning the freedom in choosing the angle of rotations. Furthermore, although the HR method can eliminate the parallel joint axes' singularity, it has its own singularities when the z_i-axis is parallel to either the x_{i-1} or y_{i-1} axes, or when z_i intersects the origin of the B_{i-1} frame.

Example 205 ★ Parametrically Continuous Convention method.

There exists another method for coordinate transformation called parametrically continuous convention (PC). The PC method defines the z_i-axis by two steps:

1. Direction of the z_i-axis is defined by two direction cosines, α_i and β_i, with respect to the axes x_{i-1} and y_{i-1}.
2. Position of the z_i-axis is defined by two distance parameters, l_i and m_i, to indicate the x_{i-1} and y_{i-1} coordinates of the intersection of z_i in the $x_{i-1}y_{i-1}$-plane from the origin and perpendicular to the z_i-axis.

The PC homogenous matrix to transform B_{i-1} coordinates into B_i is

$$
{}^{i-1}T_i = \begin{bmatrix} 1 - \frac{\alpha_i^2}{1+\gamma_i} & -\frac{\alpha_i\beta_i}{1+\gamma_i} & \alpha_i & l_i \\ -\frac{\alpha_i\beta_i}{1+\gamma_i} & 1 - \frac{\beta_i^2}{1+\gamma_i} & \beta_i & m_i \\ -\alpha_i & -\beta_i & \gamma_i & 0 \\ 0 & 0 & 0 & 1 \end{bmatrix} \tag{5.347}
$$

where

$$\gamma_i = \sqrt{1 - \alpha_i^2 - \beta_i^2}$$

(5.348)

The PC notation uses four parameters to indicate the position and orientation of the B_{i-1} frame in the B_i frame. Since we need six parameters in general, there must be two conditions:

1. The origin of the B_i frame must lie on the common normal and be along the joint axis.
2. The x_i-axis must lie along the common normal.

5.8   Summary

Forward kinematics is determination of the configuration of every link, specially the end-effector, coordinate frame in the base coordinate frame of a robot when the joint variables are given.

$$^0\mathbf{r}_P = {}^0T_n \, {}^n\mathbf{r}_P$$

(5.349)

For an n-link serial robot, it is equivalent to finding the transformation matrix ^0T_n as a function of joint variables q_i.

$$^0T_n = {}^0T_1(q_1) \, {}^1T_2(q_2) \, {}^2T_3(q_3) \, {}^3T_4(q_4) \cdots {}^{n-1}T_n(q_n)$$

(5.350)

There is a special rule for installing the coordinate frames attached to each robot's link called the standard Denavit–Hartenberg convention. Based on the DH rule, each transformation matrix $^{i-1}T_i$ from the coordinate frame B_i to B_{i-1} can be expressed by four parameters: link length a_i, link twist α_i, joint distance d_i, and joint angle θ_i.

$$^{i-1}T_i = \begin{bmatrix} \cos\theta_i & -\sin\theta_i \cos\alpha_i & \sin\theta_i \sin\alpha_i & a_i \cos\theta_i \\ \sin\theta_i & \cos\theta_i \cos\alpha_i & -\cos\theta_i \sin\alpha_i & a_i \sin\theta_i \\ 0 & \sin\alpha_i & \cos\alpha_i & d_i \\ 0 & 0 & 0 & 1 \end{bmatrix}$$

(5.351)

However, for most industrial robots, the link transformation matrix $^{i-1}T_i$ can be classified into 12 simple types.

1	R ∥ R(0)	or R ∥ P(0)
2	R ∥ R(180)	or R ∥ P(180)
3	R ⊥ R(90)	or R ⊥ P(90)
4	R ⊥ R(−90)	or R ⊥ P(−90)
5	R ⊢ R(90)	or R ⊢ P(90)
6	R ⊢ R(−90)	or R ⊢ P(−90)
7	P ∥ R(0)	or P ∥ P(0)
8	P ∥ R(180)	or P ∥ P(180)
9	P ⊥ R(90)	or P ⊥ P(90)
10	P ⊥ R(−90)	or P ⊥ P(−90)
11	P ⊢ R(90)	or P ⊢ P(90)
12	P ⊢ R(−90)	or P ⊢ P(−90)

Most industrial robots are made of a 3 DOF manipulator equipped with a 3 DOF spherical wrist. The transformation matrix $^0T_7$ can be decomposed into three submatrices $^0T_3$, $^3T_6$ and $^6T_7$.

$$^0T_6 = {}^0T_3 \, {}^3T_6 \, {}^6T_7$$

(5.352)

The matrix $^0T_3$ positions the wrist point and depends only on the manipulator joints' variables. The matrix $^3T_6$ is the wrist transformation and depends only on the manipulator wrist's variables. The constant matrix $^6T_7$ is the tools transformation matrix. Decomposing $^0T_7$ into submatrices enables us to make the forward kinematics modular.

## 5.9	Key Symbols

a	Kinematic link length, kinematic distance between z and w
$\mathbf{a}$	Turn vector of end-effector frame
$a_i, \alpha_i, d_i, \theta_i$	DH parameters of link (i)
b	Elevation of w-axis
B	Body coordinate frame
HR	Hayati–Roberts
c	cos, elevation of z-axis
d	Joint distance
$\mathbf{d}$	Translation vector, displacement vector
D	Displacement transformation matrix
DH	Denavit–Hartenberg
DOF	Degree of freedom
F	Goal frame
G, B_0	Global coordinate frame, base coordinate frame
h	Translation of a screw
$\mathbf{I} = [I]$	Identity matrix
$\hat{i}, \hat{j}, \hat{k}$	Coordinate axes unit vectors
l	Length
n	The number of links of a robot, the number of joints of a robot
$\mathbf{n}$	Tilt vector of end-effector frame
$\hat{n}$	Common normal of joint axes in Sheth method
p	Pitch of a screw, a body point, a fixed point in B
$\mathbf{r}$	Position vectors, homogenous position vector
q	Joint variable
r_i	The element i of $\mathbf{r}$
r_{ij}	The element of row i and column j of a matrix
R	Rotation transformation matrix
s	sin
$\mathbf{s}$	Location vector of a screw, twist vector of end-effector frame
$\check{s}$	Screw
S	Station frame
$SSRMS$	Space station remote manipulator system
T	Homogenous transformation matrix, tool frame
$\hat{u}$	Unit vector on axis of rotation
$\tilde{u}$	Skew symmetric matrix of the vector $\hat{u}$
W	Wrist frames
x, y, z	Local coordinate axes
X, Y, Z	Global coordinate axes
z and w	Joint axes of Sheth method

Greek

α	Link twist, roll angle of HR frame, angle from z to w about $\hat{n}$
β	Pitch angle of HR frame, angle from $\hat{n}_i$ to u_i about w_i
γ	Angle from x_i to $\hat{n}_i$ about z_i
θ	Joint angle
$\boldsymbol{\xi}$	Moment vector of a Plücker line
$\boldsymbol{\rho}$	Moment vector of $\hat{u}$ about origin
ϕ	Angle of rotation about $\hat{u}$, rotation of a screw

Symbol

vers	$1 - \cos$
$[\ ]^{-1}$	Inverse of the matrix $[\ ]$
$[\ ]^{T}$	Transpose of the matrix $[\ ]$
$\equiv$	Equivalent
$\vdash$	Orthogonal
(i)	Link number i
$\parallel$	Parallel sign
$\perp$	Perpendicular
$\times$	Vector cross product

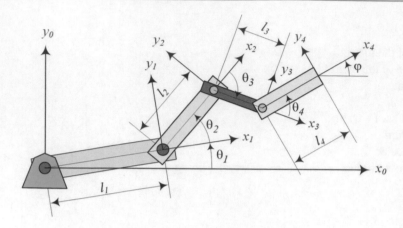

Fig. 5.57 A $4R$ planar manipulator

Exercises

1. Notation and symbols.
   Describe the meaning of

   $$\text{a} - \theta_i \quad \text{b} - \alpha_i \quad \text{c- R} \parallel \text{R}(180) \quad \text{d- P} \perp \text{R}(-90) \quad \text{e- R} \vdash \text{R}(-90)$$

   $$\text{f} -{}^{i-1}T_i \quad \text{g} - \otimes \quad \text{h} - \check{s}(a_i, \alpha_i, \hat{\imath}_i) \quad \text{i} - \check{s}(h, \phi, \hat{u}) \quad \text{j} - \check{s}(d_i, \theta_i, \hat{k}_{i-1})$$

   $$\text{k} - d_i \quad \text{l} - a_i \quad \text{m} - \begin{bmatrix} \alpha_i \, \hat{\imath}_i \\ a_i \, \hat{\imath}_i \end{bmatrix} \quad \text{n} - \begin{bmatrix} \phi \hat{u} \\ h \hat{u} \end{bmatrix} \quad \text{o} - \begin{bmatrix} \theta_i \, \hat{k}_{i-1} \\ d_i \, \hat{k}_{i-1} \end{bmatrix}$$

2. A $4R$ planar manipulator kinematics.
   For the $4R$ planar manipulator, shown in Fig. 5.57, find the followings:
   (a) DH table.
   (b) Link-type table.
   (c) Individual frame transformation matrices ${}^{i-1}T_i, i = 1, 2, 3, 4$.
   (d) Global coordinates of the end-effector.
   (e) Orientation of the end-effector φ.
3. A one-link $R \vdash R(-90)$ arm. Forward kinematics for different frame installments.
   A one-link $R \vdash R(-90)$ manipulator is shown in Fig. 5.58a and b, with different frame installments.
   (a) Find the transformation matrices ${}^0T_1, {}^1T_2, {}^0T_2$.
   (b) Compare the transformation matrix ${}^1T_2$ for both frame installations.
4. A $2R$ planar manipulator. Forward kinematics for different frame arrangements.
   Determine the link's homogenous transformation matrices ${}^1T_2, {}^2T_3, {}^1T_3$ for the $2R$ planar manipulator shown in:
   (a) Figure 5.59.
   (b) Figure 5.60.
   (c) Determine the transformation matrices ${}^3T_2, {}^2T_1, {}^3T_1$ for the manipulator of Fig. 5.59 by moving from a frame to the other, not by matrix inversion.
5. DH coordinate frame setup. It is on engineers to set up coordinate frame the way suit them.
   (a) Set up the required link coordinate frames for the manipulators in Fig. 5.11a and b.
   (b) Set up the required link coordinate frames for the manipulators in Fig. 5.61a and b using l_1, l_2, l_3 for the length of the links.

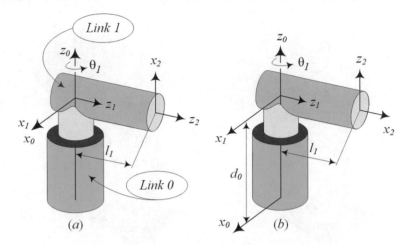

Fig. 5.58 A one-link R⊢R(−90) manipulator

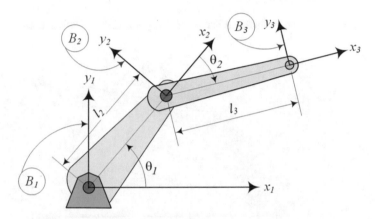

Fig. 5.59 A 2R planar manipulator with DH coordinate frames

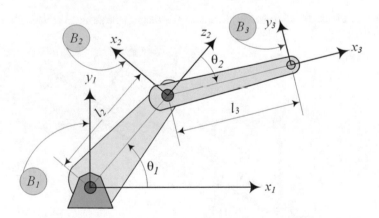

Fig. 5.60 A 2R planar manipulator with arbitrary coordinate frames

(c) Determine the forward kinematics transformation matrix of the manipulator in Fig. 5.61a and b and find their rest positions.

(d) Determine the global coordinates of the tip point of the manipulator in Fig. 5.61a and b at the position shown.

6. Frame at center. A practice on link coordinate frames not installed at either end.

Let us attach the link's coordinate frame at the geometric center of the link, $a_i/2$. Using the rigid motion and homogenous matrices, develop the transformation matrices ${}^0T_1$, ${}^1T_2$, ${}^0T_2$ for the manipulator of Fig. 5.62.

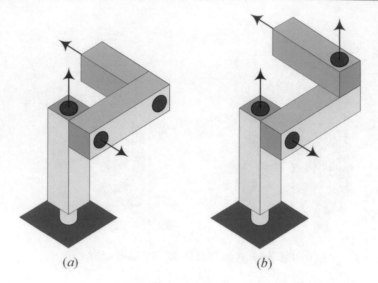

Fig. 5.61 Two manipulators that are made by connecting industrial links

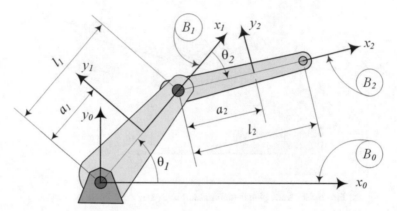

Fig. 5.62 A $2R$ planar manipulator with a coordinate frame at the geometric center of each link

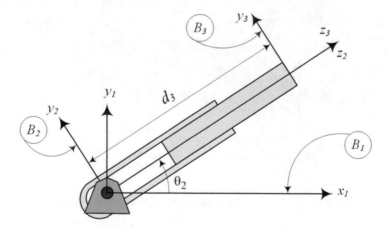

Fig. 5.63 A $2\,DOF$ polar manipulator

7. A polar manipulator. A planar two degrees of freedom robot working based on polar coordinate system.
   (a) Determine the link's transformation matrices $^{1}T_{2}$, $^{2}T_{3}$, $^{1}T_{3}$ for the polar manipulator shown in Fig. 5.63.
   (b) Determine the transformation matrices $^{3}T_{2}$, $^{2}T_{1}$, $^{3}T_{1}$ for the manipulator of Fig. 5.63 by moving from a frame to the other, not by matrix inversion.

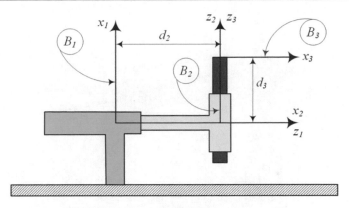

Fig. 5.64 A 2 DOF Cartesian manipulator

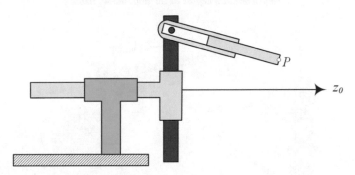

Fig. 5.65 A planar manipulator with 4 DOFs

8. A planar Cartesian manipulator. The simplest and strongest planar robot is the Cartesian manipulator that is working based on Cartesian coordinate system.
   Determine the link's transformation matrices $^1T_2, \, ^2T_3, \, ^1T_3$ for the planar Cartesian manipulator shown in Fig. 5.64. Hint: The coordinate frames are not based on DH rules.

9. ★ Manipulator designing. Industrial robots are designed modular by assembling the 12 standard links. Here is a practice. Use the industrial robot links to make a manipulator to have:
   (a) Three prismatic joints and reach every point in a three dimensional Cartesian space.
   (b) Two prismatic and one revolute joints and reach every point in a three dimensional Cartesian space.
   (c) One prismatic and two revolute joints and reach every point in a three dimensional Cartesian space.
   (d) Three revolute joints and reach every point in a three dimensional Cartesian space.

10. ★ Special manipulator design. In case special paths are supposed to be traced by the end-effector repeatedly, it is better to design the robot such that the path can be traced with the minimum number of actuators, ideally one.
    Use the industrial robot links to make a manipulator with three DOFs such that:
    (a) The tip point of the manipulator traces a circular path about a center point when two joints are lucked.
    (b) The tip point of the manipulator traces a circular path about the origin of global frame when two joints are lucked.
    (c) The tip point of the manipulator traces a straight path when two joints are lucked.
    (d) The tip point of the manipulator traces a straight path passing through the origin of global frame.

11. Coordinate frame assigning.
    Figure 5.65 depicts a planar manipulator with 4 DOFs.
    (a) Follow the DH rules and assign the link coordinate frames.
    (b) Determine the link-joint table for the manipulator.
    (c) Determine the DH transformation matrices.
    (d) Determine the coordinates of point P as functions of the joint coordinates.
    (e) Attach a tool coordinate frame at P and solve the forward kinematics to determine the orientation of the frame.
    (d) Determine the rest configuration and transformation matrix of the manipulator.

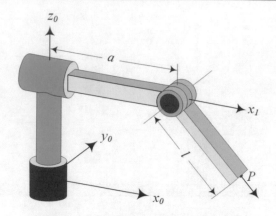

Fig. 5.66 A 3 degree of freedom manipulator

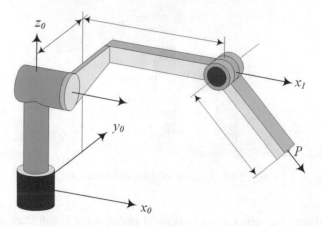

Fig. 5.67 A design of an articulated manipulator

12. Modular articulated manipulators. How to build a $3D$ manipulator by attaching a two DOF planar robot to a rotating base link.

Most industrial robots are modular. Some are manufactured by attaching a $2\,DOF$ manipulator to a one-link R⊢R(-90) arm. Articulated manipulators are made by attaching a $2R$ planar manipulator, such as the one shown in Fig. 5.59, to a one-link R⊢R(-90) manipulator shown in Fig. 5.58a. Attach the $2R$ manipulator to the one-link R⊢R(-90) arm and make an articulated manipulator. Make the required changes into the coordinate frames of Exercises 3 and 4 to find the link's transformation matrices of the articulated manipulator. Examine the rest position of the manipulator.

13. Coordinate frame completing. A practice of coordinate frame setup.

Figure 5.66 shows a 3 degree of freedom manipulator.

(a) Follow the DH rules and complete the link coordinate frames.

(b) Determine the link-joint table for the manipulator.

(c) Determine the DH transformation matrices.

(d) Determine the coordinates of point P as functions of the joint coordinates.

(e) Attach a tool coordinate frame at P and solve the forward kinematics to determine the orientation of the frame.

(f) Determine the rest configuration and transformation matrix of the manipulator.

14. Articulated manipulator. Modular robot with strange shape.

Figure 5.67 illustrates an articulated manipulator.

(a) Follow the DH rules and complete the link coordinate frames.

(b) Determine the link-joint table for the manipulator.

(c) Determine the DH transformation matrices.

(d) Determine the coordinates of point P as functions of the joint coordinates.

(e) Attach a tool coordinate frame at P and solve the forward kinematics to determine the orientation of the frame.

(f) Determine the rest configuration and transformation matrix of the manipulator.

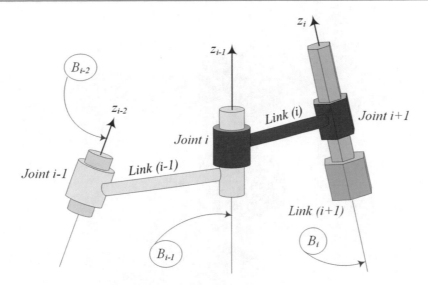

Fig. 5.68 A set of non-industrial connected links

15. Modular spherical manipulators. How to build a $3D$ manipulator by attaching a two DOF polar robot to a rotating base link.

    Spherical manipulators are made by attaching a polar manipulator shown in Fig. 5.63 to a one-link R⊢R(-90) manipulator shown in Fig. 5.58b. Attach the polar manipulator to the one-link R⊢R(-90) arm to make a spherical manipulator. Make the required changes to the coordinate frames of figures in Exercises 3 and 7 to find the link's transformation matrices of the spherical manipulator. Examine the rest position of the manipulator.

16. ★ Non-industrial links and DH parameters. In special design we may use non-standard industrial links. However, DH method works for all binary links with one joint at each end. This is an example of three general links joined to each other.

    Figure 5.68 illustrates a set of non-industrial connected links. Complete the DH coordinate frames and assign the DH parameters and transformation matrices.

17. Modular cylindrical manipulators. A three DOF manipulator that simulates cylindrical coordinate system is designed in this example.

    Cylindrical manipulators are made by attaching a 2 DOF Cartesian manipulator, shown in Fig. 5.64, to a one-link R⊢R(-90) manipulator shown in Fig. 5.58a. Attach the 2 DOF Cartesian manipulator to the one-link R⊢R(-90) arm and make a cylindrical manipulator. Make the required changes into the coordinate frames of Exercises 3 and 8 to find the link's transformation matrices of the cylindrical manipulator. Examine the rest position of the manipulator.

18. Disassembled spherical wrist. Practice on defining coordinate frame on a spherical wrist as three links.

    A spherical wrist has three revolute joints in such a way that their joint axes intersect at a common wrist point. Every revolute joint of the wrist attaches two links. Disassembled links of a spherical wrist are shown in Fig. 5.69. Define the required DH coordinate frames to the links in (a), (b), and (c) consistently. Find the transformation matrices $^3T_4$ for (a), $^4T_5$ for (b), and $^5T_6$ for (c).

19. ★ Assembled spherical wrist. There are three links in a spherical wrist, as shown in previous exercise. Here, the three links are supposed to be kinematically assembled to make a wrist.

    Label the coordinate frames attached to the spherical wrist in Fig. 5.70 according to the frames that you installed in Exercise 18. Determine the transformation matrices $^3T_6$ and $^3T_7$ for the wrist.

20. ★ A 5 DOF robot. Attachment of a 3 DOF spherical wrist to a 2 DOF manipulator.

    Figure 5.71 illustrates a five DOF robot having a spherical wrist.

    (a) Follow the DH rules and complete the link coordinate frames such that the hand of the robot at the rest position stays straight with the forearm.
    (b) Determine the link-joint table for the manipulator.
    (c) Determine the DH transformation matrices.
    (d) Determine the forward kinematics final transformation matrix.
    (e) Determine the rest configuration and transformation matrix of the manipulator.

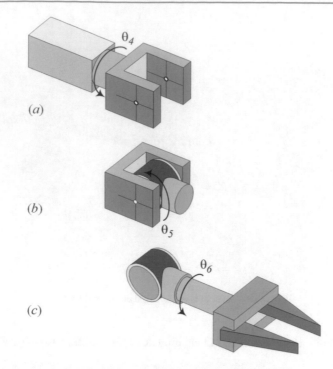

Fig. 5.69 Disassembled links of a spherical wrist

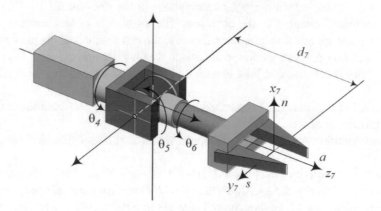

Fig. 5.70 Assembled spherical wrist

21. Articulated robots. Assembling a wrist to an articulated manipulator to make a robot.
    Attach the spherical wrist of Exercise 19 to the articulated manipulator of Exercise 12 and make a 6 DOF articulated robot. Change your DH coordinate frames in the exercises accordingly and solve the forward kinematics problem of the robot.

22. Spherical robots. Assembling a wrist to a spherical manipulator to make a robot.
    Attach the spherical wrist of Exercise 19 to the spherical manipulator of Exercise 15 and make a 6 DOF spherical robot. Change your DH coordinate frames in the exercises accordingly and solve the forward kinematics problem of the robot.

23. Cylindrical robots. Assembling a wrist to a cylindrical manipulator to make a robot.
    Attach the spherical wrist of Exercise 19 to the cylindrical manipulator of Exercise 17 and make a 6 DOF cylindrical robot. Change your DH coordinate frames in the exercises accordingly and solve the forward kinematics problem of the robot.

24. An R∥P manipulator. Kinematic analysis of the end part of a $SCARA$ manipulator.
    Figure 5.72 depicts a 2 DOF R∥P manipulator. The end-effector of the manipulator can slide on a line and rotate about the same line. Label the installed coordinate frames on the links of the manipulator and determine the transformation matrix of the end-effector to the base link.

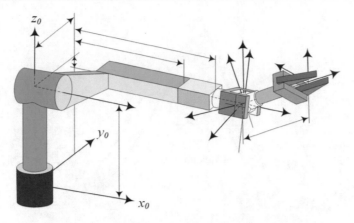

Fig. 5.71 A five DOF robot having a spherical wrist

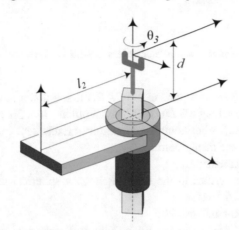

Fig. 5.72 A 2 DOF R‖P manipulator

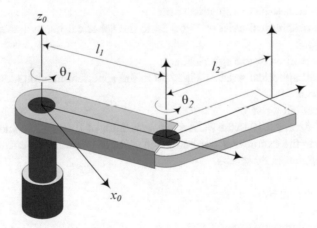

Fig. 5.73 A 2R manipulator, acting in a horizontal plane

25. Horizontal 2R manipulator. Kinematic analysis of the beginning part of a $SCARA$ manipulator.

Figure 5.73 illustrates a 2R planar manipulator that acts in a horizontal plane. Label the coordinate frames and determine the transformation matrix of the end-effector in the base frame.

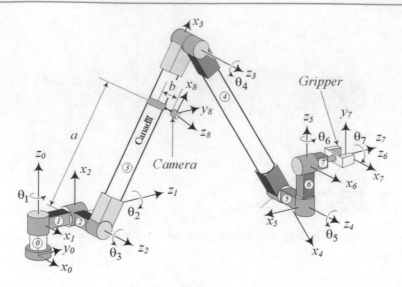

Fig. 5.74 The space shuttle remote manipulator system ($SSRMS$) with a camera attached to the link (3)

26. $SCARA$ manipulator. Attachment of the two parts of a $SCARA$ manipulator studied in the previous exercises.
    A $SCARA$ robot can be made by attaching a 2 DOF R‖P manipulator to a $2R$ planar manipulator. Attach the 2 DOF R‖P manipulator of Exercise 24 to the $2R$ horizontal manipulator of Exercise 25 and make a $SCARA$ manipulator. Solve the forward kinematics problem for the manipulator.

27. ★ Roll–Pitch–Yaw spherical wrist kinematics.
    Attach the required DH coordinate frames to the Roll–Pitch–Yaw spherical wrist of Fig. 5.34, similar to 5.32, and determine the forward kinematics of the wrist.

28. ★ Pitch–Yaw–Roll spherical wrist kinematics.
    Attach the required coordinate DH frames to the Pitch–Yaw–Roll spherical wrist of Fig. 5.35, similar to 5.32, and determine the forward kinematics of the wrist.

29. ★ Assembling a Roll–Pitch–Yaw wrist to a spherical arm.
    Assemble the Roll–Pitch–Yaw spherical wrist of Fig. 5.34 to the spherical manipulator of Fig. 5.47 and determine the forward kinematics of the robot.

30. ★ Assembling a Pitch–Yaw–Roll wrist to a spherical arm.
    Assemble the Pitch–Yaw–Roll spherical wrist of Fig. 5.50 to the spherical manipulator of Fig. 5.47 and determine the forward kinematics of the robot.

31. ★ $SCARA$ robot with a spherical wrist.
    Attach the spherical wrist of Exercise 19 to the $SCARA$ manipulator of Exercise 26 and make a 7 DOF robot. Change your DH coordinate frames in the exercises accordingly and solve the forward kinematics problem of the robot.

32. ★ Modular articulated manipulators by screws.
    Solve Exercise 12 by screws.

33. ★ Modular spherical manipulators by screws.
    Solve Exercise 15 by screws.

34. ★ Modular cylindrical manipulators by screws.
    Solve Exercise 17 by screws.

35. ★ Spherical wrist kinematics by screws.
    Solve Exercise 19 by screws.

36. ★ Modular $SCARA$ manipulator by screws.
    Solve Exercises 24, 25, and 26 by screws.

37. ★ Space station remote manipulator system.
    Attach a spherical wrist to the $SSRMS$ and make a 10 DOF robot. Solve the forward kinematics of the robot by matrix and screw methods.

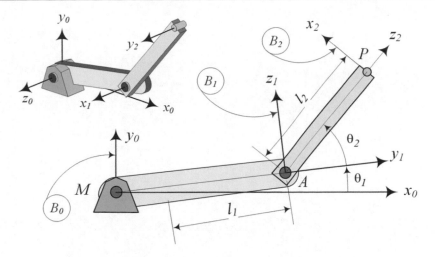

Fig. 5.75 A planar $2R$ robot with arbitrary coordinate frame setup

38. ★ Camera on a space station remote manipulator.
    Assume that we attach an inspection camera to link (3) of the space station remote manipulator system, as shown in Fig. 5.74.
    Determine the matrix $^8T_7$ such that x_8 points the origin of the gripper frame B_7 and z_8 be in (x_3, x_4)-plane and perpendicular to x_4. Then, determine the matrices $^0T_8, {}^4T_8, {}^3T_8, {}^6T_8, {}^1T_8$.
39. Forward kinematics of non-DH frames.
    Figure 5.75 illustrates a planar $2R$ robot with arbitrary coordinate frame setup. Determine its forward kinematics.
40. ★Project: Joint angles of a $2R$ robot for a given path.
    Consider the $2R$ robot of Fig. 5.75 with the following dimensions.

$$l_1 = 1\,\text{m} \qquad l_2 = 1\,\text{m} \tag{5.353}$$

The tip point P is moving from (X_1, Y_1) to (X_2, Y_2) on a straight line.

$$(X_1, Y_1) = (1, 1.5) \qquad (X_2, Y_2) = (-1, 1.5) \tag{5.354}$$

(a) Assume an elbow up configuration. Determine the location of joint A when P is at (X_1, Y_1) by intersecting a circle with radius l_1 and center M, and another circle with radius l_2 and center P. Then determine the location of joint A when P is at (X_2, Y_2) by intersecting a circle with radius l_1 and center M and another circle with radius l_2 and center P. Calculate θ_1 at the beginning and end of the motion. Divide the sweep angle of θ_1 into 10 equal angles. Position A at the 11 angular positions of θ_1. Calculate θ_2 to locate P at a point on the path of motion by trial and error and apply forward kinematics collations.

(b) Assume an elbow down configuration. Determine the location of joint A when P is at (X_1, Y_1) by intersecting a circle with radius l_1 and center M, and another circle with radius l_2 and center P. Then determine the location of joint A when P is at (X_2, Y_2) by intersecting a circle with radius l_1 and center M, and another circle with radius l_2 and center P. Calculate θ_1 at the beginning and end of the motion. Divide the sweep angle of θ_1 into 10 equal angles. Position A at the 11 angular positions of θ_1. Calculate θ_2 to locate P at a point on the path of motion by trial and error and apply forward kinematics collations.

(c) Which configuration, elbow up or elbows down, will provide minmax deviation of θ_2 from zero.

Inverse Kinematics

6

What are the joint variables for a given configuration of a robot? This is the problem to be answered by inverse kinematic analysis. In this chapter, you will see that determination of the joint variables reduces to solving a set of nonlinear coupled algebraic equations. Although there is no standard and generally applicable method to solve the inverse kinematic problem, there are a few analytic and numerical methods to solve the problem. The main difficulty of inverse kinematic is the multiple solutions such as the one that is shown in Fig. 6.1 for a planar $2R$ manipulator.

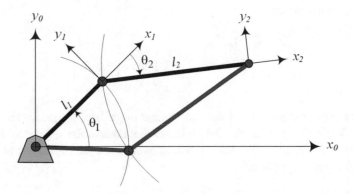

Fig. 6.1 Multiple solution for inverse kinematic problem of a planar $2R$ manipulator

---

6.1 Decoupling Technique

Determination of joint variables in terms of the end-effector position and orientation is called *inverse kinematics*. Mathematically, inverse kinematics is searching for the elements of joint variable vector $\mathbf{q}$,

$$\mathbf{q} = \begin{bmatrix} q_1 & q_2 & q_3 & \cdots & q_n \end{bmatrix}^T \tag{6.1}$$

when a transformation ^0T_n is given as a function of the joint variables $q_1, q_2, q_3, \cdots, q_n$.

$$^0T_n = {^0T_1(q_1)}\, {^1T_2(q_2)}\, {^2T_3(q_3)}\, {^3T_4(q_4)} \cdots {^{n-1}T_n(q_n)} \tag{6.2}$$

Computer-controlled robots are usually actuated in the joint variable space; however, objects to be manipulated are usually expressed in the global Cartesian coordinate frame. Therefore, carrying kinematic information, back and forth, between joint space and Cartesian space, is a need in robotics. To control the configuration of the end-effector to reach an object, the inverse kinematics problem must be solved. Hence, we need to know what the required values of joint variables are to reach a desired point in a desired orientation.

© The Author(s), under exclusive license to Springer Nature Switzerland AG 2022
R. N. Jazar, *Theory of Applied Robotics*, https://doi.org/10.1007/978-3-030-93220-6_6

Consider a robot with n prismatic or revolute joints. Such a robot has n degrees-of-freedom (DOF). To have a robot capable of reaching a particular point at a particular orientation, the robot needs six DOF, three to position the point and three to adjust the orientation of the end-effector. The result of forward kinematics of such a six DOF multibody is a 4×4 transformation matrix.

$$
\begin{aligned}
{}^0T_6 &= {}^0T_1\,{}^1T_2\,{}^2T_3\,{}^3T_4\,{}^4T_5\,{}^5T_6 \\[4pt]
&= \begin{bmatrix} {}^0R_6 & {}^0\mathbf{d}_6 \\ 0 & 1 \end{bmatrix} = \begin{bmatrix} r_{11} & r_{12} & r_{13} & r_{14} \\ r_{21} & r_{22} & r_{23} & r_{24} \\ r_{31} & r_{32} & r_{33} & r_{34} \\ 0 & 0 & 0 & 1 \end{bmatrix}
\end{aligned}
\tag{6.3}
$$

The 12 elements of ${}^0T_6$ are trigonometric functions of 6 unknown joint variables. However, because the upper left 3×3 submatrix of (6.3) is a rotation matrix, only 3 elements of them are independent. This is because of the orthogonality condition (2.285). Hence, only six equations out of the 12 equations of (6.3) are independent. Theoretically, we should be able to use the 6 independent equations and determine the 6 joint variables. The inverse kinematic analysis leads to a set of trigonometric equations. Trigonometric functions inherently provide multiple solutions. Therefore, multiple configurations of the robot are expected when the 6 equations are solved for the unknown joint variables.

It is possible to decouple the inverse kinematics problem into two subproblems, known as *inverse position* and *inverse orientation* kinematics. The practical consequence of such decoupling is to break the problem into two independent problems, each with only three unknown variables. Following the decoupling principle, the overall transformation matrix of a robot can be decomposed to a translation and a rotation.

$$
\begin{aligned}
{}^0T_6 &= {}^0D_6\,{}^0R_6 \\[4pt]
&= \begin{bmatrix} {}^0R_6 & {}^0\mathbf{d}_6 \\ 0 & 1 \end{bmatrix} = \begin{bmatrix} \mathbf{I} & {}^0\mathbf{d}_6 \\ 0 & 1 \end{bmatrix} \begin{bmatrix} {}^0R_6 & \mathbf{0} \\ 0 & 1 \end{bmatrix}
\end{aligned}
\tag{6.4}
$$

The translation matrix ${}^0D_6$ indicates the position of the end-effector in the base frame B_0 and involves only the three joint variables of the manipulator. We will solve ${}^0\mathbf{d}_6$ for the variables that control the wrist position. The rotation matrix ${}^0R_6$ indicates the orientation of the end-effector in B_0 and involves only the three joint variables of the wrist. We will solve ${}^0R_6$ for the variables that control the wrist orientation.

Proof Most robots have a wrist made of three revolute joints with three intersecting orthogonal axes at the wrist point. Taking advantage of having such wrists, we may decouple the kinematics of the wrist and manipulator by decomposing the overall forward kinematics transformation matrix ${}^0T_6$ into the wrist orientation and wrist position:

$$
{}^0T_6 = {}^0T_3\,{}^3T_6 = \begin{bmatrix} {}^0R_3 & {}^0\mathbf{d}_3 \\ 0 & 1 \end{bmatrix} \begin{bmatrix} {}^3R_6 & \mathbf{0} \\ 0 & 1 \end{bmatrix}
\tag{6.5}
$$

where the wrist orientation matrix is:

$$
{}^3R_6 = {}^0R_3^T\,{}^0R_6 = {}^0R_3^T \begin{bmatrix} r_{11} & r_{12} & r_{13} \\ r_{21} & r_{22} & r_{23} \\ r_{31} & r_{32} & r_{33} \end{bmatrix}
\tag{6.6}
$$

and the wrist position vector is:

$$
{}^0\mathbf{d}_6 = \begin{bmatrix} r_{14} \\ r_{24} \\ r_{34} \end{bmatrix}
\tag{6.7}
$$

The wrist position vector ${}^0\mathbf{d}_6 \equiv {}^0\mathbf{d}_3$ includes the manipulator joint variables only. Hence, to solve the inverse kinematics of such a robot, we must solve ${}^0\mathbf{d}_3$ for position of the wrist point and then solve ${}^3R_6$ for orientation of the wrist. The components of the wrist position vector ${}^0\mathbf{d}_6 = {}^0\mathbf{d}_{wrist}$ provide three equations for the three unknown manipulator joint variables. Solving

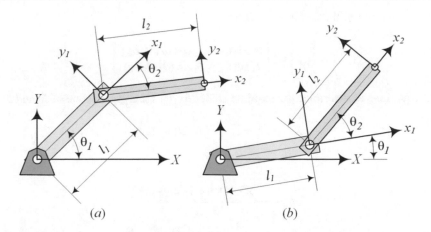

Fig. 6.2 Illustration of a $2R$ planar manipulator in two possible configurations: (**a**) elbow up and (**b**) elbow down

$^0\mathbf{d}_6$ for the manipulator joint variables and substituting them in $^3R_6$ makes $^3R_6$ of (6.6) to be only a function of wrist joint variables. Then, the wrist orientation matrix $^3R_6$ will be solved for the wrist joint variables.

In case we include the tool coordinate frame in forward kinematics, the decomposition must be done according to the following equation to exclude the effect of tool distance d_7 from the robot's kinematics.

$$^0T_7 = {}^0T_3\,{}^3T_7 = {}^0T_3\,{}^3T_6\,{}^6T_7$$

$$= \begin{bmatrix} {}^0R_3 & \mathbf{d}_w \\ 0 & 1 \end{bmatrix} \begin{bmatrix} {}^3R_6 & \mathbf{0} \\ 0 & 1 \end{bmatrix} \begin{bmatrix} \mathbf{I} & \begin{matrix} 0 \\ 0 \\ d_7 \end{matrix} \\ 0 & 1 \end{bmatrix} \tag{6.8}$$

In this case, inverse kinematics starts from determination of $^0T_6$, which can be found as:

$$^0T_6 = {}^0T_7\,{}^6T_7^{-1} \tag{6.9}$$

$$= {}^0T_7 \begin{bmatrix} 1 & 0 & 0 & 0 \\ 0 & 1 & 0 & 0 \\ 0 & 0 & 1 & d_7 \\ 0 & 0 & 0 & 1 \end{bmatrix}^{-1} = {}^0T_7 \begin{bmatrix} 1 & 0 & 0 & 0 \\ 0 & 1 & 0 & 0 \\ 0 & 0 & 1 & -d_7 \\ 0 & 0 & 0 & 1 \end{bmatrix}$$

∎

Example 207 Inverse kinematics for $2R$ planar manipulator. This is the simplest multi DOF robotic system and this example will show the inverse kinematic problems to have multiple solutions. Although, all solutions are possible, only one of the solutions will be the practical one. It is also the first time students are introduced to the trigonometric equations and the function atan2(y, x).

Figure 6.2 illustrates a $2R$ planar manipulator with two R‖R links. Employing the transformation matrix of two links R‖R(0), we find the forward kinematics of the manipulator as:

$$^0T_2 = {}^0T_1\,{}^1T_2 \tag{6.10}$$

$$= \begin{bmatrix} c\,(\theta_1+\theta_2) & -s\,(\theta_1+\theta_2) & 0 & l_1c\theta_1+l_2c\,(\theta_1+\theta_2) \\ s\,(\theta_1+\theta_2) & c\,(\theta_1+\theta_2) & 0 & l_1s\theta_1+l_2s\,(\theta_1+\theta_2) \\ 0 & 0 & 1 & 0 \\ 0 & 0 & 0 & 1 \end{bmatrix}$$

The global position of the tip point of the manipulator is at:

$$\begin{bmatrix} X \\ Y \end{bmatrix} = \begin{bmatrix} l_1 \cos\theta_1 + l_2 \cos(\theta_1 + \theta_2) \\ l_1 \sin\theta_1 + l_2 \sin(\theta_1 + \theta_2) \end{bmatrix} \tag{6.11}$$

We must use these two equations to determine the required angles θ_1, θ_2 for a given value of X and Y. To find θ_2, we use

$$X^2 + Y^2 = l_1^2 + l_2^2 + 2l_1l_2 \cos\theta_2 \tag{6.12}$$

and

$$\cos\theta_2 = \frac{X^2 + Y^2 - l_1^2 - l_2^2}{2l_1l_2} \tag{6.13}$$

$$\theta_2 = \cos^{-1} \frac{X^2 + Y^2 - l_1^2 - l_2^2}{2l_1l_2} \tag{6.14}$$

The first joint variable θ_1 of an elbow up configuration can geometrically be found from:

$$\theta_1 = \arctan\frac{Y}{X} + \arctan\frac{l_2 \sin\theta_2}{l_1 + l_2 \cos\theta_2} \tag{6.15}$$

and for an elbow down configuration from:

$$\theta_1 = \arctan\frac{Y}{X} - \arctan\frac{l_2 \sin\theta_2}{l_1 + l_2 \cos\theta_2} \tag{6.16}$$

However, we usually avoid using functions arcsin and arccos because of their inaccuracy. It is more exact if we are able to calculate angles by function arctan. Let us employ the half angle formula,

$$\tan^2\frac{\theta}{2} = \frac{1 - \cos\theta}{1 + \cos\theta} \tag{6.17}$$

to find θ_2 using an atan2 function.

$$\theta_2 = \pm 2 \, \mathrm{atan2} \sqrt{\frac{(l_1 + l_2)^2 - (X^2 + Y^2)}{(X^2 + Y^2) - (l_1 - l_2)^2}} \tag{6.18}$$

The meaning of the function $\mathrm{atan2}(y, x)$ is explained in Example 208. At the moment assume:

$$\mathrm{atan2}(y, x) \equiv \arctan\frac{y}{x} \tag{6.19}$$

The $\pm$ in Eq. (6.18) is because of the square root, which generates two solutions for θ_2. These two solutions are called elbow up and elbow down configurations, as shown in Fig. 6.2a,b respectively.

The angle θ_1 can also be found from the following alternative equation.

$$\theta_1 = \mathrm{atan2} \frac{-Xl_2 \sin\theta_2 + Y(l_1 + l_2 \cos\theta_2)}{Yl_2 \sin\theta_2 + X(l_1 + l_2 \cos\theta_2)} \tag{6.20}$$

Most of the time, the value of θ_1 should be corrected by adding or subtracting π depending on the sign of X. The two different sets of solutions for θ_1 and θ_2 correspond to the elbow up and elbow down configurations.

It is also possible to combine Equations of (6.11) and determine a trigonometric equation for θ_1.

$$2Xl_1 \cos\theta_1 + 2Yl_1 \sin\theta_1 = X^2 + Y^2 + l_1^2 - l_2^2 \tag{6.21}$$

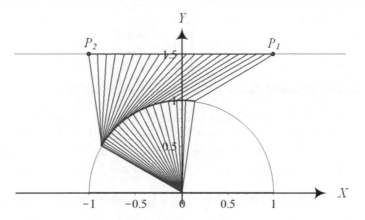

Fig. 6.3 A $2R$ planar manipulator with $l_1 = 1$ m, $l_1 = 1$ m moving from P_1 to P_2 on a straight line

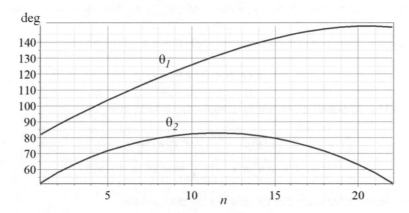

Fig. 6.4 The variation of the angles θ_1 and θ_2 of the $2R$ planar manipulator

The trigonometric equation appears in inverse kinematic problems frequently. The method of trigonometric equations will be explained in Example 209.

As an example, let us consider the motion of a $2R$ planar manipulator with $l_1 = 1$ m, $l_1 = 1$ m that its tip point is moving from P_1 $(1.2, 1.5)$ to P_2 $(-1.2, 1.5)$ on a straight line. Employing the inverse kinematic equations (6.14) and (6.15), we determine the configuration of the manipulator at any point of the path. Figure 6.3 illustrates the manipulator at $n = 22$ equally spaced points between P_1 and P_2. The variation of the angles θ_1 and θ_2 is as shown in Fig. 6.4.

Example 208 ★ Meaning of the function atan2$(y, x) \equiv \arctan \frac{y}{x}$. Inverse tangent is the most exact way to determine the value of an angle. It provides more exact solution than other inverse trigonometric functions. The argument of inverse tangent usually appears as a fraction y/x. The problem of inverse tangent is that $\arctan \frac{-y}{x} = \arctan \frac{y}{-x}$. To overcome this problem, a modified inverse tangent function will be introduced in this example.

In robotic kinematics calculation, especially in solving inverse kinematic problems, we need to find angles based on the sin and cos functions of the angles. However, arctan cannot show the effect of the individual sign for the numerator and denominator. It always represents an angle in the first or fourth quadrant. To overcome this problem and determine the joint angles in the correct quadrant, the atan2 function is introduced as:

$$\text{atan2}(y, x) = \begin{cases} \text{sgn } y \, \arctan \left| \frac{y}{x} \right| & \text{if } x > 0, y \neq 0 \\ \frac{\pi}{2} \text{sgn } y & \text{if } x = 0, y \neq 0 \\ \text{sgn } y \left(\pi - \arctan \left| \frac{y}{x} \right| \right) & \text{if } x < 0, y \neq 0 \\ \pi - \pi \text{sgn } x & \text{if } x \neq 0, y = 0 \end{cases} \tag{6.22}$$

The sgn represents the signum function.

$$\operatorname{sgn}(x) = \begin{cases} 1 & \text{if } x > 0 \\ 0 & \text{if } x = 0 \\ -1 & \text{if } x < 0 \end{cases} \tag{6.23}$$

As an example, let us compare the arctan and atan2 for four points in four quadrants.

$$
\begin{array}{llll}
x = 1, \ y = 1 & \text{then } \arctan \frac{1}{1} = 0.785 & \text{atan2}(1, 1) = 0.785 \\
x = -1, \ y = 1 & \text{then } \arctan \frac{1}{-1} = -0.785 & \text{atan2}(1, -1) = 2.356 \\
x = -1, \ y = -1 & \text{then } \arctan \frac{-1}{-1} = 0.785 & \text{atan2}(-1, -1) = -2.356 \\
x = 1, \ y = -1 & \text{then } \arctan \frac{-1}{1} = -0.785 & \text{atan2}(-1, 1) = -0.785
\end{array}
$$

In robotic kinematics calculations, wherever $\arctan \frac{y}{x}$ is needed, it is recommended to be calculated based on $\text{atan2}(y, x)$.

Example 209 ★ Solution of first type of trigonometric equation $a \cos \theta + b \sin \theta = c$. Trigonometric equations frequently appear in kinematic analysis of robots. The method of solution of the first type of trigonometric equation will be introduced in this example.

The first type of trigonometric equation for the unknown angle θ is a linear combination of $\cos \theta$ and $\sin \theta$.

$$a \cos \theta + b \sin \theta = c \tag{6.24}$$

This equation can be solved by introducing two new variables r and ϕ such that:

$$a = r \sin \phi \qquad b = r \cos \phi \tag{6.25}$$

$$r = \sqrt{a^2 + b^2} \qquad \phi = \text{atan2}(a, b) \tag{6.26}$$

We may verify these identities by expanding and substituting the new variables.

$$\sin(\phi + \theta) = \frac{c}{r} \tag{6.27}$$

$$\cos(\phi + \theta) = \pm \sqrt{1 - \frac{c^2}{r^2}} \tag{6.28}$$

Hence, the unknown angle θ of the trigonometric equation (6.24) is:

$$\theta = \text{atan2}\left(\frac{c}{r}, \pm \sqrt{1 - \frac{c^2}{r^2}}\right) - \text{atan2}(a, b) \tag{6.29}$$

$$= \text{atan2}\left(c, \pm \sqrt{r^2 - c^2}\right) - \text{atan2}(a, b) \tag{6.30}$$

$$\equiv \arctan \frac{c}{\pm \sqrt{r^2 - c^2}} - \arctan \frac{a}{b} \tag{6.31}$$

To solve the first type of trigonometric equation (6.24), we may employ new variables (6.25) to rewrite the equation in a new form,

$$c = r \sin \phi \cos \theta + r \cos \phi \sin \theta = r \sin(\phi + \theta) \tag{6.32}$$

and determine θ as:

$$\theta = \arcsin \frac{c}{r} - \arctan \frac{a}{b} \tag{6.33}$$

$$|c| < |r| \tag{6.34}$$

Solutions (6.29) and (6.33) are convertible. Therefore, Eq. (6.24) has two solutions if $c^2 < r^2$, one solution if $r^2 = c^2$, and no solution if $r^2 < c^2$.

As an example, let us solve the following equation.

$$1.5 \cos \theta + 2.5 \sin \theta = 2.549 \tag{6.35}$$

Having $a = 1.5$ and $b = 2.5$, we find r and ϕ.

$$r = \sqrt{a^2 + b^2} = 2.915475947 \tag{6.36}$$

$$\phi = \text{atan2}(a, b) = 0.5404195 \, \text{rad} \tag{6.37}$$

Therefore,

$$\begin{aligned}
\theta &= \text{atan2}(c, \pm\sqrt{r^2 - c^2}) - \text{atan2}(a, b) \\
&= \text{atan2}(2.549, \pm\sqrt{2}) - \phi \\
&= 0.5235718477 \, \text{rad}, \ 1.537181805 \, \text{rad} \\
&\approx 30 \, \text{deg}, \ 88.07 \, \text{deg}
\end{aligned} \tag{6.38}$$

Example 210 ★ Second method to solve the first type of trigonometric equation $a \cos \theta + b \sin \theta = c$. Converging a trigonometric equation using tangent function allows us to solve the equation in an alternative method, as exact and as applied as the first method of Example 209.

An alternative method to solve the first type of trigonometric equation for the unknown angle θ

$$a \cos \theta + b \sin \theta = c \tag{6.39}$$

is to convert the equation to $\tan \theta / 2$.

$$\frac{c - a + (a + c) \tan^2 \frac{1}{2}\theta - 2b \tan \frac{1}{2}\theta}{\tan^2 \frac{1}{2}\theta + 1} = 0 \tag{6.40}$$

If $\theta \neq \pi/2$, then we have a quadratic equation for $\tan^2 \frac{1}{2}\theta$.

$$(a + c) \tan^2 \frac{1}{2}\theta - 2b \tan \frac{1}{2}\theta + c - a = 0 \tag{6.41}$$

$$\theta = 2 \arctan \frac{b \pm \sqrt{a^2 + b^2 - c^2}}{a + c} \tag{6.42}$$

Example 211 An articulated manipulator. Articulated manipulator is one of the most important and practical industrial robots. This example will be a good reference for detailed analytic inverse kinematics of this manipulator.

Consider an articulated manipulator as is shown in Fig. 6.5. The links of the manipulator are R⊢R(90), R∥R(0), R⊢R(90), and their associated transformation matrices between coordinate frames are:

$$^{0}T_{1} = \begin{bmatrix} \cos\theta_1 & 0 & \sin\theta_1 & 0 \\ \sin\theta_1 & 0 & -\cos\theta_1 & 0 \\ 0 & 1 & 0 & l_1 \\ 0 & 0 & 0 & 1 \end{bmatrix} \tag{6.43}$$

$$^{1}T_{2} = \begin{bmatrix} \cos\theta_2 & -\sin\theta_2 & 0 & l_2\cos\theta_2 \\ \sin\theta_2 & \cos\theta_2 & 0 & l_2\sin\theta_2 \\ 0 & 0 & 1 & 0 \\ 0 & 0 & 0 & 1 \end{bmatrix} \tag{6.44}$$

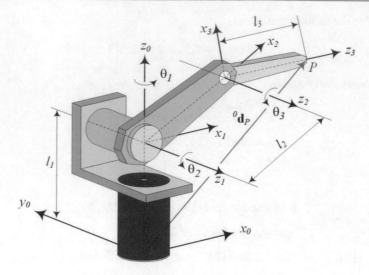

Fig. 6.5 An R⊢R∥R articulated manipulator

$$
{}^2T_3 = \begin{bmatrix} \cos\theta_3 & 0 & \sin\theta_3 & 0 \\ \sin\theta_3 & 0 & -\cos\theta_3 & 0 \\ 0 & 1 & 0 & 0 \\ 0 & 0 & 0 & 1 \end{bmatrix} \tag{6.45}
$$

The forward kinematics of the manipulator is:

$$
{}^0T_3 = {}^0T_1\,{}^1T_2\,{}^2T_3 \tag{6.46}
$$

$$
= \begin{bmatrix} c\theta_1 c(\theta_2+\theta_3) & s\theta_1 & c\theta_1 s(\theta_2+\theta_3) & l_2 c\theta_1 c\theta_2 \\ s\theta_1 c(\theta_2+\theta_3) & -c\theta_1 & s\theta_1 s(\theta_2+\theta_3) & l_2 c\theta_2 s\theta_1 \\ s(\theta_2+\theta_3) & 0 & -c(\theta_2+\theta_3) & l_1 + l_2 s\theta_2 \\ 0 & 0 & 0 & 1 \end{bmatrix}
$$

and therefore, the tip point P is at:

$$
{}^0\mathbf{d}_P = \begin{bmatrix} X \\ Y \\ Z \\ 1 \end{bmatrix} = {}^0T_3 \begin{bmatrix} 0 \\ 0 \\ l_3 \\ 1 \end{bmatrix}
$$

$$
= \begin{bmatrix} l_3 \sin(\theta_2+\theta_3)\cos\theta_1 + l_2 \cos\theta_1 \cos\theta_2 \\ l_3 \sin(\theta_2+\theta_3)\sin\theta_1 + l_2 \sin\theta_1 \cos\theta_2 \\ l_1 - l_3 \cos(\theta_2+\theta_3) + l_2 \sin\theta_2 \\ 1 \end{bmatrix} \tag{6.47}
$$

Point P is supposed to be the point at which we attach a spherical wrist. Therefore, ${}^0\mathbf{d}_P$ is the decoupled position vector of the wrist point that will not be affected by the wrist attachment. The vector ${}^0\mathbf{d}_P$ provides three equations for the three joint variables $\theta_1, \theta_2, \theta_3$ of the manipulator. In a practical problem, the left-hand side of Eq. (6.47) is a known vector and hence, X, Y, Z are given numbers. The right-hand sides are three equations with all parameters given and including only three unknowns $\theta_1, \theta_2, \theta_3$. The first angle can be found from

$$
X \sin\theta_1 - Y \cos\theta_1 = 0 \tag{6.48}
$$

that is:

$$
\theta_1 = \text{atan2}\,(Y, X) \tag{6.49}
$$

We may combine the first and second elements of $^0\mathbf{d}_P$ to find:

$$X \cos \theta_1 + Y \sin \theta_1 = l_3 \sin (\theta_2 + \theta_3) + l_2 \cos \theta_2 \tag{6.50}$$

Now, let us rewrite this equation as:

$$X \cos \theta_1 + Y \sin \theta_1 - l_2 \cos \theta_2 = l_3 \sin (\theta_2 + \theta_3) \tag{6.51}$$

and rewrite the third component of (6.47) as:

$$Z - l_1 - l_2 \sin \theta_2 = l_3 \cos (\theta_2 + \theta_3) \tag{6.52}$$

Then, combining Eqs. (6.51) and (6.52) provides:

$$(Z - l_1 - l_2 \sin \theta_2)^2 + (X \cos \theta_1 + Y \sin \theta_1 - l_2 \cos \theta_2)^2 = l_3^2 \tag{6.53}$$

or

$$-2l_2 (X \cos \theta_1 + Y \sin \theta_1) \cos \theta_2 + 2l_2 (l_1 - Z) \sin \theta_2 = l_3^2$$
$$- (l_1 - Z)^2 - l_2^2 - Y^2 - \left(X^2 - Y^2 \right) \cos^2 \theta_1 - XY \sin 2\theta_1 \tag{6.54}$$

That is a trigonometric equation of the form (6.24).

$$a \cos \theta_2 + b \sin \theta_2 = c \tag{6.55}$$

$$a = -2l_2 (X \cos \theta_1 + Y \sin \theta_1) \tag{6.56}$$

$$b = 2l_2 (l_1 - Z) \tag{6.57}$$

$$c = l_3^2 - (l_1 - Z)^2 - l_2^2 - Y^2 - \left(X^2 - Y^2 \right) \cos^2 \theta_1 - XY \sin 2\theta_1 \tag{6.58}$$

We solve this equation for θ_2 considering θ_1 has been calculated in (6.49).

$$\theta_2 = \arcsin \left(\frac{c}{\sqrt{a^2 + b^2}} \right) - \arctan \frac{a}{b} \tag{6.59}$$

Dividing (6.50) by the third element of $^0\mathbf{d}_P$ determines θ_3.

$$\tan (\theta_2 + \theta_3) = \frac{X \cos \theta_1 + Y \sin \theta_1 - l_2 \cos \theta_2}{l_1 + l_2 \sin \theta_2 - Z} \tag{6.60}$$

$$\theta_3 = \operatorname{atan2} \left(\frac{X \cos \theta_1 + Y \sin \theta_1 - l_2 \cos \theta_2}{l_1 + l_2 \sin \theta_2 - Z} \right) - \theta_2 \tag{6.61}$$

Therefore, Eqs. (6.49), (6.59), and (6.61) are the solutions of the inverse kinematics of the articulated manipulator.

There is no standard method to solve the three equations (6.47) for the three unknown angles θ_1, θ_2, θ_3, and this is a weakness of inverse kinematics. However, when a set of solution is set, they will be applicable to all similar manipulators. The method described in this example is not the only method. The readers may manipulate the equations to come up with other ways to calculate the unknowns.

Example 212 A set of equations for inverse kinematic equations of articulated manipulator. This example introduces another set of equations for inverse kinematics of articulated manipulators, suitable for computerization. Although the final equations of Example 211 are convertible to the equations introduced here.

Figure 6.5 illustrates an articulated manipulator. The forward kinematics of the manipulator ends up with three equations with three unknown joint variables θ_1, θ_2, θ_3, indicating the position of the tip point of the robot in the base Cartesian coordinate frame.

$$^0\mathbf{d}_P = \begin{bmatrix} l_3 \sin(\theta_2 + \theta_3) \cos\theta_1 + l_2 \cos\theta_1 \cos\theta_2 \\ l_3 \sin(\theta_2 + \theta_3) \sin\theta_1 + l_2 \sin\theta_1 \cos\theta_2 \\ l_1 - l_3 \cos(\theta_2 + \theta_3) + l_2 \sin\theta_2 \end{bmatrix} = \begin{bmatrix} X \\ Y \\ Z \end{bmatrix} \tag{6.62}$$

In forward kinematics, the joint variables θ_1, θ_2, θ_3 are given and the coordinates of the tip point X, Y, Z are requested. Forward kinematics is usually a simple and straightforward problem that will be solved by substitution numerical values. In inverse kinematics, the coordinates of the tip point X, Y, Z are given and the joint variables θ_1, θ_2, θ_3 are requested. Hence, in the following analysis we assume X, Y, Z are given numerical values.

Interestingly, we have a good relationship between X and Y as $Y/X = \tan\theta_1$. Therefore, calculating θ_1 is reduced to an inverse tangent calculation.

$$\theta_1 = \begin{cases} \arctan \dfrac{Y}{X} & X \geq 0 \\ \arctan \dfrac{Y}{X} + \pi & X < 0 \end{cases} \tag{6.63}$$

To combine the two possible solutions in a single equation, we may employ the *Heaviside* function H and rewrite Eqs. (6.63) in a new form.

$$\theta_1 = \arctan \frac{Y}{X} + \pi \, H(-X) \tag{6.64}$$

Heaviside is an on-off switching function. It was introduced by the English mathematician Oliver Heaviside (1850–1925) as:

$$H(x - x_0) = \begin{cases} 1 & x > x_0 \\ 0 & x < x_0 \end{cases} \tag{6.65}$$

A combination of X, Y, Z eliminates θ_3 and provides us with the following identity.

$$(Z - l_1 - l_2 \sin\theta_2)^2 + (X \cos\theta_1 + Y \sin\theta_1 - l_2 \cos\theta_2)^2 = l_3^2 \tag{6.66}$$

Considering θ_1 is a known value from (6.63), Eq. (6.66) will be a first kind of trigonometric equation for θ_2.

$$a \cos\theta_2 + b \sin\theta_2 = c \tag{6.67}$$

Its solution from Example 210 is:

$$\theta_2 = 2 \arctan \frac{b \pm \sqrt{b^2 + a^2 - c^2}}{a + c} \tag{6.68}$$

where

$$a = -2l_2 (X \cos\theta_1 + Y \sin\theta_1) \tag{6.69}$$

$$b = 2l_2 (l_1 - Z) \tag{6.70}$$

$$c = l_3^2 - (l_1 - Z)^2 - l_2^2 - Y^2 - \left(X^2 - Y^2\right) \cos^2\theta_1$$
$$- XY \sin 2\theta_1 \tag{6.71}$$

Introducing new parameters,

$$C_1 = -(a + c) \qquad C_2 = -b \qquad C_3 = a - c \tag{6.72}$$

we have:

$$\theta_2 = 2 \arctan \frac{-C_2 \pm \sqrt{C_2^2 - C_1 C_3}}{C_1} \tag{6.73}$$

where

$$C_1 = l_1^2 - 2l_1 Z + l_2^2 + \frac{2l_2 X}{\cos\theta_1} - l_3^2 + \frac{X^2}{\cos^2\theta_1} + Z^2 \tag{6.74}$$

$$C_2 = 2l_1 l_2 - 2l_2 Z \tag{6.75}$$

$$C_3 = l_1^2 - 2l_1 Z + l_2^2 - \frac{2l_2 X}{\cos\theta_1} - l_3^2 + \frac{X^2}{\cos^2\theta_1} + Z^2 \tag{6.76}$$

Finally the third component of $^0\mathbf{d}_P$ is enough to find θ_3.

$$\theta_3 = \arccos\left(\frac{l_1 - Z + l_2\sin\theta_2}{l_3}\right) - \theta_2 \tag{6.77}$$

Therefore, Eqs. (6.63), (6.73), and (6.77) are the solutions of the inverse kinematics of the articulated manipulator.

Example 213 Numerical case of an articulated manipulator. Analytic solutions must eventually provide numerical solutions to be practical. Here we use the analytic solutions of inverse kinematics of articulated manipulator to show how we recognize the correct solution out of multiple solutions.

To check the inverse kinematic equations of Example 211, let us examine an articulated manipulator with the following dimensions

$$l_1 = 1\,\text{m} \qquad l_2 = 1.05\,\text{m} \qquad l_3 = 0.89\,\text{m} \tag{6.78}$$

when its tip point is at:

$$^0\mathbf{d}_P = \begin{bmatrix} 1 & 1.1 & 1.2 \end{bmatrix}^T \equiv \begin{bmatrix} X & Y & Z \end{bmatrix}^T \tag{6.79}$$

Equation (6.49) provides θ_1.

$$\theta_1 = \text{atan2}\,(Y, X) = \arctan\frac{1.1}{1}$$
$$= 0.83298\,\text{rad} \approx 47.726\,\text{deg} \tag{6.80}$$

To determine θ_2, we should solve Eq. (6.55)

$$a\cos\theta_2 + b\sin\theta_2 = c \tag{6.81}$$

where

$$a = -2l_2\,(X\cos\theta_1 + Y\sin\theta_1) = -3.12187 \tag{6.82}$$

$$b = 2l_2\,(l_1 - Z) = -0.42 \tag{6.83}$$

$$c = l_3^2 - \left((X\cos\theta_1 + Y\sin\theta_1)^2 + l_1^2 - 2l_1 Z + l_2^2 + Z^2\right)$$
$$= -2.56 \tag{6.84}$$

We find two values for θ_2

$$\theta_2 = 0.7555\,\text{rad} \approx 43.289\,\text{deg} \tag{6.85}$$

$$\theta_2 = -0.488\,\text{rad} \approx -27.9648\,\text{deg} \tag{6.86}$$

θ_3 comes from (6.61). If $\theta_2 = 0.7555\,\text{rad}$ we have:

$$\theta_3 = \text{atan2}\left(\frac{X\cos\theta_1 + Y\sin\theta_1 - l_2\cos\theta_2}{l_1 + l_2\sin\theta_2 - Z}\right) - \theta_2$$
$$= 0.19132\,\text{rad} \approx 10.9618\,\text{deg} \tag{6.87}$$

and if $\theta_2 = -0.488$ rad, then we have:

$$\theta_3 = -0.1913 \, \text{rad} \approx -10.9618 \, \text{deg} \tag{6.88}$$

There are two sets of solutions that only one of them will match with the current position of the manipulator.

$$\theta_1 = 0.83298 \quad \theta_2 = 0.7555 \quad \theta_3 = 0.1913 \tag{6.89}$$

$$\theta_1 = 0.83298 \quad \theta_2 = -0.488 \quad \theta_3 = -0.1913 \tag{6.90}$$

To determine the correct set, we need to use them in direct kinematic analysis and calculate the position of the end-effector of the manipulator.

$$^0\mathbf{d}_P = \begin{bmatrix} l_3 \sin(\theta_2 + \theta_3) \cos\theta_1 + l_2 \cos\theta_1 \cos\theta_2 \\ l_3 \sin(\theta_2 + \theta_3) \sin\theta_1 + l_2 \sin\theta_1 \cos\theta_2 \\ l_1 - l_3 \cos(\theta_2 + \theta_3) + l_2 \sin\theta_2 \end{bmatrix} \equiv \begin{bmatrix} X \\ Y \\ Z \end{bmatrix} \tag{6.91}$$

Substituting the first set (6.89) provides us with the correct position of the end-effector.

$$^0\mathbf{d}_P = \begin{bmatrix} 1.000 \\ 1.100 \\ 1.1999 \end{bmatrix} \tag{6.92}$$

The second set (6.90) provides us with the incorrect position of the end-effector.

$$^0\mathbf{d}_P = \begin{bmatrix} 0.24774 \\ 0.27252 \\ -0.18473 \end{bmatrix} \tag{6.93}$$

Therefore, the correct solutions of the inverse kinematic for the manipulator are (6.89).

Let us also use the alternative Eqs. (6.63), (6.73), and (6.77) to solve this example to compare. Equation (6.63) gives θ_1.

$$\theta_1 = \arctan \frac{Y}{X} = \arctan \frac{1.1}{1} = 0.83298 \, \text{rad} \tag{6.94}$$

Equation (6.73) provides θ_2.

$$C_1 = l_1^2 - 2l_1 Z + l_2^2 + \frac{2l_2 X}{\cos\theta_1} - l_3^2 + \frac{X^2}{\cos^2\theta_1} + Z^2 = 5.6822 \tag{6.95}$$

$$C_2 = 2l_1 l_2 - 2l_2 Z = -0.42 \tag{6.96}$$

$$C_3 = l_1^2 - 2l_1 Z + l_2^2 - \frac{2l_2 X}{\cos\theta_1} - l_3^2 + \frac{X^2}{\cos^2\theta_1} + Z^2 = -0.56147 \tag{6.97}$$

$$\theta_2 = 2 \arctan \frac{-C_2 \pm \sqrt{C_2^2 - C_1 C_3}}{C_1} = \begin{cases} 0.7555 \, \text{rad} \\ -0.4881 \, \text{rad} \end{cases} \tag{6.98}$$

Equation (6.77) provides θ_3. It will give two values, one for every θ_2.

$$\theta_3 = \text{atan2} \left(\frac{X \cos\theta_1 + Y \sin\theta_1 - l_2 \cos\theta_2}{l_1 + l_2 \sin\theta_2 - Z} \right) - \theta_2$$

$$= \begin{cases} 0.19132 \, \text{rad} \\ 2.95027 \, \text{rad} \end{cases} \tag{6.99}$$

The alternative method also provides two sets of solutions, where the first set matches with the correct solutions (6.89).

$$\theta_1 = 0.83298 \quad \theta_2 = 0.7555 \quad \theta_3 = 0.1913 \tag{6.100}$$

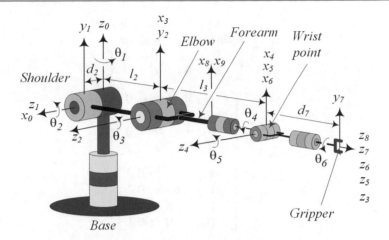

Fig. 6.6 A 6 DOF articulated manipulator

$$\theta_1 = 0.83298 \quad \theta_2 = -0.488 \quad \theta_3 = 2.95027 \tag{6.101}$$

Example 214 Inverse kinematics of an articulated robot. The decoupling method will be reviewed in this example for a 6 DOF reboot to determine manipulator joint variables and wrist orientation joint variables separately.

The forward kinematics of the articulated robot, illustrated in Fig. 6.6, was solved in Example 192, where the overall transformation matrix of the end-effector was found, based on the wrist and arm transformation matrices.

$$^0T_7 = T_{arm}T_{wrist} = \, ^0T_3 \, ^3T_7 \tag{6.102}$$

The wrist transformation matrix T_{wrist} is described in (5.231) and the manipulator transformation matrix, T_{arm} is found in (5.103). However, according to a new setup coordinate frame, as shown in Fig. 6.6, we have a $6R$ robot with a six links configurations and a displacement T_{Z,d_7}.

1	$R \vdash R(90)$
2	$R \parallel R(0)$
3	$R \vdash R(90)$
4	$R \vdash R(-90)$
5	$R \vdash R(90)$
6	$R \parallel R(0)$

Therefore, the individual links' transformation matrices are:

$$^0T_1 = \begin{bmatrix} \cos\theta_1 & 0 & \sin\theta_1 & 0 \\ \sin\theta_1 & 0 & -\cos\theta_1 & 0 \\ 0 & 1 & 0 & 0 \\ 0 & 0 & 0 & 1 \end{bmatrix} \tag{6.103}$$

$$^1T_2 = \begin{bmatrix} \cos\theta_2 & -\sin\theta_2 & 0 & l_2\cos\theta_2 \\ \sin\theta_2 & \cos\theta_2 & 0 & l_2\sin\theta_2 \\ 0 & 0 & 1 & d_2 \\ 0 & 0 & 0 & 1 \end{bmatrix} \tag{6.104}$$

$$^2T_3 = \begin{bmatrix} \cos\theta_3 & 0 & \sin\theta_3 & 0 \\ \sin\theta_3 & 0 & -\cos\theta_3 & 0 \\ 0 & 1 & 0 & 0 \\ 0 & 0 & 0 & 1 \end{bmatrix} \tag{6.105}$$

$$^3T_4 = \begin{bmatrix} \cos\theta_4 & 0 & -\sin\theta_4 & 0 \\ \sin\theta_4 & 0 & \cos\theta_4 & 0 \\ 0 & -1 & 0 & l_3 \\ 0 & 0 & 0 & 1 \end{bmatrix} \tag{6.106}$$

$$^4T_5 = \begin{bmatrix} \cos\theta_5 & 0 & \sin\theta_5 & 0 \\ \sin\theta_5 & 0 & -\cos\theta_5 & 0 \\ 0 & 1 & 0 & 0 \\ 0 & 0 & 0 & 1 \end{bmatrix} \tag{6.107}$$

$$^5T_6 = \begin{bmatrix} \cos\theta_6 & -\sin\theta_6 & 0 & 0 \\ \sin\theta_6 & \cos\theta_6 & 0 & 0 \\ 0 & 0 & 1 & 0 \\ 0 & 0 & 0 & 1 \end{bmatrix} \tag{6.108}$$

$$^6T_7 = \begin{bmatrix} 1 & 0 & 0 & 0 \\ 0 & 1 & 0 & 0 \\ 0 & 0 & 1 & d_6 \\ 0 & 0 & 0 & 1 \end{bmatrix} \tag{6.109}$$

Hence, the tool transformation matrix in the base coordinate frame is:

$$^0T_7 = {}^0T_1\,{}^1T_2\,{}^2T_3\,{}^3T_4\,{}^4T_5\,{}^5T_6\,{}^6T_7 \tag{6.110}$$

$$= {}^0T_3\,{}^3T_6\,{}^6T_7$$

$$= \begin{bmatrix} t_{11} & t_{12} & t_{13} & t_{14} \\ t_{21} & t_{22} & t_{23} & t_{24} \\ t_{31} & t_{32} & t_{33} & t_{34} \\ 0 & 0 & 0 & 1 \end{bmatrix}$$

where

$$^0T_3 = \begin{bmatrix} c\theta_1 c(\theta_2+\theta_3) & s\theta_1 & c\theta_1 s(\theta_2+\theta_3) & l_2 c\theta_1 c\theta_2 + d_2 s\theta_1 \\ s\theta_1 c(\theta_2+\theta_3) & -c\theta_1 & s\theta_1 s(\theta_2+\theta_3) & l_2 c\theta_2 s\theta_1 - d_2 c\theta_1 \\ s(\theta_2+\theta_3) & 0 & -c(\theta_2+\theta_3) & l_2 s\theta_2 \\ 0 & 0 & 0 & 1 \end{bmatrix} \tag{6.111}$$

$$^3T_6 = \begin{bmatrix} c\theta_4 c\theta_5 c\theta_6 - s\theta_4 s\theta_6 & -c\theta_6 s\theta_4 - c\theta_4 c\theta_5 s\theta_6 & c\theta_4 s\theta_5 & 0 \\ c\theta_5 c\theta_6 s\theta_4 + c\theta_4 s\theta_6 & c\theta_4 c c\theta_6 - c\theta_5 s\theta_4 s\theta_6 & s\theta_4 s\theta_5 & 0 \\ -c\theta_6 s\theta_5 & s\theta_5 s\theta_6 & c\theta_5 & l_3 \\ 0 & 0 & 0 & 1 \end{bmatrix} \tag{6.112}$$

and

$$t_{11} = c\theta_1 \left(c\left(\theta_2+\theta_3\right) \left(c\theta_4 c\theta_5 c\theta_6 - s\theta_4 s\theta_6\right) - c\theta_6 s\theta_5 s\left(\theta_2+\theta_3\right)\right)$$

$$+ s\theta_1 \left(c\theta_4 s\theta_6 + c\theta_5 c\theta_6 s\theta_4\right) \tag{6.113}$$

$$t_{21} = s\theta_1 \left(c\left(\theta_2+\theta_3\right) \left(-s\theta_4 s\theta_6 + c\theta_4 c\theta_5 c\theta_6\right) - c\theta_6 s\theta_5 s\left(\theta_2+\theta_3\right)\right)$$

$$- c\theta_1 \left(c\theta_4 s\theta_6 + c\theta_5 c\theta_6 s\theta_4\right) \tag{6.114}$$

$$t_{31} = s\left(\theta_2+\theta_3\right) \left(c\theta_4 c\theta_5 c\theta_6 - s\theta_4 s\theta_6\right) + c\theta_6 s\theta_5 c\left(\theta_2+\theta_3\right) \tag{6.115}$$

$$t_{12} = c\theta_1 \left(s\theta_5 s\theta_6 s\left(\theta_2+\theta_3\right) - c\left(\theta_2+\theta_3\right)\left(c\theta_6 s\theta_4 + c\theta_4 c\theta_5 s\theta_6\right)\right)$$

$$+ s\theta_1 \left(c\theta_4 c\theta_6 - c\theta_5 s\theta_4 s\theta_6\right) \tag{6.116}$$

$$t_{22} = s\theta_1 \left(s\theta_5 s\theta_6 s\left(\theta_2+\theta_3\right) - c\left(\theta_2+\theta_3\right)\left(c\theta_6 s\theta_4 + c\theta_4 c\theta_5 s\theta_6\right)\right)$$

$$+ c\theta_1 \left(-c\theta_4 c\theta_6 + c\theta_5 s\theta_4 s\theta_6\right) \tag{6.117}$$

$$t_{32} = -s\theta_5 s\theta_6 c\,(\theta_2 + \theta_3) - s\,(\theta_2 + \theta_3)\,(c\theta_6 s\theta_4 + c\theta_4 c\theta_5 s\theta_6) \tag{6.118}$$

$$t_{13} = s\theta_1 s\theta_4 s\theta_5 + c\theta_1\,(c\theta_5 s\,(\theta_2 + \theta_3) + c\theta_4 s\theta_5 c\,(\theta_2 + \theta_3)) \tag{6.119}$$

$$t_{23} = -c\theta_1 s\theta_4 s\theta_5 + s\theta_1\,(c\theta_5 s\,(\theta_2 + \theta_3) + c\theta_4 s\theta_5 c\,(\theta_2 + \theta_3)) \tag{6.120}$$

$$t_{33} = c\theta_4 s\theta_5 s\,(\theta_2 + \theta_3) - c\theta_5 c\,(\theta_2 + \theta_3) \tag{6.121}$$

$$t_{14} = d_6\,(s\theta_1 s\theta_4 s\theta_5 + c\theta_1\,(c\theta_4 s\theta_5 c\,(\theta_2 + \theta_3) + c\theta_5 s\,(\theta_2 + \theta_3)))$$
$$+ l_3 c\theta_1 s\,(\theta_2 + \theta_3) + d_2 s\theta_1 + l_2 c\theta_1 c\theta_2 \tag{6.122}$$

$$t_{24} = d_6\,(-c\theta_1 s\theta_4 s\theta_5 + s\theta_1\,(c\theta_4 s\theta_5 c\,(\theta_2 + \theta_3) + c\theta_5 s\,(\theta_2 + \theta_3)))$$
$$+ s\theta_1 s\,(\theta_2 + \theta_3) l_3 - d_2 c\theta_1 + l_2 c\theta_2 s\theta_1 \tag{6.123}$$

$$t_{34} = d_6\,(c\theta_4 s\theta_5 s\,(\theta_2 + \theta_3) - c\theta_5 c\,(\theta_2 + \theta_3))$$
$$+ l_2 s\theta_2 + l_3 c\,(\theta_2 + \theta_3) \tag{6.124}$$

Solution of the inverse kinematics problem begins with the wrist position vector $\mathbf{d} = \begin{bmatrix} X & Y & Z \end{bmatrix}^T$, which is $\begin{bmatrix} t_{14} & t_{24} & t_{34} \end{bmatrix}^T$ of $^0T_7$ for $d_7 = 0$, and (X, Y, Z) are coordinates of the position of the wrist point.

$$\mathbf{d} = \begin{bmatrix} (l_3 \sin(\theta_2 + \theta_3) + l_2 \cos\theta_2)\cos\theta_1 + d_2 \sin\theta_1 \\ (l_3 \sin(\theta_2 + \theta_3) + l_2 \cos\theta_2)\sin\theta_1 - d_2 \cos\theta_1 \\ l_3 \cos(\theta_2 + \theta_3) + l_2 \sin\theta_2 \end{bmatrix} = \begin{bmatrix} X \\ Y \\ Z \end{bmatrix} \tag{6.125}$$

Theoretically, we must be able to solve Eq. (6.125) for the three joint variables $\theta_1, \theta_2, \theta_3$. It can be seen that

$$X \sin\theta_1 - Y \cos\theta_1 = d_2 \tag{6.126}$$

which provides:

$$\theta_1 = 2 \operatorname{atan2}(X \pm \sqrt{X^2 + Y^2 - d_2^2}, d_2 - Y) \tag{6.127}$$

Equation (6.127) has two solutions for $X^2 + Y^2 > d_2^2$, one solution for $X^2 + Y^2 = d_2^2$, and no real solution for $X^2 + Y^2 < d_2^2$.
   Combining the first two elements of $\mathbf{d}$ gives

$$l_3 \sin(\theta_2 + \theta_3) = \pm\sqrt{X^2 + Y^2 - d_2^2} - l_2 \cos\theta_2 \tag{6.128}$$

then the third element of $\mathbf{d}$ may be utilized to find

$$l_3^2 = \left(\pm\sqrt{X^2 + Y^2 - d_2^2} - l_2 \cos\theta_2\right)^2 + (Z - l_2 \sin\theta_2)^2 \tag{6.129}$$

which can be rearranged to the following form

$$a \cos\theta_2 + b \sin\theta_2 = c \tag{6.130}$$

$$a = 2l_2\sqrt{X^2 + Y^2 - d_2^2} \qquad b = 2l_2 Z \tag{6.131}$$

$$c = X^2 + Y^2 + Z^2 - d_2^2 + l_2^2 - l_3^2 \tag{6.132}$$

with two solutions:

$$\theta_2 = \text{atan2}(\frac{c}{r}, \pm\sqrt{1 - \frac{c^2}{r^2}}) - \text{atan2}(a, b) \tag{6.133}$$

$$r^2 = a^2 + b^2 \tag{6.134}$$

Summing the squares of the elements of **d** gives

$$X^2 + Y^2 + Z^2 = d_2^2 + l_2^2 + l_3^2 + 2l_2l_3 \sin(2\theta_2 + \theta_3) \tag{6.135}$$

that provides:

$$\theta_3 = \arcsin\left(\frac{X^2 + Y^2 + Z^2 - d_2^2 - l_2^2 - l_3^2}{2l_2l_3}\right) - 2\theta_2 \tag{6.136}$$

Having $\theta_1, \theta_2, \theta_3$ means we can find the wrist point in space. However, because the joint variables in $^0T_3$ and in $^3T_6$ are independent, we should find the orientation of the end-effector by solving $^3T_6$ or $^3R_6$ for $\theta_4, \theta_5, \theta_6$.

$$
\begin{aligned}
^3R_6 &= \begin{bmatrix} c\theta_4c\theta_5c\theta_6 - s\theta_4s\theta_6 & -c\theta_6s\theta_4 - c\theta_4c\theta_5s\theta_6 & c\theta_4s\theta_5 \\ c\theta_5c\theta_6s\theta_4 + c\theta_4s\theta_6 & c\theta_4cc\theta_6 - c\theta_5s\theta_4s\theta_6 & s\theta_4s\theta_5 \\ -c\theta_6s\theta_5 & s\theta_5s\theta_6 & c\theta_5 \end{bmatrix} \\
&= \begin{bmatrix} s_{11} & s_{12} & s_{13} \\ s_{21} & s_{22} & s_{23} \\ s_{31} & s_{32} & s_{33} \end{bmatrix}
\end{aligned} \tag{6.137}
$$

The angles $\theta_4, \theta_5, \theta_6$ can be found by examining elements of $^3R_6$.

$$\theta_4 = \text{atan2}(s_{23}, s_{13}) \tag{6.138}$$

$$\theta_5 = \text{atan2}\left(\sqrt{s_{13}^2 + s_{23}^2}, s_{33}\right) \tag{6.139}$$

$$\theta_6 = \text{atan2}(s_{32}, -s_{31}) \tag{6.140}$$

Example 215 ★ General inverse kinematics formulas. Inverse trigonometric functions frequently appear in inverse kinematic analysis. To minimize numerical error, it is recommended to calculate angles by arctan. Here are some useful equations.

There are some general trigonometric equations that regularly appear in inverse kinematics problems. The following indicates the most frequently equations and solutions.

1. If

$$\sin\theta = a \tag{6.141}$$

then we have two answers: θ and $\pi - \theta$.

$$\theta = \text{atan2}\frac{a}{\pm\sqrt{1 - a^2}} \tag{6.142}$$

2. If

$$\cos\theta = b \tag{6.143}$$

then we have two answers: θ and $-\theta$.

$$\theta = \text{atan2}\frac{\pm\sqrt{1 - b^2}}{b} \tag{6.144}$$

3. If

$$\sin\theta = a \qquad \cos\theta = b \tag{6.145}$$

then:

$$\theta = \text{atan2}\frac{a}{b} \tag{6.146}$$

4. If

$$a \cos \theta + b \sin \theta = 0 \tag{6.147}$$

then we have two answers: θ and $\theta + \pi$.

$$\theta = \mathrm{atan2} \frac{a}{b} \qquad \theta = \mathrm{atan2} \frac{-a}{-b} \tag{6.148}$$

5. If

$$a \cos \theta + b \sin \theta = c \tag{6.149}$$

then:

$$\theta = \mathrm{atan2} \frac{a}{b} + \mathrm{atan2} \frac{\pm\sqrt{a^2 + b^2 - c^2}}{c} \tag{6.150}$$

6. If

$$a \cos \theta + b \sin \theta = c \tag{6.151}$$

$$a \cos \theta - b \sin \theta = d \tag{6.152}$$

then:

$$a^2 + b^2 = c^2 + d^2 \tag{6.153}$$

$$\theta = \mathrm{atan2} \frac{ac - bd}{ad + bc} \tag{6.154}$$

7. If

$$\sin \theta \sin \varphi = a \qquad \cos \theta \sin \varphi = b \tag{6.155}$$

then we have two answers: θ and $\theta + \pi$.

$$\theta = \mathrm{atan2} \frac{a}{b} \qquad \theta = \mathrm{atan2} \frac{-a}{-b} \tag{6.156}$$

8. If

$$\sin \theta \sin \varphi = a \qquad \cos \theta \sin \varphi = b \qquad \cos \varphi = c \tag{6.157}$$

then we have two answers for θ and φ: θ corresponds to φ, and $\theta + \pi$ corresponds to $-\varphi$.

$$\theta = \mathrm{atan2} \frac{a}{b} \qquad \theta = \mathrm{atan2} \frac{-a}{-b} \tag{6.158}$$

$$\varphi = \mathrm{atan2} \frac{\sqrt{a^2 + b^2}}{c} \qquad \varphi = \mathrm{atan2} \frac{-\sqrt{a^2 + b^2}}{c} \tag{6.159}$$

6.2 Inverse Transformation Technique

Assume we have the 4×4 transformation matrix $^0T_6$ from forward kinematics expressed by numbers. The matrix $^0T_6$ includes the global position and the orientation of the end-effector of a $6\,DOF$ robot in the base frame B_0. Also, assume the individual transformation matrices $^0T_1(q_1)$, $^1T_2(q_2)$, $^2T_3(q_3)$, $^3T_4(q_4)$, $^4T_5(q_5)$, and $^5T_6(q_6)$ are known as functions of joint variables analytically.

According to forward kinematics we have:

$$^0T_6 = {}^0T_1 \, {}^1T_2 \, {}^2T_3 \, {}^3T_4 \, {}^4T_5 \, {}^5T_6 \tag{6.160}$$

$$= \begin{bmatrix} r_{11} & r_{12} & r_{13} & r_{14} \\ r_{21} & r_{22} & r_{23} & r_{24} \\ r_{31} & r_{32} & r_{33} & r_{34} \\ 0 & 0 & 0 & 1 \end{bmatrix}$$

We can solve the inverse kinematics problem by solving the following equations for the unknown joint variable.

$$^1T_6 = {}^0T_1^{-1} \, {}^0T_6 \tag{6.161}$$

$$^2T_6 = {}^1T_2^{-1} \, {}^0T_1^{-1} \, {}^0T_6 \tag{6.162}$$

$$^3T_6 = {}^2T_3^{-1} \, {}^1T_2^{-1} \, {}^0T_1^{-1} \, {}^0T_6 \tag{6.163}$$

$$^4T_6 = {}^3T_4^{-1} \, {}^2T_3^{-1} \, {}^1T_2^{-1} \, {}^0T_1^{-1} \, {}^0T_6 \tag{6.164}$$

$$^5T_6 = {}^4T_5^{-1} \, {}^3T_4^{-1} \, {}^2T_3^{-1} \, {}^1T_2^{-1} \, {}^0T_1^{-1} \, {}^0T_6 \tag{6.165}$$

$$\mathbf{I} = {}^5T_6^{-1} \, {}^4T_5^{-1} \, {}^3T_4^{-1} \, {}^2T_3^{-1} \, {}^1T_2^{-1} \, {}^0T_1^{-1} \, {}^0T_6 \tag{6.166}$$

Proof The left-hand side of Eq. (6.160) is a numerical matrix. The right-hand side of Eq. (6.160) is multiplication of several individual transformation matrices, everyone with one unknown variable. If we multiply both sides of Eq. (6.160) by $^0T_1^{-1}$, we obtain an equation which on the left-hand side there is a matrix with only one unknown variable q_1, and on the right-hand side there is a matrix including all other variables q_2, q_3, q_4, q_5, q_6.

$$^0T_1^{-1} \, {}^0T_6 = {}^0T_1^{-1} \left({}^0T_1 \, {}^1T_2 \, {}^2T_3 \, {}^3T_4 \, {}^4T_5 \, {}^5T_6 \right) = {}^1T_6 \tag{6.167}$$

Note that $^0T_1^{-1}$ is the mathematical inverse of the 4×4 matrix $^0T_1$, and not an inverse transformation. So, $^0T_1^{-1}$ must be calculated by a mathematical matrix inversion. The elements of the matrix $^1T_6$ on the right-hand side are either zero, constant, or functions of q_2, q_3, q_4, q_5, q_6. A zero or constant element of the right-hand side provides the required algebraic equation to be solved for q_1.

Then, we multiply both sides of (6.167) by $^1T_2^{-1}$ to obtain a new equation.

$$^1T_2^{-1} \, {}^0T_1^{-1} \, {}^0T_6 = {}^1T_2^{-1} \, {}^0T_1^{-1} \left({}^0T_1 \, {}^1T_2 \, {}^2T_3 \, {}^3T_4 \, {}^4T_5 \, {}^5T_6 \right)$$

$$= {}^2T_6 \tag{6.168}$$

Assuming q_1 has been calculated from (6.167), the left-hand side of this equation is a function of q_2, while the elements of the matrix $^2T_6$, on the right-hand side, are either zero, constant, or functions of q_3, q_4, q_5, q_6. Equating the associated element, with constant or zero element on the right-hand side, provides the required algebraic equation to be solved for q_2.

Following this procedure, we can find the joint variables q_3, q_4, q_5, q_6 by using the following equalities, respectively.

$$^2T_3^{-1} \, {}^1T_2^{-1} \, {}^0T_1^{-1} \, {}^0T_6$$

$$= {}^2T_3^{-1} \, {}^1T_2^{-1} \, {}^0T_1^{-1} \left({}^0T_1 \, {}^1T_2 \, {}^2T_3 \, {}^3T_4 \, {}^4T_5 \, {}^5T_6 \right)$$

$$= {}^3T_6 \tag{6.169}$$

$$^3T_4^{-1} \, {}^2T_3^{-1} \, {}^1T_2^{-1} \, {}^0T_1^{-1} \, {}^0T_6$$

$$= {}^3T_4^{-1} \, {}^2T_3^{-1} \, {}^1T_2^{-1} \, {}^0T_1^{-1} \left({}^0T_1 \, {}^1T_2 \, {}^2T_3 \, {}^3T_4 \, {}^4T_5 \, {}^5T_6 \right)$$

$$= {}^4T_6 \tag{6.170}$$

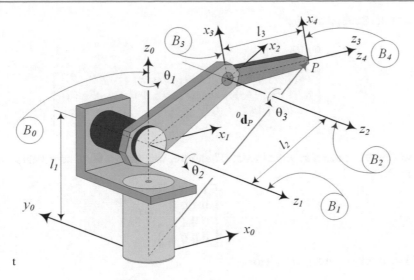

Fig. 6.7 An articulated manipulator

$$
{}^4T_5^{-1}\,{}^3T_4^{-1}\,{}^2T_3^{-1}\,{}^1T_2^{-1}\,{}^0T_1^{-1}\,{}^0T_6
$$
$$
= {}^4T_5^{-1}\,{}^3T_4^{-1}\,{}^2T_3^{-1}\,{}^1T_2^{-1}\,{}^0T_1^{-1}\left({}^0T_1\,{}^1T_2\,{}^2T_3\,{}^3T_4\,{}^4T_5\,{}^5T_6\right)
$$
$$
= {}^5T_6 \tag{6.171}
$$

$$
{}^5T_6^{-1}\,{}^4T_5^{-1}\,{}^3T_4^{-1}\,{}^2T_3^{-1}\,{}^1T_2^{-1}\,{}^0T_1^{-1}\,{}^0T_6
$$
$$
= {}^5T_6^{-1}\,{}^4T_5^{-1}\,{}^3T_4^{-1}\,{}^2T_3^{-1}\,{}^1T_2^{-1}\,{}^0T_1^{-1}\left({}^0T_1\,{}^1T_2\,{}^2T_3\,{}^3T_4\,{}^4T_5\,{}^5T_6\right)
$$
$$
= \mathbf{I} \tag{6.172}
$$

The inverse transformation technique may sometimes be called Pieper technique. ∎

Example 216 Articulated manipulator and numerical case. Articulated manipulator is a very practical industrial robot and an important educational example. Here is the use of inverse transformation technique to solve its inverse kinematics.

Consider the articulated manipulator shown in Fig. 6.7. The transformation matrices between its coordinate frames are:

$$
{}^0T_1 = \begin{bmatrix} \cos\theta_1 & 0 & \sin\theta_1 & 0 \\ \sin\theta_1 & 0 & -\cos\theta_1 & 0 \\ 0 & 1 & 0 & l_1 \\ 0 & 0 & 0 & 1 \end{bmatrix} \tag{6.173}
$$

$$
{}^1T_2 = \begin{bmatrix} \cos\theta_2 & -\sin\theta_2 & 0 & l_2\cos\theta_2 \\ \sin\theta_2 & \cos\theta_2 & 0 & l_2\sin\theta_2 \\ 0 & 0 & 1 & 0 \\ 0 & 0 & 0 & 1 \end{bmatrix} \tag{6.174}
$$

$$
{}^2T_3 = \begin{bmatrix} \cos\theta_3 & 0 & \sin\theta_3 & 0 \\ \sin\theta_3 & 0 & -\cos\theta_3 & 0 \\ 0 & 1 & 0 & 0 \\ 0 & 0 & 0 & 1 \end{bmatrix} \tag{6.175}
$$

The forward kinematics of the manipulator is:

$$
{}^0T_3 = {}^0T_1 \, {}^1T_2 \, {}^2T_3 \tag{6.176}
$$

$$
= \begin{bmatrix}
c\theta_1 c\,(\theta_2 + \theta_3) & s\theta_1 & c\theta_1 s\,(\theta_2 + \theta_3) & l_2 c\theta_1 c\theta_2 \\
s\theta_1 c\,(\theta_2 + \theta_3) & -c\theta_1 & s\theta_1 s\,(\theta_2 + \theta_3) & l_2 c\theta_2 s\theta_1 \\
s\,(\theta_2 + \theta_3) & 0 & -c\,(\theta_2 + \theta_3) & l_1 + l_2 s\theta_2 \\
0 & 0 & 0 & 1
\end{bmatrix}
$$

Point P is the point at which we attach a spherical wrist. Therefore, we attach a takht coordinate frame B_4 at P that is at a constant distance l_3 from B_3.

$$
{}^3T_4 = \begin{bmatrix}
1 & 0 & 0 & 0 \\
0 & 1 & 0 & 0 \\
0 & 0 & 1 & l_3 \\
0 & 0 & 0 & 1
\end{bmatrix} \tag{6.177}
$$

So, the overall forward kinematics of the manipulator is:

$$
{}^0T_4 = {}^0T_3 \, {}^3T_4 = \begin{bmatrix}
c\,(\theta_2 + \theta_3)\,c\theta_1 & s\theta_1 & s\,(\theta_2 + \theta_3)\,c\theta_1 & l_3 s\,(\theta_2 + \theta_3)\,c\theta_1 + l_2 c\theta_1 c\theta_2 \\
c\,(\theta_2 + \theta_3)\,s\theta_1 & -c\theta_1 & s\,(\theta_2 + \theta_3)\,s\theta_1 & l_3 s\,(\theta_2 + \theta_3)\,s\theta_1 + l_2 c\theta_2 s\theta_1 \\
s\,(\theta_2 + \theta_3) & 0 & -c\,(\theta_2 + \theta_3) & l_1 - l_3 c\,(\theta_2 + \theta_3) + l_2 s\theta_2 \\
0 & 0 & 0 & 1
\end{bmatrix} \tag{6.178}
$$

Using the following dimensions

$$
l_1 = 1\,\text{m} \qquad l_2 = 1.05\,\text{m} \qquad l_3 = 0.89\,\text{m} \tag{6.179}
$$

when its tip point is at:

$$
{}^0\mathbf{d}_P = \begin{bmatrix} 1 & 1.1 & 1.2 \end{bmatrix}^T \tag{6.180}
$$

the forward kinematics reduces to:

$$
{}^0T_4
$$
$$
= \begin{bmatrix}
\cos(\theta_2 + \theta_3)\cos\theta_1 & \sin\theta_1 & \sin(\theta_2 + \theta_3)\cos\theta_1 & 1 \\
\cos(\theta_2 + \theta_3)\sin\theta_1 & -\cos\theta_1 & \sin(\theta_2 + \theta_3)\sin\theta_1 & 1.1 \\
\sin(\theta_2 + \theta_3) & 0 & -\cos(\theta_2 + \theta_3) & 1.2 \\
0 & 0 & 0 & 1
\end{bmatrix} \tag{6.181}
$$

Let us multiply both sides of ${}^0T_4$ by ${}^0T_1^{-1}$ to have:

$$
{}^0T_1^{-1}\,{}^0T_4 = {}^0T_1^{-1}\left({}^0T_1\,{}^1T_2\,{}^2T_3\,{}^3T_4\right) = {}^1T_4 \tag{6.182}
$$

where

$$
{}^0T_1^{-1}\,{}^0T_4 = {}^1T_4
$$
$$
= \begin{bmatrix}
\cos\theta_1 & \sin\theta_1 & 0 & 0 \\
0 & 0 & 1 & -1 \\
\sin\theta_1 & -\cos\theta_1 & 0 & 0 \\
0 & 0 & 0 & 1
\end{bmatrix} {}^0T_4 \tag{6.183}
$$
$$
= \begin{bmatrix}
\cos(\theta_2 + \theta_3) & 0 & \sin(\theta_2 + \theta_3) & \cos\theta_1 + 1.1\sin\theta_1 \\
\sin(\theta_2 + \theta_3) & 0 & -\cos(\theta_2 + \theta_3) & 0.2 \\
0 & 1 & 0 & \sin\theta_1 - 1.1\cos\theta_1 \\
0 & 0 & 0 & 1
\end{bmatrix}
$$

and

$$^1T_2\,^2T_3\,^3T_4 = \tag{6.184}$$

$$\begin{bmatrix} \cos(\theta_2 + \theta_3) & 0 & \sin(\theta_2 + \theta_3) & 0.89\sin(\theta_2 + \theta_3) + 1.05\cos\theta_2 \\ \sin(\theta_2 + \theta_3) & 0 & -\cos(\theta_2 + \theta_3) & 1.05\sin\theta_2 - 0.89\cos(\theta_2 + \theta_3) \\ 0 & 1 & 0 & 0 \\ 0 & 0 & 0 & 1 \end{bmatrix}$$

The last column of the left-hand side of (6.182) is only a function of θ_1 while the right-hand side is a function of θ_2 and θ_3. Equating the element r_{24} of both sides of (6.182) provides an equation to determine θ_1.

$$\sin\theta_1 - 1.1\cos\theta_1 = 0 \tag{6.185}$$

$$\theta_1 = \mathrm{atan2}\,(1.1,\,1) = \arctan\frac{1.1}{1}$$

$$= 0.8329812667\,\text{rad} \approx 47.72631098\,\text{deg} \tag{6.186}$$

Substituting $\theta_1 = 0.83298$ rad in (6.183) provides a matrix $^1T_4$ with a numerical values in the last column.

$$^1T_4 = \begin{bmatrix} \cos(\theta_2 + \theta_3) & 0 & \sin(\theta_2 + \theta_3) & 1.4866 \\ \sin(\theta_2 + \theta_3) & 0 & -\cos(\theta_2 + \theta_3) & 0.2 \\ 0 & 1 & 0 & 0 \\ 0 & 0 & 0 & 1 \end{bmatrix} \tag{6.187}$$

We multiply both sides of (6.187) by $^1T_2^{-1}$ to have:

$$^1T_2^{-1}\,^1T_4 = {}^1T_2^{-1}\left({}^1T_2\,^2T_3\,^3T_4\right) = {}^2T_4 \tag{6.188}$$

where

$$^1T_2^{-1}\,^1T_4 = {}^2T_4 = \begin{bmatrix} \cos\theta_2 & \sin\theta_2 & 0 & -1.05 \\ -\sin\theta_2 & \cos\theta_2 & 0 & 0 \\ 0 & 0 & 1 & 0 \\ 0 & 0 & 0 & 1 \end{bmatrix} {}^1T_4 \tag{6.189}$$

$$= \begin{bmatrix} \cos\theta_3 & 0 & \sin\theta_3 & 1.4866\cos\theta_2 + 0.2\sin\theta_2 - 1.05 \\ \sin\theta_3 & 0 & -\cos\theta_3 & 0.2\cos\theta_2 - 1.4866\sin\theta_2 \\ 0 & 1 & 0 & 0 \\ 0 & 0 & 0 & 1 \end{bmatrix}$$

and

$$^2T_3\,^3T_4 = \begin{bmatrix} \cos\theta_3 & 0 & \sin\theta_3 & 0.89\sin\theta_3 \\ \sin\theta_3 & 0 & -\cos\theta_3 & -0.89\cos\theta_3 \\ 0 & 1 & 0 & 0 \\ 0 & 0 & 0 & 1 \end{bmatrix} \tag{6.190}$$

Squaring the elements r_{14} and r_{24} of the left-hand sides of (6.188) provides an equation to determine θ_2.

$$(1.4866\cos\theta_2 + 0.2\sin\theta_2 - 1.05)^2 + (0.2\cos\theta_2 - 1.4866\sin\theta_2)^2$$

$$= (0.89\sin\theta_3)^2 + (-0.89\cos\theta_3)^2 \tag{6.191}$$

$$3.1219\cos\theta_2 + 0.42\sin\theta_2 = 2.5604 \tag{6.192}$$

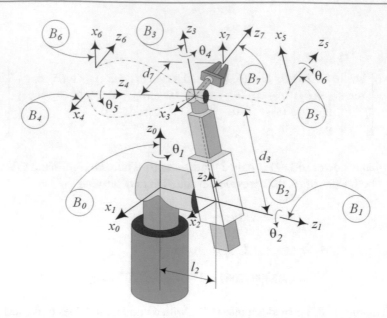

Fig. 6.8 A spherical robot made of a spherical manipulator attached to a spherical wrist

This equation has the following solutions:

$$\theta_2 = 0.7555 \, \text{rad} \approx 43.29 \, \text{deg} \tag{6.193}$$

$$\theta_2 = -0.4881 \, \text{rad} \approx -27.96 \, \text{deg} \tag{6.194}$$

Having θ_2, we can calculate θ_3 from the last column of (6.190) and (6.189).

$$\theta_3 = \arctan\left(\frac{1.486\,6\cos\theta_2 + 0.2\sin\theta_2 - 1.05}{0.2\cos\theta_2 - 1.486\,6\sin\theta_2}\right) + \pi \tag{6.195}$$

If $\theta_2 = 0.7555$ rad, then we have:arctan

$$\theta_3 = 2.95 \, \text{rad} \approx 169 \, \text{deg} \tag{6.196}$$

If $\theta_2 = -0.488$ rad, then we have:

$$\theta_3 = 0.19198 \, \text{rad} \approx 11 \, \text{deg} \tag{6.197}$$

Example 217 Inverse kinematics for a spherical robot. This robot has a long history in education and research of robotic science. Its inverse kinematics will be solved in this example by inverse transformation method.

Homogenous transformation matrices of the spherical robot shown in Fig. 6.8 are:

$$
{}^{0}T_1 = \begin{bmatrix} c\theta_1 & 0 & -s\theta_1 & 0 \\ s\theta_1 & 0 & c\theta_1 & 0 \\ 0 & -1 & 0 & 0 \\ 0 & 0 & 0 & 1 \end{bmatrix} \qquad
{}^{1}T_2 = \begin{bmatrix} c\theta_2 & 0 & s\theta_2 & 0 \\ s\theta_2 & 0 & -c\theta_2 & 0 \\ 0 & 1 & 0 & l_2 \\ 0 & 0 & 0 & 1 \end{bmatrix} \tag{6.198}
$$

$$
{}^{2}T_3 = \begin{bmatrix} 1 & 0 & 0 & 0 \\ 0 & 1 & 0 & 0 \\ 0 & 0 & 1 & d_3 \\ 0 & 0 & 0 & 1 \end{bmatrix} \qquad
{}^{3}T_4 = \begin{bmatrix} c\theta_4 & 0 & -s\theta_4 & 0 \\ s\theta_4 & 0 & c\theta_4 & 0 \\ 0 & -1 & 0 & 0 \\ 0 & 0 & 0 & 1 \end{bmatrix} \tag{6.199}
$$

$$
{}^{4}T_{5} = \begin{bmatrix} c\theta_{5} & 0 & s\theta_{5} & 0 \\ s\theta_{5} & 0 & -c\theta_{5} & 0 \\ 0 & 1 & 0 & 0 \\ 0 & 0 & 0 & 1 \end{bmatrix} \qquad {}^{5}T_{6} = \begin{bmatrix} c\theta_{6} & -s\theta_{6} & 0 & 0 \\ s\theta_{6} & c\theta_{6} & 0 & 0 \\ 0 & 0 & 1 & 0 \\ 0 & 0 & 0 & 1 \end{bmatrix} \tag{6.200}
$$

Therefore, the position and orientation of the end-effector for a set of joint variables can be found by matrix multiplication,

$$
{}^{0}T_{6} = {}^{0}T_{1} {}^{1}T_{2} {}^{2}T_{3} {}^{3}T_{4} {}^{4}T_{5} {}^{5}T_{6}
$$

$$
= \begin{bmatrix} r_{11} & r_{12} & r_{13} & r_{14} \\ r_{21} & r_{22} & r_{23} & r_{24} \\ r_{31} & r_{32} & r_{33} & r_{34} \\ 0 & 0 & 0 & 1 \end{bmatrix} \tag{6.201}
$$

where the elements of ${}^{0}T_{6}$ are the same as the elements of the matrix in Eq. (5.280).

Multiplying both sides of the (6.201) by ${}^{0}T_{1}^{-1}$ provides

$$
{}^{0}T_{1}^{-1} {}^{0}T_{6} = \begin{bmatrix} \cos\theta_{1} & \sin\theta_{1} & 0 & 0 \\ 0 & 0 & -1 & 0 \\ -\sin\theta_{1} & \cos\theta_{1} & 0 & 0 \\ 0 & 0 & 0 & 1 \end{bmatrix} \begin{bmatrix} r_{11} & r_{12} & r_{13} & r_{14} \\ r_{21} & r_{22} & r_{23} & r_{24} \\ r_{31} & r_{32} & r_{33} & r_{34} \\ 0 & 0 & 0 & 1 \end{bmatrix}
$$

$$
= \begin{bmatrix} f_{11} & f_{12} & f_{13} & f_{14} \\ f_{21} & f_{22} & f_{23} & f_{24} \\ f_{31} & f_{32} & f_{33} & f_{34} \\ 0 & 0 & 0 & 1 \end{bmatrix} \tag{6.202}
$$

where

$$
f_{1i} = r_{1i} \cos\theta_{1} + r_{2i} \sin\theta_{1} \tag{6.203}
$$

$$
f_{2i} = -r_{3i} \tag{6.204}
$$

$$
f_{3i} = r_{2i} \cos\theta_{1} - r_{1i} \sin\theta_{1} \tag{6.205}
$$

$$
i = 1, 2, 3, 4
$$

Based on the given transformation matrices, we find:

$$
{}^{1}T_{6} = {}^{1}T_{2} {}^{2}T_{3} {}^{3}T_{4} {}^{4}T_{5} {}^{5}T_{6}
$$

$$
= \begin{bmatrix} f_{11} & f_{12} & f_{13} & f_{14} \\ f_{21} & f_{22} & f_{23} & f_{24} \\ f_{31} & f_{32} & f_{33} & f_{34} \\ 0 & 0 & 0 & 1 \end{bmatrix} \tag{6.206}
$$

$$
f_{11} = -c\theta_{2}s\theta_{4}s\theta_{6} + c\theta_{6}\,(-s\theta_{2}s\theta_{5} + c\theta_{2}c\theta_{4}c\theta_{5}) \tag{6.207}
$$

$$
f_{21} = -s\theta_{2}s\theta_{4}s\theta_{6} + c\theta_{6}\,(c\theta_{2}s\theta_{5} + c\theta_{4}c\theta_{5}s\theta_{2}) \tag{6.208}
$$

$$
f_{31} = c\theta_{4}s\theta_{6} + c\theta_{5}c\theta_{6}s\theta_{4} \tag{6.209}
$$

$$
f_{12} = -c\theta_{2}c\theta_{6}s\theta_{4} - s\theta_{6}\,(-s\theta_{2}s\theta_{5} + c\theta_{2}c\theta_{4}c\theta_{5}) \tag{6.210}
$$

$$
f_{22} = -c\theta_{6}s\theta_{2}s\theta_{4} - s\theta_{6}\,(c\theta_{2}s\theta_{5} + c\theta_{4}c\theta_{5}s\theta_{2}) \tag{6.211}
$$

$$
f_{32} = c\theta_{4}c\theta_{6} - c\theta_{5}s\theta_{4}s\theta_{6} \tag{6.212}
$$

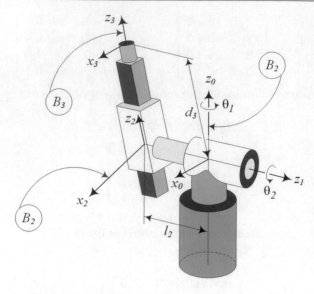

Fig. 6.9 Left shoulder configuration of a spherical robot

$$f_{13} = c\theta_5 s\theta_2 + c\theta_2 c\theta_4 s\theta_5 \tag{6.213}$$

$$f_{23} = -c\theta_2 c\theta_5 + c\theta_4 s\theta_2 s\theta_5 \tag{6.214}$$

$$f_{33} = s\theta_4 s\theta_5 \tag{6.215}$$

$$f_{14} = d_3 s\theta_2 \tag{6.216}$$

$$f_{24} = -d_3 c\theta_2 \tag{6.217}$$

$$f_{34} = l_2 \tag{6.218}$$

The only constant element of the matrix (6.206) is $f_{34} = l_2$, therefore,

$$r_{24} \cos\theta_1 - r_{14} \sin\theta_1 = l_2 \tag{6.219}$$

This is a first kind of trigonometric equations, solved in Example 209, which has a systematic method of solution.

$$\theta_1 = \arctan\frac{r_{24}}{r_{14}} - \arctan\frac{l_2}{\pm\sqrt{r_{24}^2 + r_{14}^2 - l_2^2}} \tag{6.220}$$

The $(-)$ sign corresponds to a left shoulder configuration of the robots as shown in Fig. 6.9, and the $(+)$ sign corresponds to the right shoulder configuration.

The elements f_{14} and f_{24} of matrix (6.206) are functions of θ_1 and θ_2 only.

$$f_{14} = d_3 \sin\theta_2 = r_{14} \cos\theta_1 + r_{24} \sin\theta_1 \tag{6.221}$$

$$f_{24} = -d_3 \cos\theta_2 = -r_{34} \tag{6.222}$$

Hence, it is possible to use them and find θ_2

$$\theta_2 = \tan^{-1}\frac{r_{14} \cos\theta_1 + r_{24} \sin\theta_1}{r_{34}} \tag{6.223}$$

where θ_1 must be substituted from (6.220).

In the next step, we find the third joint variable d_3 from

$$^1T_2^{-1}\,^0T_1^{-1}\,^0T_6 = \,^2T_6 \tag{6.224}$$

where

$$^1T_2^{-1} = \begin{bmatrix} \cos\theta_2 & \sin\theta_2 & 0 & 0 \\ 0 & 0 & 1 & -l_2 \\ \sin\theta_2 & -\cos\theta_2 & 0 & 0 \\ 0 & 0 & 0 & 1 \end{bmatrix} \tag{6.225}$$

and

$$^2T_6 = \begin{bmatrix} -s\theta_4s\theta_6 + c\theta_4c\theta_5c\theta_6 & -c\theta_6s\theta_4 - c\theta_4c\theta_5s\theta_6 & c\theta_4s\theta_5 & 0 \\ c\theta_4s\theta_6 + c\theta_5c\theta_6s\theta_4 & c\theta_4c\theta_6 - c\theta_5s\theta_4s\theta_6 & s\theta_4s\theta_5 & 0 \\ -c\theta_6s\theta_5 & s\theta_5s\theta_6 & c\theta_5 & d_3 \\ 0 & 0 & 0 & 1 \end{bmatrix} \tag{6.226}$$

Employing the elements of the matrices on both sides of Eq. (6.224) shows that the element (3, 4) can be utilized to find d_3.

$$d_3 = r_{34}\cos\theta_2 + r_{14}\cos\theta_1\sin\theta_2 + r_{24}\sin\theta_1\sin\theta_2 \tag{6.227}$$

There is no other element in Eq. (6.224) to be a function of another single variable. The equation $^2T_3^{-1}\,^1T_2^{-1}\,^0T_1^{-1}\,^0T_6 = \,^3T_6$ also provides no new equation. Hence, we move to the next step and evaluate θ_4 from $^4T_6$.

$$^3T_4^{-1}\,^2T_3^{-1}\,^1T_2^{-1}\,^0T_1^{-1}\,^0T_6 = \,^4T_6 \tag{6.228}$$

Evaluating $^4T_6$

$$^4T_6 = \begin{bmatrix} \cos\theta_5\cos\theta_6 & -\cos\theta_5\sin\theta_6 & \sin\theta_5 & 0 \\ \cos\theta_6\sin\theta_5 & -\sin\theta_5\sin\theta_6 & -\cos\theta_5 & 0 \\ \sin\theta_6 & \cos\theta_6 & 0 & 0 \\ 0 & 0 & 0 & 1 \end{bmatrix} \tag{6.229}$$

and the left-hand side of (6.228) utilizing

$$^2T_3^{-1} = \begin{bmatrix} 1 & 0 & 0 & 0 \\ 0 & 1 & 0 & 0 \\ 0 & 0 & 1 & -d_3 \\ 0 & 0 & 0 & 1 \end{bmatrix} \tag{6.230}$$

and

$$^3T_4^{-1} = \begin{bmatrix} \cos\theta_4 & \sin\theta_4 & 0 & 0 \\ 0 & 0 & -1 & 0 \\ -\sin\theta_4 & \cos\theta_4 & 0 & 0 \\ 0 & 0 & 0 & 1 \end{bmatrix} \tag{6.231}$$

shows that

$$^3T_4^{-1}\,^2T_3^{-1}\,^1T_2^{-1}\,^0T_1^{-1}\,^0T_6 = \begin{bmatrix} g_{11} & g_{12} & g_{13} & g_{14} \\ g_{21} & g_{22} & g_{23} & g_{24} \\ g_{31} & g_{32} & g_{33} & g_{34} \\ 0 & 0 & 0 & 1 \end{bmatrix} \tag{6.232}$$

where

$$g_{1i} = -r_{3i}c\theta_4s\theta_2 + r_{2i}\left(c\theta_1s\theta_4 + c\theta_2c\theta_4s\theta_1\right)$$
$$+ r_{1i}\left(-s\theta_1s\theta_4 + c\theta_1c\theta_2c\theta_4\right) \tag{6.233}$$
$$g_{2i} = d_3\delta_{4i} - r_{31}c\theta_2 - r_{11}c\theta_1s\theta_2 - r_{21}s\theta_1s\theta_2 \tag{6.234}$$
$$g_{3i} = r_{31}s\theta_2s\theta_4 + r_{21}\left(c\theta_1c\theta_4 - c\theta_2s\theta_1s\theta_4\right)$$
$$+ r_{11}\left(-c\theta_4s\theta_1 - c\theta_1c\theta_2s\theta_4\right) \tag{6.235}$$
$$i = 1, 2, 3, 4$$

The symbol δ_{4i} indicates the Kronecker delta and is:

$$\delta_{4i} = \begin{cases} 1 & if \ i = 4 \\ 0 & if \ i \neq 4 \end{cases} \tag{6.236}$$

Therefore, we can find θ_4 by equating the element $(3, 3)$, θ_5 by equating the elements $(1, 3)$ or $(2, 3)$, and θ_6 by equating the elements $(3, 1)$ or $(3, 2)$. Starting from element $(3, 3)$

$$r_{13} \left(-c\theta_4 s\theta_1 - c\theta_1 c\theta_2 s\theta_4\right) + r_{23} \left(c\theta_1 c\theta_4 - c\theta_2 s\theta_1 s\theta_4\right)$$
$$+r_{33} s\theta_2 s\theta_4 = 0 \tag{6.237}$$

we find θ_4

$$\theta_4 = \tan^{-1} \frac{-r_{13}s\theta_1 + r_{23}c\theta_1}{c\theta_2 \left(r_{13}c\theta_1 + r_{23}s\theta_1\right) - r_{33}s\theta_2} \tag{6.238}$$

which, based on the second value of θ_1, can also be equal to:

$$\theta_4 = \frac{\pi}{2} + \tan^{-1} \frac{-r_{13}s\theta_1 + r_{23}c\theta_1}{c\theta_2 \left(r_{13}c\theta_1 + r_{23}s\theta_1\right) - r_{33}s\theta_2} \tag{6.239}$$

Now we use elements $(1, 3)$ and $(2, 3)$, to find θ_5.

$$\sin\theta_5 = r_{23} \left(\cos\theta_1 \sin\theta_4 + \cos\theta_2 \cos\theta_4 \sin\theta_1\right) - r_{33} \cos\theta_4 \sin\theta_2$$
$$+r_{13} \left(\cos\theta_1 \cos\theta_2 \cos\theta_4 - \sin\theta_1 \sin\theta_4\right) \tag{6.240}$$

$$-\cos\theta_5 = -r_{33} \cos\theta_2 - r_{13} \cos\theta_1 \sin\theta_2 - r_{23} \sin\theta_1 \sin\theta_2 \tag{6.241}$$

$$\theta_5 = \tan^{-1} \frac{\sin\theta_5}{\cos\theta_5} \tag{6.242}$$

Finally, θ_6 can be found from the elements $(3, 1)$ and $(3, 2)$.

$$\sin\theta_6 = r_{31} \sin\theta_2 \sin\theta_4 + r_{21} \left(\cos\theta_1 \cos\theta_4 - \cos\theta_2 \sin\theta_1 \sin\theta_4\right)$$
$$+r_{11} \left(-\cos\theta_4 \sin\theta_1 - \cos\theta_1 \cos\theta_2 \sin\theta_4\right) \tag{6.243}$$

$$\cos\theta_6 = r_{32} \sin\theta_2 \sin\theta_4 + r_{22} \left(\cos\theta_1 \cos\theta_4 - \cos\theta_2 \sin\theta_1 \sin\theta_4\right)$$
$$+r_{12} \left(-\cos\theta_4 \sin\theta_1 - \cos\theta_1 \cos\theta_2 \sin\theta_4\right) \tag{6.244}$$

$$\theta_6 = \tan^{-1} \frac{\sin\theta_6}{\cos\theta_6} \tag{6.245}$$

Example 218 Inverse of parametric Euler angles transformation matrix. Any given numerical matrix that is equal to the multiplication of a finite number of matrices based on unknown variables may be solved by inverse transformation matrix method. As an example, we find Euler angles of the Euler transformation matrix.

The global rotation matrix based on Euler angles has been found in Eq. (2.174).

$$^G R_B = \left[A_{z,\psi} \ A_{x,\theta} \ A_{z,\varphi}\right]^T = R_{Z,\varphi} \ R_{X,\theta} \ R_{Z,\psi}$$

$$= \begin{bmatrix} c\varphi c\psi - c\theta s\varphi s\psi & -c\varphi s\psi - c\theta c\psi s\varphi & s\theta s\varphi \\ c\psi s\varphi + c\theta c\varphi s\psi & -s\varphi s\psi + c\theta c\varphi c\psi & -c\varphi s\theta \\ s\theta s\psi & s\theta c\psi & c\theta \end{bmatrix}$$

$$= \begin{bmatrix} r_{11} & r_{12} & r_{13} \\ r_{21} & r_{22} & r_{23} \\ r_{31} & r_{32} & r_{33} \end{bmatrix} \tag{6.246}$$

Let us assume the left-hand side matrix $^G R_B$ is given as a numerical matrix. The matrices on the right-hand side are also known, each with an unknown Euler angle. Premultiplying $^G R_B$ by $R_{Z,\varphi}^{-1}$, gives:

$$
\begin{bmatrix} \cos\varphi & \sin\varphi & 0 \\ -\sin\varphi & \cos\varphi & 0 \\ 0 & 0 & 1 \end{bmatrix} {}^G R_B
$$

$$
= \begin{bmatrix} r_{11}c\varphi + r_{21}s\varphi & r_{12}c\varphi + r_{22}s\varphi & r_{13}c\varphi + r_{23}s\varphi \\ r_{21}c\varphi - r_{11}s\varphi & r_{22}c\varphi - r_{12}s\varphi & r_{23}c\varphi - r_{13}s\varphi \\ r_{31} & r_{32} & r_{33} \end{bmatrix}
$$

$$
= \begin{bmatrix} \cos\psi & -\sin\psi & 0 \\ \cos\theta\sin\psi & \cos\theta\cos\psi & -\sin\theta \\ \sin\theta\sin\psi & \sin\theta\cos\psi & \cos\theta \end{bmatrix} \tag{6.247}
$$

Equating the elements $(1, 3)$ of both sides

$$
r_{13}\cos\varphi + r_{23}\sin\varphi = 0 \tag{6.248}
$$

gives φ.

$$
\varphi = \operatorname{atan2}(r_{13}, -r_{23}) \tag{6.249}
$$

Having φ helps us to find ψ by using elements $(1, 1)$ and $(1, 2)$

$$
\cos\psi = r_{11}\cos\varphi + r_{21}\sin\varphi \tag{6.250}
$$

$$
-\sin\psi = r_{12}\cos\varphi + r_{22}\sin\varphi \tag{6.251}
$$

therefore,

$$
\psi = \operatorname{atan2}\frac{-r_{12}\cos\varphi - r_{22}\sin\varphi}{r_{11}\cos\varphi + r_{21}\sin\varphi} \tag{6.252}
$$

In the next step, we may postmultiply $^G R_B$ by $R_{Z,\psi}^{-1}$, to provide:

$$
{}^G R_B \begin{bmatrix} \cos\psi & \sin\psi & 0 \\ -\sin\psi & \cos\psi & 0 \\ 0 & 0 & 1 \end{bmatrix}
$$

$$
= \begin{bmatrix} r_{11}c\psi - r_{12}s\psi & r_{12}c\psi + r_{11}s\psi & r_{13} \\ r_{21}c\psi - r_{22}s\psi & r_{22}c\psi + r_{21}s\psi & r_{23} \\ r_{31}c\psi - r_{32}s\psi & r_{32}c\psi + r_{31}s\psi & r_{33} \end{bmatrix}
$$

$$
= \begin{bmatrix} \cos\varphi & -\cos\theta\sin\varphi & \sin\theta\sin\varphi \\ \sin\varphi & \cos\theta\cos\varphi & -\cos\varphi\sin\theta \\ 0 & \sin\theta & \cos\theta \end{bmatrix} \tag{6.253}
$$

The elements $(3, 1)$ on both sides make an equation to find ψ.

$$
r_{31}\cos\psi - r_{31}\sin\psi = 0 \tag{6.254}
$$

Therefore, it is possible to find ψ from the following equation:

$$
\psi = \operatorname{atan2}(r_{31}, r_{31}). \tag{6.255}
$$

Finally, θ can be found using elements $(3, 2)$ and $(3, 3)$

$$r_{32}c\psi + r_{31}s\psi = \sin\theta \qquad (6.256)$$

$$r_{33} = \cos\theta \qquad (6.257)$$

which give θ.

$$\theta = \text{atan2}\,\frac{r_{32}\cos\psi + r_{31}\sin\psi}{r_{33}} \qquad (6.258)$$

As an example, let us assume the global rotation matrix based on Euler angles is given as:

$$
{}^G R_B = \begin{bmatrix} A_{z,\psi}\ A_{x,\theta}\ A_{z,\varphi} \end{bmatrix}^T = R_{Z,\varphi}\,R_{X,\theta}\,R_{Z,\psi}
$$

$$
= \begin{bmatrix} c\varphi\,c\psi - c\theta\,s\varphi\,s\psi & -c\varphi\,s\psi - c\theta\,c\psi\,s\varphi & s\theta\,s\varphi \\ c\psi\,s\varphi + c\theta\,c\varphi\,s\psi & -s\varphi\,s\psi + c\theta\,c\varphi\,c\psi & -c\varphi\,s\theta \\ s\theta\,s\psi & s\theta\,c\psi & c\theta \end{bmatrix}
$$

$$
= \begin{bmatrix} 0.126\,83 & -0.780\,33 & 0.612\,37 \\ 0.926\,78 & -0.126\,83 & -0.353\,55 \\ 0.353\,55 & 0.612\,37 & 0.707\,11 \end{bmatrix} \qquad (6.259)
$$

Premultiplying ${}^G R_B$ by $R_{Z,\varphi}^{-1}$, gives

$$
\begin{bmatrix} \cos\varphi & \sin\varphi & 0 \\ -\sin\varphi & \cos\varphi & 0 \\ 0 & 0 & 1 \end{bmatrix} {}^G R_B
$$

$$
= \begin{bmatrix} 0.126c\varphi + 0.926s\varphi & -0.780c\varphi - 0.126s\varphi & 0.612c\varphi - 0.353s\varphi \\ 0.926c\varphi - 0.126s\varphi & 0.780s\varphi - 0.126c\varphi & -0.353c\varphi - 0.612s\varphi \\ 0.353\,55 & 0.612\,37 & 0.707\,11 \end{bmatrix}
$$

$$
= \begin{bmatrix} \cos\psi & -\sin\psi & 0 \\ \cos\theta\sin\psi & \cos\theta\cos\psi & -\sin\theta \\ \sin\theta\sin\psi & \sin\theta\cos\psi & \cos\theta \end{bmatrix}. \qquad (6.260)
$$

Equating the elements $(1, 3)$ of both sides

$$0.612\,37\cos\varphi - 0.353\,55\sin\varphi = 0 \qquad (6.261)$$

gives

$$\varphi = \text{atan2}\left(\frac{0.612\,37}{0.353\,55}\right) = 1.0472\,\text{rad} = 60\,\text{deg} \qquad (6.262)$$

Having φ helps us to find ψ by using elements $(1, 1)$ and $(1, 2)$

$$\cos\psi = 0.126\cos\varphi + 0.926\sin\varphi \qquad (6.263)$$

$$-\sin\psi = -0.78\cos\varphi - 0.126\sin\varphi \qquad (6.264)$$

therefore,

$$
\psi = \text{atan2}\,\frac{0.78\cos\varphi + 0.126\sin\varphi}{0.126\cos\varphi + 0.926\sin\varphi}
$$

$$
= \text{atan2}\,\frac{0.499\,12}{0.864\,94} = 0.523\,\text{rad} = 30\,\text{deg} \qquad (6.265)
$$

Although we can find θ from elements $(2, 3)$ and $(3, 3)$, let us postmultiply $^G R_B$ by $R_{Z,\psi}^{-1}$, to follow the inverse transformation technique.

$$
\begin{aligned}
&^G R_B
\begin{bmatrix}
\cos\psi & \sin\psi & 0 \\
-\sin\psi & \cos\psi & 0 \\
0 & 0 & 1
\end{bmatrix} \\
&=
\begin{bmatrix}
0.126c\psi + 0.78s\psi & 0.126s\psi - 0.78c\psi & 0.612\,37 \\
0.926c\psi + 0.126s\psi & 0.926s\psi - 0.126c\psi & -0.353\,55 \\
0.353c\psi - 0.612s\psi & 0.612c\psi + 0.353s\psi & 0.707\,11
\end{bmatrix} \\
&=
\begin{bmatrix}
\cos\varphi & -\cos\theta\sin\varphi & \sin\theta\sin\varphi \\
\sin\varphi & \cos\theta\cos\varphi & -\cos\varphi\sin\theta \\
0 & \sin\theta & \cos\theta
\end{bmatrix}
\end{aligned}
\tag{6.266}
$$

The elements $(3, 1)$ on both sides make an equation to find ψ.

$$
0.353\,55\cos\psi - 0.612\,37\sin\psi = 0 \tag{6.267}
$$

Therefore, it is also possible to find ψ from the following equation:

$$
\psi = \text{atan2}\left(\frac{0.353\,55}{0.612\,37}\right) = 0.523\,\text{rad} = 30\,\text{deg} \tag{6.268}
$$

Finally, θ can be found using elements $(3, 2)$ and $(3, 3)$

$$
0.612\,37\cos\psi + 0.353\,55\sin\psi = \sin\theta \tag{6.269}
$$

$$
0.707\,11 = \cos\theta \tag{6.270}
$$

which gives θ.

$$
\theta = \text{atan2}\frac{0.707\,11}{0.707\,11} = 1\,\text{rad} = 45\,\text{deg} \tag{6.271}
$$

Example 219 ★ Inverse kinematics and non-standard DH frames. The inverse transformation method may equally applied on kinematics of robots, multibody kinematics, with any standard in their forward kinematics. Here the method will be applied on forward kinematics based on non-standard DH method.

Consider a 3 DOF planar manipulator shown in Fig. 5.5. The non-standard DH transformation matrices of the manipulator are:

$$
^0 T_1 =
\begin{bmatrix}
\cos\theta_1 & -\sin\theta_1 & 0 & 0 \\
\sin\theta_1 & \cos\theta_1 & 0 & 0 \\
0 & 0 & 1 & 0 \\
0 & 0 & 0 & 1
\end{bmatrix}
\tag{6.272}
$$

$$
^1 T_2 =
\begin{bmatrix}
\cos\theta_2 & -\sin\theta_2 & 0 & l_1 \\
\sin\theta_2 & \cos\theta_2 & 0 & 0 \\
0 & 0 & 1 & 0 \\
0 & 0 & 0 & 1
\end{bmatrix}
\tag{6.273}
$$

$$
^2 T_3 =
\begin{bmatrix}
\cos\theta_3 & -\sin\theta_3 & 0 & l_2 \\
\sin\theta_3 & \cos\theta_3 & 0 & 0 \\
0 & 0 & 1 & 0 \\
0 & 0 & 0 & 1
\end{bmatrix}
\tag{6.274}
$$

$$^3T_4 = \begin{bmatrix} 1 & 0 & 0 & l_3 \\ 0 & 1 & 0 & 0 \\ 0 & 0 & 1 & 0 \\ 0 & 0 & 0 & 1 \end{bmatrix} \tag{6.275}$$

The solution of the inverse kinematics problem is a mathematical problem and none of the standard or non-standard DH methods for defining link frames provides any simplicity. To calculate the inverse kinematics, we begin with calculating the forward kinematics transformation matrix $^0T_4$,

$$^0T_4 = {}^0T_1 \, {}^1T_2 \, {}^2T_3 \, {}^3T_4 \tag{6.276}$$

$$= \begin{bmatrix} \cos\theta_{123} & -\sin\theta_{123} & 0 & l_1\cos\theta_1 + l_2\cos\theta_{12} + l_3\cos\theta_{123} \\ \sin\theta_{123} & \cos\theta_{123} & 0 & l_1\sin\theta_1 + l_2\sin\theta_{12} + l_3\sin\theta_{123} \\ 0 & 0 & 1 & 0 \\ 0 & 0 & 0 & 1 \end{bmatrix}$$

$$= \begin{bmatrix} r_{11} & r_{12} & r_{13} & r_{14} \\ r_{21} & r_{22} & r_{23} & r_{24} \\ r_{31} & r_{32} & r_{33} & r_{34} \\ 0 & 0 & 0 & 1 \end{bmatrix}$$

where we used the following short notation to simplify the equation.

$$\theta_{ijk} = \theta_i + \theta_j + \theta_k \tag{6.277}$$

Examining the matrix $^0T_4$ indicates that:

$$\theta_{123} = \text{atan2}\,(r_{21}, r_{11}) \tag{6.278}$$

The next equation

$$^0T_4 \, {}^3T_4^{-1} = {}^0T_1 \, {}^1T_2 \, {}^2T_3 \tag{6.279}$$

$$\begin{bmatrix} r_{11} & r_{12} & 0 & r_{14} - l_3 r_{11} \\ r_{21} & r_{22} & 0 & r_{24} - l_3 r_{21} \\ 0 & 0 & 1 & 0 \\ 0 & 0 & 0 & 1 \end{bmatrix} = \begin{bmatrix} c\theta_{123} & -s\theta_{123} & 0 & l_1 c\theta_1 + l_2 c\theta_{12} \\ s\theta_{123} & c\theta_{123} & 0 & l_1 s\theta_1 + l_2 s\theta_{12} \\ 0 & 0 & 1 & 0 \\ 0 & 0 & 0 & 1 \end{bmatrix}$$

shows that:

$$\theta_2 = \arccos \frac{f_1^2 + f_2^2 - l_1^2 - l_2^2}{2l_1 l_2} \tag{6.280}$$

$$\theta_1 = \text{atan2}\,(f_2 f_3 - f_1 f_4, \; f_1 f_3 + f_2 f_4) \tag{6.281}$$

where

$$f_1 = r_{14} - l_3 r_{11} = c\theta_1 \,(l_2 c\theta_2 + l_1) - s\theta_1 \,(l_2 s\theta_2)$$
$$= f_3 \, c\theta_1 - f_4 \, s\theta_1 \tag{6.282}$$

$$f_2 = r_{24} - l_3 r_{21} = s\theta_1 \,(l_2 c\theta_2 + l_1) + c\theta_1 \,(l_2 s\theta_2)$$
$$= f_3 \, s\theta_1 + f_4 \, c\theta_1 \tag{6.283}$$

Finally, the angle θ_3 will be:

$$\theta_3 = \theta_{123} - \theta_1 - \theta_2 \tag{6.284}$$

6.3   ★ Iterative Technique

The inverse kinematics problem of robots can be interpreted as searching for the unknowns q_k of a set of nonlinear algebraic equations

$$^0T_n = \mathbf{T}(\mathbf{q}) \tag{6.285}$$

$$= {}^0T_1(q_1)\,{}^1T_2(q_2)\,{}^2T_3(q_3)\,{}^3T_4(q_4)\,\cdots\,{}^{n-1}T_n(q_n)$$

$$= \begin{bmatrix} r_{11} & r_{12} & r_{13} & r_{14} \\ r_{21} & r_{22} & r_{23} & r_{24} \\ r_{31} & r_{32} & r_{33} & r_{34} \\ 0 & 0 & 0 & 1 \end{bmatrix}$$

or

$$r_{ij} = r_{ij}(q_k) \qquad k = 1, 2, \cdots n \tag{6.286}$$

where n is the number of degree of freedom (DOF) of the robot. Assuming $n = 6$, maximum 6 out of the 12 equations of (6.285) are independent and can be utilized to solve for joint variables q_k. The functions $\mathbf{T}(\mathbf{q})$ are given explicitly based on forward kinematic analysis.

Numerous methods are available to find the zeros of Eq. (6.285). However, the methods are, in general, *iterative*. The most common method is known as the *Newton-Raphson method*. The iteration technique can be set in an algorithm.

Algorithm 6.1 *Inverse kinematics iteration technique.*

1. *Set the initial counter $i = 0$.*
2. *Find or guess an initial estimate $\mathbf{q}^{(0)}$.*
3. *Calculate the residue $\delta\mathbf{T}(\mathbf{q}^{(i)}) = \mathbf{J}(\mathbf{q}^{(i)})\,\delta\mathbf{q}^{(i)}$.*
   If every element of $\mathbf{T}(\mathbf{q}^{(i)})$ or its norm $\left\|\mathbf{T}(\mathbf{q}^{(i)})\right\|$ is less than a tolerance, $\left\|\mathbf{T}(\mathbf{q}^{(i)})\right\| < \epsilon$ then terminate the iteration. The $\mathbf{q}^{(i)}$ is the desired solution.
4. *Calculate $\mathbf{q}^{(i+1)} = \mathbf{q}^{(i)} + \mathbf{J}^{-1}(\mathbf{q}^{(i)})\,\delta\mathbf{T}(\mathbf{q}^{(i)})$.*
5. *Set $i = i + 1$ and return to step 3.*

The tolerance ϵ can equivalently be set up on variables

$$\mathbf{q}^{(i+1)} - \mathbf{q}^{(i)} < \epsilon \tag{6.287}$$

or on Jacobian $\mathbf{J}$.

$$\mathbf{J} - \mathbf{I} < \epsilon \tag{6.288}$$

$$\mathbf{J}(\mathbf{q}) = \left[\frac{\partial T_i}{\partial q_j}\right] \tag{6.289}$$

Proof To solve the kinematic equations for variables $\mathbf{q}$ by *iterative technique*,

$$\mathbf{T}(\mathbf{q}) = 0 \tag{6.290}$$

we begin with an initial guess $\mathbf{q}^\star$ for the joint variables $\mathbf{q}$.

$$\mathbf{q}^\star = \mathbf{q} + \delta\mathbf{q} \tag{6.291}$$

Using the forward kinematics, we determine the configuration of the end-effector frame for the guessed joint variables.

$$\mathbf{T}^\star = \mathbf{T}(\mathbf{q}^\star) \tag{6.292}$$

The difference between the calculated configuration with the forward kinematics $\mathbf{T}^{\star}$ and the desired configuration $\mathbf{T}$ represents an *error*, called *residue* $\delta\mathbf{T}$, which must be minimized.

$$\delta\mathbf{T} = \mathbf{T} - \mathbf{T}^{\star} \tag{6.293}$$

A first order Taylor expansion of the set of equations is:

$$\mathbf{T} = \mathbf{T}(\mathbf{q}^{\star} + \delta\mathbf{q}) = \mathbf{T}(\mathbf{q}^{\star}) + \frac{\partial\mathbf{T}}{\partial\mathbf{q}}\delta\mathbf{q} + O(\delta\mathbf{q}^2) \tag{6.294}$$

Assuming $\delta\mathbf{q} \ll \mathbf{I}$ allows us to work with a set of linear equations,

$$\delta\mathbf{T} = \mathbf{J}\,\delta\mathbf{q} \tag{6.295}$$

where $\mathbf{J}$ is the Jacobian matrix of the set of equations

$$\mathbf{J}(\mathbf{q}) = \left[\frac{\partial T_i}{\partial q_j}\right] \tag{6.296}$$

that implies

$$\delta\mathbf{q} = \mathbf{J}^{-1}\,\delta\mathbf{T} \tag{6.297}$$

Therefore, the unknown variables $\mathbf{q}$ are:

$$\mathbf{q} = \mathbf{q}^{\star} + \mathbf{J}^{-1}\,\delta\mathbf{T} \tag{6.298}$$

We may use the values obtained by (6.298) as a new approximation to repeat the calculations and find newer values. Repeating the methods can be summarized in the following iterative equation to converge to the exact value of the variables.

$$\mathbf{q}^{(i+1)} = \mathbf{q}^{(i)} + \mathbf{J}^{-1}(\mathbf{q}^{(i)})\,\delta\mathbf{T}(\mathbf{q}^{(i)}) \tag{6.299}$$

The main disadvantage of the iteration method is:

1− A particular solution depends on how closeness of the guess solution.
2− It cannot determine the total number of possible solutions nor can determine all of them.
3− There is no guarantee that the solution is the one matches with the current configuration of the robot.∎

Example 220 ★ Inverse kinematics for a $2R$ planar manipulator. Iterative method of inverse kinematics is based on forward kinematics, and not a depended on the type of a robot. An example on a simple robot illustrates the method which can be equally applied on any type of robots.

The position of tip point of a $2R$ planar manipulator is calculated in Example 207 as:

$$\begin{bmatrix} X \\ Y \end{bmatrix} = \begin{bmatrix} l_1\cos\theta_1 + l_2\cos(\theta_1 + \theta_2) \\ l_1 s\sin\theta_1 + l_2\sin(\theta_1 + \theta_2) \end{bmatrix} \tag{6.300}$$

To solve the inverse kinematics of the manipulator and find the joint coordinates for a known position of the tip point, we define $\mathbf{q}$, and $\mathbf{T}$.

$$\mathbf{q} = \begin{bmatrix} \theta_1 \\ \theta_2 \end{bmatrix} \qquad \mathbf{T} = \begin{bmatrix} X \\ Y \end{bmatrix} \tag{6.301}$$

Therefore, the Jacobian of the equations is:

$$\mathbf{J}(\mathbf{q}) = \left[\frac{\partial T_i}{\partial q_j}\right] = \begin{bmatrix} \dfrac{\partial X}{\partial\theta_1} & \dfrac{\partial X}{\partial\theta_2} \\ \dfrac{\partial Y}{\partial\theta_1} & \dfrac{\partial Y}{\partial\theta_2} \end{bmatrix}$$

$$= \begin{bmatrix} -l_1\sin\theta_1 - l_2\sin(\theta_1 + \theta_2) & -l_2\sin(\theta_1 + \theta_2) \\ l_1\cos\theta_1 + l_2\cos(\theta_1 + \theta_2) & l_2\cos(\theta_1 + \theta_2) \end{bmatrix} \tag{6.302}$$

The inverse of the Jacobian is

$$\mathbf{J}^{-1} = \frac{-1}{l_1 l_2 s \theta_2} \begin{bmatrix} -l_2 c \left(\theta_1 + \theta_2\right) & -l_2 s \left(\theta_1 + \theta_2\right) \\ l_1 c \theta_1 + l_2 c \left(\theta_1 + \theta_2\right) & l_1 s \theta_1 + l_2 s \left(\theta_1 + \theta_2\right) \end{bmatrix} \tag{6.303}$$

and therefore, the iterative formula (6.299) is set up as:

$$\begin{bmatrix} \theta_1 \\ \theta_2 \end{bmatrix}^{(i+1)} = \begin{bmatrix} \theta_1 \\ \theta_2 \end{bmatrix}^{(i)} + \mathbf{J}^{-1} \left(\begin{bmatrix} X \\ Y \end{bmatrix} - \begin{bmatrix} X \\ Y \end{bmatrix}^{(i)} \right) \tag{6.304}$$

Let us assume

$$l_1 = l_2 = 1 \tag{6.305}$$

$$\mathbf{T} = \begin{bmatrix} X \\ Y \end{bmatrix} = \begin{bmatrix} 1 \\ 1 \end{bmatrix} \tag{6.306}$$

and start from a guess value $\mathbf{q}^{(0)}$,

$$\mathbf{q}^{(0)} = \begin{bmatrix} \theta_1 \\ \theta_2 \end{bmatrix}^{(0)} = \begin{bmatrix} \pi/3 \\ -\pi/3 \end{bmatrix} \tag{6.307}$$

for which:

$$\delta \mathbf{T} = \begin{bmatrix} 1 \\ 1 \end{bmatrix} - \begin{bmatrix} \cos \pi/3 + \cos \left(\pi/3 + -\pi/3\right) \\ \sin \pi/3 + \sin \left(\pi/3 + -\pi/3\right) \end{bmatrix}$$

$$= \begin{bmatrix} 1 \\ 1 \end{bmatrix} - \begin{bmatrix} \frac{3}{2} \\ \frac{1}{2}\sqrt{3} \end{bmatrix} = \begin{bmatrix} -\frac{1}{2} \\ -\frac{1}{2}\sqrt{3} + 1 \end{bmatrix} \tag{6.308}$$

The Jacobian and its inverse for these values are

$$\mathbf{J} = \begin{bmatrix} -\frac{1}{2}\sqrt{3} & 0 \\ \frac{3}{2} & 1 \end{bmatrix} \qquad \mathbf{J}^{-1} = \begin{bmatrix} -\frac{2}{3}\sqrt{3} & 0 \\ \sqrt{3} & 1 \end{bmatrix} \tag{6.309}$$

and therefore,

$$\begin{bmatrix} \theta_1 \\ \theta_2 \end{bmatrix}^{(1)} = \begin{bmatrix} \theta_1 \\ \theta_2 \end{bmatrix}^{(0)} + \mathbf{J}^{-1} \delta \mathbf{T}$$

$$= \begin{bmatrix} \pi/3 \\ -\pi/3 \end{bmatrix} + \begin{bmatrix} -\frac{2}{3}\sqrt{3} & 0 \\ \sqrt{3} & 1 \end{bmatrix} \begin{bmatrix} -\frac{1}{2} \\ -\frac{1}{2}\sqrt{3} + 1 \end{bmatrix}$$

$$= \begin{bmatrix} 1.624\,5 \\ -1.779\,2 \end{bmatrix} \tag{6.310}$$

Based on the iterative technique, we can find the following values and find the solution in a few iterations.

Iteration 1.

$$\mathbf{J} = \begin{bmatrix} -\frac{1}{2}\sqrt{3} & 0 \\ \frac{3}{2} & 1 \end{bmatrix} \tag{6.311}$$

$$\delta \mathbf{T} = \begin{bmatrix} -\frac{1}{2} \\ -\frac{1}{2}\sqrt{3} + 1 \end{bmatrix} \tag{6.312}$$

$$\mathbf{q}^{(1)} = \begin{bmatrix} 1.624\,5 \\ -1.779\,2 \end{bmatrix} \tag{6.313}$$

Iteration 2.

$$\mathbf{J} = \begin{bmatrix} -0.844 & 0.154 \\ 0.934 & 0.988 \end{bmatrix} \tag{6.314}$$

$$\delta\mathbf{T} = \begin{bmatrix} 6.516 \times 10^{-2} \\ 0.155\,53 \end{bmatrix} \tag{6.315}$$

$$\mathbf{q}^{(2)} = \begin{bmatrix} 1.583 \\ -1.582 \end{bmatrix} \tag{6.316}$$

Iteration 3.

$$\mathbf{J} = \begin{bmatrix} -1.00 & -.433 \times 10^{-3} \\ .988 & .999 \end{bmatrix} \tag{6.317}$$

$$\delta\mathbf{T} = \begin{bmatrix} .119 \times 10^{-1} \\ -.362 \times 10^{-3} \end{bmatrix} \tag{6.318}$$

$$\mathbf{q}^{(3)} = \begin{bmatrix} 1.570795886 \\ -1.570867014 \end{bmatrix} \tag{6.319}$$

Iteration 4.

$$\mathbf{J} = \begin{bmatrix} -1.000 & 0.0 \\ 0.998\,50 & 1.0 \end{bmatrix} \tag{6.320}$$

$$\delta\mathbf{T} = \begin{bmatrix} -.438 \times 10^{-6} \\ .711 \times 10^{-4} \end{bmatrix} \tag{6.321}$$

$$\mathbf{q}^{(4)} = \begin{bmatrix} 1.570796329 \\ -1.570796329 \end{bmatrix} \tag{6.322}$$

The result of the fourth iteration $\mathbf{q}^{(4)}$ is close enough to the exact value $\mathbf{q} = \begin{bmatrix} \pi/2 & -\pi/2 \end{bmatrix}^T$.

Example 221 ★ Iteration technique and *n-m* relationship. A discussion on applicability and shortening of the iteration method.

1− Iteration method when $n = m$.

When the number of joint variables n is equal to the number of independent equations generated in forward kinematics m, then provided that the Jacobian matrix remains non-singular, the linearized equation

$$\delta\mathbf{T} = \mathbf{J}\,\delta\mathbf{q} \tag{6.323}$$

has a unique set of solutions and therefore, the Newton-Raphson technique may be utilized to solve the inverse kinematics problem.

The cost of the procedure depends on the number of iterations to be performed, which depends upon different parameters such as the distance between the estimated and effective solutions, and the condition number of the Jacobian matrix at the solution. Since the solution to the inverse kinematics problem is not unique, it may generate different configurations according to the choice of the estimated solution. No convergence may be observed if the initial estimate of the solution falls outside the convergence domain of the algorithm.

2− Iteration method when $n > m$.

When the number of joint variables n is more than the number of independent equations m, then the problem is an overdetermined case for which no solution exists in general because the number of joints is not enough to generate an arbitrary configuration for the end-effector. However, a solution can be generated, which minimizes the position error.

3− Iteration method when $n < m$.

When the number of joint variables n is less than the number of independent equations m, then the problem is a redundant case for which an infinite number of solutions are generally available.

6.4 ★ Comparison of the Inverse Kinematics Techniques

6.4.1 ★ Existence and Uniqueness of Solution

When the desired tool frame position $^0\mathbf{d}_7$ is outside the working space of a robot, there cannot be any real solution for the joint variables of the robot. Mathematically, in this condition, the resultant of the terms under square root signs would be negative. Furthermore, even if the tool frame position $^0\mathbf{d}_7$ is within the working space, there may be some tool orientations $^0\mathbf{R}_7$ that are not achievable without breaking joint constraints and violating one or more joint variable limits. Therefore, existing solutions for inverse kinematics problem generally depend on the geometric configuration of the robot.

In normal case when the number of joints is six, no DOF is redundant and the configuration assigned to the end-effector of the robot lies within the workspace, the inverse kinematics solution exists. The solutions are inherently multiple because of inverse trigonometric functions. Generally speaking, the different solutions correspond to possible configurations to reach the same end-effector configuration. However, physically, our robot will match with only one set of the solutions.

Generally speaking, when the solution of the inverse kinematics of a robot exists, they are not unique. Multiple solutions appear because a robot can reach a point within the working space in different configurations. Every set of solutions is usually associated with a possible configuration. The elbow up and elbow down configuration of the $2R$ manipulator in Example 207 is a simple example.

The multiplicity of the solution depends on the number of joints of the manipulator and their type. The fact that a manipulator has multiple solutions may cause problems since the system has to be able to select one of them. The criteria on which to make a decision may vary. A reasonable choice is to choose the closest solution to the current configuration.

When the number of joints is less than six, no solution exists unless freedom is reduced at the same time in the task space, for example, by constraining the tool orientation to certain directions. When the number of joints exceeds six, the structure becomes redundant and an infinite number of solutions exists to reach the same end-effector configuration within the robot workspace. Redundancy of the robot architecture is an interesting feature for systems installed in a highly constrained environment. From the kinematic point of view, the difficulty lies in formulating the environment constraints in mathematical form, to ensure the uniqueness of the solution to the inverse kinematic problem.

6.4.2 ★ Inverse Kinematics Techniques

The inverse kinematics problem of robots can be solved by several methods, such as *decoupling*, *inverse transformation*, *iterative*, *screw algebra*, *dual matrices*, *quaternions*, and *geometric techniques*. The decoupling and inverse transform technique using 4×4 homogenous transformation matrices suffers from the fact that the solution does not clearly indicate how to select the appropriate solution from multiple possible solutions for a particular configuration. Thus, these techniques rely on the skills and intuition of the engineer. The iterative solution method often requires a vast amount of computation and moreover, it does not guarantee convergence to the correct solution. It is especially weak when the robot is close to a singular and degenerate configurations. The iterative solution method also lacks a method for selecting the appropriate solution from multiple possible solutions.

Although the set of nonlinear trigonometric equations is typically not possible to be solved analytically, there are some robot structures that are *solvable* analytically. The sufficient condition of solvability is when a 6 DOF robot has three consecutive revolute joints with axes intersecting at one point. The other property of inverse kinematics is ambiguity of a solution at singular points. However, when closed-form solutions to the arm equation can be found, they are seldom unique.

6.5 ★ Singular Configuration

For any robot, redundant or not, it is possible to discover some configurations, called *singular configurations*, at which the number of DOF of the end-effector is inferior to the dimension in which it generally operates. Singular configurations happen when:

1— Two axes of prismatic joints become parallel
2— Two axes of revolute joints become identical.

At singular positions, the end-effector loses one or more degrees of freedom, because the kinematic equations become linearly dependent or certain solutions become undefined. Singular positions must be avoided as the velocities required to move the end-effector become theoretically infinite. The singular configurations can be mathematically detected from the Jacobian matrix. The Jacobian matrix $\mathbf{J}$ relates the infinitesimal displacements of the end-effector $\delta\mathbf{X}$,

$$\delta\mathbf{X} = [\delta X_1, \cdots \delta X_m] \tag{6.324}$$

to the infinitesimal joint variables $\delta\mathbf{q}$,

$$\delta\mathbf{q} = [\delta q_1, \cdots \delta q_n] \tag{6.325}$$

and has thus dimension $m \times n$, where n is the number of joints, and m is the number of end-effector DOF. When n is larger than m and $\mathbf{J}$ has full rank, then there are $m - n$ redundancies in the system to which $m - n$ arbitrary variables correspond.

The Jacobian matrix $\mathbf{J}$ also determines the relationship between end-effector velocities $\dot{\mathbf{X}}$ and joint velocities $\dot{\mathbf{q}}$.

$$\dot{\mathbf{X}} = \mathbf{J}\dot{\mathbf{q}} \tag{6.326}$$

This equation can be interpreted as a linear mapping from an m-dimensional vector space $\mathbf{X}$ to an n-dimensional vector space $\mathbf{q}$. The subspace $\mathbb{R}(\mathbf{J})$ is the *range space* of the linear mapping and represents all the possible end-effector velocities that can be generated by the n joints in the current configuration. $\mathbf{J}$ has full row-rank, which means that the system does not present any singularity in that configuration, then the range space $\mathbb{R}(\mathbf{J})$ covers the entire vector space $\mathbf{X}$. Otherwise, there exists at least one direction in which the end-effector cannot be moved.

The null space $\mathbb{N}(\mathbf{J})$ represents the solutions of $\mathbf{J}\dot{\mathbf{q}} = 0$. Therefore, any vector $\mathbf{q}^\cdot \in \mathbb{N}(\mathbf{J})$ does not generate any motion for the end-effector. If the manipulator has full rank, the dimension of the null space is then equal to the number $m - n$ of redundant DOF. When $\mathbf{J}$ is degenerate, the dimension of $\mathbb{R}(\mathbf{J})$ decreases and the dimension of the null space increases by the same amount. Therefore,

$$\dim \mathbb{R}(\mathbf{J}) + \dim \mathbb{N}(\mathbf{J}) = n \tag{6.327}$$

Configurations in which the Jacobian no longer has full rank corresponds to singularities of the robot, which are generally of two types:

1. *Workspace boundary singularities* are those occurring when the manipulator is fully stretched out or folded back on itself. In these cases, the end-effector is near or at the workspace boundary.
2. *Workspace interior singularities* are those occurring away from the boundary. In this case, generally two or more axes line up.

Mathematically, singular configurations can be found by calculating the conditions that make:

$$|\mathbf{J}| = 0 \tag{6.328}$$

or

$$\left|\mathbf{J}\mathbf{J}^T\right| = 0 \tag{6.329}$$

Identification and avoidance of singular configurations are important in robotics. Some of the main reasons are:

1. Certain directions of motion may be unattainable.
2. Some of the joint velocities are infinite.
3. Some of the joint torques are infinite.
4. There will not exist a unique solution to the inverse kinematics problem.

Detecting the singular configurations using the Jacobian determinant may be a tedious task for complex robots. However, for robots having a spherical wrist, it is possible to split the singularity detection problem into two separate problems:

1. Arm singularities resulting from the motion of the manipulator arms.
2. Wrist singularities resulting from the motion of the wrist.

6.6   Summary

Inverse kinematics refers to determining the joint variables of a robot for a given position and orientation of the end-effector frame. The forward kinematics of a 6 DOF robot generates a 4×4 transformation matrix

$$
{}^{0}T_6 = {}^{0}T_1 \, {}^{1}T_2 \, {}^{2}T_3 \, {}^{3}T_4 \, {}^{4}T_5 \, {}^{5}T_6
$$

$$
= \begin{bmatrix} {}^{0}R_6 & {}^{0}\mathbf{d}_6 \\ 0 & 1 \end{bmatrix} = \begin{bmatrix} r_{11} & r_{12} & r_{13} & r_{14} \\ r_{21} & r_{22} & r_{23} & r_{24} \\ r_{31} & r_{32} & r_{33} & r_{34} \\ 0 & 0 & 0 & 1 \end{bmatrix} \tag{6.330}
$$

where only six elements out of the 12 elements of ${}^{0}T_6$ are independent. Therefore, the inverse kinematics reduces to finding the six independent elements for a given ${}^{0}T_6$ matrix.

Decoupling, inverse transformation, and iterative techniques are three applied methods for solving the inverse kinematics problem. In decoupling technique, the inverse kinematics of a robot with a spherical wrist can be decoupled into two subproblems: inverse position and inverse orientation kinematics. Practically, the tools transformation matrix ${}^{0}T_7$ is decomposed into three submatrices ${}^{0}T_3$, ${}^{3}T_6$, and ${}^{6}T_7$.

$$
{}^{0}T_6 = {}^{0}T_3 \, {}^{3}T_6 \, {}^{6}T_7 \tag{6.331}
$$

The matrix ${}^{0}T_3$ positions the wrist point and depends on the three manipulator joints' variables. The matrix ${}^{3}T_6$ is the wrist transformation matrix and the ${}^{6}T_7$ is the tools transformation matrix.

In inverse transformation technique, we extract equations with only one unknown from the following matrix equations, step by step.

$$
{}^{1}T_6 = {}^{0}T_1^{-1} \, {}^{0}T_6 \tag{6.332}
$$

$$
{}^{2}T_6 = {}^{1}T_2^{-1} \, {}^{0}T_1^{-1} \, {}^{0}T_6 \tag{6.333}
$$

$$
{}^{3}T_6 = {}^{2}T_3^{-1} \, {}^{1}T_2^{-1} \, {}^{0}T_1^{-1} \, {}^{0}T_6 \tag{6.334}
$$

$$
{}^{4}T_6 = {}^{3}T_4^{-1} \, {}^{2}T_3^{-1} \, {}^{1}T_2^{-1} \, {}^{0}T_1^{-1} \, {}^{0}T_6 \tag{6.335}
$$

$$
{}^{5}T_6 = {}^{4}T_5^{-1} \, {}^{3}T_4^{-1} \, {}^{2}T_3^{-1} \, {}^{1}T_2^{-1} \, {}^{0}T_1^{-1} \, {}^{0}T_6 \tag{6.336}
$$

$$
\mathbf{I} = {}^{5}T_6^{-1} \, {}^{4}T_5^{-1} \, {}^{3}T_4^{-1} \, {}^{2}T_3^{-1} \, {}^{1}T_2^{-1} \, {}^{0}T_1^{-1} \, {}^{0}T_6 \tag{6.337}
$$

The iterative technique is a numerical method seeking to find the joint variable vector $\mathbf{q}$ for a set of equations $\mathbf{T}(\mathbf{q}) = 0$.

6.7   Key Symbols

0	Null vector
a, b, c	Coefficients of trigonometric equation
a	Turn vector of end-effector frame
A	Local rotation transformation matrix
B	Body coordinate frame
c	cos
d	Joint distance
d_x, d_y, d_z	Elements of **d**
d	Translation vector, displacement vector
d$_{wrist}$	Wrist position vector
D	Displacement transformation matrix
DH	Denavit–Hartenberg
DOF	Degree of freedom
f_{ij}	The element of row i and column j of a matrix
g_{ij}	The element of row i and column j of a matrix
G, B_0	Global coordinate frame, Base coordinate frame
$\mathbf{I} = [I]$	Identity matrix
J	Jacobian
l	Length
m	Number of independent equations
n	Number of links of a robot, number of joint variables
P	Point
r, ϕ	Parameters of trigonometric equation
r	Position vectors, homogenous position vector
q	Joint variable
q	Joint variables vector
r_i	The element i of **r**
r_{ij}	The element of row i and column j of a matrix
R	Rotation transformation matrix
s	sin
sgn	Signum function
$SSRMS$	Space station remote manipulator system
T	Homogenous transformation matrix
T_{arm}	Manipulator transformation matrix
T_{wrist}	Wrist transformation matrix
T	A set of nonlinear algebraic equations of **q**
x, y, z	Local coordinate axes, local coordinates
X, Y, Z	Global coordinate axes, global coordinates

Greek

δ	Kronecker function, small increment of a parameter
ϵ	Small test number to terminate a procedure
θ	Rotary joint angle
θ_{ijk}	$\theta_i + \theta_j + \theta_k$

Symbol

$[\ ]^{-1}$	Inverse of the matrix $[\ ]$
$[\ ]^T$	Transpose of the matrix $[\ ]$
$\equiv$	Equivalent
$\vdash$	Orthogonal
(i)	Link number i

$\parallel$	Parallel sign
$\perp$	Perpendicular
$\times$	Vector cross product
$\mathbf{q}^{\star}$	A guess value for $\mathbf{q}$
dim	Dimension
$\mathbb{N}$	Null space
$\mathbb{R}$	Range space

Exercises

1. Notation and symbols.
   Describe the meaning of:
   $$a - \text{atan2}\,(a, b) \quad b - {}^0 T_n \quad c - \mathbf{T(q)} \quad d - \mathbf{q} \quad e - \mathbf{J}$$

2. $3R$ planar manipulator inverse kinematics.
   Figure 5.5 illustrates an R∥R∥R planar manipulator. The forward kinematics of the manipulator generates the following matrices. Solve the inverse kinematics and find $\theta_1, \theta_2, \theta_3$ for given coordinates x_0, y_0 of the tip point and a given value of φ.

$$
{}^2 T_3 = \begin{bmatrix} \cos\theta_3 & -\sin\theta_3 & 0 & l_3 \cos\theta_3 \\ \sin\theta_3 & \cos\theta_3 & 0 & l_3 \sin\theta_3 \\ 0 & 0 & 1 & 0 \\ 0 & 0 & 0 & 1 \end{bmatrix} \tag{6.338}
$$

$$
{}^1 T_2 = \begin{bmatrix} \cos\theta_2 & -\sin\theta_2 & 0 & l_2 \cos\theta_2 \\ \sin\theta_2 & \cos\theta_2 & 0 & l_2 \sin\theta_2 \\ 0 & 0 & 1 & 0 \\ 0 & 0 & 0 & 1 \end{bmatrix} \tag{6.339}
$$

$$
{}^0 T_1 = \begin{bmatrix} \cos\theta_1 & -\sin\theta_1 & 0 & l_1 \cos\theta_1 \\ \sin\theta_1 & \cos\theta_1 & 0 & l_1 \sin\theta_1 \\ 0 & 0 & 1 & 0 \\ 0 & 0 & 0 & 1 \end{bmatrix} \tag{6.340}
$$

3. $2R$ manipulator tip point on a horizontal path.
   Consider an elbow up planar $2R$ manipulator with $l_1 = l_2 = 1$. The tip point is moving on a straight line from $P_1\,(1, 1.5)$ to $P_2\,(-1, 1.5)$.
   (a) Divide the Cartesian path in 10 equal sections and determine the joint variables at the 11 points.
   (b) ★ Calculate the joint variable θ_1 at P_1 and at P_2. Divide the range of θ_1 into 10 equal sections and determine the coordinates of the tip point and the angle θ_2 at the 11 values of θ_1.
   (c) ★ Calculate the joint variable θ_2 at P_1 and at P_2. Divide the range of θ_1 into 10 equal sections and determine the coordinates of the tip point and the angle θ_1 at the 11 values of θ_2.

4. $2R$ manipulator tip point on a tilted path.
   Consider an elbow up planar $2R$ manipulator with $l_1 = l_2 = 1$. The tip point is moving on a straight line from $P_1\,(1, 1.5)$ to $P_2\,(-1, 1)$.
   (a) Divide the Cartesian path in 10 equal sections and determine the joint variables at the 11 points.
   (b) Calculate the joint variable θ_1 at P_1 and at P_2. Divide the range of θ_1 into 10 equal sections and determine the coordinates of the tip point and the angle θ_2 at the 11 values of θ_1.
   (c) Calculate the joint variable θ_2 at P_1 and at P_2. Divide the range of θ_1 into 10 equal sections and determine the coordinates of the tip point and the angle θ_1 at the 11 values of θ_2.

5. $2R$ manipulator motion on a horizontal path.
   Consider an elbow up planar $2R$ manipulator with $l_1 = l_2 = 1$. The tip point is moving on a straight line from $P_1\,(1, 1.5)$ to $P_2\,(-1, 1.5)$ according to the following functions of time.

$$X = 1 - 6t^2 + 4t^3 \qquad Y = 1.5 \tag{6.341}$$

   (a) Calculate and plot θ_1 and θ_2 as functions of time if the time of motion is $0 \le t \le 1$.
   (b) ★ Calculate and plot $\dot{\theta}_1$ and $\dot{\theta}_2$ as functions of time.
   (c) ★ Calculate and plot $\ddot{\theta}_1$ and $\ddot{\theta}_2$ as functions of time.
   (d) ★ Calculate and plot $\dddot{\theta}_1$ and $\dddot{\theta}_2$ as functions of time.

6. A planar manipulator.
   Figure 6.10 illustrates a three DOF planar manipulator.
   (a) Determine the transformation matrices between coordinate frames.

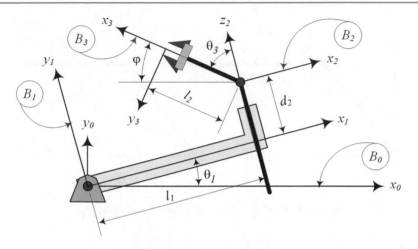

Fig. 6.10 A planar manipulator

(b) Solve the forward kinematics and determine the coordinates X, Y, and φ of the end-effector frame B_3 for a given set of joint variables θ_1, d_2, θ_3.

(c) Solve the inverse kinematics and determine the joint variables θ_1, d_2, θ_3 for a given set of end-effector coordinates X, Y, and φ.

7. $2R$ manipulator motion on a horizontal path.

Consider a planar elbow up $2R$ manipulator with $l_1 = l_2 = 1$. The tip point is moving on a straight line from P_1 $(1, 1.5)$ to P_2 $(-1, 1.5)$ with a constant speed.

$$X = 1 - vt \qquad Y = 1.5 \tag{6.342}$$

(a) Calculate v and plot θ_1 and θ_2 if the time of motion is $0 \leq t \leq 1$.

(b) Calculate v and plot θ_1 and θ_2 if the time of motion is $0 \leq t \leq 5$.

(c) Calculate v and plot θ_1 and θ_2 if the time of motion is $0 \leq t \leq 10$.

(d) ★ Plot θ_1 and θ_2 as functions of v at point $(0, 1.5)$.

8. $2R$ manipulator motion on a horizontal path.

Consider a planar elbow up $2R$ manipulator with $l_1 = l_2 = 1$. The tip point is moving on a straight line from P_1 $(1, 1.5)$ to P_2 $(-1, 1.5)$ with a constant acceleration.

$$X = 1 - \frac{1}{2}at^2 \qquad Y = 1.5 \tag{6.343}$$

(a) Calculate a and plot θ_1 and θ_2 if the time of motion is $0 \leq t \leq 1$.

(b) Calculate a and plot θ_1 and θ_2 if the time of motion is $0 \leq t \leq 5$.

(c) Calculate a and plot θ_1 and θ_2 if the time of motion is $0 \leq t \leq 10$.

(d) ★ Plot θ_1 and θ_2 as functions of a at point $(0, 1.5)$.

9. ★ $2R$ manipulator kinematics on a tilted path.

Consider a planar elbow up $2R$ manipulator with $l_1 = l_2 = 1$. The tip point is moving on a straight line from P_1 $(1, 1.5)$ to P_2 $(-1, 1.5)$ with a constant speed.

$$X = 1 - vt \qquad Y = 1.5 \tag{6.344}$$

(a) Calculate and plot θ_1 and θ_2 if the time of motion is $0 \leq t \leq 1$.

(b) Calculate and plot $\dot{\theta}_1$ and $\dot{\theta}_2$ as functions of time.

(c) Calculate and plot $\ddot{\theta}_1$ and $\ddot{\theta}_2$ as functions of time.

(d) Calculate and plot $\dddot{\theta}_1$ and $\dddot{\theta}_2$ as functions of time.

10. Acceptable lengths of a $2R$ planar manipulator.

The tip point of a $2R$ planar manipulator is at $(1, 1.1)$.

(a) Assume $l_1 = 1$. Plot θ_1 and θ_2 versus l_2 and determine the range of possible l_2 for elbow up configuration.

(b) Assume $l_2 = 1$. Plot θ_1 and θ_2 versus l_1 and determine the range of possible l_1 for elbow up configuration.

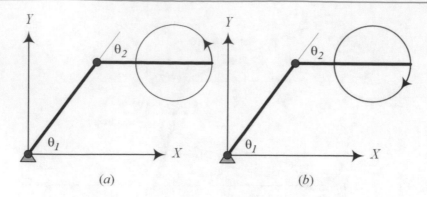

Fig. 6.11 An elbow up $2R$ manipulator on a circular path

(c) Assume $l_1 = 1$. Plot θ_1 and θ_2 versus l_2 and determine the range of possible l_2 for elbow down configuration.

(d) Assume $l_2 = 1$. Plot θ_1 and θ_2 versus l_1 and determine the range of possible l_1 for elbow down configuration.

11. $3R$ manipulator tip point on a straight path.

Consider a $3R$ articulated manipulator such as Fig. 6.5 with $l_1 = 0.5, l_2 = l_3 = 1$. The tip point is moving on a straight line from $P_1\,(1.5, 0, 1)$ to $P_2\,(-1, 1, 1.5)$.

(a) Divide the Cartesian path into 10 equal sections and determine the joint variables at the 11 points.

(b) Calculate the joint variable θ_1 at P_1 and at P_2. Divide the range of θ_1 into 10 equal sections and determine the coordinates of the tip point at the 11 values of θ_2 and θ_3.

(c) Calculate the joint variable θ_2 at P_1 and at P_2. Divide the range of θ_2 into 10 equal sections and determine the coordinates of the tip point at the 11 values of θ_1 and θ_3.

(d) Calculate the joint variable θ_3 at P_1 and at P_2. Divide the range of θ_3 into 10 equal sections and determine the coordinates of the tip point at the 11 values of θ_1 and θ_2.

12. $3R$ manipulator motion on a straight path.

Consider a $3R$ articulated manipulator such as Fig. 6.5 with $l_1 = 0.5, l_2 = l_3 = 1$. The tip point is moving on a straight line from $P_1\,(1.5, 0, 1)$ to $P_2\,(-1, 1, 1.5)$ according to the following functions of time.

$$X = 1.5 - 0.025t^3 + 0.00375t^4 - 0.00015t^5$$

$$Y = 0.01t^3 - 0.0015t^4 + 0.00006t^5 \qquad (6.345)$$

$$Z = 1 + 0.005t^3 - 0.00075t^4 + 0.00003t^5$$

(a) Calculate and plot θ_1, θ_2, and θ_3 if the time of motion is $0 \le t \le 1$.

(b) ★ Calculate and plot $\dot{\theta}_1, \dot{\theta}_2$, and $\dot{\theta}_3$ as functions of time.

(c) ★ Calculate and plot $\ddot{\theta}_1, \ddot{\theta}_2$, and $\ddot{\theta}_3$ as functions of time.

(d) ★ Calculate and plot $\dddot{\theta}_1, \dddot{\theta}_2$, and $\dddot{\theta}_3$ as functions of time.

13. An elbow up $2R$ manipulator on a circular path.

The $2R$ manipulator of Fig. 6.11 has $l_2 = l_1 = 1$. The tip point of the manipulator is supposed to move on a circular path with a radius $R = 1/3$ and centered at $(1.267, 0.8)$. Assume the manipulator starts moving when the second link is horizontal.

(a) Plot θ_1 and θ_2 if the tip point is moving counterclockwise as shown in Fig. 6.11a.

(b) Plot θ_1 and θ_2 if the tip point is moving clockwise as shown in Fig. 6.11b.

14. Acceptable lengths of a $3R$ manipulator.

The tip point of a $3R$ articulated manipulator is at $(1, 1.1, 0.5)$.

(a) Assume $l_1 = l_3 = 1$. Plot θ_1, θ_2 and θ_3 versus l_2 and determine the range of possible l_2.

(b) Assume $l_2 = l_3 = 1$. Plot θ_1, θ_2 and θ_3 versus l_1 and determine the range of possible l_1.

(c) Assume $l_2 = l_1 = 1$. Plot θ_1, θ_2 and θ_3 versus l_3 and determine the range of possible l_3.

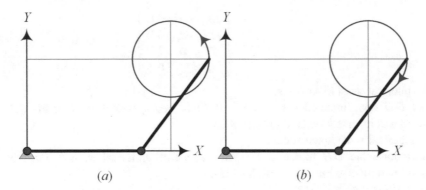

Fig. 6.12 An elbow down $2R$ manipulator on a circular path

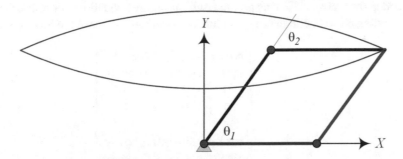

Fig. 6.13 A $2R$ manipulator on a circular path

15. An elbow down $2R$ manipulator on a circular path.

    The $2R$ manipulator of Fig. 6.12 has $l_2 = l_1 = 1$. The tip point of the manipulator is supposed to move on a circular path with a radius $R = 1/3$ and centered at $(1.267, 0.8)$. Assume the manipulator starts moving when the first link is horizontal.

    (a) Plot θ_1 and θ_2 if the tip point is moving counterclockwise as shown in Fig. 6.12a.

    (b) Plot θ_1 and θ_2 if the tip point is moving clockwise as shown in Fig. 6.12b.

16. A $2R$ manipulator on a circular path.

    The $2R$ manipulator of Fig. 6.13 has $l_2 = l_1 = 1$. The tip point of the manipulator is supposed to move on a circular path with a radius $R = 4$ and a center on Y-axis.

    (a) Assume the elbow up manipulator starts moving on the upper circular path when the second link is horizontal. Plot θ_1 and θ_2 until the first link becomes horizontal at the end of the path.

    (b) Assume the elbow down manipulator starts moving on the upper circular path when the first link is horizontal. Plot θ_1 and θ_2 until the first link becomes horizontal at the end of the path.

    (c) Assume the elbow up manipulator starts moving on the lower circular path when the second link is horizontal. Plot θ_1 and θ_2 until the first link becomes horizontal at the end of the path.

    (d) Assume the elbow down manipulator starts moving on the lower circular path when the first link is horizontal. Plot θ_1 and θ_2 until the first link becomes horizontal at the end of the path.

17. Spherical wrist inverse kinematics.

    Figure 5.30 illustrates a spherical wrist with following transformation matrices. Assume that the frame B_3 is the base frame. Solve the inverse kinematics and find $\theta_4, \theta_5, \theta_6$ for a given $^3T_6$.

$$^3T_4 = \begin{bmatrix} c\theta_4 & 0 & -s\theta_4 & 0 \\ s\theta_4 & 0 & c\theta_4 & 0 \\ 0 & -1 & 0 & 0 \\ 0 & 0 & 0 & 1 \end{bmatrix} \quad ^4T_5 = \begin{bmatrix} c\theta_5 & 0 & s\theta_5 & 0 \\ s\theta_5 & 0 & -c\theta_5 & 0 \\ 0 & 1 & 0 & 0 \\ 0 & 0 & 0 & 1 \end{bmatrix} \quad (6.346)$$

$$^5T_6 = \begin{bmatrix} c\theta_6 & -s\theta_6 & 0 & 0 \\ s\theta_6 & c\theta_6 & 0 & 0 \\ 0 & 0 & 1 & 0 \\ 0 & 0 & 0 & 1 \end{bmatrix} \tag{6.347}$$

18. ★ Roll-Pitch-Yaw spherical wrist kinematics.

   Attach the required DH coordinate frames to the Roll-Pitch-Yaw spherical wrist of Fig. 5.34, similar to 5.32, and determine the forward and inverse kinematics of the wrist.

19. ★ Pitch-Yaw-Roll spherical wrist kinematics.

   Attach the required coordinate DH frames to the Pitch-Yaw-Roll spherical wrist of Fig. 5.35, similar to 5.32, and determine the forward and inverse kinematics of the wrist.

20. $SCARA$ robot inverse kinematics.

   Consider the R∥R∥R∥P robot shown in Fig. 5.23 with the following transformation matrices. Solve the inverse kinematics and find θ_1, θ_2, θ_3, and d for a given $^0T_4$. This problem will have three equations for four unknowns. Hence, there are infinity solutions. Assuming $\theta_3 = 0$, eliminates the redundant freedom and make the inverse kinematics unique.

$$^0T_1 = \begin{bmatrix} \cos\theta_1 & -\sin\theta_1 & 0 & l_1\cos\theta_1 \\ \sin\theta_1 & \cos\theta_1 & 0 & l_1\sin\theta_1 \\ 0 & 0 & 1 & 0 \\ 0 & 0 & 0 & 1 \end{bmatrix} \tag{6.348}$$

$$^1T_2 = \begin{bmatrix} \cos\theta_2 & -\sin\theta_2 & 0 & l_2\cos\theta_2 \\ \sin\theta_2 & \cos\theta_2 & 0 & l_2\sin\theta_2 \\ 0 & 0 & 1 & 0 \\ 0 & 0 & 0 & 1 \end{bmatrix} \tag{6.349}$$

$$^2T_3 = \begin{bmatrix} \cos\theta_3 & -\sin\theta_3 & 0 & 0 \\ \sin\theta_3 & \cos\theta_3 & 0 & 0 \\ 0 & 0 & 1 & 0 \\ 0 & 0 & 0 & 1 \end{bmatrix} \qquad ^3T_4 = \begin{bmatrix} 1 & 0 & 0 & 0 \\ 0 & 1 & 0 & 0 \\ 0 & 0 & 1 & d \\ 0 & 0 & 0 & 1 \end{bmatrix} \tag{6.350}$$

$$^0T_4 = {^0T_1}\,{^1T_2}\,{^2T_3}\,{^3T_4}$$
$$= \begin{bmatrix} c\theta_{123} & -s\theta_{123} & 0 & l_1c\theta_1 + l_2c\theta_{12} \\ s\theta_{123} & c\theta_{123} & 0 & l_1s\theta_1 + l_2s\theta_{12} \\ 0 & 0 & 1 & d \\ 0 & 0 & 0 & 1 \end{bmatrix} \tag{6.351}$$
$$\theta_{123} = \theta_1 + \theta_2 + \theta_3 \qquad \theta_{12} = \theta_1 + \theta_2$$

21. R⊢R∥R articulated arm inverse kinematics.

   Figure 5.22 illustrates 3 DOF R⊢R∥R manipulator. Use the following transformation matrices and solve the inverse kinematics for $\theta_1, \theta_2, \theta_3$.

$$^0T_1 = \begin{bmatrix} \cos\theta_1 & 0 & -\sin\theta_1 & 0 \\ \sin\theta_1 & 0 & \cos\theta_1 & 0 \\ 0 & -1 & 0 & d_1 \\ 0 & 0 & 0 & 1 \end{bmatrix} \tag{6.352}$$

$$^1T_2 = \begin{bmatrix} \cos\theta_2 & -\sin\theta_2 & 0 & l_2\cos\theta_2 \\ \sin\theta_2 & \cos\theta_2 & 0 & l_2\sin\theta_2 \\ 0 & 0 & 1 & d_2 \\ 0 & 0 & 0 & 1 \end{bmatrix} \tag{6.353}$$

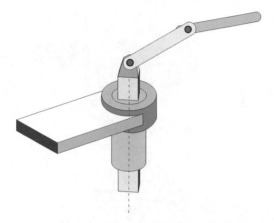

Fig. 6.14 A $PRRR$ manipulator

$$
^2T_3 = \begin{bmatrix} \cos\theta_3 & 0 & \sin\theta_3 & 0 \\ \sin\theta_3 & 0 & -\cos\theta_3 & 0 \\ 0 & 1 & 0 & 0 \\ 0 & 0 & 0 & 1 \end{bmatrix}
\tag{6.354}
$$

22. Kinematics of a $PRRR$ manipulator.

    A $PRRR$ manipulator is shown in Fig. 6.14.

    (a) Set up the links' coordinate frame according to standard DH rules.

    (b) Determine the class of each link.

    (c) Find the links' transformation matrices.

    (d) Calculate the forward kinematics of the manipulator.

    (e) Solve the inverse kinematics problem for the manipulator.

23. ★ Space station remote manipulator system inverse kinematics.

    Shuttle remote manipulator system ($SSRMS$) is shown in Fig. 5.24 schematically. The forward kinematics of the robot provides the following transformation matrices. Solve the inverse kinematics for the $SSRMS$.

$$
^0T_1 = \begin{bmatrix} c\theta_1 & 0 & -s\theta_1 & 0 \\ s\theta_1 & 0 & c\theta_1 & 0 \\ 0 & -1 & 0 & d_1 \\ 0 & 0 & 0 & 1 \end{bmatrix}
\qquad
^1T_2 = \begin{bmatrix} c\theta_2 & 0 & -s\theta_2 & 0 \\ s\theta_2 & 0 & c\theta_2 & 0 \\ 0 & -1 & 0 & d_2 \\ 0 & 0 & 0 & 1 \end{bmatrix}
\tag{6.355}
$$

$$
^2T_3 = \begin{bmatrix} c\theta_3 & -s\theta_3 & 0 & a_3c\theta_3 \\ s\theta_3 & c\theta_3 & 0 & a_3s\theta_3 \\ 0 & 0 & 1 & d_3 \\ 0 & 0 & 0 & 1 \end{bmatrix}
\qquad
^3T_4 = \begin{bmatrix} c\theta_4 & -s\theta_4 & 0 & a_4c\theta_4 \\ s\theta_4 & c\theta_4 & 0 & a_4s\theta_4 \\ 0 & 0 & 1 & d_4 \\ 0 & 0 & 0 & 1 \end{bmatrix}
\tag{6.356}
$$

$$
^4T_5 = \begin{bmatrix} c\theta_5 & 0 & s\theta_5 & 0 \\ s\theta_5 & 0 & -c\theta_5 & 0 \\ 0 & 1 & 0 & d_5 \\ 0 & 0 & 0 & 1 \end{bmatrix}
\qquad
^5T_6 = \begin{bmatrix} c\theta_6 & 0 & -s\theta_6 & 0 \\ s\theta_6 & 0 & c\theta_6 & 0 \\ 0 & -1 & 0 & d_6 \\ 0 & 0 & 0 & 1 \end{bmatrix}
\tag{6.357}
$$

$$
^6T_7 = \begin{bmatrix} c\theta_7 & -s\theta_7 & 0 & 0 \\ s\theta_7 & c\theta_7 & 0 & 0 \\ 0 & 0 & 1 & d_7 \\ 0 & 0 & 0 & 1 \end{bmatrix}
\tag{6.358}
$$

Hint: This robot is a one degree redundant robot. It has 7 joints which is one more than the required 6 DOF to reach a point at a desired orientation. To solve the inverse kinematics of this robot, we need to introduce one extra condition among the joint variables, or assign a value to one of the joint variables.

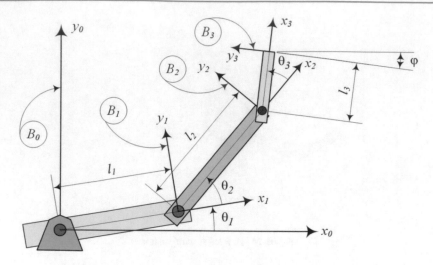

Fig. 6.15 A planar $3R$ manipulator with one DOF redundancy

(a) Assume $\theta_1 = 0$ and $^1T_7$ is given. Determine $\theta_2, \theta_3, \theta_4, \theta_5, \theta_6, \theta_7$.
(b) Assume $\theta_2 = 0$ and $^1T_7$ is given. Determine $\theta_1, \theta_3, \theta_4, \theta_5, \theta_6, \theta_7$.
(c) Assume $\theta_3 = 0$ and $^1T_7$ is given. Determine $\theta_1, \theta_2, \theta_4, \theta_5, \theta_6, \theta_7$.
(d) Assume $\theta_5 = 0$ and $^1T_7$ is given. Determine $\theta_1, \theta_2, \theta_3, \theta_4, \theta_6, \theta_7$.
(e) Assume $\theta_6 = 0$ and $^1T_7$ is given. Determine $\theta_1, \theta_2, \theta_3, \theta_4, \theta_5, \theta_7$.
(f) Assume $\theta_7 = 0$ and $^1T_7$ is given. Determine $\theta_1, \theta_2, \theta_3, \theta_4, \theta_5, \theta_6$.
(g) Determine $\theta_1, \theta_2, \theta_3, \theta_4, \theta_5, \theta_6, \theta_7$ such that f is minimized.

$$f = \theta_1 + \theta_2 + \theta_3 + \theta_4 + \theta_5 + \theta_6 + \theta_7 \tag{6.359}$$

24. ★ Project. Redundant manipulator and extra condition.
    Figure 6.15 illustrates a planar $3R$ manipulator. The manipulator has one DOF redundancy. Assume the length of the links are $l_i = 2/3$, and the end point is moving from $(1.2, 1.5)$ to $(-1.2, 1.5)$ on a straight line.
    (a) Solve the inverse kinematics if always $\theta_3 = 2\theta_2$. Divide the path into 40 increments and plot θ_i for the whole trip.
    (b) Solve the inverse kinematics if always $\theta_3 = \theta_1 - \theta_2$. Divide the path into 40 increments and plot θ_i for the whole trip.

Kinematics, the English version of the French word *cinématique* from the Greek $\kappa\acute{\iota}\upsilon\eta\mu\alpha$ ("movement"), is a branch of science that studies geometry in motion. By *motion* we mean any type of displacement, which includes changes in position and orientation. Therefore, *displacement*, and its successive derivatives with respect to time, *velocity, acceleration, jerk*, etc., all combine into kinematics.

Derivative kinematics explains how the derivatives of vectors are calculated and how they are related to each other. The simple and mixed forms of the first and second-derivative transformation formulas for a general body vector $^B\square$ are

$$\frac{^Gd}{dt}\,{}^B\square = \frac{^Bd}{dt}\,{}^B\square + {}^B_G\boldsymbol{\omega}_B \times {}^B\square$$

$$^B_A\dot{\square} = {}^B_C\dot{\square} + \left({}^B_A\boldsymbol{\omega}_B - {}^B_C\boldsymbol{\omega}_B\right) \times {}^B\square$$

$$\frac{^Gd}{dt}\frac{^Gd}{dt}\,{}^B\square = \frac{^Bd}{dt}\frac{^Bd}{dt}\,{}^B\square + {}^B_G\boldsymbol{\alpha}_B \times {}^B\square$$

$$+ 2\,{}^B_G\boldsymbol{\omega}_B \times \left(\frac{^Bd}{dt}\,{}^B\square + {}^B_G\boldsymbol{\omega}_B \times {}^B\square\right)$$

$$^B_A\ddot{\square} = {}^B_C\ddot{\square} + \left({}^B_A\boldsymbol{\alpha}_B - {}^B_C\boldsymbol{\alpha}_B\right) \times {}^B\square + 2\left({}^B_A\boldsymbol{\omega}_B - {}^B_C\boldsymbol{\omega}_B\right) \times {}^B\dot{\square}$$

$$+ {}^B_A\boldsymbol{\omega}_B \times \left({}^B_A\boldsymbol{\omega}_B \times {}^B\square\right) - {}^B_C\boldsymbol{\omega}_B \times \left({}^B_C\boldsymbol{\omega}_B \times {}^B\square\right)$$

They are the keys for relating the derivatives of vectorial quantities in different coordinate frames.

The geometric transformation formula

$$^G\mathbf{r} = {}^GR_B\,{}^B\mathbf{r}$$

is not generally correct for non-position vectors,

$$\frac{d^n}{dt^n}\,{}^G\mathbf{r} \neq {}^GR_B\frac{d^n}{dt^n}\,{}^B\mathbf{r} \qquad \mathbf{n} = 1, 2, 3, \cdots$$

unless the B and G-frames are fixed relatively.

Angular velocity is a vectorial quantity. Angular velocity of a rotating body B in a global frame G is the instantaneous rotation of the body with respect to G about an axis. Using the analytic description of angular velocity, we introduce the velocity and time derivative of homogenous transformation matrices.

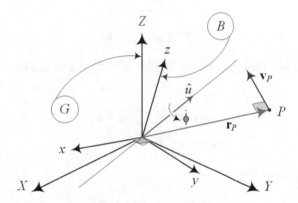

Fig. 7.1 A rotating rigid body $B(Oxyz)$ with a fixed point O in a global frame $G(OXYZ)$

7.1 Angular Velocity Vector and Matrix

Consider a continuously rotating rigid body $B(Oxyz)$ with a fixed point O in a reference frame $G(OXYZ)$ as shown in Fig. 7.1. We express the motion of the body B by a time varying rotation transformation matrix $^{G}R_B(t)$ between B and G frames. The transformation matrix maps the instantaneous coordinates of any fixed point in body frame B into their coordinates in the global frame G.

$$^{G}\mathbf{r}(t) = {}^{G}R_B(t)\, {}^{B}\mathbf{r} \tag{7.1}$$

The velocity of a body point in the global frame is $^{G}\mathbf{v}(t)$,

$$^{G}\mathbf{v}(t) = {}^{G}\dot{\mathbf{r}}(t) = {}^{G}\dot{R}_B(t)\, {}^{B}\mathbf{r} = {}_{G}\tilde{\omega}_B\, {}^{G}\mathbf{r}(t) = {}_{G}\boldsymbol{\omega}_B \times {}^{G}\mathbf{r}(t) \tag{7.2}$$

where $_{G}\boldsymbol{\omega}_B$ is the *angular velocity vector* of B with respect to G. It is equal to a rotation with *angular rate* ϕ about an *instantaneous axis of rotation* $\hat{u}$.

$$_{G}\boldsymbol{\omega}_B = \begin{bmatrix} \omega_1 \\ \omega_2 \\ \omega_3 \end{bmatrix} = \dot{\phi}\,\hat{u} \tag{7.3}$$

R. N. Jazar, *Theory of Applied Robotics*, https://doi.org/10.1007/978-3-030-93220-6_7

The angular velocity vector $\boldsymbol{\omega}$ is associated with a skew symmetric matrix $_G\tilde{\omega}_B$ called the *angular velocity matrix,*

$$_G\tilde{\omega}_B = \begin{bmatrix} 0 & -\omega_3 & \omega_2 \\ \omega_3 & 0 & -\omega_1 \\ -\omega_2 & \omega_1 & 0 \end{bmatrix} \tag{7.4}$$

where

$$_G\tilde{\omega}_B = {}^G\dot{R}_B \, {}^G R_B^T \tag{7.5}$$

$$= \dot{\phi}\,\tilde{u} \tag{7.6}$$

The B-expression of the angular velocity is similarly defined:

$$_G^B\tilde{\omega}_B = {}^G R_B^T \, {}^G\dot{R}_B \tag{7.7}$$

Employing the global and body expressions of the angular velocity of the body relative to the global coordinate frame, $_G\tilde{\omega}_B$ and $_G^B\tilde{\omega}_B$, we determine the global and body expressions of the velocity of a body point as

$$_G^G\mathbf{v}_P = {}_G^G\boldsymbol{\omega}_B \times {}^G\mathbf{r}_P \tag{7.8}$$

$$_G^B\mathbf{v}_P = {}_G^B\boldsymbol{\omega}_B \times {}^B\mathbf{r}_P \tag{7.9}$$

The G-expression $_G\tilde{\omega}_B$ and B-expression $_G^B\tilde{\omega}_B$ of the angular velocity matrix can be transformed to each other using the rotation matrix $^G R_B$:

$$_G\tilde{\omega}_B = {}^G R_B \, {}_G^B\tilde{\omega}_B \, {}^G R_B^T \tag{7.10}$$

$$_G^B\tilde{\omega}_B = {}^G R_B^T \, {}_G^G\tilde{\omega}_B \, {}^G R_B \tag{7.11}$$

They are also related to each other directly as

$$_G\tilde{\omega}_B \, {}^G R_B = {}^G R_B \, {}_G^B\tilde{\omega}_B^T \tag{7.12}$$

$$^G R_B^T \, {}_G\tilde{\omega}_B = {}_G^B\tilde{\omega}_B \, {}^G R_B^T \tag{7.13}$$

The relative angular velocity vectors of relatively moving rigid bodies can be calculated only if all the angular velocities are expressed in one coordinate frame:

$$_0\boldsymbol{\omega}_n = {}_0\boldsymbol{\omega}_1 + {}_1^0\boldsymbol{\omega}_2 + {}_2^0\boldsymbol{\omega}_3 + \cdots + {}_{n-1}^0\boldsymbol{\omega}_n = \sum_{i=1}^{n} {}_{i-1}^0\boldsymbol{\omega}_i \tag{7.14}$$

The transpose of the angular velocity matrices $_G\tilde{\omega}_B$ and $_G^B\tilde{\omega}_B$ are

$$_G\tilde{\omega}_B^T = {}^G R_B \, {}^G\dot{R}_B^T \tag{7.15}$$

$$_G^B\tilde{\omega}_B^T = {}^G\dot{R}_B^T \, {}^G R_B \tag{7.16}$$

Proof Consider a rigid body with an attached frame $B(Oxyz)$ and a fixed point O. The body frame B is initially coincident with the global frame G. Therefore, the position vector of a body point P is constant in B-frame:

$$^G\mathbf{r}(t_0) = {}^B\mathbf{r} \tag{7.17}$$

$$^G\mathbf{r}(t) = {}^G R_B(t) \, {}^B\mathbf{r} \tag{7.18}$$

The global time derivative of $^G\mathbf{r}$ is

$$
^G\mathbf{v} = {}^G\dot{\mathbf{r}} = \frac{{}^Gd}{dt}\,{}^G\mathbf{r}(t) = \frac{{}^Gd}{dt}\left[\,{}^GR_B(t)\,{}^B\mathbf{r}\right]
$$

$$
= \frac{{}^Gd}{dt}\left[\,{}^GR_B(t)\,{}^G\mathbf{r}(t_0)\right]
$$

$$
= {}^G\dot{R}_B(t)\,{}^G\mathbf{r}(t_0) = {}^G\dot{R}_B(t)\,{}^B\mathbf{r} \tag{7.19}
$$

Eliminating $^B\mathbf{r}$ between (7.18) and (7.19) determines the velocity of the point in global frame.

$$
^G\mathbf{v} = {}^G\dot{R}_B(t)\,{}^GR_B^T(t)\,{}^G\mathbf{r}(t) \tag{7.20}
$$

We denote the coefficient of $^G\mathbf{r}(t)$ by ${}_G\tilde{\omega}_B$

$$
{}_G\tilde{\omega}_B = {}^G\dot{R}_B\,{}^GR_B^T \tag{7.21}
$$

and write Eq. (7.20) as

$$
^G\mathbf{v} = {}_G\tilde{\omega}_B\,{}^G\mathbf{r}(t) \tag{7.22}
$$

or as

$$
^G\mathbf{v} = {}_G\boldsymbol{\omega}_B \times {}^G\mathbf{r}(t) \tag{7.23}
$$

The time derivative of the orthogonality condition, $^GR_B\,{}^GR_B^T = \mathbf{I}$, introduces an identity

$$
^G\dot{R}_B\,{}^GR_B^T + {}^GR_B\,{}^G\dot{R}_B^T = 0 \tag{7.24}
$$

which can be utilized to show that ${}_G\tilde{\omega}_B = [{}^G\dot{R}_B\,{}^GR_B^T]$ is a skew symmetric matrix.

$$
^G\dot{R}_B\,{}^GR_B^T = -\left[{}^GR_B\,{}^G\dot{R}_B^T\right]
$$

$$
= -\left[{}^G\dot{R}_B\,{}^GR_B^T\right]^T \tag{7.25}
$$

The vector ${}_G^G\boldsymbol{\omega}_B = {}_G\boldsymbol{\omega}_B$ is called *instantaneous angular velocity* of the body B relative to the global frame G as seen from the G-frame, and the vector ${}_G^B\boldsymbol{\omega}_B$ is the *instantaneous angular velocity* of the body B relative to the global frame G as seen from the B-frame.

A vectorial equation can be transformed to and expressed in any coordinate frame, and we may use any of the following expressions for the velocity of a body point in body or global frames,

$$
^G_G\mathbf{v}_P = {}^G_G\boldsymbol{\omega}_B \times {}^G\mathbf{r}_P \tag{7.26}
$$

$$
^B_G\mathbf{v}_P = {}^B_G\boldsymbol{\omega}_B \times {}^B\mathbf{r}_P \tag{7.27}
$$

where ${}^G_G\mathbf{v}_P$ is the global velocity of point P expressed in global frame G, and ${}^B_G\mathbf{v}_P$ is the global velocity of point P expressed in body frame B.

Generally speaking, an angular velocity vector is the instantaneous rotation of a coordinate frame A with respect to another frame B, which can be expressed in or seen from a third coordinate frame C. We indicate the first coordinate frame A by a right subscript, the second frame B by a left subscript, and the third frame C by a left superscript, ${}^C_B\boldsymbol{\omega}_A$. If the left super- and subscripts are the same, we only show the left subscript.

We can transform the G-expression of the global velocity $^G\mathbf{v}_P$ of a body point P and the B-expression of the global velocity ${}^B_G\mathbf{v}_P$ of the point P, to each other using their rotation matrix:

$$
^B_G\mathbf{v}_P = {}^BR_G\,{}^G\mathbf{v}_P = {}^BR_G\,{}_G\tilde{\omega}_B\,{}^G\mathbf{r}_P = {}^BR_G\,{}_G\tilde{\omega}_B\,{}^GR_B\,{}^B\mathbf{r}_P
$$

$$
= {}^BR_G\,{}^G\dot{R}_B\,{}^GR_B^T\,{}^GR_B\,{}^B\mathbf{r}_P = {}^BR_G\,{}^G\dot{R}_B\,{}^B\mathbf{r}_P
$$

$$
= {}^GR_B^T\,{}^G\dot{R}_B\,{}^B\mathbf{r}_P = {}^B_G\tilde{\omega}_B\,{}^B\mathbf{r}_P = {}^B_G\boldsymbol{\omega}_B \times {}^B\mathbf{r}_P \tag{7.28}
$$

$$^G\mathbf{v}_P = {}^GR_B\,{}^B_G\mathbf{v}_P = {}^GR_B\,{}^B_G\tilde{\omega}_B\,{}^B\mathbf{r}_P = {}^GR_B\,{}^B_G\tilde{\omega}_B\,{}^GR_B^T\,{}^G\mathbf{r}_P$$

$$= {}^GR_B\,{}^GR_B^T\,{}^G\dot{R}_B\,{}^GR_B^T\,{}^G\mathbf{r}_P = {}^G\dot{R}_B\,{}^GR_B^T\,{}^G\mathbf{r}_P$$

$$= {}_G\tilde{\omega}_B\,{}^G\mathbf{r}_P = {}_G\boldsymbol{\omega}_B \times {}^G\mathbf{r}_P = {}^GR_B\left({}^B_G\boldsymbol{\omega}_B \times {}^B\mathbf{r}_P\right) \tag{7.29}$$

It shows that

$$^B_G\tilde{\omega}_B = {}^GR_B^T\,{}^G\dot{R}_B \tag{7.30}$$

which is B-expression of the *instantaneous angular velocity* of B relative to the global frame G. From the definitions of ${}_G\tilde{\omega}_B$ and ${}^B_G\tilde{\omega}_B$, we are able to transform the two angular velocity matrices to each other by

$$_G\tilde{\omega}_B = {}^GR_B\,{}^B_G\tilde{\omega}_B\,{}^GR_B^T \tag{7.31}$$

$$^B_G\tilde{\omega}_B = {}^GR_B^T\,{}^G_G\tilde{\omega}_B\,{}^GR_B \tag{7.32}$$

or equivalently

$$^G\dot{R}_B = {}_G\tilde{\omega}_B\,{}^GR_B \tag{7.33}$$

$$^G\dot{R}_B = {}^GR_B\,{}^B_G\tilde{\omega}_B \tag{7.34}$$

$$_G\tilde{\omega}_B\,{}^GR_B = {}^GR_B\,{}^B_G\tilde{\omega}_B \tag{7.35}$$

The angular velocity of B in G is negative of the angular velocity of G in B if both are expressed in the same coordinate frame.

$$^G_G\tilde{\omega}_B = -{}^G_B\tilde{\omega}_G \qquad {}^G_G\boldsymbol{\omega}_B = -{}^G_B\boldsymbol{\omega}_G \tag{7.36}$$

$$^B_G\tilde{\omega}_B = -{}^B_B\tilde{\omega}_G \qquad {}^B_G\boldsymbol{\omega}_B = -{}^B_B\boldsymbol{\omega}_G \tag{7.37}$$

The angular velocity ${}_G\boldsymbol{\omega}_B$ can always be expressed in the form of

$$_G\boldsymbol{\omega}_B = \omega\hat{u} \tag{7.38}$$

where ω is the magnitude of $\boldsymbol{\omega}$ and $\hat{u}$ is a unit vector parallel to ${}_G\boldsymbol{\omega}_B$, indicating the *instantaneous axis of rotation*.

Using the Rodriguez rotation formula (3.5), we can show that

$$\dot{R}_{\hat{u},\phi} = \dot{\phi}\,\tilde{u}\,R_{\hat{u},\phi} \tag{7.39}$$

because

$$R_{\hat{u},\phi} = \mathbf{I}\cos\phi + \hat{u}\,\hat{u}^T\,\mathrm{vers}\,\phi + \tilde{u}\sin\phi \tag{7.40}$$

$$\dot{R}_{\hat{u},\phi} = \dot{\phi}\left[-\sin\phi\,\mathbf{I} + \hat{u}\,\hat{u}^T\sin\phi + \tilde{u}\cos\phi\right] \tag{7.41}$$

$$\tilde{u}\,R_{\hat{u},\phi} = \tilde{u}\left[\mathbf{I}\cos\phi + \hat{u}\,\hat{u}^T\,\mathrm{vers}\,\phi + \tilde{u}\sin\phi\right]$$

$$= \cos\phi\,\tilde{u} + \left(\hat{u}\,\hat{u}^T - \mathbf{I}\right)\sin\phi \tag{7.42}$$

$$= -\sin\phi\,\mathbf{I} + \hat{u}\,\hat{u}^T\sin\phi + \tilde{u}\cos\phi \tag{7.43}$$

and therefore,

$$_G\tilde{\omega}_B = \dot{R}_{\hat{u},\phi}\,R_{\hat{u},\phi}^T = \dot{\phi}\,\tilde{u}\,R_{\hat{u},\phi}\,R_{\hat{u},\phi}^T = \dot{\phi}\,\tilde{u} \tag{7.44}$$

$$\tilde{\omega} = \dot{\phi}\,\tilde{u} \tag{7.45}$$

$$\boldsymbol{\omega} = \dot{\phi}\,\hat{u} \tag{7.46}$$

To show the addition of relative angular velocities in Eq. (7.14), we start from a combination of rotations

$$^0R_2 = {}^0R_1 \, {}^1R_2 \tag{7.47}$$

and take a time derivative:

$$^0\dot{R}_2 = {}^0\dot{R}_1 \, {}^1R_2 + {}^0R_1 \, {}^1\dot{R}_2 \tag{7.48}$$

Substituting the derivative of the rotation matrices with

$$^0\dot{R}_2 = {}_0\tilde{\omega}_2 \, {}^0R_2 \tag{7.49}$$

$$^0\dot{R}_1 = {}_0\tilde{\omega}_1 \, {}^0R_1 \tag{7.50}$$

$$^1\dot{R}_2 = {}_1\tilde{\omega}_2 \, {}^1R_2 \tag{7.51}$$

results in

$$\begin{aligned}
{}_0\tilde{\omega}_2 \, {}^0R_2 &= {}_0\tilde{\omega}_1 \, {}^0R_1 \, {}^1R_2 + {}^0R_1 \, {}_1\tilde{\omega}_2 \, {}^1R_2 \\
&= {}_0\tilde{\omega}_1 \, {}^0R_2 + {}^0R_1 \, {}_1\tilde{\omega}_2 \, {}^0R_1^T \, {}^0R_1 \, {}^1R_2 \\
&= {}_0\tilde{\omega}_1 \, {}^0R_2 + {}_1^0\tilde{\omega}_2 \, {}^0R_2 \tag{7.52}
\end{aligned}$$

where

$$^0R_1 \, {}_1\tilde{\omega}_2 \, {}^0R_1^T = {}_1^0\tilde{\omega}_2 \tag{7.53}$$

Therefore, we find

$$_0\tilde{\omega}_2 = {}_0\tilde{\omega}_1 + {}_1^0\tilde{\omega}_2 \tag{7.54}$$

which indicates two angular velocities may be added when they are expressed in the same frame:

$$_0\boldsymbol{\omega}_2 = {}_0\boldsymbol{\omega}_1 + {}_1^0\boldsymbol{\omega}_2 \tag{7.55}$$

The expansion of this equation for any number of angular velocities would be Eq. (7.14).

Employing the relative angular velocity formula (7.55), we can find the relative velocity formula of a point P in B_2 at $^0\mathbf{r}_P$:

$$\begin{aligned}
{}_0\mathbf{v}_2 &= {}_0\boldsymbol{\omega}_2 \, {}^0\mathbf{r}_P = \left({}_0\boldsymbol{\omega}_1 + {}_1^0\boldsymbol{\omega}_2\right) {}^0\mathbf{r}_P = {}_0\boldsymbol{\omega}_1 \, {}^0\mathbf{r}_P + {}_1^0\boldsymbol{\omega}_2 \, {}^0\mathbf{r}_P \\
&= {}_0\mathbf{v}_1 + {}_1^0\mathbf{v}_2 \tag{7.56}
\end{aligned}$$

The angular velocity matrices $_G\tilde{\omega}_B$ and $_G^B\tilde{\omega}_B$ are skew symmetric and not invertible. However, we can define their transpose by the rules:

$$_G\tilde{\omega}_B^T = {}^GR_B \, {}^G\dot{R}_B^T \tag{7.57}$$

$$_G^B\tilde{\omega}_B^T = {}^G\dot{R}_B^T \, {}^GR_B \tag{7.58}$$

to get

$$_G\tilde{\omega}_B^T \, {}_G\tilde{\omega}_B = {}_G\tilde{\omega}_B \, {}_G\tilde{\omega}_B^T \tag{7.59}$$

$$_G^B\tilde{\omega}_B^T \, {}_G^B\tilde{\omega}_B = {}_G^B\tilde{\omega}_B \, {}_G^B\tilde{\omega}_B^T \tag{7.60}$$

∎

Example 222 Rotation of a body point about a global axis. An example of a continuously uniform rotation of a rigid body about the Z-axis and calculating the position and velocity of a point at a particular time.

The slab shown in Fig. 2.5 is turning about the Z-axis with $\dot{\alpha} = 10 \deg/s$. The global velocity of the corner point $P(5, 30, 10)$, after 3 s when the slab is at $\alpha = \frac{\dot{\alpha}\pi}{180}t = 30$ deg, is

$$
{}^{G}\mathbf{v}_{P} = {}^{G}\dot{R}_{B}(t)\,{}^{B}\mathbf{r}_{P} \tag{7.61}
$$

$$
= \frac{{}^{G}d}{dt}\left(\begin{bmatrix} \cos\frac{\dot{\alpha}\pi}{180}t & -\sin\frac{\dot{\alpha}\pi}{180}t & 0 \\ \sin\frac{\dot{\alpha}\pi}{180}t & \cos\frac{\dot{\alpha}\pi}{180}t & 0 \\ 0 & 0 & 1 \end{bmatrix}\right)\begin{bmatrix} 5 \\ 30 \\ 10 \end{bmatrix}
$$

$$
= \frac{\dot{\alpha}\pi}{180}\begin{bmatrix} -\sin\frac{\dot{\alpha}\pi}{180}t & -\cos\frac{\dot{\alpha}\pi}{180}t & 0 \\ \cos\frac{\dot{\alpha}\pi}{180}t & -\sin\frac{\dot{\alpha}\pi}{180}t & 0 \\ 0 & 0 & 0 \end{bmatrix}\begin{bmatrix} 5 \\ 30 \\ 10 \end{bmatrix}
$$

$$
= \frac{10\pi}{180}\begin{bmatrix} -\sin\frac{\pi}{6} & -\cos\frac{\pi}{6} & 0 \\ \cos\frac{\pi}{6} & -\sin\frac{\pi}{6} & 0 \\ 0 & 0 & 0 \end{bmatrix}\begin{bmatrix} 5 \\ 30 \\ 10 \end{bmatrix} = \begin{bmatrix} -4.97 \\ -1.86 \\ 0 \end{bmatrix}
$$

at this moment, the point P is at

$$
{}^{G}\mathbf{r}_{P} = {}^{G}R_{B}\,{}^{B}\mathbf{r}_{P} \tag{7.62}
$$

$$
= \begin{bmatrix} \cos\frac{\pi}{6} & -\sin\frac{\pi}{6} & 0 \\ \sin\frac{\pi}{6} & \cos\frac{\pi}{6} & 0 \\ 0 & 0 & 1 \end{bmatrix}\begin{bmatrix} 5 \\ 30 \\ 10 \end{bmatrix} = \begin{bmatrix} -10.67 \\ 28.48 \\ 10 \end{bmatrix}
$$

The global velocity of the point P would be

$$
{}^{G}\mathbf{v}_{P} = {}^{G}\dot{R}_{B}\,{}^{G}R_{B}^{T}\,{}^{G}\mathbf{r}_{P} \tag{7.63}
$$

$$
= \frac{10\pi}{180}\begin{bmatrix} -s\frac{\pi}{6} & -c\frac{\pi}{6} & 0 \\ c\frac{\pi}{6} & -s\frac{\pi}{6} & 0 \\ 0 & 0 & 0 \end{bmatrix}\begin{bmatrix} c\frac{\pi}{6} & -s\frac{\pi}{6} & 0 \\ s\frac{\pi}{6} & c\frac{\pi}{6} & 0 \\ 0 & 0 & 1 \end{bmatrix}^{T}\begin{bmatrix} -10.67 \\ 28.48 \\ 10 \end{bmatrix}
$$

$$
= \begin{bmatrix} -4.97 \\ -1.86 \\ 0 \end{bmatrix}
$$

Example 223 Angular velocity based on variable transformation matrix. Angular velocity exists whenever a time varying transformation matrix exists. Here, we see how to calculate angular velocity from a time varying transformation matrix.

Let us assume a time varying rotation transformation matrix about the Z-axis.

$$
{}^{G}R_{B} = \begin{bmatrix} \cos at^{3} & -\sin at^{3} & 0 \\ \sin at^{3} & \cos at^{3} & 0 \\ 0 & 0 & 1 \end{bmatrix} \tag{7.64}
$$

Its associated angular velocity will be

$$
{}_{G}\tilde{\omega}_{B} = {}^{G}\dot{R}_{B}\,{}^{G}R_{B}^{T} = \begin{bmatrix} 0 & -3at^{2} & 0 \\ 3at^{2} & 0 & 0 \\ 0 & 0 & 0 \end{bmatrix} \tag{7.65}
$$

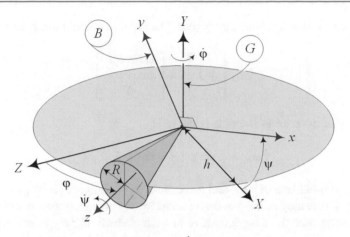

Fig. 7.2 A rotating cone with angular speeds $\dot{\psi}$ about the z-axis and $\dot{\varphi}$ about the x-axis

because

$$
{}^{G}\dot{R}_{B} = \frac{{}^{G}d}{dt}\left(\begin{bmatrix} \cos at^3 & -\sin at^3 & 0 \\ \sin at^3 & \cos at^3 & 0 \\ 0 & 0 & 1 \end{bmatrix}\right)
$$

$$
= \begin{bmatrix} -3at^2\sin at^3 & -3at^2\cos at^3 & 0 \\ 3at^2\cos at^3 & -3at^2\sin at^3 & 0 \\ 0 & 0 & 0 \end{bmatrix} \tag{7.66}
$$

To calculate ${}^{B}_{G}\tilde{\omega}_{B}$, we may either transform ${}_{G}\tilde{\omega}_{B}$ from G-frame to B-frame by (7.11) or directly calculate it by (7.7). Direct substitution of ${}^{G}R_{B}$, ${}^{G}_{G}\tilde{\omega}_{B}$, ${}^{G}\dot{R}_{B}$, and matrix multiplication will show that

$$
{}^{B}_{G}\tilde{\omega}_{B} = {}^{G}R_{B}^{T}\,{}^{G}_{G}\tilde{\omega}_{B}\,{}^{G}R_{B} = \begin{bmatrix} 0 & -3at^2 & 0 \\ 3at^2 & 0 & 0 \\ 0 & 0 & 0 \end{bmatrix} \tag{7.67}
$$

$$
{}^{B}_{G}\tilde{\omega}_{B} = {}^{G}R_{B}^{T}\,{}^{G}\dot{R}_{B} = \begin{bmatrix} 0 & -3at^2 & 0 \\ 3at^2 & 0 & 0 \\ 0 & 0 & 0 \end{bmatrix} \tag{7.68}
$$

It can also be checked that instead of coordinate transformation of skew symmetric expression of angular velocity (7.11), we may transform the angular velocity vector as a vector transformation.

$$
{}^{B}_{G}\boldsymbol{\omega}_{B} = {}^{B}R_{G}\,{}^{G}_{G}\boldsymbol{\omega}_{B} = {}^{G}R_{B}^{T}\,{}^{G}\begin{bmatrix} 0 \\ 0 \\ 3at^2 \end{bmatrix} = {}^{B}\begin{bmatrix} 0 \\ 0 \\ 3at^2 \end{bmatrix} \tag{7.69}
$$

Example 224 A rotating cone. Here is a rotating rigid body and definition of its angular velocity.

Figure 7.2 illustrates a cone with angular speeds $\dot{\psi}$ about the z-axis and $\dot{\varphi}$ about the Y-axis. It moves such that the z-axis remains in the (X, Z)-plane. The transformation matrix between frames G and B is

$$
{}^{B}R_{G} = R_{z,\psi}R_{y,\varphi}
$$

$$
= \begin{bmatrix} \cos\psi\cos\varphi & \sin\psi & -\cos\psi\sin\varphi \\ -\cos\varphi\sin\psi & \cos\psi & \sin\psi\sin\varphi \\ \sin\varphi & 0 & \cos\varphi \end{bmatrix} \tag{7.70}
$$

The rotation $\dot{\psi}\hat{k}$ about the z-axis is already about a body axis. The rotation $\dot{\varphi}$ about the Y-axis can be found in the B-frame:

$$
{}^B R_G \begin{bmatrix} 0 \\ \dot{\varphi}\,\hat{j} \\ 0 \end{bmatrix} = \begin{bmatrix} \dot{\varphi}\sin\psi\,\hat{\imath} \\ \dot{\varphi}\cos\psi\,\hat{\jmath} \\ 0 \end{bmatrix}
\tag{7.71}
$$

Therefore, the angular velocity of the cone in B-frame is

$$
{}^B_G\boldsymbol{\omega}_B = \dot{\varphi}\sin\psi\,\hat{\imath} + \dot{\varphi}\cos\psi\,\hat{\jmath} + \dot{\psi}\hat{k}
\tag{7.72}
$$

Example 225 Time derivative of elements of ${}^G R_B$ and determination of components of ${}_G\boldsymbol{\omega}_B$. General expressions of ${}^G R_B$ and ${}_G\boldsymbol{\omega}_B$ with unknown components may be used to derive some identities among their components.

Expansion of Eq. (7.33) shows how the time derivative of each element of ${}^G R_B$ is related to components of ${}_G\boldsymbol{\omega}_B$ and elements of ${}^G R_B$.

$$
{}^G\dot{R}_B = {}_G\tilde{\omega}_B\,{}^G R_B \begin{bmatrix} \dot{r}_{11} & \dot{r}_{12} & \dot{r}_{13} \\ \dot{r}_{21} & \dot{r}_{22} & \dot{r}_{23} \\ \dot{r}_{31} & \dot{r}_{32} & \dot{r}_{33} \end{bmatrix}
\tag{7.73}
$$

$$
= \begin{bmatrix} 0 & -\omega_Z & \omega_Y \\ \omega_Z & 0 & -\omega_X \\ -\omega_Y & \omega_X & 0 \end{bmatrix} \begin{bmatrix} r_{11} & r_{12} & r_{13} \\ r_{21} & r_{22} & r_{23} \\ r_{31} & r_{32} & r_{33} \end{bmatrix}
$$

$$
= \begin{bmatrix} \omega_Y r_{31} - \omega_Z r_{21} & \omega_Y r_{32} - \omega_Z r_{22} & \omega_Y r_{33} - \omega_Z r_{23} \\ \omega_Z r_{11} - \omega_X r_{31} & \omega_Z r_{12} - \omega_X r_{32} & \omega_Z r_{13} - \omega_X r_{33} \\ \omega_X r_{21} - \omega_Y r_{11} & \omega_X r_{22} - \omega_Y r_{12} & \omega_X r_{23} - \omega_Y r_{13} \end{bmatrix}
$$

$$
\tag{7.74}
$$

Expansion of Eq. (7.34) shows how the elements of ${}^G\dot{R}_B$ are related to components of ${}^B_G\boldsymbol{\omega}_B$ and elements of ${}^G R_B$:

$$
{}^G\dot{R}_B = {}^G R_B\,{}^B_G\tilde{\omega}_B \begin{bmatrix} \dot{r}_{11} & \dot{r}_{12} & \dot{r}_{13} \\ \dot{r}_{21} & \dot{r}_{22} & \dot{r}_{23} \\ \dot{r}_{31} & \dot{r}_{32} & \dot{r}_{33} \end{bmatrix}
\tag{7.75}
$$

$$
= \begin{bmatrix} r_{11} & r_{12} & r_{13} \\ r_{21} & r_{22} & r_{23} \\ r_{31} & r_{32} & r_{33} \end{bmatrix} \begin{bmatrix} 0 & -\omega_z & \omega_y \\ \omega_z & 0 & -\omega_x \\ -\omega_y & \omega_x & 0 \end{bmatrix}
$$

$$
= \begin{bmatrix} \omega_z r_{12} - \omega_y r_{13} & \omega_x r_{13} - \omega_z r_{11} & \omega_y r_{11} - \omega_x r_{12} \\ \omega_z r_{22} - \omega_y r_{23} & \omega_x r_{23} - \omega_z r_{21} & \omega_y r_{21} - \omega_x r_{22} \\ \omega_z r_{32} - \omega_y r_{33} & \omega_x r_{33} - \omega_z r_{31} & \omega_y r_{31} - \omega_x r_{32} \end{bmatrix}
\tag{7.76}
$$

Employing these expanded forms, we may determine the angular velocity when a rotation transformation matrix ${}^G R_B$ is given. As an example, assume the following matrix is known:

$$
{}^G R_B = \begin{bmatrix} \cos\theta\cos\varphi & -\sin\varphi & \cos\varphi\sin\theta \\ \cos\theta\sin\varphi & \cos\varphi & \sin\theta\sin\varphi \\ -\sin\theta & 0 & \cos\theta \end{bmatrix}
\tag{7.77}
$$

A time derivative yields

$$
{}^G\dot{R}_B = \begin{bmatrix} -\dot{\theta}c\varphi s\theta - \dot{\varphi}c\theta s\varphi & -\dot{\varphi}c\varphi & \dot{\theta}c\theta c\varphi - \dot{\varphi}s\theta s\varphi \\ \dot{\varphi}c\theta c\varphi - \dot{\theta}s\theta s\varphi & -\dot{\varphi}s\varphi & \dot{\theta}c\theta s\varphi + \dot{\varphi}c\varphi s\theta \\ -\dot{\theta}c\theta & 0 & -\dot{\theta}s\theta \end{bmatrix}
\tag{7.78}
$$

Let us use any set of three independent elements of $^G\dot{R}_B$ to determine components of $_G\boldsymbol{\omega}_B$:

$$\dot{r}_{11} = \omega_Y r_{31} - \omega_Z r_{21} = -\dot{\theta}\cos\varphi\sin\theta - \dot{\varphi}\cos\theta\sin\varphi \tag{7.79}$$

$$\dot{r}_{21} = \omega_Z r_{11} - \omega_X r_{31} = \dot{\varphi}\cos\theta\cos\varphi - \dot{\theta}\sin\theta\sin\varphi \tag{7.80}$$

$$\dot{r}_{12} = \omega_Y r_{32} - \omega_Z r_{22} = -\dot{\varphi}\cos\varphi \tag{7.81}$$

Writing these in the matrix form

$$\begin{bmatrix} \dot{r}_{11} \\ \dot{r}_{21} \\ \dot{r}_{12} \end{bmatrix} = \begin{bmatrix} 0 & r_{31} & -r_{21} \\ -r_{31} & 0 & r_{11} \\ 0 & r_{32} & -r_{22} \end{bmatrix} \begin{bmatrix} \omega_X \\ \omega_Y \\ \omega_Z \end{bmatrix} \tag{7.82}$$

provides the solutions

$$\begin{bmatrix} \omega_X \\ \omega_Y \\ \omega_Z \end{bmatrix} = \begin{bmatrix} 0 & r_{31} & -r_{21} \\ -r_{31} & 0 & r_{11} \\ 0 & r_{32} & -r_{22} \end{bmatrix}^{-1} \begin{bmatrix} \dot{r}_{11} \\ \dot{r}_{21} \\ \dot{r}_{12} \end{bmatrix}$$

$$= \begin{bmatrix} -\dot{\theta}\sin\varphi \\ \dot{\theta}\cos\varphi \\ \dot{\varphi} \end{bmatrix} \tag{7.83}$$

Example 226 Principal angular velocities. Time varying rotation of a rigid body about the global and body principal axes will be used to define principal angular velocities. These principal angular velocities will be used to decompose a general angular velocity.

The principal rotational matrices about the axes X, Y, and Z arc

$$R_{X,\gamma} = \begin{bmatrix} 1 & 0 & 0 \\ 0 & \cos\gamma & -\sin\gamma \\ 0 & \sin\gamma & \cos\gamma \end{bmatrix} \tag{7.84}$$

$$R_{Y,\beta} = \begin{bmatrix} \cos\beta & 0 & \sin\beta \\ 0 & 1 & 0 \\ -\sin\beta & 0 & \cos\beta \end{bmatrix} \tag{7.85}$$

$$R_{Z,\alpha} = \begin{bmatrix} \cos\alpha & -\sin\alpha & 0 \\ \sin\alpha & \cos\alpha & 0 \\ 0 & 0 & 1 \end{bmatrix} \tag{7.86}$$

and hence, their time derivatives are

$$\dot{R}_{X,\gamma} = \dot{\gamma}\begin{bmatrix} 0 & 0 & 0 \\ 0 & -\sin\gamma & -\cos\gamma \\ 0 & \cos\gamma & -\sin\gamma \end{bmatrix} \tag{7.87}$$

$$\dot{R}_{Y,\beta} = \dot{\beta}\begin{bmatrix} -\sin\beta & 0 & \cos\beta \\ 0 & 0 & 0 \\ -\cos\beta & 0 & -\sin\beta \end{bmatrix} \tag{7.88}$$

$$\dot{R}_{Z,\alpha} = \dot{\alpha}\begin{bmatrix} -\sin\alpha & -\cos\alpha & 0 \\ \cos\alpha & -\sin\alpha & 0 \\ 0 & 0 & 0 \end{bmatrix} \tag{7.89}$$

Therefore, the principal angular velocity matrices about the axes X, Y, and Z are

$$_G\tilde{\omega}_X = \dot{R}_{X,\gamma}R_{X,\gamma}^T = \dot{\gamma}\begin{bmatrix} 0 & 0 & 0 \\ 0 & 0 & -1 \\ 0 & 1 & 0 \end{bmatrix} \tag{7.90}$$

$$_G\tilde{\omega}_Y = \dot{R}_{Y,\beta} R_{Y,\beta}^T = \dot{\beta} \begin{bmatrix} 0 & 0 & 1 \\ 0 & 0 & 0 \\ -1 & 0 & 0 \end{bmatrix} \tag{7.91}$$

$$_G\tilde{\omega}_Z = \dot{R}_{Z,\alpha} R_{Z,\alpha}^T = \dot{\alpha} \begin{bmatrix} 0 & -1 & 0 \\ 1 & 0 & 0 \\ 0 & 0 & 0 \end{bmatrix} \tag{7.92}$$

which are equivalent to

$$_G\tilde{\omega}_X = \dot{\gamma}\,\tilde{I} \tag{7.93}$$

$$_G\tilde{\omega}_Y = \dot{\beta}\,\tilde{J} \tag{7.94}$$

$$_G\tilde{\omega}_Z = \dot{\alpha}\,\tilde{K} \tag{7.95}$$

and therefore, the principal angular velocity vectors are

$$_G\boldsymbol{\omega}_X = \omega_X\,\hat{I} = \dot{\gamma}\,\hat{I} \tag{7.96}$$

$$_G\boldsymbol{\omega}_Y = \omega_Y\,\hat{J} = \dot{\beta}\,\hat{J} \tag{7.97}$$

$$_G\boldsymbol{\omega}_Z = \omega_Z\,\hat{K} = \dot{\alpha}\,\hat{K} \tag{7.98}$$

Utilizing the same technique, we can find the following principal angular velocity matrices about the body axes:

$$_G^B\tilde{\omega}_x = R_{x,\psi}^T\,\dot{R}_{x,\psi} = -\dot{\psi}\begin{bmatrix} 0 & 0 & 0 \\ 0 & 0 & -1 \\ 0 & 1 & 0 \end{bmatrix} = -\dot{\psi}\,\tilde{\imath} \tag{7.99}$$

$$_G^B\tilde{\omega}_y = R_{y,\theta}^T\,\dot{R}_{y,\theta} = -\dot{\theta}\begin{bmatrix} 0 & 0 & 1 \\ 0 & 0 & 0 \\ -1 & 0 & 0 \end{bmatrix} = -\dot{\theta}\,\tilde{\jmath} \tag{7.100}$$

$$_G^B\tilde{\omega}_z = R_{z,\varphi}^T\,\dot{R}_{z,\varphi} = -\dot{\varphi}\begin{bmatrix} 0 & -1 & 0 \\ 1 & 0 & 0 \\ 0 & 0 & 0 \end{bmatrix} = -\dot{\varphi}\,\tilde{k} \tag{7.101}$$

Every angular velocity vector can be decomposed to three principal angular velocity vectors.

$$\begin{aligned} _G\boldsymbol{\omega}_B &= \left(_G\boldsymbol{\omega}_B \cdot \hat{I}\right)\hat{I} + \left(_G\boldsymbol{\omega}_B \cdot \hat{J}\right)\hat{J} + \left(_G\boldsymbol{\omega}_B \cdot \hat{K}\right)\hat{K} \\ &= \omega_X\,\hat{I} + \omega_Y\,\hat{J} + \omega_Z\,\hat{K} = \dot{\gamma}\,\hat{I} + \dot{\beta}\,\hat{J} + \dot{\alpha}\,\hat{K} \\ &= \boldsymbol{\omega}_X + \boldsymbol{\omega}_Y + \boldsymbol{\omega}_Z \end{aligned} \tag{7.102}$$

$$\begin{aligned} _G^B\boldsymbol{\omega}_B &= \left(_G^B\boldsymbol{\omega}_B \cdot \tilde{\imath}\right)\tilde{\imath} + \left(_G^B\boldsymbol{\omega}_B \cdot \tilde{\jmath}\right)\tilde{\jmath} + \left(_G^B\boldsymbol{\omega}_B \cdot \tilde{k}\right)\tilde{k} \\ &= \omega_x\,\tilde{\imath} + \omega_y\,\tilde{\jmath} + \omega_z\,\tilde{k} = -\dot{\psi}\,\tilde{\imath} - \dot{\theta}\,\tilde{\jmath} - \dot{\varphi}\,\tilde{k} \\ &= \boldsymbol{\omega}_x + \boldsymbol{\omega}_y + \boldsymbol{\omega}_z \end{aligned} \tag{7.103}$$

Example 227 Angular velocity in terms of Euler frequencies. In rigid body dynamics and employing Euler equations of motion, we usually prefer to define the equations in terms of Euler angles and Euler frequencies. This is how to define angular velocity in terms of Euler angles.

The angular velocity vector can be expressed by Euler frequencies. The details of derivation are described in Chap. 2, Example 40. Therefore,

$$_G^B\boldsymbol{\omega}_B = \omega_x\hat{\imath} + \omega_y\hat{\jmath} + \omega_z\hat{k} = \dot{\varphi}\hat{e}_\varphi + \dot{\theta}\hat{e}_\theta + \dot{\psi}\hat{e}_\psi$$

$$= \dot{\varphi}\begin{bmatrix} \sin\theta\sin\psi \\ \sin\theta\cos\psi \\ \cos\theta \end{bmatrix} + \dot{\theta}\begin{bmatrix} \cos\psi \\ -\sin\psi \\ 0 \end{bmatrix} + \dot{\psi}\begin{bmatrix} 0 \\ 0 \\ 1 \end{bmatrix}$$

$$= \begin{bmatrix} \sin\theta\sin\psi & \cos\psi & 0 \\ \sin\theta\cos\psi & -\sin\psi & 0 \\ \cos\theta & 0 & 1 \end{bmatrix}\begin{bmatrix} \dot{\varphi} \\ \dot{\theta} \\ \dot{\psi} \end{bmatrix} \tag{7.104}$$

and

$$_G^G\boldsymbol{\omega}_B = {}^BR_G^{-1}\,{}_G^B\boldsymbol{\omega}_B = {}^BR_G^{-1}\begin{bmatrix} \dot{\varphi}\sin\theta\sin\psi + \dot{\theta}\cos\psi \\ \dot{\varphi}\sin\theta\cos\psi - \dot{\theta}\sin\psi \\ \dot{\varphi}\cos\theta + \dot{\psi} \end{bmatrix}$$

$$= \begin{bmatrix} 0 & \cos\varphi & \sin\theta\sin\varphi \\ 0 & \sin\varphi & -\cos\varphi\sin\theta \\ 1 & 0 & \cos\theta \end{bmatrix}\begin{bmatrix} \dot{\varphi} \\ \dot{\theta} \\ \dot{\psi} \end{bmatrix} \tag{7.105}$$

where the inverse of the Euler transformation matrix is

$$^BR_G^{-1} = \begin{bmatrix} c\varphi c\psi - c\theta s\varphi s\psi & -c\varphi s\psi - c\theta c\psi s\varphi & s\theta s\varphi \\ c\psi s\varphi + c\theta c\varphi s\psi & -s\varphi s\psi + c\theta c\varphi c\psi & -c\varphi s\theta \\ s\theta s\psi & s\theta c\psi & c\theta \end{bmatrix} \tag{7.106}$$

Example 228 Angular velocity in terms of Euler frequencies. Angular velocity for decomposed triple rotation matrices of a general rotation transformation matrix.

Appendices A and B show the 12 global and 12 local axes' triple rotations. Utilizing those equations, we are able to find decomposed expressions of the angular velocity matrix and vector of a rigid body in terms of rotation frequencies. As an example, consider the Euler angles transformation matrix in case 9 of the Appendix B.

$$^BR_G = R_{z,\psi}R_{x,\theta}R_{z,\varphi} \tag{7.107}$$

The angular velocity matrix is then equal to

$$_B\tilde{\omega}_G = {}^B\dot{R}_G\,{}^BR_G^T$$

$$= \left(\dot{\varphi}\,R_{z,\psi}R_{x,\theta}\frac{dR_{z,\varphi}}{d\varphi} + \dot{\theta}\,R_{z,\psi}\frac{dR_{x,\theta}}{d\theta}R_{z,\varphi} + \dot{\psi}\,\frac{dR_{z,\psi}}{d\psi}R_{x,\theta}R_{z,\varphi}\right)$$

$$\times \left(R_{z,\psi}R_{x,\theta}R_{z,\varphi}\right)^T$$

$$= \dot{\varphi}\,R_{z,\psi}R_{x,\theta}\frac{dR_{z,\varphi}}{d\varphi}R_{z,\varphi}^T R_{x,\theta}^T R_{z,\psi}^T + \dot{\theta}\,R_{z,\psi}\frac{dR_{x,\theta}}{d\theta}R_{x,\theta}^T R_{z,\psi}^T$$

$$+ \dot{\psi}\,\frac{dR_{z,\psi}}{d\psi}R_{z,\psi}^T \tag{7.108}$$

which, in matrix form, is

$$_B\tilde{\omega}_G = \dot{\varphi}\begin{bmatrix} 0 & \cos\theta & -\sin\theta\cos\psi \\ -\cos\theta & 0 & \sin\theta\sin\psi \\ \sin\theta\cos\psi & -\sin\theta\sin\psi & 0 \end{bmatrix}$$

$$+ \dot{\theta}\begin{bmatrix} 0 & 0 & \sin\psi \\ 0 & 0 & \cos\psi \\ -\sin\psi & -\cos\psi & 0 \end{bmatrix} + \dot{\psi}\begin{bmatrix} 0 & 1 & 0 \\ -1 & 0 & 0 \\ 0 & 0 & 0 \end{bmatrix} \tag{7.109}$$

or

$$
_B\tilde\omega_G = \begin{bmatrix} 0 & \dot\psi + \dot\varphi c\theta & \dot\theta s\psi - \dot\varphi s\theta c\psi \\ -\dot\psi - \dot\varphi c\theta & 0 & \dot\theta c\psi + \dot\varphi s\theta s\psi \\ -\dot\theta s\psi + \dot\varphi s\theta c\psi & -\dot\theta c\psi - \dot\varphi s\theta s\psi & 0 \end{bmatrix} \tag{7.110}
$$

The corresponding angular velocity vector is

$$
_B\boldsymbol{\omega}_G = - \begin{bmatrix} \dot\theta c\psi + \dot\varphi s\theta s\psi \\ -\dot\theta s\psi + \dot\varphi s\theta c\psi \\ \dot\psi + \dot\varphi c\theta \end{bmatrix}
$$

$$
= - \begin{bmatrix} \sin\theta \sin\psi & \cos\psi & 0 \\ \sin\theta \cos\psi & -\sin\psi & 0 \\ \cos\theta & 0 & 1 \end{bmatrix} \begin{bmatrix} \dot\varphi \\ \dot\theta \\ \dot\psi \end{bmatrix} \tag{7.111}
$$

However,

$$
_B^B\tilde\omega_G = -_G^B\tilde\omega_B \tag{7.112}
$$

$$
_B^B\boldsymbol{\omega}_G = -_G^B\boldsymbol{\omega}_B \tag{7.113}
$$

and therefore,

$$
_G^B\boldsymbol{\omega}_B = \begin{bmatrix} \sin\theta \sin\psi & \cos\psi & 0 \\ \sin\theta \cos\psi & -\sin\psi & 0 \\ \cos\theta & 0 & 1 \end{bmatrix} \begin{bmatrix} \dot\varphi \\ \dot\theta \\ \dot\psi \end{bmatrix} \tag{7.114}
$$

which is compatible with Eq. (7.104).

Example 229 Derivative of the Rodriguez formula. Angular velocity should be able to be determined by the Rodriguez rotation matrix and its derivative. Here it shows how.

Based on the Rodriguez formula, the angle-axis rotation matrix $^G R_B$ is

$$
^G R_B = R_{\hat u,\phi} = \cos\phi\, \mathbf{I} + \hat u \hat u^T \operatorname{vers}\phi + \tilde u \sin\phi \tag{7.115}
$$

Assuming ϕ to be variable with time, the time rate of the Rodriguez formula is

$$
\begin{aligned}
^G \dot R_B = \dot R_{\hat u,\phi} &= -\dot\phi \sin\phi\, \mathbf{I} + \hat u \hat u^T \dot\phi \sin\phi + \tilde u \dot\phi \cos\phi \\
&= \dot\phi \left[-\sin\phi\, \mathbf{I} + \hat u\, \hat u^T \sin\phi + \tilde u \cos\phi \right] \\
&= \dot\phi \left[\cos\phi\, \tilde u + \left(\hat u\, \hat u^T - \mathbf{I} \right) \sin\phi \right] \\
&= \dot\phi \left[\cos\phi\, \tilde u + \tilde u^2 \sin\phi + \tilde u\, \hat u\, \hat u^T \operatorname{vers}\phi \right] \\
&= \dot\phi \tilde u \left[\mathbf{I} \cos\phi + \tilde u \sin\phi + \hat u\, \hat u^T \operatorname{vers}\phi \right] \\
&= \dot\phi\, \tilde u\, R_{\hat u,\phi}
\end{aligned} \tag{7.116}
$$

and therefore,

$$
_G\tilde\omega_B = \dot R_{\hat u,\phi}\, R_{\hat u,\phi}^T = \dot\phi\, \tilde u\, R_{\hat u,\phi}\, R_{\hat u,\phi}^T = \dot\phi\, \tilde u \tag{7.117}
$$

Example 230 ★ Angular velocity coordinate transformation. Angular velocity expression in different coordinate frames is different. The easiest way to express it in another frame is to use its skew symmetric expression and use the rule of this example.

Angular velocity $_1^1\boldsymbol{\omega}_2$ of coordinate frame B_2 with respect to B_1 and expressed in B_1 can be expressed in the base coordinate frame B_0 according to the following rule:

$$
^0R_1 \, _1\tilde\omega_2 \, ^0R_1^T = \, _1^0\tilde\omega_2 \tag{7.118}
$$

To show this equation, it is enough to apply both sides on an arbitrary vector $^0\mathbf{r}$. Therefore, the left-hand side would be

$$
\begin{aligned}
{}^0R_1 \, {}_1\tilde{\omega}_2 \, {}^0R_1^T \, {}^0\mathbf{r} &= {}^0R_1 \, {}_1\tilde{\omega}_2 \, {}^1R_0 \, {}^0\mathbf{r} = {}^0R_1 \, {}_1\tilde{\omega}_2 \, {}^1\mathbf{r} \\
&= {}^0R_1 \left({}_1\boldsymbol{\omega}_2 \times {}^1\mathbf{r} \right) = {}^0R_1 \, {}_1\boldsymbol{\omega}_2 \times {}^0R_1 \, {}^1\mathbf{r} \\
&= {}^0_1\boldsymbol{\omega}_2 \times {}^0\mathbf{r}
\end{aligned}
\tag{7.119}
$$

which is equal to the right-hand side after applying on the vector ${}^0\mathbf{r}$.

$$
{}^0_1\tilde{\omega}_2 \, {}^0\mathbf{r} = {}^0_1\boldsymbol{\omega}_2 \times {}^0\mathbf{r}
\tag{7.120}
$$

The transformation rule for B and G frames will be

$$
{}^BR_G \, {}_G\tilde{\omega}_B \, {}^BR_G^T = {}^B_G\tilde{\omega}_B
\tag{7.121}
$$

It is equivalent to

$$
{}^BR_G \, {}_G\boldsymbol{\omega}_B = {}^B_G\boldsymbol{\omega}_B
\tag{7.122}
$$

Example 231 ★ Time derivative of unit vectors. Unit vectors have constant length, and therefore, their time derivative will only be measured by their rotation.

Using Eq. (7.27), we can define the time G-derivative of unit vectors of a body coordinate frame $B(\hat{\imath}, \hat{\jmath}, \hat{k})$, rotating in the global coordinate frame $G(\hat{I}, \hat{J}, \hat{K})$.

$$
\frac{{}^G d\hat{\imath}}{dt} = {}^B_G\boldsymbol{\omega}_B \times \hat{\imath}
\tag{7.123}
$$

$$
\frac{{}^G d\hat{\jmath}}{dt} = {}^B_G\boldsymbol{\omega}_B \times \hat{\jmath}
\tag{7.124}
$$

$$
\frac{{}^G d\hat{k}}{dt} = {}^B_G\boldsymbol{\omega}_B \times \hat{k}
\tag{7.125}
$$

Example 232 ★ Derivative of a quaternion. All physical quantities may be defined by quaternions, and hence, they will have derivatives. Here is to show how to take derivative of a quaternion with respect to a parameter.

Consider a quaternion q as a function of a scalar parameter s.

$$
\begin{aligned}
q &= p_0(s) + \mathbf{p}(s) \\
&= p_0(s) + p_1(s)\,i + p_2(s)\,j + p_3(s)\,k
\end{aligned}
\tag{7.126}
$$

The derivative of q with respect to s is

$$
\frac{dq}{ds} = \frac{dp_0}{ds} + \frac{dp_1}{ds}i + \frac{dp_2}{ds}j + \frac{dp_3}{ds}k
\tag{7.127}
$$

If the parameter s is time t, we have

$$
\frac{dq}{dt} = \frac{dp_0}{dt} + \frac{dp_1}{dt}i + \frac{dp_2}{dt}j + \frac{dp_3}{dt}k
\tag{7.128}
$$

Let us show the position vector of a moving particle by a quaternion as

$$
\overleftrightarrow{r} = t^2 + \mathbf{r}(t) = t^2 + r_1(t)\,i + r_2(t)\,j + r_3(t)\,k
\tag{7.129}
$$

The velocity and acceleration vectors of the particle will be

$$
\overleftrightarrow{v} = \left(\frac{d}{dt} + \mathbf{0} \right) \left(t^2 + \mathbf{r} \right) = 2t + \mathbf{v}(t)
\tag{7.130}
$$

$$\overleftrightarrow{a} = \left(\frac{d}{dt} + \mathbf{0}\right)(2t + \mathbf{v}) = 2 + \mathbf{a}(t) \tag{7.131}$$

Example 233 ★ Angular velocity in terms of quaternion and Euler parameters. Employing quaternion and Euler parameters to express rotation transformation, we may define angular velocity by quaternions.

Starting from the unit quaternion representation of a finite rotation

$$^G\mathbf{r} = e(t)\,^B\mathbf{r}\,e^*(t) = e(t)\,^B\mathbf{r}\,e^{-1}(t) \tag{7.132}$$

where

$$e = e_0 + \mathbf{e} \tag{7.133}$$
$$e^* = e^{-1} = e_0 - \mathbf{e} \tag{7.134}$$

we can find

$$^G\dot{\mathbf{r}} = \dot{e}\,^B\mathbf{r}\,e^* + e\,^B\mathbf{r}\,\dot{e}^* = \dot{e}\,e^*\,^G\mathbf{r} + \,^G\mathbf{r}\,e\,\dot{e}^* = 2\dot{e}\,e^*\,^G\mathbf{r} \tag{7.135}$$

and therefore, the angular velocity vector can also be expressed based on quaternions.

$$_G\boldsymbol{\omega}_B = 2\dot{e}\,e^* \tag{7.136}$$

We have used the orthogonality property of unit quaternion

$$e\,e^{-1} = e\,e^* = 1 \tag{7.137}$$

which provides

$$\dot{e}\,e^* + e\,\dot{e}^* = 0 \tag{7.138}$$

The angular velocity quaternion can be expanded using quaternion products to find the angular velocity components based on Euler parameters.

$$\begin{aligned}
_G\boldsymbol{\omega}_B &= 2\dot{e}\,e^* = 2(\dot{e}_0 + \dot{\mathbf{e}})(e_0 - \mathbf{e}) \\
&= 2(\dot{e}_0 e_0 + e_0\dot{\mathbf{e}} - \dot{e}_0\mathbf{e} + \dot{\mathbf{e}}\cdot\mathbf{e} - \dot{\mathbf{e}}\times\mathbf{e}) \\
&= 2\begin{bmatrix}
0 \\
e_0\dot{e}_1 - e_1\dot{e}_0 + e_2\dot{e}_3 - e_3\dot{e}_2 \\
e_0\dot{e}_2 - e_2\dot{e}_0 - e_1\dot{e}_3 + e_3\dot{e}_1 \\
e_0\dot{e}_3 + e_1\dot{e}_2 - e_2\dot{e}_1 - e_3\dot{e}_0
\end{bmatrix} \\
&= 2\begin{bmatrix}
\dot{e}_0 & -\dot{e}_1 & -\dot{e}_2 & -\dot{e}_3 \\
\dot{e}_1 & \dot{e}_0 & -\dot{e}_3 & \dot{e}_2 \\
\dot{e}_2 & \dot{e}_3 & \dot{e}_0 & -\dot{e}_1 \\
\dot{e}_3 & -\dot{e}_2 & \dot{e}_1 & \dot{e}_0
\end{bmatrix}\begin{bmatrix}
e_0 \\
-e_1 \\
-e_2 \\
-e_3
\end{bmatrix}
\end{aligned} \tag{7.139}$$

The scalar component of the angular velocity quaternion is zero.

$$\dot{e}_0 e_0 + \dot{\mathbf{e}}\cdot\mathbf{e} = \dot{e}_0 e_0 + e_1\dot{e}_1 + e_2\dot{e}_2 + e_3\dot{e}_3 = 0 \tag{7.140}$$

The angular velocity vector can also be defined as a quaternion

$$\overleftrightarrow{_G\boldsymbol{\omega}_B} = 2\,\overleftrightarrow{\dot{e}}\,\overleftrightarrow{e^*} \tag{7.141}$$

to be utilized for definition of the derivative of a rotation quaternion.

$$\overset{\leftrightarrow}{\dot{e}} = \frac{1}{2} \overset{\longleftrightarrow}{_G\boldsymbol{\omega}_B} \overset{\leftrightarrow}{e}$$ (7.142)

A coordinate transformation can transform the angular velocity into a body coordinate frame

$$
{}_G^B\boldsymbol{\omega}_B = e^* \, {}_G^G\boldsymbol{\omega}_B \, e = 2e^* \dot{e}
$$

$$
= 2 \begin{bmatrix} e_0 & e_1 & e_2 & e_3 \\ -e_1 & e_0 & e_3 & -e_2 \\ -e_2 & -e_3 & e_0 & e_1 \\ -e_3 & -e_2 & e_1 & e_0 \end{bmatrix} \begin{bmatrix} \dot{e}_0 \\ \dot{e}_1 \\ \dot{e}_2 \\ \dot{e}_3 \end{bmatrix}
$$ (7.143)

and therefore,

$$\overset{\longleftrightarrow}{_G^B\boldsymbol{\omega}_B} = 2 \overset{\leftrightarrow}{e^*} \overset{\leftrightarrow}{\dot{e}}$$ (7.144)

$$\overset{\leftrightarrow}{\dot{e}} = \frac{1}{2} \overset{\leftrightarrow}{e} \overset{\longleftrightarrow}{_G^B\boldsymbol{\omega}_B}$$ (7.145)

Example 234 ★ Differential of Euler parameters. The definition of angular velocity by $\dot{R}$ and R requires time derivatives of the components of R. Here, we derived relationships derivative of components of R based on Euler parameters.

The rotation matrix ${}^G R_B$ based on Euler parameters is given in Eq. (3.140).

$$
{}^G R_B = \begin{bmatrix} e_0^2 + e_1^2 - e_2^2 - e_3^2 & 2(e_1 e_2 - e_0 e_3) & 2(e_0 e_2 + e_1 e_3) \\ 2(e_0 e_3 + e_1 e_2) & e_0^2 - e_1^2 + e_2^2 - e_3^2 & 2(e_2 e_3 - e_0 e_1) \\ 2(e_1 e_3 - e_0 e_2) & 2(e_0 e_1 + e_2 e_3) & e_0^2 - e_1^2 - e_2^2 + e_3^2 \end{bmatrix}
$$

$$
= \begin{bmatrix} r_{11} & r_{12} & r_{13} \\ r_{21} & r_{22} & r_{23} \\ r_{31} & r_{32} & r_{33} \end{bmatrix}
$$ (7.146)

The individual parameters can be found from any set of Eqs. (3.147)–(3.150). The first set indicates

$$
e_0 = \pm \frac{1}{2} \sqrt{1 + r_{11} + r_{22} + r_{33}}
$$

$$
e_1 = \frac{1}{4} \frac{r_{32} - r_{23}}{e_0} \qquad e_2 = \frac{1}{4} \frac{r_{13} - r_{31}}{e_0} \qquad e_3 = \frac{1}{4} \frac{r_{21} - r_{12}}{e_0}
$$ (7.147)

and therefore,

$$
\dot{e}_0 = \frac{\dot{r}_{11} + \dot{r}_{22} + \dot{r}_{33}}{8 e_0}
$$ (7.148)

$$
\dot{e}_1 = \frac{1}{4 e_0^2} \left((\dot{r}_{32} - \dot{r}_{23}) e_0 - (r_{32} - r_{23}) \dot{e}_0 \right)
$$ (7.149)

$$
\dot{e}_2 = \frac{1}{4 e_0^2} \left((\dot{r}_{13} - \dot{r}_{31}) e_0 - (r_{13} - r_{31}) \dot{e}_0 \right)
$$ (7.150)

$$
\dot{e}_3 = \frac{1}{4 e_0^2} \left((\dot{r}_{21} - \dot{r}_{12}) e_0 - (r_{21} - r_{12}) \dot{e}_0 \right)
$$ (7.151)

We may use the differential of the transformation matrix

$$
{}^G \dot{R}_B = {}_G\tilde{\omega}_B \, {}^G R_B
$$ (7.152)

to show

$$
\dot{e}_0 = \frac{1}{2} \left(-e_1 \omega_1 - e_2 \omega_2 - e_3 \omega_3 \right)
$$ (7.153)

$$\dot{e}_1 = \frac{1}{2}\left(e_0\omega_1 + e_2\omega_3 - e_3\omega_2\right) \tag{7.154}$$

$$\dot{e}_2 = \frac{1}{2}\left(e_0\omega_2 - e_1\omega_3 - e_3\omega_1\right) \tag{7.155}$$

$$\dot{e}_3 = \frac{1}{2}\left(e_0\omega_3 + e_1\omega_2 - e_2\omega_1\right) \tag{7.156}$$

Similarly, we can find $\dot{e}_1$, $\dot{e}_2$, and $\dot{e}_3$ and set the final result in a matrix form as

$$\begin{bmatrix} \dot{e}_0 \\ \dot{e}_1 \\ \dot{e}_2 \\ \dot{e}_3 \end{bmatrix} = \frac{1}{2} \begin{bmatrix} 0 & -\omega_1 & -\omega_2 & -\omega_3 \\ \omega_1 & 0 & \omega_3 & -\omega_2 \\ \omega_2 & -\omega_3 & 0 & \omega_1 \\ \omega_3 & \omega_2 & -\omega_1 & 0 \end{bmatrix} \begin{bmatrix} e_0 \\ e_1 \\ e_2 \\ e_3 \end{bmatrix} \tag{7.157}$$

or

$$\begin{bmatrix} \dot{e}_0 \\ \dot{e}_1 \\ \dot{e}_2 \\ \dot{e}_3 \end{bmatrix} = \frac{1}{2} \begin{bmatrix} e_0 & -e_3 & -e_2 & -e_1 \\ e_1 & e_0 & -e_3 & e_2 \\ e_2 & e_1 & e_0 & -e_3 \\ e_3 & -e_2 & e_1 & e_0 \end{bmatrix} \begin{bmatrix} 0 \\ \omega_1 \\ \omega_2 \\ \omega_3 \end{bmatrix} \tag{7.158}$$

Example 235 ★ Elements of the angular velocity matrix. Employing index permutation index notation, we may show the angular velocity components in a compact form.

Utilizing the permutation symbol introduced in (3.207)

$$\epsilon_{ijk} = \frac{1}{2}(i-j)(j-k)(k-i) \qquad i,j,k = 1,2,3 \tag{7.159}$$

allows us to find the elements of the angular velocity matrix, $\tilde{\omega}$, when the angular velocity vector, $\boldsymbol{\omega} = \begin{bmatrix} \omega_1 & \omega_2 & \omega_3 \end{bmatrix}^T$, is given.

$$\tilde{\omega}_{ij} = \epsilon_{ijk}\,\omega_k \tag{7.160}$$

7.2   ★ Time Derivative and Coordinate Frames

Time derivative of a vector quantity depends on the coordinate frame in which we are taking the derivative and the coordinate frame in which the vector is expressed. The time derivative of a vector $\mathbf{r}$ in the global frame is called *G-derivative* and is denoted by

$$\frac{{}^G d}{dt}\mathbf{r} \tag{7.161}$$

and the time derivative of the vector in the body frame is called the *B-derivative* and is denoted by

$$\frac{{}^B d}{dt}\mathbf{r} \tag{7.162}$$

The left superscript on the derivative symbol indicates the coordinate frame in which the derivative is taken. The unit vectors of the derivative frame are considered constant. The derivative of a vector in the same frame in which it is expressed is called a *simple derivative*. The derivative of the G-vector, ${}^G\mathbf{r}$ in G, and the derivative of the B-vector, ${}^B\mathbf{r}$ in B, are

$$ {}^B\mathbf{r} = x\hat{i} + y\hat{j} + z\hat{k} \tag{7.163}$$

$$\frac{{}^B d}{dt}\,{}^B\mathbf{r} = {}^B\dot{\mathbf{r}} = {}^B\mathbf{v} = \dot{x}\,\hat{i} + \dot{y}\,\hat{j} + \dot{z}\,\hat{k} \tag{7.164}$$

$$^G\mathbf{r} = X\hat{I} + Y\hat{J} + Z\hat{K} \tag{7.165}$$

$$\frac{^Gd}{dt}\,^G\mathbf{r} = \,^G\dot{\mathbf{r}} = \,^G\mathbf{v} = \dot{X}\hat{I} + \dot{Y}\hat{J} + \dot{Z}\hat{K} \tag{7.166}$$

It is also possible to calculate the G-derivative of $^B\mathbf{r}$ and the B-derivative of $^G\mathbf{r}$. The derivative of a vector in a frame other than the frame in which it is expressed is called a *mixed derivative*. The result of a mixed derivative would be a vector indicated by two left indices. Its left subscript indicates the frame in which the derivative is taken, and its left superscript indicates the frame it is expressed in. Derivative operation will not change the expression frame. We define the G-derivative of a body vector $^B\mathbf{r}$ by

$$^B_G\mathbf{v} = \,^B_G\dot{\mathbf{r}} = \frac{^Gd}{dt}\,^B\mathbf{r} \tag{7.167}$$

and call $^B_G\mathbf{v}$ the B-expression of the G-velocity. Similarly, the B-derivative of a global vector $^G\mathbf{r}$ is given as

$$^G_B\mathbf{v} = \,^G_B\dot{\mathbf{r}} = \frac{^Bd}{dt}\,^G\mathbf{r} \tag{7.168}$$

and we call $^G_B\mathbf{v}$ the G-expression of the B-velocity.

Whenever it is clear that we are working with velocity vectors, we may use the signs $^B_G\mathbf{v}$ and $^G_B\mathbf{v}$, which are short notations for $\frac{^Gd}{dt}\,^B\mathbf{r}$ and $\frac{^Bd}{dt}\,^G\mathbf{r}$ respectively. The left superscript of $^B_G\mathbf{v}$ indicates the frame in which $\mathbf{v}$ is expressed and the left subscript indicates the frame in which the derivative is taken. If the left superscript and subscript of a derivative vector are the same, we only keep the superscript. To read $^B_G\mathbf{v}$ and $^G_B\mathbf{v}$, we may use the expressions *G-velocity of a B-vector* for $^B_G\mathbf{v}$ and *B-velocity of a G-vector* for $^G_B\mathbf{v}$.

When the interested point P is not a fixed point in B, then, P is moving in frame B and $^B\mathbf{r}_P = \,^B\mathbf{r}_P(t)$ is not a constant vector. The G-derivative of $^B\mathbf{r}(t)$ is defined by

$$\frac{^Gd}{dt}\,^B\mathbf{r}(t) = \,^B_G\mathbf{v} = \,^B_G\dot{\mathbf{r}} = \frac{^Bd}{dt}\,^B\mathbf{r} + \,^B_G\boldsymbol{\omega}_B \times \,^B\mathbf{r}$$

$$= \,^B\mathbf{v} + \,^B_G\boldsymbol{\omega}_B \times \,^B\mathbf{r} \tag{7.169}$$

and the B-derivative of $^G\mathbf{r}(t)$ is defined by

$$\frac{^Bd}{dt}\,^G\mathbf{r}(t) = \,^G_B\mathbf{v} = \,^G_B\dot{\mathbf{r}} = \frac{^Gd}{dt}\,^G\mathbf{r}(t) - \,_G\boldsymbol{\omega}_B \times \,^G\mathbf{r}$$

$$= \,^G\mathbf{v} - \,_G\boldsymbol{\omega}_B \times \,^G\mathbf{r} \tag{7.170}$$

The G-derivative of a B-vector is a B-vector and the B-derivative of a G-vector is a G-vector.

Proof Let $G(OXYZ)$ with unit vectors $\hat{I}$, $\hat{J}$, and $\hat{K}$ to be the global coordinate frame and $B(Oxyz)$ with unit vectors $\hat{i}$, $\hat{j}$, and $\hat{k}$ to be a body coordinate frame. A point P is moving in B-frame, indicated by a time varying position vector, as shown in Fig. 7.3. The position vector of P can be expressed in the body and global frames.

$$^B\mathbf{r}_P(t) = x(t)\hat{i} + y(t)\hat{j} + z(t)\hat{k} \tag{7.171}$$

$$^G\mathbf{r}_P(t) = X(t)\hat{I} + Y(t)\hat{J} + Z(t)\hat{K} \tag{7.172}$$

The time derivative of $^B\mathbf{r}_P$ in B and $^G\mathbf{r}_P$ in G are

$$\frac{^Bd}{dt}\,^B\mathbf{r}_P = \,^B\dot{\mathbf{r}}_P = \,^B\mathbf{v}_P = \dot{x}\hat{i} + \dot{y}\hat{j} + \dot{z}\hat{k} \tag{7.173}$$

$$\frac{^Gd}{dt}\,^G\mathbf{r}_P = \,^G\dot{\mathbf{r}}_P = \,^G\mathbf{v}_P = \dot{X}\hat{I} + \dot{Y}\hat{J} + \dot{Z}\hat{K} \tag{7.174}$$

because the unit vectors of B in Eq. (7.173) and the unit vectors of G in Eq. (7.174) are considered constant.

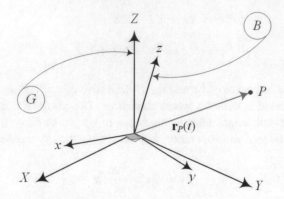

Fig. 7.3 A moving body point P at $^B\mathbf{r}(t)$ in the rotating body frame B

Using the definition (7.167), we find the G-derivative of the B-vector $^B\mathbf{r}_P$.

$$\frac{^Gd}{dt}\,^B\mathbf{r} = \frac{^Gd}{dt}\left(x\hat{\imath} + y\hat{\jmath} + z\hat{k}\right)$$

$$= \dot{x}\hat{\imath} + \dot{y}\hat{\jmath} + \dot{z}\hat{k} + x\frac{^Gd\hat{\imath}}{dt} + y\frac{^Gd\hat{\jmath}}{dt} + z\frac{^Gd\hat{k}}{dt}$$

$$= \dot{x}\hat{\imath} + \dot{y}\hat{\jmath} + \dot{z}\hat{k} + x\left(^B_G\boldsymbol{\omega}_B \times \hat{\imath}\right) + y\left(^B_G\boldsymbol{\omega}_B \times \hat{\jmath}\right) + z\left(^B_G\boldsymbol{\omega}_B \times \hat{k}\right)$$

$$= \dot{x}\hat{\imath} + \dot{y}\hat{\jmath} + \dot{z}\hat{k} + {}^B_G\boldsymbol{\omega}_B \times \left(x\hat{\imath} + y\hat{\jmath} + z\hat{k}\right)$$

$$= \frac{^Bd}{dt}\,^B\mathbf{r} + {}^B_G\boldsymbol{\omega}_B \times {}^B\mathbf{r} = {}^B\mathbf{v} + {}^B_G\boldsymbol{\omega}_B \times {}^B\mathbf{r} \tag{7.175}$$

We achieved this result because the x-, y-, and z-components of $^B\mathbf{r}_P$ are scalar. Scalars are invariant with respect to coordinate frame transformations. Therefore, if x is a scalar, then

$$\frac{^Gd}{dt}x = \frac{^Bd}{dt}x = \dot{x} \tag{7.176}$$

The B-derivative of $^G\mathbf{r}_P$ is

$$\frac{^Bd}{dt}\,^G\mathbf{r} = \frac{^Bd}{dt}\left(X\hat{I} + Y\hat{J} + Z\hat{K}\right)$$

$$= \dot{X}\hat{I} + \dot{Y}\hat{J} + \dot{Z}\hat{K} + X\frac{^Bd\hat{I}}{dt} + Y\frac{^Bd\hat{J}}{dt} + Z\frac{^Bd\hat{K}}{dt}$$

$$= \dot{X}\hat{I} + \dot{Y}\hat{J} + \dot{Z}\hat{K}$$

$$\quad + X\left(^G_B\boldsymbol{\omega}_G \times \hat{I}\right) + Y\left(^G_B\boldsymbol{\omega}_G \times \hat{J}\right) + Z\left(^G_B\boldsymbol{\omega}_G \times \hat{K}\right)$$

$$= \dot{X}\hat{I} + \dot{Y}\hat{J} + \dot{Z}\hat{K} + {}^G_B\boldsymbol{\omega}_G \times \left(X\hat{I} + Y\hat{J} + Z\hat{K}\right)$$

$$= \frac{^Gd}{dt}\,^G\mathbf{r} + {}^G_B\boldsymbol{\omega}_G \times {}^G\mathbf{r} = {}^G\mathbf{v} + {}^G_B\boldsymbol{\omega}_G \times {}^G\mathbf{r}$$

$$= {}^G\mathbf{v} - {}_G\boldsymbol{\omega}_B \times {}^G\mathbf{r} \tag{7.177}$$

The angular velocity of B relative to G is a vector quantity and can be expressed in either frames, which can be transformed from one frame to the other according to (7.10) and (7.11).

$$_G^G\boldsymbol{\omega}_B = \omega_X \,\hat{I} + \omega_Y \,\hat{J} + \omega_Z \,\hat{K} \tag{7.178}$$

$$_G^B\boldsymbol{\omega}_B = \omega_x \hat{\imath} + \omega_y \hat{\jmath} + \omega_z \hat{k} \tag{7.179}$$

Using equations of simple and mixed derivatives (7.164)–(7.170), we define the simple integrals as

$$^G\!\int {}^G\mathbf{v}dt = {}^G\!\int \left(\dot{X}\,\hat{I} + \dot{Y}\,\hat{J} + \dot{Z}\,\hat{K}\right)dt = X\,\hat{\imath} + Y\,\hat{\jmath} + Z\,\hat{k} \tag{7.180}$$

$$^B\!\int {}^B\mathbf{v}dt = {}^B\!\int \left(\dot{x}\,\hat{\imath} + \dot{y}\,\hat{\jmath} + \dot{z}\,\hat{k}\right)dt = x\,\hat{\imath} + y\,\hat{\jmath} + z\,\hat{k}$$

$$\tag{7.181}$$

Similarly, the mixed integrals of a body vector are

$$^G\!\int {}_G^B\mathbf{v}dt = {}^B\mathbf{r} = {}_G^B\tilde{\omega}_B^T \, {}_G^B\mathbf{v} \tag{7.182}$$

$$^B\!\int {}_B^G\mathbf{v}dt = {}^G\mathbf{r} = -{}_G\tilde{\omega}_B^T \, {}_B^G\mathbf{v}$$

$$\tag{7.183}$$

If a point P is moving in B, then the mixed integrals would be

$$^G\!\int {}_G^B\mathbf{v}dt = {}^B\mathbf{r} = {}_G^B\tilde{\omega}_B^T \left({}_G^B\mathbf{v} - {}^B\mathbf{v}\right) \tag{7.184}$$

$$\frac{^Gd}{dt}\,{}^B\mathbf{r}\,(t) = {}_G^B\mathbf{v} = {}_G^B\dot{\mathbf{r}} = \frac{^Bd}{dt}\,{}^B\mathbf{r} + {}_G^B\boldsymbol{\omega}_B \times {}^B\mathbf{r}$$

$$= {}^B\mathbf{v} + {}_G^B\boldsymbol{\omega}_B \times {}^B\mathbf{r} \tag{7.185}$$

$$\frac{^Bd}{dt}\,{}^G\mathbf{r}\,(t) = {}_B^G\mathbf{v} = {}_B^G\dot{\mathbf{r}} = \frac{^Gd}{dt}\,{}^G\mathbf{r}\,(t) - {}_G\boldsymbol{\omega}_B \times {}^G\mathbf{r}$$

$$= {}^G\mathbf{v} - {}_G\boldsymbol{\omega}_B \times {}^G\mathbf{r} \tag{7.186}$$

∎

Example 236 Rotation of B about Z-axis. A B-frame is rotating in G-frame with a given angular velocity. The velocity of a fixed point in B-frame is calculated by a time derivative of the position vector of the point.

A body frame B is rotating in G with angular velocity $\dot{\alpha}$ about the Z-axis. Therefore, a B-fixed point at $^B\mathbf{r}$ will be seen by an observer in G-frame at

$$^G\mathbf{r}_P = {}^GR_B\,{}^B\mathbf{r} = R_{Z,\alpha}(t)\,{}^B\mathbf{r} \tag{7.187}$$

$$- \begin{bmatrix} \cos\alpha & -\sin\alpha & 0 \\ \sin\alpha & \cos\alpha & 0 \\ 0 & 0 & 1 \end{bmatrix} \begin{bmatrix} x \\ y \\ z \end{bmatrix} = \begin{bmatrix} x\cos\alpha - y\sin\alpha \\ y\cos\alpha + x\sin\alpha \\ z \end{bmatrix}$$

The angular velocity matrix of B in G is

$$_G\tilde{\omega}_B = {}^G\dot{R}_B\,{}^GR_B^T = \dot{\alpha}\tilde{K} \tag{7.188}$$

$$_G\boldsymbol{\omega}_B = \dot{\alpha}\hat{K} \tag{7.189}$$

The B-expression of $_G\tilde{\omega}_B$ and $_G\boldsymbol{\omega}_B$ are

$$_G^B\tilde{\omega}_B = {}^GR_B^T\,{}_G^G\tilde{\omega}_B\,{}^GR_B = \dot{\alpha}\tilde{k} \tag{7.190}$$

$$_G^B\boldsymbol{\omega}_B = \dot{\alpha}\hat{k} \tag{7.191}$$

Now, we can calculate the following simple derivatives:

$$\frac{^Bd}{dt}\,{}^B\mathbf{r} = {}^B\dot{\mathbf{r}} = 0 \tag{7.192}$$

$$\frac{^Gd}{dt}\,{}^G\mathbf{r} = {}^G\dot{\mathbf{r}} = \frac{^Gd}{dt}\begin{bmatrix} x\cos\alpha - y\sin\alpha \\ y\cos\alpha + x\sin\alpha \\ z \end{bmatrix} \tag{7.193}$$

$$= (-x\dot{\alpha}\sin\alpha - y\dot{\alpha}\cos\alpha)\,\hat{I} + (x\dot{\alpha}\cos\alpha - y\dot{\alpha}\sin\alpha)\,\hat{J} + \dot{z}\hat{K}$$

For the mixed derivatives, we begin with the global velocity expressed in B.

$$\frac{^Gd}{dt}\,{}^B\mathbf{r} = {}_G^B\mathbf{v} = {}_G^B\dot{\mathbf{r}} = {}_G^B\boldsymbol{\omega}_B \times {}^B\mathbf{r}$$

$$= \dot{\alpha}\begin{bmatrix} 0 \\ 0 \\ 1 \end{bmatrix} \times \begin{bmatrix} x \\ y \\ z \end{bmatrix} = \dot{\alpha}\begin{bmatrix} -y \\ x \\ 0 \end{bmatrix}$$

$$= -y\dot{\alpha}\hat{i} + x\dot{\alpha}\hat{j} = {}_G^B\dot{\mathbf{r}} \tag{7.194}$$

We may transform the velocity vector $_G^B\dot{\mathbf{r}}$ to the global frame to find the global expression velocity, $^G\dot{\mathbf{r}}$, which is in agreement with (7.193).

$$^G\dot{\mathbf{r}} = {}^GR_B\,{}_G^B\dot{\mathbf{r}}$$

$$= \dot{\alpha}\begin{bmatrix} \cos\alpha & -\sin\alpha & 0 \\ \sin\alpha & \cos\alpha & 0 \\ 0 & 0 & 1 \end{bmatrix}\begin{bmatrix} -y \\ x \\ 0 \end{bmatrix} = \dot{\alpha}\begin{bmatrix} -y\cos\alpha - x\sin\alpha \\ x\cos\alpha - y\sin\alpha \\ 0 \end{bmatrix}$$

$$= \dot{\alpha}\left(-y\cos\alpha - x\sin\alpha\right)\hat{I} + \dot{\alpha}\left(x\cos\alpha - y\sin\alpha\right)\hat{J} \tag{7.195}$$

The next derivative is the velocity of body points relative to B and expressed in G.

$$\frac{^Bd}{dt}\,{}^G\mathbf{r} = {}_B^G\mathbf{v} = {}_B^G\dot{\mathbf{r}} = {}^G\dot{\mathbf{r}} - {}_G\boldsymbol{\omega}_B \times {}^G\mathbf{r} \tag{7.196}$$

$$= \dot{\alpha}\begin{bmatrix} -y\cos\alpha - x\sin\alpha \\ x\cos\alpha - y\sin\alpha \\ 0 \end{bmatrix} - \dot{\alpha}\begin{bmatrix} 0 \\ 0 \\ 1 \end{bmatrix}$$

$$\times \begin{bmatrix} x\cos\alpha - y\sin\alpha \\ y\cos\alpha + x\sin\alpha \\ z \end{bmatrix} = \begin{bmatrix} 0 \\ 0 \\ 0 \end{bmatrix}$$

Example 237 Time derivative of a moving point in B. A B-frame is rotating in G-frame with a given angular velocity. The velocity of a moving point in B-frame is calculated by a time derivative of the variable position vector of the point.

Consider a local frame B, rotating in G by $\dot\alpha$ about the Z-axis, and a moving point in B with variable position vector $^B\mathbf{r}_P(t) = t\hat\imath$. Therefore,

$$^G\mathbf{r}_P = {}^G R_B \, {}^B\mathbf{r}_P = R_{Z,\alpha}(t) \, {}^B\mathbf{r}_P \tag{7.197}$$

$$= \begin{bmatrix} \cos\alpha & -\sin\alpha & 0 \\ \sin\alpha & \cos\alpha & 0 \\ 0 & 0 & 1 \end{bmatrix} \begin{bmatrix} t \\ 0 \\ 0 \end{bmatrix} = t\cos\alpha\,\hat{I} + t\sin\alpha\,\hat{J}$$

The angular velocity matrix is

$$_G\tilde\omega_B = {}^G\dot{R}_B \, {}^G R_B^T = \dot\alpha\tilde{K} \tag{7.198}$$

$$_G\boldsymbol\omega_B = \dot\alpha\hat{K} \tag{7.199}$$

It can also be verified that

$$_G^B\tilde\omega_B = {}^G R_B^T \, _G\tilde\omega_B \, {}^G R_B = \dot\alpha\tilde{k} \tag{7.200}$$

$$_G^B\boldsymbol\omega_B = \dot\alpha\hat{k} \tag{7.201}$$

The simple derivatives of $^B\mathbf{r}_P$ and $^G\mathbf{r}_P$ are

$$\frac{^Bd}{dt}{}^B\mathbf{r}_P = {}^B\dot{\mathbf{r}}_P = \hat\imath \tag{7.202}$$

$$\frac{^Gd}{dt}{}^G\mathbf{r}_P = {}^G\dot{\mathbf{r}}_P$$

$$= (\cos\alpha - t\dot\alpha\sin\alpha)\,\hat{I} + (\sin\alpha + t\dot\alpha\cos\alpha)\,\hat{J} \tag{7.203}$$

To calculate the mixed derivatives, we begin with $\frac{^Gd}{dt}{}^B\mathbf{r}_P$.

$$\frac{^Gd}{dt}{}^B\mathbf{r}_P = \frac{^Bd}{dt}{}^B\mathbf{r}_P + {}_G^B\boldsymbol\omega_B \times {}^B\mathbf{r}_P$$

$$= \begin{bmatrix} 1 \\ 0 \\ 0 \end{bmatrix} + \dot\alpha \begin{bmatrix} 0 \\ 0 \\ 1 \end{bmatrix} \times \begin{bmatrix} t \\ 0 \\ 0 \end{bmatrix} = \begin{bmatrix} 1 \\ t\dot\alpha \\ 0 \end{bmatrix}$$

$$= \hat\imath + t\dot\alpha\hat\jmath = {}_G^B\dot{\mathbf{r}}_P \tag{7.204}$$

It is the global velocity of P expressed in B. We may, however, transform $_G^B\dot{\mathbf{r}}_P$ to the global frame and find the global velocity expressed in G.

$$^G\dot{\mathbf{r}}_P = {}^G R_B \, _G^B\dot{\mathbf{r}}_P$$

$$= \begin{bmatrix} \cos\alpha & -\sin\alpha & 0 \\ \sin\alpha & \cos\alpha & 0 \\ 0 & 0 & 1 \end{bmatrix} \begin{bmatrix} 1 \\ t\dot\alpha \\ 0 \end{bmatrix} = \begin{bmatrix} \cos\alpha & t\dot\alpha\sin\alpha \\ \sin\alpha + t\dot\alpha\cos\alpha \\ 0 \end{bmatrix}$$

$$= (\cos\alpha - t\dot\alpha\sin\alpha)\,\hat{I} + (\sin\alpha + t\dot\alpha\cos\alpha)\,\hat{J} \tag{7.205}$$

The next mixed derivative is $\frac{^Bd}{dt}\,{}^G\mathbf{r}_P$.

$$\frac{^Bd}{dt}\,{}^G\mathbf{r}_P = {}^G\dot{\mathbf{r}}_P - {}_G\boldsymbol{\omega}_B \times {}^G\mathbf{r}_P$$

$$= \begin{bmatrix} \cos\alpha - t\dot{\alpha}\sin\alpha \\ \sin\alpha + t\dot{\alpha}\cos\alpha \\ 0 \end{bmatrix} - \dot{\alpha}\begin{bmatrix} 0 \\ 0 \\ 1 \end{bmatrix} \times \begin{bmatrix} t\cos\alpha \\ t\sin\alpha \\ 0 \end{bmatrix}$$

$$= \begin{bmatrix} \cos\alpha \\ \sin\alpha \\ 0 \end{bmatrix} = (\cos\alpha)\,\hat{I} + (\sin\alpha)\,\hat{J} = {}^G_B\dot{\mathbf{r}}_P \tag{7.206}$$

It is the velocity of P relative to B and expressed in G. To express this velocity in B, we apply a frame transformation.

$$^B\dot{\mathbf{r}}_P = {}^GR_B^T\,{}^G_B\dot{\mathbf{r}}_P$$

$$= \begin{bmatrix} \cos\alpha & -\sin\alpha & 0 \\ \sin\alpha & \cos\alpha & 0 \\ 0 & 0 & 1 \end{bmatrix}^T \begin{bmatrix} \cos\alpha \\ \sin\alpha \\ 0 \end{bmatrix} = \begin{bmatrix} 1 \\ 0 \\ 0 \end{bmatrix} = \hat{\imath} \tag{7.207}$$

It might be more applied if we first transform the vector to the same frame in which we are taking the derivative, then apply the differential operator as a simple derivative, and then transform. the result back to the frame it was.

$$\frac{^Gd}{dt}\,{}^B\mathbf{r}_P = {}^GR_B^T\frac{^Gd}{dt}\left({}^GR_B\,{}^B\mathbf{r}_P\right) = {}^GR_B^T\frac{^Gd}{dt}\begin{bmatrix} t\cos\alpha \\ t\sin\alpha \\ 0 \end{bmatrix}$$

$$= {}^GR_B^T\begin{bmatrix} \cos\alpha - t\dot{\alpha}\sin\alpha \\ \sin\alpha + t\dot{\alpha}\cos\alpha \\ 0 \end{bmatrix} = \begin{bmatrix} 1 \\ t\dot{\alpha} \\ 0 \end{bmatrix} \tag{7.208}$$

$$\frac{^Bd}{dt}\,{}^G\mathbf{r}_P = {}^GR_B\frac{^Bd}{dt}\left({}^GR_B^T\,{}^G\mathbf{r}_P\right)$$

$$= {}^GR_B\frac{^Bd}{dt}\begin{bmatrix} t \\ 0 \\ 0 \end{bmatrix} = {}^GR_B\begin{bmatrix} 1 \\ 0 \\ 0 \end{bmatrix} = \begin{bmatrix} \cos\alpha \\ \sin\alpha \\ 0 \end{bmatrix} \tag{7.209}$$

Example 238 Orthogonality of position and velocity vectors. A velocity vector is time derivative of a position vector. If the position vector is a constant length vector, then velocity and position vectors are orthogonal.

If the position vector of a body point in global frame has a constant length and is denoted by $\mathbf{r}$, then

$$\frac{d\mathbf{r}}{dt} \cdot \mathbf{r} = 0 \tag{7.210}$$

To show this property, we may take a derivative from r^2,

$$\mathbf{r} \cdot \mathbf{r} = r^2 \tag{7.211}$$

to find

$$\frac{d}{dt}(\mathbf{r} \cdot \mathbf{r}) = \frac{d\mathbf{r}}{dt} \cdot \mathbf{r} + \mathbf{r} \cdot \frac{d\mathbf{r}}{dt} = 2\frac{d\mathbf{r}}{dt} \cdot \mathbf{r} = 0 \tag{7.212}$$

Equation (7.210) is correct in every coordinate frame and for every constant length vector, as long as the vector and the derivative are expressed in the same coordinate frame.

Example 239 ★ Derivative transformation formula. The most important equation in derivative kinematics is the derivative transformation formula. It shows how the global derivative of a vector quantity must be calculated in another rotating frame.

Consider a point P that is moving in a body frame B ($Oxyz$). The body position vector $^B\mathbf{r}_P$ is not constant and therefore, the global velocity of such a point expressed in B is

$$\frac{^G d}{dt} {}^B\mathbf{r}_P = \frac{^B d}{dt} {}^B\mathbf{r}_P + {}^B_G\boldsymbol{\omega}_B \times {}^B\mathbf{r}_P = {}^B_G\dot{\mathbf{r}}_P \tag{7.213}$$

The result of Eq. (7.213) is utilized to define transformation of the differential operator from a body to a global coordinate frame, where the box $\square$ indicates a vector quantity.

$$\frac{^G d}{dt}\square = \frac{^B d}{dt}\square + {}^B_G\boldsymbol{\omega}_B \times {}^B_G\square = {}^B_G\dot{\square} \tag{7.214}$$

Special attention must be paid to the coordinate frame in which the vector quantity $\square$ and the final result are expressed. The final result of $^B_G\square$ indicates the global (G) time derivative expressed in body frame (B). The vector quantity $\square$ might be any vector such as position, velocity, angular velocity, momentum, angular velocity, or even a time varying force vector.

Equation (7.214) is called the **derivative transformation formula** and relates the time derivative of a vector as it would be seen from frame G to its derivative as seen in frame B. The derivative transformation formula (7.214) is more general and can be applied to every vector for derivative transformation between every two relatively moving coordinate frames. The derivative transformation formula was derived by Leonhard Euler (1707–1783) for the first time.

Example 240 ★ Differential equation for rotation matrix. The definition of angular velocity is based on derivative of rotation transformation matrix. The definition may also be expressed as a differential equation with the rotation transformation matrix as the solution.

Equation (7.5) for defining the angular velocity matrix may be written as a first-order differential equation.

$$_G\tilde{\omega}_B = {}^G\dot{R}_B \, {}^G R_B^T \tag{7.215}$$

$$\frac{d}{dt} {}^G R_B - {}_G\tilde{\omega}_B \, {}^G R_B = 0 \tag{7.216}$$

The solution of the equation confirms the exponential definition of the rotation matrix as

$$^G R_B = e^{\tilde{\omega}t} \tag{7.217}$$

$$\tilde{\omega}t = \dot{\phi}\,\tilde{u} = \ln\left({}^G R_B\right) \tag{7.218}$$

We may also consider a variable rotation matrix $^G R_B(t)$ between frames B and G and the time derivative of the rotation transformation equation.

$$^G\mathbf{r}(t) = {}^G R_B(t)\, {}^G\mathbf{r}(0) \tag{7.219}$$

$$\frac{^G d}{dt} {}^G\mathbf{r}(t) = {}^G\dot{R}_B(t)\, {}^G\mathbf{r}(0) \tag{7.220}$$

If we define $^G\dot{R}_B(t)$ as

$$^G\dot{R}_B(t) = {}^G\dot{R}_B(0)\, {}^G R_B(t) \tag{7.221}$$

then we find the following equation called a first-order vectorial kinematic differential equation:

$$\frac{^G d}{dt} {}^G\mathbf{r}(t) = {}^G\dot{R}_B(0)\, {}^G R_B(t)\, {}^G\mathbf{r}(0) = {}^G\dot{R}_B(0)\, {}^G\mathbf{r}(t) \tag{7.222}$$

We should be able to determine the solution of this equation in a series form:

$$^G\mathbf{r}(t) = {}^G\mathbf{r}(0) + \frac{^G d\, {}^G\mathbf{r}(0)}{dt}t + \frac{1}{2!}\frac{^G d^2\, {}^G\mathbf{r}(0)}{dt^2}t^2 + \cdots \tag{7.223}$$

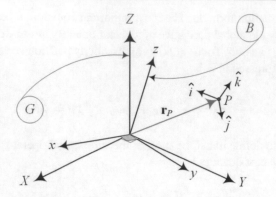

Fig. 7.4 A body coordinate frame moving with a fixed point in the global coordinate frame

Substituting the higher derivatives as

$$\frac{{}^{G}d^2}{dt^2}\,{}^{G}\mathbf{r}(t) = {}^{G}\dot{R}_B^2(0)\,{}^{G}\mathbf{r}(t) \tag{7.224}$$

$$\frac{{}^{G}d^3}{dt^3}\,{}^{G}\mathbf{r}(t) = {}^{G}\dot{R}_B^3(0)\,{}^{G}\mathbf{r}(t) \tag{7.225}$$

the series solution simplifies to an exponential solution:

$$
\begin{aligned}
{}^{G}\mathbf{r}(t) &= \left(\mathbf{I} + {}^{G}\dot{R}_B(0)\,t + \frac{1}{2!}\,{}^{G}\dot{R}_B^2(0)\,t^2 + \cdots \right)\,{}^{G}\mathbf{r}(0) \\
&= e^{{}^{G}\dot{R}_B(0)\,t}\,{}^{G}\mathbf{r}(0)
\end{aligned}
\tag{7.226}
$$

Example 241 ★ Acceleration of a body point in the global frame. Taking time derivative of angular velocity is the easiest way to derive angular acceleration.

The angular acceleration vector of a rigid body $B(Oxyz)$ in the global frame $G(OXYZ)$ is denoted by ${}_{G}\boldsymbol{\alpha}_B$ and is defined as the global time derivative of ${}_{G}\boldsymbol{\omega}_B$.

$$
{}_{G}\boldsymbol{\alpha}_B = \frac{{}^{G}d}{dt}\,{}_{G}\boldsymbol{\omega}_B \tag{7.227}
$$

Using this definition, the acceleration of a fixed body point P in the global frame is

$$
\begin{aligned}
{}^{G}\mathbf{a}_P &= \frac{{}^{G}d}{dt}\left({}_{G}\boldsymbol{\omega}_B \times {}^{G}\mathbf{r}_P\right) \\
&= {}_{G}\boldsymbol{\alpha}_B \times {}^{G}\mathbf{r}_P + {}_{G}\boldsymbol{\omega}_B \times \left({}_{G}\boldsymbol{\omega}_B \times {}^{G}\mathbf{r}_P\right)
\end{aligned}
\tag{7.228}
$$

Example 242 ★ Alternative definition of angular velocity vector. Here is a different and important way of definition of angular velocity based on unit vectors of two coordinate frames and their time derivative. This definition needs to be proven.

The angular velocity vector of a rigid body $B(\hat{\imath}, \hat{\jmath}, \hat{k})$ in global frame $G(\hat{I}, \hat{J}, \hat{K})$ can also be defined by

$$
{}_{G}^{B}\boldsymbol{\omega}_B = \hat{\imath}\left(\frac{{}^{G}d\hat{\jmath}}{dt}\cdot\hat{k}\right) + \hat{\jmath}\left(\frac{{}^{G}d\hat{k}}{dt}\cdot\hat{\imath}\right) + \hat{k}\left(\frac{{}^{G}d\hat{\imath}}{dt}\cdot\hat{\jmath}\right) \tag{7.229}
$$

Proof Consider a body coordinate frame B rotating with a fixed point in the global coordinate frame G. The fixed point of the body is taken as the origin of both coordinate frames, as is shown in Fig. 7.4. To express the motion of the body, it is sufficient to describe the motion of the body unit vectors $\hat{\imath}$, $\hat{\jmath}$, and $\hat{k}$. Let $\mathbf{r}_P$ be the position vector of a body point P. Then, $^B\mathbf{r}_P$ is a vector with constant components.

$$^B\mathbf{r}_P = x\hat{\imath} + y\hat{\jmath} + z\hat{k} \tag{7.230}$$

When the body moves, it is only the unit vectors $\hat{\imath}$, $\hat{\jmath}$, and $\hat{k}$ that vary relative to the global coordinate frame. Therefore, the vector of differential displacement is

$$d\mathbf{r}_P = x\,d\hat{\imath} + y\,d\hat{\jmath} + z\,d\hat{k} \tag{7.231}$$

which can also be expressed as

$$d\mathbf{r}_P = \left(d\mathbf{r}_P \cdot \hat{\imath}\right)\hat{\imath} + \left(d\mathbf{r}_P \cdot \hat{\jmath}\right)\hat{\jmath} + \left(d\mathbf{r}_P \cdot \hat{k}\right)\hat{k} \tag{7.232}$$

Substituting (7.231) in the right-hand side of (7.232)

$$\begin{aligned}
d\mathbf{r}_P = &\left(x\hat{\imath} \cdot d\hat{\imath} + y\hat{\imath} \cdot d\hat{\jmath} + z\hat{\imath} \cdot d\hat{k}\right)\hat{\imath} \\
&+ \left(x\hat{\jmath} \cdot d\hat{\imath} + y\hat{\jmath} \cdot d\hat{\jmath} + z\hat{\jmath} \cdot d\hat{k}\right)\hat{\jmath} \\
&+ \left(x\hat{k} \cdot d\hat{\imath} + y\hat{k} \cdot d\hat{\jmath} + z\hat{k} \cdot d\hat{k}\right)\hat{k}
\end{aligned} \tag{7.233}$$

and utilizing the unit vectors' relationships

$$\hat{\jmath} \cdot d\hat{\imath} = -\hat{\imath} \cdot d\hat{\jmath} \tag{7.234}$$

$$\hat{k} \cdot d\hat{\jmath} = -\hat{\jmath} \cdot d\hat{k} \tag{7.235}$$

$$\hat{\imath} \cdot d\hat{k} = -\hat{k} \cdot d\hat{\imath} \tag{7.236}$$

$$\hat{\imath} \cdot d\hat{\imath} = \hat{\jmath} \cdot d\hat{\jmath} = \hat{k} \cdot d\hat{k} = 0 \tag{7.237}$$

$$\hat{\imath} \cdot \hat{\jmath} = \hat{\jmath} \cdot \hat{k} = \hat{k} \cdot \hat{\imath} = 0 \tag{7.238}$$

$$\hat{\imath} \cdot \hat{\imath} = \hat{\jmath} \cdot \hat{\jmath} = \hat{k} \cdot \hat{k} = 1 \tag{7.239}$$

the $d\mathbf{r}_P$ reduces to

$$\begin{aligned}
d\mathbf{r}_P = &\left(z\hat{\imath} \cdot d\hat{k} - y\hat{\jmath} \cdot d\hat{\imath}\right)\hat{\imath} + \left(x\hat{\jmath} \cdot d\hat{\imath} - z\hat{k} \cdot d\hat{\jmath}\right)\hat{\jmath} \\
&+ \left(y\hat{k} \cdot d\hat{\jmath} - x\hat{\imath} \cdot d\hat{k}\right)\hat{k}
\end{aligned} \tag{7.240}$$

This equation can be rearranged to be expressed as a vector product.

$$d\mathbf{r}_P = \left((\hat{k} \cdot d\hat{\jmath})\hat{\imath} + (\hat{\imath} \cdot d\hat{k})\hat{\jmath} + (\hat{\jmath} \cdot d\hat{\imath})\hat{k}\right) \times \left(x\hat{\imath} + y\hat{\jmath} + z\hat{k}\right) \tag{7.241}$$

If the derivatives are respect to time, then velocity of the body fixed point will be calculated in which the angular velocity appears.

$$^B_G\mathbf{v}_P = \left((\hat{k} \cdot \frac{^Gd\hat{\jmath}}{dt})\hat{\imath} + (\hat{\imath} \cdot \frac{^Gd\hat{k}}{dt})\hat{\jmath} + (\hat{\jmath} \cdot \frac{^Gd\hat{\imath}}{dt})\hat{k}\right) \times \left(x\hat{\imath} + y\hat{\jmath} + z\hat{k}\right) \tag{7.242}$$

Comparing this result with

$$\dot{\mathbf{r}}_P = {}_G\boldsymbol{\omega}_B \times \mathbf{r}_P \tag{7.243}$$

shows that

$$^B_G\boldsymbol{\omega}_B = \hat{\imath}\left(\frac{^Gd\hat{\jmath}}{dt} \cdot \hat{k}\right) + \hat{\jmath}\left(\frac{^Gd\hat{k}}{dt} \cdot \hat{\imath}\right) + \hat{k}\left(\frac{^Gd\hat{\imath}}{dt} \cdot \hat{\jmath}\right) \tag{7.244}$$

■

Example 243 ★ Alternative proof for angular velocity definition (7.229). There are multiple methods for defining or proving the angular velocity. Here is an alternative proof of definition (7.229).

The angular velocity definition presented in Eq. (7.229) can also be shown by direct substitution for ^GR_B in the angular velocity matrix $^B_G\tilde{\omega}_B$.

$$^B_G\tilde{\omega}_B = {}^GR_B^T \, {}^G\dot{R}_B \tag{7.245}$$

Therefore,

$$
\begin{aligned}
^B_G\tilde{\omega}_B &= \begin{bmatrix} \hat{\imath}\cdot\hat{I} & \hat{\imath}\cdot\hat{J} & \hat{\imath}\cdot\hat{K} \\ \hat{\jmath}\cdot\hat{I} & \hat{\jmath}\cdot\hat{J} & \hat{\jmath}\cdot\hat{K} \\ \hat{k}\cdot\hat{I} & \hat{k}\cdot\hat{J} & \hat{k}\cdot\hat{K} \end{bmatrix} \cdot \frac{^Gd}{dt} \begin{bmatrix} \hat{I}\cdot\hat{\imath} & \hat{I}\cdot\hat{\jmath} & \hat{I}\cdot\hat{k} \\ \hat{J}\cdot\hat{\imath} & \hat{J}\cdot\hat{\jmath} & \hat{J}\cdot\hat{k} \\ \hat{K}\cdot\hat{\imath} & \hat{K}\cdot\hat{\jmath} & \hat{K}\cdot\hat{k} \end{bmatrix} \\
&= \begin{bmatrix} \hat{\imath}\cdot\dfrac{^Gd\hat{\imath}}{dt} & \hat{\imath}\cdot\dfrac{^Gd\hat{\jmath}}{dt} & \hat{\imath}\cdot\dfrac{^Gd\hat{k}}{dt} \\[2mm] \hat{\jmath}\cdot\dfrac{^Gd\hat{\imath}}{dt} & \hat{\jmath}\cdot\dfrac{^Gd\hat{\jmath}}{dt} & \hat{\jmath}\cdot\dfrac{^Gd\hat{k}}{dt} \\[2mm] \hat{k}\cdot\dfrac{^Gd\hat{\imath}}{dt} & \hat{k}\cdot\dfrac{^Gd\hat{\jmath}}{dt} & \hat{k}\cdot\dfrac{^Gd\hat{k}}{dt} \end{bmatrix}
\end{aligned} \tag{7.246}
$$

which shows the definition of angular velocity.

$$
^B_G\boldsymbol{\omega}_B = \begin{bmatrix} \dfrac{^Gd\hat{\jmath}}{dt}\cdot\hat{k} \\[2mm] \dfrac{^Gd\hat{k}}{dt}\cdot\hat{\imath} \\[2mm] \dfrac{^Gd\hat{\imath}}{dt}\cdot\hat{\jmath} \end{bmatrix} \tag{7.247}
$$

Example 244 ★ Second derivative. Using definition of derivative and coordinate frame, here we see how to calculate first and second derivatives in different coordinate frames.

In general, $^Gd\,\mathbf{r}/dt$ is a variable vector in $G(OXYZ)$ and in any other coordinate frame such as $B\,(oxyz)$. Therefore, it can be differentiated in either coordinate frame G or B. However, the order of differentiating is important.

$$\frac{^Bd}{dt}\frac{^Gd\mathbf{r}}{dt} \neq \frac{^Gd}{dt}\frac{^Bd\mathbf{r}}{dt} \tag{7.248}$$

As an example, consider a rotating body coordinate frame about the Z-axis and a variable vector such as

$$^G\mathbf{r} = t\,\hat{I} \tag{7.249}$$

Therefore,

$$\frac{^Gd\mathbf{r}}{dt} = {}^G\dot{\mathbf{r}} = \hat{I} \tag{7.250}$$

and hence,

$$
\begin{aligned}
{}^B\left(\frac{^Gd\mathbf{r}}{dt}\right) &= {}^B_G\dot{\mathbf{r}} = R_{Z,\varphi}^T\left[\hat{I}\right] = \begin{bmatrix} \cos\varphi & \sin\varphi & 0 \\ -\sin\varphi & \cos\varphi & 0 \\ 0 & 0 & 1 \end{bmatrix}\begin{bmatrix} 1 \\ 0 \\ 0 \end{bmatrix} \\
&= \cos\varphi\,\hat{\imath} - \sin\varphi\,\hat{\jmath}
\end{aligned} \tag{7.251}
$$

which provides

$$\frac{^Bd}{dt}\frac{^Gd\mathbf{r}}{dt} = -\dot{\varphi}\sin\varphi\,\hat{\imath} - \dot{\varphi}\cos\varphi\,\hat{\jmath} \tag{7.252}$$

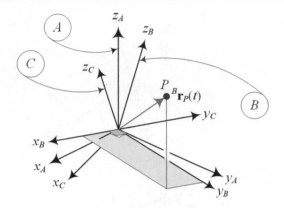

Fig. 7.5 Three relatively rotating coordinate frames, A, B, and C

and

$$^G \left(\frac{^B d}{dt} \frac{^G d\mathbf{r}}{dt} \right) = -\dot\varphi \, \hat{J} \tag{7.253}$$

Now,

$$^B \mathbf{r} = R_{Z,\varphi}^T \left[t\hat{I} \right] = t \cos\varphi \hat{\imath} - t \sin\varphi \hat{\jmath} \tag{7.254}$$

which provides

$$\frac{^B d\mathbf{r}}{dt} = (-t\dot\varphi \sin\varphi + \cos\varphi) \, \hat{\imath} - (\sin\varphi + t\dot\varphi \cos\varphi) \, \hat{\jmath} \tag{7.255}$$

and

$$^G \left(\frac{^B d\mathbf{r}}{dt} \right) = {}^G_B \dot{\mathbf{r}} = R_{Z,\varphi} \left((-t\dot\varphi \sin\varphi + \cos\varphi) \, \hat{\imath} - (\sin\varphi + t\dot\varphi \cos\varphi) \, \hat{\jmath} \right)$$

$$= \begin{bmatrix} \cos\varphi & -\sin\varphi & 0 \\ \sin\varphi & \cos\varphi & 0 \\ 0 & 0 & 1 \end{bmatrix} \begin{bmatrix} -t\dot\varphi \sin\varphi + \cos\varphi \\ -\sin\varphi - t\dot\varphi \cos\varphi \\ 0 \end{bmatrix}$$

$$= \hat{I} - t\dot\varphi \hat{J} \tag{7.256}$$

which shows

$$\frac{^G d}{dt} \frac{^B d\mathbf{r}}{dt} = - (\dot\varphi + t\ddot\varphi) \, \hat{J} \neq \frac{^B d}{dt} \frac{^G d\mathbf{r}}{dt} \tag{7.257}$$

Example 245 ★ Mixed-derivative transformation formula.

Consider three relatively rotating coordinate frames, A, B, and C, as shown in Fig. 7.5. The B-expression of the A-velocity of a moving point P in the body coordinate frame B ($Oxyz$) is

$$\frac{^A d}{dt} {}^B \mathbf{r}_P = {}^B_A \mathbf{v}_P = \frac{^B d}{dt} {}^B \mathbf{r}_P + {}^B_A \boldsymbol{\omega}_B \times {}^B \mathbf{r}_P \tag{7.258}$$

and the B-expression of the C-velocity of a moving point in the body coordinate frame B ($Oxyz$) is

$$\frac{^C d}{dt} {}^B \mathbf{r}_P = {}^B_C \mathbf{v}_P = \frac{^B d}{dt} {}^B \mathbf{r}_P + {}^B_C \boldsymbol{\omega}_B \times {}^B \mathbf{r}_P \tag{7.259}$$

Combining Eqs. (7.258) and (7.259), we find

$$^B_A \mathbf{v}_P - {}^B_A \boldsymbol{\omega}_B \times {}^B \mathbf{r}_P = {}^B_C \mathbf{v}_P - {}^B_C \boldsymbol{\omega}_B \times {}^B \mathbf{r}_P \tag{7.260}$$

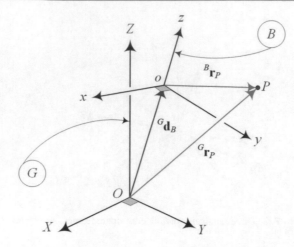

Fig. 7.6 A rigid body with an attached coordinate frame B $(oxyz)$ moving freely in a global coordinate frame $G(OXYZ)$

Rearranging (7.260) and changing the frame C in which we have taken a derivative of ${}^B\mathbf{r}_P$ to the frame A in which we need the derivative to be taken yield

$$\,{}_A^B\mathbf{v}_P = \,{}_C^B\mathbf{v}_P + \left(\,{}_A^B\boldsymbol{\omega}_B - \,{}_C^B\boldsymbol{\omega}_B\right) \times \,{}^B\mathbf{r}_P \tag{7.261}$$

We may equivalently show it as

$$\,{}_A^B\dot{\square} = \,{}_C^B\dot{\square} + \left(\,{}_A^B\boldsymbol{\omega}_B - \,{}_C^B\boldsymbol{\omega}_B\right) \times \,{}^B\square \tag{7.262}$$

We call Eq. (7.262) the *mixed-derivative transformation formula*. It presents the method used to change the frame in which the derivative of a vector ${}^B\square$ is taken. Interestingly, mixed-derivative transformation does not involve local derivatives directly.

The mixed-derivative transformation formula (7.262) is more general than the simple-derivative transformation formula (7.214). Equation (7.214) is a special case of (7.262) when $B \equiv C$ or $\,{}_C^B\boldsymbol{\omega}_B = 0$.

7.3   Rigid Body Velocity

Consider a rigid body with an attached coordinate frame B $(oxyz)$ moving freely in a fixed global coordinate frame $G(OXYZ)$, as shown in Fig. 7.6. The rigid body can rotate in the global frame, while the origin of the body frame B can translate relative to the origin of G. The coordinates of a body point P in local and global frames are related by the following equation:

$$\,{}^G\mathbf{r}_P = \,{}^GR_B\,\,{}^B\mathbf{r}_P + \,{}^G\mathbf{d}_B \tag{7.263}$$

where ${}^G\mathbf{d}_B$ indicates the position of the moving origin o relative to the fixed origin O.

The velocity of the point P in G is

$$
\begin{aligned}
{}^G\mathbf{v}_P = \,{}^G\dot{\mathbf{r}}_P &= \,{}^G\dot{R}_B\,\,{}^B\mathbf{r}_P + \,{}^G\dot{\mathbf{d}}_B \\
&= \,{}_G\tilde{\omega}_B\,\,{}^G_B\mathbf{r}_P + \,{}^G\dot{\mathbf{d}}_B = \,{}_G\tilde{\omega}_B\left(\,{}^G\mathbf{r}_P - \,{}^G\mathbf{d}_B\right) + \,{}^G\dot{\mathbf{d}}_B \\
&= \,{}_G\boldsymbol{\omega}_B \times \left(\,{}^G\mathbf{r}_P - \,{}^G\mathbf{d}_B\right) + \,{}^G\dot{\mathbf{d}}_B
\end{aligned}
\tag{7.264}
$$

Proof Direct differentiation of the position vector shows

$$
\begin{aligned}
{}^G\mathbf{v}_P = \frac{{}^Gd}{dt}\,\,{}^G\mathbf{r}_P = \,{}^G\dot{\mathbf{r}}_P &= \frac{{}^Gd}{dt}\left(\,{}^GR_B\,\,{}^B\mathbf{r}_P + \,{}^G\mathbf{d}_B\right) \\
&= \,{}^G\dot{R}_B\,\,{}^B\mathbf{r}_P + \,{}^G\dot{\mathbf{d}}_B
\end{aligned}
\tag{7.265}
$$

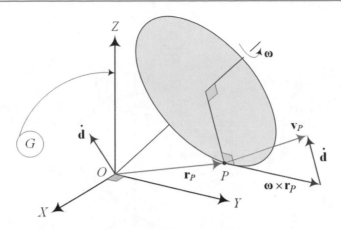

Fig. 7.7 Geometric interpretation of rigid body velocity

The local position vector ${}^B\mathbf{r}_P$ can be substituted from (7.263) to obtain:

$$
\begin{aligned}
{}^G\mathbf{v}_P &= {}^G\dot{R}_B \, {}^G R_B^T \left({}^G\mathbf{r}_P - {}^G\mathbf{d}_B \right) + {}^G\dot{\mathbf{d}}_B \\
&= {}_G\tilde{\omega}_B \left({}^G\mathbf{r}_P - {}^G\mathbf{d}_B \right) + {}^G\dot{\mathbf{d}}_B \\
&= {}_G\boldsymbol{\omega}_B \times \left({}^G\mathbf{r}_P - {}^G\mathbf{d}_B \right) + {}^G\dot{\mathbf{d}}_B
\end{aligned}
\tag{7.266}
$$

It may also be written using relative position vectors.

$$
{}^G\mathbf{v}_P = {}_G\boldsymbol{\omega}_B \times {}^G_B\mathbf{r}_P + {}^G\dot{\mathbf{d}}_B
\tag{7.267}
$$

∎

Example 246 Geometric interpretation of rigid body velocity. Kinematic characteristics of moving points can be illustrated geometrically. Illustrating velocity vectors helps to learn relative direction of different velocity components.

Figure 7.7 illustrates a body point P of a moving rigid body. The global velocity of the point P is a vector addition of rotational and translational velocities, both expressed in the global frame.

$$
{}^G\mathbf{v}_P = {}_G\boldsymbol{\omega}_B \times {}^G_B\mathbf{r}_P + {}^G\dot{\mathbf{d}}_B
\tag{7.268}
$$

At the moment, the body frame is assumed to be coincident with the global frame, and the body frame has a velocity ${}^G\dot{\mathbf{d}}_B$ with respect to the global frame. The translational velocity component ${}^G\dot{\mathbf{d}}_B$ is a common property for every point of the body, but the rotational velocity component ${}_G\boldsymbol{\omega}_B \times {}^G_B\mathbf{r}_P$ differs for different points of the body. The velocity vector ${}^G\dot{\mathbf{d}}_B$ is parallel to the velocity vector of the body at its mass center. The velocity vector ${}_G\boldsymbol{\omega}_B \times {}^G_B\mathbf{r}_P$ is perpendicular to both, ${}_G\boldsymbol{\omega}_B$ and ${}^G_B\mathbf{r}_P$. The vectors ${}_G\boldsymbol{\omega}_B$ and ${}^G_B\mathbf{r}_P + {}^G\dot{\mathbf{d}}_B$ may have any relative angle from 0 to 2π.

Example 247 The velocity of a moving point in a moving body frame. Here are the components of the velocity vector of a free particle moving in a body frame B, while B is moving and rotating in a global frame G.

Assume that point P in Fig. 7.6 is moving in frame B, indicating by a time varying position vector ${}^B\mathbf{r}_P(t)$. The global velocity of P is a composition of the velocity of P in B, rotation of B relative to G, and velocity of B relative to G.

$$
\begin{aligned}
\frac{{}^Gd}{dt} {}^G\mathbf{r}_P = {}^G\mathbf{v}_P &= \frac{{}^Gd}{dt} \left({}^G\mathbf{d}_B + {}^G R_B \, {}^B\mathbf{r}_P \right) \\
&= \frac{{}^Gd}{dt} {}^G\mathbf{d}_B + \frac{{}^Gd}{dt} \left({}^G R_B \, {}^B\mathbf{r}_P \right) \\
&= {}^G\dot{\mathbf{d}}_B + {}^G_B\dot{\mathbf{r}}_P + {}_G\boldsymbol{\omega}_B \times {}^G_B\mathbf{r}_P
\end{aligned}
\tag{7.269}
$$

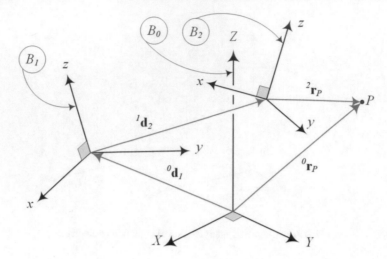

Fig. 7.8 A rigid body coordinate frame B_2 is moving in a frame B_1 that is moving in the base coordinate frame B_0

Example 248 The velocity of a body point in multiple coordinate frames. Here we look at velocity components of a fixed point in B_2-frame, while B_2 is moving in a B_1-frame, which itself is moving in another b_0-frame. The result of this example may be expanded to any number of relatively moving frames.

Consider three relatively moving frames, B_0, B_1, and B_2, as are shown in Fig. 7.8. The velocity of a body point P may be measured and expressed in any coordinate frame. If the point is stationary in a frame, say B_2, then the time derivative of $^2\mathbf{r}_P$ in B_2 is zero. If the frame B_2 is moving relative to the frame B_1, then, the time derivative of $^1\mathbf{r}_P$ is a combination of the rotational component due to rotation of B_2 relative to B_1 and the velocity of B_2 relative to B_1. In forward velocity kinematics of robots, the velocities must be measured in the base frame B_0. Therefore, the velocity of a point P in the base frame is a combination of the velocity of B_2 relative to B_1 and the velocity of B_1 relative to B_0.

The global coordinate of the body point P is

$$^0\mathbf{r}_P = {}^0\mathbf{d}_1 + {}^0_1\mathbf{d}_2 + {}^0_2\mathbf{r}_P \tag{7.270}$$

$$= {}^0\mathbf{d}_1 + {}^0R_1\,{}^1\mathbf{d}_2 + {}^0R_2\,{}^2\mathbf{r}_P \tag{7.271}$$

Therefore, the velocity of the point P can be found by combining the relative velocities.

$$^0\dot{\mathbf{r}}_P = {}^0\dot{\mathbf{d}}_1 + ({}^0\dot{R}_1\,{}^1\mathbf{d}_2 + {}^0R_1\,{}^1\dot{\mathbf{d}}_2) + {}^0\dot{R}_2\,{}^2\mathbf{r}_P$$

$$= {}^0\dot{\mathbf{d}}_1 + {}^0_0\boldsymbol{\omega}_1 \times {}^0_1\mathbf{d}_2 + {}^0R_1\,{}^1\dot{\mathbf{d}}_2 + {}^0_0\boldsymbol{\omega}_2 \times {}^0_2\mathbf{r}_P \tag{7.272}$$

Most of the time, it is better to use a relative velocity method and write

$$^0_0\mathbf{v}_P = {}^0_0\mathbf{v}_1 + {}^0_1\mathbf{v}_2 + {}^0_2\mathbf{v}_P \tag{7.273}$$

because

$$^0_0\mathbf{v}_1 = {}^0_0\dot{\mathbf{d}}_1 \tag{7.274}$$

$$^0_1\mathbf{v}_2 = {}^0_0\boldsymbol{\omega}_1 \times {}^0_1\mathbf{d}_2 + {}^0R_1\,{}^1\dot{\mathbf{d}}_2 \tag{7.275}$$

$$^0_2\mathbf{v}_P = {}^0_0\boldsymbol{\omega}_2 \times {}^0_2\mathbf{r}_P \tag{7.276}$$

and therefore,

$$^0\mathbf{v}_P = {}^0\dot{\mathbf{d}}_1 + {}^0_0\boldsymbol{\omega}_1 \times {}^0_1\mathbf{d}_2 + {}^0R_1\,{}^1\dot{\mathbf{d}}_2 + {}^0_0\boldsymbol{\omega}_2 \times {}^0_2\mathbf{r}_P \tag{7.277}$$

Example 249 Velocity vectors are free vectors. Free vectors transform from a coordinate frame to the other frame by premultiplying a transformation matrix. Hence, we may calculate velocity vector in any coordinate frame that is the simplest one and transform the result to any other coordinate frame we need them to be expressed in.

Velocity vectors are free vectors. To express them in different coordinate frames we only need to premultiply them by a proper rotation transformation matrix. Hence, considering ${}_{j}^{k}\mathbf{v}_i$ as the velocity of the origin of the coordinate frame B_i with respect to the origin of the frame B_j expressed in frame B_k, and knowing that velocity with respect to j-frame, expressed in k-frame, is negative of the velocity with respect to k-frame, expressed in j-frame,

$$
{}_{j}^{k}\mathbf{v}_i = - {}_{i}^{k}\mathbf{v}_j \tag{7.278}
$$

and

$$
{}_{j}^{k}\mathbf{v}_i = {}^{k}R_m \, {}_{j}^{m}\mathbf{v}_i \tag{7.279}
$$

we have

$$
\frac{{}^{i}d}{dt} {}_{i}^{i}\mathbf{r}_P = {}^{i}\mathbf{v}_P = {}_{j}^{i}\mathbf{v}_P + {}_{i}^{i}\boldsymbol{\omega}_j \times {}_{j}^{i}\mathbf{r}_P \tag{7.280}
$$

Example 250 ★ Zero velocity points. There is always a point with zero velocity in $2D$ rigid body motion. The point is called pole and its loci makes centroid with some applications in mechanisms, biomechanics, kinesiology, etc.

To answer whether there is a point with zero velocity, we utilize Eq. (7.264) and we set ${}^{G}\mathbf{v}_P = 0$.

$$
{}_{G}\tilde{\omega}_B \left({}^{G}\mathbf{r}_0 - {}^{G}\mathbf{d}_B \right) + {}^{G}\dot{\mathbf{d}}_B = 0 \tag{7.281}
$$

We solve the equation for ${}^{G}\mathbf{r}_0$ which refers to a point with zero velocity.

$$
{}^{G}\mathbf{r}_0 = {}^{G}\mathbf{d}_B - {}_{G}\tilde{\omega}_B^{-1} \, {}^{G}\dot{\mathbf{d}}_B \tag{7.282}
$$

The skew symmetric matrix ${}_{G}\tilde{\omega}_B$ is singular and has no inverse. In other words, there is no general solution for Eq. (7.281) in $3D$ motion.

If we restrict ourselves to planar motions, say (X, Y)-plane, then ${}_{G}\boldsymbol{\omega}_B = \omega\hat{K}$ and ${}_{G}\tilde{\omega}_B^{-1} = 1/\omega$. Hence, in $2D$ space, there is, at any time, a point with zero velocity at position ${}^{G}\mathbf{r}_0$.

$$
{}^{G}\mathbf{r}_0(t) = {}^{G}\mathbf{d}_B(t) - \frac{1}{\omega} \, {}^{G}\dot{\mathbf{d}}_B(t) \tag{7.283}
$$

The zero velocity point is called the pole or instantaneous center of rotation. The position of the pole is moving on the plane and so is a function of time. The path of the motion of pole is called centroid.

Example 251 ★ Eulerian and Lagrangian view points. Lagrangian (stationary) and Eulerian (moving) are two classical methods in modeling dynamic systems. Here is an example of their mathematical different definition.

When a variable quantity is measured within the stationary global coordinate frame, it is called absolute or the *Lagrangian viewpoint*. On the other hand, when the variable is measured within a moving body coordinate frame, it is called relative or the *Eulerian viewpoint*. In $2D$ planar motion of a rigid body, there is always a pole of zero velocity at ${}^{G}\mathbf{r}_0$, where

$$
{}^{G}\mathbf{r}_0 = {}^{G}\mathbf{d}_B - \frac{1}{\omega} \, {}^{G}\dot{\mathbf{d}}_B \tag{7.284}
$$

The position of the pole in the body coordinate frame can be found by substituting for ${}^{G}\mathbf{r}$ from (7.263)

$$
{}^{G}R_B \, {}^{B}\mathbf{r}_0 + {}^{G}\mathbf{d}_B = {}^{G}\mathbf{d}_B - {}_{G}\tilde{\omega}_B^{-1} \, {}^{G}\dot{\mathbf{d}}_B \tag{7.285}
$$

and solving for the position of the zero velocity point in the body coordinate frame ${}^{B}\mathbf{r}_0$.

$$
\begin{aligned}
{}^{B}\mathbf{r}_0 &= -{}^{G}R_B^T \, {}_{G}\tilde{\omega}_B^{-1} \, {}^{G}\dot{\mathbf{d}}_B = -{}^{G}R_B^T \left[{}^{G}\dot{R}_B \, {}^{G}R_B^T \right]^{-1} {}^{G}\dot{\mathbf{d}}_B \\
&= -{}^{G}R_B^T \left[{}^{G}R_B \, {}^{G}\dot{R}_B^{-1} \right] {}^{G}\dot{\mathbf{d}}_B = -{}^{G}\dot{R}_B^{-1} \, {}^{G}\dot{\mathbf{d}}_B
\end{aligned} \tag{7.286}
$$

Therefore, $^G\mathbf{r}_0$ indicates the path of motion of the pole in the global frame, while $^B\mathbf{r}_0$ indicates the same path in the body frame. The $^G\mathbf{r}_0$ refers to Lagrangian centroid and $^B\mathbf{r}_0$ refers to Eulerian centroid.

Example 252 ★ Screw axis and screw motion. Any rigid body motion may be expressed by a screw motion. Similarly, any rigid body velocity is a rigid body motion in time and may be expressed by a screw. This example shows how.

The screw axis is defined as a line for a moving rigid body B whose points P have velocity parallel to the angular velocity vector $_G\boldsymbol{\omega}_B = \omega\hat{u}$. Such points satisfy the following equation in which, p is a scalar:

$$^G\mathbf{v}_P = {_G\boldsymbol{\omega}_B} \times \left(^G\mathbf{r}_P - {^G\mathbf{d}_B}\right) + {^G\dot{\mathbf{d}}_B} = p\,{_G\boldsymbol{\omega}_B} \tag{7.287}$$

Because $_G\boldsymbol{\omega}_B$ is perpendicular to $_G\boldsymbol{\omega}_B \times \left(^G\mathbf{r} - {^G\mathbf{d}}\right)$, a dot product of Eq. (7.287) by $_G\boldsymbol{\omega}_B$ yields

$$p = \frac{1}{\omega^2}\left({_G\boldsymbol{\omega}_B} \cdot {^G\dot{\mathbf{d}}_B}\right) \tag{7.288}$$

Introducing a parameter k to indicate different points of the line, the equation of the screw axis is defined as

$$^G\mathbf{r}_P = {^G\mathbf{d}_B} + \frac{1}{\omega^2}\left({_G\boldsymbol{\omega}_B} \times {^G\dot{\mathbf{d}}_B}\right) + k\,{_G\boldsymbol{\omega}_B} \tag{7.289}$$

because if we have $\mathbf{a} \times \mathbf{x} = \mathbf{b}$, and $\mathbf{a} \cdot \mathbf{b} = 0$, then $\mathbf{x} = -\mathbf{a}^{-2}(\mathbf{a} \times \mathbf{b}) + k\mathbf{a}$. In our case,

$$_G\boldsymbol{\omega}_B \times \left(^G\mathbf{r}_P - {^G\mathbf{d}_B}\right) = p\,{_G\boldsymbol{\omega}_B} - {^G\dot{\mathbf{d}}_B} \tag{7.290}$$

$\left(^G\mathbf{r}_P - {^G\mathbf{d}_B}\right)$ is perpendicular to $_G\boldsymbol{\omega}_B \times \left(^G\mathbf{r}_P - {^G\mathbf{d}_B}\right)$ and hence is perpendicular to $(p\,{_G\boldsymbol{\omega}_B} - {^G\dot{\mathbf{d}}_B})$ too. Therefore, there exists at any time a line s in space, parallel to $_G\boldsymbol{\omega}_B$, which is the locus of points whose velocity is parallel to $_G\boldsymbol{\omega}_B$.

If $\mathbf{s}$ is the position vector of a point on s, then,

$$_G\boldsymbol{\omega}_B \times \left(^G\mathbf{s} - {^G\mathbf{d}_B}\right) = p\,{_G\boldsymbol{\omega}_B} - {^G\dot{\mathbf{d}}_B} \tag{7.291}$$

and the velocity of any point out of $\mathbf{s}$ is

$$^G\mathbf{v} = {_G\boldsymbol{\omega}_B} \times \left(^G\mathbf{r} - {^G\mathbf{s}}\right) + p\,{_G\boldsymbol{\omega}_B} \tag{7.292}$$

This equation expresses that at any time the velocity of a body point can be decomposed into perpendicular and parallel components to the angular velocity vector $_G\boldsymbol{\omega}_B$. Therefore, the motion of any point of a rigid body is a screw. The parameter p is the ratio of translation velocity to rotation velocity and is called **pitch**. In general, $\mathbf{s}$, $_G\boldsymbol{\omega}_B$, and p may be functions of time.

7.4   ★ Velocity Transformation Matrix

Consider the motion of a rigid body B in a global coordinate frame G, as shown in Fig. 7.6. Assume the body frame $B(oxyz)$ is coincident with the global frame $G(OXYZ)$ at an initial time t_0. At any time $t \neq t_0$, the B-frame is not necessarily coincident with G, and therefore, the homogenous transformation matrix $^GT_B(t)$ is time varying.

The global position vector $^G\mathbf{r}_P(t)$ of a point P of the rigid body is a function of time, but its local position vector $^B\mathbf{r}_P$ is a constant, which is equal to $^G\mathbf{r}_P(t_0)$.

$$^B\mathbf{r}_P \equiv {^G\mathbf{r}_P(t_0)} \tag{7.293}$$

The velocity of point P of the rigid body B as seen in the reference frame G is obtained by differentiating the position vector $^G\mathbf{r}(t)$ in the reference frame G,

$$^G\mathbf{v}_P = \frac{d}{dt}{^G\mathbf{r}_P(t)} = {^G\dot{\mathbf{r}}_P} \tag{7.294}$$

where $^G\dot{\mathbf{r}}_P$ denotes the differentiation of the position vector $^G\mathbf{r}_P(t)$ with respect to time t in the reference frame G.

The velocity of a body point in global coordinate frame can be found by applying a homogenous transformation matrix ^GV_B on the position vector $^G\mathbf{r}_P(t)$,

$$^G\mathbf{v}(t) = {^GV_B}\,{^G\mathbf{r}(t)} \tag{7.295}$$

where ${}^{G}V_{B}$ is the *velocity transformation matrix*.

$$
{}^{G}V_{B} = {}^{G}\dot{T}_{B}\,{}^{G}T_{B}^{-1}
$$

$$
= \begin{bmatrix} {}^{G}\dot{R}_{B}\,{}^{G}R_{B}^{T}\,{}^{G}\dot{\mathbf{d}}_{B} - {}^{G}\dot{R}_{B}\,{}^{G}R_{B}^{T}\,{}^{G}\mathbf{d}_{B} \\ 0 \qquad\qquad 0 \end{bmatrix}
$$

$$
= \begin{bmatrix} {}_{G}\tilde{\omega}_{B}\,{}^{G}\dot{\mathbf{d}}_{B} - {}_{G}\tilde{\omega}_{B}\,{}^{G}\mathbf{d}_{B} \\ 0 \qquad 0 \end{bmatrix} = \begin{bmatrix} {}_{G}\tilde{\omega}_{B}\,{}^{G}\mathbf{v}_{B} \\ 0 \quad 0 \end{bmatrix} \tag{7.296}
$$

Proof Based on homogenous coordinate transformation, we have

$$
{}^{G}\mathbf{r}_{P}(t) = {}^{G}T_{B}(t)\,{}^{B}\mathbf{r}_{P} = {}^{G}T_{B}(t)\,{}^{G}\mathbf{r}_{P}(t_{0}) \tag{7.297}
$$

and therefore,

$$
{}^{G}\mathbf{v}_{P} = \frac{{}^{G}d}{dt}\left[{}^{G}T_{B}\,{}^{B}\mathbf{r}_{P}\right] = {}^{G}\dot{T}_{B}\,{}^{B}\mathbf{r}_{P} = \begin{bmatrix} \frac{{}^{G}d}{dt}\,{}^{G}R_{B} & \frac{{}^{G}d}{dt}\,{}^{G}\mathbf{d}_{B} \\ 0 & 0 \end{bmatrix}{}^{B}\mathbf{r}_{P}
$$

$$
= \begin{bmatrix} {}^{G}\dot{R}_{B} & {}^{G}\dot{\mathbf{d}}_{B} \\ 0 & 0 \end{bmatrix}{}^{B}\mathbf{r}_{P} \tag{7.298}
$$

Substituting for ${}^{B}\mathbf{r}_{P}$ from Eq. (7.297) gives

$$
{}^{G}\mathbf{v}_{P} = {}^{G}\dot{T}_{B}\,{}^{G}T_{B}^{-1}\,{}^{G}\mathbf{r}_{P}(t)
$$

$$
= \begin{bmatrix} {}^{G}\dot{R}_{B} & {}^{G}\dot{\mathbf{d}}_{B} \\ 0 & 0 \end{bmatrix}\begin{bmatrix} {}^{G}R_{B}^{T} & -{}^{G}R_{B}^{T}\,{}^{G}\mathbf{d}_{B} \\ 0 & 1 \end{bmatrix}{}^{G}\mathbf{r}_{P}(t)
$$

$$
= \begin{bmatrix} {}^{G}\dot{R}_{B}\,{}^{G}R_{B}^{T}\,{}^{G}\dot{\mathbf{d}}_{B} - {}^{G}\dot{R}_{B}\,{}^{G}R_{B}^{T}\,{}^{G}\mathbf{d}_{B} \\ 0 \qquad\qquad 0 \end{bmatrix}{}^{G}\mathbf{r}_{P}(t)
$$

$$
= \begin{bmatrix} {}_{G}\tilde{\omega}_{B}\,{}^{G}\dot{\mathbf{d}}_{B} - {}_{G}\tilde{\omega}_{B}\,{}^{G}\mathbf{d}_{B} \\ 0 \qquad 0 \end{bmatrix}{}^{G}\mathbf{r}_{P}(t) \tag{7.299}
$$

Thus, the velocity of any point P of the rigid body B in the reference frame G can be obtained by premultiplying the position vector of the point P in G with the *velocity transformation matrix*, ${}^{G}V_{B}$,

$$
{}^{G}\mathbf{v}_{P}(t) = {}^{G}V_{B}\,{}^{G}\mathbf{r}_{P}(t) \tag{7.300}
$$

where

$$
{}^{G}V_{B} = {}^{G}\dot{T}_{B}\,{}^{G}T_{B}^{-1} = \begin{bmatrix} {}_{G}\tilde{\omega}_{B}\,{}^{G}\dot{\mathbf{d}}_{B} - {}_{G}\tilde{\omega}_{B}\,{}^{G}\mathbf{d}_{B} \\ 0 \qquad 0 \end{bmatrix}
$$

$$
= \begin{bmatrix} {}_{G}\tilde{\omega}_{B}\,{}^{G}\mathbf{v}_{B} \\ 0 \quad 0 \end{bmatrix} \tag{7.301}
$$

and

$$
{}_{G}\tilde{\omega}_{B} = {}^{G}\dot{R}_{B}\,{}^{G}R_{B}^{T} \tag{7.302}
$$

$$
{}^{G}\mathbf{v}_{B} = {}^{G}\dot{\mathbf{d}}_{B} - {}^{G}\dot{R}_{B}\,{}^{G}R_{B}^{T}\,{}^{G}\mathbf{d}_{B} = {}^{G}\dot{\mathbf{d}}_{B} - {}_{G}\tilde{\omega}_{B}\,{}^{G}\mathbf{d}_{B}
$$

$$
= {}^{G}\dot{\mathbf{d}}_{B} - {}_{G}\boldsymbol{\omega}_{B} \times {}^{G}\mathbf{d}_{B} \tag{7.303}
$$

The *velocity transformation matrix* ^GV_B may be assumed as a matrix operator that provides the global velocity of any point attached to $B(oxyz)$. It consists of the angular velocity matrix $_G\tilde{\omega}_B$ and the B-frame velocity $^G\dot{\mathbf{d}}_B$ both expressed in the global frame $G(OXYZ)$. The matrix ^GV_B depends on six parameters: the three components of the angular velocity vector $_G\boldsymbol{\omega}_B$ and the three components of the frame velocity $^G\dot{\mathbf{d}}_B$.

It may also be convenient to introduce a 6×1 vector called *velocity transformation vector* to simplify numerical calculations.

$$_G\mathbf{t}_B = \begin{bmatrix} ^G\mathbf{v}_B \\ _G\boldsymbol{\omega}_B \end{bmatrix} = \begin{bmatrix} ^G\dot{\mathbf{d}}_B - {}_G\tilde{\omega}_B\,^G\mathbf{d}_B \\ _G\boldsymbol{\omega}_B \end{bmatrix} \tag{7.304}$$

In analogy to the two representations of the angular velocity, the velocity of body B in reference frame G can be represented either as the velocity transformation matrix ^GV_B in (7.301) or as the velocity transformation vector $_G\mathbf{t}_B$ in (7.304). The velocity transformation vector represents a noncommensurate vector because the dimension of $_G\boldsymbol{\omega}_B$ and $^G\mathbf{v}_B$ differs.   ∎

Example 253 ★ Velocity transformation matrix based on coordinate transformation matrix. This is to show how the velocity transformation matrix is built, starting from homogenous transformation matrix.

The velocity transformation matrix can be found based on a coordinate transformation matrix.

$$^G\mathbf{r}(t) = {}^GT_B\,{}^B\mathbf{r} = \begin{bmatrix} ^GR_B & {}^G\mathbf{d}_B \\ 0 & 1 \end{bmatrix} {}^B\mathbf{r} \tag{7.305}$$

Taking a time derivative shows that

$$^G\mathbf{v} = \frac{^Gd}{dt}\left[{}^GT_B\,{}^B\mathbf{r}\right] = {}^G\dot{T}_B\,{}^B\mathbf{r} = \begin{bmatrix} ^G\dot{R}_B & {}^G\dot{\mathbf{d}}_B \\ 0 & 0 \end{bmatrix} {}^B\mathbf{r} \tag{7.306}$$

however,

$$^B\mathbf{r} = {}^GT_B^{-1}\,{}^G\mathbf{r} \tag{7.307}$$

and therefore,

$$\begin{aligned} ^G\mathbf{v} &= \begin{bmatrix} ^G\dot{R}_B & {}^G\dot{\mathbf{d}}_B \\ 0 & 0 \end{bmatrix} {}^GT_B^{-1}\,{}^G\mathbf{r} \\[2mm] &= \begin{bmatrix} ^G\dot{R}_B & {}^G\dot{\mathbf{d}}_B \\ 0 & 0 \end{bmatrix} \begin{bmatrix} ^GR_B^T & -{}^GR_B^T\,{}^G\mathbf{d}_B \\ 0 & 1 \end{bmatrix} {}^G\mathbf{r} \\[2mm] &= \begin{bmatrix} ^G\dot{R}_B\,{}^GR_B^T & {}^G\dot{\mathbf{d}}_B - {}^G\dot{R}_B\,{}^GR_B^T\,{}^G\mathbf{d}_B \\ 0 & 0 \end{bmatrix} {}^G\mathbf{r} \\[2mm] &= {}^GV_B\,{}^G\mathbf{r} \end{aligned} \tag{7.308}$$

Example 254 ★ Inverse of a velocity transformation matrix. Similar to the inverse of homogenous transformation matrices, we may determine the inverse of velocity transformation matrices by manipulating its components.

Transformation from a body frame to a global frame is given by Eq. (4.76).

$$^GT_B^{-1} = \begin{bmatrix} ^GR_B^T & -{}^GR_B^T\,{}^G\mathbf{d}_B \\ 0 & 1 \end{bmatrix} \tag{7.309}$$

Following the same principle, we may introduce the inverse velocity transformation matrix by

$$
\begin{aligned}
{}^B V_G &= {}^G V_B^{-1} \\
&= \begin{bmatrix} \left({}^G \dot{R}_B \, {}^G R_B^T\right)^{-1} & -\left({}^G \dot{R}_B \, {}^G R_B^T\right)^{-1} \left({}^G \dot{\mathbf{d}}_B - {}^G \dot{R}_B \, {}^G R_B^T \, {}^G \mathbf{d}_B\right) \\ 0 & 0 \end{bmatrix} \\
&= \begin{bmatrix} {}^G R_B \, {}^G \dot{R}_B^{-1} & -{}^G R_B \, {}^G \dot{R}_B^{-1} \left({}^G \dot{\mathbf{d}}_B - {}^G \dot{R}_B \, {}^G R_B^T \, {}^G \mathbf{d}_B\right) \\ 0 & 0 \end{bmatrix} \\
&= \begin{bmatrix} {}^G R_B \, {}^G \dot{R}_B^{-1} & -{}^G R_B \, {}^G \dot{R}_B^{-1} \, {}^G \dot{\mathbf{d}}_B + {}^G \mathbf{d}_B \\ 0 & 0 \end{bmatrix}
\end{aligned}
\tag{7.310}
$$

to have

$$
{}^G V_B \, {}^G V_B^{-1} = \mathbf{I}
\tag{7.311}
$$

Therefore, having the velocity vector of a body point ${}^G \mathbf{v}_P$ and the velocity transformation matrix ${}^G V_B$ we can find the global position of the point by

$$
{}^G \mathbf{r}_P = {}^G V_B^{-1} \, {}^G \mathbf{v}_P
\tag{7.312}
$$

Example 255 ★ Velocity transformation matrix in body frame. The velocity transformation matrix of B-frame with respect to G-frame may be expressed in either B-frame or G-frame. Here shows how to relate those expressions.

The velocity transformation matrix ${}^G V_B$ defined in the global frame G is

$$
{}^G V_B = \begin{bmatrix} {}^G \dot{R}_B \, {}^G R_B^T & {}^G \dot{\mathbf{d}}_B - {}^G \dot{R}_B \, {}^G R_B^T \, {}^G \mathbf{d}_B \\ 0 & 0 \end{bmatrix}
\tag{7.313}
$$

However, the velocity transformation matrix can be expressed in the body coordinate frame B as well,

$$
\begin{aligned}
{}^B_G V_B &= {}^G T_B^{-1} \, {}^G \dot{T}_B \\
&= \begin{bmatrix} {}^G R_B^T & -{}^G R_B^T \, {}^G \mathbf{d}_B \\ 0 & 1 \end{bmatrix} \begin{bmatrix} {}^G \dot{R}_B & {}^G \dot{\mathbf{d}}_B \\ 0 & 0 \end{bmatrix} \\
&= \begin{bmatrix} {}^G R_B^T \, {}^G \dot{R}_B & {}^G R_B^T \, {}^G \dot{\mathbf{d}}_B \\ 0 & 0 \end{bmatrix} = \begin{bmatrix} {}^B_G \omega_B & {}^B_G \dot{\mathbf{d}}_B \\ 0 & 0 \end{bmatrix}
\end{aligned}
\tag{7.314}
$$

where ${}^B_G \omega_B$ is the angular velocity vector of B with respect to G expressed in B, and ${}^B \mathbf{d} = {}^B_G \dot{\mathbf{d}}_B$ is the velocity of the origin of B in G expressed in B.

It is also possible to use a matrix multiplication to find the velocity transformation matrix in the body coordinate frame.

$$
{}^B_G \mathbf{v}_P = {}^G T_B^{-1} \, {}^G \mathbf{v}_P = {}^G T_B^{-1} \, {}^G \dot{T}_B \, {}^B \mathbf{r}_P = {}^B_G V_B \, {}^B \mathbf{r}_P
\tag{7.315}
$$

Using the definition of (7.296) and (7.314), we are able to transform the velocity transformation matrices between the B and G frames.

$$
{}^G V_B = {}^G T_B \, {}^B_G V_B \, {}^G T_B^{-1}
\tag{7.316}
$$

It can also be useful if we define the time derivative of the transformation matrix by

$$
{}^G \dot{T}_B = {}^G V_B \, {}^G T_B
\tag{7.317}
$$

or

$$
{}^G \dot{T}_B = {}^G T_B \, {}^B_G V_B
\tag{7.318}
$$

Similarly, we may define a velocity transformation matrix from link (i) to $(i-1)$ by

$$
{}^{i-1}V_i = \begin{bmatrix} {}^{i-1}\dot{R}_i\,{}^{i-1}R_i^T & {}^{i-1}\dot{\mathbf{d}} - {}^{i-1}\dot{R}_i\,{}^{i-1}R_i^T\,{}^{i-1}\mathbf{d} \\ 0 & 0 \end{bmatrix} \tag{7.319}
$$

and

$$
{}^{i}_{i-1}V_i = \begin{bmatrix} {}^{i-1}R_i^T\,{}^{i-1}\dot{R}_i & {}^{i-1}R_i^T\,{}^{i-1}\dot{\mathbf{d}} \\ 0 & 0 \end{bmatrix} \tag{7.320}
$$

where ${}^{i-1}\mathbf{d} \equiv {}^{i-1}_{i}\mathbf{d}_{i-a}$.

Example 256 Motion with a fixed point. If a point of the B-frame is fixed in G-frame, then motion of the rigid body reduces to rotation about the fixed point. Here is the simplification of the general equations for this special case.

When a point of a rigid body is fixed to the global frame, it is convenient to set the origins of the moving coordinate frame $B(Oxyz)$ and the global coordinate frame $G(OXYZ)$ on the fixed point. Under these conditions, we have

$$
{}^{G}\mathbf{d}_B = 0 \qquad {}^{G}\dot{\mathbf{d}}_B = 0 \tag{7.321}
$$

and Eq. (7.299) reduces to

$$
{}^{G}\mathbf{v}_P = {}_{G}\tilde{\omega}_B\,{}^{G}\mathbf{r}_P(t) = {}_{G}\boldsymbol{\omega}_B \times {}^{G}\mathbf{r}_P(t) \tag{7.322}
$$

Example 257 Velocity in spherical coordinates. The Cartesian coordinate system is only an option out of many other orthogonal coordinate systems. Here is a velocity analysis in spherical coordinate system.

The homogenous transformation matrix from spherical coordinates $S(Or\theta\varphi)$ to Cartesian coordinates $G(OXYZ)$ may be found by matrix transformations.

$$
\begin{aligned}
{}^{G}T_S &= R_{Z,\varphi}\,R_{Y,\theta}\,D_{Z,r} = \begin{bmatrix} {}^{G}R_B & {}^{G}\mathbf{d} \\ 0 & 1 \end{bmatrix} \\
&= \begin{bmatrix} \cos\theta\cos\varphi & -\sin\varphi & \cos\varphi\sin\theta & r\cos\varphi\sin\theta \\ \cos\theta\sin\varphi & \cos\varphi & \sin\theta\sin\varphi & r\sin\theta\sin\varphi \\ -\sin\theta & 0 & \cos\theta & r\cos\theta \\ 0 & 0 & 0 & 1 \end{bmatrix}
\end{aligned} \tag{7.323}
$$

Time derivative of ${}^{G}T_S$ determines velocity transformation matrix in spherical coordinate system.

$$
\begin{aligned}
{}^{G}\dot{T}_S &= {}^{G}V_S\,{}^{G}T_S = \begin{bmatrix} {}_{G}\tilde{\omega}_S & {}^{G}\mathbf{v}_S \\ 0 & 0 \end{bmatrix}{}^{G}T_S \\
&= \begin{bmatrix} 0 & -\dot{\varphi} & \dot{\theta}\cos\varphi & \dot{r}\cos\varphi\sin\theta \\ \dot{\varphi} & 0 & \dot{\theta}\sin\varphi & \dot{r}\sin\theta\sin\varphi \\ -\dot{\theta}\cos\varphi & -\dot{\theta}\sin\varphi & 0 & \dot{r}\cos\theta \\ 0 & 0 & 0 & 0 \end{bmatrix}{}^{G}T_B
\end{aligned} \tag{7.324}
$$

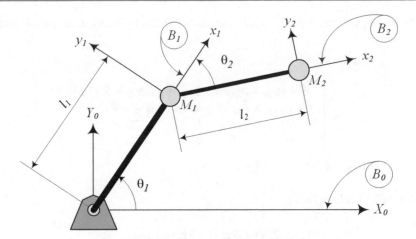

Fig. 7.9 An R‖R planar manipulator

Example 258 ★ Velocity analysis of a planar R‖R manipulator. Here is a detailed application of velocity transformation matrix and velocity analysis of a $2R$ planar manipulator. Velocity analysis will be calculated in several methods to show alternative ways and their consistency.

Figure 7.9 illustrates an R‖R planar manipulator with joint variables θ_1 and θ_2. The links (1) and (2) are both R‖R(0), and therefore the transformation matrices $^0T_1$, $^1T_2$, and $^0T_2$ are

$$
^0T_1 = \begin{bmatrix} \cos\theta_1 & -\sin\theta_1 & 0 & l_1\cos\theta_1 \\ \sin\theta_1 & \cos\theta_1 & 0 & l_1\sin\theta_1 \\ 0 & 0 & 1 & 0 \\ 0 & 0 & 0 & 1 \end{bmatrix}
\tag{7.325}
$$

$$
^1T_2 = \begin{bmatrix} \cos\theta_2 & -\sin\theta_2 & 0 & l_2\cos\theta_2 \\ \sin\theta_2 & \cos\theta_2 & 0 & l_2\sin\theta_2 \\ 0 & 0 & 1 & 0 \\ 0 & 0 & 0 & 1 \end{bmatrix}
\tag{7.326}
$$

$$
^0T_2 = {^0T_1}\,{^1T_2}
\tag{7.327}
$$

$$
= \begin{bmatrix} c\,(\theta_1+\theta_2) & -s\,(\theta_1+\theta_2) & 0 & l_2 c\,(\theta_1+\theta_2)+l_1 c\theta_1 \\ s\,(\theta_1+\theta_2) & c\,(\theta_1+\theta_2) & 0 & l_2 s\,(\theta_1+\theta_2)+l_1 s\theta_1 \\ 0 & 0 & 1 & 0 \\ 0 & 0 & 0 & 1 \end{bmatrix}
$$

The points M_1 and M_2 are at

$$
^0\mathbf{r}_{M_1} = \begin{bmatrix} l_1\cos\theta_1 \\ l_1\sin\theta_1 \\ 0 \\ 1 \end{bmatrix} \qquad
^1\mathbf{r}_{M_2} = \begin{bmatrix} l_2\cos\theta_2 \\ l_2\sin\theta_2 \\ 0 \\ 1 \end{bmatrix}
\tag{7.328}
$$

$$
^0\mathbf{r}_{M_2} = {^0T_1}\,{^1\mathbf{r}_{M_2}} = \begin{bmatrix} l_2\cos(\theta_1+\theta_2)+l_1\cos\theta_1 \\ l_2\sin(\theta_1+\theta_2)+l_1\sin\theta_1 \\ 0 \\ 1 \end{bmatrix}
\tag{7.329}
$$

To determine the velocity of M_2, we calculate $^0\dot{T}_2$ which can be calculated by direct differentiation of $^0T_2$.

$$^0\dot{T}_2 = \frac{d}{dt}\,^0T_2 \tag{7.330}$$

$$= \begin{bmatrix} -\dot{\theta}_{12}\,s\theta_{12} & -\dot{\theta}_{12}\,c\theta_{12} & 0 & -l_2\dot{\theta}_{12}\,s\theta_{12} - \dot{\theta}_1 l_1\,s\theta_1 \\ \dot{\theta}_{12}\,c\theta_{12} & -\dot{\theta}_{12}\,s\theta_{12} & 0 & l_2\dot{\theta}_{12}\,c\theta_{12} + \dot{\theta}_1 l_1\,c\theta_1 \\ 0 & 0 & 0 & 0 \\ 0 & 0 & 0 & 0 \end{bmatrix}$$

$$\theta_{12} = \theta_1 + \theta_2 \qquad \dot{\theta}_{12} = \dot{\theta}_1 + \dot{\theta}_2 \tag{7.331}$$

We may also use the chain rule to calculate $^0\dot{T}_2$

$$^0\dot{T}_2 = \frac{d}{dt}\left(^0T_1\,^1T_2\right) = {}^0\dot{T}_1\,^1T_2 + {}^0T_1\,^1\dot{T}_2 \tag{7.332}$$

where

$$^0\dot{T}_1 = \dot{\theta}_1 \begin{bmatrix} -\sin\theta_1 & -\cos\theta_1 & 0 & -l_1\sin\theta_1 \\ \cos\theta_1 & -\sin\theta_1 & 0 & l_1\cos\theta_1 \\ 0 & 0 & 0 & 0 \\ 0 & 0 & 0 & 0 \end{bmatrix} \tag{7.333}$$

$$^1\dot{T}_2 = \dot{\theta}_2 \begin{bmatrix} -\sin\theta_2 & -\cos\theta_2 & 0 & -l_2\sin\theta_2 \\ \cos\theta_2 & -\sin\theta_2 & 0 & l_2\cos\theta_2 \\ 0 & 0 & 0 & 0 \\ 0 & 0 & 0 & 0 \end{bmatrix} \tag{7.334}$$

Having $^0\dot{T}_1$ and $^1\dot{T}_2$, we can find the velocity transformation matrices $^0V_1$ and $^1V_2$ by using $^0T_1^{-1}$ and $^1T_2^{-1}$.

$$^0T_1^{-1} = \begin{bmatrix} \cos\theta_1 & \sin\theta_1 & 0 & -l_1 \\ -\sin\theta_1 & \cos\theta_1 & 0 & 0 \\ 0 & 0 & 1 & 0 \\ 0 & 0 & 0 & 1 \end{bmatrix} \tag{7.335}$$

$$^1T_2^{-1} = \begin{bmatrix} \cos\theta_2 & \sin\theta_2 & 0 & -l_2 \\ -\sin\theta_2 & \cos\theta_2 & 0 & 0 \\ 0 & 0 & 1 & 0 \\ 0 & 0 & 0 & 1 \end{bmatrix} \tag{7.336}$$

$$^0V_1 = {}^0\dot{T}_1\,^0T_1^{-1} = \dot{\theta}_1 \begin{bmatrix} 0 & -1 & 0 & 0 \\ 1 & 0 & 0 & 0 \\ 0 & 0 & 0 & 0 \\ 0 & 0 & 0 & 0 \end{bmatrix} \tag{7.337}$$

$$^1V_2 = {}^1\dot{T}_2\,^1T_2^{-1} = \dot{\theta}_2 \begin{bmatrix} 0 & -1 & 0 & 0 \\ 1 & 0 & 0 & 0 \\ 0 & 0 & 0 & 0 \\ 0 & 0 & 0 & 0 \end{bmatrix} \tag{7.338}$$

Therefore, the velocity of points M_1 and M_2 in B_0 and B_1 are

$$^0\mathbf{v}_{M_1} = {}^0V_1\,^0\mathbf{r}_{M_1} = \dot{\theta}_1 \begin{bmatrix} -l_1\sin\theta_1 \\ l_1\cos\theta_1 \\ 0 \\ 0 \end{bmatrix} \tag{7.339}$$

$$
{}^1\mathbf{v}_{M_2} = {}^1V_2\,{}^1\mathbf{r}_{M_2} = \dot\theta_2
\begin{bmatrix}
-l_2\sin\theta_2 \\
l_2\cos\theta_2 \\
0 \\
0
\end{bmatrix}
\tag{7.340}
$$

To determine the velocity of the tip point M_2 in the base frame, we can use the velocity vector addition.

$$
\begin{aligned}
{}^0\mathbf{v}_{M_2} &= {}^0\mathbf{v}_{M_1} + {}^0_1\mathbf{v}_{M_2} = {}^0\mathbf{v}_{M_1} + {}^0T_1\,{}^1\mathbf{v}_{M_2} \\
&= \begin{bmatrix}
-\left(\dot\theta_1+\dot\theta_2\right)l_2\sin\left(\theta_1+\theta_2\right) - \dot\theta_1 l_1\sin\theta_1 \\
\left(\dot\theta_1+\dot\theta_2\right)l_2\cos\left(\theta_1+\theta_2\right) + \dot\theta_1 l_1\cos\theta_1 \\
0 \\
0
\end{bmatrix}
\end{aligned}
\tag{7.341}
$$

We can also determine ${}^0\mathbf{v}_{M_2}$ by using the velocity transformation matrix ${}^0V_2$

$$
{}^0\mathbf{v}_{M_2} = {}^0V_2\,{}^0\mathbf{r}_{M_2}
\tag{7.342}
$$

where ${}^0V_2$ is

$$
{}^0V_2 = {}^0\dot T_2\,{}^0T_2^{-1} =
\begin{bmatrix}
0 & -\dot\theta_1-\dot\theta_2 & 0 & \dot\theta_2 l_1\sin\theta_1 \\
\dot\theta_1+\dot\theta_2 & 0 & 0 & -\dot\theta_2 l_1\cos\theta_1 \\
0 & 0 & 0 & 0 \\
0 & 0 & 0 & 0
\end{bmatrix}
\tag{7.343}
$$

$$
\begin{aligned}
{}^0T_2^{-1} &= {}^2T_1\,{}^1T_0 = {}^1T_2^{-1}\,{}^0T_1^{-1} \\
&= \begin{bmatrix}
\cos\left(\theta_1+\theta_2\right) & \sin\left(\theta_1+\theta_2\right) & 0 & -l_2-l_1\cos\theta_2 \\
-\sin\left(\theta_1+\theta_2\right) & \cos\left(\theta_1+\theta_2\right) & 0 & l_1\sin\theta_2 \\
0 & 0 & 1 & 0 \\
0 & 0 & 0 & 1
\end{bmatrix}
\end{aligned}
\tag{7.344}
$$

We can also determine the velocity transformation matrix ${}^0V_2$ using their addition rule ${}^0V_2 = {}^0V_1 + {}^0_1V_2$,

$$
\begin{aligned}
{}^0V_2 &= {}^0V_1 + {}^0_1V_2 \\
&= \begin{bmatrix}
0 & -\dot\theta_1-\dot\theta_2 & 0 & \dot\theta_2 l_1\sin\theta_1 \\
\dot\theta_1+\dot\theta_2 & 0 & 0 & -\dot\theta_2 l_1\cos\theta_1 \\
0 & 0 & 0 & 0 \\
0 & 0 & 0 & 0
\end{bmatrix}
\end{aligned}
\tag{7.345}
$$

where

$$
{}^0_1V_2 = {}^0T_1\,{}^1V_2\,{}^0T_1^{-1} =
\begin{bmatrix}
0 & -\dot\theta_2 & 0 & \dot\theta_2 l_1\sin\theta_1 \\
\dot\theta_2 & 0 & 0 & -\dot\theta_2 l_1\cos\theta_1 \\
0 & 0 & 0 & 0 \\
0 & 0 & 0 & 0
\end{bmatrix}
\tag{7.346}
$$

Therefore, ${}^0\mathbf{v}_{M_2}$ would be

$$
\begin{aligned}
{}^0\mathbf{v}_{M_2} &= {}^0V_2\,{}^0\mathbf{r}_{M_2} \\
&= \begin{bmatrix}
-\left(\dot\theta_1+\dot\theta_2\right)l_2\sin\left(\theta_1+\theta_2\right) - \dot\theta_1 l_1\sin\theta_1 \\
\left(\dot\theta_1+\dot\theta_2\right)l_2\cos\left(\theta_1+\theta_2\right) + \dot\theta_1 l_1\cos\theta_1 \\
0 \\
0
\end{bmatrix}
\end{aligned}
\tag{7.347}
$$

7.5 Derivative of a Homogenous Transformation Matrix

Time derivative of homogenous transformation matrix will be used to derive the velocity transformation matrix. The 4×4 homogenous transformation matrix is to move between two coordinate frames.

$$
{}^{G}T_{B} = \begin{bmatrix} {}^{G}R_{B} & {}^{G}\mathbf{d} \\ 0 & 1 \end{bmatrix} = \begin{bmatrix} r_{11} & r_{12} & r_{13} & r_{14} \\ r_{21} & r_{22} & r_{23} & r_{24} \\ r_{31} & r_{32} & r_{33} & r_{34} \\ 0 & 0 & 0 & 1 \end{bmatrix} \tag{7.348}
$$

When the elements of the transformation matrix are time varying, its derivative is

$$
\frac{{}^{G}dT}{dt} = {}^{G}\dot{T}_{B} = \begin{bmatrix} \dfrac{dr_{11}}{dt} & \dfrac{dr_{12}}{dt} & \dfrac{dr_{13}}{dt} & \dfrac{dr_{14}}{dt} \\ \dfrac{dr_{21}}{dt} & \dfrac{dr_{22}}{dt} & \dfrac{dr_{23}}{dt} & \dfrac{dr_{24}}{dt} \\ \dfrac{dr_{31}}{dt} & \dfrac{dr_{32}}{dt} & \dfrac{dr_{33}}{dt} & \dfrac{dr_{34}}{dt} \\ 0 & 0 & 0 & 0 \end{bmatrix} \tag{7.349}
$$

The time derivative of the transformation matrix can be arranged to be proportional to the transformation matrix itself

$$
{}^{G}\dot{T}_{B} = {}^{G}V_{B}\,{}^{G}T_{B} \tag{7.350}
$$

where ${}^{G}V_{B}$ is a 4×4 homogenous matrix called *velocity transformation matrix* or *velocity operator matrix* which is equal to

$$
\begin{aligned}
{}^{G}V_{B} &= {}^{G}\dot{T}_{B}\,{}^{G}T_{B}^{-1} \\
&= \begin{bmatrix} {}^{G}\dot{R}_{B}\,{}^{G}R_{B}^{T} & {}^{G}\dot{\mathbf{d}} - {}^{G}\dot{R}_{B}\,{}^{G}R_{B}^{T}\,{}^{G}\mathbf{d} \\ 0 & 0 \end{bmatrix}
\end{aligned} \tag{7.351}
$$

The homogenous matrix and its derivative based on the velocity transformation matrix are useful in forward velocity kinematics. The homogenous transformation matrix ${}^{i-1}\dot{T}_{i}$ for two links connected by a revolute joint is

$$
{}^{i-1}\dot{T}_{i} = \dot{\theta}_{i} \begin{bmatrix} -\sin\theta_{i} & -\cos\theta_{i}\cos\alpha_{i} & \cos\theta_{i}\sin\alpha_{i} & -a_{i}\sin\theta_{i} \\ \cos\theta_{i} & -\sin\theta_{i}\cos\alpha_{i} & \sin\theta_{i}\sin\alpha_{i} & a_{i}\cos\theta_{i} \\ 0 & 0 & 0 & 0 \\ 0 & 0 & 0 & 0 \end{bmatrix} \tag{7.352}
$$

and for two links connected by a prismatic joint is

$$
{}^{i-1}\dot{T}_{i} = \begin{bmatrix} 0 & 0 & 0 & 0 \\ 0 & 0 & 0 & 0 \\ 0 & 0 & 0 & \dot{d}_{i} \\ 0 & 0 & 0 & 0 \end{bmatrix} \tag{7.353}
$$

The associated velocity transformation matrix for a revolute joint is

$$
{}^{i-1}V_{i} = \dot{\theta}_{i}\,\Delta_{R} = \dot{\theta}_{i} \begin{bmatrix} 0 & -1 & 0 & 0 \\ 1 & 0 & 0 & 0 \\ 0 & 0 & 0 & 0 \\ 0 & 0 & 0 & 0 \end{bmatrix} \tag{7.354}
$$

and for a prismatic joint is

$$
{}^{i-1}V_i = \dot{d}_i \; \Delta_P = \dot{d}_i \begin{bmatrix} 0 & 0 & 0 & 0 \\ 0 & 0 & 0 & 0 \\ 0 & 0 & 0 & 1 \\ 0 & 0 & 0 & 0 \end{bmatrix}
\tag{7.355}
$$

Proof Any transformation matrix $[T]$ can be decomposed into a rotation and translation. Employing the decomposed form of $[T]$,

$$
[T] = \begin{bmatrix} R_{\hat{u},\phi} & \mathbf{d} \\ 0 & 1 \end{bmatrix} = \begin{bmatrix} \mathbf{I} & \mathbf{d} \\ 0 & 1 \end{bmatrix} \begin{bmatrix} R_{\hat{u},\phi} & 0 \\ 0 & 1 \end{bmatrix}
$$

$$
= [D]\,[R]
\tag{7.356}
$$

we can find $[\dot{T}]$ as

$$
[\dot{T}] = \begin{bmatrix} \dot{R}_{\hat{u},\phi} & \dot{\mathbf{d}} \\ 0 & 0 \end{bmatrix} = \begin{bmatrix} \mathbf{I} & \dot{\mathbf{d}} \\ 0 & 1 \end{bmatrix} \begin{bmatrix} \dot{R}_{\hat{u},\phi} & 0 \\ 0 & 1 \end{bmatrix} - \mathbf{I}
$$

$$
= [\mathbf{I} + \dot{D}][\mathbf{I} + \dot{R}] - \mathbf{I} = [V][T]
\tag{7.357}
$$

where $[V]$ is the velocity transformation matrix described as

$$
[V] = \dot{T} \; T^{-1} = \begin{bmatrix} \dot{R}_{\hat{u},\phi} & \dot{\mathbf{d}} \\ 0 & 0 \end{bmatrix} \begin{bmatrix} R_{\hat{u},\phi}^T & -R_{\hat{u},\phi}^T \mathbf{d} \\ 0 & 1 \end{bmatrix}
$$

$$
= \begin{bmatrix} \dot{R}_{\hat{u},\phi}\, R_{\hat{u},\phi}^T & \dot{\mathbf{d}} - \dot{R}_{\hat{u},\phi}\, R_{\hat{u},\phi}^T \mathbf{d} \\ 0 & 0 \end{bmatrix}
$$

$$
= \begin{bmatrix} \tilde{\omega} & \dot{\mathbf{d}} - \tilde{\omega}\,\mathbf{d} \\ 0 & 0 \end{bmatrix}
\tag{7.358}
$$

The transformation matrix between two neighbor coordinate frames of a robot is given by Eq. (5.11) based on the Denavit–Hartenberg, DH, parameters.

$$
{}^{i-1}T_i = \begin{bmatrix} \cos\theta_i & -\sin\theta_i \cos\alpha_i & \sin\theta_i \sin\alpha_i & a_i \cos\theta_i \\ \sin\theta_i & \cos\theta_i \cos\alpha_i & -\cos\theta_i \sin\alpha_i & a_i \sin\theta_i \\ 0 & \sin\alpha_i & \cos\alpha_i & d_i \\ 0 & 0 & 0 & 1 \end{bmatrix}
\tag{7.359}
$$

In case two links are connected via a revolute joint, then, θ_i is the only variable of DH matrix, and therefore direct differentiation shows

$$
{}^{i-1}\dot{T}_i = \dot{\theta}_i \begin{bmatrix} -\sin\theta_i & -\cos\theta_i \cos\alpha_i & \cos\theta_i \sin\alpha_i & -a_i \sin\theta_i \\ \cos\theta_i & -\sin\theta_i \cos\alpha_i & \sin\theta_i \sin\alpha_i & a_i \cos\theta_i \\ 0 & 0 & 0 & 0 \\ 0 & 0 & 0 & 0 \end{bmatrix}
$$

$$
= \dot{\theta}_i \begin{bmatrix} 0 & -1 & 0 & 0 \\ 1 & 0 & 0 & 0 \\ 0 & 0 & 0 & 0 \\ 0 & 0 & 0 & 0 \end{bmatrix} {}^{i-1}T_i = \dot{\theta}_i \; \Delta_R \; {}^{i-1}T_i
\tag{7.360}
$$

Hence, the *revolute velocity transformation matrix* is

$$
{}^{i-1}V_i = \dot{\theta}_i \; \Delta_R = \dot{\theta}_i \begin{bmatrix} 0 & -1 & 0 & 0 \\ 1 & 0 & 0 & 0 \\ 0 & 0 & 0 & 0 \\ 0 & 0 & 0 & 0 \end{bmatrix} \tag{7.361}
$$

However, when the two links are connected via a prismatic joint, d_i is the only variable of the DH matrix, and therefore,

$$
{}^{i-1}\dot{T}_i = \dot{d}_i \begin{bmatrix} 0 & 0 & 0 & 0 \\ 0 & 0 & 0 & 0 \\ 0 & 0 & 0 & 1 \\ 0 & 0 & 0 & 0 \end{bmatrix} {}^{i-1}T_i = \dot{d}_i \; \Delta_P \; {}^{i-1}T_i \tag{7.362}
$$

and hence, the *prismatic velocity transformation matrix* will be

$$
{}^{i-1}V_i = \dot{d}_i \; \Delta_P = \dot{d}_i \begin{bmatrix} 0 & 0 & 0 & 0 \\ 0 & 0 & 0 & 0 \\ 0 & 0 & 0 & 1 \\ 0 & 0 & 0 & 0 \end{bmatrix} \tag{7.363}
$$

The Δ_R and Δ_P are revolute and prismatic *velocity coefficient matrices* with some application in velocity analysis of robots. ∎

Example 259 Differential of a transformation matrix. A numerical example to calculate dT when differential rotation and translation, dR and dD, are given.

Assume a transformation matrix is given as

$$
T = \begin{bmatrix} 0 & 0 & 1 & 4 \\ 1 & 0 & 0 & 4 \\ 0 & 1 & 0 & 4 \\ 0 & 0 & 0 & 1 \end{bmatrix} \tag{7.364}
$$

subject to a differential rotation and a differential translation given by

$$
d\phi\hat{u} = \begin{bmatrix} 0.1 & 0.2 & 0.3 \end{bmatrix} \tag{7.365}
$$

$$
d\mathbf{d} = \begin{bmatrix} 0.6 & 0.4 & 0.2 \end{bmatrix} \tag{7.366}
$$

Then, the differential transformation matrix dT will be

$$
\begin{aligned}
dT &= [\mathbf{I} + dD][\mathbf{I} + dR] - \mathbf{I} \\
&= \begin{bmatrix} 0 & -0.3 & 0.2 & 0.6 \\ 0.3 & 0 & -0.1 & 0.4 \\ -0.2 & 0.1 & 0 & 0.2 \\ 0 & 0 & 0 & 0 \end{bmatrix}
\end{aligned} \tag{7.367}
$$

Example 260 Differential rotation and translation. It is to calculate the differential transformation matrix dT for small values of rotation $d\phi$ and translation $d\mathbf{d}$.

Assume the angle of rotation about the axis $\hat{u}$ is too small indicated by $d\phi$, and then the differential rotation matrix will be

$$
\mathbf{I} + dR_{\hat{u},\phi} = \mathbf{I} + R_{\hat{u},d\phi} = \begin{bmatrix} 1 & -u_3 d\phi & u_2 d\phi & 0 \\ u_3 d\phi & 1 & -u_1 d\phi & 0 \\ -u_2 d\phi & +u_1 d\phi & 1 & 0 \\ 0 & 0 & 0 & 1 \end{bmatrix} \tag{7.368}
$$

because when $\phi \ll 1$, we have

$$\sin \phi \simeq d\phi \tag{7.369}$$

$$\cos \phi \simeq 1 \tag{7.370}$$

$$\text{vers} \, \phi \simeq 0 \tag{7.371}$$

Differential translation $d\mathbf{d} = d(d_x \hat{I} + d_y \hat{J} + d_z \hat{K})$ is shown by a differential translation matrix

$$\mathbf{I} + dD = \begin{bmatrix} 1 & 0 & 0 & dd_x \\ 0 & 1 & 0 & dd_y \\ 0 & 0 & 1 & dd_z \\ 0 & 0 & 0 & 1 \end{bmatrix} \tag{7.372}$$

and therefore,

$$
\begin{aligned}
dT &= [\mathbf{I} + dD][\mathbf{I} + dR] - \mathbf{I} \\
&= \begin{bmatrix} 0 & -d\phi \, u_3 & d\phi \, u_2 & dd_x \\ d\phi \, u_3 & 0 & -d\phi \, u_1 & dd_y \\ -d\phi \, u_2 & d\phi \, u_1 & 0 & dd_z \\ 0 & 0 & 0 & 0 \end{bmatrix}
\end{aligned} \tag{7.373}
$$

Example 261 Combination of principal differential rotations. Assuming small rotations about the principal global axes, we calculate the combined differential rotation matrix.

The differential rotations about X, Y, and Z are

$$R_{X,d\gamma} = \begin{bmatrix} 1 & 0 & 0 & 0 \\ 0 & 1 & -d\gamma & 0 \\ 0 & d\gamma & 1 & 0 \\ 0 & 0 & 0 & 1 \end{bmatrix} \tag{7.374}$$

$$R_{Y,d\beta} = \begin{bmatrix} 1 & 0 & d\beta & 0 \\ 0 & 1 & 0 & 0 \\ -d\beta & 0 & 1 & 0 \\ 0 & 0 & 0 & 1 \end{bmatrix} \tag{7.375}$$

$$R_{Z,d\alpha} = \begin{bmatrix} 1 & -d\alpha & 0 & 0 \\ d\alpha & 1 & 0 & 0 \\ 0 & 0 & 1 & 0 \\ 0 & 0 & 0 & 1 \end{bmatrix} \tag{7.376}$$

The combination of the principal differential rotation matrices about the axes X, then Y, and then Z is

$$
\begin{aligned}
&\left[\mathbf{I} + R_{X,d\gamma}\right]\left[\mathbf{I} + R_{Y,d\beta}\right]\left[\mathbf{I} + R_{Z,d\alpha}\right] \\
&= \begin{bmatrix} 1 & 0 & 0 & 0 \\ 0 & 1 & -d\gamma & 0 \\ 0 & d\gamma & 1 & 0 \\ 0 & 0 & 0 & 1 \end{bmatrix} \begin{bmatrix} 1 & 0 & d\beta & 0 \\ 0 & 1 & 0 & 0 \\ -d\beta & 0 & 1 & 0 \\ 0 & 0 & 0 & 1 \end{bmatrix} \begin{bmatrix} 1 & -d\alpha & 0 & 0 \\ d\alpha & 1 & 0 & 0 \\ 0 & 0 & 1 & 0 \\ 0 & 0 & 0 & 1 \end{bmatrix} \\
&= \begin{bmatrix} 1 & -d\alpha & d\beta & 0 \\ d\alpha & 1 & -d\gamma & 0 \\ -d\beta & d\gamma & 1 & 0 \\ 0 & 0 & 0 & 1 \end{bmatrix} \\
&= \left[\mathbf{I} + R_{Z,d\alpha}\right]\left[\mathbf{I} + R_{Y,d\beta}\right]\left[\mathbf{I} + R_{X,d\gamma}\right]
\end{aligned} \tag{7.377}
$$

The combination of differential rotations is commutative.

$$\begin{aligned}
&\left[\mathbf{I} + R_{X,d\gamma}\right]\left[\mathbf{I} + R_{Y,d\beta}\right]\left[\mathbf{I} + R_{Z,d\alpha}\right] \\
&= \left[\mathbf{I} + R_{X,d\gamma}\right]\left[\mathbf{I} + R_{Z,d\alpha}\right]\left[\mathbf{I} + R_{Y,d\beta}\right] \\
&= \left[\mathbf{I} + R_{Y,d\beta}\right]\left[\mathbf{I} + R_{X,d\gamma}\right]\left[\mathbf{I} + R_{Z,d\alpha}\right] \\
&= \left[\mathbf{I} + R_{Y,d\beta}\right]\left[\mathbf{I} + R_{Z,d\alpha}\right]\left[\mathbf{I} + R_{X,d\gamma}\right] \\
&= \left[\mathbf{I} + R_{Z,d\alpha}\right]\left[\mathbf{I} + R_{X,d\gamma}\right]\left[\mathbf{I} + R_{Y,d\beta}\right] \\
&= \left[\mathbf{I} + R_{Z,d\alpha}\right]\left[\mathbf{I} + R_{Y,d\beta}\right]\left[\mathbf{I} + R_{X,d\gamma}\right]
\end{aligned} \tag{7.378}$$

Example 262 ★ Velocity of frame B_i in B_0. The velocity of the origin of B_i-frame in the base frame may be expressed by derivatives of all homogenous transformation matrices between B_i-frame and B_0-frame with respect to their associated joint variable.

The velocity of the frame B_i attached to the link (i) with respect to the base coordinate frame B_0 can be found by differentiating $^0\mathbf{d}_i$ in the base frame.

$$\begin{aligned}
^0\mathbf{v}_i &= \frac{^0d}{dt}\,^0\mathbf{d}_i = \frac{^0d}{dt}\left(^0T_i\,^i\mathbf{d}_i\right) \\
&= ^0\dot{T}_1\,^1T_2 \cdots\,^{i-1}T_i\,^i\mathbf{d}_i + ^0T_1\,^1\dot{T}_2\,^2T_3 \cdots\,^{i-1}T_i\,^i\mathbf{d}_i \\
&\quad + ^0T_1 \cdots\,^{i-1}\dot{T}_i\,^i\mathbf{d}_i \\
&= \left[\sum_{j=1}^{i} \frac{\partial\,^0T_i}{\partial q_j}\dot{q}_j\right]\,^i\mathbf{d}_i
\end{aligned} \tag{7.379}$$

However, the partial derivatives $\partial\,^{i-1}T_i/\partial q_i$ can be found by utilizing the velocity coefficient matrices Δ_i, which is either Δ_R or Δ_P depending on the revolute or prismatic joint between links (i) and $(i-1)$.

$$\frac{\partial\,^{i-1}T_i}{\partial q_i} = \Delta_i\,^{i-1}T_i \tag{7.380}$$

$$\Delta_R = \frac{1}{\dot{d}_i}\,^{i-1}V_i = \frac{1}{\dot{d}_i}\,^{i-1}\dot{T}_i\,^{i-1}T_i^{-1} = \begin{bmatrix} 0 & -1 & 0 & 0 \\ 1 & 0 & 0 & 0 \\ 0 & 0 & 0 & 0 \\ 0 & 0 & 0 & 0 \end{bmatrix} \tag{7.381}$$

$$\Delta_P = \frac{1}{\dot{\theta}_i}\,^{i-1}V_i = \frac{1}{\dot{\theta}_i}\,^{i-1}\dot{T}_i\,^{i-1}T_i^{-1} = \begin{bmatrix} 0 & 0 & 0 & 0 \\ 0 & 0 & 0 & 0 \\ 0 & 0 & 0 & 1 \\ 0 & 0 & 0 & 0 \end{bmatrix} \tag{7.382}$$

Hence,

$$\frac{\partial\,^0T_i}{\partial q_j} = \begin{cases} ^0T_1\,^1T_2 \cdots\,^{j-2}T_{j-1}\,\Delta_j\,^{j-1}T_j \cdots\,^{i-1}T_i & \text{for } j \leq i \\ 0 & \text{for } j > i \end{cases} \tag{7.383}$$

Example 263 V reduces to $\tilde{\omega}$, and T reduces to R if $\mathbf{d} = 0$. If displacement of a rigid body motion is zero, then T will become R, and V becomes $\tilde{\omega}$. This example shows that T and V are equivalent with R and $\tilde{\omega}$ if the only motion of body B in G is rotating.

Consider B and G coordinate frames with a common origin. In this case, $\mathbf{d} = 0$ and (7.356) will be

$$[T] = \begin{bmatrix} R_{\hat{u},\phi} & \mathbf{0} \\ 0 & 1 \end{bmatrix} = \begin{bmatrix} \mathbf{I} & \mathbf{0} \\ 0 & 1 \end{bmatrix} \begin{bmatrix} R_{\hat{u},\phi} & 0 \\ 0 & 1 \end{bmatrix}$$

$$= [\mathbf{I}]\,[R] = [R] \tag{7.384}$$

and hence, $\dot{T}$ will be

$$\dot{T} = \dot{R} \tag{7.385}$$

Therefore, the velocity transformation matrix $[V]$ will be equivalent to $\tilde{\omega}$.

$$[V] = \dot{T}\,T^{-1} = \dot{R}\,R^T = \tilde{\omega} \tag{7.386}$$

Example 264 *DH* matrix between two co-origin coordinate frames. Here, we see how the *DH* transformation matrix will be simplified if links (i) and $(i - 1)$ have the same origin.

If two neighbor coordinate frames have the same origin, then a_i and d_i of *DH* transformation matrix (5.11) will be zero. It simplifies the *DH* matrix.

$$^{i-1}T_i = \begin{bmatrix} \cos\theta_i & -\sin\theta_i\cos\alpha_i & \sin\theta_i\sin\alpha_i & 0 \\ \sin\theta_i & \cos\theta_i\cos\alpha_i & -\cos\theta_i\sin\alpha_i & 0 \\ 0 & \sin\alpha_i & \cos\alpha_i & 0 \\ 0 & 0 & 0 & 1 \end{bmatrix} \tag{7.387}$$

We can eliminate the last column and row of this matrix and show it by a rotation transformation matrix.

$$^{i-1}R_i = \begin{bmatrix} \cos\theta_i & -\sin\theta_i\cos\alpha_i & \sin\theta_i\sin\alpha_i \\ \sin\theta_i & \cos\theta_i\cos\alpha_i & -\cos\theta_i\sin\alpha_i \\ 0 & \sin\alpha_i & \cos\alpha_i \end{bmatrix} \tag{7.388}$$

When a_i and d_i are zero, the two adjacent links are connected by a revolute joint. So, θ_i is the only variable of *DH* matrix, and therefore,

$$^{i-1}\dot{R}_i = \dot{\theta}_i \begin{bmatrix} -\sin\theta_i & -\cos\theta_i\cos\alpha_i & \cos\theta_i\sin\alpha_i \\ \cos\theta_i & -\sin\theta_i\cos\alpha_i & \sin\theta_i\sin\alpha_i \\ 0 & 0 & 0 \end{bmatrix} = {}_{i-1}\omega_i\,{}^{i-1}R_i$$

$$= \dot{\theta}_i \begin{bmatrix} 0 & -1 & 0 \\ 1 & 0 & 0 \\ 0 & 0 & 0 \end{bmatrix} {}^{i-1}R_i = \dot{\theta}_i\,{}^{i-1}\tilde{k}_{i-1}\,{}^{i-1}R_i \tag{7.389}$$

which shows that the *revolute angular velocity matrix* is

$$_{i-1}\omega_i = \dot{\theta}_i\,{}^{i-1}\tilde{k}_{i-1} = \dot{\theta}_i \begin{bmatrix} 0 & -1 & 0 \\ 1 & 0 & 0 \\ 0 & 0 & 0 \end{bmatrix} \tag{7.390}$$

7.6   Summary

The transformation matrix $^G R_B$ is time dependent if a body coordinate frame B rotates continuously with respect to frame G with a common origin.

$$^G\mathbf{r}(t) = {}^G R_B(t)\,{}^B\mathbf{r} \tag{7.391}$$

Then, the global velocity of a point in B is

$$
{}^G\dot{\mathbf{r}}(t) = {}^G\mathbf{v}(t) = {}^G\dot{R}_B(t)\,{}^B\mathbf{r} = {}_G\tilde{\omega}_B\,{}^G\mathbf{r}(t) \tag{7.392}
$$

where ${}_G\tilde{\omega}_B$ is the skew symmetric angular velocity matrix.

$$
{}_G\tilde{\omega}_B = {}^G\dot{R}_B\,{}^G R_B^T = \begin{bmatrix} 0 & -\omega_3 & \omega_2 \\ \omega_3 & 0 & -\omega_1 \\ -\omega_2 & \omega_1 & 0 \end{bmatrix} \tag{7.393}
$$

The matrix ${}_G\tilde{\omega}_B$ is associated with the angular velocity vector ${}_G\omega_B = \dot{\phi}\,\hat{u}$, which is equal to an angular rate $\dot{\phi}$ about the instantaneous axis of rotation $\hat{u}$. Angular velocities of connected links of a robot may be added relatively to find the angular velocity of the link (n) in the base frame B_0.

$$
{}_0\omega_n = {}_0\omega_1 + {}_1^0\omega_2 + {}_2^0\omega_3 + \cdots + {}_{n-1}^0\omega_n = \sum_{i=1}^{n} {}_{i-1}^0\omega_i \tag{7.394}
$$

To work with angular velocities of relatively moving links, we need to follow the rules of relative derivatives in body and global coordinate frames. Derivative is a frame-dependent operation. Derivative of a vector quantity in the same frame in which the vector is expressed is a simple derivative. Derivative of a vector quantity in another frame other than the frame in which the vector is expressed is a mixed derivative. Considering two frames, we will have two simple and two mixed derivatives.

$$
\frac{{}^B d}{dt}\,{}^B\mathbf{r}_P = {}^B\dot{\mathbf{r}}_P = {}^B\mathbf{v}_P = \dot{x}\,\hat{\imath} + \dot{y}\,\hat{\jmath} + \dot{z}\,\hat{k} \tag{7.395}
$$

$$
\frac{{}^G d}{dt}\,{}^G\mathbf{r}_P = {}^G\dot{\mathbf{r}}_P = {}^G\mathbf{v}_P = \dot{X}\,\hat{I} + \dot{Y}\,\hat{J} + \dot{Z}\,\hat{K}
$$

$$
\frac{{}^G d}{dt}\,{}^B\mathbf{r}_P(t) = {}^B\dot{\mathbf{r}}_P + {}^B_G\omega_B \times {}^B\mathbf{r}_P = {}^B_G\dot{\mathbf{r}}_P \tag{7.396}
$$

$$
\frac{{}^B d}{dt}\,{}^G\mathbf{r}_P(t) = {}^G\dot{\mathbf{r}}_P - {}_G\omega_B \times {}^G\mathbf{r}_P = {}^G_B\dot{\mathbf{r}}_P \tag{7.397}
$$

The global velocity of a point P in a moving frame B at ${}^G\mathbf{r}_P$

$$
{}^G\mathbf{r}_P = {}^G R_B\,{}^B\mathbf{r}_P + {}^G\mathbf{d}_B \tag{7.398}
$$

is

$$
\begin{aligned}
{}^G\mathbf{v}_P = {}^G\dot{\mathbf{r}}_P &= {}_G\tilde{\omega}_B \left({}^G\mathbf{r}_P - {}^G\mathbf{d}_B\right) + {}^G\dot{\mathbf{d}}_B \\
&= {}_G\omega_B \times \left({}^G\mathbf{r}_P - {}^G\mathbf{d}_B\right) + {}^G\dot{\mathbf{d}}_B
\end{aligned} \tag{7.399}
$$

The velocity relationship for a body B having a continues rigid motion in G may also be expressed by a homogenous velocity transformation matrix ${}^G V_B$,

$$
{}^G\mathbf{v}(t) = {}^G V_B\,{}^G\mathbf{r}(t) \tag{7.400}
$$

where ${}^G V_B$ includes both the translational and rotational velocities of B in G.

$$
{}^G V_B = {}^G\dot{T}_B\,{}^G T_B^{-1} = \begin{bmatrix} {}_G\tilde{\omega}_B & {}^G\dot{\mathbf{d}}_B - {}_G\tilde{\omega}_B\,{}^G\mathbf{d}_B \\ 0 & 0 \end{bmatrix} \tag{7.401}
$$

7.7 Key Symbols

$\mathbf{a}$	Turn vector of end-effector frame
B	Body coordinate frame
c	cos
d	Differential, prismatic joint variable
d_x, d_y, d_z	Elements of $\mathbf{d}$
$\mathbf{d}$	Translation vector, displacement vector
D	Displacement transformation matrix
e	Rotation quaternion
e_0, e_1, e_2, e_3	Euler parameters, components of e
G, B_0	Global coordinate frame, base coordinate frame
$\hat{\imath}, \hat{\jmath}, \hat{k}$	Local coordinate axes unit vectors
$\tilde{\imath}, \tilde{\jmath}, \tilde{k}$	Skew symmetric matrices of the unit vectors $\hat{\imath}, \hat{\jmath}, \hat{k}$
$\hat{I}, \hat{J}, \hat{K}$	Global coordinate axes unit vectors
$\mathbf{I} = [I]$	Identity matrix
$\mathbf{J}$	Jacobian
l	Length
p	Pitch of a screw
q	Joint coordinate
$\mathbf{q}$	Joint coordinate vector
$\mathbf{r}$	Position vectors, homogenous position vector
r_i	The element i of $\mathbf{r}$
r_{ij}	The element of row i and column j of a matrix
R	Rotation transformation matrix
s	sin
$\mathbf{s}$	Location vector of a screw
sgn	Signum function
$SSRMS$	Space station remote manipulator system
T	Homogenous transformation matrix
T_{arm}	Manipulator transformation matrix
T_{wrist}	Wrist transformation matrix
$\mathbf{T}$	A set of nonlinear algebraic equations of $\mathbf{q}$
$\mathbf{v}$	Velocity vector
V	Velocity transformation matrix
$\hat{u}$	Unit vector along the axis of $\boldsymbol{\omega}$
$\tilde{u}$	Skew symmetric matrix of the vector $\hat{u}$
u_1, u_2, u_3	Components of $\hat{u}$
x, y, z	Local coordinate axes
X, Y, Z	Global coordinate axes

Greek

α, β, γ	Angles of rotation about the axes of global frame
δ	Kronecker function, small increment of a parameter
ϵ	Small test number to terminate a procedure
θ	Rotary joint angle
θ_{ijk}	$\theta_i + \theta_j + \theta_k$
φ, θ, ψ	Angles of rotation about the axes of body frame
ϕ	Angle of rotation about $\hat{u}$
$\boldsymbol{\omega}$	Angular velocity vector
$\tilde{\omega}$	Skew symmetric matrix of the vector $\boldsymbol{\omega}$
$\omega_1, \omega_2, \omega_3$	Components of $\boldsymbol{\omega}$

Symbol

$[\ \ ]^{-1}$	Inverse of the matrix $[\ \ ]$
$[\ \ ]^{T}$	Transpose of the matrix $[\ \ ]$
$\vdash$	Orthogonal
(i)	Link number i
$\parallel$	Parallel
$\perp$	Perpendicular
e^{*}	Conjugate of e
Δ_{P}	Prismatic velocity coefficient matrices
Δ_{R}	Revolute velocity coefficient matrices

Exercises

1. Notation and symbols.
   Describe the meaning of the following notations:

   $$
   \begin{array}{llllll}
   \text{a} - {}_{G}\boldsymbol{\omega}_B & \text{b} - {}_{B}\boldsymbol{\omega}_G & \text{c} - {}_{G}^{G}\boldsymbol{\omega}_B & \text{d} - {}_{G}^{B}\boldsymbol{\omega}_B & \text{e} - {}_{B}^{B}\boldsymbol{\omega}_G & \text{f} - {}_{B}^{G}\boldsymbol{\omega}_G \\
   \text{g} - {}_{2}^{0}\boldsymbol{\omega}_1 & \text{h} - {}_{2}^{2}\boldsymbol{\omega}_1 & \text{i} - {}_{2}^{3}\boldsymbol{\omega}_1 & \text{j} - {}^{G}\dot{R}_B & \text{k} - {}_{2}^{0}\tilde{\omega}_1 & \text{l} - {}_{j}^{k}\boldsymbol{\omega}_i \\
   \text{m} - {}^{G}\mathbf{r}_P(t) & \text{n} - {}_{2}^{G}\mathbf{v}_P & \text{o} - \Delta_R & \text{p} - \Delta_P & \text{q} - \dfrac{{}^{G}d}{dt} & \text{r} - \dfrac{{}^{B}d}{dt} \\
   \end{array}
   $$

   $$
   \text{s} - \dfrac{{}^{G}d}{dt}\,{}^{G}\mathbf{r}_P \quad \text{t} - \dfrac{{}^{G}d}{dt}\,{}^{B}\mathbf{r}_P \quad \text{u} - \dfrac{{}^{B}d}{dt}\,{}^{B}\mathbf{r}_P \quad \text{v} - {}^{G}\dot{\mathbf{r}}_P \quad \text{w} - {}^{G}\dot{\mathbf{d}}_P \quad \text{x} - {}^{G}V_B
   $$

2. Angular velocity identities.
   (a) Show that if

   $$\omega^2 = \omega_1^2 + \omega_2^2 + \omega_3^2 \tag{7.402}$$

   then

   $$\tilde{\omega}^T\,\tilde{\omega}\,\tilde{\omega}^T = \omega\tilde{\omega} \tag{7.403}$$

   $$\tilde{\omega}\,\tilde{\omega}^T\,\tilde{\omega} = \omega\tilde{\omega} \tag{7.404}$$

   $$\tilde{\omega}^3 = \omega\tilde{\omega}^T \tag{7.405}$$

   $$\tilde{\omega}^{3T} = \left(\tilde{\omega}^T\right)^3 = \left(\tilde{\omega}^3\right)^T = \omega\tilde{\omega} \tag{7.406}$$

   (b) Prove that

   $$^{G}\dot{R}_B^T\,{}^{G}R_B = -\left[{}^{G}\dot{R}_B^T\,{}^{G}R_B\right]^T \tag{7.407}$$

   (c) Using ${}^{G}R_B$ and ${}^{G}\dot{R}_B$, show that there is no $\tilde{\omega}^{-1}$.

3. Local position, global velocity.
   A body is turning about a global principal axis at a constant angular speed. Find the global velocity of a point at ${}^{B}\mathbf{r}$.

   $$^{B}\mathbf{r} = \begin{bmatrix} 5 & 30 & 10 \end{bmatrix}^T \tag{7.408}$$

   (a) The axis is Z-axis, and the angular rate is $\dot{\alpha} = 2\,\text{rad/s}$ when $\alpha = 30$ deg.
   (b) The axis is Y-axis, and the angular rate is $\dot{\beta} = 2\,\text{rad/s}$ when $\beta = 30$ deg.
   (c) The axis is X-axis, and the angular rate is $\dot{\gamma} = 2\,\text{rad/s}$ when $\gamma = 30$ deg.

4. Parametric angular velocity, global principal rotations.
   A body B is turning in a global frame G. The rotation transformation matrix can be decomposed into principal axes rotations. Determine the angular velocity ${}_{G}\tilde{\omega}_B$ and ${}_{G}\boldsymbol{\omega}_B$ for the following cases:
   (a) ${}^{G}R_B$ is the result of a rotation α about Z-axis followed by β about Y-axis.
   (b) ${}^{G}R_B$ is the result of a rotation β about Y-axis followed by α about Z-axis.
   (c) ${}^{G}R_B$ is the result of a rotation α about Z-axis followed by γ about X-axis.
   (d) ${}^{G}R_B$ is the result of a rotation γ about X-axis followed by α about Z-axis.
   (e) ${}^{G}R_B$ is the result of a rotation γ about X-axis followed by β about Y-axis.
   (f) ${}^{G}R_B$ is the result of a rotation β about Y-axis followed by γ about X-axis.

5. Numeric angular velocity, global principal rotations.
   A body B is turning in a global frame G. The rotation transformation matrix can be decomposed into principal axes rotations. Determine the angular velocity ${}_{G}\tilde{\omega}_B$ and ${}_{G}\boldsymbol{\omega}_B$ for Exercises $4(a) - -(f)$ using $\dot{\alpha} = 2\,\text{rad/s}$, $\dot{\beta} = 2\,\text{rad/s}$, $\dot{\gamma} = 2\,\text{rad/s}$ and $\alpha = 30$ deg, $\beta = 30$ deg, and $\gamma = 30$ deg.

6. Global position, constant angular velocity.

A body is turning about the global principal axis at a constant angular rate. Find the global position of a point at $^B\mathbf{r}$ after $t = 3$ s if the body and global coordinate frames were coincident at $t = 0$ s.

$$^B\mathbf{r} = \begin{bmatrix} 5 & 30 & 10 \end{bmatrix}^T \tag{7.409}$$

(a) The axis is Z-axis, and the angular rate is $\dot{\alpha} = 2$ rad/s.
(b) The axis is Y-axis, and the angular rate is $\dot{\beta} = 2$ rad/s.
(c) The axis is X-axis, and the angular rate is $\dot{\gamma} = 2$ rad/s.

7. Turning about x-axis.

Find the angular velocity matrix when the body coordinate frame is turning about a body axis.

(a) The axis is x-axis, the angular rate is $\dot{\varphi} = 2$ rad/s, and the angle is $\varphi = 45$ deg.
(b) The axis is y-axis, the angular rate is $\dot{\theta} = 2$ rad/s, and the angle is $\theta = 45$ deg.
(c) The axis is z-axis, the angular rate is $\dot{\psi} = 2$ rad/s, and the angle is $\psi = 45$ deg.

8. Combined rotation and angular velocity.

Find the rotation matrix for a body frame that turns about the global axes with given rates, and calculate the angular velocity of B in G.

(a) The axes are Z, then X, and then Y. The angles are 30 deg about the Z-axis, 30 deg about the X-axis, and 90 deg about the Y-axis. The angular rates are $\dot{\alpha} = 20$ deg/s, $\dot{\beta} = -40$ deg/s, and $\dot{\gamma} = 55$ deg/s about the Z, X, and Y axes, respectively.
(b) The axes are X, then Y, and then Z. The angles are 30 deg about X-axis, 30 deg about the Y-axis, and 90 deg about the Z-axis. The angular rates are $\dot{\alpha} = 20$ deg/s, $\dot{\beta} = -40$ deg/s, and $\dot{\gamma} = 55$ deg/s about the X, Y, and Z axes, respectively.
(c) The axes are Y, then Z, and then X. The angles are 30 deg about X-axis, 30 deg about the Y-axis, and 90 deg about the Z-axis. The angular rates are $\dot{\alpha} = 20$ deg/s, $\dot{\beta} = -40$ deg/s, and $\dot{\gamma} = 55$ deg/s about the X, Y, and Z axes, respectively.

9. ★ Global triple angular velocity matrix.

Determine the angular velocities $_G\tilde{\omega}_B$ and $_G\boldsymbol{\omega}_B$ for the global triple rotations of the Appendix A.

10. ★ Local triple angular velocity matrix.

Determine the angular velocities $_G\tilde{\omega}_B$ and $_G\boldsymbol{\omega}_B$ for the local triple rotations of the Appendix B.

11. Angular velocity, expressed in body frame.

A point P is at $\mathbf{r}_P = [1, 2, 1]^T$ in a body coordinate $B(Oxyz)$.

(a) Find $_G^B\tilde{\omega}_B$ when the body frame is turned 30 deg about the X-axis at a rate $\dot{\gamma} = 75$ deg/s, followed by 45 deg about the Z-axis at a rate $\dot{\alpha} = 25$ deg/s.
(b) Find $_G^B\tilde{\omega}_B$ when the body frame is turned 45 deg about the Z-axis at a rate $\dot{\alpha} = 25$ deg/s, followed by 30 deg about the X-axis at a rate $\dot{\gamma} = 75$ deg/s.

12. Global roll–pitch–yaw angular velocity.

Calculate the angular velocity $_G\tilde{\omega}_B$ for a global roll–pitch–yaw rotation of the following cases:

(a) $\alpha = 30$ deg, $\beta = 30$ deg, and $\gamma = 30$ deg with $\dot{\alpha} = 20$ deg/s, $\dot{\beta} = -20$ deg/s, and $\dot{\gamma} = 20$ deg/s.
(b) $\alpha = 30$ deg, $\beta = 30$ deg, and $\gamma = 30$ deg with $\dot{\alpha} = 0$ deg/s, $\dot{\beta} = -20$ deg/s, and $\dot{\gamma} = 20$ deg/s.
(c) $\alpha = 30$ deg, $\beta = 30$ deg, and $\gamma = 30$ deg with $\dot{\alpha} = 20$ deg/s, $\dot{\beta} = 0$ deg/s, and $\dot{\gamma} = 20$ deg/s.
(d) $\alpha = 30$ deg, $\beta = 30$ deg, and $\gamma = 30$ deg with $\dot{\alpha} = 20$ deg/s, $\dot{\beta} = -20$ deg/s, and $\dot{\gamma} = 0$ deg/s.
(e) $\alpha = 30$ deg, $\beta = 30$ deg, and $\gamma = 30$ deg with $\dot{\alpha} = 0$ deg/s, $\dot{\beta} = 0$ deg/s, and $\dot{\gamma} = 20$ deg/s.

13. Roll–pitch–yaw angular velocity.

Find $_G^B\tilde{\omega}_B$ and $_G\boldsymbol{\omega}_B$ for the global roll–pitch–yaw rates equal to $\dot{\alpha} = 20$ deg/s, $\dot{\beta} = -20$ deg/s, and $\dot{\gamma} = 20$ deg/s, respectively, and having the following rotation matrix:

(a)

$$^B R_G = \begin{bmatrix} 0.53 & -0.84 & 0.13 \\ 0.0 & 0.15 & 0.99 \\ -0.85 & -0.52 & 0.081 \end{bmatrix} \tag{7.410}$$

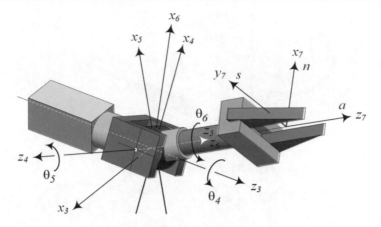

Fig. 7.10 An Eulerian wrist

(b)

$$
{}^{G}R_B = \begin{bmatrix} 0.53 & -0.84 & 0.13 \\ 0.0 & 0.15 & 0.99 \\ -0.85 & -0.52 & 0.081 \end{bmatrix}
\tag{7.411}
$$

14. Eulerian spherical wrist.

    Figure 7.10 illustrates An Eulerian wrist in motion. Assume B_3 is a globally fixed frame at the wrist point. Determine the angular velocity ${}_3\tilde{\omega}_7$ of the end-effector frame B_7 for the following cases.
    (a) Only the first motor is turning with $\dot\theta_4$ about z_3.
    (b) Only the second motor is turning with $\dot\theta_5$ about z_4.
    (c) Only the third motor is turning with $\dot\theta_6$ about z_5.
    (d) The first motor is turning with $\dot\theta_4$ about z_3 and the second motor is turning with $\dot\theta_5$ about z_4.
    (e) The first motor is turning with $\dot\theta_4$ about z_3 and the third motor is turning with $\dot\theta_6$ about z_5.
    (f) The first motor is turning with $\dot\theta_4$ about z_3 and the second motor is turning with $\dot\theta_5$ about z_4.
    (g) The first, second, and third motors are turning with $\dot\theta_4$, $\dot\theta_5$, and $\dot\theta_6$ about z_3, z_4, and z_5.

15. Angular velocity from the Rodriguez formula.

    We may find the time derivative of ${}^{G}R_B = R_{\hat{u},\phi}$ by

    $$
    {}^{G}\dot{R}_B = \frac{d}{dt}\,{}^{G}R_B = \dot\phi \frac{d}{d\phi}\,{}^{G}R_B
    \tag{7.412}
    $$

    Use the Rodriguez rotation formula and find ${}_G\tilde{\omega}_B$ and ${}_G^B\tilde{\omega}_B$.

16. Skew symmetric matrix

    Show that any square matrix can be expressed as the sum of a symmetric and a skew symmetric matrix.

    $$
    A = B + C
    \tag{7.413}
    $$

    $$
    B = \frac{1}{2}\left(A + A^{T}\right)
    \tag{7.414}
    $$

    $$
    C = \frac{1}{2}\left(A - A^{T}\right)
    \tag{7.415}
    $$

17. ★ A rotating slider robot.

    Figure 7.11 illustrates a slider link on a rotating arm. Calculate the following derivatives:

    $$
    \frac{{}^{G}d\hat{\imath}}{dt} \quad \frac{{}^{G}d\hat{\jmath}}{dt} \quad \frac{{}^{G}d\hat{k}}{dt} \quad \frac{{}^{G}d^2\hat{\imath}}{dt^2} \quad \frac{{}^{G}d^2\hat{\jmath}}{dt^2} \quad \frac{{}^{G}d^2\hat{k}}{dt^2}
    \tag{7.416}
    $$

    and find ${}^{B}\mathbf{v}$ and ${}^{B}\mathbf{a}$ of m at mass center C of the slider to calculate ${}_G^B\mathbf{a}_m = \frac{{}_G^d}{dt}\,{}^{B}\mathbf{v}_m$ using the rule of mixed derivative.

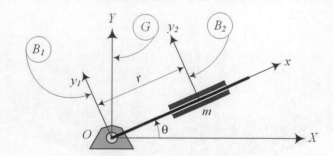

Fig. 7.11 A slider on a rotating bar

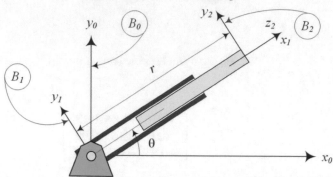

Fig. 7.12 A planar polar manipulator

$$\frac{^Gd}{dt}\left(\frac{^Bd}{dt}\mathbf{r}\right) = \frac{^Bd}{dt}\left(\frac{^Bd}{dt}\mathbf{r}\right) + {}_G^B\boldsymbol{\omega}_B \times \left(\frac{^Bd}{dt}\mathbf{r}\right) \tag{7.417}$$

18. ★ Differentiating in local and global frames.
    Consider a local point at ${}^B\mathbf{r}_P$. The local frame B is rotating in G by $\dot{\alpha}$ about the Z-axis. Calculate $\frac{^Bd}{dt}{}^B\mathbf{r}_P$, $\frac{^Gd}{dt}{}^G\mathbf{r}_P$, $\frac{^Bd}{dt}{}^G\mathbf{r}_P$, and $\frac{^Gd}{dt}{}^B\mathbf{r}_P$.
    (a) ${}^B\mathbf{r}_P = t\hat{\imath} + \hat{\jmath}$
    (b) ${}^B\mathbf{r}_P = t\hat{\imath} + t\hat{\jmath}$
    (c) ${}^B\mathbf{r}_P = t^2\hat{\imath} + \hat{\jmath}$
    (d) ${}^B\mathbf{r}_P = t\hat{\imath} + t^2\hat{\jmath}$
    (e) ${}^B\mathbf{r}_P = t\hat{\imath} + t\hat{\jmath} + t\hat{k}$
    (f) ${}^B\mathbf{r}_P = t\hat{\imath} + t^2\hat{\jmath} + t\hat{k}$
    (g) ${}^B\mathbf{r}_P = \hat{\imath}\sin t$
    (h) ${}^B\mathbf{r}_P = \hat{\imath}\sin\hat{\imath} + \hat{\jmath}\cos t + \hat{k}$

19. ★ Velocity analysis of a polar manipulator.
    Figure 7.12 illustrates a planar polar manipulator with joint variables θ and r.
    Determine ${}^0T_1$, ${}^1T_2$, ${}^0T_2$, ${}^0V_1$, ${}^1V_2$, ${}^0V_2$, and the velocity of the tip point of the manipulator.

20. ★ Velocity analysis of a spherical manipulator.
    Figure 7.13 illustrates a spherical manipulator with joint variables θ_1, θ_2, and d.
    Determine ${}^0V_1$, ${}^1V_2$, ${}^2V_3$, ${}^0V_2$, ${}^0V_3$, and the velocity of the tip point of the manipulator.

21. ★ Skew symmetric identity for angular velocity.
    Show that

$$R\tilde{\omega}R^T = \widetilde{R\boldsymbol{\omega}} \tag{7.418}$$

22. ★ Transformation of angular velocity exponents.
    Show that

$$ {}_G^B\tilde{\omega}_B^n = {}^GR_B^T\, {}_G\tilde{\omega}_B^n\, {}^GR_B \tag{7.419}$$

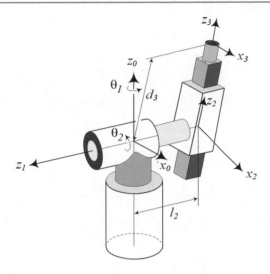

Fig. 7.13 A spherical manipulator

23. ★ An angular velocity matrix identity.
Show that

$$\tilde{\omega}^{2k+1} = (-1)^k \, \omega^{2k} \, \tilde{\omega} \qquad (7.420)$$

and

$$\tilde{\omega}^{2k} = (-1)^k \, \omega^{2(k-1)} \left(\omega^2 \, \mathbf{I} - \boldsymbol{\omega}\boldsymbol{\omega}^T \right) \qquad (7.421)$$

24. *SCARA* robot velocity kinematics analysis.
The robot R‖R‖R‖P shown in Fig. 5.23 simulates a *SCARA* manipulator. Assume $\theta_3 = 0$ and determine $^0V_1$, $^1V_2$, $^2V_3$, $^0V_2$, $^0V_3$, and the velocity of the tip point of the manipulator.

25. ★ Project: Rotating the axis of rotation.
Assume a body coordinate frame B is coincident with the global frame G. The B-frame is turning about $\mathbf{u}$ with angular speed $\dot{\phi}$. Determine the angular velocity $_G\boldsymbol{\omega}_B$.

(a)

$$\mathbf{u} = \begin{bmatrix} 1 & 1 & 1 \end{bmatrix}^T \quad \dot{\phi} = \sin t \qquad (7.422)$$

(b)

$$\mathbf{u} = \begin{bmatrix} \sin^2 t & \cos t & \sin^3 t \end{bmatrix}^T \quad \dot{\phi} = 1 \qquad (7.423)$$

(c)

$$\mathbf{u} = \begin{bmatrix} \sin t & \cos^2 t & \sin^3 t \end{bmatrix}^T \quad \dot{\phi} = t \qquad (7.424)$$

Velocity analysis of a robot is divided into forward and inverse velocity kinematics. Having the time rate of joint variables and calculating the Cartesian velocity of end-effector in the global coordinate frame is the *forward velocity kinematics*. Determination of the time rate of joint variables based on velocity of end-effector is the *inverse velocity kinematics*.

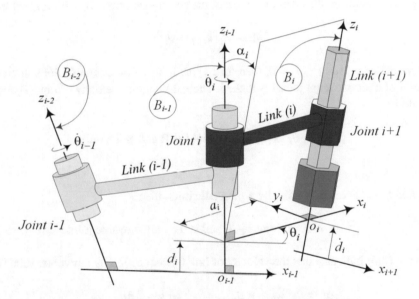

Fig. 8.1 Three connected moving links with their relative velocities

8.1 ★ Rigid Link Velocity

Every link of a robot has an angular velocity ω or a translational velocity $\dot{\mathbf{d}}$ with respect to its neighbor links as shown in Fig. 8.1. The angular velocity of link (i) in the global coordinate frame B_0 is a summation of global angular velocities of the links (j), for $j \leq i$.

$$
{}_{0}^{0}\boldsymbol{\omega}_i = \sum_{j=1}^{i} {}_{j-1}^{0}\boldsymbol{\omega}_j \tag{8.1}
$$

$$
{}_{j-1}^{0}\boldsymbol{\omega}_j = \begin{cases} \dot{\theta}_j \, {}^{0}\hat{k}_{j-1} & \text{if joint } i \text{ is R} \\ 0 & \text{if joint } i \text{ is P.} \end{cases} \tag{8.2}
$$

The velocity of the origin of B_i-frame attached to link (i) in the base coordinate frame is

$$
{}_{i-1}^{0}\dot{\mathbf{d}}_i = \begin{cases} {}_{0}^{0}\boldsymbol{\omega}_i \times {}_{i-1}^{0}\mathbf{d}_i & \text{if joint } i \text{ is R} \\ \dot{d}_i {}^{0}\hat{k}_{i-1} + {}_{0}^{0}\boldsymbol{\omega}_i \times {}_{i-1}^{0}\mathbf{d}_i & \text{if joint } i \text{ is P} \end{cases} \tag{8.3}
$$

where θ and d are Denavit–Hartenberg (DH) parameters, and ${}_{i-1}^{0}\mathbf{d}_i$ is the origin position vector of the B_i-frame with respect to B_{i-1}-frame, expressed in the base B_0-frame.

If a robot has n links, the global angular velocity of the final coordinate frame is

$$
{}_{0}^{0}\boldsymbol{\omega}_n = \sum_{i=1}^{n} {}_{i-1}^{0}\boldsymbol{\omega}_i \tag{8.4}
$$

and the global velocity vector of the last link's coordinate frame is

$$
{}_{0}^{0}\dot{\mathbf{d}}_n = \sum_{i=1}^{n} {}_{i-1}^{0}\dot{\mathbf{d}}_i \tag{8.5}
$$

Proof According to the DH definition, the position vector of the coordinate frame B_i with respect to B_{i-1} is

$$
{}_{i-1}^{0}\mathbf{d}_i = d_i {}^{0}\hat{k}_{i-1} + a_i {}^{0}\hat{\imath}_i \tag{8.6}
$$

which depends on joint variables q_j, for $j \leq i$, and, therefore, ${}_{i-1}^{0}\dot{\mathbf{d}}_i$ is also a function of joint velocities $\dot{q}_j$, for $j \leq i$.

Assume that every joint of a robot except joint i is locked. Then, the angular velocity of link (i) connecting via a revolute joint to link $(i-1)$ would be

$$
{}_{i-1}^{0}\boldsymbol{\omega}_i = \dot{\theta}_i {}^{0}\hat{k}_{i-1} \qquad \text{if joint } i \text{ is R and is the only}
$$
$$
\text{movable joint} \tag{8.7}
$$

However, if the link (i) and $(i-1)$ are connecting by a prismatic joint, then

$$
{}_{i-1}^{0}\boldsymbol{\omega}_i = 0 \quad \text{if joint } i \text{ is P and is the only movable joint} \tag{8.8}
$$

The relative position vector (8.6) indicates that the velocity of link (i) connecting by a revolute joint to link $(i-1)$ is

$$
{}_{i-1}^{0}\dot{\mathbf{d}}_i = \dot{\theta}_i {}^{0}\hat{k}_{i-1} \times a_i {}^{0}\hat{\imath}_i = {}_{i-1}^{0}\boldsymbol{\omega}_i \times {}_{i-1}^{0}\mathbf{d}_i
$$
$$
\text{if joint } i \text{ is R and is the only movable joint} \tag{8.9}
$$

We may substitute $a_i {}^{0}\hat{\imath}_i$ with ${}_{i-1}^{0}\mathbf{d}_i$ because the x_i-axis is turning about the z_{i-1}-axis with angular velocity $\dot{\theta}_i$ and, therefore,

$$
\dot{\theta}_i {}^{0}\hat{k}_{i-1} \times d_i {}^{0}\hat{k}_{i-1} = 0 \tag{8.10}
$$

However, if links (i) and $(i-1)$ are connecting by a prismatic joint, then

$$
{}_{i-1}^{0}\dot{\mathbf{d}}_i = \dot{d}_i {}^{0}\hat{k}_{i-1} \qquad \text{if joint } i \text{ is P and is the only movable joint} \tag{8.11}
$$

Now assume that all lower joints $j \leq i$ are moving. Then, the angular velocity of link (i) in the base coordinate frame is

$$
{}_{0}^{0}\boldsymbol{\omega}_i = \sum_{j=1}^{i} {}_{j-1}^{0}\boldsymbol{\omega}_j \tag{8.12}
$$

$$
{}_0^0\boldsymbol{\omega}_i = \begin{cases} \displaystyle\sum_{j=1}^{i} \dot{\theta}_i \,{}^0\hat{k}_{i-1} & \text{if joint } j \text{ is R} \\ 0 & \text{if joint } j \text{ is P} \end{cases} \tag{8.13}
$$

which can be written in a recursive form.

$$
{}_0^0\boldsymbol{\omega}_i = {}_0^0\boldsymbol{\omega}_{i-1} + {}_{i-1}^{\;0}\boldsymbol{\omega}_i \tag{8.14}
$$

$$
{}_0^0\boldsymbol{\omega}_i = \begin{cases} {}_0^0\boldsymbol{\omega}_{i-1} + \dot{\theta}_i \,{}^0\hat{k}_{i-1} & \text{if joint } i \text{ is R} \\ {}_0^0\boldsymbol{\omega}_{i-1} & \text{if joint } i \text{ is P} \end{cases} \tag{8.15}
$$

The velocity of the origin of link (i) in the base coordinate frame is

$$
{}_{i-1}^{\;0}\dot{\mathbf{d}}_i = \begin{cases} {}_0^0\boldsymbol{\omega}_i \times {}_{i-1}^{\;0}\mathbf{d}_i & \text{if joint } i \text{ is R} \\ \dot{d}_i \,{}^0\hat{k}_{i-1} + {}_0^0\boldsymbol{\omega}_i \times {}_{i-1}^{\;0}\mathbf{d}_i & \text{if joint } i \text{ is P} \end{cases} \tag{8.16}
$$

The translation and angular velocities of the last link of an n link robot is then a direct application of these results. ∎

Example 265 ★ Serial rigid links angular velocity. Many serial robots are built only with revolute joints. Here are relative velocities between adjacent links of such robot.

Consider a serial manipulator with n links and n revolute joints. The global angular velocity of link (i) in terms of the angular velocity of its previous links is

$$
{}^0\boldsymbol{\omega}_i = {}^0\boldsymbol{\omega}_{i-1} + \dot{\theta}_i \,{}^0\hat{k}_{i-1} \tag{8.17}
$$

$$
= \sum_{j=1}^{i} \dot{\theta}_j \,{}^0\hat{k}_{j-1} \tag{8.18}
$$

because

$$
{}_{i-1}^{\;0}\boldsymbol{\omega}_i = \dot{\theta}_i \,{}^0\hat{k}_{i-1} \tag{8.19}
$$

$$
{}_{i-2}^{\;0}\boldsymbol{\omega}_{i-1} = \dot{\theta}_{i-1} \,{}^0\hat{k}_{i-2} \tag{8.20}
$$

$$
{}_{i-2}^{\;0}\boldsymbol{\omega}_i = {}_{i-2}^{\;0}\boldsymbol{\omega}_{i-1} + {}_{i-1}^{\;0}\boldsymbol{\omega}_i = {}_{i-2}^{\;0}\boldsymbol{\omega}_{i-1} + \dot{\theta}_i \,{}^0\hat{k}_{i-1}
$$
$$
= \dot{\theta}_{i-1} \,{}^0\hat{k}_{i-2} + \dot{\theta}_i \,{}^0\hat{k}_{i-1} \tag{8.21}
$$

and, therefore,

$$
{}^0\boldsymbol{\omega}_i = \sum_{j=1}^{i-1} {}_{j-1}^{\;0}\boldsymbol{\omega}_j + \dot{\theta}_i \,{}^0\hat{k}_{i-1} = \sum_{j=1}^{i-1} \dot{\theta}_j \,{}^0\hat{k}_{j-1} + \dot{\theta}_i \,{}^0\hat{k}_{i-1}
$$
$$
= \sum_{j=1}^{i} \dot{\theta}_j \,{}^0\hat{k}_{j-1} \tag{8.22}
$$

Example 266 ★ Serial rigid links translational velocity. Very few serial robots are built only with translational joints. Here are relative velocities between adjacent links of such robot.

The global angular velocity of link (i) in a serial manipulator in terms of the angular velocity of its previous links is

$$
{}^0\mathbf{v}_i = {}^0\mathbf{v}_{i-1} + {}_{i-1}^{\;0}\boldsymbol{\omega}_i \times {}_{i-1}^{\;0}\mathbf{d}_i \tag{8.23}
$$

where

$$^0\mathbf{v}_i = {}^0\dot{\mathbf{d}}_i \tag{8.24}$$

or in general

$$^0\mathbf{v}_i = \sum_{j=1}^{i} \left({}^0\hat{k}_{j-1} \times {}^0_{i-1}\mathbf{d}_i\right) \dot{\theta}_j \tag{8.25}$$

because

$$^{0}_{i-1}\mathbf{v}_i = {}_0\boldsymbol{\omega}_i \times {}^{0}_{i-1}\mathbf{d}_i \tag{8.26}$$

$$^{0}_{i-2}\mathbf{v}_{i-1} = {}_0\boldsymbol{\omega}_{i-1} \times {}^{0}_{i-2}\mathbf{d}_{i-1} \tag{8.27}$$

$$\begin{aligned}
^{0}_{i-2}\mathbf{v}_i &= {}^{0}_{i-2}\mathbf{v}_{i-1} + {}^{0}_{i-1}\mathbf{v}_i = {}^{0}_{i-2}\mathbf{v}_{i-1} + {}_0\boldsymbol{\omega}_i \times {}^{0}_{i-1}\mathbf{d}_i \\
&= {}_0\boldsymbol{\omega}_{i-1} \times {}^{0}_{i-2}\mathbf{d}_{i-1} + {}_0\boldsymbol{\omega}_i \times {}^{0}_{i-1}\mathbf{d}_i \\
&= \dot{\theta}_{i-1}\,{}^0\hat{k}_{i-2} \times {}^{0}_{i-2}\mathbf{d}_{i-1} + \dot{\theta}_i\,{}^0\hat{k}_{i-1} \times {}^{0}_{i-1}\mathbf{d}_i
\end{aligned} \tag{8.28}$$

and, therefore,

$$^0\mathbf{v}_i = \sum_{j=1}^{i-1} {}^{0}_{j-1}\mathbf{v}_j + {}^{0}_{i-1}\mathbf{v}_i = \sum_{j=1}^{i} \left({}^0\hat{k}_{j-1} \times {}^{0}_{i-1}\mathbf{d}_i\right) \dot{\theta}_j \tag{8.29}$$

Example 267 ★ Recursive velocity in base frame. The velocity transformation matrix may be expressed by recursive formulae. Here is how.

The time derivative of a homogenous transformation matrix $[T]$ can be arranged in the form

$$\left[\dot{T}\right] = [V]\,[T] \tag{8.30}$$

where $[T]$ may be a link transformation matrix or the result of the forward kinematics of a robot.

$$^0T_6 = {}^0T_1\,{}^1T_2\,{}^2T_3\,{}^3T_4\,{}^4T_5\,{}^5T_6 \tag{8.31}$$

The time derivative of a transformation matrix can be computed when the velocity transformation matrix $[V]$ is calculated. The transformation matrix ^0T_i is

$$^0T_i = {}^0T_1\,{}^1T_2\,{}^2T_3 \cdots {}^{i-1}T_i \tag{8.32}$$

and the matrices ^0V_i and $^0V_{i+1}$ are

$$^0V_i = {}^0\dot{T}_i\,{}^0T_i^{-1} \tag{8.33}$$

$$\begin{aligned}
^0V_{i+1} &= {}^0\dot{T}_{i+1}\,{}^0T_{i+1}^{-1} = \frac{d}{dt}\left({}^0T_i\,{}^iT_{i+1}\right) {}^0T_{i+1}^{-1} \\
&= \left({}^0T_i\,{}^i\dot{T}_{i+1} + {}^0\dot{T}_i\,{}^iT_{i+1}\right) {}^0T_{i+1}^{-1}
\end{aligned} \tag{8.34}$$

These two equations can be combined as a recursive formula.

$$^0V_{i+1} = {}^0V_i + {}^0\dot{T}_i\,{}^iV_{i+1}\,{}^0T_i^{-1} \tag{8.35}$$

The recursive velocity transformation matrix formula can be simplified according to the type of joints connecting two links.

For a revolute joint, the velocity transformation matrix is

$$^iV_{i+1} = \dot{q}_{i+1}\,\Delta_R = \dot{\theta}_{i+1}\,\Delta_R$$

$$= \dot{\theta}_{i+1}\begin{bmatrix} 0 & -1 & 0 & 0 \\ 1 & 0 & 0 & 0 \\ 0 & 0 & 0 & 0 \\ 0 & 0 & 0 & 0 \end{bmatrix} = \dot{\theta}_{i+1}\begin{bmatrix} ^{i-1}\tilde{k}_{i-1} & 0 \\ 0 & 0 \end{bmatrix} \tag{8.36}$$

then we have

$$_0\omega_{i+1} = {}_0\omega_i + {}_i^0\omega_{i+1} = {}_0\omega_i + \dot{\theta}_{i+1}\,{}_i^0\hat{k}_{i+1} \tag{8.37}$$

which shows that the angular velocity of frame B_{i+1} is the angular velocity of frame B_i plus the relative angular velocity produced by joint variable q_{i+1}. Furthermore,

$$^0\dot{\mathbf{d}}_{i+1} = {}^0\dot{\mathbf{d}}_i + {}_i^0\dot{\mathbf{d}}_{i+1} = {}^0\dot{\mathbf{d}}_i + {}_0\tilde{\omega}_{i+1}\left({}^0\mathbf{d}_{i+1} - {}^0\mathbf{d}_i\right) \tag{8.38}$$

which shows the translational velocity is obtained by adding the translational velocity of frame B_i to the contribution of rotation of link B_{i+1}.

For a prismatic joint, the velocity coefficient matrix formula is

$$^iV_{i+1} = \dot{q}_{i+1}\,\Delta_P = \dot{d}_{i+1}\,\Delta_R$$

$$= \dot{d}_{i+1}\begin{bmatrix} 0 & 0 & 0 & 0 \\ 0 & 0 & 0 & 0 \\ 0 & 0 & 0 & 1 \\ 0 & 0 & 0 & 0 \end{bmatrix} = \dot{d}_{i+1}\begin{bmatrix} 0 & ^{i-1}\hat{k}_{i-1} \\ 0 & 0 \end{bmatrix} \tag{8.39}$$

and, therefore, we have

$$\omega_{i+1} = \omega_i \tag{8.40}$$

and

$$^0\dot{\mathbf{d}}_{i+1} = {}^0\dot{\mathbf{d}}_i + {}_0\tilde{\omega}_{i+1}\left({}^0\mathbf{d}_{i+1} - {}^0\mathbf{d}_i\right) + \dot{d}_{i+1}\,{}_i^0\hat{k}_{i+1} \tag{8.41}$$

which shows that the angular velocity of frame B_{i+1} is the same as the angular velocity of frame B_i. Furthermore, the translational velocity is obtained by adding the translational velocity due to $\dot{d}_{i+1}$ to the relative velocity due to rotation of link B_{i+1}.

8.2 Forward Velocity Kinematics

The *forward velocity kinematics* of a robot is to solve the problem of relating *joint speeds*, $\dot{\mathbf{q}}$, to the *end-effector speeds* $\dot{\mathbf{X}}$. The joint speed vector $\dot{\mathbf{q}}$ of an n degree-of-freedom (DOF) robot is an $n \times 1$ vector $\dot{\mathbf{q}}$, where $\dot{q}_i$ is the joint variable at joint i.

$$\dot{\mathbf{q}} = \begin{bmatrix} \dot{q}_1 & \dot{q}_2 & \dot{q}_3 & \cdots & \dot{q}_n \end{bmatrix}^T \tag{8.42}$$

The end-effector speed vector $\dot{\mathbf{X}}$, in the most general case, is a 6×1 vector indicating translational and rotational speeds of the end-effector.

$$\dot{\mathbf{X}} = \begin{bmatrix} \dot{X}_n & \dot{Y}_n & \dot{Z}_n & \omega_{Xn} & \omega_{Yn} & \omega_{Zn} \end{bmatrix}^T$$

$$= \begin{bmatrix} {}^0\dot{\mathbf{d}}_n \\ {}_0\boldsymbol{\omega}_n \end{bmatrix} = \begin{bmatrix} {}^0\mathbf{v}_n \\ {}_0\boldsymbol{\omega}_n \end{bmatrix} \tag{8.43}$$

The elements of *end-effector speed vector* $\dot{\mathbf{X}}$ are linearly proportional to the elements of *joint speed vector*, $\dot{\mathbf{q}}$,

$$\dot{\mathbf{X}} = \mathbf{J}\,\dot{\mathbf{q}} \tag{8.44}$$

where the $6 \times n$ proportionality matrix $\mathbf{J}(\mathbf{q})$ is called the *Jacobian matrix* of the robot.

The global expression of velocity ${}^0\mathbf{v}_n$ of the origin of B_n-frame is proportional to the manipulator joint speeds $\dot{\mathbf{q}}_D$.

$$ {}^0\mathbf{v}_n = \mathbf{J}_D\,\dot{\mathbf{q}}_D \qquad \dot{\mathbf{q}}_D \in \dot{\mathbf{q}} \tag{8.45}$$

The $3 \times n$ proportionality matrix $\mathbf{J}_D(\mathbf{q})$ is the *displacement Jacobian matrix* of the manipulator.

$$\mathbf{J}_D = \frac{\partial \mathbf{d}_n\,(\dot{\mathbf{q}}_D)}{\partial \dot{\mathbf{q}}_D} = \frac{\partial T\,(\mathbf{q})}{\partial \mathbf{q}} \tag{8.46}$$

The global expression of angular velocity ${}_0\boldsymbol{\omega}_n$ of B_n-frame is proportional to the rotational components of $\dot{\mathbf{q}}$.

$$ {}_0\boldsymbol{\omega}_n = \mathbf{J}_R\,\dot{\mathbf{q}} \tag{8.47}$$

The $3 \times n$ proportionality matrix $\mathbf{J}_R(\mathbf{q})$ is the *rotational Jacobian matrix* of the robot.

$$\mathbf{J}_R = \frac{\partial\,{}_0\boldsymbol{\omega}_n}{\partial \mathbf{q}} \tag{8.48}$$

We may combine Eqs. (8.45) and (8.47) to show the forward velocity kinematics of a robot by (8.43).

Proof Consider a robot with 6 DOF that is made of a 3 DOF manipulator to position the wrist point, and a spherical wrist with 3 DOF to orient the end-effector. The coordinate transformation of a point in the end-effector coordinate frame B_6 and the base coordinate frame B_0 is

$$ {}^0\mathbf{r} = {}^0T_6(\mathbf{q})\;{}^6\mathbf{r} = {}^0D_6\;{}^0R_6\;{}^6\mathbf{r} = \begin{bmatrix} \mathbf{I} & {}^0\mathbf{d}_6 \\ 0 & 1 \end{bmatrix} \begin{bmatrix} {}^0R_6 & \mathbf{0} \\ 0 & 1 \end{bmatrix} {}^6\mathbf{r}$$

$$ = \begin{bmatrix} {}^0R_6 & {}^0\mathbf{d}_6 \\ 0 & 1 \end{bmatrix} {}^6\mathbf{r} \tag{8.49}$$

where the transformation matrix ${}^0T_6$ is a function of 6 joint variables q_i, $i = 1, 2, \cdots, 6$. We can always divide the 6 joint variables into the end-effector position variables q_1, q_2, q_3 and end-effector orientation variables q_4, q_5, q_6. The end-effector position variables are the only variables in the position vector ${}^0\mathbf{d}_6$, and the end-effector orientation variables are the only variables in the rotation transformation matrix ${}^0R_6$.

$$ {}^0\mathbf{d}_6 = {}^0\mathbf{d}_6\,(q_1, q_2, q_3) \tag{8.50}$$
$$ {}^0R_6 = {}^0R_6\,(q_4, q_5, q_6) \tag{8.51}$$

The origin of end-effector frame B_6 is at ${}^6\mathbf{r} = \mathbf{0}$ which is globally at:

$$ {}^0\mathbf{r} = \begin{bmatrix} {}^0R_6 & {}^0\mathbf{d}_6 \\ 0 & 1 \end{bmatrix} \begin{bmatrix} \mathbf{0} \\ 1 \end{bmatrix} = \begin{bmatrix} \mathbf{I} & {}^0\mathbf{d}_6 \\ 0 & 1 \end{bmatrix} \begin{bmatrix} \mathbf{0} \\ 1 \end{bmatrix} = {}^0D_6 \begin{bmatrix} \mathbf{0} \\ 1 \end{bmatrix}$$

$$ = {}^0\mathbf{d}_6 \tag{8.52}$$

The components of end-effector displacement vector ${}^0\mathbf{d}_6 = \begin{bmatrix} X & Y & Z \end{bmatrix}$ are functions of manipulator joint variables q_1, q_2, q_3.

$$ \begin{bmatrix} X \\ Y \\ Z \end{bmatrix} = \begin{bmatrix} d_1 \\ d_2 \\ d_3 \end{bmatrix} = \begin{bmatrix} d_1\,(q_1, q_2, q_3) \\ d_2\,(q_1, q_2, q_3) \\ d_3\,(q_1, q_2, q_3) \end{bmatrix} \tag{8.53}$$

Taking a derivative of both sides indicates that each component of $^0\mathbf{v}_6 = {}^0\dot{\mathbf{r}}$ is a linear combination of $\dot{q}_1, \dot{q}_2, \dot{q}_3$.

$$\dot{X} = \frac{\partial d_1}{\partial q_1}\dot{q}_1 + \frac{\partial d_1}{\partial q_2}\dot{q}_2 + \frac{\partial d_1}{\partial q_3}\dot{q}_3$$

$$\dot{Y} = \frac{\partial d_2}{\partial q_1}\dot{q}_1 + \frac{\partial d_2}{\partial q_2}\dot{q}_2 + \frac{\partial d_2}{\partial q_3}\dot{q}_3$$

$$\dot{Z} = \frac{\partial d_3}{\partial q_1}\dot{q}_1 + \frac{\partial d_3}{\partial q_2}\dot{q}_2 + \frac{\partial d_3}{\partial q_3}\dot{q}_3 \tag{8.54}$$

It indicates that $^0\mathbf{v}_6$ is a linear combination of joint speeds q_1, q_2, q_3.

$$^0\mathbf{v}_6 = \mathbf{J}_D\,\dot{\mathbf{q}}_D = \dot{q}_1\frac{\partial\, ^0\mathbf{d}_6}{\partial q_1} + \dot{q}_2\frac{\partial\, ^0\mathbf{d}_6}{\partial q_2} + \dot{q}_3\frac{\partial\, ^0\mathbf{d}_6}{\partial q_3} \tag{8.55}$$

We may show these relations by vector and matrix expressions.

$$^0\mathbf{v}_6 = \frac{\partial\mathbf{d}_6}{\partial\mathbf{q}}\dot{\mathbf{q}}_D = \mathbf{J}_D\,\dot{\mathbf{q}}_D \tag{8.56}$$

$$\begin{bmatrix} \dot{X} \\ \dot{Y} \\ \dot{Z} \end{bmatrix} = \begin{bmatrix} \dfrac{\partial d_1}{\partial q_1} & \dfrac{\partial d_1}{\partial q_2} & \dfrac{\partial d_1}{\partial q_3} \\[2mm] \dfrac{\partial d_2}{\partial q_1} & \dfrac{\partial d_2}{\partial q_2} & \dfrac{\partial d_2}{\partial q_3} \\[2mm] \dfrac{\partial d_3}{\partial q_1} & \dfrac{\partial d_3}{\partial q_2} & \dfrac{\partial d_3}{\partial q_3} \end{bmatrix} \begin{bmatrix} \dot{q}_1 \\ \dot{q}_2 \\ \dot{q}_3 \end{bmatrix} \tag{8.57}$$

The displacement Jacobian $\mathbf{J}_D$ is equivalent to the derivative of homogenous transformation matrix T with respect to the manipulator joint coordinates.

$$\mathbf{J}_D = \frac{\partial\mathbf{d}_6}{\partial\mathbf{q}} = \frac{\partial\, ^0D_6}{\partial\mathbf{q}} = \frac{\partial\, ^0T_6}{\partial\mathbf{q}} = \frac{\partial T\,(\mathbf{q})}{\partial\mathbf{q}} \tag{8.58}$$

The angular velocity of the end-effector is

$$_0\boldsymbol{\omega}_6 = {}^0\dot{R}_6\,^0R_6^T \tag{8.59}$$

However, the time derivative of the rotational transformation matrix is

$$^0\dot{R}_6 = \frac{d}{dt}\left[{}^0R_1\,^1R_2\,^2R_3\,^3R_4\,^4R_5\,^5R_6\right] \tag{8.60}$$

$$= \dot{q}_1\frac{\partial\, ^0R_1}{\partial q_1}\,^1R_2\,^2R_3\,^3R_4\,^4R_5\,^5R_6$$

$$+\dot{q}_2\,^0R_1\frac{\partial\, ^1R_2}{\partial q_2}\,^2R_3\,^3R_4\,^4R_5\,^5R_6$$

$$+\dot{q}_3\,^0R_1\,^1R_2\frac{\partial\, ^2R_3}{\partial q_3}\,^3R_4\,^4R_5\,^5R_6$$

$$+\dot{q}_4\,^0R_1\,^1R_2\,^2R_3\frac{\partial\, ^3R_4}{\partial q_4}\,^4R_5\,^5R_6$$

$$+\dot{q}_5\,^0R_1\,^1R_2\,^2R_3\,^3R_4\frac{\partial\, ^4R_5}{\partial q_5}\,^5R_6$$

$$+\dot{q}_6\,^0R_1\,^1R_2\,^2R_3\,^3R_4\,^4R_5\frac{\partial\, ^5R_6}{\partial q_6}$$

and the transpose of $^0R_6$ is

$$
\begin{aligned}
^0R_6^T &= \left[{^0R_1} \, {^1R_2} \, {^2R_3} \, {^3R_4} \, {^4R_5} \, {^5R_6} \right]^T \\
&= {^5R_6^T} \, {^4R_5^T} \, {^3R_4^T} \, {^2R_3^T} \, {^1R_2^T} \, {^0R_1^T}
\end{aligned}
\tag{8.61}
$$

Therefore, $_0\boldsymbol{\omega}_6 = {^0\dot{R}_6} \, {^0R_6^T}$ is

$$
\begin{aligned}
_0\boldsymbol{\omega}_6 = {} & \dot{q}_1 \frac{\partial {^0R_1}}{\partial q_1} {^0R_1^T} + \dot{q}_2 \, {^0R_1} \frac{\partial {^1R_2}}{\partial q_2} {^0R_2^T} \\
& + \dot{q}_3 \, {^0R_2} \frac{\partial {^2R_3}}{\partial q_3} {^0R_3^T} + \dot{q}_4 \, {^0R_3} \frac{\partial {^3R_4}}{\partial q_4} {^0R_4^T} \\
& + \dot{q}_5 \, {^0R_4} \frac{\partial {^4R_5}}{\partial q_5} {^0R_5^T} + \dot{q}_6 \, {^0R_5} \frac{\partial {^5R_6}}{\partial q_6} {^0R_6^T}
\end{aligned}
\tag{8.62}
$$

$$
= {_0\boldsymbol{\omega}_1} + {_1^0\boldsymbol{\omega}_2} + {_2^0\boldsymbol{\omega}_3} + {_3^0\boldsymbol{\omega}_4} + {_4^0\boldsymbol{\omega}_5} + {_5^0\boldsymbol{\omega}_6}
\tag{8.63}
$$

It indicates that $^0\boldsymbol{\omega}_6$ is a linear combination of joint speeds q_i, $i = 1, 2, \cdots, 6$,

$$
\begin{aligned}
_0\boldsymbol{\omega}_6 = \mathbf{J}_R \, \dot{\mathbf{q}} = {} & \dot{q}_1 \frac{\partial {^0\boldsymbol{\omega}_6}}{\partial q_1} + \dot{q}_2 \frac{\partial {^0\boldsymbol{\omega}_6}}{\partial q_2} + \dot{q}_3 \frac{\partial {^0\boldsymbol{\omega}_6}}{\partial q_3} \\
& + \dot{q}_4 \frac{\partial {^0\boldsymbol{\omega}_6}}{\partial q_4} + \dot{q}_5 \frac{\partial {^0\boldsymbol{\omega}_6}}{\partial q_5} + \dot{q}_6 \frac{\partial {^0\boldsymbol{\omega}_6}}{\partial q_6}
\end{aligned}
\tag{8.64}
$$

where

$$
\frac{\partial {_0\boldsymbol{\omega}_6}}{\partial q_1} = \frac{\partial {^0R_1}}{\partial q_1} {^0R_1^T}
\tag{8.65}
$$

$$
\frac{\partial {_0\boldsymbol{\omega}_6}}{\partial q_2} = {^0R_1} \frac{\partial {^1R_2}}{\partial q_2} {^0R_2^T}
\tag{8.66}
$$

$$
\frac{\partial {_0\boldsymbol{\omega}_6}}{\partial q_3} = {^0R_2} \frac{\partial {^2R_3}}{\partial q_3} {^0R_3^T}
\tag{8.67}
$$

$$
\frac{\partial {_0\boldsymbol{\omega}_6}}{\partial q_4} = {^0R_3} \frac{\partial {^3R_4}}{\partial q_4} {^0R_4^T}
\tag{8.68}
$$

$$
\frac{\partial {_0\boldsymbol{\omega}_6}}{\partial q_5} = {^0R_4} \frac{\partial {^4R_5}}{\partial q_5} {^0R_5^T}
\tag{8.69}
$$

$$
\frac{\partial {_0\boldsymbol{\omega}_6}}{\partial q_6} = {^0R_5} \frac{\partial {^5R_6}}{\partial q_6} {^0R_6^T}.
\tag{8.70}
$$

Combination of the translational and rotational velocities makes the Eq. (8.76) for the velocity kinematics of the robot.

$$
\dot{\mathbf{X}} = \begin{bmatrix} {^0\mathbf{v}_n} \\ {_0\boldsymbol{\omega}_n} \end{bmatrix} = \begin{bmatrix} \mathbf{J}_D \\ \mathbf{J}_R \end{bmatrix} \dot{\mathbf{q}} = \mathbf{J} \, \dot{\mathbf{q}}
\tag{8.71}
$$

Therefore the systematic method to calculate the Jacobian matrix of the robot is

$$
\mathbf{J} = \begin{bmatrix} \dfrac{\partial {^0\mathbf{d}_6}}{\partial q_1} & \dfrac{\partial {^0\mathbf{d}_6}}{\partial q_2} & \dfrac{\partial {^0\mathbf{d}_6}}{\partial q_3} & \mathbf{0} & \mathbf{0} & \mathbf{0} \\[2ex] \dfrac{\partial {_0\boldsymbol{\omega}_6}}{\partial q_1} & \dfrac{\partial {_0\boldsymbol{\omega}_6}}{\partial q_2} & \dfrac{\partial {_0\boldsymbol{\omega}_6}}{\partial q_3} & \dfrac{\partial {_0\boldsymbol{\omega}_6}}{\partial q_4} & \dfrac{\partial {_0\boldsymbol{\omega}_6}}{\partial q_5} & \dfrac{\partial {_0\boldsymbol{\omega}_6}}{\partial q_6} \end{bmatrix}
\tag{8.72}
$$

In case the robot has n links and joints, the above equations will go from 1 to n instead of 1 to 6. So in general case, the $6 \times n$ Jacobian matrix $\mathbf{J}$ becomes

$$\mathbf{J} = \begin{bmatrix} \dfrac{\partial\,{}^0\mathbf{d}_n}{\partial q_1} & \dfrac{\partial\,{}^0\mathbf{d}_n}{\partial q_2} & \dfrac{\partial\,{}^0\mathbf{d}_n}{\partial q_3} & \cdots & \dfrac{\partial\,{}^0\mathbf{d}_n}{\partial q_n} \\ \dfrac{\partial\,{}_0\boldsymbol{\omega}_n}{\partial q_1} & \dfrac{\partial\,{}_0\boldsymbol{\omega}_n}{\partial q_2} & \dfrac{\partial\,{}_0\boldsymbol{\omega}_n}{\partial q_3} & \cdots & \dfrac{\partial\,{}_0\boldsymbol{\omega}_n}{\partial q_n} \end{bmatrix} \tag{8.73}$$

In summary, the forward velocity kinematics is: determination of the end-effector translational and angular velocities in the base frame, ${}^0\mathbf{v}_n, {}^0\boldsymbol{\omega}_n$, for a given set of joint speeds $\dot{q}_i, i = 1, 2, \cdots, n$. The components of velocity vectors ${}^0\mathbf{v}_n$ and ${}^0\boldsymbol{\omega}_n$ are proportional to the joint speeds $\dot{q}_i, i = 1, 2, \cdots, n$.

$$ {}^0\mathbf{v}_n = \mathbf{J}_D\,\dot{\mathbf{q}} \tag{8.74}$$

$$ {}_0\boldsymbol{\omega}_n = \mathbf{J}_R\,\dot{\mathbf{q}} \tag{8.75}$$

The proportionality matrices $\mathbf{J}_D$ and $\mathbf{J}_R$ are the displacement and rotational Jacobians of the robot.

We may combine Eqs. (8.74) and (8.75) as

$$ \dot{\mathbf{X}} = \mathbf{J}\,\dot{\mathbf{q}} \tag{8.76}$$

by defining the Jacobian matrix $\mathbf{J}$ and the vectors $\dot{\mathbf{X}}$ and $\dot{\mathbf{q}}$, known as *end-effector speed vector* and *joint speed vector*, respectively.

$$ \mathbf{J} = \begin{bmatrix} \mathbf{J}_D \\ \mathbf{J}_R \end{bmatrix} \tag{8.77}$$

$$ \dot{\mathbf{X}} = \begin{bmatrix} {}^0\mathbf{v}_n \\ {}_0\boldsymbol{\omega}_n \end{bmatrix} \tag{8.78}$$

$$ \dot{\mathbf{q}} = \begin{bmatrix} \dot{q}_1 & \dot{q}_2 & \dot{q}_3 & \cdots & \dot{q}_n \end{bmatrix}^T \tag{8.79}$$

We may also show $\mathbf{J}_D$ by $\mathbf{J}$, whenever we analyze the velocity kinematics of a manipulator without a wrist. ∎

Example 268 Jacobian matrix for a planar polar manipulator. Forward kinematics is the only information we need to calculate velocity kinematics, and the Jacobian matrix is the key point of the velocity kinematics. Here is to show how we calculate velocity kinematics of a 2 DOF manipulator.

Figure 8.2 illustrates a planar polar manipulator with the following forward kinematics.

$$ {}^0T_2 = {}^0T_1\,{}^1T_2 $$

$$ = \begin{bmatrix} \cos\theta & -\sin\theta & 0 & 0 \\ \sin\theta & \cos\theta & 0 & 0 \\ 0 & 0 & 1 & 0 \\ 0 & 0 & 0 & 1 \end{bmatrix} \begin{bmatrix} 1 & 0 & 0 & r \\ 0 & 1 & 0 & 0 \\ 0 & 0 & 1 & 0 \\ 0 & 0 & 0 & 1 \end{bmatrix} $$

$$ = \begin{bmatrix} \cos\theta & -\sin\theta & 0 & r\cos\theta \\ \sin\theta & \cos\theta & 0 & r\sin\theta \\ 0 & 0 & 1 & 0 \\ 0 & 0 & 0 & 1 \end{bmatrix} \tag{8.80}$$

The tip point of the manipulator is at $\begin{bmatrix} X & Y \end{bmatrix}^T$,

$$ \begin{bmatrix} X \\ Y \end{bmatrix} = \begin{bmatrix} r\cos\theta \\ r\sin\theta \end{bmatrix} \tag{8.81}$$

and, therefore, its velocity is

$$ \begin{bmatrix} \dot{X} \\ \dot{Y} \end{bmatrix} = \begin{bmatrix} \cos\theta & -r\sin\theta \\ \sin\theta & r\cos\theta \end{bmatrix} \begin{bmatrix} \dot{r} \\ \dot{\theta} \end{bmatrix} \tag{8.82}$$

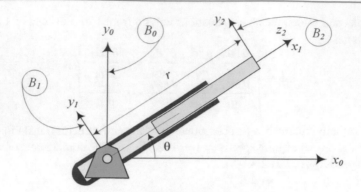

Fig. 8.2 A planar polar manipulator

which shows that

$$\mathbf{J}_D = \begin{bmatrix} \dfrac{\partial X}{\partial r} & \dfrac{\partial X}{\partial \theta} \\ \dfrac{\partial Y}{\partial r} & \dfrac{\partial Y}{\partial \theta} \end{bmatrix} = \begin{bmatrix} \cos\theta & -r\sin\theta \\ \sin\theta & r\cos\theta \end{bmatrix} \tag{8.83}$$

There is only one rotational joint coordinate, θ. The rotation matrix $^0R_2$ indicates that:

$$_0\tilde{\omega}_2 = {}^0\dot{R}_2\,{}^0R_2^T = \dot{\theta}\tilde{k} \tag{8.84}$$

$$_0\boldsymbol{\omega}_2 = \begin{bmatrix} \omega_1 \\ \omega_2 \\ \omega_3 \end{bmatrix} = \begin{bmatrix} 0 \\ 0 \\ \dot{\theta} \end{bmatrix} \tag{8.85}$$

Therefore,

$$\omega_3 = \mathbf{J}_R\,\dot{\theta} \tag{8.86}$$

$$\mathbf{J}_R = 1 \tag{8.87}$$

$$\begin{bmatrix} \dot{X} \\ \dot{Y} \\ \omega_3 \end{bmatrix} = \begin{bmatrix} \cos\theta & -r\sin\theta \\ \sin\theta & r\cos\theta \\ 0 & 1 \end{bmatrix} \begin{bmatrix} \dot{r} \\ \dot{\theta} \end{bmatrix} \tag{8.88}$$

Example 269 Jacobian matrix for the $2R$ planar manipulator. The $2R$ planar manipulator with two revolute joints has a very good educational advantage to study principles of robotics. Here we derive its Jacobian systematically as well as direct differentiation method. This is a good example for velocity analysis and calculating Jacobian matrix using systematic method.

A $2R$ planar manipulator with two R‖R links is illustrated in Fig. 8.3. The manipulator has been analyzed in Example 165 for forward kinematics, and in Example 207 for inverse kinematics.

The angular velocity of links (1) and (2) are

$$_0\boldsymbol{\omega}_1 = \dot{\theta}_1\,{}^0\hat{k}_0 \tag{8.89}$$

$$_1^0\boldsymbol{\omega}_2 = \dot{\theta}_2\,{}^0\hat{k}_1 \tag{8.90}$$

$$_0\boldsymbol{\omega}_2 = {}_0\boldsymbol{\omega}_1 + {}_1^0\boldsymbol{\omega}_2 = \left(\dot{\theta}_1 + \dot{\theta}_2\right){}^0\hat{k}_0 \tag{8.91}$$

and the global velocity of the tip position of the manipulator is

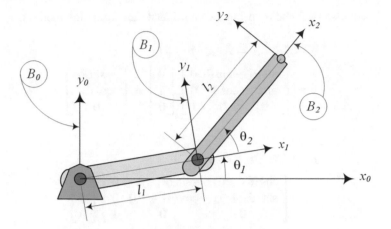

Fig. 8.3 A $2R$ planar manipulator

$$
{}^0\dot{\mathbf{d}}_2 = {}^0\dot{\mathbf{d}}_1 + {}^0_1\dot{\mathbf{d}}_2 = {}_0\boldsymbol{\omega}_1 \times {}^0\mathbf{d}_1 + {}_0\boldsymbol{\omega}_2 \times {}^0_1\mathbf{d}_2
$$

$$
= \dot{\theta}_1 {}^0\hat{k}_0 \times l_1 {}^0\hat{i}_1 + \left(\dot{\theta}_1 + \dot{\theta}_2\right) {}^0\hat{k}_0 \times l_2 {}^0\hat{i}_2
$$

$$
= \dot{\theta}_1 {}^0\hat{k}_0 \times l_1 {}^0R_1 {}^1\hat{i}_1 + \left(\dot{\theta}_1 + \dot{\theta}_2\right) {}^0\hat{k}_0 \times l_2 {}^0R_2 {}^2\hat{i}_2
$$

$$
= l_1 \dot{\theta}_1 {}^0\hat{j}_1 \times l_2 \left(\dot{\theta}_1 + \dot{\theta}_2\right) {}^0\hat{j}_2 \tag{8.92}
$$

Unit vectors must always be expressed in their own coordinate frame. It means ${}^0\hat{i}_1$ and ${}^0\hat{i}_2$ have meaning when they are transformed into 0-frame and expressed by unit vectors of the 0-frame.

$$
{}^0\hat{i}_1 = {}^0R_1 {}^1\hat{i}_1 = R_{Z,\theta_1} {}^1\hat{i}_1
$$

$$
= \begin{bmatrix} \cos\theta_1 & -\sin\theta_1 & 0 \\ \sin\theta_1 & \cos\theta_1 & 0 \\ 0 & 0 & 1 \end{bmatrix} \begin{bmatrix} 1 \\ 0 \\ 0 \end{bmatrix} = \begin{bmatrix} \cos\theta_1 \\ \sin\theta_1 \\ 0 \end{bmatrix} \tag{8.93}
$$

$$
{}^0\hat{i}_2 = {}^0R_2 {}^2\hat{i}_2 = R_{Z,\theta_1+\theta_2} {}^2\hat{i}_2
$$

$$
= \begin{bmatrix} \cos(\theta_1+\theta_2) & -\sin(\theta_1+\theta_2) & 0 \\ \sin(\theta_1+\theta_2) & \cos(\theta_1+\theta_2) & 0 \\ 0 & 0 & 1 \end{bmatrix} \begin{bmatrix} 1 \\ 0 \\ 0 \end{bmatrix} \tag{8.94}
$$

$$
= \begin{bmatrix} \cos(\theta_1+\theta_2) \\ \sin(\theta_1+\theta_2) \\ 0 \end{bmatrix} \tag{8.95}
$$

$$
{}^0\hat{k}_0 \times {}^0R_1 {}^1\hat{i}_1 = \begin{bmatrix} 0 \\ 0 \\ 1 \end{bmatrix} \times \begin{bmatrix} \cos\theta_1 \\ \sin\theta_1 \\ 0 \end{bmatrix} = \begin{bmatrix} -\sin\theta_1 \\ \cos\theta_1 \\ 0 \end{bmatrix} = {}^0\hat{j}_1 \tag{8.96}
$$

$$
{}^0\hat{k}_0 \times {}^0R_2 {}^2\hat{i}_2 = \begin{bmatrix} 0 \\ 0 \\ 1 \end{bmatrix} \times \begin{bmatrix} \cos(\theta_1+\theta_2) \\ \sin(\theta_1+\theta_2) \\ 0 \end{bmatrix}
$$

$$
= \begin{bmatrix} -\sin(\theta_1+\theta_2) \\ \cos(\theta_1+\theta_2) \\ 0 \end{bmatrix} = {}^0\hat{j}_2 \tag{8.97}
$$

The unit vectors ${}^0\hat{j}_1$ and ${}^0\hat{j}_2$ will also be found by using the coordinate transformation method.

$$
\begin{aligned}
{}^0\hat{j}_1 &= {}^0R_1\,{}^1\hat{j}_1 = R_{Z,\theta_1}\,{}^1\hat{j}_1 \\
&= \begin{bmatrix} \cos\theta_1 & -\sin\theta_1 & 0 \\ \sin\theta_1 & \cos\theta_1 & 0 \\ 0 & 0 & 1 \end{bmatrix} \begin{bmatrix} 0 \\ 1 \\ 0 \end{bmatrix} = \begin{bmatrix} -\sin\theta_1 \\ \cos\theta_1 \\ 0 \end{bmatrix}
\end{aligned}
\tag{8.98}
$$

$$
\begin{aligned}
{}^0\hat{j}_2 &= {}^0R_2\,{}^2\hat{j}_2 = R_{Z,\theta_1+\theta_2}\,{}^2\hat{j}_2 \\
&= \begin{bmatrix} \cos(\theta_1+\theta_2) & -\sin(\theta_1+\theta_2) & 0 \\ \sin(\theta_1+\theta_2) & \cos(\theta_1+\theta_2) & 0 \\ 0 & 0 & 1 \end{bmatrix} \begin{bmatrix} 0 \\ 1 \\ 0 \end{bmatrix} \\
&= \begin{bmatrix} -\sin(\theta_1+\theta_2) \\ \cos(\theta_1+\theta_2) \\ 0 \end{bmatrix}
\end{aligned}
\tag{8.99}
$$

Substituting back shows that

$$
{}^0\dot{\mathbf{d}}_2 = l_1\dot{\theta}_1 \begin{bmatrix} -\sin\theta_1 \\ \cos\theta_1 \\ 0 \end{bmatrix} \times l_2\left(\dot{\theta}_1+\dot{\theta}_2\right) \begin{bmatrix} -\sin(\theta_1+\theta_2) \\ \cos(\theta_1+\theta_2) \\ 0 \end{bmatrix}
\tag{8.100}
$$

which can be rearranged to have

$$
\begin{aligned}
\begin{bmatrix} \dot{X} \\ \dot{Y} \end{bmatrix} &= \begin{bmatrix} -l_1 s\theta_1 - l_2 s(\theta_1+\theta_2) & -l_2 s(\theta_1+\theta_2) \\ l_1 c\theta_1 + l_2 c(\theta_1+\theta_2) & l_2 c(\theta_1+\theta_2) \end{bmatrix} \begin{bmatrix} \dot{\theta}_1 \\ \dot{\theta}_2 \end{bmatrix} \\
&= \mathbf{J}_D \begin{bmatrix} \dot{\theta}_1 \\ \dot{\theta}_2 \end{bmatrix}
\end{aligned}
\tag{8.101}
$$

Taking advantage of the structural simplicity of the $2R$ manipulator, we may find its Jacobian matrix $\mathbf{J}_D$ simpler. The forward kinematics of the manipulator is

$$
{}^0T_2 = {}^0T_1\,{}^1T_2
\tag{8.102}
$$

$$
= \begin{bmatrix} c(\theta_1+\theta_2) & -s(\theta_1+\theta_2) & 0 & l_1 c\theta_1 + l_2 c(\theta_1+\theta_2) \\ s(\theta_1+\theta_2) & c(\theta_1+\theta_2) & 0 & l_1 s\theta_1 + l_2 s(\theta_1+\theta_2) \\ 0 & 0 & 1 & 0 \\ 0 & 0 & 0 & 1 \end{bmatrix}
$$

which shows the tip position ${}^0_0\mathbf{d}_2$ of the manipulator is at:

$$
{}^0_0\mathbf{d}_2 = \begin{bmatrix} X \\ Y \end{bmatrix} = \begin{bmatrix} l_1 \cos\theta_1 + l_2 \cos(\theta_1+\theta_2) \\ l_1 \sin\theta_1 + l_2 \sin(\theta_1+\theta_2) \end{bmatrix}
\tag{8.103}
$$

Direct differentiating gives

$$
\begin{bmatrix} \dot{X} \\ \dot{Y} \end{bmatrix} = \begin{bmatrix} -l_1\dot{\theta}_1 \sin\theta_1 - l_2\left(\dot{\theta}_1+\dot{\theta}_2\right)\sin(\theta_1+\theta_2) \\ l_1\dot{\theta}_1 \cos\theta_1 + l_2\left(\dot{\theta}_1+\dot{\theta}_2\right)\cos(\theta_1+\theta_2) \end{bmatrix}
\tag{8.104}
$$

which can be rearranged in a matrix form

$$\begin{bmatrix} \dot{X} \\ \dot{Y} \end{bmatrix} = \begin{bmatrix} -l_1 s\theta_1 - l_2 s (\theta_1 + \theta_2) & -l_2 s (\theta_1 + \theta_2) \\ l_1 c\theta_1 + l_2 c (\theta_1 + \theta_2) & l_2 c (\theta_1 + \theta_2) \end{bmatrix} \begin{bmatrix} \dot{\theta}_1 \\ \dot{\theta}_2 \end{bmatrix} \tag{8.105}$$

or

$$\dot{\mathbf{X}} = \mathbf{J}_D \, \dot{\boldsymbol{\theta}} \tag{8.106}$$

$\mathbf{J}_D$ is the Jacobian of the $2R$ manipulator.

$$\mathbf{J}_D = \begin{bmatrix} \dfrac{\partial X}{\partial \theta_1} & \dfrac{\partial X}{\partial \theta_2} \\ \dfrac{\partial Y}{\partial \theta_1} & \dfrac{\partial Y}{\partial \theta_2} \end{bmatrix} \tag{8.107}$$

$$= \begin{bmatrix} -l_1 \sin \theta_1 - l_2 \sin (\theta_1 + \theta_2) & -l_2 \sin (\theta_1 + \theta_2) \\ l_1 \cos \theta_1 + l_2 \cos (\theta_1 + \theta_2) & l_2 \cos (\theta_1 + \theta_2) \end{bmatrix}$$

Employing the absolute orientation angles of the links, $\theta_1, \theta_1 + \theta_2$, we can also write the velocity equation of a $2R$ manipulator in a more practical method.

$$\begin{bmatrix} \dot{X} \\ \dot{Y} \end{bmatrix} = \begin{bmatrix} -l_1 \sin \theta_1 & -l_2 \sin (\theta_1 + \theta_2) \\ l_1 \cos \theta_1 & l_2 \cos (\theta_1 + \theta_2) \end{bmatrix} \begin{bmatrix} \dot{\theta}_1 \\ \dot{\theta}_1 + \dot{\theta}_2 \end{bmatrix} \tag{8.108}$$

Example 270 Columns of the Jacobian for the $2R$ manipulator. Here we find the Jacobian matrix of the $2R$ planar manipulator using systematic method column by column.

The Jacobian of the $2R$ planar manipulator can be found systematically by using the column-by-column method. The global position vector of the coordinate frames are

$$^0_1\mathbf{d}_2 = l_2 \, ^0\hat{\imath}_2 \tag{8.109}$$

$$^0\mathbf{d}_2 = l_1 \, ^0\hat{\imath}_1 + l_2 \, ^0\hat{\imath}_2 \tag{8.110}$$

and, therefore,

$$^0\dot{\mathbf{d}}_2 = {}_0\boldsymbol{\omega}_1 \times {}^0\mathbf{d}_2 + {}^0_1\boldsymbol{\omega}_2 \times {}^0_1\mathbf{d}_2$$

$$= \dot{\theta}_1 \, ^0\hat{k}_0 \times \left(l_1 \, ^0\hat{\imath}_1 + l_2 \, ^0\hat{\imath}_2 \right) + \dot{\theta}_2 \, ^0\hat{k}_1 \times l_2 \, ^0\hat{\imath}_2$$

$$= \begin{bmatrix} ^0\hat{k}_0 \times \left(l_1 \, ^0\hat{\imath}_1 + l_2 \, ^0\hat{\imath}_2 \right) & ^0\hat{k}_1 \times l_2 \, ^0\hat{\imath}_2 \end{bmatrix} \begin{bmatrix} \dot{\theta}_1 \\ \dot{\theta}_2 \end{bmatrix} \tag{8.111}$$

which can be sct in the following form.

$$\begin{bmatrix} ^0\dot{\mathbf{d}}_2 \\ {}_0\boldsymbol{\omega}_2 \end{bmatrix} = \begin{bmatrix} ^0\hat{k}_0 \times {}^0\mathbf{d}_2 & ^0\hat{k}_1 \times {}^0_1\mathbf{d}_2 \\ ^0\hat{k}_0 & ^0\hat{k}_1 \end{bmatrix} \begin{bmatrix} \dot{\theta}_1 \\ \dot{\theta}_2 \end{bmatrix}$$

$$= \mathbf{J} \, \dot{\boldsymbol{\theta}} \tag{8.112}$$

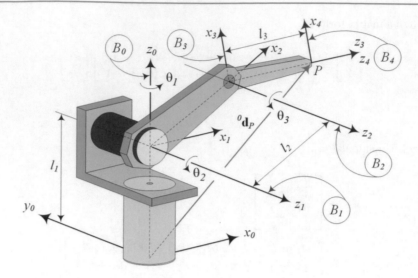

Fig. 8.4   An R⊢R∥R articulated manipulator

Example 271   An articulated manipulator. Jacobian matrix is the key point in forward velocity analysis of robots. Here we derive Jacobian matrix for a very practical industrial robot, the $3R$ articulated manipulator.

Figure 8.4 illustrates an articulated manipulator. The links of the manipulator from link (1) to (3) are R⊢R(90), R∥R(0), R⊢R(90), and their associated homogenous transformation matrices between coordinate frames are

$$
{}^0T_1 = \begin{bmatrix} \cos\theta_1 & 0 & \sin\theta_1 & 0 \\ \sin\theta_1 & 0 & -\cos\theta_1 & 0 \\ 0 & 1 & 0 & l_1 \\ 0 & 0 & 0 & 1 \end{bmatrix} \tag{8.113}
$$

$$
{}^1T_2 = \begin{bmatrix} \cos\theta_2 & -\sin\theta_2 & 0 & l_2\cos\theta_2 \\ \sin\theta_2 & \cos\theta_2 & 0 & l_2\sin\theta_2 \\ 0 & 0 & 1 & 0 \\ 0 & 0 & 0 & 1 \end{bmatrix} \tag{8.114}
$$

$$
{}^2T_3 = \begin{bmatrix} \cos\theta_3 & 0 & \sin\theta_3 & 0 \\ \sin\theta_3 & 0 & -\cos\theta_3 & 0 \\ 0 & 1 & 0 & 0 \\ 0 & 0 & 0 & 1 \end{bmatrix} \tag{8.115}
$$

Let us show the tip point of the manipulator by P. The global coordinates of P are

$$
{}^0\mathbf{r}_P = \begin{bmatrix} X_P \\ Y_P \\ Z_P \\ 1 \end{bmatrix} = {}^0T_3 \begin{bmatrix} l_3\,{}^0k_3 \end{bmatrix} = {}^0T_3 \begin{bmatrix} 0 \\ 0 \\ l_3 \\ 1 \end{bmatrix}
$$

$$
= \begin{bmatrix} \cos\theta_1 \left(l_2\cos\theta_2 + l_3\sin(\theta_2 + \theta_3)\right) \\ \sin\theta_1 \left(l_2\cos\theta_2 + l_3\sin(\theta_2 + \theta_3)\right) \\ l_1 - l_3\cos(\theta_2 + \theta_3) + l_2\sin\theta_2 \\ 1 \end{bmatrix} \tag{8.116}
$$

The coordinates of $^0\mathbf{r}_P$ will be used to determine the Jacobian matrix of the manipulator.

$$\mathbf{J} = \begin{bmatrix} \dfrac{\partial X_P}{\partial \theta_1} & \dfrac{\partial X_P}{\partial \theta_2} & \dfrac{\partial X_P}{\partial \theta_3} \\[2mm] \dfrac{\partial Y_P}{\partial \theta_1} & \dfrac{\partial Y_P}{\partial \theta_2} & \dfrac{\partial Y_P}{\partial \theta_3} \\[2mm] \dfrac{\partial Z_P}{\partial \theta_1} & \dfrac{\partial Z_P}{\partial \theta_2} & \dfrac{\partial Z_P}{\partial \theta_3} \end{bmatrix} \tag{8.117}$$

$$\frac{\partial X_P}{\partial \theta_1} = -\left(l_3 \sin(\theta_2 + \theta_3) + l_2 \cos\theta_2\right)\sin\theta_1$$

$$\frac{\partial Y_P}{\partial \theta_1} = \left(l_3 \sin(\theta_2 + \theta_3) + l_2 \cos\theta_2\right)\cos\theta_1$$

$$\frac{\partial Z_P}{\partial \theta_1} = 0 \tag{8.118}$$

$$\frac{\partial X_P}{\partial \theta_2} = \left(l_3 \cos(\theta_2 + \theta_3) - l_2 \sin\theta_2\right)\cos\theta_1$$

$$\frac{\partial Y_P}{\partial \theta_2} = \left(l_3 \cos(\theta_2 + \theta_3) - l_2 \sin\theta_2\right)\sin\theta_1$$

$$\frac{\partial Z_P}{\partial \theta_2} = l_3 \sin(\theta_2 + \theta_3) + l_2 \cos\theta_2 \tag{8.119}$$

$$\frac{\partial X_P}{\partial \theta_3} = l_3 \cos(\theta_2 + \theta_3) \cos\theta_1$$

$$\frac{\partial Y_P}{\partial \theta_3} = l_3 \cos(\theta_2 + \theta_3) \sin\theta_1$$

$$\frac{\partial Z_P}{\partial \theta_3} = l_3 \sin(\theta_2 + \theta_3) \tag{8.120}$$

$$\mathbf{J} = \begin{bmatrix} -(l_3 s\theta_{23} + l_2 c\theta_2)\, s\theta_1 & (l_3 c\theta_{23} - l_2 s\theta_2)\, c\theta_1 & l_3 c\theta_{23} c\theta_1 \\ (l_3 s\theta_{23} + l_2 c\theta_2)\, c\theta_1 & (l_3 c\theta_{23} - l_2 s\theta_2)\, s\theta_1 & l_3 c\theta_{23} s\theta_1 \\ 0 & l_3 s\theta_{23} + l_2 c\theta_2 & l_3 s\theta_{23} \end{bmatrix} \tag{8.121}$$

$$\theta_{23} = \theta_2 + \theta_3 \tag{8.122}$$

8.3 Jacobian Generating Vectors

The Jacobian matrix $\mathbf{J}$ is a linear transformation, mapping joint speeds to Cartesian speeds.

$$\dot{\mathbf{X}} = \mathbf{J}\dot{\mathbf{q}} \tag{8.123}$$

where $\mathbf{J}$ is equal to:

$$\mathbf{J} = \begin{bmatrix} ^0\hat{k}_0 \times {}^0_0\mathbf{d}_n & ^0\hat{k}_1 \times {}^0_1\mathbf{d}_n & \cdots & ^0\hat{k}_{n-1} \times {}^0_{n-1}\mathbf{d}_n \\ ^0\hat{k}_0 & ^0\hat{k}_1 & \cdots & ^0\hat{k}_{n-1} \end{bmatrix} \tag{8.124}$$

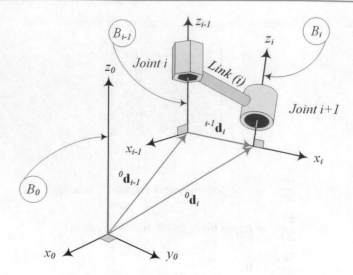

Fig. 8.5 Link (i) and associated coordinate frames

$\mathbf{J}$ can be calculated column by column. The ith column of $\mathbf{J}$ is called *Jacobian generating vector* and is denoted by $\mathbf{c}_i(\mathbf{q})$.

$$\mathbf{c}_i(\mathbf{q}) = \begin{bmatrix} {}^0\hat{k}_{i-1} \times {}^0_{i-1}\mathbf{d}_n \\ {}^0\hat{k}_{i-1} \end{bmatrix} \tag{8.125}$$

To calculate the ith column of Jacobian matrix, we need to find two vectors ${}^0_{i-1}\mathbf{d}_n$ and ${}^0\hat{k}_{i-1}$ in the base frame. The vector ${}^0_{i-1}\mathbf{d}_n$ indicates the position of origin of the n-frame with respect to $i-1$-frame, expressed in the base frame. The vector ${}^0\hat{k}_{i-1}$ is the joint axis unit vector of the frame attached to link ($i-1$), expressed in the base frame.

Calculating $\mathbf{J}$, based on the Jacobian generating vectors, shows that forward velocity kinematics is a consequence of the forward kinematics of robots.

Proof Let ${}^0\mathbf{d}_i$ and ${}^0\mathbf{d}_{i-1}$ be the global position vector of the frames B_i and B_{i-1}, while ${}^{i-1}\mathbf{d}_i$ is the position vector of the frame B_i in B_{i-1} as shown in Fig. 8.5. These three position vectors make a triangle and are related by a vector addition equation

$$\begin{aligned} {}^0\mathbf{d}_i &= {}^0\mathbf{d}_{i-1} + {}^0R_{i-1}\,{}^{i-1}\mathbf{d}_i \\ &= {}^0\mathbf{d}_{i-1} + d_i\,{}^0\hat{k}_{i-1} + a_i\,{}^0\hat{\imath}_i \end{aligned} \tag{8.126}$$

in which we have used Eq. (8.6). Taking a time derivative of ${}^0\mathbf{d}_i$,

$$\begin{aligned} {}^0\dot{\mathbf{d}}_i &= {}^0\dot{\mathbf{d}}_{i-1} + {}^0\dot{R}_{i-1}\,{}^{i-1}\mathbf{d}_i + {}^0R_{i-1}\,{}^{i-1}\dot{\mathbf{d}}_i \\ &= {}^0\dot{\mathbf{d}}_{i-1} + {}^0\dot{R}_{i-1}\left(d_i\,{}^{i-1}\hat{k}_{i-1} + a_i\,{}^{i-1}\hat{\imath}_i\right) \\ &\quad + {}^0R_{i-1}\,\dot{d}_i\,{}^{i-1}\hat{k}_{i-1} \end{aligned} \tag{8.127}$$

shows that the global velocity of the origin of B_i is a function of the translational and angular velocities of link B_{i-1}. However,

$$ {}^0_{i-1}\dot{\mathbf{d}}_i = {}^0\dot{\mathbf{d}}_i - {}^0\dot{\mathbf{d}}_{i-1} \tag{8.128}$$

$$\begin{aligned} {}^0\dot{R}_{i-1}\,{}^{i-1}\mathbf{d}_i &= {}_0\boldsymbol{\omega}_{i-1} \times {}^0R_{i-1}\,{}^{i-1}\mathbf{d}_i = {}_0\boldsymbol{\omega}_{i-1} \times {}^0_{i-1}\mathbf{d}_i \\ &= \dot{\theta}_i\,{}^0\hat{k}_{i-1} \times {}^0_{i-1}\mathbf{d}_i \end{aligned} \tag{8.129}$$

and

$$
{}^0R_{i-1}\,\dot{d}_i\,{}^{i-1}\hat{k}_{i-1} = \dot{d}_i\,{}^0R_{i-1}\,{}^{i-1}\hat{k}_{i-1} = \dot{d}_i\,{}^0\hat{k}_{i-1}
\tag{8.130}
$$

therefore,

$$
{}^0_{i-1}\dot{\mathbf{d}}_i = \dot{\theta}_i\,{}^0\hat{k}_{i-1} \times {}^0_{i-1}\mathbf{d}_i + \dot{d}_i\,{}^0\hat{k}_{i-1}
\tag{8.131}
$$

Because at each joint, either θ or d is variable, we conclude that

$$
{}^0_{i-1}\dot{\mathbf{d}}_i = {}^0_0\boldsymbol{\omega}_i \times {}^0_{i-1}\mathbf{d}_i \qquad \text{if joint } i \text{ is R}
\tag{8.132}
$$

or

$$
{}^0_{i-1}\dot{\mathbf{d}}_i = \dot{d}_i\,{}^0\hat{k}_{i-1} + {}^0_0\boldsymbol{\omega}_i \times {}^0_{i-1}\mathbf{d}_i \quad \text{if joint } i \text{ is P}
\tag{8.133}
$$

The end-effector velocity is then expressed by

$$
{}^0_0\dot{\mathbf{d}}_n = \sum_{i=1}^{n} {}^0_{i-1}\dot{\mathbf{d}}_i = \sum_{i=1}^{n} \dot{\theta}_i\,{}^0\hat{k}_{i-1} \times {}^0_{i-1}\mathbf{d}_n
\tag{8.134}
$$

and

$$
{}^0_0\boldsymbol{\omega}_n = \sum_{i=1}^{n} {}^0_{i-1}\boldsymbol{\omega}_i = \sum_{i=1}^{n} \dot{\theta}_i\,{}^0\hat{k}_{i-1}
\tag{8.135}
$$

They can be rearranged in a matrix form.

$$
\begin{bmatrix} {}^0_0\dot{\mathbf{d}}_n \\ {}^0_0\boldsymbol{\omega}_n \end{bmatrix} = \sum_{i=1}^{n} \dot{\theta}_i \begin{bmatrix} {}^0\hat{k}_{i-1} \times {}^0_{i-1}\mathbf{d}_n \\ {}^0\hat{k}_{i-1} \end{bmatrix}
$$

$$
= \begin{bmatrix} {}^0\hat{k}_0 \times {}^0_0\mathbf{d}_n & {}^0\hat{k}_1 \times {}^0_1\mathbf{d}_n & \cdots & {}^0\hat{k}_{n-1} \times {}^0_{n-1}\mathbf{d}_n \\ {}^0\hat{k}_0 & {}^0\hat{k}_1 & \cdots & {}^0\hat{k}_{n-1} \end{bmatrix} \begin{bmatrix} \dot{\theta}_1 \\ \dot{\theta}_2 \\ \vdots \\ \dot{\theta}_n \end{bmatrix}
\tag{8.136}
$$

We usually show this equation by a short notation as Eq. (8.123),

$$
\dot{\mathbf{X}} = \mathbf{J}\,\dot{\mathbf{q}}
\tag{8.137}
$$

where the vector $\dot{\mathbf{q}} = \begin{bmatrix} \dot{q}_1 & \dot{q}_2 & \cdots & \dot{q}_n \end{bmatrix}^T$ is the *joint speeds vector* and $\mathbf{J}$ is the *Jacobian* matrix of the robot.

$$
\mathbf{J} = \begin{bmatrix} {}^0\hat{k}_0 \times {}^0_0\mathbf{d}_n & {}^0\hat{k}_1 \times {}^0_1\mathbf{d}_n & \cdots & {}^0\hat{k}_{n-1} \times {}^0_{n-1}\mathbf{d}_n \\ {}^0\hat{k}_0 & {}^0\hat{k}_1 & \cdots & {}^0\hat{k}_{n-1} \end{bmatrix}
\tag{8.138}
$$

Practically, we find the Jacobian matrix column by column. Each column is a *Jacobian generating vector*, $\mathbf{c}_i(\mathbf{q})$, and is associated to joint i. If joint i is revolute, then

$$
\mathbf{c}_i(\mathbf{q}) = \begin{bmatrix} {}^0\hat{k}_{i-1} \times {}^0_{i-1}\mathbf{d}_n \\ {}^0\hat{k}_{i-1} \end{bmatrix}
\tag{8.139}
$$

and if joint i is prismatic, then $\mathbf{c}_i(\mathbf{q})$ simplifies to:

$$
\mathbf{c}_i(\mathbf{q}) = \begin{bmatrix} {}^0\hat{k}_{i-1} \\ 0 \end{bmatrix}
\tag{8.140}
$$

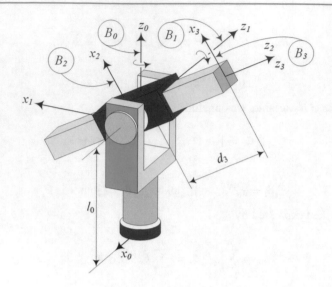

Fig. 8.6 A spherical manipulator

Equation (8.123) provides a 6×1 vector as a set of six equations. The first three equations relate the translational velocity of the end-effector joint speeds. The last three equations relate the angular velocity of the end-effector frame to the joint speeds.   ∎

Example 272 Jacobian matrix for a spherical manipulator. Spherical manipulator is an applied industrial robot and hence its kinematics analysis is important as well as educational. Here is its Jacobian matrix by generating vector method.

Figure 8.6 depicts a spherical manipulator. To find its Jacobian matrix by generating vector method, we start with determining the $^0\hat{k}_{i-1}$ axes for $i = 1, 2, 3$. It would be easier if we use the homogenous definitions to calculate $^0\hat{k}_0$, $^0\hat{k}_1$, $^0\hat{k}_2$.

$$^0\hat{k}_0 = \begin{bmatrix} 0 \\ 0 \\ 1 \\ 0 \end{bmatrix} \tag{8.141}$$

$$^0\hat{k}_1 = {^0T_1}\, {^1\hat{k}_1} \tag{8.142}$$

$$= \begin{bmatrix} \cos\theta_1 & 0 & -\sin\theta_1 & 0 \\ \sin\theta_1 & 0 & \cos\theta_1 & 0 \\ 0 & -1 & 0 & l_0 \\ 0 & 0 & 0 & 1 \end{bmatrix} \begin{bmatrix} 0 \\ 0 \\ 1 \\ 0 \end{bmatrix} = \begin{bmatrix} -\sin\theta_1 \\ \cos\theta_1 \\ 0 \\ 0 \end{bmatrix}$$

$$^0\hat{k}_2 = {^0T_2}\, {^2\hat{k}_2} \tag{8.143}$$

$$= \begin{bmatrix} c\theta_1 c\theta_2 & -s\theta_1 & c\theta_1 s\theta_2 & 0 \\ c\theta_2 s\theta_1 & c\theta_1 & s\theta_1 s\theta_2 & 0 \\ -s\theta_2 & 0 & c\theta_2 & l_0 \\ 0 & 0 & 0 & 1 \end{bmatrix} \begin{bmatrix} 0 \\ 0 \\ 1 \\ 0 \end{bmatrix} = \begin{bmatrix} \cos\theta_1 \sin\theta_2 \\ \sin\theta_1 \sin\theta_2 \\ \cos\theta_2 \\ 0 \end{bmatrix}$$

Then, the vectors $_{i-1}^{\;0}\mathbf{d}_n$ must be evaluated.

$$_0^0\mathbf{d}_3 = l_0\, {^0\hat{k}_0} + d_3\, {^0\hat{k}_2} = \begin{bmatrix} d_3 \cos\theta_1 \sin\theta_2 \\ d_3 \sin\theta_1 \sin\theta_2 \\ l_0 + d_3 \cos\theta_2 \\ 0 \end{bmatrix} \tag{8.144}$$

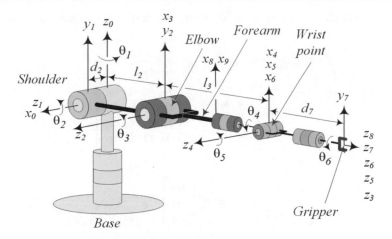

Fig. 8.7 A 6 DOF articulated manipulator

$$
{}^0_1\mathbf{d}_3 = d_3\,{}^0\hat{k}_2 = \begin{bmatrix} d_3\cos\theta_1\sin\theta_2 \\ d_3\sin\theta_1\sin\theta_2 \\ d_3\cos\theta_2 \\ 0 \end{bmatrix} \tag{8.145}
$$

Therefore, the Jacobian of the manipulator will be

$$
\begin{aligned}
\mathbf{J} &= \begin{bmatrix} {}^0\hat{k}_0 \times {}^0_0\mathbf{d}_3 & {}^0\hat{k}_1 \times {}^0_1\mathbf{d}_3 & {}^0\hat{k}_2 \times {}^0_2\mathbf{d}_3 \\ {}^0\hat{k}_0 & {}^0\hat{k}_1 & {}^0\hat{k}_2 \end{bmatrix} \\[2mm]
&= \begin{bmatrix} {}^0\hat{k}_0 \times {}^0_0\mathbf{d}_3 & {}^0\hat{k}_1 \times {}^0_1\mathbf{d}_3 & {}^0\hat{k}_2 \\ {}^0\hat{k}_0 & {}^0\hat{k}_1 & 0 \end{bmatrix} \\[2mm]
&= \begin{bmatrix} -d_3\sin\theta_1\sin\theta_2 & d_3\cos\theta_1\cos\theta_2 & \cos\theta_1\sin\theta_2 \\ d_3\cos\theta_1\sin\theta_2 & d_3\cos\theta_2\sin\theta_1 & \sin\theta_1\sin\theta_2 \\ 0 & -d_3\sin\theta_2 & \cos\theta_2 \\ 0 & -\sin\theta_1 & 0 \\ 0 & \cos\theta_1 & 0 \\ 1 & 0 & 0 \end{bmatrix}
\end{aligned} \tag{8.146}
$$

Example 273 Jacobian matrix for an articulated manipulator. Here we determine the Jacobian matrix of an articulated manipulator by generating vector method.

The Jacobian matrix of a 6 DOF articulated manipulator is a 6×6 matrix. The manipulator was shown in Fig. 8.7 and its transformation matrices were calculated in Example 214.

The ith column of the Jacobian is the generating vector $\mathbf{c}_i(\mathbf{q})$.

$$
\mathbf{c}_i(\mathbf{q}) = \begin{bmatrix} {}^0\hat{k}_{i-1} \times {}^0_{i-1}\mathbf{d}_6 \\ {}^0\hat{k}_{i-1} \end{bmatrix} \tag{8.147}
$$

For the first column of the Jacobian matrix, we need to find ${}^0\hat{k}_0$ and ${}^0\mathbf{d}_6$. The direction of the z_0-axis in the base coordinate frame is

$$
{}^0\hat{k}_0 = \begin{bmatrix} 0 \\ 0 \\ 1 \end{bmatrix} \tag{8.148}
$$

and the position vector of the end-effector frame B_6 is $^0\mathbf{d}_6$, which can be directly determined from the fourth column of the transformation matrix, $^0T_6$,

$$^0T_6 = {}^0T_1\,{}^1T_2\,{}^2T_3\,{}^3T_4\,{}^4T_5\,{}^5T_6$$

$$= \begin{bmatrix} ^0R_6 & {}^0\mathbf{d}_6 \\ 0 & 1 \end{bmatrix} = \begin{bmatrix} t_{11} & t_{12} & t_{13} & t_{14} \\ t_{21} & t_{22} & t_{23} & t_{24} \\ t_{31} & t_{32} & t_{33} & t_{34} \\ 0 & 0 & 0 & 1 \end{bmatrix} \tag{8.149}$$

which is

$$^0\mathbf{d}_6 = \begin{bmatrix} t_{14} \\ t_{24} \\ t_{34} \end{bmatrix} \tag{8.150}$$

where

$$\begin{aligned} t_{14} &= d_6\,(s\theta_1 s\theta_4 s\theta_5 + c\theta_1\,(c\theta_4 s\theta_5 c\,(\theta_2 + \theta_3) \\ &\quad + c\theta_5 s\,(\theta_2 + \theta_3))) \\ &\quad + l_3 c\theta_1 s\,(\theta_2 + \theta_3) + d_2 s\theta_1 + l_2 c\theta_1 c\theta_2 \\ t_{24} &= d_6\,(-c\theta_1 s\theta_4 s\theta_5 + s\theta_1\,(c\theta_4 s\theta_5 c\,(\theta_2 + \theta_3) \\ &\quad + c\theta_5 s\,(\theta_2 + \theta_3))) \\ &\quad + s\theta_1 s\,(\theta_2 + \theta_3)\,l_3 - d_2 c\theta_1 + l_2 c\theta_2 s\theta_1 \\ t_{34} &= d_6\,(c\theta_4 s\theta_5 s\,(\theta_2 + \theta_3) - c\theta_5 c\,(\theta_2 + \theta_3)) \\ &\quad + l_2 s\theta_2 + l_3 c\,(\theta_2 + \theta_3)\,. \end{aligned} \tag{8.151}$$

Therefore,

$$^0\hat{k}_0 \times {}^0\mathbf{d}_6 = \begin{bmatrix} 0 \\ 0 \\ 1 \end{bmatrix} \times \begin{bmatrix} t_{14} \\ t_{24} \\ t_{34} \end{bmatrix} = \begin{bmatrix} -t_{24} \\ t_{14} \\ 0 \end{bmatrix} \tag{8.152}$$

and the first Jacobian generating vector is

$$\mathbf{c}_1 = \begin{bmatrix} ^0\hat{k}_0 \times {}^0\mathbf{d}_6 \\ ^0\hat{k}_0 \end{bmatrix} = \begin{bmatrix} -t_{24} \\ t_{14} \\ 0 \\ 0 \\ 0 \\ 1 \end{bmatrix} \tag{8.153}$$

For the 2nd column,

$$\mathbf{c}_2 = \begin{bmatrix} ^0\hat{k}_1 \times {}^0_1\mathbf{d}_6 \\ ^0\hat{k}_1 \end{bmatrix} \tag{8.154}$$

we need to find $^0\hat{k}_1$ and $^0_1\mathbf{d}_6$. The z_1-axis in the base frame can be found by transforming $^1\hat{k}_1$ to the base frame.

$$^0\hat{k}_1 = {}^0R_1\,{}^1\hat{k}_1 = {}^0R_1 \begin{bmatrix} 0 \\ 0 \\ 1 \end{bmatrix}$$

$$= \begin{bmatrix} c\theta_1 & 0 & s\theta_1 \\ s\theta_1 & 0 & -c\theta_1 \\ 0 & 1 & 0 \end{bmatrix} \begin{bmatrix} 0 \\ 0 \\ 1 \end{bmatrix} = \begin{bmatrix} \sin\theta_1 \\ -\cos\theta_1 \\ 0 \end{bmatrix} \tag{8.155}$$

This determines the second half of $\mathbf{c}_2$. The first half of generating vector $\mathbf{c}_2$ is $^0\hat{k}_1 \times {}_1^0\mathbf{d}_6$. The vector $_1^0\mathbf{d}_6$ is the position of the end-effector in the coordinate frame B_1; however, it must be expressed in the base frame to be able to perform the cross product. An easy method is to find $^1\hat{k}_1 \times {}^1\mathbf{d}_6$ and transform the resultant into the base frame. The vector $^1\mathbf{d}_6$ is the fourth column of $^1T_6 = {}^1T_2\,{}^2T_3\,{}^3T_4\,{}^4T_5\,{}^5T_6$, which, from Example 214, is

$$^1\mathbf{d}_6 = \begin{bmatrix} l_2\cos\theta_2 + l_3\sin(\theta_2+\theta_3) \\ l_2\sin\theta_2 - l_3\cos(\theta_2+\theta_3) \\ d_2 \end{bmatrix} \tag{8.156}$$

Therefore, the first half of $\mathbf{c}_2$ will be

$$
\begin{aligned}
^0\hat{k}_1 \times {}_1^0\mathbf{d}_6 &= {}^0R_1\left({}^1\hat{k}_1 \times {}^1\mathbf{d}_6\right) \\
&= {}^0R_1\left(\begin{bmatrix} 0 \\ 0 \\ 1 \end{bmatrix} \times \begin{bmatrix} l_2\cos\theta_2 + l_3\sin(\theta_2+\theta_3) \\ l_2\sin\theta_2 - l_3\cos(\theta_2+\theta_3) \\ d_2 \end{bmatrix}\right) \\
&= \begin{bmatrix} \cos\theta_1\,(-l_2\sin\theta_2 + l_3\cos(\theta_2+\theta_3)) \\ \sin\theta_1\,(-l_2\sin\theta_2 + l_3\cos(\theta_2+\theta_3)) \\ l_2\cos\theta_2 + l_3\sin(\theta_2+\theta_3) \end{bmatrix}
\end{aligned} \tag{8.157}
$$

Hence, $\mathbf{c}_2$ is

$$\mathbf{c}_2 = \begin{bmatrix} ^0\hat{k}_1 \times {}_1^0\mathbf{d}_6 \\ ^0\hat{k}_1 \end{bmatrix} = \begin{bmatrix} \cos\theta_1\,(-l_2\sin\theta_2 + l_3\cos(\theta_2+\theta_3)) \\ \sin\theta_1\,(-l_2\sin\theta_2 + l_3\cos(\theta_2+\theta_3)) \\ l_2\cos\theta_2 + l_3\sin(\theta_2+\theta_3) \\ \sin\theta_1 \\ -\cos\theta_1 \\ 0 \end{bmatrix} \tag{8.158}$$

The 3rd column is made by $^0\hat{k}_2$ and $_2^0\mathbf{d}_6$. The vector $^2\mathbf{d}_6$ is position of the end-effector in B_2-frame, which will be the fourth column of $^2T_6 = {}^2T_3\,{}^3T_4\,{}^4T_5\,{}^5T_6$.

$$^2\mathbf{d}_6 = \begin{bmatrix} l_3\sin\theta_3 \\ -l_3\cos\theta_3 \\ 0 \end{bmatrix} \tag{8.159}$$

The z_2-axis in the base frame can be found by

$$^0\hat{k}_2 = {}^0R_2\,{}^2\hat{k}_2 = {}^0R_1\,{}^1R_2 \begin{bmatrix} 0 \\ 0 \\ 1 \end{bmatrix} = \begin{bmatrix} \sin\theta_1 \\ -\cos\theta_1 \\ 0 \end{bmatrix} \tag{8.160}$$

and the cross product $^0\hat{k}_2 \times {}_2^0\mathbf{d}_6$ can be found by transforming the resultant of $^2\hat{k}_2 \times {}^2\mathbf{d}_6$ into the base coordinate frame.

$$^2\hat{k}_2 \times {}^2\mathbf{d}_6 = \begin{bmatrix} l_3\cos\theta_3 \\ l_3\sin\theta_3 \\ 0 \end{bmatrix} \tag{8.161}$$

$$^0\hat{k}_2 \times {}_2^0\mathbf{d}_6 = {}^0R_2\left({}^2\hat{k}_2 \times {}^2\mathbf{d}_6\right) = \begin{bmatrix} l_3\cos\theta_1\sin(\theta_2+\theta_3) \\ l_3\sin\theta_1\sin(\theta_2+\theta_3) \\ -l_3\cos(\theta_2+\theta_3) \end{bmatrix} \tag{8.162}$$

Therefore, $\mathbf{c}_3$ is

$$
\mathbf{c}_3 = \begin{bmatrix} {}^0\hat{k}_2 \times {}^0_2\mathbf{d}_6 \\ {}^0\hat{k}_2 \end{bmatrix} = \begin{bmatrix} l_3 \cos\theta_1 \sin(\theta_2 + \theta_3) \\ l_3 \sin\theta_1 \sin(\theta_2 + \theta_3) \\ -l_3 \cos(\theta_2 + \theta_3) \\ \sin\theta_1 \\ -\cos\theta_1 \\ 0 \end{bmatrix}
\tag{8.163}
$$

The 4th column needs ${}^0\hat{k}_3$ and ${}^0_3\mathbf{d}_6$. The vector ${}^3\mathbf{d}_6$ is position of the end-effector in B_3-frame that will be found as the fourth column of ${}^3T_6 = {}^3T_4\,{}^4T_5\,{}^5T_6$.

$$
{}^3\mathbf{d}_6 = \begin{bmatrix} 0 \\ 0 \\ l_3 \end{bmatrix}
\tag{8.164}
$$

The vector ${}^0\hat{k}_3$ can be found by transforming ${}^3\hat{k}_3$ to the base frame

$$
{}^0\hat{k}_3 = {}^0R_3 \begin{bmatrix} 0 \\ 0 \\ 1 \end{bmatrix} = \begin{bmatrix} \cos\theta_1 (\cos\theta_2 \sin\theta_3 + \cos\theta_3 \sin\theta_2) \\ \sin\theta_1 (\cos\theta_2 \sin\theta_3 + \sin\theta_2 \cos\theta_3) \\ -\cos(\theta_2 + \theta_3) \end{bmatrix}
\tag{8.165}
$$

and the first half of $\mathbf{c}_4$ can be found by calculating ${}^3\hat{k}_3 \times {}^3\mathbf{d}_6$ and transforming the resultant into the base coordinate frame.

$$
{}^0R_3 \left({}^3\hat{k}_3 \times {}^3\mathbf{d}_6 \right) = {}^0R_3 \left(\begin{bmatrix} 0 \\ 0 \\ 1 \end{bmatrix} \times \begin{bmatrix} 0 \\ 0 \\ l_3 \end{bmatrix} \right) = \begin{bmatrix} 0 \\ 0 \\ 0 \end{bmatrix}
\tag{8.166}
$$

Therefore, $\mathbf{c}_4$ is

$$
\mathbf{c}_4 = \begin{bmatrix} {}^0\hat{k}_3 \times {}^0_3\mathbf{d}_6 \\ {}^0\hat{k}_3 \end{bmatrix} = \begin{bmatrix} 0 \\ 0 \\ 0 \\ \cos\theta_1 (\cos\theta_2 \sin\theta_3 + \cos\theta_3 \sin\theta_2) \\ \sin\theta_1 (\cos\theta_2 \sin\theta_3 + \sin\theta_2 \cos\theta_3) \\ -\cos(\theta_2 + \theta_3) \end{bmatrix}
\tag{8.167}
$$

The 5th column needs ${}^0\hat{k}_4$ and ${}^0_4\mathbf{d}_6$. The vector ${}^4\mathbf{d}_6$ is position of the end-effector in B_4-frame, which is the fourth column of ${}^4T_6 = {}^4T_5\,{}^5T_6$.

$$
{}^4\mathbf{d}_6 = \begin{bmatrix} 0 \\ 0 \\ 0 \end{bmatrix}
\tag{8.168}
$$

We will find the vector ${}^0\hat{k}_4$ by transforming ${}^4\hat{k}_4$ to the base frame.

$$
{}^0\hat{k}_4 = {}^0R_4 \begin{bmatrix} 0 \\ 0 \\ 1 \end{bmatrix} = \begin{bmatrix} c\theta_4 s\theta_1 - c\theta_1 s\theta_4 c(\theta_2 + \theta_3) \\ -c\theta_1 c\theta_4 - s\theta_1 s\theta_4 c(\theta_2 + \theta_3) \\ -s\theta_4 s(\theta_2 + \theta_3) \end{bmatrix}
\tag{8.169}
$$

The first half of $\mathbf{c}_5$ is ${}^4\hat{k}_4 \times {}^4\mathbf{d}_6$, expressed in the base coordinate frame.

$$
{}^0R_4 \left({}^4\hat{k}_4 \times {}^4\mathbf{d}_6 \right) = {}^0R_4 \left(\begin{bmatrix} 0 \\ 0 \\ 1 \end{bmatrix} \times \begin{bmatrix} 0 \\ 0 \\ 0 \end{bmatrix} \right) = \begin{bmatrix} 0 \\ 0 \\ 0 \end{bmatrix}
\tag{8.170}
$$

Therefore, $\mathbf{c}_5$ is

$$
\mathbf{c}_5 = \begin{bmatrix} {}^0\hat{k}_4 \times {}^0_4\mathbf{d}_6 \\ {}^0\hat{k}_4 \end{bmatrix}
$$

$$
= \begin{bmatrix} 0 \\ 0 \\ 0 \\ \cos\theta_4\sin\theta_1 - \cos\theta_1\sin\theta_4\cos(\theta_2+\theta_3) \\ -\cos\theta_1\cos\theta_4 - \sin\theta_1\sin\theta_4\cos(\theta_2+\theta_3) \\ -\sin\theta_4\sin(\theta_2+\theta_3) \end{bmatrix} \tag{8.171}
$$

The 6th column is found by calculating ${}^0\hat{k}_5$ and ${}^0\hat{k}_5 \times {}^0_5\mathbf{d}_6$. The vector ${}^5\mathbf{d}_6$ is position of the end-effector in B_5-frame, which is the fourth column of ${}^5T_6$.

$$
{}^5\mathbf{d}_6 = \begin{bmatrix} 0 \\ 0 \\ 0 \end{bmatrix} \tag{8.172}
$$

The vector ${}^0\hat{k}_5$ is

$$
{}^0\hat{k}_5 = {}^0R_5 \begin{bmatrix} 0 \\ 0 \\ 1 \end{bmatrix} \tag{8.173}
$$

$$
= \begin{bmatrix} -c\theta_1 c\theta_4 s(\theta_2+\theta_3) - s\theta_4(s\theta_1 s\theta_4 + c\theta_1 c\theta_4 c(\theta_2+\theta_3)) \\ -s\theta_1 c\theta_4 s(\theta_2+\theta_3) - s\theta_4(-c\theta_1 s\theta_4 + s\theta_1 c\theta_4 c(\theta_2+\theta_3)) \\ c\theta_4 c(\theta_2+\theta_3) - \frac{1}{2}s(\theta_2+\theta_3)s2\theta_4 \end{bmatrix}
$$

and the first half of $\mathbf{c}_6$ is ${}^5\hat{k}_5 \times {}^5\mathbf{d}_6$, expressed in the base coordinate frame.

$$
{}^0R_5\left({}^5\hat{k}_5 \times {}^5\mathbf{d}_6\right) = {}^0R_5 \left(\begin{bmatrix} 0 \\ 0 \\ 1 \end{bmatrix} \times \begin{bmatrix} 0 \\ 0 \\ 0 \end{bmatrix} \right) = \begin{bmatrix} 0 \\ 0 \\ 0 \end{bmatrix} \tag{8.174}
$$

Therefore, $\mathbf{c}_6$ is

$$
\mathbf{c}_6 = \begin{bmatrix} {}^0\hat{k}_5 \times {}^0_5\mathbf{d}_6 \\ {}^0\hat{k}_5 \end{bmatrix} \tag{8.175}
$$

$$
= \begin{bmatrix} 0 \\ 0 \\ 0 \\ -c\theta_1 c\theta_4 s(\theta_2+\theta_3) - s\theta_4(s\theta_1 s\theta_4 + c\theta_1 c\theta_4 c(\theta_2+\theta_3)) \\ -s\theta_1 c\theta_4 s(\theta_2+\theta_3) - s\theta_4(-c\theta_1 s\theta_4 + s\theta_1 c\theta_4 c(\theta_2+\theta_3)) \\ c\theta_4 c(\theta_2+\theta_3) - \frac{1}{2}s(\theta_2+\theta_3)s2\theta_4 \end{bmatrix}
$$

and the Jacobian matrix for the articulated robot is calculated.

$$
\mathbf{J} = \begin{bmatrix} \mathbf{c}_1 & \mathbf{c}_2 & \mathbf{c}_3 & \mathbf{c}_4 & \mathbf{c}_5 & \mathbf{c}_6 \end{bmatrix} \tag{8.176}
$$

Example 274 The effect of a spherical wrist on Jacobian matrix. Spherical wrist makes the last three columns of Jacobian matrix of a robot to be simple with the upper right 3×3 elements zero.

The Jacobian matrix for a robot having a spherical wrist is always of the form

$$\mathbf{J} = \begin{bmatrix} {}^0\hat{k}_0 \times {}^0_0\mathbf{d}_6 & {}^0\hat{k}_1 \times {}^0_1\mathbf{d}_6 & {}^0\hat{k}_2 \times {}^0_2\mathbf{d}_6 & 0 & 0 & 0 \\ {}^0\hat{k}_0 & {}^0\hat{k}_1 & {}^0\hat{k}_2 & {}^0\hat{k}_3 & {}^0\hat{k}_6 & {}^0\hat{k}_5 \end{bmatrix} \tag{8.177}$$

which shows the upper 3×3 submatrix is zero. This is because of the spherical wrist structure and having a wrist point as the origin of the wrist coordinate frames B_4, B_5, and B_6.

Example 275 ★ Jacobian matrix for an articulated manipulator using the direct differentiating method. Although generating vector method is the classical, general, and systematic method for calculating Jacobian, direct differentiating of the global coordinates of the end-effector is a simpler method to apply.

Figure 8.7 illustrates an articulated robot with transformation matrices given in Example 214. Using the result of the forward kinematics

$$ {}^0T_6 = \begin{bmatrix} {}^0R_6 & {}^0\mathbf{d}_6 \\ 0 & 1 \end{bmatrix} = \begin{bmatrix} t_{11} & t_{12} & t_{13} & t_{14} \\ t_{21} & t_{22} & t_{23} & t_{24} \\ t_{31} & t_{32} & t_{33} & t_{34} \\ 0 & 0 & 0 & 1 \end{bmatrix} \tag{8.178}$$

we know that the position of the end-effector is at ${}^0\mathbf{d}_6$

$$ {}^0\mathbf{d}_6 = \begin{bmatrix} X_6 \\ Y_6 \\ Z_6 \end{bmatrix} = \begin{bmatrix} t_{14} \\ t_{24} \\ t_{34} \end{bmatrix} \tag{8.179}$$

where

$$
\begin{aligned}
t_{14} &= d_6 \left(s\theta_1 s\theta_4 s\theta_5 + c\theta_1 \left(c\theta_4 s\theta_5 c \left(\theta_2 + \theta_3 \right) \right.\right. \\
&\quad \left.\left. + c\theta_5 s \left(\theta_2 + \theta_3 \right) \right) \right) \\
&\quad + l_3 c\theta_1 s \left(\theta_2 + \theta_3 \right) + d_2 s\theta_1 + l_2 c\theta_1 c\theta_2 \\
t_{24} &= d_6 \left(-c\theta_1 s\theta_4 s\theta_5 + s\theta_1 \left(c\theta_4 s\theta_5 c \left(\theta_2 + \theta_3 \right) \right.\right. \\
&\quad \left.\left. + c\theta_5 s \left(\theta_2 + \theta_3 \right) \right) \right) \\
&\quad + s\theta_1 s \left(\theta_2 + \theta_3 \right) l_3 - d_2 c\theta_1 + l_2 c\theta_2 s\theta_1 \\
t_{34} &= d_6 \left(c\theta_4 s\theta_5 s \left(\theta_2 + \theta_3 \right) - c\theta_5 c \left(\theta_2 + \theta_3 \right) \right) \\
&\quad + l_2 s\theta_2 + l_3 c \left(\theta_2 + \theta_3 \right)
\end{aligned} \tag{8.180}
$$

Taking the derivative of X_6 yields

$$
\begin{aligned}
\dot{X}_6 &= \frac{\partial X_6}{\partial \theta_1} \dot{\theta}_1 + \frac{\partial X_6}{\partial \theta_2} \dot{\theta}_2 + \cdots + \frac{\partial X_6}{\partial \theta_6} \dot{\theta}_6 \\
&= J_{11} \dot{\theta}_1 + J_{12} \dot{\theta}_2 + \cdots + J_{16} \dot{\theta}_6 \\
&= -t_{24} \dot{\theta}_1 + \cos\theta_1 \left(-l_2 \sin\theta_2 + l_3 \cos \left(\theta_2 + \theta_3 \right) \right) \dot{\theta}_2 \\
&\quad + l_3 \cos\theta_1 \sin \left(\theta_2 + \theta_3 \right) \dot{\theta}_3
\end{aligned} \tag{8.181}
$$

that shows

$$J_{11} = -t_{24}$$
$$J_{12} = \cos\theta_1 \left(-l_2 \sin\theta_2 + l_3 \cos(\theta_2 + \theta_3)\right)$$
$$J_{13} = l_3 \cos\theta_1 \sin(\theta_2 + \theta_3)$$
$$J_{14} = 0$$
$$J_{15} = 0$$
$$J_{16} = 0 \qquad (8.182)$$

Similarly, the derivatives of Y_6 and Z_6 yield

$$\dot{Y}_6 = \frac{\partial Y_6}{\partial \theta_1}\dot{\theta}_1 + \frac{\partial Y_6}{\partial \theta_2}\dot{\theta}_2 + \cdots + \frac{\partial Y_6}{\partial \theta_6}\dot{\theta}_6$$
$$= J_{21}\dot{\theta}_1 + J_{22}\dot{\theta}_2 + \cdots + J_{26}\dot{\theta}_6$$
$$= t_{14}\dot{\theta}_1 + \sin\theta_1 \left(-l_2 \sin\theta_2 + l_3 \cos(\theta_2 + \theta_3)\right)\dot{\theta}_2$$
$$+ l_3 \sin\theta_1 \sin(\theta_2 + \theta_3)\dot{\theta}_3 \qquad (8.183)$$

$$\dot{Z}_6 = \frac{\partial Z_6}{\partial \theta_1}\dot{\theta}_1 + \frac{\partial Z_6}{\partial \theta_2}\dot{\theta}_2 + \cdots + \frac{\partial Z_6}{\partial \theta_6}\dot{\theta}_6$$
$$= J_{31}\dot{\theta}_1 + J_{32}\dot{\theta}_2 + \cdots + J_{36}\dot{\theta}_6$$
$$= \left(l_2 \cos\theta_2 + l_3 \sin(\theta_2 + \theta_3)\right)\dot{\theta}_2$$
$$- l_3 \cos(\theta_2 + \theta_3)\dot{\theta}_3 \qquad (8.184)$$

and show that

$$J_{21} = t_{14}$$
$$J_{22} = \sin\theta_1 \left(-l_2 \sin\theta_2 + l_3 \cos(\theta_2 + \theta_3)\right)$$
$$J_{23} = l_3 \sin\theta_1 \sin(\theta_2 + \theta_3)$$
$$J_{24} = 0$$
$$J_{25} = 0$$
$$J_{26} = 0 \qquad (8.185)$$

$$J_{31} = 0$$
$$J_{32} = l_2 \cos\theta_2 + l_3 \sin(\theta_2 + \theta_3)$$
$$J_{33} = -l_3 \cos(\theta_2 + \theta_3)$$
$$J_{34} = 0$$
$$J_{35} = 0$$
$$J_{36} = 0 \qquad (8.186)$$

There is no explicit equation for describing the rotations of the end-effector's frame about the global axes. So, there is no equation to find differential rotations about the three axes by differentiating. This is a reason for searching indirect or more systematic methods for evaluating the rotational part of Jacobian matrix. However, the next three rows of the Jacobian matrix can be found by calculating the angular velocity vector based on the angular velocity matrix

$$
{}_0\tilde{\omega}_6 = {}^0\dot{R}_6\,{}^0R_6^T = \begin{bmatrix} 0 & -\omega_Z & \omega_Y \\ \omega_Z & 0 & -\omega_X \\ -\omega_Y & \omega_X & 0 \end{bmatrix} \tag{8.187}
$$

$$
{}_0\boldsymbol{\omega}_6 = \begin{bmatrix} \omega_X \\ \omega_Y \\ \omega_Z \end{bmatrix} \tag{8.188}
$$

and then rearranging the components to show the Jacobian elements.

$$
\omega_X = \frac{\partial \omega_X}{\partial \theta_1}\dot{\theta}_1 + \frac{\partial \omega_X}{\partial \theta_2}\dot{\theta}_2 + \cdots + \frac{\partial \omega_X}{\partial \theta_6}\dot{\theta}_6 \tag{8.189}
$$

$$
\omega_Y = \frac{\partial \omega_Y}{\partial \theta_1}\dot{\theta}_1 + \frac{\partial \omega_Y}{\partial \theta_2}\dot{\theta}_2 + \cdots + \frac{\partial \omega_Y}{\partial \theta_6}\dot{\theta}_6 \tag{8.190}
$$

$$
\omega_Z = \frac{\partial \omega_Z}{\partial \theta_1}\dot{\theta}_1 + \frac{\partial \omega_Z}{\partial \theta_2}\dot{\theta}_2 + \cdots + \frac{\partial \omega_Z}{\partial \theta_6}\dot{\theta}_6 \tag{8.191}
$$

Expanding (8.187) for the articulated manipulator shows that the angular velocity vector of the end-effector frame is

$$
\begin{aligned}
\omega_X = & \sin\theta_1\dot{\theta}_2 + \sin\theta_1\dot{\theta}_3 + \cos\theta_1\sin\theta_{23}\dot{\theta}_4 \\
& + (\cos\theta_4\sin\theta_1 - \cos\theta_1\sin\theta_4\cos\theta_{23})\,\dot{\theta}_5 \\
& - (c\theta_1 c\theta_4 s\theta_{23} + s\theta_4\,(s\theta_1 s\theta_4 + c\theta_1 c\theta_4 c\theta_{23}))\,\dot{\theta}_6
\end{aligned} \tag{8.192}
$$

$$
\begin{aligned}
\omega_Y = & -\cos\theta_1\dot{\theta}_2 - \cos\theta_1\dot{\theta}_3 + \sin\theta_1\sin\theta_{23}\dot{\theta}_4 \\
& + (-\cos\theta_1\cos\theta_4 - \sin\theta_1\sin\theta_4\cos\theta_{23})\,\dot{\theta}_5 \\
& + (-s\theta_1 c\theta_4 s\theta_{23} - s\theta_4\,(-c\theta_1 s\theta_4 + s\theta_1 c\theta_4 c\theta_{23}))\,\dot{\theta}_6
\end{aligned} \tag{8.193}
$$

$$
\begin{aligned}
\omega_Z = & \,\dot{\theta}_1 - \cos(\theta_2 + \theta_3)\,\dot{\theta}_4 - \sin\theta_4\sin(\theta_2 + \theta_3)\,\dot{\theta}_5 \\
& + \left(\cos\theta_4\cos\theta_{23} - \frac{1}{2}\sin\theta_{23}\sin2\theta_4\right)\dot{\theta}_6
\end{aligned} \tag{8.194}
$$

and, therefore,

$$
\begin{aligned}
J_{41} &= 0 \\
J_{42} &= \sin\theta_1 \\
J_{43} &= \sin\theta_1 \\
J_{44} &= \cos\theta_1\,(\cos\theta_2\sin\theta_3 + \cos\theta_3\sin\theta_2) \\
J_{45} &= \cos\theta_4\sin\theta_1 - \cos\theta_1\sin\theta_4\cos(\theta_2 + \theta_3) \\
J_{46} &= -c\theta_1 c\theta_4 s\,(\theta_2 + \theta_3) \\
&\quad -s\theta_4\,(s\theta_1 s\theta_4 + c\theta_1 c\theta_4 c\,(\theta_2 + \theta_3))
\end{aligned} \tag{8.195}
$$

$$J_{51} = 0$$

$$J_{52} = -\cos\theta_1$$

$$J_{53} = -\cos\theta_1$$

$$J_{54} = \sin\theta_1 \left(\cos\theta_2 \sin\theta_3 + \sin\theta_2 \cos\theta_3\right)$$

$$J_{55} = -\cos\theta_1 \cos\theta_4 - \sin\theta_1 \sin\theta_4 \cos\left(\theta_2 + \theta_3\right)$$

$$J_{56} = -s\theta_1 c\theta_4 s \left(\theta_2 + \theta_3\right)$$

$$-s\theta_4 \left(-c\theta_1 s\theta_4 + s\theta_1 c\theta_4 c \left(\theta_2 + \theta_3\right)\right) \tag{8.196}$$

$$J_{61} = 1$$

$$J_{62} = 0$$

$$J_{63} = 0$$

$$J_{64} = -\cos\left(\theta_2 + \theta_3\right)$$

$$J_{65} = -\sin\theta_4 \sin\left(\theta_2 + \theta_3\right)$$

$$J_{66} = \cos\theta_4 \cos\left(\theta_2 + \theta_3\right) - \frac{1}{2}\sin\left(\theta_2 + \theta_3\right)\sin 2\theta_4 \tag{8.197}$$

Example 276 ★ Analytical Jacobian and geometrical Jacobian. Here we define a new concept of the analytical Jacobian, displacement Jacobian, and angular Jacobian to be used to calculate the Jacobian matrix.

Assume the global position and orientation of the end-effector frames are specified by a set of six parameters

$$\mathbf{X} = \begin{bmatrix} {}^0\mathbf{r}_n \\ {}^0\boldsymbol{\phi}_n \end{bmatrix} \tag{8.198}$$

where

$$ {}^0\mathbf{r}_n = {}^0\mathbf{r}_n(\mathbf{q}) \tag{8.199}$$

$$ {}^0\boldsymbol{\phi}_n = {}^0\boldsymbol{\phi}_n(\mathbf{q}) \tag{8.200}$$

${}^0\boldsymbol{\phi}_n$ are based on three independent rotational parameters such as Euler angles and ${}^0\mathbf{r}_n$ is the Cartesian position of the end-effector frame, both functions of the joint variable vector, $\mathbf{q}$.

The translational velocity of the end-effector frame can be expressed by

$$ {}^0\dot{\mathbf{r}}_n = \frac{\partial \mathbf{r}}{\partial \mathbf{q}}\dot{\mathbf{q}} = \mathbf{J}_D(\mathbf{q})\,\dot{\mathbf{q}} \tag{8.201}$$

and the rotational velocity of the end-effector frame can be expressed by

$$ {}^0\dot{\boldsymbol{\phi}}_n = \frac{\partial \boldsymbol{\phi}}{\partial \mathbf{q}}\dot{\mathbf{q}} = \mathbf{J}_\phi(\mathbf{q})\,\dot{\mathbf{q}} \tag{8.202}$$

The rotational velocity vector $\dot{\boldsymbol{\phi}}$ in general differs from the angular velocity vector $\boldsymbol{\omega}$. The combination of the *displacement Jacobian* matrices $\mathbf{J}_D$ and *angular Jacobian* $\mathbf{J}_\phi$ in the form of

$$ \mathbf{J}_A = \begin{bmatrix} \mathbf{J}_D \\ \mathbf{J}_\phi \end{bmatrix} \tag{8.203}$$

is called *analytical Jacobian* to indicate its difference with *geometrical Jacobian* $\mathbf{J}$.

Having a set of orientation angles, ϕ, it is possible to find the relationship between the angular velocity ω and the rotational velocity $\dot{\phi}$. As an example, consider the Euler angles φ, θ, ψ about z, x, z axes defined in Sect. 6.3. The global angular velocity, in terms of Euler frequencies, is found in Eq. (2.215).

$$\begin{bmatrix} \omega_X \\ \omega_Y \\ \omega_Z \end{bmatrix} = \begin{bmatrix} 0 & \cos\varphi & \sin\theta\sin\varphi \\ 0 & \sin\varphi & -\cos\varphi\sin\theta \\ 1 & 0 & \cos\theta \end{bmatrix} \begin{bmatrix} \dot{\varphi} \\ \dot{\theta} \\ \dot{\psi} \end{bmatrix} \tag{8.204}$$

$$\omega = {}^G R_E \, \dot{\phi} \tag{8.205}$$

The Eulerian frequencies $\dot{\varphi}, \dot{\theta}, \dot{\psi}$ are functions of joint speeds

$$\begin{bmatrix} \dot{\varphi} \\ \dot{\theta} \\ \dot{\psi} \end{bmatrix} = \mathbf{J}_\phi \, \dot{\mathbf{q}} \tag{8.206}$$

and, therefore,

$$\mathbf{J}_R = {}^G R_E \, \mathbf{J}_\phi \tag{8.207}$$

$$\mathbf{J} = \begin{bmatrix} \mathbf{J}_D \\ \mathbf{J}_R \end{bmatrix} \tag{8.208}$$

When the angular velocity ω of the end-effector is expressed in Cartesian frequencies as

$$\omega = \begin{bmatrix} \omega_X \\ \omega_Y \\ \omega_Z \end{bmatrix} \tag{8.209}$$

then Jacobian matrix is called geometric (8.208). When the angular velocity of the end-effector is expressed in non-Cartesian frequencies such as Eulerian, then Jacobian matrix is called analytic Jacobian (8.203).

8.4 Inverse Velocity Kinematics

The *inverse velocity kinematics problem*, also known as the *resolved rates problem*, is searching for the joint speeds vector $\dot{\mathbf{q}}$ associated to known end-effector speeds vector $\dot{\mathbf{X}}$. Six DOF are needed to be able to move the end-effector in an arbitrary direction with an arbitrary angular velocity. The speeds vector of the end-effector $\dot{\mathbf{X}}$ is related to the joint speeds vector $\dot{\mathbf{q}}$ by the Jacobian matrix $\mathbf{J}$.

$$\dot{\mathbf{X}} = \begin{bmatrix} {}^0\mathbf{v}_n \\ {}^0\boldsymbol{\omega}_n \end{bmatrix} = \begin{bmatrix} \mathbf{J}_D \\ \mathbf{J}_R \end{bmatrix} \dot{\mathbf{q}} = \mathbf{J} \, \dot{\mathbf{q}} \tag{8.210}$$

Consequently, for the inverse velocity kinematics, we require the differential change in joint coordinates expressed in terms of the Cartesian translation and angular velocities of the end-effector. If the Jacobian matrix is non-singular at the moment of calculation, the inverse Jacobian $\mathbf{J}^{-1}$ exists and we are able to find the required joint speeds vector by $\mathbf{J}^{-1}$.

$$\dot{\mathbf{q}} = \mathbf{J}^{-1} \, \dot{\mathbf{X}} \tag{8.211}$$

Singular configuration is where the determinant of the Jacobian matrix is zero and, therefore, $\mathbf{J}^{-1}$ is indeterminate. Equation (8.211) determines the speeds required at the individual joints to produce a desired end-effector speeds vector $\dot{\mathbf{X}}$.

Because the inverse velocity kinematics is a consequence of the forward velocity and needs a matrix inversion, the problem is equivalent to the solution of a set of linear algebraic equations. To find $\mathbf{J}^{-1}$, every matrix inversion method may be employed.

Example 277 Inverse velocity of a planar polar manipulator. Forward velocity kinematics is all we need to calculate inverse velocity kinematics, and the Jacobian matrix is the key point of the forward and inverse velocity kinematics. Here is to show how we calculate inverse velocity kinematics of a 2 *DOF* manipulator.

Figure 8.2 illustrates a planar polar manipulator with the following forward velocity equation.

$$\begin{bmatrix} \dot{X} \\ \dot{Y} \end{bmatrix} = \begin{bmatrix} \cos\theta & -r\sin\theta \\ \sin\theta & r\cos\theta \end{bmatrix} \begin{bmatrix} \dot{r} \\ \dot{\theta} \end{bmatrix} \tag{8.212}$$

To determine the inverse velocity, we need to determine the inverse of the Jacobian matrix $\mathbf{J}$.

$$\mathbf{J} = \begin{bmatrix} \dfrac{\partial X}{\partial r} & \dfrac{\partial X}{\partial \theta} \\ \dfrac{\partial Y}{\partial r} & \dfrac{\partial Y}{\partial \theta} \end{bmatrix} = \begin{bmatrix} \cos\theta & -r\sin\theta \\ \sin\theta & r\cos\theta \end{bmatrix} \tag{8.213}$$

$$\mathbf{J}^{-1} = \frac{1}{\dfrac{\partial X}{\partial r}\dfrac{\partial Y}{\partial \theta} - \dfrac{\partial X}{\partial \theta}\dfrac{\partial Y}{\partial r}} \begin{bmatrix} \dfrac{\partial Y}{\partial \theta} & -\dfrac{\partial X}{\partial \theta} \\ -\dfrac{\partial Y}{\partial r} & \dfrac{\partial X}{\partial r} \end{bmatrix}$$

$$= \frac{1}{r}\begin{bmatrix} \cos\theta & \sin\theta \\ -\sin\theta & \cos\theta \end{bmatrix} \tag{8.214}$$

This $\mathbf{J}^{-1}$ is indeterminate only if $r = 0$. Therefore, the joint speeds for a given velocity of the end-effector will be

$$\begin{bmatrix} \dot{r} \\ \dot{\theta} \end{bmatrix} = \frac{1}{r}\begin{bmatrix} \cos\theta & \sin\theta \\ -\sin\theta & \cos\theta \end{bmatrix}\begin{bmatrix} \dot{X} \\ \dot{Y} \end{bmatrix} \tag{8.215}$$

Example 278 Inverse velocity of a $2R$ planar manipulator. The $2R$ planar manipulator has important role in robotic education and research. Every step of robotic analysis of this manipulator must be studied in detail as a good sample to learn how to deal with more complicated robots. Here is the inverse kinematics analysis of the $2R$ planar manipulator, which is nothing but calculating $\mathbf{J}^{-1}$.

Forward and inverse kinematics of a $2R$ planar manipulator have been analyzed in Example 165 and Example 207. Its Jacobian and forward velocity kinematics are also found in Example 269 as:

$$\dot{\mathbf{X}} = \mathbf{J}\dot{\mathbf{q}} \tag{8.216}$$

$$\begin{bmatrix} \dot{X} \\ \dot{Y} \end{bmatrix} = \begin{bmatrix} -l_1 s\theta_1 - l_2 s(\theta_1 + \theta_2) & -l_2 s(\theta_1 + \theta_2) \\ l_1 c\theta_1 + l_2 c(\theta_1 + \theta_2) & l_2 c(\theta_1 + \theta_2) \end{bmatrix}\begin{bmatrix} \dot{\theta}_1 \\ \dot{\theta}_2 \end{bmatrix} \tag{8.217}$$

The inverse velocity kinematics needs to calculate the inverse of the Jacobian.

$$\dot{\mathbf{q}} = \mathbf{J}^{-1}\dot{\mathbf{X}} \tag{8.218}$$

$$\begin{bmatrix} \dot{\theta}_1 \\ \dot{\theta}_2 \end{bmatrix} = \begin{bmatrix} -l_1 s\theta_1 - l_2 s(\theta_1 + \theta_2) & -l_2 s(\theta_1 + \theta_2) \\ l_1 c\theta_1 + l_2 c(\theta_1 + \theta_2) & l_2 c(\theta_1 + \theta_2) \end{bmatrix}^{-1}\begin{bmatrix} \dot{X} \\ \dot{Y} \end{bmatrix}$$

$$\mathbf{J}^{-1} = \frac{-1}{l_1 l_2 s\theta_2}\begin{bmatrix} -l_2 c(\theta_1 + \theta_2) & -l_2 s(\theta_1 + \theta_2) \\ l_1 c\theta_1 + l_2 c(\theta_1 + \theta_2) & l_1 s\theta_1 + l_2 s(\theta_1 + \theta_2) \end{bmatrix} \tag{8.219}$$

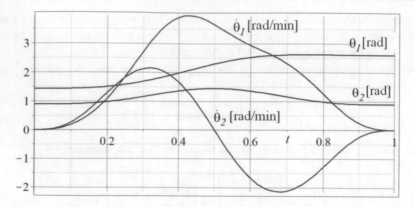

Fig. 8.8 Inverse kinematics and inverse velocity of joint angle and joint angular velocity as function of time for a $2R$ robot

Therefore, the joint angular speeds of the manipulator, $(\dot{\theta}_1, \dot{\theta}_2)$, in terms of the Cartesian velocity components of the end point, $(\dot{X}, \dot{Y})$, are

$$\dot{\theta}_1 = \frac{\dot{X}\,\cos(\theta_1 + \theta_2) + \dot{Y}\,\sin(\theta_1 + \theta_2)}{l_1\,\sin\theta_2} \tag{8.220}$$

$$\dot{\theta}_2 = \frac{\dot{X}\,(l_1\,c\theta_1 + l_2\,c\,(\theta_1 + \theta_2)) + \dot{Y}\,(l_1\,s\theta_1 + l_2\,s\,(\theta_1 + \theta_2))}{-l_1 l_2\,\sin\theta_2} \tag{8.221}$$

If we use the following forward velocity equations,

$$\begin{bmatrix} \dot{X} \\ \dot{Y} \end{bmatrix} = \begin{bmatrix} -l_1\sin\theta_1 & -l_2\sin(\theta_1 + \theta_2) \\ l_1\cos\theta_1 & l_2\cos(\theta_1 + \theta_2) \end{bmatrix} \begin{bmatrix} \dot{\theta}_1 \\ \dot{\theta}_1 + \dot{\theta}_2 \end{bmatrix} \tag{8.222}$$

then the inverse velocity kinematics of $2R$ manipulator may be written simpler.

$$\begin{bmatrix} \dot{\theta}_1 \\ \dot{\theta}_1 + \dot{\theta}_2 \end{bmatrix} = \begin{bmatrix} -l_1\sin\theta_1 & -l_2\sin(\theta_1 + \theta_2) \\ l_1\cos\theta_1 & l_2\cos(\theta_1 + \theta_2) \end{bmatrix}^{-1} \begin{bmatrix} \dot{X} \\ \dot{Y} \end{bmatrix}$$

$$= \frac{1}{l_1 l_2\sin\theta_2} \begin{bmatrix} l_2\cos(\theta_1 + \theta_2) & l_2\sin(\theta_1 + \theta_2) \\ -l_1\cos\theta_1 & -l_1\sin\theta_1 \end{bmatrix} \begin{bmatrix} \dot{X} \\ \dot{Y} \end{bmatrix} \tag{8.223}$$

$$\dot{\theta}_1 = \frac{\dot{X}\,\cos(\theta_1 + \theta_2) + \dot{Y}\,\sin(\theta_1 + \theta_2)}{l_1\,\sin\theta_2} \tag{8.224}$$

$$\dot{\theta}_2 = \frac{\dot{X}\,\cos\theta_1 + \dot{Y}\,\sin\theta_1}{-l_2\,\sin\theta_2} - \dot{\theta}_1 \tag{8.225}$$

As an example, consider a planar $2R$ manipulator with the following dimensions and a straight path of motion one unit of time, say one minute, $0 \le t \le 1$ min.

$$l_1 = l_2 = 1\,\text{m} \tag{8.226}$$

$$X(0) = 1\,\text{m} \quad Y(0) = 0 \quad X(1) = -1\,\text{m} \quad Y(1) = 0 \tag{8.227}$$

A septic time function for X will move the end-effector from $X(0) = 1$ to $X(1) = -1$ with zero velocity, zero acceleration, and zero jerk at both ends.

$$X = 1 - 70t^4 + 168t^5 - 140t^6 + 40t^7 \tag{8.228}$$

Employing inverse kinematics and inverse velocity analysis, Fig. 8.8 depicts the joint angle and joint angular velocity as function of time.

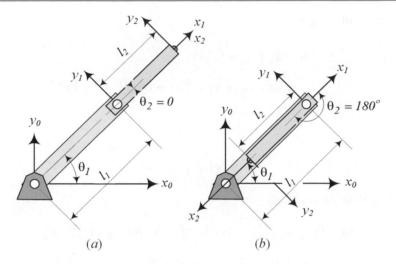

Fig. 8.9 Singular configurations of a $2R$ planar manipulator

Example 279 Singular configuration of a $2R$ manipulator. A $2R$ manipulator goes into singular configuration whenever the two arms are colinear. Here is the mathematical explanation.

Singularity of a $2R$ manipulator occurs when determinant of the Jacobian (8.217) is zero. From Example 278, we have

$$\mathbf{J} = \begin{bmatrix} -l_1 s\theta_1 - l_2 s\,(\theta_1 + \theta_2) & -l_2 s\,(\theta_1 + \theta_2) \\ l_1 c\theta_1 + l_2 c\,(\theta_1 + \theta_2) & l_2 c\,(\theta_1 + \theta_2) \end{bmatrix} \tag{8.229}$$

The determinant of $\mathbf{J}$ is

$$|\mathbf{J}| = l_1 l_2 \sin\theta_2 \tag{8.230}$$

Therefore, the singular configurations of the manipulator are

$$\theta_2 = 0 \qquad \theta_2 = 180 \text{ deg} \tag{8.231}$$

These are corresponding to the fully extended or fully contracted configurations of the links, as shown in Fig. 8.9a and b, respectively. At the singular configurations, the value of θ_1 is indeterminate and may have any real value. The two columns of the Jacobian matrix become parallel because Eq. (8.217) will be

$$\begin{bmatrix} \dot{X} \\ \dot{Y} \end{bmatrix} = 2l_1 \begin{bmatrix} -s\theta_1 \\ c\theta_1 \end{bmatrix} \dot{\theta}_1 + l_2 \begin{bmatrix} -s\theta_1 \\ c\theta_1 \end{bmatrix} \dot{\theta}_2$$

$$= \left(2l_1 \dot{\theta}_1 + l_2 \dot{\theta}_2 \right) \begin{bmatrix} -s\theta_1 \\ c\theta_1 \end{bmatrix} \tag{8.232}$$

In singular configurations, the end point can only move in the direction perpendicular to the arm links.

Example 280 ★ Analytic method for inverse velocity kinematics. The velocity vector of the tip point of a robot is a G-vector with components based on joint variables and speeds. Theoretically, we must be able to solve for joint speeds by vector calculus, although this is not a classical and recommended method.

Theoretically, we must be able to calculate the joint velocities from the equations describing the forward velocities; however, such a calculation is not easy in a general case.

As an example, consider a $2R$ planar manipulator shown in Fig. 8.3. The end point velocity of the $2R$ manipulator was expressed in Eq. (8.111) as:

$${}^0\dot{\mathbf{d}}_2 = \dot{\theta}_1 \, {}^0\hat{k}_0 \times \left(l_1 \, {}^0\hat{\imath}_1 + l_2 \, {}^0\hat{\imath}_2 \right) + \dot{\theta}_2 \, {}^0\hat{k}_1 \times l_2 \, {}^0\hat{\imath}_2 \tag{8.233}$$

A dot product of this equation with $^0\hat{i}_2$ gives

$$
\begin{aligned}
{}^0\dot{\mathbf{d}}_2 \cdot {}^0\hat{i}_2 &= \dot{\theta}_1 \left({}^0\hat{k}_0 \times l_1 {}^0\hat{i}_1 \right) \cdot {}^0\hat{i}_2 \\
&= l_1 \dot{\theta}_1 {}^0\hat{k}_0 \cdot \left({}^0\hat{i}_1 \times {}^0\hat{i}_2 \right) = l_1 \dot{\theta}_1 {}^0\hat{k}_0 \cdot {}^0\hat{i}_2 \sin\theta_2 \\
&= l_1 \dot{\theta}_1 \sin\theta_2
\end{aligned}
\tag{8.234}
$$

and, therefore,

$$
\dot{\theta}_1 = \frac{{}^0\dot{\mathbf{d}}_2 \cdot {}^0\hat{i}_2}{l_1 \sin\theta_2}
\tag{8.235}
$$

Now a dot product of (8.233) with $^0\hat{i}_1$ reduces to

$$
\begin{aligned}
{}^0\dot{\mathbf{d}}_2 \cdot {}^0\hat{i}_1 &= \dot{\theta}_1 \left({}^0\hat{k}_0 \times l_2 {}^0\hat{i}_2 \right) \cdot {}^0\hat{i}_1 + \dot{\theta}_2 \left({}^0\hat{k}_1 \times l_2 {}^0\hat{i}_2 \right) \cdot {}^0\hat{i}_1 \\
&= l_2 \left(\dot{\theta}_1 + \dot{\theta}_2 \right) {}^0\hat{k}_0 \cdot \left({}^0\hat{i}_2 \times {}^0\hat{i}_1 \right) \\
&= -l_2 \left(\dot{\theta}_1 + \dot{\theta}_2 \right) \sin\theta_2
\end{aligned}
\tag{8.236}
$$

and, therefore,

$$
\dot{\theta}_2 = -\dot{\theta}_1 - \frac{{}^0\dot{\mathbf{d}}_2 \cdot {}^0\hat{i}_1}{l_2 \sin\theta_2}
\tag{8.237}
$$

Therefore, we can determine the joint speeds $\dot{\theta}_1$ and $\dot{\theta}_2$ when the speeds of the end point $^0\dot{\mathbf{d}}_2 = \begin{bmatrix} \dot{X} & \dot{Y} \end{bmatrix}$ are given.

Example 281 ★ Inverse Jacobian matrix for a robot with spherical wrist. Spherical wrist is the most common industrial wrist. Its inverse velocity kinematics will be employed in many types of robotics velocity analysis.

The Jacobian matrix for an articulated robot with a spherical wrist is calculated in Example 273.

$$
\mathbf{J} = \begin{bmatrix} {}^0\hat{k}_0 \times {}^0_0\mathbf{d}_6 & {}^0\hat{k}_1 \times {}^0_1\mathbf{d}_6 & {}^0\hat{k}_2 \times {}^0_2\mathbf{d}_6 & 0 & 0 & 0 \\ {}^0\hat{k}_0 & {}^0\hat{k}_1 & {}^0\hat{k}_2 & {}^0\hat{k}_3 & {}^0\hat{k}_4 & {}^0\hat{k}_5 \end{bmatrix}
\tag{8.238}
$$

The upper right 3×3 submatrix of $\mathbf{J}$ is zero. This is a result of having spherical wrist structure and having the last three position vectors to be zero.

Let us split the Jacobian matrix into four 3×3 submatrices and write it as:

$$
\mathbf{J} = \begin{bmatrix} A & B \\ C & D \end{bmatrix} = \begin{bmatrix} A & 0 \\ C & D \end{bmatrix}
\tag{8.239}
$$

where

$$
[A] = \begin{bmatrix} {}^0\hat{k}_0 \times {}^0_0\mathbf{d}_6 & {}^0\hat{k}_1 \times {}^0_1\mathbf{d}_6 & {}^0\hat{k}_2 \times {}^0_2\mathbf{d}_6 \end{bmatrix}
\tag{8.240}
$$

$$
[C] = \begin{bmatrix} {}^0\hat{k}_0 & {}^0\hat{k}_1 & {}^0\hat{k}_2 \end{bmatrix}
\tag{8.241}
$$

$$
[D] = \begin{bmatrix} {}^0\hat{k}_3 & {}^0\hat{k}_4 & {}^0\hat{k}_5 \end{bmatrix}
\tag{8.242}
$$

Inversion of such a Jacobian is simpler if we take advantage of $B = 0$. The forward velocity kinematics of the robot can be written as:

$$
\dot{\mathbf{X}} = \mathbf{J}\dot{\mathbf{q}}
\tag{8.243}
$$

$$\begin{bmatrix} {}^{0}\dot{\mathbf{d}}_2 \\ {}_0\boldsymbol{\omega}_2 \end{bmatrix} = \begin{bmatrix} A & 0 \\ C & D \end{bmatrix} \begin{bmatrix} \dot{\theta}_1 \\ \dot{\theta}_2 \\ \dot{\theta}_3 \\ \dot{\theta}_4 \\ \dot{\theta}_5 \\ \dot{\theta}_6 \end{bmatrix} \tag{8.244}$$

The upper half of the equation is

$$ {}^{0}\dot{\mathbf{d}}_2 = [A] \begin{bmatrix} \dot{\theta}_1 \\ \dot{\theta}_2 \\ \dot{\theta}_3 \end{bmatrix} \tag{8.245}$$

which can be inverted as:

$$\begin{bmatrix} \dot{\theta}_1 \\ \dot{\theta}_2 \\ \dot{\theta}_3 \end{bmatrix} = A^{-1}\,{}^{0}\dot{\mathbf{d}}_2 \tag{8.246}$$

The lower half of the equation is

$$ {}_0\boldsymbol{\omega}_2 = \begin{bmatrix} C & D \end{bmatrix} \begin{bmatrix} \dot{\theta}_1 \\ \dot{\theta}_2 \\ \dot{\theta}_3 \\ \dot{\theta}_4 \\ \dot{\theta}_5 \\ \dot{\theta}_6 \end{bmatrix} = [C] \begin{bmatrix} \dot{\theta}_1 \\ \dot{\theta}_2 \\ \dot{\theta}_3 \end{bmatrix} + [D] \begin{bmatrix} \dot{\theta}_4 \\ \dot{\theta}_5 \\ \dot{\theta}_6 \end{bmatrix} \tag{8.247}$$

and, therefore,

$$\begin{aligned} \begin{bmatrix} \dot{\theta}_4 \\ \dot{\theta}_5 \\ \dot{\theta}_6 \end{bmatrix} &= D^{-1} \left({}_0\boldsymbol{\omega}_2 - [C] \begin{bmatrix} \dot{\theta}_1 \\ \dot{\theta}_2 \\ \dot{\theta}_3 \end{bmatrix} \right) \\ &= D^{-1} \left({}_0\boldsymbol{\omega}_2 - [C]\, A^{-1}\,{}^{0}\dot{\mathbf{d}}_2 \right) \end{aligned} \tag{8.248}$$

Example 282 Inverse velocity of an articulated manipulator. Inverse velocity kinematics of articulated manipulator will be calculated by inverse Jacobian method. This is a necessary step in completing study of this applied industrial manipulator.

The end point of the articulated manipulator of Fig. 8.4 is found in Exercise 271.

$$ {}^{0}\mathbf{r}_P = \begin{bmatrix} X_P \\ Y_P \\ Z_P \\ 1 \end{bmatrix} = \begin{bmatrix} \cos\theta_1\,(l_2\cos\theta_2 + l_3\sin(\theta_2+\theta_3)) \\ \sin\theta_1\,(l_2\cos\theta_2 + l_3\sin(\theta_2+\theta_3)) \\ l_1 - l_3\cos(\theta_2+\theta_3) + l_2\sin\theta_2 \\ 1 \end{bmatrix} \tag{8.249}$$

Using the components of ${}^{0}\mathbf{r}_P$, we calculated the Jacobian matrix of the manipulator by differentiating

$$ \mathbf{J} = \begin{bmatrix} \dfrac{\partial X_P}{\partial\theta_1} & \dfrac{\partial X_P}{\partial\theta_2} & \dfrac{\partial X_P}{\partial\theta_3} \\[2mm] \dfrac{\partial Y_P}{\partial\theta_1} & \dfrac{\partial Y_P}{\partial\theta_2} & \dfrac{\partial Y_P}{\partial\theta_3} \\[2mm] \dfrac{\partial Z_P}{\partial\theta_1} & \dfrac{\partial Z_P}{\partial\theta_2} & \dfrac{\partial Z_P}{\partial\theta_3} \end{bmatrix} \tag{8.250}$$

to solve the forward kinematics of the manipulator.

$$\begin{bmatrix} \dot{X}_P \\ \dot{Y}_P \\ \dot{Z}_P \end{bmatrix} = \mathbf{J} \begin{bmatrix} \dot{\theta}_1 \\ \dot{\theta}_2 \\ \dot{\theta}_3 \end{bmatrix} \tag{8.251}$$

To solve the inverse velocity kinematics of the manipulator, we need to calculate $\mathbf{J}^{-1}$.

$$\mathbf{J}^{-1} = \begin{bmatrix} \dfrac{\partial X_P}{\partial \theta_1} & \dfrac{\partial X_P}{\partial \theta_2} & \dfrac{\partial X_P}{\partial \theta_3} \\ \dfrac{\partial Y_P}{\partial \theta_1} & \dfrac{\partial Y_P}{\partial \theta_2} & \dfrac{\partial Y_P}{\partial \theta_3} \\ \dfrac{\partial Z_P}{\partial \theta_1} & \dfrac{\partial Z_P}{\partial \theta_2} & \dfrac{\partial Z_P}{\partial \theta_3} \end{bmatrix}^{-1} = \begin{bmatrix} a_{11} & a_{12} & a_{13} \\ a_{21} & a_{22} & a_{23} \\ a_{31} & a_{32} & a_{33} \end{bmatrix} \tag{8.252}$$

$$a_{11} = -\frac{\sin\theta_1}{l_3 \sin(\theta_2+\theta_3) + l_2 \cos\theta_2}$$

$$a_{21} = -\frac{1}{l_2}(\sin(\theta_2+\theta_3))\frac{\cos\theta_1}{\cos\theta_3}$$

$$a_{31} = \frac{1}{l_2 l_3}\frac{\cos\theta_1}{\cos\theta_3}(l_3 \sin(\theta_2+\theta_3) + l_2 \cos\theta_2) \tag{8.253}$$

$$a_{12} = \frac{\cos\theta_1}{l_3 \sin(\theta_2+\theta_3) + l_2 \cos\theta_2}$$

$$a_{22} = -\frac{1}{l_2}\frac{\sin(\theta_2+\theta_3)}{\cos\theta_3}\sin\theta_1$$

$$a_{32} = \frac{\sin\theta_1}{l_2 l_3 \cos\theta_3}(l_3 \sin(\theta_2+\theta_3) + l_2 \cos\theta_2) \tag{8.254}$$

$$a_{13} = 0$$

$$a_{23} = \frac{1}{l_2}\frac{\cos(\theta_2+\theta_3)}{\cos\theta_3}$$

$$a_{33} = -\frac{1}{l_2 l_3 \cos\theta_3}(l_3 \cos(\theta_2+\theta_3) - l_2 \sin\theta_2) \tag{8.255}$$

Therefore, the joint speeds of the manipulator $\dot\theta_1, \dot\theta_2, \dot\theta_3$ are

$$\begin{bmatrix} \dot\theta_1 \\ \dot\theta_2 \\ \dot\theta_3 \end{bmatrix} = \mathbf{J}^{-1}\begin{bmatrix} \dot X_P \\ \dot Y_P \\ \dot Z_P \end{bmatrix} \tag{8.256}$$

8.5 ★ Linear Algebraic Equations

By increasing the number of links, the analytic calculation in robotics becomes a tedious task and numerical calculations are needed. We review the most frequent needed numerical analysis in robotics.

In robotic analysis, there exist problems and situations, such as inverse kinematics, and inverse velocity kinematics, that we need to solve a set of coupled linear or nonlinear algebraic equations. Here we review a few applied numerical methods of solving algebraic equations to be used in robotics.

Consider a system of n linear algebraic equations with real constant coefficients,

$$a_{11}x_1 + a_{12}x_2 + \cdots + a_{1n}x_n = b_1$$
$$a_{21}x_1 + a_{22}x_2 + \cdots + a_{2n}x_n = b_2$$
$$\cdots = \cdots$$
$$a_{n1}x_1 + a_{n2}x_2 + \cdots + a_{nn}x_n = b_n \tag{8.257}$$

which can also be written in a matrix form.

$$[A]\mathbf{x} = \mathbf{b} \tag{8.258}$$

$[A]$ is the $n \times n$ coefficient matrix, $\mathbf{x}$ is the $n \times 1$ unknown vector, and $\mathbf{b}$ is the $n \times 1$ known vector. There are numerous methods for solving this set of equations. Among the most efficient methods is the *LU factorization method*.

For every non-singular matrix $[A]$, there exists an upper triangular matrix $[U]$ with nonzero diagonal elements and a lower triangular matrix $[L]$ with unit diagonal elements.

$$[A] = [L][U] \tag{8.259}$$

$$[A] = \begin{bmatrix} a_{11} & a_{12} & \cdots & a_{1n} \\ a_{21} & a_{22} & \cdots & a_{2n} \\ \cdots & \cdots & \cdots & \cdots \\ a_{n1} & a_{n2} & \cdots & a_{nn} \end{bmatrix} \tag{8.260}$$

$$[L] = \begin{bmatrix} 1 & 0 & \cdots & 0 \\ l_{21} & 1 & \cdots & 0 \\ \cdots & \cdots & \cdots & \cdots \\ l_{n1} & l_{n2} & \cdots & 1 \end{bmatrix} \tag{8.261}$$

$$[U] = \begin{bmatrix} u_{11} & u_{12} & \cdots & u_{1n} \\ 0 & u_{22} & \cdots & u_{2n} \\ \cdots & \cdots & \cdots & \cdots \\ 0 & 0 & \cdots & u_{nn} \end{bmatrix} \tag{8.262}$$

The process of factoring $[A]$ into $[L][U]$ is called *LU* factorization. Once the $[L]$ and $[U]$ matrices are obtained, the set of equations will get a new form.

$$[L][U]\mathbf{x} = \mathbf{b} \tag{8.263}$$

The set of equations can be transforming into

$$[L]\mathbf{y} = \mathbf{b} \tag{8.264}$$

and

$$[U]\mathbf{x} = \mathbf{y} \tag{8.265}$$

Equations (8.264) and (8.265) are both a triangular set of equations and their solutions are easy to obtain by forward and backward substitution.

$$y_i = b_i - \sum_{j=1}^{i-1} y_j l_{ij} \tag{8.266}$$

$$x_i = \frac{1}{u_{ii}} \left(y_i - \sum_{j=i+1}^{n} x_j u_{ij} \right) \tag{8.267}$$

Proof To show how $[A]$ can be transformed into $[L][U]$, let us consider a 4×4 matrix.

$$\begin{bmatrix} a_{11} & a_{12} & a_{13} & a_{14} \\ a_{21} & a_{22} & a_{23} & a_{24} \\ a_{31} & a_{32} & a_{33} & a_{34} \\ a_{41} & a_{42} & a_{43} & a_{44} \end{bmatrix} = \begin{bmatrix} 1 & 0 & 0 & 0 \\ l_{21} & 1 & 0 & 0 \\ l_{31} & l_{32} & 1 & 0 \\ l_{41} & l_{42} & l_{43} & 1 \end{bmatrix} \begin{bmatrix} u_{11} & u_{12} & u_{13} & u_{14} \\ 0 & u_{22} & u_{23} & u_{24} \\ 0 & 0 & u_{33} & u_{34} \\ 0 & 0 & 0 & u_{44} \end{bmatrix} \tag{8.268}$$

Employing a dummy matrix $[B]$, we may combine the elements of $[L]$ and $[U]$ to make $[B]$.

$$[B] = \begin{bmatrix} u_{11} & u_{12} & u_{13} & u_{14} \\ l_{21} & u_{22} & u_{23} & u_{24} \\ l_{31} & l_{32} & u_{33} & u_{34} \\ l_{41} & l_{42} & l_{43} & u_{44} \end{bmatrix} \tag{8.269}$$

The elements of $[B]$ will be calculated one by one, in the following order:

$$[B] = \begin{bmatrix} (1) & (2) & (3) & (4) \\ (5) & (8) & (9) & (10) \\ (6) & (11) & (13) & (14) \\ (7) & (12) & (15) & (16) \end{bmatrix} \tag{8.270}$$

The process for generating a matrix $[B]$, associated to an $n \times n$ matrix $[A]$, is performed in $n - 1$ iterations. After $i - 1$ iterations, the matrix is in the following form:

$$[B] = \begin{bmatrix} u_{1,1} & u_{1,2} & \cdots & u_{1,i-1} & \cdots & \cdots & u_{1,n} \\ l_{2,1} & u_{2,2} & \cdots & \cdots & \cdots & \cdots & u_{2,n} \\ \cdots & \cdots & \cdots & \cdots & \cdots & & \cdots \\ \cdots & \cdots & \cdots & \cdots & \cdots & \cdots & u_{i-1,n} \\ \cdots & \cdots & \cdots & \cdots & \lceil & & \rceil \\ \cdots & \cdots & \cdots & \cdots & | & D_i & | \\ l_{n,1} & l_{n,2} & \cdots & l_{n,i-1} & \lfloor & & \rfloor \end{bmatrix} \tag{8.271}$$

The unprocessed $(n - i + 1) \times (n - i + 1)$ submatrix in the lower right corner is denoted by $[D_i]$ and has the same elements as $[A]$. In the ith step, the LU factorization method converts $[D_i]$

$$[D_i] = \begin{bmatrix} d_{ii} & \mathbf{r}_i^T \\ \mathbf{s}_i & [H_{i+1}] \end{bmatrix} \tag{8.272}$$

to a new form.

$$[D_i] = \begin{bmatrix} u_{ii} & \mathbf{u}_i^T \\ \mathbf{l}_i & [D_{i+1}] \end{bmatrix} \tag{8.273}$$

Direct multiplication shows that

$$u_{11} = a_{11} \qquad u_{12} = a_{12} \qquad u_{13} = a_{13} \qquad u_{14} = a_{14} \tag{8.274}$$

$$l_{21} = \frac{a_{21}}{u_{11}} \qquad l_{31} = \frac{a_{31}}{u_{11}} \qquad l_{41} = \frac{a_{41}}{u_{11}} \tag{8.275}$$

$$u_{22} = a_{22} - l_{21}u_{12} \qquad u_{23} = a_{23} - l_{21}u_{13} \qquad u_{24} = a_{24} - l_{21}u_{14} \tag{8.276}$$

$$l_{32} = \frac{a_{32} - l_{31}u_{12}}{u_{22}} \qquad l_{42} = \frac{a_{42} - l_{41}u_{12}}{u_{22}} \tag{8.277}$$

$$u_{33} = a_{33} - (l_{31}u_{13} + l_{32}u_{23}) \qquad u_{34} = a_{34} - (l_{31}u_{14} + l_{32}u_{24}) \tag{8.278}$$

$$l_{43} = \frac{a_{43} - (l_{41}u_{13} + l_{42}u_{23})}{u_{33}} \tag{8.279}$$

$$u_{44} = a_{44} - (l_{41}u_{14} + l_{42}u_{24} + l_{43}u_{34}) \tag{8.280}$$

Therefore, the general formula for getting elements of $[L]$ and $[U]$ corresponding to an $n \times n$ coefficients matrix $[A]$ can be written as

$$u_{ij} = a_{ij} - \sum_{k=1}^{i-1} l_{ik}u_{kj} \qquad i \le j \qquad j = 1, \cdots n \tag{8.281}$$

$$l_{ij} = \frac{a_{ij} - \sum_{k=1}^{j-1} l_{ik}u_{kj}}{u_{jj}} \qquad j \le i \qquad i = 1, \cdots n \tag{8.282}$$

For $i = 1$, the rule for u reduces to

$$u_{1j} = a_{1j} \tag{8.283}$$

and for $j = 1$, the rule for l reduces to

$$l_{i1} = \frac{a_{i1}}{u_{11}} \tag{8.284}$$

The calculation of element (k) of the dummy matrix $[B]$, which is an element of $[L]$ or $[U]$, involves only the elements of $[A]$ in the same position and some previously calculated elements of $[B]$.

The LU factorization technique can be set up in an algorithm for easier numerical calculations.

Algorithm 8.1 *LU factorization technique for an $n \times n$ matrix $[A]$.*

1. *Set the initial counter $i = 1$.*
2. *Set $[D_1] = [A]$.*
3. *Calculate $[D_{i+1}]$ from $[D_i]$ according to*

$$u_{ii} = d_{ii} \tag{8.285}$$

$$\mathbf{u}_i^T = \mathbf{r}_i^T \tag{8.286}$$

$$\mathbf{l}_i = \frac{1}{u_{ii}} \mathbf{s}_i \tag{8.287}$$

$$[D_{i+1}] = [H_{i+1}] - \mathbf{l}_i \mathbf{u}_i^T \tag{8.288}$$

4. *Set $i = i + 1$. If $i = n$, then LU factorization is completed. Otherwise return to step 3.*

After decomposing the matrix $[A]$ into the matrices $[L]$ and $[U]$, the set of equations can be solved based on the following algorithm.

Algorithm 8.2 *LU solution technique.*

1. *Calculate $\mathbf{y}$ from $[L]\mathbf{y} = \mathbf{b}$ by*

$$y_1 = b_1$$
$$y_2 = b_2 - y_1 l_{21}$$
$$y_3 = b_3 - y_1 l_{31} - y_2 l_{32}$$
$$\cdots$$
$$y_i = b_i - \sum_{j=1}^{i-1} y_j l_{ij} \tag{8.289}$$

2. *Calculate $\mathbf{x}$ from $[U]\mathbf{x} = \mathbf{y}$ by*

$$x_n = \frac{y_n}{u_{n,n}}$$
$$x_{n-1} = \frac{y_{n-1} - x_n u_{n-1,n}}{u_{n-1,n-1}}$$
$$\cdots$$
$$x_i = \frac{1}{u_{ii}} \left(y_i - \sum_{j=i+1}^{n} x_j u_{ij} \right) \tag{8.290}$$

■

Example 283 Solution of a set of four equations. Assuming a set of equations are given, here we go through *LU* factorization algorithms to show how it works.

Consider a set of four linear algebraic equations.

$$[A]\mathbf{x} = \mathbf{b} \tag{8.291}$$

$$[A] = \begin{bmatrix} 2 & 1 & 3 & -3 \\ 1 & 0 & -1 & -2 \\ 0 & 2 & 2 & 1 \\ 3 & 1 & 0 & -2 \end{bmatrix} \qquad \mathbf{b} = \begin{bmatrix} 1 \\ 2 \\ 0 \\ -2 \end{bmatrix} \tag{8.292}$$

Following the *LU* factorization algorithm, we first set

$$i = 1 \qquad [D_1] = [A] \tag{8.293}$$

to find

$$d_{11} = 2 \qquad \mathbf{r}_1^T = \begin{bmatrix} 1 & 3 & -3 \end{bmatrix} \tag{8.294}$$

$$\mathbf{s}_1 = \begin{bmatrix} 1 \\ 0 \\ 3 \end{bmatrix} \qquad [H_2] = \begin{bmatrix} 0 & -1 & -2 \\ 2 & 2 & 1 \\ 1 & 0 & -2 \end{bmatrix} \tag{8.295}$$

and calculate $[D_2]$.

$$u_{11} = d_{11} = 2 \tag{8.296}$$

$$\mathbf{u}_1^T = \mathbf{r}_1^T = \begin{bmatrix} 1 & 3 & -3 \end{bmatrix} \qquad \mathbf{l}_1 = \frac{1}{u_{11}}\mathbf{s}_1 = \begin{bmatrix} \frac{1}{2} \\ 0 \\ \frac{3}{2} \end{bmatrix} \tag{8.297}$$

$$[D_2] = [H_2] - \mathbf{l}_1 \mathbf{u}_1^T = \begin{bmatrix} -\frac{1}{2} & -\frac{5}{2} & -\frac{1}{2} \\ 2 & 2 & 1 \\ -\frac{1}{2} & -\frac{9}{2} & \frac{5}{2} \end{bmatrix} \tag{8.298}$$

In the second step we have

$$i = 2 \tag{8.299}$$

$$d_{22} = -\frac{1}{2} \qquad \mathbf{r}_2^T = \begin{bmatrix} -\frac{5}{2} & -\frac{1}{2} \end{bmatrix} \tag{8.300}$$

$$\mathbf{s}_2 = \begin{bmatrix} 2 \\ -\frac{1}{2} \end{bmatrix} \qquad [H_3] = \begin{bmatrix} 2 & 1 \\ -\frac{9}{2} & \frac{5}{2} \end{bmatrix} \tag{8.301}$$

and calculate $[D_3]$.

$$u_{22} = d_{22} = -\frac{1}{2} \tag{8.302}$$

$$\mathbf{u}_2^T = \mathbf{r}_2^T = \begin{bmatrix} -\frac{5}{2} & -\frac{1}{2} \end{bmatrix} \qquad \mathbf{l}_2 = \frac{1}{u_{22}}\mathbf{s}_2 = \begin{bmatrix} -4 \\ 1 \end{bmatrix} \tag{8.303}$$

$$[D_3] = [H_3] - \mathbf{l}_2 \mathbf{u}_2^T = \begin{bmatrix} -8 & -1 \\ -2 & 3 \end{bmatrix} \tag{8.304}$$

In the third step we set

$$i = 3 \tag{8.305}$$

and find

$$d_{33} = -8 \qquad \mathbf{r}_3^T = \begin{bmatrix} -1 \end{bmatrix} \tag{8.306}$$

$$\mathbf{s}_3 = \begin{bmatrix} -2 \end{bmatrix} \qquad [H_4] = \begin{bmatrix} 3 \end{bmatrix} \tag{8.307}$$

and, therefore, $[D_4]$ is calculated.

$$u_{33} = d_{33} = -8 \tag{8.308}$$

$$\mathbf{u}_3^T = \mathbf{r}_3^T = [-1] \qquad \mathbf{l}_3 = \frac{1}{u_{33}} \mathbf{s}_3 = \left[\frac{1}{4}\right] \tag{8.309}$$

$$[D_4] = [H_4] - \mathbf{l}_3 \, \mathbf{u}_3^T = \left[\frac{13}{4}\right] \tag{8.310}$$

After these calculations, the matrices $[B]$, $[L]$, and $[U]$ are found as following.

$$[B] = \begin{bmatrix} 2 & 1 & 3 & -3 \\ \frac{1}{2} & -\frac{1}{2} & -\frac{5}{2} & -\frac{1}{2} \\ 0 & 4 & -8 & -1 \\ \frac{3}{2} & 1 & \frac{1}{4} & \frac{13}{4} \end{bmatrix} \tag{8.311}$$

$$[L] = \begin{bmatrix} 1 & 0 & 0 & 0 \\ \frac{1}{2} & 1 & 0 & 0 \\ 0 & -4 & 1 & 0 \\ \frac{3}{2} & 1 & \frac{1}{4} & 1 \end{bmatrix} \tag{8.312}$$

$$[U] = \begin{bmatrix} 2 & 1 & 3 & -3 \\ 0 & -\frac{1}{2} & -\frac{5}{2} & -\frac{1}{2} \\ 0 & 0 & -8 & -1 \\ 0 & 0 & 0 & \frac{13}{4} \end{bmatrix} \tag{8.313}$$

Now a vector $\mathbf{y}$ can be found to satisfy

$$[L]\mathbf{y} = \mathbf{b} \qquad \mathbf{y} = \begin{bmatrix} 1 \\ 3/2 \\ 6 \\ -13/2 \end{bmatrix} \tag{8.314}$$

and finally the unknown vector $\mathbf{x}$ should be found.

$$[U]\mathbf{x} = \mathbf{y} \qquad \mathbf{x} = \begin{bmatrix} -5/2 \\ 3/2 \\ -1/2 \\ -2 \end{bmatrix} \tag{8.315}$$

Example 284 LU factorization with **pivoting**. Pivoting is a method to minimize numerical errors in *LU* factorization. It is reordering equations such that the maximum numbers appear on the main diagonal of the coefficient matrix.

In the process of *LU* factorization, the situation of $u_{ii} = 0$ generates a division by zero, which must be avoided. In this situation, pivoting must be applied. By pivoting, we change the order of equations to have a coefficient matrix with the largest elements, in absolute value, as diagonal elements. The largest element is called the **pivot element**.

As an example, consider the following set of equations:

$$[A]\mathbf{x} = \mathbf{b} \tag{8.316}$$

$$\begin{bmatrix} 2 & 1 & 3 & -3 \\ 1 & 0 & -1 & 2 \\ 0 & 2 & 0 & 1 \\ 3 & 1 & 4 & -2 \end{bmatrix} \begin{bmatrix} x_1 \\ x_2 \\ x_3 \\ x_4 \end{bmatrix} = \begin{bmatrix} 1 \\ 2 \\ 0 \\ -2 \end{bmatrix} \tag{8.317}$$

We move rows such that the largest first element goes to d_{11} by interchanging row 1 with 4, and then column 1 with 3.

$$
\begin{bmatrix} 3 & 1 & 4 & -2 \\ 1 & 0 & -1 & 2 \\ 0 & 2 & 0 & 1 \\ 2 & 1 & 3 & -3 \end{bmatrix} \begin{bmatrix} x_1 \\ x_2 \\ x_3 \\ x_4 \end{bmatrix} = \begin{bmatrix} -2 \\ 2 \\ 0 \\ 1 \end{bmatrix}
\tag{8.318}
$$

$$
\begin{bmatrix} 4 & 1 & 3 & -2 \\ -1 & 0 & 1 & 2 \\ 0 & 2 & 0 & 1 \\ 3 & 1 & 2 & -3 \end{bmatrix} \begin{bmatrix} x_3 \\ x_2 \\ x_1 \\ x_4 \end{bmatrix} = \begin{bmatrix} -2 \\ 2 \\ 0 \\ 1 \end{bmatrix}
\tag{8.319}
$$

Then the largest element in the 3×3 submatrix in the lower right corner will move to d_{22}

$$
\begin{bmatrix} 4 & -2 & 3 & 1 \\ -1 & 2 & 1 & 0 \\ 0 & 1 & 0 & 2 \\ 3 & -3 & 2 & 1 \end{bmatrix} \begin{bmatrix} x_3 \\ x_4 \\ x_1 \\ x_2 \end{bmatrix} = \begin{bmatrix} -2 \\ 2 \\ 0 \\ 1 \end{bmatrix}
\tag{8.320}
$$

$$
\begin{bmatrix} 4 & -2 & 3 & 1 \\ 3 & -3 & 2 & 1 \\ 0 & 1 & 0 & 2 \\ -1 & 2 & 1 & 0 \end{bmatrix} \begin{bmatrix} x_3 \\ x_4 \\ x_1 \\ x_2 \end{bmatrix} = \begin{bmatrix} -2 \\ 1 \\ 0 \\ 2 \end{bmatrix}
\tag{8.321}
$$

and finally the largest element in the 2×2 in the lower right corner will move to d_{33}.

$$
\begin{bmatrix} 4 & -2 & 1 & 3 \\ 3 & -3 & 1 & 2 \\ 0 & 1 & 2 & 0 \\ -1 & 2 & 0 & 1 \end{bmatrix} \begin{bmatrix} x_3 \\ x_4 \\ x_2 \\ x_1 \end{bmatrix} = \begin{bmatrix} -2 \\ 1 \\ 0 \\ 2 \end{bmatrix}
\tag{8.322}
$$

To apply the LU factorization and LU solution algorithm, we define a new set of equations.

$$
\left[A' \right] \mathbf{x}' = \mathbf{b}'
\tag{8.323}
$$

$$
\begin{bmatrix} 4 & -2 & 1 & 3 \\ 3 & -3 & 1 & 2 \\ 0 & 1 & 2 & 0 \\ -1 & 2 & 0 & 1 \end{bmatrix} \begin{bmatrix} x_1' \\ x_2' \\ x_3' \\ x_4' \end{bmatrix} = \begin{bmatrix} -2 \\ 1 \\ 0 \\ 2 \end{bmatrix}
\tag{8.324}
$$

Based on the LU factorization algorithm, in the first step we set

$$
i = 1
\tag{8.325}
$$

and find $[D_1]$

$$
[D_1] = \left[A' \right] \qquad d_{11} = 4 \qquad \mathbf{r}_1^T = \begin{bmatrix} -2 & 1 & 3 \end{bmatrix}
\tag{8.326}
$$

$$
\mathbf{s}_1 = \begin{bmatrix} 3 \\ 0 \\ -1 \end{bmatrix} \qquad [H_2] = \begin{bmatrix} -3 & 1 & 2 \\ 1 & 2 & 0 \\ 2 & 0 & 1 \end{bmatrix}
\tag{8.327}
$$

to calculate $[D_2]$.

$$
u_{11} = d_{11} = 4 \qquad \mathbf{u}_1^T = \mathbf{r}_1^T = \begin{bmatrix} -2 & 1 & 3 \end{bmatrix}
\tag{8.328}
$$

$$\mathbf{l}_1 = \frac{1}{u_{11}} \mathbf{s}_1 = \begin{bmatrix} \frac{3}{4} \\ 0 \\ -\frac{1}{4} \end{bmatrix} \tag{8.329}$$

$$[D_2] = [H_2] - \mathbf{l}_1 \mathbf{u}_1^T = \begin{bmatrix} -\frac{3}{2} & \frac{1}{4} & -\frac{1}{4} \\ 1 & 2 & 0 \\ \frac{3}{2} & \frac{1}{4} & \frac{7}{4} \end{bmatrix} \tag{8.330}$$

For the second step, we have

$$i = 2 \tag{8.331}$$

and

$$d_{22} = -\frac{3}{2} \qquad \mathbf{r}_2^T = \begin{bmatrix} \frac{1}{4} & -\frac{1}{4} \end{bmatrix} \tag{8.332}$$

$$\mathbf{s}_2 = \begin{bmatrix} 1 \\ \frac{3}{2} \end{bmatrix} \qquad [H_3] = \begin{bmatrix} 2 & 0 \\ \frac{1}{4} & \frac{7}{4} \end{bmatrix} \tag{8.333}$$

and then we calculate $[D_3]$.

$$u_{22} = d_{22} = -\frac{3}{2} \tag{8.334}$$

$$\mathbf{u}_2^T = \mathbf{r}_2^T = \begin{bmatrix} \frac{1}{4} & -\frac{1}{4} \end{bmatrix} \qquad \mathbf{l}_2 = \frac{1}{u_{22}} \mathbf{s}_2 = \begin{bmatrix} -\frac{2}{3} \\ -1 \end{bmatrix} \tag{8.335}$$

$$[D_3] = [H_3] - \mathbf{l}_2 \mathbf{u}_2^T = \begin{bmatrix} \frac{13}{6} & -\frac{1}{6} \\ \frac{1}{2} & \frac{3}{2} \end{bmatrix} \tag{8.336}$$

In the third step, we set

$$i = 3 \tag{8.337}$$

and find

$$d_{33} = \frac{13}{6} \qquad \mathbf{r}_3^T = \begin{bmatrix} -\frac{1}{6} \end{bmatrix} \tag{8.338}$$

$$\mathbf{s}_3 = \begin{bmatrix} \frac{1}{2} \end{bmatrix} \qquad [H_4] = \begin{bmatrix} \frac{3}{2} \end{bmatrix} \tag{8.339}$$

and calculate $[D_4]$.

$$u_{33} = d_{33} = \frac{13}{6} \tag{8.340}$$

$$\mathbf{u}_3^T = \mathbf{r}_3^T = \begin{bmatrix} -\frac{1}{6} \end{bmatrix} \qquad \mathbf{l}_3 = \frac{1}{u_{33}} \mathbf{s}_3 = \begin{bmatrix} \frac{3}{13} \end{bmatrix} \tag{8.341}$$

$$[D_4] = [H_4] - \mathbf{l}_3 \mathbf{u}_3^T = \begin{bmatrix} \frac{20}{13} \end{bmatrix} \tag{8.342}$$

Therefore, the matrices $[L]$ and $[U]$ are

$$[L] = \begin{bmatrix} 1 & 0 & 0 & 0 \\ \frac{3}{4} & 1 & 0 & 0 \\ 0 & -\frac{2}{3} & 1 & 0 \\ -\frac{1}{4} & -1 & \frac{3}{13} & 1 \end{bmatrix} \tag{8.343}$$

$$[U] = \begin{bmatrix} 4 & -2 & 1 & 3 \\ 0 & -\frac{3}{2} & \frac{1}{4} & -\frac{1}{4} \\ 0 & 0 & \frac{13}{6} & -\frac{1}{6} \\ 0 & 0 & 0 & \frac{20}{13} \end{bmatrix} \tag{8.344}$$

and now we can find the vector **y**.

$$[L]\,\mathbf{y} = \mathbf{b}' \tag{8.345}$$

$$\mathbf{y} = \begin{bmatrix} -2 \\ 5/2 \\ 5/3 \\ 47/13 \end{bmatrix} \tag{8.346}$$

The unknown vector **x**′ can then be calculated,

$$[U]\,\mathbf{x}' = \mathbf{y} \tag{8.347}$$

$$\mathbf{x}' = \begin{bmatrix} -69/29 \\ -19/10 \\ 19/20 \\ 47/20 \end{bmatrix} = \begin{bmatrix} x_3 \\ x_4 \\ x_2 \\ x_1 \end{bmatrix} \tag{8.348}$$

and, therefore, the unknowns will be found.

$$\mathbf{x} = \begin{bmatrix} 47/20 \\ 19/20 \\ -69/20 \\ -19/10 \end{bmatrix} \tag{8.349}$$

Example 285 ★ Uniqueness of solution. Here we study the conditions that a set of linear equations provide unique or multiple situations.

Consider a set of n linear equations, $[A]\,\mathbf{x} = \mathbf{b}$. If $[A]$ is square and non-singular, then there exists a unique solution $\mathbf{x} = [A]^{-1}\,\mathbf{b}$. However, if the linear system of equations involves n variables and m equations

$$a_{11}x_1 + a_{12}x_2 + \cdots + a_{1n}x_n = b_1$$
$$a_{21}x_1 + a_{22}x_2 + \cdots + a_{2n}x_n = b_2$$
$$\cdots \quad = \cdots$$
$$a_{m1}x_1 + a_{m2}x_2 + \cdots + a_{mn}x_n = b_m \tag{8.350}$$

then three classes of solutions are possible.

1. A unique solution exists and the system is called consistent.
2. No solution exists and the system is called inconsistent.
3. Multiple solutions exist and the system is called undetermined.

Example 286 ★ Ill conditioned and well conditioned. Stability of solution of a set of equation depends on how much they change when there is a small change in coefficient matrix or the known vector. Here is their classifications and conditions.

A system of equations, $[A]\,\mathbf{x} = \mathbf{b}$, is considered to be **well conditioned** if a small change in $[A]$ or **b** results in a small change in the solution vector **x**. A system of equations, $[A]\,\mathbf{x} = \mathbf{b}$, is considered to be **ill conditioned** if a small change in $[A]$ or **b** results in a big change in the solution vector **x**. The system of equations is ill conditioned when $[A]$ has rows or columns so nearly dependent on each other.

Consider the following set of equations:

$$[A]\,\mathbf{x} = \mathbf{b} \tag{8.351}$$

$$\begin{bmatrix} 2 & 3.99 \\ 1 & 2 \end{bmatrix} \begin{bmatrix} x_1 \\ x_2 \end{bmatrix} = \begin{bmatrix} 1.99 \\ 1 \end{bmatrix} \tag{8.352}$$

The solution of this set of equations is

$$\begin{bmatrix} x_1 \\ x_2 \end{bmatrix} = \begin{bmatrix} -1.0 \\ 1.0 \end{bmatrix} \tag{8.353}$$

Let us make a small change in the **b** vector

$$\begin{bmatrix} 2 & 3.99 \\ 1 & 2 \end{bmatrix} \begin{bmatrix} x_1 \\ x_2 \end{bmatrix} = \begin{bmatrix} 1.98 \\ 1.01 \end{bmatrix} \tag{8.354}$$

and see how the solution will change.

$$\begin{bmatrix} x_1 \\ x_2 \end{bmatrix} = \begin{bmatrix} -6.99 \\ 4.0 \end{bmatrix} \tag{8.355}$$

Now we make a small change in $[A]$ matrix

$$\begin{bmatrix} 2.01 & 3.98 \\ 0.99 & 2.01 \end{bmatrix} \begin{bmatrix} x_1 \\ x_2 \end{bmatrix} = \begin{bmatrix} 1.99 \\ 1 \end{bmatrix} \tag{8.356}$$

and solve the equations

$$\begin{bmatrix} x_1 \\ x_2 \end{bmatrix} = \begin{bmatrix} 0.1988 \\ 0.3993 \end{bmatrix} \tag{8.357}$$

Therefore, the set of Eqs. (8.352) is ill conditioned and is sensitive to perturbation in $[A]$ and **b**. However, the set of equations

$$\begin{bmatrix} 2 & 3 \\ 1 & 2 \end{bmatrix} \begin{bmatrix} x_1 \\ x_2 \end{bmatrix} = \begin{bmatrix} 1 \\ 1 \end{bmatrix} \tag{8.358}$$

is well conditioned because small changes in $[A]$ or **b** cannot change the solution drastically.

The sensitivity of the solution **x** to small perturbations in $[A]$ and **b** is measured in terms of the **condition number** of $[A]$ by

$$\frac{\|\triangle \mathbf{x}\|}{\|\mathbf{x}\|} \leq con\,(A)\,\frac{\|\triangle A\|}{\|A\|} \tag{8.359}$$

where

$$con\,(A) = \|A^{-1}\|\,\|A\| \tag{8.360}$$

and $\|A\|$ is a norm of $[A]$. If $con\,(A) = 1$, then $[A]$ is called perfectly conditioned. The matrix $[A]$ is well conditioned if $con\,(A) < 1$ and it is ill conditioned if $con\,(A) > 1$. In fact, the relative change in the norm of the coefficient matrix, $[A]$, can be amplified by $con\,(A)$ to make the upper limit of the relative change in the norm of the solution vector **x**.

Proof Start with a set of equations

$$[A]\,\mathbf{x} = \mathbf{b} \tag{8.361}$$

and change the matrix $[A]$ to $[A']$. Then the solution **x** will change to **x**'.

$$[A']\,\mathbf{x}' = \mathbf{b} \tag{8.362}$$

Therefore,

$$[A]\,\mathbf{x} = [A']\,\mathbf{x}' = ([A] + \triangle A)\,(\mathbf{x} + \triangle \mathbf{x}) \tag{8.363}$$

where

$$\triangle A = [A'] - [A] \tag{8.364}$$
$$\triangle \mathbf{x} = \mathbf{x}' - \mathbf{x} \tag{8.365}$$

Expanding (8.363)

$$[A]\,\mathbf{x} = [A]\,\mathbf{x} + [A]\,\triangle \mathbf{x} + \triangle A\,(\mathbf{x} + \triangle \mathbf{x}) \tag{8.366}$$

and simplifying

$$\triangle \mathbf{x} = -A^{-1}\triangle A\,(\mathbf{x} + \triangle \mathbf{x}) \tag{8.367}$$

shows that

$$\|\triangle \mathbf{x}\| \leq \|A^{-1}\|\,\|\triangle A\|\,\|\mathbf{x} + \triangle \mathbf{x}\| \tag{8.368}$$

Multiplying both sides of (8.368) by the norm $\|A\|$ leads to Eq. (8.359). ∎

Example 287 ★ Norm of a matrix. Norm of coefficient matrix is a number to be used in calculating condition number of [A]. Here is how to calculate norm of a matrix.

The norm of a matrix is a scalar positive number and is defined for every kind of matrices including square, rectangular, invertible, and non-invertible. There are several definitions for the norm of a matrix. The most important ones are

$$\|A\|_1 = \max_{1 \le j \le n} \sum_{i=1}^{n} |a_{ij}| \tag{8.369}$$

$$\|A\|_2 = \lambda_{Max}\left(A^T A\right) \tag{8.370}$$

$$\|A\|_\infty = \max_{1 \le i \le n} \sum_{j=1}^{n} |a_{ij}| \tag{8.371}$$

$$\|A\|_F = \sum_{i=1}^{n} \sum_{j=1}^{n} a_{ij}^2 \tag{8.372}$$

The norm-infinity, $\|A\|_\infty$, is the one we accept to calculate the con (A) in Eq. (8.360). The norm-infinity, $\|A\|_\infty$, is also called the **row sum norm** and **uniform norm**. To calculate $\|A\|_\infty$, we find the sum of the absolute of the elements of each row of the matrix [A] and pick the largest sum.

As an example, the norm of

$$[A] = \begin{bmatrix} 1 & 3 & -3 \\ -1 & -1 & 2 \\ 2 & 4 & -2 \end{bmatrix} \tag{8.373}$$

is

$$\begin{aligned}
\|A\|_\infty &= \max_{1 \le i \le n} \sum_{j=1}^{n} |a_{ij}| \\
&= Max \{(|1| + |3| + |-3|), (|-1| + |-1| + |2|), \\
&\quad (|2| + |4| + |-2|)\} \\
&= Max \{7, 4, 8\} = 8
\end{aligned} \tag{8.374}$$

We may check the following relations between norms of matrices.

$$\|[A] + [B]\| \le \|[A]\| + \|[B]\| \tag{8.375}$$

$$\|[A][B]\| \le \|[A]\| \|[B]\| \tag{8.376}$$

8.6 Matrix Inversion

There are numerous techniques for matrix inversion. However the method based on the LU factorization can simplify our numerical calculations employing the method we have already applied for solving a set of linear algebraic equations.

Assume a matrix [A] could be decomposed into [L][U]

$$[A] = [L][U] \tag{8.377}$$

where

$$[A] = \begin{bmatrix} a_{11} & a_{12} & \cdots & a_{1n} \\ a_{21} & a_{22} & \cdots & a_{2n} \\ \cdots & \cdots & \cdots & \cdots \\ a_{n1} & a_{n2} & \cdots & a_{nn} \end{bmatrix} \tag{8.378}$$

$$[L] = \begin{bmatrix} 1 & 0 & \cdots & 0 \\ l_{21} & 1 & \cdots & 0 \\ \cdots & \cdots & \cdots & \cdots \\ l_{n1} & l_{n2} & \cdots & 1 \end{bmatrix} \tag{8.379}$$

$$[U] = \begin{bmatrix} u_{11} & u_{12} & \cdots & u_{1n} \\ 0 & u_{22} & \cdots & u_{2n} \\ \cdots & \cdots & \cdots & \cdots \\ 0 & 0 & \cdots & u_{nn} \end{bmatrix} \tag{8.380}$$

then its inverse matrix, $[A]^{-1}$, would be

$$[A]^{-1} = [U]^{-1}[L]^{-1} \tag{8.381}$$

Proof Because $[L]$ and $[U]$ are triangular matrices, their inverses are also triangular. The elements of the matrix $[M] = [L]^{-1}$ are

$$m_{ij} = -l_{ij} - \sum_{k=j+1}^{i-1} l_{ik} m_{kj} \qquad j < i \qquad i = 2, 3, \cdots n - 1 \tag{8.382}$$

$$[M] = [L]^{-1} = \begin{bmatrix} 1 & 0 & \cdots & 0 \\ m_{21} & 1 & \cdots & 0 \\ \cdots & \cdots & \cdots & \cdots \\ m_{n1} & m_{n2} & \cdots & 1 \end{bmatrix} \tag{8.383}$$

and the elements of the matrix $[V] = [U]^{-1}$ are

$$v_{ij} = \begin{cases} \dfrac{1}{u_{ij}} & j = i \ \ i = n, n - 1, \cdots, 1 \\ \dfrac{-1}{u_{ii}} \displaystyle\sum_{k=i+1}^{j} u_{ik} v_{kj} & j \geq i \ \ i = n - 1, \cdots, 2. \end{cases} \tag{8.384}$$

$$[V] = [U]^{-1} = \begin{bmatrix} v_{11} & v_{12} & v_{13} & v_{14} \\ 0 & v_{22} & v_{23} & v_{24} \\ 0 & 0 & v_{33} & v_{34} \\ 0 & 0 & 0 & v_{44} \end{bmatrix} \tag{8.385}$$

∎

Example 288 Solution of a set of equations by matrix inversion. *LU* factorization may be employed to calculate inversion of the coefficient matrix to solve a set of linear algebraic equations. Here is to show how.

Consider a set of four linear algebraic equations.

$$[A]\mathbf{x} = \mathbf{b} \tag{8.386}$$

$$\begin{bmatrix} 2 & 1 & 3 & -3 \\ 1 & 0 & -1 & -2 \\ 0 & 2 & 2 & 1 \\ 3 & 1 & 0 & -2 \end{bmatrix} \begin{bmatrix} x_1 \\ x_2 \\ x_3 \\ x_4 \end{bmatrix} = \begin{bmatrix} 1 \\ 2 \\ 0 \\ -2 \end{bmatrix} \tag{8.387}$$

Following the *LU* factorization algorithm, we can decompose the coefficient matrix.

$$[A] = [L][U] \tag{8.388}$$

$$[L] = \begin{bmatrix} 1 & 0 & 0 & 0 \\ \frac{1}{2} & 1 & 0 & 0 \\ 0 & -4 & 1 & 0 \\ \frac{3}{2} & 1 & \frac{1}{4} & 1 \end{bmatrix} \qquad [U] = \begin{bmatrix} 2 & 1 & 3 & -3 \\ 0 & -\frac{1}{2} & -\frac{5}{2} & -\frac{1}{2} \\ 0 & 0 & -8 & -1 \\ 0 & 0 & 0 & \frac{13}{4} \end{bmatrix} \tag{8.389}$$

The inverse of matrices $[L]$ and $[U]$ are

$$[L]^{-1} = \begin{bmatrix} 1 & 0 & 0 & 0 \\ -\frac{1}{2} & 1 & 0 & 0 \\ -2 & 4 & 1 & 0 \\ -\frac{1}{2} & -2 & -\frac{1}{4} & 1 \end{bmatrix} \qquad [U]^{-1} = \begin{bmatrix} \frac{1}{2} & 1 & -\frac{1}{8} & \frac{15}{26} \\ 0 & -2 & \frac{5}{8} & -\frac{3}{26} \\ 0 & 0 & -\frac{1}{8} & -\frac{1}{26} \\ 0 & 0 & 0 & \frac{4}{13} \end{bmatrix} \tag{8.390}$$

and, therefore, the solution of the equations is

$$\mathbf{x} = [U]^{-1} [L]^{-1} \mathbf{b} = \begin{bmatrix} -5/2 \\ 3/2 \\ -1/2 \\ -2 \end{bmatrix} \tag{8.391}$$

Example 289 ★ *LU* factorization method compared to other methods. The effectiveness of the *LU* factorization depends on the number of equations. The more the number of equations, the more efficient *LU* factorization compared to other methods.

Every non-singular matrix $[A]$ can be decomposed into lower and upper triangular matrices $[A] = [L][U]$. Then, the solution of a set of equations $[A]\mathbf{x} = \mathbf{b}$ will be found.

$$[L][U]\mathbf{x} = \mathbf{b} \tag{8.392}$$

Multiplying both sides by L^{-1} shows that

$$[U]\mathbf{x} = [L]^{-1}\mathbf{b} \tag{8.393}$$

and the problem is broken into two new sets of equations.

$$[L]\mathbf{y} = \mathbf{b} \qquad [U]\mathbf{x} = \mathbf{y} \tag{8.394}$$

The computational time required to decompose $[A]$ into $[L][U]$ is proportional to $n^3/3$, where n is the number of equations. Then, the computational time for solving each set of $[L]\mathbf{y} = \mathbf{b}$ and $[U]\mathbf{x} = \mathbf{y}$ is proportional to $n^2/2$. Therefore, the total computational time for solving a set of equations by the *LU* factorization method is proportional to $n^2 + n^3/3$. The Gaussian elimination (G) method takes a computational time proportional to $n^2/2 + n^3/3$, forward elimination takes a time proportional to $n^3/3$, and back substitution takes a time proportional to $n^2/2$.

On the other hand, the total computational time required to inverse a matrix using the *LU* factorization method is proportional to $4n^3/3$. However, the Gaussian elimination method needs $n^4/3 + n^3/2$. Their comparison indicates that Gaussian elimination method needs more operation than *LU* factorization method for $n > 2$.

$$\frac{n^4}{3} + \frac{n^3}{2} > \frac{4n^3}{3} \qquad n > 2 \tag{8.395}$$

The deference of required operations for the G and LU methods makes the function $G - LU = \frac{n^4}{3} + \frac{n^3}{2} - \frac{4n^3}{3}$ showing a rapid increase in the number of calculations for the Gaussian elimination, compared to the *LU* factorization. As an example, for a 6×6 matrix inversion, we need 540 calculations for the Gaussian elimination method, compared to 288 calculations for the *LU* factorization method.

Example 290 ★ Analytic inversion method. Matrix inversion may be calculated by an analytic method based on adjoint and cofactor method. Analytic inversion is not efficient for numerical calculations.

If the $n \times n$ matrix $[A] = [a_{ij}]$ is non-singular, that is $\det(A) \neq 0$. We may compute the inverse, $[A]^{-1}$, by dividing the adjoint matrix A_a by the determinant of $[A]$.

$$A^{-1} = \frac{A_a}{\det(A)} \tag{8.396}$$

The adjoint or adjugate matrix of the matrix $[A]$ is the transpose of the cofactor matrix of $[A]$.

$$A^a = A_c^T \tag{8.397}$$

The cofactor matrix, denoted by A_c, for a matrix $[A]$, is made of the matrix $[A]$ by replacing each of its elements by its cofactor. The cofactor associated with the element a_{ij} is defined by

$$A_{c,ij} = (-1)^{i+j} A_{ij} \tag{8.398}$$

where A_{ij} is the ij-minor of $[A]$. Associated with each element a_{ij} of the matrix $[A]$, there exists a minor A_{ij}, which is a number equal to the value of the determinant of the submatrix obtained by deleting row i and column j of the matrix $[A]$.

The determinant of $[A]$ is calculated by

$$\det(A) = \sum_{j=1}^{n} a_{ij} A_{c,ij} \tag{8.399}$$

therefore, if we have

$$[A] = \begin{bmatrix} a_{11} & a_{12} & a_{13} \\ a_{21} & a_{22} & a_{23} \\ a_{31} & a_{32} & a_{33} \end{bmatrix} \tag{8.400}$$

then the elements of adjoint matrix A_a are

$$A_{a,11} = A_{c,11} = (-1)^2 \begin{vmatrix} a_{22} & a_{23} \\ a_{32} & a_{33} \end{vmatrix} \tag{8.401}$$

$$A_{a,21} = A_{c12} = (-1)^3 \begin{vmatrix} a_{21} & a_{23} \\ a_{31} & a_{33} \end{vmatrix} \tag{8.402}$$

$$\vdots$$

$$A_{a,33} = A_{c,33} = (-1)^6 \begin{vmatrix} a_{11} & a_{12} \\ a_{21} & a_{22} \end{vmatrix} \tag{8.403}$$

and the determinant of $[A]$ is

$$\det(A) = a_{11}a_{22}a_{33} - a_{11}a_{23}a_{32} - a_{12}a_{21}a_{33}$$
$$+ a_{12}a_{31}a_{23} + a_{21}a_{13}a_{32} - a_{13}a_{22}a_{31} \tag{8.404}$$

As an example, consider a 3×3 matrix as follows.

$$[A] = \begin{bmatrix} 3 & 4 & 8 \\ 7 & 2 & 5 \\ 9 & 6 & 1 \end{bmatrix} \tag{8.405}$$

The associated adjoint matrix for $[A]$ is

$$A_a = A_c^T = \begin{bmatrix} -28 & 38 & 24 \\ 44 & -69 & 18 \\ 4 & 41 & -22 \end{bmatrix}^T = \begin{bmatrix} -28 & 44 & 4 \\ 38 & -69 & 41 \\ 24 & 18 & -22 \end{bmatrix} \tag{8.406}$$

and the determinant of $[A]$ is

$$\det[A] = 260 \tag{8.407}$$

and, therefore,

$$[A]^{-1} = \frac{A_a}{\det(A)} = \begin{bmatrix} -\frac{7}{65} & \frac{11}{65} & \frac{1}{65} \\ \frac{19}{130} & -\frac{69}{260} & \frac{41}{260} \\ \frac{6}{65} & \frac{9}{130} & -\frac{11}{130} \end{bmatrix} \qquad (8.408)$$

Example 291 ★ Cayley–Hamilton matrix inversion. Another smart and efficient method of matrix inversion is the Cayley–Hamilton, which is working based on the fact that a matrix will satisfy its own characteristic equation.

The Cayley–Hamilton theorem says: Every non-singular matrix satisfies its own characteristic equation. The characteristic equation of an $n \times n$ matrix $[A] = \begin{bmatrix} a_{ij} \end{bmatrix}$ is

$$\det(A - \lambda I) = |A - \lambda I|$$
$$= P(\lambda) = \lambda^n + a_{n-1}\lambda^{n-1} + \cdots + a_1\lambda + a_0 = 0 \qquad (8.409)$$

Hence, the characteristic equation of an $n \times n$ matrix is a polynomial of degree n. The Cayley–Hamilton theorem yields

$$P(A) = A^n + a_{n-1}A^{n-1} + \cdots + a_1 A + a_0 = 0 \qquad (8.410)$$

Multiplying both sides of this polynomial by A^{-1} and solving for A^{-1} provides

$$A^{-1} = -\frac{1}{a_0}\left[A^{n-1} + a_{n-1}A^{n-2} + \cdots + a_2 A + a_1 I\right]. \qquad (8.411)$$

Therefore, if

$$[A] = \begin{bmatrix} a_{11} & a_{12} & a_{13} \\ a_{21} & a_{22} & a_{23} \\ a_{31} & a_{32} & a_{33} \end{bmatrix} \qquad (8.412)$$

and

$$\det(A) = \sum_{j=1}^{n} a_{ij} A_{c,ij} \qquad (8.413)$$

then the characteristic equation of $[A]$ is

$$\begin{aligned} P(\lambda) = \det(A - \lambda I) &= \lambda^3 + (-a_{11} - a_{22} - a_{33})\lambda^2 \\ &+ (a_{11}a_{22} - a_{12}a_{21} + a_{11}a_{33} - a_{13}a_{31} \\ &\quad + a_{22}a_{33} - a_{23}a_{32})\lambda \\ &+ a_{11}a_{23}a_{32} + a_{12}a_{21}a_{33} + a_{13}a_{22}a_{31} \\ &- a_{11}a_{22}a_{33} - a_{12}a_{31}a_{23} - a_{21}a_{13}a_{32} \end{aligned} \qquad (8.414)$$

As an example, consider a 3×3 matrix

$$[A] = \begin{bmatrix} 1 & 2 & 3 \\ 4 & 6 & 7 \\ 5 & 8 & 9 \end{bmatrix} \qquad (8.415)$$

with following characteristic equation:

$$\lambda^3 - 16\lambda^2 - 10\lambda - 2 = 0 \qquad (8.416)$$

Because $[A]$ satisfies its own characteristic equation, we have

$$A^3 - 16A^2 - 10A - 2 = 0 \qquad (8.417)$$

Multiplying both sides by A^{-1}

$$A^{-1}A^3 - 16A^{-1}A^2 - 10A^{-1}A = 2A^{-1} \qquad (8.418)$$

provides us with the inverse matrix.

$$A^{-1} = \frac{1}{2}\left(A^2 - 16A - 10I\right)$$

$$= \frac{1}{2}\left[\begin{bmatrix} 1 & 2 & 3 \\ 4 & 6 & 7 \\ 5 & 8 & 9 \end{bmatrix}^2 - 16\begin{bmatrix} 1 & 2 & 3 \\ 4 & 6 & 7 \\ 5 & 8 & 9 \end{bmatrix} - 10\begin{bmatrix} 1 & 0 & 0 \\ 0 & 1 & 0 \\ 0 & 0 & 1 \end{bmatrix}\right]$$

$$= \begin{bmatrix} -1 & 3 & -2 \\ -\frac{1}{2} & -3 & \frac{5}{2} \\ 1 & 1 & -1 \end{bmatrix} \tag{8.419}$$

8.7 Nonlinear Algebraic Equations

Inverse kinematic problem ends up to a set of nonlinear coupled algebraic equations. Consider a set of nonlinear algebraic equations

$$\mathbf{f}(\mathbf{q}) = 0 \tag{8.420}$$

or

$$
\begin{aligned}
f_1(q_1, q_2, \cdots, q_n) &= 0 \\
f_2(q_1, q_2, \cdots, q_n) &= 0 \\
&\cdots \\
f_n(q_1, q_2, \cdots, q_n) &= 0
\end{aligned}
\tag{8.421}
$$

where the function and variable vectors are

$$\mathbf{f} = \begin{bmatrix} f_1(\mathbf{q}) \\ f_2(\mathbf{q}) \\ \cdots \\ f_n(\mathbf{q}) \end{bmatrix} \qquad \mathbf{q} = \begin{bmatrix} q_1 \\ q_2 \\ \cdots \\ q_n \end{bmatrix} \tag{8.422}$$

To solve the set of Eqs. (8.420), we begin with a guess solution vector $\mathbf{q}^{(0)}$ and employ the following *iteration formula* to search for a better solution,

$$\mathbf{q}^{(i+1)} = \mathbf{q}^{(i)} - \mathbf{J}^{-1}(\mathbf{q}^{(i)})\,\mathbf{f}(\mathbf{q}^{(i)}) \tag{8.423}$$

where $\mathbf{J}^{-1}(\mathbf{q}^{(i)})$ is the Jacobian matrix of the system of equations evaluated at step (i), $\mathbf{q} = \mathbf{q}^{(i)}$.

$$[\mathbf{J}] = \left[\frac{\partial f_i}{\partial q_j}\right] \tag{8.424}$$

Utilizing the iteration formula (8.423), we can approach an exact solution. The iteration method based on a guess solution is called *Newton–Raphson method*, which is the most common method for solving a set of nonlinear algebraic equations.

A set of nonlinear equations usually has multiple solutions and the main disadvantage of the Newton–Raphson method for solving a set of nonlinear equations is that the solution it provides may not be the solution of interest. The solution that the method will approach depends highly on the initial estimation. Hence, having a correct estimate helps to detect the proper solution.

Proof Let us define the increment $\boldsymbol{\delta}^{(i)}$ as

$$\boldsymbol{\delta}^{(i)} = \mathbf{q}^{(i+1)} - \mathbf{q}^{(i)} \tag{8.425}$$

and expand the set of equations around $\mathbf{q}^{(i+1)}$.

$$\mathbf{f}(\mathbf{q}^{(i+1)}) = \mathbf{f}(\mathbf{q}^{(i)}) + \mathbf{J}(\mathbf{q}^{(i)})\,\boldsymbol{\delta}^{(i)} \tag{8.426}$$

Assume that $\mathbf{q}^{(i+1)}$ is the exact solution of Eq. (8.420). Therefore, $\mathbf{f}(\mathbf{q}^{(i+1)}) = 0$ and we may use

$$\mathbf{J}(\mathbf{q}^{(i)}) \, \boldsymbol{\delta}^{(i)} = -\mathbf{f}(\mathbf{q}^{(i)}) \tag{8.427}$$

to find the increment $\boldsymbol{\delta}^{(i)}$

$$\boldsymbol{\delta}^{(i)} = -\mathbf{J}^{-1}(\mathbf{q}^{(i)}) \, \mathbf{f}(\mathbf{q}^{(i)}) \tag{8.428}$$

and determine the solution.

$$\begin{aligned} \mathbf{q}^{(i+1)} &= \mathbf{q}^{(i)} + \boldsymbol{\delta}^{(i)} \\ &= \mathbf{q}^{(i)} - \mathbf{J}^{-1}(\mathbf{q}^{(i)}) \, \mathbf{f}(\mathbf{q}^{(i)}) \end{aligned} \tag{8.429}$$

The Newton–Raphson iteration method can be set up as an algorithm for better application.

Algorithm 8.3 *Newton–Raphson iteration method for* $\mathbf{f}(\mathbf{q}) = 0$.

1. *Set the initial counter* $i = 0$.
2. *Evaluate an estimate solution* $\mathbf{q} = \mathbf{q}^{(i)}$.
3. *Calculate the Jacobian matrix* $[\mathbf{J}] = \left[\frac{\partial f_i}{\partial q_j}\right]$ *at* $\mathbf{q} = \mathbf{q}^{(i)}$.
4. *Solve for* $\boldsymbol{\delta}^{(i)}$ *from the set of linear equations* $\mathbf{J}(\mathbf{q}^{(i)}) \, \boldsymbol{\delta}^{(i)} = -\mathbf{f}(\mathbf{q}^{(i)})$.
5. *If* $\left|\boldsymbol{\delta}^{(i)}\right| < \epsilon$, *where* ϵ *is an arbitrary tolerance, then* $\mathbf{q}^{(i)}$ *is the solution. Otherwise calculate* $\mathbf{q}^{(i+1)} = \mathbf{q}^{(i)} + \boldsymbol{\delta}^{(i)}$.
6. *Set* $i = i + 1$ *and return to step* 3.

$\blacksquare$

Example 292 Inverse kinematics problem for a $2R$ planar robot. Employing the Newton–Raphson iteration method and the Algorithm 8.3, here is how to calculate inverse kinematic of a manipulator for a given forward kinematics.

The end point of a $2R$ planar manipulator can be expressed by two nonlinear algebraic equations.

$$\begin{bmatrix} f_1(\theta_1, \theta_2) \\ f_2(\theta_1, \theta_2) \end{bmatrix} = \begin{bmatrix} l_1 \cos\theta_1 + l_2 \cos(\theta_1 + \theta_2) - X \\ l_1 \sin\theta_1 + l_2 \sin(\theta_1 + \theta_2) - Y \end{bmatrix} = 0 \tag{8.430}$$

Assuming

$$l_1 = l_2 = 1 \tag{8.431}$$

and the end point at

$$\begin{bmatrix} X \\ Y \end{bmatrix} = \begin{bmatrix} 1 \\ 1 \end{bmatrix} \tag{8.432}$$

we are looking for the associated variables that provide the desired position of the end point.

$$\boldsymbol{\theta} = \begin{bmatrix} \theta_1 \\ \theta_2 \end{bmatrix} \tag{8.433}$$

Due to simplicity of the system of equations, the Jacobian of the equations and its inverse can be found in closed form.

$$\begin{aligned} \mathbf{J}(\boldsymbol{\theta}) = \left[\frac{\partial f_i}{\partial \theta_j}\right] &= \begin{bmatrix} \dfrac{\partial f_1}{\partial \theta_1} & \dfrac{\partial f_1}{\partial \theta_2} \\ \dfrac{\partial f_2}{\partial \theta_1} & \dfrac{\partial f_2}{\partial \theta_2} \end{bmatrix} \\ &= \begin{bmatrix} -l_1 \sin\theta_1 - l_2 \sin(\theta_1 + \theta_2) & -l_2 \sin(\theta_1 + \theta_2) \\ l_1 \cos\theta_1 + l_2 \cos(\theta_1 + \theta_2) & l_2 \cos(\theta_1 + \theta_2) \end{bmatrix} \end{aligned} \tag{8.434}$$

$$\mathbf{J}^{-1} = \frac{-1}{l_1 l_2 s\theta_2} \begin{bmatrix} -l_2 c\,(\theta_1 + \theta_2) & -l_2 s\,(\theta_1 + \theta_2) \\ l_1 c\theta_1 + l_2 c\,(\theta_1 + \theta_2) & l_1 s\theta_1 + l_2 s\,(\theta_1 + \theta_2) \end{bmatrix} \tag{8.435}$$

The Newton–Raphson iteration algorithm may now be started by setting $i = 0$ and evaluating an estimate solution.

$$\mathbf{q}^{(0)} = \begin{bmatrix} \theta_1 \\ \theta_2 \end{bmatrix}^{(0)} = \begin{bmatrix} \pi/3 \\ -\pi/3 \end{bmatrix} \tag{8.436}$$

Therefore,

$$\mathbf{J}(\frac{\pi}{3}, \frac{\pi}{3}) = \begin{bmatrix} -\sqrt{3} & -\frac{1}{2}\sqrt{3} \\ 0 & -\frac{1}{2} \end{bmatrix} \tag{8.437}$$

$$\mathbf{f}(\frac{\pi}{3}, \frac{\pi}{3}) = \begin{bmatrix} -1 \\ \sqrt{3} - 1 \end{bmatrix} \tag{8.438}$$

$$\boldsymbol{\delta}^{(0)} = -\mathbf{J}^{-1}(\boldsymbol{\theta}^{(0)})\,\mathbf{f}(\boldsymbol{\theta}^{(0)}) = \begin{bmatrix} -1.3094 \\ 1.4641 \end{bmatrix} \tag{8.439}$$

and a better solution would be

$$\begin{bmatrix} \theta_1 \\ \theta_2 \end{bmatrix}^{(1)} = \begin{bmatrix} \pi/3 \\ \pi/3 \end{bmatrix} + \begin{bmatrix} -1.3094 \\ 1.4641 \end{bmatrix} = \begin{bmatrix} -0.2622 \\ 2.5113 \end{bmatrix} \tag{8.440}$$

In the next iterations we find

$$\begin{bmatrix} \theta_1 \\ \theta_2 \end{bmatrix}^{(2)} = \begin{bmatrix} -0.2622 \\ 2.5113 \end{bmatrix} + \begin{bmatrix} -0.06952 \\ -0.80337 \end{bmatrix} = \begin{bmatrix} -0.3317 \\ 1.7079 \end{bmatrix} \tag{8.441}$$

$$\begin{bmatrix} \theta_1 \\ \theta_2 \end{bmatrix}^{(3)} = \begin{bmatrix} -0.3317 \\ 1.7079 \end{bmatrix} + \begin{bmatrix} 0.31414 \\ -0.068348 \end{bmatrix} = \begin{bmatrix} -0.0176 \\ 1.63958 \end{bmatrix} \tag{8.442}$$

$$\begin{bmatrix} \theta_1 \\ \theta_2 \end{bmatrix}^{(4)} = \begin{bmatrix} -0.0176 \\ 1.63958 \end{bmatrix} + \begin{bmatrix} 0.016275 \\ -0.06739 \end{bmatrix} = \begin{bmatrix} -0.0013 \\ 1.5722 \end{bmatrix} \tag{8.443}$$

$$\begin{bmatrix} \theta_1 \\ \theta_2 \end{bmatrix}^{(5)} = \begin{bmatrix} -0.0013 \\ 1.5722 \end{bmatrix} + \begin{bmatrix} 0.1304 \\ -0.139 \end{bmatrix} = \begin{bmatrix} -0.295 \times 10^{-8} \\ 1.571 \end{bmatrix} \tag{8.444}$$

$$\begin{bmatrix} \theta_1 \\ \theta_2 \end{bmatrix}^{(6)} = \begin{bmatrix} -0.3 \times 10^{-8} \\ 1.571 \end{bmatrix} + \begin{bmatrix} 0.29 \times 10^{-8} \\ -0.85 \times 10^{-6} \end{bmatrix}$$
$$= \begin{bmatrix} -0.49 \times 10^{-10} \\ 1.571 \end{bmatrix} \tag{8.445}$$

$$\begin{bmatrix} \theta_1 \\ \theta_2 \end{bmatrix}^{(7)} = \begin{bmatrix} -0.49 \times 10^{-10} \\ 1.571 \end{bmatrix} + \begin{bmatrix} -0.41 \times 10^{-19} \\ -0.2 \times 10^{-9} \end{bmatrix}$$
$$= \begin{bmatrix} -0.49 \times 10^{-10} \\ 1.571 \end{bmatrix} \tag{8.446}$$

and this answer is close enough to the exact elbow down configuration.

$$\begin{bmatrix} \theta_1 \\ \theta_2 \end{bmatrix} = \begin{bmatrix} 0 \\ \pi/2 \end{bmatrix} \tag{8.447}$$

Example 293 ★ Alternative and expanded proof for Newton–Raphson iteration method. Here is introducing a more general and more applied Newton–Raphson iteration method.

Consider the following set of equations in which we are searching for the exact solutions q_j where j is the number of unknowns and i is the number of equations.

$$y_i = f_i(q_j) \qquad i = 1, \cdots, n \qquad j = 1, \cdots, m \tag{8.448}$$

Assume that for a given y_i an approximate solution $q_j^\star$ is available. The difference between the exact solution q_j and the approximate solution $q_j^\star$ is

$$\delta_j = q_j - q_j^\star \tag{8.449}$$

where the value of equations for the approximate solution $q_j^\star$ is denoted by Y_i.

$$Y_i = f_i(q_j^\star) \tag{8.450}$$

The iteration method is based on the minimization of δ_j to make the solution of y_i to be as close as possible to the exact solution.

$$y_i = f_i(q_j^\star + \delta_j) \tag{8.451}$$

A first-order Taylor expansion of this equation is

$$y_i = f_i(q_j^\star) + \sum_{i=1}^{m} \frac{\partial f_i}{\partial q_j} \delta_j + O(\delta_j^2) \tag{8.452}$$

We may define

$$Y_i = f_i(q_j^\star) \tag{8.453}$$

and the residual quantity

$$r_i = y_i - Y_i \tag{8.454}$$

to write

$$\mathbf{r} = \mathbf{J}\delta + O(\delta^2) \tag{8.455}$$

where $\mathbf{J}$ is the Jacobian matrix of the set of equations.

$$\mathbf{J} = \left[\frac{\partial f_i}{\partial q_j} \right] \tag{8.456}$$

The method of solution depends on the relative value of m and n. Therefore, three cases must be considered.

1. $m = n$

Provided that the Jacobian matrix remains non-singular, the linearized equation

$$\mathbf{r} = \mathbf{J}\delta\mathbf{q} \tag{8.457}$$

possesses a unique solution, and the Newton–Raphson technique may then be utilized to solve Eq. (8.448). The stepwise procedure is illustrated in Fig. 8.10.

The effectiveness of the procedure depends on the number of iterations to be performed, which depends on the initial estimate of $q_j^\star$ and on the dimension of the Jacobian matrix. Since the solution to nonlinear equations is not unique, it may generate different sets of solutions depending on the initial guess. Furthermore, convergence may not occur if the initial estimate of the solution falls outside the convergence domain of the algorithm. In this case, much effort is needed to attain a numerical solution.

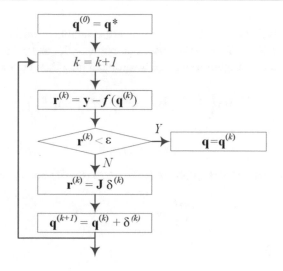

Fig. 8.10 Newton–Raphson iteration method for solving a set of nonlinear algebraic equations

2. $m < n$

This is the *overdetermined case* for which no solution exists, because the number of unknowns (such as the number of joints in robots) are not sufficient enough to generate a solution (such as an arbitrary configuration of the end-effector). A solution can, however, be generated that minimizes an error.

Consider the problem

$$\min\left(F = \frac{1}{2}\sum_{i=1}^{n} w_i \left[y_i - f_i(q_j)\right]^2\right) \tag{8.458}$$

or, in matrix form,

$$\min\left(F = \frac{1}{2}\sum_{i=1}^{n} [\mathbf{y} - \mathbf{f}(\mathbf{q})]^T \mathbf{W} [\mathbf{y} - \mathbf{f}(\mathbf{q})]\right) \tag{8.459}$$

where $\mathbf{W}$ is a set of weighting factors giving a relative importance to each of the kinematic equations.

$$\mathbf{W} = diag(w_1 \cdots w_n) \tag{8.460}$$

The error is minimum when

$$\frac{\partial F}{\partial q_j} = -\sum_i \frac{\partial f_j}{\partial q_j} w_i \left[y_i - f_i(q_j)\right] = 0 \tag{8.461}$$

or, in matrix form,

$$\mathbf{J}^T \mathbf{W} [\mathbf{y} - \mathbf{f}(\mathbf{q})] = 0 \tag{8.462}$$

A Taylor expansion of the third factor shows that the linear correction to an estimated solution $\mathbf{q}^\star$ is

$$\mathbf{J}^T \mathbf{W} \left[\mathbf{y} - \mathbf{f}(\mathbf{q}^\star)\right] - \mathbf{J} \delta = 0 \tag{8.463}$$

The correction equation is

$$\mathbf{J}^T \mathbf{W} \mathbf{J} \delta = \mathbf{J}^T \mathbf{W} \mathbf{r} \tag{8.464}$$

where $\mathbf{r}$ is the residual vector defined by Eq. (8.454).

The weighting factor $\mathbf{W}$ is a positive definite diagonal matrix, and, therefore, the matrix $\mathbf{J}^T \mathbf{W} \mathbf{J}$ is always symmetric and invertible. It provides the generalized inverse to the Jacobian matrix.

$$\mathbf{J}^{-1} = \left[\mathbf{J}^T \mathbf{W} \mathbf{J}\right]^{-1} \mathbf{J}^T \mathbf{W} \tag{8.465}$$

When the Jacobian is revertible, the solution for (8.459) is the solution to the nonlinear system (8.448).

3. $m > n$

This is the *redundant* case for which an infinity of solutions are available. Selection of an appropriate solution can be made under the condition that it is optimal in some sense. For example, let us find a solution for (8.448) that minimizes the deviation from a given reference configuration $q^{(0)}$. The problem may then be formulated as of finding the minimum of a constrained function

$$\min\left(F = \frac{1}{2}\left[\mathbf{q} - \mathbf{q}^{(0)}\right]^T \mathbf{W}\left[\mathbf{q} - \mathbf{q}^{(0)}\right]\right) \tag{8.466}$$

subject to a constraint equation.

$$\mathbf{y} - \mathbf{f}(\mathbf{q}) = 0 \tag{8.467}$$

Using the technique of Lagrangian multipliers, problem (8.466) and (8.467) may be replaced by an equivalent problem

$$\frac{\partial G}{\partial \mathbf{q}} = 0 \qquad \frac{\partial G}{\partial \lambda} = 0 \tag{8.468}$$

with the definition of the functional.

$$G(\mathbf{q}, \lambda) = \frac{1}{2}\left[\mathbf{q} - \mathbf{q}^{(0)}\right]^T \mathbf{W}\left[\mathbf{q} - \mathbf{q}^{(0)}\right] + \lambda^T \left[\mathbf{y} - \mathbf{f}(\mathbf{q})\right] \tag{8.469}$$

It leads to a system of $m + n$ equations with $m + n$ unknowns.

$$\mathbf{W}\left[\mathbf{q} - \mathbf{q}^{(0)}\right] - \mathbf{J}^T\lambda = 0 \tag{8.470}$$

$$\mathbf{y} - \mathbf{f}(\mathbf{q}) = 0 \tag{8.471}$$

Linearization of Eqs. (8.470) provides the system of equations for the displacement corrections and variations of Lagrangian multipliers

$$\mathbf{W}\,\delta - \mathbf{J}^T\,\lambda' = 0 \tag{8.472}$$

$$\mathbf{W}\,\delta = \mathbf{r} \tag{8.473}$$

where λ' is the increment of λ.

Substitution of the solution δ obtained from the first equation of (8.470) into the second one yields

$$\mathbf{J}\mathbf{W}^{-1}\mathbf{J}^T\delta\lambda = \mathbf{r} \tag{8.474}$$

or, in terms of the displacement correction.

$$\delta\mathbf{q} = \mathbf{W}^{-1}\mathbf{J}^T\left(\mathbf{J}\mathbf{W}^{-1}\mathbf{J}^T\right) \tag{8.475}$$

The matrix $\mathbf{J}^+$ has the meaning of a pseudo-inverse to the singular Jacobian matrix $\mathbf{J}$.

$$\mathbf{J}^+ = \mathbf{W}^{-1}\mathbf{J}^T\left(\mathbf{J}\mathbf{W}^{-1}\mathbf{J}^T\right)^{-1} \tag{8.476}$$

It verifies the identity

$$\mathbf{J}\mathbf{J}^+ = \mathbf{I} \tag{8.477}$$

and, whenever $\mathbf{J}$ is invertible, we have

$$\mathbf{J}^+ = \mathbf{J}^{-1} \tag{8.478}$$

8.8 ★ Jacobian Matrix From Link Transformation Matrices

In robot motion, we need to calculate the Jacobian matrix in a very short time for every configuration of the robot. The Jacobian matrix of a robot can be found easier in an algorithmic way by evaluating columns of the Jacobian $\mathbf{J}$

$$\mathbf{J} = \begin{bmatrix} \mathbf{c}_1 & \mathbf{c}_2 & \cdots & \mathbf{c}_n \end{bmatrix}$$

$$= \begin{bmatrix} {}^0\tilde{k}_0\,{}^0_0\mathbf{d}_n & {}^0\tilde{k}_1\,{}^0_1\mathbf{d}_n & \cdots & {}^0\tilde{k}_{n-1}\,{}^0_{n-1}\mathbf{d}_n \\ {}^0\hat{k}_0 & {}^0\hat{k}_1 & \cdots & {}^0\hat{k}_{n-1} \end{bmatrix} \tag{8.479}$$

$$= \begin{bmatrix} {}^0\hat{k}_0 \times {}^0_0\mathbf{d}_n & {}^0\hat{k}_1 \times {}^0_1\mathbf{d}_n & \cdots & {}^0\hat{k}_{n-1} \times {}^0_{n-1}\mathbf{d}_n \\ {}^0\hat{k}_0 & {}^0\hat{k}_1 & \cdots & {}^0\hat{k}_{n-1} \end{bmatrix} \tag{8.480}$$

where $\mathbf{c}_i$ is called the *Jacobian generating vector* and ${}^0\hat{k}_{i-1}$ is the vector associated to the skew matrix ${}^0\tilde{k}_{i-1}$.

$$\mathbf{c}_i = \begin{bmatrix} {}^0\tilde{k}_{i-1}\,{}^0_{i-1}\mathbf{d}_n \\ {}^0\hat{k}_{i-1} \end{bmatrix} = \begin{bmatrix} {}^0\hat{k}_{i-1} \times {}^0_{i-1}\mathbf{d}_n \\ {}^0\hat{k}_{i-1} \end{bmatrix} \tag{8.481}$$

This method is solely based on link transformation matrices found in forward kinematics and does not involve differentiation.

The matrix ${}^0\tilde{k}_{i-1}$ is

$$ {}^0\tilde{k}_{i-1} = {}^0R_{i-1}\,{}^{i-1}\tilde{k}_{i-1}\,{}^0R_{i-1}^T \tag{8.482}$$

which means ${}^0\hat{k}_{i-1}$ to be a unit vector in the direction of joint axis i in the global coordinate frame. For a revolute joint, we have

$$\mathbf{c}_i = \begin{bmatrix} {}^0\tilde{k}_{i-1}\,{}^0_{i-1}\mathbf{d}_n \\ {}^0\hat{k}_{i-1} \end{bmatrix} \tag{8.483}$$

and for a prismatic joint we have

$$\mathbf{c}_i = \begin{bmatrix} {}^0\hat{k}_{i-1} \\ 0 \end{bmatrix} \tag{8.484}$$

Proof Transformation between two coordinate frames is based on a transformation matrix that is a combination of rotation matrix R and the position vector $\mathbf{d}$.

$$ {}^G\mathbf{r} = {}^GT_B\,{}^B\mathbf{r} \tag{8.485}$$

$$T = \begin{bmatrix} R & \mathbf{d} \\ 0 & 1 \end{bmatrix} \tag{8.486}$$

Introducing the infinitesimal transformation matrix

$$\delta T = \begin{bmatrix} \delta R & \delta\mathbf{d} \\ 0 & 0 \end{bmatrix} \tag{8.487}$$

leads to

$$\delta T\ T^{-1} = \begin{bmatrix} \tilde{\delta\theta} & \delta\mathbf{v} \\ 0 & 0 \end{bmatrix} \tag{8.488}$$

where

$$T^{-1} = \begin{bmatrix} R^T & -R^T\mathbf{d} \\ 0 & 1 \end{bmatrix} \tag{8.489}$$

and, therefore, $\tilde{\delta\theta}$ is the infinitesimal rotations matrix,

$$\widetilde{\delta\theta} = \delta R \ R^T = \begin{bmatrix} 0 & -\delta\theta_z & \delta\theta_y \\ \delta\theta_z & 0 & -\delta\theta_x \\ -\delta\theta_y & \delta\theta_x & 0 \end{bmatrix} \tag{8.490}$$

and $\delta\mathbf{v}$ is a vector related to infinitesimal displacements.

$$\delta\mathbf{v} = \delta\mathbf{d} - \widetilde{\delta\theta} \ \mathbf{d} \tag{8.491}$$

Let us define a 6×1 coordinate vector describing the rotational and translational coordinates of the end-effector

$$\mathbf{X} = \begin{bmatrix} \mathbf{d} \\ \boldsymbol{\theta} \end{bmatrix} \tag{8.492}$$

and its variation is $\delta\mathbf{X}$.

$$\delta\mathbf{X} = \begin{bmatrix} \delta\mathbf{d} \\ \delta\boldsymbol{\theta} \end{bmatrix} \tag{8.493}$$

The Jacobian matrix $\mathbf{J}$ is then a matrix that maps differential joint variables to differential end-effector motion.

$$\delta\mathbf{X} = \frac{\partial \mathbf{T}(\mathbf{q})}{\partial \mathbf{q}} \delta\mathbf{q} = \mathbf{J} \ \delta\mathbf{q} \tag{8.494}$$

The transformation matrix T, generated in forward kinematics, is a function of joint coordinates

$$^0T_n = \mathbf{T}(\mathbf{q}) \tag{8.495}$$
$$= {}^0T_1(q_1) \, {}^1T_2(q_2) \, {}^2T_3(q_3) \, {}^3T_4(q_4) \cdots {}^{n-1}T_n(q_n)$$

therefore, the infinitesimal transformation matrix is

$$\delta T = \sum_{i=1}^{n} {}^0T_1(q_1) \, {}^1T_2(q_2) \cdots \frac{\delta\left({}^{i-1}T_i\right)}{\delta q_i} \cdots {}^{n-1}T_n(q_n) \cdot \delta q_i \tag{8.496}$$

Interestingly, the partial derivative of the transformation matrix can be arranged in the form

$$\frac{\delta\left({}^{i-1}T_i\right)}{\delta q_i} = {}^{i-1}\Delta_{i-1} \, {}^{i-1}T_i \tag{8.497}$$

where according to DH transformation matrix ${}^{i-1}T_i$ is

$$^{i-1}T_i = \begin{bmatrix} \cos\theta_i & -\sin\theta_i\cos\alpha_i & \sin\theta_i\sin\alpha_i & a_i\cos\theta_i \\ \sin\theta_i & \cos\theta_i\cos\alpha_i & -\cos\theta_i\sin\alpha_i & a_i\sin\theta_i \\ 0 & \sin\alpha_i & \cos\alpha_i & d_i \\ 0 & 0 & 0 & 1 \end{bmatrix}$$
$$= \begin{bmatrix} {}^{i-1}R_i & {}^{i-1}\mathbf{d}_i \\ 0 & 1 \end{bmatrix} \tag{8.498}$$

The *velocity coefficient matrices* Δ_i for a revolute joint is

$$^{i-1}\Delta_{i-1} = \Delta_R = \begin{bmatrix} {}^{i-1}\tilde{k}_{i-1} & 0 \\ 0 & 0 \end{bmatrix} = \begin{bmatrix} 0 & -1 & 0 & 0 \\ 1 & 0 & 0 & 0 \\ 0 & 0 & 0 & 0 \\ 0 & 0 & 0 & 0 \end{bmatrix} \tag{8.499}$$

and for a prismatic joint is

$$^{i-1}\Delta_{i-1} = \Delta_P = \begin{bmatrix} 0 & {}^{i-1}\hat{k}_{i-1} \\ 0 & 0 \end{bmatrix} = \begin{bmatrix} 0 & 0 & 0 & 0 \\ 0 & 0 & 0 & 0 \\ 0 & 0 & 0 & 1 \\ 0 & 0 & 0 & 0 \end{bmatrix} \tag{8.500}$$

We may now express each term of (8.496) in the following form.

$$^0T_1(q_1)\,{}^1T_2(q_2) \cdots \frac{\delta\left({}^{i-1}T_i\right)}{\delta q_i} \cdots {}^{n-1}T_n(q_n) = C_i\,T \tag{8.501}$$

$$\begin{aligned} C_i &= \begin{bmatrix} {}^0T_1\,{}^1T_2 \cdots {}^{i-2}T_{i-1} \end{bmatrix} \frac{\delta\left({}^{i-1}T_i\right)}{\delta q_i} \begin{bmatrix} {}^0T_1\,{}^1T_2 \cdots {}^{i-1}T_i \end{bmatrix}^{-1} \\ &= \begin{bmatrix} {}^0T_1\,{}^1T_2 \cdots {}^{i-2}T_{i-1} \end{bmatrix} {}^{i-1}\Delta_{i-1}\,{}^{i-1}T_i \begin{bmatrix} {}^0T_1\,{}^1T_2 \cdots {}^{i-1}T_i \end{bmatrix}^{-1} \\ &= \begin{bmatrix} {}^0T_1\,{}^1T_2 \cdots {}^{i-2}T_{i-1} \end{bmatrix} {}^{i-1}\Delta_{i-1} \begin{bmatrix} {}^0T_1\,{}^1T_2 \cdots {}^{i-2}T_{i-1} \end{bmatrix}^{-1} \\ &= {}^0T_1\,{}^1T_2 \cdots {}^{i-2}T_{i-1}\,{}^{i-1}\Delta_{i-1}\,{}^{i-2}T_{i-1}^{-1} \cdots {}^1T_2^{-1}\,{}^0T_1^{-1} \end{aligned} \tag{8.502}$$

The matrix C_i can be rearranged for a revolute joint,

$$C_i = \begin{bmatrix} {}^0\hat{k}_{i-1} & {}^0\tilde{k}_{i-1}\,{}^0\mathbf{d}_n - {}^0\tilde{k}_{i-1}\,{}^0\mathbf{d}_{i-1} \\ 0 & 0 \end{bmatrix} \tag{8.503}$$

and for a prismatic joint.

$$C_i = \begin{bmatrix} {}^0\hat{k}_{i-1} & 0 \\ 0 & 0 \end{bmatrix} \tag{8.504}$$

C_i has six independent terms that can be combined in a 6×1 vector. This is the generating vector $\mathbf{c}_i$ that makes the ith column of the Jacobian matrix. The Jacobian generating vector for a revolute joint is

$$\mathbf{c}_i = \begin{bmatrix} {}^0\tilde{k}_{i-1}\,{}^0_{i-1}\mathbf{d}_n \\ {}^0\hat{k}_{i-1} \end{bmatrix} \tag{8.505}$$

and for a prismatic joint is

$$\mathbf{c}_i = \begin{bmatrix} 0 \\ {}^0\hat{k}_{i-1} \end{bmatrix} \tag{8.506}$$

The position vector ${}^0\mathbf{d}_i$ indicated the origin of the B_i-frame in the base B_0-frame. Hence, ${}^0_{i-1}\mathbf{d}_n$ indicated the origin of the end-effector coordinate B_n-frame with respect to coordinate B_{i-1}-frame and expressed in the base frame B_0.

Therefore, the Jacobian matrix describing the instantaneous kinematics of the robot can be obtained.

$$\mathbf{J} = \begin{bmatrix} {}^0\tilde{k}_0\,{}^0_0\mathbf{d}_n & {}^0\tilde{k}_1\,{}^0_1\mathbf{d}_n & \cdots & {}^0\tilde{k}_{n-1}\,{}^0_{n-1}\mathbf{d}_n \\ {}^0\hat{k}_0 & {}^0\hat{k}_1 & \cdots & {}^0\hat{k}_{n-1} \end{bmatrix} \tag{8.507}$$

■

Example 294 Jacobian matrix for articulated robots.

The forward and inverse kinematics of the articulated robot has been analyzed in Example 214 with the following individual transformation matrices:

$$
{}^0T_1 = \begin{bmatrix} c\theta_1 & 0 & s\theta_1 & 0 \\ s\theta_1 & 0 & -c\theta_1 & 0 \\ 0 & 1 & 0 & 0 \\ 0 & 0 & 0 & 1 \end{bmatrix} \quad {}^1T_2 = \begin{bmatrix} c\theta_2 & -s\theta_2 & 0 & l_2 c\theta_2 \\ s\theta_2 & c\theta_2 & 0 & l_2 s\theta_2 \\ 0 & 0 & 1 & d_2 \\ 0 & 0 & 0 & 1 \end{bmatrix}
$$

$$
{}^2T_3 = \begin{bmatrix} c\theta_3 & 0 & s\theta_3 & 0 \\ s\theta_3 & 0 & -c\theta_3 & 0 \\ 0 & 1 & 0 & 0 \\ 0 & 0 & 0 & 1 \end{bmatrix} \quad {}^3T_4 = \begin{bmatrix} c\theta_4 & 0 & -s\theta_4 & 0 \\ s\theta_4 & 0 & c\theta_4 & 0 \\ 0 & -1 & 0 & l_3 \\ 0 & 0 & 0 & 1 \end{bmatrix}
$$

$$
{}^4T_5 = \begin{bmatrix} c\theta_5 & 0 & s\theta_5 & 0 \\ s\theta_5 & 0 & -c\theta_5 & 0 \\ 0 & 1 & 0 & 0 \\ 0 & 0 & 0 & 1 \end{bmatrix} \quad {}^5T_6 = \begin{bmatrix} c\theta_6 & -s\theta_6 & 0 & 0 \\ s\theta_6 & c\theta_6 & 0 & 0 \\ 0 & 0 & 1 & 0 \\ 0 & 0 & 0 & 1 \end{bmatrix} \tag{8.508}
$$

The articulated robot has 6 DOF and, therefore, its Jacobian matrix is a 6×6 matrix

$$
\mathbf{J}(\mathbf{q}) = \begin{bmatrix} \mathbf{c}_1(\mathbf{q}) & \mathbf{c}_2(\mathbf{q}) & \cdots & \mathbf{c}_6(\mathbf{q}) \end{bmatrix} \tag{8.509}
$$

that relates the translational and angular velocities of the end-effector to the joints' velocities $\dot{\mathbf{q}}$.

$$
\begin{bmatrix} \mathbf{v} \\ \boldsymbol{\omega} \end{bmatrix} = \mathbf{J}(\mathbf{q})\,\dot{\mathbf{q}} \tag{8.510}
$$

The ith column vector $\mathbf{c}_i(\mathbf{q})$ for a revolute joint is given by

$$
\mathbf{c}_i(\mathbf{q}) = \begin{bmatrix} {}^0\hat{k}_{i-1} \times {}^0_{i-1}\mathbf{d}_6 \\ {}^0\hat{k}_{i-1} \end{bmatrix} \tag{8.511}
$$

and for a prismatic joint is given as:

$$
\mathbf{c}_i(\mathbf{q}) = \begin{bmatrix} {}^0\hat{k}_{i-1} \\ 0 \end{bmatrix} \tag{8.512}
$$

Column 1 The first column of the Jacobian matrix has the simplest calculation, because it is based on the contribution of the z_0-axis and the position of the end-effector frame ${}^0\mathbf{d}_6$ and these are the result of forward kinematics. The direction of the z_0-axis in the base coordinate frame always is

$$
{}^0\hat{k}_0 = \begin{bmatrix} 0 \\ 0 \\ 1 \end{bmatrix} \tag{8.513}
$$

and the position vector of the end-effector frame B_6 is given by ${}^0\mathbf{d}_6$ directly determined from ${}^0T_6$,

$$
{}^0T_6 = {}^0T_1\,{}^1T_2\,{}^2T_3\,{}^3T_4\,{}^4T_5\,{}^5T_6
$$

$$
= \begin{bmatrix} {}^0R_6 & {}^0\mathbf{d}_6 \\ 0 & 1 \end{bmatrix} = \begin{bmatrix} t_{11} & t_{12} & t_{13} & t_{14} \\ t_{21} & t_{22} & t_{23} & t_{24} \\ t_{31} & t_{32} & t_{33} & t_{34} \\ 0 & 0 & 0 & 1 \end{bmatrix} \tag{8.514}
$$

$$
{}^0\mathbf{d}_6 = \begin{bmatrix} t_{14} \\ t_{24} \\ t_{34} \end{bmatrix} \tag{8.515}
$$

where

$$
\begin{aligned}
t_{14} = & \, d_6 \left(s\theta_1 s\theta_4 s\theta_5 + c\theta_1 \left(c\theta_4 s\theta_5 c \left(\theta_2 + \theta_3 \right) \right. \right. \\
& \left. \left. + c\theta_5 s \left(\theta_2 + \theta_3 \right) \right) \right) \\
& + l_3 c\theta_1 s \left(\theta_2 + \theta_3 \right) + d_2 s\theta_1 + l_2 c\theta_1 c\theta_2
\end{aligned}
\tag{8.516}
$$

$$
\begin{aligned}
t_{24} = & \, d_6 \left(-c\theta_1 s\theta_4 s\theta_5 + s\theta_1 \left(c\theta_4 s\theta_5 c \left(\theta_2 + \theta_3 \right) \right. \right. \\
& \left. \left. + c\theta_5 s \left(\theta_2 + \theta_3 \right) \right) \right) \\
& + s\theta_1 s \left(\theta_2 + \theta_3 \right) l_3 - d_2 c\theta_1 + l_2 c\theta_2 s\theta_1
\end{aligned}
\tag{8.517}
$$

$$
\begin{aligned}
t_{34} = & \, d_6 \left(c\theta_4 s\theta_5 s \left(\theta_2 + \theta_3 \right) - c\theta_5 c \left(\theta_2 + \theta_3 \right) \right) \\
& + l_2 s\theta_2 + l_3 c \left(\theta_2 + \theta_3 \right).
\end{aligned}
\tag{8.518}
$$

Therefore,

$$
{}^0\hat{k}_0 \times {}^0\mathbf{d}_6 = {}^0\tilde{k}_0 \, {}^0\mathbf{d}_6
$$

$$
= \begin{bmatrix} 0 & -1 & 0 \\ 1 & 0 & 0 \\ 0 & 0 & 0 \end{bmatrix} \begin{bmatrix} t_{14} \\ t_{24} \\ t_{34} \end{bmatrix} = \begin{bmatrix} -t_{24} \\ t_{14} \\ 0 \end{bmatrix}
\tag{8.519}
$$

and the first Jacobian generating vector is

$$
\mathbf{c}_1 = \begin{bmatrix} -t_{24} \\ t_{14} \\ 0 \\ 0 \\ 0 \\ 1 \end{bmatrix}
\tag{8.520}
$$

Column 2 The z_1-axis in the base frame can be found by transformation.

$$
{}^0\hat{k}_1 = {}^0R_1 \begin{bmatrix} 0 \\ 0 \\ 1 \end{bmatrix} = \begin{bmatrix} c\theta_1 & 0 & s\theta_1 \\ s\theta_1 & 0 & -c\theta_1 \\ 0 & 1 & 0 \end{bmatrix} \begin{bmatrix} 0 \\ 0 \\ 1 \end{bmatrix} = \begin{bmatrix} \sin\theta_1 \\ -\cos\theta_1 \\ 0 \end{bmatrix}
\tag{8.521}
$$

The second half of $\mathbf{c}_2$ needs the cross product of ${}^0\hat{k}_1$ and position vector ${}^1\mathbf{d}_6$. The vector ${}^1\mathbf{d}_6$ is the position of the end-effector in the coordinate frame B_1; however, it must be described in the base frame to be able to perform the cross product. An easier method is to find ${}^1\hat{k}_1 \times {}^1\mathbf{d}_6$ and transform the resultant into the base frame.

$$
{}^0\hat{k}_1 \times {}^1\mathbf{d}_6 = {}^0R_1 \left({}^1\hat{k}_1 \times {}^1\mathbf{d}_6 \right)
$$

$$
= {}^0R_1 \left(\begin{bmatrix} 0 \\ 0 \\ 1 \end{bmatrix} \times \begin{bmatrix} l_2 \cos\theta_2 + l_3 \sin(\theta_2 + \theta_3) \\ l_2 \sin\theta_2 - l_3 \cos(\theta_2 + \theta_3) \\ d_2 \end{bmatrix} \right)
$$

$$
= \begin{bmatrix} \cos\theta_1 \left(-l_2 \sin\theta_2 + l_3 \cos(\theta_2 + \theta_3) \right) \\ \sin\theta_1 \left(-l_2 \sin\theta_2 + l_3 \cos(\theta_2 + \theta_3) \right) \\ l_2 \cos\theta_2 + l_3 \sin(\theta_2 + \theta_3) \end{bmatrix}
\tag{8.522}
$$

Therefore, $\mathbf{c}_2$ is found as a 6×1 vector.

$$\mathbf{c}_2 = \begin{bmatrix} \cos\theta_1 \left(-l_2 \sin\theta_2 + l_3 \cos(\theta_2 + \theta_3)\right) \\ \sin\theta_1 \left(-l_2 \sin\theta_2 + l_3 \cos(\theta_2 + \theta_3)\right) \\ l_2 \cos\theta_2 + l_3 \sin(\theta_2 + \theta_3) \\ \sin\theta_1 \\ -\cos\theta_1 \\ 0 \end{bmatrix} \tag{8.523}$$

Column 3 The z_2-axis in the base frame can be found using the same method.

$$^0\hat{k}_2 = {}^0R_2 \, {}^2\hat{k}_2 = {}^0R_1 \, {}^1R_2 \begin{bmatrix} 0 \\ 0 \\ 1 \end{bmatrix} = \begin{bmatrix} \sin\theta_1 \\ -\cos\theta_1 \\ 0 \end{bmatrix} \tag{8.524}$$

The second half of $\mathbf{c}_3$ can be found by finding $^2\hat{k}_2 \times {}^2\mathbf{d}_6$ and transforming the resultant into the base coordinate frame.

$$^2\hat{k}_2 \times {}^2\mathbf{d}_6 = \begin{bmatrix} l_3 \cos\theta_3 \\ l_3 \sin\theta_3 \\ 0 \end{bmatrix} \tag{8.525}$$

$$^0R_2 \left({}^2\hat{k}_2 \times {}^2\mathbf{d}_6 \right) = \begin{bmatrix} l_3 \cos\theta_1 \sin(\theta_2 + \theta_3) \\ l_3 \sin\theta_1 \sin(\theta_2 + \theta_3) \\ -l_3 \cos(\theta_2 + \theta_3) \end{bmatrix} \tag{8.526}$$

Therefore, $\mathbf{c}_3$ is

$$\mathbf{c}_3 = \begin{bmatrix} l_3 \cos\theta_1 \sin(\theta_2 + \theta_3) \\ l_3 \sin\theta_1 \sin(\theta_2 + \theta_3) \\ -l_3 \cos(\theta_2 + \theta_3) \\ \sin\theta_1 \\ -\cos\theta_1 \\ 0 \end{bmatrix} \tag{8.527}$$

Column 4 The z_3-axis in the base frame is

$$^0\hat{k}_3 = {}^0R_1 \, {}^1R_2 \, {}^2R_3 \begin{bmatrix} 0 \\ 0 \\ 1 \end{bmatrix}$$

$$= \begin{bmatrix} \cos\theta_1 \left(\cos\theta_2 \sin\theta_3 + \cos\theta_3 \sin\theta_2\right) \\ \sin\theta_1 \left(\cos\theta_2 \sin\theta_3 + \sin\theta_2 \cos\theta_3\right) \\ -\cos(\theta_2 + \theta_3) \end{bmatrix} \tag{8.528}$$

and the second half of $\mathbf{c}_4$ can be found by finding $^3\hat{k}_3 \times {}^3\mathbf{d}_6$ and transforming the resultant into the base coordinate frame.

$$^0R_3 \left({}^3\hat{k}_3 \times {}^3\mathbf{d}_6 \right) = {}^0R_3 \left(\begin{bmatrix} 0 \\ 0 \\ 1 \end{bmatrix} \times \begin{bmatrix} 0 \\ 0 \\ l_3 \end{bmatrix} \right) = \begin{bmatrix} 0 \\ 0 \\ 0 \end{bmatrix} \tag{8.529}$$

Therefore, $\mathbf{c}_4$ is

$$\mathbf{c}_4 = \begin{bmatrix} 0 \\ 0 \\ 0 \\ \cos\theta_1 \left(\cos\theta_2 \sin\theta_3 + \cos\theta_3 \sin\theta_2\right) \\ \sin\theta_1 \left(\cos\theta_2 \sin\theta_3 + \sin\theta_2 \cos\theta_3\right) \\ -\cos(\theta_2 + \theta_3) \end{bmatrix} \tag{8.530}$$

Column 5 The z_4-axis in the base frame is

$$
{}^0\hat{k}_4 = {}^0R_4 \begin{bmatrix} 0 \\ 0 \\ 1 \end{bmatrix} = \begin{bmatrix} c\theta_4 s\theta_1 - c\theta_1 s\theta_4 c\,(\theta_2 + \theta_3) \\ -c\theta_1 c\theta_4 - s\theta_1 s\theta_4 c\,(\theta_2 + \theta_3) \\ -s\theta_4 s\,(\theta_2 + \theta_3) \end{bmatrix}
\tag{8.531}
$$

and the second half of $\mathbf{c}_5$ can be found by finding ${}^4\hat{k}_4 \times {}^4\mathbf{d}_6$ and transforming the resultant into the base coordinate frame.

$$
{}^0R_4 \left({}^4\hat{k}_4 \times {}^4\mathbf{d}_6 \right) = {}^0R_4 \left(\begin{bmatrix} 0 \\ 0 \\ 1 \end{bmatrix} \times \begin{bmatrix} 0 \\ 0 \\ 0 \end{bmatrix} \right) = \begin{bmatrix} 0 \\ 0 \\ 0 \end{bmatrix}
\tag{8.532}
$$

Therefore, $\mathbf{c}_5$ is

$$
\mathbf{c}_5 = \begin{bmatrix} 0 \\ 0 \\ 0 \\ \cos\theta_4 \sin\theta_1 - \cos\theta_1 \sin\theta_4 \cos\,(\theta_2 + \theta_3) \\ -\cos\theta_1 \cos\theta_4 - \sin\theta_1 \sin\theta_4 \cos\,(\theta_2 + \theta_3) \\ -\sin\theta_4 \sin\,(\theta_2 + \theta_3) \end{bmatrix}
\tag{8.533}
$$

Column 6 The z_5-axis in the base frame is

$$
{}^0\hat{k}_5 = {}^0R_5 \begin{bmatrix} 0 \\ 0 \\ 1 \end{bmatrix}
\tag{8.534}
$$

$$
= \begin{bmatrix} -c\theta_1 c\theta_4 s\,(\theta_2 + \theta_3) - s\theta_4\,(s\theta_1 s\theta_4 + c\theta_1 c\theta_4 c\,(\theta_2 + \theta_3)) \\ -s\theta_1 c\theta_4 s\,(\theta_2 + \theta_3) - s\theta_4\,(-c\theta_1 s\theta_4 + s\theta_1 c\theta_4 c\,(\theta_2 + \theta_3)) \\ c\theta_4 c\,(\theta_2 + \theta_3) - \frac{1}{2}s\,(\theta_2 + \theta_3)\,s2\theta_4 \end{bmatrix}
$$

and the second half of $\mathbf{c}_6$ can be found by finding ${}^5\hat{k}_5 \times {}^5\mathbf{d}_6$ and transforming the resultant into the base coordinate frame.

$$
{}^0R_5 \left({}^5\hat{k}_5 \times {}^5\mathbf{d}_6 \right) = {}^0R_5 \left(\begin{bmatrix} 0 \\ 0 \\ 1 \end{bmatrix} \times \begin{bmatrix} 0 \\ 0 \\ 0 \end{bmatrix} \right) = \begin{bmatrix} 0 \\ 0 \\ 0 \end{bmatrix}
\tag{8.535}
$$

Therefore, $\mathbf{c}_6$ is

$$
\mathbf{c}_6 = \begin{bmatrix} 0 \\ 0 \\ 0 \\ -c\theta_1 c\theta_4 s\,(\theta_2 + \theta_3) - s\theta_4\,(s\theta_1 s\theta_4 + c\theta_1 c\theta_4 c\,(\theta_2 + \theta_3)) \\ -s\theta_1 c\theta_4 s\,(\theta_2 + \theta_3) - s\theta_4\,(-c\theta_1 s\theta_4 + s\theta_1 c\theta_4 c\,(\theta_2 + \theta_3)) \\ c\theta_4 c\,(\theta_2 + \theta_3) - \frac{1}{2}s\,(\theta_2 + \theta_3)\,s2\theta_4 \end{bmatrix}
\tag{8.536}
$$

8.9   Summary

Each link of a serial robot has an angular and a translational velocity. The angular velocity of link (i) in the global coordinate frame can be found as a summation of the global angular velocities of its lower links

$$
{}_0^0\boldsymbol{\omega}_i = \sum_{j=1}^{i} {}_{j-1}^{0}\boldsymbol{\omega}_j .
$$

(8.537)

Using DH parameters, the angular velocity of link (j) with respect to link $(j-1)$ is

$$
{}_{j-1}^{0}\boldsymbol{\omega}_j = \begin{cases} \dot{\theta}_j \ {}^{0}\hat{k}_{j-1} & \text{if joint } i \text{ is R} \\ 0 & \text{if joint } i \text{ is P} \end{cases}
$$

(8.538)

The translational velocity of link (i) is the global velocity of the origin of coordinate frame B_i attached to link (i)

$$
{}_{i-1}^{0}\dot{\mathbf{d}}_i = \begin{cases} {}_0^0\boldsymbol{\omega}_i \times {}_{i-1}^{0}\mathbf{d}_i & \text{if joint } i \text{ is R} \\ \dot{d}_i \ {}^{0}\hat{k}_{i-1} + {}_0^0\boldsymbol{\omega}_i \times {}_{i-1}^{0}\mathbf{d}_i & \text{if joint } i \text{ is P} \end{cases}
$$

(8.539)

where θ and d are DH parameters and $\mathbf{d}$ is the frame's origin position vector.

The velocity kinematics of a robot is defined by the relationship between joint speeds $\dot{\mathbf{q}}$

$$
\dot{\mathbf{q}} = \begin{bmatrix} \dot{q}_n \ \dot{q}_n \ \dot{q}_n \ \cdots \ \dot{q}_n \end{bmatrix}^T
$$

(8.540)

and global speeds of the end-effector $\dot{\mathbf{X}}$.

$$
\dot{\mathbf{X}} = \begin{bmatrix} \dot{X}_n \ \dot{Y}_n \ \dot{Z}_n \ \omega_{Xn} \ \omega_{Yn} \ \omega_{Zn} \end{bmatrix}^T
$$

(8.541)

Such a relationship introduces Jacobian matrix $\mathbf{J}$.

$$
\dot{\mathbf{X}} = \mathbf{J}\,\dot{\mathbf{q}}
$$

(8.542)

Having $\mathbf{J}$, we are able to find the end-effector speeds for a given set of joint speeds and vice versa. Jacobian is a function of joint coordinates and is the main tool in velocity kinematics of robots.

We practically calculate $\mathbf{J}$ by Jacobian generating vectors denoted by $\mathbf{c}_i(\mathbf{q})$

$$
\mathbf{c}_i(\mathbf{q}) = \begin{bmatrix} {}^{0}\hat{k}_{i-1} \times {}_{i-1}^{0}\mathbf{d}_n \\ {}^{0}\hat{k}_{i-1} \end{bmatrix}
$$

(8.543)

where $\mathbf{c}_i(\mathbf{q})$ makes the column $i-1$ of the Jacobian matrix.

$$
\mathbf{J} = \begin{bmatrix} \mathbf{c}_0 \ \mathbf{c}_1 \ \mathbf{c}_2 \ \cdots \ \mathbf{c}_{n-1} \end{bmatrix}
$$

(8.544)

There are some general numerical calculations needed in robot kinematics. Solutions to a set of linear and nonlinear algebraic equations are the most important ones for calculating a matrix inversion and a Jacobian matrix. An applied solution for a set of linear equations is LU factorization, and a practical method for a set of nonlinear equations is the Newton–Raphson method. Both of these methods are cast in applied algorithms.

8.10 Key Symbols

a	Kinematic length of a link
$\mathbf{a}$	Turn vector of end-effector frame
A	Coefficient matrix
A, B, C, D	Submatrices of $\mathbf{J}$
a_{ij}	The element of row i and column j of A
b	The vector of known values in a set of linear equations
B	Body coordinate frame
	Dummy matrix with upper U and lower L
c	cos
$\mathbf{c}$	Jacobian generating vector
con	Condition number
d	Differential, prismatic joint variable
d_x, d_y, d_z	Elements of $\mathbf{d}$
$\mathbf{d}$	Translation vector, displacement vector
det	Determinant
D	Displacement transformation matrix
	Lower right submatrix of B
e	Rotation quaternion
$\mathbf{f}$	A set of nonlinear algebraic equations
G, B_0	Global coordinate frame, base coordinate frame
H	Dummy matrix to calculate D
$\hat{\imath}, \hat{\jmath}, \hat{k}$	Local coordinate axes unit vectors
$\tilde{\imath}, \tilde{\jmath}, \tilde{k}$	Skew symmetric matrices of the unit vector $\hat{\imath}, \hat{\jmath}, \hat{k}$
$\hat{I}, \hat{J}, \hat{K}$	Global coordinate axes unit vectors
$\mathbf{I} = [\mathbf{I}]$	Identity matrix
$\mathbf{J}$	Jacobian, geometric Jacobian
$\mathbf{J}_D$	Displacement Jacobian
$\mathbf{J}_R$	Rotational Jacobian
$\mathbf{J}_\phi$	Angular Jacobian
$\mathbf{J}_A$	Analytic Jacobian
l	Length
l_{ij}	The element of row i and column j of L
L	Lower triangle submatrix of A
m	Number of independent equations
n	Number of rows and columns of A
P	Prismatic joint, point
q	The vector of unknowns of $\mathbf{f}$, vector of joint variables
	Joint coordinate
$\mathbf{q}$	Vector joint coordinates
r_i	The element i of $\mathbf{r}$
r_{ij}	The element of row i and column j of a matrix
$\mathbf{r}$	Position vectors, homogenous position vector
R	Rotation transformation matrix, revolute joint
s	sin
$\mathbf{s}^T, \mathbf{l}^T$	Nondiagonal first column of D
sgn	Signum function
t_{ij}	The element of row i and column j of T
T	Homogenous transformation matrix
T_{arm}	Manipulator transformation matrix
T_{wrist}	Wrist transformation matrix
$\mathbf{T}$	A set of nonlinear algebraic equations of $\mathbf{q}$

u_{ij}	The element of row i and column j of U
$\hat{u}$	Unit vector along the axis of $\boldsymbol{\omega}$
$\tilde{u}$	Skew symmetric matrix of the vector $\hat{u}$
u_1, u_2, u_3	Components of $\hat{u}$
U	Upper triangle submatrix of A
$\mathbf{u}^T, \mathbf{r}^T$	Nondiagonal first row of D
$\mathbf{v}$	Velocity vector
V	Velocity transformation matrix
$\mathbf{W}$	Weight factor matrix
$\mathbf{x}$	Vector of unknowns
X, Y, Z	Global coordinate axes
$\mathbf{y}$	Dummy vector of unknowns
x, y, z	Local coordinate axes
X, Y, Z	Global coordinate axes, coordinates of end-effector

Greek

α, β, γ	Angles of rotation about the axes of global frame
δ	Kronecker function, small increment of a parameter
δ	Small increment of a parameter
$\boldsymbol{\delta}$	Difference in $\mathbf{q}$ for in two steps of iteration
ϵ	Small test number to terminate a procedure
θ	Rotary joint angle
θ_{ijk}	$\theta_i + \theta_j + \theta_k$
$\boldsymbol{\theta}$	Vector of θ_i
φ, θ, ψ	Angles of rotation about the axes of body frame
ϕ	Angle of rotation about $\hat{u}$
$\boldsymbol{\omega}$	Angular velocity vector
$\tilde{\omega}$	Skew symmetric matrix of the vector $\boldsymbol{\omega}$
$\omega_1, \omega_2, \omega_3$	Components of $\boldsymbol{\omega}$

Symbol

DOF	Degree of freedom
$[\quad]^{-1}$	Inverse of the matrix $[\quad]$
$[\quad]^T$	Transpose of the matrix $[\quad]$
$\vdash$	Orthogonal
(i)	Link number i
$\|$	Parallel
$\perp$	Perpendicular
e^*	Conjugate of e
$\|\quad\|$	Norm of the matrix $[\quad]$
$[\quad]^{-1}$	Inverse of the matrix $[\quad]$
$[\quad]^T$	Transpose of the matrix $[\quad]$
$[\quad]^+$	Pseudo-inverse of the matrix $[\quad]$
$\equiv$	Equivalent
$\vdash$	Orthogonal
(i)	Link number i
$\|$	Parallel sign
$\perp$	perpendicular
$\times$	Vector cross product
$\mathbf{q}^\star$	A guess value for $\mathbf{q}$
$\triangle$	Perturbation in a vector or a matrix

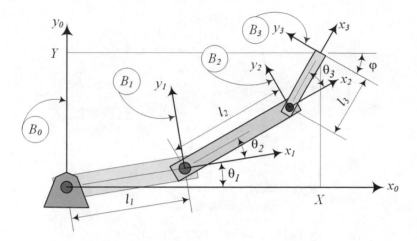

Fig. 8.11 A $3R$, R‖R‖R planar manipulator

Exercises

1. Notation and symbols.
   Describe the meaning of following notations.

$$a -^{i-1}\mathbf{d}_i b -_{i-1}^{0}\mathbf{d}_i c -_{i-1}^{i}\mathbf{d}_i d -_{i}^{i-1}\mathbf{d}_{i-1} e -_{i-1}^{i-1}\mathbf{d}_i f -_{0}^{i-1}\mathbf{d}_i$$

$$g -^{i-1}\dot{\mathbf{d}}_i h -_{i-1}^{0}\dot{\mathbf{d}}_i i -_{i-1}^{i}\dot{\mathbf{d}}_i j -_{i}^{i-1}\dot{\mathbf{d}}_{i-1} k -_{i-1}^{i-1}\dot{\mathbf{d}}_i l -_{0}^{i-1}\dot{\mathbf{d}}_i$$

$$m -^{0}\hat{k}_i n -^{0}\hat{k}_{i-1} o -_{i-1}^{0}\hat{k}_{i-2} p -_{i-1}^{0}\mathbf{v}_i q -_{i-1}^{i}\mathbf{v}_i r -^{i}\mathbf{v}_i$$

$$s - [L]\, t - [U]\, u - [B]\, v - [D_i]\, w - u_{ii} x - d_{ii}$$

$$y - \mathrm{con}\,(A)\, z - \|A\|_\infty\, A - \|A\|_1\, B - \|A\|_2\, C - \mathbf{c}_i D - \mathbf{X}$$

$$E - \mathbf{J} F - \dot{\mathbf{q}} G -^{i-1}\Delta_{i-1} H - \dot{T} I - \dot{V} J - \boldsymbol{\theta}$$

2. $3R$ planar manipulator velocity kinematics.
   Figure 8.11 illustrates an R‖R‖R planar manipulator. The forward kinematics of the manipulator provides the following transformation matrices:

$$
{}^{2}T_3 = \begin{bmatrix} c\theta_3 & -s\theta_3 & 0 & l_3 c\theta_3 \\ s\theta_3 & c\theta_3 & 0 & l_3 s\theta_3 \\ 0 & 0 & 1 & 0 \\ 0 & 0 & 0 & 1 \end{bmatrix}
\qquad
{}^{1}T_2 = \begin{bmatrix} c\theta_2 & -s\theta_2 & 0 & l_2 c\theta_2 \\ s\theta_2 & c\theta_2 & 0 & l_2 s\theta_2 \\ 0 & 0 & 1 & 0 \\ 0 & 0 & 0 & 1 \end{bmatrix}
$$

$$
{}^{0}T_1 = \begin{bmatrix} c\theta_1 & -s\theta_1 & 0 & l_1 c\theta_1 \\ s\theta_1 & c\theta_1 & 0 & l_1 s\theta_1 \\ 0 & 0 & 1 & 0 \\ 0 & 0 & 0 & 1 \end{bmatrix}
\tag{8.545}
$$

Calculate the Jacobian matrix, $\mathbf{J}$, using direct differentiating, and find the Cartesian velocity vector of the end point for numerical values.

$$\theta_1 = 56\ \mathrm{deg} \qquad \theta_2 = -28\ \mathrm{deg} \qquad \theta_3 = -10\ \mathrm{deg}$$

$$l_1 = 100\ \mathrm{cm} \qquad l_2 = 55\ \mathrm{cm} \qquad l_3 = 30\ \mathrm{cm} \tag{8.546}$$

$$\dot{\theta}_1 = 30\ \mathrm{deg}/\mathrm{s} \qquad \dot{\theta}_2 = 10\ \mathrm{deg}/\mathrm{s} \qquad \dot{\theta}_3 = -10\ \mathrm{deg}/\mathrm{s}$$

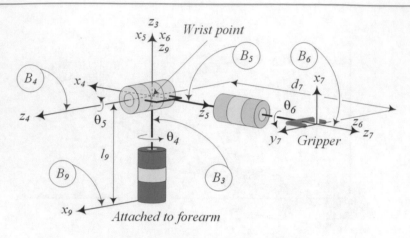

Fig. 8.12 Schematic of a spherical wrist

3. Spherical wrist velocity kinematics.

   Figure 8.12 illustrates a schematic of a spherical wrist. The associated transformation matrices are given below. Assume the frame B_3 is the base frame. Find the angular velocity vector of the coordinate frame B_6.

$$
{}^3T_4 = \begin{bmatrix} c\theta_4 & 0 & -s\theta_4 & 0 \\ s\theta_4 & 0 & c\theta_4 & 0 \\ 0 & -1 & 0 & 0 \\ 0 & 0 & 0 & 1 \end{bmatrix}
\qquad
{}^4T_5 = \begin{bmatrix} c\theta_5 & 0 & s\theta_5 & 0 \\ s\theta_5 & 0 & -c\theta_5 & 0 \\ 0 & 1 & 0 & 0 \\ 0 & 0 & 0 & 1 \end{bmatrix}
$$

$$
{}^5T_6 = \begin{bmatrix} c\theta_6 & -s\theta_6 & 0 & 0 \\ s\theta_6 & c\theta_6 & 0 & 0 \\ 0 & 0 & 1 & 0 \\ 0 & 0 & 0 & 1 \end{bmatrix}
\tag{8.547}
$$

4. Spherical wrist and tool's frame velocity kinematics.

   Assume we attach a tool's coordinate frame, with the following transformation matrix, to the last coordinate frame B_6 of the spherical wrist of Fig. 8.12.

$$
{}^6T_7 = \begin{bmatrix} 1 & 0 & 0 & 0 \\ 0 & 1 & 0 & 0 \\ 0 & 0 & 1 & d_6 \\ 0 & 0 & 0 & 1 \end{bmatrix}
\tag{8.548}
$$

   The wrist transformation matrices are given in Exercise 3. Assume that the frame B_3 is the base frame and find the translational and angular velocities of the tool's coordinate frame B_7.

5. *SCARA* manipulator velocity kinematics.

   An R‖R‖R‖P *SCARA* manipulator is shown in Fig. 8.17 with the following transformation matrices. Calculate the Jacobian matrix of the robot using
   (a) the Jacobian generating vector technique.
   (b) the direct differentiating method.

$$
{}^0T_1 = \begin{bmatrix} c\theta_1 & -s\theta_1 & 0 & l_1c\theta_1 \\ s\theta_1 & c\theta_1 & 0 & l_1s\theta_1 \\ 0 & 0 & 1 & 0 \\ 0 & 0 & 0 & 1 \end{bmatrix}
\qquad
{}^2T_3 = \begin{bmatrix} c\theta_3 & -s\theta_3 & 0 & 0 \\ s\theta_3 & c\theta_3 & 0 & 0 \\ 0 & 0 & 1 & 0 \\ 0 & 0 & 0 & 1 \end{bmatrix}
\tag{8.549}
$$

$$
{}^1T_2 = \begin{bmatrix} c\theta_2 & -s\theta_2 & 0 & l_2c\theta_2 \\ s\theta_2 & c\theta_2 & 0 & l_2s\theta_2 \\ 0 & 0 & 1 & 0 \\ 0 & 0 & 0 & 1 \end{bmatrix}
\qquad
{}^3T_4 = \begin{bmatrix} 1 & 0 & 0 & 0 \\ 0 & 1 & 0 & 0 \\ 0 & 0 & 1 & d \\ 0 & 0 & 0 & 1 \end{bmatrix}
\tag{8.550}
$$

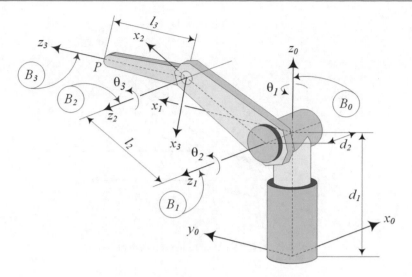

Fig. 8.13 An R⊢R∥R articulated arm

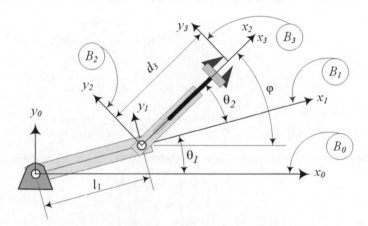

Fig. 8.14 A RRP planar redundant manipulator

6. R⊢R∥R articulated arm velocity kinematics.

Figure 8.13 illustrates a 3 DOF R⊢R∥R manipulator with the following transformation matrices. Find the Jacobian matrix using direct differentiating, and Jacobian-generating vector methods.

$$
{}^{0}T_1 = \begin{bmatrix} c\theta_1 & 0 & -s\theta_1 & 0 \\ s\theta_1 & 0 & c\theta_1 & 0 \\ 0 & -1 & 0 & d_1 \\ 0 & 0 & 0 & 1 \end{bmatrix}
\qquad
{}^{2}T_3 = \begin{bmatrix} c\theta_3 & 0 & s\theta_3 & 0 \\ s\theta_3 & 0 & -c\theta_3 & 0 \\ 0 & 1 & 0 & 0 \\ 0 & 0 & 0 & 1 \end{bmatrix}
\tag{8.551}
$$

$$
{}^{1}T_2 = \begin{bmatrix} c\theta_2 & -s\theta_2 & 0 & l_2c\theta_2 \\ s\theta_2 & c\theta_2 & 0 & l_2s\theta_2 \\ 0 & 0 & 1 & d_2 \\ 0 & 0 & 0 & 1 \end{bmatrix}
\qquad
{}^{0}\mathbf{r}_P = {}^{0}T_3 \begin{bmatrix} 0 \\ 0 \\ l_3 \\ 1 \end{bmatrix}
\tag{8.552}
$$

7. A RRP planar redundant manipulator. Redundant manipulator is the one that has more degrees of freedom than needed. A planar manipulator theoretically needs only two DOF. Here we have a planar manipulator with three DOF.

Figure 8.14 illustrates a 3 DOF planar manipulator with joint variables θ_1, θ_2, d_3.

(a) Determine the link transformation matrices and calculate ${}^{0}T_3$.

(b) Solve the inverse kinematics of the manipulator for a given values of X, Y, φ, where X, Y are global coordinates of the end-effector frame B_3, and φ is the angular coordinate of B_3.

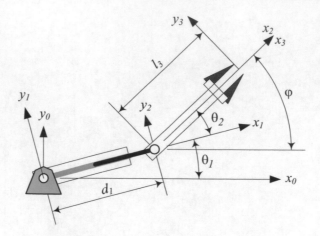

Fig. 8.15 A RPR planar redundant manipulator

(c) Show that the following equation can be a set of solution for inverse kinematic problem.

$$\theta_2 = \tan^{-1} \frac{\beta}{\pm\sqrt{1 - \beta^2}} \qquad \beta = \frac{\sqrt{X^2 + Y^2}}{l_1} \sin\left(\theta_2 - \alpha\right) \tag{8.553}$$

$$\theta_1 = \tan^{-1} \frac{Y}{X} - \left(\theta_2 - \alpha\right) \qquad \alpha = \varphi - \tan^{-1} \frac{Y}{X} \tag{8.554}$$

$$d_3 = \sqrt{l_1^2 + X^2 + Y^2 - 2l_1\sqrt{X^2 + Y^2} \cos\left(\theta_2 - \alpha\right)} \tag{8.555}$$

(d) Determine the Jacobian matrix of the manipulator and show that the following equation solves the forward velocity kinematics.

$$\begin{bmatrix} \dot{X} \\ \dot{Y} \\ \dot{\varphi} \end{bmatrix} = \mathbf{J} \begin{bmatrix} \dot{\theta}_1 \\ \dot{\theta}_2 \\ \dot{d}_3 \end{bmatrix} \tag{8.556}$$

$$\mathbf{J} = \begin{bmatrix} -l_1 s\theta_1 - d_3 s\left(\theta_1 + \theta_2\right) & -d_3 s\left(\theta_1 + \theta_2\right) & c\left(\theta_1 + \theta_2\right) \\ -l_1 c\theta_1 + d_3 c\left(\theta_1 + \theta_2\right) & d_3 c\left(\theta_1 + \theta_2\right) & s\left(\theta_1 + \theta_2\right) \\ 1 & 1 & 0 \end{bmatrix} \tag{8.557}$$

(e) Determine $\mathbf{J}^{-1}$ and solve the inverse velocity kinematics.

8. A RPR planar redundant manipulator.
   (a) Figure 8.15 illustrates a 3 DOF planar manipulator with joint variables θ_1, d_2, and θ_2.
   (b) Determine the link transformation matrices and calculate $^0T_3$.
   (c) Solve the inverse kinematics of the manipulator for a given values of X, Y, φ, where X, Y are global coordinates of the end-effector frame B_3, and φ is the angular coordinate of B_3.
   (d) Determine the Jacobian matrix of the manipulator to solve the forward velocity kinematics.

$$\begin{bmatrix} \dot{X} \\ \dot{Y} \\ \dot{\varphi} \end{bmatrix} = \mathbf{J} \begin{bmatrix} \dot{\theta}_1 \\ \dot{\theta}_2 \\ \dot{d}_3 \end{bmatrix} \tag{8.558}$$

   (e) Determine $\mathbf{J}^{-1}$ and solve the inverse velocity kinematics.

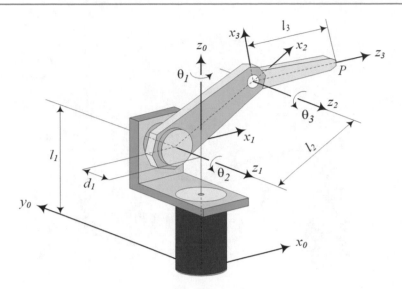

Fig. 8.16 An offset articulated manipulator

9. An offset articulated manipulator.
   Figure 8.16 illustrates an offset articulated manipulator.
   (a) Determine the forward kinematics of the manipulator.
   (b) Determine the global coordinates of the tip point P.
   (c) Determine the Jacobian of the manipulator, using direct differentiating.
   (d) Determine the Jacobian of the manipulator, using generating vectors.
   (e) Determine the inverse Jacobian matrix to solve the inverse velocity kinematics.

10. Articulated robots.
    Attach the spherical wrist of Exercise 19 to the articulated manipulator of Fig. 8.16 and make a 6 DOF articulated robot. Determine the Jacobian of the robot, using generating vectors.

11. Spherical robots.
    Attach the spherical wrist of Exercise 19 to the spherical manipulator of Exercise 15 and make a 6 DOF spherical robot. Determine the Jacobian of the robot, using generating vectors.

12. Cylindrical robots.
    Attach the spherical wrist of Exercise 19 to the cylindrical manipulator of Exercise 17 and make a 6 DOF cylindrical robot. Determine the Jacobian of the robot, using generating vectors.

13. $SCARA$ robot inverse velocity kinematics.
    Figure 8.17 illustrates a $SCARA$ robot. Assume $\theta_3 = 0$.
    (a) Determine the coordinates of the origin of B_4 in $G \equiv B_0$.
    (b) Determine the Jacobian of the manipulator, using direct differentiating.
    (c) Determine the Jacobian of the manipulator, using generating vectors.
    (d) Determine the inverse Jacobian matrix to solve the inverse velocity kinematics.

14. ★ $SCARA$ robot with B_0 on the ground.
    Figure 8.18 illustrates a $SCARA$ robot. Assume $\theta_3 = 0$.
    (a) Determine the coordinates of the origin of B_4 in $G \equiv B_0$.
    (b) Determine the Jacobian of the manipulator, using direct differentiating.
    (c) Determine the Jacobian of the manipulator, using generating vectors.
    (d) Determine the inverse Jacobian matrix to solve the inverse velocity kinematics.

15. Rigid link velocity.
    Figure 8.19 illustrates the coordinate frames and velocity vectors of a rigid link (i). Find
    (a) velocity $^0\mathbf{v}_i$ of the link at C_i in terms of $\dot{\mathbf{d}}_i$ and $\dot{\mathbf{d}}_{i-1}$.
    (b) angular velocity of the link $^0\boldsymbol{\omega}_i$ in terms of $\dot{\mathbf{d}}_i$ and $\dot{\mathbf{d}}_{i-1}$.
    (c) velocity $^0\mathbf{v}_i$ of the link at C_i in terms of proximal joint i velocity.
    (d) velocity $^0\mathbf{v}_i$ of the link at C_i in terms of distal joint $i + 1$ velocity.
    (e) velocity of proximal joint i in terms of distal joint $i + 1$ velocity.
    (f) velocity of distal joint $i + 1$ in terms of proximal joint i velocity.

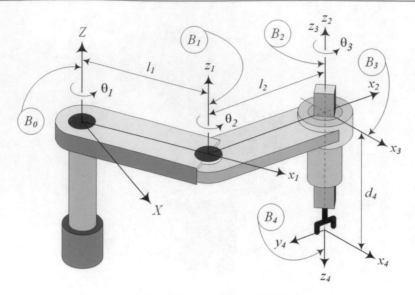

Fig. 8.17 A *SCARA* robot

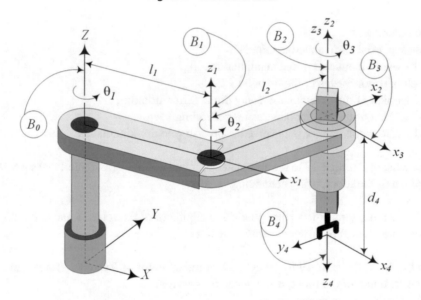

Fig. 8.18 A *SCARA* robot with B_0 on the ground

16. ★ Spherical robot velocity kinematics.

A spherical manipulator R⊢R⊢P, equipped with a spherical wrist, is shown in Fig. 5.50. The transformation matrices of the robot are given in Example 195. Find the Jacobian matrix of the robot.

17. ★ Space station remote manipulator system velocity kinematics.

The transformation matrices for the shuttle remote manipulator system (*SRMS*), shown in Fig. 5.24, are given in the Example 182. Solve the velocity kinematics of the *SRMS* by calculating the Jacobian matrix.

18. *LU* factorization method.

Use the *LU* factorization method and find the associated $[L]$ and $[U]$ for the following matrices.

(a)

$$[A] = \begin{bmatrix} 1 & 4 & 8 \\ 5 & 2 & 7 \\ 9 & 6 & 3 \end{bmatrix} \tag{8.559}$$

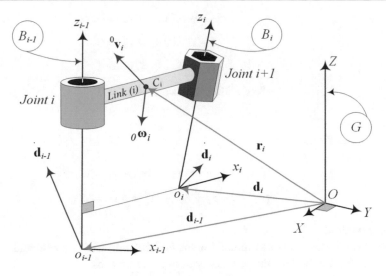

Fig. 8.19 Rigid link velocity vectors

(b)

$$[B] = \begin{bmatrix} 2 & -1 & 3 & -3 \\ 1 & 3 & -1 & -2 \\ 0 & 2 & 2 & 4 \\ 3 & 1 & 5 & -2 \end{bmatrix} \tag{8.560}$$

(c)

$$[C] = \begin{bmatrix} -2 & -1 & 3 & -3 & 6 \\ 1 & 3 & -1 & -2 & 0 \\ 1 & 2 & 2 & 4 & -2 \\ 3 & 1 & 5 & -2 & -1 \\ 7 & -5 & 2 & 1 & 1 \end{bmatrix} \tag{8.561}$$

19. *LU* inversion method.

Use the *LU* inversion method and find the inverse of the matrices in Exercises 18.

20. *LU* calculations.

Use the *LU* inversion method and calculate the inversion of the following matrices based on the matrices in Exercises 18.

$$D = AB \qquad E = AB^{-1} \qquad F = A^{-1}B \qquad G = A^{-1}B^{-1} \tag{8.562}$$

21. A set of liner equations.

Use the *LU* factorization method and solve the following set of equations and show that the solutions are $x_1 = 4$, $x_2 = 1$, $x_3 = 2$.

$$-3x_1 + 8x_2 + 5x_3 = 6$$
$$2x_1 - 7x_2 + 4x_3 = 9 \tag{8.563}$$
$$x_1 + 9x_2 - 6x_3 = 1$$

22. A set of six equations.

Use the *LU* factorization method and solve the following set of equations and show that the solutions are $x_1 = 75$, $x_2 = 52$, $x_3 = 40$, $x_4 = 31$, $x_5 = 22$, $x_6 = 10$.

$$11x_1 - 5x_2 - x_6 = 500$$
$$-20x_1 + 41x_2 - 15x_3 - 6x_5 = 0$$

$$-3x_2 + 7x_3 - 4x_4 = 0$$

$$-x_3 + 2x_4 - x_5 = 0 \qquad (8.564)$$

$$-2x_1 - 15x_5 + 47x_6 = 0$$

$$-3x_2 - 10x_4 + 28x_5 - 15x_6 = 0$$

23. A set of nonlinear equations.
    Solve the following set of equations.

$$x_1x_2 - 2x_1 - x_2 = 0$$

$$x_1^2x_2 - 2x_1x_2 + x_2 - 2x_1^2 + 4x_1 = 2 \qquad (8.565)$$

24. ★ Gaussian elimination method.
    There are two situations where the Gaussian elimination method fails: division by zero and round-off errors.
    Examine the LU factorization method for the possibility of division by zero.
25. ★ Number of subtractions as a source of round-off error.
    Round-off error is common in numerical techniques; however, it increases by increasing the number of subtractions.
    Apply the Gaussian elimination and LU factorization methods for solving a set of four equations

$$\begin{bmatrix} 2 & 1 & 3 & -3 \\ 1 & 0 & -1 & -2 \\ 0 & 2 & 2 & 1 \\ 3 & 1 & 0 & -2 \end{bmatrix} \begin{bmatrix} x_1 \\ x_2 \\ x_3 \\ x_4 \end{bmatrix} = \begin{bmatrix} 1 \\ 2 \\ 0 \\ -2 \end{bmatrix} \qquad (8.566)$$

and count the number of subtractions in each method.
26. ★ Jacobian matrix from transformation matrices.
    Use the Jacobian matrix technique from links' transformation matrices and find the Jacobian matrix of the R‖R‖R planar manipulator shown in Fig. 5.21. Choose a set of sample data for the dimensions and kinematics of the manipulator and find the inverse of the Jacobian matrix.
27. ★ Jacobian matrix for a spherical wrist.
    Use the Jacobian matrix technique from links' transformation matrices and find the Jacobian matrix of the spherical wrist shown in Fig. 5.30. Assume that the frame B_3 is the base frame.
28. Jacobian matrix for a $SCARA$ manipulator.
    Use the Jacobian matrix technique from links' transformation matrices and find the Jacobian matrix of the R‖R‖R‖P robot shown in Fig. 5.23. Assume $\theta_3 = 0$.
29. Jacobian matrix for an R⊢R‖R articulated manipulator.
    Figure 5.22 illustrates a 3 DOF R⊢R‖R manipulator. Use the Jacobian matrix technique from links' transformation matrices and find the Jacobian matrix for the manipulator.
30. ★ Partitioning inverse method.
    Calculate the matrix inversion for the matrices in Exercise 18 using the partitioning inverse method.
31. ★ Analytic matrix inversion.
    Use the analytic and LU factorization methods and find the inverse of $[A]$ or an arbitrary 3×3 matrix.

$$[A] = \begin{bmatrix} 1 & 4 & 8 \\ 5 & 2 & 7 \\ 9 & 6 & 3 \end{bmatrix} \qquad (8.567)$$

Count and compare the number of arithmetic operations.

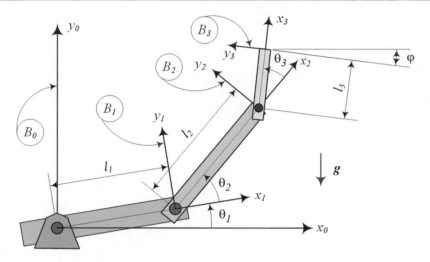

Fig. 8.20 A planar $3R$ manipulator with one DOF redundancy

32. ★ Cayley–Hamilton matrix inversion.
Use the Cayley–Hamilton and LU factorization methods and find the inverse of $[A]$ or an arbitrary 3×3 matrix.

$$[A] = \begin{bmatrix} 1 & 4 & 8 \\ 5 & 2 & 7 \\ 9 & 6 & 3 \end{bmatrix} \tag{8.568}$$

Count and compare the number of arithmetic operations.

33. ★ Norms of matrices.
Calculate the following norms of the matrices in Exercise 18.

$$\|A\|_1 = \underset{1 \le j \le n}{Max} \sum_{i=1}^{n} |a_{ij}| \tag{8.569}$$

$$\|A\|_2 = \lambda_{Max} \left(A^T A \right) \tag{8.570}$$

$$\|A\|_\infty = \underset{1 \le i \le n}{Max} \sum_{j=1}^{n} |a_{ij}| \tag{8.571}$$

$$\|A\|_F = \sum_{i=1}^{n} \sum_{j=1}^{n} a_{ij}^2 \tag{8.572}$$

34. ★ Project. Redundant manipulator and extra condition.
Figure 8.20 illustrates a planar $3R$ manipulator with one DOF redundancy. Assume the length of the links is $l_i = 2/3$, and the end point is moving from $(1.2, 1.5)$ to $(-1.2, 1.5)$ on a straight lime according to the following time functions.

$$X = 1 - 6t^2 + 4t^3 \qquad Y = 1.5 \tag{8.573}$$

(a) Solve the inverse kinematics if always $\dot{\theta}_3 = 2\dot{\theta}_2$. Plot $\dot{\theta}_i$ for the whole trip.
(b) Solve the inverse kinematics if always $\dot{\theta}_3 = \dot{\theta}_1 - \dot{\theta}_2$. Plot $\dot{\theta}_i$ for the whole trip.

Acceleration Kinematics

9

Acceleration kinematics of robots is divided into forward and inverse acceleration kinematics. To study and develop the acceleration kinematics of robots, we need to introduce the concept of angular acceleration and rigid body acceleration. Links of a robot are considered rigid bodies in relative motion. They will have rotational and translational accelerations. Angular acceleration of a rigid body with respect to a global frame is the time derivative of instantaneous angular velocity of the body. In general, it is a vectorial quantity that is in a different direction than angular velocity. In this chapter we learn acceleration kinematics of robots.

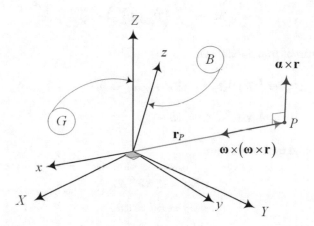

Fig. 9.1 A rotating rigid body $B(Oxyz)$ with a fixed point O in a reference frame $G(OXYZ)$

9.1 Angular Acceleration Vector and Matrix

Consider a rotating rigid body $B(Oxyz)$ with a fixed point O in a reference frame $G(OXYZ)$ as shown in Fig. 9.1. The velocity vector of a fixed point in the body frame is

$$
{}^G\mathbf{v}(t) = {}^G\dot{\mathbf{r}}(t) = {}_G\tilde{\omega}_B\,{}^G\mathbf{r}(t) = {}_G\boldsymbol{\omega}_B \times {}^G\mathbf{r}(t) \tag{9.1}
$$

This equation can be utilized to find the acceleration vector of a body point at ${}^G\mathbf{r}(t)$.

$$
{}^G\mathbf{a} = {}^G\dot{\mathbf{v}} = \frac{{}^Gd}{dt}\,{}^G\dot{\mathbf{r}}(t) = {}^G\ddot{\mathbf{r}} = {}_GS_B\,{}^G\mathbf{r} \tag{9.2}
$$

$$
= {}_G\boldsymbol{\alpha}_B \times {}^G\mathbf{r} + {}_G\boldsymbol{\omega}_B \times \left({}_G\boldsymbol{\omega}_B \times {}^G\mathbf{r}\right) \tag{9.3}
$$

$$= \left({}_G\tilde{\alpha}_B + {}_G\tilde{\omega}_B^2 \right) {}^G\mathbf{r} = \left({}_G\dot{\tilde{\omega}}_B + {}_G\tilde{\omega}_B^2 \right) {}^G\mathbf{r} \tag{9.4}$$

$$= \left[\ddot{\phi}\tilde{u} + \dot{\phi}\dot{\tilde{u}} + \dot{\phi}^2\tilde{u}^2 \right] {}^G\mathbf{r} \tag{9.5}$$

$$= \left(\ddot{\phi}\hat{u} + \dot{\phi}\dot{\hat{\mathbf{u}}} \right) \times {}^G\mathbf{r} + \dot{\phi}^2\hat{u} \times \left(\hat{u} \times {}^G\mathbf{r} \right) \tag{9.6}$$

$$= {}^G\ddot{R}_B {}^G R_B^T {}^G\mathbf{r} \tag{9.7}$$

${}_G\boldsymbol{\alpha}_B$ is the *angular acceleration vector* of the body point with respect to and expressed in the *G*-frame.

$$_G\boldsymbol{\alpha}_B = \frac{{}^G d}{dt} {}_G\boldsymbol{\omega}_B \tag{9.8}$$

$$_G\tilde{\alpha}_B = {}_G\dot{\tilde{\omega}}_B \tag{9.9}$$

The acceleration ${}^G\ddot{\mathbf{r}}$ can also be defined using the *angular acceleration matrix* ${}_G\tilde{\alpha}_B$ and *rotational acceleration transformation* ${}_G S_B$

$$^G\ddot{\mathbf{r}} = {}_G S_B {}^G\mathbf{r} = \left[{}_G\tilde{\alpha}_B + {}_G\tilde{\omega}_B^2 \right] {}^G\mathbf{r} \tag{9.10}$$

where ${}_G\tilde{\alpha}_B$ is the *angular acceleration matrix*,

$$_G\tilde{\alpha}_B = {}_G\dot{\tilde{\omega}}_B = \frac{{}^G d}{dt} \left(\dot{R}_{\hat{u},\phi} \, R_{\hat{u},\phi}^T \right) = \ddot{\phi}\,\tilde{u} + \dot{\phi}\dot{\tilde{u}} \tag{9.11}$$

and ${}_G S_B$ is the *rotational acceleration transformation*.

$$_G S_B = {}^G\ddot{R}_B {}^G R_B^T = {}_G\tilde{\alpha}_B + {}_G\tilde{\omega}_B^2 = {}_G\tilde{\alpha}_B - {}_G\tilde{\omega}_B \, {}_G\tilde{\omega}_B^T$$

$$= \ddot{\phi}\,\tilde{u} + \dot{\phi}\dot{\tilde{u}} + \dot{\phi}^2\tilde{u}\tilde{u} \tag{9.12}$$

The angular velocity vector ${}_G\boldsymbol{\omega}_B$ and matrix ${}_G\tilde{\omega}_B$ are

$$_G\tilde{\omega}_B = {}^G\dot{R}_B {}^G R_B^T \tag{9.13}$$

$$_G\boldsymbol{\omega}_B = \dot{\phi}\hat{u} = \dot{\phi}\hat{u}_\omega \tag{9.14}$$

The relative angular acceleration of two bodies B_1, B_2 in the global frame G can be combined.

$$_G\boldsymbol{\alpha}_2 = \frac{{}^G d}{dt} {}_G\boldsymbol{\omega}_2 = {}_G\boldsymbol{\alpha}_1 + {}_1^G\boldsymbol{\alpha}_2 \tag{9.15}$$

$$_G S_2 = {}_G S_1 + {}_1^G S_2 + 2 {}_G\tilde{\omega}_1 \, {}_1^G\tilde{\omega}_2 \tag{9.16}$$

The *B*-expression of ${}^G\mathbf{a}$ and ${}_G S_B$ are

$$_G^B\mathbf{a} = {}_G^B\boldsymbol{\alpha}_B \times {}^B\mathbf{r} + {}_G^B\boldsymbol{\omega}_B \times \left({}_G^B\boldsymbol{\omega}_B \times {}^B\mathbf{r} \right) \tag{9.17}$$

$$_G^B S_B = {}^B R_G {}^G\ddot{R}_B = {}_G^B\tilde{\alpha}_B + {}_G^B\tilde{\omega}_B^2 \tag{9.18}$$

The global and body expressions of the rotational acceleration transformations $_GS_B$ and $_G^BS_B$ can be transformed to each other by the following rules:

$$_GS_B = {}^GR_B \, _G^BS_B \, {}^GR_B^T \tag{9.19}$$

$$_G^BS_B = {}^GR_B^T \, _GS_B \, {}^GR_B \tag{9.20}$$

Proof The global position and velocity vectors of a body point are

$$^G\mathbf{r} = {}^GR_B \, {}^B\mathbf{r} \tag{9.21}$$

$$^G\mathbf{v} = {}^G\dot{\mathbf{r}} = {}^G\dot{R}_B \, {}^B\mathbf{r} = {}_G\tilde{\omega}_B \, {}^G\mathbf{r} = {}_G\boldsymbol{\omega}_B \times {}^G\mathbf{r} \tag{9.22}$$

Differentiating Equation (9.22) gives

$$^G\mathbf{a} = {}^G\ddot{\mathbf{r}} = {}_G\dot{\boldsymbol{\omega}}_B \times {}^G\mathbf{r} + {}_G\boldsymbol{\omega}_B \times {}^G\dot{\mathbf{r}}$$

$$= {}_G\boldsymbol{\alpha}_B \times {}^G\mathbf{r} + {}_G\boldsymbol{\omega}_B \times \left({}_G\boldsymbol{\omega}_B \times {}^G\mathbf{r} \right) \tag{9.23}$$

Employing the axis–angle expression of angular velocity,

$$\boldsymbol{\omega} = \dot{\phi}\hat{u} = \dot{\phi} \begin{bmatrix} u_1 \\ u_2 \\ u_3 \end{bmatrix} = \begin{bmatrix} \omega_1 \\ \omega_2 \\ \omega_3 \end{bmatrix} \tag{9.24}$$

$$\tilde{\omega} = \dot{\phi}\tilde{u} = \dot{\phi} \begin{bmatrix} 0 & -u_3 & u_2 \\ u_3 & 0 & -u_1 \\ -u_2 & u_1 & 0 \end{bmatrix} = \begin{bmatrix} 0 & -\omega_3 & \omega_2 \\ \omega_3 & 0 & -\omega_1 \\ -\omega_2 & \omega_1 & 0 \end{bmatrix}$$

$$\tag{9.25}$$

we find the angular acceleration vector $\boldsymbol{\alpha}$ and matrix $\tilde{\alpha}$ in terms of the instantaneous axis and angle of rotation:

$$\boldsymbol{\alpha} = \dot{\boldsymbol{\omega}} = \ddot{\phi}\hat{u} + \dot{\phi}\dot{\hat{u}} \tag{9.26}$$

$$\tilde{\alpha} = \dot{\tilde{\omega}} = \ddot{\phi}\tilde{u} + \dot{\phi}\dot{\tilde{u}} \tag{9.27}$$

We may substitute the matrix expressions of angular velocity and acceleration in (9.23) to derive Eq. (9.6).

$$^G\ddot{\mathbf{r}} = {}_G\boldsymbol{\alpha}_B \times {}^G\mathbf{r} + {}_G\boldsymbol{\omega}_B \times \left({}_G\boldsymbol{\omega}_B \times {}^G\mathbf{r} \right)$$

$$= \left(\ddot{\phi}\hat{u} + \dot{\phi}\dot{\hat{u}} \right) \times {}^G\mathbf{r} + \dot{\phi}^2\hat{u} \times \left(\hat{u} \times {}^G\mathbf{r} \right) \tag{9.28}$$

$$= {}_G\tilde{\alpha}_B \, {}^G\mathbf{r} + {}_G\tilde{\omega}_B \, {}_G\tilde{\omega}_B \, {}^G\mathbf{r} \tag{9.29}$$

$$= \left[{}_G\tilde{\alpha}_B + {}_G\tilde{\omega}_B^2 \right] {}^G\mathbf{r} = \left[{}_G\dot{\tilde{\omega}}_B + {}_G\tilde{\omega}_B^2 \right] {}^G\mathbf{r} \tag{9.30}$$

$$= \left[\ddot{\phi}\tilde{u} + \dot{\phi}\dot{\tilde{u}} + \dot{\phi}^2\tilde{u}^2 \right] {}^G\mathbf{r} = {}_GS_B \, {}^G\mathbf{r} \tag{9.31}$$

Recalling that

$$_G\tilde{\omega}_B = {}^G\dot{R}_B \, {}^GR_B^T \tag{9.32}$$

$$^G\dot{\mathbf{r}}(t) = {}_G\tilde{\omega}_B \, {}^G\mathbf{r}(t) \tag{9.33}$$

we find Eqs. (9.7) and (9.11):

$$
\begin{aligned}
{}^{G}\ddot{\mathbf{r}} &= \frac{{}^{G}d}{dt}\left[{}^{G}\dot{R}_{B}\,{}^{G}R_{B}^{T}\,{}^{G}\mathbf{r}\right] \\
&= {}^{G}\ddot{R}_{B}\,{}^{G}R_{B}^{T}\,{}^{G}\mathbf{r} + {}^{G}\dot{R}_{B}\,{}^{G}\dot{R}_{B}^{T}\,{}^{G}\mathbf{r} + \left[{}^{G}\dot{R}_{B}\,{}^{G}R_{B}^{T}\right]\left[{}^{G}\dot{R}_{B}\,{}^{G}R_{B}^{T}\right]{}^{G}\mathbf{r} \\
&= \left[{}^{G}\ddot{R}_{B}\,{}^{G}R_{B}^{T} + {}^{G}\dot{R}_{B}\,{}^{G}\dot{R}_{B}^{T} + \left[{}^{G}\dot{R}_{B}\,{}^{G}R_{B}^{T}\right]^{2}\right]{}^{G}\mathbf{r} \\
&= \left[{}^{G}\ddot{R}_{B}\,{}^{G}R_{B}^{T} - \left[{}^{G}\dot{R}_{B}\,{}^{G}R_{B}^{T}\right]^{2} + \left[{}^{G}\dot{R}_{B}\,{}^{G}R_{B}^{T}\right]^{2}\right]{}^{G}\mathbf{r} \\
&= {}^{G}\ddot{R}_{B}\,{}^{G}R_{B}^{T}\,{}^{G}\mathbf{r}
\end{aligned}
\tag{9.34}
$$

$$
\begin{aligned}
{}_{G}\tilde{\alpha}_{B} = {}_{G}\dot{\tilde{\omega}}_{B} &= {}^{G}\ddot{R}_{B}\,{}^{G}R_{B}^{T} + {}^{G}\dot{R}_{B}\,{}^{G}\dot{R}_{B}^{T} \\
&= {}^{G}\ddot{R}_{B}\,{}^{G}R_{B}^{T} + {}^{G}\dot{R}_{B}\,{}^{G}R_{B}^{T}\,{}^{G}R_{B}\,{}^{G}\dot{R}_{B}^{T} \\
&= {}^{G}\ddot{R}_{B}\,{}^{G}R_{B}^{T} + \left[{}^{G}\dot{R}_{B}\,{}^{G}R_{B}^{T}\right]\left[{}^{G}\dot{R}_{B}\,{}^{G}R_{B}^{T}\right]^{T} \\
&= {}^{G}\ddot{R}_{B}\,{}^{G}R_{B}^{T} + {}_{G}\tilde{\omega}_{B}\,{}_{G}\tilde{\omega}_{B}^{T} = {}^{G}\ddot{R}_{B}\,{}^{G}R_{B}^{T} - {}_{G}\tilde{\omega}_{B}^{2}
\end{aligned}
\tag{9.35}
$$

which indicates that

$$
{}^{G}\ddot{R}_{B}\,{}^{G}R_{B}^{T} = {}_{G}\tilde{\alpha}_{B} + {}_{G}\tilde{\omega}_{B}^{2} = {}_{G}S_{B}
\tag{9.36}
$$

The expanded forms of the angular accelerations ${}_{G}\boldsymbol{\alpha}_{B}$, ${}_{G}\tilde{\alpha}_{B}$ and *rotational acceleration transformation* ${}_{G}S_{B}$ are

$$
\begin{aligned}
{}_{G}\tilde{\alpha}_{B} = {}_{G}\dot{\tilde{\omega}}_{B} &= \ddot{\phi}\tilde{u} + \dot{\phi}\dot{\tilde{u}} = \begin{bmatrix} 0 & -\dot{\omega}_{3} & \dot{\omega}_{2} \\ \dot{\omega}_{3} & 0 & -\dot{\omega}_{1} \\ -\dot{\omega}_{2} & \dot{\omega}_{1} & 0 \end{bmatrix} \\
&= \begin{bmatrix} 0 & -\dot{u}_{3}\dot{\phi} - u_{3}\ddot{\phi} & \dot{u}_{2}\dot{\phi} + u_{2}\ddot{\phi} \\ \dot{u}_{3}\dot{\phi} + u_{3}\ddot{\phi} & 0 & -\dot{u}_{1}\dot{\phi} - u_{1}\ddot{\phi} \\ -\dot{u}_{2}\dot{\phi} - u_{2}\ddot{\phi} & \dot{u}_{1}\dot{\phi} + u_{1}\ddot{\phi} & 0 \end{bmatrix}
\end{aligned}
\tag{9.37}
$$

$$
{}_{G}\boldsymbol{\alpha}_{B} = \begin{bmatrix} \dot{\omega}_{1} \\ \dot{\omega}_{2} \\ \dot{\omega}_{3} \end{bmatrix} = \begin{bmatrix} \dot{u}_{1}\dot{\phi} + u_{1}\ddot{\phi} \\ \dot{u}_{2}\dot{\phi} + u_{2}\ddot{\phi} \\ \dot{u}_{3}\dot{\phi} + u_{3}\ddot{\phi} \end{bmatrix}
\tag{9.38}
$$

$$
\begin{aligned}
{}_{G}S_{B} &= {}_{G}\dot{\tilde{\omega}}_{B} + {}_{G}\tilde{\omega}_{B}^{2} = {}_{G}\tilde{\alpha}_{B} + {}_{G}\tilde{\omega}_{B}^{2} \\
&= \begin{bmatrix} -\omega_{2}^{2} - \omega_{3}^{2} & \omega_{1}\omega_{2} - \dot{\omega}_{3} & \dot{\omega}_{2} + \omega_{1}\omega_{3} \\ \dot{\omega}_{3} + \omega_{1}\omega_{2} & -\omega_{1}^{2} - \omega_{3}^{2} & \omega_{2}\omega_{3} - \dot{\omega}_{1} \\ \omega_{1}\omega_{3} - \dot{\omega}_{2} & \dot{\omega}_{1} + \omega_{2}\omega_{3} & -\omega_{1}^{2} - \omega_{2}^{2} \end{bmatrix}
\end{aligned}
\tag{9.39}
$$

$$
\begin{aligned}
{}_{G}S_{B} &= \ddot{\phi}\tilde{u} + \dot{\phi}\dot{\tilde{u}} + \dot{\phi}^{2}\tilde{u}^{2} \\
&= \begin{bmatrix} -\dot{\phi}^{2}\left(u_{2}^{2} + u_{3}^{2}\right) & u_{1}u_{2}\dot{\phi}^{2} - \dot{u}_{3}\dot{\phi} - u_{3}\ddot{\phi} & u_{1}u_{3}\dot{\phi}^{2} + \dot{u}_{2}\dot{\phi} + u_{2}\ddot{\phi} \\ u_{1}u_{2}\dot{\phi}^{2} + \dot{u}_{3}\dot{\phi} + u_{3}\ddot{\phi} & -\dot{\phi}^{2}\left(u_{1}^{2} + u_{3}^{2}\right) & u_{2}u_{3}\dot{\phi}^{2} - \dot{u}_{1}\dot{\phi} - u_{1}\ddot{\phi} \\ u_{1}u_{3}\dot{\phi}^{2} - \dot{u}_{2}\dot{\phi} - u_{2}\ddot{\phi} & u_{2}u_{3}\dot{\phi}^{2} - \dot{u}_{1}\dot{\phi} + u_{1}\ddot{\phi} & -\dot{\phi}^{2}\left(u_{1}^{2} + u_{2}^{2}\right) \end{bmatrix} \\
&= \begin{bmatrix} (u_{1}^{2} - 1)\dot{\phi}^{2} & u_{1}u_{2}\dot{\phi}^{2} - \dot{u}_{3}\dot{\phi} - u_{3}\ddot{\phi} & u_{1}u_{3}\dot{\phi}^{2} + \dot{u}_{2}\dot{\phi} + u_{2}\ddot{\phi} \\ u_{1}u_{2}\dot{\phi}^{2} + \dot{u}_{3}\dot{\phi} + u_{3}\ddot{\phi} & (u_{2}^{2} - 1)\dot{\phi}^{2} & u_{2}u_{3}\dot{\phi}^{2} - \dot{u}_{1}\dot{\phi} - u_{1}\ddot{\phi} \\ u_{1}u_{3}\dot{\phi}^{2} - \dot{u}_{2}\dot{\phi} - u_{2}\ddot{\phi} & u_{2}u_{3}\dot{\phi}^{2} - \dot{u}_{1}\dot{\phi} + u_{1}\ddot{\phi} & (u_{3}^{2} - 1)\dot{\phi}^{2} \end{bmatrix}
\end{aligned}
\tag{9.40}
$$

Therefore, the position, velocity, and acceleration vectors of a body point are

$$^B\mathbf{r}_P = x\hat{\imath} + y\hat{\jmath} + z\hat{k} \tag{9.41}$$

$$^G\mathbf{v}_P = {}^G\dot{\mathbf{r}}_P = \frac{^Gd}{dt}{}^G\mathbf{r}_P = {}_G\boldsymbol{\omega}_B \times {}^G\mathbf{r} \tag{9.42}$$

$$\begin{aligned}
^G\mathbf{a}_P = {}^G\dot{\mathbf{v}}_P = {}^G\ddot{\mathbf{r}}_P &= \frac{^Gd^2}{dt^2}{}^G\mathbf{r}_P \\
&= {}_G\boldsymbol{\alpha}_B \times {}^G\mathbf{r} + {}_G\boldsymbol{\omega}_B \times {}^G\dot{\mathbf{r}} \\
&= {}_G\boldsymbol{\alpha}_B \times {}^G\mathbf{r} + {}_G\boldsymbol{\omega}_B \times ({}_G\boldsymbol{\omega}_B \times {}^G\mathbf{r})
\end{aligned} \tag{9.43}$$

The angular acceleration expressed in the body frame is the body derivative of the angular velocity vector. To show this, we use the derivative transport formula (7.214).

$$\begin{aligned}
{}_G^B\boldsymbol{\alpha}_B = \frac{^Gd}{dt}{}_G^B\boldsymbol{\omega}_B &= \frac{^Bd}{dt}{}_G^B\boldsymbol{\omega}_B + {}_G^B\boldsymbol{\omega}_B \times {}_G^B\boldsymbol{\omega}_B = \frac{^Bd}{dt}{}_G^B\boldsymbol{\omega}_B \\
&= {}_G^B\dot{\boldsymbol{\omega}}_B
\end{aligned} \tag{9.44}$$

The angular acceleration of B in G can always be expressed in the form

$$_G\boldsymbol{\alpha}_B = {}_G\alpha_B\,\hat{u}_\alpha \tag{9.45}$$

where $\hat{u}_\alpha$ is a unit vector parallel to $_G\boldsymbol{\alpha}_B$. The angular velocity and angular acceleration vectors are not parallel in general, and therefore,

$$\hat{u}_\alpha \neq \hat{u}_\omega \tag{9.46}$$

$$_G\boldsymbol{\alpha}_B \neq {}_G\dot{\boldsymbol{\omega}}_B \tag{9.47}$$

However, the only special case is when the axis of rotation is fixed in both G and B frames, $\hat{u} = \hat{u}_\alpha = \hat{u}_\omega$.

$$_G\boldsymbol{\alpha}_B = \alpha\,\hat{u} = \dot{\omega}\,\hat{u} = \ddot{\phi}\,\hat{u} \tag{9.48}$$

The angular velocity of several bodies rotating relative to each other are related according to relative velocity equation:

$$_0\boldsymbol{\omega}_n = {}_0\boldsymbol{\omega}_1 + {}_1^0\boldsymbol{\omega}_2 + {}_2^0\boldsymbol{\omega}_3 + \cdots + {}_{n-1}^0\boldsymbol{\omega}_n \tag{9.49}$$

The angular accelerations of several relatively rotating rigid bodies follow the same rule:

$$_0\boldsymbol{\alpha}_n = {}_0\boldsymbol{\alpha}_1 + {}_1^0\boldsymbol{\alpha}_2 + {}_2^0\boldsymbol{\alpha}_3 + \cdots + {}_{n-1}^0\boldsymbol{\alpha}_n \tag{9.50}$$

To show this fact and develop the relative acceleration formula, we consider two relatively rotating rigid links, B_1, B_2 in a base coordinate frame B_0 with a fixed point at O. The angular velocities of the links are related by

$$_0\boldsymbol{\omega}_2 = {}_0\boldsymbol{\omega}_1 + {}_1^0\boldsymbol{\omega}_2 \tag{9.51}$$

Their angular accelerations are

$$_0\boldsymbol{\alpha}_1 = \frac{^0d}{dt}{}_0\boldsymbol{\omega}_1 \tag{9.52}$$

$$_0\boldsymbol{\alpha}_2 = \frac{^0d}{dt}\,_0\boldsymbol{\omega}_2 = {_0\boldsymbol{\alpha}_1} + {_1^0\boldsymbol{\alpha}_2} \tag{9.53}$$

and therefore,

$$
\begin{aligned}
_0S_2 &= {_0\tilde{\alpha}_2} + {_0\tilde{\omega}_2^2} = {_0\tilde{\alpha}_1} + {_1^0\tilde{\alpha}_2} + \left({_0\tilde{\omega}_1} + {_1^0\tilde{\omega}_2}\right)^2 \\
&= {_0\tilde{\alpha}_1} + {_1^0\tilde{\alpha}_2} + {_0\tilde{\omega}_1^2} + {_1^0\tilde{\omega}_2^2} + 2\,{_0\tilde{\omega}_1}\,{_1^0\tilde{\omega}_2} \\
&= {_0S_1} + {_1^0S_2} + 2\,{_0\tilde{\omega}_1}\,{_1^0\tilde{\omega}_2}
\end{aligned}
\tag{9.54}
$$

Equation (9.54) is the required *relative acceleration transformation formula*. It indicates the method of calculation of relative accelerations for a multibody. As a more general case, consider a six-link multibody. The angular acceleration of link (6) in the base frame would be

$$
\begin{aligned}
_0S_6 = {}&{_0S_1} + {_1^0S_2} + {_2^0S_3} + {_3^0S_4} + {_4^0S_5} + {_5^0S_6} \\
&+ 2\,{_0\tilde{\omega}_1}\left({_1^0\tilde{\omega}_2} + {_2^0\tilde{\omega}_3} + {_3^0\tilde{\omega}_4} + {_4^0\tilde{\omega}_5} + {_5^0\tilde{\omega}_6}\right) \\
&+ 2\,{_1^0\tilde{\omega}_2}\left({_2^0\tilde{\omega}_3} + {_3^0\tilde{\omega}_4} + {_4^0\tilde{\omega}_5} + {_5^0\tilde{\omega}_6}\right) \\
&\;\;\vdots \\
&+ 2\,{_4^0\tilde{\omega}_5}\left({_5^0\tilde{\omega}_6}\right)
\end{aligned}
\tag{9.55}
$$

We can transform the G and B-expression of the global acceleration of a body point P to each other using a rotation matrix:

$$
\begin{aligned}
_G^B\mathbf{a}_P &= {}^B R_G\,{}^G\mathbf{a}_P = {}^B R_G\,{}_G S_B\,{}^G\mathbf{r}_P = {}^B R_G\,{}_G S_B\,{}^G R_B\,{}^B\mathbf{r}_P \\
&= {}^B R_G\,{}^G\ddot{R}_B\,{}^G R_B^T\,{}^G R_B\,{}^B\mathbf{r}_P = {}^B R_G\,{}^G\ddot{R}_B\,{}^B\mathbf{r}_P \\
&= {}^G R_B^T\,{}^G\ddot{R}_B\,{}^B\mathbf{r}_P = {}_G^B S_B\,{}^B\mathbf{r}_P = \left({}_G^B\tilde{\alpha}_B + {}_G^B\tilde{\omega}_B^2\right){}^B\mathbf{r}_P \\
&= {}_G^B\boldsymbol{\alpha}_B \times {}^B\mathbf{r} + {}_G^B\boldsymbol{\omega}_B \times \left({}_G^B\boldsymbol{\omega}_B \times {}^B\mathbf{r}\right)
\end{aligned}
\tag{9.56}
$$

$$
\begin{aligned}
^G\mathbf{a}_P &= {}^G R_B\,{}_G^B\mathbf{a}_P = {}^G R_B\,{}_G^B S_B\,{}^B\mathbf{r}_P = {}^G R_B\,{}_G^B S_B\,{}^G R_B^T\,{}^G\mathbf{r}_P \\
&= {}^G R_B\,{}^G R_B^T\,{}^G\ddot{R}_B\,{}^G R_B^T\,{}^G\mathbf{r}_P = {}^G\ddot{R}_B\,{}^G R_B^T\,{}^G\mathbf{r}_P \\
&= {}_G S_B\,{}^G\mathbf{r}_P = \left({}_G\tilde{\alpha}_B + {}_G\tilde{\omega}_B^2\right){}^G\mathbf{r} \\
&= {}_G\boldsymbol{\alpha}_B \times {}^G\mathbf{r} + {}_G\boldsymbol{\omega}_B \times \left({}_G\boldsymbol{\omega}_B \times {}^G\mathbf{r}\right)
\end{aligned}
\tag{9.57}
$$

From the definitions of ${}_G S_B$ and ${}_G^B S_B$ in (9.12) and (9.18) and comparing with (9.56) and (9.57), we are able to transform the two rotational acceleration transformations by

$$_G S_B = {}^G R_B\,{}_G^B S_B\,{}^G R_B^T \tag{9.58}$$

$$_G^B S_B = {}^G R_B^T\,{}_G S_B\,{}^G R_B \tag{9.59}$$

and derive the following useful equations:

$$^G\ddot{R}_B = {}_G S_B\,{}^G R_B \tag{9.60}$$

$$^G\ddot{R}_B = {}^G R_B\,{}_G^B S_B \tag{9.61}$$

$$_G S_B\,{}^G R_B = {}^G R_B\,{}_G^B S_B \tag{9.62}$$

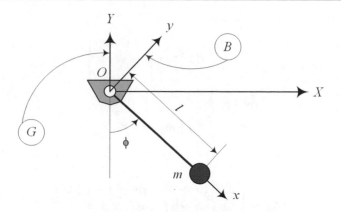

Fig. 9.2 A simple pendulum

The angular acceleration of B in G is negative of the angular acceleration of G in B if both are expressed in the same coordinate frame:

$$_{G}\tilde{\alpha}_{B} = -\,_{B}^{G}\tilde{\alpha}_{G} \qquad _{G}\alpha_{B} = -\,_{B}^{G}\alpha_{G} \tag{9.63}$$

$$_{G}^{B}\tilde{\alpha}_{B} = -\,_{B}\tilde{\alpha}_{G} \qquad _{G}^{B}\alpha_{B} = -\,_{B}\alpha_{G} \tag{9.64}$$

The term $_{G}\alpha_{B} \times {}^{G}\mathbf{r}$ in (9.43) is called the *tangential acceleration*, which is a function of the angular acceleration of B in G. The term $_{G}\omega_{B} \times \left(_{G}\omega_{B} \times {}^{G}\mathbf{r} \right)$ in ${}^{G}\mathbf{a}$ is called *centripetal acceleration* and is a function of the angular velocity of B in G. Using the Rodriguez rotation formula, it can also be shown that

$$\begin{aligned}
_{G}\tilde{\alpha}_{B} = _{G}\dot{\tilde{\omega}}_{B} &= \frac{^{G}d}{dt}\left(\dot{R}_{\hat{u},\phi}\, R_{\hat{u},\phi}^{T} \right) \\
&= \ddot{R}_{\hat{u},\phi}\, R_{\hat{u},\phi}^{T} - \dot{R}_{\hat{u},\phi}\left[R_{\hat{u},\phi}^{T}\, \dot{R}_{\hat{u},\phi}\, R_{\hat{u},\phi}^{T} \right] \tag{9.65} \\
&= \ddot{R}_{\hat{u},\phi}\, R_{\hat{u},\phi}^{T} - \left[\dot{R}_{\hat{u},\phi}\, R_{\hat{u},\phi}^{T} \right]\left[\dot{R}_{\hat{u},\phi}\, R_{\hat{u},\phi}^{T} \right] \tag{9.66} \\
&= \ddot{R}_{\hat{u},\phi}\, R_{\hat{u},\phi}^{T} - _{G}\tilde{\omega}_{B}^{2} \tag{9.67} \\
&= \ddot{\phi}\,\tilde{u} + \dot{\phi}\dot{\tilde{u}} \tag{9.68}
\end{aligned}$$

because

$$\dot{R}_{\hat{u},\phi} = \dot{\phi}\,\tilde{u}\, R_{\hat{u},\phi} \tag{9.69}$$

$$\begin{aligned}
\ddot{R}_{\hat{u},\phi} &= \frac{^{G}d}{dt}\left(\dot{\phi}\,\tilde{u}\, R_{\hat{u},\phi} \right) \\
&= \ddot{\phi}\,\tilde{u}\, R_{\hat{u},\phi} + \dot{\phi}\,\dot{\tilde{u}}\, R_{\hat{u},\phi} + \dot{\phi}\,\tilde{u}\, \dot{R}_{\hat{u},\phi} \\
&= \left[\ddot{\phi}\,\tilde{u} + \dot{\phi}\dot{\tilde{u}} + \dot{\phi}^{2}\tilde{u}\tilde{u} \right] R_{\hat{u},\phi} \tag{9.70}
\end{aligned}$$

∎

Example 295 Velocity and acceleration of a simple pendulum. Pendulum is a good example to examine acceleration of a point mass in a rotating coordinate frame. This is a good example to show the components of acceleration vector in body and global frames. Tangential and centripetal accelerations are only B-frame observation.

A point mass attached to a massless rod and hanging from a revolute joint is called a *simple pendulum*. Figure 9.2 illustrates a simple pendulum. A local coordinate frame B is attached to the pendulum that rotates in a global frame G. The position vector of the bob and the angular velocity vector $_{G}\omega_{B}$ are

$$^B\mathbf{r} = l\hat{\imath} \tag{9.71}$$

$$^G\mathbf{r} = {}^GR_B\,{}^B\mathbf{r} = \begin{bmatrix} l\sin\phi \\ -l\cos\phi \\ 0 \end{bmatrix} \tag{9.72}$$

$$^B_G\boldsymbol{\omega}_B = \dot{\phi}\hat{k} \tag{9.73}$$

$$_G\boldsymbol{\omega}_B = {}^GR_B^T\,{}^B_G\boldsymbol{\omega}_B = \dot{\phi}\,\hat{K} \tag{9.74}$$

where,

$$^GR_B = \begin{bmatrix} \cos\left(\frac{3}{2}\pi+\phi\right) & -\sin\left(\frac{3}{2}\pi+\phi\right) & 0 \\ \sin\left(\frac{3}{2}\pi+\phi\right) & \cos\left(\frac{3}{2}\pi+\phi\right) & 0 \\ 0 & 0 & 1 \end{bmatrix}$$

$$= \begin{bmatrix} \sin\phi & \cos\phi & 0 \\ -\cos\phi & \sin\phi & 0 \\ 0 & 0 & 1 \end{bmatrix} \tag{9.75}$$

Its velocity is

$$^B_G\mathbf{v} = {}^B\dot{\mathbf{r}} + {}^B_G\boldsymbol{\omega}_B \times {}^B_G\mathbf{r} = 0 + \dot{\phi}\hat{k} \times l\hat{\imath} = l\,\dot{\phi}\hat{\jmath} \tag{9.76}$$

$$^G\mathbf{v} = {}^GR_B\,{}^B\mathbf{v} = \begin{bmatrix} l\,\dot{\phi}\cos\phi \\ l\,\dot{\phi}\sin\phi \\ 0 \end{bmatrix} \tag{9.77}$$

The acceleration of the bob is then equal to

$$^B_G\mathbf{a} = {}^B_G\dot{\mathbf{v}} + {}^B_G\boldsymbol{\omega}_B \times {}^B_G\mathbf{v}$$

$$= l\,\ddot{\phi}\hat{\jmath} + \dot{\phi}\hat{k} \times l\,\dot{\phi}\hat{\jmath} = l\,\ddot{\phi}\hat{\jmath} - l\,\dot{\phi}^2\hat{\imath} \tag{9.78}$$

$$^G\mathbf{a} = {}^GR_B\,{}^B\mathbf{a} = \begin{bmatrix} l\,\ddot{\phi}\cos\phi - l\,\dot{\phi}^2\sin\phi \\ l\,\ddot{\phi}\sin\phi + l\,\dot{\phi}^2\cos\phi \\ 0 \end{bmatrix} \tag{9.79}$$

Example 296 Rotation of a body point about a global axis. A numerical example of how to determine acceleration of a body point while the body is rotating in a global frame.

Consider a turning rigid body about the Z-axis with a constant angular acceleration $\ddot{\alpha} = 2\,\text{rad/s}^2$. The global acceleration of a body point at $P(5, 30, 10)$ cm when the body is at $\dot{\alpha} = 10\,\text{rad/s}$ and $\alpha = 30$ deg is

$$^G\mathbf{a}_P = {}^G\ddot{R}_B(t)\,{}^B\mathbf{r}_P \tag{9.80}$$

$$= \begin{bmatrix} -87.6 & 48.27 & 0 \\ -48.27 & -87.6 & 0 \\ 0 & 0 & 0 \end{bmatrix} \begin{bmatrix} 5 \\ 30 \\ 10 \end{bmatrix} = \begin{bmatrix} 1010 \\ -2869.4 \\ 0 \end{bmatrix} \text{cm/s}$$

where,

$$^G\ddot{R}_B = \frac{{}^Gd^2}{dt^2}\,{}^GR_B = \dot{\alpha}\frac{{}^Gd}{d\alpha}\,{}^GR_B = \ddot{\alpha}\frac{{}^Gd}{d\alpha}\,{}^GR_B + \dot{\alpha}^2\frac{{}^Gd^2}{d\alpha^2}\,{}^GR_B$$

$$= \ddot{\alpha}\begin{bmatrix} -\sin\alpha & -\cos\alpha & 0 \\ \cos\alpha & -\sin\alpha & 0 \\ 0 & 0 & 0 \end{bmatrix} + \dot{\alpha}^2\begin{bmatrix} -\cos\alpha & \sin\alpha & 0 \\ -\sin\alpha & -\cos\alpha & 0 \\ 0 & 0 & 0 \end{bmatrix} \tag{9.81}$$

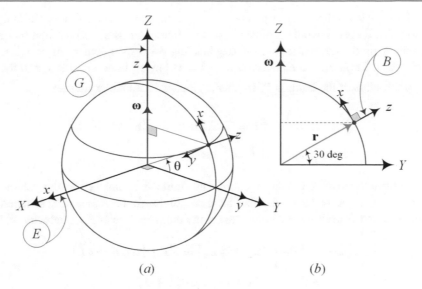

Fig. 9.3 Motion of a vehicle at latitude 30 deg and heading north on the Earth

At this moment, the point P is at

$$
{}^G\mathbf{r}_P = {}^G R_B \, {}^B\mathbf{r}_P \tag{9.82}
$$

$$
= \begin{bmatrix} \cos\dfrac{\pi}{6} & -\sin\dfrac{\pi}{6} & 0 \\ \sin\dfrac{\pi}{6} & \cos\dfrac{\pi}{6} & 0 \\ 0 & 0 & 1 \end{bmatrix} \begin{bmatrix} 5 \\ 30 \\ 10 \end{bmatrix} = \begin{bmatrix} -10.67 \\ 28.48 \\ 10 \end{bmatrix} \text{cm}
$$

What if the body point P was at ${}^B\mathbf{r}_P = \begin{bmatrix} 5 & 30 & 10 \end{bmatrix}^T$ cm while the body was turning with a constant angular acceleration $\ddot{\alpha} = 2\,\text{rad/s}^2$ about the Z-axis? Assume the body frame is at $\alpha = 30$ deg, its angular speed $\dot{\alpha} = 10$ deg/s. The transformation matrix ${}^G R_B$ between the B and G-frames will be

$$
{}^G R_B = \begin{bmatrix} \cos\dfrac{\pi}{6} & -\sin\dfrac{\pi}{6} & 0 \\ \sin\dfrac{\pi}{6} & \cos\dfrac{\pi}{6} & 0 \\ 0 & 0 & 1 \end{bmatrix} \approx \begin{bmatrix} 0.866 & -0.5 & 0 \\ 0.5 & 0.866 & 0 \\ 0 & 0 & 1 \end{bmatrix} \tag{9.83}
$$

and therefore, the acceleration of point P is

$$
{}^G\mathbf{a}_P = {}^G\ddot{R}_B \, {}^G R_B^T \, {}^G\mathbf{r}_P = \begin{bmatrix} 1010 \\ -2869.4 \\ 0 \end{bmatrix} \text{cm/s}^2 \tag{9.84}
$$

where

$$
\frac{{}^G d^2}{dt^2} \, {}^G R_B = \ddot{\alpha} \frac{{}^G d}{d\alpha} \, {}^G R_B - \dot{\alpha}^2 \frac{{}^G d^2}{d\alpha^2} \, {}^G R_B \tag{9.85}
$$

is the same as (9.81).

Example 297 Motion of a vehicle on the Earth. A practical example to show acceleration of a moving particle in a rotating coordinate frame. It also indicates relative order of magnitude of different terms of acceleration on Earth.

Consider a moving vehicle on the Earth at latitude 30 deg heading north, as shown in Fig. 9.3. The vehicle has a velocity of $v = {}^B_E\dot{\mathbf{r}} = 80\,\text{km/h} = 22.22\,\text{m/s}$ and acceleration $a = {}^B_E\ddot{\mathbf{r}} = 0.1\,\text{m/s}^2$, both with respect to the road expressed in the vehicle B-frame, $\left(\hat{\imath}, \hat{\jmath}, \hat{k}\right)$. Radius of the Earth is R, and hence, the vehicle's kinematics are

$$
{}^B_E\mathbf{r} = R\hat{k}\,\text{m} \qquad {}^B_E\dot{\mathbf{r}} = 22.22\hat{\imath}\,\text{m/s} \qquad {}^B_E\ddot{\mathbf{r}} = 0.1\hat{\imath}\,\text{m/s}^2
$$
$$
\dot{\theta} = \frac{v}{R}\text{rad/s} \qquad \ddot{\theta} = \frac{a}{R}\text{rad/s}^2 \tag{9.86}
$$

There are three coordinate frames involved. A body coordinate frame B is attached to the vehicle. A global motionless coordinate G is set up at the center of the Earth. Another local coordinate frame E is rigidly attached to the Earth and turns with the Earth. The frames E and G are assumed coincident at the moment. The angular velocity of B is

$$
{}^B_G\boldsymbol{\omega}_B = {}^BR_G \left({}_G\boldsymbol{\omega}_E + {}^G_E\boldsymbol{\omega}_B \right) = {}^BR_G \left(\omega_E\,\hat{K} + \dot{\theta}\,\hat{I} \right)
$$
$$
= (\omega_E\cos\theta)\,\hat{\imath} + (\omega_E\sin\theta)\,\hat{k} + \dot{\theta}\,\hat{\jmath}
$$
$$
= (\omega_E\cos\theta)\,\hat{\imath} + (\omega_E\sin\theta)\,\hat{k} + \frac{v}{R}\,\hat{\jmath} \tag{9.87}
$$

Therefore, the velocity and acceleration of the vehicle are

$$
{}^B_G\mathbf{v} = {}^B\dot{\mathbf{r}} + {}^B_G\boldsymbol{\omega}_B \times {}^B_G\mathbf{r} = 0 + {}^B_G\boldsymbol{\omega}_B \times R\hat{k} = v\hat{\imath} - (R\omega_E\cos\theta)\,\hat{\jmath} \tag{9.88}
$$

$$
{}^B_G\mathbf{a} = {}^B_G\dot{\mathbf{v}} + {}^B_G\boldsymbol{\omega}_B \times {}^B_G\mathbf{v}
$$
$$
= a\hat{\imath} + \left(R\omega_E\dot{\theta}\sin\theta\right)\hat{\jmath} + \begin{bmatrix} \omega_E\cos\theta \\ v/R \\ \omega_E\sin\theta \end{bmatrix} \times \begin{bmatrix} v \\ -R\omega_E\cos\theta \\ 0 \end{bmatrix}
$$
$$
= a\hat{\imath} + \left(R\omega_E\dot{\theta}\sin\theta\right)\hat{\jmath} + \begin{bmatrix} R\omega_E^2\cos\theta\sin\theta \\ v\omega_E\sin\theta \\ -\dfrac{1}{R}v^2 - R\omega_E^2\cos^2\theta \end{bmatrix}
$$
$$
= \begin{bmatrix} a + R\omega_E^2\cos\theta\sin\theta \\ 2R\omega_E\dot{\theta}\sin\theta \\ -\dfrac{1}{R}v^2 - R\omega_E^2\cos^2\theta \end{bmatrix} \tag{9.89}
$$

The term $a\hat{\imath}$ is the acceleration of B relative to Earth, $(2R\omega_E\dot{\theta}\sin\theta)\hat{\jmath}$ is the Coriolis acceleration, $-\frac{v^2}{R}\hat{k}$ is the centripetal acceleration due to traveling, and $-R\omega_E^2$ is the centripetal acceleration due to Earth's rotation.

Substituting the numerical values and accepting $R \approx 6.3677 \times 10^6$ m yield

$$
{}^B_G\mathbf{v} = 22.22\hat{\imath} - 6.3677 \times 10^6 \left(\frac{2\pi}{24 \times 3600} \frac{366.25}{365.25} \right) \cos\frac{\pi}{6}\,\hat{\jmath}
$$
$$
= 22.22\hat{\imath} - 402.13\hat{\jmath}\,\text{m/s} \tag{9.90}
$$

$$
{}^B_G\mathbf{a} = 1.5662 \times 10^{-2}\hat{\imath} + 1.6203 \times 10^{-3}\hat{\jmath} - 2.5473 \times 10^{-2}\hat{k}\,\text{m/s}^2 \tag{9.91}
$$

Example 298 ★ Combination of rotational acceleration transformation. Here, relative angular acceleration and relative rotational acceleration transformation equations are being derived analytically.

Let us consider a pair of relatively rotating rigid links in a base coordinate frame B_0 with a fixed point at O. The angular velocities of the links are related as

$$_0\boldsymbol{\omega}_2 = {_0}\boldsymbol{\omega}_1 + {_1^0}\boldsymbol{\omega}_2 \tag{9.92}$$

$$_0\tilde{\omega}_2 = {_0}\tilde{\omega}_1 + {_1^0}\tilde{\omega}_2 \tag{9.93}$$

So, their angular accelerations are

$$_0\boldsymbol{\alpha}_1 = \frac{^0 d}{dt} {_0}\boldsymbol{\omega}_1 \tag{9.94}$$

$$_0\boldsymbol{\alpha}_2 = \frac{^0 d}{dt} {_0}\boldsymbol{\omega}_2 = {_0}\boldsymbol{\alpha}_1 + {_1^0}\boldsymbol{\alpha}_2 \tag{9.95}$$

$$_0\tilde{\alpha}_2 = {_0}\tilde{\alpha}_1 + {_1^0}\tilde{\alpha}_2 \tag{9.96}$$

Similarly we have

$$_0 S_1 = {_0}\tilde{\alpha}_1 + {_0}\tilde{\omega}_1^2 \tag{9.97}$$

$$_0 S_2 = {_0}\tilde{\alpha}_2 + {_0}\tilde{\omega}_2^2 \tag{9.98}$$

and therefore,

$$\begin{aligned}
_0 S_2 &= {_0}\tilde{\alpha}_2 + {_0}\tilde{\omega}_2^2 = {_0}\tilde{\alpha}_1 + {_1^0}\tilde{\alpha}_2 + \left({_0}\tilde{\omega}_1 + {_1^0}\tilde{\omega}_2\right)^2 \\
&= {_0}\tilde{\alpha}_1 + {_1^0}\tilde{\alpha}_2 + {_0}\tilde{\omega}_1^2 + {_1^0}\tilde{\omega}_2^2 + 2\,{_0}\tilde{\omega}_1 {_1^0}\tilde{\omega}_2 \\
&= {_0} S_1 + {_1^0} S_2 + 2\,{_0}\tilde{\omega}_1 {_1^0}\tilde{\omega}_2
\end{aligned} \tag{9.99}$$

and

$$_0 S_2 \neq {_0} S_1 + {_1^0} S_2 \tag{9.100}$$

Equation (9.99) is the relative rotational acceleration transformation equation. It expresses the relative accelerations for a multi-link robot. As an example, consider a $6R$ articulated robot with six revolute joints. The angular acceleration of the end-effector frame in the base frame would be

$$_0\boldsymbol{\alpha}_6 = {_0}\boldsymbol{\alpha}_1 + {_1^0}\boldsymbol{\alpha}_2 + {_2^0}\boldsymbol{\alpha}_3 + {_3^0}\boldsymbol{\alpha}_4 + {_4^0}\boldsymbol{\alpha}_5 + {_5^0}\boldsymbol{\alpha}_6 \tag{9.101}$$

$$\begin{aligned}
_0 S_6 &= {_0} S_1 + {_1^0} S_2 + {_2^0} S_3 + {_3^0} S_4 + {_4^0} S_5 + {_5^0} S_6 \\
&\quad + 2\,{_0}\tilde{\omega}_1 \left({_1^0}\tilde{\omega}_2 + {_2^0}\tilde{\omega}_3 + {_3^0}\tilde{\omega}_4 + {_4^0}\tilde{\omega}_5 + {_5^0}\tilde{\omega}_6\right) \\
&\quad + 2\,{_1^0}\tilde{\omega}_2 \left({_2^0}\tilde{\omega}_3 + {_3^0}\tilde{\omega}_4 + {_4^0}\tilde{\omega}_5 + {_5^0}\tilde{\omega}_6\right) \\
&\qquad \vdots \\
&\quad + 2\,{_4^0}\tilde{\omega}_5 \left({_5^0}\tilde{\omega}_6\right)
\end{aligned} \tag{9.102}$$

We can transform the G-expression and B-expression of the global acceleration of a body point P, to each other using a rotation matrix.

$$\begin{aligned}
_G^B \mathbf{a}_P &= {^B R_G}\, {^G}\mathbf{a}_P = {^B R_G}\, {_G} S_B\, {^G}\mathbf{r}_P = {^B R_G}\, {_G} S_B\, {^G R_B}\, {^B}\mathbf{r}_P \\
&= {^B R_G}\, {^G}\ddot{R}_B\, {^G R_B^T}\, {^G R_B}\, {^B}\mathbf{r}_P = {^B R_G}\, {^G}\ddot{R}_B\, {^B}\mathbf{r}_P \\
&= {^G R_B^T}\, {^G}\ddot{R}_B\, {^B}\mathbf{r}_P = {_G^B} S_B\, {^B}\mathbf{r}_P = \left({_G^B}\tilde{\alpha}_B + {_G^B}\tilde{\omega}_B^2\right) {^B}\mathbf{r}_P \\
&= {_G^B}\boldsymbol{\alpha}_B \times {^B}\mathbf{r} + {_G^B}\boldsymbol{\omega}_B \times \left({_G^B}\boldsymbol{\omega}_B \times {^B}\mathbf{r}\right)
\end{aligned} \tag{9.103}$$

$$
\begin{aligned}
{}^{G}\mathbf{a}_{P} &= {}^{G}R_{B}\,{}^{B}_{G}\mathbf{a}_{P} = {}^{G}R_{B}\,{}^{B}_{G}S_{B}\,{}^{B}\mathbf{r}_{P} = {}^{G}R_{B}\,{}^{B}_{G}S_{B}\,{}^{G}R_{B}^{T}\,{}^{G}\mathbf{r}_{P} \\
&= {}^{G}R_{B}\,{}^{G}R_{B}^{T}\,{}^{G}\ddot{R}_{B}\,{}^{G}R_{B}^{T}\,{}^{G}\mathbf{r}_{P} = {}^{G}\ddot{R}_{B}\,{}^{G}R_{B}^{T}\,{}^{G}\mathbf{r}_{P} \\
&= {}_{G}S_{B}\,{}^{G}\mathbf{r}_{P} = \left({}_{G}\tilde{\alpha}_{B} + {}_{G}\tilde{\omega}_{B}^{2}\right)\,{}^{G}\mathbf{r} \\
&= {}_{G}\boldsymbol{\alpha}_{B} \times {}^{G}\mathbf{r} + {}_{G}\boldsymbol{\omega}_{B} \times \left({}_{G}\boldsymbol{\omega}_{B} \times {}^{G}\mathbf{r}\right)
\end{aligned}
\tag{9.104}
$$

The term ${}_{G}\boldsymbol{\alpha}_{B} \times {}^{G}\mathbf{r}$ is called the tangential acceleration, which is a function of the angular acceleration of B in G. The term ${}_{G}\boldsymbol{\omega}_{B} \times \left({}_{G}\boldsymbol{\omega}_{B} \times {}^{G}\mathbf{r}\right)$ in ${}^{G}\mathbf{a}$ is called the centripetal acceleration that is a function of the angular velocity of B in G.

From the definitions of ${}_{G}S_{B}$ and ${}^{B}_{G}S_{B}$ and comparing with (9.103) and (9.104), we can transform the two rotational acceleration transformations,

$$
{}_{G}S_{B} = {}^{G}R_{B}\,{}^{B}_{G}S_{B}\,{}^{G}R_{B}^{T}
\tag{9.105}
$$

$$
{}^{B}_{G}S_{B} = {}^{G}R_{B}^{T}\,{}_{G}S_{B}\,{}^{G}R_{B}
\tag{9.106}
$$

and derive the following equations:

$$
{}^{G}\ddot{R}_{B} = {}_{G}S_{B}\,{}^{G}R_{B}
\tag{9.107}
$$

$$
{}^{G}\ddot{R}_{B} = {}^{G}R_{B}\,{}^{B}_{G}S_{B}
\tag{9.108}
$$

$$
{}_{G}S_{B}\,{}^{G}R_{B} = {}^{G}R_{B}\,{}^{B}_{G}S_{B}
\tag{9.109}
$$

The angular acceleration of B in G is negative of the angular acceleration of G in B if both are expressed in the same coordinate frame.

$$
{}_{G}\tilde{\alpha}_{B} = -{}^{G}_{B}\tilde{\alpha}_{G} \qquad {}_{G}\boldsymbol{\alpha}_{B} = -{}^{G}_{B}\boldsymbol{\alpha}_{G}
\tag{9.110}
$$

$$
{}^{B}_{G}\tilde{\alpha}_{B} = -{}_{B}\tilde{\alpha}_{G} \qquad {}^{B}_{G}\boldsymbol{\alpha}_{B} = -{}_{B}\boldsymbol{\alpha}_{G}
\tag{9.111}
$$

The relative angular acceleration formula,

$$
{}_{0}\boldsymbol{\alpha}_{2} = {}_{0}\boldsymbol{\alpha}_{1} + {}^{0}_{1}\boldsymbol{\alpha}_{2}
\tag{9.112}
$$

$$
{}_{0}\boldsymbol{\alpha}_{n} = {}_{0}\boldsymbol{\alpha}_{1} + {}^{0}_{1}\boldsymbol{\alpha}_{2} + {}^{0}_{2}\boldsymbol{\alpha}_{3} + \cdots + {}^{0}_{n-1}\boldsymbol{\alpha}_{n} = \sum_{i=1}^{n} {}^{0}_{i-1}\boldsymbol{\alpha}_{i}
\tag{9.113}
$$

are correct if and only if all of the angular accelerations are expressed in the B_{0}-frame. Therefore, any equation of the form

$$
{}_{0}\boldsymbol{\alpha}_{2} \neq {}_{0}\boldsymbol{\alpha}_{1} + {}_{1}\boldsymbol{\alpha}_{2}
\tag{9.114}
$$

$$
\boldsymbol{\alpha}_{0} \neq \boldsymbol{\alpha}_{1} + \boldsymbol{\alpha}_{2}
\tag{9.115}
$$

$$
{}_{0}\boldsymbol{\alpha}_{3} \neq {}_{0}\boldsymbol{\alpha}_{1} + {}_{0}\boldsymbol{\alpha}_{2}
\tag{9.116}
$$

is wrong or is not completely expressed.

Example 299 ★ Alternative proof of relative acceleration formula. Here is another approach to show the relative angular acceleration formula starting from rotational transformation matrix.

To show addition of the relative angular accelerations in Eqs. (9.54) and (9.55), we may start from a combination of rotations,

$$
{}^{0}R_{2} = {}^{0}R_{1}\,{}^{1}R_{2}
\tag{9.117}
$$

and take their time derivatives.

$$
{}^{0}\dot{R}_{2} = {}^{0}\dot{R}_{1}\,{}^{1}R_{2} + {}^{0}R_{1}\,{}^{1}\dot{R}_{2}
\tag{9.118}
$$

$$
{}^{0}\ddot{R}_{2} = {}^{0}\ddot{R}_{1}\,{}^{1}R_{2} + 2\,{}^{0}\dot{R}_{1}\,{}^{1}\dot{R}_{2} + {}^{0}R_{1}\,{}^{1}\ddot{R}_{2}
\tag{9.119}
$$

Substituting the derivatives of rotation matrices with

$$
{}^0\ddot{R}_2 = {}_0S_2\,{}^0R_2 \qquad {}^0\ddot{R}_1 = {}_0S_1\,{}^0R_1 \qquad {}^1\ddot{R}_2 = {}_1S_2\,{}^1R_2 \tag{9.120}
$$

$$
{}^0\dot{R}_2 = {}_0\tilde{\omega}_2\,{}^0R_2 \qquad {}^0\dot{R}_1 = {}_0\tilde{\omega}_1\,{}^0R_1 \qquad {}^1\dot{R}_2 = {}_1\tilde{\omega}_2\,{}^1R_2 \tag{9.121}
$$

yields

$$
\begin{aligned}
{}_0S_2\,{}^0R_2 &= {}_0S_1\,{}^0R_1\,{}^1R_2 + 2\,{}_0\tilde{\omega}_1\,{}^0R_1\,{}_1\tilde{\omega}_2\,{}^1R_2 + {}^0R_1\,{}_1S_2\,{}^1R_2 \\
&= {}_0S_1\,{}^0R_2 + 2\,{}_0\tilde{\omega}_1\,{}^0R_1\,{}_1\tilde{\omega}_2\,{}^0R_1^T\,{}^0R_1\,{}^1R_2 + {}^0R_1\,{}_1S_2\,{}^1R_2 \\
&= {}_0S_1\,{}^0R_2 + 2\,{}_0\tilde{\omega}_1\,{}_1^0\tilde{\omega}_2\,{}^0R_2 + {}^0R_1\,{}_1S_2\,{}^0R_1^T\,{}^0R_1\,{}^1R_2 \\
&= {}_0S_1\,{}^0R_2 + 2\,{}_0\tilde{\omega}_1\,{}_1^0\tilde{\omega}_2\,{}^0R_2 + {}_1^0S_2\,{}^0R_2
\end{aligned} \tag{9.122}
$$

Therefore, we find

$$
{}_0S_2 = {}_0S_1 + {}_1^0S_2 + 2\,{}_0\tilde{\omega}_1\,{}_1^0\tilde{\omega}_2 \tag{9.123}
$$

which is equivalent to

$$
{}_0\tilde{\alpha}_2 + {}_0\tilde{\omega}_2^2 = {}_0\tilde{\alpha}_1 + {}_0\tilde{\omega}_1^2 + {}_1^0\tilde{\alpha}_2 + {}_1^0\tilde{\omega}_2^2 + 2\,{}_0\tilde{\omega}_1\,{}_1^0\tilde{\omega}_2 \tag{9.124}
$$

Simplifying this equation shows that

$$
\begin{aligned}
{}_0\tilde{\alpha}_2 &= {}_0\tilde{\alpha}_1 + {}_1^0\tilde{\alpha}_2 + {}_0\tilde{\omega}_1^2 + {}_1^0\tilde{\omega}_2^2 + 2\,{}_0\tilde{\omega}_1\,{}_1^0\tilde{\omega}_2 - {}_0\tilde{\omega}_2^2 \\
&= {}_0\tilde{\alpha}_1 + {}_1^0\tilde{\alpha}_2 + \left({}_0\tilde{\omega}_1 + {}_1^0\tilde{\omega}_2\right)^2 - {}_0\tilde{\omega}_2^2 \\
&= {}_0\tilde{\alpha}_1 + {}_1^0\tilde{\alpha}_2 + {}_0\tilde{\omega}_2^2 - {}_0\tilde{\omega}_2^2 = {}_0\tilde{\alpha}_1 + {}_1^0\tilde{\alpha}_2
\end{aligned} \tag{9.125}
$$

which indicates two angular accelerations may be added when they are expressed in the same frame:

$$
{}_0\boldsymbol{\alpha}_2 = {}_0\boldsymbol{\alpha}_1 + {}_1^0\boldsymbol{\alpha}_2 \tag{9.126}
$$

Example 300 ★ Angular acceleration and Euler angles. Expression of angular velocity and acceleration in terms of Euler angles and their time rate is practical method for rigid body dynamics. Here is how to derive angular acceleration.

The angular velocity ${}_G\boldsymbol{\omega}_B$ in terms of Euler angles is

$$
\begin{aligned}
{}_G^G\boldsymbol{\omega}_B &= \begin{bmatrix} \omega_X \\ \omega_Y \\ \omega_Z \end{bmatrix} = \begin{bmatrix} 0 & \cos\varphi & \sin\theta\sin\varphi \\ 0 & \sin\varphi & -\cos\varphi\sin\theta \\ 1 & 0 & \cos\theta \end{bmatrix} \begin{bmatrix} \dot{\varphi} \\ \dot{\theta} \\ \dot{\psi} \end{bmatrix} \\
&= \begin{bmatrix} \dot{\theta}\cos\varphi + \dot{\psi}\sin\theta\sin\varphi \\ \dot{\theta}\sin\varphi - \dot{\psi}\cos\varphi\sin\theta \\ \dot{\varphi} + \dot{\psi}\cos\theta \end{bmatrix}
\end{aligned} \tag{9.127}
$$

The angular acceleration is then equal to a simple derivative of ${}_G^G\boldsymbol{\omega}_B$.

$$
\begin{aligned}
{}_G^G\boldsymbol{\alpha}_B &= \frac{{}^Gd}{dt}\,{}_G^G\boldsymbol{\omega}_B \\
&= \begin{bmatrix} \cos\varphi\left(\ddot{\theta} + \dot{\varphi}\dot{\psi}\sin\theta\right) + \sin\varphi\left(\ddot{\psi}\sin\theta + \dot{\theta}\dot{\psi}\cos\theta - \dot{\theta}\dot{\varphi}\right) \\ \sin\varphi\left(\ddot{\theta} + \dot{\varphi}\dot{\psi}\sin\theta\right) + \cos\varphi\left(\dot{\theta}\dot{\varphi} - \ddot{\psi}\sin\theta - \dot{\theta}\dot{\psi}\cos\theta\right) \\ \ddot{\varphi} + \ddot{\psi}\cos\theta - \dot{\theta}\dot{\psi}\sin\theta \end{bmatrix}
\end{aligned} \tag{9.128}
$$

The angular acceleration vector in the body coordinate frame is then equal to

$$
{}^B_G\boldsymbol{\alpha}_B = {}^G R^T_B \, {}^G_G\boldsymbol{\alpha}_B \tag{9.129}
$$

$$
= \begin{bmatrix} c\varphi c\psi - c\theta s\varphi s\psi & c\psi s\varphi + c\theta c\varphi s\psi & s\theta s\psi \\ -c\varphi s\psi - c\theta c\psi s\varphi & -s\varphi s\psi + c\theta c\varphi c\psi & s\theta c\psi \\ s\theta s\varphi & -c\varphi s\theta & c\theta \end{bmatrix} {}^G_G\boldsymbol{\alpha}_B
$$

$$
= \begin{bmatrix} \cos\psi\left(\ddot{\theta} + \dot{\varphi}\dot{\psi}\sin\theta\right) + \sin\psi\left(\ddot{\varphi}\sin\theta + \dot{\theta}\dot{\varphi}\cos\theta - \dot{\theta}\dot{\psi}\right) \\ \cos\psi\left(\ddot{\varphi}\sin\theta + \dot{\theta}\dot{\varphi}\cos\theta - \dot{\theta}\dot{\psi}\right) - \sin\psi\left(\ddot{\theta} + \dot{\varphi}\dot{\psi}\sin\theta\right) \\ \ddot{\varphi}\cos\theta - \ddot{\psi} - \dot{\theta}\dot{\varphi}\sin\theta \end{bmatrix}
$$

Example 301 ★ *B-expression of angular acceleration.* Angular acceleration is a vectorial quantity and hence, it can be expressed in B or G frames. Here is to show how to express angular acceleration in B-frame.

The angular acceleration expressed in the body frame is the body derivative of the angular velocity vector. To show this, we use the derivative transport formula (7.214):

$$
{}^B_G\boldsymbol{\alpha}_B = {}^B_G\dot{\boldsymbol{\omega}}_B = \frac{{}^G d}{dt}\,{}^B_G\boldsymbol{\omega}_B
$$

$$
= \frac{{}^B d}{dt}\,{}^B_G\boldsymbol{\omega}_B + {}^B_G\boldsymbol{\omega}_B \times {}^B_G\boldsymbol{\omega}_B = \frac{{}^B d}{dt}\,{}^B_G\boldsymbol{\omega}_B \tag{9.130}
$$

Interestingly, the global and body derivatives of ${}^B_G\boldsymbol{\omega}_B$ are equal, because ${}_G\boldsymbol{\omega}_B$ is about an axis $\hat{u}$ that is instantaneously fixed in both, B and G.

$$
\frac{{}^G d}{dt}\,{}^B_G\boldsymbol{\omega}_B = \frac{{}^B d}{dt}\,{}^B_G\boldsymbol{\omega}_B = {}^B_G\boldsymbol{\alpha}_B \tag{9.131}
$$

A vector $\boldsymbol{\alpha}$ can generally indicate the angular acceleration of a coordinate frame A with respect to another frame B. It can be expressed in or seen from a third coordinate frame C. We indicate the first coordinate frame A by a right subscript, the second frame B by a left subscript, and the third frame C by a left superscript, ${}^C_B\boldsymbol{\alpha}_A$. If the left super and subscripts are the same, we only show the subscript. So, the angular acceleration of A with respect to B as seen from C is the C-expression of ${}_B\boldsymbol{\alpha}_A$:

$$
{}^C_B\boldsymbol{\alpha}_A = {}^C R_B \, {}_B\boldsymbol{\alpha}_A \tag{9.132}
$$

Transforming ${}^G\mathbf{a}$ to the body frame provides the body expression of the acceleration vector:

$$
{}^B_G\mathbf{a}_P = {}^G R^T_B \, {}^G\mathbf{a} = {}^G R^T_B \, {}_G S_B \, {}^G\mathbf{r} = {}^G R^T_B \, {}^G\ddot{R}_B \, {}^G R^T_B \, {}^G\mathbf{r}
$$

$$
= {}^G R^T_B \, {}^G\ddot{R}_B \, {}^B\mathbf{r} \tag{9.133}
$$

We denote the coefficient of ${}^B\mathbf{r}$ by ${}^B_G S_B$

$$
{}^B_G S_B = {}^G R^T_B \, {}^G\ddot{R}_B \tag{9.134}
$$

and rewrite Eq. (9.133) as

$$
{}^B_G\mathbf{a}_P = {}^B_G S_B \, {}^B\mathbf{r}_P \tag{9.135}
$$

where, ${}^B_G S_B$ is the rotational acceleration transformation of the B-frame relative to G-frame as seen from the B-frame.

Example 302 Principal angular accelerations. Angular acceleration for rotation about principal global axes is calculated here.

The principal rotational matrices about the axes X, Y, and Z are given by $R_{X,\gamma}$, $R_{Y,\beta}$, and $R_{Z,\alpha}$. Based on their time derivatives, the principal angular velocities about the axes X, Y, and Z are

$$
\tilde{\omega}_X = \dot{R}_{X,\gamma} R^T_{X,\gamma} = \dot{\gamma}\tilde{I} \qquad \boldsymbol{\omega}_X = \omega_X \hat{I} = \dot{\gamma}\hat{I} \tag{9.136}
$$

$$
\tilde{\omega}_Y = \dot{R}_{Y,\beta} R^T_{Y,\beta} = \dot{\beta}\tilde{J} \qquad \boldsymbol{\omega}_Y = \omega_Y \hat{J} = \dot{\beta}\hat{J} \tag{9.137}
$$

$$
\tilde{\omega}_Z = \dot{R}_{Z,\alpha} R^T_{Z,\alpha} = \dot{\alpha}\tilde{K} \qquad \boldsymbol{\omega}_Z = \omega_Z \hat{K} = \dot{\alpha}\hat{K} \tag{9.138}
$$

Taking another derivative shows that the principal angular accelerations about the axes X, Y, and Z are

$$\tilde{\alpha}_X = \ddot{R}_{X,\gamma} R_{X,\gamma}^T + \dot{R}_{X,\gamma} \dot{R}_{X,\gamma}^T = \ddot{\gamma} \tilde{I} \quad \boldsymbol{\alpha}_X = \alpha_X \hat{I} = \ddot{\gamma} \hat{I} \tag{9.139}$$

$$\tilde{\alpha}_Y = \ddot{R}_{Y,\beta} R_{Y,\beta}^T + \dot{R}_{Y,\beta} \dot{R}_{Y,\beta}^T = \ddot{\beta} \tilde{J} \quad \boldsymbol{\alpha}_Y = \alpha_Y \hat{J} = \ddot{\beta} \hat{J} \tag{9.140}$$

$$\tilde{\alpha}_Z = \ddot{R}_{Z,\alpha} R_{Z,\alpha}^T + \dot{R}_{Z,\alpha} \dot{R}_{Z,\alpha}^T = \ddot{\alpha} \tilde{K} \quad \boldsymbol{\alpha}_Z = \alpha_Z \hat{K} = \ddot{\alpha} \hat{K} \tag{9.141}$$

and therefore,

$$S_{X,\ddot{\gamma}} = \ddot{R}_{X,\gamma} R_{X,\gamma}^T = \tilde{\alpha}_X + \tilde{\omega}_X^2 = \ddot{\gamma} \tilde{I} + \dot{\gamma}^2 \tilde{I} \tilde{I} \tag{9.142}$$

$$S_{Y,\ddot{\beta}} = \ddot{R}_{Y,\beta} R_{Y,\beta}^T = \tilde{\alpha}_Y + \tilde{\omega}_Y^2 = \ddot{\beta} \tilde{J} + \dot{\beta}^2 \tilde{J} \tilde{J} \tag{9.143}$$

$$S_{Z,\ddot{\alpha}} = \ddot{R}_{Z,\alpha} R_{Z,\alpha}^T = \tilde{\alpha}_Z + \tilde{\omega}_Z^2 = \ddot{\alpha} \tilde{K} + \dot{\alpha}^2 \tilde{K} \tilde{K} \tag{9.144}$$

$$\ddot{R}_{X,\gamma} = \left(\ddot{\gamma} \tilde{I} + \dot{\gamma}^2 \tilde{I} \tilde{I} \right) R_{X,\gamma} \tag{9.145}$$

$$\ddot{R}_{Y,\beta} = \left(\ddot{\beta} \tilde{J} + \dot{\beta}^2 \tilde{J} \tilde{J} \right) R_{Y,\beta} \tag{9.146}$$

$$\ddot{R}_{Z,\alpha} = \left(\ddot{\alpha} \tilde{K} + \dot{\alpha}^2 \tilde{K} \tilde{K} \right) R_{Z,\alpha} \tag{9.147}$$

Example 303 ★ Alternative definition of angular acceleration vector. Angular acceleration may also be defined by orthogonality condition and time derivative of the unit vectors of B-frame.

Similar to the definition of the angular velocity vector in Eq. (7.229),

$$_G^B \boldsymbol{\omega}_B = \hat{\imath} \left(\frac{^G d\hat{\jmath}}{dt} \cdot \hat{k} \right) + \hat{\jmath} \left(\frac{^G d\hat{k}}{dt} \cdot \hat{\imath} \right) + \hat{k} \left(\frac{^G d\hat{\imath}}{dt} \cdot \hat{\jmath} \right) \tag{9.148}$$

we define the angular acceleration vector $_G^B \boldsymbol{\alpha}_B$ of a rigid body $B(\hat{\imath}, \hat{\jmath}, \hat{k})$ in the global frame $G(\hat{I}, \hat{J}, \hat{K})$.

$$
\begin{aligned}
_G^B \boldsymbol{\alpha}_B &= \frac{^G d\hat{\imath}}{dt} \left(\frac{^G d\hat{\jmath}}{dt} \cdot \hat{k} \right) + \frac{1}{2} \left(\frac{^G d^2\hat{\jmath}}{dt^2} \cdot \hat{k} - \frac{^G d^2\hat{k}}{dt^2} \cdot \hat{\jmath} \right) \hat{\imath} \\
&+ \frac{^G d\hat{\jmath}}{dt} \left(\frac{^G d\hat{k}}{dt} \cdot \hat{\imath} \right) + \frac{1}{2} \left(\frac{^G d^2\hat{k}}{dt^2} \cdot \hat{\imath} - \frac{^G d^2\hat{\imath}}{dt^2} \cdot \hat{k} \right) \hat{\jmath} \\
&+ \frac{^G d\hat{k}}{dt} \left(\frac{^G d\hat{\imath}}{dt} \cdot \hat{\jmath} \right) + \frac{1}{2} \left(\frac{^G d^2\hat{\imath}}{dt^2} \cdot \hat{\jmath} - \frac{^G d^2\hat{\jmath}}{dt^2} \cdot \hat{\imath} \right) \hat{k}
\end{aligned} \tag{9.149}
$$

To prove (9.149), we take a G-derivative from (9.148).

$$
\begin{aligned}
_G^B \boldsymbol{\alpha}_B &= \frac{^G d\hat{\imath}}{dt} \left(\frac{^G d\hat{\jmath}}{dt} \cdot \hat{k} \right) + \frac{^G d\hat{\jmath}}{dt} \left(\frac{^G d\hat{k}}{dt} \cdot \hat{\imath} \right) + \frac{^G d\hat{k}}{dt} \left(\frac{^G d\hat{\imath}}{dt} \cdot \hat{\jmath} \right) \\
&+ \hat{\imath} \left(\frac{^G d^2\hat{\jmath}}{dt^2} \cdot \hat{k} + \frac{^G d\hat{\jmath}}{dt} \cdot \frac{^G d\hat{k}}{dt} \right) \\
&+ \hat{\jmath} \left(\frac{^G d^2\hat{k}}{dt^2} \cdot \hat{\imath} + \frac{^G d\hat{k}}{dt} \cdot \frac{^G d\hat{\imath}}{dt} \right) \\
&+ \hat{k} \left(\frac{^G d^2\hat{\imath}}{dt^2} \cdot \hat{\jmath} + \frac{^G d\hat{\imath}}{dt} \cdot \frac{^G d\hat{\jmath}}{dt} \right)
\end{aligned} \tag{9.150}
$$

Employing the unit vectors relationships

$$\hat{\imath} \cdot \hat{\imath} = \hat{\jmath} \cdot \hat{\jmath} = \hat{k} \cdot \hat{k} = 1 \tag{9.151}$$

$$\hat{\imath} \cdot \hat{\jmath} = \hat{\jmath} \cdot \hat{k} = \hat{k} \cdot \hat{\imath} = 0 \tag{9.152}$$

$$\hat{\imath} \cdot d\hat{\imath} = \hat{\jmath} \cdot d\hat{\jmath} = \hat{k} \cdot d\hat{k} = 0 \tag{9.153}$$

$$\hat{\jmath} \cdot d\hat{\imath} = -\hat{\imath} \cdot d\hat{\jmath} \quad \hat{k} \cdot d\hat{\jmath} = -\hat{\jmath} \cdot d\hat{k} \quad \hat{\imath} \cdot d\hat{k} = -\hat{k} \cdot d\hat{\imath} \tag{9.154}$$

$$\hat{\imath} \cdot d^2\hat{\imath} = -d\hat{\imath} \cdot d\hat{\imath}$$
$$\hat{\jmath} \cdot d^2\hat{\jmath} = -d\hat{\jmath} \cdot d\hat{\jmath} \tag{9.155}$$
$$\hat{k} \cdot d^2\hat{k} = -d\hat{k} \cdot d\hat{k}$$

$$\hat{\imath} \cdot d^2\hat{\jmath} + \hat{\jmath} \cdot d^2\hat{\imath} = -2d\hat{\imath} \cdot d\hat{\jmath}$$
$$\hat{\jmath} \cdot d^2\hat{k} + \hat{k} \cdot d^2\hat{\jmath} = -2d\hat{\jmath} \cdot d\hat{k} \tag{9.156}$$
$$\hat{k} \cdot d^2\hat{\imath} + \hat{\imath} \cdot d^2\hat{k} = -2d\hat{k} \cdot d\hat{\imath}$$

we can simplify (9.150)–(9.149).

Example 304 ★ Technical point on $_G\tilde{\omega}_B, {}^B_G S_B, {}_G\tilde{\alpha}_B$. These three derivative quantities are acting as operator and transformers. Here their technical points are described.

The derivative kinematics of a rigid body with a fixed point begins by differentiating the kinematic transformation between the body B and global G coordinate frames:

$$^G\mathbf{r} = {}^G R_B \, {}^B\mathbf{r} \tag{9.157}$$

The kinematic transformation matrix $^G R_B$ takes the coordinates of a point in the B-frame and determines the coordinates of the point in the G-frame. The matrix $^G R_B$ is a kinematic or geometric transformation because the dimension of what it takes and what it provides are the same.

The first derivative of (9.157) indicates the time rate of motion of a body point and introduces the angular velocity,

$$^G\mathbf{v} = {}^G\dot{R}_B \, {}^B\mathbf{r} = {}_G\tilde{\omega}_B \, {}^G\mathbf{r} \tag{9.158}$$

where $_G\tilde{\omega}_B$ is the skew symmetric angular velocity matrix of B in G. This matrix is skew symmetric and its associated vector is the angular velocity vector. The matrix $_G\tilde{\omega}_B$ also acts as an operator and a transformer. It takes the global position vector of a body point, $^G\mathbf{r}$, and determines its global velocity vector, $^G\mathbf{v}$. So, it is also called the rotational velocity transformation.

The second derivative of (9.157) indicates the time rate of velocity and introduces the rotational acceleration transformation $_G S_B$,

$$^G\mathbf{a} = {}^G\ddot{R}_B \, {}^B\mathbf{r} = {}_G S_B \, {}^G\mathbf{r} \tag{9.159}$$

where $_G S_B$ is not skew symmetric, so it does not indicate a vector. However, $_G S_B$ acts as an operator and a transformer. It takes the global position vector of a body point, $^G\mathbf{r}$, and determines its global acceleration vector, $^G\mathbf{a}$. It is called the rotational acceleration transformation. The matrix $_G S_B$ is the sum of two matrices:

$$_G S_B = {}_G\tilde{\alpha}_B + {}_G\tilde{\omega}_B \, {}_G\tilde{\omega}_B = {}_G\tilde{\alpha}_B + {}_G\tilde{\omega}_B^2 \tag{9.160}$$

The first matrix, $_G\tilde{\alpha}_B$, is the time derivative of $_G\tilde{\omega}_B$. So, it is a skew symmetric matrix and indicates the angular acceleration matrix and vector. However, $_G\tilde{\alpha}_B$ cannot transform a position vector to its acceleration vector. The second matrix, $_G\tilde{\omega}_B^2$, is the square of the angular velocity matrix. It is also not a skew symmetric matrix and indicates no vector.

We may consider $_G\tilde{\alpha}_B$ and $_G\tilde{\omega}_B^2$ as transformers because when they operate on $^G\mathbf{r}$ they respectively provide the tangential and centripetal components of the acceleration vector $^G\mathbf{a}$.

Example 305 ★ Angular jerk. The next derivative of angular velocity and angular acceleration will be angular jerk. Jerk is needed when shaking is a problem to study and measure.

Angular second acceleration matrix $\tilde{\chi} = \dot{\tilde{\alpha}} = \ddot{\tilde{\omega}}$ is a skew symmetric matrix associated to the angular jerk $\chi = \dot{\alpha} = \ddot{\omega}$.

$$
{}_G\tilde{\chi}_B = {}_G\dot{\tilde{\alpha}}_B = {}_G\ddot{\tilde{\omega}}_B = \frac{{}^Gd}{dt}\left({}^G\ddot{R}_B\,{}^GR_B^T + {}^G\dot{R}_B\,{}^G\dot{R}_B^T\right)
$$

$$
= {}^G\dddot{R}_B\,{}^GR_B^T + 2\,{}^G\ddot{R}_B\,{}^G\dot{R}_B^T + {}^G\dot{R}_B\,{}^G\ddot{R}_B^T
$$

$$
= {}_GU_B + 2\,{}_GS_B\,{}_G\tilde{\omega}_B^T + {}_G\tilde{\omega}_B\,{}_GS_B^T \tag{9.161}
$$

$$
= \frac{{}^Gd}{dt}\left(\ddot{\phi}\tilde{u} + \dot{\phi}\dot{\tilde{u}}\right) = \dddot{\phi}\tilde{u} + 2\ddot{\phi}\dot{\tilde{u}} + \dot{\phi}\ddot{\tilde{u}} \tag{9.162}
$$

The global jerk, ${}^G\mathbf{j}$, of a body point P at ${}^G\mathbf{r}$ is

$$
{}^G\mathbf{j} = {}^G\dddot{\mathbf{r}} = \frac{{}^Gd}{dt}\left({}_GS_B\,{}^G\mathbf{r}\right) = {}^G\dddot{R}_B\,{}^GR_B^T\,{}^G\mathbf{r}
$$

$$
= \left[\dddot{\phi}\tilde{u} + 2\ddot{\phi}\dot{\tilde{u}} + \dot{\phi}\ddot{\tilde{u}} + 3\dot{\phi}\ddot{\phi}\tilde{u}^2 + 2\dot{\phi}^2\dot{\tilde{u}}\tilde{u} + \dot{\phi}^2\tilde{u}\dot{\tilde{u}} + \dot{\phi}^3\tilde{u}^3\right]{}^G\mathbf{r}
$$

$$
= {}_GU_B\,{}^G\mathbf{r} \tag{9.163}
$$

${}_GU_B = {}^G\dddot{R}_B\,{}^GR_B^T$ is the rotational jerk transformation between B and G. Hence, the angular jerk matrix would be

$$
\tilde{\chi} = \dot{\tilde{\alpha}} = \ddot{\tilde{\omega}} = \begin{bmatrix} j_{11} & j_{12} & j_{13} \\ j_{21} & j_{22} & j_{23} \\ j_{31} & j_{32} & j_{33} \end{bmatrix} \tag{9.164}
$$

where,

$$
j_{11} = 3u_1\dot{u}_1\dot{\phi}^2 + 3\left(u_1^2 - 1\right)\dot{\phi}\ddot{\phi}
$$

$$
j_{21} = (2u_2\dot{u}_1 + \dot{u}_2 u_1)\dot{\phi}^2 + 3u_2 u_1\dot{\phi}\ddot{\phi}
$$
$$
+ \left(\ddot{u}_3\dot{\phi} + 2\dot{u}_3\ddot{\phi} + u_3\dddot{\phi} - u_3\dot{\phi}^3\right)
$$

$$
j_{31} = (2u_3\dot{u}_1 + \dot{u}_3 u_1)\dot{\phi}^2 + 3u_3 u_1\dot{\phi}\ddot{\phi}
$$
$$
+ \left(\ddot{u}_2\dot{\phi} + 2\dot{u}_2\ddot{\phi} + u_2\dddot{\phi} - u_2\dot{\phi}^3\right)
$$

$$
j_{12} = (2u_1\dot{u}_2 + \dot{u}_1 u_2)\dot{\phi}^2 + 3u_1 u_2\dot{\phi}\ddot{\phi}
$$
$$
+ \left(\ddot{u}_3\dot{\phi} + 2\dot{u}_3\ddot{\phi} + u_3\dddot{\phi} - u_3\dot{\phi}^3\right)
$$

$$
j_{22} = 3u_2\dot{u}_2\dot{\phi}^2 + 3\left(u_2^2 - 1\right)\dot{\phi}\ddot{\phi}
$$

$$
j_{32} = (2u_3\dot{u}_2 + \dot{u}_3 u_2)\dot{\phi}^2 + 3u_3 u_2\dot{\phi}\ddot{\phi}
$$
$$
+ \left(\ddot{u}_1\dot{\phi} + 2\dot{u}_1\ddot{\phi} + u_1\dddot{\phi} - u_1\dot{\phi}^3\right)
$$

$$j_{13} = (2u_1\dot{u}_3 + \dot{u}_1 u_3)\,\dot{\phi}^2 + 3u_1 u_3\dot{\phi}\ddot{\phi}$$
$$+ \left(\ddot{u}_2\dot{\phi} + 2\dot{u}_2\ddot{\phi} + u_2\dddot{\phi} - u_2\dot{\phi}^3\right)$$
$$j_{23} = (2u_2\dot{u}_3 + \dot{u}_2 u_3)\,\dot{\phi}^2 + 3u_2 u_3\dot{\phi}\ddot{\phi}$$
$$+ \left(\ddot{u}_1\dot{\phi} + 2\dot{u}_1\ddot{\phi} + u_1\dddot{\phi} - u_1\dot{\phi}^3\right)$$
$$j_{33} = 3u_3\dot{u}_3\dot{\phi}^2 + 3\left(u_3^2 - 1\right)\dot{\phi}\ddot{\phi} \tag{9.165}$$

Example 306 ★ Rotational jerk transformation. It is a matrix that applies on global position vector and provides global jerk vector.

Consider a body coordinate frame B with a fixed point in a global frame G. The B-frame is turning in G with angular velocity ${}_G\boldsymbol{\omega}_B$ and acceleration ${}_G\boldsymbol{\alpha}_B$. The global jerk ${}^G\mathbf{j}$ of a body point at ${}^G\mathbf{r}$ is

$$
{}^G\mathbf{j} = {}^G\dddot{\mathbf{r}} = \frac{{}^Gd}{dt}\left({}_G S_B\,{}^G\mathbf{r}\right) = \frac{{}^Gd^2}{dt^2}\left({}_G\tilde{\omega}_B\,{}^G\mathbf{r}\right) = \frac{{}^Gd^3}{dt^3}\,{}^G\mathbf{r}
$$
$$
= \frac{{}^Gd}{dt}\left(\left[\ddot{\phi}\tilde{u} + \dot{\phi}\dot{\tilde{u}} + \dot{\phi}^2\tilde{u}^2\right]{}^G\mathbf{r}\right)
$$
$$
= \left[\dddot{\phi}\tilde{u} + 2\ddot{\phi}\dot{\tilde{u}} + \dot{\phi}\ddot{\tilde{u}} + 3\dot{\phi}\ddot{\phi}\tilde{u}^2 + 2\dot{\phi}^2\tilde{u}\dot{\tilde{u}} + \dot{\phi}^2\tilde{u}\dot{\tilde{u}} + \dot{\phi}^3\tilde{u}^3\right]{}^G\mathbf{r}
$$
$$
= {}_G U_B\,{}^G\mathbf{r} \tag{9.166}
$$

where, ${}_G U_B$ is the *rotational jerk transformation*.

$$
{}_G U_B = \dddot{\phi}\tilde{u} + 2\ddot{\phi}\dot{\tilde{u}} + \dot{\phi}\ddot{\tilde{u}} + 3\dot{\phi}\ddot{\phi}\tilde{u}^2 + 2\dot{\phi}^2\tilde{u}\dot{\tilde{u}} + \dot{\phi}^2\tilde{u}\dot{\tilde{u}} + \dot{\phi}^3\tilde{u}^3 \tag{9.167}
$$

Employing ${}^G\mathbf{a} = {}^G\ddot{R}_B\,{}^G R_B^T\,{}^G\mathbf{r}$, we can find the jerk of the body point and jerk transformation, based on the rotational transformation matrix ${}^G R_B$:

$$
{}^G\mathbf{j} = \frac{{}^Gd}{dt}\left({}^G\mathbf{a}\right) = \frac{{}^Gd}{dt}\left({}_G S_B\,{}^G\mathbf{r}\right) = \frac{{}^Gd}{dt}\left({}^G\ddot{R}_B\,{}^G R_B^T\,{}^G\mathbf{r}\right)
$$
$$
= \left({}^G\dddot{R}_B\,{}^G R_B^T + {}^G\ddot{R}_B\,{}^G\dot{R}_B^T + {}^G\ddot{R}_B\,{}^G R_B^T\,{}^G\dot{R}_B\,{}^G R_B^T\right){}^G\mathbf{r}
$$
$$
= \left({}^G\dddot{R}_B\,{}^G R_B^T + {}^G\ddot{R}_B\,{}^G R_B^T\,{}^G R_B\,{}^G\dot{R}_B^T + {}^G\ddot{R}_B\,{}^G R_B^T\,{}^G\dot{R}_B\,{}^G R_B^T\right){}^G\mathbf{r}
$$
$$
= \left({}^G\dddot{R}_B\,{}^G R_B^T + {}^G\ddot{R}_B\,{}^G R_B^T\left({}^G R_B\,{}^G\dot{R}_B^T + \left[{}^G R_B\,{}^G\dot{R}_B^T\right]^T\right)\right){}^G\mathbf{r}
$$
$$
= {}^G\dddot{R}_B\,{}^G R_B^T\,{}^G\mathbf{r} \tag{9.168}
$$

$$
{}_G U_B = {}^G\dddot{R}_B\,{}^G R_B^T \tag{9.169}
$$

Using ${}^G\mathbf{a} = {}_G\boldsymbol{\alpha}_B \times {}^G\mathbf{r} + {}_G\boldsymbol{\omega}_B \times \left({}_G\boldsymbol{\omega}_B \times {}^G\mathbf{r}\right)$, we can find the vectorial expression formula for the jerk of a body point:

$$
{}^G\mathbf{j} = \frac{{}^Gd}{dt}\left({}^G\mathbf{a}\right) = \frac{{}^Gd}{dt}\left({}_G\boldsymbol{\alpha}_B \times {}^G\mathbf{r} + {}_G\boldsymbol{\omega}_B \times \left({}_G\boldsymbol{\omega}_B \times {}^G\mathbf{r}\right)\right)
$$
$$
= {}_G\boldsymbol{\chi}_B \times {}^G\mathbf{r} + 2\,{}_G\boldsymbol{\alpha}_B \times \left({}_G\boldsymbol{\omega}_B \times {}^G\mathbf{r}\right)
$$
$$
+ {}_G\boldsymbol{\omega}_B \times \left({}_G\boldsymbol{\alpha}_B \times {}^G\mathbf{r}\right)
$$
$$
+ {}_G\boldsymbol{\omega}_B \times \left({}_G\boldsymbol{\omega}_B \times \left({}_G\boldsymbol{\omega}_B \times {}^G\mathbf{r}\right)\right) \tag{9.170}
$$

where ${}_G\boldsymbol{\chi}_B$ is the angular jerk vector of B relative to G.

The matrix form of the jerk is

$$
{}^G\mathbf{j} = \frac{{}^G d}{dt} \left(\left({}_G\tilde{\alpha}_B + {}_G\tilde{\omega}_B^2 \right) {}^G\mathbf{r} \right)
$$

$$
= \left({}_G\dot{\tilde{\alpha}}_B + 2\,{}_G\tilde{\omega}_B\,{}_G\tilde{\alpha}_B + \left({}_G\tilde{\alpha}_B + {}_G\tilde{\omega}_B^2 \right) {}_G\tilde{\omega}_B \right) {}^G\mathbf{r}
$$

$$
= \left({}_G\tilde{\chi}_B + 2\,{}_G\tilde{\omega}_B\,{}_G\tilde{\alpha}_B + {}_G\tilde{\alpha}_B\,{}_G\tilde{\omega}_B + {}_G\tilde{\omega}_B^3 \right) {}^G\mathbf{r} \tag{9.171}
$$

and therefore the rotational jerk transformation ${}_G U_B$ is

$$
{}_G U_B = {}_G\tilde{\chi}_B + 2\,{}_G\tilde{\omega}_B\,{}_G\tilde{\alpha}_B + {}_G\tilde{\alpha}_B\,{}_G\tilde{\omega}_B + {}_G\tilde{\omega}_B^3 \tag{9.172}
$$

Example 307 ★ Angular acceleration in terms of quaternion and Euler parameters. Quaternions are close to Euler parameters and both may be used to express angular accelerations.

To express the acceleration of a body point by quaternions and Euler parameters, we consider the velocity equation

$$
{}^G\dot{\mathbf{r}} = 2\dot{e}\,e^*\,{}^G\mathbf{r} \tag{9.173}
$$

and take a time derivative

$$
{}^G\ddot{\mathbf{r}} = 2\ddot{e}\,e^*\,{}^G\mathbf{r} + 2\dot{e}\,\dot{e}^*\,{}^G\mathbf{r} + 4\dot{e}^2\,e^{*2}\,{}^G\mathbf{r}
$$

$$
= 2\left(\ddot{e}\,e^* + \dot{e}\,\dot{e}^* + 2\dot{e}^2\,e^{*2} \right) {}^G\mathbf{r} \tag{9.174}
$$

Therefore, the quaternion expression of the acceleration transformation matrix is

$$
{}_G S_B = 2\left(\ddot{e}\,e^* + \dot{e}\,\dot{e}^* + 2\dot{e}^2\,e^{*2} \right) \tag{9.175}
$$

To determine the angular acceleration quaternion, let us use the angular velocity

$$
{}_G\boldsymbol{\omega}_B = 2\dot{e}\,e^* \tag{9.176}
$$

and take a derivative.

$$
{}_G\boldsymbol{\alpha}_B = 2\ddot{e}\,e^* + 2\dot{e}\,\dot{e}^* \tag{9.177}
$$

Employing the definition of the angular velocity based on rotational quaternion,

$$
\overleftrightarrow{{}_G\overrightarrow{\omega}_B} = 2\,\overleftrightarrow{\dot{e}}\,\overleftrightarrow{e^*} \tag{9.178}
$$

$$
\overleftrightarrow{{}_G^B\overrightarrow{\omega}_B} = 2\,\overleftrightarrow{e^*}\,\overleftrightarrow{\dot{e}} \tag{9.179}
$$

we are able to define the angular acceleration quaternion.

$$
\overleftrightarrow{{}_G\overrightarrow{\alpha}_B} = 2\,\overleftrightarrow{\ddot{e}}\,\overleftrightarrow{e^*} + 2\,\overleftrightarrow{\dot{e}}\,\overleftrightarrow{\dot{e}^*} \tag{9.180}
$$

$$
\overleftrightarrow{{}_G^B\overrightarrow{\alpha}_B} = 2\,\overleftrightarrow{e^*}\,\overleftrightarrow{\ddot{e}} + 2\,\overleftrightarrow{\dot{e}}\,\overleftrightarrow{\dot{e}^*} \tag{9.181}
$$

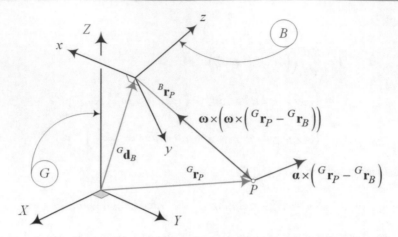

Fig. 9.4 A rigid body with coordinate frame $B\ (oxyz)$ moving freely in a fixed global coordinate frame $G(OXYZ)$

9.2   Rigid Body Acceleration

Consider a rigid body with an attached local coordinate frame $B\ (oxyz)$ moving freely in a fixed global coordinate frame $G(OXYZ)$. The rigid body can rotate in the global frame, while the origin of the body frame B can translate relative to the origin of G. The coordinates of a body point P in local and global frames, as shown in Fig. 9.4, are related by the following equation:

$$^G\mathbf{r}_P = {}^GR_B\ {}^B\mathbf{r}_P + {}^G\mathbf{d}_B \tag{9.182}$$

where $^G\mathbf{d}_B$ indicates the position of the moving origin o relative to the fixed origin O.

The acceleration of point P in G is

$$\begin{aligned}
^G\mathbf{a}_P &= {}^G\dot{\mathbf{v}}_P = {}^G\ddot{\mathbf{r}}_P \\
&= {}_G\boldsymbol{\alpha}_B \times \left({}^G\mathbf{r}_P - {}^G\mathbf{d}_B\right) \\
&\quad + {}_G\boldsymbol{\omega}_B \times \left({}_G\boldsymbol{\omega}_B \times \left({}^G\mathbf{r}_P - {}^G\mathbf{d}_B\right)\right) + {}^G\ddot{\mathbf{d}}_B
\end{aligned} \tag{9.183}$$

Proof The acceleration of point P is a consequence of differentiating the velocity equation (7.266) or (7.267).

$$^G\mathbf{v}_P = {}_G\boldsymbol{\omega}_B \times {}^G_B\mathbf{r}_P + {}^G\dot{\mathbf{d}}_B \tag{9.184}$$

$$\begin{aligned}
^G\mathbf{a}_P &= \frac{^Gd}{dt}\,{}^G\mathbf{v}_P = {}_G\boldsymbol{\alpha}_B \times {}^G_B\mathbf{r}_P + {}_G\boldsymbol{\omega}_B \times {}^G_B\dot{\mathbf{r}}_P + {}^G\ddot{\mathbf{d}}_B \\
&= {}_G\boldsymbol{\alpha}_B \times {}^G_B\mathbf{r}_P + {}_G\boldsymbol{\omega}_B \times \left({}_G\boldsymbol{\omega}_B \times {}^G_B\mathbf{r}_P\right) + {}^G\ddot{\mathbf{d}}_B \\
&= {}_G\boldsymbol{\alpha}_B \times \left({}^G\mathbf{r}_P - {}^G\mathbf{d}_B\right) \\
&\quad + {}_G\boldsymbol{\omega}_B \times \left({}_G\boldsymbol{\omega}_B \times \left({}^G\mathbf{r}_P - {}^G\mathbf{d}_B\right)\right) + {}^G\ddot{\mathbf{d}}_B.
\end{aligned} \tag{9.185}$$

The term $_G\boldsymbol{\omega}_B \times \left({}_G\boldsymbol{\omega}_B \times {}^G_B\mathbf{r}_P\right)$ is called *centripetal acceleration*, which is independent of the angular acceleration. The term $_G\boldsymbol{\alpha}_B \times {}^G_B\mathbf{r}_P$ is the *tangential acceleration* and is perpendicular to $^G_B\mathbf{r}_P$. ∎

Example 308 Acceleration of a body point. A review of acceleration equation of a point of a rigid body, starting from its position vector and taking derivatives step by step.

Consider a rigid body is moving and rotating in a global frame. The acceleration of a body point can be found by taking twice time derivative of its position vector.

$$^G\mathbf{r}_P = {}^GR_B\ {}^B\mathbf{r}_P + {}^G\mathbf{d}_B \tag{9.186}$$

$$^{G}\dot{\mathbf{r}}_P = {}^{G}\dot{R}_B\,{}^{B}\mathbf{r}_P + {}^{G}\dot{\mathbf{d}}_B \tag{9.187}$$

$$
\begin{aligned}
^{G}\ddot{\mathbf{r}}_P &= {}^{G}\ddot{R}_B\,{}^{B}\mathbf{r}_P + {}^{G}\ddot{\mathbf{d}}_B \\
&= {}^{G}\ddot{R}_B\,{}^{G}R_B^{T}\left({}^{G}\mathbf{r}_P - {}^{G}\mathbf{d}_B\right) + {}^{G}\ddot{\mathbf{d}}_B
\end{aligned}
\tag{9.188}
$$

Differentiating the angular velocity matrix

$$_{G}\tilde{\omega}_B = {}^{G}\dot{R}_B\,{}^{G}R_B^{T} \tag{9.189}$$

shows that

$$
\begin{aligned}
_{G}\dot{\tilde{\omega}}_B &= \frac{^{G}d}{dt}\,_{G}\tilde{\omega}_B = {}^{G}\ddot{R}_B\,{}^{G}R_B^{T} + {}^{G}\dot{R}_B\,{}^{G}\dot{R}_B^{T} \\
&= {}^{G}\ddot{R}_B\,{}^{G}R_B^{T} + {}_{G}\tilde{\omega}_B\,{}_{G}\tilde{\omega}_B^{T}
\end{aligned}
\tag{9.190}
$$

and therefore,

$$^{G}\ddot{R}_B\,{}^{G}R_B^{T} = {}_{G}\dot{\tilde{\omega}}_B - {}_{G}\tilde{\omega}_B\,{}_{G}\tilde{\omega}_B^{T} \tag{9.191}$$

Hence, the acceleration vector of the body point becomes

$$^{G}\ddot{\mathbf{r}}_P = \left({}_{G}\dot{\tilde{\omega}}_B - {}_{G}\tilde{\omega}_B\,{}_{G}\tilde{\omega}_B^{T}\right)\left({}^{G}\mathbf{r}_P - {}^{G}\mathbf{d}_B\right) + {}^{G}\ddot{\mathbf{d}}_B \tag{9.192}$$

where,

$$_{G}\dot{\tilde{\omega}}_B = {}_{G}\tilde{\alpha}_B = \begin{bmatrix} 0 & -\dot{\omega}_3 & \dot{\omega}_2 \\ \dot{\omega}_3 & 0 & -\dot{\omega}_1 \\ -\dot{\omega}_2 & \dot{\omega}_1 & 0 \end{bmatrix} \tag{9.193}$$

and

$$_{G}\tilde{\omega}_B\,{}_{G}\tilde{\omega}_B^{T} = \begin{bmatrix} \omega_2^2 + \omega_3^2 & -\omega_1\omega_2 & -\omega_1\omega_3 \\ -\omega_1\omega_2 & \omega_1^2 + \omega_3^2 & -\omega_2\omega_3 \\ -\omega_1\omega_3 & -\omega_2\omega_3 & \omega_1^2 + \omega_2^2 \end{bmatrix} \tag{9.194}$$

Example 309 Acceleration of joint 2 of a $2R$ planar manipulator. Second joint of a $2R$ manipulator is moving on a circle about the first joint. Acceleration of a rotating point about a center point is a good example to show analytically the acceleration has two components of centripetal and tangential.

A $2R$ planar manipulator is illustrated in Fig. 9.5. The elbow joint has a circular motion about the base joint. Knowing that

$$_{0}\boldsymbol{\omega}_1 = \dot{\theta}_1\,{}^{0}\hat{k}_0 \tag{9.195}$$

we can write

$$_{0}\boldsymbol{\alpha}_1 = {}_{0}\dot{\boldsymbol{\omega}}_1 = \ddot{\theta}_1\,{}^{0}\hat{k}_0 \tag{9.196}$$

$$_{0}\dot{\boldsymbol{\omega}}_1 \times {}^{0}\mathbf{r}_1 = \ddot{\theta}_1\,{}^{0}\hat{k}_0 \times {}^{0}\mathbf{r}_1 = \ddot{\theta}_1\,R_{Z,\theta+90}\,{}^{0}\mathbf{r}_1 \tag{9.197}$$

$$_{0}\boldsymbol{\omega}_1 \times \left({}_{0}\boldsymbol{\omega}_1 \times {}^{0}\mathbf{r}_1\right) = -\dot{\theta}_1^2\,{}^{0}\mathbf{r}_1 \tag{9.198}$$

and calculate the acceleration of the elbow joint.

$$^{0}\ddot{\mathbf{r}}_1 = \ddot{\theta}_1\,R_{Z,\theta+90}\,{}^{0}\mathbf{r}_1 - \dot{\theta}_1^2\,{}^{0}\mathbf{r}_1 \tag{9.199}$$

The first term, $\ddot{\theta}_1\,R_{Z,\theta+90}\,{}^{0}\mathbf{r}_1$, is tangential acceleration and is perpendicular to the radial position vector, $^{0}\mathbf{r}_1$, and the second term, $-\dot{\theta}_1^2\,{}^{0}\mathbf{r}_1$, is centripetal acceleration and is in opposite direction of $^{0}\mathbf{r}_1$.

Example 310 Acceleration of a moving point in a moving body frame. The kinematics of a moving point in a rotating and translating coordinate frame B is the most general case of kinematic analysis of a point. This example shows different components of the acceleration of such a point.

Assume the point P in Fig. 9.4 is indicated by a time varying body position vector ${}^B\mathbf{r}_P(t)$. Then, the velocity and acceleration of P can be found by applying the derivative transformation formula (7.214).

$$\frac{{}^G d}{dt}{}^B\Box = \frac{{}^B d}{dt}{}^B\Box + {}^B_G\boldsymbol{\omega}_B \times {}^B_G\Box = {}^B_G\dot{\Box} \tag{9.200}$$

$$\begin{aligned}
{}^B_G\mathbf{v}_P &= \frac{{}^G d}{dt}{}^B_G\mathbf{r}_P = {}^B_G\dot{\mathbf{d}}_B + {}^B\dot{\mathbf{r}}_P + {}^B_G\boldsymbol{\omega}_B \times {}^B\mathbf{r}_P \\
&= {}^B_G\dot{\mathbf{d}}_B + {}^B\mathbf{v}_P + {}^B_G\boldsymbol{\omega}_B \times {}^B\mathbf{r}_P
\end{aligned} \tag{9.201}$$

$$\begin{aligned}
{}^B_G\mathbf{a}_P &= {}^B_G\ddot{\mathbf{d}}_B + {}^B\ddot{\mathbf{r}}_P + {}^B_G\boldsymbol{\omega}_B \times {}^B\dot{\mathbf{r}}_P + {}^B_G\dot{\boldsymbol{\omega}}_B \times {}^B\mathbf{r}_P \\
&\quad + {}^B_G\boldsymbol{\omega}_B \times \left({}^B\dot{\mathbf{r}}_P + {}^B_G\boldsymbol{\omega}_B \times {}^B\mathbf{r}_P\right) \\
&= {}^G\ddot{\mathbf{d}}_B + {}^B\mathbf{a}_P + 2{}^B_G\boldsymbol{\omega}_B \times {}^B\mathbf{v}_P + {}^B_G\dot{\boldsymbol{\omega}}_B \times {}^B\mathbf{r}_P \\
&\quad + {}^B_G\boldsymbol{\omega}_B \times \left({}^B_G\boldsymbol{\omega}_B \times {}^B\mathbf{r}_P\right)
\end{aligned} \tag{9.202}$$

It is also possible to take the derivative from Eq. (7.263) with the assumption ${}^B\dot{\mathbf{r}}_P \neq 0$ and find the acceleration of P.

$$ {}^G\mathbf{r}_P = {}^G R_B \, {}^B\mathbf{r}_P + {}^G\mathbf{d}_B \tag{9.203}$$

$$\begin{aligned}
{}^G\dot{\mathbf{r}}_P &= {}^G\dot{R}_B \, {}^B\mathbf{r}_P + {}^G R_B \, {}^B\dot{\mathbf{r}}_P + {}^G\dot{\mathbf{d}}_B \\
&= {}_G\boldsymbol{\omega}_B \times {}^G R_B \, {}^B\mathbf{r}_P + {}^G R_B \, {}^B\dot{\mathbf{r}}_P + {}^G\dot{\mathbf{d}}_B
\end{aligned} \tag{9.204}$$

$$\begin{aligned}
{}^G\ddot{\mathbf{r}}_P &= {}_G\dot{\boldsymbol{\omega}}_B \times {}^G R_B \, {}^B\mathbf{r}_P + {}_G\boldsymbol{\omega}_B \times {}^G\dot{R}_B \, {}^B\mathbf{r}_P + {}_G\boldsymbol{\omega}_B \times {}^G R_B \, {}^B\dot{\mathbf{r}}_P \\
&\quad + {}^G\dot{R}_B \, {}^B\dot{\mathbf{r}}_P + {}^G R_B \, {}^B\ddot{\mathbf{r}}_P + {}^G\ddot{\mathbf{d}}_B \\
&= {}_G\dot{\boldsymbol{\omega}}_B \times {}^G_B\mathbf{r}_P + {}_G\boldsymbol{\omega}_B \times \left({}_G\boldsymbol{\omega}_B \times {}^G\mathbf{r}_P\right) + 2\,{}_G\boldsymbol{\omega}_B \times {}^G_B\dot{\mathbf{r}}_P \\
&\quad + {}^G_B\ddot{\mathbf{r}}_P + {}^G\ddot{\mathbf{d}}_B
\end{aligned} \tag{9.205}$$

The first term, ${}_G\dot{\boldsymbol{\omega}}_B \times {}^G_B\mathbf{r}_P$, is the tangential acceleration; the second term, ${}_G\boldsymbol{\omega}_B \times \left({}_G\boldsymbol{\omega}_B \times {}^G\mathbf{r}_P\right)$, is the centripetal acceleration. The third term on the right-hand side, $2\,{}_G\boldsymbol{\omega}_B \times {}^G_B\dot{\mathbf{r}}_P$, is called the Coriolis acceleration. The Coriolis acceleration is perpendicular to both, ${}_G\boldsymbol{\omega}_B$ and ${}^B\dot{\mathbf{r}}_P$.

9.3   ★ Acceleration Transformation Matrix

Consider the motion of a rigid body B in the global coordinate frame G, as shown in Fig. 9.4. Assume the body fixed frame $B(oxyz)$ is coincident at an initial time t_0 with the global frame $G(OXYZ)$. At any time $t \neq t_0$, B is not necessarily coincident with G, and therefore, the homogenous transformation matrix ${}^G T_B(t)$ is time varying.

The acceleration of a body point in the global coordinate frame can be found by applying a homogenous acceleration transformation matrix

$$ {}^G\mathbf{a}_P(t) = {}^G A_B \, {}^G\mathbf{r}_P(t) \tag{9.206}$$

where, hereafter ${}^G A_B$ is the *acceleration transformation matrix*

$$ {}^G A_B = \begin{bmatrix} {}_G\tilde{\alpha}_B - {}_G\tilde{\omega}_B \, {}_G\tilde{\omega}_B^T \, {}^G\ddot{\mathbf{d}}_B - \left({}_G\tilde{\alpha}_B - {}_G\tilde{\omega}_B \, {}_G\tilde{\omega}_B^T \right) {}^G\mathbf{d}_B \\ 0 \qquad\qquad\qquad\qquad 0 \end{bmatrix} $$

$$ = \begin{bmatrix} {}_G S_B \, {}^G\ddot{\mathbf{d}}_B - {}_G S_B \, {}^G\mathbf{d}_B \\ 0 \qquad\qquad 0 \end{bmatrix} \tag{9.207} $$

Proof Based on homogenous coordinate transformation we have

$$ {}^G\mathbf{r}_P(t) = {}^G T_B \, {}^B\mathbf{r}_P $$

$$ = \begin{bmatrix} {}^G R_B & {}^G\mathbf{d}_B \\ 0 & 1 \end{bmatrix} {}^B\mathbf{r}_P = {}^G T_B \, {}^G\mathbf{r}_P(t_0) \tag{9.208} $$

$$ {}^G\mathbf{v}_P = {}^G\dot{T}_B \, {}^G T_B^{-1} \, {}^G\mathbf{r}_P(t) $$

$$ = \begin{bmatrix} {}^G\dot{R}_B \, {}^G R_B^T \, {}^G\dot{\mathbf{d}}_B - {}^G\dot{R}_B \, {}^G R_B^T \, {}^G\mathbf{d}_B \\ 0 \qquad\qquad\qquad 0 \end{bmatrix} {}^G\mathbf{r}_P(t) $$

$$ = \begin{bmatrix} {}_G\tilde{\omega}_B \, {}^G\dot{\mathbf{d}}_B - {}_G\tilde{\omega}_B \, {}^G\mathbf{d}_B \\ 0 \qquad\qquad 0 \end{bmatrix} {}^G\mathbf{r}_P(t) $$

$$ = {}^G V_B \, {}^G\mathbf{r}_P(t) \tag{9.209} $$

To find the acceleration of a body point in the global frame, we take twice the time derivative from ${}^G\mathbf{r}_P(t) = {}^G T_B \, {}^B\mathbf{r}_P$,

$$ {}^G\mathbf{a}_P(t) = \frac{d^2}{dt^2} \, {}^G T_B \, {}^B\mathbf{r}_P = {}^G\ddot{T}_B \, {}^B\mathbf{r}_P \tag{9.210} $$

and substitute for ${}^B\mathbf{r}_P$.

$$ {}^G\mathbf{a}_P(t) = {}^G\ddot{T}_B \, {}^G T_B^{-1} \, {}^G\mathbf{r}_P(t) = {}^G A_B \, {}^G\mathbf{r}_P(t) \tag{9.211} $$

$$ {}^G A_B = {}^G\ddot{T}_B \, {}^G T_B^{-1} \tag{9.212} $$

Substituting for ${}^G\ddot{T}_B$ and ${}^G T_B^{-1}$ provides

$$ {}^G\mathbf{a}_P(t) = \begin{bmatrix} {}^G\ddot{R}_B & {}^G\ddot{\mathbf{d}}_B \\ 0 & 0 \end{bmatrix} \begin{bmatrix} {}^G R_B^T & -{}^G R_B^T \, {}^G\mathbf{d}_B \\ 0 & 1 \end{bmatrix} {}^G\mathbf{r}_P(t) $$

$$ = \begin{bmatrix} {}^G\ddot{R}_B \, {}^G R_B^T \, {}^G\ddot{\mathbf{d}}_B - {}^G\ddot{R}_B \, {}^G R_B^T \, {}^G\mathbf{d}_B \\ 0 \qquad\qquad\qquad 0 \end{bmatrix} {}^G\mathbf{r}_P(t) $$

$$ = \begin{bmatrix} {}_G\tilde{\alpha}_B - \tilde{\omega}\,\tilde{\omega}^T \, {}^G\ddot{\mathbf{d}}_B - \left({}_G\tilde{\alpha}_B - \tilde{\omega}\,\tilde{\omega}^T \right) {}^G\mathbf{d}_B \\ 0 \qquad\qquad\qquad\qquad 0 \end{bmatrix} {}^G\mathbf{r}_P(t) $$

$$ = \begin{bmatrix} {}_G S_B \, {}^G\ddot{\mathbf{d}}_B - {}_G S_B \, {}^G\mathbf{d}_B \\ 0 \qquad\qquad 0 \end{bmatrix} = {}^G A_B \, {}^G\mathbf{r}_P(t) \tag{9.213} $$

where,

$$ {}^G\ddot{R}_B \, {}^G R_B^T = {}_G S_B = {}_G\dot{\tilde{\omega}}_B - \tilde{\omega}\,\tilde{\omega}^T = {}_G\tilde{\alpha}_B - \tilde{\omega}\,\tilde{\omega}^T \tag{9.214} $$

∎

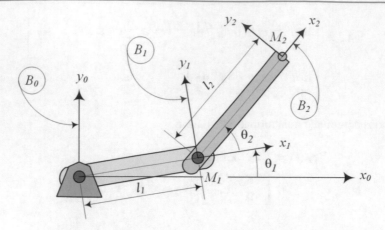

Fig. 9.5 An R∥R planar manipulator with joint variables θ_1 and θ_2

Example 311 ★ Kinematics of the gripper of a planar R∥R manipulator. $2R$ manipulator is the main component of many multi- degree of freedom and hence, its kinematics will appear in robotic analysis very often. Here, its kinematics will be solved from position to acceleration in different methods.

Figure 9.5 illustrates an R∥R planar manipulator with joint variables θ_1 and θ_2. The links (1) and (2) are both R∥R(0) and therefore, the transformation matrices $^0T_1$, $^1T_2$, and $^0T_2$ are

$$
^0T_1 =
\begin{bmatrix}
\cos\theta_1 & -\sin\theta_1 & 0 & l_1\cos\theta_1 \\
\sin\theta_1 & \cos\theta_1 & 0 & l_1\sin\theta_1 \\
0 & 0 & 1 & 0 \\
0 & 0 & 0 & 1
\end{bmatrix}
\tag{9.215}
$$

$$
^1T_2 =
\begin{bmatrix}
\cos\theta_2 & -\sin\theta_2 & 0 & l_2\cos\theta_2 \\
\sin\theta_2 & \cos\theta_2 & 0 & l_2\sin\theta_2 \\
0 & 0 & 1 & 0 \\
0 & 0 & 0 & 1
\end{bmatrix}
\tag{9.216}
$$

$$
^0T_2 = {}^0T_1\,{}^1T_2
\tag{9.217}
$$

$$
=
\begin{bmatrix}
c\,(\theta_1+\theta_2) & -s\,(\theta_1+\theta_2) & 0 & l_2c\,(\theta_1+\theta_2)+l_1c\theta_1 \\
s\,(\theta_1+\theta_2) & c\,(\theta_1+\theta_2) & 0 & l_2s\,(\theta_1+\theta_2)+l_1s\theta_1 \\
0 & 0 & 1 & 0 \\
0 & 0 & 0 & 1
\end{bmatrix}
$$

The points M_1 and M_2 are at

$$
^0\mathbf{r}_{M_1} =
\begin{bmatrix}
l_1\cos\theta_1 \\
l_1\sin\theta_1 \\
0 \\
1
\end{bmatrix}
\qquad
^1\mathbf{r}_{M_2} =
\begin{bmatrix}
l_2\cos\theta_2 \\
l_2\sin\theta_2 \\
0 \\
1
\end{bmatrix}
\tag{9.218}
$$

$$
^0\mathbf{r}_{M_2} = {}^0T_1\,{}^1\mathbf{r}_{M_2} =
\begin{bmatrix}
l_2\cos\,(\theta_1+\theta_2)+l_1\cos\theta_1 \\
l_2\sin\,(\theta_1+\theta_2)+l_1\sin\theta_1 \\
0 \\
1
\end{bmatrix}
\tag{9.219}
$$

To determine the velocity and acceleration of M_2, we need to calculate $^0\dot{T}_2$, which can be calculated by direct differentiation of $^0T_2$.

$$^0\dot{T}_2 = \frac{d}{dt}\,^0T_2 \tag{9.220}$$

$$= \begin{bmatrix} -\dot{\theta}_{12}s\theta_{12} & -\dot{\theta}_{12}c\theta_{12} & 0 & -l_2\dot{\theta}_{12}s\theta_{12} - \dot{\theta}_1 l_1 s\theta_1 \\ \dot{\theta}_{12}c\theta_{12} & -\dot{\theta}_{12}s\theta_{12} & 0 & l_2\dot{\theta}_{12}c\theta_{12} + \dot{\theta}_1 l_1 c\theta_1 \\ 0 & 0 & 0 & 0 \\ 0 & 0 & 0 & 0 \end{bmatrix}$$

$$\theta_{12} = \theta_1 + \theta_2 \qquad \dot{\theta}_{12} = \dot{\theta}_1 + \dot{\theta}_2 \tag{9.221}$$

We can also calculate $^0\dot{T}_2$ from $^0T_2 = {}^0T_1\,{}^1T_2$ by chain rule

$$^0\dot{T}_2 = \frac{d}{dt}\left({}^0T_1\,{}^1T_2\right) = {}^0\dot{T}_1\,{}^1T_2 + {}^0T_1\,{}^1\dot{T}_2 \tag{9.222}$$

where,

$$^0\dot{T}_1 = \dot{\theta}_1 \begin{bmatrix} -\sin\theta_1 & -\cos\theta_1 & 0 & -l_1\sin\theta_1 \\ \cos\theta_1 & -\sin\theta_1 & 0 & l_1\cos\theta_1 \\ 0 & 0 & 0 & 0 \\ 0 & 0 & 0 & 0 \end{bmatrix} \tag{9.223}$$

$$^1\dot{T}_2 = \dot{\theta}_2 \begin{bmatrix} -\sin\theta_2 & -\cos\theta_2 & 0 & -l_2\sin\theta_2 \\ \cos\theta_2 & -\sin\theta_2 & 0 & l_2\cos\theta_2 \\ 0 & 0 & 0 & 0 \\ 0 & 0 & 0 & 0 \end{bmatrix} \tag{9.224}$$

Having $^0\dot{T}_1$ and $^1\dot{T}_2$, we can find the velocity transformation matrices $^0V_1$ and $^1V_2$ by using $^0T_1^{-1}$ and $^1T_2^{-1}$.

$$^0T_1^{-1} = \begin{bmatrix} \cos\theta_1 & \sin\theta_1 & 0 & -l_1 \\ -\sin\theta_1 & \cos\theta_1 & 0 & 0 \\ 0 & 0 & 1 & 0 \\ 0 & 0 & 0 & 1 \end{bmatrix} \tag{9.225}$$

$$^1T_2^{-1} = \begin{bmatrix} \cos\theta_2 & \sin\theta_2 & 0 & -l_2 \\ -\sin\theta_2 & \cos\theta_2 & 0 & 0 \\ 0 & 0 & 1 & 0 \\ 0 & 0 & 0 & 1 \end{bmatrix} \tag{9.226}$$

$$^0V_1 = {}^0\dot{T}_1\,{}^0T_1^{-1} = \dot{\theta}_1\,{}^0_1\tilde{k} \tag{9.227}$$

$$^1V_2 = {}^1\dot{T}_2\,{}^1T_2^{-1} = \dot{\theta}_2\,{}^1_2\tilde{k} \tag{9.228}$$

Now, we can determine the velocity of points M_1 and M_2 in B_0 and B_1, respectively.

$$^0\mathbf{v}_{M_1} = {}^0V_1\,{}^0\mathbf{r}_{M_1} = \dot{\theta}_1 \begin{bmatrix} -l_1\sin\theta_1 \\ l_1\cos\theta_1 \\ 0 \\ 0 \end{bmatrix} \tag{9.229}$$

$$^1\mathbf{v}_{M_2} = {}^1V_2\,{}^1\mathbf{r}_{M_2} = \dot{\theta}_2 \begin{bmatrix} -l_2\sin\theta_2 \\ l_2\cos\theta_2 \\ 0 \\ 0 \end{bmatrix} \tag{9.230}$$

To determine the velocity of the tip point M_2 in the base frame, we can use the velocity vector addition.

$$
{}^0\mathbf{v}_{M_2} = {}^0\mathbf{v}_{M_1} + {}^0_1\mathbf{v}_{M_2} = {}^0\mathbf{v}_{M_1} + {}^0T_1 \, {}^1\mathbf{v}_{M_2}
$$

$$
= \begin{bmatrix} -\left(\dot{\theta}_1 + \dot{\theta}_2\right) l_2 \sin\left(\theta_1 + \theta_2\right) - \dot{\theta}_1 l_1 \sin\theta_1 \\ \left(\dot{\theta}_1 + \dot{\theta}_2\right) l_2 \cos\left(\theta_1 + \theta_2\right) + \dot{\theta}_1 l_1 \cos\theta_1 \\ 0 \\ 0 \end{bmatrix} \tag{9.231}
$$

We can also determine ${}^0\mathbf{v}_{M_2}$ by using the velocity transformation matrix ${}^0V_2$

$$
{}^0\mathbf{v}_{M_2} = {}^0V_2 \, {}^0\mathbf{r}_{M_2} \tag{9.232}
$$

where, the velocity transformation matrix ${}^0V_2$ is

$$
{}^0V_2 = {}^0\dot{T}_2 \, {}^0T_2^{-1} = \begin{bmatrix} 0 & -\dot{\theta}_1 - \dot{\theta}_2 & 0 & \dot{\theta}_2 l_1 \sin\theta_1 \\ \dot{\theta}_1 + \dot{\theta}_2 & 0 & 0 & -\dot{\theta}_2 l_1 \cos\theta_1 \\ 0 & 0 & 0 & 0 \\ 0 & 0 & 0 & 0 \end{bmatrix} \tag{9.233}
$$

$$
{}^0T_2^{-1} = {}^2T_1 \, {}^1T_0 = {}^1T_2^{-1} \, {}^0T_1^{-1} \tag{9.234}
$$

$$
= \begin{bmatrix} \cos\left(\theta_1 + \theta_2\right) & \sin\left(\theta_1 + \theta_2\right) & 0 & -l_2 - l_1 \cos\theta_2 \\ -\sin\left(\theta_1 + \theta_2\right) & \cos\left(\theta_1 + \theta_2\right) & 0 & l_1 \sin\theta_2 \\ 0 & 0 & 1 & 0 \\ 0 & 0 & 0 & 1 \end{bmatrix}
$$

Furthermore, we can determine the velocity transformation matrix ${}^0V_2$ using their addition rule,

$$
{}^0V_2 = {}^0V_1 + {}^0_1V_2 \tag{9.235}
$$

where,

$$
{}^0_1V_2 = {}^0T_1 \, {}^1V_2 \, {}^0T_1^{-1} = \begin{bmatrix} 0 & -\dot{\theta}_2 & 0 & \dot{\theta}_2 l_1 \sin\theta_1 \\ \dot{\theta}_2 & 0 & 0 & -\dot{\theta}_2 l_1 \cos\theta_1 \\ 0 & 0 & 0 & 0 \\ 0 & 0 & 0 & 0 \end{bmatrix} \tag{9.236}
$$

Therefore, ${}^0\mathbf{v}_{M_2}$ will be

$$
{}^0\mathbf{v}_{M_2} = {}^0V_2 \, {}^0\mathbf{r}_{M_2} \tag{9.237}
$$

To determine the acceleration of M_2, we need to calculate ${}^0\ddot{T}_2$, which can be calculated by direct differentiation of ${}^0\dot{T}_2$.

$$
{}^0\ddot{T}_2 = \frac{d}{dt}\,{}^0\dot{T}_2 = \frac{d}{dt}\frac{d}{dt}\left({}^0T_1 \, {}^1T_2\right) = \frac{d}{dt}\left({}^0\dot{T}_1 \, {}^1T_2 + {}^0T_1 \, {}^1\dot{T}_2\right)
$$

$$
= {}^0\ddot{T}_1 \, {}^1T_2 + 2\,{}^0\dot{T}_1 \, {}^1\dot{T}_2 + {}^0T_1 \, {}^1\ddot{T}_2 \tag{9.238}
$$

We have,

$$
{}^0\ddot{T}_1 = \frac{d}{dt}\,{}^0\dot{T}_1 \tag{9.239}
$$

$$
= \begin{bmatrix}
-\dot{\theta}_1^2 c\theta_1 - \ddot{\theta}_1 s\theta_1 & \dot{\theta}_1^2 s\theta_1 - \ddot{\theta}_1 c\theta_1 & 0 & -\ddot{\theta}_1 l_1 s\theta_1 - l_1\dot{\theta}_1^2 c\theta_1 \\
\ddot{\theta}_1 c\theta_1 - \dot{\theta}_1^2 s\theta_1 & -\dot{\theta}_1^2 c\theta_1 - \ddot{\theta}_1 s\theta_1 & 0 & l_1\ddot{\theta}_1 c\theta_1 - l_1\dot{\theta}_1^2 s\theta_1 \\
0 & 0 & 0 & 0 \\
0 & 0 & 0 & 0
\end{bmatrix}
$$

$$
{}^1\ddot{T}_2 = \frac{d}{dt}\,{}^1\dot{T}_2 \tag{9.240}
$$

$$
= \begin{bmatrix}
-\dot{\theta}_2^2 c\theta_2 - \ddot{\theta}_2 s\theta_2 & \dot{\theta}_2^2 s\theta_2 - \ddot{\theta}_2 c\theta_2 & 0 & -\ddot{\theta}_2 l_2 s\theta_2 - l_2\dot{\theta}_2^2 c\theta_2 \\
\ddot{\theta}_2 c\theta_2 - \dot{\theta}_2^2 s\theta_2 & -\dot{\theta}_2^2 c\theta_2 - \ddot{\theta}_2 s\theta_2 & 0 & \ddot{\theta}_2 l_2 c\theta_2 - \dot{\theta}_2^2 l_2 s\theta_2 \\
0 & 0 & 0 & 0 \\
0 & 0 & 0 & 0
\end{bmatrix}
$$

and therefore,

$$
{}^0\ddot{T}_2 = {}^0\ddot{T}_1\,{}^1T_2 + 2\,{}^0\dot{T}_1\,{}^1\dot{T}_2 + {}^0T_1\,{}^1\ddot{T}_2
$$

$$
= \begin{bmatrix}
r_{11} & r_{12} & 0 & r_{14} \\
r_{21} & r_{22} & 0 & r_{24} \\
0 & 0 & 0 & 0 \\
0 & 0 & 0 & 0
\end{bmatrix} \tag{9.241}
$$

$$
r_{11} = -\dot{\theta}_{12}^2 \cos\theta_{12} - \ddot{\theta}_{12} \sin\theta_{12}
$$

$$
r_{21} = -\dot{\theta}_{12}^2 \sin\theta_{12} + \ddot{\theta}_{12} \cos\theta_{12}
$$

$$
r_{12} = \dot{\theta}_{12}^2 \sin\theta_{12} - \ddot{\theta}_{12} \cos\theta_{12}
$$

$$
r_{22} = -\dot{\theta}_{12}^2 \cos\theta_{12} - \ddot{\theta}_{12} \sin\theta_{12}
$$

$$
r_{14} = -\dot{\theta}_{12}^2 l_2 \cos\theta_{12} - \ddot{\theta}_{12} l_2 \sin\theta_{12}
$$
$$
-\dot{\theta}_1^2 l_1 \cos\theta_1 - \ddot{\theta}_1 l_1 \sin\theta_1
$$

$$
r_{24} = \ddot{\theta}_{12} l_2 \cos\theta_{12} - \dot{\theta}_{12}^2 l_2 \sin\theta_{12}
$$
$$
-\dot{\theta}_1^2 l_1 \sin\theta_1 + \ddot{\theta}_1 l_1 \cos\theta_1 \tag{9.242}
$$

Having ${}^0\ddot{T}_1$, ${}^1\ddot{T}_2$, ${}^0\ddot{T}_2$, we can find the acceleration transformation matrices ${}^0A_1$, ${}^1A_2$, and ${}^0A_2$ by using ${}^0T_1^{-1}$, ${}^1T_2^{-1}$, and ${}^0T_2^{-1}$.

$$
{}^0A_1 = {}^0\ddot{T}_1\,{}^0T_1^{-1} = \begin{bmatrix}
-\dot{\theta}_1^2 & -\ddot{\theta}_1 & 0 & 0 \\
\ddot{\theta}_1 & -\dot{\theta}_1^2 & 0 & 0 \\
0 & 0 & 0 & 0 \\
0 & 0 & 0 & 0
\end{bmatrix} \tag{9.243}
$$

$$
{}^1A_2 = {}^1\ddot{T}_2\,{}^1T_2^{-1} = \begin{bmatrix}
-\dot{\theta}_2^2 & -\ddot{\theta}_2 & 0 & 0 \\
\ddot{\theta}_2 & -\dot{\theta}_2^2 & 0 & 0 \\
0 & 0 & 0 & 0 \\
0 & 0 & 0 & 0
\end{bmatrix} \tag{9.244}
$$

$$^0A_2 = {}^0\ddot{T}_2 \, {}^0T_2^{-1} \tag{9.245}$$

$$= \begin{bmatrix} -\dot{\theta}_{12}^2 & -\ddot{\theta}_{12} & 0 & l_1\dot{\theta}_2^2\cos\theta_1 + 2\theta_1\dot{\theta}_2 l_1\cos\theta_1 + \ddot{\theta}_2 l_1\sin\theta_1 \\ \ddot{\theta}_{12} & -\dot{\theta}_{12}^2 & 0 & l_1\dot{\theta}_2^2\sin\theta_1 + 2\theta_1\dot{\theta}_2 l_1\sin\theta_1 - \ddot{\theta}_2 l_1\cos\theta_1 \\ 0 & 0 & 0 & 0 \\ 0 & 0 & 0 & 0 \end{bmatrix}$$

Now, we can determine the acceleration of points M_1 and M_2 in B_0 and B_1, respectively.

$$^0\mathbf{a}_{M_1} = {}^0A_1 \, {}^0\mathbf{r}_{M_1} = \begin{bmatrix} -l_1\dot{\theta}_1^2\cos\theta_1 - \ddot{\theta}_1 l_1\sin\theta_1 \\ \ddot{\theta}_1 l_1\cos\theta_1 - \dot{\theta}_1^2 l_1\sin\theta_1 \\ 0 \\ 0 \end{bmatrix} \tag{9.246}$$

$$^1\mathbf{a}_{M_2} = {}^1A_2 \, {}^1\mathbf{r}_{M_2} = \begin{bmatrix} -l_2\dot{\theta}_2^2\cos\theta_2 - \ddot{\theta}_2 l_2\sin\theta_2 \\ \ddot{\theta}_2 l_2\cos\theta_2 - \dot{\theta}_2^2 l_2\sin\theta_2 \\ 0 \\ 0 \end{bmatrix} \tag{9.247}$$

$$^0\mathbf{a}_{M_2} = {}^0A_2 \, {}^0\mathbf{r}_{M_2}$$

$$= \begin{bmatrix} -\ddot{\theta}_{12}l_2(\cos\theta_{12} + \sin\theta_{12}) - \dot{\theta}_1^2 l_1\cos\theta_1 - \ddot{\theta}_1 l_1\sin\theta_1 \\ \ddot{\theta}_{12}l_2(\cos\theta_{12} - \sin\theta_{12}) - \dot{\theta}_1^2 l_1\sin\theta_1 + \ddot{\theta}_1 l_1\cos\theta_1 \\ 0 \\ 0 \end{bmatrix} \tag{9.248}$$

Example 312 ★ Jerk transformation matrix. Jerk is the next derivative after acceleration and it will be useful whenever shaking of the carrying object by the gripper is needed to be controlled. Here jerk transformation matrix is calculated.

Following the same pattern of acceleration transformation matrix, we may define a jerk transformation matrix $^G J_B$ as

$$^G\mathbf{j}_P(t) = {}^G J_B \, {}^G\mathbf{r}_P(t) \tag{9.249}$$

where,

$$^G J_B = \begin{bmatrix} {}^G\dddot{R}_B & {}^G\dddot{\mathbf{d}}_B \\ 0 & 0 \end{bmatrix} \begin{bmatrix} {}^G R_B^T & -{}^G R_B^T \, {}^G\mathbf{d}_B \\ 0 & 1 \end{bmatrix}$$

$$= \begin{bmatrix} {}^G\dddot{R}_B \, {}^G R_B^T & {}^G\dddot{\mathbf{d}}_B - {}^G\dddot{R}_B \, {}^G R_B^T \, {}^G\mathbf{d}_B \\ 0 & 0 \end{bmatrix} \tag{9.250}$$

and

$$^G\dddot{R}_B \, {}^G R_B^T = {}_G U_B$$

$$= {}_G\dddot{\tilde{\omega}}_B - 2\left({}_G\dot{\tilde{\omega}}_B - \tilde{\omega}\,\tilde{\omega}^T\right)\tilde{\omega}^T - \tilde{\omega}\left({}_G\dot{\tilde{\omega}}_B - \tilde{\omega}\,\tilde{\omega}^T\right)^T$$

$$= {}_G\tilde{\chi}_B - 2\left({}_G\tilde{\alpha}_B - \tilde{\omega}\,\tilde{\omega}^T\right)\tilde{\omega}^T - \tilde{\omega}\left({}_G\tilde{\alpha}_B - \tilde{\omega}\,\tilde{\omega}^T\right)^T \tag{9.251}$$

Example 313 Velocity, acceleration, and jerk transformation matrices. This is to compare the transformation matrices of velocity, acceleration, and jerk.

The velocity transformation matrix is to map a position vector to its velocity vector. Assume $\mathbf{p}$ and $\mathbf{q}$ denote the position of two body points P and Q.

$$
{}^G\mathbf{q} - {}^G\mathbf{p} = {}^G R_B \left({}^B\mathbf{q} - {}^B\mathbf{p}\right) \tag{9.252}
$$

Assuming we have the kinematics of one of the points, say P, then the position of the other point is

$$
{}^G\mathbf{q} = {}^G\mathbf{p} + {}^G R_B \left({}^B\mathbf{q} - {}^B\mathbf{p}\right) \tag{9.253}
$$

This equation is similar to

$$
{}^G\mathbf{r} = {}^G\mathbf{d} + {}^G R_B \, {}^B\mathbf{r} \tag{9.254}
$$

where the origin of the B-frame is at point P and ${}^B\mathbf{r}$ indicates the position of Q relative to P. Taking a time derivative shows that

$$
{}^G\dot{\mathbf{q}} - {}^G\dot{\mathbf{p}} = {}_G\tilde{\omega}_B \left({}^G\mathbf{q} - {}^G\mathbf{p}\right) \tag{9.255}
$$

which can be converted to

$$
\begin{bmatrix} {}^G\dot{\mathbf{q}} \\ 0 \end{bmatrix} = \begin{bmatrix} {}_G\tilde{\omega}_B & {}^G\dot{\mathbf{p}} - {}_G\tilde{\omega}_B \, {}^G\mathbf{p} \\ 0 & 0 \end{bmatrix} \begin{bmatrix} {}^G\mathbf{q} \\ 1 \end{bmatrix}
$$

$$
= {}^G V_B \begin{bmatrix} {}^G\mathbf{q} \\ 1 \end{bmatrix} \tag{9.256}
$$

The matrix $[V]$ is velocity transformation matrix. Similarly, we obtain the acceleration equation

$$
\begin{aligned}
{}^G\ddot{\mathbf{q}} - {}^G\ddot{\mathbf{p}} &= {}_G\tilde{\alpha}_B \left({}^G\mathbf{q} - {}^G\mathbf{p}\right) + {}_G\tilde{\omega}_B \left({}^G\dot{\mathbf{q}} - {}^G\dot{\mathbf{p}}\right) \\
&= {}_G\tilde{\alpha}_B \left({}^G\mathbf{q} - {}^G\mathbf{p}\right) + {}_G\tilde{\omega}_B \, {}_G\tilde{\omega}_B \left({}^G\mathbf{q} - {}^G\mathbf{p}\right) \\
&= {}_G\tilde{\alpha}_B \left({}^G\mathbf{q} - {}^G\mathbf{p}\right) - {}_G\tilde{\omega}_B \, {}_G\tilde{\omega}_B^T \left({}^G\mathbf{q} - {}^G\mathbf{p}\right) \\
&= \left({}_G\tilde{\alpha}_B - {}_G\tilde{\omega}_B \, {}_G\tilde{\omega}_B^T\right) \left({}^G\mathbf{q} - {}^G\mathbf{p}\right) \\
&= {}_G S_B \left({}^G\mathbf{q} - {}^G\mathbf{p}\right)
\end{aligned} \tag{9.257}
$$

which can be converted to

$$
\begin{bmatrix} {}^G\ddot{\mathbf{q}} \\ 0 \end{bmatrix} = {}^G A_B \begin{bmatrix} {}^G\mathbf{q} \\ 1 \end{bmatrix} \tag{9.258}
$$

$$
\begin{aligned}
[A] &= \begin{bmatrix} {}_G\tilde{\alpha}_B - {}_G\tilde{\omega}_B \, {}_G\tilde{\omega}_B^T & {}^G\ddot{\mathbf{p}} - \left({}_G\tilde{\alpha}_B - {}_G\tilde{\omega}_B \, {}_G\tilde{\omega}_B^T\right) {}^G\mathbf{p} \\ 0 & 0 \end{bmatrix} \\
&= \begin{bmatrix} {}_G S_B & {}^G\ddot{\mathbf{p}} - {}_G S_B \, {}^G\mathbf{p} \\ 0 & 0 \end{bmatrix}
\end{aligned} \tag{9.259}
$$

where $[A]$ is the acceleration transformation matrix for rigid motion.

The jerk matrix can be found after another differentiation

$$
\begin{aligned}
{}^G\dddot{\mathbf{q}} - {}^G\dddot{\mathbf{p}} &= \\
&= \left(\left({}_G\dot{\tilde{\alpha}}_B + 2\,{}_G\tilde{\omega}_B \, {}_G\tilde{\alpha}_B\right) + \left({}_G\tilde{\alpha}_B + {}_G\tilde{\omega}_B^2\right) {}_G\tilde{\omega}_B\right) \left({}^G\mathbf{q} - {}^G\mathbf{p}\right) \\
&= \left({}_G\dot{\tilde{\alpha}}_B + 2\,{}_G\tilde{\omega}_B \, {}_G\tilde{\alpha}_B + \left({}_G\tilde{\alpha}_B + {}_G\tilde{\omega}_B^2\right) {}_G\tilde{\omega}_B\right) \left({}^G\mathbf{q} - {}^G\mathbf{p}\right) \\
&= \left({}_G\tilde{\chi}_B + 2\,{}_G\tilde{\omega}_B \, {}_G\tilde{\alpha}_B + {}_G\tilde{\alpha}_B \, {}_G\tilde{\omega}_B + {}_G\tilde{\omega}_B^3\right) \left({}^G\mathbf{q} - {}^G\mathbf{p}\right)
\end{aligned} \tag{9.260}
$$

which can be converted to

$$
\begin{bmatrix} {}^G\dddot{\mathbf{q}} \\ 0 \end{bmatrix} = {}^G J_B \begin{bmatrix} \mathbf{q} \\ 1 \end{bmatrix} \tag{9.261}
$$

where $[J]$ is the jerk transformation matrix.

$$[J] = \begin{bmatrix} J_{11} & J_{12} \\ 0 & 0 \end{bmatrix} \tag{9.262}$$

$$J_{11} = {}_G\tilde{\chi}_B + 2\,{}_G\tilde{\omega}_B\,{}_G\tilde{\alpha}_B + {}_G\tilde{\alpha}_B\,{}_G\tilde{\omega}_B + {}_G\tilde{\omega}_B^3 \tag{9.263}$$

$$J_{12} = {}^G\ddot{\mathbf{p}} - \left({}_G\tilde{\chi}_B + 2\,{}_G\tilde{\omega}_B\,{}_G\tilde{\alpha}_B + {}_G\tilde{\alpha}_B\,{}_G\tilde{\omega}_B + {}_G\tilde{\omega}_B^3 \right)\,{}^G\mathbf{p}$$

$$\tag{9.264}$$

9.4   Forward Acceleration Kinematics

The forward acceleration kinematics problem is the method of relating the end-effector accelerations to $\ddot{\mathbf{X}}$ the joint accelerations, $\ddot{\mathbf{q}}$. It is

$$\ddot{\mathbf{X}} = \mathbf{J}\,\ddot{\mathbf{q}} + \dot{\mathbf{J}}\,\dot{\mathbf{q}} \tag{9.265}$$

where, $[\mathbf{J}]$ is the Jacobian matrix, $[\dot{\mathbf{J}}]$ is time derivative of Jacobian matrix, $\mathbf{q}$ is the joint variable vector, $\dot{\mathbf{q}}$ is the joint velocity vector, and $\ddot{\mathbf{q}}$ is the *joint acceleration vector.*

$$\mathbf{q} = \begin{bmatrix} q_1 & q_2 & q_3 & \cdots & q_n \end{bmatrix}^T \tag{9.266}$$

$$\dot{\mathbf{q}} = \begin{bmatrix} \dot{q}_1 & \dot{q}_2 & \dot{q}_3 & \cdots & \dot{q}_n \end{bmatrix}^T \tag{9.267}$$

$$\ddot{\mathbf{q}} = \begin{bmatrix} \ddot{q}_1 & \ddot{q}_2 & \ddot{q}_3 & \cdots & \ddot{q}_n \end{bmatrix}^T \tag{9.268}$$

However, $\dot{\mathbf{X}}$ and $\ddot{\mathbf{X}}$ are the *end-effector configuration velocity* and *acceleration vectors,* respectively.

$$\mathbf{X} = \begin{bmatrix} X_n & Y_n & Z_n & \varphi_n & \theta_n & \psi_n \end{bmatrix}^T \tag{9.269}$$

$$\dot{\mathbf{X}} = \begin{bmatrix} {}^0\mathbf{v}_n \\ {}^0\boldsymbol{\omega}_n \end{bmatrix} = \begin{bmatrix} {}^0\dot{\mathbf{d}}_n \\ {}^0\boldsymbol{\omega}_n \end{bmatrix} \tag{9.270}$$

$$= \begin{bmatrix} \dot{X}_n & \dot{Y}_n & \dot{Z}_n & \omega_{Xn} & \omega_{Yn} & \omega_{Zn} \end{bmatrix}^T \tag{9.271}$$

$$\ddot{\mathbf{X}} = \begin{bmatrix} {}^0\mathbf{a}_n \\ {}^0\boldsymbol{\alpha}_n \end{bmatrix} = \begin{bmatrix} {}^0\ddot{\mathbf{d}}_n \\ {}^0\dot{\boldsymbol{\omega}}_n \end{bmatrix}$$

$$= \begin{bmatrix} \ddot{X}_n & \ddot{Y}_n & \ddot{Z}_n & \dot{\omega}_{Xn} & \dot{\omega}_{Yn} & \dot{\omega}_{Zn} \end{bmatrix}^T \tag{9.272}$$

To calculate the time derivative of the Jacobian matrix $[\dot{\mathbf{J}}]$, we use Eqs. (8.132)–(8.135).

$$_{i-1}^{\ 0}\dot{\mathbf{d}}_i = \begin{cases} {}_0^0\boldsymbol{\omega}_i \times {}_{i-1}^{\ 0}\mathbf{d}_i & \text{if joint } i \text{ is R} \\ \dot{d}_i\,{}^0\hat{k}_{i-1} + {}_0^0\boldsymbol{\omega}_i \times {}_{i-1}^{\ 0}\mathbf{d}_i & \text{if joint } i \text{ is P} \end{cases}$$

$$\tag{9.273}$$

$$_{i-1}^{\ 0}\boldsymbol{\omega}_i = \begin{cases} \dot{\theta}_i\,{}^0\hat{k}_{i-1} & \text{if joint } i \text{ is R} \\ 0 & \text{if joint } i \text{ is P} \end{cases}$$

$$\tag{9.274}$$

Taking a derivative to find the acceleration of link (i) with respect to its previous link $(i-1)$, yields

$$
{}_{i-1}^{0}\ddot{\mathbf{d}}_i = \begin{cases} {}_0^0\dot{\boldsymbol{\omega}}_i \times {}_{i-1}^{0}\mathbf{d}_i + {}_0^0\boldsymbol{\omega}_i \times \left({}_0^0\boldsymbol{\omega}_i \times {}_{i-1}^{0}\mathbf{d}_i \right) & \text{if joint } i \text{ is R} \\[2mm] {}_0^0\dot{\boldsymbol{\omega}}_i \times {}_{i-1}^{0}\mathbf{d}_i + {}_0^0\boldsymbol{\omega}_i \times \left({}_0^0\boldsymbol{\omega}_i \times {}_{i-1}^{0}\mathbf{d}_i \right) \\ \quad + \ddot{d}_i \, {}^0\hat{k}_{i-1} + 2\dot{d}_i \, {}_0^0\boldsymbol{\omega}_{i-1} \times {}^0\hat{k}_{i-1} & \text{if joint } i \text{ is P} \end{cases}
$$

$$(9.275)$$

$$
{}_{i-1}^{0}\dot{\boldsymbol{\omega}}_i = \begin{cases} \ddot{\theta}_i \, {}^0\hat{k}_{i-1} + \dot{\theta}_i \, {}_0^0\boldsymbol{\omega}_{i-1} \times {}^0\hat{k}_{i-1} & \text{if joint } i \text{ is R} \\ 0 & \text{if joint } i \text{ is P} \end{cases}
$$

$$(9.276)$$

Therefore, the acceleration vectors of the end-effector frame are

$$
{}_0^0\ddot{\mathbf{d}}_n = \sum_{i=1}^{n} {}_{i-1}^{0}\ddot{\mathbf{d}}_i
$$

$$(9.277)$$

and

$$
{}_0^0\dot{\boldsymbol{\omega}}_n = \sum_{i=1}^{n} {}_{i-1}^{0}\boldsymbol{\omega}_i.
$$

$$(9.278)$$

The acceleration relationships can also be rearranged in a recursive form.

$$
{}_{i-1}^{0}\dot{\boldsymbol{\omega}}_i = \begin{cases} \ddot{\theta}_i \, {}^0\hat{k}_{i-1} + \dot{\theta}_i \, {}_0^0\boldsymbol{\omega}_{i-1} \times {}^0\hat{k}_{i-1} & \text{if joint } i \text{ is R} \\ 0 & \text{if joint } i \text{ is P} \end{cases}
$$

$$(9.279)$$

Example 314 Forward acceleration of the $2R$ planar manipulator. An example to show how $[\mathbf{J}]$ will be calculated.
   The forward velocity of the $2R$ planar manipulator is found as

$$
\dot{\mathbf{X}} = \mathbf{J}\,\dot{\mathbf{q}}
$$

$$(9.280)$$

$$
\begin{bmatrix} \dot{X} \\ \dot{Y} \end{bmatrix} = \begin{bmatrix} -l_1 s\theta_1 - l_2 s\left(\theta_1 + \theta_2\right) & -l_2 s\left(\theta_1 + \theta_2\right) \\ l_1 c\theta_1 + l_2 c\left(\theta_1 + \theta_2\right) & l_2 c\left(\theta_1 + \theta_2\right) \end{bmatrix} \begin{bmatrix} \dot{\theta}_1 \\ \dot{\theta}_2 \end{bmatrix}
$$

The differential of the Jacobian matrix is

$$
\dot{\mathbf{J}} = \begin{bmatrix} \dot{J}_{11} & \dot{J}_{12} \\ \dot{J}_{21} & \dot{J}_{22} \end{bmatrix}
$$

$$(9.281)$$

where,

$$
\dot{J}_{11} = \left(-l_1 \cos\theta_1 - l_2 \cos\left(\theta_1 + \theta_2\right)\right)\dot{\theta}_1 - l_2 \cos\left(\theta_1 + \theta_2\right)\dot{\theta}_2
$$

$$
\dot{J}_{12} = -l_2 \cos\left(\theta_1 + \theta_2\right)\dot{\theta}_1 - l_2 \cos\left(\theta_1 + \theta_2\right)\dot{\theta}_2
$$

$$
\dot{J}_{21} = \left(-l_1 \sin\theta_1 - l_2 \sin\left(\theta_1 + \theta_2\right)\right)\dot{\theta}_1 - l_2 \sin\left(\theta_1 + \theta_2\right)\dot{\theta}_2
$$

$$
\dot{J}_{22} = -l_2 \sin\left(\theta_1 + \theta_2\right)\dot{\theta}_1 - l_2 \sin\left(\theta_1 + \theta_2\right)\dot{\theta}_2
$$

$$(9.282)$$

and therefore, the forward acceleration kinematics of the manipulator can be rearranged.

$$
\ddot{\mathbf{X}} = \mathbf{J}\,\ddot{\mathbf{q}} + \dot{\mathbf{J}}\,\dot{\mathbf{q}}
$$

$$(9.283)$$

For the $2R$ manipulator it is easier to show the acceleration in the following form:

$$\begin{bmatrix} \ddot{X} \\ \ddot{Y} \end{bmatrix} = \begin{bmatrix} -l_1 \sin\theta_1 & -l_2 \sin(\theta_1+\theta_2) \\ l_1 \cos\theta_1 & l_2 \cos(\theta_1+\theta_2) \end{bmatrix} \begin{bmatrix} \ddot{\theta}_1 \\ \ddot{\theta}_1 + \ddot{\theta}_2 \end{bmatrix}$$

$$- \begin{bmatrix} l_1 \cos\theta_1 & l_2 \cos(\theta_1+\theta_2) \\ l_1 \sin\theta_1 & l_2 \sin(\theta_1+\theta_2) \end{bmatrix} \begin{bmatrix} \dot{\theta}_1^2 \\ (\dot{\theta}_1 + \dot{\theta}_2)^2 \end{bmatrix} \tag{9.284}$$

Example 315 ★ Acceleration based on position vector. It is an analytic calculation of acceleration when the position vector of the end-effector is given as function of joint variables.

Assume that the position vector $\mathbf{r}$ of the end-effector of a manipulator is given as function of its joint coordinates $\mathbf{q}$.

$$\mathbf{r} = \begin{bmatrix} r_1(\mathbf{q}) \\ r_2(\mathbf{q}) \\ r_3(\mathbf{q}) \end{bmatrix} = \begin{bmatrix} r_1(q_1, q_2, q_3) \\ r_2(q_1, q_2, q_3) \\ r_3(q_1, q_2, q_3) \end{bmatrix} \tag{9.285}$$

The velocity of the end-effector is

$$\dot{\mathbf{r}} = \frac{\partial \mathbf{r}}{\partial \mathbf{q}} \dot{\mathbf{q}} = \mathbf{J}_D \, \dot{\mathbf{q}} \tag{9.286}$$

where,

$$\mathbf{J}_D = \frac{\partial \mathbf{r}}{\partial \mathbf{q}} = \begin{bmatrix} \partial r_1/\partial q_1 & \partial r_1/\partial q_2 & \partial r_1/\partial q_3 \\ \partial r_2/\partial q_1 & \partial r_2/\partial q_2 & \partial r_2/\partial q_3 \\ \partial r_3/\partial q_1 & \partial r_3/\partial q_2 & \partial r_3/\partial q_3 \end{bmatrix} \tag{9.287}$$

The second derivative of $\mathbf{r}$ is

$$\ddot{\mathbf{r}} = \frac{\partial \mathbf{r}}{\partial \mathbf{q}} \ddot{\mathbf{q}} + \frac{d}{dt}\left(\frac{\partial \mathbf{r}}{\partial \mathbf{q}}\right) \dot{\mathbf{q}} \tag{9.288}$$

where,

$$\frac{d}{dt}\left(\frac{\partial \mathbf{r}}{\partial \mathbf{q}}\right) = \frac{d}{dt}\mathbf{J}_D = \sum_{i=1}^{3} \frac{d\mathbf{J}_D}{dq_i} \dot{q}_i$$

$$= \frac{d\mathbf{J}_D}{dq_1}\dot{q}_1 + \frac{d\mathbf{J}_D}{dq_2}\dot{q}_2 + \frac{d\mathbf{J}_D}{dq_3}\dot{q}_3 \tag{9.289}$$

9.5   Inverse Acceleration Kinematics

The inverse kinematics acceleration is to calculate the joint acceleration vector $\ddot{\mathbf{q}}$ for a given end-effector acceleration $\ddot{\mathbf{X}}$. The inverse acceleration kinematics will be solved from the forward acceleration kinematics equation.

$$\ddot{\mathbf{X}} = \begin{bmatrix} {}^0\mathbf{a}_n \\ {}^0\boldsymbol{\alpha}_n \end{bmatrix} = \mathbf{J}\ddot{\mathbf{q}} + \dot{\mathbf{J}}\dot{\mathbf{q}} \tag{9.290}$$

Assuming that the Jacobian matrix, $\mathbf{J}$, is square and non-singular, the joint acceleration vector $\ddot{\mathbf{q}}$ can be found by matrix inversion.

$$\ddot{\mathbf{q}} = \mathbf{J}^{-1}\left(\ddot{\mathbf{X}} - \dot{\mathbf{J}}\dot{\mathbf{q}}\right) \tag{9.291}$$

Calculating the joint acceleration vector $\ddot{\mathbf{q}}$ is a matrix operation, realizing that all vector and matrices of $\mathbf{X}$, $\dot{\mathbf{X}}$, $\ddot{\mathbf{X}}$, $\mathbf{q}$, $\dot{\mathbf{q}}$, $[\mathbf{J}]$, $[\dot{\mathbf{J}}]$ are already calculated in position and velocity kinematics analysis.

Proof As long as the Jacobian matrix $\mathbf{J}$ is square and non-singular, it has an inverse $\mathbf{J}^{-1}$ that can be used to solve the set of linear algebraic Equations (9.290) for the joint acceleration vector $\ddot{\mathbf{q}}$.

However, calculating $\dot{\mathbf{J}}$ and $\mathbf{J}^{-1}$ becomes more tedious by increasing the DOF of robots. The alternative technique is to write the equation in a new form

$$\ddot{\mathbf{X}} - \dot{\mathbf{J}}\dot{\mathbf{q}} = \mathbf{J}\ddot{\mathbf{q}} \tag{9.292}$$

which is similar to Eq. (8.239) for a robot with a spherical wrist

$$\ddot{\mathbf{X}} - \dot{\mathbf{J}}\dot{\mathbf{q}} = \begin{bmatrix} A & 0 \\ C & D \end{bmatrix} \ddot{\mathbf{q}} \tag{9.293}$$

or

$$\begin{bmatrix} \mathbf{m} \\ \mathbf{n} \end{bmatrix} = \begin{bmatrix} A & 0 \\ C & D \end{bmatrix} \begin{bmatrix} \ddot{q}_1 \\ \ddot{q}_2 \\ \ddot{q}_3 \\ \ddot{q}_4 \\ \ddot{q}_5 \\ \ddot{q}_6 \end{bmatrix} \tag{9.294}$$

where,

$$\begin{bmatrix} \mathbf{m} \\ \mathbf{n} \end{bmatrix} = \ddot{\mathbf{X}} - \dot{\mathbf{J}}\dot{\mathbf{q}} \tag{9.295}$$

Therefore, the inverse acceleration kinematics problem can be solved as

$$\begin{bmatrix} \ddot{\theta}_1 \\ \ddot{\theta}_2 \\ \ddot{\theta}_3 \end{bmatrix} = A^{-1} [\mathbf{m}] \tag{9.296}$$

and

$$\begin{bmatrix} \ddot{q}_4 \\ \ddot{q}_5 \\ \ddot{q}_6 \end{bmatrix} = D^{-1} \left([\mathbf{n}] - [C] \begin{bmatrix} \ddot{q}_1 \\ \ddot{q}_2 \\ \ddot{q}_3 \end{bmatrix} \right) \tag{9.297}$$

The matrix $\dot{\mathbf{J}}\dot{\mathbf{q}} = \ddot{\mathbf{X}} - \mathbf{J}\ddot{\mathbf{q}}$ is called the *acceleration bias vector* and can be calculated by differentiating from

$$_0^0\dot{\mathbf{d}}_6 = \sum_{i=1}^{3} {}_0^0\boldsymbol{\omega}_i \times {}_{i-1}^{0}\mathbf{d}_i \tag{9.298}$$

$$_0^0\boldsymbol{\omega}_6 = \sum_{i=1}^{6} \dot{\theta}_i \, {}^0\hat{k}_{i-1} \tag{9.299}$$

to find

$$_0^0\mathbf{a}_6 = {}_0^0\ddot{\mathbf{d}}_6 = \sum_{i=1}^{3} \left({}_0^0\dot{\boldsymbol{\omega}}_i \times {}_{i-1}^{0}\mathbf{d}_i + {}_0^0\boldsymbol{\omega}_i \times \left({}_0^0\boldsymbol{\omega}_i \times {}_{i-1}^{0}\mathbf{d}_i \right) \right) \tag{9.300}$$

and

$$_0^0\boldsymbol{\alpha}_6 = {}_0^0\dot{\boldsymbol{\omega}}_6 = \sum_{i-1}^{6} \left(\ddot{\theta}_i \, {}^0\hat{k}_{i-1} + {}_0^0\boldsymbol{\omega}_i \times \dot{\theta}_i \, {}^0\hat{k}_{i-1} \right) \tag{9.301}$$

The angular acceleration vector $_0^0\boldsymbol{\alpha}_6$ is the second half of $\ddot{\mathbf{X}}$. Then, subtracting the second half of $\mathbf{J}\ddot{\mathbf{q}}$ from $_0^0\boldsymbol{\alpha}_6$ provides the second half of the bias vector.

$$\sum_{i=1}^{6} {}_0^0\boldsymbol{\omega}_i \times \dot{\theta}_i \, {}^0\hat{k}_{i-1} \tag{9.302}$$

We substitute $_0^0\boldsymbol{\omega}_i$ and $_0^0\dot{\boldsymbol{\omega}}_i$

$$_0^0\boldsymbol{\omega}_i = \sum_{j=1}^{i} \dot{\theta}_j\, {}^0\hat{k}_{j-1} \tag{9.303}$$

$$_0^0\dot{\boldsymbol{\omega}}_i = \sum_{j=1}^{i} \left(\ddot{\theta}_j\, {}^0\hat{k}_{j-1} + {}_0^0\boldsymbol{\omega}_{j-1} \times \dot{\theta}_j\, {}^0\hat{k}_{j-1} \right) \tag{9.304}$$

in Eq. (9.300)

$$\begin{aligned}
_0^0\ddot{\mathbf{d}}_6 &= \sum_{i=1}^{3}\sum_{j=1}^{i} \left(\ddot{\theta}_j\, {}^0\hat{k}_{j-1} + {}_0^0\boldsymbol{\omega}_{j-1} \times \dot{\theta}_j\, {}^0\hat{k}_{j-1} \right) \times {}_{i-1}^{0}\mathbf{d}_i \\
&\quad + \sum_{i=1}^{3}\sum_{j=1}^{i} \dot{\theta}_j\, {}^0\hat{k}_{j-1} \times \left({}_0^0\boldsymbol{\omega}_i \times {}_{i-1}^{0}\mathbf{d}_i \right) \\
&= \sum_{i=1}^{3}\sum_{j=1}^{i} \ddot{\theta}_j\, {}^0\hat{k}_{j-1} \times {}_{i-1}^{0}\mathbf{d}_i \\
&\quad + \sum_{i=1}^{3}\sum_{j=1}^{i} \left({}_0^0\boldsymbol{\omega}_{j-1} \times \dot{\theta}_j\, {}^0\hat{k}_{j-1} \right) \times {}_{i-1}^{0}\mathbf{d}_i \\
&\quad + \sum_{i=1}^{3}\sum_{j=1}^{i} \dot{\theta}_j\, {}^0\hat{k}_{j-1} \times \left({}_0^0\boldsymbol{\omega}_i \times {}_{i-1}^{0}\mathbf{d}_i \right)
\end{aligned} \tag{9.305}$$

to find the first half of the bias vector.

$$\sum_{i=1}^{3}\sum_{j=1}^{i} \left({}_0^0\boldsymbol{\omega}_{j-1} \times \dot{\theta}_j\, {}^0\hat{k}_{j-1} \right) \times {}_{i-1}^{0}\mathbf{d}_i$$

$$+ \sum_{i=1}^{3}\sum_{j=1}^{i} \dot{\theta}_j\, {}^0\hat{k}_{j-1} \times \left({}_0^0\boldsymbol{\omega}_i \times {}_{i-1}^{0}\mathbf{d}_i \right) \tag{9.306}$$

∎

Example 316 Inverse acceleration of a $2R$ planar manipulator. Inverse acceleration kinematics of $2R$ manipulator is a good example to review how to calculate joint accelerations.

The analytic calculation of forward velocity and forward acceleration of the $2R$ planar manipulator is

$$\dot{\mathbf{X}} = \mathbf{J}\,\dot{\mathbf{q}} \tag{9.307}$$

$$\begin{bmatrix} \dot{X} \\ \dot{Y} \end{bmatrix} = \begin{bmatrix} -l_1\sin\theta_1 & -l_2\sin(\theta_1+\theta_2) \\ l_1\cos\theta_1 & l_2\cos(\theta_1+\theta_2) \end{bmatrix} \begin{bmatrix} \dot{\theta}_1 \\ \dot{\theta}_1+\dot{\theta}_2 \end{bmatrix}$$

$$\ddot{\mathbf{X}} = \mathbf{J}\,\ddot{\mathbf{q}} + \dot{\mathbf{J}}\,\dot{\mathbf{q}} = \mathbf{J}\,\ddot{\mathbf{q}} + \mathbf{J}'\,\dot{\mathbf{q}}^2 \tag{9.308}$$

$$\begin{bmatrix} \ddot{X} \\ \ddot{Y} \end{bmatrix} = \begin{bmatrix} -l_1\sin\theta_1 & -l_2\sin(\theta_1+\theta_2) \\ l_1\cos\theta_1 & l_2\cos(\theta_1+\theta_2) \end{bmatrix} \begin{bmatrix} \ddot{\theta}_1 \\ \ddot{\theta}_1+\ddot{\theta}_2 \end{bmatrix}$$

$$\quad - \begin{bmatrix} l_1\cos\theta_1 & l_2\cos(\theta_1+\theta_2) \\ l_1\sin\theta_1 & l_2\sin(\theta_1+\theta_2) \end{bmatrix} \begin{bmatrix} \dot{\theta}_1^2 \\ (\dot{\theta}_1+\dot{\theta}_2)^2 \end{bmatrix}$$

Also, the derivative and inverse of its Jacobian matrices are calculated as follows:

$$\dot{\mathbf{J}} = \begin{bmatrix} -l_1\dot{\theta}_1 c\theta_1 - l_2\left(\dot{\theta}_1 + \dot{\theta}_2\right) c\left(\theta_1 + \theta_2\right) & -l_2\left(\dot{\theta}_1 + \dot{\theta}_2\right) c\left(\theta_1 + \theta_2\right) \\ -l_1\dot{\theta}_1 s\theta_1 - l_2\left(\dot{\theta}_1 + \dot{\theta}_2\right) s\left(\theta_1 + \theta_2\right) & -l_2\left(\dot{\theta}_1 + \dot{\theta}_2\right) s\left(\theta_1 + \theta_2\right) \end{bmatrix} \tag{9.309}$$

$$\mathbf{J}^{-1} = \frac{-1}{l_1 l_2 s\theta_2} \begin{bmatrix} -l_2 c\left(\theta_1 + \theta_2\right) & -l_2 s\left(\theta_1 + \theta_2\right) \\ l_1 c\theta_1 + l_2 c\left(\theta_1 + \theta_2\right) & l_1 s\theta_1 + l_2 s\left(\theta_1 + \theta_2\right) \end{bmatrix} \tag{9.310}$$

The inverse acceleration kinematics of the manipulator $\ddot{\mathbf{q}}$ will be calculated by substituting the matrices and performing matrix operations.

$$\ddot{\mathbf{q}} = \mathbf{J}^{-1}\left(\ddot{\mathbf{X}} - \dot{\mathbf{J}}\dot{\mathbf{q}}\right) = \mathbf{J}^{-1}\left(\ddot{\mathbf{X}} - \mathbf{J}'\dot{\mathbf{q}}^2\right) \tag{9.311}$$

$$\begin{bmatrix} \ddot{\theta}_1 \\ \ddot{\theta}_1 + \ddot{\theta}_2 \end{bmatrix} =$$

$$\frac{1}{l_1 l_2 s\theta_2} \begin{bmatrix} l_2 \cos\left(\theta_1 + \theta_2\right) & l_2 \sin\left(\theta_1 + \theta_2\right) \\ -l_1 \cos\theta_1 & -l_1 \sin\theta_1 \end{bmatrix} \begin{bmatrix} \ddot{X} \\ \ddot{Y} \end{bmatrix}$$

$$+ \frac{1}{l_1 l_2 s\theta_2} \begin{bmatrix} l_1 l_2 \cos\theta_1 & l_2^2 \\ -l_1^2 & -l_1 l_2 \cos\theta_1 \end{bmatrix} \begin{bmatrix} \dot{\theta}_1^2 \\ \left(\dot{\theta}_1 + \dot{\theta}_2\right)^2 \end{bmatrix} \tag{9.312}$$

If we use the direct forward velocity equations,

$$\begin{bmatrix} \dot{X} \\ \dot{Y} \end{bmatrix} = \begin{bmatrix} -l_1\dot{\theta}_1 \sin\theta_1 - l_2\left(\dot{\theta}_1 + \dot{\theta}_2\right) \sin\left(\theta_1 + \theta_2\right) \\ l_1\dot{\theta}_1 \cos\theta_1 + l_2\left(\dot{\theta}_1 + \dot{\theta}_2\right) \cos\left(\theta_1 + \theta_2\right) \end{bmatrix} \tag{9.313}$$

we will have the forward acceleration equations by another differentiating.

$$\ddot{\mathbf{X}} = \begin{bmatrix} -l_1 s\theta_1 - l_2 s\left(\theta_1 + \theta_2\right) & -l_2 s\left(\theta_1 + \theta_2\right) \\ l_1 c\theta_1 + l_2 c\left(\theta_1 + \theta_2\right) & l_2 c\left(\theta_1 + \theta_2\right) \end{bmatrix} \begin{bmatrix} \ddot{\theta}_1 \\ \ddot{\theta}_2 \end{bmatrix}$$

$$- \begin{bmatrix} l_1 \cos\theta_1 & l_2 \cos\left(\theta_1 + \theta_2\right) \\ l_1 \sin\theta_1 & l_2 \sin\left(\theta_1 + \theta_2\right) \end{bmatrix} \begin{bmatrix} \dot{\theta}_1^2 \\ \left(\dot{\theta}_1 + \dot{\theta}_2\right)^2 \end{bmatrix} \tag{9.314}$$

The joint acceleration vector $\ddot{\mathbf{q}}$ can be found by matrix calculation.

$$\begin{bmatrix} \ddot{\theta}_1 \\ \ddot{\theta}_2 \end{bmatrix} = \begin{bmatrix} -l_1 s\theta_1 - l_2 s\left(\theta_1 + \theta_2\right) & -l_2 s\left(\theta_1 + \theta_2\right) \\ l_1 c\theta_1 + l_2 c\left(\theta_1 + \theta_2\right) & l_2 c\left(\theta_1 + \theta_2\right) \end{bmatrix}^{-1}$$

$$\times \left(\ddot{\mathbf{X}} + \begin{bmatrix} l_1 \cos\theta_1 & l_2 \cos\left(\theta_1 + \theta_2\right) \\ l_1 \sin\theta_1 & l_2 \sin\left(\theta_1 + \theta_2\right) \end{bmatrix} \begin{bmatrix} \dot{\theta}_1^2 \\ \left(\dot{\theta}_1 + \dot{\theta}_2\right)^2 \end{bmatrix} \right)$$

$$= \frac{-1}{l_1 l_2 s\theta_2} \begin{bmatrix} -l_2 \cos\left(\theta_1 + \theta_2\right) & -l_2 \sin\left(\theta_1 + \theta_2\right) \\ l_1 c\theta_1 + l_2 c\left(\theta_1 + \theta_2\right) & l_1 s\theta_1 + l_2 s\left(\theta_1 + \theta_2\right) \end{bmatrix} \begin{bmatrix} \ddot{X} \\ \ddot{Y} \end{bmatrix}$$

$$+ \frac{-1}{l_1 l_2 s\theta_2} \begin{bmatrix} -l_1 l_2 \cos\theta_2 & -l_2^2 \\ l_1\left(l_1 + l_2 c\theta_2\right) & l_2\left(l_2 + l_1 c\theta_2\right) \end{bmatrix} \begin{bmatrix} \dot{\theta}_1^2 \\ \left(\dot{\theta}_1 + \dot{\theta}_2\right)^2 \end{bmatrix} \tag{9.315}$$

As an example, let us consider a planar $2R$ manipulator with the following dimensions such that its end-effector moves on a straight line in one minute, $0 \leq t \leq 1\,\text{min}$.

$$l_1 = l_2 = 1\,\text{m} \tag{9.316}$$

$$X(0) = 1\,\text{m} \quad Y(0) = 0 \quad X(1) = -1\,\text{m} \quad Y(1) = 0 \tag{9.317}$$

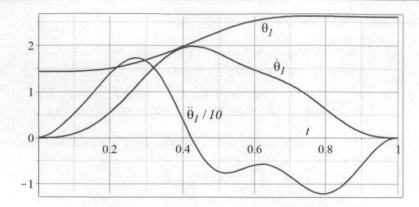

Fig. 9.6 Inverse kinematics of the joint 1 as functions of time for a 2*R* robot

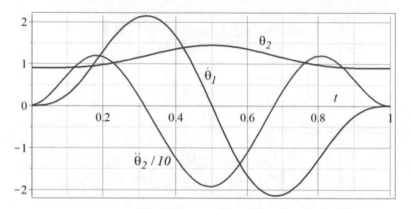

Fig. 9.7 Inverse kinematics of the joint 2 as functions of time for a 2*R* robot

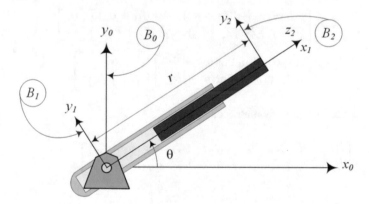

Fig. 9.8 A planar polar manipulator

A septic time function for X will move the end-effector from $X(0) = 1$ to $X(1) = -1$ with zero velocity, zero acceleration, zero jerk at both ends.

$$X = 1 - 70t^4 + 168t^5 - 140t^6 + 40t^7 \tag{9.318}$$

Employing inverse kinematics analysis, Figs. 9.6 and 9.7 depict the angle, angular velocity, and angular acceleration of joint 1 and 2 as function of time, respectively.

Example 317 Inverse acceleration of a polar planar manipulator. The polar manipulator is also a planar manipulator, which is the main arm of many industrial robots. Its inverse acceleration kinematics is calculated here.

Figure 9.8 illustrates a planar polar manipulator with the following forward velocity kinematics:

$$\dot{\mathbf{X}} = \mathbf{J}\dot{\mathbf{q}} \tag{9.319}$$

$$\begin{bmatrix} \dot{X} \\ \dot{Y} \end{bmatrix} = \begin{bmatrix} \cos\theta & -r\sin\theta \\ \sin\theta & r\cos\theta \end{bmatrix} \begin{bmatrix} \dot{r} \\ \dot{\theta} \end{bmatrix} \tag{9.320}$$

The $\dot{X}$, and $\dot{Y}$ are components of the global velocity of the tip point and $[\mathbf{J}]$ is the displacement Jacobian matrix of the manipulator.

$$\mathbf{J} = \begin{bmatrix} \cos\theta & -r\sin\theta \\ \sin\theta & r\cos\theta \end{bmatrix} \tag{9.321}$$

To determine the acceleration of the end-effector, we take a derivative of (9.319).

$$\ddot{\mathbf{X}} = \mathbf{J}\ddot{\mathbf{q}} + \dot{\mathbf{J}}\dot{\mathbf{q}} \tag{9.322}$$

The time derivative of Jacobian is

$$\dot{\mathbf{J}} = \frac{d}{dt}\begin{bmatrix} \cos\theta & -r\sin\theta \\ \sin\theta & r\cos\theta \end{bmatrix}$$

$$= \begin{bmatrix} -\dot{\theta}\sin\theta & -\dot{r}\sin\theta - r\dot{\theta}\cos\theta \\ \dot{\theta}\cos\theta & \dot{r}\cos\theta - r\dot{\theta}\sin\theta \end{bmatrix} \tag{9.323}$$

and therefore, the forward acceleration of the manipulator is

$$\begin{bmatrix} \ddot{X} \\ \ddot{Y} \end{bmatrix} = \begin{bmatrix} \cos\theta & -r\sin\theta \\ \sin\theta & r\cos\theta \end{bmatrix} \begin{bmatrix} \ddot{r} \\ \ddot{\theta} \end{bmatrix}$$

$$+ \begin{bmatrix} -\dot{\theta}\sin\theta & -\dot{r}\sin\theta - r\dot{\theta}\cos\theta \\ \dot{\theta}\cos\theta & \dot{r}\cos\theta - r\dot{\theta}\sin\theta \end{bmatrix} \begin{bmatrix} \dot{r} \\ \dot{\theta} \end{bmatrix} \tag{9.324}$$

To determine $\ddot{\mathbf{q}}$, we solve Eq. (9.324) for $\begin{bmatrix} \ddot{r} & \ddot{\theta} \end{bmatrix}^T$.

$$\ddot{\mathbf{q}} = \mathbf{J}^{-1}\left(\ddot{\mathbf{X}} - \dot{\mathbf{J}}\dot{\mathbf{q}}\right) \tag{9.325}$$

The inverse of Jacobian is

$$\mathbf{J}^{-1} = \begin{bmatrix} \cos\theta & \sin\theta \\ -\dfrac{1}{r}\sin\theta & \dfrac{1}{r}\cos\theta \end{bmatrix} \tag{9.326}$$

and therefore, the inverse acceleration of the manipulator would be

$$\ddot{\mathbf{q}} = \mathbf{J}^{-1}\ddot{\mathbf{X}} - \left[\mathbf{J}^{-1}\dot{\mathbf{J}}\right]\dot{\mathbf{q}} \tag{9.327}$$

$$\begin{bmatrix} \ddot{r} \\ \ddot{\theta} \end{bmatrix} = \begin{bmatrix} \cos\theta & \sin\theta \\ -\dfrac{1}{r}\sin\theta & \dfrac{1}{r}\cos\theta \end{bmatrix} \begin{bmatrix} \ddot{X} \\ \ddot{Y} \end{bmatrix}$$

$$- \begin{bmatrix} 0 & -r\dot{\theta} \\ \dfrac{1}{r}\dot{\theta} & \dfrac{1}{r}\dot{r} \end{bmatrix} \begin{bmatrix} \dot{r} \\ \dot{\theta} \end{bmatrix} \tag{9.328}$$

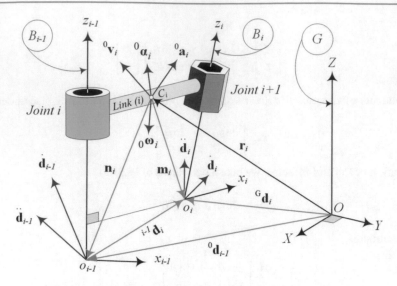

Fig. 9.9 Illustration of vectorial kinematics information of a link (i)

9.6 ★ Rigid Link Recursive Acceleration

Figure 9.9 illustrates a link (i) of a manipulator and shows its velocity and acceleration vectorial characteristics. Based on velocity and acceleration kinematics of rigid links we may write a set of recursive equations to relate the kinematics information of each link (i) in the coordinate frame attached to the previous link ($i-1$). We may then generalize the method of analysis to be suitable for a robot with any number of links.

Translational acceleration of the link (i) is denoted by $^0\mathbf{a}_i$ and is measured at the mass center C_i. Angular acceleration of the link (i) is denoted by $^0\boldsymbol{\alpha}_i$ and is usually shown at the mass center C_i, although $^0\boldsymbol{\alpha}_i$ is the same for all points of a rigid link. We calculate the translational acceleration of the link (i) at C_i by referring to the acceleration at the distal end of the link. Hence,

$$
\begin{aligned}
^0\mathbf{a}_i = {}&^0\ddot{\mathbf{d}}_i + {}_0\boldsymbol{\alpha}_i \times \left(^0\mathbf{r}_i - {}^0\mathbf{d}_i \right) \\
&+ {}_0\boldsymbol{\omega}_i \times \left({}_0\boldsymbol{\omega}_i \times \left(^0\mathbf{r}_i - {}^0\mathbf{d}_i \right) \right)
\end{aligned}
\tag{9.329}
$$

which can also be expressed in the body frame B_i.

$$
\begin{aligned}
{}^i_0\mathbf{a}_i = {}&^i\ddot{\mathbf{d}}_i + {}^i_0\boldsymbol{\alpha}_i \times \left(^i\mathbf{r}_i - {}^i\mathbf{d}_i \right) \\
&+ {}^i_0\boldsymbol{\omega}_i \times \left({}^i_0\boldsymbol{\omega}_i \times \left(^i\mathbf{r}_i - {}^i\mathbf{d}_i \right) \right)
\end{aligned}
\tag{9.330}
$$

The angular acceleration of the link (i) is

$$
{}_0\boldsymbol{\alpha}_i =
\begin{cases}
{}_0\boldsymbol{\alpha}_{i-1} + \ddot{\theta}_i\, {}^0\hat{k}_{i-1} + {}_0\boldsymbol{\omega}_{i-1} \times \dot{\theta}_i\, {}^0\hat{k}_{i-1} & \text{if joint } i \text{ is R} \\
{}_0\boldsymbol{\alpha}_{i-1} & \text{if joint } i \text{ is P}
\end{cases}
\tag{9.331}
$$

that can also be expressed in the body frame B_i.

$$
{}^i_0\boldsymbol{\alpha}_i =
\begin{cases}
\begin{aligned}
&{}^iT_{i-1}\left({}^{i-1}_0\boldsymbol{\alpha}_{i-1} + \ddot{\theta}_i\, {}^{i-1}\hat{k}_{i-1} \right) \\
&+ {}^iT_{i-1}\left({}^{i-1}_0\boldsymbol{\omega}_{i-1} \times \dot{\theta}_i\, {}^{i-1}\hat{k}_{i-1} \right)
\end{aligned} & \text{if joint } i \text{ is R} \\[2ex]
{}^iT_{i-1}\, {}^{i-1}_0\boldsymbol{\alpha}_{i-1} & \text{if joint } i \text{ is P}
\end{cases}
\tag{9.332}
$$

Proof According to rigid body acceleration in Eq. (9.183), the acceleration of the point C_i is

$$^0\mathbf{a}_i = \frac{^0d}{dt}\,^0\mathbf{v}_i \tag{9.333}$$

$$= {}_0\boldsymbol{\alpha}_i \times \left(^0\mathbf{r}_i - {}^0\mathbf{d}_i\right) + {}_0\boldsymbol{\omega}_i \times \left({}_0\boldsymbol{\omega}_i \times \left(^0\mathbf{r}_i - {}^0\mathbf{d}_i\right)\right) + {}^0\ddot{\mathbf{d}}_i$$

We can transform the acceleration of C_i to the body frame B_i.

$$^i\mathbf{a}_i = {}^0T_i^{-1}\,^0\mathbf{a}_i \tag{9.334}$$

$$= {}_0^i\boldsymbol{\alpha}_i \times \left(^i\mathbf{r}_i - {}^i\mathbf{d}_i\right) + {}_0^i\boldsymbol{\omega}_i \times \left({}_0^i\boldsymbol{\omega}_i \times \left(^i\mathbf{r}_i - {}^i\mathbf{d}_i\right)\right) + {}^i\ddot{\mathbf{d}}_i$$

Let us introduce the vectors $^0\mathbf{n}_i$ and $^0\mathbf{m}_i$ to define the position of o_{i-1} and o_i with respect to the mass center C_i.

$$^0\mathbf{n}_i = {}^0\mathbf{d}_{i-1} - {}^0\mathbf{r}_i \tag{9.335}$$

$$^0\mathbf{m}_i = {}^0\mathbf{d}_i - {}^0\mathbf{r}_i \tag{9.336}$$

The vectors $^0\mathbf{n}_i$ and $^0\mathbf{m}_i$ simplify the dynamic equations of robots.

Using Eq. (9.99) or (9.101) as a rule for adding relative angular accelerations, we find

$$_0\boldsymbol{\alpha}_i = {}_0\boldsymbol{\alpha}_{i-1} + {}_{i-1}^0\boldsymbol{\alpha}_i + {}_0\boldsymbol{\omega}_{i-1} \times {}_{i-1}^0\boldsymbol{\omega}_i \tag{9.337}$$

however, the angular velocity $_{i-1}^0\boldsymbol{\omega}_i$ and angular acceleration $_{i-1}^0\boldsymbol{\alpha}_i$ are

$$_{i-1}^0\boldsymbol{\omega}_i = \dot{\theta}_i\,^0\hat{k}_{i-1} \quad \text{if joint } i \text{ is R} \tag{9.338}$$

$$_{i-1}^0\boldsymbol{\alpha}_i = \ddot{\theta}_i\,^0\hat{k}_{i-1} \quad \text{if joint } i \text{ is R} \tag{9.339}$$

or

$$_{i-1}^0\boldsymbol{\omega}_i = 0 \quad \text{if joint } i \text{ is P} \tag{9.340}$$

$$_{i-1}^0\boldsymbol{\alpha}_i = 0 \quad \text{if joint } i \text{ is P} \tag{9.341}$$

Therefore,

$$_0\boldsymbol{\alpha}_i = \begin{cases} {}_0\boldsymbol{\alpha}_{i-1} + \ddot{\theta}_i\,^0\hat{k}_{i-1} + {}_0\boldsymbol{\omega}_{i-1} \times \dot{\theta}_i\,^0\hat{k}_{i-1} & \text{if joint } i \text{ is R} \\ {}_0\boldsymbol{\alpha}_{i-1} & \text{if joint } i \text{ is P} \end{cases} \tag{9.342}$$

and it can be transformed to any coordinate frame, including B_i, to find Eq. (9.332).

$$_0^i\boldsymbol{\alpha}_i = {}^0T_i^{-1}\,_0\boldsymbol{\alpha}_i = {}_0^i\boldsymbol{\alpha}_{i-1} + \ddot{\theta}_i\,_0^i\hat{k}_{i-1} + {}_0^i\boldsymbol{\omega}_{i-1} \times \dot{\theta}_i\,_0^i\hat{k}_{i-1}$$

$$= {}^iT_{i-1}\left(_0^{i-1}\boldsymbol{\alpha}_{i-1} + \ddot{\theta}_i\,_0^{i-1}\hat{k}_{i-1} + {}_0^{i-1}\boldsymbol{\omega}_{i-1} \times \dot{\theta}_i\,_0^{i-1}\hat{k}_{i-1}\right) \tag{9.343}$$

∎

Example 318 ★ Recursive angular velocity equation for link (i). Recursive equations are very applied because they can be computerized and calculate kinematics of links one by one, each one related to the next or previous link. Here we derive angular velocity recursive equations for links of a robot.

The recursive global angular velocity equation for a link (i) is

$$_0\boldsymbol{\omega}_i = \begin{cases} {}_0\boldsymbol{\omega}_{i-1} + \dot{\theta}_i\,^0\hat{k}_{i-1} & \text{if joint } i \text{ is R} \\ {}_0\boldsymbol{\omega}_{i-1} & \text{if joint } i \text{ is P} \end{cases} \tag{9.344}$$

We may transform this equation to the body frame B_i.

$$
{}_0^i\boldsymbol{\omega}_i = {}^0T_i^{-1}\,{}_0\boldsymbol{\omega}_i = {}^0T_i^{-1}\left({}_0\boldsymbol{\omega}_{i-1} + \dot{\theta}_i\,{}^0\hat{k}_{i-1}\right) = {}_0^i\boldsymbol{\omega}_{i-1} + \dot{\theta}_i\,{}^i\hat{k}_{i-1}
$$

$$
= {}^iT_{i-1}\left({}_0^{i-1}\boldsymbol{\omega}_{i-1} + \dot{\theta}_i\,{}^{i-1}\hat{k}_{i-1}\right) \tag{9.345}
$$

Therefore, we can use a recursive equation to find the locally expressed angular velocity of link (i) by having the angular velocity of its lower link $(i-1)$. In this equation, every vector is expressed in its own coordinate frame.

$$
{}_0^i\boldsymbol{\omega}_i = \begin{cases} {}^iT_{i-1}\left({}_0^{i-1}\boldsymbol{\omega}_{i-1} + \dot{\theta}_i\,{}^{i-1}\hat{k}_{i-1}\right) & \text{if joint } i \text{ is R} \\ {}^iT_{i-1}\,{}_0^{i-1}\boldsymbol{\omega}_{i-1} & \text{if joint } i \text{ is P} \end{cases} \tag{9.346}
$$

Example 319 ★ Recursive translational velocity equation for link (i). Here we derive translational velocity recursive equations for links of a robot.

The recursive global translational velocity equation for a link (i) is

$$
{}^0\dot{\mathbf{d}}_i = \begin{cases} {}^0\dot{\mathbf{d}}_{i-1} + {}_0\boldsymbol{\omega}_i \times {}_{i-1}^0\mathbf{d}_i & \text{if joint } i \text{ is R} \\ {}^0\dot{\mathbf{d}}_{i-1} + {}_0\boldsymbol{\omega}_i \times {}_{i-1}^0\mathbf{d}_i + \dot{d}_i\,{}^0\hat{k}_{i-1} & \text{if joint } i \text{ is P} \end{cases} \tag{9.347}
$$

We may transform this equation to the body frame B_i.

$$
{}_0^i\dot{\mathbf{d}}_i = {}^0T_i^{-1}\,{}_0\dot{\mathbf{d}}_i = {}^0T_i^{-1}\left({}^0\dot{\mathbf{d}}_{i-1} + {}_0\boldsymbol{\omega}_i \times {}_{i-1}^0\mathbf{d}_i + \dot{d}_i\,{}^0\hat{k}_{i-1}\right)
$$

$$
= {}_0^i\mathbf{d}_{i-1} + {}_0^i\boldsymbol{\omega}_i \times {}_{i-1}^i\mathbf{d}_i + \dot{d}_i\,{}^i\hat{k}_{i-1}
$$

$$
= {}^iT_{i-1}\left({}_0^{i-1}\mathbf{d}_{i-1} + \dot{d}_i\,{}^{i-1}\hat{k}_{i-1}\right) + {}_0^i\boldsymbol{\omega}_i \times {}_{i-1}^i\mathbf{d}_i \tag{9.348}
$$

Therefore, we may use a recursive equation to find the locally expressed translational velocity of link (i) by having the translational velocity of its lower link $(i-1)$. In this equation, every vector is expressed in its own coordinate frame.

$$
{}^i\dot{\mathbf{d}}_i = \begin{cases} {}^iT_{i-1}\left({}_0^{i-1}\mathbf{d}_{i-1} + \dot{d}_i\,{}^{i-1}\hat{k}_{i-1}\right) \\ \quad + {}_0^i\boldsymbol{\omega}_i \times {}_{i-1}^i\mathbf{d}_i & \text{if joint } i \text{ is R} \\ \\ {}^iT_{i-1}\,{}_0^{i-1}\mathbf{d}_{i-1} + {}_0^i\boldsymbol{\omega}_i \times {}_{i-1}^i\mathbf{d}_i & \text{if joint } i \text{ is P} \end{cases} \tag{9.349}
$$

The angular velocity equation is a consequence of the relative velocity Eq. (7.55) and the rigid link's angular velocity based on DH parameters (8.2). The translational velocity equation also comes from rigid link velocity analysis (8.3).

Example 320 ★ Recursive joints' translational acceleration. Kinematics of joints of a robot may also be expressed by recursive equations. Here is the recursive acceleration analysis of joint i of a robot.

Equations (9.329) and (9.331) determine the translational and angular accelerations of link (i) in the base coordinate frame B_0. We may similarly determine the recursive translational acceleration for a link (i) at joint i, using the translational acceleration of link $(i-)$ at joint $i-1$.

$$
{}^0\ddot{\mathbf{d}}_i = \begin{cases} {}^0\ddot{\mathbf{d}}_{i-1} + {}_0\dot{\boldsymbol{\omega}}_i \times {}_{i-1}^0\mathbf{d}_i \\ \quad + {}_0\boldsymbol{\omega}_i \times \left({}_0\boldsymbol{\omega}_i \times {}_{i-1}^0\mathbf{d}_i\right) & \text{if joint } i \text{ is R} \\ \\ {}^0\ddot{\mathbf{d}}_{i-1} + {}_0\dot{\boldsymbol{\omega}}_i \times {}_{i-1}^0\mathbf{d}_i \\ \quad + {}_0\boldsymbol{\omega}_i \times \left({}_0\boldsymbol{\omega}_i \times {}_{i-1}^0\mathbf{d}_i\right) \\ \quad + \ddot{d}_i\,{}^0\hat{k}_{i-1} + 2\,{}_0\boldsymbol{\omega}_i \times \dot{d}_i\,{}^0\hat{k}_{i-1} & \text{if joint } i \text{ is P} \end{cases} \tag{9.350}
$$

where,

$$
{}_{i-1}^0\mathbf{d}_i = {}^0\mathbf{d}_i - {}^0\mathbf{d}_{i-1} \tag{9.351}
$$

We may also transform this equation to the local frame B_i.

$$
\begin{aligned}
{}^i_0\ddot{\mathbf{d}}_i &= {}^0T_i^{-1}\,{}^0\ddot{\mathbf{d}}_i \\
&= {}^i\ddot{\mathbf{d}}_{i-1} + {}^i_0\dot{\boldsymbol{\omega}}_i \times {}^i_{i-1}\mathbf{d}_i + {}^i_0\boldsymbol{\omega}_i \times \left({}^i_0\boldsymbol{\omega}_i \times {}^i_{i-1}\mathbf{d}_i\right) \\
&\quad + \ddot{d}_i\,{}^i\hat{k}_{i-1} + 2\,{}^i_0\boldsymbol{\omega}_i \times \dot{d}_i\,{}^i\hat{k}_{i-1} \\
&= {}^iT_{i-1}\left({}^{i-1}_0\ddot{\mathbf{d}}_{i-1} + \ddot{d}_i\,{}^{i-1}\hat{k}_{i-1}\right) + 2\,{}^i_0\boldsymbol{\omega}_i \times \dot{d}_i\,{}^iT_{i-1}\,{}^{i-1}\hat{k}_{i-1} \\
&\quad + {}^i_0\boldsymbol{\omega}_i \times \left({}^i_0\boldsymbol{\omega}_i \times {}^i_{i-1}\mathbf{d}_i\right) + {}^i_0\dot{\boldsymbol{\omega}}_i \times {}^i_{i-1}\mathbf{d}_i
\end{aligned}
\tag{9.352}
$$

Therefore, we may use a recursive equation to find the locally expressed translational acceleration of link (i) by having the translational acceleration of its lower link $(i-1)$. In this equation, every vector is expressed in its own coordinate frame.

$$
{}^i_0\ddot{\mathbf{d}}_i =
\begin{cases}
\begin{aligned}
&{}^iT_{i-1}\,{}^{i-1}_0\ddot{\mathbf{d}}_{i-1} \\
&\quad + {}^i_0\boldsymbol{\omega}_i \times \left({}^i_0\boldsymbol{\omega}_i \times {}^i_{i-1}\mathbf{d}_i\right) \\
&\quad + {}^i_0\dot{\boldsymbol{\omega}}_i \times {}^i_{i-1}\mathbf{d}_i
\end{aligned} & \text{if joint } i \text{ is R} \\[2ex]
\begin{aligned}
&{}^iT_{i-1}\left({}^{i-1}_0\ddot{\mathbf{d}}_{i-1} + \ddot{d}_i\,{}^{i-1}\hat{k}_{i-1}\right) \\
&\quad + 2\,{}^i_0\boldsymbol{\omega}_i \times \dot{d}_i\,{}^iT_{i-1}\,{}^{i-1}\hat{k}_{i-1} \\
&\quad + {}^i_0\boldsymbol{\omega}_i \times \left({}^i_0\boldsymbol{\omega}_i \times {}^i_{i-1}\mathbf{d}_i\right) \\
&\quad + {}^i_0\dot{\boldsymbol{\omega}}_i \times {}^i_{i-1}\mathbf{d}_i
\end{aligned} & \text{if joint } i \text{ is P}
\end{cases}
\tag{9.353}
$$

Equation (9.350) is the result of differentiating (9.347), and Eq. (9.353) is the result of transforming (9.350) to the local frame B_i.

Example 321 ★ Recursive accelerations for revolute joints. There are many robots with only revolute joint. Kinematics of such robot will be simpler as reviewed in this example.

If all joints of a robot are revolute, then Eq. (9.350) for translational acceleration of joint i becomes

$$
{}^0\ddot{\mathbf{d}}_i = {}^0\ddot{\mathbf{d}}_{i-1} + {}_0\boldsymbol{\alpha}_i \times {}_{i-1}^0\mathbf{d}_i + {}_0\boldsymbol{\omega}_i \times \left({}_0\boldsymbol{\omega}_i \times {}_{i-1}^0\mathbf{d}_i\right)
\tag{9.354}
$$

and the angular acceleration of link (i) becomes

$$
{}_0\boldsymbol{\alpha}_i = {}_0\boldsymbol{\alpha}_{i-1} + \ddot{\theta}_i\,{}^0\hat{k}_{i-1} + {}_0\boldsymbol{\omega}_{i-1} \times \dot{\theta}_i\,{}^0\hat{k}_{i-1}
\tag{9.355}
$$

Equation (9.344) also simplifies to the following equation for global angular velocity of link (i):

$$
{}_0\boldsymbol{\omega}_i = {}_0\boldsymbol{\omega}_{i-1} + \dot{\theta}_i\,{}^0\hat{k}_{i-1}
\tag{9.356}
$$

Starting from the first link, we find the angular velocity of the first four links as

$$
{}_0\boldsymbol{\omega}_1 = \dot{\theta}_1\,{}^0\hat{k}_0
\tag{9.357}
$$

$$
{}_0\boldsymbol{\omega}_2 = {}_0\boldsymbol{\omega}_1 + \dot{\theta}_2\,{}^0\hat{k}_1 = \dot{\theta}_1\,{}^0\hat{k}_0 + \dot{\theta}_2\,{}^0\hat{k}_1
\tag{9.358}
$$

$$
{}_0\boldsymbol{\omega}_3 = {}_0\boldsymbol{\omega}_2 + \dot{\theta}_3\,{}^0\hat{k}_2 = \dot{\theta}_1\,{}^0\hat{k}_0 + \dot{\theta}_2\,{}^0\hat{k}_1 + \dot{\theta}_3\,{}^0\hat{k}_2
\tag{9.359}
$$

$$
{}_0\boldsymbol{\omega}_4 = {}_0\boldsymbol{\omega}_3 + \dot{\theta}_4\,{}^0\hat{k}_3 = \dot{\theta}_1\,{}^0\hat{k}_0 + \dot{\theta}_2\,{}^0\hat{k}_1 + \dot{\theta}_3\,{}^0\hat{k}_2 + \dot{\theta}_4\,{}^0\hat{k}_3
\tag{9.360}
$$

and the angular velocity of the link (i) as

$$
{}_0\boldsymbol{\omega}_i = {}_0\boldsymbol{\omega}_{i-1} + \dot{\theta}_i\,{}^0\hat{k}_{i-1} = \sum_{j=1}^i \dot{\theta}_j\,{}^0\hat{k}_{j-1}
\tag{9.361}
$$

The first four angular accelerations will be

$$_0\boldsymbol{\alpha}_1 = \ddot{\theta}_1 \, {}^0\hat{k}_0 \tag{9.362}$$

$$\begin{aligned} _0\boldsymbol{\alpha}_2 &= {}_0\boldsymbol{\alpha}_1 + \ddot{\theta}_2 \, {}^0\hat{k}_1 + {}_0\boldsymbol{\omega}_1 \times \dot{\theta}_2 \, {}^0\hat{k}_1 \\ &= \ddot{\theta}_1 \, {}^0\hat{k}_0 + \ddot{\theta}_2 \, {}^0\hat{k}_1 + \dot{\theta}_1 \, {}^0\hat{k}_0 \times \dot{\theta}_2 \, {}^0\hat{k}_1 \end{aligned} \tag{9.363}$$

$$\begin{aligned} _0\boldsymbol{\alpha}_3 &= {}_0\boldsymbol{\alpha}_2 + \ddot{\theta}_3 \, {}^0\hat{k}_2 + {}_0\boldsymbol{\omega}_2 \times \dot{\theta}_3 \, {}^0\hat{k}_2 \\ &= \ddot{\theta}_1 \, {}^0\hat{k}_0 + \ddot{\theta}_2 \, {}^0\hat{k}_1 + \dot{\theta}_1 \, {}^0\hat{k}_0 \times \dot{\theta}_2 \, {}^0\hat{k}_1 \\ &\quad + \ddot{\theta}_3 \, {}^0\hat{k}_2 + \left(\dot{\theta}_1 \, {}^0\hat{k}_0 + \dot{\theta}_2 \, {}^0\hat{k}_1 \right) \times \dot{\theta}_3 \, {}^0\hat{k}_2 \\ &= \ddot{\theta}_1 \, {}^0\hat{k}_0 + \ddot{\theta}_2 \, {}^0\hat{k}_1 + \ddot{\theta}_3 \, {}^0\hat{k}_2 \\ &\quad + \dot{\theta}_1 \, {}^0\hat{k}_0 \times \dot{\theta}_2 \, {}^0\hat{k}_1 + \left(\dot{\theta}_1 \, {}^0\hat{k}_0 + \dot{\theta}_2 \, {}^0\hat{k}_1 \right) \times \dot{\theta}_3 \, {}^0\hat{k}_2 \end{aligned} \tag{9.364}$$

$$\begin{aligned} _0\boldsymbol{\alpha}_4 &= {}_0\boldsymbol{\alpha}_3 + \ddot{\theta}_4 \, {}^0\hat{k}_3 + {}_0\boldsymbol{\omega}_3 \times \dot{\theta}_4 \, {}^0\hat{k}_3 \\ &= \ddot{\theta}_1 \, {}^0\hat{k}_0 + \ddot{\theta}_2 \, {}^0\hat{k}_1 + \ddot{\theta}_3 \, {}^0\hat{k}_2 + \ddot{\theta}_4 \, {}^0\hat{k}_3 \\ &\quad + \dot{\theta}_1 \, {}^0\hat{k}_0 \times \dot{\theta}_2 \, {}^0\hat{k}_1 + \left(\dot{\theta}_1 \, {}^0\hat{k}_0 + \dot{\theta}_2 \, {}^0\hat{k}_1 \right) \times \dot{\theta}_3 \, {}^0\hat{k}_2 \\ &\quad + \left(\dot{\theta}_1 \, {}^0\hat{k}_0 + \dot{\theta}_2 \, {}^0\hat{k}_1 + \dot{\theta}_3 \, {}^0\hat{k}_2 \right) \times \dot{\theta}_4 \, {}^0\hat{k}_3 \end{aligned} \tag{9.365}$$

or

$$\begin{aligned} _0\boldsymbol{\alpha}_4 &= \dot{\theta}_1 \, {}^0\hat{k}_0 \times \left(\dot{\theta}_2 \, {}^0\hat{k}_1 + \dot{\theta}_3 \, {}^0\hat{k}_2 + \dot{\theta}_4 \, {}^0\hat{k}_3 \right) \\ &\quad + \dot{\theta}_2 \, {}^0\hat{k}_1 \times \left(\dot{\theta}_3 \, {}^0\hat{k}_2 + \dot{\theta}_4 \, {}^0\hat{k}_3 \right) + \dot{\theta}_3 \, {}^0\hat{k}_2 \times \dot{\theta}_4 \, {}^0\hat{k}_3 \end{aligned} \tag{9.366}$$

So, we can find the angular acceleration of link (i) as

$$\begin{aligned} _0\boldsymbol{\alpha}_i &= {}_0\boldsymbol{\alpha}_{i-1} + \ddot{\theta}_i \, {}^0\hat{k}_{i-1} + {}_0\boldsymbol{\omega}_{i-1} \times \dot{\theta}_i \, {}^0\hat{k}_{i-1} \\ &= \sum_{j=1}^{i} \ddot{\theta}_j \, {}^0\hat{k}_{j-1} + \sum_{j=1}^{i-1} \left(\sum_{k=1}^{j} \dot{\theta}_k \, {}^0\hat{k}_{k-1} \times \dot{\theta}_{j+1} \, {}^0\hat{k}_j \right) \end{aligned} \tag{9.367}$$

or

$$_0\boldsymbol{\alpha}_i = \sum_{j=1}^{i} \ddot{\theta}_j \, {}^0\hat{k}_{j-1} + \sum_{j=1}^{i-1} \left(\dot{\theta}_j \, {}^0\hat{k}_{j-1} \times \sum_{k=j+1}^{i} \dot{\theta}_k \, {}^0\hat{k}_{k-1} \right) \tag{9.368}$$

However, because of ${}^0\hat{k}_i \times {}^0\hat{k}_i = 0$, we can rearrange the equation as

$$\begin{aligned} _0\boldsymbol{\alpha}_i &= {}_0\boldsymbol{\alpha}_{i-1} + \ddot{\theta}_i \, {}^0\hat{k}_{i-1} + {}_0\boldsymbol{\omega}_{i-1} \times \dot{\theta}_i \, {}^0\hat{k}_{i-1} \\ &= \sum_{j=1}^{i} \ddot{\theta}_j \, {}^0\hat{k}_{j-1} + \sum_{j=1}^{i} \left(\sum_{k=1}^{j} \dot{\theta}_k \, {}^0\hat{k}_{k-1} \times \dot{\theta}_j \, {}^0\hat{k}_{j-1} \right) \end{aligned} \tag{9.369}$$

Using the above equations, we can find the translational acceleration of frame B_i.

$$
\begin{aligned}
{}^0\ddot{\mathbf{d}}_i &= {}^0\ddot{\mathbf{d}}_{i-1} + {}_0\boldsymbol{\alpha}_i \times {}_{i-1}^{\ 0}\mathbf{d}_i + {}_0\boldsymbol{\omega}_i \times \left({}_0\boldsymbol{\omega}_i \times {}_{i-1}^{\ 0}\mathbf{d}_i \right) \\
&= \sum_{j=1}^{i} {}_0\boldsymbol{\alpha}_j \times {}_{j-1}^{\ 0}\mathbf{d}_j + \sum_{j=1}^{i} {}_0\boldsymbol{\omega}_j \times \left({}_0\boldsymbol{\omega}_j \times {}_{j-1}^{\ 0}\mathbf{d}_j \right) \\
&= \sum_{j=1}^{i} \left(\sum_{m=1}^{j} \ddot{\theta}_m \, {}^0\hat{k}_{m-1} + \sum_{m=1}^{j} \left(\sum_{k=1}^{m} \dot{\theta}_k \, {}^0\hat{k}_{k-1} \times \dot{\theta}_m \, {}^0\hat{k}_{m-1} \right) \right) \\
&\quad \times {}_{j-1}^{\ 0}\mathbf{d}_j + \sum_{j=1}^{i} \left(\sum_{m=1}^{j} \dot{\theta}_m \, {}^0\hat{k}_{m-1} \times \left(\sum_{m=1}^{j} \dot{\theta}_m \, {}^0\hat{k}_{m-1} \times {}_{j-1}^{\ 0}\mathbf{d}_j \right) \right)
\end{aligned}
\tag{9.370}
$$

Example 322 ★ Jacobian rate generating vector. Jacobian matrix is a key point for derivative kinematics of a robot. The elements of Jacobian matrix are related to velocity kinematics and hence, Jacobian can also be expressed by recursive equations.

The forward acceleration kinematic needs differential of Jacobian matrix.

$$
\begin{bmatrix} {}^0\mathbf{a}_n \\ {}^0\boldsymbol{\alpha}_n \end{bmatrix} = \begin{bmatrix} {}^0\ddot{\mathbf{d}}_n \\ {}^0\dot{\boldsymbol{\omega}}_n \end{bmatrix} = \mathbf{J}\ddot{\mathbf{q}} + \dot{\mathbf{J}}\dot{\mathbf{q}}
\tag{9.371}
$$

From Eq. (8.136) we know that

$$
\begin{bmatrix} {}^0\dot{\mathbf{d}}_n \\ {}_0\boldsymbol{\omega}_n \end{bmatrix} = \sum_{i=1}^{n} \begin{bmatrix} {}^0\hat{k}_{i-1} \times {}_{i-1}^{\ 0}\mathbf{d}_n \\ {}^0\hat{k}_{i-1} \end{bmatrix} \dot{\theta}_i
\tag{9.372}
$$

Taking a derivative, we have

$$
\begin{bmatrix} {}^0\ddot{\mathbf{d}}_n \\ {}^0\dot{\boldsymbol{\omega}}_n \end{bmatrix} = \begin{bmatrix} {}^0\mathbf{a}_n \\ {}^0\boldsymbol{\alpha}_n \end{bmatrix} = \sum_{i=1}^{n} \begin{bmatrix} {}^0\hat{k}_{i-1} \times {}_{i-1}^{\ 0}\mathbf{d}_n \\ {}^0\hat{k}_{i-1} \end{bmatrix} \ddot{\theta}_i + \sum_{i=1}^{n} \begin{bmatrix} \dfrac{d}{dt}\left({}^0\hat{k}_{i-1} \times {}_{i-1}^{\ 0}\mathbf{d}_n \right) \\ \dfrac{d}{dt}\, {}^0\hat{k}_{i-1} \end{bmatrix} \dot{\theta}_i
\tag{9.373}
$$

To expand this equation, let us begin with $\dfrac{d}{dt}\, {}^0\hat{k}_{i-1}$.

$$
\frac{d}{dt}\, {}^0\hat{k}_{i-1} = {}_0\boldsymbol{\omega}_{i-1} \times {}^0\hat{k}_{i-1}
\tag{9.374}
$$

Using (9.361), we have

$$
\frac{d}{dt}\, {}^0\hat{k}_{i-1} = \sum_{j=1}^{i-1} \dot{\theta}_j \, {}^0\hat{k}_{j-1} \times {}^0\hat{k}_{i-1}
\tag{9.375}
$$

The first row of (9.373) is

$$
\frac{d}{dt}\left({}^0\hat{k}_{i-1} \times {}_{i-1}^{\ 0}\mathbf{d}_n \right) = \frac{d\, {}^0\hat{k}_{i-1}}{dt} \times {}_{i-1}^{\ 0}\mathbf{d}_n + {}^0\hat{k}_{i-1} \times \frac{d}{dt}\, {}_{i-1}^{\ 0}\mathbf{d}_n
\tag{9.376}
$$

Let us use

$$
{}_{i-1}^{\ 0}\mathbf{d}_i = {}^0\mathbf{d}_i - {}^0\mathbf{d}_{i-1}
\tag{9.377}
$$

$$
{}_{i-1}^{\ 0}\mathbf{d}_n = \sum_{j=i}^{n} {}_{j-1}^{\ 0}\mathbf{d}_j
\tag{9.378}
$$

to find the first term as

$$\frac{d\,^0\hat{k}_{i-1}}{dt} \times\,_{i-1}^0\mathbf{d}_n = \left(\sum_{j=1}^{i-1} \dot{\theta}_j\,^0\hat{k}_{j-1} \times\,^0\hat{k}_{i-1}\right) \times \sum_{k=i}^n\,_{k-1}^0\mathbf{d}_k = \sum_{j=1}^{i-1}\sum_{k=i}^n \left(^0\hat{k}_{j-1} \times\,^0\hat{k}_{i-1}\right) \times\,_{k-1}^0\mathbf{d}_k\,\dot{\theta}_j \qquad (9.379)$$

The second term is

$$^0\hat{k}_{i-1} \times \frac{d}{dt}\,_{i-1}^0\mathbf{d}_n = {}^0\hat{k}_{i-1} \times \frac{d}{dt}\sum_{k=i}^n\,_{k-1}^0\mathbf{d}_k$$

$$= {}^0\hat{k}_{i-1} \times \sum_{k=i}^n \frac{d}{dt}\,_{k-1}^0\mathbf{d}_k$$

$$= {}^0\hat{k}_{i-1} \times \sum_{k=i}^n\,_0\boldsymbol{\omega}_k \times\,_{k-1}^0\mathbf{d}_k$$

$$= \sum_{k=i}^n\,^0\hat{k}_{i-1} \times \left(_0\boldsymbol{\omega}_k \times\,_{k-1}^0\mathbf{d}_k\right)$$

$$= \sum_{k=i}^n\,^0\hat{k}_{i-1} \times \left(\sum_{j=1}^k \dot{\theta}_j\,^0\hat{k}_{j-1} \times\,_{k-1}^0\mathbf{d}_k\right)$$

$$= \sum_{k=i}^n\sum_{j=1}^k\,^0\hat{k}_{i-1} \times \left(^0\hat{k}_{j-1} \times\,_{k-1}^0\mathbf{d}_k\right)\dot{\theta}_j \qquad (9.380)$$

Therefore,

$$\frac{d}{dt}\left(^0\hat{k}_{i-1} \times\,_{i-1}^0\mathbf{d}_n\right) = \sum_{j=1}^{i-1}\left(^0\hat{k}_{j-1} \times\,^0\hat{k}_{i-1}\right) \times\,_{j-1}^0\mathbf{d}_j\,\dot{\theta}_j$$

$$+ \sum_{k=i}^n\sum_{j=1}^k\,^0\hat{k}_{i-1} \times \left(^0\hat{k}_{j-1} \times\,_{k-1}^0\mathbf{d}_k\right)\dot{\theta}_j \qquad (9.381)$$

and Eq. (9.373) becomes

$$\begin{bmatrix} ^0\dot{\mathbf{v}}_n \\ ^0\dot{\boldsymbol{\omega}}_n \end{bmatrix} = \sum_{i=1}^n \begin{bmatrix} ^0\hat{k}_{i-1} \times\,_{i-1}^0\mathbf{d}_n \\ ^0\hat{k}_{i-1} \end{bmatrix}\ddot{\theta}_i$$

$$+ \sum_{i=1}^n \begin{bmatrix} \sum_{j=1}^{i-1}\left(^0\hat{k}_{j-1} \times\,^0\hat{k}_{i-1}\right) \times\,_{j-1}^0\mathbf{d}_j\,\dot{\theta}_j \\ \quad + \sum_{k=i}^n\sum_{j=1}^k\,^0\hat{k}_{i-1} \times \left(^0\hat{k}_{j-1} \times\,_{k-1}^0\mathbf{d}_k\right)\dot{\theta}_j \\ \sum_{j=1}^{i-1}\dot{\theta}_j\,^0\hat{k}_{j-1} \times\,^0\hat{k}_{i-1} \end{bmatrix}\dot{\theta}_i \qquad (9.382)$$

The first bracket is the Jacobian matrix,

$$\mathbf{J} = \sum_{i=1}^n \begin{bmatrix} ^0\hat{k}_{i-1} \times\,_{i-1}^0\mathbf{d}_n \\ ^0\hat{k}_{i-1} \end{bmatrix}$$

$$= \begin{bmatrix} ^0\hat{k}_0 \times\,_0^0\mathbf{d}_n & ^0\hat{k}_1 \times\,_1^0\mathbf{d}_n & \cdots & ^0\hat{k}_{n-1} \times\,_{n-1}^0\mathbf{d}_n \\ ^0\hat{k}_0 & ^0\hat{k}_1 & \cdots & ^0\hat{k}_{n-1} \end{bmatrix} \qquad (9.383)$$

and the second bracket is the time rate of Jacobian matrix.

$$
\mathbf{\dot{J}} = \sum_{i=1}^{n} \begin{bmatrix} \sum_{j=1}^{i-1} \left({}^0\hat{k}_{j-1} \times {}^0\hat{k}_{i-1} \right) \times {}_{j-1}{}^0\mathbf{d}_j \, \dot{\theta}_j \\ + \sum_{k=i}^{n} \sum_{j=1}^{k} {}^0\hat{k}_{i-1} \times \left({}^0\hat{k}_{j-1} \times {}_{k-1}{}^0\mathbf{d}_k \right) \dot{\theta}_j \\ \sum_{j=1}^{i-1} \dot{\theta}_j \, {}^0\hat{k}_{j-1} \times {}^0\hat{k}_{i-1} \end{bmatrix} \tag{9.384}
$$

Similar to the Jacobian generating vector $\mathbf{c}_i$ in (8.125), we can define a Jacobian generating rate vector $\dot{\mathbf{c}}_i$,

$$
\dot{\mathbf{c}}_i(\mathbf{q}) = \begin{bmatrix} \sum_{j=1}^{i-1} \left({}^0\hat{k}_{j-1} \times {}^0\hat{k}_{i-1} \right) \times {}_{j-1}{}^0\mathbf{d}_j \, \dot{\theta}_j \\ + \sum_{k=i}^{n} \sum_{j=1}^{k} {}^0\hat{k}_{i-1} \times \left({}^0\hat{k}_{j-1} \times {}_{k-1}{}^0\mathbf{d}_k \right) \dot{\theta}_j \\ \sum_{j=1}^{i-1} {}^0\hat{k}_{j-1} \times {}^0\hat{k}_{i-1} \, \dot{\theta}_j \end{bmatrix} \tag{9.385}
$$

to determine $\dot{\mathbf{J}}$ column by column.

$$
\dot{\mathbf{J}} = \begin{bmatrix} \dot{\mathbf{c}}_1 & \dot{\mathbf{c}}_2 & \dot{\mathbf{c}}_3 & \cdots & \dot{\mathbf{c}}_n \end{bmatrix} \tag{9.386}
$$

Example 323 ★ Recursive acceleration in base frame. The acceleration of a link can be calculated by its velocity transformation matrix.

The acceleration of connected links can also be calculated recursively. Let us start with computing ${}^0\dot{V}_i$, for the absolute accelerations of link (i),

$$
{}^0\dot{V}_i = {}^0\ddot{T}_i \; {}^0T_i^{-1} \tag{9.387}
$$

and then, calculating the acceleration matrix for link $(i+1)$.

$$
\begin{aligned}
{}^0\dot{V}_{i+1} &= {}^0\ddot{T}_{i+1} \; {}^0T_{i+1}^{-1} \\
&= \left({}^0\ddot{T}_i \; {}^iT_{i+1} + 2\,{}^0\dot{T}_i \; {}^i\dot{T}_{i+1} + {}^0T_i \; {}^i\ddot{T}_{i+1} \right) {}^0T_{i+1}^{-1} \\
&= {}^0\ddot{T}_i \; {}^0T_i^{-1} + 2\,{}^0\dot{T}_i \; {}^0T_i^{-1}\,{}^0T_i \; {}^i\dot{T}_{i+1}\,{}^iT_{i+1}^{-1} \; {}^0T_i^{-1} \\
&\quad + {}^0T_i \; {}^i\ddot{T}_{i+1}\,{}^iT_{i+1}^{-1} \; {}^0T_i^{-1}
\end{aligned} \tag{9.388}
$$

These two equations can be put in a recursive form.

$$
{}^0\dot{V}_{i+1} = {}^0\dot{V}_i + 2\,{}^0\dot{V}_i \; {}^0T_i \; {}^iV_{i+1}\,{}^0T_i^{-1} + {}^0T_i \; {}^i\dot{V}_{i+1}\,{}^0T_i^{-1} \tag{9.389}
$$

For a revolute joint, we have

$$
{}^iV_{i+1} = \dot{q}_{i+1}\,\Delta_R = \dot{\theta}_{i+1}\,\Delta_R = \dot{\theta}_{i+1} \begin{bmatrix} 0 & -1 & 0 & 0 \\ 1 & 0 & 0 & 0 \\ 0 & 0 & 0 & 0 \\ 0 & 0 & 0 & 0 \end{bmatrix} \tag{9.390}
$$

and therefore,

$$
\begin{aligned}
{}^i\dot{V}_{i+1} &= \ddot{\theta}_{i+1}\,\Delta_R - \dot{\theta}_{i+1}^2\,\Delta_R\,\Delta_R^T \\
&= \ddot{\theta}_{i+1} \begin{bmatrix} 0 & -1 & 0 & 0 \\ 1 & 0 & 0 & 0 \\ 0 & 0 & 0 & 0 \\ 0 & 0 & 0 & 0 \end{bmatrix} - \dot{\theta}_{i+1}^2 \begin{bmatrix} 1 & 0 & 0 & 0 \\ 0 & 1 & 0 & 0 \\ 0 & 0 & 0 & 0 \\ 0 & 0 & 0 & 0 \end{bmatrix}
\end{aligned} \tag{9.391}
$$

9.7 ★ Second Derivative and Coordinate Frames

The time derivative of a vector depends on the coordinate frame in which it is expressed and the frame in which we are taking the derivative. Consider a global frame $G\ (OXYZ)$ and a body frame $B\ (Oxyz)$. The second derivative of a vector follows the same rule of the first G and B-derivative of the G and B-vector. The derivative of a B-vector $^B\mathbf{v}$ in B and the derivative of a G-vector $^G\mathbf{v}$ in G are

$$^B\mathbf{v} = \dot{x}\hat{\imath} + \dot{y}\hat{\jmath} + \dot{z}\hat{k} \tag{9.392}$$

$$^G\mathbf{v} = \dot{X}\hat{I} + \dot{Y}\hat{J} + \dot{Z}\hat{K} \tag{9.393}$$

$$^B\mathbf{a} = \frac{^Bd^2}{dt^2}\,^B\mathbf{r} = \frac{^Bd}{dt}\,^B\mathbf{v} = \,^B\ddot{\mathbf{r}} = \ddot{x}\hat{\imath} + \ddot{y}\hat{\jmath} + \ddot{z}\hat{k} \tag{9.394}$$

$$^G\mathbf{a} = \frac{^Gd^2}{dt^2}\,^G\mathbf{r} = \frac{^Gd}{dt}\,^G\mathbf{v} = \,^G\ddot{\mathbf{r}} = \ddot{X}\hat{I} + \ddot{Y}\hat{J} + \ddot{Z}\hat{K} \tag{9.395}$$

We call $^G\mathbf{a}$ and $^B\mathbf{a}$ *simple accelerations* because they are *simple derivatives* of the *simple velocities* $^G\mathbf{v}$ and $^B\mathbf{v}$. We may also calculate the *mixed derivatives* and find the G-derivative of $^B_G\mathbf{v}$ and the B-derivative of $^G_B\mathbf{v}$. The velocity $^B_G\mathbf{v}$ is B-expression of G-velocity of a point, and $^G_B\mathbf{v}$ is the G-expression of B-velocity of a point.

The G-derivative of $^B_G\mathbf{v}$ is

$$^B_G\mathbf{a} = \frac{^Gd}{dt}\,^B_G\mathbf{v} = \,^B_G\boldsymbol{\alpha}_B \times \,^B\mathbf{r} + \,^B_G\boldsymbol{\omega}_B \times \left(^B_G\boldsymbol{\omega}_B \times \,^B\mathbf{r}\right) \tag{9.396}$$

and the B-derivative of $^G_B\mathbf{v}$ is

$$^G_B\mathbf{a} = \frac{^Bd}{dt}\,^G_B\mathbf{v} = -\,_G\boldsymbol{\alpha}_B \times \,^G\mathbf{r} + \,_G\boldsymbol{\omega}_B \times \left(_G\boldsymbol{\omega}_B \times \,^G\mathbf{r}\right) \tag{9.397}$$

We call $^B_G\mathbf{a}$ the B-expression of the G-acceleration and $^G_B\mathbf{a}$ the G-expression of the B-acceleration. The left superscript of $^B_G\mathbf{a}$ indicates the frame in which $\mathbf{a}$ is expressed, and the left subscript indicates the frame in which derivative is taken. If the left super and subscripts of an acceleration vector are the same, it is a simple acceleration vector and we only keep the superscript. To read the mixed accelerations $^B_G\mathbf{a}$ and $^G_B\mathbf{a}$, we may use the *G-acceleration of a B-velocity* and *B-acceleration of a G-velocity*, respectively.

When the interested body point P is not a fixed point in B, then P is moving in frame B with a variable body position $^B\mathbf{r}_P = \,^B\mathbf{r}_P\,(t)$ and velocity $^B\mathbf{v}_P = \,^B\mathbf{v}_P\,(t)$. The mixed derivatives of $\mathbf{v}\,(t)$ are defined by

$$\begin{aligned}
^B_G\mathbf{a} &= \frac{^Gd}{dt}\,^B_G\mathbf{v}\,(t) \\
&= \,^B\mathbf{a} + \,^B_G\boldsymbol{\alpha}_B \times \,^B\mathbf{r} + 2\,^B_G\boldsymbol{\omega}_B \\
&\quad \times \,^B\mathbf{v} + \,^B_G\boldsymbol{\omega}_B \times \left(^B_G\boldsymbol{\omega}_B \times \,^B\mathbf{r}\right)
\end{aligned} \tag{9.398}$$

$$\begin{aligned}
^G_B\mathbf{a} &= \frac{^Bd}{dt}\,^G_B\mathbf{v}\,(t) \\
&= \,^G\mathbf{a} - \,_G\boldsymbol{\alpha}_B \times \,^G\mathbf{r} - 2\,_G\boldsymbol{\omega}_B \\
&\quad \times \,^G\mathbf{v} + \,_G\boldsymbol{\omega}_B \times \left(_G\boldsymbol{\omega}_B \times \,^G\mathbf{r}\right)
\end{aligned} \tag{9.399}$$

Proof A vector is called a B-vector if it is expressed in the B-frame, and similarly it is a G-vector if it is expressed in the G-frame. If B represents a rigid body, then P is a body point and hence is fixed in B. For a body point, the body position vector $^B\mathbf{r}$ is constant in B and its body derivatives would be zero.

Having a rotating B-frame in a fixed global G-frame, we can define eight different accelerations as the second time derivatives of a position vector $\mathbf{r}$:

$$1.\quad \frac{^Gd}{dt}\frac{^Gd}{dt}\,^G\mathbf{r} = \frac{^Gd}{dt}\,^G\mathbf{v} = \,^G\mathbf{a} \tag{9.400}$$

$$2. \quad \frac{^{G}d}{dt}\frac{^{G}d}{dt}{}^{B}\mathbf{r} = \frac{^{G}d}{dt}{}^{B}_{G}\mathbf{v} = {}^{BB}_{GG}\mathbf{a} = {}^{B}_{G}\mathbf{a} \tag{9.401}$$

$$3. \quad \frac{^{G}d}{dt}\frac{^{B}d}{dt}{}^{G}\mathbf{r} = \frac{^{G}d}{dt}{}^{G}_{B}\mathbf{v} = {}^{GG}_{GB}\mathbf{a} \tag{9.402}$$

$$4. \quad \frac{^{G}d}{dt}\frac{^{B}d}{dt}{}^{B}\mathbf{r} = \frac{^{G}d}{dt}{}^{B}\mathbf{v} = {}^{BB}_{GB}\mathbf{a} \tag{9.403}$$

$$5. \quad \frac{^{B}d}{dt}\frac{^{G}d}{dt}{}^{G}\mathbf{r} = \frac{^{B}d}{dt}{}^{G}\mathbf{v} = {}^{GG}_{BG}\mathbf{a} \tag{9.404}$$

$$6. \quad \frac{^{B}d}{dt}\frac{^{G}d}{dt}{}^{B}\mathbf{r} = \frac{^{B}d}{dt}{}^{B}_{G}\mathbf{v} = {}^{BB}_{BG}\mathbf{a} \tag{9.405}$$

$$7. \quad \frac{^{B}d}{dt}\frac{^{B}d}{dt}{}^{G}\mathbf{r} = \frac{^{B}d}{dt}{}^{G}_{B}\mathbf{v} = {}^{GG}_{BB}\mathbf{a} = {}^{G}_{B}\mathbf{a} \tag{9.406}$$

$$8. \quad \frac{^{B}d}{dt}\frac{^{B}d}{dt}{}^{B}\mathbf{r} = \frac{^{B}d}{dt}{}^{B}\mathbf{v} = {}^{B}\mathbf{a} \tag{9.407}$$

Only the first and eighth second derivatives are simple accelerations. The two simple accelerations of (9.394) and (9.395), can be found by simple differentiation of a vector in the same frame in which they are expressed. Therefore, the G-derivative of the G-velocity and the B-derivative of the B-velocity provide the simple G and B-accelerations, respectively:

$$\frac{^{G}d^{2}}{dt^{2}}{}^{G}\mathbf{r} = \frac{^{G}d}{dt}{}^{G}\mathbf{v} = \frac{^{G}d}{dt}\frac{^{G}d}{dt}{}^{G}\mathbf{r} = {}^{G}\mathbf{a} = {}^{G}\ddot{\mathbf{r}}$$
$$= \ddot{X}\hat{I} + \ddot{Y}\hat{J} + \ddot{Z}\hat{K} \tag{9.408}$$

$$\frac{^{B}d^{2}}{dt^{2}}{}^{B}\mathbf{r} = \frac{^{B}d}{dt}{}^{B}\mathbf{v} = \frac{^{B}d}{dt}\frac{^{B}d}{dt}{}^{B}\mathbf{r} = {}^{B}\mathbf{a} = {}^{B}\ddot{\mathbf{r}}$$
$$= \ddot{x}\hat{\imath} + \ddot{y}\hat{\jmath} + \ddot{z}\hat{k} \tag{9.409}$$

Recalling the derivative transfer formula (7.214) and the mixed velocities,

$$\frac{^{G}d}{dt}{}^{B}\mathbf{r} = {}^{B}_{G}\mathbf{v} = {}^{B}_{G}\dot{\mathbf{r}} = \frac{^{B}d}{dt}{}^{B}\mathbf{r} + {}^{B}_{G}\boldsymbol{\omega}_{B} \times {}^{B}\mathbf{r}$$
$$= {}^{B}\mathbf{v} + {}^{B}_{G}\boldsymbol{\omega}_{B} \times {}^{B}\mathbf{r} \tag{9.410}$$

$$\frac{^{B}d}{dt}{}^{G}\mathbf{r} = {}^{G}_{B}\mathbf{v} = {}^{G}_{B}\dot{\mathbf{r}} = \frac{^{G}d}{dt}{}^{G}\mathbf{r}(t) - {}_{G}\boldsymbol{\omega}_{B} \times {}^{G}\mathbf{r}$$
$$= {}^{G}\mathbf{v} - {}_{G}\boldsymbol{\omega}_{B} \times {}^{G}\mathbf{r} \tag{9.411}$$

we can find the mixed accelerations in (9.401)–(9.406). The simple acceleration cases of 1 and 8, and the mixed acceleration cases of 2 and 7 are being used in robotic analysis.

The second case, ${}^{B}_{G}\mathbf{a} = {}^{BB}_{GG}\mathbf{a}$, is the B-expression of a G-acceleration. It happens when the position vector of a point is given in B while derivatives are taken in G.

$$
\begin{aligned}
{}^{B}_{G}\mathbf{a} = {}^{BB}_{GG}\mathbf{a} &= \frac{{}^{G}d}{dt} {}^{B}_{G}\mathbf{v} = \frac{{}^{G}d}{dt}\left({}^{B}\mathbf{v} + {}^{B}_{G}\boldsymbol{\omega}_B \times {}^{B}\mathbf{r} \right) \\
&= {}^{B}\mathbf{a} + {}^{B}_{G}\boldsymbol{\omega}_B \times {}^{B}\mathbf{v} + \left({}^{B}_{G}\boldsymbol{\alpha}_B + {}^{B}_{G}\boldsymbol{\omega}_B \times {}^{B}_{G}\boldsymbol{\omega}_B \right) \times {}^{B}\mathbf{r} \\
&\quad + {}^{B}_{G}\boldsymbol{\omega}_B \times \left({}^{B}\mathbf{v} + {}^{B}_{G}\boldsymbol{\omega}_B \times {}^{B}\mathbf{r} \right) \\
&= {}^{B}\mathbf{a} + {}^{B}_{G}\boldsymbol{\alpha}_B \times {}^{B}\mathbf{r} + 2\,{}^{B}_{G}\boldsymbol{\omega}_B \times {}^{B}\mathbf{v} \\
&\quad + {}^{B}_{G}\boldsymbol{\omega}_B \times \left({}^{B}_{G}\boldsymbol{\omega}_B \times {}^{B}\mathbf{r} \right)
\end{aligned}
\tag{9.412}
$$

The first term, ${}^{B}\mathbf{a}$, is the body acceleration of the moving point P in B, regardless of the rotation of B in G. So, ${}^{B}\mathbf{a}$ can be assumed as the acceleration of the moving point in B with respect to a body fixed point that is coincident with P at the moment. The second term, ${}^{B}_{G}\boldsymbol{\alpha}_B \times {}^{B}\mathbf{r}$, is the *tangential acceleration* of a body point that is coincident with P at the moment. The third term, $2\,{}^{B}_{G}\boldsymbol{\omega}_B \times {}^{B}\mathbf{v}$, is the *Coriolis acceleration* and is the result of a moving point in B while B is rotating in G. The fourth term, ${}^{B}_{G}\boldsymbol{\omega}_B \times \left({}^{B}_{G}\boldsymbol{\omega}_B \times {}^{B}\mathbf{r} \right)$, is the *centripetal acceleration* of a body point that is coincident with P at the moment.

The B-expression of the G-acceleration, ${}^{B}_{G}\mathbf{a}$, is the most applied and practical acceleration in the dynamics of robots and rigid bodies. In robotics, the point P is a fixed point in B and the acceleration ${}^{B}_{G}\mathbf{a}$ simplifies to

$$
{}^{B}_{G}\mathbf{a} = \frac{{}^{G}d}{dt} {}^{B}_{G}\mathbf{v} = {}^{B}_{G}\boldsymbol{\alpha}_B \times {}^{B}\mathbf{r} + {}^{B}_{G}\boldsymbol{\omega}_B \times \left({}^{B}_{G}\boldsymbol{\omega}_B \times {}^{B}\mathbf{r} \right)
\tag{9.413}
$$

A body point will have only a tangential and a centripetal acceleration. Using ${}^{B}_{G}\mathbf{a}$ is a practical method of acceleration analysis as we usually need to measure the kinematic information of a moving link in the body coordinate frame.

The third case, ${}^{GG}_{GB}\mathbf{a}$, is the G-expression of the G-derivative of ${}^{G}_{B}\mathbf{v}$ that is the B-derivative of the G-expression of a position vector.

$$
\begin{aligned}
{}^{GG}_{GB}\mathbf{a} &= \frac{{}^{G}d}{dt} {}^{G}_{B}\mathbf{v} = \frac{{}^{G}d}{dt}\frac{{}^{B}d}{dt} {}^{G}\mathbf{r} = \frac{{}^{G}d}{dt}\left({}^{G}\mathbf{v} - {}_{G}\boldsymbol{\omega}_B \times {}^{G}\mathbf{r} \right) \\
&= {}^{G}\mathbf{a} - {}_{G}\boldsymbol{\alpha}_B \times {}^{G}\mathbf{r} - {}_{G}\boldsymbol{\omega}_B \times {}^{G}\mathbf{v}
\end{aligned}
\tag{9.414}
$$

For a fixed point in B, we have ${}^{G}_{B}\mathbf{v} = 0$ and ${}^{GG}_{GB}\mathbf{a} = 0$ and Eq. (9.414) reduces to the G-expression of (9.413).

The fourth case, ${}^{BB}_{GB}\mathbf{a}$, is the B-expression of the G-derivative of ${}^{B}_{B}\mathbf{v}$ that is the B-derivative of the B-expression of a position vector.

$$
\begin{aligned}
{}^{BB}_{GB}\mathbf{a} &= \frac{{}^{G}d}{dt} {}^{B}\mathbf{v} = \frac{{}^{G}d}{dt}\frac{{}^{B}d}{dt} {}^{B}\mathbf{r} = \frac{{}^{G}d}{dt}\left({}^{B}\mathbf{v} \right) \\
&= {}^{B}\mathbf{a} + {}^{B}_{G}\boldsymbol{\omega}_B \times {}^{B}\mathbf{v}
\end{aligned}
\tag{9.415}
$$

The fifth case, ${}^{GG}_{BG}\mathbf{a}$, is the G-expression of the B-derivative of ${}^{G}_{G}\mathbf{v}$ that is the G-derivative of the G-expression of a position vector.

$$
\begin{aligned}
{}^{GG}_{BG}\mathbf{a} &= \frac{{}^{B}d}{dt} {}^{G}\mathbf{v} = \frac{{}^{B}d}{dt}\frac{{}^{G}d}{dt} {}^{G}\mathbf{r} = \frac{{}^{B}d}{dt}\left({}^{G}\mathbf{v} \right) \\
&= {}^{G}\mathbf{a} - {}_{G}\boldsymbol{\omega}_B \times {}^{G}\mathbf{v}
\end{aligned}
\tag{9.416}
$$

The sixth case, ${}^{BB}_{BG}\mathbf{a}$, is the B-expression of the B-derivative of ${}^{G}_{B}\mathbf{v}$ that is the B-derivative of the G-expression of a position vector:

$$
\begin{aligned}
{}^{BB}_{BG}\mathbf{a} &= \frac{{}^{B}d}{dt} {}^{B}_{G}\mathbf{v} = \frac{{}^{B}d}{dt}\frac{{}^{G}d}{dt} {}^{B}\mathbf{r} = \frac{{}^{B}d}{dt}\left({}^{B}\mathbf{v} + {}^{B}_{G}\boldsymbol{\omega}_B \times {}^{B}\mathbf{r} \right) \\
&= {}^{B}\mathbf{a} + {}^{B}_{G}\boldsymbol{\alpha}_B \times {}^{B}\mathbf{r} + {}^{B}_{G}\boldsymbol{\omega}_B \times {}^{B}\mathbf{v}
\end{aligned}
\tag{9.417}
$$

The seventh case, $_{BB}^{GG}\mathbf{a}$, is the G-expression of the B-derivative of $_B^G\mathbf{v}$ that is the B-derivative of the G-expression of a position vector:

$$
\begin{aligned}
_B^G\mathbf{a} = {}_{BB}^{GG}\mathbf{a} &= \frac{^Bd}{dt}\,{}_B^G\mathbf{v} = \frac{^Bd}{dt}\frac{^Bd}{dt}\,{}^G\mathbf{r} = \frac{^Bd}{dt}\left({}^G\mathbf{v} - {}_G\boldsymbol{\omega}_B \times {}^G\mathbf{r}\right)\\
&= {}^G\mathbf{a} - {}_G\boldsymbol{\omega}_B \times {}^G\mathbf{v} - \left({}_G\boldsymbol{\alpha}_B - {}_G\boldsymbol{\omega}_B \times {}_G\boldsymbol{\omega}_B\right) \times {}^G\mathbf{r}\\
&\quad - {}_G\boldsymbol{\omega}_B \times \left({}^G\mathbf{v} - {}_G\boldsymbol{\omega}_B \times {}^G\mathbf{r}\right)\\
&= {}^G\mathbf{a} - {}_G\boldsymbol{\alpha}_B \times {}^G\mathbf{r} - 2\,{}_G\boldsymbol{\omega}_B \times {}^G\mathbf{v}\\
&\quad + {}_G\boldsymbol{\omega}_B \times \left({}_G\boldsymbol{\omega}_B \times {}^G\mathbf{r}\right)
\end{aligned}
\tag{9.418}
$$

The seventh acceleration, $_B^G\mathbf{a}$, is the same as the second acceleration, $_G^B\mathbf{a}$, if we switch the name of the coordinate frames B and G. So, $_G^B\mathbf{a}$ is the G-acceleration of a moving point in B while the observer is in the B-frame, and $_B^G\mathbf{a}$ is the G-acceleration of a moving point in B when the observer is in G.

These accelerations for a point of a rigid link are

$$
^G\mathbf{a} = \frac{^Gd}{dt}\frac{^Gd}{dt}\,{}^G\mathbf{r} = \ddot{X}\hat{I} + \ddot{Y}\hat{J} + \ddot{Z}\hat{K}
\tag{9.419}
$$

$$
_G^B\mathbf{a} = \frac{^Gd}{dt}\frac{^Gd}{dt}\,{}^B\mathbf{r} = {}_G^B\boldsymbol{\alpha}_B \times {}^B\mathbf{r} + {}_G^B\boldsymbol{\omega}_B \times \left({}_G^B\boldsymbol{\omega}_B \times {}^B\mathbf{r}\right)
\tag{9.420}
$$

$$
_{GB}^{GG}\mathbf{a} = \frac{^Gd}{dt}\frac{^Bd}{dt}\,{}^G\mathbf{r} = {}^G\mathbf{a} - {}_G\boldsymbol{\alpha}_B \times {}^G\mathbf{r} - {}_G\boldsymbol{\omega}_B \times {}^G\mathbf{v}
\tag{9.421}
$$

$$
_{GB}^{BB}\mathbf{a} = \frac{^Gd}{dt}\frac{^Bd}{dt}\,{}^B\mathbf{r} = 0
\tag{9.422}
$$

$$
_{BG}^{GG}\mathbf{a} = \frac{^Bd}{dt}\frac{^Gd}{dt}\,{}^G\mathbf{r} = {}^G\mathbf{a} - {}_G\boldsymbol{\omega}_B \times {}^G\mathbf{v}
\tag{9.423}
$$

$$
_{BG}^{BB}\mathbf{a} = \frac{^Bd}{dt}\frac{^Gd}{dt}\,{}^B\mathbf{r} = {}_G^B\boldsymbol{\alpha}_B \times {}^B\mathbf{r}
\tag{9.424}
$$

$$
\begin{aligned}
_B^G\mathbf{a} &= \frac{^Bd}{dt}\frac{^Bd}{dt}\,{}^G\mathbf{r}\\
&= {}^G\mathbf{a} - {}_G\boldsymbol{\alpha}_B \times {}^G\mathbf{r} - 2\,{}_G\boldsymbol{\omega}_B \times {}^G\mathbf{v}\\
&\quad + {}_G\boldsymbol{\omega}_B \times \left({}_G\boldsymbol{\omega}_B \times {}^G\mathbf{r}\right)
\end{aligned}
\tag{9.425}
$$

$$
^B\mathbf{a} = \frac{^Bd}{dt}\frac{^Bd}{dt}\,{}^B\mathbf{r} = 0
\tag{9.426}
$$

Considering the Newton equation of motion, $\mathbf{F} = m\mathbf{a}$ for a particle of mass m. The equation of motion in the global coordinate frame G is

$$
^G\mathbf{F} = m\,{}^G\mathbf{a} = m\frac{^Gd}{dt}\frac{^Gd}{dt}\,{}^G\mathbf{r}
\tag{9.427}
$$

It means that both derivatives of acceleration must be taken in G. If the position vector $\mathbf{r}$ was expressed in the body coordinate frame B, then the Newton equation of motion would be

$$
^B\mathbf{F} = m\,{}_G^B\mathbf{a} = m\frac{^Gd}{dt}\frac{^Gd}{dt}\,{}^B\mathbf{r}
\tag{9.428}
$$

■

Example 324 Transformation method to find mixed derivatives. The mixed derivatives are easier to find if we calculate simple derivatives and then transform the result to the coordinate we need. Here is how.

It is more applied if we transform the position vector to the same frame in which we are taking the derivative and then apply the differential operator. Let us assume a point is moving in a body coordinate frame with the position vector $^B\mathbf{r}$,

$$
^B\mathbf{r} = \begin{bmatrix} \cos\alpha t \\ \sin\alpha t \\ 0 \end{bmatrix} \qquad \alpha = const.
\tag{9.429}
$$

while the body coordinate frame is turning about the X-axis with angular velocity and acceleration $\dot{\gamma}$ and $\ddot{\gamma}$.

$$
_G\boldsymbol{\omega}_B = \dot{\gamma}\,\hat{I} \qquad _G\boldsymbol{\alpha}_B = \ddot{\gamma}\,\hat{I}
\tag{9.430}
$$

To determine $_G^B\mathbf{v}$ and $_G^B\mathbf{a}$, we transform $^B\mathbf{r}$ to the G-frame.

$$
\begin{aligned}
^G\mathbf{r} &= {}^G R_B \, {}^B\mathbf{r} \\
&= \begin{bmatrix} 1 & 0 & 0 \\ 0 & \cos\gamma & -\sin\gamma \\ 0 & \sin\gamma & \cos\gamma \end{bmatrix} \begin{bmatrix} \cos\alpha t \\ \sin\alpha t \\ 0 \end{bmatrix} = \begin{bmatrix} \cos\alpha t \\ \sin\alpha t \cos\gamma \\ \sin\alpha t \sin\gamma \end{bmatrix}
\end{aligned}
\tag{9.431}
$$

Taking derivatives provides the global velocity and acceleration:

$$
^G\mathbf{v} = \frac{{}^G d}{dt}\,{}^G\mathbf{r} = \begin{bmatrix} -\alpha\sin\alpha t \\ \alpha\cos\alpha t\cos\gamma - \dot{\gamma}\sin\alpha t\sin\gamma \\ \alpha\cos\alpha t\sin\gamma + \dot{\gamma}\sin\alpha t\cos\gamma \end{bmatrix}
\tag{9.432}
$$

$$
\begin{aligned}
^G\mathbf{a} &= \frac{{}^G d}{dt}\,{}^G\mathbf{v} \\[4pt]
&= \begin{bmatrix} -\alpha^2\cos\alpha t \\ -\left(\alpha^2 + \dot{\gamma}^2\right)\sin\alpha t\cos\gamma - (\ddot{\gamma}\sin\alpha t + 2\alpha\dot{\gamma}\cos\alpha t)\sin\gamma \\ -\left(\alpha^2 + \dot{\gamma}^2\right)\sin\alpha t\sin\gamma + (\ddot{\gamma}\sin\alpha t + 2\alpha\dot{\gamma}\cos\alpha t)\cos\gamma \end{bmatrix}
\end{aligned}
\tag{9.433}
$$

Using the kinematic transformation matrix $^G R_B$, we are able to calculate $_G^B\mathbf{v}$ and $_G^B\mathbf{a}$:

$$
_G^B\mathbf{v} = {}^G R_B^T \, {}^G\mathbf{v} = \begin{bmatrix} -\alpha\sin\alpha t \\ \alpha\cos\alpha t \\ \dot{\gamma}\sin\alpha t \end{bmatrix}
\tag{9.434}
$$

$$
_G^B\mathbf{a} = {}^G R_B^T \, {}^G\mathbf{a} = \begin{bmatrix} -\alpha^2\cos\alpha t \\ -\left(\alpha^2 + \dot{\gamma}^2\right)\sin\alpha t \\ \ddot{\gamma}\sin\alpha t + 2\alpha\dot{\gamma}\cos\alpha t \end{bmatrix}
\tag{9.435}
$$

Example 325 Mixed velocity and simple acceleration of a moving point in B. This is to show how velocity and acceleration of point is calculated in different frames and expressed in different frames.

Consider a body frame $B(Oxyz)$ that is rotating in a global frame $G(OXYZ)$ with an angular velocity $\dot{\alpha}$ about the Z-axis, and a moving point P in B at $^B\mathbf{r}_P(t)$.

$$
^B\mathbf{r}_P(t) = t^2\,\hat{\imath}
\tag{9.436}
$$

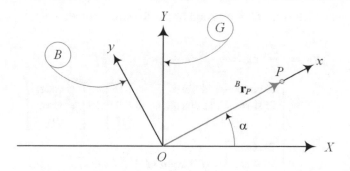

Fig. 9.10 Top view of a rotating body frame B in G with an angular velocity $\dot{\alpha}$ about the Z-axis and a moving point P in B

Figure 9.10 illustrates a top view of the system when B is at an angle α. The global position vector of the point is

$$
{}^G\mathbf{r}_P = {}^G_{\ }R_B \, {}^B\mathbf{r}_P = R_{Z,\alpha}(t) \, {}^B\mathbf{r}_P = \begin{bmatrix} \cos\alpha & -\sin\alpha & 0 \\ \sin\alpha & \cos\alpha & 0 \\ 0 & 0 & 1 \end{bmatrix} \begin{bmatrix} t^2 \\ 0 \\ 0 \end{bmatrix}
$$

$$
= t^2 \cos\alpha \, \hat{I} + t^2 \sin\alpha \, \hat{J} \tag{9.437}
$$

The angular velocity of ${}_G\boldsymbol{\omega}_B$ is

$$
{}_G\tilde{\omega}_B = {}^G\dot{R}_B \, {}^G R_B^T = \dot{\alpha}\tilde{K} \qquad {}_G\boldsymbol{\omega}_B = \dot{\alpha}\hat{K} \tag{9.438}
$$

It may also be verified that the body expression of the angular velocity is

$$
{}^B_G\tilde{\omega}_B = {}^G R_B^T \, {}_G\tilde{\omega}_B \, {}^G R_B = \dot{\alpha}\tilde{k} \qquad {}^B_G\boldsymbol{\omega}_B = \dot{\alpha}\hat{k} \tag{9.439}
$$

The simple velocities of P in both frames are

$$
{}^B\mathbf{v}_P = \frac{{}^B d}{dt} \, {}^B\mathbf{r}_P = {}^B\dot{\mathbf{r}}_P = 2t\hat{i} \tag{9.440}
$$

$$
{}^G\mathbf{v}_P = \frac{{}^G d}{dt} \, {}^G\mathbf{r}_P = {}^G\dot{\mathbf{r}}_P
$$

$$
= \left(2t\cos\alpha - t^2\dot{\alpha}\sin\alpha\right)\hat{I} + \left(2t\sin\alpha + t^2\dot{\alpha}\cos\alpha\right)\hat{J} \tag{9.441}
$$

For the mixed velocities we start with ${}^B_G\mathbf{v}$, which is the B-expression of the G-velocity of P.

$$
{}^B_G\mathbf{v}_P = \frac{{}^G d}{dt} \, {}^B\mathbf{r}_P = \frac{{}^B d}{dt} \, {}^B\mathbf{r}_P + {}^B_G\boldsymbol{\omega}_B \times {}^B\mathbf{r}_P
$$

$$
= \begin{bmatrix} 2t \\ 0 \\ 0 \end{bmatrix} + \dot{\alpha} \begin{bmatrix} 0 \\ 0 \\ 1 \end{bmatrix} \times \begin{bmatrix} t^2 \\ 0 \\ 0 \end{bmatrix} = \begin{bmatrix} 2t \\ t^2\dot{\alpha} \\ 0 \end{bmatrix}
$$

$$
= 2t\hat{i} + t^2\dot{\alpha}\hat{j} \tag{9.442}
$$

It can also be found by a coordinate frame transformation.

$$
{}^B_G\mathbf{v}_P = {}^G R_B^T \, {}^G\mathbf{v}_P = 2t\,\hat{i} + t^2\dot{\alpha}\,\hat{j} \tag{9.443}
$$

The next mixed velocity is ${}^{G}_{B}\mathbf{v}_P$, which is the G-expression of the B-velocity of P.

$$
\begin{aligned}
{}^{G}_{B}\mathbf{v}_P &= \frac{{}^{B}d}{dt}{}^{G}\mathbf{r}_P = \frac{{}^{G}d}{dt}{}^{G}\mathbf{r}_P - {}_{G}\boldsymbol{\omega}_B \times {}^{G}\mathbf{r}_P \\[2mm]
&= \begin{bmatrix} 2t\cos\alpha - t^2\dot{\alpha}\sin\alpha \\ 2t\sin\alpha + t^2\dot{\alpha}\cos\alpha \\ 0 \end{bmatrix} - \dot{\alpha}\begin{bmatrix} 0 \\ 0 \\ 1 \end{bmatrix} \times \begin{bmatrix} t^2\cos\alpha \\ t^2\sin\alpha \\ 0 \end{bmatrix} \\[2mm]
&= \begin{bmatrix} 2t\cos\alpha \\ 2t\sin\alpha \\ 0 \end{bmatrix} = (2t\cos\alpha)\,\hat{I} + (2t\sin\alpha)\,\hat{J}
\end{aligned}
\tag{9.444}
$$

To find this velocity, we can also apply a kinematic transformation ${}^{G}R_B$ to ${}^{B}\mathbf{v}_P$.

$$
{}^{G}_{B}\mathbf{v}_P = {}^{G}R_B\,{}^{B}\mathbf{v}_P = (2t\cos\alpha)\,\hat{I} + (2t\sin\alpha)\,\hat{J}
\tag{9.445}
$$

The simple accelerations of P are

$$
{}^{B}\mathbf{a}_P = \frac{{}^{B}d}{dt}{}^{B}\mathbf{v}_P = {}^{B}\ddot{\mathbf{r}}_P = 2\hat{\imath}
\tag{9.446}
$$

$$
\begin{aligned}
{}^{G}\mathbf{a}_P &= \frac{{}^{G}d}{dt}{}^{G}\mathbf{v}_P = {}^{G}\ddot{\mathbf{r}}_P \\[2mm]
&= \left(2\cos\alpha - 4t\dot{\alpha}\sin\alpha - t^2\ddot{\alpha}\sin\alpha - t^2\dot{\alpha}^2\cos\alpha\right)\hat{I} \\[2mm]
&\quad + \left(2\sin\alpha + 4t\dot{\alpha}\cos\alpha + t^2\ddot{\alpha}\cos\alpha - t^2\dot{\alpha}^2\sin\alpha\right)\hat{J}
\end{aligned}
\tag{9.447}
$$

These velocities and simple accelerations of the moving point P in B help us to determine the mixed accelerations of P. The acceleration ${}^{B}_{G}\mathbf{a} = {}^{BB}_{GG}\mathbf{a}$ is

$$
\begin{aligned}
{}^{B}_{G}\mathbf{a} &= {}^{B}\mathbf{a} + {}^{B}_{G}\boldsymbol{\alpha}_B \times {}^{B}\mathbf{r} + 2{}^{B}_{G}\boldsymbol{\omega}_B \times {}^{B}\mathbf{v} + {}^{B}_{G}\boldsymbol{\omega}_B \times \left({}^{B}_{G}\boldsymbol{\omega}_B \times {}^{B}\mathbf{r}\right) \\[2mm]
&= \begin{bmatrix} 2 - t^2\dot{\alpha}^2 \\ t\left(4\dot{\alpha} + t\ddot{\alpha}\right) \\ 0 \end{bmatrix}
\end{aligned}
\tag{9.448}
$$

We must also be able to determine the mixed accelerations ${}^{B}_{G}\mathbf{a}$ by a kinematic transformation:

$$
\begin{aligned}
{}^{B}_{G}\mathbf{a} &= {}^{G}R_B^{T}\,{}^{G}\mathbf{a} \\[2mm]
&= \begin{bmatrix} \cos\alpha & -\sin\alpha & 0 \\ \sin\alpha & \cos\alpha & 0 \\ 0 & 0 & 1 \end{bmatrix}^{T} \begin{bmatrix} 2t\cos\alpha - t^2\dot{\alpha}\sin\alpha \\ 2t\sin\alpha + t^2\dot{\alpha}\cos\alpha \\ 0 \end{bmatrix} \\[2mm]
&= \begin{bmatrix} 2 - t^2\dot{\alpha}^2 \\ t\left(4\dot{\alpha} + t\ddot{\alpha}\right) \\ 0 \end{bmatrix}
\end{aligned}
\tag{9.449}
$$

The acceleration $_{GB}^{GG}\mathbf{a}$ is

$$
\begin{aligned}
_{GB}^{GG}\mathbf{a} &= {}^G\mathbf{a} - {}_G\boldsymbol{\alpha}_B \times {}^G\mathbf{r} - {}_G\boldsymbol{\omega}_B \times {}^G\mathbf{v} \\
&= \begin{bmatrix} 2\cos\alpha - 2t\dot\alpha\sin\alpha \\ 2\sin\alpha + 2t\dot\alpha\cos\alpha \\ 0 \end{bmatrix}
\end{aligned}
\tag{9.450}
$$

the acceleration $_{GB}^{BB}\mathbf{a}$ is

$$
_{GB}^{BB}\mathbf{a} = {}^B\mathbf{a} + {}_G^B\boldsymbol{\omega}_B \times {}^B\mathbf{v} = \begin{bmatrix} 2 \\ 2t\dot\alpha \\ 0 \end{bmatrix}
\tag{9.451}
$$

the acceleration $_{BG}^{GG}\mathbf{a}$ is

$$
_{BG}^{GG}\mathbf{a} = {}^G\mathbf{a} - {}_G\boldsymbol{\omega}_B \times {}^G\mathbf{v} = \begin{bmatrix} 2\cos\alpha - (\ddot\alpha t + 2\dot\alpha)\,t\sin\alpha \\ 2\sin\alpha + (\ddot\alpha t + 2\dot\alpha)\,t\cos\alpha \\ 0 \end{bmatrix}
\tag{9.452}
$$

the acceleration $_{BG}^{BB}\mathbf{a}$ is

$$
_{BG}^{BB}\mathbf{a} = {}^B\mathbf{a} + {}_G^B\boldsymbol{\alpha}_B \times {}^B\mathbf{r} + {}_G^B\boldsymbol{\omega}_B \times {}^B\mathbf{v} = \begin{bmatrix} 2 \\ (\ddot\alpha t + 2\dot\alpha)\,t \\ 0 \end{bmatrix}
\tag{9.453}
$$

and the acceleration $_{BG}^{BB}\mathbf{a}$ is

$$
\begin{aligned}
_B^G\mathbf{a} &= {}_{BB}^{GG}\mathbf{a} \\
&= {}^G\mathbf{a} - {}_G\boldsymbol{\alpha}_B \times {}^G\mathbf{r} - 2\,{}_G\boldsymbol{\omega}_B \times {}^G\mathbf{v} + {}_G\boldsymbol{\omega}_B \times \left({}_G\boldsymbol{\omega}_B \times {}^G\mathbf{r}\right) \\
&= \begin{bmatrix} 2\cos\alpha \\ 2\sin\alpha \\ 0 \end{bmatrix}
\end{aligned}
\tag{9.454}
$$

We can also determine the mixed accelerations $_B^G\mathbf{a}$ by a kinematic transformation.

$$
\begin{aligned}
_B^G\mathbf{a} &= {}^GR_B\,{}^B\mathbf{a} \\
&= \begin{bmatrix} \cos\alpha & -\sin\alpha & 0 \\ \sin\alpha & \cos\alpha & 0 \\ 0 & 0 & 1 \end{bmatrix} \begin{bmatrix} 2 \\ 0 \\ 0 \end{bmatrix} = \begin{bmatrix} 2\cos\alpha \\ 2\sin\alpha \\ 0 \end{bmatrix}
\end{aligned}
\tag{9.455}
$$

Example 326 ★ Second-derivative transformation formula. When a position vector is given in the B-frame and its acceleration is needed to be used in Newton equation of motion, two G-derivatives need to be taken. The result is the acceleration of the global acceleration of the point, expressed in B-frame. Such derivative is needed to express equation of motion of moving objects such as vehicle, ship, airplane, etc.

Consider a point P that can move in the body coordinate frame $B\,(Oxyz)$. Using the derivative transformation formula (7.214) we find the B-expression of the G-velocity.

$$
\frac{^Gd}{dt}\,{}^B\mathbf{r}_P = {}_G^B\mathbf{v}_P = \frac{^Bd}{dt}\,{}^B\mathbf{r}_P + {}_G^B\boldsymbol{\omega}_B \times {}^B\mathbf{r}_P
\tag{9.456}
$$

Another G-derivative of this equation provides the B-expression for the global acceleration of P as Eq. (9.412).

$$
_G^B\mathbf{a} = {}^B\mathbf{a} + {}_G^B\boldsymbol{\alpha}_B \times {}^B\mathbf{r} + 2\,{}_G^B\boldsymbol{\omega}_B \times {}^B\mathbf{v} + {}_G^B\boldsymbol{\omega}_B \times \left({}_G^B\boldsymbol{\omega}_B \times {}^B\mathbf{r}\right)
\tag{9.457}
$$

Using this result, we can define the *second-derivative transformation formula* of a B-vector $^B\square$ from the body to the global coordinate frame:

$$\frac{^Gd}{dt}\frac{^Gd}{dt}\,^B\square = {}^B_G\ddot{\square}$$

$$= \frac{^Bd}{dt}\frac{^Bd}{dt}\,^B\square + {}^B_G\boldsymbol{\alpha}_B \times {}^B\square$$

$$+ 2\,^B_G\boldsymbol{\omega}_B \times \left(\frac{^Bd}{dt}\,^B\square + {}^B_G\boldsymbol{\omega}_B \times {}^B\square\right) \tag{9.458}$$

The final result $^B_G\ddot{\square}$ shows the second global time derivative expressed in the body frame, or simply the B-expression of the second G-derivative of a B-vector $^B\square$. The vector $^B\square$ may be any vector quantity such as position, velocity, angular velocity, momentum, angular momentum, a time varying force vector, etc.

Example 327 ★ *Rāzī acceleration* $\mathbf{a}_{Ra}$. If an A-frame is moving in a B-frame, while B-frame is moving in C-frame, we may have the first derivative of an A-vector in the B-frame and the second derivative in the C-frame. This would be a mixed second derivative, which was the base to discover a new acceleration term called Razi acceleration.

Consider three relatively rotating frames A, B, C. The mixed double acceleration $^{AA}_{CB}\mathbf{a}$ is the A-expression of the second derivative of a position vector $^A\mathbf{r}$ when the first derivative is taken in the B-frame and the second derivative is taken in the C-frame.

$$^{AA}_{CB}\mathbf{a} = {}^A\mathbf{a} + {}^A_C\boldsymbol{\omega}_A \times {}^A\mathbf{v} + {}^A_B\boldsymbol{\alpha}_A \times {}^A\mathbf{r} + \left(^A_C\boldsymbol{\omega}_A \times {}^A_B\boldsymbol{\omega}_A\right) \times {}^A\mathbf{r}$$

$$+ {}^A_B\boldsymbol{\omega}_A \times {}^A\mathbf{v} + {}^A_B\boldsymbol{\omega}_A \times \left(^A_C\boldsymbol{\omega}_A \times {}^A\mathbf{r}\right) \tag{9.459}$$

The mixed double derivative is made when we take the first and second derivatives of $^A\mathbf{r}$ in different coordinate frames B and C.

Using the derivative transformation formula, the B-derivative of $^A\mathbf{r}$ is

$$\frac{^Bd}{dt}\,^A\mathbf{r} = {}^A_B\mathbf{v} = \frac{^Ad}{dt}\,^A\mathbf{r} + {}^A_B\boldsymbol{\omega}_A \times {}^A\mathbf{r} \tag{9.460}$$

A time derivative of this equation in a third frame C would be

$$\frac{^Cd}{dt}\frac{^Bd}{dt}\,^A\mathbf{r} = \frac{^Cd}{dt}\,^A_B\mathbf{v} = {}^{AA}_{CB}\mathbf{a} = \frac{^Cd}{dt}\left(\frac{^Ad}{dt}\,^A\mathbf{r}\right) + \frac{^Cd}{dt}\left(^A_B\boldsymbol{\omega}_A \times {}^A\mathbf{r}\right)$$

$$= \frac{^Ad}{dt}\frac{^Ad}{dt}\,^A\mathbf{r} + {}^A_C\boldsymbol{\omega}_A \times \frac{^Ad}{dt}\,^A\mathbf{r}$$

$$+ \left(\frac{^Ad}{dt}\,^A_B\boldsymbol{\omega}_A + {}^A_C\boldsymbol{\omega}_A \times {}^A_B\boldsymbol{\omega}_A\right) \times {}^A\mathbf{r}$$

$$+ {}^A_B\boldsymbol{\omega}_A \times \left(\frac{^Ad}{dt}\,^A\mathbf{r} + {}^A_C\boldsymbol{\omega}_A \times {}^A\mathbf{r}\right)$$

$$= {}^A\mathbf{a} + {}^A_C\boldsymbol{\omega}_A \times {}^A\mathbf{v} + {}^A_B\boldsymbol{\alpha}_A \times {}^A\mathbf{r} + \left(^A_C\boldsymbol{\omega}_A \times {}^A_B\boldsymbol{\omega}_A\right) \times {}^A\mathbf{r}$$

$$+ {}^A_B\boldsymbol{\omega}_A \times {}^A\mathbf{v} + {}^A_B\boldsymbol{\omega}_A \times \left(^A_C\boldsymbol{\omega}_A \times {}^A\mathbf{r}\right) \tag{9.461}$$

We call the acceleration $^{AA}_{CB}\mathbf{a}$ the *mixed double acceleration*. The first term $^A\mathbf{a}$ is the local A-acceleration of a moving point P in the A-frame. The combined terms $^A_C\boldsymbol{\omega}_A \times {}^A\mathbf{v} + {}^A_B\boldsymbol{\omega}_A \times {}^A\mathbf{v}$ is the *mixed Coriolis acceleration*.

$$\mathbf{a}_{Co} = {}^{AA}_{CB}\mathbf{a}_{Ra} = {}^A_C\boldsymbol{\omega}_A \times {}^A\mathbf{v} + {}^A_B\boldsymbol{\omega}_A \times {}^A\mathbf{v} \tag{9.462}$$

The term $_B^A\boldsymbol{\alpha}_A \times {}^A\mathbf{r}$ is the *tangential acceleration* of P. The term $_B^A\boldsymbol{\omega}_A \times \left(_C^A\boldsymbol{\omega}_A \times {}^A\mathbf{r}\right)$ is the mixed centripetal acceleration. The term $\left(_C^A\boldsymbol{\omega}_A \times {}_B^A\boldsymbol{\omega}_A\right) \times {}^A\mathbf{r}$ is a new term in the acceleration of P that cannot be seen in the simple acceleration $_G^B\mathbf{a}$. This term is called the *Rāzī acceleration* $\mathbf{a}_{Ra}$.

$$\mathbf{a}_{Ra} = {}_{CB}^{AA}\mathbf{a}_{Ra} = \left(_C^A\boldsymbol{\omega}_A \times {}_B^A\boldsymbol{\omega}_A\right) \times {}^A\mathbf{r} \tag{9.463}$$

The difference between the two terms in the Coriolis acceleration is clearer when we consider three coordinate frames B_1, B_2, and G. The frame B_2 is turning in B_1, and B_1 is turning in G. There is also a moving point P in a coordinate frame B_2. The B_2-expression of the velocity of P in B_1 is

$$_1^2\mathbf{v} = \frac{^1d}{dt}{}^2\mathbf{r} = \frac{^2d}{dt}{}^2\mathbf{r} + {}_1^2\boldsymbol{\omega}_2 \times {}^2\mathbf{r} = {}^2\mathbf{v} + {}_1^2\boldsymbol{\omega}_2 \times {}^2\mathbf{r} \tag{9.464}$$

The derivative of $_1^2\mathbf{v}$ in G provides a double mixed acceleration $_{G1}^{22}\mathbf{a}$,

$$\frac{^Gd}{dt}\frac{^1d}{dt}{}^2\mathbf{r} = {}_{G1}^{22}\mathbf{a} = \frac{^Gd}{dt}\left(\frac{^2d}{dt}{}^2\mathbf{r}\right) + \frac{^Gd}{dt}\left(_1^2\boldsymbol{\omega}_2 \times {}^2\mathbf{r}\right)$$

$$= \frac{^2d}{dt}\frac{^2d}{dt}{}^2\mathbf{r} + {}_G^2\boldsymbol{\omega}_2 \times \frac{^2d}{dt}{}^2\mathbf{r}$$

$$+ \left(\frac{^2d}{dt}{}_1^2\boldsymbol{\omega}_2 + {}_G^2\boldsymbol{\omega}_2 \times {}_1^2\boldsymbol{\omega}_2\right) \times {}^2\mathbf{r}$$

$$+ {}_1^2\boldsymbol{\omega}_2 \times \left(\frac{^2d}{dt}{}^2\mathbf{r} + {}_G^2\boldsymbol{\omega}_2 \times {}^2\mathbf{r}\right) \tag{9.465}$$

that can be simplified to

$$\frac{^Gd}{dt}\frac{^1d}{dt}{}^2\mathbf{r} = {}^2\mathbf{a} + {}_G^2\boldsymbol{\omega}_2 \times {}^2\mathbf{v} + {}_1^2\boldsymbol{\alpha}_2 \times {}^2\mathbf{r} + \left(_G^2\boldsymbol{\omega}_2 \times {}_1^2\boldsymbol{\omega}_2\right)$$

$$\times {}^2\mathbf{r} + {}_1^2\boldsymbol{\omega}_2 \times {}^2\mathbf{v} + {}_1^2\boldsymbol{\omega}_2 \times \left(_G^2\boldsymbol{\omega}_2 \times {}^2\mathbf{r}\right) \tag{9.466}$$

which is equivalent to (9.459) and indicates that, if $B_1 \neq G$, then $_G^2\boldsymbol{\omega}_2 \times {}^2\mathbf{v} \neq {}_1^2\boldsymbol{\omega}_2 \times {}^2\mathbf{v}$, and there exists a mixed Coriolis acceleration:

$$\mathbf{a}_{Co} = {}_{G1}^{22}\mathbf{a}_{Co} = {}_1^2\boldsymbol{\omega}_2 \times {}^2\mathbf{v} + {}_G^2\boldsymbol{\omega}_2 \times {}^2\mathbf{v} \tag{9.467}$$

Furthermore, when $B_1 \neq G$, there exists a Rāzī acceleration term $\mathbf{a}_{Ra}$ in the acceleration of P that cannot be seen in $_G^B\mathbf{a}$:

$$\mathbf{a}_{Ra} = {}_{G1}^{22}\mathbf{a}_{Ra} = \left(_G^2\boldsymbol{\omega}_2 \times {}_1^2\boldsymbol{\omega}_2\right) \times {}^2\mathbf{r} \tag{9.468}$$

Zakariyā Rāzī (Rhazes, 865 − 925) was a Persian mathematician, chemist, and physician who is considered the father of pediatrics.

9.8 Summary

When a body coordinate frame B and a global frame G have a common origin, the global acceleration of a point P in frame B is

$$^G\ddot{\mathbf{r}} = \frac{^Gd}{dt}{}^G\mathbf{v}_P = {}_G\boldsymbol{\alpha}_B \times {}^G\mathbf{r} + {}_G\boldsymbol{\omega}_B \times \left(_G\boldsymbol{\omega}_B \times {}^G\mathbf{r}\right) \tag{9.469}$$

where, $_G\boldsymbol{\alpha}_B$ is the angular acceleration of B with respect to G

$$_G\boldsymbol{\alpha}_B = \frac{^Gd}{dt}{}_G\boldsymbol{\omega}_B. \tag{9.470}$$

However, when the body coordinate frame B has a rigid motion with respect to G, then

$$
{}^G\mathbf{a}_P = \frac{{}^G d}{dt} {}^G\mathbf{v}_P = {}_G\boldsymbol{\alpha}_B \times \left({}^G\mathbf{r}_P - {}^G\mathbf{d}_B\right)
$$
$$
+ {}_G\boldsymbol{\omega}_B \times \left({}_G\boldsymbol{\omega}_B \times \left({}^G\mathbf{r}_P - {}^G\mathbf{d}_B\right)\right) + {}^G\ddot{\mathbf{d}}_B \tag{9.471}
$$

where ${}^G\mathbf{d}_B$ indicates the position of the origin of B with respect to the origin of G.

Angular accelerations of two links are related according to

$$
{}_0\boldsymbol{\alpha}_2 = {}_0\boldsymbol{\alpha}_1 + {}_1^0\boldsymbol{\alpha}_2. \tag{9.472}
$$

The acceleration relationship for a body B having a rigid motion in G may also be expressed by a homogenous acceleration transformation matrix ${}^G A_B$

$$
{}^G\mathbf{a}_P(t) = {}^G A_B \, {}^G\mathbf{r}_P(t) \tag{9.473}
$$

where,

$$
{}^G A_B = {}^G\ddot{T}_B \, {}^G T_B^{-1}
$$
$$
= \begin{bmatrix} {}_G\tilde{\alpha}_B - \tilde{\omega}\,\tilde{\omega}^T & {}^G\ddot{\mathbf{d}}_B - \left({}_G\tilde{\alpha}_B - \tilde{\omega}\,\tilde{\omega}^T\right) {}^G\mathbf{d}_B \\ 0 & 0 \end{bmatrix}. \tag{9.474}
$$

The forward acceleration kinematics of a robot is defined by

$$
\ddot{\mathbf{X}} = \mathbf{J}\,\ddot{\mathbf{q}} + \dot{\mathbf{J}}\,\dot{\mathbf{q}} \tag{9.475}
$$

that is a relationship between joint coordinate acceleration

$$
\ddot{\mathbf{q}} = \begin{bmatrix} \ddot{q}_1 & \ddot{q}_2 & \ddot{q}_3 & \cdots & \ddot{q}_n \end{bmatrix}^T \tag{9.476}
$$

and global acceleration of the end-effector

$$
\ddot{\mathbf{X}} = \begin{bmatrix} \ddot{X}_n & \ddot{Y}_n & \ddot{Z}_n & \dot{\omega}_{Xn} & \dot{\omega}_{Yn} & \dot{\omega}_{Zn} \end{bmatrix}^T. \tag{9.477}
$$

Such a relationship introduces the time derivative of Jacobian matrix $\begin{bmatrix} \dot{\mathbf{J}} \end{bmatrix}$.

Having the forward acceleration kinematics of a robot as (9.475), we can determine the inverse acceleration kinematics of the robot algebraically and determine the joint accelerations for a given set of end-effector acceleration, speed, and configuration.

$$
\ddot{\mathbf{q}} = \mathbf{J}^{-1}\left(\ddot{\mathbf{X}} - \dot{\mathbf{J}}\,\dot{\mathbf{q}}\right) \tag{9.478}
$$

9.9   Key Symbols

$\mathbf{a}$	Acceleration vector
A	Acceleration transformation matrix
B	Body coordinate frame
c	cos
C	Mass center
d_x, d_y, d_z	Elements of $\mathbf{d}$
$\mathbf{d}$	Translation vector, displacement vector
D	Displacement transformation matrix
e	Rotation quaternion
G, B_0	Global coordinate frame, Base coordinate frame
$\hat{\imath}, \hat{\jmath}, \hat{k}$	Local coordinate axes unit vectors
$\tilde{\imath}, \tilde{\jmath}, \tilde{k}$	Skew symmetric matrices of the unit vector $\hat{\imath}, \hat{\jmath}, \hat{k}$
$\hat{I}, \hat{J}, \hat{K}$	Global coordinate axes unit vectors
$\mathbf{I} = [\mathbf{I}]$	Identity matrix
$\mathbf{j}$	Jerk vector
j_{ij}	The element of row i and column j of $\tilde{\chi}$
$\mathbf{J} = [J]$	Jacobian
J_{ij}	The element of row i and column j of $[J]$
l	Length
m	Number of independent equations
$\mathbf{n}_i$	Position vector of o_{i-1} with respect to C_i
$\mathbf{m}_i$	Position vector of o_i with respect to C_i
$\mathbf{p}, \mathbf{q}$	Position vectors of P, Q
$\mathbf{q}$	Joint variable vector
P, Q	Points
$\mathbf{r}$	Position vectors, homogenous position vector
r_i	The element i of $\mathbf{r}$
r_{ij}	The element of row i and column j of a matrix
R	Rotation transformation matrix, radius
s	sin
S	Rotational acceleration transformation
T	Homogenous transformation matrix
$\hat{u}$	Unit vector along the axis of $\boldsymbol{\omega}$
$\tilde{u}$	Skew symmetric matrix of the vector $\hat{u}$
u_1, u_2, u_3	Components of $\hat{u}$
U	Rotational jerk transformation
v	Velocity
$\mathbf{v}$	Velocity vector
V	Velocity transformation matrix
x, y, z	Local coordinate axes
X, Y, Z	Global coordinate axes

Greek

α	Angular acceleration
$\boldsymbol{\alpha}$	Angular acceleration vector
$\tilde{\alpha} = \dot{\tilde{\omega}}$	Angular acceleration matrix
$\alpha_1, \alpha_2, \alpha_3$	Components of $\boldsymbol{\alpha}$
θ	Rotary joint angle
θ_{ijk}	$\theta_i + \theta_j + \theta_k\, \varphi, \theta, \psi]$ Euler angles
ϕ	Angle of rotation about $\hat{u}$

ω	Angular velocity
$\boldsymbol{\omega}$	Angular velocity vector
$\tilde{\omega}$	Skew symmetric matrix of the vector $\boldsymbol{\omega}$
$\omega_1, \omega_2, \omega_3$	Components of $\boldsymbol{\omega}$
$\tilde{\chi} = \dot{\tilde{\alpha}}$	Angular jerk matrix
$\boldsymbol{\chi} = \dot{\boldsymbol{\alpha}}$	Angular jerk vector

Symbol

$[\ \ ]^{-1}$	Inverse of the matrix $[\ \ ]$
$[\ \ ]^{T}$	TRANSPOSE of the matrix $[\ \ ]$
$\equiv$	Equivalent
$\vdash$	Orthogonal
(i)	Link number i
$\parallel$	Parallel sign
$\perp$	Perpendicular
$\times$	Vector cross product
$\overleftrightarrow{e}$	Matrix form of a quaternion e
E	Earth
lim	Limit function
sgn	Signum function

Exercises

1. Notation and symbols.

    Describe the meaning of the following notations:

    $$a - {}_G\alpha_B \quad b - {}_B\alpha_G \quad c - {}^G_G\alpha_B \quad d - {}^B_G\alpha_B \quad e - {}^B_B\alpha_G \quad f - {}^G_B\alpha_G$$

    $$g - {}^0_2\alpha_1 \quad h - {}^0_1\alpha_2 \quad i - {}^1_2\alpha_1 \quad j - {}^2_2\alpha_1 \quad k - {}^3_2\alpha_1 \quad l - {}^k_j\alpha_i$$

    $$m - {}^0_2\tilde{\alpha}_1 \quad n - {}^2_1\dot{\tilde{\omega}}_2 \quad o - {}^G A_B \quad p - {}^G\dot{V}_B \quad q - \dot{\mathbf{j}} \quad r - \ddot{\mathbf{X}}$$

2. Local position, global acceleration.

    A body is turning about a global principal axis at a constant angular acceleration of $2\,\text{rad/s}^2$. Find the global velocity and acceleration of a point P at ${}^B\mathbf{r}$,

    $$ {}^B\mathbf{r} = \begin{bmatrix} 5 & 30 & 10 \end{bmatrix}^T \tag{9.479}$$

    (a) If the axis is the Z-axis, the angular velocity is $2\,\text{rad/s}$, and the angle of rotation is $\pi/3\,\text{rad}$.
    (b) If the axis is the X-axis, the angular velocity is $1\,\text{rad/s}$, and the angle of rotation is $\pi/4\,\text{rad}$.
    (c) If the axis is the Y-axis, the angular velocity is $3\,\text{rad/s}$, and the angle of rotation is $\pi/6\,\text{rad}$.

3. Global position, constant angular acceleration.

    A body is turning about the Z-axis at a constant angular acceleration $\ddot{\alpha} = 0.2\,\text{rad/s}^2$. Find the global position of a point at ${}^B\mathbf{r}$,

    $$ {}^B\mathbf{r} = \begin{bmatrix} 5 & 30 & 10 \end{bmatrix}^T \tag{9.480}$$

    after $t = 3\,\text{s}$ when $\dot{\alpha} = 2\,\text{rad/s}$, if the body and global coordinate frames were coincident at $t = 0\,\text{s}$.

4. ★ Angular velocity and acceleration matrices.

    A body B is turning continuously in the global frame G. The transformation matrix can be simulated by a rotation α about Z-axis followed by a rotation β about X-axis.
    (a) Determine the axis and angle of rotation $\hat{u}$ and ϕ as functions of α and β.
    (b) Determine if α and β are changing with constant rate, while $\hat{u}$ and ϕ also change. Determine the conditions that keep $\hat{u}$ constant and the conditions that keep ϕ constant.
    (c) Assume $\hat{u}$ is constant and $\dot{\phi} = 5\,\text{rad/s}$. Determine $\dot{\alpha}$ and $\dot{\beta}$ when $\phi = 30\,\text{deg}$.
    (d) Determine $\dot{\phi}$ when $\alpha = 30\,\text{deg}$, $\beta = 45\,\text{deg}$, $\dot{\alpha} = 5\,\text{rad/s}$, and $\dot{\beta} = 5\,\text{rad/s}$.

5. Turning about x-axis.

    Find the angular acceleration matrix when the body coordinate frame is turning $-5\,\text{deg/s}^2$, $35\,\text{deg/s}$ at $45\,\text{deg}$ about the x-axis.

6. A roller in a circle.

    Figure 9.11 shows a roller in a circular path. Find the velocity and acceleration of a point P on the circumference of the roller.

7. Angular acceleration and Euler angles.

    Calculate the angular velocity and acceleration vectors in body and global coordinate frames if the Euler angles and their derivatives are as below.

    $$\begin{array}{lll} \varphi = 0.25\,\text{rad} & \dot{\varphi} = 2.5\,\text{rad/s} & \ddot{\varphi} = 25\,\text{rad/s}^2 \\ \theta = -0.25\,\text{rad} & \dot{\theta} = -4.5\,\text{rad/s} & \ddot{\theta} = 35\,\text{rad/s}^2 \\ \psi = 0.5\,\text{rad} & \dot{\psi} = 3\,\text{rad/s} & \ddot{\psi} = 25\,\text{rad/s}^2 \end{array} \tag{9.481}$$

8. Angular acceleration by Euler angles.

    Employing the Euler angles transformation matrix,
    (a) Determine the linear relations between the Cartesian angular velocity ${}_G\boldsymbol{\omega}_B$ and $\dot{\varphi}, \dot{\theta}, \dot{\psi}$.
    (b) Determine the linear relations between the Cartesian angular acceleration ${}_G\boldsymbol{\alpha}_B$ and $\ddot{\varphi}, \ddot{\theta}, \ddot{\psi}$.
    (c) ★ Determine the linear relations between the Cartesian angular jerk ${}_G\mathbf{j}_B$ and $\dddot{\varphi}, \dddot{\theta}, \dddot{\psi}$.

9. Combined rotation and angular acceleration.

    Find the rotation matrix for a body frame after 30 deg rotation about the Z-axis, followed by 30 deg about the X-axis, and then 90 deg about the Y-axis. Then calculate the angular velocity of the body if it is turning with $\dot{\alpha} = 20\,\text{deg/s}$,

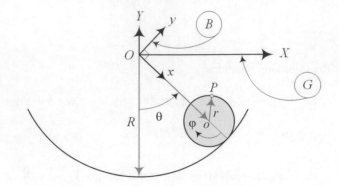

Fig. 9.11 A roller in a circular path

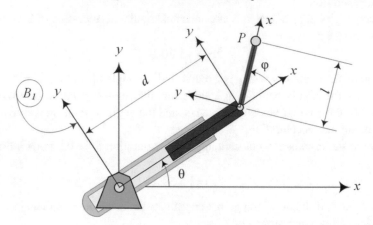

Fig. 9.12 A planar RPR manipulator

$\dot{\beta} = -40$ deg /s, and $\dot{\gamma} = 55$ deg /s about the Z, Y, and X axes, respectively. Finally, calculate the angular acceleration of the body if it is turning with $\ddot{\alpha} = 2$ deg /s^2, $\ddot{\beta} = 4$ deg /s^2, and $\ddot{\gamma} = -6$ deg /s^2 about the Z, Y, and X axes.

10. An RPR manipulator.

Label the coordinate frames and find the velocity and acceleration of point P at the endpoint of the manipulator shown in Fig. 9.12.

11. ★ Differentiation and coordinate frame.

How can we define these derivatives?

$$\frac{{}^G d}{dt}\frac{{}^G d}{dt}{}^G\mathbf{r} \qquad \frac{{}^G d}{dt}\frac{{}^G d}{dt}{}^B\mathbf{r} \qquad \frac{{}^G d}{dt}\frac{{}^B d}{dt}{}^G\mathbf{r}$$

$$\frac{{}^B d}{dt}\frac{{}^B d}{dt}{}^B\mathbf{r} \qquad \frac{{}^B d}{dt}\frac{{}^B d}{dt}{}^G\mathbf{r}\frac{{}^B d}{dt}\frac{{}^G d}{dt}{}^G\mathbf{r} \qquad\qquad (9.482)$$

$$\frac{{}^G d}{dt}\frac{{}^B d}{dt}{}^B\mathbf{r} \qquad \frac{{}^B d}{dt}\frac{{}^G d}{dt}{}^B\mathbf{r}$$

12. ★ Third derivative and coordinate frames.

Consider a global frame G $(OXYZ)$, a body frame B $(Oxyz)$, and a body point P that is moving in the frame B with a variable body position ${}^B\mathbf{r}_P = {}^B\mathbf{r}_P(t)$ velocity, ${}^B\mathbf{v}_P = {}^B\mathbf{v}_P(t)$, and acceleration ${}^B\mathbf{a}_P = {}^B\mathbf{a}_P(t)$. Determine how many possible simple and mixed jerks we can define.

13. A RRP planar redundant manipulator.

Figure 9.13 illustrates a 3 DOF planar manipulator with joint variables θ_1, θ_2, d_3.

(a) Solve the forward kinematics of the manipulator and calculate the position and orientation of the end-effector $X, Y,$ φ for a given set of joint variables θ_1, θ_2, d_3, where, X, Y are global coordinates of the end-effector frame B_3, and φ is the angular coordinate of B_3.

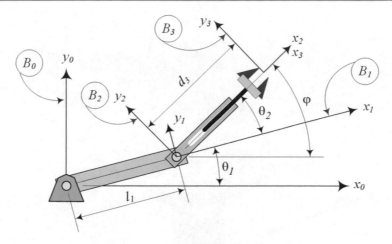

Fig. 9.13 A RRP planar redundant manipulator

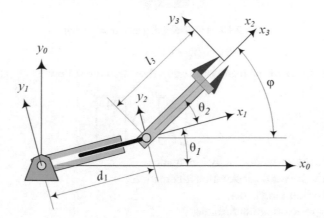

Fig. 9.14 A RPR planar redundant manipulator

(b) Solve the inverse kinematics of the manipulator and determine θ_1, θ_2, d_3 for given values of X, Y, φ.

(c) Determine the Jacobian matrix of the manipulator and show that the following equation solves the forward velocity kinematics:

$$\begin{bmatrix} \dot{X} \\ \dot{Y} \\ \dot{\varphi} \end{bmatrix} = \mathbf{J} \begin{bmatrix} \dot{\theta}_1 \\ \dot{\theta}_2 \\ \dot{d}_3 \end{bmatrix} \tag{9.483}$$

(d) Determine $\mathbf{J}^{-1}$ and solve the inverse velocity kinematics.

(e) Determine $\dot{\mathbf{J}}$ and solve the forward acceleration kinematics.

(f) Solve the inverse acceleration kinematics.

14. ★ Mixed velocity and simple acceleration.

Consider a local frame $B(Oxyz)$, which is rotating in $G(OXYZ)$ with an angular velocity $\dot{\alpha} = 10\,\text{rad/s}^2$ about the Z-axis and a moving point P in B at

$$^B\mathbf{r}_P(t) = \sin 2t\,\hat{\imath} \tag{9.484}$$

Determine $_G\tilde{\omega}_D, {}^B_G\tilde{\omega}_D, {}^B\mathbf{v}, {}^B_G\mathbf{v}, {}^G_D\mathbf{v}, {}^B\mathbf{a}, {}^G\mathbf{a}, {}^B_G\mathbf{a}, {}^G_R\mathbf{a}, {}^{GG}_{GR}\mathbf{a}, {}^{BB}_{GB}\mathbf{a}, {}^{GG}_{BG}\mathbf{a}, {}^{BB}_{BG}\mathbf{a}$.

15. A RPR planar redundant manipulator.

(a) Figure 9.14 illustrates a 3 DOF planar manipulator with joint variables θ_1, d_2, θ_2.

(b) Solve the forward kinematics of the manipulator and calculate the position and orientation of the end-effector X, Y, φ for a given set of joint variables θ_1, θ_2, d_3, where, X, Y are global coordinates of the end-effector frame B_3, and φ is the angular coordinate of B_3.

(c) Solve the inverse kinematics of the manipulator and determine θ_1, θ_2, d_3 for given values of X, Y, φ.

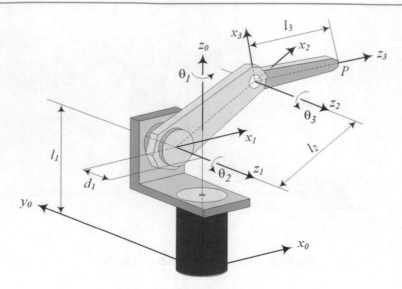

Fig. 9.15 An offset articulated manipulator

(d) Determine the Jacobian matrix of the manipulator and show that the following equation solves the forward velocity kinematics:

$$
\begin{bmatrix} \dot{X} \\ \dot{Y} \\ \dot{\varphi} \end{bmatrix} = \mathbf{J} \begin{bmatrix} \dot{\theta}_1 \\ \dot{\theta}_2 \\ \dot{d}_3 \end{bmatrix}
\tag{9.485}
$$

(e) Determine $\mathbf{J}^{-1}$ and solve the inverse velocity kinematics.
(f) Determine $\dot{\mathbf{J}}$ and solve the forward acceleration kinematics.
(g) Solve the inverse acceleration kinematics.

16. ★ Mixed second-derivative transformation formula.

Consider three relatively rotating coordinate frames A, B, and C. The frame C is turning about the y_B-axis of the B frame with angular velocity 8 rad/s and angular acceleration 10 rad/s^2, while B is turning about the x_A-axis with angular velocity 5 rad/s and turning about the z_A-axis with angular acceleration 3 rad/s^2. Determine the mixed double acceleration $^{AA}_{CB}\mathbf{a}$ if:

(a) A point P is at $^C\mathbf{r} = \begin{bmatrix} 1 & 1 & 1 \end{bmatrix}$ and moving with velocity $^C\mathbf{v} = \begin{bmatrix} -10 & 0 & 10 \end{bmatrix}$.
(b) A point P is at $^C\mathbf{r} = \begin{bmatrix} \sin t & 0 & 0 \end{bmatrix}$.
(c) A point P is at $^C\mathbf{r} = \begin{bmatrix} \cot s & \sin t & 0 \end{bmatrix}$.
(d) A point P is at $^C\mathbf{r} = \begin{bmatrix} \cot s & \sin t & t \end{bmatrix}$.

17. ★ An offset articulated manipulator.

Figure 9.15 illustrates an offset articulated manipulator.

(a) Solve the forward kinematics of the manipulator.
(b) Solve the inverse kinematics of the manipulator.
(c) Solve the forward velocity kinematics of the manipulator.
(d) Solve the inverse velocity kinematics of the manipulator.
(e) Solve the forward acceleration kinematics of the manipulator.
(f) Solve the inverse acceleration kinematics of the manipulator.

18. ★ Coriolis acceleration.

A disc with radius $R = 1$ m is turning in a horizontal plane with $\boldsymbol{\omega} = \omega \hat{K}$ and $\boldsymbol{\alpha} = \alpha \hat{K}$. We shoot a particle m with speed $\mathbf{v} = 10\hat{I}$ m/s from the center of the disc, which is at the origin of the disc coordinate frame $B\,(Oxyz)$ and global coordinate frame $G\,(OXYZ)$. If we ignore any friction between m and the disc, determine the local coordinate of the point where m reaches the periphery of the disc if:

(a) $\omega = 10$ rad/s and $\alpha = 0$
(b) $\omega = 10$ rad/s and $\alpha = 1$ rad/s^2
(c) $\omega = 10$ rad/s and $\alpha = -1$ rad/s^2
(d) $\omega = \sin t$ rad/s.

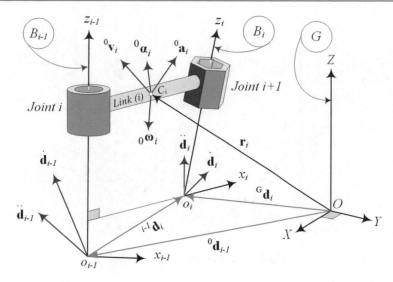

Fig. 9.16 Kinematics of a rigid link

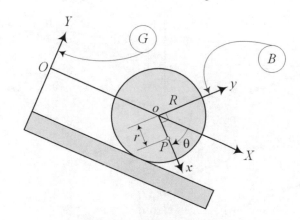

Fig. 9.17 A rolling disc on an inclined flat

19. Rigid link acceleration.

   Figure 9.16 illustrates the coordinate frames and kinematics of a rigid link (i). Assume that the angular velocity of the link, as well as the velocity and acceleration at proximal and distal joints, are given.

   (a) Find velocity and acceleration of the link at C_i in terms of proximal joint i velocity and acceleration.

   (b) Find velocity and acceleration of the link at C_i in terms of distal joint $i + 1$ velocity and acceleration.

   (c) Find velocity and acceleration of the proximal joint i in terms of distal joint $i + 1$ velocity and acceleration.

   (d) Find velocity and acceleration of the distal joint $i + 1$ in terms of proximal joint i velocity and acceleration.

20. ★ Coriolis and effective forces.

   A disc with radius $R = 1$ m is turning in a horizontal plane with $\boldsymbol{\omega} = \omega \hat{K}$ and $\boldsymbol{\alpha} = \alpha \hat{K}$. We shoot a particle m in a radial channel with speed $\mathbf{v} = 10\hat{I}$ m/s from the center of the disc, which is at the origin of the disc coordinate frame $B\,(Oxyz)$ and global coordinate frame $G\,(OXYZ)$. If we ignore any friction between m and the channel, determine the Coriolis and effective forces on m during its motion if:

   (a) $\omega = 10$ rad/s and $\alpha = 0$.

   (b) $\omega = 10$ rad/s and $\alpha = 1$ rad/s^2.

   (c) $\omega = 10$ rad/s and $\alpha = -1$ rad/s^2.

   (d) $\omega = \sin t$ rad/s.

21. ★ Jerk of a point in a roller.

   Figure 9.17 shows a rolling disc on an inclined flat surface. Find the velocity, acceleration, and jerk of point P in body and global coordinate frames.

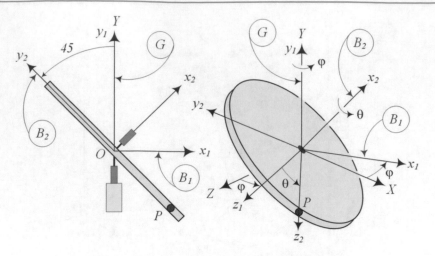

Fig. 9.18 A rotating disc on a needle

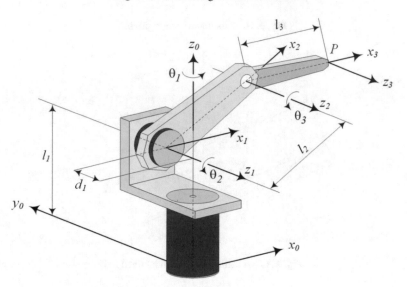

Fig. 9.19 A modified offset articulated manipulator

22. ★ Rāzī acceleration.

Consider a point P on the disc with radius $R = 1$ m in Fig. 9.18 and the following information:

$$
{}_1^2\boldsymbol{\omega}_2 = \dot{\theta}\hat{\imath}_2 = \begin{bmatrix} \dot{\theta} \\ 0 \\ 0 \end{bmatrix} \quad {}_G^1\boldsymbol{\omega}_1 = \dot{\varphi}\hat{J} = \begin{bmatrix} 0 \\ \dot{\varphi} \\ 0 \end{bmatrix} \quad {}^2\mathbf{r}_P = \begin{bmatrix} 0 \\ 0 \\ r \end{bmatrix}
\tag{9.486}
$$

Determine ${}^G\mathbf{v}_P$, ${}^1\mathbf{v}$, ${}_1^2\mathbf{v}$, ${}_G^2\mathbf{v}$, ${}_1^2\mathbf{a}$, ${}^1\mathbf{a}$, ${}_G^2\mathbf{a}$, ${}_{G1}^{22}\mathbf{a}$, tangential acceleration, mixed centripetal acceleration, mixed Coriolis acceleration, and Rāzī acceleration if:

(a) $r = R$, $\dot{\theta} = 10$ rad/s, $\dot{\varphi} = 2$ rad/s, and all other variables are zero

(b) $r = R$, $\dot{\theta} = \cos 10t$ rad/s, $\dot{\varphi} = \cos 2t$ rad/s, and all other variables are zero

(c) $r = R$, $\dot{\theta} = 10$ rad/s, $\dot{\varphi} = 2$ rad/s, $\ddot{\theta} = 1$ rad/s, $\ddot{\varphi} = -0.2$ rad/s, and all other variables are zero

(d) $r = t$, $\dot{\theta} = \cos 10t$ rad/s, $\dot{\varphi} = \cos 2t$ rad/s, and all other variables are zero

(e) $r = R \sin t$, $\dot{\theta} = \cos 10t$ rad/s, $\dot{\varphi} = \cos 2t$ rad/s, and all other variables are zero

23. ★ Project. A modified offset articulated manipulator.

Figure 9.19 illustrates an offset articulated manipulator with a different end-effector coordinate frame. Determine the inverse acceleration kinematics of the robot.

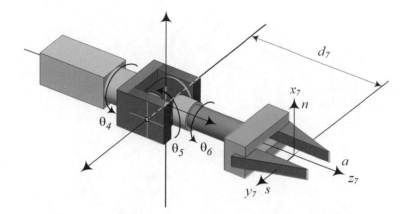

Fig. 9.20 Assembled spherical wrist

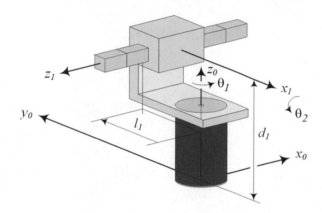

Fig. 9.21 A cylindrical manipulator

24. ★ Project. Articulated robots.

    Attach the spherical wrist of Fig. 9.20 to the articulated manipulator of Fig. 9.15 and make a 6 DOF articulated robot. Determine the inverse acceleration kinematics of the robot.

25. ★ Project-Spherical robots.

    Attach the spherical wrist of Fig. 9.20 to the spherical manipulator of Fig. 7.13 and make a 6 DOF spherical robot. Determine the inverse acceleration kinematics of the robot.

26. ★ Project. Cylindrical robots.

    Attach the spherical wrist of Fig. 9.20 to the cylindrical manipulator of Fig. 9.21 and make a 6 DOF cylindrical robot. Determine the inverse acceleration kinematics of the robot.

27. ★ Project. $SCARA$ robot inverse acceleration kinematics.

    Figure 9.22 illustrates a $SCARA$ robot. Assume $\theta_3 = 0$.

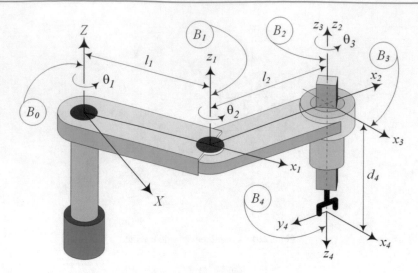

Fig. 9.22 A *SCARA* robot

Dynamics is the science of *motion*. It describes why and how a motion occurs when forces and moments are applied on massive particles and bodies. The motion can be considered as evolution of the position, orientation, and their time derivatives. In robotics, the dynamic equations of motion for manipulators are utilized to set up the fundamental equations for control. The links and arms in a robotic system are modeled as rigid bodies. Therefore, the dynamic properties of the rigid body take a central place in robot dynamics. As arms of a robot may rotate or translate with respect to each other, translational and rotational equations of motion must be developed and expressed in body-attached coordinate frames $B_1, B_2, B_3, \cdots$ or in the global base reference frame $B_0 = G$.

There are basically two problems in robot dynamics: *direct dynamics* and *inverse dynamics*.

Problem 1. We want the links of a robot to move in a specified manner. What forces and moments are required to achieve the motion?

Problem 1 is called *direct dynamics* and is easier to solve when the equations of motion are in hand because it needs differentiating of kinematics equations. The first problem includes robots statics to keep a robot at a given position such as the rest position of a robot. In such condition, the problem reduces to finding forces such that no motion takes place when they act. However, there are many meaningful problems of the first type that involve robot motion rather than rest. An important example is that of calculating the required forces that must act on a robot such that its end-effector moves on a given path and with a prescribed time history from the start configuration to the final configuration.

Problem 2. The applied forces and moments on a robot are completely specified. How will the robot move?

The second problem is called *inverse dynamics* and is more difficult to solve because it needs integration of equations of motion. The variety of the applied problems of the second type is interesting. Problem 2 is essentially a prediction since we wish to find the robot motion for all future times when the initial state of each link is given.

In this part, we develop techniques to derive the equations of motion for a robot.

Applied Dynamics

Relation between kinematics and the cause of change of kinematics is called equations of motion. Deriving the equations of motion and the expression of their solution is called dynamics. Dynamics of a robot may be considered as the motion of a rigid link with respect to a fixed global coordinate frame. The principles of dynamics as well as derivation methods of Newton and Euler equations of motion that express the translational and rotational motion of rigid bodies are reviewed in this chapter.

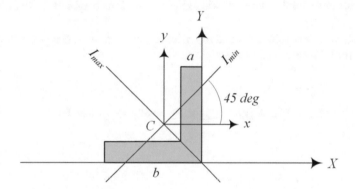

Fig. 10.1 Principal coordinate frame for a symmetric L-section

10.1 Force and Moment

In this section, we review the definition, elements, and principal equations of motion in dynamics.

10.1.1 Force and Moment

In Newtonian dynamics, the acting forces on a system of connected rigid bodies can be divided into *internal* and *external* *forces*. Internal forces are acting between connected bodies and appear as action and reaction forces. External forces are acting from outside of the system and appear as applied driving forces. An external force can be a *contact force*, such as traction force at tireprint of a driving wheel, or a *body force*, such as gravitational force on a robot link. The external forces and moments are called *load*, and a set of forces and moments acting on a rigid body are called a *force system*. The *resultant* **F** is the vectorial sum of all the external forces acting on a body, and the *resultant* **M** is the vectorial sum of all the moments of the external forces acting on the body.

$$\mathbf{F} = \sum_i \mathbf{F}_i \tag{10.1}$$

$$\mathbf{M} = \sum_i \mathbf{M}_i \tag{10.2}$$

Consider an acting force $\mathbf{F}$ on a point P at $\mathbf{r}_P$. The moment of $\mathbf{F}$ about the origin is

$$\mathbf{M} = \mathbf{r}_P \times \mathbf{F} \tag{10.3}$$

The moment of the force $\mathbf{F}$, about a point Q at $\mathbf{r}_Q$, is

$$\mathbf{M}_Q = (\mathbf{r}_P - \mathbf{r}_Q) \times \mathbf{F} \tag{10.4}$$

The *moment of the force* about a directional line l with unit vector $\hat{u}$ passing through the origin is

$$\mathbf{M}_l = \hat{u} \cdot (\mathbf{r}_P \times \mathbf{F}) \tag{10.5}$$

The moment of a force may also be called *torque* or *moment* although they are conceptually different.

The effect of a force system is equivalent to the effect of the resultant force and resultant moment of the force system. Any two force systems are equivalent if their resultant force and resultant moment are equal. If the resultant force of a force system is zero, the resultant moment of the force system is independent of the origin of the coordinate frame. Such a resultant moment is called *couple*.

When a force system is reduced to a resultant $\mathbf{F}_P$ and $\mathbf{M}_P$ with respect to a reference point P, we may change the reference point to another point Q and find the new resultants as

$$\mathbf{F}_Q = \mathbf{F}_P \tag{10.6}$$

$$\mathbf{M}_Q = \mathbf{M}_P + (\mathbf{r}_P - \mathbf{r}_Q) \times \mathbf{F}_P = \mathbf{M}_P + {}_Q\mathbf{r}_P \times \mathbf{F}_P \tag{10.7}$$

10.1.2 Momentum

The *momentum* of a moving rigid body is a vector quantity equal to the total mass of the body times the translational velocity of its mass center C.

$$\mathbf{p} = m\mathbf{v} \tag{10.8}$$

The momentum $\mathbf{p}$ may also be called *translational momentum* or *linear momentum*.

Consider a rigid body with momentum $\mathbf{p}$. The *moment of momentum*, $\mathbf{L}$, about a directional line l with directional unit vector $\hat{u}$ passing through the origin is

$$\mathbf{L}_l = \hat{u} \cdot (\mathbf{r}_C \times \mathbf{p}) \tag{10.9}$$

where $\mathbf{r}_C$ is the position vector of the mass center C. The moment of momentum about the origin is

$$\mathbf{L} = \mathbf{r}_C \times \mathbf{p} \tag{10.10}$$

The moment of momentum $\mathbf{L}$ may also be called *angular momentum*.

10.1.3 Equation of Motion

The application of a force system is emphasized by *Newton's second and third laws of motion*. The second law of motion, also called the *Newton's equation of motion*, states that the global rate of change of *linear momentum* is proportional to the global *applied force*.

$${}^G\mathbf{F} = \frac{{}^Gd}{dt}\,{}^G\mathbf{p} = \frac{{}^Gd}{dt}\left(m\,{}^G\mathbf{v}\right) \tag{10.11}$$

The third Newton's law of motion states that the action and reaction forces acting between two bodies are equal and opposite.

The second law of motion can be expanded to include rotational motions. Hence, the second law of motion also states that the global rate of change of *angular momentum* is proportional to the global *applied moment*.

$$^G\mathbf{M} = \frac{^Gd}{dt}\,{}^G\mathbf{L} \tag{10.12}$$

Proof Differentiating from moment of momentum (10.10) shows that

$$\frac{^Gd}{dt}\,{}^G\mathbf{L} = \frac{^Gd}{dt}\,(\mathbf{r}_C \times \mathbf{p}) = \left(\frac{^Gd\mathbf{r}_C}{dt} \times \mathbf{p} + \mathbf{r}_C \times \frac{^Gd\mathbf{p}}{dt}\right)$$

$$= {}^G\mathbf{r}_C \times \frac{^Gd\mathbf{p}}{dt} = {}^G\mathbf{r}_C \times {}^G\mathbf{F} = {}^G\mathbf{M} \tag{10.13}$$

∎

10.1.4 Work and Energy

The *kinetic energy K* of a moving point P with mass m at a position $^G\mathbf{r}_P$ and velocity $^G\mathbf{v}_P$ is

$$K = \frac{1}{2}m\left(^G\mathbf{v} \cdot {}^G\mathbf{v}\right) \tag{10.14}$$

where G indicates the global coordinate frame and B indicates body coordinate frame in which the velocity vector $\mathbf{v}_P$ is expressed. The work done by the applied force $^G\mathbf{F}$ on m in moving from point 1 to point 2 on a path, indicated by a vector $^G\mathbf{r}$, is

$$_1W_2 = \int_1^2 {}^G\mathbf{F} \cdot d\,{}^G\mathbf{r} \tag{10.15}$$

However,

$$\int_1^2 {}^G\mathbf{F} \cdot d\,{}^G\mathbf{r} = m\int_1^2 \frac{^Gd}{dt}\,{}^G\mathbf{v} \cdot {}^G\mathbf{v}\,dt = \frac{1}{2}m\int_1^2 \frac{d}{dt}v^2\,dt$$

$$= \frac{1}{2}m\left(v_2^2 - v_1^2\right) = K_2 - K_1 \tag{10.16}$$

that shows $_1W_2$ is equal to the difference of the kinetic energy between the terminal and initial points.

$$_1W_2 = K_2 - K_1 \tag{10.17}$$

Equation (10.17) is called the *principle of work and energy*.

If there is a scalar *potential field function* $V = V(x, y, z)$ such that

$$\mathbf{F} = -\nabla V = -\frac{dV}{d\mathbf{r}} = -\left(\frac{\partial V}{\partial x}\hat{\imath} + \frac{\partial V}{\partial y}\hat{\jmath} + \frac{\partial V}{\partial z}\hat{k}\right) \tag{10.18}$$

then the principle of work and energy simplifies to the principle of *conservation of energy*,

$$K_1 + V_1 = K_2 + V_2 \tag{10.19}$$

The value of the potential force function $V = V(x, y, z)$ is the *potential energy* of the system.

Example 328 Position of center of mass. Mass center of every link of a robot must be known to locate the point at which the acceleration and velocity of the link are referred to. Here is an example to show analytic method of calculating the location of mass center for objects their mass centers are not geometrically obvious.

The position of the mass center of a rigid body in a coordinate frame is indicated by $\mathbf{r}_C$ and is usually measured in the body coordinate frame. The location of mass center will be calculated by integrals over the volume of the body.

$$^B\mathbf{r}_C = \frac{1}{m} \int_B \mathbf{r}\, dm \tag{10.20}$$

$$\begin{bmatrix} x_C \\ y_C \\ z_C \end{bmatrix} = \begin{bmatrix} \frac{1}{m} \int_B x\, dm \\ \frac{1}{m} \int_B y\, dm \\ \frac{1}{m} \int_B z\, dm \end{bmatrix} \tag{10.21}$$

Applying the mass center integral on the symmetric L-section rigid body with $\rho = 1$ shown in Fig. 10.1 provides the location of C of the section. The x position of C is

$$x_C = \frac{1}{m} \int_B x\, dm = \frac{1}{A} \int_B x\, dA = -\frac{b^2 + ab - a^2}{4ab + 2a^2} \tag{10.22}$$

and because of symmetry, we have

$$y_C = -x_C = \frac{b^2 + ab - a^2}{4ab + 2a^2} \qquad z_C = 0 \tag{10.23}$$

Example 329 Exponential decaying force. Having the force function, determination of position as a function of time is the outcome of inverse dynamics problem. Here is an example.

Consider a point mass m that is under an exponentially decaying force $F(t) = ce^{-t}$, where c is a constant:

$$m\frac{dv}{dt} = ce^{-t} \tag{10.24}$$

The velocity of the mass will be

$$m \int_{v_0}^{v} dv = \int_{t_0}^{t} ce^{-t}\, dt$$

$$v = v_0 + \frac{c}{m}\left(e^{-t_0} - e^{-t}\right) \tag{10.25}$$

which can be used to find the position.

$$\int_{x_0}^{x} dx = \int_{t_0}^{t} \left(v_0 + \frac{c}{m}\left(e^{-t_0} - e^{-t}\right)\right) dt$$

$$x = x_0 - \frac{c}{m}(1 + t_0 - t)\, e^{-t_0} + v_0(t - t_0) + \frac{c}{m}e^{-t} \tag{10.26}$$

If the initial time t_0 is assumed to be zero, then the position and velocity of the mass are simplified to

$$x = x_0 - \frac{c}{m}(1 - t) + v_0 t + \frac{c}{m}e^{-t} \tag{10.27}$$

$$v = v_0 + \frac{c}{m}\left(1 - e^{-t}\right) \tag{10.28}$$

Example 330 Motion equation of a system of particles. There are situations in which there are multiple particles involved. Here is what would be the equation of motion for multiple particles.

Consider a group of n particles $m_i, i = 1, 2, 3, \cdots, n$, with position vectors $\mathbf{r}_i$ in a global coordinate frame G. The position vector of the mass center C of the particles is at $\mathbf{r}_C$, where m_C is the total mass of the system.

$$\mathbf{r}_C = \frac{1}{m_C} \sum_{i=1}^{n} m_i \mathbf{r}_i \qquad m_C = \sum_{i=1}^{n} m_i \qquad (10.29)$$

The force acting on each particle m_i can be decomposed into an external force $\mathbf{F}_i$ and an internal force $\sum_{j=1}^{n} \mathbf{f}_{ij}$. The internal force $\mathbf{f}_{ij}$, with the condition $\mathbf{f}_{ii} = 0$, is the force that particle m_j applies on m_i. The motion equation of the particle m_i would be

$$m_i \frac{d^2 \mathbf{r}_i}{dt^2} = \mathbf{F}_i + \sum_{j=1}^{n} \mathbf{f}_{ij} \qquad i = 1, 2, 3, \cdots, n \qquad (10.30)$$

By adding the n equations of motion of all particles we have

$$\sum_{i=1}^{n} m_i \frac{d^2 \mathbf{r}_i}{dt^2} = \sum_{i=1}^{n} \mathbf{F}_i + \sum_{i=1}^{n} \sum_{j=1}^{n} \mathbf{f}_{ij} \qquad (10.31)$$

Because of the third law of Newton,

$$\mathbf{f}_{ij} = -\mathbf{f}_{ji} \qquad (10.32)$$

the summation of the internal forces is zero:

$$\sum_{i=1}^{n} \sum_{j=1}^{n} \mathbf{f}_{ij} = 0 \qquad (10.33)$$

Therefore, the equation of motion of all particles reduces to

$$m_C \frac{d^2 \mathbf{r}_0}{dt^2} = \mathbf{F}_C \qquad (10.34)$$

where $\mathbf{F}_C$ is the resultant of all the external forces.

$$\mathbf{F}_C = \sum_{i=1}^{n} \mathbf{F}_i \qquad (10.35)$$

Equation (10.34) states that the motion of the mass center C of a system of particles is the same as if all the masses m were concentrated at that point and were acted upon by the resultant of all the external forces $\mathbf{F}_C$.

Example 331 Work, force, and kinetic energy in a unidirectional motion. An example to show how we use work and kinetic energy principle.

A mass $m = 2$ kg has an initial kinetic energy $K = 12$ J. The mass is under a constant force $\mathbf{F} = F\hat{I} = 4\hat{I}$ and moves from $X(0) = 1$ to $X(t_f) = 22$ m at a terminal time t_f. The work done by the force during this motion is

$$W = \int_{\mathbf{r}(0)}^{\mathbf{r}(t_f)} \mathbf{F} \cdot d\mathbf{r} = \int_{1}^{22} 4 \, dX = 21 \, \text{Nm} = 21 \, \text{J} \qquad (10.36)$$

The kinetic energy at the terminal time is

$$K(t_f) = W + K(0) = 33 \, \text{J} \qquad (10.37)$$

which shows that the terminal speed of the mass is

$$v_2 = \sqrt{\frac{2K(t_f)}{m}} = \sqrt{33} \, \text{m/s} \qquad (10.38)$$

As a more complete work calculation, consider a planar force $\mathbf{F}$,

$$\mathbf{F} = 2xy\hat{I} + 3x^2\hat{J} \, \text{N} \qquad (10.39)$$

that moves a mass m on a planar curve

$$y = x^2 \tag{10.40}$$

from $(0, 0)$ to $(3, 1)$ m. Using

$$dy = 2x \, dx \tag{10.41}$$

we can calculate the work done by the force:

$$
{}_1W_2 = \int_{P_1}^{P_2} {}^G\mathbf{F} \cdot d\mathbf{r} = \int_{(0,0)}^{(3,1)} \left(2xy \, dx + 3x^2 \, dy\right)
$$

$$
= \int_0^3 \left(2x^3 \, dx + 6x^3 \, dx\right) = \int_0^3 8x^3 \, dx = 162 \, \text{Nm} \tag{10.42}
$$

To see an example of potential field, consider a force field $\mathbf{F}$,

$$\mathbf{F} = 2xy\hat{\imath} + x^2\hat{\jmath} + 2z\hat{k} \, \text{N} \tag{10.43}$$

The work done by the force when it moves from $P_1(0, 0, 0)$ to $P_2(1, 3, 5)$ m on a line

$$x = t \qquad y = 3t \qquad z = 5t \tag{10.44}$$

is

$$
{}_1W_2 = \int_{P_1}^{P_2} {}^G\mathbf{F} \cdot d\mathbf{r} = \int_0^1 \left(10z + 2xy + 3x^2\right) dt
$$

$$
= \int_0^1 \left(50t + 6t^2 + 3t^2\right) dt = 28 \, \text{Nm} \tag{10.45}
$$

We may change the path from a straight line to a curve

$$x = t \qquad y = 3t^2 \qquad z = 5t^3 \tag{10.46}$$

and calculate the work ${}_1W_2$ again:

$$
{}_1W_2 = \int_{P_1}^{P_2} {}^G\mathbf{F} \cdot d\mathbf{r} = \int_0^1 \begin{bmatrix} 2xy \\ x^2 \\ 2z \end{bmatrix} \cdot \begin{bmatrix} dt \\ 6t \, dt \\ 15t^2 \, dt \end{bmatrix}
$$

$$
= \int_0^1 2\left(3tx^2 + 15t^2z + xy\right) dt
$$

$$
= 2\int_0^1 \left(75t^5 + 6t^3\right) dt = 28 \, \text{Nm} \tag{10.47}
$$

Because ${}_1W_2$ is independent of the path from P_1 to P_2, the force field (10.43) might be a potential field. To check, we may calculate the curl of the field because curl of a potential force is zero: $\nabla \times \mathbf{F} = 0$.

$$
\nabla \times \mathbf{F} = \begin{bmatrix} \partial/\partial x \\ \partial/\partial y \\ \partial/\partial z \end{bmatrix} \times \begin{bmatrix} 2xy \\ x^2 \\ 2z \end{bmatrix} = \begin{bmatrix} 0 \\ 0 \\ 0 \end{bmatrix} \tag{10.48}
$$

So, the force field is a potential.

Example 332 ★ Newton equation in a rotating frame. Analysis of motion of objects in a rotating coordinate frame is very applied in engineering as well as study of every moving objects on the Earth. Such expression has an interesting term called the Coriolis acceleration. Here is the expression of Newton equation of motion in a rotating frame.

Consider a spherical rigid body (such as the Earth) with a fixed point that is rotating with a constant angular velocity. The equation of motion for a moving point P on the rigid body is

$$
{}^{B}\mathbf{F} = m \, {}^{B}\mathbf{a}_P + m \, {}^{B}_{G}\boldsymbol{\omega}_B \times \left({}^{B}_{G}\boldsymbol{\omega}_B \times {}^{B}\mathbf{r}_P \right) + 2m \, {}^{B}_{G}\boldsymbol{\omega}_B \times {}^{B}\dot{\mathbf{r}}_P
$$
$$
\neq m \, {}^{B}\mathbf{a}_P \tag{10.49}
$$

It shows that the Newton equation of motion $\mathbf{F} = m\,\mathbf{a}$, which is only correct in a globally stationary frame, will have extra terms when expressed in the rotating frame. The equation of motion of a moving point on the surface of the Earth can be rearranged to

$$
{}^{B}\mathbf{F} - m \, {}^{B}_{G}\boldsymbol{\omega}_B \times \left({}^{B}_{G}\boldsymbol{\omega}_B \times {}^{B}\mathbf{r}_P \right) - 2m \, {}^{B}_{G}\boldsymbol{\omega}_B \times {}^{B}\mathbf{v}_P = m \, {}^{B}\mathbf{a}_P \tag{10.50}
$$

Equation (10.50) is the equation of motion to an observer in the rotating frame, which in this case is an observer on the Earth. The left-hand side of this equation is called the effective force,

$$
\mathbf{F}_{eff} = {}^{B}\mathbf{F} - m \, {}^{B}_{G}\boldsymbol{\omega}_B \times \left({}^{B}_{G}\boldsymbol{\omega}_B \times {}^{B}\mathbf{r}_P \right) - 2m \, {}^{B}_{G}\boldsymbol{\omega}_B \times {}^{B}\mathbf{v}_P \tag{10.51}
$$

because it seems that the particle is moving in the Earth frame B under the influence of this force. The second term is the negative of the centrifugal force and pointing outward. The maximum value of this force on the Earth surface is on the equator that is about 0.3% of the acceleration of gravity.

$$
r\omega^2 = 6378.388 \times 10^3 \times \left(\frac{2\pi}{24 \times 3600} \frac{366.25}{365.25} \right)^2
$$
$$
= 3.3917 \times 10^{-2}\,\text{m/s}^2 \tag{10.52}
$$

If we add the variation of the gravitational acceleration because of a change of radius from $R = 6356912\,\text{m}$ at the pole to $R = 6378388\,\text{m}$ on the equator, then the variation of the acceleration of gravity becomes 0.53%. So, generally speaking, a sportsman such as a pole-vaulter can show a better record in a competition held on the equator.

The third term is called the Coriolis force, F_C, which is perpendicular to both $\boldsymbol{\omega}$ and ${}^{B}\mathbf{v}_P$. For a mass m moving on the north hemisphere at a latitude θ toward the equator, we should provide a lateral eastward force equal to the Coriolis effect to force the mass, keeping its direction relative to the ground.

$$
F_C = 2m \, {}^{B}_{G}\boldsymbol{\omega}_B \times {}^{B}\mathbf{v}_m = 1.4584 \times 10^{-4}\, {}^{B}\mathbf{p}_m \cos\theta \,\,\text{kg m/s}^2 \tag{10.53}
$$

The Coriolis effect is the reason of wearing out the west side of railways, roads, and rivers. The lack of providing Coriolis force is the reason for turning the direction of winds, projectiles, and shooting objects westward.

Gaspard-Gustave de Coriolis (1792–1843) is the French mathematician who showed the importance of the Coriolis term in a rotating coordinate frame.

Example 333 ★ Force function in equation of motion. In Newtonian mechanics, applied forces may only be functions of time, position, and velocity and nothing else. Here is a proof.

The first and second Newton laws of motion provide no predication or expectation of the force function and its arguments. Qualitatively, force is whatever changes the motion, and quantitatively, force is whatever is equal to mass times acceleration. Mathematically, equation of motion provides a vectorial second order differential equation:

$$
m\ddot{\mathbf{r}} = \mathbf{F}\left(\dot{\mathbf{r}}, \mathbf{r}, t \right) \tag{10.54}
$$

We assume that the force function may generally be a function of time t, position $\mathbf{r}$, and velocity $\dot{\mathbf{r}}$. In other words, the Newton equation of motion is correct as long as we can show that force is only a function of $\dot{\mathbf{r}}, \mathbf{r}, t$. If there is a force that depends on the acceleration, jerk, or other variables that cannot be reduced to $\dot{\mathbf{r}}, \mathbf{r}, t$, we do not know the equation of motion because we do not have an equation of motion for those forces.

$$\mathbf{F}(\mathbf{r}, \dot{\mathbf{r}}, \ddot{\mathbf{r}}, \dddot{\mathbf{r}}, \cdots, t) \neq m\ddot{\mathbf{r}} \tag{10.55}$$

So, in Newtonian mechanics, we assume that force can only be a function of $\dot{\mathbf{r}}$, $\mathbf{r}$, t and nothing else. In the real world, force may be a function of many things; however, we always ignore any other variables than $\dot{\mathbf{r}}$, $\mathbf{r}$, t, or make some approximations accordingly.

Because Eq. (10.54) is a linear equation of force $\mathbf{F}$, it accepts the superposition principle. When a mass m is affected by several forces $\mathbf{F}_1$, $\mathbf{F}_2$, $\mathbf{F}_3$, $\cdots$, we may calculate their summation vectorially and apply the resultant force on m.

$$\mathbf{F} = \mathbf{F}_1 + \mathbf{F}_2 + \mathbf{F}_3 + \cdots \tag{10.56}$$

So, if a force $\mathbf{F}_1$ provides acceleration $\ddot{\mathbf{r}}_1$ and $\mathbf{F}_2$ provides $\ddot{\mathbf{r}}_2$,

$$m\ddot{\mathbf{r}}_1 = \mathbf{F}_1 \qquad m\ddot{\mathbf{r}}_2 = \mathbf{F}_2 \tag{10.57}$$

then the resultant force $\mathbf{F}_3 = \mathbf{F}_1 + \mathbf{F}_2$ provides the acceleration $\ddot{\mathbf{r}}_3$ such that

$$\ddot{\mathbf{r}}_3 = \ddot{\mathbf{r}}_1 + \ddot{\mathbf{r}}_2 \tag{10.58}$$

To see that the Newton equation of motion is not correct when the force is not only a function of $\dot{\mathbf{r}}$, $\mathbf{r}$, t, let us assume that a particle with mass m is under two acceleration-dependent forces $F_1(\ddot{x})$ and $F_2(\ddot{x})$ on the x-axis:

$$m\ddot{x}_1 = F_1(\ddot{x}_1) \qquad m\ddot{x}_2 = F_2(\ddot{x}_2) \tag{10.59}$$

The acceleration of m under the action of both forces would be $\ddot{x}_3$,

$$m\ddot{x}_3 = F_1(\ddot{x}_3) + F_2(\ddot{x}_3) \tag{10.60}$$

however, we must have

$$\ddot{x}_3 = \ddot{x}_1 + \ddot{x}_2 \tag{10.61}$$

but we have

$$m(\ddot{x}_1 + \ddot{x}_2) = F_1(\ddot{x}_1 + \ddot{x}_2) + F_2(\ddot{x}_1 + \ddot{x}_2) \neq F_1(\ddot{x}_1) + F_2(\ddot{x}_2) \tag{10.62}$$

Example 334 ★ Every force system is equivalent to a wrench. Wrench in dynamics is similar to screw motion in kinematics. It is an interesting concept but not very applied. Here is to show how a wrench is calculated to be equivalent to a force system.

The Poinsot theorem states: Every force system is equivalent to a single force, plus a moment parallel to the force. Let $\mathbf{F}$ and $\mathbf{M}$ be the resultant force and moment of a force system. We decompose the moment into parallel and perpendicular components, $\mathbf{M}_\parallel$ and $\mathbf{M}_\perp$, to the force axis. The force $\mathbf{F}$ and the perpendicular moment $\mathbf{M}_\perp$ can be replaced by a single force $\mathbf{F}'$ parallel to $\mathbf{F}$. Therefore, the force system is reduced to a force $\mathbf{F}'$ and a moment $\mathbf{M}_\parallel$ parallel to each other. A force and a moment about the force axis are the equivalent wrenches to the given force system.

Poinsot theorem is similar to the Chasles theorem that states every rigid body motion is equivalent to a screw, which is a translation plus a rotation about the axis of translation. Louis Poinsot (1777–1859) is a French mathematician and physicist who showed how a system of forces can be resolved into a wrench.

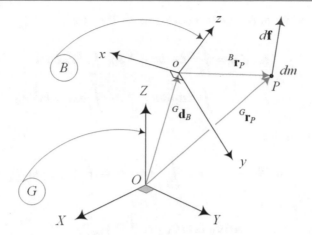

Fig. 10.2 A body point mass moving with velocity $^G\mathbf{v}_P$ and acted on by force $d\mathbf{f}$

10.2   Rigid Body Translational Kinetics

Figure 10.2 depicts a moving body B in a global frame G. Assume the body frame is attached at the center of mass C of the body. Point P indicates an infinitesimal sphere of the body with a very small mass dm. The point mass dm is acted on by an infinitesimal force $d\mathbf{f}$ and has a global velocity $^G\mathbf{v}_P$. According to Newton law of motion we have

$$d\mathbf{f} = {}^G\mathbf{a}_P\, dm \tag{10.63}$$

however, the equation of motion for the whole body in global coordinate frame is

$$^G\mathbf{F} = m\, {}^G\mathbf{a}_B \tag{10.64}$$

which can be expressed in the body coordinate frame as

$$^B\mathbf{F} = m\, {}^B_G\mathbf{a}_B + m\, {}^B_G\boldsymbol{\omega}_B \times {}^B\mathbf{v}_B \tag{10.65}$$

$$\begin{bmatrix} F_x \\ F_y \\ F_z \end{bmatrix} = \begin{bmatrix} ma_x + m\left(\omega_y v_z - \omega_z v_y\right) \\ ma_y - m\left(\omega_x v_z - \omega_z v_x\right) \\ ma_z + m\left(\omega_x v_y - \omega_y v_x\right) \end{bmatrix} \tag{10.66}$$

In these equations, $^G\mathbf{a}_B$ is the acceleration vector of the body C in global frame, m is the total mass of the body, and $\mathbf{F}$ is the resultant of the external forces acted on the body at C.

Proof A body coordinate frame at the center of mass is called a *central frame*. If the frame B is a central frame, then the *center of mass*, C, yields

$$\int_B {}^B\mathbf{r}_{dm}\, dm = 0 \tag{10.67}$$

The global position vector of dm is related to its local position vector by $^G\mathbf{r}_{dm}$,

$$^G\mathbf{r}_{dm} = {}^G\mathbf{d}_B + {}^GR_B\, {}^B\mathbf{r}_{dm} \tag{10.68}$$

where $^G\mathbf{d}_B$ is the global position vector of the central body frame B. Therefore,

$$\int_B {}^G\mathbf{r}_{dm}\, dm = \int_B {}^G\mathbf{d}_B\, dm + {}^GR_B \int_m {}^B\mathbf{r}_{dm}\, dm$$

$$= \int_B {}^G\mathbf{d}_B\, dm = {}^G\mathbf{d}_B \int_B dm = m\, {}^G\mathbf{d}_B \qquad (10.69)$$

A time derivative of both sides shows that

$$m\, {}^G\dot{\mathbf{d}}_B = m\, {}^G\mathbf{v}_B = \int_B {}^G\dot{\mathbf{r}}_{dm}\, dm = \int_B {}^G\mathbf{v}_{dm}\, dm \qquad (10.70)$$

and another derivative indicates that

$$m\, {}^G\dot{\mathbf{v}}_B = m\, {}^G\mathbf{a}_B = \int_B {}^G\dot{\mathbf{v}}_{dm}\, dm \qquad (10.71)$$

However, we have $d\mathbf{f} = {}^G\dot{\mathbf{v}}_P\, dm$, and hence,

$$m\, {}^G\mathbf{a}_B = \int_B d\mathbf{f} \qquad (10.72)$$

The integral on the right-hand side accounts for all the forces acting on all particles in the body. The internal forces cancel one another out, so the net result is the vector sum of all the externally applied forces, $\mathbf{F}$, and therefore,

$$^G\mathbf{F} = m\, {}^G\mathbf{a}_B = m\, {}^G\dot{\mathbf{v}}_B \qquad (10.73)$$

In the body coordinate frame we have

$$^B\mathbf{F} = {}^BR_G\, {}^G\mathbf{F} = m\, {}^BR_G\, {}^G\mathbf{a}_B = m\, {}^B_G\mathbf{a}_B$$

$$= m\, {}^B\mathbf{a}_B + m\, {}^B_G\boldsymbol{\omega}_B \times {}^B\mathbf{v}_B \qquad (10.74)$$

The expanded form of the Newton's equation in the body coordinate frame is then equal to

$$^B\mathbf{F} = m\, {}^B\mathbf{a}_B + m\, {}^B_G\boldsymbol{\omega}_B \times {}^B\mathbf{v}_B \qquad (10.75)$$

$$\begin{bmatrix} F_x \\ F_y \\ F_z \end{bmatrix} = m \begin{bmatrix} a_x \\ a_y \\ a_z \end{bmatrix} + m \begin{bmatrix} \omega_x \\ \omega_y \\ \omega_z \end{bmatrix} \times \begin{bmatrix} v_x \\ v_y \\ v_z \end{bmatrix}$$

$$= \begin{bmatrix} ma_x + m\left(\omega_y v_z - \omega_z v_y\right) \\ ma_y - m\left(\omega_x v_z - \omega_z v_x\right) \\ ma_z + m\left(\omega_x v_y - \omega_y v_x\right) \end{bmatrix} \qquad (10.76)$$

$\blacksquare$

Example 335 A wound ribbon. Translational motion of a rigid is always along with rigid body rotation; otherwise, the rigid body may be treated as a particle. Here is an example of a planar rotating and translating rigid body motion.

Figure 10.3 illustrates a ribbon of negligible weight and thickness that is wound tightly around a uniform massive disc of radius R and mass m. The ribbon is fastened to a rigid support, and the disc is released to roll down vertically. There are two forces acting on the disc during the motion, its weight mg and the tension of the ribbon T. The translational equation of motion of the disc is expressed easier in the global coordinate frame:

$$\sum F_Y = -mg + T = m\ddot{Y} \qquad (10.77)$$

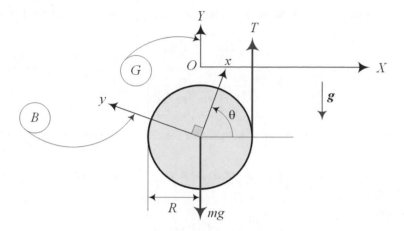

Fig. 10.3 A ribbon of negligible weight and thickness that is wound tightly around a uniform massive disc

The rotational equation of motion is simpler if expressed in the body coordinate frame:

$$\sum M_z = TR = {}^B I \, {}^B_G \dot{\boldsymbol{\omega}}_B + {}^B_G \boldsymbol{\omega}_B \times {}^B I \, {}^B_G \boldsymbol{\omega}_B = I \ddot{\theta} \tag{10.78}$$

There is a constraint between the coordinates Y and θ.

$$Y = Y_0 - R\theta \tag{10.79}$$

To solve the motion, let us eliminate T between (10.77) and (10.78) to obtain a new equation,

$$m\ddot{Y} = -mg + \frac{I}{R}\ddot{\theta} \tag{10.80}$$

and use the constraint to eliminate $\ddot{Y}$.

$$\ddot{\theta} = \frac{mg}{\left(\dfrac{I}{R} + mR\right)} \tag{10.81}$$

Now, we can find T, $\ddot{Y}$, and Y:

$$T = \frac{I}{R}\ddot{\theta} = \frac{I}{mR^2 + I}mg \tag{10.82}$$

$$\ddot{Y} = -\frac{mR^2}{mR^2 + I}g \quad Y = -\frac{mR^2}{mR^2 + I}gt^2 + \dot{Y}(0)\,t + Y(0) \tag{10.83}$$

For a point mass with $I = 0$, the falling acceleration is the same as the free fall of a particle. However, being a rigid body and having $I \neq 0$, the falling acceleration of the disc will be less. This is because the kinetic energy of the disc splits between rotation and translation.

10.3 Rigid Body Rotational Kinetics

The rigid body rotational equation of motion is expressed by the *Euler equation*

$$
\begin{aligned}
{}^B\mathbf{M} = \frac{{}^G d}{dt} {}^B\mathbf{L} &= {}^B\dot{\mathbf{L}} + {}^B_G\boldsymbol{\omega}_B \times {}^B\mathbf{L} \\
&= {}^B I \, {}^B_G\dot{\boldsymbol{\omega}}_B + {}^B_G\boldsymbol{\omega}_B \times \left({}^B I \, {}^B_G\boldsymbol{\omega}_B\right)
\end{aligned} \tag{10.84}
$$

where **L** is the *angular momentum* and $[I]$ is the *mass moment* or *moment of inertia* of the rigid body.

$$^B\mathbf{L} = \, ^B I \, ^B_G \boldsymbol{\omega}_B \tag{10.85}$$

$$[I] = \begin{bmatrix} I_{xx} & I_{xy} & I_{xz} \\ I_{yx} & I_{yy} & I_{yz} \\ I_{zx} & I_{zy} & I_{zz} \end{bmatrix} \tag{10.86}$$

The elements of $[I]$ are only functions of the mass distribution of the rigid body and may be defined by geometric integrals,

$$I_{ij} = \int_B \left(r_i^2 \delta_{mn} - x_{im} x_{jn} \right) dm \qquad i, j = 1, 2, 3 \tag{10.87}$$

where δ_{ij} is Kronecker's delta.

$$\delta_{mn} = \begin{cases} 1 \; if \; m = n \\ 0 \; if \; m \neq n \end{cases} \tag{10.88}$$

The expanded form of the Euler equation is

$$M_x = I_{xx}\dot{\omega}_x + I_{xy}\dot{\omega}_y + I_{xz}\dot{\omega}_z - \left(I_{yy} - I_{zz} \right) \omega_y \omega_z$$
$$- I_{yz} \left(\omega_z^2 - \omega_y^2 \right) - \omega_x \left(\omega_z I_{xy} - \omega_y I_{xz} \right) \tag{10.89}$$

$$M_y = I_{yx}\dot{\omega}_x + I_{yy}\dot{\omega}_y + I_{yz}\dot{\omega}_z - (I_{zz} - I_{xx}) \omega_z \omega_x$$
$$- I_{xz} \left(\omega_x^2 - \omega_z^2 \right) - \omega_y \left(\omega_x I_{yz} - \omega_z I_{xy} \right) \tag{10.90}$$

$$M_z = I_{zx}\dot{\omega}_x + I_{zy}\dot{\omega}_y + I_{zz}\dot{\omega}_z - \left(I_{xx} - I_{yy} \right) \omega_x \omega_y$$
$$- I_{xy} \left(\omega_y^2 - \omega_x^2 \right) - \omega_z \left(\omega_y I_{xz} - \omega_x I_{yz} \right) \tag{10.91}$$

which can be reduced to a set of simpler equations in the *principal coordinate frame*.

$$M_1 = I_1\dot{\omega}_1 - (I_2 - I_2) \, \omega_2 \omega_3$$
$$M_2 = I_2\dot{\omega}_2 - (I_3 - I_1) \, \omega_3 \omega_1 \tag{10.92}$$
$$M_3 = I_3\dot{\omega}_3 - (I_1 - I_2) \, \omega_1 \omega_2$$

The principal coordinate frame is denoted by numbers 123 instead of xyz to indicate the first, second, and third *principal axes*. The parameters I_{ij}, $i \neq j$ are zero in the principal frame. The body and principal coordinate frame are assumed to sit at C of the body. .

The kinetic energy of a rotating rigid body is

$$K = \frac{1}{2} \left(I_{xx}\omega_x^2 + I_{yy}\omega_y^2 + I_{zz}\omega_z^2 \right)$$
$$- I_{xy}\omega_x\omega_y - I_{yz}\omega_y\omega_z - I_{zx}\omega_z\omega_x$$
$$= \frac{1}{2}\boldsymbol{\omega} \cdot \mathbf{L} = \frac{1}{2}\boldsymbol{\omega}^T I \, \boldsymbol{\omega} \tag{10.93}$$

that in the principal coordinate frame reduces to

$$K = \frac{1}{2} \left(I_1\omega_1^2 + I_2\omega_2^2 + I_3\omega_3^2 \right) \tag{10.94}$$

Proof Let m_i be the mass of the ith particle of a rigid body B, which is made of n particles and $\mathbf{r}_i$ is the Cartesian position vector of m_i in a central body fixed coordinate frame $Oxyz$.

$$\mathbf{r}_i = \, ^B\mathbf{r}_i = \begin{bmatrix} x_i & y_i & z_i \end{bmatrix}^T \tag{10.95}$$

Assume $_G^B\boldsymbol{\omega}_B$ to be the angular velocity of the rigid body with respect to the ground, expressed in the body coordinate frame.

$$\boldsymbol{\omega} = {}_G^B\boldsymbol{\omega}_B = \begin{bmatrix} \omega_x & \omega_y & \omega_z \end{bmatrix}^T \tag{10.96}$$

The angular momentum of m_i is

$$
\begin{aligned}
\mathbf{L}_i &= \mathbf{r}_i \times m_i \dot{\mathbf{r}}_i = m_i \left[\mathbf{r}_i \times (\boldsymbol{\omega} \times \mathbf{r}_i) \right] \\
&= m_i \left[(\mathbf{r}_i \cdot \mathbf{r}_i)\,\boldsymbol{\omega} - (\mathbf{r}_i \cdot \boldsymbol{\omega})\,\mathbf{r}_i \right] \\
&= m_i r_i^2 \boldsymbol{\omega} - m_i (\mathbf{r}_i \cdot \boldsymbol{\omega})\,\mathbf{r}_i
\end{aligned}
\tag{10.97}
$$

Hence, the angular momentum of the rigid body would be

$$\mathbf{L} = \boldsymbol{\omega} \sum_{i=1}^{n} m_i r_i^2 - \sum_{i=1}^{n} m_i (\mathbf{r}_i \cdot \boldsymbol{\omega})\,\mathbf{r}_i \tag{10.98}$$

Substitution for $\mathbf{r}_i$ and $\boldsymbol{\omega}$ provides us with

$$
\begin{aligned}
\mathbf{L} = {}&\left(\omega_x \hat{\imath} + \omega_y \hat{\jmath} + \omega_z \hat{k} \right) \sum_{i=1}^{n} m_i \left(x_i^2 + y_i^2 + z_i^2 \right) \\
&- \sum_{i=1}^{n} m_i \left(x_i \omega_x + y_i \omega_y + z_i \omega_z \right) \cdot \left(x_i \hat{\imath} + y_i \hat{\jmath} + z_i \hat{k} \right)
\end{aligned}
\tag{10.99}
$$

and therefore,

$$
\begin{aligned}
\mathbf{L} = {}&\sum_{i=1}^{n} m_i \left(x_i^2 + y_i^2 + z_i^2 \right) \omega_x \hat{\imath} + \sum_{i=1}^{n} m_i \left(x_i^2 + y_i^2 + z_i^2 \right) \omega_y \hat{\jmath} \\
&+ \sum_{i=1}^{n} m_i \left(x_i^2 + y_i^2 + z_i^2 \right) \omega_z \hat{k} \\
&- \sum_{i=1}^{n} m_i \left(x_i \omega_x + y_i \omega_y + z_i \omega_z \right) x_i \hat{\imath} \\
&- \sum_{i=1}^{n} m_i \left(x_i \omega_x + y_i \omega_y + z_i \omega_z \right) y_i \hat{\jmath} \\
&- \sum_{i=1}^{n} m_i \left(x_i \omega_x + y_i \omega_y + z_i \omega_z \right) z_i \hat{k}
\end{aligned}
\tag{10.100}
$$

or

$$
\begin{aligned}
\mathbf{L} = {}&\sum_{i=1}^{n} m_i \left[\left(x_i^2 + y_i^2 + z_i^2 \right) \omega_x - \left(x_i \omega_x + y_i \omega_y + z_i \omega_z \right) x_i \right] \hat{\imath} \\
&+ \sum_{i=1}^{n} m_i \left[\left(x_i^2 + y_i^2 + z_i^2 \right) \omega_y - \left(x_i \omega_x + y_i \omega_y + z_i \omega_z \right) y_i \right] \hat{\jmath} \\
&+ \sum_{i=1}^{n} m_i \left[\left(x_i^2 + y_i^2 + z_i^2 \right) \omega_z - \left(x_i \omega_x + y_i \omega_y + z_i \omega_z \right) z_i \right] \hat{k}
\end{aligned}
\tag{10.101}
$$

which can be rearranged as

$$\mathbf{L} = \sum_{i=1}^{n} \left[m_i \left(y_i^2 + z_i^2 \right) \right] \omega_x \hat{\imath} + \sum_{i=1}^{n} \left[m_i \left(z_i^2 + x_i^2 \right) \right] \omega_y \hat{\jmath}$$

$$+ \sum_{i=1}^{n} \left[m_i \left(x_i^2 + y_i^2 \right) \right] \omega_z \hat{k}$$

$$- \left(\sum_{i=1}^{n} \left(m_i x_i y_i \right) \omega_y + \sum_{i=1}^{n} \left(m_i x_i z_i \right) \omega_z \right) \hat{\imath}$$

$$- \left(\sum_{i=1}^{n} \left(m_i y_i z_i \right) \omega_z + \sum_{i=1}^{n} \left(m_i y_i x_i \right) \omega_x \right) \hat{\jmath}$$

$$- \left(\sum_{i=1}^{n} \left(m_i z_i x_i \right) \omega_x + \sum_{i=1}^{n} \left(m_i z_i y_i \right) \omega_y \right) \hat{k} \tag{10.102}$$

By introducing the mass moment matrix $[I]$ with the following elements,

$$I_{xx} = \sum_{i=1}^{n} \left[m_i \left(y_i^2 + z_i^2 \right) \right] \tag{10.103}$$

$$I_{yy} = \sum_{i=1}^{n} \left[m_i \left(z_i^2 + x_i^2 \right) \right] \tag{10.104}$$

$$I_{zz} = \sum_{i=1}^{n} \left[m_i \left(x_i^2 + y_i^2 \right) \right] \tag{10.105}$$

$$I_{xy} = I_{yx} = - \sum_{i=1}^{n} \left(m_i x_i y_i \right) \tag{10.106}$$

$$I_{yz} = I_{zy} = - \sum_{i=1}^{n} \left(m_i y_i z_i \right) \tag{10.107}$$

$$I_{zx} = I_{xz} = - \sum_{i=1}^{n} \left(m_i z_i x_i \right) \tag{10.108}$$

we can write the angular momentum $\mathbf{L}$ in a concise form,

$$L_x = I_{xx}\omega_x + I_{xy}\omega_y + I_{xz}\omega_z \tag{10.109}$$

$$L_y = I_{yx}\omega_x + I_{yy}\omega_y + I_{yz}\omega_z \tag{10.110}$$

$$L_z = I_{zx}\omega_x + I_{zy}\omega_y + I_{zz}\omega_z \tag{10.111}$$

or in a matrix form

$$\begin{bmatrix} L_x \\ L_y \\ L_z \end{bmatrix} = \begin{bmatrix} I_{xx} & I_{xy} & I_{xz} \\ I_{yx} & I_{yy} & I_{yz} \\ I_{zx} & I_{zy} & I_{zz} \end{bmatrix} \begin{bmatrix} \omega_x \\ \omega_y \\ \omega_z \end{bmatrix} \tag{10.112}$$

$$\mathbf{L} = I \cdot \boldsymbol{\omega} \tag{10.113}$$

For a rigid body that is a continuous compact solid body, the summations will be replaced by integrations over the volume of the body as in Eq. (10.87).

The Euler equation of motion for a rigid body is

$$^{B}\mathbf{M} = \frac{^{G}d}{dt} \, ^{B}\mathbf{L} \tag{10.114}$$

where $^{B}\mathbf{M}$ is the resultant of the external moments applied on the rigid body. The angular momentum vector $^{B}\mathbf{L}$ is defined in the body coordinate frame B. Hence, its time derivative in the global coordinate frame is

$$\frac{^{G}d \, ^{B}\mathbf{L}}{dt} = \, ^{B}\dot{\mathbf{L}} + \, ^{B}_{G}\boldsymbol{\omega}_{B} \times \, ^{B}\mathbf{L} \tag{10.115}$$

Therefore, $^{B}\mathbf{M}$ is

$$^{B}\mathbf{M} = \frac{d\mathbf{L}}{dt} = \dot{\mathbf{L}} + \dot{\omega} \times L = I\dot{\omega} + \omega \times (I\omega) \tag{10.116}$$

$$\begin{aligned}
^{B}\mathbf{M} = &\left(I_{xx}\dot{\omega}_x + I_{xy}\dot{\omega}_y + I_{xz}\dot{\omega}_z \right)\hat{\imath} \\
&+ \omega_y \left(I_{xz}\omega_x + I_{yz}\omega_y + I_{zz}\omega_z \right)\hat{\imath} \\
&- \omega_z \left(I_{xy}\omega_x + I_{yy}\omega_y + I_{yz}\omega_z \right)\hat{\imath} \\
&+ \left(I_{yx}\dot{\omega}_x + I_{yy}\dot{\omega}_y + I_{yz}\dot{\omega}_z \right)\hat{\jmath} \\
&+ \omega_z \left(I_{xx}\omega_x + I_{xy}\omega_y + I_{xz}\omega_z \right)\hat{\jmath} \\
&- \omega_x \left(I_{xz}\omega_x + I_{yz}\omega_y + I_{zz}\omega_z \right)\hat{\jmath} \\
&+ \left(I_{zx}\dot{\omega}_x + I_{zy}\dot{\omega}_y + I_{zz}\dot{\omega}_z \right)\hat{k} \\
&+ \omega_x \left(I_{xy}\omega_x + I_{yy}\omega_y + I_{yz}\omega_z \right)\hat{k} \\
&- \omega_y \left(I_{xx}\omega_x + I_{xy}\omega_y + I_{xz}\omega_z \right)\hat{k}
\end{aligned} \tag{10.117}$$

Therefore, the most general form of the Euler equations of motion for a rigid body in a body frame attached to C is

$$M_x = I_{xx}\dot{\omega}_x + I_{xy}\dot{\omega}_y + I_{xz}\dot{\omega}_z - \left(I_{yy} - I_{zz} \right)\omega_y\omega_z$$
$$- I_{yz}\left(\omega_z^2 - \omega_y^2 \right) - \omega_x\left(\omega_z I_{xy} - \omega_y I_{xz} \right) \tag{10.118}$$

$$M_y = I_{yx}\dot{\omega}_x + I_{yy}\dot{\omega}_y + I_{yz}\dot{\omega}_z - \left(I_{zz} - I_{xx} \right)\omega_z\omega_x$$
$$- I_{xz}\left(\omega_x^2 - \omega_z^2 \right) - \omega_y\left(\omega_x I_{yz} - \omega_z I_{xy} \right) \tag{10.119}$$

$$M_z = I_{zx}\dot{\omega}_x + I_{zy}\dot{\omega}_y + I_{zz}\dot{\omega}_z - \left(I_{xx} - I_{yy} \right)\omega_x\omega_y$$
$$- I_{xy}\left(\omega_y^2 - \omega_x^2 \right) - \omega_z\left(\omega_y I_{xz} - \omega_x I_{yz} \right) \tag{10.120}$$

Assume that we are able to rotate the body frame about its origin to find an orientation that makes $I_{ij} = 0$, for $i \neq j$. In such a principal coordinate frame, the Euler equations reduce to

$$M_1 = I_1\dot{\omega}_1 - \left(I_2 - I_3 \right)\omega_2\omega_3 \tag{10.121}$$

$$M_2 = I_2\dot{\omega}_2 - \left(I_3 - I_1 \right)\omega_3\omega_1 \tag{10.122}$$

$$M_3 = I_3\dot{\omega}_3 - \left(I_1 - I_2 \right)\omega_1\omega_2 \tag{10.123}$$

The kinetic energy of a rigid body may be found by the integral of the kinetic energy of a mass element dm, over the whole body.

$$
K = \frac{1}{2} \int_B \dot{\mathbf{v}}^2 dm = \frac{1}{2} \int_B (\boldsymbol{\omega} \times \mathbf{r}) \cdot (\boldsymbol{\omega} \times \mathbf{r}) \, dm
$$

$$
= \frac{\omega_x^2}{2} \int_B \left(y^2 + z^2 \right) dm + \frac{\omega_y^2}{2} \int_B \left(z^2 + x^2 \right) dm
$$

$$
+ \frac{\omega_z^2}{2} \int_B \left(x^2 + y^2 \right) dm
$$

$$
- \omega_x \omega_y \int_B xy \, dm - \omega_y \omega_z \int_B yz \, dm - \omega_z \omega_x \int_B zx \, dm
$$

$$
= \frac{1}{2} \left(I_{xx} \omega_x^2 + I_{yy} \omega_y^2 + I_{zz} \omega_z^2 \right)
$$

$$
- I_{xy} \omega_x \omega_y - I_{yz} \omega_y \omega_z - I_{zx} \omega_z \omega_x \tag{10.124}
$$

The kinetic energy can be rearranged and expressed in a matrix multiplication form.

$$
K = \frac{1}{2} \boldsymbol{\omega}^T I \, \boldsymbol{\omega} = \frac{1}{2} \boldsymbol{\omega} \cdot \mathbf{L} \tag{10.125}
$$

When the body frame is principal, the kinetic energy will be simplified.

$$
K = \frac{1}{2} \left(I_1 \omega_1^2 + I_2 \omega_2^2 + I_3 \omega_3^2 \right) \tag{10.126}
$$

∎

Example 336 Spherical pendulum. Analysis of physical systems is the best way to understand analytic methods. Here is deriving angular acceleration of a spherical pendulum.

A pendulum free to oscillate in any plane is called a spherical pendulum. The tip mass of such pendulum will be free to move on the surface of a sphere around the fulcrum. Consider a pendulum with a point mass m at the tip point of a long, massless, and straight bar with length l. The pendulum is hanging from a point $A\,(0, 0, 0)$ in a local coordinate frame $B_1\,(x_1, y_1, z_1)$.

To indicate the mass m, we attach a coordinate frame $B_2\,(x_2, y_2, z_2)$ to the pendulum at point A as is shown in Fig. 10.4. The pendulum makes an angle β with the vertical z_1-axis and swings in the plane (x_2, z_2) that makes an angle γ with the plane (x_1, z_1). Therefore, the transformation matrix between B_2 and B_1 is

$$
{}^2R_1 = R_{y_2, -\beta} \, R_{z_2, \gamma}
$$

$$
= \begin{bmatrix} \cos\gamma\cos\beta & \cos\beta\sin\gamma & \sin\beta \\ -\sin\gamma & \cos\gamma & 0 \\ -\cos\gamma\sin\beta & -\sin\gamma\sin\beta & \cos\beta \end{bmatrix} \tag{10.127}
$$

The position vectors of m are

$$
{}^2\mathbf{r} = \begin{bmatrix} 0 \\ 0 \\ -l \end{bmatrix} \qquad {}^1\mathbf{r} = {}^1R_2 \, {}^2\mathbf{r} = \begin{bmatrix} l\cos\gamma\sin\beta \\ l\sin\beta\sin\gamma \\ -l\cos\beta \end{bmatrix} \tag{10.128}
$$

The equation of motion of m is

$$
{}^1\mathbf{M} = I \, {}_1\boldsymbol{\alpha}_2 \tag{10.129}
$$

$$
{}^1\mathbf{r} \times m \, {}^1\mathbf{g} = ml^2 \, {}_1\boldsymbol{\alpha}_2 \tag{10.130}
$$

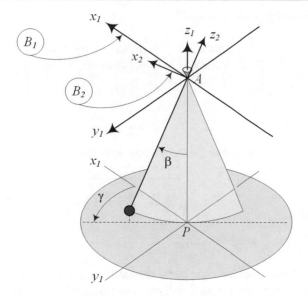

Fig. 10.4 A spherical pendulum

$$\begin{bmatrix} l\cos\gamma\sin\beta \\ l\sin\beta\sin\gamma \\ -l\cos\beta \end{bmatrix} \times m \begin{bmatrix} 0 \\ 0 \\ -g_0 \end{bmatrix} = ml^2\,{}_1\boldsymbol{\alpha}_2 \tag{10.131}$$

Therefore,

$$_1\boldsymbol{\alpha}_2 = \frac{g_0}{l}\begin{bmatrix} -\sin\beta\sin\gamma \\ \cos\gamma\sin\beta \\ 0 \end{bmatrix} \tag{10.132}$$

To find the angular acceleration of B_2 in B_1, we use ${}^2R_1$:

$$\begin{aligned}
{}^1\dot{R}_2 &= \dot{\beta}\frac{d}{d\beta}\,{}^2R_1 + \dot{\gamma}\frac{d}{d\gamma}\,{}^2R_1 \\
&= \begin{bmatrix} -\dot{\beta}c\gamma s\beta - \dot{\gamma}c\beta s\gamma & -\dot{\gamma}c\gamma & \dot{\gamma}s\beta s\gamma - \dot{\beta}c\beta c\gamma \\ \dot{\gamma}c\beta c\gamma - \dot{\beta}s\beta s\gamma & -\dot{\gamma}s\gamma & -\dot{\beta}c\beta s\gamma - \dot{\gamma}c\gamma s\beta \\ \dot{\beta}c\beta & 0 & -\dot{\beta}s\beta \end{bmatrix}
\end{aligned} \tag{10.133}$$

$$_1\tilde{\omega}_2 = {}^1\dot{R}_2\,{}^1R_2^T = \begin{bmatrix} 0 & -\dot{\gamma} & -\dot{\beta}\cos\gamma \\ \dot{\gamma} & 0 & -\dot{\beta}\sin\gamma \\ \dot{\beta}\cos\gamma & \dot{\beta}\sin\gamma & 0 \end{bmatrix} \tag{10.134}$$

$$\begin{aligned}
{}^1\ddot{R}_2 &= \ddot{\beta}\frac{d}{d\beta}\,{}^2R_1 + \dot{\beta}^2\frac{d^2}{d\beta^2}\,{}^2R_1 + \dot{\beta}\dot{\gamma}\frac{d^2}{d\gamma d\beta}\,{}^2R_1 \\
&\quad + \ddot{\gamma}\frac{d}{d\gamma}\,{}^2R_1 + \dot{\gamma}\dot{\beta}\frac{d^2}{d\beta d\gamma}\,{}^2R_1 + \dot{\gamma}^2\frac{d^2}{d\gamma^2}\,{}^2R_1
\end{aligned} \tag{10.135}$$

$$\begin{aligned}
{}_1\tilde{\alpha}_2 &= {}^1\ddot{R}_2\,{}^1R_2^T - {}_1\tilde{\omega}_2^2 \\
&= \begin{bmatrix} 0 & -\ddot{\gamma} & -\ddot{\beta}c\gamma + \dot{\beta}\dot{\gamma}s\gamma \\ \ddot{\gamma} & 0 & -\ddot{\beta}s\gamma - \dot{\beta}\dot{\gamma}c\gamma \\ \ddot{\beta}c\gamma - \dot{\beta}\dot{\gamma}s\gamma & \ddot{\beta}s\gamma + \dot{\beta}\dot{\gamma}c\gamma & 0 \end{bmatrix}
\end{aligned} \tag{10.136}$$

Therefore, the equation of motion of the pendulum would be

$$\frac{g_0}{l} \begin{bmatrix} -\sin\beta\sin\gamma \\ \cos\gamma\sin\beta \\ 0 \end{bmatrix} = \begin{bmatrix} \ddot{\beta}\sin\gamma + \dot{\beta}\dot{\gamma}\cos\gamma \\ -\ddot{\beta}\cos\gamma + \dot{\beta}\dot{\gamma}\sin\gamma \\ \ddot{\gamma} \end{bmatrix} \tag{10.137}$$

The third equation indicates that

$$\dot{\gamma} = \dot{\gamma}_0 \qquad \gamma = \dot{\gamma}_0 t + \gamma_0 \tag{10.138}$$

The second and third equations can be combined to the form a new equation,

$$\ddot{\beta} = -\sqrt{\frac{g_0^2}{l^2}\sin^2\beta + \dot{\beta}^2\dot{\gamma}_0^2} \tag{10.139}$$

which reduces to the equation of a simple pendulum if $\dot{\gamma}_0 = 0$.

Example 337 Steady rotation of a freely rotating rigid body. Free motion of a rigid body is a situation where all externally applied forces and moments are zero. Such a rigid body will move in a steady state behavior determined by initial condition.

The Newton–Euler equations of motion for a rigid body are

$$^G\mathbf{F} = m\,{}^G\dot{\mathbf{v}} \tag{10.140}$$

$$^B\mathbf{M} = I\,{}^B_G\dot{\boldsymbol\omega}_B + {}^B_G\boldsymbol\omega_B \times {}^B\mathbf{L} \tag{10.141}$$

Consider a situation where the resultant applied force and moment on the body are zero.

$$\mathbf{F} = 0 \qquad \mathbf{M} = 0 \tag{10.142}$$

Based on the Newton equation, the velocity of the mass center will be constant in the global coordinate frame. The Euler equation reduces to

$$\dot{\omega}_1 = \frac{I_2 - I_3}{I_1}\omega_2\omega_3 \tag{10.143}$$

$$\dot{\omega}_2 = \frac{I_3 - I_1}{I_{22}}\omega_3\omega_1 \tag{10.144}$$

$$\dot{\omega}_3 = \frac{I_1 - I_2}{I_3}\omega_1\omega_2 \tag{10.145}$$

showing that the angular velocity can be constant if

$$I_1 = I_2 = I_3 \tag{10.146}$$

or if two principal moments of inertia, say I_1 and I_2, are zero and the third angular velocity, in this case ω_3, is initially zero, or if the angular velocity vector is initially parallel to a principal axis.

Example 338 Angular momentum of a two-link manipulator. An example of a two arm manipulator treated as connected rigid bodies.

A two-link manipulator is shown in Fig. 10.5. Link A rotates with angular velocity $\dot{\varphi}$ about the z-axis of its local coordinate frame. Link B is attached to link A and has angular velocity $\dot{\psi}$ with respect to A about the x_A-axis. We assume A and G to be coincident at $\varphi = 0$; therefore, the rotation matrix between A and G is

$$^G R_A = \begin{bmatrix} \cos\varphi(t) & -\sin\varphi(t) & 0 \\ \sin\varphi(t) & \cos\varphi(t) & 0 \\ 0 & 0 & 1 \end{bmatrix} \tag{10.147}$$

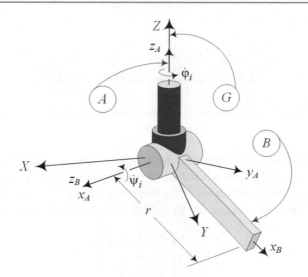

Fig. 10.5 A two-link manipulator

The frame B is related to the frame A by Euler angles $\varphi = 90$ deg, $\theta = 90$ deg, and $\psi = \psi$; hence,

$$
{}^A R_B = \begin{bmatrix} c\pi c\psi - c\pi s\pi s\psi & -c\pi s\psi - c\pi c\psi s\pi & s\pi s\pi \\ c\psi s\pi + c\pi c\pi s\psi & -s\pi s\psi + c\pi c\pi c\psi & -c\pi s\pi \\ s\pi s\psi & s\pi c\psi & c\pi \end{bmatrix}
$$

$$
\begin{bmatrix} -\cos\psi & \sin\psi & 0 \\ \sin\psi & \cos\psi & 0 \\ 0 & 0 & -1 \end{bmatrix} \tag{10.148}
$$

and therefore,

$$
{}^G R_B = {}^G R_A \, {}^A R_B \tag{10.149}
$$

$$
= \begin{bmatrix} -\cos\varphi\cos\psi - \sin\varphi\sin\psi & \cos\varphi\sin\psi - \cos\psi\sin\varphi & 0 \\ \cos\varphi\sin\psi - \cos\psi\sin\varphi & \cos\varphi\cos\psi + \sin\varphi\sin\psi & 0 \\ 0 & 0 & -1 \end{bmatrix}
$$

The angular velocities of A in G and B in A are

$$
{}_G\boldsymbol{\omega}_A = \dot\varphi \hat{K} \qquad {}_A\boldsymbol{\omega}_B = \dot\psi \hat{\imath}_A \tag{10.150}
$$

Moment of inertia matrices for the arms A and B can be defined as

$$
{}^A I_A = \begin{bmatrix} I_{A1} & 0 & 0 \\ 0 & I_{A2} & 0 \\ 0 & 0 & I_{A3} \end{bmatrix} \qquad {}^B I_B = \begin{bmatrix} I_{B1} & 0 & 0 \\ 0 & I_{B2} & 0 \\ 0 & 0 & I_{B3} \end{bmatrix} \tag{10.151}
$$

These moments of inertia must be transformed into the global frame.

$$
{}^G I_A = {}^G R_B \, {}^A I_A \, {}^G R_A{}^T \qquad {}^G I_B = {}^G R_B \, {}^B I_B \, {}^G R_B^T \tag{10.152}
$$

The total angular momentum of the manipulator is

$$
{}^G \mathbf{L} = {}^G \mathbf{L}_A + {}^G \mathbf{L}_B \tag{10.153}
$$

where

$$^G \mathbf{L}_A = {}^G I_A \, {}_G \boldsymbol{\omega}_A \tag{10.154}$$

$$^G \mathbf{L}_B = {}^G I_B \, {}_G \boldsymbol{\omega}_B = {}^G I_B \left({}^G_A \boldsymbol{\omega}_B + {}_G \boldsymbol{\omega}_A \right) \tag{10.155}$$

10.4 Mass Moment Matrix

In analysis of the motion of rigid bodies, two types of integrals appear that belong to the geometry of the body. The first type defines the center of mass and is important when the translation motion of the body is considered. The second is the *mass moment* that appears when the rotational motion of the body is considered. The mass moment is also called *moment of inertia*, *centrifugal moments*, or *deviation moments*. Every rigid body has a 3×3 moment of inertia matrix $[I]$.

$$[I] = \begin{bmatrix} I_{xx} & I_{xy} & I_{xz} \\ I_{yx} & I_{yy} & I_{yz} \\ I_{zx} & I_{zy} & I_{zz} \end{bmatrix} \tag{10.156}$$

The diagonal elements I_{ij}, $i = j$ are called *polar mass moments*,

$$I_{xx} = I_x = \int_B \left(y^2 + z^2 \right) dm \tag{10.157}$$

$$I_{yy} = I_y = \int_B \left(z^2 + x^2 \right) dm \tag{10.158}$$

$$I_{zz} = I_z = \int_B \left(x^2 + y^2 \right) dm \tag{10.159}$$

and the off-diagonal elements I_{ij}, $i \neq j$ are called *products mass moments*

$$I_{xy} = I_{yx} = - \int_B xy \, dm \tag{10.160}$$

$$I_{yz} = I_{zy} = - \int_B yz \, dm \tag{10.161}$$

$$I_{zx} = I_{xz} = - \int_B zx \, dm \tag{10.162}$$

The elements of $[I]$ are second moments of inertia about a body coordinate frame attached to the mass center C of the body. Therefore, $[I]$ is a frame-dependent quantity and must be written as $^B I$ to show the frame it is computed in.

$$
\begin{aligned}
^B I &= \int_B \begin{bmatrix} y^2 + z^2 & -xy & -zx \\ -xy & z^2 + x^2 & -yz \\ -zx & -yz & x^2 + y^2 \end{bmatrix} dm \\
&= \int_B \left(r^2 \mathbf{I} - \mathbf{r} \mathbf{r}^T \right) dm = \int_B -\tilde{r} \, \tilde{r} \, dm
\end{aligned} \tag{10.163}
$$

Mass moments can be transformed from a coordinate frame B_1 into another coordinate frame B_2, both installed at the mass center of the body, according to the rule of the *rotated-axes theorem*.

$$^2 I = {}^2 R_1 \, {}^1 I \, {}^2 R_1^T \tag{10.164}$$

Transformation of the mass moment from a central frame B_1 located at $^{B_2} \mathbf{r}_C$ into another frame B_2, which is parallel to B_1, is, according to the rule of *parallel-axes theorem*.

$$^2 I = {}^1 I + m \, \tilde{r}_C \, \tilde{r}_C^T \tag{10.165}$$

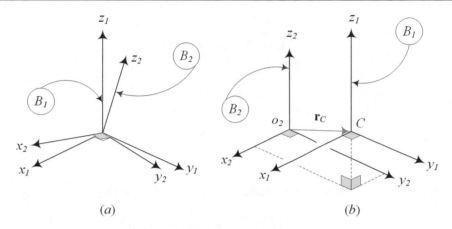

Fig. 10.6 (a) Two coordinate frames with a common origin at the mass center of a rigid body. (b) A central coordinate frame B_1 and a translated frame B_2

If the body coordinate frame $Oxyz$ is located such that the products of inertia vanish, the body coordinate frame is the *principal coordinate frame* and the associated mass moments are *principal mass moments*. Principal axes and principal mass moments can be found by solving the following equation for unknown eigenvalues, λ.

$$\begin{vmatrix} I_{xx} - \lambda & I_{xy} & I_{xz} \\ I_{yx} & I_{yy} - \lambda & I_{yz} \\ I_{zx} & I_{zy} & I_{zz} - \lambda \end{vmatrix} = 0 \tag{10.166}$$

$$\det\left(\left[I_{ij}\right] - \lambda\left[\delta_{ij}\right]\right) = 0 \tag{10.167}$$

Equation (10.167) is a cubic equation in λ and we obtain three eigenvalues that are the principal mass moments.

$$I_1 = I_x \qquad I_2 = I_y \qquad I_3 = I_z \tag{10.168}$$

Proof Consider two coordinate frames with a common origin at the mass center of a rigid body as shown in Fig. 10.6a. The angular velocity vector and angular momentum vector of a rigid body transform from the frame B_1 into B_2 by vector transformation rule.

$$^2\boldsymbol{\omega} = {}^2R_1\,{}^1\boldsymbol{\omega} \tag{10.169}$$

$$^2\mathbf{L} = {}^2R_1\,{}^1\mathbf{L} \tag{10.170}$$

However, $\mathbf{L}$ and $\boldsymbol{\omega}$ are related according to the mass moment matrix,

$$^1\mathbf{L} = {}^1I\,{}^1\boldsymbol{\omega} \tag{10.171}$$

and therefore,

$$^2\mathbf{L} = {}^2R_1\,{}^1I\,{}^2R_1^T\,{}^2\boldsymbol{\omega} = {}^2I\,{}^2\boldsymbol{\omega} \tag{10.172}$$

which shows how to transfer the mass moment matrix from the coordinate frame B_1 into a rotated frame B_2,

$$^2I = {}^2R_1\,{}^1I\,{}^2R_1^T \tag{10.173}$$

Now consider a central frame B_1, shown in Fig. 10.6b, at $^2\mathbf{r}_C$, which translates into a fixed frame B_2 such that their axes remain parallel. The angular velocity and angular momentum of the rigid body transform from frame B_1 into frame B_2 by the following relationships.

$$^2\boldsymbol{\omega} = {}^1\boldsymbol{\omega} \tag{10.174}$$

$$^2\mathbf{L} = {}^1\mathbf{L} + (\mathbf{r}_C \times m\mathbf{v}_C) \tag{10.175}$$

Therefore,

$$\begin{aligned}
^2\mathbf{L} &= {}^1\mathbf{L} + m\,{}^2\mathbf{r}_C \times \left({}^2\boldsymbol{\omega} \times {}^2\mathbf{r}_C\right) \\
&= {}^1\mathbf{L} + \left(m\,{}^2\tilde{r}_C\,{}^2\tilde{r}_C^T\right){}^2\boldsymbol{\omega} \\
&= \left({}^1 I + m\,{}^2\tilde{r}_C\,{}^2\tilde{r}_C^T\right){}^2\boldsymbol{\omega}
\end{aligned} \tag{10.176}$$

which shows how to transfer the mass moment from frame B_1 into a parallel frame B_2

$$^2 I = {}^{B_1} I + m\,\tilde{r}_C\,\tilde{r}_C^T \tag{10.177}$$

This is the *parallel-axes theorem*, which is also called the *Huygens–Steiner theorem*.

Referring to Eq. (10.173) for transformation of the mass moment into a rotated frame, we can always find a frame B_2 in which $^2 I$ is diagonal. In such a frame, we have

$$^2 R_1\,{}^1 I = {}^2 I\,{}^2 R_1 \tag{10.178}$$

$$\begin{bmatrix} r_{11} & r_{12} & r_{13} \\ r_{21} & r_{22} & r_{23} \\ r_{31} & r_{32} & r_{33} \end{bmatrix} \begin{bmatrix} I_{xx} & I_{xy} & I_{xz} \\ I_{yx} & I_{yy} & I_{yz} \\ I_{zx} & I_{zy} & I_{zz} \end{bmatrix}$$
$$= \begin{bmatrix} I_1 & 0 & 0 \\ 0 & I_2 & 0 \\ 0 & 0 & I_3 \end{bmatrix} \begin{bmatrix} r_{11} & r_{12} & r_{13} \\ r_{21} & r_{22} & r_{23} \\ r_{31} & r_{32} & r_{33} \end{bmatrix} \tag{10.179}$$

which shows that I_1, I_2, and I_3 are eigenvalues of $^1 I$. These eigenvalues can be found by solving the following equation for λ:

$$\begin{vmatrix} I_{xx} - \lambda & I_{xy} & I_{xz} \\ I_{yx} & I_{yy} - \lambda & I_{yz} \\ I_{zx} & I_{zy} & I_{zz} - \lambda \end{vmatrix} = 0 \tag{10.180}$$

The eigenvalues I_1, I_2, and I_3 are *principal mass moments*, and their associated eigenvectors are called *principal directions*. The coordinate frame made by the eigenvectors is the *principal body coordinate frame*. In the principal coordinate frame, the rigid body angular momentum will be

$$\begin{bmatrix} L_1 \\ L_2 \\ L_3 \end{bmatrix} = \begin{bmatrix} I_1 & 0 & 0 \\ 0 & I_2 & 0 \\ 0 & 0 & I_3 \end{bmatrix} \begin{bmatrix} \omega_1 \\ \omega_2 \\ \omega_3 \end{bmatrix} \tag{10.181}$$

∎

Example 339 Principal mass moments. An easy example of how to determine principal mass moments and principal axes of a mass moment matrix associated to a symmetric rigid body with respect to the z-axis.

Consider a mass moment matrix $[I]$ for a rigid body that is symmetric with respect to the z-axis.

$$[I] = \begin{bmatrix} 20 & -2 & 0 \\ -2 & 30 & 0 \\ 0 & 0 & 40 \end{bmatrix} \tag{10.182}$$

To determine its principal mass moments, we set up the determinant (10.167)

$$\begin{vmatrix} 20 - \lambda & -2 & 0 \\ -2 & 30 - \lambda & 0 \\ 0 & 0 & 40 - \lambda \end{vmatrix} = 0 \tag{10.183}$$

which leads to a third degree characteristic equation.

$$(20 - \lambda)(30 - \lambda)(40 - \lambda) - 4(40 - \lambda) = 0 \tag{10.184}$$

Three roots of Eq. (10.184) are

$$I_1 = 30.385 \qquad I_2 = 19.615 \qquad I_3 = 40 \tag{10.185}$$

and therefore, the principal moment of inertia matrix is

$$I = \begin{bmatrix} 30.385 & 0 & 0 \\ 0 & 19.615 & 0 \\ 0 & 0 & 40 \end{bmatrix} \tag{10.186}$$

The direction of a principal axis x_i is determined by solving the following set of algebraic equations for direction cosines.

$$\begin{bmatrix} I_{xx} - I_i & I_{xy} & I_{xz} \\ I_{yx} & I_{yy} - I_i & I_{yz} \\ I_{zx} & I_{zy} & I_{zz} - I_i \end{bmatrix} \begin{bmatrix} \cos \alpha_i \\ \cos \beta_i \\ \cos \gamma_i \end{bmatrix} = \begin{bmatrix} 0 \\ 0 \\ 0 \end{bmatrix} \tag{10.187}$$

The direction cosines must satisfy a constraint equation.

$$\cos^2 \alpha_i + \cos^2 \beta_i + \cos^2 \gamma_i = 1 \tag{10.188}$$

For the first principal moment of inertia $I_1 = 30.385$ we have

$$\begin{bmatrix} 20 - 30.385 & -2 & 0 \\ -2 & 30 - 30.385 & 0 \\ 0 & 0 & 40 - 30.385 \end{bmatrix} \begin{bmatrix} \cos \alpha_1 \\ \cos \beta_1 \\ \cos \gamma_1 \end{bmatrix} = \begin{bmatrix} 0 \\ 0 \\ 0 \end{bmatrix} \tag{10.189}$$

or

$$-10.385 \cos \alpha_1 - 2 \cos \beta_1 + 0 = 0 \tag{10.190}$$

$$-2 \cos \alpha_1 - 0.385 \cos \beta_1 + 0 = 0 \tag{10.191}$$

$$0 + 0 + 9.615 \cos \gamma_1 = 0 \tag{10.192}$$

and we obtain

$$\alpha_1 = 79.1 \text{ deg} \qquad \beta_1 = 169.1 \text{ deg} \qquad \gamma_1 = 90.0 \text{ deg} \tag{10.193}$$

Using $I_2 = 19.615$ for the second principal axis,

$$\begin{bmatrix} 20 - 19.62 & -2 & 0 \\ -2 & 30 - 19.62 & 0 \\ 0 & 0 & 40 - 19.62 \end{bmatrix} \begin{bmatrix} \cos \alpha_2 \\ \cos \beta_2 \\ \cos \gamma_2 \end{bmatrix} = \begin{bmatrix} 0 \\ 0 \\ 0 \end{bmatrix} \tag{10.194}$$

we obtain

$$\alpha_2 = 10.9 \text{ deg} \qquad \beta_2 = 79.1 \text{ deg} \qquad \gamma_2 = 90.0 \text{ deg} \tag{10.195}$$

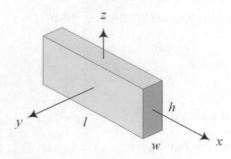

Fig. 10.7 A homogenous rectangular link

The third principal axis is for $I_3 = 40$,

$$
\begin{bmatrix}
20 - 40 & -2 & 0 \\
-2 & 30 - 40 & 0 \\
0 & 0 & 40 - 40
\end{bmatrix}
\begin{bmatrix}
\cos \alpha_3 \\
\cos \beta_3 \\
\cos \gamma_3
\end{bmatrix}
=
\begin{bmatrix}
0 \\
0 \\
0
\end{bmatrix}
\tag{10.196}
$$

which leads to

$$
\alpha_3 = 90.0 \text{ deg} \qquad \beta_3 = 90.0 \text{ deg} \qquad \gamma_3 = 0.0 \text{ deg}
\tag{10.197}
$$

Example 340 Mass moment of rigid rectangular link. A simple example to show how mass moment volume integral will be calculated.

Consider a homogenous rectangular link with mass m, length l, width w, and height h, as shown in Fig. 10.7. The body central coordinate frame is attached to the link at its mass center. To calculate the mass moments matrix of the link by integral method, we begin with calculating I_{xx}.

$$
\begin{aligned}
I_{xx} &= \int_B \left(y^2 + z^2 \right) dm = \int_v \left(y^2 + z^2 \right) \rho \, dv \\
&= \frac{m}{lwh} \int_v \left(y^2 + z^2 \right) dv \\
&= \frac{m}{lwh} \int_{-h/2}^{h/2} \int_{-w/2}^{w/2} \int_{-l/2}^{l/2} \left(y^2 + z^2 \right) dx \, dy \, dz \\
&= \frac{m}{12} \left(w^2 + h^2 \right)
\end{aligned}
\tag{10.198}
$$

The I_{yy} and I_{zz} can be calculated similarly.

$$
I_{yy} = \frac{m}{12} \left(h^2 + l^2 \right) \qquad I_{zz} = \frac{m}{12} \left(l^2 + w^2 \right)
\tag{10.199}
$$

Because the coordinate frame is central, the products of inertia must be zero. To show this, we examine I_{xy}.

$$
\begin{aligned}
I_{xy} = I_{yx} &= -\int_B xy \, dm = \int_v xy \rho \, dv \\
&= \frac{m}{lwh} \int_{-h/2}^{h/2} \int_{-w/2}^{w/2} \int_{-l/2}^{l/2} xy \, dx \, dy \, dz = 0
\end{aligned}
\tag{10.200}
$$

Therefore, the mass moment matrix for the rigid rectangular link in its principal frame is

$$
I =
\begin{bmatrix}
\frac{m}{12} \left(w^2 + h^2 \right) & 0 & 0 \\
0 & \frac{m}{12} \left(h^2 + l^2 \right) & 0 \\
0 & 0 & \frac{m}{12} \left(l^2 + w^2 \right)
\end{bmatrix}
\tag{10.201}
$$

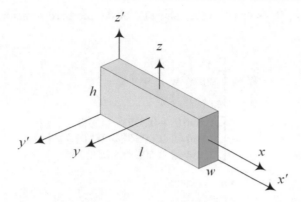

Fig. 10.8 A rigid rectangular link in the principal and non-principal frames

Example 341 Translation of the mass moment matrix. An example for expressing a given mass moment matrix in another translated coordinate frame.

The mass moment matrix of the rigid link shown in Fig. 10.8, in the principal frame $B(oxyz)$, is given in Eq. (10.201). The mass moment matrix in the parallel and non-principal frame $B'(ox'y'z')$ can be found by applying the parallel-axes transformation formula (10.165).

$$^{B'}I = \,^{B}I + m \,^{B'}\tilde{r}_C \,^{B'}\tilde{r}_C^T \tag{10.202}$$

The center of mass position vector in B'-frame is

$$^{B'}\mathbf{r}_C = \frac{1}{2}\begin{bmatrix} l \\ w \\ h \end{bmatrix} \tag{10.203}$$

and therefore,

$$^{B'}\tilde{r}_C = \frac{1}{2}\begin{bmatrix} 0 & -h & w \\ h & 0 & -l \\ -w & l & 0 \end{bmatrix} \tag{10.204}$$

that provides mass moment matrix in the B'-frame.

$$^{B'}I = \begin{bmatrix} \frac{1}{3}h^2m + \frac{1}{3}mw^2 & -\frac{1}{4}lmw & -\frac{1}{4}hlm \\ -\frac{1}{4}lmw & \frac{1}{3}h^2m + \frac{1}{3}l^2m & -\frac{1}{4}hmw \\ -\frac{1}{4}hlm & -\frac{1}{4}hmw & \frac{1}{3}l^2m + \frac{1}{3}mw^2 \end{bmatrix} \tag{10.205}$$

Example 342 Principal rotation matrix.

Consider a body inertia matrix as

$$[I] = \begin{bmatrix} 2/3 & -1/2 & -1/2 \\ -1/2 & 5/3 & -1/4 \\ -1/2 & -1/4 & 5/3 \end{bmatrix} \tag{10.206}$$

The eigenvalues and eigenvectors of $[I]$ are

$$I_1 = 0.2413 \qquad \begin{bmatrix} 2.351 \\ 1 \\ 1 \end{bmatrix} \equiv \begin{bmatrix} 0.8569 \\ 0.36448 \\ 0.36448 \end{bmatrix} = \mathbf{w}_1 \tag{10.207}$$

$$I_2 = 1.8421 \qquad \begin{bmatrix} -0.851 \\ 1 \\ 1 \end{bmatrix} \equiv \begin{bmatrix} -0.5156 \\ 0.60588 \\ 0.60588 \end{bmatrix} = \mathbf{w}_2 \tag{10.208}$$

$$I_3 = 1.9167 \qquad \begin{bmatrix} 0 \\ -1 \\ 1 \end{bmatrix} \equiv \begin{bmatrix} 0.0 \\ -0.70711 \\ 0.70711 \end{bmatrix} = \mathbf{w}_3 \tag{10.209}$$

The normalized eigenvector matrix W is equal to the transpose of the required transformation matrix to make the mass moment matrix diagonal.

$$
W = \begin{bmatrix} | & | & | \\ \mathbf{w}_1 & \mathbf{w}_2 & \mathbf{w}_3 \\ | & | & | \end{bmatrix} = {}^2R_1^T
$$

$$
= \begin{bmatrix} 0.856\,9 & -0.515\,6 & 0.0 \\ 0.364\,48 & 0.605\,88 & -0.707\,11 \\ 0.364\,48 & 0.605\,88 & 0.707\,11 \end{bmatrix} \tag{10.210}
$$

We may verify that

$$
{}^2I \approx {}^2R_1\,{}^1I\,{}^2R_1^T = W^T\,{}^1I\,W
$$

$$
= \begin{bmatrix} 0.2413 & -1 \times 10^{-4} & 0.0 \\ -1 \times 10^{-4} & 1.842\,1 & -1 \times 10^{-19} \\ 0.0 & 0.0 & 1.916\,7 \end{bmatrix} \tag{10.211}
$$

Example 343 ★ Coefficients of the characteristic equation. The characteristic equation for all mass moment matrices is a third degree algebraic equation. The coefficients of the equation are particular functions of elements of $[I]$. Here are those functions.

The determinant (10.180) for calculating the principal moments of inertia,

$$
\begin{vmatrix} I_{xx} - \lambda & I_{xy} & I_{xz} \\ I_{yx} & I_{yy} - \lambda & I_{yz} \\ I_{zx} & I_{zy} & I_{zz} - \lambda \end{vmatrix} = 0 \tag{10.212}
$$

leads to a third degree equation of λ, called the characteristic equation.

$$
\lambda^3 - a_1 \lambda^2 + a_2 \lambda - a_3 = 0 \tag{10.213}
$$

The coefficients of the characteristic equation are called the principal invariants of $[I]$. The coefficients of the characteristic equation can directly be found from the following equations of the mass moments.

$$
a_1 = I_{xx} + I_{yy} + I_{zz} = \operatorname{tr}[I] \tag{10.214}
$$

$$
a_2 = I_{xx}I_{yy} + I_{yy}I_{zz} + I_{zz}I_{xx} - I_{xy}^2 - I_{yz}^2 - I_{zx}^2
$$

$$
= \begin{vmatrix} I_{xx} & I_{xy} \\ I_{yx} & I_{yy} \end{vmatrix} + \begin{vmatrix} I_{yy} & I_{yz} \\ I_{zy} & I_{zz} \end{vmatrix} + \begin{vmatrix} I_{xx} & I_{xz} \\ I_{zx} & I_{zz} \end{vmatrix}
$$

$$
= \frac{1}{2}\left(a_1^2 - \operatorname{tr}[I^2]\right) \tag{10.215}
$$

$$
a_3 = I_{xx}I_{yy}I_{zz} + I_{xy}I_{yz}I_{zx} + I_{zy}I_{yx}I_{xz}
$$

$$
\quad - \left(I_{xx}I_{yz}I_{zy} + I_{yy}I_{zx}I_{xz} + I_{zz}I_{xy}I_{yx}\right)
$$

$$
= I_{xx}I_{yy}I_{zz} + 2I_{xy}I_{yz}I_{zx} - \left(I_{xx}I_{yz}^2 + I_{yy}I_{zx}^2 + I_{zz}I_{xy}^2\right)
$$

$$
= \det[I] \tag{10.216}
$$

Example 344 ★ The principal mass moments are coordinate invariants. Mass moment matrix in different coordinate frames has different values; however, the principal values of all of them are the same. Such parameters that remain the same regardless of the coordinate frame are called coordinate invariant. Here is a proof.

The roots of the mass moment characteristic equation are the principal mass moments and all are real numbers. The principal mass moments are extreme values. That is, the principal mass moments determine the smallest and largest values of I_{ii} for a rigid body. Because the smallest and largest values of I_{ii} do not depend on the choice of the body coordinate frame, the solution of the characteristic equation is not dependent on the coordinate frame. In other words, if I_1, I_2, I_3 are the principal mass moments for ^1I in B_1-frame, the principal moments of inertia for ^2I in B_2-frame are also I_1, I_2, I_3, when

$$^2I = {}^2R_1 \, {}^1I \, {}^2R_1^T \tag{10.217}$$

We conclude that I_1, I_2, I_3 are coordinate invariants of the matrix $[I]$, and therefore any quantity that depends on I_1, I_2, I_3 is also coordinate invariant. The matrix $[I]$ has only three independent invariants and every other invariant can be expressed in terms of I_1, I_2, I_3.

Because I_1, I_2, I_3 are the solutions of the characteristic equation of $[I]$ given in (10.213), we may write the determinant (10.180) in the following form.

$$(\lambda - I_1)(\lambda - I_2)(\lambda - I_3) = 0 \tag{10.218}$$

Expanded form of this equation is

$$\lambda^3 - (I_1 + I_2 + I_3)\lambda^2 + (I_1 I_2 + I_2 I_3 + I_3 I_1)\lambda - I_1 I_2 I_3 = 0 \tag{10.219}$$

By comparing (10.219) and (10.213) we conclude that

$$a_1 = I_{xx} + I_{yy} + I_{zz} = I_1 + I_2 + I_3 \tag{10.220}$$

$$a_2 = I_{xx} I_{yy} + I_{yy} I_{zz} + I_{zz} I_{xx} - I_{xy}^2 - I_{yz}^2 - I_{zx}^2$$

$$= I_1 I_2 + I_2 I_3 + I_3 I_1 \tag{10.221}$$

$$a_3 = I_{xx} I_{yy} I_{zz} + 2 I_{xy} I_{yz} I_{zx} - \left(I_{xx} I_{yz}^2 + I_{yy} I_{zx}^2 + I_{zz} I_{xy}^2 \right)$$

$$= I_1 I_2 I_3 \tag{10.222}$$

Being able to express the coefficients a_1, a_2, a_3 as functions of I_1, I_2, I_3 determines that the coefficients of the characteristic equation are also coordinate invariant.

Example 345 ★ Mass moment with respect to a plane, a line, and a point. In robotics, we are only interested in mass moments matrix of rigid bodies as defined in this section. However, mass moments may also be defined with respect to a plane or a line or a point. Here is their definition.

The mass moment of a system of particles may be defined with respect to a plane, a line, or a point as the sum of the products of the mass of the particles into the square of the perpendicular distance from the particle to the plane, the line, or the point. For a continuous body, the sum would be replaced by definite integral over the volume of the body.

The mass moments with respect to the xy, yz, and zx planes are

$$I_{z^2} = \int_B z^2 dm \qquad I_{y^2} = \int_B y^2 dm \qquad I_{x^2} = \int_B x^2 dm \tag{10.223}$$

The mass moments with respect to the x-, y-, and z-axes are

$$I_x = \int_B \left(y^2 + z^2 \right) dm \tag{10.224}$$

$$I_y = \int_B \left(z^2 + x^2 \right) dm \tag{10.225}$$

$$I_z = \int_B \left(x^2 + y^2 \right) dm \tag{10.226}$$

and therefore,

$$I_x = I_{y^2} + I_{z^2} \qquad I_y = I_{z^2} + I_{x^2} \qquad I_z = I_{x^2} + I_{y^2} \tag{10.227}$$

The mass moment with respect to the origin is

$$I_o = \int_B \left(x^2 + y^2 + z^2\right) dm = I_{x^2} + I_{y^2} + I_{z^2}$$

$$= \frac{1}{2} \left(I_x + I_y + I_z\right) \tag{10.228}$$

Because the choice of the coordinate frame is arbitrary, we can say that the mass moment with respect to a line is the sum of the mass moments with respect to any two mutually orthogonal planes that pass through the line. The mass moment with respect to a point has similar meaning for three mutually orthogonal planes intersecting at the point.

10.5   Lagrange's Form of Newton's Equations

Newton's equation of motion can be transformed into

$$\frac{d}{dt}\left(\frac{\partial K}{\partial \dot{q}_r}\right) - \frac{\partial K}{\partial q_r} = F_r \qquad r = 1, 2, \cdots n \tag{10.229}$$

where

$$F_r = \sum_{i=1}^n \left(F_{ix}\frac{\partial f_i}{\partial q_1} + F_{iy}\frac{\partial g_i}{\partial q_2} + F_{iz}\frac{\partial h_i}{\partial q_n}\right) \tag{10.230}$$

Equation (10.229) is called the *Lagrange equation of motion*, where K is the kinetic energy of the n DOF system, q_r, $r = 1, 2, \cdots, n$, are the generalized coordinates of the system, $\mathbf{F} = \begin{bmatrix} F_{ix} & F_{iy} & F_{iz} \end{bmatrix}^T$ is the external force acting on the ith particle of the system, and F_r is the generalized force associated to q_r.

Proof Let m_i be the mass of one of the particles of a system, and let (x_i, y_i, z_i) be its Cartesian coordinates in a globally fixed coordinate frame. Assume that the coordinates of every individual particle are functions of another set of coordinates $q_1, q_2, q_3, \cdots, q_n$ and possibly time t.

$$x_i = f_i(q_1, q_2, q_3, \cdots, q_n, t) \tag{10.231}$$

$$y_i = g_i(q_1, q_2, q_3, \cdots, q_n, t) \tag{10.232}$$

$$z_i = h_i(q_1, q_2, q_3, \cdots, q_n, t) \tag{10.233}$$

If F_{xi}, F_{yi}, F_{zi} are components of the total acting force on the particle m_i, then the Newton equations of motion for the particle would be

$$F_{xi} = m_i\ddot{x}_i \tag{10.234}$$

$$F_{yi} = m_i\ddot{y}_i \tag{10.235}$$

$$F_{zi} = m_i\ddot{z}_i \tag{10.236}$$

We multiply both sides of these equations, respectively, by

$$\frac{\partial f_i}{\partial q_r} \qquad \frac{\partial g_i}{\partial q_r} \qquad \frac{\partial h_i}{\partial q_r} \tag{10.237}$$

and add them up for all the n particles.

$$\sum_{i=1}^n m_i \left(\ddot{x}_i\frac{\partial f_i}{\partial q_r} + \ddot{y}_i\frac{\partial g_i}{\partial q_r} + \ddot{z}_i\frac{\partial h_i}{\partial q_r}\right)$$

$$= \sum_{i=1}^n \left(F_{xi}\frac{\partial f_i}{\partial q_r} + F_{yi}\frac{\partial g_i}{\partial q_r} + F_{zi}\frac{\partial h_i}{\partial q_r}\right) \tag{10.238}$$

Taking a time derivative of Eq. (10.231),

$$\dot{x}_i = \frac{\partial f_i}{\partial q_1}\dot{q}_1 + \frac{\partial f_i}{\partial q_2}\dot{q}_2 + \frac{\partial f_i}{\partial q_3}\dot{q}_3 + \cdots + \frac{\partial f_i}{\partial q_n}\dot{q}_n + \frac{\partial f_i}{\partial t} \tag{10.239}$$

we find

$$\frac{\partial \dot{x}_i}{\partial \dot{q}_r} = \frac{\partial}{\partial \dot{q}_r}\left(\frac{\partial f_i}{\partial q_1}\dot{q}_1 + \frac{\partial f_i}{\partial q_2}\dot{q}_2 + \cdots + \frac{\partial f_i}{\partial q_n}\dot{q}_n + \frac{\partial f_i}{\partial t}\right)$$

$$= \frac{\partial f_i}{\partial q_r} \tag{10.240}$$

and therefore,

$$\ddot{x}_i\frac{\partial f_i}{\partial q_r} = \ddot{x}_i\frac{\partial \dot{x}_i}{\partial \dot{q}_r} = \frac{d}{dt}\left(\dot{x}_i\frac{\partial \dot{x}_i}{\partial \dot{q}_r}\right) - \dot{x}_i\frac{d}{dt}\left(\frac{\partial \dot{x}_i}{\partial \dot{q}_r}\right) \tag{10.241}$$

However,

$$\dot{x}_i\frac{d}{dt}\left(\frac{\partial \dot{x}_i}{\partial \dot{q}_r}\right) = \dot{x}_i\frac{d}{dt}\left(\frac{\partial f_i}{\partial q_r}\right)$$

$$= \dot{x}_i\left(\frac{\partial^2 f_i}{\partial q_1 \partial q_r}\dot{q}_1 + \cdots + \frac{\partial^2 f_i}{\partial q_n \partial q_r}\dot{q}_n + \frac{\partial^2 f_i}{\partial t \partial q_r}\right)$$

$$= \dot{x}_i\frac{\partial}{\partial q_r}\left(\frac{\partial f_i}{\partial q_1}\dot{q}_1 + \frac{\partial f_i}{\partial q_2}\dot{q}_2 + \cdots + \frac{\partial f_i}{\partial q_n}\dot{q}_n + \frac{\partial f_i}{\partial t}\right)$$

$$= \dot{x}_i\frac{\partial \dot{x}_i}{\partial q_r} \tag{10.242}$$

and we have

$$\ddot{x}_i\frac{\partial \dot{x}_i}{\partial \dot{q}_r} = \frac{d}{dt}\left(\dot{x}_i\frac{\partial \dot{x}_i}{\partial \dot{q}_r}\right) - \dot{x}_i\frac{\partial \dot{x}_i}{\partial q_r} \tag{10.243}$$

which is equal to

$$\ddot{x}_i\frac{\dot{x}_i}{\dot{q}_r} = \frac{d}{dt}\left[\frac{\partial}{\partial \dot{q}_r}\left(\frac{1}{2}\dot{x}_i^2\right)\right] - \frac{\partial}{\partial q_r}\left(\frac{1}{2}\dot{x}_i^2\right) \tag{10.244}$$

Now substituting (10.241) and (10.244) into the left-hand side of (10.238) leads to

$$\sum_{i=1}^{n} m_i\left(\ddot{x}_i\frac{\partial f_i}{\partial q_r} + \ddot{y}_i\frac{\partial g_i}{\partial q_r} + \ddot{z}_i\frac{\partial h_i}{\partial q_r}\right)$$

$$= \sum_{i=1}^{n} m_i\frac{d}{dt}\left[\frac{\partial}{\partial \dot{q}_r}\left(\frac{1}{2}\dot{x}_i^2 + \frac{1}{2}\dot{y}_i^2 + \frac{1}{2}\dot{z}_i^2\right)\right]$$

$$- \sum_{i=1}^{n} m_i\frac{\partial}{\partial q_r}\left(\frac{1}{2}\dot{x}_i^2 + \frac{1}{2}\dot{y}_i^2 + \frac{1}{2}\dot{z}_i^2\right)$$

$$- \frac{1}{2}\sum_{i=1}^{n} m_i\frac{d}{dt}\left[\frac{\partial}{\partial \dot{q}_r}\left(\dot{x}_i^2 + \dot{y}_i^2 + \dot{z}_i^2\right)\right]$$

$$- \frac{1}{2}\sum_{i=1}^{n} m_i\frac{\partial}{\partial q_r}\left(\dot{x}_i^2 + \dot{y}_i^2 + \dot{z}_i^2\right)$$

$$= \frac{d}{dt}\frac{\partial K}{\partial \dot{q}_r} - \frac{\partial K}{\partial q_r} \tag{10.245}$$

where

$$\frac{1}{2} \sum_{i=1}^{n} m_i \left(\dot{x}_i^2 + \dot{y}_i^2 + \dot{z}_i^2 \right) = K \tag{10.246}$$

K is the *kinetic energy* of the system. Therefore, the Newton equations of motion (10.234), (10.235), (10.236) are converted into

$$\frac{d}{dt} \left(\frac{\partial K}{\partial \dot{q}_r} \right) - \frac{\partial K}{\partial q_r} = \sum_{i=1}^{n} \left(F_{xi} \frac{\partial f_i}{\partial q_r} + F_{yi} \frac{\partial g_i}{\partial q_r} + F_{zi} \frac{\partial h_i}{\partial q_r} \right) \tag{10.247}$$

Because of (10.231), (10.232), and (10.233), the kinetic energy is a function of $q_1, q_2, q_3, \cdots, q_n$ and time t. The left-hand side of Eq. (10.247) includes the kinetic energy of the whole system, and the right-hand side is a generalized force that shows the effect of changing coordinates from x_i to q_j on the external forces. Let us assume that the coordinate q_r alters to $q_r + \delta q_r$, while the other coordinates $q_1, q_2, q_3, \cdots, q_{r-1}, q_{r+1}, \cdots, q_n$ and time t remain unaltered. So, the coordinates of m_i are changed into

$$x_i + \frac{\partial f_i}{\partial q_r} \delta q_r \tag{10.248}$$

$$y_i + \frac{\partial g_i}{\partial q_r} \delta q_r \tag{10.249}$$

$$z_i + \frac{\partial h_i}{\partial q_r} \delta q_r \tag{10.250}$$

and the work done in this virtual displacement by all forces acting on the particles of the system is

$$\delta W = \sum_{i=1}^{n} \left(F_{xi} \frac{\partial f_i}{\partial q_r} + F_{yi} \frac{\partial g_i}{\partial q_r} + F_{zi} \frac{\partial h_i}{\partial q_r} \right) \delta q_r \tag{10.251}$$

As the work done by internal forces appears in opposite pairs, only the work by external forces remains in Eq. (10.251). Let us denote the virtual work by

$$\delta W = F_r (q_1, q_2, q_3, \cdots, q_n, t) \, \delta q_r \tag{10.252}$$

Then we have

$$\frac{d}{dt} \left(\frac{\partial K}{\partial \dot{q}_r} \right) - \frac{\partial K}{\partial q_r} = F_r \tag{10.253}$$

where

$$F_r = \sum_{i=1}^{n} \left(F_{xi} \frac{\partial f_i}{\partial q_r} + F_{yi} \frac{\partial g_i}{\partial q_r} + F_{zi} \frac{\partial h_i}{\partial q_r} \right) \tag{10.254}$$

Equation (10.253) is the Lagrange form of equations of motion. This equation is true for all values of r from 1 to n. We thus have n second-order ordinary differential equations in which $q_1, q_2, q_3, \cdots, q_n$ are the dependent variables and t is the independent variable. The coordinates $q_1, q_2, q_3, \cdots, q_n$ are called *generalized coordinates* and can be any measurable variables to provide the configuration of the system. Because the number of equations and the number of dependent variables are equal, the equations are theoretically sufficient to determine the motion of all m_i.   ∎

Example 346 A simple pendulum. Simple pendulum means a pendulum made on a massless bar and a pint mass, oscillating in a vertical plane. Here is to find its equation of motion using Lagrange method.

A pendulum is shown in Fig. 10.9. Using x and y to indicate Cartesian position of m and using $\theta = q$ as the generalized coordinate, we have

$$x = f(\theta) = l \sin \theta \tag{10.255}$$

$$y = g(\theta) = l \cos \theta \tag{10.256}$$

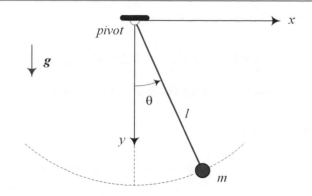

Fig. 10.9 A simple pendulum

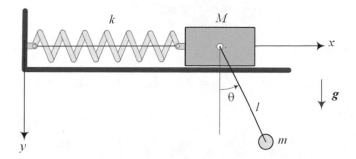

Fig. 10.10 A vibration absorber

$$K = \frac{1}{2}m\left(\dot{x}^2 + \dot{y}^2\right) = \frac{1}{2}ml^2\dot{\theta}^2 \tag{10.257}$$

$$\frac{d}{dt}\left(\frac{\partial K}{\partial \dot{\theta}}\right) - \frac{\partial K}{\partial \theta} = \frac{d}{dt}(ml^2\dot{\theta}) = ml^2\ddot{\theta} \tag{10.258}$$

The external force components, acting on m, are

$$F_x = 0 \qquad F_y = mg \tag{10.259}$$

and therefore the generalized force on m can be calculated.

$$F_\theta = F_x\frac{\partial f}{\partial \theta} + F_y\frac{\partial g}{\partial \theta} = -mgl\sin\theta \tag{10.260}$$

Hence, the equation of motion for the pendulum is

$$ml^2\ddot{\theta} = -mgl\sin\theta \tag{10.261}$$

Example 347 A pendulum attached to an oscillating mass. Attaching a pendulum to a vibrating mass makes a two *DOF* dynamic system. It is a good example to see how Lagrange method is being applied to derive equations of motion of multi-*DOF* systems.

Figure 10.10 illustrates a vibrating mass with a hanging pendulum. The pendulum can act as a vibration absorber if designed properly. Starting with coordinate relationships

$$x_M = f_M = x \qquad\qquad y_M = g_M = 0 \tag{10.262}$$

$$x_m = f_m = x + l\sin\theta \qquad y_m = g_m = l\cos\theta \tag{10.263}$$

we may find the kinetic energy in terms of the generalized coordinates x and θ.

$$
\begin{aligned}
K &= \frac{1}{2}M\left(\dot{x}_M^2 + \dot{y}_M^2\right) + \frac{1}{2}m\left(\dot{x}_m^2 + \dot{y}_m^2\right) \\
&= \frac{1}{2}M\dot{x}^2 + \frac{1}{2}m\left(\dot{x}^2 + l^2\dot{\theta}^2 + 2l\dot{x}\dot{\theta}\cos\theta\right)
\end{aligned}
\tag{10.264}
$$

Then, the left-hand sides of Lagrange equations are

$$
\frac{d}{dt}\left(\frac{\partial K}{\partial \dot{x}}\right) - \frac{\partial K}{\partial x} = (M+m)\ddot{x} + ml\ddot{\theta}\cos\theta - ml\dot{\theta}^2\sin\theta
$$

$$
\tag{10.265}
$$

$$
\frac{d}{dt}\left(\frac{\partial K}{\partial \dot{\theta}}\right) - \frac{\partial K}{\partial \theta} = ml^2\ddot{\theta} + ml\ddot{x}\cos\theta
\tag{10.266}
$$

The external forces acting on M and m are

$$
F_{x_M} = -kx \qquad F_{y_M} = 0 \qquad F_{x_m} = 0 \qquad F_{y_m} = mg
\tag{10.267}
$$

Therefore, the generalized forces are

$$
\begin{aligned}
F_x &= F_{x_M}\frac{\partial f_M}{\partial x} + F_{y_M}\frac{\partial g_M}{\partial x} + F_{x_m}\frac{\partial f_m}{\partial x} + F_{y_m}\frac{\partial g_m}{\partial x} \\
&= -kx
\end{aligned}
\tag{10.268}
$$

$$
\begin{aligned}
F_\theta &= F_{x_M}\frac{\partial f_M}{\partial \theta} + F_{y_M}\frac{\partial g_M}{\partial \theta} + F_{x_m}\frac{\partial f_m}{\partial \theta} + F_{y_m}\frac{\partial g_m}{\partial \theta} \\
&= -mgl\sin\theta
\end{aligned}
\tag{10.269}
$$

and finally the Lagrange equations of motion will be found.

$$
(M+m)\ddot{x} + ml\ddot{\theta}\cos\theta - ml\dot{\theta}^2\sin\theta = -kx
\tag{10.270}
$$

$$
ml^2\ddot{\theta} + ml\ddot{x}\cos\theta = -mgl\sin\theta
\tag{10.271}
$$

Example 348 ★ Potential force field. In mechanics, gravitational and spring forces are the most common potential forces. In case there is no nonpotential forces, the Lagrange equation reduces to energy conservation law. Here is an analytic proof.

If a system of masses m_i are moving in a potential force field, $\nabla_i V$, their Newton equations of motion will be

$$
\mathbf{F}_{m_i} = -\nabla_i V
\tag{10.272}
$$

$$
m_i\ddot{\mathbf{r}}_i = -\nabla_i V \qquad i = 1, 2, \cdots n
\tag{10.273}
$$

Inner product of equations of motion with $\dot{\mathbf{r}}_i$ and adding the equations,

$$
\sum_{i=1}^{n} m_i\dot{\mathbf{r}}_i \cdot \ddot{\mathbf{r}}_i = -\sum_{i=1}^{n}\dot{\mathbf{r}}_i \cdot \nabla_i V
\tag{10.274}
$$

and then integrating over time

$$
\frac{1}{2}\sum_{i=1}^{n} m_i\dot{\mathbf{r}}_i \cdot \dot{\mathbf{r}}_i = -\int \sum_{i=1}^{n}\mathbf{r}_i \cdot \nabla_i V
\tag{10.275}
$$

show that

$$K = - \int \sum_{i=1}^{n} \left(\frac{\partial V}{\partial x_i} x_i + \frac{\partial V}{\partial y_i} y_i + \frac{\partial V}{\partial z_i} z_i \right) = -V + E \tag{10.276}$$

where E is the constant of integration. E is called mechanical energy of the system and is equal to kinetic plus potential energies.

Example 349 Kinetic energy of the Earth. To have a sense of numerical value of kinetic energy, we will compare the kinetic energy of the Earth for its spin and for its rotation about the Sun, separately.

The Earth is approximately a rotating rigid body about a fixed axis. The two motions of the Earth are called revolution about the Sun, and rotation about an axis approximately fixed in the Earth. The kinetic energy of the Earth due to its rotation is

$$\begin{aligned} K_1 &= \frac{1}{2} I \omega_1^2 \\ &= \frac{1}{2} \frac{2}{5} \left(5.9742 \times 10^{24} \right) \left(\frac{6356912 + 6378388}{2} \right)^2 \\ &\quad \times \left(\frac{2\pi}{24 \times 3600} \frac{366.25}{365.25} \right)^2 \\ &= 2.5762 \times 10^{29} \, \text{J} \end{aligned} \tag{10.277}$$

and the kinetic energy of the Earth due to its revolution is

$$\begin{aligned} K_2 &= \frac{1}{2} M r^2 \omega_2^2 \\ &= \frac{1}{2} \left(5.9742 \times 10^{24} \right) \left(1.49475 \times 10^{11} \right)^2 \\ &\quad \times \left(\frac{2\pi}{24 \times 3600} \frac{1}{365.25} \right)^2 \\ &= 2.6457 \times 10^{33} \, \text{J} \end{aligned} \tag{10.278}$$

The r is the distance from the Sun, ω_1 is the angular speed about its axis, and ω_2 is the angular speed about the Sun. The total kinetic energy of the Earth is $K = K_1 + K_2$. The ratio of the revolutionary to rotational kinetic energies is a large number.

$$\frac{K_2}{K_1} = \frac{2.6457 \times 10^{33}}{2.5762 \times 10^{29}} \approx 10000 \tag{10.279}$$

Example 350 ★ Non-Cartesian coordinate system. Lagrange method works for any type of coordinate systems. Here is an example of a non-Cartesian coordinate system.

The parabolic coordinate system and Cartesian coordinate systems are related according to

$$x = \eta \xi \cos \varphi \qquad y = \eta \xi \sin \varphi \qquad z = \frac{\left(\xi^2 - \eta^2 \right)}{2} \tag{10.280}$$

$$\xi^2 = \sqrt{x^2 + y^2 + z^2} + z \qquad \eta^2 = \sqrt{x^2 + y^2 + z^2} - z \tag{10.281}$$

$$\varphi = \tan^{-1} \frac{y}{x} \tag{10.282}$$

An electron in a uniform electric field along the positive z-axis is also under the action of an attractive central force field due to the nuclei of the atom.

$$\mathbf{F} = -\frac{k}{r^2}\hat{e}_r = -\nabla\left(-\frac{k}{r}\right) \tag{10.283}$$

The influence of a uniform electric field on the motion of the electrons in atoms is called the *Stark effect*, and it is easier to analyze its motion in a parabolic coordinate system.

The kinetic energy in a parabolic coordinate system is

$$
\begin{aligned}
K &= \frac{1}{2}m\left(\dot{x}^2 + \dot{y}^2 + \dot{z}^2\right) \\
&= \frac{1}{2}m\left[\left(\eta^2 + \xi^2\right)\left(\dot{\eta}^2 + \dot{\xi}^2\right) + \eta^2\xi^2\dot{\varphi}^2\right]
\end{aligned} \tag{10.284}
$$

and the force acting on the electron is

$$
\begin{aligned}
\mathbf{F} &= -\nabla\left(-\frac{k}{r} + eEz\right) \\
&= -\nabla\left(-\frac{2k}{\xi^2 + \eta^2} + \frac{eE}{2}\left(\xi^2 - \eta^2\right)\right)
\end{aligned} \tag{10.285}
$$

They lead to the following generalized forces:

$$F_\eta = \mathbf{F}\cdot\mathbf{b}_\eta = -\frac{4k\eta}{\left(\xi^2 + \eta^2\right)^2} + eE\eta \tag{10.286}$$

$$F_\xi = \mathbf{F}\cdot\mathbf{b}_\xi = -\frac{4k\xi}{\left(\xi^2 + \eta^2\right)^2} - eE\eta \tag{10.287}$$

$$F_\varphi = 0 \tag{10.288}$$

where $\mathbf{b}_\xi$, $\mathbf{b}_\eta$, $\mathbf{b}_\varphi$ are base vectors of the coordinate system

$$\mathbf{b}_\xi = \frac{\partial\mathbf{r}}{\partial\xi} = \eta\cos\varphi\,\hat{i} + \eta\sin\varphi\,\hat{j} + \xi\hat{k} \tag{10.289}$$

$$\mathbf{b}_\eta = \frac{\partial\mathbf{r}}{\partial\eta} = \xi\cos\varphi\,\hat{i} + \xi\sin\varphi\,\hat{j} - \eta\hat{k} \tag{10.290}$$

$$\mathbf{b}_\varphi = \frac{\partial\mathbf{r}}{\partial\varphi} = -\eta\xi\sin\varphi\,\hat{i} + \eta\xi\cos\varphi\,\hat{j} \tag{10.291}$$

Therefore, following the Lagrange method, the equations of motion of the electron are

$$F_\eta = \frac{d}{dt}\left[m\dot{\eta}\left(\xi^2 + \eta^2\right)\right] - m\eta\left(\dot{\eta}^2 + \dot{\xi}^2\right) - m\eta\xi^2\dot{\varphi}^2 \tag{10.292}$$

$$F_\xi = \frac{d}{dt}\left[m\dot{\xi}\left(\xi^2 + \eta^2\right)\right] - m\eta\left(\dot{\eta}^2 + \dot{\xi}^2\right) - m\xi\eta^2\dot{\varphi}^2 \tag{10.293}$$

$$F_\varphi = \frac{d}{dt}\left(m\eta^2\xi^2\dot{\varphi}^2\right) \tag{10.294}$$

10.6 Lagrangian Mechanics

Assume for some forces $\mathbf{F}$ there is a function V, called *potential energy*, such that the force is derivable from V.

$$\mathbf{F} = \begin{bmatrix} F_{ix} & F_{iy} & F_{iz} \end{bmatrix}^T = -\nabla V \tag{10.295}$$

Such a force is called *potential* or *conservative force*. Then, the Lagrange equation of motion can be written as

$$\frac{d}{dt}\left(\frac{\partial \mathcal{L}}{\partial \dot{q}_r}\right) - \frac{\partial \mathcal{L}}{\partial q_r} = Q_r \qquad r = 1, 2, \cdots n \tag{10.296}$$

where $\mathcal{L}$ is the *Lagrangian* of the system

$$\mathcal{L} = K - V \tag{10.297}$$

and Q_r is the nonpotential generalized force.

Proof Consider an external conservative force $\mathbf{F}$ acting on a system.

$$\mathbf{F} = \begin{bmatrix} F_{xi} & F_{yi} & F_{zi} \end{bmatrix}^T = -\nabla V \tag{10.298}$$

Then the Lagrange equation can be expressed simpler.

$$\frac{d}{dt}\left(\frac{\partial K}{\partial \dot{q}_r}\right) - \frac{\partial K}{\partial q_r} = -\frac{\partial V}{\partial q_1} \qquad r = 1, 2, \cdots n \tag{10.299}$$

The virtual work ∂W done by the forces in an arbitrary virtual displacement $\delta q_1, \delta q_2, \delta q_3, \cdots, \delta q_n$ is

$$\partial W = -\frac{\partial V}{\partial q_1}\delta q_1 - \frac{\partial V}{\partial q_2}\delta q_2 - \cdots - \frac{\partial V}{\partial q_n}\delta q_n \tag{10.300}$$

Introducing the Lagrangian function $\mathcal{L}$ converts the Lagrange equation for the conservative system into a compact expression. The Lagrangian is also called *kinetic potential*.

$$\mathcal{L} = K - V \tag{10.301}$$

$$\frac{d}{dt}\left(\frac{\partial \mathcal{L}}{\partial \dot{q}_r}\right) - \frac{\partial \mathcal{L}}{\partial q_r} = 0 \qquad r = 1, 2, \cdots n \tag{10.302}$$

If a force is not conservative, then the work done by the force is

$$\delta W = \sum_{i=1}^{n}\left(F_{xi}\frac{\partial f_i}{\partial q_r} + F_{yi}\frac{\partial g_i}{\partial q_r} + F_{zi}\frac{\partial h_i}{\partial q_r}\right)\delta q_r$$

$$= Q_r\,\delta q_r \tag{10.303}$$

and the equation of motion would be

$$\frac{d}{dt}\left(\frac{\partial \mathcal{L}}{\partial \dot{q}_r}\right) - \frac{\partial \mathcal{L}}{\partial q_r} = Q_r \qquad r = 1, 2, \cdots n \tag{10.304}$$

where Q_r is the nonpotential generalized force doing work in a virtual displacement of the rth generalized coordinate q_r. ∎

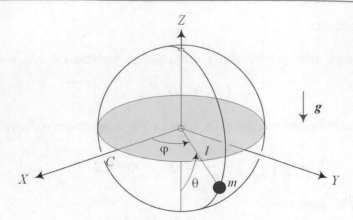

Fig. 10.11 A spherical pendulum

Example 351 Spherical pendulum. Spherical pendulum is a nonlinear conservative two DOF dynamic system. A good example to learn how Lagrange method derives equations of motion.

A pendulum analogy is utilized in modeling of many dynamic problems. Figure 10.11 illustrates a spherical pendulum with a tip mass m and a massless bar of length l. The angles φ and θ may be used as describing coordinates of the system. At the first step, we determine Cartesian coordinates of the mass as a function of the generalized coordinates.

$$\begin{bmatrix} X \\ Y \\ Z \end{bmatrix} = \begin{bmatrix} r \cos \varphi \sin \theta \\ r \sin \theta \sin \varphi \\ -r \cos \theta \end{bmatrix} \tag{10.305}$$

Then, the kinetic and potential energies of the pendulum must be determined.

$$K = \frac{1}{2} m \left(l^2 \dot{\theta}^2 + l^2 \dot{\varphi}^2 \sin^2 \theta \right) \tag{10.306}$$

$$V = -mgl \cos \theta \tag{10.307}$$

The kinetic potential function of this system then determines the Lagrangian $\mathcal{L}$ of the system.

$$\mathcal{L} = \frac{1}{2} m \left(l^2 \dot{\theta}^2 + l^2 \dot{\varphi}^2 \sin^2 \theta \right) + mgl \cos \theta \tag{10.308}$$

Having the Lagrangian is equivalent to having the equations of motion. They will be derived by differentiation and substitution in Eq. (10.302).

$$\ddot{\theta} - \dot{\varphi}^2 \sin \theta \cos \theta + \frac{g}{l} \sin \theta = 0 \tag{10.309}$$

$$\ddot{\varphi} \sin^2 \theta + 2 \dot{\varphi} \dot{\theta} \sin \theta \cos \theta = 0 \tag{10.310}$$

Example 352 A one-link manipulator. Deriving equation of motion of an example on a one DOF dynamic system similar to a robotic arm with actuator.

A one-link manipulator is illustrated in Fig. 10.12. Assume that there is viscous friction in the joint where an ideal motor applies the torque Q to move the arm. The rotor of an ideal motor has no mass moment by assumption. The kinetic and potential energies of the manipulator are

$$K = \frac{1}{2} I \dot{\theta}^2 = \frac{1}{2} \left(I_C + ml^2 \right) \dot{\theta}^2 \tag{10.311}$$

$$V = -mg \cos \theta \tag{10.312}$$

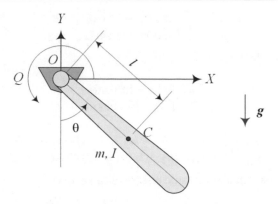

Fig. 10.12 A compound pendulum with a motor at O

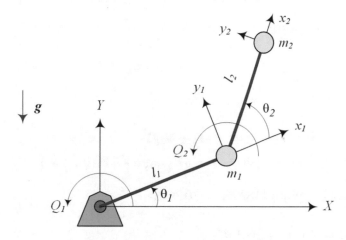

Fig. 10.13 An ideal model of a 2R planar manipulator

where m is the mass and I is the mass moment of the pendulum about O. The Lagrangian of the manipulator will be

$$\mathcal{L} = K - V = \frac{1}{2}I\dot{\theta}^2 + mg\cos\theta \tag{10.313}$$

and therefore, the equation of motion of the manipulator can be found.

$$M = \frac{d}{dt}\left(\frac{\partial\mathcal{L}}{\partial\dot{\theta}}\right) - \frac{\partial\mathcal{L}}{\partial\theta} = I\ddot{\theta} + mgl\sin\theta \tag{10.314}$$

The generalized force M is the contribution of the motor torque Q and the viscous friction torque $-c\dot{\theta}$. Hence, the equation of motion of the manipulator will be

$$Q = I\ddot{\theta} + c\dot{\theta} + mgl\sin\theta \tag{10.315}$$

Example 353 The ideal $2R$ planar manipulator dynamics. An example of applying Lagrange method on a manipulator and deriving its equations of motion. Ideal $2R$ manipulator is the one with massless arm and point mass motors and payload

An ideal model of a $2R$ planar manipulator is illustrated in Fig. 10.13. It is called ideal because we assumed the links are massless and there is no friction. The masses m_1 and m_2 are the second motor to run the second link and the load at the end point. We take the absolute angle θ_1 and the relative angle θ_2 as the generalized coordinates to express the configuration of the manipulator. The global positions of m_1 and m_2 are

$$\begin{bmatrix} X_1 \\ Y_2 \end{bmatrix} = \begin{bmatrix} l_1\cos\theta_1 \\ l_1\sin\theta_1 \end{bmatrix} \tag{10.316}$$

$$\begin{bmatrix} X_2 \\ Y_2 \end{bmatrix} = \begin{bmatrix} l_1 \cos\theta_1 + l_2 \cos(\theta_1 + \theta_2) \\ l_1 \sin\theta_1 + l_2 \sin(\theta_1 + \theta_2) \end{bmatrix} \tag{10.317}$$

and therefore, the global velocities of the masses are

$$\begin{bmatrix} \dot{X}_1 \\ \dot{Y}_1 \end{bmatrix} = \begin{bmatrix} -l_1\dot{\theta}_1 \sin\theta_1 \\ l_1\dot{\theta}_1 \cos\theta_1 \end{bmatrix} \tag{10.318}$$

$$\begin{bmatrix} \dot{X}_2 \\ \dot{Y}_2 \end{bmatrix} = \begin{bmatrix} -l_1\dot{\theta}_1 \sin\theta_1 - l_2\left(\dot{\theta}_1 + \dot{\theta}_2\right)\sin(\theta_1 + \theta_2) \\ l_1\dot{\theta}_1 \cos\theta_1 + l_2\left(\dot{\theta}_1 + \dot{\theta}_2\right)\cos(\theta_1 + \theta_2) \end{bmatrix} \tag{10.319}$$

The kinetic energy of this manipulator is made of kinetic energy of the masses.

$$\begin{aligned}
K &= K_1 + K_2 = \frac{1}{2}m_1\left(\dot{X}_1^2 + \dot{Y}_1^2\right) + \frac{1}{2}m_2\left(\dot{X}_2^2 + \dot{Y}_2^2\right) \\
&= \frac{1}{2}m_1 l_1^2\dot{\theta}_1^2 \\
&\quad + \frac{1}{2}m_2\left(l_1^2\dot{\theta}_1^2 + l_2^2\left(\dot{\theta}_1 + \dot{\theta}_2\right)^2 + 2l_1 l_2\dot{\theta}_1\left(\dot{\theta}_1 + \dot{\theta}_2\right)\cos\theta_2\right)
\end{aligned} \tag{10.320}$$

The potential energy of the manipulator is

$$\begin{aligned}
V &= V_1 + V_2 = m_1 g Y_1 + m_2 g Y_2 \\
&= m_1 g l_1 \sin\theta_1 + m_2 g\left(l_1 \sin\theta_1 + l_2 \sin(\theta_1 + \theta_2)\right)
\end{aligned} \tag{10.321}$$

The Lagrangian is then obtained from Eqs. (10.320) and (10.321).

$$\begin{aligned}
\mathcal{L} &= K - V = \frac{1}{2}m_1 l_1^2\dot{\theta}_1^2 \\
&\quad + \frac{1}{2}m_2\left(l_1^2\dot{\theta}_1^2 + l_2^2\left(\dot{\theta}_1 + \dot{\theta}_2\right)^2 + 2l_1 l_2\dot{\theta}_1\left(\dot{\theta}_1 + \dot{\theta}_2\right)\cos\theta_2\right) \\
&\quad - \left(m_1 g l_1 \sin\theta_1 + m_2 g\left(l_1 \sin\theta_1 + l_2 \sin(\theta_1 + \theta_2)\right)\right)
\end{aligned} \tag{10.322}$$

Lagrangian provides us with the required partial derivatives.

$$\frac{\partial\mathcal{L}}{\partial\theta_1} = -\left(m_1 + m_2\right)g l_1 \cos\theta_1 - m_2 g l_2 \cos(\theta_1 + \theta_2) \tag{10.323}$$

$$\begin{aligned}
\frac{\partial\mathcal{L}}{\partial\dot{\theta}_1} &= \left(m_1 + m_2\right)l_1^2\dot{\theta}_1 + m_2 l_2^2\left(\dot{\theta}_1 + \dot{\theta}_2\right) \\
&\quad + m_2 l_1 l_2\left(2\dot{\theta}_1 + \dot{\theta}_2\right)\cos\theta_2
\end{aligned} \tag{10.324}$$

$$\begin{aligned}
\frac{d}{dt}\left(\frac{\partial\mathcal{L}}{\partial\dot{\theta}_1}\right) &= \left(m_1 + m_2\right)l_1^2\ddot{\theta}_1 + m_2 l_2^2\left(\ddot{\theta}_1 + \ddot{\theta}_2\right) \\
&\quad + m_2 l_1 l_2\left(2\ddot{\theta}_1 + \ddot{\theta}_2\right)\cos\theta_2 \\
&\quad - m_2 l_1 l_2\dot{\theta}_2\left(2\dot{\theta}_1 + \dot{\theta}_2\right)\sin\theta_2
\end{aligned} \tag{10.325}$$

$$\frac{\partial\mathcal{L}}{\partial\theta_2} = -m_2 l_1 l_2\dot{\theta}_1\left(\dot{\theta}_1 + \dot{\theta}_2\right)\sin\theta_2 - m_2 g l_2 \cos(\theta_1 + \theta_2) \tag{10.326}$$

$$\frac{\partial\mathcal{L}}{\partial\dot{\theta}_2} = m_2 l_2^2\left(\dot{\theta}_1 + \dot{\theta}_2\right) + m_2 l_1 l_2\dot{\theta}_1 \cos\theta_2 \tag{10.327}$$

$$\frac{d}{dt}\left(\frac{\partial \mathcal{L}}{\partial \dot{\theta}_2}\right) = m_2 l_2^2 \left(\ddot{\theta}_1 + \ddot{\theta}_2\right) + m_2 l_1 l_2 \ddot{\theta}_1 \cos\theta_2 - m_2 l_1 l_2 \dot{\theta}_1 \dot{\theta}_2 \sin\theta_2 \tag{10.328}$$

Therefore, the equations of motion for the $2R$ manipulator are

$$
\begin{aligned}
Q_1 &= \frac{d}{dt}\left(\frac{\partial \mathcal{L}}{\partial \dot{\theta}_1}\right) - \frac{\partial \mathcal{L}}{\partial \theta_1} \\
&= (m_1 + m_2)\, l_1^2 \ddot{\theta}_1 + m_2 l_2^2 \left(\ddot{\theta}_1 + \ddot{\theta}_2\right) \\
&\quad + m_2 l_1 l_2 \left(2\ddot{\theta}_1 + \ddot{\theta}_2\right)\cos\theta_2 - m_2 l_1 l_2 \dot{\theta}_2 \left(2\dot{\theta}_1 + \dot{\theta}_2\right)\sin\theta_2 \\
&\quad + (m_1 + m_2)\, g l_1 \cos\theta_1 + m_2 g l_2 \cos(\theta_1 + \theta_2)
\end{aligned}
\tag{10.329}
$$

$$
\begin{aligned}
Q_2 &= \frac{d}{dt}\left(\frac{\partial \mathcal{L}}{\partial \dot{\theta}_2}\right) - \frac{\partial \mathcal{L}}{\partial \theta_2} \\
&= m_2 l_2^2 \left(\ddot{\theta}_1 + \ddot{\theta}_2\right) + m_2 l_1 l_2 \ddot{\theta}_1 \cos\theta_2 - m_2 l_1 l_2 \dot{\theta}_1 \dot{\theta}_2 \sin\theta_2 \\
&\quad + m_2 l_1 l_2 \dot{\theta}_1 \left(\dot{\theta}_1 + \dot{\theta}_2\right)\sin\theta_2 + m_2 g l_2 \cos(\theta_1 + \theta_2)
\end{aligned}
\tag{10.330}
$$

The generalized forces Q_1 and Q_2 are the required forces to vary the generalized coordinates. In this case, Q_1 is the torque at the base motor and Q_2 is the torque of the motor at m_1.

The equations of motion can be rearranged to have a more systematic form.

$$
\begin{aligned}
Q_1 &= \left((m_1 + m_2)\, l_1^2 + m_2 l_2 \left(l_2 + 2l_1 \cos\theta_2\right)\right)\ddot{\theta}_1 \\
&\quad + m_2 l_2 \left(l_2 + l_1 \cos\theta_2\right)\ddot{\theta}_2 \\
&\quad - 2m_2 l_1 l_2 \sin\theta_2\, \dot{\theta}_1 \dot{\theta}_2 - m_2 l_1 l_2 \sin\theta_2\, \dot{\theta}_2^2 \\
&\quad + (m_1 + m_2)\, g l_1 \cos\theta_1 + m_2 g l_2 \cos(\theta_1 + \theta_2)
\end{aligned}
\tag{10.331}
$$

$$
\begin{aligned}
Q_2 &= m_2 l_2 \left(l_2 + l_1 \cos\theta_2\right)\ddot{\theta}_1 + m_2 l_2^2 \ddot{\theta}_2 \\
&\quad + m_2 l_1 l_2 \sin\theta_2\, \dot{\theta}_1^2 + m_2 g l_2 \cos(\theta_1 + \theta_2)
\end{aligned}
\tag{10.332}
$$

Example 354 Trebuchet. Trebuchet is an interesting mechanical device similar to a robotic manipulator. Here is how to derive the equations of motion of a sample trebuchet.

A trebuchet is shown schematically in Fig. 10.14. It is a shooting weapon of war for the mass m_2 that is powered by a falling massive counterweight m_1. A beam AB is pivoted to the chassis with two unequal sections a and b. The figure shows a trebuchet at its initial configuration. The origin of a global coordinate frame is set at the pivot point. The counterweight m_1 is at (x_1, y_1) and is hinged at the shorter arm of the beam at a distance c from the end B. The mass of the projectile is m_2 and it is at the end of a massless sling with a length l attached to the end of the longer arm of the beam. The three independent variable angles α, θ, γ express the motion of the device. We consider the parameters a, b, c, d, l, m_1, m_2 to be constant and determine the equations of motion by the Lagrange method.

Figure 10.15 illustrates the trebuchet when it is in motion. The position coordinates of masses m_1 and m_2 are

$$x_1 = b\sin\theta - c\sin(\theta + \gamma) \tag{10.333}$$

$$y_1 = -b\cos\theta + c\cos(\theta + \gamma) \tag{10.334}$$

$$x_2 = -a\sin\theta - l\sin(-\theta + \alpha) \tag{10.335}$$

$$y_2 = -a\cos\theta - l\cos(-\theta + \alpha) \tag{10.336}$$

Taking a time derivative provides the velocity components.

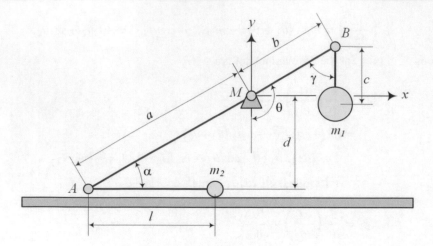

Fig. 10.14 A trebuchet at starting position

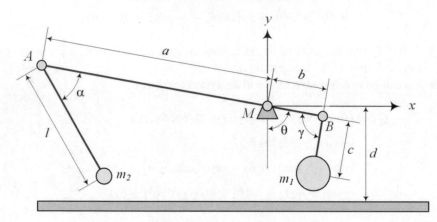

Fig. 10.15 A trebuchet in motion

$$\dot{x}_1 = b\dot{\theta}\cos\theta - c\left(\dot{\theta} + \dot{\gamma}\right)\cos\left(\theta + \gamma\right) \tag{10.337}$$

$$\dot{y}_1 = b\dot{\theta}\sin\theta - c\left(\dot{\theta} + \dot{\gamma}\right)\sin\left(\theta + \gamma\right) \tag{10.338}$$

$$\dot{x}_2 = l\left(c - \dot{\alpha}\right)\cos\left(\alpha - \theta\right) - a\dot{\theta}\cos\left(\theta\right) \tag{10.339}$$

$$\dot{y}_2 = a\dot{\theta}\sin\theta - l\left(\dot{\theta} - \dot{\alpha}\right)\sin\left(\alpha - \theta\right) \tag{10.340}$$

The kinetic energy of the system is

$$
\begin{aligned}
K &= \frac{1}{2}m_1 v_1^2 + \frac{1}{2}m_2 v_2^2 = \frac{1}{2}m_1\left(\dot{x}_1^2 + \dot{y}_1^2\right) + \frac{1}{2}m_2\left(\dot{x}_2^2 + \dot{y}_2^2\right) \\
&= \frac{1}{2}m_1\left(\left(b^2 + c^2\right)\dot{\theta}^2 + c^2\dot{\gamma}^2 + 2c^2\dot{\theta}\dot{\gamma}\right) \\
&\quad - m_1 bc\dot{\theta}\left(\dot{\theta} + \dot{\gamma}\right)\cos\gamma \\
&\quad + \frac{1}{2}m_2\left(\left(a^2 + l^2\right)\dot{\theta}^2 + l^2\dot{\alpha}^2 - 2l^2\dot{\theta}\dot{\alpha}\right) \\
&\quad - m_2 al\dot{\theta}\left(\dot{\theta} - \dot{\alpha}\right)\cos\left(2\theta - \alpha\right)
\end{aligned}
\tag{10.341}
$$

The potential energy of the system can be calculated by y position of the masses.

$$
\begin{aligned}
V &= m_1 g y_1 + m_2 g y_2 \\
&= m_1 g \left(-b \cos\theta + c \cos\left(\theta + \gamma\right) \right) \\
&\quad + m_2 g \left(-a \cos\theta - l \cos\left(-\theta + \alpha\right) \right)
\end{aligned}
\tag{10.342}
$$

Having the energies K and V, we can set up the Lagrangian $\mathcal{L}$.

$$
\mathcal{L} = K - V
\tag{10.343}
$$

Using the Lagrangian, we are able to find the three equations of motion.

$$
\frac{d}{dt}\left(\frac{\partial \mathcal{L}}{\partial \dot{\theta}}\right) - \frac{\partial \mathcal{L}}{\partial \theta} = 0
\tag{10.344}
$$

$$
\frac{d}{dt}\left(\frac{\partial \mathcal{L}}{\partial \dot{\alpha}}\right) - \frac{\partial \mathcal{L}}{\partial \alpha} = 0
\tag{10.345}
$$

$$
\frac{d}{dt}\left(\frac{\partial \mathcal{L}}{\partial \dot{\gamma}}\right) - \frac{\partial \mathcal{L}}{\partial \gamma} = 0
\tag{10.346}
$$

The trebuchet appeared in 500–400 B.C. in China and was developed by Persian armies around 300 B.C. It was used by the Arabs against the Romans during 600–1200 A.D. The trebuchet may also be called the manganic, manjaniq, catapults, or onager. The Persian word "Manganic" is the root of the words "mechanic" and "machine."

10.7 Summary

The translational and rotational equations of motion for a rigid body, expressed in the global coordinate frame, are

$$
{}^G\mathbf{F} = \frac{{}^G d}{dt}\, {}^G\mathbf{p}
\tag{10.347}
$$

$$
{}^G\mathbf{M} = \frac{{}^G d}{dt}\, {}^G\mathbf{L}
\tag{10.348}
$$

where ${}^G\mathbf{F}$ and ${}^G\mathbf{M}$ indicate the resultant of the external forces and moments applied on the rigid body and measured at C. The vector ${}^G\mathbf{p}$ is the momentum and ${}^G\mathbf{L}$ is the moment of momentum for the rigid body at C

$$
\mathbf{p} = m\,\mathbf{v}
\tag{10.349}
$$

$$
\mathbf{L} = \mathbf{r}_C \times \mathbf{p}
\tag{10.350}
$$

The expressions of the equations of motion in the body coordinate frame are

$$
\begin{aligned}
{}^B\mathbf{F} &= {}^G\dot{\mathbf{p}} + {}^B_G\boldsymbol{\omega}_B \times {}^B\mathbf{p} \\
&= m\,{}^B\mathbf{a}_B + m\,{}^B_G\boldsymbol{\omega}_B \times {}^B\mathbf{v}_B
\end{aligned}
\tag{10.351}
$$

$$
\begin{aligned}
{}^B\mathbf{M} &= {}^B\dot{\mathbf{L}} + {}^B_G\boldsymbol{\omega}_B \times {}^B\mathbf{L} \\
&= {}^B I\,{}^B_G\dot{\boldsymbol{\omega}}_B + {}^B_G\boldsymbol{\omega}_B \times \left({}^B I\,{}^B_G\boldsymbol{\omega}_B\right)
\end{aligned}
\tag{10.352}
$$

where I is the moment of inertia for the rigid body.

$$I = \begin{bmatrix} I_{xx} & I_{xy} & I_{xz} \\ I_{yx} & I_{yy} & I_{yz} \\ I_{zx} & I_{zy} & I_{zz} \end{bmatrix}$$

(10.353)

The elements of I are only functions of the mass distribution of the rigid body and are defined by

$$I_{ij} = \int_B \left(r_i^2 \delta_{mn} - x_{im} x_{jn} \right) dm \ , \ i, j = 1, 2, 3$$

(10.354)

where δ_{ij} is Kronecker's delta.

Every rigid body has a principal body coordinate frame in which the moment of inertia is in the form.

$$^B I = \begin{bmatrix} I_1 & 0 & 0 \\ 0 & I_2 & 0 \\ 0 & 0 & I_3 \end{bmatrix}$$

(10.355)

The rotational equation of motion in the principal coordinate frame simplifies to

$$M_1 = I_1 \dot{\omega}_1 - (I_2 - I_2)\,\omega_2 \omega_3$$
$$M_2 = I_2 \dot{\omega}_2 - (I_3 - I_1)\,\omega_3 \omega_1$$
$$M_3 = I_3 \dot{\omega}_3 - (I_1 - I_2)\,\omega_1 \omega_2$$

(10.356)

Utilizing homogenous position vectors we also define pseudo-inertia matrix $\bar{I}$ with application in robot dynamics as

$$^B \bar{I} = \int_B \mathbf{r}\,\mathbf{r}^T \, dm$$

(10.357)

$$= \begin{bmatrix} \frac{-I_{xx}+I_{yy}+I_{zz}}{2} & I_{xy} & I_{xz} & mx_C \\ I_{yx} & \frac{I_{xx}-I_{yy}+I_{zz}}{2} & I_{yz} & my_C \\ I_{zx} & I_{zy} & \frac{I_{xx}+I_{yy}-I_{zz}}{2} & mz_C \\ mx_C & my_C & mz_C & m \end{bmatrix}$$

where $^B \mathbf{r}_C$ is the position of the mass center in the body frame.

$$^B \mathbf{r}_C = \begin{bmatrix} x_C \\ y_C \\ z_C \end{bmatrix} = \begin{bmatrix} \frac{1}{m}\int_B x\,dm \\ \frac{1}{m}\int_B y\,dm \\ \frac{1}{m}\int_B z\,dm \end{bmatrix}$$

(10.358)

The equations of motion for a mechanical system having $n\,DOF$ can also be found by the Lagrange equation

$$\frac{d}{dt}\left(\frac{\partial \mathcal{L}}{\partial \dot{q}_r} \right) - \frac{\partial \mathcal{L}}{\partial q_r} = Q_r \ \ r = 1, 2, \cdots n$$

(10.359)

$$\mathcal{L} = K - V$$

(10.360)

where $\mathcal{L}$ is the *Lagrangian* of the system, K is the kinetic energy, V is the potential energy, and Q_r is the nonpotential generalized force.

$$Q_r = \sum_{i=1}^n \left(Q_{ix} \frac{\partial f_i}{\partial q_1} + Q_{iy} \frac{\partial g_i}{\partial q_2} + Q_{iz} \frac{\partial h_i}{\partial q_n} \right)$$

(10.361)

The parameters $q_r, r = 1, 2, \cdots, n$, are the generalized coordinates of the system, $\mathbf{Q} = \begin{bmatrix} Q_{ix} & Q_{iy} & Q_{iz} \end{bmatrix}^T$ is the external force acting on the ith particle of the system, and Q_r is the generalized force associated to q_r. When (x_i, y_i, z_i) are the Cartesian coordinates in a globally fixed coordinate frame for the particle m_i, then its coordinates may be functions of another set of coordinates $q_1, q_2, q_3, \cdots, q_n$ and possibly time t.

$$x_i = f_i(q_1, q_2, q_3, \cdots, q_n, t) \tag{10.362}$$

$$y_i = g_i(q_1, q_2, q_3, \cdots, q_n, t) \tag{10.363}$$

$$z_i = h_i(q_1, q_2, q_3, \cdots, q_n, t) \tag{10.364}$$

10.8   Key Symbols

$\mathbf{a}$	Acceleration vector
A	Acceleration transformation matrix
B	Body coordinate frame
C	Mass center
$\mathbf{d}$	Translation vector, displacement vector
D	Displacement transformation matrix
e	Rotation quaternion
$\mathbf{f}, \mathbf{F}$	Force vector
g	Gravitational acceleration
G, B_0	Global coordinate frame, base coordinate frame
h	Height
$\hat{\imath}, \hat{\jmath}, \hat{k}$	Local coordinate axes unit vectors
$\hat{I}, \hat{J}, \hat{K}$	Global coordinate axes unit vectors
$I = [I]$	Mass moment matrix
$\bar{I} = [\bar{I}]$	Pseudo-inertia matrix
$\mathbf{I} = [\mathbf{I}]$	Identity matrix
k	Spring stiffness
K	Kinetic energy
l	Length
$\mathbf{L}$	Angular moment vector, moment of moment
$\mathcal{L}$	Lagrangian
m	The number of independent equations
$\mathbf{M}$	Moment vector, torque vector
M_i	The element i of $\mathbf{M}$
$\mathbf{p}$	Momentum
$\mathbf{p}, \mathbf{q}$	Position vectors of P, Q
$\mathbf{q}$	Joint variable vector
P, Q	Points
r	Radius
$\mathbf{r}$	Position vectors, homogenous position vector
r_i	The element i of $\mathbf{r}$
R	Rotation transformation matrix, radius
$\hat{u}$	Unit vector along the axis of $\boldsymbol{\omega}$
u_1, u_2, u_3	Components of $\hat{u}$
v	Velocity
$\mathbf{v}$	Velocity vector
V	Potential energy
w	Width
W	Work, normalized eigenvector matrix
x, y, z	Local coordinate axes
X, Y, Z	Global coordinate axes

Greek

α	Angular acceleration
$\boldsymbol{\alpha}$	Angular acceleration vector
$\alpha_1, \alpha_2, \alpha_3$	Components of $\boldsymbol{\alpha}$
δ	Kronecker delta
θ	Rotary joint angle
φ, θ, ψ	Euler angles
ϕ	Angle of rotation about $\hat{u}$
ω	Angular velocity

$\boldsymbol{\omega}$ Angular velocity vector
$\tilde{\omega}$ Skew symmetric matrix of the vector $\boldsymbol{\omega}$
$\omega_1, \omega_2, \omega_3$ Components of $\boldsymbol{\omega}$

Symbol
$[\ \ ]^{-1}$ Inverse of the matrix $[\ \ ]$
$[\ \ ]^{T}$ Transpose of the matrix $[\ \ ]$
∇ Gradient
(i) Link number i
$\parallel$ Parallel
$\perp$ Perpendicular
$\times$ Vector cross product
$\overset{\leftrightarrow}{e}$ Matrix form of a quaternion e
E The Earth
lim   Limit function
sgn   Signum function
tr   Trace

Exercises

1. Notation and symbols.

   Describe the meaning of these notations.

$$a - {}^{G}\mathbf{p}_P \quad b - \mathbf{L}_l \quad c - {}^{G}\mathbf{F} \quad d - {}_{1}W_2 \quad e - K \quad f - V$$

$$g - dm \quad h - [I] \quad i - I_{ij} \quad j - {}^{G}_{B}\mathbf{F} \quad k - {}^{G}\mathbf{M} \quad l - I_1$$

$$m - {}^{0}_{2}\tilde{\alpha}_1 \quad n - {}^{2}_{1}\tilde{\omega}_2 \quad o - {}^{G}A_B \quad p - {}^{G}\dot{V}_B \quad q - \dot{\mathbf{J}} \quad r - \ddot{\mathbf{X}}$$

2. Kinetic energy of a rigid link.

   Consider a straight and uniform bar as a rigid link of a manipulator. The link has a mass m. Show that the kinetic energy of the bar can be expressed as

$$K = \frac{1}{6}m\left(\mathbf{v}_1 \cdot \mathbf{v}_1 + \mathbf{v}_1 \cdot \mathbf{v}_2 + \mathbf{v}_2 \cdot \mathbf{v}_2\right) \tag{10.365}$$

   where $\mathbf{v}_1$ and $\mathbf{v}_2$ are the velocity vectors of the end points of the link.

3. Position and velocity of mass center C of system of discrete particles.

   There are three particles $m_1 = 1\,\text{kg}$, $m_2 = 2\,\text{kg}$, $m_3 = 3\,\text{kg}$, at

$$\mathbf{r}_1 = \begin{bmatrix} 1 \\ -1 \\ 1 \end{bmatrix} \quad \mathbf{r}_2 = \begin{bmatrix} -1 \\ -3 \\ 2 \end{bmatrix} \quad \mathbf{r}_3 = \begin{bmatrix} 2 \\ -1 \\ -3 \end{bmatrix} \tag{10.366}$$

   Their velocities are

$$\mathbf{v}_1 = \begin{bmatrix} 2 \\ 1 \\ 1 \end{bmatrix} \quad \mathbf{v}_2 = \begin{bmatrix} -1 \\ 0 \\ 2 \end{bmatrix} \quad \mathbf{v}_3 = \begin{bmatrix} 3 \\ -2 \\ -1 \end{bmatrix} \tag{10.367}$$

   (a) Find the position and velocity of the system at C.

   (b) Calculate the system's momentum and moment of momentum.

   (c) Calculate the system's kinetic energy and determine the rotational and translational parts of the kinetic energy.

4. Newton's equation of motion in the body frame.

   Show that Newton's equation of motion in the body frame is

$$\begin{bmatrix} F_x \\ F_y \\ F_z \end{bmatrix} = m \begin{bmatrix} a_x \\ a_y \\ a_z \end{bmatrix} + \begin{bmatrix} 0 & -\omega_z & \omega_y \\ \omega_z & 0 & -\omega_x \\ -\omega_y & \omega_x & 0 \end{bmatrix} \begin{bmatrix} v_x \\ v_y \\ v_z \end{bmatrix} \tag{10.368}$$

5. Work on a curved path.

   A particle of mass m is moving on a circular path given by ${}^{G}\mathbf{r}_P$.

$$^{G}\mathbf{r}_P = \cos\theta\,\hat{I} + \sin\theta\,\hat{J} + 4\,\hat{K} \tag{10.369}$$

Calculate the work of a force ${}^{G}\mathbf{F}$ when the particle moves from $\theta = 0$ to $\theta = \frac{\pi}{2}$.

   (a)

$$^{G}\mathbf{F} = \frac{z^2 - y^2}{(x + y)^2}\,\hat{I} + \frac{y^2 - x^2}{(x + y)^2}\,\hat{J} + \frac{x^2 - y^2}{(x + z)^2}\,\hat{K} \tag{10.370}$$

   (b)

$$^{G}\mathbf{F} = \frac{z^2 - y^2}{(x + y)^2}\,\hat{I} + \frac{2y}{x + y}\,\hat{J} + \frac{x^2 - y^2}{(x + z)^2}\,\hat{K} \tag{10.371}$$

6. Newton's equation of motion.

   Find the equations of motion for the system shown in Fig. 10.9 based on Newton's method.

7. Acceleration in the body frame.

   Find the acceleration vector for the end point of the two-link manipulator shown in Fig. 10.5, expressed in frame A.

8. Principal moments of inertia.

   Find the principal mass moments and axes of the following mass moments matrices:

   (a)

$$[I] = \begin{bmatrix} 3 & 2 & 2 \\ 2 & 2 & 0 \\ 2 & 0 & 4 \end{bmatrix} \tag{10.372}$$

   (b)

$$[I] = \begin{bmatrix} 3 & 2 & 4 \\ 2 & 0 & 2 \\ 4 & 2 & 3 \end{bmatrix} \tag{10.373}$$

   (c)

$$[I] = \begin{bmatrix} 100 & 20\sqrt{3} & 0 \\ 20\sqrt{3} & 60 & 0 \\ 0 & 0 & 10 \end{bmatrix} \tag{10.374}$$

9. ★ The solutions of mass moment characteristic equation are real.

   The solution of a cubic equation where $a \neq 0$ can be found in a systematic way.

$$ax^3 + bx^2 + cx + d = 0 \tag{10.375}$$

Transform the equation into a new form with discriminant $4p^3 + q^2$,

$$y^3 + 3py + q = 0 \tag{10.376}$$

Use the transformation $x = y - \frac{b}{3a}$, where

$$p = \frac{3ac - b^2}{9a^2} \qquad q = \frac{2b^3 - 9abc + 27a^2d}{27a^3} \tag{10.377}$$

The solutions are then

$$y_1 = \sqrt[3]{\alpha} - \sqrt[3]{\beta} \tag{10.378}$$

$$y_2 = e^{\frac{2\pi i}{3}}\sqrt[3]{\alpha} - e^{\frac{4\pi i}{3}}\sqrt[3]{\beta} \qquad y_3 = e^{\frac{4\pi i}{3}}\sqrt[3]{\alpha} - e^{\frac{2\pi i}{3}}\sqrt[3]{\beta}$$

where

$$\alpha = \frac{-q + \sqrt{q^2 + 4p^3}}{2} \qquad \beta = \frac{-q + \sqrt{q^2 + 4p^3}}{2} \tag{10.379}$$

For real values of p and q, if the discriminant is positive, then one root is real, and two roots are complex conjugates. If the discriminant is zero, then there are three real roots, of which at least two are equal. If the discriminant is negative, then there are three unequal real roots.

Apply this theory for the characteristic equation of the matrix $[I]$ and show that the principal moments of inertia are real.

10. Moment of inertia.

    Calculate the mass moments of shown objects in Fig. 10.16, in a principal Cartesian coordinate frame at C.

    (a) A cylinder similar to a uniform arm with a circular cross section, Fig. 10.16a.

    (b) A rectangular box similar to a uniform arm with a rectangular cross section, Fig. 10.16b.

    (c) A house similar to a prismatic bar with a nonsymmetric polygon cross section, Fig. 10.16c.

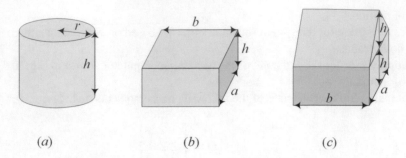

Fig. 10.16 Solid objects to calculate mass moments

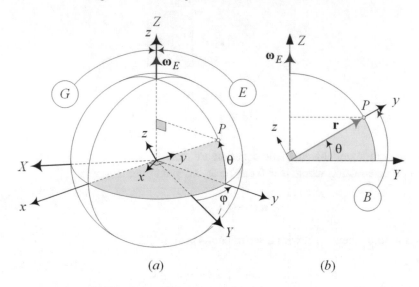

Fig. 10.17 The location on the Earth is defined by longitude φ and latitude θ

11. Global differential of angular momentum.

Convert the mass moment $^B I$ and the angular velocity $^B_G \omega_B$ into the global coordinate frame and then find the differential of angular momentum. It is an alternative method to show that

$$
\frac{^G d}{dt} \, ^B\mathbf{L} = \frac{^G d}{dt} \left(^B I \, ^B_G \omega_B \right)
$$
$$
= {}^B\dot{\mathbf{L}} + {}^B_G \omega_B \times {}^B\mathbf{L} = I\dot{\omega} + \omega \times (I\omega) \tag{10.380}
$$

12. Rotated mass moment matrix.

A principal mass moment matrix $^2 I$ in B_2-frame is given.

$$
^2 I = \begin{bmatrix} 3 & 0 & 0 \\ 0 & 5 & 0 \\ 0 & 0 & 4 \end{bmatrix} \tag{10.381}
$$

The principal frame was achieved by rotating the initial body coordinate frame 30 deg about the x-axis, followed by 45 deg about the z-axis. Find the initial moment of inertia matrix $^1 I$.

13. Rotation of moment of inertia matrix.

Find the required rotation matrix that transforms the mass moment matrix $[I]$ into a diagonal matrix.

$$
[I] = \begin{bmatrix} 3 & 2 & 2 \\ 2 & 2 & 0.1 \\ 2 & 0.1 & 4 \end{bmatrix} \tag{10.382}
$$

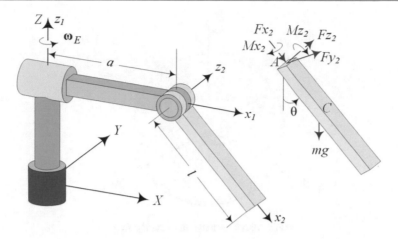

Fig. 10.18 A two-link manipulator

14. Kinematics of a moving car on the Earth.

A vehicle on the Earth is located by its longitude φ from a fixed meridian, say, the Greenwich meridian, and its latitude θ from the equator, as shown in Fig. 10.17. We attach a coordinate frame B at the center of the Earth with x-axis on the equator's plane and y-axis pointing the vehicle. There are also two coordinate frames E and G where E is attached to the Earth and G is the global coordinate frame. Show that the angular velocity of B and the velocity of the vehicle are

$$\,^B_G\boldsymbol{\omega}_B = \dot{\theta}\,\hat{\imath}_B + (\omega_E + \dot{\varphi})\sin\theta\,\hat{\jmath}_B + (\omega_E + \dot{\varphi})\cos\theta\,\hat{k} \tag{10.383}$$

$$\,^B_G\mathbf{v}_P = -r\,(\omega_E + \dot{\varphi})\cos\theta\,\hat{\imath}_B + r\dot{\theta}\,\hat{k} \tag{10.384}$$

Calculate the acceleration of the vehicle.

15. Equations of motion for a rotating arm.

Find the equations of motion for the rotating links shown in Fig. 10.18 based on the Lagrange method.

16. Equations of motion from Lagrangian.

Consider a physical system with a Lagrangian $\mathcal{L}$.

$$\mathcal{L} = \frac{1}{2}m\,(a\dot{x} + b\dot{y})^2 - \frac{1}{2}k\,(ax + by)^2 \tag{10.385}$$

Determine the equations of motion of the system. The coefficients m, k, a, and b are constant.

17. Lagrangian from equation of motion.

Find the Lagrangian associated to the following equations of motions:

(a)

$$mr^2\ddot{\theta} + k_1l_1\theta + k_2l_2\theta + mgl = 0 \tag{10.386}$$

(b)

$$\ddot{r} - r\dot{\theta}^2 = 0 \qquad r^2\ddot{\theta} + 2r\dot{r}\dot{\theta} = 0 \tag{10.387}$$

18. A mass on a rotating ring.

A particle of mass m is free to slide on a rotating vertical ring as shown in Fig. 10.19. The ring is turning with a constant angular velocity $\omega = \dot{\varphi}$ about the Z-axis. Determine the equation of motion of the particle. The local coordinate frame is set up with the x-axis pointing the particle and the z-axis in the plane of the ring and parallel with the tangent to the ring at the position of the mass.

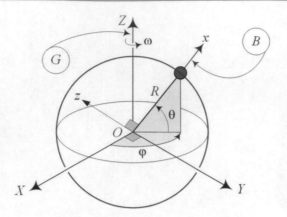

Fig. 10.19 A mass on a rotating ring

19. ★ Particle in electromagnetic field.

Show that equations of motion of a particle with mass m with a Lagrangian $\mathcal{L}$

$$\mathcal{L} = \frac{1}{2}m\dot{\mathbf{r}}^2 - e\Phi + e\dot{\mathbf{r}} \cdot \mathbf{A} \tag{10.388}$$

are

$$m\ddot{q}_i = e\left(-\frac{\partial\Phi}{\partial q_i} - \frac{\partial A_i}{\partial t}\right) + e\sum_{j=1}^{3}\dot{q}_j\left(\frac{\partial A_j}{\partial q_i} - \frac{\partial A_i}{\partial q_j}\right) \tag{10.389}$$

where

$$\mathbf{q} = \begin{bmatrix} q_1 \\ q_2 \\ q_3 \end{bmatrix} = \begin{bmatrix} x \\ y \\ z \end{bmatrix} \tag{10.390}$$

Then convert the equations of motion into a vectorial form

$$m\ddot{\mathbf{r}} = e\mathbf{E}(\mathbf{r}, t) + e\dot{\mathbf{r}} \times \mathbf{B}(\mathbf{r}, t) \tag{10.391}$$

where $\mathbf{E}$ and $\mathbf{B}$ are electric and magnetic fields.

$$\mathbf{E} = -\nabla\Phi - \frac{\partial\mathbf{A}}{\partial t} \qquad \mathbf{B} = \nabla \times \mathbf{A} \tag{10.392}$$

20. Inverted pendulum on a moving cart.

Figure 10.20 illustrates an inverted pendulum on a moving cart. If there is a friction between the cart and the ground surface with coefficient μ_s, and a friction at the pivot of the pendulum with coefficient μ_p, show that the equations of motion of the system are

$$\ddot{\theta} = \frac{g\sin\theta + \left(\mu_s\,\mathrm{sgn}\,\dot{x} - ml\dot{\theta}^2\sin\theta - F\right)\cos\theta - \dfrac{\mu_p}{ml}\dot{\theta}}{l\left(\dfrac{4}{3} - \dfrac{m\cos^2\theta}{M+m}\right)} \tag{10.393}$$

$$\ddot{x} = \frac{F + ml\left(\dot{\theta}^2\sin\theta - \ddot{\theta}\cos\theta\right) - \mu_s\,\mathrm{sgn}\,\dot{x}}{M+m} \tag{10.394}$$

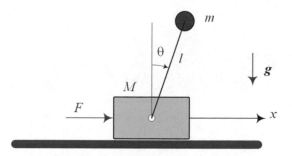

Fig. 10.20 An inverted pendulum on a moving cart

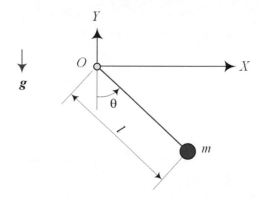

Fig. 10.21 A pendulum with a vibrating pivot

where

$$\operatorname{sgn} x = \begin{cases} 1 & x > 0 \\ 0 & x = 0 \\ -1 & x < 0 \end{cases} \tag{10.395}$$

21. Forced vibration of a pendulum.

Figure 10.21 illustrates a simple pendulum having a length l and a bob with mass m. Find the equation of motion if:

(a) The pivot O has a dictated motion in X direction.

$$X_O = a \sin \omega t \tag{10.396}$$

(b) The pivot O has a dictated motion in Y direction.

$$Y_O = b \sin \omega t \tag{10.397}$$

(c) The pivot O has a uniform motion on a circle.

$$\mathbf{r}_O = R \cos \omega t \ \hat{I} + R \sin \omega t \ \hat{J} \tag{10.398}$$

We find the dynamics equations of motion of robots by two methods: *Newton–Euler* and *Lagrange*. The Newton–Euler method is more fundamental and finds the dynamic equations to determine the required actuators' force and torque to move a robot, as well as the joint forces. Lagrange method provides only the required differential equations that determines the actuators' force and torque.

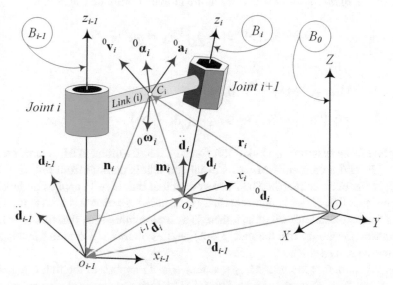

Fig. 11.1 A link (i) and its vectorial kinematic characteristics

11.1 Rigid Link Newton–Euler Dynamics

Figure 11.1 illustrates the link number (i) of a manipulator and its velocity and acceleration vectorial characteristics. Figure 11.2 illustrates free body diagram of the link (i). The force $\mathbf{F}_{i-1}$ and moment $\mathbf{M}_{i-1}$ are the resultant force and moment that link ($i-1$) applies to link (i) at joint i. Similarly, $\mathbf{F}_i$ and $\mathbf{M}_i$ are the resultant force and moment that link (i) applies to link ($i+1$) at joint $i+1$. We measure and show the force systems ($\mathbf{F}_{i-1}$, $\mathbf{M}_{i-1}$) and ($\mathbf{F}_i$, $\mathbf{M}_i$) at the origin of the coordinate frames B_{i-1} and B_i, respectively. The sum of the external loads acting on the link (i) are shown by $\sum \mathbf{F}_{e_i}$ and $\sum \mathbf{M}_{e_i}$.

R. N. Jazar, *Theory of Applied Robotics*, https://doi.org/10.1007/978-3-030-93220-6_11

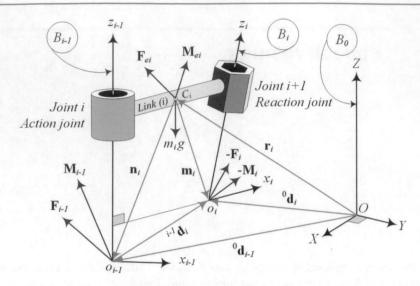

Fig. 11.2 Force system on link (i)

The *Newton–Euler equations of motion* for the link (i) in the global coordinate frame are:

$$^0\mathbf{F}_{i-1} - {}^0\mathbf{F}_i + \sum {}^0\mathbf{F}_{e_i} = m_i \, {}^0\mathbf{a}_i \tag{11.1}$$

$$\begin{aligned}
&^0\mathbf{M}_{i-1} - {}^0\mathbf{M}_i + \sum {}^0\mathbf{M}_{e_i} \\
&+ \left({}^0\mathbf{d}_{i-1} - {}^0\mathbf{r}_i\right) \times {}^0\mathbf{F}_{i-1} - \left({}^0\mathbf{d}_i - {}^0\mathbf{r}_i\right) \times {}^0\mathbf{F}_i = {}^0I_i \, {}_0\boldsymbol{\alpha}_i
\end{aligned} \tag{11.2}$$

Proof There are three applied force systems on a link (i): A force $\mathbf{F}_{i-1}$ and a moment $\mathbf{M}_{i-1}$ at its proximal end that is coming from link $(i-1)$; a force $-\mathbf{F}_i$ and a moment $-\mathbf{M}_i$ at its distal end that is coming from link $(i+1)$; the resultant external force system $\sum \mathbf{F}_{e_i}$, $\sum \mathbf{M}_{e_i}$ at the mass center C_i. The force system that link $(i-1)$ applies on link (i) is called *action* force system, and the force system that link $(i+1)$ applies on link (i) is called *reaction* force system.

First, let us only look at forces on link (i). At joint i, there is an *action force* $\mathbf{F}_{i-1}$ that link $(i-1)$ applies on link (i), and at joint $i+1$ there is a *reaction force* $-\mathbf{F}_i$ that the link $(i+1)$ applies on link (i). Action force is also called *driving force*, and reaction force is also called *driven force*.

Similarly, at joint i there is an *action moment* $\mathbf{M}_{i-1}$ that link $(i-1)$ applies to the link (i), and at joint $i+1$ there is a *reaction moment* $-\mathbf{M}_i$ that link $(i+1)$ applies to the link (i). The action moment is also called *driving moment*, and the reaction moment is also called *driven moment*.

Therefore, there is a driving force system $(\mathbf{F}_{i-1}, \mathbf{M}_{i-1})$ at the origin of the coordinate frame B_{i-1}, and a driven force system $(-\mathbf{F}_i, -\mathbf{M}_i)$ at the origin of the coordinate frame B_i. The driving force system $(\mathbf{F}_{i-1}, \mathbf{M}_{i-1})$ gives motion to link (i) and the driven force system $(\mathbf{F}_i, \mathbf{M}_i)$ gives motion to link $(i+1)$.

In addition to the action and reaction force systems, there might be some external forces acting on the link (i) that their resultant makes a force system $(\sum \mathbf{F}_{e_i}, \sum \mathbf{M}_{e_i})$ at the mass center C_i. In robotic application, weight is usually the only external load on middle links, and reactions from the environment are extra external force systems on the base and end-effector links. The force and moment that the base actuator applies to the first link are $\mathbf{F}_0$ and $\mathbf{M}_0$, and the force and moment that the end-effector applies to the environment are $\mathbf{F}_n$ and $\mathbf{M}_n$. If weight is the only external load on link (i) and it is in $-{}^0\hat{k}_0$ direction, and $\mathbf{g}$ is the gravitational acceleration vector, then we have the B_0-expression of the external forces on link (i) as:

$$\sum {}^0\mathbf{F}_{e_i} = m_i \, {}^0\mathbf{g} = -m_i \, g \, {}^0\hat{k}_0 \tag{11.3}$$

$$\sum {}^0\mathbf{M}_{e_i} = {}^0\mathbf{r}_i \times m_i \, {}^0\mathbf{g} = -{}^0\mathbf{r}_i \times m_i \, g \, {}^0\hat{k}_0 \tag{11.4}$$

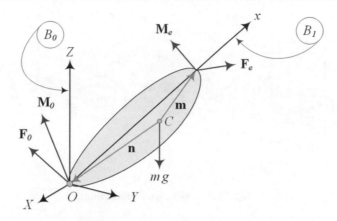

Fig. 11.3 A one-link manipulator

As showing in Fig. 11.2, we indicate the global position of the mass center of the link by $^0\mathbf{r}_i$, and the global position of the origin of body frames B_i and B_{i-1} by $^0\mathbf{d}_i$ and $^0\mathbf{d}_{i-1}$ respectively. The link's velocities $^0\mathbf{v}_i$, $_0\boldsymbol{\omega}_i$ and accelerations $^0\mathbf{a}_i$, $_0\boldsymbol{\alpha}_i$ are measured and shown at C_i. The physical properties of the link (i) are specified by its mass m_i and mass moment matrix ^0I_i about the link's mass center C_i.

The Newton's equation of motion determines that the sum of forces applied to the link (i) is equal to the mass of the link times its acceleration at C_i.

$$^0\mathbf{F}_{i-1} - {}^0\mathbf{F}_i + \sum {}^0\mathbf{F}_{e_i} = m_i \, {}^0\mathbf{a}_i \tag{11.5}$$

For the Euler equation, in addition to the action and reaction moments, we must add the moments of the action and reaction forces about C_i. The moment of $-\mathbf{F}_i$ and $\mathbf{F}_{i-1}$ are equal to $-\mathbf{m}_i \times \mathbf{F}_i$ and $\mathbf{n}_i \times \mathbf{F}_{i-1}$ where $\mathbf{m}_i$ is the position vector of o_i from C_i and $\mathbf{n}_i$ is the position vector of o_{i-1} from C_i. Therefore, the link's Euler equation of motion is:

$$\begin{aligned} {}^0\mathbf{M}_{i-1} - {}^0\mathbf{M}_i + \sum {}^0\mathbf{M}_{e_i} \\ + {}^0\mathbf{n}_i \times {}^0\mathbf{F}_{i-1} - {}^0\mathbf{m}_i \times {}^0\mathbf{F}_i = {}^0I_i \, {}_0\boldsymbol{\alpha}_i \end{aligned} \tag{11.6}$$

The position vectors $\mathbf{n}_i$ and $\mathbf{m}_i$ can be expressed in terms of $^0\mathbf{d}_i$, $^0\mathbf{r}_i$, $^0\mathbf{d}_{i-1}$,

$$^0\mathbf{n}_i = {}^0\mathbf{d}_{i-1} - {}^0\mathbf{r}_i \tag{11.7}$$

$$^0\mathbf{m}_i = {}^0\mathbf{d}_i - {}^0\mathbf{r}_i \tag{11.8}$$

$$_{i-1}^0\mathbf{d}_i = {}^0\mathbf{m}_i - {}^0\mathbf{n}_i \tag{11.9}$$

to derive Eq. (11.2).

Because there is one translational and one rotational equation of motion for each link of a robot, there are $2n$ vectorial equations of motion for an n link robot. However, there are $2(n+1)$ forces and moments involved. Therefore, one set of force systems (usually $\mathbf{F}_n$ and $\mathbf{M}_n$) must be specified to solve the equations and find the joints' force and moment. ∎

Example 355 One-link manipulator. One-link manipulator is an ideal and simple enough example to learn how to develop equations of motion in robotics.

Figure 11.3 depicts a link attached to the ground via a spherical joint at O. The free body diagram (FBD) of the link is made of an external force $^0\mathbf{F}_e$ and moment $^0\mathbf{M}_e$ at the endpoint, gravity mg, and the driving force $^0\mathbf{F}_0$ and moment $^0\mathbf{M}_0$ at the joint O. The Newton–Euler equations for the link are:

$$^0\mathbf{F}_0 + {}^0\mathbf{F}_e + mg \, \hat{K} = m \, {}^0\mathbf{a}_C \tag{11.10}$$

$$^0\mathbf{M}_0 + {}^0\mathbf{M}_e + {}^0\mathbf{n} \times {}^0\mathbf{F}_0 + {}^0\mathbf{m} \times {}^0\mathbf{F}_e = {}^0I \, {}_0\boldsymbol{\alpha}_1 \tag{11.11}$$

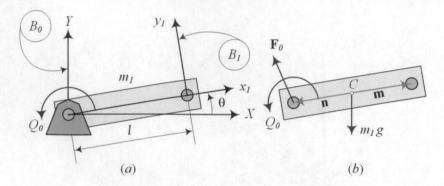

Fig. 11.4 A turning uniform beam

To derive the equations of motion and study the application of Newton–Euler equations, let us consider the uniform beam of Fig. 11.4a. Figure 11.4b illustrates the FBD of the beam and its relative position vectors $\mathbf{m}$ and $\mathbf{n}$.

$$^0\mathbf{m} = \begin{bmatrix} \dfrac{l}{2}\cos\theta \\ \dfrac{l}{2}\sin\theta \\ 0 \end{bmatrix} \qquad ^0\mathbf{n} = \begin{bmatrix} -\dfrac{l}{2}\cos\theta \\ -\dfrac{l}{2}\sin\theta \\ 0 \end{bmatrix} \tag{11.12}$$

The kinematics of the beam are:

$$^0\mathbf{r} = -\,^0\mathbf{n} \tag{11.13}$$

$$^0\mathbf{d} = -\,^0\mathbf{n} + \,^0\mathbf{m} \tag{11.14}$$

where $^0\mathbf{r}$ indicates the position of C, and $^0\mathbf{d}$ indicates the position of the tip point, both in B_0.

$$_0\boldsymbol{\omega}_1 = \dot{\theta}\,\hat{K} \tag{11.15}$$

$$_0\boldsymbol{\alpha}_1 = \,_0\dot{\boldsymbol{\omega}}_1 = \ddot{\theta}\,\hat{K} \tag{11.16}$$

$$\mathbf{g} = -g\,\hat{J} \tag{11.17}$$

$$^0\mathbf{a}_C = \,_0\boldsymbol{\alpha}_1 \times \,^0\mathbf{r} - \,_0\boldsymbol{\omega}_1 \times \left(\,_0\boldsymbol{\omega}_1 \times \,^0\mathbf{r}\right)$$

$$= \begin{bmatrix} -\dfrac{l}{2}\ddot{\theta}\sin\theta + \dfrac{l}{2}\dot{\theta}^2\,(\cos\theta) \\ \dfrac{l}{2}\ddot{\theta}\cos\theta + \dfrac{l}{2}\dot{\theta}^2\sin\theta \\ 0 \end{bmatrix} \tag{11.18}$$

The forces on the beam are:

$$^0\mathbf{F}_0 = \begin{bmatrix} F_X \\ F_Y \\ F_Z \end{bmatrix} \qquad ^0\mathbf{F}_e = \begin{bmatrix} 0 \\ 0 \\ 0 \end{bmatrix} \tag{11.19}$$

$$^0\mathbf{M}_0 = \begin{bmatrix} Q_X \\ Q_Y \\ Q_Z \end{bmatrix} \qquad ^0\mathbf{M}_e = \begin{bmatrix} 0 \\ 0 \\ 0 \end{bmatrix} \tag{11.20}$$

Let us assume that $^1I_1$ to be the principal mass moment matrix of the beam about its mass center expressed in B_1.

$$^1I_1 = \begin{bmatrix} I_x & 0 & 0 \\ 0 & I_y & 0 \\ 0 & 0 & I_z \end{bmatrix} \tag{11.21}$$

The transformation matrix $^0R_1$ can be used to calculate $^0I_1$.

$$^0R_1 = R_{Z,\theta} = \begin{bmatrix} \cos\theta & -\sin\theta & 0 \\ \sin\theta & \cos\theta & 0 \\ 0 & 0 & 1 \end{bmatrix} \tag{11.22}$$

$$^0I_1 = R_{Z,\theta}\,^1I_1\,R_{Z,\theta}^T = \,^0R_1 \begin{bmatrix} I_x & 0 & 0 \\ 0 & I_y & 0 \\ 0 & 0 & I_z \end{bmatrix} \,^0R_1^T$$

$$= \begin{bmatrix} I_x\cos^2\theta + I_y\sin^2\theta & (I_x - I_y)\cos\theta\sin\theta & 0 \\ (I_x - I_y)\cos\theta\sin\theta & I_y\cos^2\theta + I_x\sin^2\theta & 0 \\ 0 & 0 & I_z \end{bmatrix} \tag{11.23}$$

Substituting the above information in Eqs. (11.10) and (11.11) provides the following equations of motion.

$$^0\mathbf{F}_0 + \,^0\mathbf{F}_e + m_1\mathbf{g} = m_1\,^0\mathbf{a}_C \tag{11.24}$$

$$\begin{bmatrix} F_X \\ F_Y \\ F_Z \end{bmatrix} = \begin{bmatrix} -\dfrac{1}{2}m_1l\left(\ddot{\theta}\sin\theta - \dot{\theta}^2\cos\theta\right) \\ \dfrac{1}{2}m_1l\left(\ddot{\theta}\cos\theta + \dot{\theta}^2\sin\theta\right) + m_1g \\ 0 \end{bmatrix} \tag{11.25}$$

$$^0\mathbf{M}_0 + \,^0\mathbf{M}_e + \,^0\mathbf{n} \times \,^0\mathbf{F}_0 + \,^0\mathbf{m} \times \,^0\mathbf{F}_e = I_0\boldsymbol{\alpha}_1 \tag{11.26}$$

$$^0\mathbf{M}_0 = \begin{bmatrix} Q_X \\ Q_Y \\ Q_Z \end{bmatrix} = \begin{bmatrix} \dfrac{l}{2}F_Z\sin\theta \\ -\dfrac{l}{2}F_Z\cos\theta \\ I_z\ddot{\theta} + \dfrac{l}{2}F_Y\cos\theta - \dfrac{l}{2}F_X\sin\theta \end{bmatrix} \tag{11.27}$$

Let us substitute the force components from (11.25) to determine the components of the driving moment $^0\mathbf{M}_0$.

$$\begin{bmatrix} Q_X \\ Q_Y \\ Q_Z \end{bmatrix} = \begin{bmatrix} 0 \\ 0 \\ \left(I_z + \dfrac{m_1l^2}{4}\right)\ddot{\theta} + \dfrac{1}{2}m_1gl\cos\theta \end{bmatrix} \tag{11.28}$$

Example 356 A turning uniform beam with a tip mass. Hanging a load to the tip point of a one-link manipulator makes a one-link system with payload. Here is to show how the equations of motion of this system will be found.

Let us consider the uniform beam of Fig. 11.5a with a hanging mass m_2 at the tip point. Figure 11.5b illustrates the *FBD* of the beam. The mass center of the beam is at $^1\mathbf{r}_1$

$$^1\mathbf{r}_1 = \frac{m_1}{m_1 + m_2}\begin{bmatrix} l/2 \\ 0 \\ 0 \end{bmatrix} + \frac{m_2}{m_1 + m_2}\begin{bmatrix} l \\ 0 \\ 0 \end{bmatrix}$$

$$= \begin{bmatrix} \dfrac{m_1 + 2m_2}{2(m_1 + m_2)}l \\ 0 \\ 0 \end{bmatrix} = \begin{bmatrix} r_x \\ 0 \\ 0 \end{bmatrix} \tag{11.29}$$

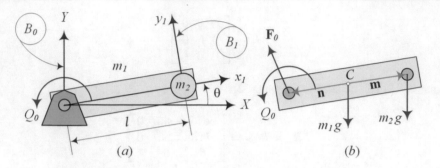

Fig. 11.5 A uniform beam with a hanging weight m_2 at the tip point

and its relative position vectors **m** and **n** are:

$$^1\mathbf{n}_1 = -\,^1\mathbf{r}_1 = -r_x\hat{\imath} \tag{11.30}$$

$$^1\mathbf{m}_1 = l\hat{\imath} - \,^1\mathbf{r}_1 = (l - r_x)\,\hat{\imath} \tag{11.31}$$

$$^0\mathbf{d}_1 = -\,^1\mathbf{n}_1 + \,^1\mathbf{m}_1 = l\hat{\imath} \tag{11.32}$$

$$^0\mathbf{m} = \begin{bmatrix} (l - r_x)\cos\theta \\ (l - r_x)\sin\theta \\ 0 \end{bmatrix} \qquad ^0\mathbf{n} = \begin{bmatrix} -r_x\cos\theta \\ -r_x\sin\theta \\ 0 \end{bmatrix} \tag{11.33}$$

The kinematics of the beam are:

$$_0\boldsymbol{\omega}_1 = \dot{\theta}\,\hat{K} \tag{11.34}$$

$$_0\boldsymbol{\alpha}_1 = \,_0\dot{\boldsymbol{\omega}}_1 = \ddot{\theta}\,\hat{K} \tag{11.35}$$

$$\mathbf{g} = -g\,\hat{J} \tag{11.36}$$

$$^0\mathbf{a}_C = \,_0\boldsymbol{\alpha}_1 \times \,^0\mathbf{r}_1 + \,_0\boldsymbol{\omega}_1 \times \left(_0\boldsymbol{\omega}_1 \times \,^0\mathbf{r}_1\right)$$

$$= \begin{bmatrix} -r_x\ddot{\theta}\sin\theta + r_x\dot{\theta}^2\,(\cos\theta) \\ r_x\ddot{\theta}\cos\theta + r_x\dot{\theta}^2\sin\theta \\ 0 \end{bmatrix} \tag{11.37}$$

The forces on the beam are:

$$^0\mathbf{F}_0 = \begin{bmatrix} F_X \\ F_Y \\ F_Z \end{bmatrix} \qquad ^0\mathbf{F}_e = \begin{bmatrix} 0 \\ 0 \\ 0 \end{bmatrix} \tag{11.38}$$

$$^0\mathbf{M}_0 = \begin{bmatrix} Q_X \\ Q_Y \\ Q_Z \end{bmatrix} \qquad ^0\mathbf{M}_e = \begin{bmatrix} 0 \\ 0 \\ 0 \end{bmatrix} \tag{11.39}$$

Let us assume that $^1I_1$ is the mass moment matrix of the beam about its center.

$$^1I_1 = \begin{bmatrix} I_x & 0 & 0 \\ 0 & I_y & 0 \\ 0 & 0 & I_z \end{bmatrix} \tag{11.40}$$

The mass moment matrix of the manipulator about the common mass center at $^1\mathbf{r}_1$ is:

$$^1I_1 = \begin{bmatrix} I_x & 0 & 0 \\ 0 & I_y & 0 \\ 0 & 0 & I_3 \end{bmatrix} \tag{11.41}$$

$$I_3 = I_z + m_1 \left(r_x - \frac{l}{2} \right)^2 + m_2 \left(l - r_x \right)^2 \tag{11.42}$$

Knowing the transformation matrix $^0R_1$, we can determine $^0I_1$.

$$^0R_1 = R_{Z,\theta} = \begin{bmatrix} \cos\theta & -\sin\theta & 0 \\ \sin\theta & \cos\theta & 0 \\ 0 & 0 & 1 \end{bmatrix} \tag{11.43}$$

$$^0I_1 = R_{Z,\theta} \, ^1I_1 \, R_{Z,\theta}^T = \, ^0R_1 \begin{bmatrix} I_x & 0 & 0 \\ 0 & I_y & 0 \\ 0 & 0 & I_3 \end{bmatrix} \, ^0R_1^T$$

$$= \begin{bmatrix} I_x \cos^2\theta + I_y \sin^2\theta & (I_x - I_y)\cos\theta\sin\theta & 0 \\ (I_x - I_y)\cos\theta\sin\theta & I_y \cos^2\theta + I_x \sin^2\theta & 0 \\ 0 & 0 & I_3 \end{bmatrix} \tag{11.44}$$

Substituting the above information in Eqs. (11.10) and (11.11) provides the following equations of motion.

$$^0\mathbf{F}_0 + \, ^0\mathbf{F}_e + m_1 g \, \hat{K} = m_1 \, ^0\mathbf{a}_C + m_2 \frac{1}{2} \, ^0\mathbf{a}_C \tag{11.45}$$

$$\begin{bmatrix} F_X \\ F_Y \\ F_Z \end{bmatrix} = \begin{bmatrix} -(m_1 + m_2) \, r_x \left(\ddot{\theta} \sin\theta - \dot{\theta}^2 \cos\theta \right) \\ (m_1 + m_2) \, r_x \left(\ddot{\theta} \cos\theta + \dot{\theta}^2 \sin\theta \right) + (m_2 + m_1) \, g \\ 0 \end{bmatrix} \tag{11.46}$$

$$^0\mathbf{M}_0 + \, ^0\mathbf{M}_e + \, ^0\mathbf{n} \times \, ^0\mathbf{F}_0 + \, ^0\mathbf{m} \times \, ^0\mathbf{F}_e = I \, _0\boldsymbol{\alpha}_1 \tag{11.47}$$

$$^0\mathbf{M}_0 = \begin{bmatrix} Q_X \\ Q_Y \\ Q_Z \end{bmatrix} = \begin{bmatrix} r_x F_Z \sin\theta \\ -r_x F_Z \cos\theta \\ I_3 \ddot{\theta} + r_x F_Y \cos\theta - r_x F_X \sin\theta \end{bmatrix} \tag{11.48}$$

Let us substitute the force components from (11.46) to determine the components of the driving moment $^0\mathbf{M}_0$.

$$\begin{bmatrix} Q_X \\ Q_Y \\ Q_Z \end{bmatrix} = \begin{bmatrix} 0 \\ 0 \\ \left(I_z + (m_1 + m_2) \, r_x^2 \right) \ddot{\theta} + (m_1 + m_2) \, r_x g \cos\theta \end{bmatrix} \tag{11.49}$$

Substituting r_x provides the required torque Q_0.

$$Q_0 = Q_Z = \left(\frac{1}{4} m_1 l^2 + m_2 l^2 + I_z \right) \ddot{\theta} + \left(\frac{1}{2} m_1 + m_2 \right) g l \cos\theta \tag{11.50}$$

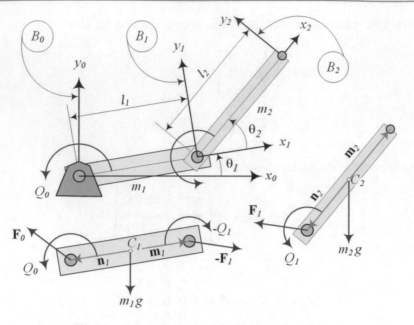

Fig. 11.6 Free body diagram of a $2R$ planar manipulator

Example 357 $2R$ planar manipulator Newton–Euler dynamics. To derive equations of motion of a robot, we only need to draw free body diagram (FBD) of every link and write the Newton and Euler equation for every link in the base coordinate frame. Here is the equations of motion of a $2R$ manipulator.

A $2R$ planar manipulator and its free body diagram are shown in Fig. 11.6 along with the FBD of every link. The torques of actuators are parallel to the Z-axis and are indicated by Q_0 and Q_1. The Newton–Euler equations of motion for the first link are:

$$^0\mathbf{F}_0 - {}^0\mathbf{F}_1 + m_1 g\,\hat{J} = m_1\,{}^0\mathbf{a}_1 \tag{11.51}$$

$$^0\mathbf{Q}_0 - {}^0\mathbf{Q}_1 + {}^0\mathbf{n}_1 \times {}^0\mathbf{F}_0 - {}^0\mathbf{m}_1 \times {}^0\mathbf{F}_1 = {}^0I_1\,{}_0\boldsymbol{\alpha}_1 \tag{11.52}$$

and the equations of motion for the second link are:

$$^0\mathbf{F}_1 + m_2 g\,\hat{J} = m_2\,{}^0\mathbf{a}_2 \tag{11.53}$$

$$^0\mathbf{Q}_1 + {}^0\mathbf{n}_2 \times {}^0\mathbf{F}_1 = {}^0I_2\,{}_0\boldsymbol{\alpha}_2 \tag{11.54}$$

There are four equations for four unknowns $\mathbf{F}_0, \mathbf{F}_1, \mathbf{Q}_0, \mathbf{Q}_1$. These equations can be set in a matrix form

$$[A]\mathbf{x} = \mathbf{b} \tag{11.55}$$

where

$$[A] = \begin{bmatrix} 1 & 0 & 0 & -1 & 0 & 0 \\ 0 & 1 & 0 & 0 & -1 & 0 \\ n_{1y} & -n_{1x} & 1 & -m_{1y} & m_{1x} & -1 \\ 0 & 0 & 0 & 1 & 0 & 0 \\ 0 & 0 & 0 & 0 & 1 & 0 \\ 0 & 0 & 0 & n_{2y} & -n_{2x} & 1 \end{bmatrix} \tag{11.56}$$

$$\mathbf{x} = \begin{bmatrix} F_{0x} \\ F_{0y} \\ Q_0 \\ F_{1x} \\ F_{1y} \\ Q_1 \end{bmatrix} \qquad \mathbf{b} = \begin{bmatrix} m_1 a_{1x} \\ m_1 a_{1y} - m_1 g \\ {}^0I_1\alpha_1 \\ m_2 a_{2x} \\ m_2 a_{2y} - m_2 g \\ {}^0I_2\alpha_2 \end{bmatrix} \tag{11.57}$$

Example 358 Equations for joint actuators. Newton–Euler equations provide with joint torque and joint force equations. Torque equations are those we need for robot control. To separate the equations, there should be some manipulations. Here is an example for $2R$ manipulators.

In robot dynamics, we usually do not need to find joint forces. Actuator torques are much more important as they are used to control a robot. In Example 357 we derived four equations for dynamics of the $2R$ manipulator that is shown in Fig. 11.6.

$$
{}^0\mathbf{F}_0 - {}^0\mathbf{F}_1 + m_1 g\,\hat{J} = m_1\,{}^0\mathbf{a}_1 \tag{11.58}
$$

$$
{}^0\mathbf{Q}_0 - {}^0\mathbf{Q}_1 + {}^0\mathbf{n}_1 \times {}^0\mathbf{F}_0 - {}^0\mathbf{m}_1 \times {}^0\mathbf{F}_1 = {}^0I_1\,{}_0\boldsymbol{\alpha}_1 \tag{11.59}
$$

$$
{}^0\mathbf{F}_1 + m_2 g\,\hat{J} = m_2\,{}^0\mathbf{a}_2 \tag{11.60}
$$

$$
{}^0\mathbf{Q}_1 + {}^0\mathbf{n}_2 \times {}^0\mathbf{F}_1 = {}^0I_2\,{}_0\boldsymbol{\alpha}_2 \tag{11.61}
$$

However, we may eliminate the joint forces $\mathbf{F}_0$, $\mathbf{F}_1$ and reduce the number of equations to two for the two torques $\mathbf{Q}_0$ and $\mathbf{Q}_1$. Eliminating $\mathbf{F}_1$ between (11.60) and (11.61) provides

$$
{}^0\mathbf{Q}_1 = {}^0I_2\,{}_0\boldsymbol{\alpha}_2 - {}^0\mathbf{n}_2 \times \left(m_2\,{}^0\mathbf{a}_2 - m_2 g\,\hat{J} \right) \tag{11.62}
$$

and eliminating $\mathbf{F}_0$ and $\mathbf{F}_1$ between (11.58) and (11.59) gives:

$$
\begin{aligned}
{}^0\mathbf{Q}_0 = {}^0\mathbf{Q}_1 + {}^0I_1\,{}_0\boldsymbol{\alpha}_1 + {}^0\mathbf{m}_1 \times \left(m_2\,{}^0\mathbf{a}_2 - m_2 g\,\hat{J} \right) \\
- {}^0\mathbf{n}_1 \times \left(m_1\,{}^0\mathbf{a}_1 - m_1 g\,\hat{J} + m_2\,{}^0\mathbf{a}_2 - m_2 g\,\hat{J} \right)
\end{aligned} \tag{11.63}
$$

The joint forces $\mathbf{F}_0$ and $\mathbf{F}_1$, if we are interested, are equal to:

$$
{}^0\mathbf{F}_1 = m_2\,{}^0\mathbf{a}_2 - m_2 g\,\hat{J} \tag{11.64}
$$

$$
{}^0\mathbf{F}_0 = m_1\,{}^0\mathbf{a}_1 + m_2\,{}^0\mathbf{a}_2 - (m_1 + m_2)\,g\,\hat{J} \tag{11.65}
$$

Example 359 $2R$ planar manipulator with massive arms and joints. Here is an example of a $2R$ manipulator with payload and massive motors, as well as massive links.

In a real situation for a $2R$ planar manipulators, we generally have a massive motor at joint 0 to turn the link (1) and a massive motor at joint 1 to turn the link (2). We may also carry a massive object by the gripper at the tip point. The motor at joint 0 is siting on the ground and its weight will not effect the dynamics of the manipulator. The FBD of the manipulator is shown in Fig. 11.7.

The massive joints will displace the position of C_i and changes the relative position vectors $\mathbf{m}$ and $\mathbf{n}$. We will have the same equations of motion (11.51)–(11.53) provided we determine $\mathbf{m}$ and $\mathbf{n}$ for the new position of C_i as suggested in Fig. 11.8.

$$
{}^0\mathbf{F}_0 - {}^0\mathbf{F}_1 + (m_{11} + m_{12})\,g\,\hat{J} = m_1\,{}^0\mathbf{a}_1 \tag{11.66}
$$

$$
{}^0\mathbf{Q}_0 - {}^0\mathbf{Q}_1 + {}^0\mathbf{n}_1 \times {}^0\mathbf{F}_0 - {}^0\mathbf{m}_1 \times {}^0\mathbf{F}_1 = {}^0I_1\,{}_0\boldsymbol{\alpha}_1 \tag{11.67}
$$

$$
{}^0\mathbf{F}_1 + (m_{21} + m_{22})\,g\,\hat{J} = (m_{21} + m_{22})\,{}^0\mathbf{a}_2 \tag{11.68}
$$

$$
{}^0\mathbf{Q}_1 + {}^0\mathbf{n}_2 \times {}^0\mathbf{F}_1 = {}^0I_2\,{}_0\boldsymbol{\alpha}_2 \tag{11.69}
$$

We may show the masses as

$$
m_1 = m_{11} + m_{12} \tag{11.70}
$$

$$
m_2 = m_{21} + m_{22} \tag{11.71}
$$

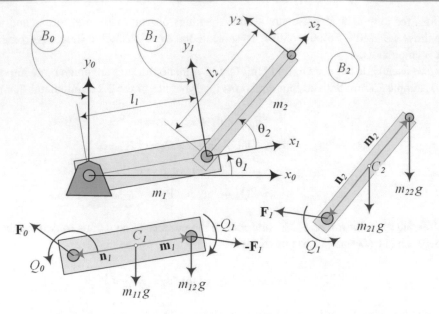

Fig. 11.7 A $2R$ planar manipulator with massive arms and massive joints

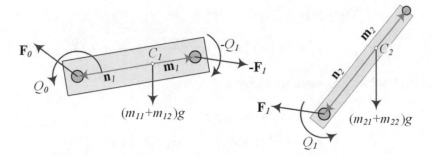

Fig. 11.8 Determination of the vectors **m** and **n** for new positions of mass center C_i

and use the same Eqs. (11.60)–(11.61).

Example 360 $2R$ planar manipulator general equations. There are many details and information in the equations of motion. This is a complete example for a general $2R$ manipulator to show every steps and all information we take from equations of motion.

Let us analyze a general $2R$ manipulator that has massive arms and carries a payload m_0 as is shown in Fig. 11.9. The equations of motion are:

$$^0\mathbf{F}_0 - {}^0\mathbf{F}_1 + m_1 g\, \hat{J} = m_1\, {}^0\mathbf{a}_1 \tag{11.72}$$

$$^0\mathbf{Q}_0 - {}^0\mathbf{Q}_1 + {}^0\mathbf{n}_1 \times {}^0\mathbf{F}_0 - {}^0\mathbf{m}_1 \times {}^0\mathbf{F}_1 = {}^0I_1\, {}_0\boldsymbol{\alpha}_1 \tag{11.73}$$

$$^0\mathbf{F}_1 + (m_0 + m_2)\, g\, \hat{J} = m_2\, {}^0\mathbf{a}_2 \tag{11.74}$$

$$^0\mathbf{Q}_1 + {}^0\mathbf{n}_2 \times {}^0\mathbf{F}_1 + {}^0\mathbf{m}_2 \times m_0 g\, \hat{J} = {}^0I_2\, {}_0\boldsymbol{\alpha}_2 \tag{11.75}$$

Elimination the joint forces $\mathbf{F}_0$, $\mathbf{F}_1$ provides the following equations for the torques $\mathbf{Q}_0$ and $\mathbf{Q}_1$.

$$^0\mathbf{Q}_1 = {}^0I_2\, {}_0\boldsymbol{\alpha}_2 - {}^0\mathbf{n}_2 \times \left(m_2\, {}^0\mathbf{a}_2 - (m_0 + m_2)\, g\, \hat{J} \right) - {}^0\mathbf{m}_2 \times m_0 g\, \hat{J} \tag{11.76}$$

$$^0\mathbf{Q}_0 = {}^0\mathbf{Q}_1 + {}^0I_1\, {}_0\boldsymbol{\alpha}_1 + {}^0\mathbf{m}_1 \times \left(m_2\, {}^0\mathbf{a}_2 - (m_0 + m_2)\, g\, \hat{J} \right)$$
$$- {}^0\mathbf{n}_1 \times \left(m_1\, {}^0\mathbf{a}_1 + m_2\, {}^0\mathbf{a}_2 - (m_0 + m_1 + m_2)\, g\, \hat{J} \right) \tag{11.77}$$

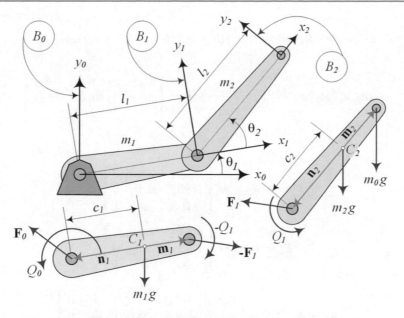

Fig. 11.9 A $2R$ manipulator that has massive arms and carries a payload m_0

The forces $\mathbf{F}_0$ and $\mathbf{F}_1$ are equal to:

$$^0\mathbf{F}_1 = m_2\,{}^0\mathbf{a}_2 - (m_0 + m_2)\,g\,\hat{J} \tag{11.78}$$

$$^0\mathbf{F}_0 = m_1\,{}^0\mathbf{a}_1 + m_2\,{}^0\mathbf{a}_2 - (m_0 + m_1 + m_2)\,g\,\hat{J} \tag{11.79}$$

where

$$^0\mathbf{r}_1 = -\,{}^0\mathbf{n}_1 \tag{11.80}$$

$$^0\mathbf{r}_2 = -\,{}^0\mathbf{n}_1 + {}^0\mathbf{m}_1 - {}^0\mathbf{n}_2 \tag{11.81}$$

$$^0\mathbf{d}_1 = -\,{}^0\mathbf{n}_1 + {}^0\mathbf{m}_1 \tag{11.82}$$

$$^0_1\mathbf{d}_2 = -\,{}^0\mathbf{n}_2 + {}^0\mathbf{m}_2 \tag{11.83}$$

$$^0\mathbf{d}_2 = -\,{}^0\mathbf{n}_1 + {}^0\mathbf{m}_1 - {}^0\mathbf{n}_2 + {}^0\mathbf{m}_2 \tag{11.84}$$

In a general case, the local position vectors of C_i are:

$$^0\mathbf{n}_1 = {}^0R_1\,{}^1\mathbf{n}_1 = -R_{Z,\theta_1}\,c_1\,{}^1\hat{\imath}_1 = \begin{bmatrix} -c_1 \cos\theta_1 \\ -c_1 \sin\theta_1 \\ 0 \end{bmatrix} \tag{11.85}$$

$$\begin{aligned}
^0\mathbf{n}_2 &= -\,{}^0R_2\,{}^2\mathbf{n}_2 = -\,{}^0R_1\,{}^1R_2\,{}^2\mathbf{n}_2 \\
&= -R_{Z,\theta_1}\,R_{Z,\theta_2}\,c_2\,{}^2\hat{\imath}_2 = \begin{bmatrix} -c_2 \cos(\theta_1 + \theta_2) \\ -c_2 \sin(\theta_1 + \theta_2) \\ 0 \end{bmatrix}
\end{aligned} \tag{11.86}$$

$$^1\mathbf{n}_2 = -\,{}^1R_2\,{}^2\mathbf{n}_2 = -R_{Z,\theta_2}\,c_2\,{}^2\hat{\imath}_2 = \begin{bmatrix} -c_2 \cos\theta_2 \\ -c_2 \sin\theta_2 \\ 0 \end{bmatrix} \tag{11.87}$$

$$^0\mathbf{m}_1 = {}^0R_1 \, {}^1\mathbf{m}_1 = R_{Z,\theta_1} \, (l_1 - c_1) \, {}^1\hat{\imath}_1 = \begin{bmatrix} (l_1 - c_1) \cos\theta_1 \\ (l_1 - c_1) \sin\theta_1 \\ 0 \end{bmatrix} \tag{11.88}$$

$$^0\mathbf{m}_2 = {}^0R_2 \, {}^2\mathbf{m}_2 = {}^0R_2 \, (l_2 - c_2) \, {}^2\hat{\imath}_2 = \begin{bmatrix} (l_2 - c_2) \cos(\theta_1 + \theta_2) \\ (l_2 - c_2) \sin(\theta_1 + \theta_2) \\ 0 \end{bmatrix} \tag{11.89}$$

where

$$^0R_1 = R_{Z,\theta_1} = \begin{bmatrix} \cos\theta_1 & -\sin\theta_1 & 0 \\ \sin\theta_1 & \cos\theta_1 & 0 \\ 0 & 0 & 1 \end{bmatrix} \tag{11.90}$$

$$^1R_2 = R_{Z,\theta_2} = \begin{bmatrix} \cos\theta_2 & -\sin\theta_2 & 0 \\ \sin\theta_2 & \cos\theta_2 & 0 \\ 0 & 0 & 1 \end{bmatrix} \tag{11.91}$$

$$^0R_2 = R_{z,\theta_1+\theta_2} = \begin{bmatrix} \cos(\theta_1 + \theta_2) & -\sin(\theta_1 + \theta_2) & 0 \\ \sin(\theta_1 + \theta_2) & \cos(\theta_1 + \theta_2) & 0 \\ 0 & 0 & 1 \end{bmatrix} \tag{11.92}$$

The position vectors are as follows.

$$^0\mathbf{r}_1 = -{}^0\mathbf{n}_1 = {}^0R_1 \, {}^1\mathbf{r}_1 = {}^0R_1 \, c_1\hat{\imath}_1 = \begin{bmatrix} c_1 \cos\theta_1 \\ c_1 \sin\theta_1 \\ 0 \end{bmatrix} \tag{11.93}$$

$$^0\mathbf{r}_2 = -{}^0\mathbf{n}_1 + {}^0\mathbf{m}_1 - {}^0\mathbf{n}_2 = {}^0\mathbf{d}_1 + {}^0R_2 \, {}^2\mathbf{r}_2$$
$$= \begin{bmatrix} l_1 \cos\theta_1 + c_2 \cos(\theta_1 + \theta_2) \\ l_1 \sin\theta_1 + c_2 \sin(\theta_1 + \theta_2) \\ 0 \end{bmatrix} \tag{11.94}$$

$$^0\mathbf{d}_1 = -{}^0\mathbf{n}_1 + {}^0\mathbf{m}_1 = \begin{bmatrix} l_1 \cos\theta_1 \\ l_1 \sin\theta_1 \\ 0 \end{bmatrix} \tag{11.95}$$

$$^0_1\mathbf{d}_2 = -{}^0\mathbf{n}_2 + {}^0\mathbf{m}_2 = \begin{bmatrix} l_2 \cos(\theta_1 + \theta_2) \\ l_2 \sin(\theta_1 + \theta_2) \\ 0 \end{bmatrix} \tag{11.96}$$

$$^0\mathbf{d}_2 = -{}^0\mathbf{n}_1 + {}^0\mathbf{m}_1 - {}^0\mathbf{n}_2 + {}^0\mathbf{m}_2$$
$$= {}^0\mathbf{d}_1 + {}^0_1\mathbf{d}_2 = \begin{bmatrix} l_2 \cos(\theta_1 + \theta_2) + l_1 \cos\theta_1 \\ l_2 \sin(\theta_1 + \theta_2) + l_1 \sin\theta_1 \\ 0 \end{bmatrix} \tag{11.97}$$

The links' angular velocities and accelerations are:

$$_0\boldsymbol{\omega}_1 = \dot{\theta}_1 \, \hat{K} \tag{11.98}$$

$$_0\boldsymbol{\alpha}_1 = {}_0\dot{\boldsymbol{\omega}}_1 = \ddot{\theta}_1 \, \hat{K} \tag{11.99}$$

$$_0\boldsymbol{\omega}_2 = \left(\dot{\theta}_1 + \dot{\theta}_2\right) \hat{K} \tag{11.100}$$

$$_0\boldsymbol{\alpha}_2 = {_0}\dot{\boldsymbol{\omega}}_2 = \left(\ddot{\theta}_1 + \ddot{\theta}_2\right)\hat{K} \tag{11.101}$$

The translational acceleration of C_i are:

$$
\begin{aligned}
{}^0\mathbf{a}_1 &= {_0}\boldsymbol{\alpha}_1 \times {}^0\mathbf{r}_1 - {_0}\boldsymbol{\omega}_1 \times \left({_0}\boldsymbol{\omega}_1 \times {}^0\mathbf{r}_1\right) \\
&= \begin{bmatrix} -c_1\ddot{\theta}_1 \sin\theta_1 + c_1\dot{\theta}_1^2 \cos\theta_1 \\ c_1\ddot{\theta}_1 \cos\theta_1 + c_1\dot{\theta}_1^2 \sin\theta_1 \\ 0 \end{bmatrix}
\end{aligned} \tag{11.102}
$$

$$
\begin{aligned}
{}^0\ddot{\boldsymbol{d}}_1 &= {_0}\boldsymbol{\alpha}_1 \times {}^0\mathbf{d}_1 - {_0}\boldsymbol{\omega}_1 \times \left({_0}\boldsymbol{\omega}_1 \times {}^0\mathbf{d}_1\right) \\
&= \begin{bmatrix} -l_1\ddot{\theta}_1 \sin\theta_1 + l_1\dot{\theta}_1^2 \cos\theta_1 \\ l_1\ddot{\theta}_1 \cos\theta_1 + l_1\dot{\theta}_1^2 \sin\theta_1 \\ 0 \end{bmatrix}
\end{aligned} \tag{11.103}
$$

$$
\begin{aligned}
{}^0\ddot{\boldsymbol{d}}_2 &= \frac{{}^G d^2\, {}^0\mathbf{d}_2}{dt^2} = {}^0\ddot{\boldsymbol{d}}_1 + {_1^0}\dot{\boldsymbol{\omega}}_2 \times {_1^0}\mathbf{d}_2 - {_1^0}\boldsymbol{\omega}_2 \times \left({_1^0}\boldsymbol{\omega}_2 \times {_1^0}\mathbf{d}_2\right) \\
&= \begin{bmatrix} {}^0\ddot{d}_{2x} \\ {}^0\ddot{d}_{2y} \\ 0 \end{bmatrix}
\end{aligned} \tag{11.104}
$$

$$
\begin{aligned}
{}^0\ddot{d}_{2x} = &-l_1\ddot{\theta}_1 \sin\theta_1 - l_2\ddot{\theta}_2 \sin(\theta_1 + \theta_2) \\
&+ l_1\dot{\theta}_1^2 \cos\theta_1 + l_2\dot{\theta}_2^2 \cos(\theta_1 + \theta_2)
\end{aligned} \tag{11.105}
$$

$$
\begin{aligned}
{}^0\ddot{d}_{2y} = &\; l_1\ddot{\theta}_1 \cos\theta_1 + l_2\ddot{\theta}_2 \cos(\theta_1 + \theta_2) \\
&+ l_1\dot{\theta}_1^2 \sin\theta_1 + l_2\dot{\theta}_2^2 \sin(\theta_1 + \theta_2)
\end{aligned} \tag{11.106}
$$

$$
\begin{aligned}
{}^0\mathbf{a}_2 &= {}^0\ddot{\boldsymbol{d}}_2 + {_0}\boldsymbol{\alpha}_2 \times \left({}^0\mathbf{r}_2 - {}^0\mathbf{d}_2\right) - {_0}\boldsymbol{\omega}_2 \times \left({_0}\boldsymbol{\omega}_2 \times \left({}^0\mathbf{r}_2 - {}^0\mathbf{d}_2\right)\right) \\
&= {}^0\ddot{\boldsymbol{d}}_2 - {_0}\boldsymbol{\alpha}_2 \times {}^0\mathbf{m}_2 + {_0}\boldsymbol{\omega}_2 \times \left({_0}\boldsymbol{\omega}_2 \times {}^0\mathbf{m}_2\right) \\
&= \begin{bmatrix} {}^0a_{2x} \\ {}^0a_{2y} \\ 0 \end{bmatrix}
\end{aligned} \tag{11.107}
$$

$$
\begin{aligned}
{}^0a_{2x} = &\left((l_2 - c_2)\left(\ddot{\theta}_1 + \ddot{\theta}_2\right) - l_2\ddot{\theta}_2\right)\sin(\theta_1 + \theta_2) \\
&- l_1\ddot{\theta}_1 \sin\theta_1 + l_1\dot{\theta}_1^2 \cos\theta_1 \\
&- \left((l_2 - c_2)\left(\dot{\theta}_1 + \dot{\theta}_2\right)^2 - l_2\dot{\theta}_2^2\right)\cos(\theta_1 + \theta_2)
\end{aligned} \tag{11.108}
$$

$$
\begin{aligned}
{}^0a_{2y} = &-\left((l_2 - c_2)\left(\ddot{\theta}_1 + \ddot{\theta}_2\right) - l_2\ddot{\theta}_2\right)\cos(\theta_1 + \theta_2) \\
&+ l_1\ddot{\theta}_1 \cos\theta_1 + l_1\dot{\theta}_1^2 \sin\theta_1 \\
&- \left((l_2 - c_2)\left(\dot{\theta}_1 + \dot{\theta}_2\right)^2 - l_2\dot{\theta}_2^2\right)\sin(\theta_1 + \theta_2)
\end{aligned} \tag{11.109}
$$

The moment of inertia matrices in the global coordinate frame are:

$$^0I_1 = R_{Z,\theta_1} \, ^1I_1 \, R_{Z,\theta_1}^T = \, ^0R_1 \begin{bmatrix} I_{x_1} & 0 & 0 \\ 0 & I_{y_1} & 0 \\ 0 & 0 & I_{z_1} \end{bmatrix} \, ^0R_1^T$$

$$= \begin{bmatrix} I_{x_1}c^2\theta_1 + I_{y_1}s^2\theta_1 & \left(I_{x_1} - I_{y_1}\right)c\theta_1 s\theta_1 & 0 \\ \left(I_{x_1} - I_{y_1}\right)c\theta_1 s\theta_1 & I_{y_1}c^2\theta_1 + I_{x_1}s^2\theta_1 & 0 \\ 0 & 0 & I_{z_1} \end{bmatrix} \tag{11.110}$$

$$^0I_2 = \, ^0R_2 \, ^2I_2 \, ^0R_2^T = \, ^0R_2 \begin{bmatrix} I_{x_2} & 0 & 0 \\ 0 & I_{y_2} & 0 \\ 0 & 0 & I_{z_2} \end{bmatrix} \, ^0R_2^T$$

$$= \begin{bmatrix} I_{x_2}c^2\theta_{12} + I_{y_2}s^2\theta_{12} & \left(I_{x_2} - I_{y_2}\right)c\theta_{12} s\theta_{12} & 0 \\ \left(I_{x_2} - I_{y_2}\right)c\theta_{12} s\theta_{12} & I_{y_2}c^2\theta_{12} + I_{x_2}s^2\theta_{12} & 0 \\ 0 & 0 & I_{z_2} \end{bmatrix} \tag{11.111}$$

$$\theta_{12} = \theta_1 + \theta_2 \tag{11.112}$$

Substituting these results in Eqs. (11.76) and (11.77), and solving for Q_0 and Q_1, provides the dynamic equations for the $2R$ manipulator.

$$^0\mathbf{Q}_1 = \, ^0I_2 \, _0\boldsymbol{\alpha}_2 - \, ^0\mathbf{n}_2 \times \left(m_2 \, ^0\mathbf{a}_2 - (m_0 + m_2)\, g \, \hat{J}\right) - \, ^0\mathbf{m}_2 \times m_0 g \, \hat{J}$$

$$= \begin{bmatrix} 0 \\ 0 \\ ^0Q_{1z} \end{bmatrix} \tag{11.113}$$

$$^0Q_{1z} = \left(I_{z_2} + m_2 c_2^2 - m_2 l_2 c_2 + m_2 l_1 c_2 \cos\theta_2\right)\ddot{\theta}_1$$
$$+ \left(I_{z_2} + m_2 c_2^2\right)\ddot{\theta}_2 - m_2 c_2 l_1 \dot{\theta}_1^2 \sin\theta_2$$
$$- (m_2 c_2 + m_0 l_2)\, g \cos(\theta_1 + \theta_2) \tag{11.114}$$

$$^0\mathbf{Q}_0 = \, ^0\mathbf{Q}_1 + \, ^0I_1 \, _0\boldsymbol{\alpha}_1 + \, ^0\mathbf{m}_1 \times \left(m_2 \, ^0\mathbf{a}_2 - (m_0 + m_2)\, g \, \hat{J}\right)$$
$$- \, ^0\mathbf{n}_1 \times \left(m_1 \, ^0\mathbf{a}_1 - m_1 g \, \hat{J} + \left(m_2 \, ^0\mathbf{a}_2 - (m_0 + m_2)\, g \, \hat{J}\right)\right)$$

$$= \begin{bmatrix} 0 \\ 0 \\ ^0Q_{0z} \end{bmatrix} \tag{11.115}$$

$$^0Q_{0z} = \left(I_{z_1} + I_{z_2} + m_1 c_1^2 + m_2 \left(l_1^2 + c_2^2 - l_2 c_2 + l_1 \left(2c_2 - l_2\right)\cos\theta_2\right)\right)\ddot{\theta}_1$$
$$+ \left(I_{z_2} + m_2 c_2 \left(c_2 + l_1 \cos\theta_2\right)\right)\ddot{\theta}_2 - m_2 l_1 l_2 \dot{\theta}_1^2 \sin\theta_2$$
$$+ m_2 l_1 c_2 \dot{\theta}_2^2 \sin\theta_2 - 2 m_2 l_1 \left(l_2 - c_2\right)\dot{\theta}_1 \dot{\theta}_2 \sin\theta_2$$
$$- (m_0 l_1 + m_1 c_1 + m_2 l_1)\, g \cos\theta_1$$
$$- (m_0 l_2 + m_2 c_2)\, g \cos(\theta_1 + \theta_2) \tag{11.116}$$

Substituting the vectorial information of (11.85)–(11.111) in (11.78) and (11.79), we find the joint forces of the general $2R$ manipulator. The manipulator has massive arms with mass center at C_i and carries a payload m_0.

$$^0\mathbf{F}_1 = m_2\,{}^0\mathbf{a}_2 - (m_0 + m_2)\,g\,\hat{J} = \begin{bmatrix} {}^0F_{1x} \\ {}^0F_{1y} \\ 0 \end{bmatrix} \qquad (11.117)$$

$$\begin{aligned}
^0F_{1x} &= (m_2\,(l_2 - c_2)\sin(\theta_1 + \theta_2) - m_2 l_1 \sin\theta_1)\,\ddot{\theta}_1 \\
&\quad - m_2 c_2 \ddot{\theta}_2 \sin(\theta_1 + \theta_2) + m_2 c_2 \dot{\theta}_2^2 \cos(\theta_1 + \theta_2) \\
&\quad + (m_2\,(l_2 - c_2)\cos(\theta_1 + \theta_2) + m_2 l_1 \cos\theta_1)\,\dot{\theta}_1^2 \\
&\quad + 2 m_2 l_1\,(l_2 - c_2)\,\dot{\theta}_1 \dot{\theta}_2 \cos(\theta_1 + \theta_2)
\end{aligned} \qquad (11.118)$$

$$\begin{aligned}
^0F_{1y} &= (-m_2\,(l_2 - c_2)\cos(\theta_1 + \theta_2) + m_2 l_1 \cos\theta_1)\,\ddot{\theta}_1 \\
&\quad + m_2 c_2 \ddot{\theta}_2 \cos(\theta_1 + \theta_2) + m_2 c_2 \dot{\theta}_2^2 \sin(\theta_1 + \theta_2) \\
&\quad + (-m_2\,(l_2 - c_2)\sin(\theta_1 + \theta_2) + m_2 l_1 \sin\theta_1)\,\dot{\theta}_1^2 \\
&\quad - 2 m_2 l_1\,(l_2 - c_2)\,\dot{\theta}_1 \dot{\theta}_2 \sin(\theta_1 + \theta_2) \\
&\quad - (m_0 + m_2)\,g
\end{aligned} \qquad (11.119)$$

$$^0\mathbf{F}_0 = m_1\,{}^0\mathbf{a}_1 + m_2\,{}^0\mathbf{a}_2 - (m_0 + m_1 + m_2)\,g\,\hat{J} = \begin{bmatrix} {}^0F_{0x} \\ {}^0F_{0y} \\ 0 \end{bmatrix} \qquad (11.120)$$

$$\begin{aligned}
^0F_{0x} &= (m_2\,(l_2 - c_2)\sin(\theta_1 + \theta_2) - (m_1 c_1 + m_2 l_1)\sin\theta_1)\,\ddot{\theta}_1 \\
&\quad - m_2 c_2 \ddot{\theta}_2 \sin(\theta_1 + \theta_2) - m_2 c_2 \dot{\theta}_2^2 \cos(\theta_1 + \theta_2) \\
&\quad + (-m_2\,(l_2 - c_2)\cos(\theta_1 + \theta_2) + (m_2 l_1 + m_1 c_1)\cos\theta_1)\,\dot{\theta}_1^2 \\
&\quad - 2 m_2\,(l_2 - c_2)\,\dot{\theta}_1 \dot{\theta}_2 \cos(\theta_1 + \theta_2)
\end{aligned} \qquad (11.121)$$

$$\begin{aligned}
^0F_{0y} &= (-m_2\,(l_2 - c_2)\cos(\theta_1 + \theta_2) + (m_2 l_1 + m_1 c_1)\cos\theta_1)\,\ddot{\theta}_1 \\
&\quad + m_2 c_2 \ddot{\theta}_2 \cos(\theta_1 + \theta_2) + m_2 c_2 \dot{\theta}_2^2 \sin(\theta_1 + \theta_2) \\
&\quad + (-m_2\,(l_2 - c_2)\sin(\theta_1 + \theta_2) + (m_2 l_1 + m_1 c_1)\sin\theta_1)\,\dot{\theta}_1^2 \\
&\quad - 2 m_2 l_1\,(l_2 - c_2)\,\dot{\theta}_1 \dot{\theta}_2 \sin(\theta_1 + \theta_2) \\
&\quad - (m_0 + m_1 + m_2)\,g
\end{aligned} \qquad (11.122)$$

Example 361 ★ Matrix form of equations of motion. It is always possible and useful to convert the equations of motion of a robot into a matrix form of $\mathbf{D}(\mathbf{q})\,\ddot{q} + \mathbf{C}(\mathbf{q},\dot{q})\dot{q} + \mathbf{G}(\mathbf{q}) = \mathbf{Q}$. It is a better form for computer calculations. Here is an example how to convert equations of motion of $2R$ planar manipulator into matrix form.

Let us rearrange the equations of motion (11.114) and (11.116) of $2R$ planar manipulator in a matrix form.

$$\mathbf{D}(\mathbf{q})\,\ddot{q} + \mathbf{C}(\mathbf{q},\dot{q})\dot{q} + \mathbf{G}(\mathbf{q}) = \mathbf{Q} \qquad (11.123)$$

$$\mathbf{q} = \begin{bmatrix} \theta_1 \\ \theta_2 \end{bmatrix} \qquad \mathbf{Q} = \begin{bmatrix} {}^0Q_{1z} \\ {}^0Q_{2z} \end{bmatrix} \qquad (11.124)$$

$$\mathbf{D}(\mathbf{q}) = \begin{bmatrix} Z_1 - Z_2 + Z_3 \cos\theta_2 & Z_1 \\ Z_1 + Z_4 - Z_2 + Z_5 + Z_6 \cos\theta_2 & Z_1 + Z_3 \cos\theta_2 \end{bmatrix} \qquad (11.125)$$

$$\mathbf{C}(\mathbf{q},\dot{q}) = \begin{bmatrix} -Z_3 \dot{\theta}_1 \sin\theta_2 & 0 \\ (-Z_7 \dot{\theta}_1 - Z_8 \dot{\theta}_2)\sin\theta_2 & (Z_3 \dot{\theta}_2 - Z_8 \dot{\theta}_1)\sin\theta_2 \end{bmatrix} \qquad (11.126)$$

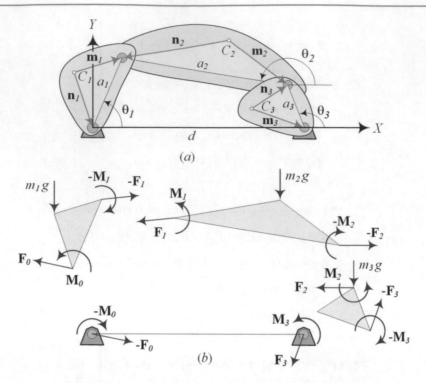

Fig. 11.10 A four-bar linkage, and free body diagram of each link

$$\mathbf{G}(\mathbf{q}) = \begin{bmatrix} -Z_9 \cos\left(\theta_1 + \theta_2\right) \\ -Z_9 \cos\left(\theta_1 + \theta_2\right) + Z_{10} \cos\theta_1 \end{bmatrix} \tag{11.127}$$

$$Z_1 = I_{z_2} + m_2 c_2^2 \tag{11.128}$$

$$Z_2 = m_2 l_2 c_2 \tag{11.129}$$

$$Z_3 = m_2 l_1 c_2 \tag{11.130}$$

$$Z_4 = I_{z_1} + m_1 c_1^2 \tag{11.131}$$

$$Z_5 = m_2 l_1^2 \tag{11.132}$$

$$Z_6 = m_2 l_1 \left(2c_2 - l_2\right) \tag{11.133}$$

$$Z_7 = m_2 l_1 l_2 \tag{11.134}$$

$$Z_8 = m_2 l_1 \left(l_2 - c_2\right) \tag{11.135}$$

$$Z_9 = \left(m_2 c_2 + m_0 l_2\right) g \tag{11.136}$$

$$Z_{10} = \left(m_0 l_1 + m_1 c_1 + m_2 l_1\right) g \tag{11.137}$$

Example 362 A four-bar linkage dynamics. A planar mechanism is a multi-link with two dimensional forces and one dimensional moments. This example illustrates how to define the equations of motion of a multi-link mechanism similar to a robotic manipulator.

Figure 11.10a illustrates a closed-loop four-bar linkage and Fig. 11.10b shows the free body diagrams of the links. The position of the mass centers is given, and therefore the vectors $^0\mathbf{n}_i$ and $^0\mathbf{m}_i$ for each link are known. The Newton–Euler equations for the link (i) are:

$$^0\mathbf{F}_{i-1} - {}^0\mathbf{F}_i + m_i g\, \hat{J} = m_i\, {}^0\mathbf{a}_i \tag{11.138}$$

$$^0\mathbf{M}_{i-1} - {}^0\mathbf{M}_i + {}^0\mathbf{n}_i \times {}^0\mathbf{F}_{i-1} - {}^0\mathbf{m}_i \times {}^0\mathbf{F}_i = I_i\, {}_0\boldsymbol{\alpha}_i \tag{11.139}$$

and therefore, we have three sets of equations for links (1), (2), (3). Link (0) is the ground link and motionless.

$$^0\mathbf{F}_0 - {}^0\mathbf{F}_1 + m_1 g\,\hat{J} = m_1\,{}^0\mathbf{a}_1 \tag{11.140}$$

$$^0\mathbf{M}_0 - {}^0\mathbf{M}_1 + {}^0\mathbf{n}_1 \times {}^0\mathbf{F}_0 - {}^0\mathbf{m}_1 \times {}^0\mathbf{F}_1 = I_1\,{}_0\boldsymbol{\alpha}_1 \tag{11.141}$$

$$^0\mathbf{F}_1 - {}^0\mathbf{F}_2 + m_2 g\,\hat{J} = m_2\,{}^0\mathbf{a}_2 \tag{11.142}$$

$$^0\mathbf{M}_1 - {}^0\mathbf{M}_2 + {}^0\mathbf{n}_2 \times {}^0\mathbf{F}_1 - {}^0\mathbf{m}_2 \times {}^0\mathbf{F}_2 = I_2\,{}_0\boldsymbol{\alpha}_2 \tag{11.143}$$

$$^0\mathbf{F}_2 - {}^0\mathbf{F}_3 + m_3 g\,\hat{J} = m_2\,{}^0\mathbf{a}_2 \tag{11.144}$$

$$^0\mathbf{M}_2 - {}^0\mathbf{M}_3 + {}^0\mathbf{n}_3 \times {}^0\mathbf{F}_2 - {}^0\mathbf{m}_3 \times {}^0\mathbf{F}_3 = I_3\,{}_0\boldsymbol{\alpha}_3 \tag{11.145}$$

Let us assume that there is no friction in joints and the mechanism is planar. Therefore, the force vectors are in the (X, Y)-plane, and the moments are parallel to Z-axis. So, the equations of motion simplify to:

$$^0\mathbf{F}_0 - {}^0\mathbf{F}_1 + m_1 g\,\hat{J} = m_1\,{}^0\mathbf{a}_1 \tag{11.146}$$

$$^0\mathbf{M}_0 + {}^0\mathbf{n}_1 \times {}^0\mathbf{F}_0 - {}^0\mathbf{m}_1 \times {}^0\mathbf{F}_1 = I_1\,{}_0\boldsymbol{\alpha}_1 \tag{11.147}$$

$$^0\mathbf{F}_1 - {}^0\mathbf{F}_2 + m_2 g\,\hat{J} = m_2\,{}^0\mathbf{a}_2 \tag{11.148}$$

$$^0\mathbf{n}_2 \times {}^0\mathbf{F}_1 - {}^0\mathbf{m}_2 \times {}^0\mathbf{F}_2 = I_2\,{}_0\boldsymbol{\alpha}_2 \tag{11.149}$$

$$^0\mathbf{F}_2 - {}^0\mathbf{F}_3 + m_3 g\,\hat{J} = m_2\,{}^0\mathbf{a}_2 \tag{11.150}$$

$$^0\mathbf{n}_3 \times {}^0\mathbf{F}_2 - {}^0\mathbf{m}_3 \times {}^0\mathbf{F}_3 = I_3\,{}_0\boldsymbol{\alpha}_3 \tag{11.151}$$

where $^0\mathbf{M}_0$ is the driving torque of the mechanism. The number of equations reduces to 9 and the unknowns of the equations are:

$$F_{0x}, F_{0y}, F_{1x}, F_{1y}, F_{2x}, F_{2y}, F_{3x}, F_{3y}, M_0 \tag{11.152}$$

We can rearrange the set of equations in a matrix form

$$[A]\,\mathbf{x} = \mathbf{b} \tag{11.153}$$

where

$$[A] = \begin{bmatrix} 1 & 0 & -1 & 0 & 0 & 0 & 0 & 0 & 0 \\ 0 & 1 & 0 & -1 & 0 & 0 & 0 & 0 & 0 \\ -n_{1y} & n_{1x} & m_{1y} & -m_{1x} & 0 & 0 & 0 & 0 & 1 \\ 0 & 0 & 1 & 0 & -1 & 0 & 0 & 0 & 0 \\ 0 & 0 & 0 & 1 & 0 & -1 & 0 & 0 & 0 \\ 0 & 0 & -n_{2y} & n_{2x} & m_{2y} & -m_{2x} & 0 & 0 & 0 \\ 0 & 0 & 0 & 0 & 1 & 0 & -1 & 0 & 0 \\ 0 & 0 & 0 & 0 & 0 & 1 & 0 & -1 & 0 \\ 0 & 0 & 0 & 0 & -n_{3y} & n_{3x} & m_{3y} & -m_{3x} & 0 \end{bmatrix} \tag{11.154}$$

$$\mathbf{x} = \begin{bmatrix} F_{0x} \\ F_{0y} \\ F_{1x} \\ F_{1y} \\ F_{2x} \\ F_{2y} \\ F_{3x} \\ F_{3y} \\ M_0 \end{bmatrix} \quad \mathbf{b} = \begin{bmatrix} m_1 a_{1x} \\ m_1 a_{1y} - m_1 g \\ I_1 \alpha_1 \\ m_2 a_{2x} \\ m_2 a_{2y} - m_2 g \\ I_2 \alpha_2 \\ m_3 a_{3x} \\ m_3 a_{3y} - m_3 g \\ I_3 \alpha_3 \end{bmatrix} \tag{11.155}$$

The coefficient matrix $[A]$ describes the geometry of the mechanism, the vector $\mathbf{x}$ is the unknown forces, and the vector $\mathbf{b}$ indicates the dynamic terms. To solve the dynamics of the four-bar mechanism, we must calculate the accelerations ${}^0\mathbf{a}_i$ and ${}_0\boldsymbol{\alpha}_i$ and then find the required driving moment ${}^0\mathbf{M}_0$ and the joints' forces.

The force $\mathbf{F}_s$ is called the shaking force and shows the reaction of the mechanism on the ground.

$$\mathbf{F}_s = \mathbf{F}_3 - \mathbf{F}_0 \tag{11.156}$$

11.2 ★ Recursive Newton–Euler Dynamics

An advantage of the Newton–Euler equations of motion in robotic application is that we can calculate the joint forces of one link at a time. Therefore, starting from the end-effector link, we can analyze the links one by one and end up at the base link or vice versa. For such an analysis, we need to reform the Newton–Euler equations of motion to work in a *recursive* form in the link's frame.

The *backward recursive Newton–Euler equations of motion* for the link (i) in its body coordinate frame B_i are:

$$ {}^i\mathbf{F}_{i-1} = {}^i\mathbf{F}_i - \sum {}^i\mathbf{F}_{e_i} + m_i\, {}^i_0\mathbf{a}_i \tag{11.157}$$

$$ {}^i\mathbf{M}_{i-1} = {}^i\mathbf{M}_i - \sum {}^i\mathbf{M}_{e_i} - \left({}^i\mathbf{d}_{i-1} - {}^i\mathbf{r}_i \right) \times {}^i\mathbf{F}_{i-1} $$
$$ + \left({}^i\mathbf{d}_i - {}^i\mathbf{r}_i \right) \times {}^i\mathbf{F}_i + {}^iI_i\, {}^i_0\boldsymbol{\alpha}_i + {}^i_0\boldsymbol{\omega}_i \times {}^iI_i\, {}^i_0\boldsymbol{\omega}_i \tag{11.158}$$

$$ {}^i\mathbf{n}_i = {}^i\mathbf{d}_{i-1} - {}^i\mathbf{r}_i \tag{11.159}$$

$$ {}^i\mathbf{m}_i = {}^i\mathbf{d}_i - {}^i\mathbf{r}_i \tag{11.160}$$

When the driving force system $({}^i\mathbf{F}_{i-1}, {}^i\mathbf{M}_{i-1})$ is found in frame B_i, we can transform them to the frame B_{i-1} and apply the Newton–Euler equation for link $(i-1)$.

$$ {}^{i-1}\mathbf{F}_{i-1} = {}^{i-1}T_i\, {}^i\mathbf{F}_{i-1} \tag{11.161}$$

$$ {}^{i-1}\mathbf{M}_{i-1} = {}^{i-1}T_i\, {}^i\mathbf{M}_{i-1} \tag{11.162}$$

The negative of the converted force system acts as the driven force system $(-{}^{i-1}\mathbf{F}_{i-1}, -{}^{i-1}\mathbf{M}_{i-1})$ for the link $(i-1)$.

The *forward recursive Newton–Euler equations of motion* for the link (i) in its body coordinate frame B_i are:

$$ {}^i\mathbf{F}_i = {}^i\mathbf{F}_{i-1} + \sum {}^i\mathbf{F}_{e_i} - m_i\, {}^i_0\mathbf{a}_i \tag{11.163}$$

$$ {}^i\mathbf{M}_i = {}^i\mathbf{M}_{i-1} + \sum {}^i\mathbf{M}_{e_i} + \left({}^i\mathbf{d}_{i-1} - {}^i\mathbf{r}_i \right) \times {}^i\mathbf{F}_{i-1} $$
$$ - \left({}^i\mathbf{d}_i - {}^i\mathbf{r}_i \right) \times {}^i\mathbf{F}_i - {}^iI_i\, {}^i_0\boldsymbol{\alpha}_i - {}^i_0\boldsymbol{\omega}_i \times {}^iI_i\, {}^i_0\boldsymbol{\omega}_i \tag{11.164}$$

$$ {}^i\mathbf{n}_i = {}^i\mathbf{d}_{i-1} - {}^i\mathbf{r}_i \tag{11.165}$$

$$ {}^i\mathbf{m}_i = {}^i\mathbf{d}_i - {}^i\mathbf{r}_i \tag{11.166}$$

When the reaction force system $({}^i\mathbf{F}_i, {}^i\mathbf{M}_i)$ is found in frame B_i, we can transform them to frame B_{i+1}.

$$^{i+1}\mathbf{F}_i = {}^i T_{i+1}^{-1} \, {}^i\mathbf{F}_i \tag{11.167}$$

$$^{i+1}\mathbf{M}_i = {}^i T_{i+1}^{-1} \, {}^i\mathbf{M}_i \tag{11.168}$$

The negative of the converted force system acts as the action force system $(-{}^{i+1}\mathbf{F}_i, -{}^{i+1}\mathbf{M}_i)$ for the link $(i + 1)$.

Proof The Euler equation for a rigid link in body coordinate frame is:

$$^B\mathbf{M} = \frac{{}^G d}{dt} \, {}^B\mathbf{L} = {}^B\dot{\mathbf{L}} + {}^B_G\boldsymbol{\omega}_B \times {}^B\mathbf{L}$$

$$= {}^i I_i \, {}_i\boldsymbol{\alpha}_i + {}^B_G\boldsymbol{\omega}_B \times {}^i I_i \, {}_i\boldsymbol{\omega}_i \tag{11.169}$$

where $\mathbf{L}$ is the angular momentum of the link.

$$^B\mathbf{L} = {}^B I \, {}^B_G\boldsymbol{\omega}_B \tag{11.170}$$

We may solve the Newton–Euler equations of motion (11.1) and (11.2) for the action force system

$$^0\mathbf{F}_{i-1} = {}^0\mathbf{F}_i - \sum {}^0\mathbf{F}_{e_i} + m_i \, {}^0\mathbf{a}_i \tag{11.171}$$

$$^0\mathbf{M}_{i-1} = {}^0\mathbf{M}_i - \sum {}^0\mathbf{M}_{e_i} - \left({}^0\mathbf{d}_{i-1} - {}^0\mathbf{r}_i\right) \times {}^0\mathbf{F}_{i-1}$$

$$+ \left({}^0\mathbf{d}_i - {}^0\mathbf{r}_i\right) \times {}^0\mathbf{F}_i + \frac{{}^0 d}{dt} \, {}^0\mathbf{L}_i \tag{11.172}$$

and then transform the equations to the coordinate frame B_i attached to the link (i) to make the recursive form of the Newton–Euler equations of motion.

$$^i\mathbf{F}_{i-1} = {}^0 T_i^{-1} \, {}^0\mathbf{F}_{i-1} = {}^i\mathbf{F}_i - \sum {}^i\mathbf{F}_{e_i} + m_i \, {}^i_0\mathbf{a}_i \tag{11.173}$$

$$^i\mathbf{M}_{i-1} = {}^0 T_i^{-1} \, {}^0\mathbf{M}_{i-1}$$

$$= {}^i\mathbf{M}_i - \sum {}^i\mathbf{M}_{e_i} - \left({}^i\mathbf{d}_{i-1} - {}^i\mathbf{r}_i\right) \times {}^i\mathbf{F}_{i-1} \tag{11.174}$$

$$+ \left({}^i\mathbf{d}_i - {}^i\mathbf{r}_i\right) \times {}^i\mathbf{F}_i + \frac{{}^0 d}{dt} \, {}^i\mathbf{L}_i$$

$$= {}^i\mathbf{M}_i - \sum {}^i\mathbf{M}_{e_i} - \left({}^i\mathbf{d}_{i-1} - {}^i\mathbf{r}_i\right) \times {}^i\mathbf{F}_{i-1}$$

$$+ \left({}^i\mathbf{d}_i - {}^i\mathbf{r}_i\right) \times {}^i\mathbf{F}_i + {}^i I_i \, {}^i_0\boldsymbol{\alpha}_i + {}^i_0\boldsymbol{\omega}_i \times {}^i I_i \, {}^i_0\boldsymbol{\omega}_i \tag{11.175}$$

Starting from link (i) and deriving the equations of motion of the previous link $(i-1)$ is called the *backward Newton–Euler equations of motion*.

We may also start from link (i) and derive the equations of motion of the next link $(i + 1)$. This method is called the *forward Newton–Euler equations of motion*. Employing the Newton–Euler equations of motion (11.157) and (11.158), we can write them in a *forward recursive* form in coordinate frame B_i attached to the link (i).

$$^i\mathbf{F}_i = {}^i\mathbf{F}_{i-1} + \sum {}^i\mathbf{F}_{e_i} - m_i \, {}^i_0\mathbf{a}_i \tag{11.176}$$

$$^i\mathbf{M}_i = {}^i\mathbf{M}_{i-1} + \sum {}^i\mathbf{M}_{e_i} + \left({}^i\mathbf{d}_{i-1} - {}^i\mathbf{r}_i\right) \times {}^i\mathbf{F}_{i-1}$$

$$- \left({}^i\mathbf{d}_i - {}^i\mathbf{r}_i\right) \times {}^i\mathbf{F}_i - {}^i I_i \, {}^i_0\boldsymbol{\alpha}_i - {}^i_0\boldsymbol{\omega}_i \times {}^i I_i \, {}^i_0\boldsymbol{\omega}_i \tag{11.177}$$

$$^i\mathbf{n}_i = {}^i\mathbf{d}_{i-1} - {}^i\mathbf{r}_i \tag{11.178}$$

$$^i\mathbf{m}_i = {}^i\mathbf{d}_i - {}^i\mathbf{r}_i \tag{11.179}$$

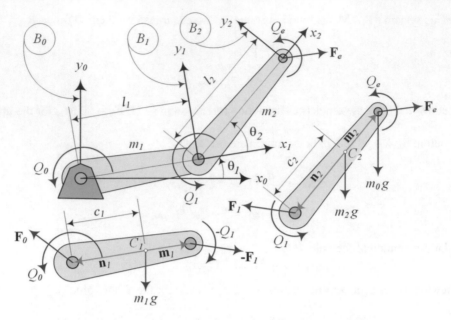

Fig. 11.11   A $2R$ planar manipulator carrying a load at the endpoint

Using the forward Newton–Euler equations of motion (11.176) and (11.177), we can calculate the reaction force system ($^i\mathbf{F}_i$, $^i\mathbf{M}_i$) by having the action force system ($^i\mathbf{F}_{i-1}$, $^i\mathbf{M}_{i-1}$). When the reaction force system ($^i\mathbf{F}_i$, $^i\mathbf{M}_i$) is found in frame B_i, we can transform them to frame B_{i+1}.

$$^{i+1}\mathbf{F}_i = {}^iT_{i+1}^{-1}\,{}^i\mathbf{F}_i \tag{11.180}$$

$$^{i+1}\mathbf{M}_i = {}^iT_{i+1}^{-1}\,{}^i\mathbf{M}_i \tag{11.181}$$

The negative of the converted force system acts as the action force system ($-^{i+1}\mathbf{F}_i$, $-^{i+1}\mathbf{M}_i$) for the link ($i+1$) and we can apply the Newton–Euler equation to the link ($i+1$).

The forward Newton–Euler equations of motion allows us to start from a known action force system ($^1\mathbf{F}_0$, $^1\mathbf{M}_0$), that the base link applies to the link (1), and calculate the action force of the next link. Therefore, analyzing the links of a robot, one by one, we end up with the force system that the end-effector applies to the environment.

Using the forward or backward recursive Newton–Euler equations of motion depends on the measurement and sensory system of the robot.   ∎

Example 363   ★ Recursive dynamics of a $2R$ planar manipulator. The $2R$ planar manipulator is a good example to practice how to derive recursive dynamic equations.

Consider the $2R$ planar manipulator shown in Fig. 11.11. The manipulator is carrying a force system at the endpoint. We use this manipulator to show how we can, step by step, develop the dynamic equations for a robot.

The backward recursive Newton–Euler equations of motion for the first link are:

$$
\begin{aligned}
^1\mathbf{F}_0 &= {}^1\mathbf{F}_1 - \sum {}^1\mathbf{F}_{e_1} + m_1\,{}_0^1\mathbf{a}_1 \\
&= {}^1\mathbf{F}_1 - m_1\,{}^1\mathbf{g} + m_1\,{}_0^1\mathbf{a}_1
\end{aligned}
\tag{11.182}
$$

$$
\begin{aligned}
^1\mathbf{M}_0 &= {}^1\mathbf{M}_1 - \sum {}^1\mathbf{M}_{e_1} - \left({}^1\mathbf{d}_0 - {}^1\mathbf{r}_1\right) \times {}^1\mathbf{F}_0 \\
&\quad + \left({}^1\mathbf{d}_1 - {}^1\mathbf{r}_1\right) \times {}^1\mathbf{F}_1 + {}^1I_1\,{}_0^1\boldsymbol{\alpha}_1 + {}_0^1\boldsymbol{\omega}_1 \times {}^1I_1\,{}_0^1\boldsymbol{\omega}_1 \\
&= {}^1\mathbf{M}_1 - {}^1\mathbf{n}_1 \times {}^1\mathbf{F}_0 + {}^1\mathbf{m}_1 \times {}^1\mathbf{F}_1 \\
&\quad + {}^1I_1\,{}_0^1\boldsymbol{\alpha}_1 + {}_0^1\boldsymbol{\omega}_1 \times {}^1I_1\,{}_0^1\boldsymbol{\omega}_1
\end{aligned}
\tag{11.183}
$$

and the backward recursive equations of motion for the second link are:

$$^2\mathbf{F}_1 = {}^2\mathbf{F}_2 - \sum {}^2\mathbf{F}_{e_2} + m_2 \, {}^2_0\mathbf{a}_2$$

$$= -m_2 \, {}^2\mathbf{g} - {}^2\mathbf{F}_e + m_2 \, {}^2_0\mathbf{a}_2 \tag{11.184}$$

$$^2\mathbf{M}_1 = {}^2\mathbf{M}_2 - \sum {}^2\mathbf{M}_{e_2} - \left({}^2\mathbf{d}_1 - {}^2\mathbf{r}_2\right) \times {}^2\mathbf{F}_1$$

$$+ \left({}^2\mathbf{d}_2 - {}^2\mathbf{r}_2\right) \times {}^2\mathbf{F}_2 + {}^2I_2 \, {}^2_0\boldsymbol{\alpha}_2 + {}^2_0\boldsymbol{\omega}_2 \times {}^2I_2 \, {}^2_0\boldsymbol{\omega}_2$$

$$= -{}^2\mathbf{M}_e - {}^2\mathbf{m}_2 \times {}^2\mathbf{F}_e - {}^2\mathbf{n}_2 \times {}^2\mathbf{F}_1$$

$$+ {}^2I_2 \, {}^2_0\boldsymbol{\alpha}_2 + {}^2_0\boldsymbol{\omega}_2 \times {}^2I_2 \, {}^2_0\boldsymbol{\omega}_2 \tag{11.185}$$

The manipulator consists of two R‖R(0) links, therefore their transformation matrices $^{i-1}T_i$ are of class (5.32). Substituting $d_i = 0$ and $a_i = l_i$ produces the following transformation matrices.

$$^0T_1 = \begin{bmatrix} \cos\theta_1 & -\sin\theta_1 & 0 & l_1\cos\theta_1 \\ \sin\theta_1 & \cos\theta_1 & 0 & l_1\sin\theta_1 \\ 0 & 0 & 1 & 0 \\ 0 & 0 & 0 & 1 \end{bmatrix} \tag{11.186}$$

$$^1T_2 = \begin{bmatrix} \cos\theta_2 & -\sin\theta_2 & 0 & l_2\cos\theta_2 \\ \sin\theta_2 & \cos\theta_2 & 0 & l_2\sin\theta_2 \\ 0 & 0 & 1 & 0 \\ 0 & 0 & 0 & 1 \end{bmatrix} \tag{11.187}$$

The homogenous mass moments matrices are:

$$^1I_1 = \frac{m_1 l_1^2}{12} \begin{bmatrix} 0&0&0&0 \\ 0&1&0&0 \\ 0&0&1&0 \\ 0&0&0&0 \end{bmatrix} \qquad ^2I_2 = \frac{m_2 l_2^2}{12} \begin{bmatrix} 0&0&0&0 \\ 0&1&0&0 \\ 0&0&1&0 \\ 0&0&0&0 \end{bmatrix} \tag{11.188}$$

The homogenous mass moment of inertia matrix is obtained by appending a zero row and column to the I matrix.

The position vectors involved are:

$$^1\mathbf{n}_1 = \begin{bmatrix} -l_1/2 \\ 0 \\ 0 \\ 0 \end{bmatrix} \qquad ^2\mathbf{n}_2 = \begin{bmatrix} -l_2/2 \\ 0 \\ 0 \\ 0 \end{bmatrix} \tag{11.189}$$

$$^1\mathbf{m}_1 = \begin{bmatrix} l_1/2 \\ 0 \\ 0 \\ 0 \end{bmatrix} \qquad ^2\mathbf{m}_2 = \begin{bmatrix} l_2/2 \\ 0 \\ 0 \\ 0 \end{bmatrix} \tag{11.190}$$

$$^1\mathbf{r}_1 = -{}^1\mathbf{n}_1 \qquad ^2\mathbf{r}_2 = -{}^2\mathbf{n}_1 + {}^2\mathbf{m}_2 - {}^2\mathbf{n}_2 \tag{11.191}$$

The angular velocities and accelerations are:

$$^1_0\boldsymbol{\omega}_1 = \begin{bmatrix} 0 \\ 0 \\ \dot\theta_1 \\ 0 \end{bmatrix} \qquad ^2_0\boldsymbol{\omega}_2 = \begin{bmatrix} 0 \\ 0 \\ \dot\theta_1 + \dot\theta_2 \\ 0 \end{bmatrix} \tag{11.192}$$

$$
{}_0^1\boldsymbol{\alpha}_1 = \begin{bmatrix} 0 \\ 0 \\ \ddot{\theta}_1 \\ 0 \end{bmatrix} \qquad {}_0^2\boldsymbol{\alpha}_2 = \begin{bmatrix} 0 \\ 0 \\ \ddot{\theta}_1 + \ddot{\theta}_2 \\ 0 \end{bmatrix}
\tag{11.193}
$$

The translational acceleration of C_1 is:

$$
{}_0^1\mathbf{a}_1 = {}_0^1\boldsymbol{\alpha}_1 \times \left(-{}^1\mathbf{m}_1\right) + {}_0^1\boldsymbol{\omega}_1 \times \left({}_0^1\boldsymbol{\omega}_1 \times \left(-{}^1\mathbf{m}_1\right)\right) + {}_0^1\ddot{\boldsymbol{d}}_1
$$

$$
= \begin{bmatrix} -\frac{1}{2}l_1\dot{\theta}_1^2 \\ \frac{1}{2}l_1\ddot{\theta}_1 \\ 0 \\ 0 \end{bmatrix}
\tag{11.194}
$$

because,

$$
{}^1\ddot{\boldsymbol{d}}_1 = 2\,{}^1\mathbf{a}_1
\tag{11.195}
$$

The translational acceleration of C_2 is:

$$
{}_0^2\mathbf{a}_2 = {}_0^2\boldsymbol{\alpha}_2 \times \left(-{}^2\mathbf{m}_2\right) + {}_0^2\boldsymbol{\omega}_2 \times \left({}_0^2\boldsymbol{\omega}_2 \times \left(-{}^2\mathbf{m}_2\right)\right) + {}_0^2\ddot{\boldsymbol{d}}_2
$$

$$
= \begin{bmatrix} -\frac{1}{2}l_2\left(\dot{\theta}_1 + \dot{\theta}_2\right)^2 \\ \frac{1}{2}l_2\left(\ddot{\theta}_1 + \ddot{\theta}_2\right) \\ 0 \\ 0 \end{bmatrix}
\tag{11.196}
$$

because,

$$
{}^2\ddot{\boldsymbol{d}}_2 = 2\,{}^2\mathbf{a}_2
\tag{11.197}
$$

The gravitational acceleration vector in the links' frame is:

$$
{}^1\mathbf{g} = {}^0T_1^{-1}\,{}^0\mathbf{g} = \begin{bmatrix} -g\sin\theta_2 \\ g\cos\theta_2 \\ 0 \\ 0 \end{bmatrix}
\tag{11.198}
$$

$$
{}^2\mathbf{g} = {}^0T_2^{-1}\,{}^0\mathbf{g} = \begin{bmatrix} -g\sin(\theta_1 + \theta_2) \\ g\cos(\theta_1 + \theta_2) \\ 0 \\ 0 \end{bmatrix}
\tag{11.199}
$$

The external load is usually given in the global coordinate frame. We must transform them to the interested link's frame to apply the recursive equations of motion. Therefore, the external force system expressed in B_2 is:

$$
{}^2\mathbf{F}_e = {}^0T_2^{-1}\,{}^0\mathbf{F}_e = \begin{bmatrix} F_{ex}\cos(\theta_1 + \theta_2) + F_{ey}\sin(\theta_1 + \theta_2) \\ F_{ey}\cos(\theta_1 + \theta_2) - F_{ex}\sin(\theta_1 + \theta_2) \\ 0 \\ 0 \end{bmatrix}
\tag{11.200}
$$

$$
{}^2\mathbf{M}_e = {}^0T_2^{-1}\,{}^0\mathbf{M}_e = \begin{bmatrix} 0 \\ 0 \\ M_e \\ 0 \end{bmatrix}
\tag{11.201}
$$

Now, we start from the final link and calculate its action force system. The backward Newton equation for link (2) is:

$$
{}^2\mathbf{F}_1 = -m_2\,{}^2\mathbf{g} - {}^2\mathbf{F}_e + m_2\,{}^2_0\mathbf{a}_2 = \begin{bmatrix} {}^2F_{1x} \\ {}^2F_{1y} \\ 0 \\ 0 \end{bmatrix} \tag{11.202}
$$

$$
{}^2F_{1x} = -\frac{1}{2}l_2 m_2 \left(\dot\theta_1 + \dot\theta_2\right)^2 - F_{ex}\cos\left(\theta_1 + \theta_2\right)
$$
$$
- \left(F_{ey} - gm_2\right)\sin\left(\theta_1 + \theta_2\right) \tag{11.203}
$$

$$
{}^2F_{1y} = \frac{1}{2}l_2 m_2 \left(\ddot\theta_1 + \ddot\theta_2\right) + F_{ex}\sin\left(\theta_1 + \theta_2\right)
$$
$$
- \left(F_{ey} + gm_2\right)\cos\left(\theta_1 + \theta_2\right) \tag{11.204}
$$

and the backward Euler equation for link (2) is:

$$
{}^2\mathbf{M}_1 = -{}^2\mathbf{M}_e - {}^2\mathbf{m}_2 \times {}^2\mathbf{F}_e - {}^2\mathbf{n}_2 \times {}^2\mathbf{F}_1
$$
$$
+ {}^2I_2\,{}^2_0\boldsymbol{\alpha}_2 + {}^2_0\boldsymbol{\omega}_2 \times {}^2I_2\,{}^2_0\boldsymbol{\omega}_2 = \begin{bmatrix} 0 \\ 0 \\ {}^2M_{1z} \\ 0 \end{bmatrix} \tag{11.205}
$$

where

$$
{}^2M_{1z} = -M_e + l_2 F_{ex}\sin\left(\theta_1 + \theta_2\right) - l_2 F_{ey}\cos\left(\theta_1 + \theta_2\right)
$$
$$
+ \frac{1}{3}l_2^2 m_2 \left(\ddot\theta_1 + \ddot\theta_2\right) - \frac{1}{2}gl_2 m_2 \cos\left(\theta_1 + \theta_2\right) \tag{11.206}
$$

Finally the action force on link (1) is:

$$
{}^1\mathbf{F}_0 = {}^1\mathbf{F}_1 - m_1\,{}^1\mathbf{g} + m_1\,{}^1_0\mathbf{a}_1
$$
$$
= {}^1T_2\,{}^2\mathbf{F}_1 - m_1\,{}^1\mathbf{g} + m_1\,{}^1_0\mathbf{a}_1 = \begin{bmatrix} {}^1F_{0x} \\ {}^1F_{0y} \\ 0 \\ 0 \end{bmatrix} \tag{11.207}
$$

where

$$
{}^1F_{0x} = -F_{ex}\cos\theta_1 - \left(F_{ey} - gm_1\right)\sin\theta_1
$$
$$
- \frac{1}{2}l_2 m_2 \left(\ddot\theta_1 + \ddot\theta_2\right)\sin\theta_2 - \frac{1}{2}l_2 m_2 \left(\dot\theta_1 + \dot\theta_2\right)^2 \cos\theta_2
$$
$$
+ gm_2\sin\left(2\theta_2 + \theta_1\right) - \frac{1}{2}l_1 m_1 \dot\theta_1^2 \tag{11.208}
$$

$$
{}^1F_{0y} = F_{ex}\sin\theta_1 - \left(F_{ey} + gm_1\right)\cos\theta_1
$$
$$
+ \frac{1}{2}l_2 m_2 \left(\ddot\theta_1 + \ddot\theta_2\right)\cos\theta_2 - \frac{1}{2}l_2 m_2 \left(\dot\theta_1 + \dot\theta_2\right)^2 \sin\theta_2
$$
$$
- gm_2\cos\left(2\theta_2 + \theta_1\right) + \frac{1}{2}l_1 m_1 \ddot\theta_1 \tag{11.209}
$$

and the action moment on link (1) is:

$$
\begin{aligned}
{}^1\mathbf{M}_0 &= {}^1\mathbf{M}_1 - {}^1\mathbf{n}_1 \times {}^1\mathbf{F}_0 + {}^1\mathbf{m}_1 \times {}^1\mathbf{F}_1 \\
&\quad + {}^1I_1\,{}^1_0\boldsymbol{\alpha}_1 + {}^1_0\boldsymbol{\omega}_1 \times {}^1I_1\,{}^1_0\boldsymbol{\omega}_1 \\
&= {}^1T_2\,{}^2\mathbf{M}_1 - {}^1\mathbf{n}_1 \times {}^1\mathbf{F}_0 + {}^1\mathbf{m}_1 \times {}^1T_2\,{}^2\mathbf{F}_1 \\
&\quad + {}^1I_1\,{}^1_0\boldsymbol{\alpha}_1 + {}^1_0\boldsymbol{\omega}_1 \times {}^1I_1\,{}^1_0\boldsymbol{\omega}_1 =
\begin{bmatrix} 0 \\ 0 \\ {}^1M_{0z} \\ 0 \end{bmatrix}
\end{aligned}
\tag{11.210}
$$

where

$$
\begin{aligned}
{}^1M_{0z} &= -M_e + \frac{1}{3}l_2^2 m_2\left(\ddot{\theta}_1 + \ddot{\theta}_2\right) + \frac{1}{3}l_1^2 m_1\ddot{\theta}_1 \\
&\quad - \left(F_{ey}l_2 + \frac{1}{2}gl_2 m_2\right)\cos\left(\theta_1 + \theta_2\right) \\
&\quad - \frac{1}{2}l_1 m_1 g\cos\theta_1 + F_{ex}l_2\sin\left(\theta_1 + \theta_2\right)
\end{aligned}
\tag{11.211}
$$

Example 364 ★ Actuator's force and torque. Actuator forces and torques are the required to derive a robot. Their expression for recursive equations are shown here.

Applying a backward recursive force analysis ends up with a set of known force systems at joints. Each joint is driven by a motor known as an actuator that applies a force in a P joint, or a torque in an R joint. When the joint i is prismatic, the force of the driving actuator is along the z_{i-1}-axis showing that $\hat{k}_{i-1}$ component of the joint force $\mathbf{F}_i$ is supported by the actuator.

$$
F_m = {}^0\hat{k}_{i-1}^T\,{}^0\mathbf{F}_i
\tag{11.212}
$$

The $\hat{i}_{i-1}$ and $\hat{j}_{i-1}$ components of $\mathbf{F}_i$ must be supported by the bearings of the joint. Similarly, when the joint i is revolute, the torque of the driving actuator is along the z_{i-1}-axis showing that $\hat{k}_{i-1}$ component of the joint torque $\mathbf{M}_i$ is supported by the actuator.

$$
M_m = {}^0\hat{k}_{i-1}^T\,{}^0\mathbf{M}_i
\tag{11.213}
$$

The $\hat{i}_{i-1}$ and $\hat{j}_{i-1}$ components of $\mathbf{M}_i$ must be supported by the bearings of the joint.

11.3　Robot Lagrange Dynamics

The Lagrange equation of motion provides a systematic approach to obtain the dynamics equations of robots. The Lagrangian $\mathcal{L}$ is defined as the difference between the kinetic K and potential V energies.

$$
\mathcal{L} = K - V
\tag{11.214}
$$

The Lagrange equation of motion for a robotic system can be found by applying the Lagrange equation

$$
\frac{d}{dt}\left(\frac{\partial\mathcal{L}}{\partial\dot{q}_i}\right) - \frac{\partial\mathcal{L}}{\partial q_i} = Q_i \qquad i = 1, 2, \cdots n
\tag{11.215}
$$

where q_i is the coordinates by which the energies are expressed, and Q_i is the corresponding generalized nonpotential force that makes q_i to vary.

The equations of motion for an n link serial manipulator can be set in a matrix form

$$
\mathbf{D}(\mathbf{q})\,\ddot{\mathbf{q}} + \mathbf{H}(\mathbf{q},\dot{q}) + \mathbf{G}(\mathbf{q}) = \mathbf{Q}
\tag{11.216}
$$

or

$$\mathbf{D(q)}\,\ddot{q} + \mathbf{C(q,\dot{q})}\dot{q} + \mathbf{G(q)} = \mathbf{Q} \tag{11.217}$$

or in a summation form:

$$\sum_{j=1}^{n} D_{ij}(q)\,\ddot{q}_j + \sum_{k=1}^{n}\sum_{m=1}^{n} H_{ikm}\dot{q}_k\dot{q}_m + G_i = Q_i \tag{11.218}$$

D_{ij} is an $n \times n$ inertial-type symmetric matrix,

$$D_{ij} = \sum_{k=1}^{n}\left(\mathbf{J}_{Dk}^{T}\,m_k\,\mathbf{J}_{Dk} + \frac{1}{2}\mathbf{J}_{Rk}^{T}\,{}^0I_k\,\mathbf{J}_{Rk}\right) \tag{11.219}$$

H_{ikm} is the velocity coupling vector,

$$H_{ijk} = \sum_{j=1}^{n}\sum_{k=1}^{n}\left(\frac{\partial D_{ij}}{\partial q_k} - \frac{1}{2}\frac{\partial D_{jk}}{\partial q_i}\right) \tag{11.220}$$

and G_i is the gravitational vector.

$$G_i = \sum_{j=1}^{n} m_j \mathbf{g}^T \mathbf{J}_{Dj}^{(i)} \tag{11.221}$$

Proof Kinetic energy of link (i) is:

$$K_i = \frac{1}{2}\,{}^0\mathbf{v}_i^T\,m_i\,{}^0\mathbf{v}_i + \frac{1}{2}\,{}_0\boldsymbol{\omega}_i^T\,{}^0I_i\,{}_0\boldsymbol{\omega}_i \tag{11.222}$$

where m_i is the mass of the link, ${}^iI_i$ is the mass moment matrix of the link in the link's frame B_i, ${}^0\mathbf{v}_i$ is the global velocity of the link at its mas center C, and ${}_0\boldsymbol{\omega}_i$ is the global angular velocity of the link with respect to the base frame B_0.

The translational and angular velocity vectors can be expressed based on the joint coordinate velocities, utilizing the *Jacobian of the link* $\mathbf{J}_i$.

$$\mathbf{R}_i = \begin{bmatrix} {}^0\mathbf{v}_i \\ {}^0\boldsymbol{\omega}_i \end{bmatrix} = \begin{bmatrix} \mathbf{J}_{Di} \\ \mathbf{J}_{Ri} \end{bmatrix} \dot{q} = \mathbf{J}_i\,\dot{q} \tag{11.223}$$

The link's Jacobian $\mathbf{J}_i$ is a $6 \times n$ matrix that transforms the instantaneous joint coordinate velocities into the instantaneous link's translational and angular velocities. The jth column of $\mathbf{J}_i$ is made of $\mathbf{c}_{Di}^{(j)}$ and $\mathbf{c}_{Ri}^{(j)}$. For $j \leq i$, they are:

$$\mathbf{c}_{Di}^{(j)} = \begin{cases} \hat{k}_{j-1} \times {}_{j-1}^{0}\mathbf{r}_i & \text{for a } R \text{ joint} \\ \hat{k}_{j-1} & \text{for a } P \text{ joint} \end{cases} \tag{11.224}$$

$$\mathbf{c}_{Ri}^{(j)} = \begin{cases} \hat{k}_{j-1} & \text{for a } R \text{ joint} \\ 0 & \text{for a } P \text{ joint} \end{cases} \tag{11.225}$$

where ${}_{j-1}^{0}\mathbf{r}_i$ is the position vector of C of the link (i) in the coordinate frame B_{j-1} expressed in the base frame. The columns of $\mathbf{J}_i$ are zero for $j > i$.

The kinetic energy K of the whole robot is then equal to:

$$\begin{aligned} K &= \sum_{i=1}^{n} K_i = \frac{1}{2}\sum_{i=1}^{n}\left({}^0\mathbf{v}_i^T\,m_i\,{}^0\mathbf{v}_i + \frac{1}{2}\,{}_0\boldsymbol{\omega}_i^T\,{}^0I_i\,{}_0\boldsymbol{\omega}_i\right) \\ &= \frac{1}{2}\sum_{i=1}^{n}\left(\left(\mathbf{J}_{Di}\,\dot{q}_i\right)^T m_i\,\left(\mathbf{J}_{Di}\,\dot{q}_i\right) + \frac{1}{2}\left(\mathbf{J}_{Ri}\,\dot{q}_i\right)^T {}^0I_i\,\left(\mathbf{J}_{Ri}\,\dot{q}_i\right)\right) \\ &= \frac{1}{2}\dot{q}_i^T\left(\sum_{i=1}^{n}\left(\mathbf{J}_{Di}^T\,m_i\,\mathbf{J}_{Di} + \frac{1}{2}\mathbf{J}_{Ri}^T\,{}^0I_i\,\mathbf{J}_{Ri}\right)\right)\dot{q}_i \end{aligned} \tag{11.226}$$

where ^0I_i is the mass moment matrix of the link (i) about its C and expressed in the base frame.

$$^0I_i = {^0R_i} \, {^iI_i} \, {^0R_i^T} \tag{11.227}$$

The kinetic energy K may be written in a more convenient form as

$$K = \frac{1}{2} \dot{q}_i^T \, D \, \dot{q}_i \tag{11.228}$$

where D is an $n \times n$ matrix called the *manipulator inertia matrix*.

$$D = \sum_{i=1}^{n} \left(\mathbf{J}_{Di}^T \, m_i \, \mathbf{J}_{Di} + \frac{1}{2} \mathbf{J}_{Ri}^T \, {^0I_i} \, \mathbf{J}_{Ri} \right) \tag{11.229}$$

The potential energy V_i of the link (i) is due to gravity

$$V_i = -m_i \, {^0\mathbf{g}} \cdot {^0\mathbf{r}_i} \tag{11.230}$$

and therefore, the total potential energy of the manipulator is:

$$V = \sum_{i=1}^{n} V_i = - \sum_{i=1}^{n} m_i \, {^0\mathbf{g}^T} \, {^0\mathbf{r}_i} \tag{11.231}$$

where $^0\mathbf{g}$ is the gravitational acceleration vector expressed in the base frame.

The Lagrangian of the manipulator is:

$$\mathcal{L} = K - V = \frac{1}{2} \dot{q}_i^T \, D \, \dot{q}_i + \sum_{i=1}^{n} m_i \, {^0\mathbf{g}^T} \, {^0\mathbf{r}_i}$$

$$= \frac{1}{2} \sum_{i=1}^{n} \sum_{j=1}^{n} D_{ij} \, \dot{q}_i \dot{q}_j + \sum_{i=1}^{n} m_i \, {^0\mathbf{g}^T} \, {^0\mathbf{r}_i} \tag{11.232}$$

Based on the Lagrangian $\mathcal{L}$ we find the partial differentials.

$$\frac{\partial \mathcal{L}}{\partial q_i} = \frac{1}{2} \frac{\partial}{\partial q_i} \left(\sum_{j=1}^{n} \sum_{k=1}^{n} D_{jk} \, \dot{q}_j \dot{q}_k \right) + \sum_{j=1}^{n} m_j \, {^0\mathbf{g}^T} \, \frac{\partial \, {^0\mathbf{r}_j}}{\partial q_i}$$

$$= \frac{1}{2} \sum_{j=1}^{n} \sum_{k=1}^{n} \frac{\partial D_{jk}}{\partial q_i} \, \dot{q}_j \dot{q}_k + \sum_{j=1}^{n} m_j \, {^0\mathbf{g}^T} \, \mathbf{J}_{Dj}^{(i)} \tag{11.233}$$

$$\frac{\partial \mathcal{L}}{\partial \dot{q}_i} = \sum_{j=1}^{n} D_{ij} \, \dot{q}_j \tag{11.234}$$

$$\frac{d}{dt} \frac{\partial \mathcal{L}}{\partial \dot{q}_i} = \sum_{j=1}^{n} D_{ij} \, \ddot{q}_j + \sum_{j=1}^{n} \frac{dD_{ij}}{dt} \, \dot{q}_j$$

$$= \sum_{j=1}^{n} D_{ij} \, \ddot{q}_j + \sum_{j=1}^{n} \sum_{k=1}^{n} \frac{\partial D_{ij}}{\partial q_k} \, \dot{q}_k \dot{q}_j \tag{11.235}$$

The generalized force Q_i of the Lagrange equations is:

$$Q_i = M_i + \mathbf{J}^T \mathbf{F}_e \tag{11.236}$$

$$\mathbf{F}_e = \begin{bmatrix} -\mathbf{F}_{en}^T & -\mathbf{M}_{en}^T \end{bmatrix}^T \tag{11.237}$$

where M_i is the ith actuator force at joint i, and $\mathbf{F}_e$ is the external force system applied on the end-effector.

Finally, the Lagrange equations of motion for an n-link manipulator are:

$$\sum_{j=1}^{n} D_{ij}(q)\, \ddot{q}_j + H_{ikm}\dot{q}_k\dot{q}_m + G_i = Q_i \tag{11.238}$$

where

$$H_{ijk} = \sum_{j=1}^{n}\sum_{k=1}^{n} \left(\frac{\partial D_{ij}}{\partial q_k} - \frac{1}{2}\frac{\partial D_{jk}}{\partial q_i} \right) \tag{11.239}$$

$$G_i = \sum_{j=1}^{n} m_j \mathbf{g}^T \mathbf{J}_{Dj}^{(i)} \tag{11.240}$$

We can show the equations of motion for a manipulator in a more concise form to simplify matrix calculations.

$$\mathbf{D(q)}\, \ddot{q} + \mathbf{H(q, \dot{q})} + \mathbf{G(q)} = \mathbf{Q} \tag{11.241}$$

The term $\mathbf{G(q)}$ is called the *gravitational force vector* and the term $\mathbf{H(q, \dot{q})}$ is called the *velocity coupling vector*. The velocity coupling vector may sometimes be written in another form.

$$\mathbf{H(q, \dot{q})} = \mathbf{C(q, \dot{q})\dot{q}} \tag{11.242}$$

∎

Example 365 A prismatic-revolute planar manipulator. A PR planar robotic manipulator with a prismatic and a revolute joints will be examined to derive its equations of motion by Lagrange method. It is a good example to find the quietens of motion in matrix form, step by step.

Figure 11.12 illustrates a prismatic-revolute planar manipulator with massless links and two massive points m_1 and m_2. The joint variables are indicated by q_1 and q_2. To determine the equations of motion, we begin with calculating its kinetic energy.

$$K_1 = \frac{1}{2}m_1\dot{q}_1^2 \tag{11.243}$$

$$\begin{aligned} K_2 &= \frac{1}{2}m_2\dot{X}_2^2 + \frac{1}{2}m_2\dot{Y}_2^2 \\ &= \frac{1}{2}m_2\left(\frac{d}{dt}(q_1 + l\cos q_2)\right)^2 + \frac{1}{2}m_2\left(\frac{d}{dt}(l\sin q_2)\right)^2 \\ &= \frac{1}{2}m_2\left(\dot{q}_1 - l\dot{q}_2\sin q_2\right)^2 + \frac{1}{2}m_2\left(l\dot{q}_2\cos q_2\right) \\ &= \frac{1}{2}m_2\left(\dot{q}_1^2 + l^2\dot{q}_2^2 - 2l\dot{q}_1\dot{q}_2\sin q_2\right) \end{aligned} \tag{11.244}$$

The potential energy of the manipulator is:

$$V = m_2 g Y_2 = m_2 g l \sin q_2 \tag{11.245}$$

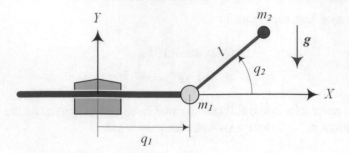

Fig. 11.12 A prismatic-revolute planar manipulator

Therefore, the Lagrangian is:

$$\mathcal{L} = K - V = K_1 + K_2 - V$$

$$= \frac{1}{2}m_1\dot{q}_1^2 + \frac{1}{2}m_2\left(\dot{q}_1^2 + l^2\dot{q}_2^2 - 2l\dot{q}_1\dot{q}_2\sin q_2\right) - m_2 gl \sin q_2 \qquad (11.246)$$

Applying the Lagrange equation

$$\frac{d}{dt}\left(\frac{\partial \mathcal{L}}{\partial \dot{q}_i}\right) - \frac{\partial \mathcal{L}}{\partial q_i} = Q_i \qquad i = 1, 2 \qquad (11.247)$$

provides the following equations of motion.

$$(m_1 + m_2)\ddot{q}_1 - m_2 l \ddot{q}_2 \sin q_2 - m_2 l \dot{q}_2^2 \cos q_2 = Q_1 \qquad (11.248)$$

$$m_2 l^2 \ddot{q}_1 - m_2 l \ddot{q}_1 \sin q_2 + m_2 gl \cos q_2 = Q_2 \qquad (11.249)$$

We can rearrange these equations to the matrix form of (11.217)

$$\mathbf{D(q)}\begin{bmatrix} \ddot{q}_1 \\ \ddot{q}_2 \end{bmatrix} + \mathbf{C(q, \dot{q})}\begin{bmatrix} \dot{q}_1 \\ \dot{q}_2 \end{bmatrix} + \mathbf{G(q)} = \begin{bmatrix} Q_1 \\ Q_2 \end{bmatrix} \qquad (11.250)$$

where

$$\mathbf{D(q)} = \begin{bmatrix} m_1 + m_2 & -m_2 l \sin q_2 \\ m_2 l^2 & -m_2 l \sin q_2 \end{bmatrix} \qquad (11.251)$$

$$\mathbf{C(q, \dot{q})} = \begin{bmatrix} 0 & -m_2 l \dot{q}_2 \cos q_2 \\ 0 & 0 \end{bmatrix} \qquad (11.252)$$

$$\mathbf{G(q)} = \begin{bmatrix} 0 \\ m_2 gl \cos q_2 \end{bmatrix} \qquad (11.253)$$

Example 366 A planar polar manipulator. A RP planar robotic manipulator with a revolute and a prismatic joints will be used to determine its equations of motion by Lagrange method and set them in matrix form.

Figure 11.13 illustrates a revolute-prismatic planar polar manipulator with massless link and a massive point m. The kinetic energy K of the manipulator is:

$$K = \frac{1}{2}m_2 \dot{X}_2^2 + \frac{1}{2}m_2 \dot{Y}_2^2$$

$$= \frac{1}{2}m\left(\frac{d}{dt}(q_1 \cos q_2)\right)^2 + \frac{1}{2}m\left(\frac{d}{dt}(q_1 \sin q_2)\right)^2$$

$$= \frac{1}{2}m\left(\dot{q}_1^2 + q_1^2 \dot{q}_2^2\right) \qquad (11.254)$$

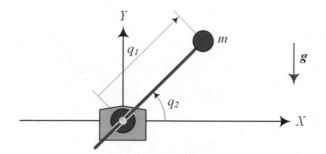

Fig. 11.13 A planar polar manipulator

The potential energy V of the manipulator is:

$$V = mgY_2 = mgq_1 \sin q_2 \tag{11.255}$$

and therefore, the Lagrangian $\mathcal{L}$ of the manipulator is:

$$\mathcal{L} = K - V = \frac{1}{2}m\left(\dot{q}_1^2 + q_1^2\dot{q}_2^2\right) - mgq_1\sin q_2 \tag{11.256}$$

Applying the Lagrange equation

$$\frac{d}{dt}\left(\frac{\partial \mathcal{L}}{\partial \dot{q}_i}\right) - \frac{\partial \mathcal{L}}{\partial q_i} = Q_i \qquad i = 1,2 \tag{11.257}$$

provides the following equations of motion.

$$m\ddot{q}_1 - mq_1\dot{q}_2^2 + mg\sin q_2 = Q_1 \tag{11.258}$$
$$mq_1^2\ddot{q}_2 + 2mq_1\dot{q}_1\dot{q}_2 + mgq_1\cos q_2 = Q_2 \tag{11.259}$$

Let us rearrange these equations to the matrix form of (11.217).

$$\mathbf{D}(\mathbf{q})\begin{bmatrix}\ddot{q}_1\\\ddot{q}_2\end{bmatrix} + \mathbf{C}(\mathbf{q},\dot{q})\begin{bmatrix}\dot{q}_1\\\dot{q}_2\end{bmatrix} + \mathbf{G}(\mathbf{q}) = \begin{bmatrix}Q_1\\Q_2\end{bmatrix} \tag{11.260}$$

$$\mathbf{D}(\mathbf{q}) = \begin{bmatrix}m & 0\\0 & mq_1^2\end{bmatrix} \tag{11.261}$$

$$\mathbf{C}(\mathbf{q},\dot{q}) = \begin{bmatrix}0 & -mq_1\dot{q}_2\\mq_1\dot{q}_2 & mq_1\dot{q}_1\end{bmatrix} \tag{11.262}$$

$$\mathbf{G}(\mathbf{q}) = \begin{bmatrix}mg\sin q_2\\mgq_1\cos q_2\end{bmatrix} \tag{11.263}$$

Example 367 General model of $2R$ planar manipulator. Deriving equations of motion of a $2R$ manipulator by Lagrange method and arranging the result in matrix form.

Consider a general $2R$ manipulator with massive arms and joints while carrying a payload m_0 as is shown in Fig. 11.14. The first motor that drives link (1) is on the ground. The second motor with mass m_{12} drives link (2) and is mounted on link (1). The mass of first and second links are m_{11} and m_{21}, respectively. The global position vectors of the mass centers C_i and massive joints are:

$${}^0\mathbf{r}_1 = {}^0R_1\,{}^1\mathbf{r}_1 = R_{Z,\theta_1}\,c_1\,{}^1\hat{\imath}_1 = \begin{bmatrix}c_1\cos\theta_1\\c_1\sin\theta_1\\0\end{bmatrix} \tag{11.264}$$

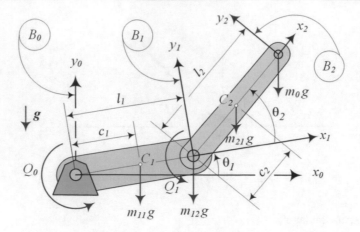

Fig. 11.14   A $2R$ manipulator with massive arms and a carrying payload m_0

$$^0\mathbf{r}_2 = {}^0\mathbf{d}_1 + {}^0R_2\,{}^2\mathbf{r}_2 = {}^0\mathbf{d}_1 + R_{Z,\theta_1}\,R_{Z,\theta_2}\,c_2\,{}^2\hat{\imath}_2$$

$$= \begin{bmatrix} l_1\cos\theta_1 + c_2\cos(\theta_1+\theta_2) \\ l_1\sin\theta_1 + c_2\sin(\theta_1+\theta_2) \\ 0 \end{bmatrix} \tag{11.265}$$

$$^0\mathbf{d}_1 = {}^0R_1\,{}^1\mathbf{r}_1 = R_{Z,\theta_1}\,l_1\,{}^1\hat{\imath}_1 = \begin{bmatrix} l_1\cos\theta_1 \\ l_1\sin\theta_1 \\ 0 \end{bmatrix} \tag{11.266}$$

$$^0\mathbf{d}_2 = {}^0\mathbf{d}_1 + {}^0R_2\,{}^2\mathbf{d}_2 = \begin{bmatrix} l_2\cos(\theta_1+\theta_2) + l_1\cos\theta_1 \\ l_2\sin(\theta_1+\theta_2) + l_1\sin\theta_1 \\ 0 \end{bmatrix} \tag{11.267}$$

where

$$^0R_1 = R_{Z,\theta_1} = \begin{bmatrix} \cos\theta_1 & -\sin\theta_1 & 0 \\ \sin\theta_1 & \cos\theta_1 & 0 \\ 0 & 0 & 1 \end{bmatrix} \tag{11.268}$$

$$^1R_2 = R_{Z,\theta_2} = \begin{bmatrix} \cos\theta_2 & -\sin\theta_2 & 0 \\ \sin\theta_2 & \cos\theta_2 & 0 \\ 0 & 0 & 1 \end{bmatrix} \tag{11.269}$$

$$^0R_2 = {}^0R_1\,{}^1R_2 = \begin{bmatrix} \cos(\theta_1+\theta_2) & -\sin(\theta_1+\theta_2) & 0 \\ \sin(\theta_1+\theta_2) & \cos(\theta_1+\theta_2) & 0 \\ 0 & 0 & 1 \end{bmatrix} \tag{11.270}$$

The links' angular velocity is:

$$_0\boldsymbol{\omega}_1 = \dot{\theta}_1\,\hat{K} \qquad _0\boldsymbol{\omega}_2 = \left(\dot{\theta}_1 + \dot{\theta}_2\right)\hat{K} \tag{11.271}$$

The mass moment matrices in the global coordinate frame are:

$$^0I_1 = R_{Z,\theta_1}\,{}^1I_1\,R_{Z,\theta_1}^T = {}^0R_1 \begin{bmatrix} I_{x_1} & 0 & 0 \\ 0 & I_{y_1} & 0 \\ 0 & 0 & I_{z_1} \end{bmatrix} {}^0R_1^T$$

$$= \begin{bmatrix} I_{x_1}c^2\theta_1 + I_{y_1}s^2\theta_1 & \left(I_{x_1}-I_{y_1}\right)c\theta_1 s\theta_1 & 0 \\ \left(I_{x_1}-I_{y_1}\right)c\theta_1 s\theta_1 & I_{y_1}c^2\theta_1 + I_{x_1}s^2\theta_1 & 0 \\ 0 & 0 & I_{z_1} \end{bmatrix} \tag{11.272}$$

$$
{}^0I_2 = {}^0R_2 \, {}^2I_2 \, {}^0R_2^T = {}^0R_2 \begin{bmatrix} I_{x_2} & 0 & 0 \\ 0 & I_{y_2} & 0 \\ 0 & 0 & I_{z_2} \end{bmatrix} {}^0R_2^T
$$

$$
= \begin{bmatrix} I_{x_2}c^2\theta_{12} + I_{y_2}s^2\theta_{12} & \left(I_{x_2} - I_{y_2}\right)c\theta_{12}s\theta_{12} & 0 \\ \left(I_{x_2} - I_{y_2}\right)c\theta_{12}s\theta_{12} & I_{y_2}c^2\theta_{12} + I_{x_2}s^2\theta_{12} & 0 \\ 0 & 0 & I_{z_2} \end{bmatrix} \tag{11.273}
$$

$$
\theta_{12} = \theta_1 + \theta_2 \tag{11.274}
$$

The velocity of C_i and the masses are:

$$
{}^0\mathbf{v}_1 = \frac{{}^0d}{dt}{}^0\mathbf{r}_1 = \begin{bmatrix} -c_1\dot{\theta}_1 \sin\theta_1 \\ c_1\dot{\theta}_1 \cos\theta_1 \\ 0 \end{bmatrix} \tag{11.275}
$$

$$
{}^0\mathbf{v}_2 = \frac{{}^0d}{dt}{}^0\mathbf{r}_2
$$

$$
= \begin{bmatrix} -l_1\dot{\theta}_1 \sin\theta_1 - c_2\left(\dot{\theta}_1 + \dot{\theta}_2\right)\sin\left(\theta_1 + \theta_2\right) \\ l_1\dot{\theta}_1 \cos\theta_1 + c_2\left(\dot{\theta}_1 + \dot{\theta}_2\right)\cos\left(\theta_1 + \theta_2\right) \\ 0 \end{bmatrix} \tag{11.276}
$$

$$
{}^0\dot{\boldsymbol{d}}_1 = \begin{bmatrix} -l_1\dot{\theta}_1 \sin\theta_1 \\ l_1\dot{\theta}_1 \cos\theta_1 \\ 0 \end{bmatrix} \tag{11.277}
$$

$$
{}^0\dot{\boldsymbol{d}}_2 = \begin{bmatrix} -l_1\dot{\theta}_1 \sin\theta_1 - l_2\left(\dot{\theta}_1 + \dot{\theta}_2\right)\sin\left(\theta_1 + \theta_2\right) \\ l_1\dot{\theta}_1 \cos\theta_1 + l_2\left(\dot{\theta}_1 + \dot{\theta}_2\right)\cos\left(\theta_1 + \theta_2\right) \\ 0 \end{bmatrix} \tag{11.278}
$$

To calculate Lagrangian $\mathcal{L} = K - V$, we determine the energies of the manipulator. The kinetic energy of the manipulator is:

$$
K = \frac{1}{2}m_{12} \, {}^0\dot{\boldsymbol{d}}_1 \cdot {}^0\dot{\boldsymbol{d}}_1 + \frac{1}{2}m_{11} \, {}^0\mathbf{v}_1 \cdot {}^0\mathbf{v}_1
$$

$$
+ \frac{1}{2}m_0 \, {}^0\dot{\boldsymbol{d}}_2 \cdot {}^0\dot{\boldsymbol{d}}_2 + \frac{1}{2}m_{21} \, {}^0\mathbf{v}_2 \cdot {}^0\mathbf{v}_2
$$

$$
+ \frac{1}{2}{}_0\boldsymbol{\omega}_1^T \, {}^0I_{1\,0}\boldsymbol{\omega}_1 + \frac{1}{2}{}_0\boldsymbol{\omega}_2^T \, {}^0I_{2\,0}\boldsymbol{\omega}_2 \tag{11.279}
$$

which after substituting (11.268)–(11.278) would be:

$$
K = \frac{1}{2}\left(m_{11}c_1^2 + m_{12}l_1^2 + I_{z_1}\right)\dot{\theta}_1^2
$$

$$
+ \frac{1}{2}m_{21}\left(-l_1\dot{\theta}_1 \sin\theta_1 - c_2\left(\dot{\theta}_1 + \dot{\theta}_2\right)\sin\left(\theta_1 + \theta_2\right)\right)^2
$$

$$
+ \frac{1}{2}m_{21}\left(l_1\dot{\theta}_1 \cos\theta_1 + c_2\left(\dot{\theta}_1 + \dot{\theta}_2\right)\cos\left(\theta_1 + \theta_2\right)\right)^2
$$

$$
+ \frac{1}{2}m_0\left(-l_1\dot{\theta}_1 \sin\theta_1 - l_2\left(\dot{\theta}_1 + \dot{\theta}_2\right)\sin\left(\theta_1 + \theta_2\right)\right)^2
$$

$$
+ \frac{1}{2}m_0\left(l_1\dot{\theta}_1 \cos\theta_1 + l_2\left(\dot{\theta}_1 + \dot{\theta}_2\right)\cos\left(\theta_1 + \theta_2\right)\right)^2
$$

$$
+ \frac{1}{2}I_{z_2}\left(\dot{\theta}_1 + \dot{\theta}_2\right)^2 \tag{11.280}
$$

The potential energy of the manipulator is:

$$V = m_{11}gc_1 \sin\theta_1 + m_{12}gl_1 \sin\theta_1$$
$$+ m_{21}g \left(l_1 \sin\theta_1 + c_2 \sin(\theta_1 + \theta_2)\right)$$
$$+ m_0 g \left(l_1 \sin\theta_1 + l_2 \sin(\theta_1 + \theta_2)\right) \tag{11.281}$$

Applying the Lagrange equation

$$\frac{d}{dt}\left(\frac{\partial \mathcal{L}}{\partial \dot{\theta}_1}\right) - \frac{\partial \mathcal{L}}{\partial \theta_1} = Q_0 \tag{11.282}$$

$$\frac{d}{dt}\left(\frac{\partial \mathcal{L}}{\partial \dot{\theta}_2}\right) - \frac{\partial \mathcal{L}}{\partial \theta_2} = Q_1 \tag{11.283}$$

determines the general equations of motion which can be rearranged in a matrix form.

$$\begin{bmatrix} D_{11} & D_{12} \\ D_{21} & D_{22} \end{bmatrix} \begin{bmatrix} \ddot{\theta}_1 \\ \ddot{\theta}_2 \end{bmatrix} + \begin{bmatrix} C_{11} & C_{12} \\ C_{21} & C_{22} \end{bmatrix} \begin{bmatrix} \dot{\theta}_1 \\ \dot{\theta}_2 \end{bmatrix}$$
$$+ \begin{bmatrix} G_1 \\ G_2 \end{bmatrix} = \begin{bmatrix} Q_o \\ Q_1 \end{bmatrix} \tag{11.284}$$

$$D_{11} = 2l_1 \left(m_{21}c_2 + m_0 l_2\right) \cos\theta_2 + I_{z_1} + I_{z_2}$$
$$+ m_{11}c_1^2 + m_{12}l_1^2 + m_{21}\left(c_2^2 + l_1^2\right) + m_0\left(l_1^2 + l_2^2\right) \tag{11.285}$$

$$D_{12} = l_1 \left(m_{21}c_2 + m_0 l_2\right) \cos\theta_2 + I_{z_2} + m_0 l_2^2 + m_{21}c_2^2 \tag{11.286}$$

$$D_{21} = l_1 \left(m_{21}c_2 + m_0 l_2\right) \cos\theta_2 + I_{z_2} + m_{21}c_2^2 + m_0 l_2^2 \tag{11.287}$$

$$D_{22} = I_{z_2} + m_{21}c_2^2 + m_0 l_2^2 \tag{11.288}$$

$$C_{11} = -l_1 \left(m_{21}c_2 + m_0 l_2\right) \dot{\theta}_2 \sin\theta_2 \tag{11.289}$$

$$C_{12} = -l_1 \left(m_{21}c_2 + m_0 l_2\right) \left(\dot{\theta}_1 + \dot{\theta}_2\right) \sin\theta_2 \tag{11.290}$$

$$C_{21} = l_1 \left(m_{21}c_2 + m_0 l_2\right) \dot{\theta}_1 \sin\theta_2 \tag{11.291}$$

$$C_{22} = 0 \tag{11.292}$$

$$G_1 = \left((m_{21} + m_{12} + m_0) l_1 + m_{11}c_1\right) g \cos\theta_1$$
$$+ \left(m_{21}c_2 + m_0 l_2\right) g \cos(\theta_1 + \theta_2) \tag{11.293}$$

$$G_2 = \left(m_{21}c_2 + m_0 l_2\right) g \cos(\theta_1 + \theta_2) \tag{11.294}$$

To have all equations that we need for control, path planning, and computer analysis of $2R$ manipulator, here we explain dynamics of the most common special cases of the manipulator. Figure 11.14 illustrates a general $2R$ manipulator with massive arms and joints while carrying a payload of mass m_0. The second motor has a mass m_{12} and is mounted on link (1). The mass of first and second links are m_{11} and m_{21}, respectively, and their mass centers are at distance c_1 and c_2 from o_{i-1} on x_i-axis. The general equations of motion for the $2R$ planar manipulator are given in Eqs. (11.284).

1. Massless arms.

   When the mass of the links of the manipulator are much less than the masses of its motors and the carrying load, we may use a massless arm model. The equations of motion for a massless arm $2R$ planar manipulator are calculated by substituting $m_{11} = 0$, $m_{21} = 0$, $I_{z_1} = 0$, $I_{z_2} = 0$ in Eqs. (11.284).

$$D_{11} = 2m_0 l_1 l_2 \cos\theta_2 + m_{12} l_1^2 + m_0 \left(l_1^2 + l_2^2\right) \tag{11.295}$$

$$D_{12} = m_0 l_1 l_2 \cos\theta_2 + m_0 l_2^2 \tag{11.296}$$

$$D_{21} = m_0 l_1 l_2 \cos\theta_2 + m_0 l_2^2 \tag{11.297}$$

$$D_{22} = m_0 l_2^2 \tag{11.298}$$

$$C_{11} = -m_0 l_1 l_2 \dot{\theta}_2 \sin\theta_2 \tag{11.299}$$

$$C_{12} = -m_0 l_1 l_2 \left(\dot{\theta}_1 + \dot{\theta}_2\right) \sin\theta_2 \tag{11.300}$$

$$C_{21} = m_0 l_1 l_2 \dot{\theta}_1 \sin\theta_2 \qquad C_{22} = 0 \tag{11.301}$$

$$G_1 = (m_{12} + m_0)\, g l_1 \cos\theta_1 + m_0 g l_2 \cos(\theta_1 + \theta_2) \tag{11.302}$$

$$G_2 = m_0 g l_2 \cos(\theta_1 + \theta_2) \tag{11.303}$$

$$Q_o = \left(m_0 \left(l_1^2 + l_2^2 + 2 l_1 l_2 \cos\theta_2\right) + m_{12} l_1^2\right) \ddot{\theta}_1$$
$$+ m_0 \left(l_2^2 + l_1 l_2 \cos\theta_2\right) \ddot{\theta}_2$$
$$- 2 m_0 l_1 l_2 \dot{\theta}_1 \dot{\theta}_2 \sin\theta_2 - m_0 l_1 l_2 \dot{\theta}_2^2 \sin\theta_2$$
$$+ (m_{12} + m_0)\, g l_1 \cos\theta_1 + m_0 g l_2 \cos(\theta_1 + \theta_2) \tag{11.304}$$

$$Q_1 = m_0 \left(l_2^2 + l_1 l_2 \cos\theta_2\right) \ddot{\theta}_1 + m_0 l_2^2 \ddot{\theta}_2$$
$$+ m_0 l_1 l_2 \dot{\theta}_1^2 \sin\theta_2 + m_0 g l_2 \cos(\theta_1 + \theta_2) \tag{11.305}$$

2. Massless joints.

When the mass of the links of the manipulator are much more than the masses of its motors and the carrying load, we may use a massless joint model. The equations of motion for a massless joints $2R$ planar manipulator are calculated by substituting $m_{12} = 0$, $m_0 = 0$ in Eqs. (11.284).

$$D_{11} = 2m_{21} l_1 c_2 \cos\theta_2 + I_{z_1} + I_{z_2}$$
$$+ m_{11} c_1^2 + m_{21} \left(c_2^2 + l_1^2\right) \tag{11.306}$$

$$D_{12} = m_{21} c_2 \left(l_1 \cos\theta_2 + c_2\right) + I_{z_2} \tag{11.307}$$

$$D_{21} = m_{21} c_2 \left(l_1 \cos\theta_2 + c_2\right) + I_{z_2} \tag{11.308}$$

$$D_{22} = I_{z_2} + m_{21} c_2^2 \tag{11.309}$$

$$C_{11} = -m_{21} l_1 c_2 \dot{\theta}_2 \sin\theta_2 \tag{11.310}$$

$$C_{12} = -m_{21} l_1 c_2 \left(\dot{\theta}_1 + \dot{\theta}_2\right) \sin\theta_2 \tag{11.311}$$

$$C_{21} = m_{21} l_1 c_2 \dot{\theta}_1 \sin\theta_2 \qquad C_{22} = 0 \tag{11.312}$$

$$G_1 = (m_{21} l_1 + m_{11} c_1)\, g \cos\theta_1 + m_{21} c_2 g \cos(\theta_1 + \theta_2) \tag{11.313}$$

$$G_2 = m_{21} c_2 g \cos(\theta_1 + \theta_2) \tag{11.314}$$

If the links of the manipulator are uniform and symmetric, then

$$c_1 = l_1/2 \qquad c_2 = l_2/2 \tag{11.315}$$

As an example, let us determine the torques of motors of a planar $2R$ manipulator with massless arms and the following dimensions. The end-effector moves on a straight line in one minute, $0 \leq t \leq 1$ min.

$$l_1 = l_2 = 1\,\text{m} \qquad m_0 = 1\,\text{kg} \qquad m_{12} = 1\,\text{kg} \tag{11.316}$$

$$X(0) = 1\,\text{m} \qquad Y(0) = 0 \tag{11.317}$$

$$X(1) = -1\,\text{m} \qquad Y(1) = 0 \tag{11.318}$$

We assume a septic time function for X to move the end-effector from $X(0) = 1$ to $X(1) = -1$ with zero velocity, zero acceleration, zero jerk at both ends.

$$X = 1 - 70t^4 + 168t^5 - 140t^6 + 40t^7 \tag{11.319}$$

To make the dynamic equations of motion suitable for numerical calculations, we express the equations of motion (11.304) and (11.305) for $2R$ robotic manipulators in the following form.

$$Q_o = A\ddot{\theta}_1 + B\ddot{\theta}_2 + C\dot{\theta}_1\dot{\theta}_2 + D\dot{\theta}_2^2 + M \tag{11.320}$$

$$Q_1 = E\ddot{\theta}_1 + F\ddot{\theta}_2 + G\dot{\theta}_1^2 + N \tag{11.321}$$

$$A = A(\theta_2) = m_0 \left(l_1^2 + l_2^2 + 2l_1 l_2 \cos\theta_2 \right) + m_{12} l_1^2 \tag{11.322}$$

$$B = B(\theta_2) = m_0 \left(l_2^2 + l_1 l_2 \cos\theta_2 \right) \tag{11.323}$$

$$C = C(\theta_2) = -2m_0 l_1 l_2 \sin\theta_2 \tag{11.324}$$

$$D = D(\theta_2) = -m_0 l_1 l_2 \sin\theta_2 \tag{11.325}$$

$$M = M(\theta_1, \theta_2) = (m_{12} + m_0)\, g l_1 \cos\theta_1 + m_0 g l_2 \cos(\theta_1 + \theta_2) \tag{11.326}$$

$$E = E(\theta_2) = B = m_0 \left(l_2^2 + l_1 l_2 \cos\theta_2 \right) \tag{11.327}$$

$$F = m_0 l_2^2 \tag{11.328}$$

$$G = G(\theta_2) = -D = m_0 l_1 l_2 \sin\theta_2 \tag{11.329}$$

$$N = N(\theta_1, \theta_2) = m_0 g l_2 \cos(\theta_1 + \theta_2)$$

Employing the inverse kinematics analysis, and dynamic equations (11.320) and (11.321), Fig. 11.15 illustrates the torques Q_0, and Q_1 and joints 1 and 2 as function of time to move the robot's end-effector from point $(1, 1.5)$ to $(-1, 1.5)$ on a straight line in one unit of time.

Having torques and the results of inverse kinematic analysis enable us to calculate the consumed energy E and needed power P of motors.

$$E = \int_{\theta_{1i}}^{\theta_{1f}} Q_0 d\theta_1 + \int_{\theta_{2i}}^{\theta_{2f}} Q_1 d\theta_2 \equiv \sum_{k=1}^{n} Q_{0_k} d\theta_{1_k} + \sum_{k=1}^{n} Q_{1_k} d\theta_{2_k} \tag{11.330}$$

$$P_0 = Q_0(t)\, \dot{\theta}_1(t) \qquad P_1 = Q_1(t)\, \dot{\theta}_2(t) \tag{11.331}$$

Figure 11.16 illustrates time history of the power of motors of this example. Assuming motors will be using energy regardless of torque direction, the absolute value of the powers must be considered as required powers. Hence, consumed energy will be the area under the curve of absolute value of power.

Example 368 Lagrange equation for $2R$ manipulators with massive arms. A step by step deriving the matrix form of equations of motion based on Lagrange method for a $2R$ manipulator. Although the name of the actuation force and torques are better to be indexed by the link they are installed on, such as Fig. 11.14, we may name there as our preference. Here is a traditional naming.

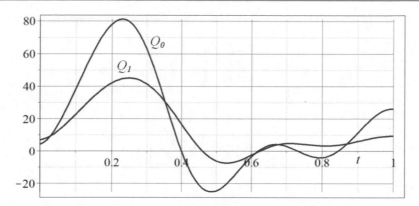

Fig. 11.15 The torques Q_0, and Q_1 of joints 1 and 2 of a 2R robot as function of time. The path of motion of the robot is from point (1, 1.5) to $(-1, 1.5)$ in one unit of time

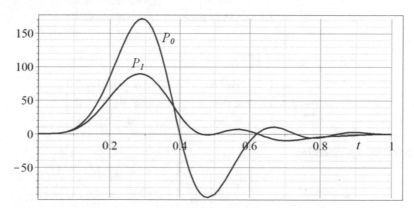

Fig. 11.16 Time history of the power of motors

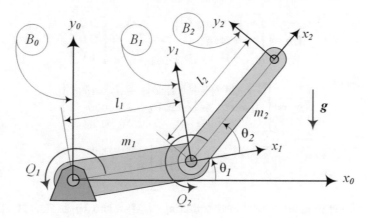

Fig. 11.17 A 2R planar manipulator with massive links

A 2R planar manipulator is shown in Fig. 11.17. Its homogenous transformation matrices are given in Eqs. (5.29) and (5.30).

$$
{}^0R_1 = \begin{bmatrix} \cos\theta_1 & -\sin\theta_1 & 0 \\ \sin\theta_1 & \cos\theta_1 & 0 \\ 0 & 0 & 1 \end{bmatrix} \tag{11.332}
$$

$$
{}^1R_2 = \begin{bmatrix} \cos\theta_2 & -\sin\theta_2 & 0 \\ \sin\theta_2 & \cos\theta_2 & 0 \\ 0 & 0 & 1 \end{bmatrix} \tag{11.333}
$$

Assuming the links are made of homogenous material in a bar shape, the position vectors of the mass center C_i are:

$$^i\mathbf{r}_i = \begin{bmatrix} -l_i/2 \\ 0 \\ 0 \end{bmatrix} \qquad i = 1, 2 \tag{11.334}$$

and the mass moment matrices are:

$$^iI_i = \frac{1}{12} m_i l_i^2 \begin{bmatrix} 0 & 0 & 0 \\ 0 & 1 & 0 \\ 0 & 0 & 1 \end{bmatrix} \tag{11.335}$$

Therefore,

$$^0I_1 = {}^0R_1 \, {}^1I_1 \, {}^0R_1^T$$
$$= \frac{1}{12} m_1 l_1^2 \begin{bmatrix} \sin^2 \theta_1 & -\cos\theta_1 \sin\theta_1 & 0 \\ -\cos\theta_1 \sin\theta_1 & \cos^2\theta_1 & 0 \\ 0 & 0 & 1 \end{bmatrix} \tag{11.336}$$

$$^0I_2 = {}^0R_2 \, {}^2I_2 \, {}^0R_2^T$$
$$= \frac{1}{12} m_2 l_2^2 \begin{bmatrix} \sin^2 \theta_{12} & -\cos\theta_{12} \sin\theta_{12} & 0 \\ -\cos\theta_{12} \sin\theta_{12} & \cos^2\theta_{12} & 0 \\ 0 & 0 & 1 \end{bmatrix} \tag{11.337}$$

The gravity is assumed to be in $-\hat{\jmath}_0$ direction

$$^0\mathbf{g} = \begin{bmatrix} 0 \\ -g \\ 0 \end{bmatrix} \tag{11.338}$$

and the link Jacobian matrices are:

$$\mathbf{J}_{D1} = \begin{bmatrix} -\frac{1}{2} l_1 \sin\theta_1 & 0 \\ \frac{1}{2} l_1 \cos\theta_1 & 0 \\ 0 & 0 \end{bmatrix} \qquad \mathbf{J}_{R1} = \begin{bmatrix} 0 & 0 \\ 0 & 0 \\ 1 & 0 \end{bmatrix} \tag{11.339}$$

$$\mathbf{J}_{D2} = \begin{bmatrix} -l_1 \sin\theta_1 - \frac{1}{2} l_2 \sin\theta_{12} & -\frac{1}{2} l_2 \sin\theta_{12} \\ l_1 \cos\theta_1 + \frac{1}{2} l_2 \cos\theta_{12} & \frac{1}{2} l_2 \cos\theta_{12} \\ 0 & 0 \end{bmatrix} \tag{11.340}$$

$$\mathbf{J}_{R2} = \begin{bmatrix} 0 & 0 \\ 0 & 0 \\ 1 & 0 \end{bmatrix} \tag{11.341}$$

We calculate the manipulator mass moment matrix by substituting ^0I_i, $\mathbf{J}_{Di}$, and $\mathbf{J}_{Ri}$ in Eq. (11.229).

$$D = \sum_{i=1}^{2} \left(\mathbf{J}_{Di}^T \, m_i \, \mathbf{J}_{Di} + \frac{1}{2} \mathbf{J}_{Ri}^T \, {}^0I_i \, \mathbf{J}_{Ri} \right) \tag{11.342}$$

$$= \mathbf{J}_{D1}^T \, m_1 \, \mathbf{J}_{D1} + \frac{1}{2} \mathbf{J}_{R1}^T \, {}^0I_1 \, \mathbf{J}_{R1} + \mathbf{J}_{D2}^T \, m_2 \, \mathbf{J}_{D2} + \frac{1}{2} \mathbf{J}_{R2}^T \, {}^0I_2 \, \mathbf{J}_{R2}$$

$$= \begin{bmatrix} \frac{1}{3} m_1 l_1^2 + m_2 \left(l_1^2 + l_1 l_2 c\theta_2 + \frac{1}{3} l_2^2 \right) & m_2 \left(\frac{1}{2} l_1 l_2 c\theta_2 + \frac{1}{3} l_2^2 \right) \\ m_2 \left(\frac{1}{2} l_1 l_2 c\theta_2 + \frac{1}{3} l_2^2 \right) & \frac{1}{3} m_2 l_2^2 \end{bmatrix}$$

The velocity coupling vector $\mathbf{H}$ has two elements that are:

$$H_1 = \sum_{j=1}^{1} \sum_{k=1}^{1} \left(\frac{\partial D_{1j}}{\partial q_k} - \frac{1}{2} \frac{\partial D_{jk}}{\partial q_1} \right) \dot{q}_j \dot{q}_k$$

$$= -m_2 l_1 l_2 \left(\dot{\theta}_1 + \frac{1}{2} \dot{\theta}_2 \right) \dot{\theta}_2 \sin \theta_2 \tag{11.343}$$

$$H_2 = \sum_{j=1}^{2} \sum_{k=1}^{2} \left(\frac{\partial D_{2j}}{\partial q_k} - \frac{1}{2} \frac{\partial D_{jk}}{\partial q_2} \right) \dot{q}_j \dot{q}_k$$

$$= \frac{1}{2} m_2 l_1 l_2 \dot{\theta}_1^2 \sin \theta_2. \tag{11.344}$$

The elements of the gravitational force vector $\mathbf{G}$ are:

$$G_1 = \frac{1}{2} m_1 g l_1 \cos \theta_1 + m_2 g l_1 \cos \theta_1 + \frac{1}{2} m_2 g l_2 \cos \theta_{12} \tag{11.345}$$

$$G_2 = \frac{1}{2} m_2 g l_2 \cos \theta_{12} \tag{11.346}$$

Now we can assemble the equations of motion for the $2R$ planar manipulator. Assuming no external force on the end-effector, the equations of motion are:

$$Q_1 = \left(\frac{1}{3} m_1 l_1^2 + m_2 \left(l_1^2 + l_1 l_2 c\theta_2 + \frac{1}{3} l_2^2 \right) \right) \ddot{\theta}_1$$

$$+ m_2 l_2 \left(\frac{1}{2} l_1 c\theta_2 + \frac{1}{3} l_2 \right) \ddot{\theta}_2 - m_2 l_1 l_2 \left(\dot{\theta}_1 + \frac{1}{2} \dot{\theta}_2 \right) \dot{\theta}_2 \sin \theta_2$$

$$+ \left(\frac{1}{2} m_1 + m_2 \right) g l_1 \cos \theta_1 + \frac{1}{2} m_2 g l_2 \cos \theta_{12} \tag{11.347}$$

$$Q_2 = m_2 \left(\frac{1}{2} l_1 l_2 c\theta_2 + \frac{1}{3} l_2^2 \right) \ddot{\theta}_1 + \frac{1}{3} m_2 l_2^2 \ddot{\theta}_2$$

$$+ \frac{1}{2} m_2 l_1 l_2 \dot{\theta}_1^2 \sin \theta_2 + \frac{1}{2} m_2 g l_2 \cos \theta_{12} \tag{11.348}$$

Example 369 Lagrange equation of a one-link manipulator.

To show the advantage and simplicity of the Lagrange method when compared to Newton–Euler method, let us derive the equation of motion of the uniform beam of Fig. 11.18 with a mass m_2 at the tip point. This is similar to the system of Fig. 11.5a.

The beam is uniform with a mass center at ${}^0\mathbf{r}_1$ while the tip mass is at ${}^0\mathbf{d}_1$, both in B_0.

$$
{}^0\mathbf{r}_1 = {}^0R_1 \, {}^1\mathbf{r}_1 = \begin{bmatrix} \dfrac{l}{2} \cos \theta \\ \dfrac{l}{2} \sin \theta \\ 0 \end{bmatrix} \tag{11.349}
$$

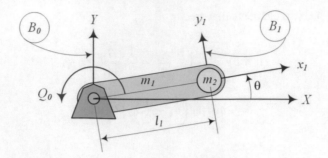

Fig. 11.18 A uniform beam with a hanging weight m_2 at the tip point

$$^0\mathbf{d}_1 = {}^0R_1\,{}^1\mathbf{d}_1 = \begin{bmatrix} l\cos\theta \\ l\sin\theta \\ 0 \end{bmatrix} \tag{11.350}$$

$$^0R_1 = R_{Z,\theta} = \begin{bmatrix} \cos\theta & -\sin\theta & 0 \\ \sin\theta & \cos\theta & 0 \\ 0 & 0 & 1 \end{bmatrix} \tag{11.351}$$

The angular velocity of the beam is:

$$_0\boldsymbol{\omega}_1 = \dot{\theta}\,\hat{K} \tag{11.352}$$

and therefore, the velocity of C and m_2 are:

$$^0\mathbf{v}_1 = {}_0\boldsymbol{\omega}_1 \times {}^0\mathbf{r}_1 = \begin{bmatrix} -\dfrac{l}{2}\dot{\theta}\sin\theta \\ \dfrac{l}{2}\dot{\theta}\cos\theta \\ 0 \end{bmatrix} \tag{11.353}$$

$$^0\dot{\mathbf{d}}_1 = {}_0\boldsymbol{\omega}_1 \times {}^0\mathbf{d}_1 = \begin{bmatrix} -l\dot{\theta}\sin\theta \\ l\dot{\theta}\cos\theta \\ 0 \end{bmatrix} \tag{11.354}$$

The kinetic energy of the manipulator is:

$$\begin{aligned} K_2 &= \frac{1}{2}m_2\,{}^0\dot{\mathbf{d}}_1 \cdot {}^0\dot{\mathbf{d}}_1 + \frac{1}{2}m_1\,{}^0\mathbf{v}_1 \cdot {}^0\mathbf{v}_1 + \frac{1}{2}\,_0\boldsymbol{\omega}_1^T\,{}^0I_1\,_0\boldsymbol{\omega}_1 \\ &= \frac{1}{8}l^2\dot{\theta}^2\,(m_1 + 4m_2) + \frac{1}{2}I_z\dot{\theta}^2 \end{aligned} \tag{11.355}$$

$$\begin{aligned} ^0I_1 &= R_{Z,\theta}\,{}^1I_1\,R_{Z,\theta}^T = {}^0R_1 \begin{bmatrix} I_x & 0 & 0 \\ 0 & I_y & 0 \\ 0 & 0 & I_z \end{bmatrix} {}^0R_1^T \\ &= \begin{bmatrix} I_x\cos^2\theta + I_y\sin^2\theta & \left(I_x - I_y\right)\cos\theta\sin\theta & 0 \\ \left(I_x - I_y\right)\cos\theta\sin\theta & I_y\cos^2\theta + I_x\sin^2\theta & 0 \\ 0 & 0 & I_z \end{bmatrix} \end{aligned} \tag{11.356}$$

The potential energy of the manipulator is:

$$\begin{aligned} V &= m_1 g Y_1 + m_2 g Y_2 = m_1 g r_Y + m_2 g d_Y \\ &= m_1 g \frac{l}{2}\sin\theta + m_2 g l \sin\theta \end{aligned} \tag{11.357}$$

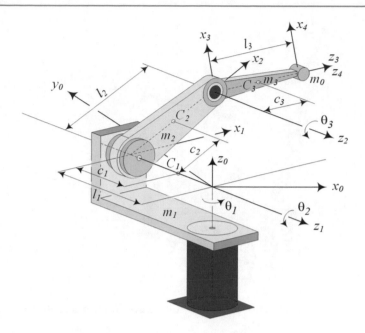

Fig. 11.19 An articulated manipulator with massive links and a massive load at the tip point

and therefore, the Lagrangian of the manipulator will be:

$$\mathcal{L} = K - V = \frac{1}{8}l^2\dot{\theta}^2 \left(m_1 + 4m_2\right) + \frac{1}{2}I_z\dot{\theta}^2$$

$$-m_1 g\frac{l}{2}\sin\theta - m_2 gl\sin\theta \tag{11.358}$$

Applying the Lagrange equation

$$\frac{d}{dt}\left(\frac{\partial\mathcal{L}}{\partial\dot{\theta}}\right) - \frac{\partial\mathcal{L}}{\partial\theta} = Q_0 \tag{11.359}$$

$$\frac{\partial\mathcal{L}}{\partial\dot{\theta}} = \frac{1}{4}l^2\left(m_1 + 4m_2\right)\dot{\theta} + I_z\dot{\theta} \tag{11.360}$$

$$\frac{d}{dt}\left(\frac{\partial\mathcal{L}}{\partial\dot{\theta}}\right) = \left(\frac{1}{4}m_1 l^2 + m_2 l^2 + I_z\right)\ddot{\theta} \tag{11.361}$$

$$\frac{\partial\mathcal{L}}{\partial\theta} = -m_1 g\frac{l}{2}\cos\theta - m_2 gl\cos\theta \tag{11.362}$$

determines the equation of motion.

$$Q_0 = \left(\frac{1}{4}m_1 l^2 + m_2 l^2 + I_z\right)\ddot{\theta} + \left(\frac{1}{2}m_1 + m_2\right)gl\cos\theta \tag{11.363}$$

It is the same equation as (11.50).

Example 370 Equations of motion of an articulated manipulator. This is very common and practical $3D$ industrial manipulator. Here is deriving its equations of motion by Lagrange method.

Figure 11.19 illustrates an articulated manipulator with massive links and a massive load at the tip point. Points C_i, $i = 1, 2, 3$ indicate the mass centers of the links with masses m_i, $i = 1, 2, 3$. The tip point has a mass of m_0. A top view of the manipulator is shown in Fig. 11.20.

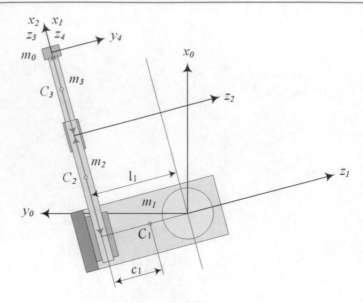

Fig. 11.20   A top view of an articulated manipulator with massive links and a massive load at the tip point

The link (1) of the manipulator is an R⊢R(90) with an extra displacement l_1 along z_1. To determine the transformation matrix $^0R_1$ we can begin from a coincident configuration of B_1 and B_0 and move B_1 to its current configuration by a sequence of proper rotations and displacements.

$$^1T_0 = D_{z_1,l_1} \, R_{x_1,\pi/2} \, R_{z_1,\theta_1}$$

$$= \begin{bmatrix} \cos\theta_1 & \sin\theta_1 & 0 & 0 \\ 0 & 0 & 1 & 0 \\ \sin\theta_1 & -\cos\theta_1 & 0 & l_1 \\ 0 & 0 & 0 & 1 \end{bmatrix} \tag{11.364}$$

$$^0T_1 = {^1T_0^{-1}} = \begin{bmatrix} \cos\theta_1 & 0 & \sin\theta_1 & -l_1\sin\theta_1 \\ \sin\theta_1 & 0 & -\cos\theta_1 & l_1\cos\theta_1 \\ 0 & 1 & 0 & 0 \\ 0 & 0 & 0 & 1 \end{bmatrix} \tag{11.365}$$

The second and third links are R‖R(0), R⊢R(90), and their associated transformation matrices between coordinate frames are:

$$^1T_2 = \begin{bmatrix} \cos\theta_2 & -\sin\theta_2 & 0 & l_2\cos\theta_2 \\ \sin\theta_2 & \cos\theta_2 & 0 & l_2\sin\theta_2 \\ 0 & 0 & 1 & 0 \\ 0 & 0 & 0 & 1 \end{bmatrix} \tag{11.366}$$

$$^2T_3 = \begin{bmatrix} \cos\theta_3 & 0 & \sin\theta_3 & 0 \\ \sin\theta_3 & 0 & -\cos\theta_3 & 0 \\ 0 & 1 & 0 & 0 \\ 0 & 0 & 0 & 1 \end{bmatrix} \tag{11.367}$$

The global position vectors of the mass centers C_i and joints are:

$$^0\mathbf{r}_1 = {^0T_1} \, {^1\mathbf{r}_1} = {^0T_1} \begin{bmatrix} 0 \\ 0 \\ c_1 \\ 1 \end{bmatrix} = \begin{bmatrix} -(l_1 - c_1)\sin\theta_1 \\ (l_1 - c_1)\cos\theta_1 \\ 0 \\ 1 \end{bmatrix} \tag{11.368}$$

$$^0\mathbf{d}_1 = {}^0T_1\,{}^1\mathbf{d}_1 = {}^0T_1 \begin{bmatrix} 0 \\ 0 \\ 0 \\ 1 \end{bmatrix} = \begin{bmatrix} -l_1\sin\theta_1 \\ l_1\cos\theta_1 \\ 0 \\ 1 \end{bmatrix} \tag{11.369}$$

$$^0\mathbf{r}_2 = {}^0\mathbf{d}_1 + {}^0T_2\,{}^2\mathbf{r}_2 = \begin{bmatrix} -2l_1\sin\theta_1 + (c_2+l_2)\cos\theta_1\cos\theta_2 \\ 2l_1\cos\theta_1 + (c_2+l_2)\cos\theta_2\sin\theta_1 \\ (c_2+l_2)\sin\theta_2 \\ 2 \end{bmatrix} \tag{11.370}$$

$$^0\mathbf{d}_2 = {}^0\mathbf{d}_1 + {}^0T_2\,{}^2\mathbf{d}_2 = \begin{bmatrix} 2l_2\cos\theta_1\cos\theta_2 - 2l_1\sin\theta_1 \\ 2l_1\cos\theta_1 + 2l_2\cos\theta_2\sin\theta_1 \\ 2l_2\sin\theta_2 \\ 2 \end{bmatrix} \tag{11.371}$$

$$^0\mathbf{r}_3 = {}^0\mathbf{d}_2 + {}^0T_3\,{}^3\mathbf{r}_3$$

$$= \begin{bmatrix} c_3\cos\theta_1\left(\sin(\theta_2+\theta_3) + 3l_2\cos\theta_2\right) - 3l_1\sin\theta_1 \\ c_3\sin\theta_1\left(\sin(\theta_2+\theta_3) + 3l_2\cos\theta_2\right) + 3l_1\cos\theta_1 \\ 3l_2\sin\theta_2 - c_3\cos(\theta_2+\theta_3) \\ 3 \end{bmatrix} \tag{11.372}$$

$$^0\mathbf{d}_3 = {}^0\mathbf{d}_2 + {}^0T_3\,{}^3\mathbf{d}_3$$

$$= \begin{bmatrix} l_3\cos\theta_1\left(\sin(\theta_2+\theta_3) + 3l_2\cos\theta_2\right) - 3l_1\sin\theta_1 \\ l_3\sin\theta_1\left(\sin(\theta_2+\theta_3) + 3l_2\cos\theta_2\right) + 3l_1\cos\theta_1 \\ 3l_2\sin\theta_2 - l_3\cos(\theta_2+\theta_3) \\ 3 \end{bmatrix} \tag{11.373}$$

The links' angular velocity is:

$$_0\boldsymbol{\omega}_1 = \dot{\theta}_1\,\hat{k}_0 \qquad _1\boldsymbol{\omega}_2 = \dot{\theta}_2\,\hat{k}_1 \qquad _2\boldsymbol{\omega}_3 = \dot{\theta}_3\,\hat{k}_2 \tag{11.374}$$

$$_0\tilde{\omega}_2 = {}_0\tilde{\omega}_1 + {}_1^0\tilde{\omega}_2 = {}_0\tilde{\omega}_1 + {}^0R_1\,{}_1\tilde{\omega}_2\,{}^0R_1^T$$

$$= \begin{bmatrix} 0 & -\dot{\theta}_1 & -\dot{\theta}_2\cos\theta_1 \\ \dot{\theta}_1 & 0 & -\dot{\theta}_2\sin\theta_1 \\ \dot{\theta}_2\cos\theta_1 & \dot{\theta}_2\sin\theta_1 & 0 \end{bmatrix} \tag{11.375}$$

$$_0\tilde{\omega}_3 = {}_0\tilde{\omega}_2 + {}_2^0\tilde{\omega}_3 = {}_0\tilde{\omega}_2 + {}^0R_2\,{}_2\tilde{\omega}_3\,{}^0R_2^T \tag{11.376}$$

$$= \begin{bmatrix} 0 & -\dot{\theta}_1 & -\left(\dot{\theta}_2+\dot{\theta}_3\right)\cos\theta_1 \\ \dot{\theta}_1 & 0 & -\left(\dot{\theta}_2+\dot{\theta}_3\right)\sin\theta_1 \\ \left(\dot{\theta}_2+\dot{\theta}_3\right)\cos\theta_1 & \left(\dot{\theta}_2+\dot{\theta}_3\right)\sin\theta_1 & 0 \end{bmatrix}$$

The mass moment matrices in the global coordinate frame are:

$$^0I_1 = {}^0R_1 \begin{bmatrix} I_{x_1} & 0 & 0 \\ 0 & I_{y_1} & 0 \\ 0 & 0 & I_{z_1} \end{bmatrix} {}^0R_1^T \tag{11.377}$$

$$= \begin{bmatrix} I_{x_1}\cos^2\theta_1 + I_{z_1}\sin^2\theta_1 & \left(I_{x_1} - I_{z_1}\right)\cos\theta_1\sin\theta_1 & 0 \\ \left(I_{x_1} - I_{z_1}\right)\cos\theta_1\sin\theta_1 & I_{z_1}\cos^2\theta_1 + I_{x_1}\sin^2\theta_1 & 0 \\ 0 & 0 & I_{y_1} \end{bmatrix}$$

$$^0I_2 = {}^0R_2 \, {}^2I_2 \, {}^0R_2^T = {}^0R_2 \begin{bmatrix} I_{x_2} & 0 & 0 \\ 0 & I_{y_2} & 0 \\ 0 & 0 & I_{z_2} \end{bmatrix} {}^0R_2^T \tag{11.378}$$

$$^0I_3 = {}^0R_3 \, {}^3I_3 \, {}^0R_3^T = {}^0R_3 \begin{bmatrix} I_{x_3} & 0 & 0 \\ 0 & I_{y_3} & 0 \\ 0 & 0 & I_{z_3} \end{bmatrix} {}^0R_3^T \tag{11.379}$$

The velocity of C_i and the joints are:

$$^0\mathbf{v}_1 = \frac{^0d}{dt} \, {}^0\mathbf{r}_1 = \begin{bmatrix} -(l_1 - c_1)\,\dot\theta_1 \cos\theta_1 \\ -(l_1 - c_1)\,\dot\theta_1 \sin\theta_1 \\ 0 \end{bmatrix} \tag{11.380}$$

$$^0\mathbf{v}_2 = \frac{^0d}{dt} \, {}^0\mathbf{r}_2 \qquad {}^0\mathbf{v}_3 = \frac{^0d}{dt} \, {}^0\mathbf{r}_3 \tag{11.381}$$

$$^0\dot{\mathbf{d}}_1 = \begin{bmatrix} -l_1\dot\theta_1 \cos\theta_1 \\ -l_1\dot\theta_1 \sin\theta_1 \\ 0 \end{bmatrix} \tag{11.382}$$

$$^0\dot{\mathbf{d}}_2 = \frac{^0d}{dt} \, {}^0\mathbf{d}_2 \qquad {}^0\dot{\mathbf{d}}_3 = \frac{^0d}{dt} \, {}^0\mathbf{d}_3 \tag{11.383}$$

The kinetic and potential energies of the manipulator are:

$$K = \frac{1}{2}m_1 \, {}^0\mathbf{v}_1 \cdot {}^0\mathbf{v}_1 + \frac{1}{2}m_{21} \, {}^0\mathbf{v}_2 \cdot {}^0\mathbf{v}_2 + \frac{1}{2}m_3 \, {}^0\mathbf{v}_3 \cdot {}^0\mathbf{v}_3$$

$$+ \frac{1}{2}m_0 \, {}^0\dot{\mathbf{d}}_3 \cdot {}^0\dot{\mathbf{d}}_3 + \frac{1}{2} \, {}_0\boldsymbol{\omega}_1^T \, {}^0I_1 \, {}_0\boldsymbol{\omega}_1$$

$$+ \frac{1}{2} \, {}_0\boldsymbol{\omega}_2^T \, {}^0I_2 \, {}_0\boldsymbol{\omega}_2 + \frac{1}{2} \, {}_0\boldsymbol{\omega}_3^T \, {}^0I_3 \, {}_0\boldsymbol{\omega}_3 \tag{11.384}$$

$$V = m_2 g r_{2z} + m_3 g r_{3z} + m_0 g d_{3z} \tag{11.385}$$

Using the Lagrangian of the manipulator $\mathcal{L} = K - V$, and applying the Lagrange equation, we determines the equations of motion.

$$\frac{d}{dt}\left(\frac{\partial\mathcal{L}}{\partial\dot\theta_1}\right) - \frac{\partial\mathcal{L}}{\partial\theta_1} = Q_0 \tag{11.386}$$

$$\frac{d}{dt}\left(\frac{\partial\mathcal{L}}{\partial\dot\theta_2}\right) - \frac{\partial\mathcal{L}}{\partial\theta_2} = Q_1 \tag{11.387}$$

$$\frac{d}{dt}\left(\frac{\partial\mathcal{L}}{\partial\dot\theta_3}\right) - \frac{\partial\mathcal{L}}{\partial\theta_3} = Q_2 \tag{11.388}$$

Example 371 Equations of motion of a Cartesian manipulator. This is mathematically the simplest type of manipulators. Here is how to find its equations of motion, ignoring wrist motions.

Figure 11.21 illustrates a Cartesian manipulator with massive motors and a massive load at the tip point. The manipulator has 3 DOF and with three mutually orthogonal rigid links. The three joints of the robot are prismatic, each one equipped with a motor to apply forces Q_1, Q_2, Q_3, respectively. The robot's joint variables are q_1, q_2, q_3. The kinetic K and potential P energies and Lagrangian $\mathcal{L}$ of this manipulator are:

$$K = \frac{1}{2}\left(m_1\dot{q}_1^2 + m_2\left(\dot{q}_1^2 + \dot{q}_2^2\right) + m_3\left(\dot{q}_1^2 + \dot{q}_2^2 + \dot{q}_3^2\right)\right)$$

$$= \frac{1}{2}\left((m_1 + m_2 + m_3)\,\dot{q}_1^2 + (m_2 + m_3)\,\dot{q}_2^2 + m_3\dot{q}_3^2\right) \tag{11.389}$$

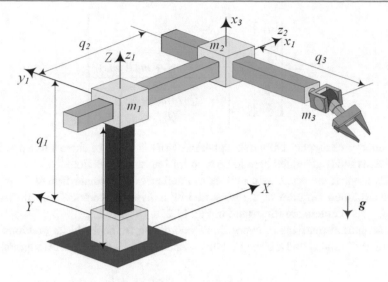

Fig. 11.21 A Cartesian manipulator with massive motors and a massive load at the tip point

$$V = (m_1 + m_2 + m_3)\, g\, q_1 \tag{11.390}$$

$$\mathcal{L} = K - V \tag{11.391}$$

So we have:

$$
\begin{aligned}
&\frac{\partial \mathcal{L}}{\partial \dot{q}_1} = (m_1 + m_2 + m_3)\, \dot{q}_1 \qquad &&\frac{d}{dt}\frac{\partial \mathcal{L}}{\partial \dot{q}_1} = (m_1 + m_2 + m_3)\, \ddot{q}_1 \\[4pt]
&\frac{\partial \mathcal{L}}{\partial \dot{q}_2} = (m_2 + m_3)\, \dot{q}_2 \qquad &&\frac{d}{dt}\frac{\partial \mathcal{L}}{\partial \dot{q}_2} = (m_2 + m_3)\, \ddot{q}_2 \\[4pt]
&\frac{\partial \mathcal{L}}{\partial \dot{q}_3} = m_3 \dot{q}_3 \qquad &&\frac{d}{dt}\frac{\partial \mathcal{L}}{\partial \dot{q}_3} = m_3 \ddot{q}_3 \\[4pt]
&\frac{\partial \mathcal{L}}{\partial q_1} = -(m_1 + m_2 + m_3)\, g \qquad &&\frac{\partial \mathcal{L}}{\partial q_2} = \frac{\partial \mathcal{L}}{\partial q_3} = 0
\end{aligned}
\tag{11.392}
$$

and the equations of motion will be as follows.

$$(m_1 + m_2 + m_3)\, \ddot{q}_1 + (m_1 + m_2 + m_3)\, g = Q_1 \tag{11.393}$$

$$(m_2 + m_3)\, \ddot{q}_2 = Q_2 \tag{11.394}$$

$$m_3 \ddot{q}_3 = Q_3 \tag{11.395}$$

$$
\begin{bmatrix} m_1 + m_2 + m_3 & 0 & 0 \\ 0 & m_2 + m_3 & 0 \\ 0 & 0 & m_3 \end{bmatrix}
\begin{bmatrix} \ddot{q}_1 \\ \ddot{q}_2 \\ \ddot{q}_3 \end{bmatrix}
$$
$$
+ \begin{bmatrix} m_1 + m_2 + m_3 \\ 0 \\ 0 \end{bmatrix}
= \begin{bmatrix} Q_1 \\ Q_2 \\ Q_3 \end{bmatrix}
\tag{11.396}
$$

Usually it is better to express the equations of motion in state-space, so we work only with first-order equations. Software usually work easier with first-order equations. Equations of motion of this Cartesian manipulator in state-space will be:

$$
\frac{d}{dt}
\begin{bmatrix}
q_1 \\ q_2 \\ q_3 \\ \dot{q}_1 \\ \dot{q}_2 \\ \dot{q}_3
\end{bmatrix}
=
\begin{bmatrix}
\dot{q}_1 \\
\dot{q}_2 \\
\dot{q}_3 \\
\dfrac{Q_1 - (m_1 + m_2 + m_3)}{m_1 + m_2 + m_3} \\
Q_2 / (m_2 + m_3) \\
Q_3 / m_3
\end{bmatrix}
\tag{11.397}
$$

Example 372 ★ Kinetic energy of a spherical wrist. Spherical wrist is made of three links and three orthogonal revolute joints. Kinetic energy of the wrist is calculated here to be used for Lagrange equation.

Figure 11.22a illustrates a spherical wrist with minimum number of coordinate frames to express relative position of its three loving links. The exploded diagram of the wrist and its individual members are illustrated in Fig. 11.22b. Their coordinate frames and their mass centers are illustrated in Fig. 11.22c.

The homogenous transformation matrices of every link's coordinate frame B_i to its previous frame B_{i-1} are only one rotation about z_{i-1}-axis with different α_i and d_i. The first link is and R⊢R(−90), as z_1 is orthogonal to z_0 with $\alpha_1 = -90$ deg and $a_1 = 0$, $d_1 = l_1$.

$$
{}^0T_1 =
\begin{bmatrix}
\cos\theta_1 & 0 & -\sin\theta_1 & 0 \\
\sin\theta_1 & 0 & \cos\theta_1 & 0 \\
0 & -1 & 0 & l_1 \\
0 & 0 & 0 & 1
\end{bmatrix}
\tag{11.398}
$$

The second link is and R⊢R(90), as z_2 is orthogonal to z_1 with $\alpha_1 = 90$ deg and $a_2 = 0$, $d_2 = 0$.

$$
{}^1T_2 =
\begin{bmatrix}
\cos\theta_2 & 0 & \sin\theta_2 & 0 \\
\sin\theta_2 & 0 & -\cos\theta_2 & 0 \\
0 & 1 & 0 & 0 \\
0 & 0 & 0 & 1
\end{bmatrix}
\tag{11.399}
$$

The second link is and R∥R(0), as z_3 is along z_2 with $a_3 = 0$ and $d_3 = l_3$.

$$
{}^2T_3 =
\begin{bmatrix}
\cos\theta_3 & -\sin\theta_3 & 0 & 0 \\
\sin\theta_3 & \cos\theta_3 & 0 & 0 \\
0 & 0 & 1 & l_3 \\
0 & 0 & 0 & 1
\end{bmatrix}
\tag{11.400}
$$

Their associate rotation transformation matrices are:

$$
{}^0R_1 =
\begin{bmatrix}
\cos\theta_1 & 0 & -\sin\theta_1 \\
\sin\theta_1 & 0 & \cos\theta_1 \\
0 & -1 & 0
\end{bmatrix}
\tag{11.401}
$$

$$
{}^1R_2 =
\begin{bmatrix}
\cos\theta_2 & 0 & \sin\theta_2 \\
\sin\theta_2 & 0 & -\cos\theta_2 \\
0 & 1 & 0
\end{bmatrix}
\tag{11.402}
$$

$$
{}^2R_3 =
\begin{bmatrix}
\cos\theta_3 & -\sin\theta_3 & 0 \\
\sin\theta_3 & \cos\theta_3 & 0 \\
0 & 0 & 1
\end{bmatrix}
\tag{11.403}
$$

$$
{}^0R_2 = {}^0R_1 \, {}^1R_2 =
\begin{bmatrix}
\cos\theta_1 \cos\theta_2 & -\sin\theta_1 & \cos\theta_1 \sin\theta_2 \\
\cos\theta_2 \sin\theta_1 & \cos\theta_1 & \sin\theta_1 \sin\theta_2 \\
-\sin\theta_2 & 0 & \cos\theta_2
\end{bmatrix}
\tag{11.404}
$$

$$
{}^0R_3 = {}^0R_1 \, {}^1R_2 \, {}^2R_3
$$

$$
=
\begin{bmatrix}
c\theta_1 c\theta_2 c\theta_3 - s\theta_1 s\theta_3 & -c\theta_3 s\theta_1 - c\theta_1 c\theta_2 s\theta_3 & c\theta_1 s\theta_2 \\
c\theta_1 s\theta_3 + c\theta_2 c\theta_3 s\theta_1 & c\theta_1 c\theta_3 - c\theta_2 s\theta_1 s\theta_3 & s\theta_1 s\theta_2 \\
-c\theta_3 s\theta_2 & s\theta_2 s\theta_3 & c\theta_2
\end{bmatrix}
\tag{11.405}
$$

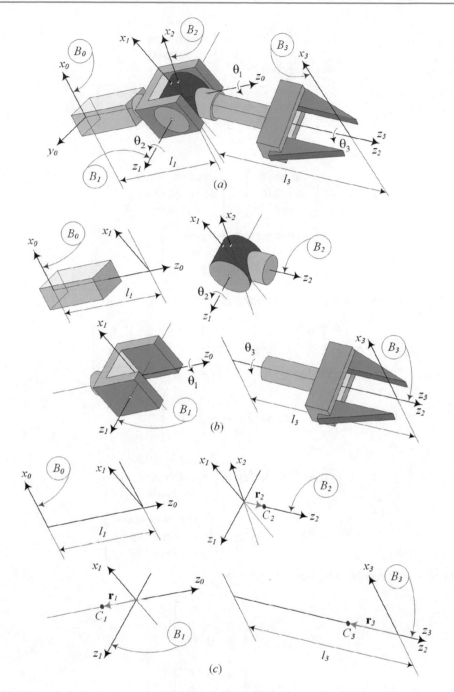

Fig. 11.22 A spherical wrist, its exploded diagram, the coordinate frame of each link, and their mass centers

The angular velocity vectors of links are:

$$
{}_0\boldsymbol{\omega}_1 = \begin{bmatrix} 0 \\ 0 \\ \dot{\theta}_1 \end{bmatrix} \qquad {}_1\boldsymbol{\omega}_2 = \begin{bmatrix} 0 \\ 0 \\ \dot{\theta}_2 \end{bmatrix} \qquad {}_2\boldsymbol{\omega}_3 = \begin{bmatrix} 0 \\ 0 \\ \dot{\theta}_3 \end{bmatrix} \tag{11.406}
$$

$$
{}_1^0\boldsymbol{\omega}_2 = {}^0R_1\,{}_1\boldsymbol{\omega}_2 = \begin{bmatrix} -\dot{\theta}_2 \sin\theta_1 \\ \dot{\theta}_2 \cos\theta_1 \\ 0 \end{bmatrix} \tag{11.407}
$$

$$
{}_2^0\boldsymbol{\omega}_3 = {}^0R_1\,{}^1R_2\,{}_2\boldsymbol{\omega}_3 = \begin{bmatrix} \dot{\theta}_3 \cos\theta_1 \sin\theta_2 \\ \dot{\theta}_3 \sin\theta_1 \sin\theta_2 \\ \dot{\theta}_3 \cos\theta_2 \end{bmatrix} \tag{11.408}
$$

$$
{}_0\boldsymbol{\omega}_2 = {}_0\boldsymbol{\omega}_1 + {}_1^0\boldsymbol{\omega}_2
$$

$$
= \begin{bmatrix} 0 \\ 0 \\ \dot{\theta}_1 \end{bmatrix} + \begin{bmatrix} -\dot{\theta}_2 \sin\theta_1 \\ \dot{\theta}_2 \cos\theta_1 \\ 0 \end{bmatrix} = \begin{bmatrix} -\dot{\theta}_2 \sin\theta_1 \\ \dot{\theta}_2 \cos\theta_1 \\ \dot{\theta}_1 \end{bmatrix} \tag{11.409}
$$

$$
{}_0\boldsymbol{\omega}_3 = {}_0\boldsymbol{\omega}_2 + {}_2^0\boldsymbol{\omega}_3
$$

$$
= \begin{bmatrix} -\dot{\theta}_2 \sin\theta_1 \\ \dot{\theta}_2 \cos\theta_1 \\ \dot{\theta}_1 \end{bmatrix} + \begin{bmatrix} \dot{\theta}_3 \cos\theta_1 \sin\theta_2 \\ \dot{\theta}_3 \sin\theta_1 \sin\theta_2 \\ \dot{\theta}_3 \cos\theta_2 \end{bmatrix}
$$

$$
= \begin{bmatrix} \dot{\theta}_3 \cos\theta_1 \sin\theta_2 - \dot{\theta}_2 \sin\theta_1 \\ \dot{\theta}_2 \cos\theta_1 + \dot{\theta}_3 \sin\theta_1 \sin\theta_2 \\ \dot{\theta}_1 + \dot{\theta}_3 \cos\theta_2 \end{bmatrix} \tag{11.410}
$$

The position vector of the mass centers of link (i) expressed in frame B_i is:

$$
{}^1\mathbf{r}_1 = \begin{bmatrix} 0 \\ r_1 \\ 0 \end{bmatrix} \qquad {}^2\mathbf{r}_2 = \begin{bmatrix} 0 \\ 0 \\ r_2 \end{bmatrix} \qquad {}^3\mathbf{r}_3 = \begin{bmatrix} 0 \\ 0 \\ -r_3 \end{bmatrix} \tag{11.411}
$$

$$
{}^0\mathbf{r}_1 = {}^0R_1\,{}^1\mathbf{r}_1 = \begin{bmatrix} 0 \\ 0 \\ -r_1 \end{bmatrix} \tag{11.412}
$$

$$
{}^0\mathbf{r}_2 = {}^0R_1\,{}^1R_2\,{}^2\mathbf{r}_2 = \begin{bmatrix} r_2 \cos\theta_1 \sin\theta_2 \\ r_2 \sin\theta_1 \sin\theta_2 \\ r_2 \cos\theta_2 \end{bmatrix} \tag{11.413}
$$

$$
{}^0\mathbf{r}_3 = {}^0R_1\,{}^1R_2\,{}^2R_3\,{}^3\mathbf{r}_3 = \begin{bmatrix} -r_3 \cos\theta_1 \sin\theta_2 \\ -r_3 \sin\theta_1 \sin\theta_2 \\ -r_3 \cos\theta_2 \end{bmatrix} \tag{11.414}
$$

The mass moment matrices of links are assumed to be principal.

$$
{}^1I_1 = \begin{bmatrix} I_{11} & 0 & 0 \\ 0 & I_{12} & 0 \\ 0 & 0 & I_{13} \end{bmatrix} \tag{11.415}
$$

$$
{}^2I_2 = \begin{bmatrix} I_{21} & 0 & 0 \\ 0 & I_{22} & 0 \\ 0 & 0 & I_{23} \end{bmatrix} \tag{11.416}
$$

$$
{}^3I_3 = \begin{bmatrix} I_{31} & 0 & 0 \\ 0 & I_{32} & 0 \\ 0 & 0 & I_{33} \end{bmatrix} \tag{11.417}
$$

$$
{}^0I_1 = {}^0R_1\,{}^1I_1\,{}^0R_1^T \tag{11.418}
$$

$$
= \begin{bmatrix} I_{11} \cos^2\theta_1 + I_{13} \sin^2\theta_1 & (I_{11} - I_{13}) \cos\theta_1 \sin\theta_1 & 0 \\ (I_{11} - I_{13}) \cos\theta_1 \sin\theta_1 & I_{13} \cos^2\theta_1 + I_{11} \sin^2\theta_1 & 0 \\ 0 & 0 & I_{12} \end{bmatrix}
$$

$$^0I_2 = {}^0R_2\, {}^2I_2\, {}^0R_2^T = \begin{bmatrix} I_{211} & I_{212} & I_{213} \\ I_{221} & I_{222} & I_{223} \\ I_{231} & I_{232} & I_{233} \end{bmatrix} \tag{11.419}$$

$$I_{211} = \left(I_{21}\cos^2\theta_2 + I_{23}\sin^2\theta_2\right)\cos^2\theta_1 + I_{22}\sin^2\theta_1 \tag{11.420}$$

$$I_{221} = \left(I_{21}\cos^2\theta_2 + I_{23}\sin^2\theta_2 - I_{22}\right)\cos\theta_1\sin\theta_1 \tag{11.421}$$

$$I_{231} = (I_{23} - I_{21})\cos\theta_1\cos\theta_2\sin\theta_2 \tag{11.422}$$

$$I_{212} = \left(I_{21}\cos^2\theta_2 + I_{23}\sin^2\theta_2 - I_{22}\right)\cos\theta_1\sin\theta_1 \tag{11.423}$$

$$I_{222} = I_{22}\cos^2\theta_1 + \left(I_{21}\cos^2\theta_2 + I_{23}\sin^2\theta_2\right)\sin^2\theta_1 \tag{11.424}$$

$$I_{232} = (I_{23} - I_{21})\cos\theta_2\sin\theta_1\sin\theta_2 \tag{11.425}$$

$$I_{213} = (I_{23} - I_{21})\cos\theta_1\cos\theta_2\sin\theta_2 \tag{11.426}$$

$$I_{223} = (I_{23} - I_{21})\cos\theta_2\sin\theta_1\sin\theta_2 \tag{11.427}$$

$$I_{233} = I_{23}\cos^2\theta_2 + I_{21}\sin^2\theta_2 \tag{11.428}$$

$$^0I_3 = {}^0R_3\, {}^3I_3\, {}^0R_3^T = \begin{bmatrix} I_{311} & I_{312} & I_{313} \\ I_{321} & I_{322} & I_{323} \\ I_{331} & I_{332} & I_{333} \end{bmatrix} \tag{11.429}$$

$$I_{311} = I_{31}\left(\sin\theta_1\sin\theta_3 - \cos\theta_1\cos\theta_2\cos\theta_3\right)^2 + I_{33}\cos^2\theta_1\sin^2\theta_2$$
$$+ I_{32}\left(\cos\theta_3\sin\theta_1 + \cos\theta_1\cos\theta_2\sin\theta_3\right)^2 \tag{11.430}$$

$$I_{321} = -I_{31}\left(\cos\theta_1\sin\theta_3 + \cos\theta_2\cos\theta_3\sin\theta_1\right)\times$$
$$\left(\sin\theta_1\sin\theta_3 - \cos\theta_1\cos\theta_2\cos\theta_3\right)$$
$$- I_{32}\left(\cos\theta_3\sin\theta_1 + \cos\theta_1\cos\theta_2\sin\theta_3\right)\times$$
$$\left(\cos\theta_1\cos\theta_3 - \cos\theta_2\sin\theta_1\sin\theta_3\right)$$
$$+ I_{33}\cos\theta_1\sin\theta_1\sin^2\theta_2 \tag{11.431}$$

$$I_{331} = I_{31}\left(\cos\theta_3\sin\theta_2\right)\left(\sin\theta_1\sin\theta_3 - \cos\theta_1\cos\theta_2\cos\theta_3\right)$$
$$- I_{32}\left(\sin\theta_2\sin\theta_3\right)\left(\cos\theta_3\sin\theta_1 + \cos\theta_1\cos\theta_2\sin\theta_3\right)$$
$$+ I_{33}\cos\theta_1\cos\theta_2\sin\theta_2 \tag{11.432}$$

$$I_{312} = -I_{31}\left(\cos\theta_1\sin\theta_3 + \cos\theta_2\cos\theta_3\sin\theta_1\right)$$
$$\times \left(\sin\theta_1\sin\theta_3 - \cos\theta_1\cos\theta_2\cos\theta_3\right)$$
$$- I_{32}\left(\cos\theta_3\sin\theta_1 + \cos\theta_1\cos\theta_2\sin\theta_3\right)$$
$$\times \left(\cos\theta_1\cos\theta_3 - \cos\theta_2\sin\theta_1\sin\theta_3\right)$$
$$+ I_{33}\cos\theta_1\sin\theta_1\sin^2\theta_2 \tag{11.433}$$

$$I_{322} = I_{31}\left(\cos\theta_1\sin\theta_3 + \cos\theta_2\cos\theta_3\sin\theta_1\right)^2$$
$$+ I_{32}\left(\cos\theta_1\cos\theta_3 - \cos\theta_2\sin\theta_1\sin\theta_3\right)^2$$
$$+ I_{33}\sin^2\theta_1\sin^2\theta_2 \tag{11.434}$$

$$I_{332} = I_{32}\left(\sin\theta_2\sin\theta_3\right)\left(\cos\theta_1\cos\theta_3 - \cos\theta_2\sin\theta_1\sin\theta_3\right)$$
$$- I_{31}\left(\cos\theta_3\sin\theta_2\right)\left(\cos\theta_1\sin\theta_3 + \cos\theta_2\cos\theta_3\sin\theta_1\right)$$
$$+ I_{33}\cos\theta_2\sin\theta_1\sin\theta_2 \tag{11.435}$$

$$I_{313} = I_{31} \left(\cos \theta_3 \sin \theta_2 \right) \left(\sin \theta_1 \sin \theta_3 - \cos \theta_1 \cos \theta_2 \cos \theta_3 \right)$$

$$- I_{32} \left(\sin \theta_2 \sin \theta_3 \right) \left(\cos \theta_3 \sin \theta_1 + \cos \theta_1 \cos \theta_2 \sin \theta_3 \right)$$

$$+ I_{33} \cos \theta_1 \cos \theta_2 \sin \theta_2 \tag{11.436}$$

$$I_{323} = I_{32} \left(\sin \theta_2 \sin \theta_3 \right) \left(\cos \theta_1 \cos \theta_3 - \cos \theta_2 \sin \theta_1 \sin \theta_3 \right)$$

$$- I_{31} \left(\cos \theta_3 \sin \theta_2 \right) \left(\cos \theta_1 \sin \theta_3 + \cos \theta_2 \cos \theta_3 \sin \theta_1 \right)$$

$$+ I_{33} \cos \theta_2 \sin \theta_1 \sin \theta_2 \tag{11.437}$$

$$I_{333} = I_{33} \cos^2 \theta_2 + I_{31} \cos^2 \theta_3 \sin^2 \theta_2 + I_{32} \sin^2 \theta_2 \sin^2 \theta_3 \tag{11.438}$$

Velocity of the mass centers are:

$$^1\mathbf{v}_1 = {}^1_0\boldsymbol{\omega}_1 \times {}^1\mathbf{r}_1 = {}^0R_1^T \, {}_0\boldsymbol{\omega}_1 \times {}^1\mathbf{r}_1 = \begin{bmatrix} 0 \\ 0 \\ 0 \end{bmatrix} \tag{11.439}$$

$$^0\mathbf{v}_1 = {}^0R_1 \, {}^1\mathbf{v}_1 = \begin{bmatrix} 0 \\ 0 \\ 0 \end{bmatrix} \tag{11.440}$$

$$^2\mathbf{v}_2 = {}^2_0\boldsymbol{\omega}_2 \times {}^2\mathbf{r}_2 = {}^2R_0 \, {}_0\boldsymbol{\omega}_2 \times {}^2\mathbf{r}_2 = \begin{bmatrix} r_2\dot{\theta}_2 \\ r_2\dot{\theta}_1 \sin \theta_2 \\ 0 \end{bmatrix} \tag{11.441}$$

$$^0\mathbf{v}_2 = {}^0R_2 \, {}^2\mathbf{v}_2 = \begin{bmatrix} r_2\dot{\theta}_2 \cos \theta_1 \cos \theta_2 - r_2\dot{\theta}_1 \sin \theta_1 \sin \theta_2 \\ r_2\dot{\theta}_1 \cos \theta_1 \sin \theta_2 + r_2\dot{\theta}_2 \cos \theta_2 \sin \theta_1 \\ -r_2\dot{\theta}_2 \sin \theta_2 \end{bmatrix} \tag{11.442}$$

$$^3\mathbf{v}_3 = {}^3_0\boldsymbol{\omega}_3 \times {}^3\mathbf{r}_3 = {}^3R_0 \, {}_0\boldsymbol{\omega}_3 \times {}^3\mathbf{r}_3$$

$$= \begin{bmatrix} -r_3\dot{\theta}_2 \cos \theta_3 - r_3\dot{\theta}_1 \sin \theta_2 \sin \theta_3 \\ r_3\dot{\theta}_2 \sin \theta_3 - r_3\dot{\theta}_1 \cos \theta_3 \sin \theta_2 \\ 0 \end{bmatrix} \tag{11.443}$$

$$^0\mathbf{v}_3 = {}^0R_3 \, {}^3\mathbf{v}_3$$

$$= \begin{bmatrix} r_3\dot{\theta}_1 \sin \theta_1 \sin \theta_2 - r_3\dot{\theta}_2 \cos \theta_1 \cos \theta_2 \\ -r_3\dot{\theta}_1 \cos \theta_1 \sin \theta_2 - r_3\dot{\theta}_2 \cos \theta_2 \sin \theta_1 \\ r_3\dot{\theta}_2 \sin \theta_2 \end{bmatrix} \tag{11.444}$$

The kinetic energy of the link (1) will be:

$$K_1 = \frac{1}{2}m \, {}^0\mathbf{v}_1^T \, {}^0\mathbf{v}_1 + \frac{1}{2} \, {}_0\boldsymbol{\omega}_1^T \, {}^0I_1 \, {}_0\boldsymbol{\omega}_1 = \frac{1}{2}\dot{\theta}_1^2 \, I_{12} \tag{11.445}$$

The kinetic energy of the link (2) will be:

$$K_2 = \frac{1}{2}m \, {}^0\mathbf{v}_2^T \, {}^0\mathbf{v}_2 + \frac{1}{2} \, {}_0\boldsymbol{\omega}_2^T \, {}^0I_2 \, {}_0\boldsymbol{\omega}_2$$

$$= \frac{1}{4}mr_2^2 \left(\dot{\theta}_1^2 + 2\dot{\theta}_2^2 - \dot{\theta}_1^2 \cos 2\theta_2 \right) + \frac{1}{2} I_{22}\dot{\theta}_2^2$$

$$+ \frac{1}{4} \left(I_{21} + I_{23} \right) \dot{\theta}_1^2 + \frac{1}{4} \left(I_{23} - I_{21} \right) \dot{\theta}_1^2 \cos 2\theta_2 \tag{11.446}$$

The kinetic energy of the link (3) will be:

$$
\begin{aligned}
K_3 &= \frac{1}{2} m \, {}^0\mathbf{v}_3^T \, {}^0\mathbf{v}_3 + \frac{1}{2} \, {}_0\boldsymbol{\omega}_3^T \, {}^0 I_3 \, {}_0\boldsymbol{\omega}_3 \\
&= \frac{1}{4} m r_3^2 \left(\dot{\theta}_1^2 + 2\dot{\theta}_2^2 - \dot{\theta}_1^2 \cos 2\theta_2 \right) \\
&\quad + \frac{1}{2} I_{31} \dot{\theta}_2^2 \sin^2 \theta_3 + \frac{1}{2} I_{33} \dot{\theta}_3^2 \\
&\quad + \frac{1}{2} I_{32} \left(\dot{\theta}_2^2 \cos^2 \theta_3 + \dot{\theta}_1^2 \sin^2 \theta_2 \right) \\
&\quad + \left(\dot{\theta}_3 + \frac{1}{2} \dot{\theta}_1 \cos \theta_2 \right) I_{33} \dot{\theta}_1 \cos \theta_2 \\
&\quad + (I_{32} - I_{31}) \left(\dot{\theta}_2 \cos \theta_3 \sin \theta_3 - \frac{1}{2} \dot{\theta}_1 \sin \theta_2 \right) \dot{\theta}_1 \sin \theta_2
\end{aligned}
\tag{11.447}
$$

Therefore, the total kinetic energy K of the spherical wrist is the sum of kinetic energy of the individual links.

$$
K = K_1 + K_2 + K_3
\tag{11.448}
$$

Example 373 General form of Lagrange equations of robots. This is to understand what would be general form of equations of motion if the robot is not indicated, so that would be common equations for all robots, as well as other dynamic systems.

Consider a robot manipulator of $n\,DOF$. The kinetic energy function K of the manipulator may always be expressed in a quadratic form of the generalized speed vector $\dot{q} = \frac{d}{dt}\mathbf{q}$ of the generalized coordinates vector $\mathbf{q}$.

$$
K\left(\mathbf{q}, \dot{q}\right) = \frac{1}{2} \dot{q}^T M\left(\mathbf{q}\right) \dot{q}
\tag{11.449}
$$

The symmetric and positive definite $M\left(\mathbf{q}\right)$ is the inertia matrix of dimension $n \times n$.

The potential energy $P\left(\mathbf{q}\right)$ does not have a specific general form but it always depends on the generalized coordinates vector $\mathbf{q}$. Hence the general form of the Lagrangian $\mathcal{L}\left(\mathbf{q}, \dot{q}\right)$ of the robot will be:

$$
\mathcal{L}\left(\mathbf{q}, \dot{q}\right) = \frac{1}{2} \dot{q}^T M\left(\mathbf{q}\right) \dot{q} - P\left(\mathbf{q}\right)
\tag{11.450}
$$

The Lagrange equations of motion will be:

$$
\frac{\partial \mathcal{L}}{\partial \dot{q}} = M\left(\mathbf{q}\right) \dot{q}
\tag{11.451}
$$

$$
\frac{d}{dt}\left(\frac{\partial \mathcal{L}}{\partial \dot{q}} \right) = M\left(\mathbf{q}\right) \ddot{q} + \dot{M}\left(\mathbf{q}\right) \dot{q}
\tag{11.452}
$$

$$
\frac{\partial \mathcal{L}}{\partial \mathbf{q}} = \frac{1}{2} \frac{\partial}{\partial \mathbf{q}} \left(\dot{q}^T M\left(\mathbf{q}\right) \dot{q} \right) - \frac{\partial}{\partial \mathbf{q}} P\left(\mathbf{q}\right)
\tag{11.453}
$$

$$
M\left(\mathbf{q}\right) \ddot{q} + \dot{M}\left(\mathbf{q}\right) \dot{q} - \frac{1}{2} \frac{\partial}{\partial \mathbf{q}} \left(\dot{q}^T M\left(\mathbf{q}\right) \dot{q} \right) + \frac{\partial}{\partial \mathbf{q}} P\left(\mathbf{q}\right) = \mathbf{Q}
\tag{11.454}
$$

It can also be rewritten in a more common form,

$$
M\left(\mathbf{q}\right) \ddot{q} + C\left(\mathbf{q}, \dot{q}\right) \dot{q} + G\left(\mathbf{q}\right) = \mathbf{Q}
\tag{11.455}
$$

where

$$G\left(\mathbf{q}\right) = \frac{\partial}{\partial \mathbf{q}} P\left(\mathbf{q}\right) \tag{11.456}$$

$$C\left(\mathbf{q}, \dot{\mathbf{q}}\right) \dot{q} = \dot{M}\left(\mathbf{q}\right) \dot{q} - \frac{1}{2} \frac{\partial}{\partial \mathbf{q}} \left(\dot{q}^T M\left(\mathbf{q}\right) \dot{q}\right) \tag{11.457}$$

The equations in state-space form will be:

$$\frac{d}{dt} \begin{bmatrix} \mathbf{q} \\ \dot{q} \end{bmatrix} = \begin{bmatrix} \dot{q} \\ M^{-1}\left(\mathbf{q}\right) \left[\mathbf{Q} - C\left(\mathbf{q}, \dot{\mathbf{q}}\right) \dot{q} - G\left(\mathbf{q}\right)\right] \end{bmatrix} \tag{11.458}$$

Example 374 ★ Christoffel operator. Christoffel symbol $\Gamma_{j,k}^i$ is a short notation for particular sum of partial derivatives. It happens frequently in $3D$ kinematics and dynamics. Here is an example to rewrite $\mathbf{H}$ with Christoffel symbol.

The symbol $\Gamma_{j,k}^i$ is one version of the Christoffel symbol or Christoffel operator that helps to shorten $\mathbf{H}$ matrix.

$$\Gamma_{j,k}^i = \frac{1}{2} \left(\frac{\partial D_{ij}}{\partial q_k} + \frac{\partial D_{ik}}{\partial q_j} - \frac{\partial D_{jk}}{\partial q_i} \right) \tag{11.459}$$

The velocity coupling vector H_{ijk} is a Christoffel symbol.

$$\begin{aligned} H_{ijk} &= \sum_{j=1}^n \sum_{k=1}^n \left(\frac{\partial D_{ij}}{\partial q_k} - \frac{1}{2} \frac{\partial D_{jk}}{\partial q_i} \right) \\ &= \frac{1}{2} \sum_{j=1}^n \sum_{k=1}^n \left(\frac{\partial D_{ij}}{\partial q_k} + \frac{\partial D_{ik}}{\partial q_j} - \frac{\partial D_{jk}}{\partial q_i} \right) \end{aligned} \tag{11.460}$$

Using Christoffel symbol, we can write the equations of motion of a robot as:

$$\sum_{j=1}^n D_{ij}(q) \ddot{q}_j + \sum_{j=1}^n \sum_{k=1}^n \Gamma_{j,k}^i \dot{q}_k \dot{q}_m + G_i = Q_i \tag{11.461}$$

Elwin Bruno Christoffel (1829–1900) is a German mathematician who introduced the Christoffel symbols and fundamental concepts of differential geometry.

Example 375 ★ No gravity and no external force.

Assume a robot is working in space where there is no gravity and there is no external force applied on the end-effector of the robot. In these conditions, the Lagrangian of the manipulator simplifies to

$$\mathcal{L} = \frac{1}{2} \sum_{i=1}^n \sum_{j=1}^n D_{ij} \dot{q}_i \dot{q}_j \tag{11.462}$$

and the equations of motion reduce to

$$\frac{d}{dt} \left(\frac{\partial \mathcal{L}}{\partial \dot{q}_i} \right) - \frac{\partial \mathcal{L}}{\partial q_i} = \sum_{i=1}^n \sum_{j=1}^n D_{ij} \left(\ddot{q}_i + \Gamma_{l,m}^j \dot{q}_l \dot{q}_m \right) \tag{11.463}$$

11.4 ★ Lagrange Equations and Link Transformation Matrices

The matrix form of the equations of motion for a robot, based on the Lagrange equations, is:

$$\mathbf{D}(\mathbf{q})\ddot{\mathbf{q}} + \mathbf{H}(\mathbf{q}, \dot{q}) + \mathbf{G}(\mathbf{q}) = \mathbf{Q} \tag{11.464}$$

which can also be written in a summation form.

$$\sum_{j=1}^{n} D_{ij}(q)\ddot{q}_j + \sum_{j=1}^{n}\sum_{k=1}^{n} H_{ikm}\dot{q}_k\dot{q}_m + G_i = Q_i \tag{11.465}$$

The matrix $\mathbf{D}(\mathbf{q})$ is an $n \times n$ inertial-type symmetric matrix.

$$D_{ij} = \sum_{r=\max i,j}^{n} \text{tr}\left(\frac{\partial\, {}^0T_r}{\partial q_i}\, {}^r\bar{I}_r\, \frac{\partial\, {}^0T_r}{\partial q_j}^T\right) \tag{11.466}$$

The H_{ikm} is the velocity coupling term,

$$H_{ijk} = \sum_{r=\max i,j,k}^{n} \text{tr}\left(\frac{\partial^2\, {}^0T_r}{\partial q_j \partial q_k}\, {}^r\bar{I}_r\, \frac{\partial\, {}^0T_r}{\partial q_i}^T\right) \tag{11.467}$$

and G_i is the gravitational vector.

$$G_i = -\sum_{r=i}^{n} m_r \mathbf{g}^T \frac{\partial\, {}^0T_r}{\partial q_i}\, {}^r\mathbf{r}_r \tag{11.468}$$

Proof Position vector of a point P of the link (i) at $^i\mathbf{r}_P$ in the body coordinate B_i can be transformed to the base frame.

$$^0\mathbf{r}_P = {}^0T_i\, {}^i\mathbf{r}_P \tag{11.469}$$

Therefore, its velocity and square of velocity in the base frame are:

$$^0\dot{\mathbf{r}}_P = \sum_{j=1}^{i} \frac{\partial\, {}^0T_i}{\partial q_j}\dot{q}_j\, {}^i\mathbf{r}_P \tag{11.470}$$

$$\begin{aligned}
{}^0\dot{r}_P^2 &= {}^0\dot{\mathbf{r}}_P \cdot {}^0\dot{\mathbf{r}}_P = \text{tr}\left({}^0\dot{\mathbf{r}}_P\, {}^0\dot{\mathbf{r}}_P^T\right) \\
&= \text{tr}\left(\sum_{j=1}^{i} \frac{\partial\, {}^0T_i}{\partial q_j}\dot{q}_j\, {}^i\mathbf{r}_P \sum_{k=1}^{i}\left[\frac{\partial\, {}^0T_i}{\partial q_k}\dot{q}_k\, {}^i\mathbf{r}_P\right]^T\right) \\
&= \text{tr}\left(\sum_{j=1}^{i}\sum_{k=1}^{i} \frac{\partial\, {}^0T_i}{\partial q_j}\, {}^i\mathbf{r}_P\, {}^i\mathbf{r}_P^T \left[\frac{\partial\, {}^0T_i}{\partial q_k}\right]^T \dot{q}_j\dot{q}_k\right)
\end{aligned} \tag{11.471}$$

The kinetic energy of point P having a small mass dm is then equal to:

$$dK_P = \frac{1}{2} \text{tr} \left(\sum_{j=1}^{i} \sum_{k=1}^{i} \frac{\partial\, ^0T_i}{\partial q_j}\, ^i\mathbf{r}_P\, ^i\mathbf{r}_P^T\, \frac{\partial\, ^0T_i}{\partial q_k}^T \dot{q}_j \dot{q}_k \right) dm$$

$$= \frac{1}{2} \text{tr} \left(\sum_{j=1}^{i} \sum_{k=1}^{i} \frac{\partial\, ^0T_i}{\partial q_j} \left(^i\mathbf{r}_P\, dm\, ^i\mathbf{r}_P^T \right) \frac{\partial\, ^0T_i}{\partial q_k}^T \dot{q}_j \dot{q}_k \right) \quad (11.472)$$

and hence, the kinetic energy of the link (i) would be:

$$K_i = \int_{B_i} dK_P$$

$$= \frac{1}{2} \text{tr} \left(\sum_{j=1}^{i} \sum_{k=1}^{i} \frac{\partial\, ^0T_i}{\partial q_j} \left(\int_{B_i} ^i\mathbf{r}_P\, ^i\mathbf{r}_P^T\, dm \right) \frac{\partial\, ^0T_i}{\partial q_k}^T \dot{q}_j \dot{q}_k \right) \quad (11.473)$$

The integral in Eq. (11.473) is the pseudo-inertia matrix for the link (i).

$$^i\bar{I}_i = \int_{B_i} ^i\mathbf{r}_P\, ^i\mathbf{r}_P^T\, dm \quad (11.474)$$

Therefore, kinetic energy of the link (i) becomes:

$$K_i = \frac{1}{2} \text{tr} \left(\sum_{j=1}^{i} \sum_{k=1}^{i} \frac{\partial\, ^0T_i}{\partial q_j}\, ^i\bar{I}_i\, \frac{\partial\, ^0T_i}{\partial q_k}^T \dot{q}_j \dot{q}_k \right) \quad (11.475)$$

The kinetic energy of a robot having n links is a summation of the kinetic energies of each link.

$$K = \sum_{i=1}^{n} K_i = \frac{1}{2} \text{tr} \sum_{i=1}^{n} \left(\sum_{j=1}^{i} \sum_{k=1}^{i} \frac{\partial\, ^0T_i}{\partial q_j}\, ^i\bar{I}_i\, \frac{\partial\, ^0T_i}{\partial q_k}^T \dot{q}_j \dot{q}_k \right) \quad (11.476)$$

We may also add the kinetic energy due to the actuating motors K_a that are installed at joints of the robot,

$$K_a = \begin{cases} \sum_{i=1}^{n} \frac{1}{2} I_i \dot{q}_i^2 & \text{if joint } i \text{ is R} \\ \sum_{i=1}^{n} \frac{1}{2} m_i \dot{q}_i^2 & \text{if joint } i \text{ is P} \end{cases} \quad (11.477)$$

where I_i is the mass moment of the rotary actuator at joint i, and m_i is the mass of the translatory actuator. However, we may assume that the motors are concentrated masses at joints and add the mass of motor at joint i to the mass of link $(i-1)$ and adjust the inertial parameters of the link. The motor at joint i will drive the link (i).

For the potential energy we assume the gravity is the only source of potential energy. Therefore, the potential energy of the link (i) with respect to the base coordinate frame is:

$$V_i = -m_i\, ^0\mathbf{g} \cdot\, ^0\mathbf{r}_i = -m_i\, ^0\mathbf{g}^T\, ^0T_i\, ^i\mathbf{r}_i \quad (11.478)$$

where $^0\mathbf{g} = \begin{bmatrix} g_x & g_y & g_z & 0 \end{bmatrix}^T$ is the gravitational acceleration, usually in the direction $-z_0$, and $^0\mathbf{r}_i$ is the position vector of C of link (i) in the base frame. The potential energy of the whole robot is then equal to:

$$V = \sum_{i=1}^{n} V_i = -\sum_{i=1}^{n} m_i \mathbf{g}^T\, ^0T_i\, ^i\mathbf{r}_i \quad (11.479)$$

The Lagrangian of a robot is found by substituting (11.476) and (11.479) in the Lagrange equation (11.214).

$$\mathcal{L} = K - V = \frac{1}{2} \sum_{i=1}^{n} \sum_{j=1}^{i} \sum_{k=1}^{i} \text{tr} \left(\frac{\partial\, {}^0T_i}{\partial q_j}\, {}^i\bar{I}_i\, \frac{\partial\, {}^0T_i}{\partial q_k}^T \right) \dot{q}_j\, \dot{q}_k$$

$$+ \sum_{i=1}^{n} m_i\, {}^0\mathbf{g}^T\, {}^0T_i\, {}^i\mathbf{r}_i \tag{11.480}$$

The dynamic equations of motion of a robot can now be found by applying the Lagrange equations (11.215) to Eq. (11.480). We develop the equations of motion term by term. Differentiating the $\mathcal{L}$ with respect to $\dot{q}_r$ is:

$$\frac{\partial\mathcal{L}}{\partial\dot{q}_r} = \frac{1}{2} \sum_{i=1}^{n} \sum_{k=1}^{i} \text{tr} \left(\frac{\partial\, {}^0T_i}{\partial q_r}\, {}^i\bar{I}_i\, \frac{\partial\, {}^0T_i}{\partial q_k}^T \right) \dot{q}_k$$

$$+ \frac{1}{2} \sum_{i=1}^{n} \sum_{j=1}^{i} \text{tr} \left(\frac{\partial\, {}^0T_i}{\partial q_j}\, {}^i\bar{I}_i\, \frac{\partial\, {}^0T_i}{\partial q_r}^T \right) \dot{q}_j$$

$$= \sum_{i=r}^{n} \sum_{j=1}^{i} \text{tr} \left(\frac{\partial\, {}^0T_i}{\partial q_j}\, {}^i\bar{I}_i\, \frac{\partial\, {}^0T_i}{\partial q_r}^T \right) \dot{q}_j \tag{11.481}$$

because

$$\frac{\partial\, {}^0T_i}{\partial q_r} = 0 \quad \text{for } r > i \tag{11.482}$$

and

$$\text{tr} \left(\frac{\partial\, {}^0T_i}{\partial q_j}\, {}^i\bar{I}_i\, \frac{\partial\, {}^0T_i}{\partial q_k}^T \right) = \text{tr} \left(\frac{\partial\, {}^0T_i}{\partial q_k}\, {}^i\bar{I}_i\, \frac{\partial\, {}^0T_i}{\partial q_j}^T \right) \tag{11.483}$$

Time derivative of $\partial\mathcal{L}/\partial\dot{q}_r$ is:

$$\frac{d}{dt}\frac{\partial\mathcal{L}}{\partial\dot{q}_r} = \sum_{i=r}^{n} \sum_{j=1}^{i} \text{tr} \left(\frac{\partial\, {}^0T_i}{\partial q_j}\, {}^i\bar{I}_i\, \frac{\partial\, {}^0T_i}{\partial q_r}^T \right) \ddot{q}_j$$

$$+ \sum_{i=r}^{n} \sum_{j=1}^{i} \sum_{k=1}^{i} \text{tr} \left(\frac{\partial^2\, {}^0T_i}{\partial q_j \partial q_k}\, {}^i\bar{I}_i\, \frac{\partial\, {}^0T_i}{\partial q_r}^T \right) \dot{q}_j \dot{q}_k$$

$$+ \sum_{i=r}^{n} \sum_{j=1}^{i} \sum_{k=1}^{i} \text{tr} \left(\frac{\partial^2\, {}^0T_i}{\partial q_r \partial q_k}\, {}^i\bar{I}_i\, \frac{\partial\, {}^0T_i}{\partial q_j}^T \right) \dot{q}_j \dot{q}_k \tag{11.484}$$

The last term of the Lagrange equation is:

$$\frac{\partial\mathcal{L}}{\partial q_r} = \frac{1}{2} \sum_{i=r}^{n} \sum_{j=1}^{i} \sum_{k=1}^{i} \text{tr} \left(\frac{\partial^2\, {}^0T_i}{\partial q_j \partial q_r}\, {}^i\bar{I}_i\, \frac{\partial\, {}^0T_i}{\partial q_k}^T \right) \dot{q}_j\, \dot{q}_k$$

$$+ \frac{1}{2} \sum_{i=r}^{n} \sum_{j=1}^{i} \sum_{k=1}^{i} \text{tr} \left(\frac{\partial^2\, {}^0T_i}{\partial q_k \partial q_r}\, {}^i\bar{I}_i\, \frac{\partial\, {}^0T_i}{\partial q_j}^T \right) \dot{q}_j\, \dot{q}_k$$

$$+ \sum_{i=r}^{n} m_i \mathbf{g}^T\, \frac{\partial\, {}^0T_i}{\partial q_r}\, {}^i\mathbf{r}_i \tag{11.485}$$

which can be simplified to:

$$\frac{\partial \mathcal{L}}{\partial q_r} = \sum_{i=r}^{n} \sum_{j=1}^{i} \sum_{k=1}^{i} \text{tr}\left(\frac{\partial^2 {}^0 T_i}{\partial q_r \partial q_j} \, {}^i \bar{I}_i \, \frac{\partial {}^0 T_i}{\partial q_k}^T \right) \dot{q}_j \dot{q}_k$$

$$+ \sum_{i=r}^{n} m_i \mathbf{g}^T \frac{\partial {}^0 T_i}{\partial q_r} \, {}^i \mathbf{r}_i \tag{11.486}$$

Interestingly, the third term in Eq. (11.484) is equal to the first term in (11.486). So, substituting these equations in the Lagrange equation can be simplified.

$$\frac{d}{dt}\left(\frac{\partial \mathcal{L}}{\partial \dot{q}_i} \right) - \frac{\partial \mathcal{L}}{\partial q_i} = \sum_{j=i}^{n} \sum_{k=1}^{j} \text{tr}\left(\frac{\partial {}^0 T_j}{\partial q_k} \, {}^j \bar{I}_j \, \frac{\partial {}^0 T_j}{\partial q_i}^T \right) \ddot{q}_k$$

$$+ \sum_{j=i}^{n} \sum_{k=1}^{j} \sum_{m=1}^{j} \text{tr}\left(\frac{\partial^2 {}^0 T_j}{\partial q_k \partial q_m} \, {}^j \bar{I}_j \, \frac{\partial {}^0 T_j}{\partial q_i}^T \right) \dot{q}_k \dot{q}_m$$

$$- \sum_{j=i}^{n} m_j \mathbf{g}^T \frac{\partial {}^0 T_j}{\partial q_i} \, {}^j \mathbf{r}_j \tag{11.487}$$

Finally, the equations of motion for an n link robot are:

$$Q_i = \sum_{j=i}^{n} \sum_{k=1}^{j} \text{tr}\left(\frac{\partial {}^0 T_j}{\partial q_k} \, {}^j \bar{I}_j \, \frac{\partial {}^0 T_j}{\partial q_i}^T \right) \ddot{q}_k$$

$$+ \sum_{j=i}^{n} \sum_{k=1}^{j} \sum_{m=1}^{j} \text{tr}\left(\frac{\partial^2 {}^0 T_j}{\partial q_k \partial q_m} \, {}^j \bar{I}_j \, \frac{\partial {}^0 T_j}{\partial q_i}^T \right) \dot{q}_k \dot{q}_m$$

$$- \sum_{j=i}^{n} m_j \mathbf{g}^T \frac{\partial {}^0 T_j}{\partial q_i} \, {}^j \mathbf{r}_j. \tag{11.488}$$

The equations of motion can be written in a more concise form

$$Q_i = \sum_{j=1}^{n} D_{ij} \ddot{q}_j + \sum_{j=1}^{n} \sum_{k=1}^{n} H_{ijk} \dot{q}_j \dot{q}_k + G_i \tag{11.489}$$

where

$$D_{ij} = \sum_{r=\max i,j}^{n} \text{tr}\left(\frac{\partial {}^0 T_r}{\partial q_j} \, {}^r \bar{I}_r \, \frac{\partial {}^0 T_r}{\partial q_i}^T \right) \tag{11.490}$$

$$H_{ijk} = \sum_{r=\max i,j,k}^{n} \text{tr}\left(\frac{\partial^2 {}^0 T_r}{\partial q_j \partial q_k} \, {}^r \bar{I}_r \, \frac{\partial {}^0 T_r}{\partial q_i}^T \right) \tag{11.491}$$

$$G_i = - \sum_{r=i}^{n} m_r \mathbf{g}^T \frac{\partial {}^0 T_r}{\partial q_i} \, {}^r \mathbf{r}_r \tag{11.492}$$

■

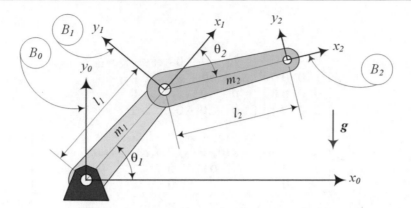

Fig. 11.23 A $2R$ planar manipulator with massive links

Example 376 $2R$ manipulator with massive links. It is to derive matrix form of equations of motion by employing homogenous transformation matrices for $2R$ manipulator.

A $2R$ planar manipulator with massive links is shown in Fig. 11.23. We assume the mass center C of each link is in the middle of the link and the motors at each joint is massless. The links' homogenous transformation matrices are:

$$
^0T_1 = \begin{bmatrix} \cos\theta_1 & -\sin\theta_1 & 0 & l_1\cos\theta_1 \\ \sin\theta_1 & \cos\theta_1 & 0 & l_1\sin\theta_1 \\ 0 & 0 & 1 & 0 \\ 0 & 0 & 0 & 1 \end{bmatrix}
\tag{11.493}
$$

$$
^1T_2 = \begin{bmatrix} \cos\theta_2 & -\sin\theta_2 & 0 & l_2\cos\theta_2 \\ \sin\theta_2 & \cos\theta_2 & 0 & l_2\sin\theta_2 \\ 0 & 0 & 1 & 0 \\ 0 & 0 & 0 & 1 \end{bmatrix}
\tag{11.494}
$$

$$
^0T_2 = {}^0T_1\,{}^1T_2
\tag{11.495}
$$

$$
= \begin{bmatrix} c\,(\theta_1+\theta_2) & -s\,(\theta_1+\theta_2) & 0 & l_1c\theta_1 + l_2c\,(\theta_1+\theta_2) \\ s\,(\theta_1+\theta_2) & c\,(\theta_1+\theta_2) & 0 & l_1s\theta_1 + l_2s\,(\theta_1+\theta_2) \\ 0 & 0 & 1 & 0 \\ 0 & 0 & 0 & 1 \end{bmatrix}
$$

Employing the velocity coefficient matrix Δ_R for revolute joints, we can write:

$$
\frac{\partial\,^0T_1}{\partial\theta_1} = \Delta_R\,^0T_1
\tag{11.496}
$$

$$
= \begin{bmatrix} 0 & -1 & 0 & 0 \\ 1 & 0 & 0 & 0 \\ 0 & 0 & 0 & 0 \\ 0 & 0 & 0 & 0 \end{bmatrix} \begin{bmatrix} \cos\theta_1 & -\sin\theta_1 & 0 & l_1\cos\theta_1 \\ \sin\theta_1 & \cos\theta_1 & 0 & l_1\sin\theta_1 \\ 0 & 0 & 1 & 0 \\ 0 & 0 & 0 & 1 \end{bmatrix}
$$

$$
= \begin{bmatrix} -\sin\theta_1 & -\cos\theta_1 & 0 & -l_1\sin\theta_1 \\ \cos\theta_1 & -\sin\theta_1 & 0 & l_1\cos\theta_1 \\ 0 & 0 & 0 & 0 \\ 0 & 0 & 0 & 0 \end{bmatrix}
$$

$$\frac{\partial\,^0T_2}{\partial\theta_1} = \Delta_R\,^0T_2 \tag{11.497}$$

$$= \begin{bmatrix} 0 & -1 & 0 & 0 \\ 1 & 0 & 0 & 0 \\ 0 & 0 & 0 & 0 \\ 0 & 0 & 0 & 0 \end{bmatrix} \begin{bmatrix} c\theta_{12} & -s\theta_{12} & 0 & l_1c\theta_1 + l_2c\theta_{12} \\ s\theta_{12} & c\theta_{12} & 0 & l_1s\theta_1 + l_2s\theta_{12} \\ 0 & 0 & 1 & 0 \\ 0 & 0 & 0 & 1 \end{bmatrix}$$

$$= \begin{bmatrix} -s\,(\theta_1 + \theta_2) & -c\,(\theta_1 + \theta_2) & 0 & -l_1s\theta_1 - l_2s\,(\theta_1 + \theta_2) \\ c\,(\theta_1 + \theta_2) & -s\,(\theta_1 + \theta_2) & 0 & l_1c\theta_1 + l_2c\,(\theta_1 + \theta_2) \\ 0 & 0 & 0 & 0 \\ 0 & 0 & 0 & 0 \end{bmatrix}$$

$$\frac{\partial\,^0T_2}{\partial\theta_2} = {}^0T_1\,\Delta_R\,{}^1T_2 \tag{11.498}$$

$$= \begin{bmatrix} -s\,(\theta_1 + \theta_2) & -c\,(\theta_1 + \theta_2) & 0 & -l_2s\,(\theta_1 + \theta_2) \\ c\,(\theta_1 + \theta_2) & -s\,(\theta_1 + \theta_2) & 0 & l_2c\,(\theta_1 + \theta_2) \\ 0 & 0 & 0 & 0 \\ 0 & 0 & 0 & 0 \end{bmatrix}$$

Assuming all the product of inertias are zero, we have:

$$^1\bar{I}_1 = \begin{bmatrix} \frac{1}{3}m_1l_1^2 & 0 & 0 & -\frac{1}{2}m_1l_1 \\ 0 & 0 & 0 & 0 \\ 0 & 0 & 0 & 0 \\ -\frac{1}{2}m_1l_1 & 0 & 0 & m_1 \end{bmatrix} \tag{11.499}$$

$$^2\bar{I}_2 = \begin{bmatrix} \frac{1}{3}m_2l_2^2 & 0 & 0 & -\frac{1}{2}m_2l_2 \\ 0 & 0 & 0 & 0 \\ 0 & 0 & 0 & 0 \\ -\frac{1}{2}m_2l_2 & 0 & 0 & m_2 \end{bmatrix} \tag{11.500}$$

Using inertia and derivative of transformation matrices we can calculate the inertial-type symmetric matrix $\mathbf{D}(\mathbf{q})$.

$$D_{11} = \mathrm{tr}\left(\frac{\partial\,^0T_1}{\partial q_1}\,{}^1\bar{I}_1\,\frac{\partial\,^0T_1}{\partial q_1}^T\right) + \mathrm{tr}\left(\frac{\partial\,^0T_2}{\partial q_1}\,{}^2\bar{I}_2\,\frac{\partial\,^0T_2}{\partial q_1}^T\right)$$

$$= \frac{1}{3}m_1l_1^2 + m_2\left(l_1^2 + \frac{1}{3}l_2^2\right) + m_2l_1l_2\cos\theta_2 \tag{11.501}$$

$$D_{12} = D_{21} = \mathrm{tr}\left(\frac{\partial\,^0T_2}{\partial q_1}\,{}^2\bar{I}_2\,\frac{\partial\,^0T_2}{\partial q_1}^T\right)$$

$$= \frac{1}{3}m_2l_2^2 + m_2l_1^2 + m_2l_1l_2\cos\theta_2 \tag{11.502}$$

$$D_{22} = \mathrm{tr}\left(\frac{\partial\,^0T_2}{\partial q_2}\,{}^2\bar{I}_2\,\frac{\partial\,^0T_2}{\partial q_2}^T\right) = \frac{1}{3}l_2^2m_2 \tag{11.503}$$

The coupling terms $\mathbf{H}(\mathbf{q}, \dot{q})$ are calculated as below

$$
\begin{aligned}
H_1 &= \sum_{k=1}^{2} \sum_{m=1}^{2} H_{1km} \dot{q}_k \dot{q}_m \\
&= H_{111} \dot{q}_1 \dot{q}_1 + H_{112} \dot{q}_1 \dot{q}_2 + H_{121} \dot{q}_2 \dot{q}_1 + H_{122} \dot{q}_2 \dot{q}_2
\end{aligned}
\tag{11.504}
$$

$$
\begin{aligned}
H_2 &= \sum_{k=1}^{2} \sum_{m=1}^{2} H_{2km} \dot{q}_k \dot{q}_m \\
&= H_{211} \dot{q}_1 \dot{q}_1 + H_{212} \dot{q}_1 \dot{q}_2 + H_{221} \dot{q}_2 \dot{q}_1 + H_{222} \dot{q}_2 \dot{q}_2
\end{aligned}
\tag{11.505}
$$

where

$$
H_{ijk} = \sum_{r=\max i,j,k}^{n} \mathrm{tr}\left(\frac{\partial^2 {}^0 T_r}{\partial q_j \partial q_k} {}^r \bar{I}_r \frac{\partial {}^0 T_r}{\partial q_i}^T \right)
\tag{11.506}
$$

These calculations yields:

$$
\mathbf{H} = \begin{bmatrix} -\frac{1}{2} m_2 l_1 l_2 \dot{\theta}_2^2 \sin\theta_2 - m_2 l_1 l_2 \dot{\theta}_1 \dot{\theta}_2 \sin\theta_2 \\ \frac{1}{2} m_2 l_1 l_2 \dot{\theta}_1^2 \sin\theta_2 \end{bmatrix}
\tag{11.507}
$$

The last terms of equations of motion are the gravitational vector $\mathbf{G}(\mathbf{q})$.

$$
\begin{aligned}
G_1 &= -m_1 \mathbf{g}^T \frac{\partial {}^0 T_1}{\partial q_1} {}^1 \mathbf{r}_1 - m_2 \mathbf{g}^T \frac{\partial {}^0 T_2}{\partial q_1} {}^2 \mathbf{r}_2 \\
&= -m_1 \begin{bmatrix} 0 \\ -g \\ 0 \\ 0 \end{bmatrix}^T \begin{bmatrix} -\sin\theta_1 & -\cos\theta_1 & 0 & -l_1 \sin\theta_1 \\ \cos\theta_1 & -\sin\theta_1 & 0 & l_1 \cos\theta_1 \\ 0 & 0 & 0 & 0 \\ 0 & 0 & 0 & 0 \end{bmatrix} \begin{bmatrix} -\frac{l_1}{2} \\ 0 \\ 0 \\ 1 \end{bmatrix} \\
&\quad -m_2 \begin{bmatrix} 0 \\ -g \\ 0 \\ 0 \end{bmatrix}^T \begin{bmatrix} -s\theta_{12} & -c\theta_{12} & 0 & -l_1 s\theta_1 - l_2 s\theta_{12} \\ c\theta_{12} & -s\theta_{12} & 0 & l_1 c\theta_1 + l_2 c\theta_{12} \\ 0 & 0 & 0 & 0 \\ 0 & 0 & 0 & 0 \end{bmatrix} \begin{bmatrix} -\frac{l_1}{2} \\ 0 \\ 0 \\ 1 \end{bmatrix} \\
&= \frac{1}{2} m_1 g l_1 \cos\theta_1 + \frac{1}{2} m_2 g l_1 \cos(\theta_1 + \theta_2) + m_2 g l_1 \cos\theta_1
\end{aligned}
\tag{11.508}
$$

$$
\begin{aligned}
G_2 &= -m_2 \mathbf{g}^T \frac{\partial {}^0 T_2}{\partial q_2} {}^2 \mathbf{r}_2 \\
&= -m_2 \begin{bmatrix} 0 \\ -g \\ 0 \\ 0 \end{bmatrix}^T \begin{bmatrix} -s\theta_{12} & -c\theta_{12} & 0 & -l_2 s\theta_{12} \\ c\theta_{12} & -s\theta_{12} & 0 & l_2 c\theta_{12} \\ 0 & 0 & 0 & 0 \\ 0 & 0 & 0 & 0 \end{bmatrix} \begin{bmatrix} -\frac{l_1}{2} \\ 0 \\ 0 \\ 1 \end{bmatrix} \\
&= \frac{1}{2} m_2 g l_2 \cos(\theta_1 + \theta_2) .
\end{aligned}
\tag{11.509}
$$

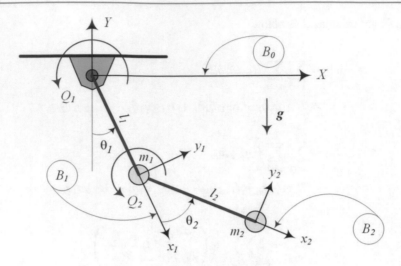

Fig. 11.24 $2R$ manipulator mounted on a ceiling

Finally the equations of motion for the $2R$ planar manipulator are found.

$$
\begin{bmatrix} Q_1 \\ Q_2 \end{bmatrix} =
$$

$$
\begin{bmatrix} \frac{1}{3}m_1 l_1^2 + m_2\left(l_1^2 + \frac{1}{3}l_2^2 + l_1 l_2 c\theta_2\right) & m_2\left(l_1^2 + \frac{1}{3}l_2^2 + l_1 l_2 c\theta_2\right) \\ \left(l_1^2 + \frac{1}{3}l_2^2\right)m_2 + m_2 l_1 l_2 c\theta_2 & \frac{1}{3}l_2^2 m_2 \end{bmatrix} \begin{bmatrix} \ddot{\theta}_1 \\ \ddot{\theta}_2 \end{bmatrix}
$$

$$
+ \begin{bmatrix} -\frac{1}{2}m_2 l_1 l_2 \dot{\theta}_2^2 \sin\theta_2 - m_2 l_1 l_2 \dot{\theta}_1 \dot{\theta}_2 \sin\theta_2 \\ \frac{1}{2}m_2 l_1 l_2 \dot{\theta}_1^2 \sin\theta_2 \end{bmatrix}
$$

$$
+ \begin{bmatrix} \frac{1}{2}m_1 g l_1 \cos\theta_1 + \frac{1}{2}m_2 g l_1 \cos(\theta_1 + \theta_2) + m_2 g l_1 \cos\theta_1 \\ \frac{1}{2}m_2 g l_2 \cos(\theta_1 + \theta_2) \end{bmatrix} \tag{11.510}
$$

Example 377 $2R$ manipulator mounted on ceiling. Here is to derive equations of motion of a ceiling attached ideal $2R$ manipulator by link transformation matrices method. Attaching robots to ceiling is very common in industry.

Figure 11.24 depicts an ideal $2R$ planar manipulator mounted on a ceiling. Ceiling mounting is an applied method in robotic operated assembly lines. The Lagrangian of the manipulator is

$$
\mathcal{L} = K - V = \frac{1}{2}m_1 l_1^2 \dot{\theta}_1^2
$$

$$
+ \frac{1}{2}m_2\left(l_1^2 \dot{\theta}_1^2 + l_2^2\left(\dot{\theta}_1 + \dot{\theta}_2\right)^2 + 2l_1 l_2 \dot{\theta}_1\left(\dot{\theta}_1 + \dot{\theta}_2\right)\cos\theta_2\right)
$$

$$
+ m_1 g l_1 \cos\theta_1 + m_2 g\left(l_1 \cos\theta_1 + l_2 \cos(\theta_1 + \theta_2)\right) \tag{11.511}
$$

which leads to the following equations of motion:

$$
Q_1 = \left((m_1 + m_2)\, l_1^2 + m_2 l_2^2 + 2m_2 l_1 l_2 \cos\theta_2\right)\ddot{\theta}_1
$$

$$
+ m_2 l_2\left(l_2 + l_1 \cos\theta_2\right)\ddot{\theta}_2
$$

$$
- 2m_2 l_1 l_2 \sin\theta_2 \dot{\theta}_1 \dot{\theta}_2 - m_2 l_1 l_2 \sin\theta_2 \dot{\theta}_2^2
$$

$$
+ (m_1 + m_2)\, g l_1 \sin\theta_1 + m_2 g l_2 \sin(\theta_1 + \theta_2) \tag{11.512}
$$

$$Q_2 = m_2 l_2 \left(l_2 + l_1 \cos \theta_2 \right) \ddot{\theta}_1 + m_2 l_2^2 \ddot{\theta}_2$$
$$-2 m_2 l_1 l_2 \sin \theta_2 \dot{\theta}_1 \left(\dot{\theta}_1 + \dot{\theta}_2 \right) - m_2 g l_2 \sin \left(\theta_1 + \theta_2 \right) . \tag{11.513}$$

The equations of motion can be rearranged to

$$Q_1 = D_{11} \ddot{\theta}_1 + D_{12} \ddot{\theta}_2 + H_{111} \dot{\theta}_1^2 + H_{122} \dot{\theta}_2^2$$
$$+ H_{112} \dot{\theta}_1 \dot{\theta}_2 + H_{121} \dot{\theta}_2 \dot{\theta}_1 + G_1 \tag{11.514}$$

$$Q_2 = D_{21} \ddot{\theta}_1 + D_{22} \ddot{\theta}_2 + H_{211} \dot{\theta}_1^2 + H_{222} \dot{\theta}_2^2$$
$$+ H_{212} \dot{\theta}_1 \dot{\theta}_2 + H_{221} \dot{\theta}_2 \dot{\theta}_1 + G_2 \tag{11.515}$$

where

$$D_{11} = (m_1 + m_2) l_1^2 + m_2 l_2^2 + 2 m_2 l_1 l_2 \cos \theta_2 \tag{11.516}$$
$$D_{12} = m_2 l_2 \left(l_2 + l_1 \cos \theta_2 \right) \tag{11.517}$$
$$D_{21} = D_{12} = m_2 l_2 \left(l_2 + l_1 \cos \theta_2 \right) \tag{11.518}$$
$$D_{22} = m_2 l_2^2 \tag{11.519}$$

$$H_{111} = 0 \tag{11.520}$$
$$H_{122} = -m_2 l_1 l_2 \sin \theta_2 \tag{11.521}$$
$$H_{211} = -m_2 l_1 l_2 \sin \theta_2 \tag{11.522}$$

$$H_{222} = 0 \tag{11.523}$$
$$H_{112} = H_{121} = -m_2 l_1 l_2 \sin \theta_2 \tag{11.524}$$
$$H_{212} = H_{221} = -m_2 l_1 l_2 \sin \theta_2 \tag{11.525}$$

$$G_1 = (m_1 + m_2) g l_1 \sin \theta_1 + m_2 g l_2 \sin \left(\theta_1 + \theta_2 \right) \tag{11.526}$$
$$G_2 = m_2 g l_2 \sin \left(\theta_1 + \theta_2 \right) \tag{11.527}$$

11.5 Robot Statics

At the beginning and at the end of a rest-to-rest mission, a robot must keep its specified configurations. To hold the position and orientation, the actuators must apply some required forces to balance the external and gravity loads applied to the robot. Calculating the required actuators' force to hold a robot in a specific configuration is called *robot statics analysis*.

In a static condition, the globally expressed Newton–Euler equations for the link (i) can be written in a recursive form,

$${}^0\mathbf{F}_{i-1} = {}^0\mathbf{F}_i - \sum {}^0\mathbf{F}_{e_i} \tag{11.528}$$

$${}^0\mathbf{M}_{i-1} = {}^0\mathbf{M}_i - \sum {}^0\mathbf{M}_{e_i} + {}_{i-1}^{0}\mathbf{d}_i \times {}^0\mathbf{F}_i \tag{11.529}$$

where

$${}_{i-1}^{0}\mathbf{d}_i = {}^0\mathbf{d}_i - {}^0\mathbf{d}_{i-1} \tag{11.530}$$

Therefore, we are able to calculate the action force system $(\mathbf{F}_{i-1}, \mathbf{M}_{i-1})$ when the reaction force system $(-\mathbf{F}_i, -\mathbf{M}_i)$ is given. The position vectors and force systems on link (i) are shown in Fig. 11.25.

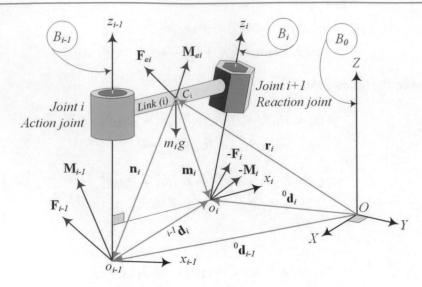

Fig. 11.25   Position vectors and force system on link (i)

Proof In a static condition, the Newton–Euler equations of motion (11.1) and (11.2) for the link (i) reduce to force and moment balance equations.

$$^0\mathbf{F}_{i-1} - {}^0\mathbf{F}_i + \sum {}^0\mathbf{F}_{e_i} = 0 \tag{11.531}$$

$$^0\mathbf{M}_{i-1} - {}^0\mathbf{M}_i + \sum {}^0\mathbf{M}_{e_i} + {}^0\mathbf{n}_i \times {}^0\mathbf{F}_{i-1} - {}^0\mathbf{m}_i \times {}^0\mathbf{F}_i = 0 \tag{11.532}$$

These equations can be rearranged into a *backward recursive* form.

$$^0\mathbf{F}_{i-1} = {}^0\mathbf{F}_i - \sum {}^0\mathbf{F}_{e_i} \tag{11.533}$$

$$^0\mathbf{M}_{i-1} = {}^0\mathbf{M}_i - \sum {}^0\mathbf{M}_{e_i} - {}^0\mathbf{n}_i \times {}^0\mathbf{F}_{i-1} + {}^0\mathbf{m}_i \times {}^0\mathbf{F}_i \tag{11.534}$$

However, we may also transform the Euler equation from C_i to O_{i-1} and find Eq. (11.529).

Practically, we measure the position of mass center $\mathbf{r}_i$ and the relative position of B_i and B_{i-1} in the coordinate frame B_i attached to the link (i). Hence, we must transform ${}^i\mathbf{r}_i$ and ${}_{i-1}{}^i\mathbf{d}_i$ to the base frame.

$$^0\mathbf{r}_i = {}^0T_i \, {}^i\mathbf{r}_i \tag{11.535}$$

$$_{i-1}{}^0\mathbf{d}_i = {}^0T_i \, _{i-1}{}^i\mathbf{d}_i \tag{11.536}$$

The external load is usually the gravitational force $m_i\mathbf{g}$.

$$\sum {}^0\mathbf{F}_{e_i} = m_i \, {}^0\mathbf{g} \tag{11.537}$$

$$\sum {}^0\mathbf{M}_{e_i} = {}^0\mathbf{r}_i \times m_i \, {}^0\mathbf{g} \tag{11.538}$$

Using the DH parameters, we may express the relative position vector $_{i-1}{}^i\mathbf{d}_i$.

$$_{i-1}{}^i\mathbf{d}_i = \begin{bmatrix} a_i \\ d_i \sin \alpha_i \\ d_i \cos \alpha_i \\ 1 \end{bmatrix} \tag{11.539}$$

The backward recursive equations (11.528) and (11.529) allow us to start with a known force system $(\mathbf{F}_n, \mathbf{M}_n)$ at B_n, applied from the end-effector to the environment, and calculate the force system at B_{n-1}.

$$^0\mathbf{F}_{n-1} = {}^0\mathbf{F}_n - \sum {}^0\mathbf{F}_{e_n} \tag{11.540}$$

$$^0\mathbf{M}_{n-1} = {}^0\mathbf{M}_n - \sum {}^0\mathbf{M}_{e_n} + {}_{n-1}^{0}\mathbf{d}_n \times {}^0\mathbf{F}_n \tag{11.541}$$

Following the same procedure and calculating force system at proximal end by having the force system at distal end of each link ends up to the force system at the base. In this procedure, the force system applied by the end-effector to the environment is assumed to be known.

It is also possible to rearrange the static equations (11.531) and (11.532) into a *forward recursive* form.

$$^0\mathbf{F}_i = {}^0\mathbf{F}_{i-1} + \sum {}^0\mathbf{F}_{e_i} \tag{11.542}$$

$$^0\mathbf{M}_i = {}^0\mathbf{M}_{i-1} + \sum {}^0\mathbf{M}_{e_i} + {}^0\mathbf{n}_i \times {}^0\mathbf{F}_{i-1} - {}^0\mathbf{m}_i \times {}^0\mathbf{F}_i \tag{11.543}$$

Transforming the Euler equation from C_i to O_i simplifies the forward recursive equations into the more practical equations.

$$^0\mathbf{F}_i = {}^0\mathbf{F}_{i-1} + \sum {}^0\mathbf{F}_{e_i} \tag{11.544}$$

$$^0\mathbf{M}_i = {}^0\mathbf{M}_{i-1} + \sum {}^0\mathbf{M}_{e_i} - {}_{i-1}^{0}\mathbf{d}_i \times {}^0\mathbf{F}_{i-1} \tag{11.545}$$

Using the forward recursive equations (11.544) and (11.545) we can start with a known force system $(\mathbf{F}_0, \mathbf{M}_0)$ at B_0, applied from the base to the link (1), and calculate the force system at B_1.

$$^0\mathbf{F}_1 = {}^0\mathbf{F}_0 + \sum {}^0\mathbf{F}_{e_1} \tag{11.546}$$

$$^0\mathbf{M}_1 = {}^0\mathbf{M}_0 + \sum {}^0\mathbf{M}_{e_1} - {}^0\mathbf{d}_1 \times {}^0\mathbf{F}_0 \tag{11.547}$$

Following this procedure and calculating force system at the distal end by having the force system at the proximal end of each link ends up at the force system applied to the environment by the end-effector. In this procedure, the force system applied by the base actuators to the first link is assumed to be known. ∎

Example 378 Statics of a $4R$ planar manipulator. Calculating static torques at four joint motors of a $4R$ planar manipulator is a practical example of how to analyze static configuration of robots.

Figure 11.26 illustrates a $4R$ planar manipulator with the DH coordinate frames set up for each link. Assume the end-effector force system applied to the environment is known.

$$^4\mathbf{F}_4 = \begin{bmatrix} F_x \\ F_y \\ 0 \end{bmatrix} \qquad {}^4\mathbf{M}_4 = \begin{bmatrix} 0 \\ 0 \\ M_z \end{bmatrix} \tag{11.548}$$

In addition, we assume the links to be uniform such that their C are located at the midpoint of each link, and the gravitational acceleration is:

$$\mathbf{g} = -g\,\hat{\jmath}_0 \tag{11.549}$$

The manipulator consists of four R‖R(0) links, therefore their transformation matrices $^{i-1}T_i$ are of class (5.32) that because $d_i = 0$ and $a_i = l_i$, simplifies to:

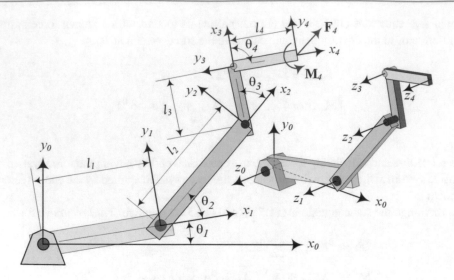

Fig. 11.26 A $4R$ planar manipulator

$$
{}^{i-1}T_i = \begin{bmatrix} \cos\theta_i & -\sin\theta_i & 0 & l_i \cos\theta_i \\ \sin\theta_i & \cos\theta_i & 0 & l_i \sin\theta_i \\ 0 & 0 & 1 & 0 \\ 0 & 0 & 0 & 1 \end{bmatrix} \tag{11.550}
$$

The C position vectors ${}^i\mathbf{r}_i$ and the relative position vectors ${}_{i-1}^{\ 0}\mathbf{d}_i$ are:

$$
{}^i\mathbf{r}_i = \begin{bmatrix} l_i/2 \\ 0 \\ 0 \\ 0 \end{bmatrix} \qquad {}_{i-1}^{\ i}\mathbf{d}_i = \begin{bmatrix} l_i \\ 0 \\ 0 \\ 0 \end{bmatrix} \tag{11.551}
$$

and therefore,

$$
{}^0\mathbf{r}_i = {}^0T_i\,{}^i\mathbf{r}_i \qquad {}_{i-1}^{\ 0}\mathbf{d}_i = {}^0T_i\,{}_{i-1}^{\ i}\mathbf{d}_i \tag{11.552}
$$

where

$$
{}^0T_i = {}^0T_1 \cdots {}^{i-1}T_i \tag{11.553}
$$

The static force at joints 3, 2, and 1 are:

$$
{}^0\mathbf{F}_3 = {}^0\mathbf{F}_4 - \sum {}^0\mathbf{F}_{e_4} = {}^0\mathbf{F}_4 + m_4 g\,{}^0\hat{j}_0 \tag{11.554}
$$

$$
= \begin{bmatrix} F_x \\ F_y \\ 0 \\ 0 \end{bmatrix} + m_4 g \begin{bmatrix} 0 \\ 1 \\ 0 \\ 0 \end{bmatrix} = \begin{bmatrix} F_x \\ F_y + m_4 g \\ 0 \\ 0 \end{bmatrix}
$$

$$
{}^0\mathbf{F}_2 = {}^0\mathbf{F}_3 - m_3 g\,{}^0\hat{j}_0 \tag{11.555}
$$

$$
= \begin{bmatrix} F_x \\ F_y + m_4 g \\ 0 \\ 0 \end{bmatrix} + m_3 g \begin{bmatrix} 0 \\ 1 \\ 0 \\ 0 \end{bmatrix} = \begin{bmatrix} F_x \\ F_y + (m_3 + m_4)\, g \\ 0 \\ 0 \end{bmatrix}
$$

$$^0\mathbf{F}_1 = {}^0\mathbf{F}_2 - m_2 g \, {}^0\hat{j}_0$$

$$= \begin{bmatrix} F_x \\ F_y + (m_3 + m_4) \, g \\ 0 \\ 0 \end{bmatrix} + m_2 g \begin{bmatrix} 0 \\ 1 \\ 0 \\ 0 \end{bmatrix}$$

$$= \begin{bmatrix} F_x \\ F_y + g \, (m_2 + m_3 + m_4) \\ 0 \\ 0 \end{bmatrix} \tag{11.556}$$

$$^0\mathbf{F}_0 = {}^0\mathbf{F}_1 - m_1 g \, {}^0\hat{j}_0$$

$$= \begin{bmatrix} F_x \\ F_y + g \, (m_2 + m_3 + m_4) \\ 0 \\ 0 \end{bmatrix} + m_1 g \begin{bmatrix} 0 \\ 1 \\ 0 \\ 0 \end{bmatrix}$$

$$= \begin{bmatrix} F_x \\ F_y + g \, (m_1 + m_2 + m_3 + m_4) \\ 0 \\ 0 \end{bmatrix} \tag{11.557}$$

The static moment at joints 3, 2, and 1 are:

$$^0\mathbf{M}_3 = {}^0\mathbf{M}_4 - \sum {}^0\mathbf{M}_{e_i} + {}^0_3\mathbf{d}_4 \times {}^0\mathbf{F}_4$$

$$= {}^0\mathbf{M}_4 + {}^0\mathbf{r}_4 \times m_4 g \, {}^0\hat{j}_0 + {}^0_3\mathbf{d}_4 \times {}^0\mathbf{F}_4$$

$$= {}^0\mathbf{M}_4 + {}^0\mathbf{r}_4 \times m_4 g \, {}^0\hat{j}_0 + {}^0_3\mathbf{d}_4 \times {}^0\mathbf{F}_4$$

$$= {}^0\mathbf{M}_4 + m_4 g \, \left({}^0T_4 \, {}^4\mathbf{r}_4 \times {}^0\hat{j}_0\right) + \left({}^0T_4 \, {}^4_3\mathbf{d}_4\right) \times {}^0\mathbf{F}_4$$

$$= \begin{bmatrix} 0 \\ 0 \\ M_{3z} \\ 0 \end{bmatrix} \tag{11.558}$$

$$M_{3z} = M_z + l_4 F_y \cos \theta_{1234} - l_4 F_x \sin \theta_{1234} + \frac{1}{2} g l_4 m_4 \cos \theta_{1234} \tag{11.559}$$

$$^0\mathbf{M}_2 = {}^0\mathbf{M}_3 + m_3 g \, \left({}^0T_3 \, {}^3\mathbf{r}_3 \times {}^0\hat{j}_0\right) + \left({}^0T_3 \, {}^3_2\mathbf{d}_3\right) \times {}^0\mathbf{F}_3$$

$$= \begin{bmatrix} 0 \\ 0 \\ M_{2z} \\ 0 \end{bmatrix} \tag{11.560}$$

$$M_{2z} = M_z + l_4 F_y \cos \theta_{1234} - l_4 F_x \sin \theta_{1234} + \frac{1}{2} g l_4 m_4 \cos \theta_{1234}$$

$$+ \frac{1}{2} g l_3 m_3 \cos \theta_{123} - l_3 F_x \sin \theta_{123}$$

$$+ l_3 \left(F_y + g m_4\right) \cos \theta_{123} \tag{11.561}$$

$$^0\mathbf{M}_1 = {}^0\mathbf{M}_2 + m_2 g \left({}^0T_2 \, {}^2\mathbf{r}_2 \times {}^0\hat{\jmath}_0\right) + \left({}^0T_2 \, {}^2_1\mathbf{d}_2\right) \times {}^0\mathbf{F}_2$$

$$= \begin{bmatrix} 0 \\ 0 \\ M_{1z} \\ 0 \end{bmatrix} \tag{11.562}$$

$$M_{1z} = M_z + l_4 F_y \cos\theta_{1234} - l_4 F_x \sin\theta_{1234} + \frac{1}{2} g l_4 m_4 \cos\theta_{1234}$$

$$+ \frac{1}{2} g l_3 m_3 \cos\theta_{123} - l_3 F_x \sin\theta_{123} + l_3 \cos\theta_{123} \left(F_y + g m_4\right)$$

$$+ \frac{1}{2} g l_2 m_2 \cos\theta_{12} - l_2 F_x \sin\theta_{12}$$

$$+ l_2 \cos\theta_{12} \left(F_y + g \left(m_3 + m_4\right)\right) \tag{11.563}$$

$$^0\mathbf{M}_0 = {}^0\mathbf{M}_1 + m_1 g \left({}^0T_1 \, {}^1\mathbf{r}_1 \times {}^0\hat{\jmath}_0\right) + \left({}^0T_1 \, {}^1_0\mathbf{d}_1\right) \times {}^0\mathbf{F}_1$$

$$= \begin{bmatrix} 0 \\ 0 \\ M_{0z} \\ 0 \end{bmatrix} \tag{11.564}$$

$$M_{0z} = M_z + l_4 F_y \cos\theta_{1234} - l_4 F_x \sin\theta_{1234} + \frac{1}{2} g l_4 m_4 \cos\theta_{1234}$$

$$+ \frac{1}{2} g l_3 m_3 \cos\theta_{123} - l_3 F_x \sin\theta_{123} + l_3 \cos\theta_{123} \left(F_y + g m_4\right)$$

$$- l_2 F_x \sin\theta_{12} + \frac{1}{2} g l_2 m_2 \cos\theta_{12} + l_2 \cos\theta_{12} \left(F_y + g \left(m_3 + m_4\right)\right)$$

$$+ \frac{1}{2} g l_1 m_1 \cos\theta_1 - l_1 F_x \sin\theta_1$$

$$+ l_1 \cos\theta_1 \left(F_y + g \left(m_2 + m_3 + m_4\right)\right) \tag{11.565}$$

where

$$\theta_{1234} = \theta_1 + \theta_2 + \theta_3 + \theta_4 \tag{11.566}$$

$$\theta_{123} = \theta_1 + \theta_2 + \theta_3 \tag{11.567}$$

$$\theta_{12} = \theta_1 + \theta_2 \tag{11.568}$$

Example 379 Recursive force equation in link's frame. The static equations can also be expressed in body coordinate frame, which may be the preferred form in computer calculation.

Practically, it is easier to measure and calculate the force systems in the kink's frame. Therefore, we may rewrite the backward recursive Equations (11.528) and (11.529) in the following form and calculate the proximal force system from the distal force system in the link's frame.

$$^i\mathbf{F}_{i-1} = {}^i\mathbf{F}_i - \sum {}^i\mathbf{F}_{e_i} \tag{11.569}$$

$$^i\mathbf{M}_{i-1} = {}^i\mathbf{M}_i - \sum {}^i\mathbf{M}_{e_i} + {}^i_{i-1}\mathbf{d}_i \times {}^i\mathbf{F}_i \tag{11.570}$$

The calculated force system, then, may be transformed to the previous link's coordinate frame by a transformation,

$$^{i-1}\mathbf{F}_{i-1} = {}^{i-1}T_i \, {}^i\mathbf{F}_{i-1} \tag{11.571}$$

$$^{i-1}\mathbf{M}_{i-1} = {}^{i-1}T_i \, {}^i\mathbf{M}_{i-1} \tag{11.572}$$

or they may be transformed to any other coordinate frame including the base frame.

$$^{0}\mathbf{F}_{i-1} = {}^{0}T_{i} \, {}^{i}\mathbf{F}_{i-1} \tag{11.573}$$

$$^{0}\mathbf{M}_{i-1} = {}^{0}T_{i} \, {}^{i}\mathbf{M}_{i-1} \tag{11.574}$$

Example 380 Actuator's force and torque. All actuators apply their actuation force or torque only in one dimensional. Here is a discussion on static force at joint and the component that will be supported by actuators.

Applying a backward or forward recursive static force analysis ends up with a set of known force systems at joints. Each joint is driven by a motor or generally an actuator that applies a force in a P joint, or a torque in an R joint. When the joint i is prismatic, the actuator force is applied along the axis of the joint i. Therefore, the force of the driving motor is along the z_{i-1}-axis,

$$F_m = {}^{0}\hat{k}_{i-1}^{T} \, {}^{0}\mathbf{F}_i \tag{11.575}$$

showing that the $\hat{k}_{i-1}$ component of the joint force $\mathbf{F}_i$ is supported by the actuator, while the $\hat{\imath}_{i-1}$ and $\hat{\jmath}_{i-1}$ components of $\mathbf{F}_i$ must be supported by the bearings of the joint.

Similarly, when joint i is revolute, the actuator torque is applied about the axis of joint i. Therefore, the torque of the driving motor is along the z_{i-1}-axis

$$M_m = {}^{0}\hat{k}_{i-1}^{T} \, {}^{0}\mathbf{M}_i \tag{11.576}$$

showing that the $\hat{k}_{i-1}$ component of the joint torque $\mathbf{M}_i$ is supported by the actuator, while the $\hat{\imath}_{i-1}$ and $\hat{\jmath}_{i-1}$ components of $\mathbf{M}_i$ must be supported by the bearings of the joint.

11.6 Summary

Dynamics equations of motion for a robot can be found by both Newton–Euler and Lagrange methods. In the Newton–Euler method, each link (i) is a rigid body and therefore, its translational and rotational equations of motion in the base coordinate frame are:

$$m_i \, {}^{0}\mathbf{a}_i = {}^{0}\mathbf{F}_{i-1} - {}^{0}\mathbf{F}_i + \sum {}^{0}\mathbf{F}_{e_i} \tag{11.577}$$

$$^{0}I_i \, {}_{0}\boldsymbol{\alpha}_i = {}^{0}\mathbf{M}_{i-1} - {}^{0}\mathbf{M}_i + \sum {}^{0}\mathbf{M}_{e_i}$$
$$+ \left({}^{0}\mathbf{d}_{i-1} - {}^{0}\mathbf{r}_i\right) \times {}^{0}\mathbf{F}_{i-1} - \left({}^{0}\mathbf{d}_i - {}^{0}\mathbf{r}_i\right) \times {}^{0}\mathbf{F}_i \tag{11.578}$$

The force $\mathbf{F}_{i-1}$ and moment $\mathbf{M}_{i-1}$ are the resultant force and moment that link $(i-1)$ applics to link (i) at joint i. Similarly, $\mathbf{F}_i$ and $\mathbf{M}_i$ are the resultant force and moment that link (i) applies to link $(i+1)$ at joint $i+1$. We measure the force systems $(\mathbf{F}_{i-1}, \mathbf{M}_{i-1})$ and $(\mathbf{F}_i, \mathbf{M}_i)$ at the origin of the coordinate frames B_{i-1} and B_i, respectively. The sum of the external loads acting on the link (i) are $\sum \mathbf{F}_{e_i}$ and $\sum \mathbf{M}_{e_i}$. The vector ${}^{0}\mathbf{r}_i$ is the global position vector of C_i and ${}^{0}\mathbf{d}_i$ is the global position vector of the origin of B_i. The vector ${}^{0}\boldsymbol{\alpha}_i$ is the angular acceleration and ${}^{0}\mathbf{a}_i$ is the translational acceleration of the link (i) measured at the mass center C_i.

$$^{0}\mathbf{a}_i = {}^{0}\ddot{\mathbf{d}}_i + {}_{0}\boldsymbol{\alpha}_i \times \left({}^{0}\mathbf{r}_i - {}^{0}\mathbf{d}_i\right) + {}_{0}\boldsymbol{\omega}_i \times \left({}_{0}\boldsymbol{\omega}_i \times \left({}^{0}\mathbf{r}_i - {}^{0}\mathbf{d}_i\right)\right) \tag{11.579}$$

$$_{0}\boldsymbol{\alpha}_i = \begin{cases} {}_{0}\boldsymbol{\alpha}_{i-1} + \ddot{\theta}_i \, {}^{0}\hat{k}_{i-1} + {}_{0}\boldsymbol{\omega}_{i-1} \times \dot{\theta}_i \, {}^{0}\hat{k}_{i-1} & \text{if joint } i \text{ is R} \\ {}_{0}\boldsymbol{\alpha}_{i-1} & \text{if joint } i \text{ is P} \end{cases} \tag{11.580}$$

Weight is usually the only external load on middle links of a robot, and reactions from the environment are extra external force systems on the base and end-effector links. The force and moment that the base actuator applies to the first link are $\mathbf{F}_0$ and $\mathbf{M}_0$, and the force and moment that the end-effector applies to the environment are $\mathbf{F}_n$ and $\mathbf{M}_n$. If weight is the only external load on link (i) and it is in $-{}^{0}\hat{k}_0$ direction, then we have

$$\sum {}^{0}\mathbf{F}_{e_i} = m_i \, {}^{0}\mathbf{g} = -m_i \, g \, {}^{0}\hat{k}_0 \tag{11.581}$$

$$\sum {}^0\mathbf{M}_{e_i} = {}^0\mathbf{r}_i \times m_i \, {}^0\mathbf{g} = -{}^0\mathbf{r}_i \times m_i \, g \, {}^0\hat{k}_0 \tag{11.582}$$

where $\mathbf{g}$ is the gravitational acceleration vector.

The Newton–Euler equation of motion can also be written in link's coordinate frame in a forward or backward method. The backward Newton–Euler equations of motion for link (i) in the local coordinate frame B_i are

$$ {}^i\mathbf{F}_{i-1} = {}^i\mathbf{F}_i - \sum {}^i\mathbf{F}_{e_i} + m_i \, {}^i_0\mathbf{a}_i \tag{11.583}$$

$$ {}^i\mathbf{M}_{i-1} = {}^i\mathbf{M}_i - \sum {}^i\mathbf{M}_{e_i} - \left({}^i\mathbf{d}_{i-1} - {}^i\mathbf{r}_i\right) \times {}^i\mathbf{F}_{i-1}$$

$$ + \left({}^i\mathbf{d}_i - {}^i\mathbf{r}_i\right) \times {}^i\mathbf{F}_i + {}^iI_i \, {}^i_0\boldsymbol{\alpha}_i + {}^i_0\boldsymbol{\omega}_i \times {}^iI_i \, {}^i_0\boldsymbol{\omega}_i \tag{11.584}$$

where

$$ {}^i\mathbf{n}_i = {}^i\mathbf{d}_{i-1} - {}^i\mathbf{r}_i \tag{11.585}$$

$$ {}^i\mathbf{m}_i = {}^i\mathbf{d}_i - {}^i\mathbf{r}_i \tag{11.586}$$

and

$$ {}^i_0\mathbf{a}_i = {}^i\ddot{\mathbf{d}}_i + {}^i_0\boldsymbol{\alpha}_i \times \left({}^i\mathbf{r}_i - {}^i\mathbf{d}_i\right) + {}^i_0\boldsymbol{\omega}_i \times \left({}^i_0\boldsymbol{\omega}_i \times \left({}^i\mathbf{r}_i - {}^i\mathbf{d}_i\right)\right) \tag{11.587}$$

$$ {}^i_0\boldsymbol{\alpha}_i = \begin{cases} {}^iT_{i-1}\left({}^{i-1}_{\,0}\boldsymbol{\alpha}_{i-1} + \ddot{\theta}_i \, {}^{i-1}\hat{k}_{i-1}\right) \\ + {}^iT_{i-1}\left({}^{i-1}_{\,0}\boldsymbol{\omega}_{i-1} \times \dot{\theta}_i \, {}^{i-1}\hat{k}_{i-1}\right) & \text{if joint } i \text{ is R} \\ {}^iT_{i-1} \, {}^{i-1}_{\,0}\boldsymbol{\alpha}_{i-1} & \text{if joint } i \text{ is P} \end{cases} \tag{11.588}$$

In this method, we search for the driving force system $({}^i\mathbf{F}_{i-1}, {}^i\mathbf{M}_{i-1})$ by having the driven force system $({}^i\mathbf{F}_i, {}^i\mathbf{M}_i)$ and the resultant external force system $({}^i\mathbf{F}_{e_i}, {}^i\mathbf{M}_{e_i})$. When the driving force system $({}^i\mathbf{F}_{i-1}, {}^i\mathbf{M}_{i-1})$ is found in frame B_i, we can transform them to the frame B_{i-1} and apply the Newton–Euler equation for link $(i-1)$.

$$ {}^{i-1}\mathbf{F}_{i-1} = {}^{i-1}T_i \, {}^i\mathbf{F}_{i-1} \tag{11.589}$$

$$ {}^{i-1}\mathbf{M}_{i-1} = {}^{i-1}T_i \, {}^i\mathbf{M}_{i-1} \tag{11.590}$$

The negative of the converted force system acts as the driven force system $(-{}^{i-1}\mathbf{F}_{i-1}, -{}^{i-1}\mathbf{M}_{i-1})$ for the link $(i-1)$.

The forward Newton–Euler equations of motion for link (i) in the local coordinate frame B_i are

$$ {}^i\mathbf{F}_i = {}^i\mathbf{F}_{i-1} + \sum {}^i\mathbf{F}_{e_i} - m_i \, {}^i_0\mathbf{a}_i \tag{11.591}$$

$$ {}^i\mathbf{M}_i = {}^i\mathbf{M}_{i-1} + \sum {}^i\mathbf{M}_{e_i} + \left({}^i\mathbf{d}_{i-1} - {}^i\mathbf{r}_i\right) \times {}^i\mathbf{F}_{i-1}$$

$$ - \left({}^i\mathbf{d}_i - {}^i\mathbf{r}_i\right) \times {}^i\mathbf{F}_i - {}^iI_i \, {}^i_0\boldsymbol{\alpha}_i - {}^i_0\boldsymbol{\omega}_i \times {}^iI_i \, {}^i_0\boldsymbol{\omega}_i. \tag{11.592}$$

$$ {}^i\mathbf{n}_i = {}^i\mathbf{d}_{i-1} - {}^i\mathbf{r}_i \tag{11.593}$$

$$ {}^i\mathbf{m}_i = {}^i\mathbf{d}_i - {}^i\mathbf{r}_i \tag{11.594}$$

Using the forward Newton–Euler equations of motion, we can calculate the reaction force system $({}^i\mathbf{F}_i, {}^i\mathbf{M}_i)$ by having the action force system $({}^i\mathbf{F}_{i-1}, {}^i\mathbf{M}_{i-1})$. When the reaction force system $({}^i\mathbf{F}_i, {}^i\mathbf{M}_i)$ is found in frame B_i, we can transform them to the frame B_{i+1}

$$ {}^{i+1}\mathbf{F}_i = {}^iT_{i+1}^{-1} \, {}^i\mathbf{F}_i \tag{11.595}$$

$$ {}^{i+1}\mathbf{M}_i = {}^iT_{i+1}^{-1} \, {}^i\mathbf{M}_i \tag{11.596}$$

The negative of the converted force system acts as the action force system $(-^{i+1}\mathbf{F}_i, -^{i+1}\mathbf{M}_i)$ for the link $(i + 1)$ and we can apply the Newton–Euler equation to the link $(i + 1)$. The forward Newton–Euler equations of motion allow us to start from a known action force system $(^1\mathbf{F}_0, {}^1\mathbf{M}_0)$, that the base link applies to the link (1), and calculate the action force of the next link. Therefore, analyzing the links of a robot, one by one, we end up with the force system that the end-effector applies to the environment.

The Lagrange equation of motion

$$\frac{d}{dt}\left(\frac{\partial \mathcal{L}}{\partial \dot{q}_i}\right) - \frac{\partial \mathcal{L}}{\partial q_i} = Q_i \quad i = 1, 2, \cdots n \tag{11.597}$$

$$\mathcal{L} = K - V \tag{11.598}$$

provides a systematic approach to obtain the dynamics equations for robots. The variables q_i are the coordinates by which the energies are expressed and the Q_i is the corresponding generalized nonpotential force.

The equations of motion for an n link serial manipulator, based on Newton–Euler or Lagrangian, can always be set in a matrix form

$$\mathbf{D}(\mathbf{q})\,\ddot{q} + \mathbf{H}(\mathbf{q}, \dot{q}) + \mathbf{G}(\mathbf{q}) = \mathbf{Q} \tag{11.599}$$

or

$$\mathbf{D}(\mathbf{q})\,\ddot{q} + \mathbf{C}(\mathbf{q}, \dot{q})\dot{q} + \mathbf{G}(\mathbf{q}) = \mathbf{Q} \tag{11.600}$$

or in a summation form

$$\sum_{j=1}^{n} D_{ij}(q)\,\ddot{q}_j + \sum_{k=1}^{n}\sum_{m=1}^{n} H_{ikm}\dot{q}_k\dot{q}_m + G_i = Q_i \tag{11.601}$$

where $\mathbf{D}(\mathbf{q})$ is an $n \times n$ inertial-type symmetric matrix

$$D = \sum_{i=1}^{n}\left(\mathbf{J}_{Di}^T\, m_i\, \mathbf{J}_{Di} + \frac{1}{2}\mathbf{J}_{Ri}^T\, {}^0 I_i\, \mathbf{J}_{Ri}\right) \tag{11.602}$$

H_{ikm} is the velocity coupling vector

$$H_{ijk} = \sum_{j=1}^{n}\sum_{k=1}^{n}\left(\frac{\partial D_{ij}}{\partial q_k} - \frac{1}{2}\frac{\partial D_{jk}}{\partial q_i}\right) \tag{11.603}$$

G_i is the gravitational vector

$$G_i = \sum_{j=1}^{n} m_j \mathbf{g}^T \mathbf{J}_{Dj}^{(i)} \tag{11.604}$$

and $\mathbf{J}_i$ is the Jacobian matrix of the robot

$$\dot{X}_i = \begin{bmatrix} {}^0\mathbf{v}_i \\ {}^0\boldsymbol{\omega}_i \end{bmatrix} = \begin{bmatrix} \mathbf{J}_{Di} \\ \mathbf{J}_{Ri} \end{bmatrix}\dot{q} = \mathbf{J}_i\,\dot{q}. \tag{11.605}$$

To hold a robot in a stationary configuration, the actuators must apply some required forces to balance the external loads applied to the robot. In the static condition, the globally expressed Newton–Euler equations for the link (i) can be written in a recursive form

$$^0\mathbf{F}_{i-1} = {}^0\mathbf{F}_i - \sum {}^0\mathbf{F}_{e_i} \tag{11.606}$$

$$^0\mathbf{M}_{i-1} = {}^0\mathbf{M}_i - \sum {}^0\mathbf{M}_{e_i} + {}_{i-1}^{0}\mathbf{d}_i \times {}^0\mathbf{F}_i. \tag{11.607}$$

Now we are able to calculate the action force system $(\mathbf{F}_{i-1}, \mathbf{M}_{i-1})$ when the reaction force system $(-\mathbf{F}_i, -\mathbf{M}_i)$ is given.

11.7   Key Symbols

a	Kinematic link length
$\mathbf{a}$	Acceleration vector
$[A]$	Coefficient matrix of a set of linear equations
$\mathbf{b}$	Vector of known values in a set of linear equations
B	Body coordinate frame
c	cos
c_i	Position of the mass center of link (i) in B_i
$\mathbf{c}$	Jacobian generating vector
C	Mass center
$\mathbf{C}(\mathbf{q}, \dot{q})$	Damping-type matrix of equation of motion
d_x, d_y, d_z	Elements of $\mathbf{d}$
$\mathbf{d}$	Translation vector, joint position vector
$\mathbf{d}_i$	Position vector of the origin of B_i
D	Displacement transformation matrix
$\mathbf{D}(\mathbf{q})$	Inertial-type matrix of equation of motion
E	Energy
$\mathbf{F}_{e_i}$	External force acting on the link(i)
$\mathbf{F}_i$	The force that link (i) applies to $(i+1)$ at joint $i+1$
$\mathbf{F}_{i-1}$	The force that link $(i-1)$ applies to link (i) at joint i
$\mathbf{F}_s$	Shaking force
$\mathbf{g}$	Gravitational acceleration vector
G, B_0	Global coordinate frame, Base coordinate frame
$\mathbf{G}(\mathbf{q})$	Gravitational vector of equation of motion
$\mathbf{H}(\mathbf{q}, \dot{q})$	Velocity coupling vector of equation of motion
$\hat{\imath}, \hat{\jmath}, \hat{k}$	Local coordinate axes unit vectors
$\hat{I}, \hat{J}, \hat{K}$	Global coordinate axes unit vectors
$I = [I]$	Mass moment matrix
$\bar{I} = [\bar{I}]$	Pseudo-inertia matrix
$\mathbf{I} = [\mathbf{I}]$	Identity matrix
$\mathbf{J}$	Jacobian
K	Kinetic energy
l	Length
$\mathbf{L}$	Angular moment vector, moment of moment
$\mathcal{L}$	Lagrangian
m	Mass
$\mathbf{m}_i$	Position vector of o_i from C_i
$\mathbf{n}_i$	Position vector of o_{i-1} from C_i
$\mathbf{M}_{e_i}$	External moment acting on the link (i)
$\mathbf{M}_i$	The moment that link (i) applies to $(i+1)$ at joint$i+1$
$\mathbf{M}_{i-1}$	The moment that link $(i-1)$ applies to link (i) at joint i

P	Power
q	Generalized coordinate
Q	Torque of an actuator, generalized nonpotential force
$\mathbf{Q}$	Moment vector at a joint
$\mathbf{r}$	Position vectors, homogenous position vector, Global position of the mass center of a link
r_i	The element i of $\mathbf{r}$
r_{ij}	The element of row i and column j of a matrix
R	Rotation transformation matrix
s	sin
T	Homogenous transformation matrix
$\mathbf{v}$	Translational velocity vector
V	Potential energy
x, y, z	Local coordinate axes
$\mathbf{x}$	Vector of unknown values in a set of linear equations
X, Y, Z	Global coordinate axes
Z_i	Short notation of an equation

Greek

α	Angular acceleration
$\boldsymbol{\alpha}$	Angular acceleration vector
$\alpha_1, \alpha_2, \alpha_3$	Components of $\boldsymbol{\alpha}$
θ	Rotary joint angle
θ_{ijk}	$\theta_i + \theta_j + \theta_k$
$\omega_1, \omega_2, \omega_3$	Components of $\boldsymbol{\omega}$
ε	Small test number to terminate a procedure
θ	Rotary joint angle
θ_{ijk}	$\theta_i + \theta_j + \theta_k$
ω	Angular velocity
$\boldsymbol{\omega}$	Angular velocity vector
$\tilde{\omega}$	Skew symmetric matrix of the vector $\boldsymbol{\omega}$

Symbol

$[\ ]^{-1}$	Inverse of the matrix $[\ ]$
$[\ ]^{T}$	Transpose of the matrix $[\ ]$
$\equiv$	Equivalent
(i)	Link number i
FBD	Free body diagram
tr	Trace

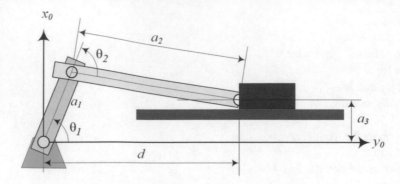

Fig. 11.27   A planar slider-crank mechanism

Exercises

1. Notation and symbols.

   Describe the meaning of these notations.

$$a - \mathbf{F}_2 \quad b - {}^0\mathbf{F}_1 \quad c - {}^1\mathbf{F}_1 \quad d - {}^2\mathbf{M}_1 \quad e - {}^2\mathbf{M}_{e1} \quad f - {}^B\mathbf{M}$$

$$g - \mathbf{m}_2 \quad h - {}^0\mathbf{n}_2 \quad i - {}_{i-1}^{0}\mathbf{d}_i \quad j - {}^0\mathbf{d}_i \quad k - {}^0\mathbf{r}_i \quad l - {}^{i-1}\mathbf{d}_i$$

$$m - {}^0\mathbf{L}_2 \quad n - {}^0\mathbf{I}_2 \quad o - {}_{i-1}^{0}\mathbf{L}_i \quad p - \mathbf{K}_i \quad q - \mathbf{V}_i \quad r - {}^{i-1}\mathbf{I}_i$$

2. *PR* manipulator dynamics.

   Find the equations of motion for the planar polar manipulator shown in Fig. 5.14. Eliminate the joints' constraint force and moment to derive the equations for the actuators' force or moment.

3. ★ Even order recursive translational velocity.

   Find an equation to relate the velocity of link (i) to the velocity of link $(i - 2)$, and the velocity of link (i) to the velocity of link $(i + 2)$.

4. ★ Even order recursive angular velocity.

   Find an equation to relate the angular velocity of link (i) to the angular velocity of link $(i - 2)$, and the angular velocity of link (i) to the angular velocity of link $(i + 2)$.

5. ★ Even order recursive translational acceleration.

   Find an equation to relate the acceleration of link (i) to the acceleration of link $(i - 2)$, and the acceleration of link (i) to the acceleration of link $(i + 2)$.

6. ★ Even order recursive angular acceleration.

   Find an equation to relate the angular acceleration of link (i) to the angular acceleration of link $(i - 2)$, and the angular acceleration of link (i) to the angular acceleration of link $(i + 2)$.

7. ★ Acceleration in different frames.

   For the $2R$ planar manipulator shown in Fig. 11.6, find ${}_1^0\mathbf{a}_2, {}_0^1\mathbf{a}_2, {}_2^0\mathbf{a}_1, {}_0^2\mathbf{a}_1, {}_0^2\mathbf{a}_2$, and ${}_1^0\mathbf{a}_1$.

8. Slider-crank mechanism dynamics.

   A planar slider-crank mechanism is shown in Fig. 11.27. Set up the link coordinate frames, develop the Newton–Euler equations of motion, and find the driving moment at the base revolute joint.

9. ★ Global differential of a link momentum.

   In recursive Newton–Euler equations of motion, why we do not use the following Newton equation?

$$ {}^i\mathbf{F} = \frac{{}^Gd}{dt}\, {}^i\mathbf{F} = \frac{{}^Gd}{dt} m\, {}^i\mathbf{v} = m\, {}^i\dot{\mathbf{v}} + {}_0^i\omega_i \times m\, {}^i\mathbf{v} \tag{11.608}$$

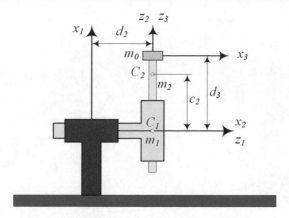

Fig. 11.28 A 2 DOF Cartesian manipulator

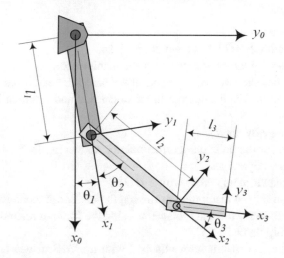

Fig. 11.29 A 3R planar manipulator attached to a wall

10. A planar Cartesian manipulator.

    Determine the equations of motion of the planar Cartesian manipulator shown in Fig. 11.28. *Hint*: The coordinate frames are not based on DH rules.

11. 3R planar manipulator dynamics.

    A 3R planar manipulator is shown in Fig. 11.29. The manipulator is attached to a wall and therefore, $\mathbf{g} = g\,{}^0\hat{\imath}_0$.

    (a) Find the Newton–Euler equations of motion for the manipulator. Do your calculations in the global frame and derive the dynamic force and moment at each joint.

    (b) Reduce the number of equations to three for moments at joints.

    (c) Substitute the vectorial quantities and calculate the moments in terms of geometry and angular variables of the manipulator.

(d) 3R planar manipulator recursive dynamics.

    The manipulator shown in Fig. 11.29 is a 3R planar manipulator attached to a wall and therefore, $\mathbf{g} = -g\,{}^0\hat{\imath}_0$.

    (a) Find the equations of motion for the manipulator utilizing the backward recursive Newton–Euler technique.

    (b) ★ Find the equations of motion for the manipulator utilizing the forward recursive Newton–Euler technique.

12. 3R planar manipulator Lagrange dynamics.

    Find the equations of motion for the 3R planar manipulator shown in Fig. 11.29 utilizing the Lagrange technique. The manipulator is attached to a wall and therefore, $\mathbf{g} = -g\,{}^0\hat{\imath}_0$.

13. An articulated manipulator.

    Figure 11.30 illustrates an articulated manipulator with massless arms and two massive points m_1 and m_2.

    (a) Follow the DH rules and complete the link coordinate frames.

    (b) Determine the DH transformation matrices.

    (c) Determine the equations of motion of the manipulator using Lagrange method.

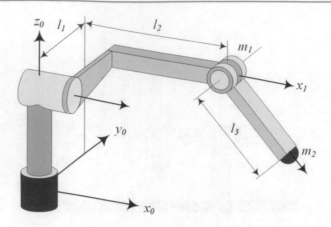

Fig. 11.30 An articulated manipulator

14. Polar planar manipulator dynamics.

    A polar planar manipulator with 2 DOF is shown in Fig. 5.14.

    (a) Determine the Newton–Euler equations of motion for the manipulator.

    (b) Reduce the number of equations to two, for moments at the base joint and force at the P joint.

    (c) Substitute the vectorial quantities and calculate the action force and moment in terms of geometry and angular variables of the manipulator.

15. Polar planar manipulator Lagrange dynamics.

    Find the equations of motion for the polar planar manipulator, shown in Fig. 5.14, utilizing the Lagrange technique.

16. Polar planar manipulator recursive dynamics.

    Figure 5.14 depicts a polar planar manipulator with 2 DOF.

    (a) Find the equations of motion for the manipulator utilizing the backward recursive Newton–Euler technique.

    (b) ★ Find the equations of motion for the manipulator utilizing the forward recursive Newton–Euler technique.

17. ★ Dynamics of a spherical manipulator.

    Figure 5.46 illustrates a spherical manipulator attached with a spherical wrist. Analyze the robot and derive the equations of motion for joints action force and moment. Assume $\mathbf{g} = -g\,{}^{0}\hat{k}_0$ and the end-effector is carrying a mass m.

18. ★ Dynamics of an articulated manipulator.

    Figure 5.22 illustrates an articulated manipulator R⊢R∥R. Use $\mathbf{g} = -g\,{}^{0}\hat{k}_0$ and find the manipulator's equations of motion.

19. A planar manipulator.

    Figure 11.31 illustrates a three DOF planar manipulator. Determine the equations of motion of the manipulator if the links are massless and there are two massive points m_1 and m_2.

20. ★ Dynamics of a $SCARA$ robot.

    Calculate the dynamic joints' force system for the $SCARA$ robot R∥R∥R∥P shown in Fig. 5.24 if $\mathbf{g} = -g\,{}^{0}\hat{k}_0$.

21. ★ Dynamics of an $SRMS$ manipulator.

    Figure 5.24 shows a model of the Shuttle remote manipulator system ($SRMS$).

    (a) Derive the equations of motion for the $SRMS$ and calculate the joints' force system for $\mathbf{g} = 0$.

    (b) Derive the equations of motion for the $SRMS$ and calculate the joints' force system for $\mathbf{g} = -g\,{}^{0}\hat{k}_0$.

    (c) Eliminate the constraint forces and reduce the number of equations equal to the number of action moments.

    (d) Assume the links are made of a uniform cylinder with radius $r = .25$ m and $m = 12$ kg/m. Use the characteristics indicated in Table 5.10 and find the equations of motion when the end-effector is holding a 24 kg mass.

22. Statics of a $3R$ planar manipulator.

    Figure 11.29 illustrates a $3R$ planar manipulator attached to a wall. Derive the static force and moment at each joint to keep the configuration of the manipulator if $\mathbf{g} = -g\,{}^{0}\hat{\imath}_0$.

23. An RPR planar redundant manipulator.

    Figure 11.32 illustrates a 3 DOF planar manipulator with joint variables θ_1, d_2, and θ_2. Determine the equations of motion of the manipulator if the links are massless and there are two massive points m_1 and m_2.

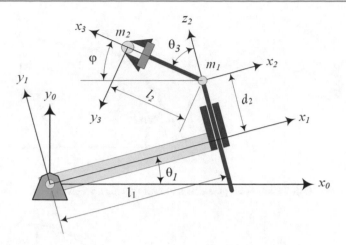

Fig. 11.31 A planar manipulator

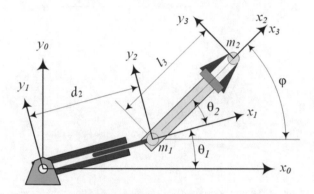

Fig. 11.32 A RPR planar redundant manipulator

24. ★ Recursive dynamics of an articulated manipulator.

Figure 11.19 illustrates an articulated manipulator R⊢R∥R. Use $\mathbf{g} = -g\,{}^0\hat{k}_0$ and find the manipulator's equations of motion

(a) utilizing the backward recursive Newton–Euler technique.

(b) utilizing the forward recursive Newton–Euler technique.

25. ★ Recursive dynamics of a $SCARA$ robot.

A $SCARA$ robot R∥R∥R∥P is shown in Fig. 9.22. If $\mathbf{g} = -g\,{}^0\hat{k}_0$ determine the dynamic equations of motion by

(a) utilizing the backward recursive Newton–Euler technique.

(b) utilizing the forward recursive Newton–Euler technique.

26. ★ Recursive dynamics of an $SRMS$ manipulator.

Figure 5.24 shows a model of the Shuttle remote manipulator system ($SRMS$).

(a) Derive the equations of motion for the $SRMS$ utilizing the backward recursive Newton–Euler technique for $\mathbf{g} = 0$.

(b) Derive the equations of motion for the $SRMS$ utilizing the forward recursive Newton–Euler technique for $\mathbf{g} = 0$.

27. ★ Work done by actuators.

Consider a $2R$ planar manipulator moving on a given path. Assume that the endpoint of a $2R$ manipulator moves with constant speed $v = 1$ m/s from P_1 to P_2, on a path made of two semi-circles as shown in Fig. 11.33. Calculate the work done by the actuators if $l_1 = l_2 = 1$ m and the manipulator is carrying a 12 kg mass. The center of the circles are at $(0.75\,\text{m}, 0.5\,\text{m})$ and $(-0.75\,\text{m}, 0.5\,\text{m})$.

28. ★ Lagrange dynamics of an articulated manipulator.

Figure 11.19 illustrates an articulated manipulator R⊢R∥R. Use $\mathbf{g} = -g\,{}^0\hat{k}_0$ and find the manipulator's equations of motion utilizing the Lagrange technique.

29. ★ Lagrange dynamics of a $SCARA$ robot.

A $SCARA$ robot R∥R∥R∥P is shown in Fig. 9.22. If $\mathbf{g} = -g\,{}^0\hat{k}_0$ determine the dynamic equations of motion by applying the Lagrange technique.

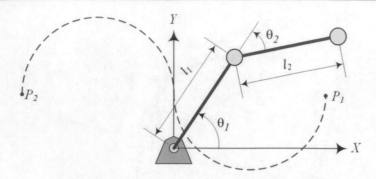

Fig. 11.33 A $2R$ planar manipulator moving on a given path

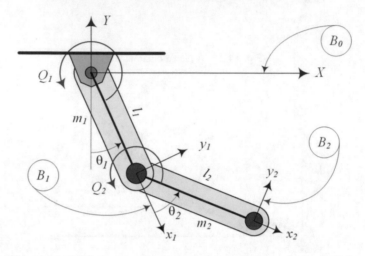

Fig. 11.34 A $2R$ planar manipulator attached to a ceiling in static condition

30. ★ Lagrange dynamics of an $SRMS$ manipulator.

Figure 9.22 shows a model of the Shuttle remote manipulator system ($SRMS$). Derive the equations of motion for the $SRMS$ utilizing the Lagrange technique for

(a) $\mathbf{g} = 0$

(b) $\mathbf{g} = -g\,{}^0\hat{k}_0$.

31. Statics of a $2R$ planar manipulator.

Figure 11.34 illustrates a $2R$ planar manipulator attached to a ceiling. The links are uniform with the following characteristics.

$$m_1 = 24\,\text{kg} \qquad m_2 = 18\,\text{kg} \qquad l_1 = 1\,\text{m} \qquad l_2 = 1\,\text{m} \tag{11.609}$$

$$\mathbf{g} = -g\,{}^0\hat{j}_0 \tag{11.610}$$

There is a load $\mathbf{F}_e = -14g\,{}^0\hat{j}_0\text{N}$ at the endpoint. Calculate the static moments Q_1 and Q_2 for $\theta_1 = 30$ deg and $\theta_2 = 45$ deg.

32. Statics of a $2R$ planar manipulator at a different base angle.

In Exercise 31 keep $\theta_2 = 45$ deg and calculate the static moments Q_1 and Q_2 as functions of θ_1. Plot Q_1 and Q_2 versus θ_1 and find the configuration that minimizes Q_1, Q_2, $Q_1 + Q_2$, and the potential energy V.

33. ★ Statics of an articulated manipulator.

An articulated manipulator R⊢R‖R is shown in Fig. 11.19. Find the static force and moment at joints for $\mathbf{g} = g\,{}^0\hat{k}_0$. The end-effector is carrying a 20 kg mass. Calculate the maximum base force moment.

34. ★ Statics of a $SCARA$ robot.

Calculate the static joints' force system for the $SCARA$ robot R‖R‖R‖P shown in Fig. 5.24 if $\mathbf{g} = -g\,{}^0\hat{k}_0$ and the end-effector is carrying a 10 kg mass.

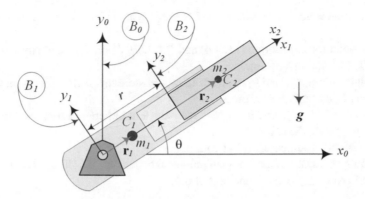

Fig. 11.35 A planar polar manipulator with massive links

35. ★ Statics of a spherical manipulator.

Figure 5.46 illustrates a spherical manipulator attached with a spherical wrist. Analyze the robot and calculate the static force system in joints for $\mathbf{g} = -g\,^0\hat{k}_0$ if the end-effector is carrying a 12 kg mass.

36. ★ Statics of an $SRMS$ manipulator.

A model of the Shuttle remote manipulator system ($SRMS$) is shown in Fig. 9.22. Analyze the static configuration of the $SRMS$ and calculate the joints' force system for $\mathbf{g} = -g\,^0\hat{k}_0$.

Assume the links are made of a uniform cylinder with radius $r = 0.25$ m and $m = 12$ kg/m. Use the characteristics indicated in Table 5.10 and find the maximum value of the base force system for a 24 kg mass held by the end-effector. The $SRMS$ is supposed to work in a no-gravity field.

37. Polar manipulator with massive links.

Figure 11.35 illustrates a planar polar manipulator with massive links. Derive the below equations of motion for Q as the torque of a motor at origin of B_0 to turn link (1), and F for actuating force of a jack between origin of B_1 and B_2 to move link (2) respect to link (1).

$$m_2\ddot{r} + m_2 g \sin\theta - m_2 (r + r_2)\,\dot{\theta}^2 = F \tag{11.611}$$

$$m_1 r_1^2 + m_2 (r + r_2)^2\,\ddot{\theta} + 2m_2 (r + r_2)\,\dot{r}\dot{\theta}$$
$$+g\,(m_1 r_1 + m_2 (r + r_2))\cos\theta = Q \tag{11.612}$$

38. Project. $2R$ robot torque calculation.

Consider a $2R$ manipulator as shown in Fig. 11.14, that its tip point is supposed to move on a line $Y = f(X)$ from $(1, 0.5)$ to $(-1, 1.5)$ in one unit of time.

$$l_1 = l_2 = 1\,\text{m} \qquad m_0 = 1\,\text{kg} \qquad m_{12} = 1\,\text{kg} \tag{11.613}$$

Assume the links of the robot are massless. A septic time function for X will move the end-effector from $X(0) = 1$ to $X(1) = -1$ with zero velocity, zero acceleration, zero jerk at both ends.

$$X = 1 - 70t^4 + 168t^5 - 140t^6 + 40t^7 \tag{11.614}$$

(a) Determine and plot $X, \dot{X}, \ddot{X}, Y, \dot{Y}, \ddot{Y}$ as functions of time.
(b) Divide the total time of motion into 100 equal segments. Calculate $X, \dot{X}, \ddot{X}, Y, \dot{Y}, \ddot{Y}$ at the 101 points and store them.
(c) Using inverse kinematic analysis, determine, and store θ_1, θ_2 at the 101 points. Plot the graphs of θ_1, θ_2 as functions of time.
(d) Using inverse velocity analysis, determine, and store $\dot{\theta}_1, \dot{\theta}_2$ at the 101 points. Plot the graphs of $\dot{\theta}_1, \dot{\theta}_2$ as functions of time.

(e) Using inverse acceleration analysis, determine, and store $\ddot{\theta}_1, \ddot{\theta}_2$ at the 101 points. Plot the graphs of $\ddot{\theta}_1, \ddot{\theta}_2$ as functions of time.

(f) Employing dynamic equations of motion (11.304) and (11.305), determine, and store Q_0, Q_1 at the 101 points. Plot the graphs of Q_0, Q_1 as functions of time.

(g) Using the stored values of Q_0, Q_1 and $\dot{\theta}_1, \dot{\theta}_2$ determine, and store the power of each motor P_1, P_2 at the 101 points. Plot the graphs of P_1, P_2 as functions of time.

(f) Using the stored values of Q_0, Q_1 and θ_1, θ_2 calculate the amount energy that each motor used E_1, E_2. Calculate the total consumed energy by the robot.

39. ★ Project. Ceiling and ground mounted $2R$ robot comparison.

Figure 11.24 illustrates an ideal $2R$ planar manipulator mounted on a ceiling. The tip point will move on a line $Y = f(X)$ from $(1, -0.5)$ to $(-1, -1.5)$ in one unit of time.

$$l_1 = l_2 = 1\,\text{m} \qquad m_0 = 1\,\text{kg} \qquad m_{12} = 1\,\text{kg} \tag{11.615}$$

Assume the links of the robot are massless and use a septic time function for X to move the tip point from $X(0) = 1$ to $X(1) = -1$ with zero velocity, zero acceleration, zero jerk at both ends.

$$X = 1 - 70t^4 + 168t^5 - 140t^6 + 40t^7 \tag{11.616}$$

(a) Calculate the time history of the torques and powers of the motors in this maneuver. Determine the total consumed energy.

(b) Consider the same robot is mounted on the ground and the tip point is to move on a line $Y = f(X)$ from $(1, 0.5)$ to $(-1, 1.5)$ in one unit of time. Use the same time function for X as (11.616) and calculate the time history of the torques and powers of the motors in this maneuver. Determine the total consumed energy.

(c) Compare torques, powers, and consumed energy of the ceiling and ground mounted $2R$ robot, and make engineering conclusion and comment.

40. ★ Project. Inverse dynamics of $2R$ robot.

Consider a $2R$ manipulator as shown in Fig. 11.14 with massless arms. Assume the robot to be at its rest position at which, $\theta_1 = \theta_2 = 0$.

$$l_1 = l_2 = 1\,\text{m} \qquad m_0 = 1\,\text{kg} \qquad m_{12} = 1\,\text{kg} \tag{11.617}$$

Apply the following torques for $0 \le t \le 8$ where t is in second.

$$Q_0 = 10 \sin \frac{\pi}{4} t\,\text{N}\,\text{m} \qquad Q_1 = 5 \sin \frac{\pi}{4} t\,\text{N}\,\text{m} \tag{11.618}$$

(a) Solve the equations of motion numerically and determine θ_1, θ_2.

(b) Using forward kinematics, determine the path of motion of the tip point of the robot.

(c) Employing numerical solution of the equations of motion, determine $\dot{\theta}_1, \dot{\theta}_2, \ddot{\theta}_1, \ddot{\theta}_2$ and the velocity and acceleration of the tip point.

(d) Having θ_1, θ_2 from part (a), determine $\dot{\theta}_1, \dot{\theta}_2, \ddot{\theta}_1, \ddot{\theta}_2$ by numerical differentiation and compare the result with part (c).

Control is the science of *desired motion*. It relates the dynamics and kinematics of a robot to a prescribed motion. It includes optimization problems to determine forces so that the system will behave optimally. A typical example is the situation in which the initial and terminal configurations of a robot are given and the forces acting on the robot must be found to have the motion in minimum time.

Path or trajectory planning is a part of control, in which we plan a path followed by the manipulator in a planned time profile. Paths can be planned in joint or Cartesian space. Joint path planning directly specifies the time evolution of the joint variables. However, Cartesian path specifies the position and orientation of the end frame. So, path includes attaining a desired target from an initial configuration. It may also include avoiding obstacles. Joint path planning is a relatively simple task because it does not involve inverse kinematics, but it is hard to digest the motion of the manipulator in Cartesian space. However, Cartesian coordinates make sense but need inverse kinematics calculation.

In this part, we develop techniques to derive the required commands to control the robot's task.

Path Planning

Path planning includes three tasks: (1) Defining a geometric curve for the end-effector between two points. (2) Defining a rotational motion between two orientations. (3) Defining a time function for variation of a coordinate between two given values. All of these three definitions are called *path planning*. Figure 12.1 illustrates a path of the tip point of a $2R$ manipulator between points P_1 and P_2 to avoid two obstacles.

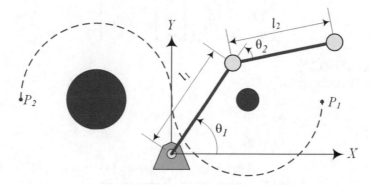

Fig. 12.1 A path of the tip point of a $2R$ manipulator to avoid two obstacles

12.1 Cubic Path

A cubic function is the simplest polynomial to determine the time behavior of a variable, such as a joint variable, between two given values, rest-to-rest. A cubic path in joint space for the joint variable $q(t)$ between two points $q(t_0)$ and $q(t_f)$ is

$$q(t) = a_0 + a_1 t + a_2 t^2 + a_3 t^3 \tag{12.1}$$

where

$$a_0 = \frac{(3t_0 - t_f)\, t_f^2}{(t_0 - t_f)^3} q_0 + \frac{(t_0 - 3t_f)\, t_0^2}{(t_0 - t_f)^3} q_f - \frac{t_f\, q_0' + t_0\, q_f'}{(t_0 - t_f)^2} t_0 t_f \tag{12.2}$$

$$a_1 = \frac{-6 t_0 t_f}{(t_0 - t_f)^3}(q_0 - q_f) + \frac{2t_0 + t_f}{(t_0 - t_f)^2} t_f\, q_0' + \frac{t_0 + 2t_f}{(t_0 - t_f)^2} t_0\, q_f' \tag{12.3}$$

$$a_2 = \frac{3(t_0 + t_f)}{(t_0 - t_f)^3}(q_0 - q_f) - \frac{t_0 + 2t_f}{(t_f - t_0)^2} q_0' - \frac{2t_0 + t_f}{(t_f - t_0)^2} q_f' \tag{12.4}$$

$$a_3 = \frac{-2\left(q_0 - q_f\right)}{\left(t_0 - t_f\right)^3} + \frac{q_0' + q_f'}{\left(t_0 - t_f\right)^2} \tag{12.5}$$

and

$$
\begin{aligned}
q(t_0) &= q_0 & \dot{q}(t_0) &= q_0' \\
q(t_f) &= q_f & \dot{q}(t_f) &= q_f'
\end{aligned} \tag{12.6}
$$

Proof A cubic polynomial has four coefficients. Therefore, it can satisfy the position and velocity constraints at the initial and final points. For simplicity, we call the value of the variable the *position* and the rate of the variable the *speed*. Assume the position and speed of a variable at the initial time t_0 and at the final time t_f are given as (12.6).

Substituting the boundary conditions (12.6) in the position and speed equations of the variable q,

$$q(t) = a_0 + a_1 t + a_2 t^2 + a_3 t^3 \tag{12.7}$$

$$\dot{q}(t) = a_1 + 2a_2 t + 3a_3 t^2 \tag{12.8}$$

generates four equations for the coefficients of the path.

$$
\begin{bmatrix}
1 & t_0 & t_0^2 & t_0^3 \\
0 & 1 & 2t_0 & 3t_0^2 \\
1 & t_f & t_f^2 & t_f^3 \\
0 & 1 & 2t_f & 3t_f^2
\end{bmatrix}
\begin{bmatrix}
a_0 \\ a_1 \\ a_2 \\ a_3
\end{bmatrix}
=
\begin{bmatrix}
q_0 \\ q_0' \\ q_f \\ q_f'
\end{bmatrix} \tag{12.9}
$$

Their solutions are given in (12.2)–(12.5).

In case that $t_0 = 0$, the coefficients simplify.

$$a_0 = q_0 \tag{12.10}$$

$$a_1 = q_0' \tag{12.11}$$

$$a_2 = \frac{3\left(q_f - q_0\right) - \left(2q_0' + q_f'\right) t_f}{t_f^2} \tag{12.12}$$

$$a_3 = \frac{-2\left(q_f - q_0\right) + \left(q_0' + q_f'\right) t_f}{t_f^3} \tag{12.13}$$

It is also possible to employ a time shift and search for a cubic polynomial of $(t - t_0)$.

$$q(t) = a_0 + a_1 (t - t_0) + a_2 (t - t_0)^2 + a_3 (t - t_0)^3 \tag{12.14}$$

Now, the boundary conditions (12.6) generate a set of equations

$$
\begin{bmatrix}
1 & 0 & 0 & 0 \\
0 & 1 & 0 & 0 \\
1 & (t_f - t_0) & (t_f - t_0)^2 & (t_f - t_0)^3 \\
0 & 1 & 2(t_f - t_0) & 3(t_f - t_0)^2
\end{bmatrix}
\begin{bmatrix}
a_0 \\ a_1 \\ a_2 \\ a_3
\end{bmatrix}
=
\begin{bmatrix}
q_0 \\ q_0' \\ q_f \\ q_f'
\end{bmatrix} \tag{12.15}
$$

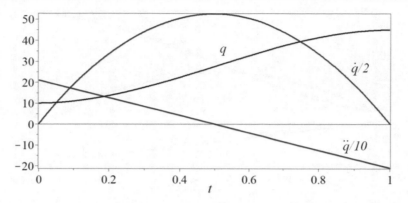

Fig. 12.2 Kinematics of a rest-to-rest cubic path

with the following solutions:

$$
\begin{bmatrix} a_0 \\ a_1 \\ a_2 \\ a_3 \end{bmatrix} = \begin{bmatrix} q_0 \\ q_0' \\ -\left(t_f - t_0\right)^{-2}\left(3q_0 - 3q_f - 2t_0q_0' - t_0q_f' + 2t_fq_0' + t_fq_f'\right) \\ \left(t_f - t_0\right)^{-3}\left(2q_0 - 2q_f - t_0q_0' - t_0q_f' + t_fq_0' + t_fq_f'\right) \end{bmatrix}
\tag{12.16}
$$

A disadvantage of cubic paths is the acceleration jump at boundaries that introduces infinite jerks. ■

Example 381 Rest-to-rest cubic path. Here we analyze a rest-to-rest cubic path to look at the graphical variation of the variable and see how cubic paths works.

Assume $q(0) = 10$ deg, $q(1) = 45$ deg, and $\dot{q}(0) = \dot{q}(1) = 0$. The set of equations and the coefficients of the cubic path will be

$$
\begin{bmatrix} 1 & 0 & 0 & 0 \\ 0 & 1 & 0 & 0 \\ 1 & 1 & 1 & 1 \\ 0 & 1 & 2 & 3 \end{bmatrix} \begin{bmatrix} a_0 \\ a_1 \\ a_2 \\ a_3 \end{bmatrix} = \begin{bmatrix} 10 \\ 0 \\ 45 \\ 0 \end{bmatrix}
\tag{12.17}
$$

$$
a_0 = 10 \qquad a_1 = 0 \qquad a_2 = 105 \qquad a_3 = -70
\tag{12.18}
$$

which generate a cubic path for the variable $q(t)$.

$$
q(t) = 10 + 105t^2 - 70t^3 \text{ deg}
\tag{12.19}
$$

$$
\dot{q}(t) = 210t - 210t^2 \text{ deg}/\text{s}
\tag{12.20}
$$

$$
\ddot{q}(t) = 210 - 420t \text{ deg}/\text{s}^2
\tag{12.21}
$$

The path information is shown in Fig. 12.2.

Example 382 To-rest cubic path. A path may start from a nonzero speed value and end to rest as zero speed. Such path may be used as the last piece of a series of paths.

Assume the angle of a revolute joint starts from $\theta(0) = 10$ deg, $\dot{\theta}(0) = 10$ deg$/$s and ends at $\theta(2) = 45$ deg, $\dot{\theta}(2) = 0$. The coefficients of a cubic path for this motion are

$$
a_0 = 10 \qquad a_1 = 10 \qquad a_2 = \frac{65}{4} \qquad a_3 = \frac{-25}{4}
\tag{12.22}
$$

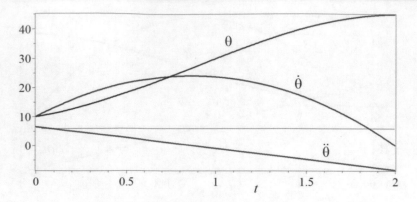

Fig. 12.3 Kinematics of a to-rest cubic path in joint space

The kinematics of the path are

$$\theta(t) = 10 + 10t + 16.25t^2 - 6.25t^3 \text{ deg} \tag{12.23}$$

$$\dot{\theta}(t) = 10 + 32.5t - 18.75t^2 \text{ deg/s} \tag{12.24}$$

$$\ddot{\theta}(t) = 32.5 - 37.5t \text{ deg/s}^2 \tag{12.25}$$

and are shown graphically in Fig. 12.3.

Example 383 Rest-to-rest path with a constant speed in the middle. The conditions on a path may be more complicated than only rest-to-rest. Here is an example of a path that needs to have a constant speed for a duration of time, as well as rest-to-rest conditions.

Assume we need a rest-to-rest path with a constant given speed $\dot{q} = \dot{q}_c$ for $t_1 < t < t_2$, where $t_0 < t_1 < t_2 < t_f$. The required conditions to satisfy are

$$q(t_0) = q_0 \qquad \dot{q}(t_0) = q_0' \tag{12.26}$$

$$\dot{q}(t) = q_c' \qquad t_1 < t < t_2 \tag{12.27}$$

$$q(t_f) = q_f \qquad \dot{q}(t_f) = q_f' \tag{12.28}$$

The path has three parts: (1) rest-to, (2) constant-speed, and (3) to-rest. We need an equation for the rest-to part of the motion to achieve the given speed. There are three boundary conditions for the first part of the path. The conditions are the initial position and speed and the final constant speed. Assuming $t_0 = 0$, the conditions are

$$q_1(0) = q_0 \qquad \dot{q}_1(0) = 0 \qquad \dot{q}_1(t_1) = q_c' \tag{12.29}$$

A quadratic path has three coefficients and may be utilized to satisfy the three initial conditions.

$$q_1(t) = a_0 + a_1 t + a_2 t^2 \tag{12.30}$$

$$\dot{q}_1(t) = a_1 + 2a_2 t \tag{12.31}$$

Employing the conditions generates the following equations:

$$q_0 = a_0 \qquad 0 = a_1 \qquad q_c' = 2a_2 t_1 \tag{12.32}$$

Therefore, the rest-to path is

$$q_1(t) = q_0 + \frac{q_c'}{2t_1} t^2 \qquad 0 < t < t_1 \tag{12.33}$$

Given the specific constant speed, q'_c shows that the path in the middle part is

$$\dot{q}_2(t) = q'_c \tag{12.34}$$

$$q_2(t) = q'_c t + C_1 \qquad t_1 < t < t_2 \tag{12.35}$$

The constant of integration can be found by utilizing the position condition at $t = t_1$.

$$q_0 + \frac{q'_c}{2t_1}t_1^2 = q'_c t_1 + C_1 \tag{12.36}$$

$$C_1 = q_0 - \frac{1}{2}t_1 q'_c \tag{12.37}$$

There are four conditions for the to-rest part of the path.

$$q_3(t_f) = q_f \qquad \dot{q}_3(t_f) = 0 \tag{12.38}$$

$$q_3(t_2) = q_2(t_2) = q_2 = q'_c t_2 + q_0 - \frac{1}{2}t_1 q'_c \tag{12.39}$$

$$\dot{q}_3(t_2) = \dot{q}_2(t_2) = q'_c \tag{12.40}$$

Therefore, it can be calculated utilizing a cubic equation.

$$q_3(t) = b_0 + b_1 t + b_2 t^2 + b_3 t^3 \tag{12.41}$$

$$\dot{q}_3(t) = b_1 + 2b_2 t + 3b_3 t^2 \tag{12.42}$$

$$t_2 < t < t_f \tag{12.43}$$

The conditions generate four equations

$$\begin{bmatrix} 1 & t_f & t_f^2 & t_f^3 \\ 0 & 1 & 2t_f & 3t_f^2 \\ 1 & t_2 & t_2^2 & t_2^3 \\ 0 & 1 & 2t_2 & 3t_2^2 \end{bmatrix} \begin{bmatrix} b_0 \\ b_1 \\ b_2 \\ b_3 \end{bmatrix} = \begin{bmatrix} q_f \\ 0 \\ q'_c t_2 + q_0 - \frac{1}{2}t_1 q'_c \\ q'_c \end{bmatrix} \tag{12.44}$$

with the following solutions:

$$b_0 = -t_2 t_f^2 \frac{q'_c}{-2t_2 t_f + t_2^2 + t_f^2} + q_2 \frac{t_f^3 - 3t_2 t_f^2}{-t_2^3 + t_f^3 - 3t_2 t_f^2 + 3t_2^2 t_f}$$

$$+ q_f \frac{-t_2^3 + 3t_2^2 t_f}{-t_2^3 + t_f^3 - 3t_2 t_f^2 + 3t_2^2 t_f} \tag{12.45}$$

$$b_1 = q'_c \frac{2t_2 t_f + t_f^2}{-2t_2 t_f + t_2^2 + t_f^2} + 6q_2 t_2 \frac{t_f}{-t_2^3 + t_f^3 - 3t_2 t_f^2 + 3t_2^2 t_f}$$

$$- 6t_2 q_f \frac{t_f}{-t_2^3 + t_f^3 - 3t_2 t_f^2 + 3t_2^2 t_f} \tag{12.46}$$

$$b_2 = q'_c \frac{-t_2 - 2t_f}{-2t_2 t_f + t_2^2 + t_f^2} + q_2 \frac{-3t_2 - 3t_f}{-t_2^3 + t_f^3 - 3t_2 t_f^2 + 3t_2^2 t_f}$$

$$+ q_f \frac{3t_2 + 3t_f}{-t_2^3 + t_f^3 - 3t_2 t_f^2 + 3t_2^2 t_f} \tag{12.47}$$

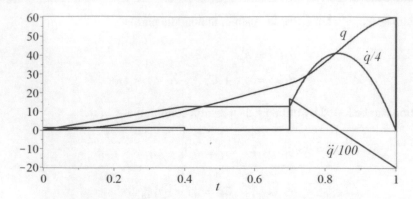

Fig. 12.4 A piecewise rest-to-rest path with a constant velocity in the middle

$$b_3 = \frac{q_c'}{-2t_2t_f + t_2^2 + t_f^2} + 2\frac{q_2}{-t_2^3 + t_f^3 - 3t_2t_f^2 + 3t_2^2t_f}$$
$$- 2\frac{q_f}{-t_2^3 + t_f^3 - 3t_2t_f^2 + 3t_2^2t_f} \tag{12.48}$$

A graph of the path for the following values is illustrated in Fig. 12.4.

$$t_1 = 0.4\,\text{s} \qquad t_2 = 0.7\,\text{s} \qquad t_f = 1\,\text{s}$$
$$q_0 = 0 \qquad q_f = 60\,\text{deg} \qquad q_c' = 50\,\text{deg}\,/\text{s} \tag{12.49}$$

12.2 Polynomial Path

Polynomial paths are the easiest functions to satisfy multiple conditions. The number of required conditions determines the degree of the polynomial for $q = q(t)$. In general, a polynomial path of degree n needs $n + 1$ conditions.

$$q(t) = a_0 + a_1t + a_2t^2 + \cdots + a_nt^n \tag{12.50}$$

The conditions may be of two types: positions at a series of points, so that the trajectory will pass through all specified points, or position, speed, acceleration, and jerk at two points, so that the smoothness of the path can be controlled.

The problem of searching for the coefficients of a polynomial reduces to a set of linear algebraic equations that may be solved numerically. However, the path planning can be simplified by splitting the whole path into a series of segments and utilizing combinations of lower order polynomials for different segments of the path. The polynomials must then be joined together to satisfy all the required boundary conditions. Examples will illustrate how to define conditions and determine associated polynomial path functions.

Example 384 Quintic path. To eliminate acceleration jump at both ends of a rest-to-rest path in quadratic path, we add the zero acceleration conditions at both ends. These new conditions require using a fifth degree path function.

Forcing a variable to have specific position, speed, and acceleration at boundaries introduces six conditions:

$$q(t_0) = q_0 \qquad \dot{q}(t_0) = q_0' \qquad \ddot{q}(t_0) = q_0''$$
$$q(t_f) = q_f \qquad \dot{q}(t_f) = q_f' \qquad \ddot{q}(t_f) = q_f'' \tag{12.51}$$

A five degree polynomial can satisfy these conditions,

$$q(t) = a_0 + a_1t + a_2t^2 + a_3t^3 + a_4t^4 + a_5t^5 \tag{12.52}$$

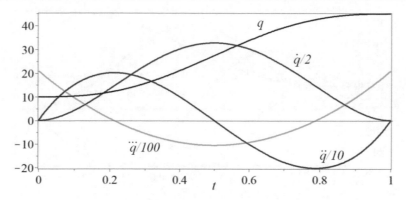

Fig. 12.5 A quintic rest-to-rest path

and generates a set of six equations.

$$
\begin{bmatrix}
1 & t_0 & t_0^2 & t_0^3 & t_0^4 & t_0^5 \\
0 & 1 & 2t_0 & 3t_0^2 & 4t_0^3 & 5t_0^4 \\
0 & 0 & 2 & 6t_0 & 12t_0^2 & 20t_0^3 \\
1 & t_f & t_f^2 & t_f^3 & t_f^4 & t_f^5 \\
0 & 1 & 2t_f & 3t_f^2 & 4t_f^3 & 5t_f^4 \\
0 & 0 & 2 & 6t_f & 12t_f^2 & 20t_f^3
\end{bmatrix}
\begin{bmatrix}
a_0 \\ a_1 \\ a_2 \\ a_3 \\ a_4 \\ a_5
\end{bmatrix}
=
\begin{bmatrix}
q_0 \\ q_0' \\ q_0'' \\ q_f \\ q_f' \\ q_f''
\end{bmatrix}
\tag{12.53}
$$

The set of equations is linear and can be solved by matrix inversion.

As an example, let us consider a rest-to-rest path with no acceleration at the rest positions with the following conditions:

$$
q(0) = 10 \text{ deg} \qquad \dot{q}(0) = 0 \qquad \ddot{q}(0) = 0
$$
$$
q(1) = 45 \text{ deg} \qquad \dot{q}(1) = 0 \qquad \ddot{q}(1) = 0
\tag{12.54}
$$

The coefficients of the quintic function can be found by solving a set of equations.

$$
\begin{bmatrix}
1 & 0 & 0 & 0 & 0 & 0 \\
0 & 1 & 0 & 0 & 0 & 0 \\
0 & 0 & 2 & 0 & 0 & 0 \\
1 & 1 & 1 & 1 & 1 & 1 \\
0 & 1 & 2 & 3 & 4 & 5 \\
0 & 0 & 2 & 6 & 12 & 20
\end{bmatrix}
\begin{bmatrix}
a_0 \\ a_1 \\ a_2 \\ a_3 \\ a_4 \\ a_5
\end{bmatrix}
=
\begin{bmatrix}
10 \\ 0 \\ 0 \\ 45 \\ 0 \\ 0
\end{bmatrix}
\tag{12.55}
$$

$$
\begin{bmatrix}
a_0 \\ a_1 \\ a_2 \\ a_3 \\ a_4 \\ a_5
\end{bmatrix}
=
\begin{bmatrix}
10 \\ 0 \\ 0 \\ 350 \\ -525 \\ 210
\end{bmatrix}
\tag{12.56}
$$

The path equation is then equal to

$$
q(t) = 10 + 350t^3 - 525t^4 + 210t^5
\tag{12.57}
$$

which is shown in Fig. 12.5.

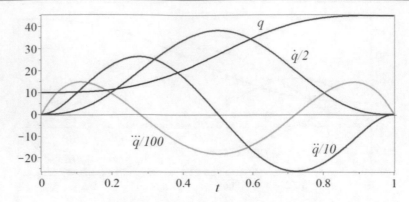

Fig. 12.6 A jerk zero at start–stop path

Example 385 Septic path. To remove the nonzero jerk at both ends of quintic path, a jerk zero at a start–stop path is needed. A seven degree polynomial can make a rest-to-rest path with no acceleration and no jerk at boundaries.

To make a path to start and stop with zero jerk, a seven degree polynomial and eight boundary conditions must be employed.

$$q(t) = a_0 + a_1 t + a_2 t^2 + a_3 t^3 + a_4 t^4 + a_5 t^5 + a_6 t^6 + a_7 t^7 \tag{12.58}$$

$$q(0) = q_0 \qquad \dot{q}(0) = 0 \qquad \ddot{q}(0) = 0 \qquad \dddot{q}(0) = 0$$

$$q(1) = q_f \qquad \dot{q}(1) = 0 \qquad \ddot{q}(1) = 0 \qquad \dddot{q}(1) = 0$$

$$\tag{12.59}$$

Such a zero jerk start–stop path for $q(0) = 10$ deg and $q(1) = 45$ deg can be found by solving the following set of equations for the unknown coefficients $a_0, a_1, \cdots, a_7$:

$$\begin{bmatrix} 1 & 0 & 0 & 0 & 0 & 0 & 0 & 0 \\ 0 & 1 & 0 & 0 & 0 & 0 & 0 & 0 \\ 0 & 0 & 2 & 0 & 0 & 0 & 0 & 0 \\ 0 & 0 & 0 & 6 & 0 & 0 & 0 & 0 \\ 1 & 1 & 1 & 1 & 1 & 1 & 1 & 1 \\ 0 & 1 & 2 & 3 & 4 & 5 & 6 & 7 \\ 0 & 0 & 2 & 6 & 12 & 20 & 30 & 42 \\ 0 & 0 & 0 & 6 & 24 & 60 & 120 & 210 \end{bmatrix} \begin{bmatrix} a_0 \\ a_1 \\ a_2 \\ a_3 \\ a_4 \\ a_5 \\ a_6 \\ a_7 \end{bmatrix} = \begin{bmatrix} 10 \\ 0 \\ 0 \\ 0 \\ 45 \\ 0 \\ 0 \\ 0 \end{bmatrix} \tag{12.60}$$

The solutions make the required seventh degree polynomial.

$$q(t) = 10 + 1225t^4 - 2940t^5 + 2450t^6 - 700t^7 \tag{12.61}$$

A graph of this path is illustrated in Fig. 12.6.

Figure 12.2 depicts the path of a rest-to-rest motion with no condition on acceleration and jerk. Figure 12.5 shows an improvement by forcing the motion to have zero accelerations at start and stop. In Fig. 12.6, the motion is forced to have zero acceleration and zero jerk at start and stop. Hence, it shows the smoothest start and stop. However, increasing the smoothness of the start and stop increases the peak value of acceleration.

Example 386 Constant acceleration path. Considering acceleration to be proportional to force, minimization of the maximum acceleration could be optimal path planning for some applications. The minimum acceleration happens when it remains constant during the motion. Here is the path planning for constant acceleration.

A rest-to-rest path with constant acceleration has two segments with positive and negative accelerations. Let us assume the absolute value of the positive and negative accelerations is given.

$$|\ddot{q}(t)| = a_c \tag{12.62}$$

For this symmetric assumption, the switching point will happen at the midway. The first half of the motion has a positive acceleration and a second degree polynomial is the simplest function to satisfy the initial conditions.

$$q_1(t) = \frac{1}{2}a_c t^2 + q_0 \qquad \dot{q}_1(t) = a_c t \tag{12.63}$$

$$0 < t < \frac{1}{2}t_f \tag{12.64}$$

The constants of integration of Eq. (12.62) or the coefficients of the path function are found based on the initial conditions.

$$q_1(0) = q_0 \qquad \dot{q}_1(0) = 0 \tag{12.65}$$

For the second half of the path, we start with a general second degree polynomial,

$$q_2(t) = a_0 + a_1 t + a_2 t^2 \qquad \frac{1}{2}t_f < t < t_f \tag{12.66}$$

and impose the following boundary conditions to determine the coefficients:

$$q_2(t_f) = q_f \qquad a_0 + a_1 t_f + a_2 t_f^2 = q_f \tag{12.67}$$

$$\dot{q}_2(t_f) = 0 \qquad a_1 + 2a_2 t_f = 0 \tag{12.68}$$

$$q_1\left(\frac{t_f}{2}\right) = q_2\left(\frac{t_f}{2}\right) \tag{12.69}$$

$$\frac{a_c}{2}\left(\frac{t_f}{2}\right)^2 + q_0 = a_0 + a_1\frac{t_f}{2} + a_2\left(\frac{t_f}{2}\right)^2 \tag{12.70}$$

$$\dot{q}_1\left(\frac{t_f}{2}\right) = \dot{q}_2\left(\frac{t_f}{2}\right) \qquad a_c\frac{t_f}{2} = a_1 + a_2 t_f$$

There are four conditions that make four equations, and there are five unknowns, a_0, a_1, a_2, a_c, and t_f. Out of these parameters, we may choose either the time of travel t_f or the value of acceleration a_c, but not both. Let us assume the total time t_f is set to generate four equations for the unknown parameters a_0, a_1, a_2, and a_c.

$$\begin{bmatrix} 1 & t_f & t_f^2 & 0 \\ 0 & 1 & 2t_f & 0 \\ 1 & \frac{1}{2}t_f & \frac{1}{4}t_f^2 & -\frac{1}{8}t_f^2 \\ 0 & 1 & t_f & -\frac{1}{2}t_f \end{bmatrix} \begin{bmatrix} a_0 \\ a_1 \\ a_2 \\ a_c \end{bmatrix} = \begin{bmatrix} q_f \\ 0 \\ q_0 \\ 0 \end{bmatrix} \tag{12.71}$$

The equations will have the following solutions:

$$\begin{bmatrix} a_0 \\ a_1 \\ a_2 \\ a_c \end{bmatrix} = \begin{bmatrix} 2q_0 - q_f \\ \dfrac{4}{t_f}(q_f - q_0) \\ -\dfrac{2}{t_f^2}(q_f - q_0) \\ \dfrac{4}{t_f^2}(q_f - q_0) \end{bmatrix} \tag{12.72}$$

Having a set of data as

$$q_0 = 10 \text{ deg} \qquad q_f = 45 \text{ deg} \qquad t_f = 1 \tag{12.73}$$

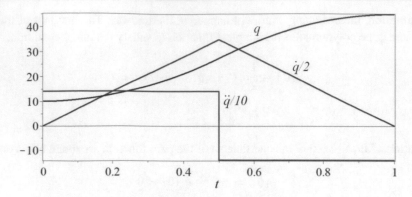

Fig. 12.7 A constant acceleration path

provides us with

$$a_0 = -25 \qquad a_1 = 140 \qquad a_2 = -70 \qquad a_c = 140 \tag{12.74}$$

Figure 12.7 illustrates kinematics of constant acceleration path.

Example 387 Point sequence path. If the whole path of motion is important, we may simulate the path with a series of points where the variable must be at particular times. Polynomial path planning is an easy way to determine the path.

A path can be assigned via a series of points that the variable must attain at specific times. The points may also be located to approximate a trajectory. Consider an example path specified by four points q_0, q_1, q_2, and q_3, such that the points are reached at times t_0, t_1, t_2, and t_3, respectively. In addition to positions, we usually impose constraint on initial and final velocities and accelerations.

$$q(t_0) = q_0 \qquad \dot{q}(t_0) = 0 \qquad \ddot{q}(t_0) = 0$$

$$q(t_1) = q_1 \qquad q(t_2) = q_2$$

$$q(t_3) = q_3 \qquad \dot{q}(t_3) = 0 \qquad \ddot{q}(t_3) = 0 \tag{12.75}$$

There are eight conditions, and hence, a seven degree polynomial can be utilized to satisfy the conditions.

$$q(t) = a_0 + a_1 t + a_2 t^2 + a_3 t^3 + a_4 t^4$$
$$+ a_5 t^5 + a_6 t^6 + a_7 t^7 \tag{12.76}$$
$$t_0 = 0 \tag{12.77}$$

The set of equations for the unknown coefficients is

$$
\begin{bmatrix}
1 & t_0 & t_0^2 & t_0^3 & t_0^4 & t_0^5 & t_0^6 & t_0^7 \\
0 & 1 & 2t_0 & 3t_0^2 & 4t_0^3 & 5t_0^4 & 6t_0^5 & 7t_0^6 \\
0 & 0 & 2 & 6t_0 & 12t_0^2 & 20t_0^3 & 30t_0^4 & 42t_0^5 \\
1 & t_1 & t_1^2 & t_1^3 & t_1^4 & t_1^5 & t_1^6 & t_1^7 \\
1 & t_2 & t_2^2 & t_2^3 & t_2^4 & t_2^5 & t_2^6 & t_2^7 \\
1 & t_3 & t_3^2 & t_3^3 & t_3^4 & t_3^5 & t_3^6 & t_3^7 \\
0 & 1 & 2t_3 & 3t_3^2 & 4t_3^3 & 5t_3^4 & 6t_3^5 & 7t_3^6 \\
0 & 0 & 2 & 6t_3 & 12t_3^2 & 20t_3^3 & 30t_3^4 & 42t_3^5
\end{bmatrix}
\begin{bmatrix}
a_0 \\ a_1 \\ a_2 \\ a_3 \\ a_4 \\ a_5 \\ a_6 \\ a_7
\end{bmatrix}
=
\begin{bmatrix}
q_0 \\ 0 \\ 0 \\ q_1 \\ q_2 \\ q_3 \\ 0 \\ 0
\end{bmatrix}
\tag{12.78}
$$

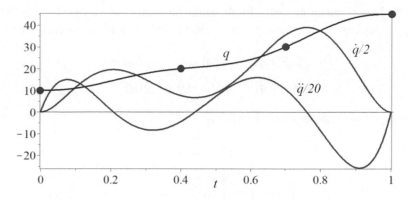

Fig. 12.8 A point sequence path

which can be simplified to

$$
\begin{bmatrix}
1 & 0 & 0 & 0 & 0 & 0 & 0 & 0 \\
0 & 1 & 0 & 0 & 0 & 0 & 0 & 0 \\
0 & 0 & 2 & 0 & 0 & 0 & 0 & 0 \\
1 & 0.4 & 0.4^2 & 0.4^3 & 0.4^4 & 0.4^5 & 0.4^6 & 0.4^7 \\
1 & 0.7 & 0.7^2 & 0.7^3 & 0.7^4 & 0.7^5 & 0.7^6 & 0.7^7 \\
1 & 1 & 1 & 1 & 1 & 1 & 1 & 1 \\
0 & 1 & 2 & 3 & 4 & 5 & 6 & 7 \\
0 & 0 & 2 & 6 & 12 & 20 & 30 & 42
\end{bmatrix}
\begin{bmatrix}
a_0 \\ a_1 \\ a_2 \\ a_3 \\ a_4 \\ a_5 \\ a_6 \\ a_7
\end{bmatrix}
=
\begin{bmatrix}
10 \\ 0 \\ 0 \\ 20 \\ 30 \\ 45 \\ 0 \\ 0
\end{bmatrix}
\tag{12.79}
$$

for the following example data:

$$q(0) = 10 \text{ deg} \qquad \dot{q}(0) = 0 \qquad \ddot{q}(0) = 0$$

$$q(0.4) = 20 \text{ deg}$$

$$q(0.7) = 30 \text{ deg}$$

$$q(1) = 45 \text{ deg} \qquad \dot{q}(1) = 0 \qquad \ddot{q}(1) = 0 \tag{12.80}$$

The solution for the coefficients is

$$
\begin{bmatrix}
a_0 \\ a_1 \\ a_2 \\ a_3 \\ a_4 \\ a_5 \\ a_6 \\ a_7
\end{bmatrix}
=
\begin{bmatrix}
10 \\ 0 \\ 0 \\ 1500.5 \\ -7053 \\ 12891 \\ -10380 \\ 3076.9
\end{bmatrix}
\tag{12.81}
$$

These coefficients generate a path as shown in Fig. 12.8.

This method provides a continuous and differentiable function for the variable q. Continuity and differentiability of $q = q(t)$ is an advantage to provide a continuous velocity, acceleration, and jerk. However, the number of equations increases by increasing the number of points, which needs larger data storage and increases the calculating time.

Example 388 Splitting a path into a series of segments. When the number of path points is too many, we may split the whole path into smaller segments with a low polynomial function in each segment. Sawing the segments is the main art of this path planning method.

Instead of using a single high degree polynomial for the entire trajectory, we may prefer to split the trajectory into some segments and use a simpler function of low degree polynomials. Consider a path for the following boundary conditions:

$$q(t_0) = q_0 \qquad \dot{q}(t_0) = 0 \qquad \ddot{q}(t_0) = 0$$

$$q(t_4) = q_3 \qquad \dot{q}(t_4) = 0 \qquad \ddot{q}(t_4) = 0 \tag{12.82}$$

which must also pass through three middle points given below:

$$q(t_1) = q_1$$

$$q(t_2) = q_2$$

$$q(t_3) = q_3 \tag{12.83}$$

Let us split the entire path into four segments, namely $q_1(t)$, $q_2(t)$, $q_3(t)$, and $q_4(t)$.

$$q_1(t) \text{ for } q(t_0) < q_1(t) < q(t_1) \text{ and } t_0 < t < t_1$$
$$q_2(t) \text{ for } q(t_1) < q_2(t) < q(t_2) \text{ and } t_1 < t < t_2$$
$$q_3(t) \text{ for } q(t_2) < q_3(t) < q(t_3) \text{ and } t_2 < t < t_3$$
$$q_4(t) \text{ for } q(t_3) < q_4(t) < q(t_4) \text{ and } t_3 < t < t_4$$

The boundary conditions for the first segment are

$$q_1(t_0) = q_0 \qquad \dot{q}_1(t_0) = 0 \qquad \ddot{q}_1(t_0) = 0$$

$$q_1(t_1) = q_1 \tag{12.84}$$

which can be satisfied by a cubic function.

$$q_1(t) = a_0 + a_1 (t - t_0) + a_2 (t - t_0)^2 + a_3 (t - t_0)^3 \tag{12.85}$$

The coefficients can be calculated by solving a set of algebraic equations.

$$\begin{bmatrix} 1 & 0 & 0 & 0 \\ 0 & 1 & 0 & 0 \\ 0 & 0 & 2 & 0 \\ 1 & (t_1 - t_0) & (t_1 - t_0)^2 & (t_1 - t_0)^3 \end{bmatrix} \begin{bmatrix} a_0 \\ a_1 \\ a_2 \\ a_3 \end{bmatrix} = \begin{bmatrix} q_0 \\ 0 \\ 0 \\ q_1 \end{bmatrix} \tag{12.86}$$

$$a_0 = q_0 \qquad a_1 = 0 \qquad a_2 = 0 \qquad a_3 = \frac{q_1 - q_0}{(t_1 - t_0)^3} \tag{12.87}$$

The path in the second segment must satisfy the following boundary conditions:

$$q_2(t_1) = q_1$$

$$\dot{q}_2(t_1) = \dot{q}_1(t_1) = a_1 + 2a_2 (t_1 - t_0)^2 + 3a_3 (t_1 - t_0)^2$$

$$= q_0 + 3\frac{q_1 - q_0}{t_1 - t_0}$$

$$q_2(t_2) = q_2 \tag{12.88}$$

A quadratic polynomial will satisfy these conditions:

$$q_2(t) = b_0 + b_1 t + b_2 t^2 \tag{12.89}$$

The coefficients are the solutions of three equations.

$$\begin{bmatrix} 1 & t_1 & t_1^2 \\ 0 & 1 & 2t_1 \\ 1 & t_2 & t_2^2 \end{bmatrix} \begin{bmatrix} b_0 \\ b_1 \\ b_2 \end{bmatrix} = \begin{bmatrix} q_1 \\ q_0 + 3\frac{q_1 - q_0}{t_1 - t_0} \\ q_2 \end{bmatrix} \tag{12.90}$$

$$b_0 = q_2 \frac{t_1^2}{-2t_1t_2 + t_1^2 + t_2^2} + q_1 \frac{-2t_1t_2 + t_2^2}{-2t_1t_2 + t_1^2 + t_2^2}$$
$$-t_1 \frac{t_2}{-t_1 + t_2} \left(q_0 + 3\frac{-q_0 + q_1}{-t_0 + t_1} \right) \tag{12.91}$$

$$b_1 = 2q_1 \frac{t_1}{-2t_1t_2 + t_1^2 + t_2^2} - 2q_2 \frac{t_1}{-2t_1t_2 + t_1^2 + t_2^2}$$
$$+\frac{t_1 + t_2}{-t_1 + t_2} \left(q_0 + 3\frac{-q_0 + q_1}{-t_0 + t_1} \right) \tag{12.92}$$

$$b_2 = -\frac{q_1}{-2t_1t_2 + t_1^2 + t_2^2} + \frac{q_2}{-2t_1t_2 + t_1^2 + t_2^2}$$
$$-\frac{1}{-t_1 + t_2} \left(q_0 + 3\frac{-q_0 + q_1}{-t_0 + t_1} \right) \tag{12.93}$$

The boundary conditions in the third segment are

$$q_3(t_2) = q_2 \qquad \dot{q}_3(t_2) = \dot{q}_2(t_2) = b_1 + 2b_2t_2$$
$$q_3(t_3) = q_3 \tag{12.94}$$

We can satisfy these conditions with a quadratic equation,

$$q_3(t) = c_0 + c_1 t + c_2 t^2 \tag{12.95}$$

which provides three equations for the unknown coefficients.

$$\begin{bmatrix} 1 & t_2 & t_2^2 \\ 0 & 1 & 2t_2 \\ 1 & t_3 & t_3^2 \end{bmatrix} \begin{bmatrix} c_0 \\ c_1 \\ c_2 \end{bmatrix} = \begin{bmatrix} q_2 \\ b_1 + 2b_2t_2 \\ q_3 \end{bmatrix} \tag{12.96}$$

$$c_0 = -t_2 \frac{t_3}{-t_2 + t_3} (b_1 + 2b_2t_2) + q_3 \frac{t_2^2}{-2t_2t_3 + t_2^2 + t_3^2}$$
$$+q_2 \frac{-2t_2t_3 + t_3^2}{-2t_2t_3 + t_2^2 + t_3^2} \tag{12.97}$$

$$c_1 = \frac{t_2 + t_3}{-t_2 + t_3} (b_1 + 2b_2t_2) + 2q_2 \frac{t_2}{-2t_2t_3 + t_2^2 + t_3^2}$$
$$-2q_3 \frac{t_2}{-2t_2t_3 + t_2^2 + t_3^2} \tag{12.98}$$

$$c_2 = -\frac{1}{-t_2 + t_3} (b_1 + 2b_2t_2) - \frac{q_2}{-2t_2t_3 + t_2^2 + t_3^2}$$
$$+\frac{q_3}{-2t_2t_3 + t_2^2 + t_3^2}. \tag{12.99}$$

The boundary conditions for the fourth segment are

$$q_4(t_3) = q_3 \qquad \dot{q}_4(t_3) = \dot{q}_3(t_3) = c_1 + 2c_2t_3$$

$$q_4(t_4) = q_4 \qquad \dot{q}_4(t_4) = 0 \qquad \ddot{q}_4(t_4) = 0 \tag{12.100}$$

which needs a fourth degree polynomial to be satisfied.

$$q_4(t) = d_0 + d_1t + d_2t^2 + d_3t^3 + d_4t^4 \tag{12.101}$$

Substituting the boundary conditions generates a set of four equations for the coefficient.

$$\begin{bmatrix} 1 & t_3 & t_3^2 & t_3^3 & t_3^4 \\ 0 & 1 & 2t_2 & 3t_2^2 & 4t_2^3 \\ 1 & t_4 & t_4^2 & t_4^3 & t_4^4 \\ 0 & 1 & 2t_4 & 3t_4^2 & 4t_4^3 \\ 0 & 0 & 2 & 6t_4 & 12t_4^2 \end{bmatrix} \begin{bmatrix} d_0 \\ d_1 \\ d_2 \\ d_3 \\ d_4 \end{bmatrix} = \begin{bmatrix} q_3 \\ c_1 + 2c_2t_3 \\ q_4 \\ 0 \\ 0 \end{bmatrix} \tag{12.102}$$

As an example, a set of conditions given by

$$t_0 = 0 \qquad t_1 = 0.4 \qquad t_2 = 0.7 \qquad t_3 = 0.9 \qquad t_4 = 1 \tag{12.103}$$

$$q(0) = 10 \text{ deg} \qquad \dot{q}(0) = 0 \qquad \ddot{q}(0) = 0$$

$$q(0.4) = 20 \text{ deg}$$

$$q(0.7) = 30 \text{ deg}$$

$$q(0.9) = 35 \text{ deg}$$

$$q(1) = 45 \text{ deg} \qquad \dot{q}_i(1) = 0 \qquad \ddot{q}(1) = 0 \tag{12.104}$$

yields

$$q_1(t) = 10 + 156.25t^3 \tag{12.105}$$

$$q_2(t) = -41.56 + 222.78t - 172.2t^2 \tag{12.106}$$

$$q_3(t) = 148.99 - 321.67t + 216.67t^2 \tag{12.107}$$

$$q_4(t) = 198545 - 827166.6672t$$
$$+1290500t^2 - 893500t^3 + 231666.67t^4 \tag{12.108}$$

It is shown in Fig. 12.9 graphically.

The main disadvantage of the segment method is the lack of a smooth overall path and having a discontinuous acceleration. To increase the smoothness of the path, we need to use higher degree polynomials and put constraints on acceleration and possibly jerk.

Equations (12.105)–(12.108) indicate that

$$\ddot{q}_1(t_1) = 375 \qquad \ddot{q}_2(t_1) = -344.4 \tag{12.109}$$

$$\ddot{q}_2(t_2) = -344.4 \qquad \ddot{q}_3(t_2) = 433.34 \tag{12.110}$$

$$\ddot{q}_3(t_3) = 433.34 \qquad \ddot{q}_4(t_3) = 7900 \tag{12.111}$$

$$\ddot{q}_1(t_1) \neq \ddot{q}_2(t_1) \qquad \ddot{q}_2(t_2) \neq \ddot{q}_3(t_2) \qquad \ddot{q}_3(t_3) \neq \ddot{q}_4(t_3) \tag{12.112}$$

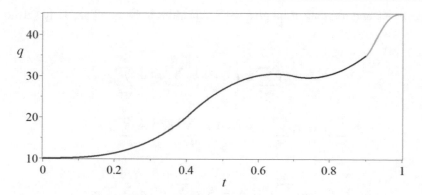

Fig. 12.9 Splitting a path into a series of segments

Therefore, the acceleration of the path is not continuous at the connection points and shows a finite jump. A jump in acceleration introduces an infinity jerk. Having continuous acceleration is the minimum requirement for smoothness of a path. A piecewise path with continuous acceleration is called spline.

Example 389 ★ Least-squares polynomial. When there are many number of points to define a path, it is easier to plan the path by a desired function and determine its coefficients such that the function move over the points or as close as possible. Least-square is a common method to determine the coefficients of a selected mathematical function for the path.

When the number of points to approximate a trajectory is too large, we may use a low degree polynomial to pass close to the points. Least-squares is an applied method to determine the coefficients of a selected polynomial to approximate the path. Consider a path with N given points, $p_1, p_2, \ldots, p_N$.

$$p_i = p(t_i) \qquad i = 1, 2, 3, \cdots, N \tag{12.113}$$

A polynomial of degree n is supposed to approximate the path. If $N = n + 1$, then the polynomial passes exactly through all given points. To work with lower degree polynomials, we choose $n < N + 1$.

$$q = a_0 + a_1 t + a_2 t^2 + \cdots + a_n t^n \tag{12.114}$$

Having the N points (12.113) and the polynomial (12.114), we define an error e_i at t_i.

$$e_i = p_i - q_i = p_i - a_0 - a_1 t_i - a_2 t_i^2 - \cdots - a_n t_i^n \tag{12.115}$$

The sum of e_i^2 for all points p_i is the total error e.

$$e = \sum_{i=1}^{N} e_i^2 = \sum_{i=1}^{N} \left(p_i - a_0 - a_1 t_i - a_2 t_i^2 - \cdots - a_n t_i^n \right)^2 \tag{12.116}$$

The minimum error e provides the best approximate polynomial (12.114). At the minimum, all the partial derivatives $\partial e / \partial a_0$, $\partial e / \partial a_1, \cdots, \partial e / \partial a_n$ vanish. These conditions generate $n + 1$ equations for the unknown coefficients, $a_0, a_1, a_2, \cdots$, and a_n.

$$\frac{\partial e}{\partial a_0} = -2 \sum_{i=1}^{N} \left(p_i - a_0 - a_1 t_i - a_2 t_i^2 - \cdots - a_n t_i^n \right) = 0$$

$$\frac{\partial e}{\partial a_1} = -2 \sum_{i=1}^{N} t_i \left(p_i - a_0 - a_1 t_i - a_2 t_i^2 - \cdots - a_n t_i^n \right) = 0$$

$$\cdots$$

$$\frac{\partial e}{\partial a_n} = -2 \sum_{i=1}^{N} t_i^n \left(p_i - a_0 - a_1 t_i - a_2 t_i^2 - \cdots - a_n t_i^n \right) = 0 \tag{12.117}$$

Dividing each equation by -2 and rearrangement give $n + 1$ equations to be simultaneously solved for the coefficients a_i, $i = 1, 2, \cdots, n$.

$$a_0 N + a_1 \sum_{i=1}^{N} t_i + \cdots + a_n \sum_{i=1}^{N} t_i^n = \sum_{i=1}^{N} p_i$$

$$a_0 \sum_{i=1}^{N} t_i + a_1 \sum_{i=1}^{N} t_i^2 + \cdots + a_n \sum_{i=1}^{N} t_i^{n+1} = \sum_{i=1}^{N} t_i p_i$$

$$\cdots$$

$$a_0 \sum_{i=1}^{N} t_i^n + a_1 \sum_{i=1}^{N} t_i^{n+1} + \cdots + a_n \sum_{i=1}^{N} t_i^{2n} = \sum_{i=1}^{N} t_i^n p_i \qquad (12.118)$$

The equations can be solved by matrix inversion when they are set in matrix form.

$$[A]\,\mathbf{a} = \mathbf{b} \qquad (12.119)$$

$$[A] = \begin{bmatrix} N & \sum_{i=1}^{N} t_i & \sum_{i=1}^{N} t_i^2 & \cdots & \sum_{i=1}^{N} t_i^n \\ \sum_{i=1}^{N} t_i & \sum_{i=1}^{N} t_i^2 & \sum_{i=1}^{N} t_i^3 & \cdots & \sum_{i=1}^{N} t_i^{n+1} \\ \cdots & \cdots & \cdots & \cdots & \cdots \\ \sum_{i=1}^{N} t_i^n & \sum_{i=1}^{N} t_i^{n+1} & \sum_{i=1}^{N} t_i^{n+2} & \cdots & \sum_{i=1}^{N} t_i^{2n} \end{bmatrix} \qquad (12.120)$$

$$\mathbf{a} = \begin{bmatrix} a_0 \\ a_1 \\ a_2 \\ \cdots \\ a_n \end{bmatrix} \qquad \mathbf{b} = \begin{bmatrix} \sum_{i=1}^{N} p_i \\ \sum_{i=1}^{N} t_i p_i \\ \sum_{i=1}^{N} t_i^2 p_i \\ \cdots \\ \sum_{i=1}^{N} t_i^n p_i \end{bmatrix} \qquad (12.121)$$

12.3 ★ Non-polynomial Path Planning

A path of motion in either joint or Cartesian spaces may be defined based on different mathematical functions. Harmonic and cycloid functions are the most common paths.

$$q(t) = a_0 + a_1 \cos a_2 t + a_3 \sin a_2 t \qquad (12.122)$$

$$q(t) = a_0 + a_1 t - a_2 \sin a_3 t \qquad (12.123)$$

However, we may also use other function approximate methods such as Fourier,

$$q(t) = \frac{A_0}{2} + \sum_{n=1}^{\infty} [A_n \cos(nx) + B_n \sin(nx)] \qquad (12.124)$$

$$A_0 = \frac{1}{\pi} \int_{-\pi}^{\pi} q(t) dt \qquad (12.125)$$

$$A_n = \frac{1}{\pi} \int_{-\pi}^{\pi} q(t) \cos(nx)\, dt \qquad (12.126)$$

$$B_n = \frac{1}{\pi} \int_{-\pi}^{\pi} q(t) \sin(nx)\, dt \qquad (12.127)$$

Legendre,

$$q_n(t) = \sum_{i=0}^{n} L_i(t)q(t_i) \tag{12.128}$$

$$L_i(t) = \prod_{j=0, j \neq i}^{n} \frac{t - t_j}{t_i - t_j} \qquad i = 0, 1, 2, \dots, n \tag{12.129}$$

and Chebyshev

$$q_{n+1}(t) = 2t q_n(t) - q_{n-1}(t) \tag{12.130}$$

$$q_0(t) = 1 \qquad q_1(t) = t \tag{12.131}$$

The main challenge of non-polynomial equations compared to polynomial functions is working with nonlinear algebraic equations to determine unknown coefficients.

Example 390 Harmonic path. A full single harmonic function can be made with four coefficients, and hence, such function can satisfy four conditions similar to cubic functions. However, harmonic functions give a smoother behavior. Here is to show how to determine coefficients of a full harmonic function for rest-to-rest path.

Consider a harmonic path between two points $q(t_0)$ and $q(t_f)$,

$$q(t) = a_0 + a_1 \cos a_2 t + a_3 \sin a_2 t \tag{12.132}$$

with the rest-to-rest boundary conditions.

$$q(t_0) = q_0 \qquad \dot{q}(t_0) = 0$$
$$q(t_f) = q_f \qquad \dot{q}(t_f) = 0 \tag{12.133}$$

Applying the conditions to the harmonic Eq. (12.132) provides the following solutions:

$$q_0 = a_0 + a_1 \cos a_2 t_0 + a_3 \sin a_2 t_0 \tag{12.134}$$

$$q_f = a_0 + a_1 \cos a_2 t_f + a_3 \sin a_2 t_f \tag{12.135}$$

$$\dot{q}_0 = -a_1 \sin a_2 t_0 + a_3 \cos a_2 t_0 \tag{12.136}$$

$$\dot{q}_f = -a_1 \sin a_2 t_f + a_3 \cos a_2 t_f \tag{12.137}$$

The equations are nonlinear and it is the main disadvantage of using harmonic functions. Assuming

$$t_0 = 0 \qquad q_0 = 10 \text{ deg} \qquad \dot{q}_0 = 0$$
$$t_f = 1 \qquad q_f = 45 \text{ deg} \qquad \dot{q}_f = 0 \tag{12.138}$$

makes the equations to be

$$10 = a_0 + a_1 \qquad 45 = a_0 + a_1 \cos a_2 + a_3 \sin a_2 \tag{12.139}$$

$$0 = a_3 a_2 \qquad 0 = -a_1 \sin a_2 + a_3 \cos a_2 \tag{12.140}$$

and their solutions are

$$a_0 = 55/2 \qquad a1 = -35/2 \qquad a2 = \pi \qquad a3 = 0 \tag{12.141}$$

These coefficients make the harmonic path.

$$q(t) = \frac{55}{2} - \frac{35}{2} \cos \pi t \tag{12.142}$$

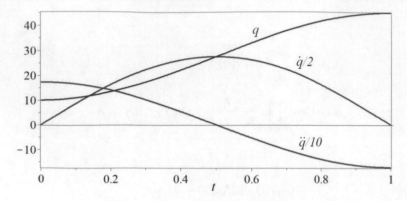

Fig. 12.10 A harmonic path

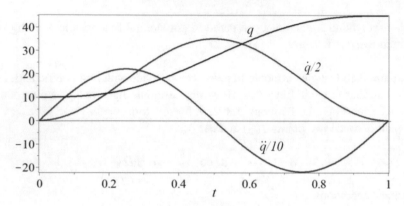

Fig. 12.11 A cycloid path

The general solution is

$$q(t) = \frac{1}{2}\left(q_f + q_0 - (q_f - q_0)\cos\frac{\pi(t - t_0)}{t_f - t_0}\right) \tag{12.143}$$

A plot of the solution is depicted in Fig. 12.10.

Example 391 Cycloid path. For the rest-to-rest maneuver, a cycloid function provides a much smoother path than polynomial and harmonic paths.

A cycloid path between two points $q(t_0)$ and $q(t_f)$ with rest-to-rest boundary conditions

$$q(t_0) = q_0 \qquad \dot{q}(t_0) = 0$$
$$q(t_f) = q_f \qquad \dot{q}(t_f) = 0 \tag{12.144}$$

is

$$q(t) = q_0 + \frac{q_f - q_0}{\pi}\left(\frac{\pi(t - t_0)}{t_f - t_0} - \frac{1}{2}\sin\frac{2\pi(t - t_0)}{t_f - t_0}\right) \tag{12.145}$$

A plot of the cycloid path is illustrated in Fig. 12.11 for the following numerical values:

$$t_0 = 0 \qquad q_0 = 10 \text{ deg} \qquad \dot{q}_0 = 0$$
$$t_f = 1 \qquad q_f = 45 \text{ deg} \qquad \dot{q}_f = 0 \tag{12.146}$$

Comparing Fig. 12.11 with 12.5 indicates that the main kinematic characteristics of a cycloid path are similar to quintic rest-to-rest path.

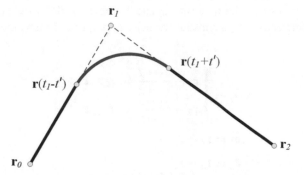

Fig. 12.12 Transition parabola between two line segments as a path in Cartesian space

12.4 ★ Spatial Path Design

Cartesian spatial path is a curve in base frame between two given points P_0 and P_f that the end-effector of a manipulator is supposed to move on during $t_0 \le t \le t_f$.

$$P_0 = P_0 (X_0, Y_0, Z_0) \qquad P_f = P_f (X_f, Y_f, Z_f) \tag{12.147}$$

The Cartesian path may be defined by a parametric function as a geometric space curve,

$$Z = Z(X) \qquad Y = Y(X) \tag{12.148}$$

$$X = X(t) \tag{12.149}$$

where

$$X(t_0) = X_0 \qquad X(t_f) = X_f \tag{12.150}$$

To move on the curve, we define a time path function for X between P_0 and P_f to relate all three coordinates to the time variable on the geometric path (12.148).

By defining a position vector $\mathbf{r}(t) = X\hat{I} + Y\hat{J} + Z\hat{K}$, designing the spatial curve will be equal to designing a path for $\mathbf{r}(t)$. A practical method is to design a spatial curve utilizing straight lines with constant velocity and avoid the sharp corners by deformed smooth transition curves. The path connecting points $\mathbf{r}_0$ to $\mathbf{r}_2$, and passing close to a sharp corner $\mathbf{r}_1$ on a transition curve, can be designed by a piecewise function.

$$
\begin{aligned}
\mathbf{r}(t) &= \mathbf{r}_1 - \frac{t_1 - t}{t_1 - t_0} (\mathbf{r}_1 - \mathbf{r}_0) & t_0 \le t \le t_1 - t' \\
\mathbf{r}(t) &= \mathbf{r}_1 - \frac{(t - t' - t_1)^2}{4t' (t_1 - t_0)} (\mathbf{r}_1 - \mathbf{r}_0) \\
&\quad + \frac{(t + t' - t_1)^2}{4t' (t_2 - t_1)} (\mathbf{r}_2 - \mathbf{r}_1) & t_1 - t' \le t \le t_1 + t' \\
\mathbf{r}(t) &= \mathbf{r}_1 - \frac{t_1 - t}{t_2 - t_1} (\mathbf{r}_2 - \mathbf{r}_1) & t_1 + t' \le t \le t_2
\end{aligned}
\tag{12.151}
$$

The path starts from $\mathbf{r}_0$ at time t_0 and moves with constant velocity $\mathbf{v}_1 = \frac{\mathbf{r}_1 - \mathbf{r}_0}{t_1 - t_0}$ along a line until the point at switching time $t_1 - t'$. At this time, the path switches to a constant acceleration parabola. At another switching point at time $t_1 + t'$, the path switches to the second line and moves with constant velocity $\mathbf{v}_2 = \frac{\mathbf{r}_2 - \mathbf{r}_1}{t_2 - t_1}$ toward the destination at point $\mathbf{r}_2$. The connecting time t' is selected by the designer. The time $t_1 - t_0$ is the required time to move from $\mathbf{r}_0$ to $\mathbf{r}_1$ and $t_2 - t_1$ is the required time to move from $\mathbf{r}_1$ to $\mathbf{r}_2$, if there were no transition path. The path is shown in Fig. 12.12 schematically.

Proof The first line segment starts from a point $\mathbf{r}_0$ at time t_0 and, without any deformation, it arrives at point $\mathbf{r}_1$ at time t_1 at a constant velocity. The second line ends with a constant velocity at point $\mathbf{r}_2$ at time t_2 and, without deformation, it would start from point $\mathbf{r}_1$ at time t_1.

$$\mathbf{r}(t) = \begin{cases} \mathbf{r}_1 - \dfrac{t_1 - t}{t_1 - t_0}\boldsymbol{\delta}_1 & t_0 \le t \le t_1 \\ \mathbf{r}_1 - \dfrac{t_1 - t}{t_2 - t_1}\boldsymbol{\delta}_2 & t_1 \le t \le t_2 \end{cases} \tag{12.152}$$

$$\boldsymbol{\delta}_1 = \mathbf{r}_1 - \mathbf{r}_0 \tag{12.153}$$

$$\boldsymbol{\delta}_2 = \mathbf{r}_2 - \mathbf{r}_1 \tag{12.154}$$

$$\dot{\mathbf{r}}(t) = \begin{cases} \dfrac{\boldsymbol{\delta}_1}{t_1 - t_0} & t_0 \le t \le t_1 \\ \dfrac{\boldsymbol{\delta}_2}{t_2 - t_1} & t_1 \le t \le t_2 \end{cases} \tag{12.155}$$

Let us introduce an interval time t' before arriving at $\mathbf{r}_1$ to switch from the line to a transition curve. The transition curve is then between times $t_1 - t'$ and $t_1 + t'$. The simplest transition curve is a parabola which, at the end points, will have the same speed as the line segments.

The boundary positions of the transition curve on the first and second lines are

$$\mathbf{r}(t_1 - t') = \mathbf{r}_1 - \frac{t'}{t_1 - t_0}\boldsymbol{\delta}_1 \tag{12.156}$$

$$\mathbf{r}(t_1 + t') = \mathbf{r}_1 + \frac{t'}{t_2 - t_1}\boldsymbol{\delta}_2 \tag{12.157}$$

The velocity at the beginning and final points of the transition curve is, respectively, equal to

$$\dot{\mathbf{r}}(t_1 - t') = \frac{1}{t_1 - t_0}\boldsymbol{\delta}_1 \tag{12.158}$$

$$\dot{\mathbf{r}}(t_1 + t') = \frac{1}{t_2 - t_1}\boldsymbol{\delta}_2 \tag{12.159}$$

Assume the acceleration of motion along the transition curve is constant

$$\ddot{\mathbf{r}}(t) = \ddot{\mathbf{r}}_c = const \tag{12.160}$$

and therefore, the transition curve after integration is equal to

$$\mathbf{r}(t) = \mathbf{r}(t_1 - t') + (t - t_1 + t')\dot{\mathbf{r}}(t_1 - t') + \frac{1}{2}(t - t_1 + t')^2\ddot{\mathbf{r}}_c \tag{12.161}$$

Substituting (12.156) and (12.158) yields

$$\mathbf{r}(t) = \mathbf{r}_1 + \frac{t - t_1}{t_1 - t_0}\boldsymbol{\delta}_1 + \frac{1}{2}\ddot{\mathbf{r}}_c(t - t_1 + t')^2 \tag{12.162}$$

The transition curve $\mathbf{r}(t)$ must be at the end point when $t = t_1 + t'$,

$$\mathbf{r}(t_1 + t') = \mathbf{r}_1 + \frac{t'}{t_2 - t_1}\boldsymbol{\delta}_2 = \mathbf{r}_1 + \frac{t'}{t_1 - t_0}\boldsymbol{\delta}_1 + 2\ddot{\mathbf{r}}_c\, t_1^2 \tag{12.163}$$

and therefore, the acceleration on the curve is equal to

$$\ddot{r}_c = \frac{1}{2t'} \left(\frac{\delta_2}{t_2 - t_1} - \frac{\delta_1}{t_1 - t_0} \right) \tag{12.164}$$

Hence, the transition curve equation becomes

$$\mathbf{r}(t) = \mathbf{r}_1 - \delta_1 \frac{(t - t' - t_1)^2}{4t'(t_1 - t_0)} + \delta_2 \frac{(t + t' - t_1)^2}{4t'(t_2 - t_1)} \tag{12.165}$$

showing that the path between $\mathbf{r}_0$ and $\mathbf{r}_2$ has a piecewise function given in (12.151).

A Cartesian path followed by the manipulator, plus the time profile along the path, specifies the position and orientation of the end frame. Issues in Cartesian path planning include attaining a specific target from an initial starting point, avoiding obstacles, and staying within manipulator capabilities. A path is usually modeled by n *control points*. The control points are connected via straight lines and the transient parabolas will be implemented to exclude the sharp corners.

An alternative method is applying an interpolating or approximating method, such as least-squared, to design a continuous path over the control points, or close to them. ∎

Example 392 A $2D$ path mathematical expression. Here is a simple example of how to connect a spatial path to be traced by a time function.

Assume the tip point of a manipulator is supposed to move on a given line $Y = f(X)$.

$$Y = -0.25998X + 0.3705 \tag{12.166}$$

The tip point moves form P_1 to P_2 in $10\,\text{s}$.

$$X_{P_1} = 0.411\,22 \qquad Y_{P_1} = 0.263\,59 \tag{12.167}$$

$$X_{P_2} = -2.818\,8 \times 10^{-2} \qquad Y_{P_2} = 0.377\,83 \tag{12.168}$$

Let us define a rest-to-rest cubic path for X.

$$X = 0.41122 - 0.01149096t^2 + 0.000766064t^3 \tag{12.169}$$

We determine the equation of Y as a function of t by substituting $X = X(t)$ in the line Eq. (12.166).

$$Y = -1.9916 \times 10^{-4}t^3 + 2.9874 \times 10^{-3}t^2 + 0.26359 \tag{12.170}$$

Example 393 A path in $2D$ Cartesian space. Here is designing a spatial curve in a two dimensional space.

Consider a line in the (X, Y)-plane connecting $(1, 0)$ and $(1, 1)$ and another line connecting $(1, 1)$ and $(0, 1)$. Assume the time is zero at $(1, 0)$, $t = 1\,\text{s}$ at $(1, 1)$, and $t = 2\,\text{s}$ at $(0, 1)$. For an interval time $t' = 0.1\,\text{s}$, the position vectors at control points are

$$\mathbf{r}_0 = \hat{\imath} \qquad \mathbf{r}_1 = \hat{\imath} + \hat{\jmath} \qquad \mathbf{r}_2 = \hat{\jmath} \tag{12.171}$$

$$\mathbf{r}(t_1 - t') = \mathbf{r}_1 - \frac{t'}{t_1}\delta_1 = \hat{\imath} + \left(1 - \frac{t'}{t_1}\right)\hat{\jmath} \tag{12.172}$$

$$\mathbf{r}(t_1 + t') = \mathbf{r}_1 + \frac{t'}{t_2}\delta_2 = \left(1 - \frac{t'}{t_2}\right)\hat{\imath} + \hat{\jmath} \tag{12.173}$$

where

$$\delta_1 = \mathbf{r}_1 - \mathbf{r}_0 = \hat{\jmath} \qquad \delta_2 = \mathbf{r}_2 - \mathbf{r}_1 = -\hat{\imath} \tag{12.174}$$

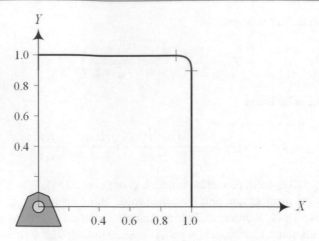

Fig. 12.13 A transition parabola connecting two lines

The path of motion is then expressed by the following piecewise function as shown in Fig. 12.13:

$$\mathbf{r}(t) = \begin{cases} \hat{\imath} + t\hat{\jmath} & 0 \le t \le 0.9 \\ \left(1 - \frac{(t-0.9)^2}{0.4}\right)\hat{\imath} + \left(1 - \frac{(t-1.1)^2}{0.4}\right)\hat{\jmath} & 0.9 \le t \le 1.1 \\ (2-t)\hat{\imath} + \hat{\jmath} & 1.1 \le t \le 2 \end{cases} \tag{12.175}$$

The velocity of motion along the path is also a piecewise function given below.

$$\dot{r}(t) = \begin{cases} \hat{\jmath} & 0 \le t \le 0.9 \\ \frac{t-0.9}{0.2}\hat{\imath} - \frac{t-1.1}{0.2}\hat{\jmath} & 0.9 \le t \le 1.1 \\ -\hat{\imath} & 1.1 \le t \le 2 \end{cases} \tag{12.176}$$

12.5 Forward Path Robot Motion

Forward path robot motion is having the joint variables as functions of time, and employing the forward kinematics of manipulators, to determine the path of motion for the end-effector of a robot.

The position of the end-effector of a manipulator, (X, Y, Z), is determined by the values of the joint variables of the manipulator q_i. If the joint variables are given functions of time $q_i(t)$, then so will be the position of the end-effector. Substituting the time varying functions for the joint variables, the forward kinematics of the manipulator determines the geometric path of the end-effector.

$$^0\mathbf{r}_P = \begin{bmatrix} X(q_i(t)) \\ Y(q_i(t)) \\ Z(q_i(t)) \end{bmatrix} \tag{12.177}$$

Proof Consider a robot with 6 joint variables $q_1(t), q_2(t), \ldots, q_6(t)$, all varying with known functions of time. The result of forward kinematics is a 4×4 homogenous transformation matrix $^0T_6$.

$$^0T_6 = {}^0T_1(q_1)\,{}^1T_2(q_2)\,{}^2T_3(q_3)\,{}^3T_4(q_4)\,{}^4T_5(q_5)\,{}^5T_6(q_6)$$

$$= \begin{bmatrix} {}^0R_6(t) & {}^0\mathbf{d}_6(t) \\ 0 & 1 \end{bmatrix} = \begin{bmatrix} r_{11} & r_{12} & r_{13} & d_X(t) \\ r_{21} & r_{22} & r_{23} & d_Y(t) \\ r_{31} & r_{32} & r_{33} & d_Z(t) \\ 0 & 0 & 0 & 1 \end{bmatrix} \tag{12.178}$$

The upper left 3×3 submatrix of $^0T_6$ is the rotation matrix, controlling the orientation of the end-effector. The last column of $^0T_6$ indicates the components of the position vector of the end-effector frame in the base frame.

$$^0\mathbf{d}_6(t) = \begin{bmatrix} d_X(t) \; d_Y(t) \; d_Z(t) \end{bmatrix}^T \tag{12.179}$$

Assuming that $^0\mathbf{d}_6 = {}^0\mathbf{r}_P$ indicates the point of interest in the base frame, its components would be the global coordinates (X, Y, Z) of the point. The coordinates X, Y, and Z are functions of three joint coordinates of the manipulator q_1, q_2, and q_3. The joint coordinates are given functions of time, and hence, the coordinates are functions of time, indicating the path of motion of the end-effector in B_0-frame.

$$^0\mathbf{d}_6(t) = \begin{bmatrix} X(q_i(t)) \\ Y(q_i(t)) \\ Z(q_i(t)) \end{bmatrix} = \begin{bmatrix} d_Y(q_i(t)) \\ d_Y(q_i(t)) \\ d_Z(q_i(t)) \end{bmatrix} \tag{12.180}$$

The forward path motion of robots is the secondary problem which will come into action after solving the inverse path robot motion. In practice, we start with defining the geometric path of motion of the end-effector and then inverse kinematics determines how joint variables would vary. Defining a time function of the geometric path of the end-effector makes the joint variable to be functions of time. Then by feeding the functions to the joint actuators, we get to the forward path motion of robots and expect the end-effector to move on the planned geometric path. Therefore, defining time function of the joint variables is usually not in the hand of engineer. ∎

Example 394 2R manipulator motion based on joints' path. This is to show how the forward path robot motion is working for a 2R planar manipulator.

Assume we have calculated the paths of the two joints of a 2R planar manipulator according to cubic functions, and they are

$$\theta_1(t) = 10 + 105t^2 - 70t^3 \text{ deg} \tag{12.181}$$

$$\theta_2(t) = 10 + 350t^3 - 525t^4 + 210t^5 \text{ deg} \tag{12.182}$$

The joints' paths satisfy the following rest-to-rest and no acceleration conditions for $0 \le t \le 1$:

$$\theta_1(0) = 10 \text{ deg} \qquad \dot{\theta}_1(0) = 0$$
$$\theta_1(1) = 45 \text{ deg} \qquad \dot{\theta}_1(1) = 0 \tag{12.183}$$

$$\theta_2(0) = 10 \text{ deg} \qquad \dot{\theta}_2(0) = 0 \qquad \ddot{\theta}_2(0) = 0$$
$$\theta_2(1) = 45 \text{ deg} \qquad \dot{\theta}_2(1) = 0 \qquad \ddot{\theta}_2(1) = 0 \tag{12.184}$$

The forward kinematics of a 2R manipulator are found in Example 165 as below:

$$^0T_2 = {}^0T_1 \, {}^1T_2$$

$$= \begin{bmatrix} c(\theta_1 + \theta_2) & -s(\theta_1 + \theta_2) & 0 & l_1 c\theta_1 + l_2 c(\theta_1 + \theta_2) \\ s(\theta_1 + \theta_2) & c(\theta_1 + \theta_2) & 0 & l_1 s\theta_1 + l_2 s(\theta_1 + \theta_2) \\ 0 & 0 & 1 & 0 \\ 0 & 0 & 0 & 1 \end{bmatrix} \tag{12.185}$$

The fourth column of $^0T_2$ indicates the Cartesian position of the tip point of the manipulator in the base frame. Therefore, the X and Y components of the tip point are

$$X = l_1 \cos\theta_1 + l_2 \cos(\theta_1 + \theta_2) \tag{12.186}$$

$$Y = l_1 \sin\theta_1 + l_2 \sin(\theta_1 + \theta_2) \tag{12.187}$$

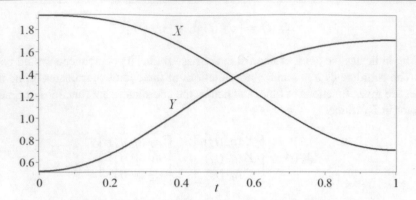

Fig. 12.14 X and Y components of the tip point position of a $2R$ planar manipulator

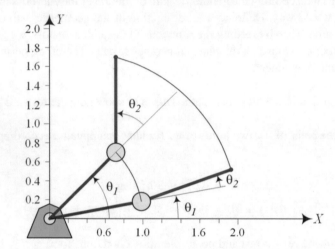

Fig. 12.15 Configuration of a $2R$ manipulator at initial and final positions

Substituting θ_1 and θ_2 from (12.181) and (12.182) provides the time variation of the position of the tip point. These variations, for $l_1 = l_2 = 1$ m, are shown in Fig. 12.14, while the configurations of the manipulator at initial and final positions are shown in Fig. 12.15.

As long as the joint variables are defined and given as functions of time, it is immaterial which joint turns first. The joint variables are relative coordinates and the final configuration of the robot would be the same. The actuators can also turn together.

Moving a robot by applying a set of joint paths is not always a proper method. In case the joint variables are not monotonic in time and are fluctuating, defining a joint path is more complicated. Furthermore, it is not easy to move the end-effector of a robot on a desired geometric path by defining joint paths.

Example 395 A $2R$ robot moving along a line. Here we show how an inverse path motion will end up to forward path motion.

Let us consider a $2R$ manipulator with

$$l_1 = l_2 = 0.25 \tag{12.188}$$

that its tip point is supposed to move on a given line $Y = f(X)$ as is shown in Fig. 12.16.

$$Y = -0.25998X + 0.3705 \tag{12.189}$$

Assume the first angle is moving between 45 deg and 135 deg in 10 s based on a cubic path.

$$\theta_1 = \frac{\pi}{4} + \frac{3\pi}{200}t^2 - \frac{\pi}{1000}t^3 \qquad 0 < t < 10\,\text{s} \tag{12.190}$$

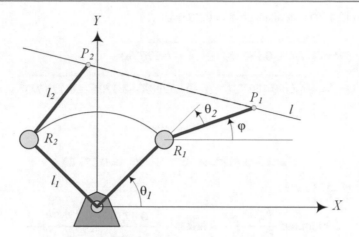

Fig. 12.16 A $2R$ robot moving along a given line

$$45 \text{ deg} < \theta_1 < 135 \text{ deg} \tag{12.191}$$

The elbow joint R will move on a circle and at the beginning is at

$$X_{R_1} = 0.25 \cos \frac{\pi}{4} = 0.176\,78 \tag{12.192}$$

$$Y_{R_1} = 0.25 \sin \frac{\pi}{4} = 0.176\,78 \tag{12.193}$$

Point P_1 must be on the line (12.189) at a distance $d = 0.25$ from R_1.

$$d = \sqrt{(X - 0.176\,78)^2 + (Y - 0.176\,78)^2}$$
$$= \sqrt{(X - 0.176\,78)^2 + (-0.25998X + 0.3705 - 0.176\,78)^2}$$
$$= 0.25 \tag{12.194}$$

Therefore, P_1 is at

$$X_{P_1} = 0.411\,22 \qquad Y_{P_1} = 0.263\,59 \tag{12.195}$$

and the initial values of angles φ and θ_2 are

$$\varphi = \arctan \frac{Y_{P_1} - Y_{R_1}}{X_{P_1} - X_{R_1}} = \arctan \frac{0.263\,59 - 0.176\,78}{0.411\,22 - 0.176\,78}$$
$$= 0.354\,63 \text{ rad} \approx 20.319 \text{ deg} \tag{12.196}$$

$$\theta_2 = \theta_1 - \varphi = \frac{\pi}{4} - 0.354\,63$$
$$= 0.430\,77 \text{ rad} \approx 24.681 \text{ deg} \tag{12.197}$$

The elbow joint R at the final position is at

$$X_{R_2} = 0.25 \cos \frac{3\pi}{4} = -0.176\,78 \tag{12.198}$$

$$Y_{R_2} = 0.25 \sin \frac{3\pi}{4} = 0.176\,78 \tag{12.199}$$

Point P_2 must be on the line (12.189) at a distance $d = 0.25$ from R_2.

$$
\begin{aligned}
d &= \sqrt{(X + 0.176\,78)^2 + (Y - 0.176\,78)^2} \\
&= \sqrt{(X + 0.176\,78)^2 + (-0.25998X + 0.3705 - 0.176\,78)^2} \\
&= 0.25
\end{aligned}
\tag{12.200}
$$

Therefore, P_2 is at

$$
X_{P_2} = -2.818\,8 \times 10^{-2} \qquad Y_{P_2} = 0.377\,83
\tag{12.201}
$$

and the final values of angles φ and θ_2 are

$$
\begin{aligned}
\varphi &= \arctan \frac{Y_{P_2} - Y_{R_2}}{X_{P_2} - X_{R_2}} = \arctan \frac{0.377\,83 - 0.176\,78}{-2.818\,8 \times 10^{-2} + 0.176\,78} \\
&= 0.934\,32\,\text{rad} \approx 53.533\,\text{deg}
\end{aligned}
\tag{12.202}
$$

$$
\begin{aligned}
\theta_2 = \theta_1 - \varphi &= \frac{3\pi}{4} - 0.934\,32 \\
&= 1.421\,9\,\text{rad} \approx 81.469\,\text{deg}
\end{aligned}
\tag{12.203}
$$

To determine θ_2 during the motion, we should follow the same procedure. Let us find the position of the elbow joint R as a function of θ_1.

$$
X_R = 0.25 \cos\theta_1 \qquad Y_R = 0.25 \sin\theta_1
\tag{12.204}
$$

The tip point P must be on the line (12.189) at a distance $d = 0.25$ from the elbow joint R.

$$
\begin{aligned}
d &= \sqrt{(X_P - 0.25 \cos\theta_1)^2 + (Y_P - 0.25 \sin\theta_1)^2} \\
&= \sqrt{(X_P - 0.25 \cos\theta_1)^2 + (-0.25998X_P + 0.3705 - 0.25 \sin\theta_1)^2} \\
&= 0.25
\end{aligned}
\tag{12.205}
$$

The solution of this equation for X_P and substitution in (12.189) provide the coordinates (X_P, Y_P) of the tip point P during the motion. Then, the angle φ and θ_2 would be

$$
\varphi = \arctan \frac{Y_P - Y_R}{X_P - X_R} = \arctan \frac{Y_P - 0.25 \sin\theta_1}{X_P - 0.25 \cos\theta_1}
\tag{12.206}
$$

$$
\theta_2 = \theta_1 - \varphi = \theta_1 - \arctan \frac{Y_P - 0.25 \sin\theta_1}{X_P - 0.25 \cos\theta_1}
\tag{12.207}
$$

Therefore, to make the point P moving along the line (12.189), while θ_1 is varying as (12.190), the angle θ_2 must vary according to (12.207).

12.6   Inverse Path Robot Motion

The Cartesian path planning of the end-effector is the most natural application of path planning. Considering the pick and place motion as the main job of industrial robots, we have to determine a desired geometric path for the end-effector in the three dimensional Cartesian space of the base frame. We may then define a time path for one of the coordinates, say X, and determine the time history of the other coordinates by using the geometric path. Having the time functions of the coordinates of the end-effector, we can determine the velocity, acceleration, and jerk behavior of the end-effector.

Inverse kinematics will determine the kinematics of joint variables. Substituting the joint variables' position, velocity, and acceleration in the dynamic equations of motion provides the required actuators' torque or force to move the end-effector on the desired path with the planned kinematics. The geometric Cartesian path is an applied method of path planning in robotics because it can control the level of force and jerk inserted by the hand of a robot to the carrying object. Path planning in Cartesian space also determines the geometric constraints of the external world. Cartesian path needs inverse kinematics to determine the time history of the joint variables.

Example 396 Joint path for a designed Cartesian path. Moving the end point of a $2R$ manipulator on a horizontal line and inverse kinematics shows how to make a robot to move on a designed Cartesian path.

Consider a rest-to-rest Cartesian path from point $(1, 1.5)$ to point $(-1, 1.5)$ on a straight line $Y = 1.5$. A cubic polynomial can satisfy the position and velocity constraints at initial and final points.

$$X(0) = X_0 = 1 \qquad \dot{X}(0) = \dot{X}_0 = 0$$
$$X(1) = X_f = -1 \qquad \dot{X}(1) = \dot{X}_f = 0 \tag{12.208}$$

The coefficients of the polynomial are

$$a_0 = 1 \qquad a_1 = 0 \qquad a_2 = -6 \qquad a_3 = 4 \tag{12.209}$$

and the Cartesian path is

$$X = 1 - 6t^2 + 4t^3 \tag{12.210}$$
$$Y = 1.5 \tag{12.211}$$

The inverse kinematics of a $2R$ planar manipulator is calculated in Example 207 as

$$\theta_2 = \pm 2 \operatorname{atan2} \sqrt{\frac{(l_1 + l_2)^2 - (X^2 + Y^2)}{(X^2 + Y^2) - (l_1 - l_2)^2}} \tag{12.212}$$

$$\theta_1 = \operatorname{atan2} \frac{X(l_1 + l_2 \cos\theta_2) + Y l_2 \sin\theta_2}{Y(l_1 + l_2 \cos\theta_2) - X l_2 \sin\theta_2} \tag{12.213}$$

where the sign $(\pm)$ indicates the elbow up and elbow down configurations of the manipulator. Depending on the initial configuration at point $(1, 1.5)$, the manipulator is supposed to stay in that configuration. Let us consider an elbow up configuration. Therefore, we accept only those values of the joint valuables that belong to the elbow up configuration. Substituting (12.210) and (12.211) in (12.212) and (12.213) provides the path in joint space.

$$\theta_2 = \pm 2 \operatorname{atan2} \sqrt{\frac{(l_1 + l_2)^2 - \left(4t^3 - 6t^2 + 2.5\right)^2}{\left(4t^3 - 6t^2 + 1\right)^2 - (l_1 - l_2)^2}} \tag{12.214}$$

$$\theta_1 = \operatorname{atan2} \frac{\left(1 - 6t^2 + 4t^3\right)(l_1 + l_2 \cos\theta_2) + 1.5 l_2 \sin\theta_2}{1.5(l_1 + l_2 \cos\theta_2) - \left(1 - 6t^2 + 4t^3\right) l_2 \sin\theta_2} \tag{12.215}$$

A graphical illustration of the manipulator at every 1/30th of the total time is shown in Fig. 12.17.

Example 397 A $2R$ manipulator on a line. Here is another example of moving the end-effector of a manipulator on a given path. Inverse kinematics is the key to determine joint variables kinematics for given motions of the end-effector.

Consider the $2R$ manipulator with

$$l_1 = l_2 = 0.25 \,\mathrm{m} \tag{12.216}$$

that its tip point is supposed to move on a given line

$$Y = -0.25998X + 0.3705 \tag{12.217}$$

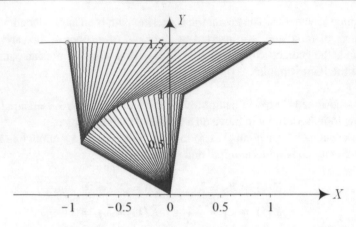

Fig. 12.17 Illustration of a $2R$ manipulator when the tip point moves on a straight line $y = 1.5$

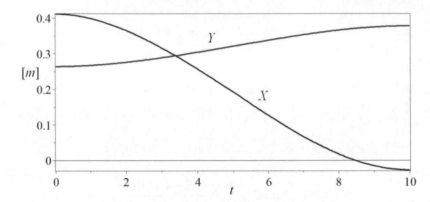

Fig. 12.18 Cartesian coordinates of the tip point versus time

between P_1 and P_2 in 10 s.

$$X_{P_1} = 0.41122 \qquad Y_{P_1} = 0.26359 \tag{12.218}$$

$$X_{P_2} = -0.0282 \qquad Y_{P_2} = 0.37783 \tag{12.219}$$

Defining a rest-to-rest cubic path for X, we determine the Cartesian path of the tip point.

$$X = 0.41122 - 0.0131826t^2 + 0.00087884t^3 \tag{12.220}$$

$$Y = -0.00022848t^3 + 0.003427t^2 + 0.26359 \tag{12.221}$$

The kinematics of the tip point are shown in Fig. 12.18. Employing the inverse kinematics of Eqs. (6.14) and (6.15), we find the variation of the joint angles as are shown in Fig. 12.19.

Let us divide the total time of the motion in $n = 40$ equal intervals. The configuration of the manipulator at each time step is shown in Fig. 12.20.

Example 398 A $2R$ manipulator on a line with no end acceleration. To have a smooth rest-to-rest maneuver, the path must have zero acceleration at boundaries. Here is an improvement to smoothness of motion of end-effector on an inclined line.

Consider the $2R$ manipulator with

$$l_1 = l_2 = 0.25 \, \text{m} \tag{12.222}$$

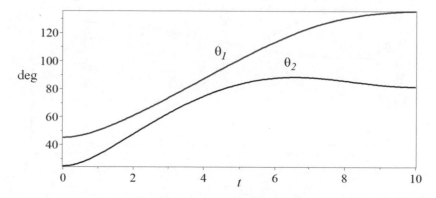

Fig. 12.19 The variation of joint angles of the 2R manipulator

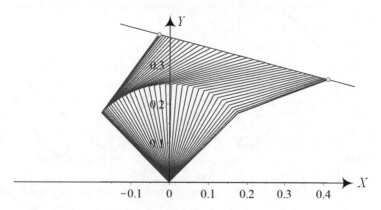

Fig. 12.20 The configuration of the 2R manipulator at 42 equal time steps

that its tip point is supposed to move on a given line between P_1 and P_2 in 10 s.

$$Y = -0.25998X + 0.3705 \tag{12.223}$$

$$X_{P_1} = 0.41122 \qquad Y_{P_1} = 0.26359 \tag{12.224}$$

$$X_{P_2} = -0.0282 \qquad Y_{P_2} = 0.37783 \tag{12.225}$$

Let us define a quintic path for X to apply a zero acceleration at both ends.

$$X = 0.41122 - 0.0043942t^3 + 0.00065913t^4$$
$$-0.0000263652t^5 \tag{12.226}$$

Substituting X in the line Eq. (12.223), we also determine the variation of Y.

$$Y = 0.26359 + 0.0011424t^3 - 0.00017136t^4$$
$$+0.0000068544t^5 \tag{12.227}$$

Using the Cartesian components (12.226) and (12.227), we determine the kinematics of the tip point as are shown in Figs. 12.21 and 12.22. Using Eqs. (6.14) and (6.15), we find the variation of the joint angles as are shown in Fig. 12.23.

For the same path of motion of Fig. 12.20, the difference of path plans of (12.220) and (12.226) cannot be seen clearly in Figs. 12.18 and 12.21 or 12.19 and 12.23 unless we put them together.

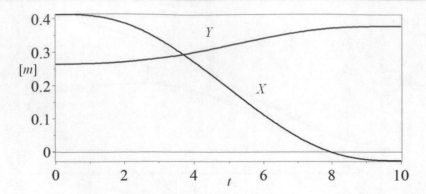

Fig. 12.21 Cartesian coordinates of the tip point versus time on a no end acceleration path

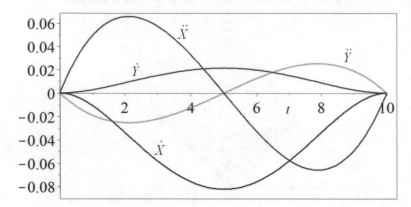

Fig. 12.22 Components of the tip point velocity and acceleration versus time on a no end acceleration path

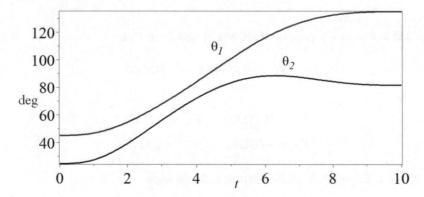

Fig. 12.23 The variation of joint angles of the $2R$ manipulator on a no end acceleration path

Example 399 Path of a $2R$ manipulator to avoid obstacle. If a robot is not equipped with vision, it should take the path blindly. It is the job of engineer to train the robot to take an obstacle free path. Here is an example.

Consider the $2R$ manipulator of Fig. 12.24 with equal arms' length.

$$l_1 = l_2 = 0.25\,\text{m} \tag{12.228}$$

The tip point is supposed to move from P_1 to P_2 in 10 s.

$$X_{P_1} = 0.41122 \qquad Y_{P_1} = 0.26359 \tag{12.229}$$

$$X_{P_2} = -0.0282 \qquad Y_{P_2} = 0.37783 \tag{12.230}$$

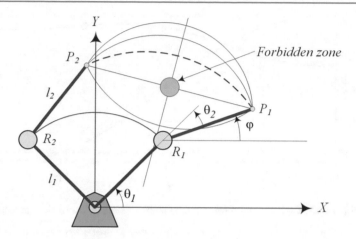

Fig. 12.24 A few circular paths between P_1 and P_2 to go around forbidden zone at P_3

However, there is an obstacle as a circular forbidden zone at point P_3, where the tip point cannot pass.

$$X_{P_3} = 0.19151 \qquad Y_{P_3} = 0.32071 \tag{12.231}$$

$$\left(X - X_{P_3}\right)^2 + \left(Y - Y_{P_3}\right)^2 = 0.025^2 \tag{12.232}$$

To find a path between P_1 and P_2 to go around the obstacle, let us choose a circular arc with a center on the bisector of $P_1 P_2$. Figure 12.24 depicts a few optional paths. The arc must be within the working space of the manipulator.

$$(l_1 - l_2)^2 < X^2 + Y^2 < (l_1 + l_2)^2 \tag{12.233}$$
$$0 < X^2 + Y^2 < 0.5^2 \tag{12.234}$$

The center of the circular path should be on the following line:

$$Y - Y_{P_3} = 3.8464 \left(X - X_{P_3}\right) \tag{12.235}$$

Let us pick a point P_C to be the center of the circular path at

$$X_C = 0.1 \qquad Y_C = -0.06 \tag{12.236}$$

Therefore, the equation of the path would be

$$(X - X_C)^2 + (Y - Y_C)^2 = 0.45^2 \tag{12.237}$$

This path is shown in Fig. 12.24 with a dashed arc.

Let us use a quintic time path for X to apply a zero acceleration at both ends.

$$X = 0.41122 - 0.0043942t^3 + 0.00065913t^4$$
$$- 0.0000263652t^5 \tag{12.238}$$

Substituting X in the path Eq. (12.237), we determine the time path of Y.

$$Y = Y_C + \sqrt{0.45^2 - (X - X_C)^2} \tag{12.239}$$

The kinematics of the tip point are shown in Figs. 12.25 and 12.26. Equations (6.14) and (6.15) provide the joint angles as are shown in Fig. 12.27. The configuration of the manipulator at 40 equal time steps is shown in Fig. 12.28.

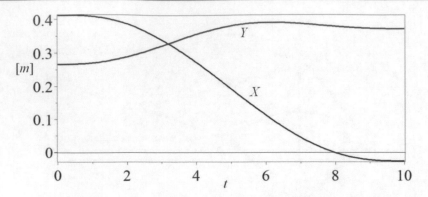

Fig. 12.25 Cartesian coordinates of the tip point versus time on a circular path

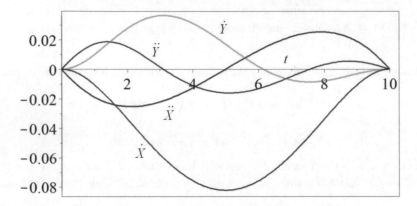

Fig. 12.26 Components of the tip point velocity and acceleration versus time on a circular path

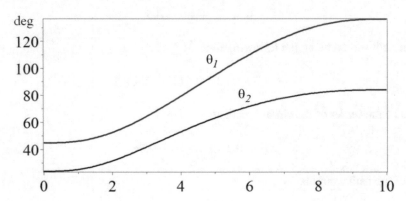

Fig. 12.27 The variation of joint angles of the $2R$ manipulator on an obstacle avoidance path

Example 400 Articulated manipulator on a line.

Figure 12.29 illustrates an articulated manipulator. The tip point of the manipulator is supposed to move from point P_1 to P_2 in 10 s.

$$l_1 = 0.5 \, \text{m} \qquad l_2 = 1.0 \, \text{m} \qquad l_3 = 1.0 \, \text{m} \tag{12.240}$$

$$\mathbf{r}_{P_1} = \begin{bmatrix} 1.5 \\ 0.0 \\ 1.0 \end{bmatrix} \qquad \mathbf{r}_{P_2} = \begin{bmatrix} -1.0 \\ 1.0 \\ 1.5 \end{bmatrix} \tag{12.241}$$

Using a quintic path for X, we find the following function to express the time variation of X:

$$X = 1.5 - 0.025t^3 + 0.00375t^4 - 0.00015t^5 \tag{12.242}$$

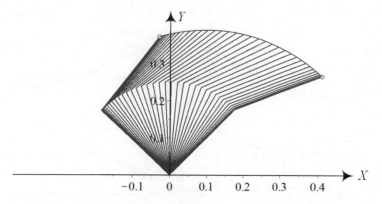

Fig. 12.28 The configuration of the $2R$ manipulator at 42 equal time steps on a circular path

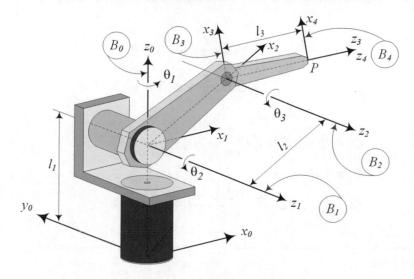

Fig. 12.29 An articulated manipulator

Let us connect P_1 and P_2 by a straight line and determine the time variation of Y and Z.

$$Y = Y_{P_1} + \frac{Y_{P_2} - Y_{P_1}}{X_{P_2} - X_{P_1}}(X - X_{P_1})$$

$$= 0.010t^3 - 0.0015t^4 + 0.00006t^5 \tag{12.243}$$

$$Z = Z_{P_1} + \frac{Z_{P_2} - Z_{P_1}}{X_{P_2} - X_{P_1}}(X - X_{P_1})$$

$$= 1 + 0.005t^3 - 0.00075t^4 + 0.00003t^5 \tag{12.244}$$

Using the inverse kinematic equations, we can determine the time history of joint variables of the manipulator as are shown in Fig. 12.30.

$$\theta_3 = \arccos\left(\frac{l_1 - Z + l_2 \sin\theta_2}{l_3}\right) - \theta_2 \tag{12.245}$$

$$\theta_2 = 2\arctan\frac{-C_2 + \sqrt{C_2^2 - C_1 C_3}}{C_1} \tag{12.246}$$

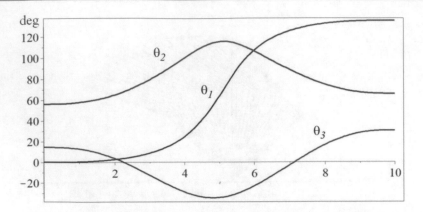

Fig. 12.30 The time history of joint variables of an articulated manipulator

$$\theta_1 = \begin{cases} \arctan \dfrac{Y}{X} & X \geq 0 \\[2mm] \arctan \dfrac{Y}{X} + \pi & X < 0 \end{cases} \tag{12.247}$$

$$C_1 = l_1^2 - 2l_1 Z + l_2^2 + \frac{2l_2 X}{\cos \theta_1} - l_3^2 + \frac{X^2}{\cos^2 \theta_1} + Z^2 \tag{12.248}$$

$$C_2 = 2l_1 l_2 - 2l_2 Z \tag{12.249}$$

$$C_3 = l_1^2 - 2l_1 Z + l_2^2 - \frac{2l_2 X}{\cos \theta_1} - l_3^2 + \frac{X^2}{\cos^2 \theta_1} + Z^2 \tag{12.250}$$

12.7 ★ Rotational Path

Consider an end-effector frame to have a rotation matrix $^G R_0$ at an initial orientation at time t_0. The end-effector must be at a final orientation $^G R_f$ at time t_f. The *rotational path* is defined by the angle-axis rotation matrix $R_{\hat{u},\phi}$

$$R_{^0\hat{u},\phi} = {^0R_f} = {^G R_0^T}\, {^G R_f} \tag{12.251}$$

which transforms the end-effector frame from the final orientation $^G R_f$ to the initial orientation $^G R_0$. The axis of rotation $^0 \hat{u}$ is defined by a unit vector expressed in the initial frame. Therefore, the desired rotation matrix for going from initial to the final orientation would be

$$R_{^0\hat{u},\phi}^T = {^G R_f^T}\, {^G R_0} \tag{12.252}$$

Keeping $^0 \hat{u}$ constant, we can define an angular path for φ to vary $R_{^0\hat{u},\phi}^T$ from $^G R_0$ to $^G R_f$ at t_f.

To control a rotation, we may define a series of control orientations $^G R_1$, $^G R_2$, $\cdots$, $^G R_n$ between the initial and final orientations and rotate the end-effector frame through the control orientations. When there is a control orientation $^G R_1$ between the initial and final orientations, then the initial orientation $^G R_0$ transforms to the control orientation $^G R_1$ using an angle–axis rotation $R_{^0\hat{u},\phi_0}$, and then it transforms from the control orientation $^G R_1$ to the final orientation using a second-angle axis rotation $R_{^1\hat{u},\phi_1}$.

$$R_{^0\hat{u},\phi_0} = {^G R_0^T}\, {^G R_1} \tag{12.253}$$

$$R_{^1\hat{u},\phi_1} = {^G R_1^T}\, {^G R_f} \tag{12.254}$$

Proof According to the Rodriguez rotation formula (3.5),

$$^0 R_f = R_{^0\hat{u},\phi} = \mathbf{I} \cos \phi + {^0\hat{u}}\, {^0\hat{u}^T} \operatorname{vers} \phi + {^0\tilde{u}} \sin \phi, \tag{12.255}$$

the angle and axis that transform a frame B_f to another frame B_0 are found from:

$$\cos \phi = \frac{1}{2} \left(\text{tr} \left({}^0 R_f \right) - 1 \right) \tag{12.256}$$

$$ {}^0 \tilde{u} = \frac{1}{2 \sin \phi} \left({}^0 R_f - {}^0 R_f^T \right) \tag{12.257}$$

If ${}^G R_0$ is the rotation matrix from B_0 to the global frame G and ${}^G R_f$ is the rotation matrix from B_f to G, then

$$ {}^G R_f = {}^G R_0 \, {}^0 R_f \tag{12.258}$$

and therefore,

$$ {}^0 R_f = R_{{}^0 \hat{u}, \phi} = {}^G R_0^T \, {}^G R_f \tag{12.259}$$

We define a linearly time dependent rotation matrix by varying the angle of rotation about the axis of rotation

$$ {}^0 R_f(t) = R_{{}^0 \hat{u}, (\frac{t-t_0}{t_f-t_0}) \phi} \tag{12.260}$$

$$= \begin{bmatrix} r_{11}(t) & r_{12}(t) & r_{13}(t) \\ r_{21}(t) & r_{22}(t) & r_{23}(t) \\ r_{31}(t) & r_{32}(t) & r_{33}(t) \end{bmatrix} \qquad t_0 \le t \le t_f$$

where t_0 is the time when the end-effector frame is at orientation ${}^G R_0$ and t_f is the time at which the end-effector frame is at orientation ${}^G R_f$.

$$ r_{11}(t) = u_1^2 \, \text{vers} \left(\frac{t-t_0}{t_f-t_0} \right) \phi + \cos \left(\frac{t-t_0}{t_f-t_0} \right) \phi$$

$$ r_{21}(t) = u_1 u_2 \, \text{vers} \left(\frac{t-t_0}{t_f-t_0} \right) \phi + u_3 \sin \left(\frac{t-t_0}{t_f-t_0} \right) \phi$$

$$ r_{31}(t) = u_1 u_3 \, \text{vers} \left(\frac{t-t_0}{t_f-t_0} \right) \phi - u_2 \sin \left(\frac{t-t_0}{t_f-t_0} \right) \phi \tag{12.261}$$

$$ r_{12}(t) = u_1 u_2 \, \text{vers} \left(\frac{t-t_0}{t_f-t_0} \right) \phi - u_3 \sin \left(\frac{t-t_0}{t_f-t_0} \right) \phi$$

$$ r_{22}(t) = u_2^2 \, \text{vers} \left(\frac{t-t_0}{t_f-t_0} \right) \phi + \cos \left(\frac{t-t_0}{t_f-t_0} \right) \phi$$

$$ r_{32}(t) = u_2 u_3 \, \text{vers} \left(\frac{t-t_0}{t_f-t_0} \right) \phi + u_1 \sin \left(\frac{t-t_0}{t_f-t_0} \right) \phi \tag{12.262}$$

$$ r_{13}(t) = u_1 u_3 \, \text{vers} \left(\frac{t-t_0}{t_f-t_0} \right) \phi + u_2 \sin \left(\frac{t-t_0}{t_f-t_0} \right) \phi$$

$$ r_{23}(t) = u_2 u_3 \, \text{vers} \left(\frac{t-t_0}{t_f-t_0} \right) \phi - u_1 \sin \left(\frac{t-t_0}{t_f-t_0} \right) \phi$$

$$ r_{33}(t) = u_3^2 \, \text{vers} \left(\frac{t-t_0}{t_f-t_0} \right) \phi + \cos \left(\frac{t-t_0}{t_f-t_0} \right) \phi \tag{12.263}$$

The matrix ${}^0 R_f(t)$ can turn the final frame about the axis of rotation ${}^0 \hat{u}$ onto the initial frame.

$$ {}^G R_f = {}^G R_0 \, {}^0 R_f(t) \tag{12.264}$$

If there is a control orientation frame ${}^G R_1$ between the initial and final orientations, then

$$
{}^G R_1 = {}^G R_0 \, {}^0 R_1 \tag{12.265}
$$

$$
{}^G R_f = {}^G R_1 \, {}^1 R_f \tag{12.266}
$$

and therefore,

$$
R_{{}^0\hat{u},\phi_0} = {}^0 R_1 = {}^G R_0^T \, {}^G R_1 \tag{12.267}
$$

$$
R_{{}^1\hat{u},\phi_1} = {}^1 R_f = {}^G R_1^T \, {}^G R_f. \tag{12.268}
$$

The rotation matrices ${}^0 R_1$ and ${}^1 R_f$ may be defined as linearly time varying rotation matrices.

$$
\begin{aligned}
{}^0 R_1(t) &= R_{{}^0\hat{u},(\frac{t-t_0}{t_1-t_0})\phi_0} & t_0 \le t \le t_1 \\
{}^1 R_f(t) &= R_{{}^1\hat{u},(\frac{t-t_1}{t_f-t_1})\phi_1} & t_1 \le t \le t_f
\end{aligned} \tag{12.269}
$$

Using these variable matrices, we can turn the end-effector frame from the initial orientation ${}^G R_0$ about ${}^0\hat{u}$ to achieve the control orientation ${}^G R_1$ and then turn the end-effector frame about ${}^1\hat{u}$ to achieve the final orientation ${}^G R_f$.

Following a parabola transition technique similar to (12.151), we may define an orientation path connecting ${}^G R_0$ and ${}^G R_f$ and passing close to the corner orientation ${}^G R_1$ on a transient rotation path. The path starts from ${}^G R_0$ at time t_0 and turns with constant angular velocity along an axis until $t = t_1 - t'$. At this time, the path switches to a rotational parabolic path with constant angular acceleration. At another switching orientation at time $t = t_1 + t'$, the path switches to the second path and turns with constant velocity toward the destination orientation ${}^G R_f$. The time $t_1 - t_0$ is the required time to move from ${}^G R_0$ to ${}^G R_1$, and $t_2 - t_1$ is the required time to move from ${}^G R_1$ to ${}^G R_f$ if there were no transition path.

We introduce an interval time t' before arriving at orientation ${}^G R_1$ to switch from the first path segment to a transition path. The transition path is then between times $t_1 - t'$ and $t_1 + t'$. At the second switching orientation, the transition path ends at the same angular velocity as the third path segment.

The boundary positions of the transition path between the first and third segments are, respectively,

$$
\begin{aligned}
{}^G R_1(t_1 - t') &= {}^G R_0 \, {}^0 R_1(t_1 - t') \\
&= {}^G R_0 \, R_{{}^0\hat{u},(1-\frac{t'}{t_1-t_0})\phi_0} & t = t_1 - t'
\end{aligned} \tag{12.270}
$$

$$
\begin{aligned}
{}^G R_f(t_1 + t') &= {}^G R_1 \, {}^1 R_f(t_1 + t') \\
&= {}^G R_1 \, R_{{}^1\hat{u},(\frac{t'}{t_f-t_1})\phi_1} & t = t_1 + t'
\end{aligned} \tag{12.271}
$$

The transition path is then equal to

$$
\begin{aligned}
R_t(t) &= {}^G R_0 \, {}^0 R_1 \left(\frac{t_1 - t' - t}{2t'} - \frac{(t - t' - t_1)^2}{4t'(t_1 - t_0)} \right) \\
&\quad \times {}^1 R_f \left(\frac{(t + t' - t_1)^2}{4t'(t_f - t_1)} \right) \\
&= {}^G R_0 \, R_{{}^0\hat{u},(\frac{t_1-t'-t}{2t'}-\frac{(t-t'-t_1)^2}{4t'(t_1-t_0)})\phi_0} \, R_{{}^1\hat{u},(\frac{(t+t'-t_1)^2}{4t'(t_f-t_1)})\phi_1}
\end{aligned} \tag{12.272}
$$

$$
t_1 - t' \le t \le t_1 + t'
$$

and the entire path is

$$
\begin{aligned}
R(t) &= {}^0 R_1(t) = R_{{}^0\hat{u},(\frac{t-t_0}{t_1-t_0})\phi_0} & t_0 \le t \le t_1 - t' \\
R(t) &= R_t(t) & t_1 - t' \le t \le t_1 + t' \\
R(t) &= {}^1 R_f(t) = R_{{}^1\hat{u},(\frac{t-t_1}{t_f-t_1})\phi_1} & t_1 + t' \le t \le t_2
\end{aligned} \tag{12.273}
$$

Example 401 Rotation about Z-axis. Rotation about an axis is showing how a rotational path is designed.

Consider a body B which is initially coincident with the global coordinate frame G at $t = 0$. Its initial transformation matrix is an identity.

$$^G R_1 = \mathbf{I} \tag{12.274}$$

B is supposed to be at $^G R_2$ after 10 s.

$$^G R_2 = \begin{bmatrix} -1 & 0 & 0 \\ 0 & -1 & 0 \\ 0 & 0 & 1 \end{bmatrix} \tag{12.275}$$

Let us consider the axis of rotation $^2 R_1$ to be the Z-axis, and the angle of rotation is π. The transformation matrix between the initial and final orientations of B_1 and B_2 is

$$^2 R_1 = {^G R_1^T}\, {^G R_2} = \begin{bmatrix} -1 & 0 & 0 \\ 0 & -1 & 0 \\ 0 & 0 & 1 \end{bmatrix} \tag{12.276}$$

Let us define a cubic rest-to-rest path for the angle of rotation α.

$$\alpha = \frac{3\pi}{100} t^2 - \frac{\pi}{500} t^3 \tag{12.277}$$

The angular path of B between B_1 an B_2 would be

$$^2 R_1 = \begin{bmatrix} \cos\alpha & -\sin\alpha & 0 \\ \sin\alpha & \cos\alpha & 0 \\ 0 & 0 & 1 \end{bmatrix} \tag{12.278}$$

$$= \begin{bmatrix} \cos\frac{3\pi}{100}t^2 - \frac{\pi}{500}t^3 & -\sin\frac{3\pi}{100}t^2 - \frac{\pi}{500}t^3 & 0 \\ \sin\frac{3\pi}{100}t^2 - \frac{\pi}{500}t^3 & \cos\frac{3\pi}{100}t^2 - \frac{\pi}{500}t^3 & 0 \\ 0 & 0 & 1 \end{bmatrix}$$

Example 402 Rotation about X-axis. Here is a cubic path of rotation about a global axis to study rotational paths.

A body B is initially at $^G R_1$.

$$^G R_1 = \begin{bmatrix} 1 & 0 & 0 \\ 0 & \cos\dfrac{\pi}{10} & -\sin\dfrac{\pi}{10} \\ 0 & \sin\dfrac{\pi}{10} & \cos\dfrac{\pi}{10} \end{bmatrix} \tag{12.279}$$

The body is supposed to be at $^G R_2$ in 10 s.

$$^G R_2 = \begin{bmatrix} 1 & 0 & 0 \\ 0 & \cos\dfrac{\pi}{2} & -\sin\dfrac{\pi}{2} \\ 0 & \sin\dfrac{\pi}{2} & \cos\dfrac{\pi}{2} \end{bmatrix} \tag{12.280}$$

The axis of rotation $^2 R_1$ is the X-axis, and the angle of rotation is $\frac{2}{5}\pi = \frac{\pi}{2} - \frac{\pi}{10}$. We define a cubic rest-to-rest path for the angle of rotation γ,

$$\gamma = \frac{\pi}{10} + \frac{3\pi t^2}{250} - \frac{\pi t^3}{1250} \tag{12.281}$$

to determine the angular path of B between G and B_2.

$$
{}^G R_2 = \begin{bmatrix} 1 & 0 & 0 \\ 0 & \cos \gamma & -\sin \gamma \\ 0 & \sin \gamma & \cos \gamma \end{bmatrix}
\tag{12.282}
$$

At any time t, the body B with respect to B_1 is at ${}^1 R_2$.

$$
{}^1 R_2 = {}^1 R_G \, {}^G R_2
\tag{12.283}
$$

$$
= \begin{bmatrix} 1 & 0 & 0 \\ 0 & 0.951 \cos \gamma - 0.309 \sin \gamma & -0.309 \cos \gamma - 0.951 \sin \gamma \\ 0 & 0.309 \cos \gamma + 0.951 \sin \gamma & 0.951 \cos \gamma - 0.309 \sin \gamma \end{bmatrix}
$$

12.8   Summary

A serial robot may be assumed as a variable geometrical chain of links that relates the configuration of its end-effector to the Cartesian coordinate frame in which the base frame is attached. Forward kinematics are mathematical–geometrical relations that provide the end-effector configuration by having the joint coordinates. On the other hand, the inverse kinematics are mathematical–geometrical relations that provide joint coordinates for a given end-effector configuration.

The Cartesian path of motion for the end-effector must be expressed as a function of time to find the links' velocity and acceleration. The first applied path function that can provide a rest-to-rest motion is a cubic path for a variable $q_i(t)$ between two given points $q_i(t_0)$ and $q_i(t_f)$

$$
q_i(t) = a_0 + a_1 t + a_2 t^2 + a_3 t^3
\tag{12.284}
$$

By increasing the requirements, such as zero acceleration or jerk at some points on the path, we need to employ higher polynomials to satisfy the conditions. An n degree polynomial can satisfy $n + 1$ conditions. It is also possible to split a multiple conditional path into some intervals with fewer conditions. The interval paths must then be connected to satisfy their boundary conditions.

A path of motion may also be defined based on different mathematical functions. Harmonic and cycloid functions are the most common paths. Non-polynomial equations introduce some advantages, due to simpler expression, and some disadvantages due to nonlinearity. When a path of motion either in joint or in Cartesian coordinates space is defined, forward and inverse kinematics must be utilized to find the path of motion in the other space.

Rotational maneuver of the end-effector about the wrist point needs a rotational path. A rotational path may mathematically be defined similar to a Cartesian path utilizing the Rodriguez formula and rotation matrices.

12.9 Key Symbols

a_c Constant acceleration

a_i, b_i, c_i Coefficient of path equation

B Body coordinate frame

C Constant of integral

G, B_0 Global coordinate frame, base coordinate frame

l Length

q Dependent variable coordinate, joint variable

Cartesian variable

$\mathbf{r}$ Position vectors, homogenous position vector

r_{ij} The element of row i and column j of a matrix

R Rotation transformation matrix

t Dependent variable, time

t_0 Initial time

t_f Final time

$\hat{u}$ Axis of rotation

x, y, z Local coordinate axes

X, Y, Z Global coordinate axes

Greek

δ Difference of position vectors

θ Rotary joint angle, joint variable

ϕ Angle of rotation

Symbol

$[\quad]^{-1}$ Inverse of the matrix $[\quad]$

$[\quad]^T$ Transpose of the matrix $[\quad]$

$\equiv$ Equivalent

$\vdash$ Orthogonal

(i) Link number i

Exercises

1. Rest-to-rest cubic path.

   Find a cubic path for a joint coordinate to satisfy the following conditions:

   (a)
   $$q(0) = -10 \text{ deg}, \ q(1) = 45 \text{ deg}, \ \dot{q}(0) = \dot{q}(1) = 0 \tag{12.285}$$

   (b)
   $$q(0) = 0 \text{ deg}, \ q(1) = 50 \text{ deg}, \ \dot{q}(0) = \dot{q}(1) = 0 \tag{12.286}$$

   (c)
   $$q(0) = 10 \text{ deg}, \ q(1) = 60 \text{ deg}, \ \dot{q}(0) = \dot{q}(1) = 0 \tag{12.287}$$

2. To-rest path.

   Find a quadratic path to satisfy the following conditions:

   $$q(0) = -10 \text{ deg}, \ q(1) = 45 \text{ deg}, \ \dot{q}(1) = 0 \tag{12.288}$$

   Calculate the initial velocity of the path using the quadratic path. Then, find a cubic path to satisfy the same boundary conditions as the quadratic path. Compare the maximum accelerations of the two paths.

3. Constant velocity path.

   Calculate a path to satisfy the following conditions:

   $$q(0) = -10 \text{ deg}, \ q(10) = 45 \text{ deg}, \ \dot{q}(0) = \dot{q}(10) = 0 \tag{12.289}$$

   and move with constant velocity $\dot{q} = 25 \text{ deg}/s$ between 12 deg and 35 deg.

4. Constant acceleration path.

   Calculate a path with constant acceleration $\ddot{q} = 25 \text{ deg}/s^2$ between 12 deg and 35 deg, and satisfy the following conditions:

   $$q(0) = -10 \text{ deg}, \ q(10) = 45 \text{ deg}, \ \dot{q}(0) = \dot{q}(10) = 0 \tag{12.290}$$

5. Zero jerk path.

   Find a path to satisfy the following boundary conditions:

   $$q(0) = 0, \ q(1) = 66 \text{ deg}, \ \dot{q}(0) = \dot{q}(1) = 0 \tag{12.291}$$

   and have zero jerk at the beginning, middle, and end points.

6. Control points.

   Find a path to satisfy these conditions

   $$q(0) = 10 \text{ deg}, \ q(1) = 95 \text{ deg}, \ \dot{q}(0) = \dot{q}(1) = 0 \tag{12.292}$$

   and pass through the following control points:

   $$q(0.25) = 30 \text{ deg}, \ q(0.5) = 65 \text{ deg} \tag{12.293}$$

7. A jerk zero at start–middle–stop path.

   To make a path have jerk as close to zero as possible, an eight degree polynomial

   $$q(t) = a_0 + a_1 t + a_2 t^2 + a_3 t^3 + a_4 t^4 + a_5 t^5 + a_6 t^6 + a_7 t^7 + a_7 t^8 \tag{12.294}$$

   and nine boundary conditions can be employed. Find the path.

$$q(0) = 0 \qquad \dot{q}(0) = 0 \; \ddot{q}(0) = 0 \; \dddot{q}(0) = 0$$
$$\dddot{q}(0.5) = 0 \qquad (12.295)$$
$$q(1) = 120 \text{ deg } \dot{q}(1) = 0 \; \ddot{q}(1) = 0 \; \dddot{q}(1) = 0$$

8. Point sequence path.

The conditions for a sequence of points are given here. Find a path to satisfy the conditions given below:

(a)

$$q(0) = 5 \text{ deg} \qquad \dot{q}(0) = 0 \; \ddot{q}(0) = 0$$
$$q(0.4) = 35 \text{ deg}$$
$$q(0.75) = 65 \text{ deg} \qquad (12.296)$$
$$q(1) = 100 \text{ deg} \quad \dot{q}(1) = 0 \; \ddot{q}(1) = 0.$$

(b)

$$q(0) = 5 \text{ deg} \qquad \dot{q}(0) = 0 \qquad \ddot{q}(0) = 0$$
$$q(2) = 15 \text{ deg} \quad q(4) = 35 \text{ deg } q(7.5) = 65 \text{ deg} \qquad (12.297)$$
$$q(10) = 100 \text{ deg} \; \dot{q}(10) = 0 \qquad \ddot{q}(10) = 0.$$

9. Splitting a path into a series of segments.

Using the splitting method, find a path for the following conditions:

$$q(0) = 5 \text{ deg} \qquad \dot{q}(0) = 0 \qquad \ddot{q}(0) = 0$$
$$q(2) = 15 \text{ deg} \quad q(4) = 35 \text{ deg } q(7.5) = 65 \text{ deg} \qquad (12.298)$$
$$q(10) = 100 \text{ deg} \; \dot{q}(10) = 0 \qquad \ddot{q}(10) = 0.$$

by breaking the entire path into four segments.

$$q_1(t) \text{ for } \quad q(0) < q_1(t) < q(2) \quad \text{and} \quad 0 < t < 2$$
$$q_2(t) \text{ for } \quad q(2) < q_2(t) < q(4) \quad \text{and} \quad 2 < t < 4$$
$$q_3(t) \text{ for } \quad q(4) < q_3(t) < q(7.5) \text{ and} \quad 4 < t < 7.5 \qquad (12.299)$$
$$q_4(t) \text{ for } q(7.5) < q_4(t) < q(10) \text{ and } 7.5 < t < 10$$

10. ★ Extra conditions.

To have a smooth overall path in the splitting method, we may add extra conditions to match the segments. Solve Exercise 9 having a zero jerk transition between segments.

11. ★ Least-squared path.

Using the least-squared method, find the best polynomial path of degree n to approximate a path given by the following points:

$$q(0) = 5 \text{ deg} \qquad q(1) = 7 \text{ deg} \qquad q(2) = 15 \text{ deg}$$
$$q(3) = 21 \text{ deg} \qquad q(4) = 35 \text{ deg} \qquad q(7.5) = 65 \text{ deg} \qquad (12.300)$$
$$q(9) = 85 \text{ deg} \qquad q(10) = 100 \text{ deg}$$

(a) $n = 2$.
(b) $n = 3$.
(c) $n = 4$.
(d) $n = 5$.

12. ★ Least-squared path and boundary conditions.

Using the least-squared method, find the best polynomial path of degree n to approximate a path given by the following points and conditions:

$$q(0) = 5 \text{ deg} \qquad \dot{q}(0) = 0 \qquad \ddot{q}(0) = 0$$
$$q(2) = 15 \text{ deg} \quad q(4) = 35 \text{ deg } q(7.5) = 65 \text{ deg} \qquad (12.301)$$
$$q(10) = 100 \text{ deg} \; \dot{q}(10) = 0 \qquad \ddot{q}(10) = 0.$$

(a) $n = 2$.
(b) $n = 3$.
(c) $n = 4$.
(d) $n = 5$.

13. $2R$ manipulator motion to follow a joint path.

Find the path of the end point of a $2R$ manipulator, with $l_1 = l_2 = 1\,\text{m}$, if the joint variables follow the following given paths:

$$\theta_1(t) = 10 + 156.25t^3 \tag{12.302}$$

$$\theta_2(t) = -41.56 + 222.78t - 172.2t^2 \tag{12.303}$$

14. $3R$ planar manipulator motion to follow a joint path.

Find the path of the end point of a $3R$ manipulator, with

$$l_1 = 1\,\text{m} \qquad l_2 = 0.65\,\text{m} \qquad l_3 = 0.35\,\text{m} \tag{12.304}$$

if the joint variables follow these given paths:

$$\theta_1(t) = -41.56 + 222.78t - 172.2t^2 \tag{12.305}$$

$$\theta_2(t) = 148.99 - 321.67t + 216.67t^2 \tag{12.306}$$

$$\theta_3(t) = 198545 - 827166.6672t \tag{12.307}$$

$$+1290500t^2 - 893500t^3 + 231666.67t^4$$

Calculate the maximum acceleration and jerk of the end point.

15. R⊢R∥R articulated arm motion.

Find the Cartesian trajectory of the end point of an articulated manipulator, shown in Fig. 12.29, if the geometric parameters are

$$d_1 = 1\,\text{m} \qquad d_2 = 0 \qquad l_2 = 1\,\text{m} \qquad l_3 = 1\,\text{m} \tag{12.308}$$

and the joints' paths are

$$\theta_1(t) = -41.56 + 222.78t - 172.2t^2 \tag{12.309}$$

$$\theta_2(t) = 148.99 - 321.67t + 216.67t^2 \tag{12.310}$$

$$\theta_3(t) = 198545 - 827166.6672t \tag{12.311}$$

$$+1290500t^2 - 893500t^3 + 231666.67t^4$$

16. Cartesian paths.

Connect the following points with a straight line. Determine the Cartesian coordinates as functions of time for rest-to-rest paths in $t = 1\,\text{s}$:

(a)

$$P_1 = (1.5, 1.5) \qquad P_2 = (-0.5, 1.5) \tag{12.312}$$

(b)

$$P_1 = (0, 0) \qquad P_2 = (1, 1.5) \tag{12.313}$$

(c)

$$P_1 = (-1.5, 1) \qquad P_2 = (0.5, 1.5) \tag{12.314}$$

(d)

$$P_1 = (-1.5, 1, 0) \qquad P_2 = (0.5, 1.5, 1) \tag{12.315}$$

(e)

$$P_1 = (-1, 0, -1) \qquad P_2 = (-0.5, 1.5.1) \tag{12.316}$$

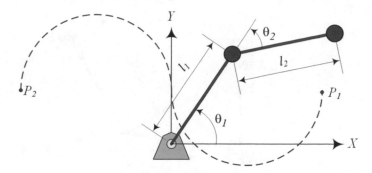

Fig. 12.31 A $2R$ manipulator moves on a path made of two semi-circles

17. Cartesian path for a $2R$ manipulator.

    Consider a $2R$ planar manipulator.

    (a) Calculate a cubic rest-to-rest path in Cartesian space to join the following points with a straight line:

    $$P_1 = (1.5, 1) \qquad P_2 = (-0.5, 1.5) \tag{12.317}$$

    (b) Calculate and plot the joint coordinates of the manipulator, with $l_1 = l_2 = 1\,\text{m}$, that follows the Cartesian path.
    (c) Calculate the maximum angular acceleration of the joint variables.

18. Cartesian path for a $3R$ manipulator.

    Consider a $3R$ articulated manipulator with $l_1 = l_2 = l_3 = 1\,\text{m}$.

    (a) Calculate a cubic rest-to-rest path in Cartesian space to join the following points with a straight line.

    $$P_1 = (-1.5, 1, 0) \qquad P_2 = (0.5, 1.5, 1) \tag{12.318}$$

    (b) Calculate and plot the joint coordinates of the manipulator that follows the Cartesian path.
    (c) Calculate the maximum angular velocity and acceleration of the joint variables.

19. ★ Joint path for a given Cartesian path.

    Assume that the end point of a $2R$ manipulator moves with constant speed $v = 1\,\text{m/s}$ from P_1 to P_2, on a path made of two semi-circles, as shown in Fig. 12.31. The center of the circles is at $(0.75\,\text{m}, 0.5\,\text{m})$ and $(-0.75\,\text{m}, 0.5\,\text{m})$.

    (a) Calculate and plot the joints' path if $l_1 = l_2 = 1\,\text{m}$.
    (b) Calculate the value and positions of the maximum angular velocity in joint variables.
    (c) Calculate the value and positions of the maximum angular acceleration in joint variables.
    (d) Calculate the value and positions of the maximum angular jerk in joint variables.

20. ★ Obstacle avoidance and path planning.

    Let us determine a path between $P_1 = (1.5, 1)$ and $P_2 = (-1, 1)$ to avoid the obstacle shown in Fig. 12.32. The path may be made of two straight lines with a transition circular path in the middle. The radius of the circle is $r = 0.5\,\text{m}$ and the center of the circle is at the lower point of the obstacle. The lines connect to the circle smoothly.

    The end point of the $2R$ manipulator, with $l_1 = l_2 = 1\,\text{m}$, starts at rest from P_1 and moves along the first line with constant acceleration. The end point keeps its speed constant $v = 1\,\text{m/s}$ on the circular path and then moves with constant acceleration on the final line segment to stop at P_2.

    (a) Calculate and plot the joints' paths.
    (b) Find the value and position of the maximum angular velocity for both joints' variable.
    (c) Find the value and position of the maximum angular acceleration for both joints' variable.
    (d) Find the value and position of the maximum angular jerk for both joints' variable.

21. ★ Joint path for a given Cartesian path.

    (a) Connect the points $P_1 = (1.1, 0.8, 0.5)$ and $P_2 = (-1, 1, 0.35)$ with a straight line.
    (b) Find a rest-to-rest cubic path and plot the Cartesian coordinates X, Y, and Z as functions of time.
    (c) Calculate the joints' path for an articulated manipulator, shown in Fig. 12.29, if the geometric parameters are

    $$d_1 = 1\,\text{m} \qquad d_2 = 0 \qquad l_2 = 1\,\text{m} \qquad l_3 = 1\,\text{m} \tag{12.319}$$

    (d) Find the value and position of the maximum angular velocity, acceleration, and jerk for the joints' variable.

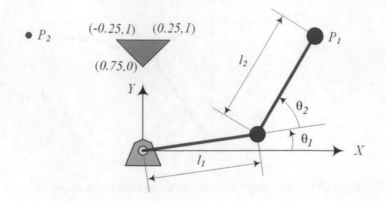

Fig. 12.32 An obstacle in the Cartesian space of motion for a $2R$ manipulator

22. ★ Transition parabola.

In Exercise 20, connect the points $P_1 = (1.5, 1)$ and $P_2 = (-1, 1)$ with two straight lines, using $P_0 = (0, 0.6)$ as a corner. Design a parabolic transition path to avoid the corner if the total time of motion is 12 s and

(a) the interval time is $t' = 1$ s.
(b) the interval time is $t' = 2$ s.
(c) the interval time is $t' = 5$ s.
(d) the interval time is $t' = 8$ s.
(e) the interval time is $t' = 10$ s.

23. ★ Rotational path.

Consider a body frame B that turns 90 deg about Z-axis. Determine the rotation transformation matrix $^G R_B(t)$ such that

(a) the rotation takes place in $t = 1$ s and the angular velocity is constant.
(b) the rotation takes place in $t = 1$ s and the rotation is rest-to-rest.

24. ★ Combined rotational path.

Consider a body frame B that turns 90 deg about Z-axis and 60 deg about X-axis.

(a) Determine the rotation transformation matrix $^G R_B(t)$ such that the body first turns about Z-axis in $t_1 = 1$ s rest-to-rest and then turns about X-axis in $t_2 = 1$ s rest-to-rest.
(b) Multiply the rotation matrices of $R_Z(t)$ and $R_X(t)$. Now $^G R_B(t)$ has only one time variable. Where would B be after $t = 1$ s?
(c) Multiply the rotation matrices of R_Z and R_X and determine $^G R_B$. Determine the angle and axis of rotation of $^G R_B$. Define a rest-to-rest path for the angle of rotation to move B from initial to final orientation in $t = 1$ s.

25. ★ Euler angles' rotational path.

Assume that the spherical wrist of a 6 DOF robot starts from rest position and turns about the axes of the final coordinate frame B_6 in order z-x-z for $\varphi = 15$ deg, $\theta = 38$ deg, and $\psi = 77$ deg. The frame B_6 is installed at the wrist point.

(a) Design a rest-to-rest cubic rotational path for the angles φ, θ, and ψ, if each rotation takes 1 s.
(b) Find the axis and angle of rotation, $(\hat{u}, \phi)$, that moves the wrist from the initial to the final orientation.
(c) Design a cubic rotational path for the axis–angle rotation if it takes 3 s.
(d) Calculate the Euler angles' path $\varphi(t)$, $\theta(t)$, and $\psi(t)$ for this motion.
(e) Calculate and compare the maximum angular velocity, acceleration, and jerk for φ, θ, and ψ in the first and second motions in parts a and c.
(f) Calculate the maximum angular velocity, acceleration, and jerk of ϕ in the second motion in part c.

★ Time Optimal Control

The main job of an industrial robot is to move an object on a pre-specified path, rest-to-rest, repeatedly. To increase productivity, the robot should do the job in minimum time. In this chapter we introduce a numerical method to solve the time optimal control problem of multi degree of freedom robots.

13.1 ★ Minimum Time and Bang-Bang Control

The most important job of industrial robots is moving between two points rest-to-rest. Minimum time control is what we need to increase industrial robots productivity. The objective of time optimal control is to transfer the end-effector of a robot from an initial position to a desired destination in minimum time. Consider a system with the following equation of motion:

$$\dot{\mathbf{x}} = \mathbf{f}\left(\mathbf{x}(t), \mathbf{Q}(t)\right) \tag{13.1}$$

where $\mathbf{Q}$ is the control input, and $\mathbf{x}$ is the state vector of the system.

$$\mathbf{x} = \begin{bmatrix} \mathbf{q} \\ \dot{\mathbf{q}} \end{bmatrix} \tag{13.2}$$

The minimum time problem is always subject to bounded input such as:

$$|\mathbf{Q}(t)| \leq \mathbf{Q}_{Max} \tag{13.3}$$

The solution of the time optimal control problem subject to bounded input is *bang-bang control*. The control in which the input variable takes either the maximum or minimum values is called bang-bang control.

Proof The goal of minimum time control is to find the trajectory $\mathbf{x}(t)$ and input $\mathbf{Q}(t)$ starting from an initial state $\mathbf{x}_0(t)$ and arriving at the final state $\mathbf{x}_f(t)$ under the condition that the whole trajectory minimizes the time integral.

$$J = \int_{t_0}^{t_f} dt \tag{13.4}$$

The input command vector $\mathbf{Q}(t)$ usually has limit constraint such as (13.3).

We define a scalar Hamiltonian function H, and a vector $\mathbf{p}$,

$$H(\mathbf{x}, \mathbf{Q}, \mathbf{p}) = \mathbf{p}^T \mathbf{f}\left(\mathbf{x}(t), \mathbf{Q}(t)\right) \tag{13.5}$$

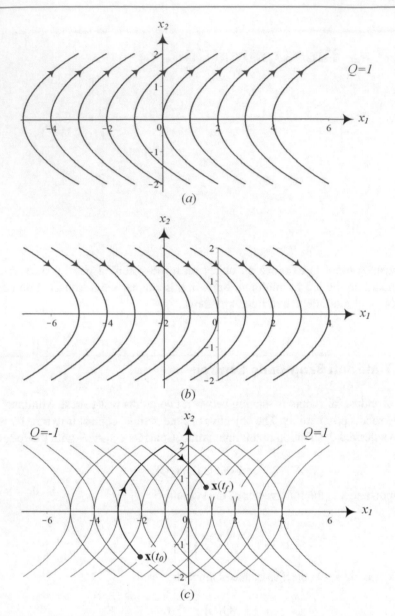

Fig. 13.1 Optimal path for $Q = \ddot{x}$ in phase plane and the mesh of optimal paths in phase plane

that provide the following two equations of motion.

$$\mathbf{\dot{R}} = \frac{\partial H}{\partial \mathbf{p}}^T \tag{13.6}$$

$$\dot{p} = -\frac{\partial H}{\partial \mathbf{x}}^T \tag{13.7}$$

Based on the *Pontryagin principle*, the optimal input $\mathbf{Q}(t)$ is the one that minimizes the function H. Such an optimal input is to apply the maximum effort, $\mathbf{Q}_{Max}$ or $-\mathbf{Q}_{Max}$, over the entire time interval. When the control command takes a value at the boundary of its admissible region, it is said to be *saturated*. The function H is *Hamiltonian*, and the vector $\mathbf{p}$ is called a *co-state*.

The components of the control command $\mathbf{Q}(t)$ are allowed to be piecewise continuous and the values they can take may be any number within the bounded region of the control space. As an example, consider a 2 DOF system $\mathbf{Q}(t) = \begin{bmatrix} Q_1 & Q_2 \end{bmatrix}^T$

with the restriction $|Q_i| < 1$, $i = 1, 2$. The control space is a circle in the (Q_1, Q_2)-plane. The control components may have any piecewise continuous value within the circle. Such controls are called admissible. ∎

Example 403 ★ A linear dynamic system. Time optimal control of a linear system is good example to show the logic and principle of bang-bang control.

Consider a linear dynamic system along with a constraint on the input variable Q given by

$$Q = \ddot{x} \qquad Q \le 1 \tag{13.8}$$

or equivalently by

$$\dot{x} = [A]\mathbf{x} + \mathbf{b}Q \tag{13.9}$$

where

$$\mathbf{x} = \begin{bmatrix} x_1 \\ x_2 \end{bmatrix} \qquad [A] = \begin{bmatrix} 0 & 1 \\ 0 & 0 \end{bmatrix} \qquad \mathbf{b} = \begin{bmatrix} 0 \\ 1 \end{bmatrix} \tag{13.10}$$

By defining a co-state vector

$$\mathbf{p} = \begin{bmatrix} p_1 \\ p_2 \end{bmatrix} \tag{13.11}$$

the Hamiltonian (13.5) becomes:

$$H(\mathbf{x}, \mathbf{Q}, \mathbf{p}) = \mathbf{p}^T \left([A]\mathbf{x} + \mathbf{b}Q\right) \tag{13.12}$$

that provides two first-order differential state equations.

$$\dot{x} = \frac{\partial H}{\partial \mathbf{p}}^T = [A]\mathbf{x} + \mathbf{b}Q \tag{13.13}$$

$$\dot{p} = -\frac{\partial H}{\partial \mathbf{x}}^T = -[A]\mathbf{p} \tag{13.14}$$

Equation (13.14) is

$$\begin{bmatrix} \dot{p}_1 \\ \dot{p}_2 \end{bmatrix} = \begin{bmatrix} 0 \\ -p_2 \end{bmatrix} \tag{13.15}$$

which can be integrated to find $\mathbf{p}$.

$$\mathbf{p} = \begin{bmatrix} p_1 \\ p_2 \end{bmatrix} = \begin{bmatrix} C_1 \\ -C_1 t + C_2 \end{bmatrix} \tag{13.16}$$

The Hamiltonian is then equal to:

$$H = Qp_2 + p_1 x_2 = (-C_1 t + C_2) Q + p_1 x_2 \tag{13.17}$$

The control command Q only appears in

$$\mathbf{p}^T \mathbf{b}Q = (-C_1 t + C_2) Q \tag{13.18}$$

which can be maximized by:

$$Q(t) = \begin{cases} 1 & if \ -C_1 t + C_2 \ge 0 \\ -1 & if \ -C_1 t + C_2 < 0 \end{cases} \tag{13.19}$$

This solution implies that $Q(t)$ has a jump point at $t = C_2/C_1$. The jump point, at which the control command suddenly changes from maximum to minimum or from minimum to maximum, is called the switching point.

Substituting the control input (13.19) into (13.9) gives us two first-order differential equations.

$$\begin{bmatrix} \dot{x}_1 \\ \dot{x}_2 \end{bmatrix} = \begin{bmatrix} x_2 \\ Q \end{bmatrix} \tag{13.20}$$

Equation (13.20) can be integrated to find the path $\mathbf{x}(t)$.

$$\begin{bmatrix} x_1 \\ x_2 \end{bmatrix} = \begin{cases} \begin{bmatrix} \frac{1}{2}\left(t + C_3\right)^2 + C_4 \\ t + C_3 \end{bmatrix} & if \quad Q = 1 \\ \begin{bmatrix} -\frac{1}{2}\left(t - C_3\right)^2 + C_4 \\ -t + C_3 \end{bmatrix} & if \quad Q = -1 \end{cases} \tag{13.21}$$

The constants of integration, C_1, C_2, C_3, C_4, must be calculated based on the following boundary conditions

$$\mathbf{x}_0 = \mathbf{x}(t_0) \qquad \mathbf{x}_f = \mathbf{x}(t_f) \tag{13.22}$$

Eliminating t between equations in (13.21) provides the relationship between the state variables x_1 and x_2.

$$x_1 = \begin{cases} \frac{1}{2}x_2^2 + C_4 & if \quad Q = 1 \\ -\frac{1}{2}x_2^2 + C_4 & if \quad Q = -1 \end{cases} \tag{13.23}$$

These equations show a series of parabolic curves in the (x_1, x_2)-plane with C_4 as a parameter. The parabolas are shown in Fig. 13.1a and b with the arrows indicating the direction of motion on the paths. The (x_1, x_2)-plane is called the phase plane.

Considering that there is one switching point in this system, the overall optimal paths are shown in Fig. 13.1c. As an example, assume the state of the system at initial and final times are $\mathbf{x}(t_0)$ and $\mathbf{x}(t_f)$, respectively. The motion starts with $Q = 1$, which forces the system to move on the control path $x_1 = \frac{1}{2}x_2^2 + \left(x_{10} - \frac{1}{2}x_{20}^2\right)$ up to the intersection point with $x_1 = -\frac{1}{2}x_2^2 + \left(x_{1f} + \frac{1}{2}x_{2f}^2\right)$. The intersection is the switching point at which the control input changes to $Q = -1$. The switching point is at:

$$x_1 = \frac{1}{4}\left(2x_{10} + 2x_f - x_{20}^2 + x_{2f}^2\right) \tag{13.24}$$

$$x_2 = \sqrt{\left(x_{1f} + \frac{1}{2}x_{2f}^2\right) - \left(x_{10} - \frac{1}{2}x_{20}^2\right)} \tag{13.25}$$

Example 404 ★ Robot equations in state equations. The best way to apply optimal control methods on dynamic systems is to express the equations of motion in state space with a set of first-order differential equations. Here is to show equations of motion of robots in state space form.

The vector form of the equations of motion of a robot is:

$$\mathbf{D}(\mathbf{q})\,\ddot{\mathbf{q}} + \mathbf{H}(\mathbf{q}, \dot{\mathbf{q}}) + \mathbf{G}(\mathbf{q}) = \mathbf{Q} \tag{13.26}$$

We can define a state vector $\mathbf{x}$,

$$\mathbf{x} = \begin{bmatrix} \mathbf{q} \\ \dot{\mathbf{q}} \end{bmatrix} \tag{13.27}$$

and transform the equations of motion to an equation in state space

$$\dot{\mathbf{x}} = \mathbf{f}\left(\mathbf{x}(t), \mathbf{Q}(t)\right) \tag{13.28}$$

where

$$\mathbf{f}\left(\mathbf{x}(t), \mathbf{Q}(t)\right) = \begin{bmatrix} \dot{\mathbf{q}} \\ \mathbf{D}^{-1}\left(\mathbf{Q} - \mathbf{H} - \mathbf{G}\right) \end{bmatrix} \tag{13.29}$$

Example 405 ★ Time optimal control for robots. The analytic time optimal control with bounded input is expressed for robotic equations.

Assume that a robot is initially at $\mathbf{x}_0$,

$$\mathbf{x}(\mathbf{t}_0) = \mathbf{x}_0 = \begin{bmatrix} \mathbf{q}_0 \\ \dot{\mathbf{q}}_0 \end{bmatrix} \tag{13.30}$$

and it is supposed to be finally at $\mathbf{x}_f$, in the shortest possible time.

$$\mathbf{x}(\mathbf{t}_f) = \mathbf{x}_f = \begin{bmatrix} \mathbf{q}_f \\ \dot{\mathbf{q}}_f \end{bmatrix} \tag{13.31}$$

The torques of the actuators at each joint are assumed to be bounded.

$$|Q_i| \le Q_{i_{Max}} \tag{13.32}$$

The optimal control problem is to minimize the time performance index J.

$$J = \int_{t_0}^{t_f} dt = t_f - t_0 \tag{13.33}$$

The Hamiltonian H is defined as:

$$H(\mathbf{x}, \mathbf{Q}, \mathbf{p}) = \mathbf{p}^T \mathbf{f}(\mathbf{x}(t), \mathbf{Q}(t)) \tag{13.34}$$

which provides the following two sets of equations.

$$\mathbf{R} = \frac{\partial H}{\partial \mathbf{p}}^T \qquad \dot{p} = -\frac{\partial H}{\partial \mathbf{x}}^T \tag{13.35}$$

The optimal control input $\mathbf{Q}_t(t)$ is the one that minimizes the function H. Hamiltonian minimization reduces the time optimal control problem to a two-point boundary value problem. The boundary conditions are the states of the robot at times t_0 and t_f. Due to nonlinearity of the robots' equations of motion, there is no analytic solution for the boundary value problem. Hence, a numerical technique must be developed.

Example 406 ★ Euler–Lagrange equation. Optimal control is a sub-problem of calculus of variation and Lagrange equation. Here is showing this fact.

To show that a path $x = x^\star(t)$ is a minimizing path for the functional J with boundary conditions $x(t_0) = x_0, x(t_f) = x_f$,

$$J(x) = \int_{t_0}^{t_f} f(x, \dot{x}, t) dt \tag{13.36}$$

we need to show that

$$J(x) \ge J(x^\star) \tag{13.37}$$

for all continuous paths $x(t)$ satisfying the boundary conditions. Any path $x(t)$ satisfying the boundary conditions $x(t_0) = x_0$, $x(t_f) = x_f$, is called *admissible*. To see if $x^\star(t)$ is the optimal path, we may examine the integral J for every admissible path. An admissible path may be defined by

$$x(t) = x^\star + \epsilon y(t) \qquad \epsilon \ll 1 \tag{13.38}$$

where

$$y(t_0) = y(t_f) = 0 \tag{13.39}$$

and ϵ is a small number. Substituting $x(t)$ in J and subtracting from (13.36) provides ΔJ.

$$\begin{aligned} \Delta J &= J\left(x^\star + \epsilon y(t)\right) - J\left(x^\star\right) \\ &= \int_{t_0}^{t_f} f(x^\star + \epsilon y, \dot{x}^\star + \epsilon \dot{y}, t) dt - \int_{t_0}^{t_f} f(x^\star, x^\star, t) dt \end{aligned} \tag{13.40}$$

Let us expand $f(x^\star + \epsilon y, \dot{x}^\star + \epsilon \dot{y}, t)$ about $(x^\star, \dot{x}^\star)$

$$f(x^\star + \epsilon y, \dot{x}^\star + \epsilon \dot{y}, t) = f(x^\star, \dot{x}^\star, t) + \epsilon \left(y\frac{\partial f}{\partial x} + \dot{y}\frac{\partial f}{\partial \dot{x}} \right)$$

$$+\epsilon^2 \left(y^2 \frac{\partial^2 f}{\partial x^2} + 2y\dot{y}\frac{\partial^2 f}{\partial x \partial \dot{x}} + \dot{y}^2 \frac{\partial^2 f}{\partial \dot{x}^2} \right) dt$$

$$+ O\left(\epsilon^3\right) \tag{13.41}$$

and find ΔJ

$$\Delta J = \epsilon V_1 + \epsilon^2 V_2 + O\left(\epsilon^3\right) \tag{13.42}$$

where

$$V_1 = \int_{t_0}^{t_f} \left(y\frac{\partial f}{\partial x} + \dot{y}\frac{\partial f}{\partial \dot{x}} \right) dt \tag{13.43}$$

$$V_2 = \int_{t_0}^{t_f} \left(y^2 \frac{\partial^2 f}{\partial x^2} + 2y\dot{y}\frac{\partial^2 f}{\partial x \partial \dot{x}} + \dot{y}^2 \frac{\partial^2 f}{\partial \dot{x}^2} \right) \tag{13.44}$$

The first integral, V_1, is called the first variation of J, and the second integral, V_2, is called the second variation of J. All the higher variations are combined and shown as $O\left(\epsilon^3\right)$. If $x^\star$ is the minimizing curve, then it is necessary that $\Delta J \geq 0$ for every admissible $y(t)$. If we divide ΔJ by ϵ and make $\epsilon \to 0$ then we find a necessary condition for $x^\star$ to be the optimal path as $V_1 = 0$. This condition is equivalent to:

$$\int_{t_0}^{t_f} \left(y\frac{\partial f}{\partial x} + \dot{y}\frac{\partial f}{\partial \dot{x}} \right) dt = 0 \tag{13.45}$$

Integrating by parts yields:

$$\int_{t_0}^{t_f} \dot{y}\frac{\partial f}{\partial \dot{x}} dt = \left(y\frac{\partial f}{\partial \dot{x}} \right) \Big|_{t_0}^{t_f} - \int_{t_0}^{t_f} y\frac{d}{dt}\left(\frac{\partial f}{\partial \dot{x}} \right) dt \tag{13.46}$$

Because $y(t_0) = y(t_f) = 0$, the first term on the right-hand side is zero. Therefore, the minimization integral condition (13.45), for every admissible $y(t)$, reduces to:

$$\int_{t_0}^{t_f} y\left(\frac{\partial f}{\partial x} - \frac{d}{dt}\frac{\partial f}{\partial \dot{x}} \right) dt = 0 \tag{13.47}$$

The terms in the parentheses are continuous functions of t, evaluated on the optimal path $x^\star$, and they do not involve $y(t)$. So, the only way that the bounded integral of the parentheses $\left(\frac{\partial f}{\partial x} - \frac{d}{dt}\frac{\partial f}{\partial \dot{x}} \right)$, multiplied by a nonzero function $y(t)$, from t_0 and t_f to be zero, is that it is zero itself.

$$\frac{\partial f}{\partial x} - \frac{d}{dt}\frac{\partial f}{\partial \dot{x}} = 0 \tag{13.48}$$

Equation (13.48) is a necessary condition for $x = x^\star(t)$ to be a solution of the minimization problem (13.36). This differential equation is called the Euler–Lagrange equation. It is the same Lagrange equation that we utilized to derive the equations of motion of a robot. The second necessary condition to have $x = x^\star(t)$ as a minimizing solution is that the second variation, evaluated on $x^\star(t)$, must be negative.

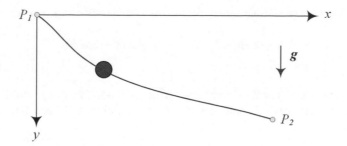

Fig. 13.2 A curve joining points P_1 and P_2, and a frictionless sliding point

Example 407 ★ The Lagrange equation for extremizing $J = \int_1^2 \dot{x}^2 dt$. Applying Lagrange equation and solving the optimal solution for an example of dynamic system.

The Lagrange equation for extremizing the functional J

$$J = \int_1^2 \dot{x}^2 dt \tag{13.49}$$

is

$$\frac{\partial f}{\partial x} - \frac{d}{dt}\frac{\partial f}{\partial \dot{x}} = -\ddot{x} = 0 \tag{13.50}$$

that shows the optimal path will be:

$$x = C_1 t + C_2 \tag{13.51}$$

Considering the boundary conditions $x(1) = 0$, $x(2) = 3$ provides the optimal solution.

$$x = 3t - 3 \tag{13.52}$$

Example 408 ★ Brachistochrone problem. Minimum time of sliding a bid on a wire is the first time optimal control in history of science that was solved mathematically. It is still a very good and practical example.

We may utilize the Lagrange equation and find the frictionless curve joining two points as shown if Fig. 13.2, along which a particle falling from rest due to gravity, travels from the higher to the lower point in a minimum time. This is the well-known brachistochrone problem. If v is the velocity of the falling point along the curve, then the time required to fall an arc length ds is ds/v. Then, the objective function J to find the curve of minimum time would be:

$$J = \int_1^2 \frac{ds}{v} \tag{13.53}$$

However,

$$ds = \sqrt{1 + y'^2}dx \tag{13.54}$$

and according to the conservation of energy, we have:

$$v = \sqrt{2gy} \tag{13.55}$$

Therefore, the objective function simplifies to:

$$J = \int_1^2 \sqrt{\frac{1 + y'^2}{2gy}}dx \tag{13.56}$$

Applying the Lagrange equations we find the optimal solution,

$$y\left(1 + y'^2\right) = 2r \tag{13.57}$$

where r is a constant. The optimal curve starting from $y(0) = 0$ can be expressed by two parametric equations.

$$x = r\,(\beta - \sin\beta) \qquad y = r\,(1 - \cos\beta) \tag{13.58}$$

The optimal curve is a cycloid.

The name of the problem is derived from the Greek word "$\beta\rho\alpha\chi\iota\sigma\tau\sigma\zeta$," meaning "shortest," and "$\chi\rho\sigma\nu\sigma\zeta$," meaning "time." The brachistochrone problem was originally discussed by Galilei in 1630 and later solved by Johann and Jacob Bernoulli in 1696.

Example 409 ★ Lagrange multiplier. When an optimal problem must satisfy some constraints, the constrains will be included in the objective function, each by a weight factor called Lagrange multiplier. Here is to show how to deal with constraint and solve the optimal control problem.

Assume $f(x)$ is defined on an open interval (a, b) and has continuous first and second-order derivatives in some neighborhood of $x_0 \in (a, b)$. The point x_0 is a local extremum of $f(x)$ if:

$$\frac{df(x_0)}{dx} = 0 \tag{13.59}$$

Assume $f(\mathbf{x}) = 0$, $\mathbf{x} \in \mathbb{R}^n$ and $g_i(\mathbf{x}) = 0$, $i = 1, 2, \cdots, m$ are functions defined on an open real region $\mathbb{R}^n$ and have continuous first and second-order derivatives in $\mathbb{R}^n$. The necessary condition that $\mathbf{x}_0$ be an extremum of $f(\mathbf{x})$, subject to the constraints $g_i(\mathbf{x}) = 0$, is that there exist m Lagrange multipliers λ_i, $i = 1, 2, \cdots, m$ such that:

$$\nabla\left(s + \sum \lambda_i g_i\right) = 0 \tag{13.60}$$

As an example, we can find the minimum of $f(x)$

$$f = 1 - x_1^2 - x_2^2 \tag{13.61}$$

subject to $g(x)$

$$g = x_1^2 + x_2 - 1 = 0 \tag{13.62}$$

by finding the gradient of $f + \lambda g$.

$$\nabla\left(1 - x_1^2 - x_2^2 + \lambda\left(x_1^2 + x_2 - 1\right)\right) = 0 \tag{13.63}$$

That leads to:

$$\frac{\partial f}{\partial x_1} = -2x_1 + 2\lambda x_1 = 0 \tag{13.64}$$

$$\frac{\partial f}{\partial x_2} = -2x_2 + \lambda = 0 \tag{13.65}$$

To find the three unknowns, x_1, x_2, and λ, we employ Eqs. (13.64), (13.65), and (13.62). There are two sets of solutions as follows:

$$\begin{array}{lll} x_1 = 0 & x_2 = 1 & \lambda = 2 \\ x_1 = \pm 1/\sqrt{2} & x_2 = 1/2 & \lambda = 1 \end{array} \tag{13.66}$$

13.2 ★ Floating Time Method

Consider a particle with mass m, as shown in Fig. 13.3, is moving according to the following equation of motion:

$$m\ddot{x} = g(x, \dot{x}) + f(t) \tag{13.67}$$

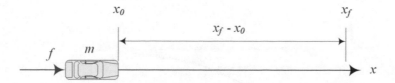

Fig. 13.3 Rest-to-rest motion of a mass on a straight line time optimally

where $g(x, \dot{x})$ is a general nonlinear external force function, and $f(t)$ is a unknown input control force function. The control command $f(t)$ is bounded and is subject to be determined.

$$|f(t)| \le F \tag{13.68}$$

The particle starts from rest at position $x(0) = x_0$ and moves on a straight line to the destination point $x(t_f) = x_f$ at which it stops.

We can solve this rest-to-rest control problem and find the required $f(t)$ to move m from $\dot{x}_0$ to $\dot{x}_f$ in minimum time utilizing the *floating time algorithm*.

Algorithm 13.1 Floating time technique

1. *Divide the preplanned path of motion $x(t)$ into $s + 1$ intervals and specify all coordinate values x_i, $(i = 0, 1, 2, 3, \ldots, s + 1)$.*
2. *Set $f_0 = +F$ and calculate τ_0.*

$$\tau_0 = \sqrt{\frac{2m\,(x_1 - x_0)}{F}} \tag{13.69}$$

3. *Set $f_{s+1} = -F$ and calculate τ_s.*

$$\tau_s = \sqrt{\frac{2m\,(x_s - x_{s-1})}{-F}} \tag{13.70}$$

4. *For i from 1 to $s - 1$, calculate τ_i such that $f_i = +F$ and*

$$
\begin{aligned}
f_i &= m\ddot{x}_i - g(x_i, \dot{x}_i) \\
&= \frac{4m}{\tau_i^2 + \tau_{i-1}^2} \left(\frac{\tau_{i-1}}{\tau_i + \tau_{i-1}} x_{i+1} + \frac{\tau_i}{\tau_i + \tau_{i-1}} x_{i-1} + x_i \right) \\
&\quad - g(x_i, \dot{x}_i)
\end{aligned}
\tag{13.71}
$$

5. *If $|f_i| \le F$, then stop, otherwise set $j = s$.*
6. *Calculate τ_{j-1} such that $f_j = -F$.*
7. *If $\left| f_{j-1} \right| \le F$, then stop, otherwise set $j = j - 1$ and return to step 6*

Proof Assume $g(x_i, \dot{x}_i) = 0$ and $x(t)$, as shown in Fig. 13.4, is the time history of motion for the point mass m. We divide the path of motion into $s + 1$ arbitrary, and not necessarily equal, segments. Hence, the coordinates x_i, $(i = 0, 1, 2, \ldots, s+1)$ are known. The *floating time $\tau_i = t_i - t_{i-1}$* is defined as the required time to move m from x_i to x_{i+1}.

Utilizing the central difference method, we may define the first and second derivatives at point i by

$$\dot{x}_i = \frac{x_{i+1} - x_{i-1}}{\tau_i + \tau_{i-1}} \tag{13.72}$$

$$\ddot{x}_i = \frac{4}{\tau_i^2 + \tau_{i-1}^2} \left(\frac{\tau_{i-1}}{\tau_i + \tau_{i-1}} x_{i+1} + \frac{\tau_i}{\tau_i + \tau_{i-1}} x_{i-1} - x_i \right). \tag{13.73}$$

These equations indicate that the velocity and acceleration at point i depend on x_i, and two adjacent points x_{i-1}, and x_{i+1}, as well as on the floating times τ_i, and τ_{i-1}. Therefore, two extra points, x_{-1} and x_{s+2}, before the initial point and after the

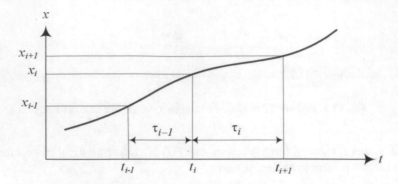

Fig. 13.4 Time history of motion for the point mass m

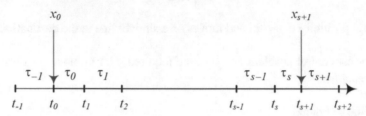

Fig. 13.5 Introducing two extra points, x_{-1} and x_{s+2}, before the initial and after the final points

final point are needed to define velocity and acceleration at x_0 and x_{s+1}. These extra points and their corresponding floating times are shown in Fig. 13.5.

The rest conditions at the beginning and at the end of motion require the following constraints.

$$x_{-1} = x_1 \qquad x_{s+2} = x_s \tag{13.74}$$

$$\tau_0 = \tau_{-1} \qquad \tau_{s+1} = \tau_s \tag{13.75}$$

Using Eq. (13.73), the equation of motion, $f_i = m\ddot{x}_i$, at the initial point is:

$$f_0 = m\ddot{x}_0 = \frac{4m}{2\tau_0^2}(x_1 - x_0) \tag{13.76}$$

The minimum value of the first floating time τ_0 is found by setting $f_0 = F$.

$$\tau_0 = \sqrt{\frac{2m(x_1 - x_0)}{F}} \tag{13.77}$$

It is the minimum value of the first floating time because if τ_0 is less than the value given by (13.77), then f_0 will be greater than F and breaks the constraint (13.68). On the other hand, if τ_0 is greater than the value given by (13.77), then f_0 will be less than F and the input is not saturated yet. The same conditions exist at the final point where the equation of motion is:

$$f_{s+1} = m\ddot{x}_{s+1} = \frac{4m}{2\tau_s^2}(x_s - x_{s-1}) \tag{13.78}$$

The minimum value of the final floating time, τ_s, is achieved by setting $f_{s+1} = -F$.

$$\tau_s = \sqrt{\frac{2m(x_s - x_{s-1})}{-F}} \tag{13.79}$$

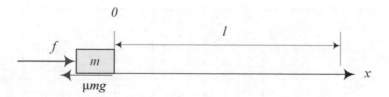

Fig. 13.6 A rectilinear motion of a rigid mass m under the influence of a control force $f(t)$ and a friction force μmg

To find the minimum value of τ_1, we develop the equation of motion at x_1.

$$f_1 = \frac{4m}{\tau_1^2 + \tau_0^2}\left(\frac{\tau_0}{\tau_1 + \tau_0}x_2 + \frac{\tau_1}{\tau_1 + \tau_0}x_0 - x_1\right) \tag{13.80}$$

It is an equation with two unknowns f_1 and τ_1. We are able to find τ_1 numerically by adjusting τ_1 to provide $f_1 = F$. Applying this procedure we are able to find the minimum floating times τ_{i+1} by applying the maximum force constraint $f_i = F$, and solving the equation of motion for τ_{i+1} numerically. When τ_i is known and the maximum force is applied to find the next floating time, τ_{i+1}, we are in the *forward path of the floating time algorithm*. In the last step of the forward path, τ_{s-1} is found at x_{s-1}. At this step, all the floating times τ_i, $(i = 0, 1, 2, \ldots, s)$ are known, while $f_i = F$ for $i = 0, 1, 2, \ldots, s-1$, and $f_i = -F$ for $i = s$. Then, the force f_s is the only variable that is not calculated during the forward path by using the equation of motion at point x_s. The value of f_s is actually dictated by the equation of motion at point x_s, because τ_{s-1} is known from the forward path procedure, and τ_s is known from Eq. (13.79) to satisfy the final point condition. Therefore, the value of f_s can be found from the equation of motion at $i = s$ by substituting τ_s, τ_{s-1}, x_{s-1}, x_s, and x_{s+1}.

$$f_s = \frac{4m}{\tau_s^2 + \tau_{s-1}^2}\left(\frac{\tau_{s-1}}{\tau_s + \tau_{s-1}}x_{s+1} + \frac{\tau_s}{\tau_s + \tau_{s-1}}x_{s-1} - x_s\right) \tag{13.81}$$

Now, if f_s does not break the constraint $|f(t)| \leq F$, the problem is solved and the minimum time motion is determined. The input signals, f_i, $(i = 0, 1, 2, \ldots, s+1)$, $i \neq s$, are always saturated, and also none of the floating times τ_i can be reduced any more. However, it is expected that f_s breaks the constraint $|f(t)| \leq F$ because accelerating in the first $s-1$ number of steps with $f = F$ produces a large amount of kinetic energy and a huge deceleration is needed to stop the mass m in the final step.

Now we reverse the procedure, and start a *backward path*. According to (13.81), f_s can be adjusted to satisfy the constraint $f_s = -F$ by tuning τ_{s-1}. Now f_{n-1} must be checked for the constraint $|f(t)| \leq F$. This is because τ_{s-2} is already found in the forward path, and τ_{s-1} in the backward path. Hence, the value of f_{n-1} is dictated by the equation of motion at point x_{s-1}. If f_{s-1} does not break the constraint $|f(t)| \leq F$, the problem is solved and the time optimal motion is achieved. Otherwise, the backward path must be continued to a point where the force constraint is satisfied. The position x_k in the backward path, where $|f_k| \leq F$, is called *switching point* because $f_j = F$ for $j < k$, $0 \leq j < k$ and $f_j = -F$ for $j > k$, $k < j \leq s+1$. ∎

Example 410 ★ Moving a mass on a rough surface. Applying the floating time method and solving time optimal control motion of a mass on a rough surface shows well how to apply the method on applied dynamic systems.

Consider a rectilinear motion of a rigid mass m under a variable force $f(t)$ and a friction force μmg, as shown in Fig. 13.6. The force is bounded by $|f(t)| \leq F$, where $\pm F$ is the limit of available force. It is necessary to find a function $f(t)$ that moves m, from the initial conditions $x(0) = 0$, $v(0) = 0$ to the final conditions $x(t_f) = l > 0$, $v(t_f) = 0$ in minimum total time $t = t_f$. The motion is described by the following equation of motion and boundary conditions.

$$f = m\ddot{x} - \mu mg \tag{13.82}$$

$$x(0) = 0 \quad v(0) = 0$$
$$x(t_f) = l \quad v(t_f) = 0 \tag{13.83}$$

Using the theory of optimal control, we know that a time optimal control solution for $\mu = 0$ is a piecewise constant function where the only discontinuity is at the switching point $t = \tau = t_f/2$ and

$$f(t) = \begin{cases} F & \text{if } t < \tau \\ -F & \text{if } t > \tau \end{cases} \tag{13.84}$$

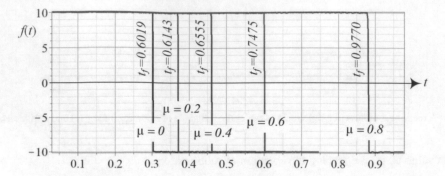

Fig. 13.7 Time history of the optimal input $f(t)$ for different friction coefficients μ

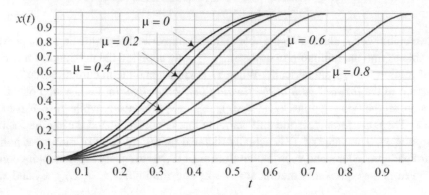

Fig. 13.8 Time history of the optimal motion $x(t)$ for different friction coefficients μ

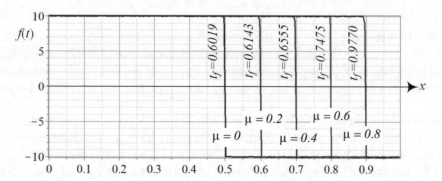

Fig. 13.9 Position history of the optimal force $f(t)$ for different friction coefficients μ

Therefore, the time optimal control solution for moving a mass m from $x(0) = x_0 = 0$ to $x(t_f) = x_f = l$ on a smooth straight line is a bang-bang control with only one switching time. The input force $f(t)$ is on its maximum, $f = F$, before the switching point $x = (x_f - x_0)/2$ at $\tau = t_f/2$, and $f = -F$ after the switching point. Any asymmetric characteristics, such as friction, will make the problem asymmetric by moving the switching point.

In applying the floating time algorithm, we assume that a particle of unit mass, $m = 1$ kg, slides under Coulomb friction on a rough horizontal surface. The magnitude of the friction force is μmg, where μ is the friction coefficient and $g = 9.81$ m/s^2. We apply the floating time algorithm using the following numerical values:

$$F = 10\,\text{N} \qquad l = 1\,\text{m} \qquad s + 1 = 200 \tag{13.85}$$

Figures 13.7, 13.8, and 13.9 show the results for some different values of μ. Figure 13.7 illustrates the time history of the optimal input force for different values of μ. Each curve is indicated by the value of μ and the corresponding minimum time of motion t_f. Time history of the optimal motions $x(t)$ are shown in Fig. 13.8, while the time history of the optimal inputs $f(t)$ are shown in Fig. 13.9. The switching times and positions are shown in Figs. 13.7 and 13.9, respectively.

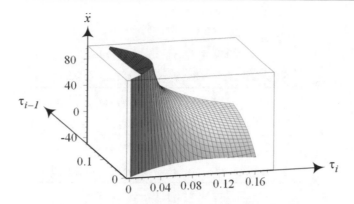

Fig. 13.10 Behavior of $\ddot{x}_i$ as a function of τ_i and τ_{i-1}

If $\mu = 0$, then switching occurs at the midpoint of the motion $x(\tau) = l/2$ and halfway through the time $\tau = t_f/2$. Increasing μ delays both the switching times and the switching positions. The total time of motion also increases by increasing μ.

Example 411 ★ First and second derivatives in 3-point central difference method. Here is showing how derivative equations are calculated for central time difference method.

Using a Taylor series, we expand x at points x_{i-1} and x_{i+1} as an extrapolation of point x_i.

$$x_{i+1} = x_i + \dot{x}_i \tau_i + \frac{1}{2}\ddot{x}_i \tau_i^2 + \cdots \tag{13.86}$$

$$x_{i-1} = x_i - \dot{x}_i \tau_{i-1} + \frac{1}{2}\ddot{x}_i \tau_{i-1}^2 - \cdots \tag{13.87}$$

Accepting the first two terms and calculating $x_{i+1} - x_{i-1}$ provides $\dot{x}_i$.

$$\dot{x}_i = \frac{x_{i+1} - x_{i-1}}{\tau_i + \tau_{i-1}} \tag{13.88}$$

Now, accepting the first three terms of the Taylor series and calculating $x_{i+1} + x_{i-1}$ provides $\ddot{x}_i$.

$$\ddot{x}_i = \frac{4}{\tau_i^2 + \tau_{i-1}^2}\left(\frac{\tau_{i-1}}{\tau_i + \tau_{i-1}}x_{i+1} + \frac{\tau_i}{\tau_i + \tau_{i-1}}x_{i-1} - x_i\right) \tag{13.89}$$

In applying the central difference method for numerical solution of differential equations, the time intervals are kept equal and hence, the equations for 3-point central difference will be simpler.

$$\dot{x}_i = \frac{x_{i+1} - x_{i-1}}{2h} \qquad \ddot{x}_i = \frac{x_{i-1} - 2x_i + x_{i+1}}{h^2} \tag{13.90}$$

$$h = \tau_i \qquad i = 0, 1, 2, \cdots, n$$

Example 412 ★ Convergence. Here is the proof of convergence of floating time method. It is to show on what conditions the method provides the correct solution.

The floating time algorithm presents an iterative method. The convergence criteria of such numerical method must be identified. In addition, a condition must be defined to terminate the iteration. In the forward path, we calculate the floating time τ_i by adjusting it to a value that provides $f_i = F$. The floating time τ_i converges to the minimum possible value, as long as $\partial \ddot{x}_i/\partial \tau_i < 0$ and $\partial \ddot{x}_i/\partial \tau_{i-1} > 0$. Figure 13.10 illustrates the behavior of $\ddot{x}_i$ as a function of τ_i and τ_{i-1}. Using Eq. (13.73), the required conditions are fulfilled within a basin of convergence,

$$Z_1 x_{s+1} + Z_2 x_s + Z_3 x_{s-1} < 0 \tag{13.91}$$

$$Z_4 x_{s+1} + Z_5 x_s + Z_6 x_{s-1} > 0 \tag{13.92}$$

where

$$Z_1 = \frac{8 \left(6\tau_i^4 \tau_{i-1} + 8\tau_i^3 \tau_{i-1}^2 + 6\tau_i^2 \tau_{i-1}^3\right)}{\left(\tau_i^2 + \tau_{i-1}^2\right)^3 (\tau_i + \tau_{i-1})^3} \tag{13.93}$$

$$Z_2 = \frac{8 \left(\tau_{i-1}^5 - 3\tau_i^5 - 8\tau_i^3 \tau_{i-1}^2 - 9\tau_i^4 \tau_{i-1} + 3\tau_i \tau_{i-1}^4\right)}{\left(\tau_i^2 + \tau_{i-1}^2\right)^3 (\tau_i + \tau_{i-1})^3} \tag{13.94}$$

$$Z_3 = \frac{8 \left(-\tau_{i-1}^5 + 3\tau_i^5 + 3\tau_i^4 \tau_{i-1} - 3\tau_i \tau_{i-1}^4 - 6\tau_i^2 \tau_{i-1}^3\right)}{\left(\tau_i^2 + \tau_{i-1}^2\right)^3 (\tau_i + \tau_{i-1})^3} \tag{13.95}$$

$$Z_4 = \frac{8 \left(3\tau_{i-1}^5 - \tau_i^5 - 6\tau_i^3 \tau_{i-1}^2 - 3\tau_i^4 \tau_{i-1} + 3\tau_i \tau_{i-1}^4\right)}{\left(\tau_i^2 + \tau_{i-1}^2\right)^3 (\tau_i + \tau_{i-1})^3} \tag{13.96}$$

$$Z_5 = \frac{8 \left(-3\tau_{i-1}^5 + \tau_i^5 + 3\tau_i^4 \tau_{i-1} - 9\tau_i \tau_{i-1}^4 - 8\tau_i^2 \tau_{i-1}^3\right)}{\left(\tau_i^2 + \tau_{i-1}^2\right)^3 (\tau_i + \tau_{i-1})^3} \tag{13.97}$$

$$Z_6 = \frac{8 \left(6\tau_i^3 \tau_{i-1}^2 + 8\tau_i^2 \tau_{i-1}^3 + 6\tau_i \tau_{i-1}^4\right)}{\left(\tau_i^2 + \tau_{i-1}^2\right)^3 (\tau_i + \tau_{i-1})^3} \tag{13.98}$$

The convergence conditions guarantee that $\ddot{x}_i$ decreases with an increase in τ_i and increases with an increase in τ_{i-1}. Therefore, if either τ_i or τ_{i-1} is fixed, we are able to find the other floating time by setting $f_i = F$. Convergence conditions for backward path are changed to:

$$Z_1 x_{s+1} + Z_2 x_s + Z_3 x_{s-1} > 0 \tag{13.99}$$

$$Z_4 x_{s+1} + Z_5 x_s + Z_6 x_{s-1} < 0 \tag{13.100}$$

A termination criterion may be defined by

$$||f_i| - F| \leq \epsilon \tag{13.101}$$

where ϵ is a user-specified number. The termination criterion provides a good method to make sure that the maximum deviation is within certain bounds.

Example 413 ★ Analytic calculating of floating times. Here is the answer to the question if we are able to calculate floating times analytically.

The rest condition at the beginning of the motion of an m on a straight line yields:

$$x_{-1} = x_1 \qquad \tau_0 = \tau_{-1} \tag{13.102}$$

The first floating time τ_0 is found by setting $f_0 = F$ and developing the equation of motion $f_i = m\ddot{x}_i$ at point x_0.

$$\tau_0 = \sqrt{\frac{2m (x_1 - x_0)}{F}} \tag{13.103}$$

Now the equation of motion at point x_1 would be:

$$f_1 = \frac{4m}{\tau_1^2 + \tau_0^2} \left(\frac{\tau_0}{\tau_1 + \tau_0} x_2 + \frac{\tau_1}{\tau_1 + \tau_0} x_0 + x_1\right) \tag{13.104}$$

Substituting τ_0 from (13.103) into (13.104) and applying $f_1 = F$ provides the following equation that must be solved for τ_1.

$$F = \frac{4mF}{2mx_1 - 2mx_0 + F\tau_1^2} \left(\frac{\tau_0 x_2 + \tau_1 x_0}{\tau_1 + \sqrt{\frac{2m(x_1-x_0)}{F}}} + x_1\right) \tag{13.105}$$

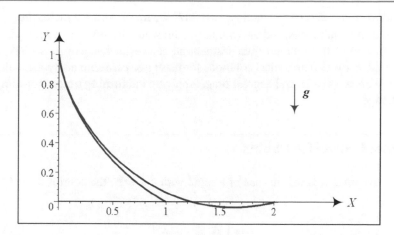

Fig. 13.11 Time optimal path for a falling unit mass from $A(0, 1)$ to two different destinations

Then substituting τ_1 from (13.105) into the equation of motion at x_2, and setting $f_2 = F$ leads to a new equation to find τ_2. This procedure can similarly be applied to the other steps. However, calculating the floating times in closed form is not straightforward and getting more complicated step by step, hence, a numerical solution is needed. The equations for calculating τ_i are nonlinear and therefore have multiple solutions. Each positive solution must be examined for the constraint $f_i = F$. Negative solutions are not acceptable.

Example 414 ★ Brachistochrone and path planning. Brachistochrone is a time optimal control problem in which the applied force is given but the path is not. This problem has analytic solution.

The floating time method can be applicable to path planning problems. As an illustrative example, let us consider the well-known brachistochrone problem. The problem as Johann Bernoulli states is: "A material particle moves without friction along a curve. This curve connects point A with point B (point A is placed above point B). No forces affect it, except the gravitational attraction. The time of travel from A to B must be the smallest. This brings up the question: what is the form of this curve?"

The classical solution of the brachistochrone problem is a cycloid and its parametric equation is:

$$x = r\left(\beta - \sin\beta\right) \qquad y = r\left(1 - \cos\beta\right) \tag{13.106}$$

where r is the radius of the corresponding cycloid and β is the angle of rotation of r. When $\beta = 0$ the particle is at the beginning point $A(0, 0)$. The particle is at the second point B when $\beta = \beta_B$. The value of β_B can be obtained from

$$x_B = r\left(\beta_B - \sin\beta_B\right) \qquad y_B = r\left(1 - \cos\beta_B\right) \tag{13.107}$$

The total time of the motion is:

$$t_f = \beta_B\sqrt{\frac{r}{g}} \tag{13.108}$$

In a path planning problem, except for the boundaries, the path of motion is not known. Hence, the position of x_i in Eqs. (13.72) and (13.73) are not given. Knowing the initial and final positions, we fix x_i coordinates while keeping y_i coordinates free. We will obtain the optimal path of motion by applying the known input force and searching for the optimum y_i that minimizes the floating times.

Consider the points $B_1(1, 0)$ and $B_2(2, 0)$ as two different destinations of motion for a unit mass, $m = 1$, falling from point $A(0, 1)$. Figure 13.11 illustrates the optimal path of motion for the two destinations, obtained by the floating time method for $s = 100$. The total time of motion is $t_{f_1} = 0.61084$ s, and $t_{f_2} = 0.8057$ s, respectively. In this calculation, the gravitational acceleration is assumed $g = 10\,\text{m/s}^2$ in $-Y$ direction.

An analytic solution shows that ;

$$\beta_{B_1} = 1.934563\,\text{rad} \qquad \beta_{B_2} = 2.554295\,\text{rad} \tag{13.109}$$

The corresponding total times are $t_{f_1} = 0.6176$ s, and $t_{f_2} = 0.8077$ s, respectively. By increasing s, the calculated minimum time would be closer to the analytical results, and the evaluated path would be closer to a cycloid.

A more interesting and more realistic problem of brachistochrone can be brachistochrone with friction and brachistochrone with linear drag. Although there are also analytical solutions for these two cases, no analytical solution has been developed for brachistochrone with nonlinear (say second degree) drag. Applying the floating time algorithm for this kind of problem can be an interesting challenge.

13.3   ★ Time Optimal Control for Robots

Robots are multiple DOF dynamic systems. In case of a robot with n DOF, the control force $\mathbf{f}$ and the output position $\mathbf{x}$ are vectors.

$$\mathbf{f} = \begin{bmatrix} f_1 & f_2 & \cdots & f_n \end{bmatrix}^T \tag{13.110}$$

$$\mathbf{x} = \begin{bmatrix} x_1 & x_2 & \cdots & x_n \end{bmatrix}^T \tag{13.111}$$

The constraint on the input force vector can be shown by

$$|\mathbf{f}_i| \leq \mathbf{F} \tag{13.112}$$

where the elements of the limit vector $\mathbf{F} \in \mathbb{R}^n$ may be different. The floating time algorithm is applied similar to the algorithm 13.2. At each step all the elements of the force vector $\mathbf{f}$ must be examined for their constraints. To attain the time optimal control, at least one element of the input vector $\mathbf{f}$ must be saturated at each step, while all the other elements are within their limits.

Algorithm 13.2 Floating time technique for n DOF systems

1. *Divide the preplanned path of motion* $\mathbf{x}(t)$ *into* $s + 1$ *intervals and specify all coordinate vectors* $\mathbf{x}_i$, ($i = 0, 1, 2, 3, \ldots$ $, s + 1$).
2. *Develop the equations of motion at* $\mathbf{x}_0$ *and calculate* τ_0 *for which only one component of the force vector* $\mathbf{f}_0$ *is saturated on its higher limit, while all the other components are within their limits.*

$$f_{0_k} = F_k \ , \ k \in \{0, 1, 2, \cdots, n\}$$
$$f_{0_r} \leq F_r \ , \ r = 0, 1, 2, \cdots, n \ , \ r \neq k$$

3. *Develop the equations of motion at* $\mathbf{x}_{s+1}$ *and calculate* τ_s *for which only one component of the force vector* $\mathbf{f}_{s+1}$ *is saturated on its higher limit, while all the other components are within their limits.*

$$f_{s+1_k} = -F_k \ , \ k \in \{0, 1, 2, \cdots, n\}$$
$$f_{s+1_r} \leq F_r \ \ , \ r = 0, 1, 2, \cdots, n \ , \ r \neq k$$

4. *For i from 1 to* $s - 1$, *calculate* τ_i *such that only one component of the force vector* $\mathbf{f}_i$ *is saturated on its higher limit, while all the other components are within their limits.*

$$f_{i_k} = -F_k \ , \ k \in \{0, 1, 2, \cdots, n\}$$
$$f_{i_r} \leq F_r \ \ , \ r = 0, 1, 2, \cdots, n \ , \ r \neq k$$

5. *If* $|\mathbf{f}_s| \leq \mathbf{F}$, *then stop, otherwise set* $j = s$.
6. *Calculate* τ_{j-1} *such that only one component of the force vector* $\mathbf{f}_j$ *is saturated on its lower limit, while all the other components are within their limits.*
7. *If* $|\mathbf{f}_{j-1}| \leq F$, *then stop, otherwise set* $j = j - 1$ *and return to step 6.*

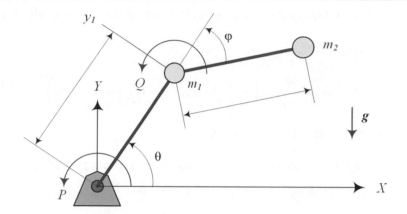

Fig. 13.12 A $2R$ planar manipulator with rigid arms

The aim of minimum time control is to guide a robot on a given path in minimum time to increase the robot's productivity. Except for low order, autonomous, and linear problems, there is no general analytic solution for the time optimal control problems of dynamic systems. The problem of time optimal control is always a bounded input problem. If there exists an admissible time optimal control for a given initial condition and final target, then, at any time, at least one of the control variables attains its maximum or minimum value. Based on Pontryagin's principle, the solution of minimum time problems with bounded inputs is a bang-bang control, indicating that at least one of the input actuators must be saturated at any time. However, determining switching points at which the saturated input signal is replaced with another saturated signal is not straightforward and is the main concern of numerical solution methods.

In a general case, the problem reduces to a two points boundary value problem. The corresponding non-singular, nonlinear two-point boundary value problem must be solved to determine the switching times. A successful approach is to assume that the configuration trajectory of the dynamic system is preplanned, and then reduce the problem to a minimum time motion along the trajectory.

Example 415 ★ $2R$ manipulator on a straight line. Applying floating time algorithm on motion of a $2R$ manipulator on a straight line and determining time optimal torques of motors.

Consider a $2R$ planar manipulator that its endpoint moves rest-to-rest from point $(1, 1.5)$ to point $(-1, 1.5)$ on a straight line $Y = 1.5$. Figure 13.12 illustrates a $2R$ planar manipulator with rigid arms. The manipulator has two rotary joints, whose angular positions are defined by the coordinates θ and φ. The joint axes are both parallel to the Z-axis of the global coordinate frame, and the robot moves in the (X, Y)-plane. Gravity acts in the $-Y$ direction and the lengths of the arms are l_1 and l_2.

We express the equations of motion for $2R$ robotic manipulators in the following form where P and Q are the motors' torques.

$$P = A\ddot{\theta} + B\ddot{\varphi} + C\dot{\theta}\dot{\varphi} + D\dot{\varphi}^2 + M \tag{13.113}$$

$$Q = E\ddot{\theta} + F\ddot{\varphi} + G\dot{\theta}^2 + N \tag{13.114}$$

$$A = A(\varphi) = m_1 l_1^2 + m_2 \left(l_1^2 + l_2^2 + 2l_1 l_2 \cos\varphi\right)$$
$$B = B(\varphi) = m_2 \left(l_2^2 + l_1 l_2 \cos\varphi\right)$$
$$C = C(\varphi) = -2m_2 l_1 l_2 \sin\varphi$$
$$D = D(\varphi) = -m_2 l_1 l_2 \sin\varphi$$
$$E = E(\varphi) = B$$
$$F = m_2 l_2^2$$
$$G = G(\varphi) = -D$$
$$M = M(\theta, \varphi) = (m_1 + m_2)gl_1 \cos\theta + m_2 gl_2 \cos(\theta + \varphi)$$
$$N = N(\theta, \varphi) = m_2 gl_2 \cos(\theta + \varphi) \tag{13.115}$$

Following Eqs. (13.72) and (13.73), we define two functions to discretize velocity v and acceleration a.

$$v(x_i) = \frac{x_{i+1} - x_{i-1}}{\tau_i + \tau_{i-1}} \tag{13.116}$$

$$a(x_i) = \frac{4}{\tau_i^2 + \tau_{i-1}^2} \left(\frac{\tau_{i-1}}{\tau_i + \tau_{i-1}} x_{i+1} + \frac{\tau_i}{\tau_i + \tau_{i-1}} x_{i-1} - x_i \right) \tag{13.117}$$

Then, the equations of motion at each instant may be written as:

$$P_i(t) = A_i a(\theta) + B_i a(\varphi) + C_i v(\theta) v(\varphi) + D_i v^2(\varphi) + M_i \tag{13.118}$$

$$Q_i(t) = E_i a(\theta) + F_i a(\varphi) + G_i v^2(\theta) + N_i \tag{13.119}$$

$$v(\theta_i) = \dot{\theta}_i = \frac{\theta_{i+1} - \theta_{i-1}}{\tau_i + \tau_{i-1}} \tag{13.120}$$

$$a(\theta_i) = \ddot{\theta}_i$$

$$= \frac{4}{\tau_i^2 + \tau_{i-1}^2} \left(\frac{\tau_{i-1}}{\tau_i + \tau_{i-1}} \theta_{i+1} + \frac{\tau_i}{\tau_i + \tau_{i-1}} \theta_{i-1} - \theta_i \right) \tag{13.121}$$

$$v(\varphi) = \dot{\varphi}_i = \frac{\varphi_{i+1} - \varphi_{i-1}}{\tau_i + \tau_{i-1}} \tag{13.122}$$

$$a(\varphi) = \ddot{\varphi}_i$$

$$= \frac{4}{\tau_i^2 + \tau_{i-1}^2} \left(\frac{\tau_{i-1}}{\tau_i + \tau_{i-1}} \varphi_{i+1} + \frac{\tau_i}{\tau_i + \tau_{i-1}} \varphi_{i-1} - \varphi_i \right) \tag{13.123}$$

where P_i and Q_i are the required actuator torques at instant i. Knowing the path of the end-effector and solving the inverse kinematics of the robot, all values of joint variables θ_i, φ_i are known at each i. Therefore, all coefficients of (13.115) are known, and the floating times τ_i are the only unknowns on the right-hand side of Equations of motion (13.118) and (13.119).

Actuators are assumed to be bounded.

$$|P_i(t)| \le P_M \qquad |Q_i(t)| \le Q_M \tag{13.124}$$

To move the manipulator time optimally along a known trajectory, motors must exert torques with a known time history. Using the floating time method, the motion starts at point $i = 0$ and ends at point $i = s + 1$. Introducing two extra points at $i = -1$ and $i = n + 2$, and applying the rest boundary conditions, yields:

$$\begin{aligned} i = 0 & \quad v(x_0) = 0 \quad \tau_0 = \tau_{-1} \quad \theta_1 = \theta_{-1} \quad \varphi_1 = \varphi_{-1} \\ i = s + 1 & \quad v(x_{s+1}) = 0 \quad \tau_s = \tau_{s+1} \quad \theta_s = \theta_{s+2} \quad \varphi_s = \varphi_{s+2} \end{aligned} \tag{13.125}$$

$$v(x_i) = \frac{x_{i+1} - x_{i-1}}{\tau_i + \tau_{i-1}} \tag{13.126}$$

$$a(x_i) = \frac{4}{\tau_i^2 + \tau_{i-1}^2} \left(\frac{\tau_{i-1}}{\tau_i + \tau_{i-1}} x_{i+1} + \frac{\tau_i}{\tau_i + \tau_{i-1}} x_{i-1} - x_i \right) \tag{13.127}$$

At $t = 0$, we have:

$$i = 0 \quad \begin{aligned} v(x_0) &= \frac{x_1 - x_{-1}}{\tau_0 + \tau_{-1}} = 0 \\ a(x_0) &= \frac{4(x_1 - x_0)}{\tau_0^2 + \tau_{-1}^2} = \frac{2(x_1 - x_0)}{\tau_0^2} \end{aligned} \tag{13.128}$$

$$P_0(t) = A_0 a(\theta_0) + B_0 a\left(\varphi_0\right) + M_0 \tag{13.129}$$

$$Q_0(t) = E_0 a(\theta_0) + F_0 a\left(\varphi_0\right) + N_0 \tag{13.130}$$

The first floating time τ_0 will be calculated from Eq. (13.129) by setting $P_0(t) = P_M$ and substituting for $a(\theta_0)$, $a\left(\varphi_0\right)$.

$$
\begin{aligned}
P_M &= A_0 a(\theta_0) + B_0 a\left(\varphi_0\right) + M_0 \\
&= A_0 \frac{2\left(\theta_1 - \theta_0\right)}{\tau_0^2} + B_0 \frac{2\left(\varphi_1 - \varphi_0\right)}{\tau_0^2} + M_0
\end{aligned}
\tag{13.131}
$$

$$\tau_0 = \sqrt{2}\sqrt{\frac{A_0\left(\theta_1 - \theta_0\right) + B_0\left(\varphi_1 - \varphi_0\right)}{P_M - M_0}} \tag{13.132}$$

$$
\begin{aligned}
Q_0(t) &= E_0 \frac{2\left(\theta_1 - \theta_0\right)}{\tau_0^2} + F_0 \frac{2\left(\varphi_1 - \varphi_0\right)}{\tau_0^2} + N_0 \\
&= E_0 \frac{\left(\theta_1 - \theta_0\right)\left(P_M - M_0\right)}{A_0\left(\theta_1 - \theta_0\right) + B_0\left(\varphi_1 - \varphi_0\right)} \\
&\quad + F_0 \frac{\left(\varphi_1 - \varphi_0\right)\left(P_M - M_0\right)}{A_0\left(\theta_1 - \theta_0\right) + B_0\left(\varphi_1 - \varphi_0\right)} + N_0
\end{aligned}
\tag{13.133}
$$

For this value of τ_0, we must have $Q_0(t) \leq Q_M$, otherwise, we must calculate τ_0 from (13.130).

$$
\begin{aligned}
Q_M &= E_0 a(\theta_0) + F_0 a\left(\varphi_0\right) + N_0 \\
&= E_0 \frac{2\left(\theta_1 - \theta_0\right)}{\tau_0^2} + F_0 \frac{2\left(\varphi_1 - \varphi_0\right)}{\tau_0^2} + N_0
\end{aligned}
\tag{13.134}
$$

$$\tau_0 = \sqrt{2}\sqrt{\frac{E_0\left(\theta_1 - \theta_0\right) + F_0\left(\varphi_1 - \varphi_0\right)}{Q_M - N_0}} \tag{13.135}$$

$$
\begin{aligned}
P_0 &= A_0 \frac{2\left(\theta_1 - \theta_0\right)}{\tau_0^2} + B_0 \frac{2\left(\varphi_1 - \varphi_0\right)}{\tau_0^2} + M_0 \\
&= A_0 \frac{\left(\theta_1 - \theta_0\right)\left(Q_M - N_0\right)}{E_0\left(\theta_1 - \theta_0\right) + F_0\left(\varphi_1 - \varphi_0\right)} \\
&\quad + B_0 \frac{\left(\varphi_1 - \varphi_0\right)\left(Q_M - N_0\right)}{E_0\left(\theta_1 - \theta_0\right) + F_0\left(\varphi_1 - \varphi_0\right)} + M_0
\end{aligned}
\tag{13.136}
$$

At $t = s + 1$, we have:

$$i = s + 1 \qquad
\begin{aligned}
v(x_{s+1}) &= \frac{x_{s+2} - x_s}{2\tau_s} = 0 \\
a(x_{s+1}) &= \frac{-2\left(x_{s+1} - x_s\right)}{\tau_s^2}
\end{aligned}
\tag{13.137}$$

$$P_{s+1}(t) = A_{s+1} a(\theta_{s+1}) + B_{s+1} a\left(\varphi_{s+1}\right) + M_{s+1} \tag{13.138}$$

$$Q_{s+1}(t) = E_{s+1} a(\theta_{s+1}) + F_{s+1} a\left(\varphi_{s+1}\right) + N_{s+1} \tag{13.139}$$

The last floating time τ_s will be calculated from Eq. (13.138) by setting $P_{s+1}(t) = -P_M$ and substituting for $a(\theta_{s+1})$, $a\left(\varphi_{s+1}\right)$.

$$-P_M = A_{s+1}a(\theta_{s+1}) + B_{s+1}a\left(\varphi_{s+1}\right) + M_{s+1}$$

$$= A_{s+1}\frac{-2\left(\theta_{s+1}-\theta_s\right)}{\tau_s^2} + B_{s+1}\frac{-2\left(\varphi_{s+1}-\varphi_s\right)}{\tau_s^2} + M_{s+1} \tag{13.140}$$

$$\tau_s = \sqrt{2}\sqrt{\frac{A_{s+1}\left(\theta_{s+1}-\theta_s\right)+B_{s+1}\left(\varphi_{s+1}-\varphi_s\right)}{P_M + M_{s+1}}} \tag{13.141}$$

$$Q_{s+1}(t)$$

$$= E_{s+1}\frac{-2\left(\theta_{s+1}-\theta_s\right)}{\tau_s^2} + F_{s+1}\frac{-2\left(\varphi_{s+1}-\varphi_s\right)}{\tau_s^2} + N_{s+1}$$

$$= E_{s+1}\frac{-\left(\theta_{s+1}-\theta_s\right)\left(P_M + M_{s+1}\right)}{A_{s+1}\left(\theta_{s+1}-\theta_s\right)+B_{s+1}\left(\varphi_{s+1}-\varphi_s\right)}$$

$$+F_{s+1}\frac{-\left(\varphi_{s+1}-\varphi_s\right)\left(P_M + M_{s+1}\right)}{A_{s+1}\left(\theta_{s+1}-\theta_s\right)+B_{s+1}\left(\varphi_{s+1}-\varphi_s\right)} + N_{s+1} \tag{13.142}$$

For this value of τ_s, we must have $Q_{s+1}(t) \geq -Q_M$, otherwise, we must calculate τ_s from (13.139).

$$-Q_M = E_{s+1}a(\theta_{s+1}) + F_{s+1}a\left(\varphi_{s+1}\right) + N_{s+1}$$

$$= E_{s+1}\frac{-2\left(\theta_{s+1}-\theta_s\right)}{\tau_s^2} + F_{s+1}\frac{-2\left(\varphi_{s+1}-\varphi_s\right)}{\tau_s^2} + N_{s+1} \tag{13.143}$$

$$\tau_s = \sqrt{2}\sqrt{\frac{E_{s+1}\left(\theta_{s+1}-\theta_s\right)+F_{s+1}\left(\varphi_{s+1}-\varphi_s\right)}{Q_M + N_{s+1}}} \tag{13.144}$$

$$P_{s+1}(t)$$

$$= A_{s+1}\frac{-2\left(\theta_{s+1}-\theta_s\right)}{\tau_s^2} + B_{s+1}\frac{-2\left(\varphi_{s+1}-\varphi_s\right)}{\tau_s^2} + M_{s+1}$$

$$= A_{s+1}\frac{-\left(\theta_{s+1}-\theta_s\right)\left(Q_M + N_{s+1}\right)}{E_{s+1}\left(\theta_{s+1}-\theta_s\right)+F_{s+1}\left(\varphi_{s+1}-\varphi_s\right)}$$

$$+B_{s+1}\frac{-\left(\varphi_{s+1}-\varphi_s\right)\left(Q_M + N_{s+1}\right)}{E_{s+1}\left(\theta_{s+1}-\theta_s\right)+F_{s+1}\left(\varphi_{s+1}-\varphi_s\right)} + M_{s+1} \tag{13.145}$$

All inputs of the manipulator at instant i are controlled by the common floating times τ_i and τ_{i-1}. In the forward path, when one of the inputs saturates at instant i, while the other is less than its limit, the minimum τ_i is achieved. Any reduction in τ_i increases the saturated input and breaks one of the constraints (13.124). The same is true in the backward path when we search for τ_{i-1}. Consider the following numerical values and the path of motion illustrated in Fig. 13.13.

$$m_1 = m_2 = 1\,\text{kg} \quad l_1 = l_2 = 1\,\text{m} \quad P_M = Q_M = 100\,\text{Nm} \tag{13.146}$$

To apply the floating time algorithm, the path of motion in Cartesian space must first be transformed into joint space using inverse kinematics. Then, the path of motion in joint space must be discretized to an arbitrary interval, say 100, and the algorithm 13.2 should be applied.

Figure 13.14 depicts the actuators' torque for minimum time motion after applying the floating time algorithm. In this maneuver, there exists one switching point, where the grounded actuator switches from maximum to minimum. The ungrounded actuator never saturates, but as expected, one of the inputs is always saturated. Calculating the floating times allows us to calculate the kinematics information of motion in joint coordinate space. Time histories of the joint coordinates can be utilized to determine the kinematics of the end-effector in Cartesian space.

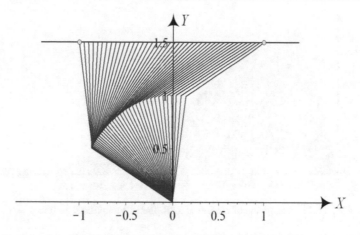

Fig. 13.13 A $2R$ planar manipulator, moving from point $(1, 1.5)$ to point $(-1, 1.5)$ on a straight line $Y = 1.5$

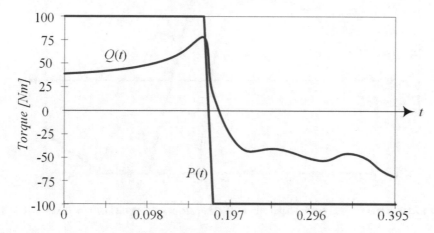

Fig. 13.14 Time optimal control inputs for a 2R manipulator moving on line $Y = 1.5$, $-1 < X < 1$

Example 416 ★ Multiple switching points. It is applying the floating time algorithm on a $2R$ manipulator for a maneuver to have multiple switching points and different motor saturations.

The $2R$ manipulator shown in Fig. 13.12 is made to follow the path illustrated in Fig. 13.15. The floating time algorithm is run for the following data:

$$m_1 = m_2 = 1\,\text{kg} \quad l_1 = l_2 = 1\,\text{m} \quad P_M = Q_M = 100\,\text{N m}$$
$$X(0) = 1.9\,\text{m} \quad X(t_f) = 0.5\,\text{m} \quad Y = 0 \tag{13.147}$$

It leads to the solution shown in Fig. 13.16. There are three switching points for this motion. It is seen that the optimal motion starts while the grounded actuator is saturated and the ungrounded actuator applies a positive torque within its limits. At the first switching point, the ungrounded actuator reaches its negative limit. The grounded actuator shows a change from positive to negative until it reaches its negative limit when the second switching occurs. Between the second and third switching points, the grounded actuator is saturated. Finally, when the ungrounded actuator touches its negative limit for the second time, the third switching occurs.

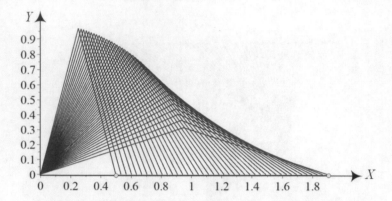

Fig. 13.15 Illustration of motion of a 2R planar manipulator on line $y = 0$

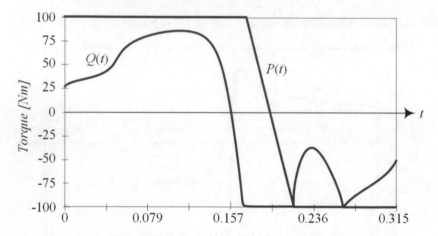

Fig. 13.16 Time optimal control inputs for a 2R manipulator moving on line $Y = 0, 0.5 < X < 1.9$

13.4 Summary

Practically, every actuator can provide only a bounded output. When an actuator is working on its limit, we call it saturated. Time optimal control of an n DOF robot has a simple solution: At every instant of time, at least one actuator must be saturated while the others are within their limits. Floating time is an applied method to find the saturated actuator, the switching points, and the output of the non-saturated actuators. Switching points are the points that the saturated actuator switches with another one.

The floating time method is based on discrete equations of motion, utilizing variable time increments. Then, following a recursive algorithm, it calculates the required output for the robot's actuators to follow a given path of motion.

13.5   Key Symbols

A	Coefficient matrix of variables
b	Coefficient vector of control commands
B	Body coordinate frame
c	cos, Air resistance coefficient
C	Constant of integral
$\mathbf{D}, \mathbf{G}, \mathbf{H}$	Coefficient matrices of robot equation of motion
f	Function, force, control command
F	Force, control command
G, B_0	Global coordinate frame, Base coordinate frame
H	Hamiltonian
J	Objective function
l	Length
m	Mass
$\mathbf{p}$	Momentum vector
P, Q	Torque, control command
$\mathbf{r}$	Position vectors, homogenous position vector
R	Rotation transformation matrix
s	sin, Arc length, number of increments
t	Time
V	Variation
x, y, z	Local coordinate axes
$\mathbf{x}$	Vector of joint states
X, Y, Z	Global coordinate axes

Greek

β	Cycloid angular variable
δ	Kronecker function, variation of a variable
ϵ	Small number
θ	Rotary joint angle
λ	Lagrange multiplier
μ	Coefficient of friction
τ	Floating time increment
$\triangle$	Difference

Symbol

$[\ \ ]^{-1}$	Inverse of the matrix $[\ \ ]$
$[\ \ ]^{T}$	Transpose of the matrix $[\ \ ]$
$\mathbf{q}^{\star}$	A guess value for $\mathbf{q}$
$\mathbb{R}$	Set of real numbers

Exercises

1. Notation and symbols.
   Describe the meaning of these notations.

$$a - \tau_0 \quad b - \tau_i \quad c - f_i \quad d - x_i \quad e - \ddot{x}_i$$

$$f - \dot{x}_i \quad g - \tau_{-1} \quad h - \tau_s \quad i - x_{s+1} \quad j - f_s$$

2. ★ Time optimal control of a 2 DOF system.
   Consider a dynamical system with the following equations of motion.

$$\dot{x}_1 = -3x_1 + 2x_2 + 5Q \tag{13.148}$$

$$\dot{x}_2 = 2x_1 - 3x_2 \tag{13.149}$$

It must start from an arbitrary initial condition and finish at $x_1 = x_2 = 0$, with a bounded control input $|Q| \le 1$.
   Show that the functions

$$f_1 = -3x_1 + 2x_2 + 5Q \tag{13.150}$$

$$f_2 = -2x_1 - 3x_2 \tag{13.151}$$

$$f_3 = 1 \tag{13.152}$$

along with the Hamiltonian function H

$$H = -1 + p_1 \left(-3x_1 + 2x_2 + 5Q\right) + p_2 \left(2x_1 - 3x_2\right) \tag{13.153}$$

and the co-state variables p_1 and p_2 can solve the problem.
3. ★ Nonlinear objective function.
   Consider a one-dimensional control problem, where Q is the control command.

$$\dot{x} = -x + Q \tag{13.154}$$

The variable $x = x(t)$ must satisfy the boundary conditions

$$x(0) = a \qquad x(t_f) = b \tag{13.155}$$

and minimize the objective function J.

$$J = \frac{1}{2} \int_0^{t_f} Q^2 dt \tag{13.156}$$

Show that the functions

$$f_1 = \frac{1}{2} Q^2 \qquad f_2 = -x + Q \qquad f_3 = 0 \tag{13.157}$$

along with the Hamiltonian function H

$$H = \frac{1}{2} p_0 Q^2 + p_1 \left(-x + Q\right) \tag{13.158}$$

and the co-state variables p_0 and p_1 can solve the problem.
4. ★ Time optimal control to origin.
   Consider a dynamical system that must start from an arbitrary initial condition and finish at the origin of the phase plane, $x_1 = x_2 = 0$, with a bounded control $|Q| \le 1$.

$$\dot{x}_1 = x_2 \qquad \dot{x}_2 = -x_1 + Q \tag{13.159}$$

Find the control command to do this motion in minimum time.

5. ★ A linear dynamical system.

Consider a linear dynamical system

$$\dot{x} = [A]x + bQ \tag{13.160}$$

where

$$x = \begin{bmatrix} x_1 \\ x_2 \end{bmatrix} \qquad [A] = \begin{bmatrix} 0 & 1 \\ 0 & 0 \end{bmatrix} \qquad b = \begin{bmatrix} 0 \\ 1 \end{bmatrix} \tag{13.161}$$

subject to a bounded constraint on the control command.

$$Q \leq 1 \tag{13.162}$$

Find the time optimal control command Q to move from the system from x_0 to x_1.

(a)

$$x_0 = \begin{bmatrix} -1 \\ -1 \end{bmatrix} \qquad x_1 = \begin{bmatrix} 1 \\ 1 \end{bmatrix} \tag{13.163}$$

(b)

$$x_0 = \begin{bmatrix} -1 \\ -1 \end{bmatrix} \qquad x_1 = \begin{bmatrix} 3 \\ 1 \end{bmatrix} \tag{13.164}$$

6. ★ Constraint minimization.

Find the local minima and maxima of $f(x)$

$$f(x) = x_1^2 + x_2^2 + x_3^2 \tag{13.165}$$

subject to the following constraints

(a)

$$g_1 = x_1 + x_2 + x_3 - 3 = 0 \tag{13.166}$$

(b)

$$g_1 = x_1^2 + x_2^2 + x_3^2 - 5 = 0 \tag{13.167}$$
$$g_2 = x_1^2 + x_2^2 + x_3^2 - 2x_1 - 3 = 0 \tag{13.168}$$

7. ★ A control command with different limits.

Consider a rectilinear motion of a point mass $m = 1$ kg under the influence of a control force $f(t)$ on a smooth surface. The force is bounded to $F_1 \leq f(t) \leq F_2$. The mass is supposed to move from the initial conditions $x(0) = 0$, $v(0) = 0$ to the final conditions $x(t_f) = 10$ m, $v(t_f) = 0$ in minimum total time $t = t_f$. Use the floating time algorithm to find the required control command $f(t) = m\ddot{x}$ and the switching time for the following data.

(a)

$$F_1 = 10\,\text{N} \quad F_2 = 10\,\text{N} \tag{13.169}$$

(b)

$$F_1 = 8\,\text{N} \quad F_2 = 10\,\text{N} \tag{13.170}$$

(c)

$$F_1 = 10\,\text{N} \quad F_2 = 8\,\text{N} \tag{13.171}$$

8. ★ A control command with different limits.

Find the time optimal control command $|f(t)| \leq 20$ N to move the mass $m = 2$ kg from rest at point P_1 to P_2, and return to stop at point P_3, as shown in Fig. 13.17. The value of μ is:

(a)

$$\mu = 0 \tag{13.172}$$

(b)

$$\mu = 0.2 \tag{13.173}$$

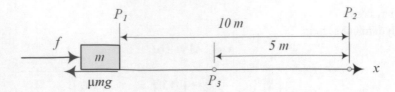

Fig. 13.17 A rectilinear motion of a mass m from rest at point P_1 to P_2, and a return to stop at point P_3

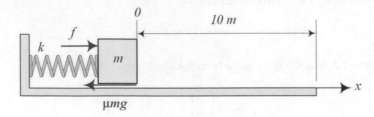

Fig. 13.18 A rectilinear motion of a mass m on a rough surface and attached to a wall with a spring

9. ★ Resistive media.

Consider the mass $m = 2\,\text{kg}$ in Fig. 13.17 that is supposed to move from P_1 to P_2 rest-to-rest in minimum time. The control command is limited to $|f(t)| \le 20\,\text{N}$. However, there is an air resistant proportional to the velocity $c\dot{x}$. Determine the optimal $f(t)$, if

(a)

$$\mu = 0 \qquad c = 0.1 \tag{13.174}$$

(b)

$$\mu = 0.2 \qquad c = 0.1 \tag{13.175}$$

10. ★ Motion of a mass under friction and spring forces.

Find the optimal control command $|f(t)| \le 100\,\text{N}$ to move the mass $m = 1\,\text{kg}$ rest-to-rest from $x(0) = 0$ to $x(t_f) = 10\,\text{m}$. The mass is moving on a rough surface with coefficient μ and is attached to a wall by a linear spring with stiffness k, as shown in Fig. 13.18. The value of μ and k are:

(a)

$$\mu = 0.1 \qquad k = 2\,\text{N/m} \tag{13.176}$$

(b)

$$\mu = 0.5 \qquad k = 5\,\text{N/m} \tag{13.177}$$

11. ★ Convergence conditions.

Verify Eqs. (13.93)–(13.98) for convergence condition of the floating time algorithm.

12. ★ 2R manipulator moving on a line and a circle.

Calculate the actuators' torque for the 2R manipulator, shown in Fig. 13.19, such that the end-point moves time optimally from $P_1(1.5\,\text{m}, 0.5\,\text{m})$ to $P_2(0, 0.5\,\text{m})$. The manipulator has the following characteristics:

$$m_1 = m_2 = 1\,\text{kg} \qquad l_1 = l_2 = 1\,\text{m} \tag{13.178}$$

$$|P(t)| \le 100\,\text{Nm} \qquad |Q(t)| \le 80\,\text{Nm} \tag{13.179}$$

The path of motion is:

(a) a straight line.

(b) a semi-circle with a center at $(0.75\,\text{m}, 0.5\,\text{m})$.

13. ★ Time optimal control for a polar manipulator.

Figure 13.20 illustrates a polar manipulator that is controlled by a torque Q and a force P. The base actuator rotates the manipulator and a force P slides the second link on the first link. Find the optimal controls to move the endpoint from $P_1(1.5\,\text{m}, 1\,\text{m})$ to $P_2(-1, 0.5\,\text{m})$ for the following data:

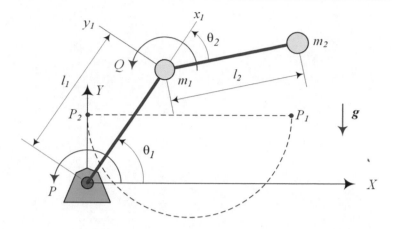

Fig. 13.19 A $2R$ manipulator moves between two points on a line and a semi-circle

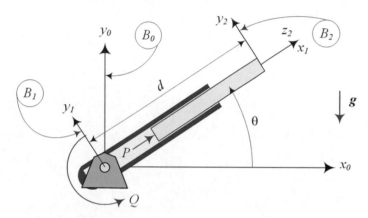

Fig. 13.20 A polar manipulator, controlled by a torque Q and a force P

$$m_1 = 5\,\text{kg} \qquad m_2 = 3\,\text{kg}$$
$$|Q(t)| \leq 100\,\text{Nm} \quad |P(t)| \leq 80\,\text{Nm}$$

(13.180)

14. ★ Control of an articulated manipulator.

Find the time optimal control of an articulated manipulator, shown in Fig. 12.29, to move from $P_1 = (1.1, 0.8, 0.5)$ to $P_2 = (-1, 1, 0.35)$ on a straight line. The geometric parameters of the manipulator are given below. Assume the links are made of uniform bars.

$$\begin{aligned} &d_1 = 1\,\text{m} &&d_2 = 0 \\ &l_2 = 1\,\text{m} &&&l_3 = 1\,\text{m} \\ &m_1 = 25\,\text{kg} &m_2 = 12\,\text{kg} &&m_3 = 8\,\text{kg} \\ &|Q_1(t)| \leq 180\,\text{Nm} \;\; &|Q_2(t)| \leq 100\,\text{Nm} \;\; &|Q_3(t)| \leq 50\,\text{Nm} \end{aligned}$$

(13.181)

Control Techniques

Using inverse kinematics, we calculate the joint kinematics for a desired geometric path of the end-effector of a robot. Substitution of the joint kinematics in equations of motion provides the actuator commands. Applying the commands will move the end-effector of the robot on the desired path ideally. However, because of perturbations and non-modeled phenomena, the robot will not follow the desired path. The techniques that minimize or remove the difference are called the *control techniques*.

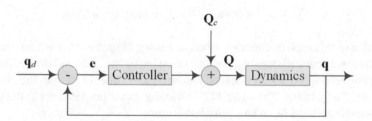

Fig. 14.1 Illustration of feedback control algorithm

14.1 Open- and Closed-Loop Control

A robot is a mechanism with an actuator at each joint i to apply a force or torque to derive the link (i). The robot is instrumented with position, velocity, and possibly acceleration sensors to measure the joint variables' kinematics. The measured values are kinematics information of the frame B_i, attached to the link (i), relative to the frame B_{i-1} or B_0.

To make each joint of the robot to follow a desired mathematical path function, we must provide the required torque command. Assume the desired path of joint variables $\mathbf{q}_d = \mathbf{q}(t)$ are given as functions of time. Then, the required torques $\mathbf{Q}_c$ that make the robot to follow the desired motion are calculated by the equations of motion.

$$\mathbf{Q}_c = \mathbf{D}(\mathbf{q}_d)\,\ddot{\mathbf{q}}_d + \mathbf{H}(\mathbf{q}_d, \dot{\mathbf{q}}_d) + \mathbf{G}(\mathbf{q}_d) \tag{14.1}$$

The subscripts d and c stand for *desired* and *controlled*, respectively.

In an ideal world, the variables can be measured exactly and the robot can perfectly work based on the equations of motion (14.1). Then, the actuators' *control command* $\mathbf{Q}_c$ can make the desired path $\mathbf{q}_d$ to happen. This is an *open-loop control algorithm* in which the control commands are calculated based on a known desired path and the equations of motion. Then, the control commands are fed to the system to generate the desired path. Therefore, in an open-loop control algorithm, we expect the robot to follow the designed path; however, there is no mechanism to compensate any possible error.

Assume we are watching the robot during its motion by measuring the joints' kinematics. At any instant there can be a difference between the actual joint variables and the desired values. The difference is called *error* and is measured by a variable **e**.

R. N. Jazar, *Theory of Applied Robotics*, https://doi.org/10.1007/978-3-030-93220-6_14

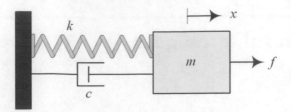

Fig. 14.2 A linear mass-spring-damper oscillator

$$\mathbf{e} = \mathbf{q} - \mathbf{q}_d \tag{14.2}$$

$$\dot{\mathbf{e}} = \dot{q} - \dot{q}_d \tag{14.3}$$

Let us define a control law and calculate a new control command vector by

$$\mathbf{Q} = \mathbf{Q}_c + \mathbf{k}_D \dot{e} + \mathbf{k}_P \mathbf{e} \tag{14.4}$$

where $\mathbf{k}_P$ and $\mathbf{k}_D$ are *constant control gains*. The control law compares the actual joint state variables $(\mathbf{q}, \dot{q})$ with the desired values $(\mathbf{q}_d, \dot{q}_d)$ and generates a command proportionally. Applying the new control command changes the dynamic equations of the robot to produce the actual joint variables $\mathbf{q}$.

$$\mathbf{Q}_c + \mathbf{k}_D \dot{e} + \mathbf{k}_P \mathbf{e} = \mathbf{D}(\mathbf{q})\ddot{q} + \mathbf{H}(\mathbf{q}, \dot{q}) + \mathbf{G}(\mathbf{q}) \tag{14.5}$$

Figure 14.1 illustrates the idea of this type of control method in a *block diagram*. This is a *closed-loop control algorithm*, in which the control commands are calculated based on the difference between actual and desired variables. Reading the actual variables and comparing with the desired values is called *feedback*, and because of that, the closed-loop control algorithm is also called a *feedback control algorithm*. The controller provides a signal proportional to the error and its time rate. This signal is added to the predicted command $\mathbf{Q}_c$ to compensate the error.

The principle of feedback control can be expressed as: *Increase the control command when the actual variable is smaller than the desired value and decrease the control command when the actual variable is larger than the desired value.*

Example 421 Mass-spring-damper oscillator. An example to illustrate the idea of feedback control.

Consider a linear oscillator made by a mass-spring-damper system shown in Fig. 14.2. The equation of motion for the oscillator under the effect of an external force f is

$$m\ddot{x} + c\dot{x} + kx = f \tag{14.6}$$

where, f is the control command, m is the mass of the oscillating object, c is the viscous damping, and k is the stiffness of the spring. The required force to achieve a desired displacement $x_d = x(t)$ is calculated from the equation of motion.

$$f_c = m\ddot{x}_d + c\dot{x}_d + kx_d \tag{14.7}$$

The open-loop control algorithm is shown in Fig. 14.3a.

To remove any possible error, we may use the difference between the desired and actual outputs $e = x - x_d$, and define a control law.

$$f = f_c + k_D \dot{e} + k_P e \qquad k_D > 0 \qquad k_P > 0 \tag{14.8}$$

The new control law uses a feedback command as shown in Fig. 14.3b. It is also possible to define a new control law only based on the error signal.

$$f = -k_D \dot{e} - k_P e \tag{14.9}$$

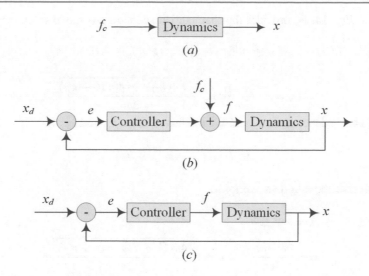

Fig. 14.3 Open-loop and closed-loop control algorithms for a linear oscillator

Employing this law, we can define a more compact feedback control algorithm and modify the equation of motion.

$$m\ddot{x} + (c + k_D)\,\dot{x} + (k + k_P)\,x = k_D\dot{x}_d + k_Px_d \tag{14.10}$$

The equation of the system can be summarized in a block diagram as shown in Fig. 14.3c.

A general scheme of a feedback control system is that a signal from the output feeds back to be compared to the input. This feedback signal closes a loop and makes it reasonable to use the words **feedback** and **close-loop**. The principle of a closed-loop control is to detect any error between the actual output and the desired. As long as the error signal is not zero, the controller keeps changing the control command so that the error signal converges to zero.

Example 422 Stability of a controlled system. How a linear control system is stable to make any error signals to approach zero? Here is the analysis of stability of a control system.

Consider a linear mass-spring-damper oscillator as shown in Fig. 14.2 with a known equation of motion.

$$m\ddot{x} + c\dot{x} + kx = f \tag{14.11}$$

We define a control law based on the actual output

$$f = -k_D\dot{x} - k_Px \qquad k_D > 0 \qquad k_P > 0 \tag{14.12}$$

and modify the equation of motion.

$$m\ddot{x} + (c + k_D)\,\dot{x} + (k + k_P)\,x = 0 \tag{14.13}$$

Comparing with the open-loop equation (14.11), the closed-loop equation shows that the oscillator acts as a free vibrating system under the action of new stiffness $k + k_P$ and damping $c + k_D$. Hence, the control law has changed the apparent stiffness and damping of the actual system. This example introduces the most basic application of control theory to improve the characteristics of a system and run the system to behave in a desired manner.

A control system must be stable when the desired output of the system changes and also be able to eliminate the effect of small disturbances. Stability of a control system is defined as: The output must remain bounded for a given input or a bounded disturbance function.

To investigate the stability of the system, we must solve the closed-loop differential equation (14.13) for its eigenvalues. The equation is linear and therefore, it has an exponential solution.

$$x = e^{\lambda t} \tag{14.14}$$

Substituting the solution into Eq. (14.13) provides the characteristic equation with two solutions.

$$m\lambda^2 + (c + k_D)\,\lambda + (k + k_P) = 0 \tag{14.15}$$

$$\lambda_{1,2} = -\frac{c + k_D}{2m} \pm \frac{\sqrt{(c + k_D)^2 - 4m\,(k + k_P)}}{2m} \tag{14.16}$$

The nature of the solution (14.14) depends on λ_1 and λ_2, and therefore on k_D and k_P. If

$$(c + k_D)^2 < 4m\,(k + k_P) \tag{14.17}$$

then the roots of the characteristic equation are complex.

$$\lambda = -a \pm bi \tag{14.18}$$

$$a = \frac{c + k_D}{2m} \qquad b = \frac{\sqrt{4m\,(k + k_P) - (c + k_D)^2}}{2m} \tag{14.19}$$

In this case, the solution of the equation of motion will be

$$x = C e^{-\xi \omega_n t} \sin\left(\omega_n \sqrt{1 - \xi^2}\, t + \varphi\right) \tag{14.20}$$

$$\omega_n = \sqrt{\frac{k + k_P}{m}} = \sqrt{a^2 + b^2} \tag{14.21}$$

$$\xi = \frac{c + k_D}{2\sqrt{m\,(k + k_P)}} = \frac{a}{\sqrt{a^2 + b^2}} \tag{14.22}$$

The parameter ω_n is called natural frequency, and ξ is the damping ratio of the system. The damping ratio controls the behavior of the system according to the following categories:

1. If $\xi = 0$, then the characteristic values are purely imaginary.

$$\lambda_{1,2} = \pm bi = \pm i\,\frac{1}{2m}\sqrt{4m\,(k + k_P)} \tag{14.23}$$

   In this case, the system has no damping, and therefore, it oscillates with a constant amplitude around the equilibrium, $x = 0$, forever.
2. If $0 < \xi < 1$, then the system is under-damped and it oscillates around the equilibrium with a decaying amplitude. The system is asymptotically stable in this case.
3. If $\xi = 1$, then the system is critically damped. A critically damped oscillator has the fastest return to the equilibrium in an unoscillatory manner.
4. If $\xi > 1$, then the system is over-damped and it slowly returns to the equilibrium in an unoscillatory manner. The characteristic values are real and the solution of an over-damped oscillator is

$$x = A e^{\lambda_1 t} + B e^{\lambda_2 t} \qquad (\lambda_{1,2}) \in \mathbb{R} \tag{14.24}$$

5. If $\xi < 0$, then the system is unstable because the solution would be

$$x = A e^{\lambda_1 t} + B e^{\lambda_2 t} \qquad \text{Re}(\lambda_{1,2}) > 0 \tag{14.25}$$

It shows a motion with an increasing amplitude.

As an example, consider a system with the following characteristic equation:

$$\lambda^2 + 6\lambda + 10 = 0 \tag{14.26}$$

$$\lambda_{1,2} = -3 \pm i \tag{14.27}$$

Solutions of this equation show a stable system because $\mathrm{Re}\,(\lambda_{1,2}) = -3 < 0$.

Characteristic equations are linear polynomials. Hence for high degree equations, it is possible to use numerical methods, such as Newton–Raphson, to find the solution and determine the stability of the system.

Example 423 ★ Complex roots. Complex roots make real response. Here is to show this fact.

In case the characteristic equation has complex roots

$$\lambda_{1,2} = a \pm bi \tag{14.28}$$

we may employ the Euler formula

$$e^{i\theta} = \cos\theta + i\sin\theta \tag{14.29}$$

and show that the solution can be written in a new form

$$\begin{aligned} x &= C_1 e^{at}\,(\cos bt + i\sin bt) + C_2 e^{at}\,(\cos bt - i\sin bt) \\ &= e^{at}\,(A\cos bt + B\sin bt) \end{aligned} \tag{14.30}$$

where, C_1 and C_2 are complex, and A and B are real numbers.

$$A = C_1 + C_2 \qquad B = (C_1 - C_2)\,i \tag{14.31}$$

Example 424 Robot Control Algorithms. Here is a fundamental classification of control algorithms that are being employed in robotics.

Robots are nonlinear dynamical systems, and there is no general method for designing a nonlinear controller to be suitable for every robot in every mission. However, there are a variety of alternative and complementary methods, each best applicable to particular class of robots in a particular mission. The most important control methods are as follows.

Feedback Linearization or Computed Torque Control Technique In feedback linearization technique, we define a control law to obtain a linear differential equation for error command, and then use the linear control design techniques. The feedback linearization technique can be applied to robots successfully; however, it does not guarantee robustness according to parameter uncertainty or disturbances. This technique is a model-based control method, because the control law is designed based on a nominal model of the robot.

Linear Control Technique The simplest technique for controlling robots is to design a linear controller based on linearization of the equations of motion about an operating point. The linearization technique locally determines the stability of the robot. Proportional, integral, and derivative, or any combination of them are the most practical linear control techniques.

Adaptive Control Technique Adaptive control is a technique for controlling uncertain or time varying robots. Adaptive control technique is more effective for low DOF robots.

Robust and Adaptive Control Technique In the robust control method, the controller is designed based on the nominal model plus some uncertainty. Uncertainty can be in any parameter, such as the load carrying by the end-effector. For example, we develop a control technique to be effective for loads in a range of 1–10 kg.

Gain-Scheduling Control Technique Gain-scheduling is a technique that tries to apply the linear control techniques to the nonlinear dynamics of robots. In gain-scheduling, we select a number of control points to cover the range of robot operation. Then at each control point, we make a linear time varying approximation to the robot dynamics and design a linear controller. The parameters of the controller are then interpolated or scheduled between control points.

14.2   Computed Torque Control

Dynamics of a robot can be expressed in the form

$$\mathbf{Q} = \mathbf{D}(\mathbf{q})\,\ddot{\mathbf{q}} + \mathbf{H}(\mathbf{q}, \dot{q}) + \mathbf{G}(\mathbf{q}) \tag{14.32}$$

where $\mathbf{q}$ is the vector of joint variables, and $\mathbf{Q}(\mathbf{q}, \dot{q}, t)$ is the torques applied at joints. Assume a desired path in joint space is given by a twice differentiable function $\mathbf{q} = \mathbf{q}_d(t) \in C^2$. Hence, the desired time history of joints' position, velocity, and acceleration are known. We can control the robot to follow the desired path, by introducing a *computed torque control law* as below

$$\mathbf{Q} = \mathbf{D}(\mathbf{q})\left(\ddot{q}_d - \mathbf{k}_D\dot{e} - \mathbf{k}_P\mathbf{e}\right) + \mathbf{H}(\mathbf{q}, \dot{q}) + \mathbf{G}(\mathbf{q}) \tag{14.33}$$

where $\mathbf{k}_D$ and $\mathbf{k}_P$ are constant gain diagonal matrices and $\mathbf{e}$ is the error vector.

$$\mathbf{e} = \mathbf{q} - \mathbf{q}_d \tag{14.34}$$

The control law is stable and applied as long as all the eigenvalues of the following matrix have negative real part:

$$[A] = \begin{bmatrix} 0 & \mathbf{I} \\ -\mathbf{k}_P & -\mathbf{k}_D \end{bmatrix} \tag{14.35}$$

Proof The required torque $\mathbf{Q}_c$ to track a desired path $\mathbf{q}_d(t)$ can directly be found by substituting the path function into the equations of motion.

$$\mathbf{Q}_c = \mathbf{D}(\mathbf{q}_d)\,\ddot{q}_d + \mathbf{H}(\mathbf{q}_d, \dot{q}_d) + \mathbf{G}(\mathbf{q}_d) \tag{14.36}$$

The calculated torques are called *control inputs*, and the control is based on the *open-loop control law*. In an open-loop control, we have the equations of motion for a robot and we need the required torques to move the robot on a given path. Open-loop control is a blind method because the current state of the robot is not used for calculating the inputs.

Due to non-modeled parameters and also errors in adjustment, there is always a difference between the desired and actual paths. To make the robot's actual path converge to the desired path, we must introduce a feedback control. Let us use the feedback signal of the actual path and apply the computed torque control law (14.33) to the robot. Substituting the control law in the equations of motion (14.32) yields

$$\ddot{e} + \mathbf{k}_D\dot{e} + \mathbf{k}_P\mathbf{e} = 0 \tag{14.37}$$

This is a linear differential equation for the error variable between the actual and desired outputs. If the $n \times n$ gain matrices $\mathbf{k}_D$ and $\mathbf{k}_P$ are assumed to be diagonal, then we may rewrite the error equation in a matrix form.

$$\frac{d}{dt}\begin{bmatrix} \mathbf{e} \\ \dot{e} \end{bmatrix} = \begin{bmatrix} 0 & \mathbf{I} \\ -\mathbf{k}_P & -\mathbf{k}_D \end{bmatrix}\begin{bmatrix} \mathbf{e} \\ \dot{e} \end{bmatrix} = [A]\begin{bmatrix} \mathbf{e} \\ \dot{e} \end{bmatrix} \tag{14.38}$$

The linear differential equation (14.38) is asymptotically stable when all the eigenvalues of $[A]$ have negative real part. The matrix $\mathbf{k}_P$ has the role of natural frequency, and $\mathbf{k}_D$ acts similar to damping.

$$\mathbf{k}_P = \begin{bmatrix} \omega_1^2 & 0 & 0 & 0 \\ 0 & \omega_2^2 & 0 & 0 \\ 0 & 0 & \cdots & 0 \\ 0 & 0 & 0 & \omega_n^2 \end{bmatrix} \tag{14.39}$$

$$\mathbf{k}_D = \begin{bmatrix} 2\xi_1\omega_1 & 0 & 0 & 0 \\ 0 & 2\xi_2\omega_2 & 0 & 0 \\ 0 & 0 & \cdots & 0 \\ 0 & 0 & 0 & 2\xi_n\omega_n \end{bmatrix} \tag{14.40}$$

Because $\mathbf{k}_D$ and $\mathbf{k}_P$ are diagonal, we can adjust the gain matrices $\mathbf{k}_D$ and $\mathbf{k}_P$ to control the response speed of the robot at each joint independently. A simple choice for the matrices is to set $\xi_i = 0$, $i = 1, 2, \cdots, n$, and make each joint response equal to the response of a critically damped linear second-order system with natural frequency ω_i.

The computed torque control law (14.33) has two components as shown below.

$$\mathbf{Q} = \underbrace{\mathbf{D(q)}\ddot{\mathbf{q}}_d + \mathbf{H(q, \dot{q})} + \mathbf{G(q)}}_{\mathbf{Q}_{ff}} + \underbrace{\mathbf{D(q)}\left(-\mathbf{k}_D\dot{\mathbf{e}} - \mathbf{k}_P\mathbf{e}\right)}_{\mathbf{Q}_{fb}} \tag{14.41}$$

The first term, $\mathbf{Q}_{ff}$, is the *feedforward* command, which is the required torques based on open-loop control law. When there is no error, the control input $\mathbf{Q}_{ff}$ makes the robot follow the desired path $\mathbf{q}_d$. The second term, $\mathbf{Q}_{fb}$, is the *feedback* command, which is the correction torques to reduce the errors in the path of the robot.

Computed torque control is also called *feedback linearization*, which is an applied technique for robots' nonlinear control design. To apply the feedback linearization technique, we develop a control law to eliminate all nonlinearities and reduce the problem to the linear second-order equation of error signal (14.37). ∎

Example 425 Computed force control for an oscillator. To illustrate how computed force control should be applied on a dynamic system.

Figure 14.2 depicts a linear mass-spring-damper oscillator under the action of a control force.

$$m\ddot{x} + c\dot{x} + kx = f \tag{14.42}$$

Applying a computed force control law

$$f = m\left(\ddot{x}_d - k_D\dot{e} - k_P e\right) + c\dot{x} + kx \tag{14.43}$$
$$e = x - x_d \tag{14.44}$$

makes a differential equation for the error signal e.

$$\ddot{e} + k_D\dot{e} + k_P e = 0 \tag{14.45}$$

The solution of the error equation is

$$e = A e^{\lambda_1 t} + B e^{\lambda_2 t} \tag{14.46}$$
$$\lambda_{1,2} = -k_D \pm \sqrt{k_D^2 - 4k_P} \tag{14.47}$$

where A and B are functions of initial conditions, and $\lambda_{1,2}$ are solutions of the characteristic equation.

$$m\lambda^2 + k_D\lambda + k_P = 0 \tag{14.48}$$

The solution (14.46) is stable and $e \to 0$ exponentially as $t \to \infty$ if $k_D > 0$.

Example 426 Inverted pendulum. To keep an inverted pendulum to stay at its unstable upright equilibrium point, a torque Q must be applied at the base joint, calculated by feedback signals.

Consider an inverted pendulum shown in Fig. 14.4. Its equation of motion is

$$ml^2\ddot{\theta} - mgl\sin\theta = Q \tag{14.49}$$

To control the pendulum and bring it from an initial angle $\theta = \theta_0$ to the vertical-up position, we may employ a feedback control law.

$$Q = -k_D\dot{\theta} - k_P\theta - mgl\sin\theta \tag{14.50}$$

The parameters k_D and k_P are positive gains and are assumed constants. The control law (14.50) transforms the dynamics of the system, showing that the system behaves as a stable mass-spring-damper.

$$ml^2\ddot{\theta} + k_D\dot{\theta} + k_P\theta = 0 \tag{14.51}$$

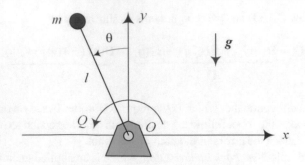

Fig. 14.4 A controlled inverted pendulum

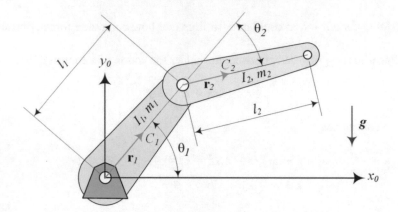

Fig. 14.5 A 2R planar manipulator with massive links

In case the desired position of the pendulum is at a nonzero angle, $\theta = \theta_d$, we may employ a feedback control law based on the error $e = \theta - \theta_d$ as below,

$$Q = ml^2\ddot{\theta}_d - k_D\dot{e} - k_P e - mgl\sin\theta \tag{14.52}$$

Substituting this control law in the equation of motion (14.49) shows that the dynamics of the controlled system is governed by an equation for the error signal.

$$ml^2\ddot{e} + k_D\dot{e} + k_P e = 0 \tag{14.53}$$

Example 427 Control of a 2R planar manipulator. To illustrate how computed force control should be applied on a robotic system.

A 2R planar manipulator is shown in Fig. 14.5 with dynamic equations given below.

$$\begin{bmatrix} Q_1 \\ Q_2 \end{bmatrix} = \begin{bmatrix} D_{11} & D_{12} \\ D_{21} & D_{22} \end{bmatrix} \begin{bmatrix} \ddot{\theta}_1 \\ \ddot{\theta}_2 \end{bmatrix}$$
$$+ \begin{bmatrix} C_{11} & C_{12} \\ C_{21} & C_{22} \end{bmatrix} \begin{bmatrix} \dot{\theta}_1 \\ \dot{\theta}_2 \end{bmatrix} + \begin{bmatrix} G_1 \\ G_2 \end{bmatrix} \tag{14.54}$$

$$D_{11} = m_1 r_1^2 + I_1 + m_2 \left(l_1^2 + l_1 r_2 \cos\theta_2 + r_2^2 \right) + I_2 \tag{14.55}$$

$$D_{21} = D_{12} = m_2 l_1 r_2 \cos\theta_2 + m_2 r_2^2 + I_2 \tag{14.56}$$

$$D_{22} = m^2 r_2^2 + I_2 \tag{14.57}$$

$$C_{11} = -m_2 l_1 r_2 \dot{\theta}_2 \sin\theta_2 \tag{14.58}$$

$$C_{21} = -m_2 l_1 r_2 (\dot{\theta}_1 + \dot{\theta}_2) \sin\theta_2 \tag{14.59}$$

$$C_{12} = m_2 l_1 r_2 \dot{\theta}_1 \sin \theta_2 \tag{14.60}$$

$$C_{22} = 0 \tag{14.61}$$

$$G_1 = m_1 g r_1 \cos \theta_1 + m_2 g \left(l_1 \cos \theta_1 + r_2 \cos \left(\theta_1 + \theta_2 \right) \right) \tag{14.62}$$

$$G_2 = m_2 g r_2 \cos \left(\theta_1 + \theta_2 \right) \tag{14.63}$$

Let us write the equations of motion in the following form:

$$\mathbf{D(q)} \, \ddot{q} + \mathbf{C(q, \dot{q})} \, \dot{q} + \mathbf{G(q)} = \mathbf{Q} \tag{14.64}$$

and multiply both sides by $\mathbf{D}^{-1}$ to reform the equations of motion.

$$\ddot{q} + \mathbf{D}^{-1} \mathbf{C} \, \dot{q} + \mathbf{D}^{-1} \mathbf{G} = \mathbf{D}^{-1} \mathbf{Q} \tag{14.65}$$

To control the manipulator to follow a desired path $\mathbf{q} = \mathbf{q}_d(t)$, we apply the following control law:

$$\mathbf{Q} = \mathbf{D(q)} \, \mathbf{U} + \mathbf{C(q, \dot{q})} \dot{q} + \mathbf{G(q)} \tag{14.66}$$

where

$$\mathbf{U} = \ddot{q}_d - 2k \, \dot{e} - k^2 \, \mathbf{e} \tag{14.67}$$

$$\mathbf{e} = \mathbf{q} - \mathbf{q}_d. \tag{14.68}$$

The vector $\mathbf{U}$ is the controller input, $\mathbf{e}$ is the position error, and k is a positive constant gain number. Substituting the control law into the equation of motion shows that the error vector satisfies a linear second-order ordinary differential equation.

$$\ddot{e} + 2k \, \dot{e} + k^2 \, \mathbf{e} = 0 \tag{14.69}$$

Error will exponentially converge to zero for $k > 0$.

14.3   Linear Control Technique

Linearization of a robot's equations of motion about an operating point while applying a linear control algorithm is a traditional and fundamental robot control technique. This technique works well in a vicinity of the operating point. Hence, it is only a locally stable method. The linear control techniques are proportional, integral, derivative, and any combination of them. The idea is to linearize the nonlinear equations of motion about some reference operating points to make a linear system, design a controller for the linear system, and then, apply the control. This technique will always result in a stable controller in some neighborhood of the operating point. However, the stable neighborhood may be quite small and hard to be determined.

A proportional-integral-derivative (PID) control algorithm employs a position error, derivative error, and integral error to develop a control law. Hence, a PID control law has the following general form for the input command:

$$Q = k_P \, e + k_I \int_0^t e \, dt + k_D \, \dot{e} \tag{14.70}$$

where $e = q - q_d$ is the error signal, and k_P, k_I, and k_D are positive constant gains associated to the proportional, integral, and derivative controllers. The control command Q is thus a sum of three terms: the P-term, which is proportional to error e, the I-term, which is proportional to the integral of the error, and the D-term, which is proportional to the derivative of the error.

14.3.1 Proportional Control

In case of proportional control, the PID control law (14.70) reduces to

$$Q = k_P\, e + Q_d \tag{14.71}$$

The variable Q_d is the desired control command, which is called a *bias* or *reset factor*. When the error signal is zero, the control command is equal to the desired value. The proportional control has a drawback that results in a constant finite error at steady state condition.

14.3.2 Integral Control

The main function of an integral control is to eliminate the steady state error and make the system follow the set point at steady state conditions. The integral controller leads to an increasing control command for a positive error, and a decreasing control command for a negative error. An integral controller is usually used with a proportional controller. The control law for a PI controller is

$$Q = k_P\, e + k_I \int_0^t e\, dt \tag{14.72}$$

14.3.3 Derivative Control

The purpose of derivative control is to improve the closed-loop stability of a system. A derivative controller has a predicting action by extrapolating the error using a tangent to the error curve. A derivative controller is usually used with a proportional controller. The PD control law is

$$Q = k_P\, e + k_D\, \dot e \tag{14.73}$$

Proof Any linear system behaves linearly if it is sufficiently near a reference operating point. Consider a nonlinear system

$$\dot q = \mathbf{f}(\mathbf{q}, \mathbf{Q}) \tag{14.74}$$

where $\mathbf{q}_d$ is a solution generated by a specific input $\mathbf{Q}_c$.

$$\dot q_d = \mathbf{f}(\mathbf{q}_d, \mathbf{Q}_c) \tag{14.75}$$

Assume $\delta\mathbf{q}$ is a small change from the reference point $\mathbf{q}_d$ because of a small change $\delta\mathbf{Q}$ from $\mathbf{Q}$.

$$\mathbf{q} = \mathbf{q}_d + \delta\mathbf{q} \tag{14.76}$$
$$\mathbf{Q} = \mathbf{Q}_c + \delta\mathbf{Q} \tag{14.77}$$

If the changes $\delta\mathbf{q}$ and $\delta\mathbf{Q}$ are assumed small at all times, then Eq. (14.74) can be approximated by its Taylor expansion and $\mathbf{q}$ be the solution of the following equation:

$$\dot q = \frac{\partial \mathbf{f}}{\partial \mathbf{q}_d}\mathbf{q} + \frac{\partial \mathbf{f}}{\partial \mathbf{Q}_d}\mathbf{Q} \tag{14.78}$$

The partial derivative matrices $\left[\frac{\partial \mathbf{f}}{\partial \mathbf{q}_d}\right]$ and $\left[\frac{\partial \mathbf{f}}{\partial \mathbf{Q}_d}\right]$ are evaluated at the reference point $(\mathbf{q}_d, \mathbf{Q}_d)$. ∎

Example 428 ★ Linear control for a pendulum. How to linearize equation of motion of a nonlinear system around a set point and define a PID control to keep the system as the set point.

Figure 14.6 illustrates a controlled pendulum as a one-arm manipulator. The equation of motion for the arm is

$$Q = I\,\ddot\theta + c\,\dot\theta + mgl\,\sin\theta \tag{14.79}$$

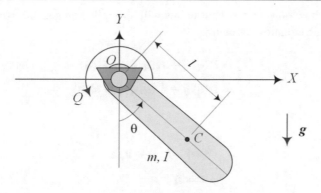

Fig. 14.6 A controlled pendulum as a one-arm manipulator

where I is the arm's mass moment about the pivot joint and m is the mass of the arm. The joint has a viscous damping c, the distance between the pivot and mass center C, is l. Introducing a new set of variables

$$\theta = x_1 \qquad \dot\theta = x_2 \tag{14.80}$$

converts the equation of motion into state-space expression.

$$\dot x_1 = x_2 \qquad \dot x_2 = \frac{Q - c x_2 - mgl \sin x_1}{I} \tag{14.81}$$

The linearized form of these equations is

$$\begin{bmatrix} \dot x_1 \\ \dot x_2 \end{bmatrix} = \begin{bmatrix} 0 & 1 \\ -mgl/I & -c \end{bmatrix} \begin{bmatrix} x_1 \\ x_2 \end{bmatrix} + \begin{bmatrix} 0 & 0 \\ 0 & 1/I \end{bmatrix} \begin{bmatrix} 0 \\ Q \end{bmatrix} \tag{14.82}$$

Assume that the reference point is

$$\mathbf{x}_d = \begin{bmatrix} x_1 \\ x_2 \end{bmatrix} = \begin{bmatrix} \pi/2 \\ 0 \end{bmatrix} \qquad Q_c = mgl \tag{14.83}$$

The coefficient matrices in Eq. (14.82) must then be evaluated at the reference point. We use a set of sample data

$$m = 1\,\text{kg} \quad l = 0.35\,\text{m} \quad I = 0.07\,\text{kg.m}^2 \quad c = 0.01\,\text{Ns/m} \tag{14.84}$$

and find the partial derivatives.

$$\frac{\partial \mathbf{f}}{\partial \mathbf{q}_d} = \begin{bmatrix} 0 & 1 \\ -49.05 & -0.01 \end{bmatrix} \tag{14.85}$$

$$\frac{\partial \mathbf{f}}{\partial \mathbf{Q}_c} = \begin{bmatrix} 0 & 0 \\ 0 & 14.286 \end{bmatrix} \tag{14.86}$$

Now we have a linear system and we may apply any control law that applies to linear systems. For instance, a PID control law can control the arm around the reference point.

$$\mathbf{Q} = \mathbf{Q}_c - \mathbf{k}_D \dot{\mathbf{e}} - \mathbf{k}_P \mathbf{e} + \mathbf{k}_I \int_0^t \mathbf{e}\,dt \tag{14.87}$$

$$\mathbf{e} = \mathbf{q} - \mathbf{q}_d = \begin{bmatrix} x_1 - \pi/2 \\ x_2 \end{bmatrix} \tag{14.88}$$

Example 429 PD control. To show how a linear control law will be applied on general dynamics of robots.

Consider a robot with dynamic equations as below.

$$\mathbf{Q} = \mathbf{D}(\mathbf{q})\,\ddot{\mathbf{q}} + \mathbf{H}(\mathbf{q}, \dot{\mathbf{q}}) + \mathbf{G}(\mathbf{q})$$

$$= \mathbf{D}(\mathbf{q})\,\ddot{q} + \mathbf{C}(\mathbf{q}, \dot{\mathbf{q}})\dot{q} + \mathbf{G}(\mathbf{q}) \tag{14.89}$$

Let us define a PD control law.

$$\mathbf{Q} = -\mathbf{k}_D\dot{e} - \mathbf{k}_P\mathbf{e} \tag{14.90}$$

$$\mathbf{e} = \mathbf{q} - \mathbf{q}_d \tag{14.91}$$

Applying the PD control will produce the following control equation.

$$\mathbf{D}(\mathbf{q})\,\ddot{q} + \mathbf{C}(\mathbf{q}, \dot{\mathbf{q}})\dot{q} + \mathbf{G}(\mathbf{q}) + \mathbf{k}_D\left(\dot{\mathbf{q}} - \dot{\mathbf{q}}_d\right) - \mathbf{k}_P\left(\mathbf{q} - \mathbf{q}_d\right) = 0 \tag{14.92}$$

This control is ideal when $\mathbf{q}_d$ is a constant vector associated with a specific configuration of a robot, and therefore $\dot{q}_d = 0$. In this case the PD controller can make the configuration $\mathbf{q}_d$ globally stable.

In case of a path given by $\mathbf{q} = \mathbf{q}_d(t)$, we define a modified PD controller,

$$\mathbf{Q} = \mathbf{D}(\mathbf{q})\ddot{\mathbf{q}}_d + \mathbf{C}(\mathbf{q}, \dot{\mathbf{q}})\dot{q}_d + \mathbf{G}(\mathbf{q}) - \mathbf{k}_D\dot{e} - \mathbf{k}_P\mathbf{e} \tag{14.93}$$

and reduce the closed-loop equation to an equation for the error signal $\mathbf{e}$.

$$\mathbf{D}(\mathbf{q})\ddot{e} + (\mathbf{C}(\mathbf{q}, \dot{\mathbf{q}}) + \mathbf{k}_D)\,\dot{e} + \mathbf{k}_P\mathbf{e} = 0 \tag{14.94}$$

The linearization of this equation about a control point $\mathbf{q} = \mathbf{q}_d = const$ provides a stable dynamics for the error signal.

$$\mathbf{D}(\mathbf{q}_d)\ddot{e} + \mathbf{k}_D\dot{e} + \mathbf{k}_P\mathbf{e} = 0 \tag{14.95}$$

14.4   Sensing and Control

Position, velocity, acceleration, and force sensors are the most common sensors used in robotics. Consider the inverted pendulum shown in Fig. 14.4 as a one DOF manipulator with the following equation of motion:

$$ml^2\ddot{\theta} - c\dot{\theta} - mgl\sin\theta = Q \tag{14.96}$$

From an open-loop control viewpoint, we need to provide a moment $Q_c(t)$ to force the manipulator to follow a desired path of motion $\theta_d(t)$.

$$Q_c = ml^2\ddot{\theta}_d - c\dot{\theta}_d - mgl\sin\theta_d \tag{14.97}$$

In robotics, we calculate Q_c from the dynamics equation and dictate it to actuators. The manipulator will respond to the applied moment and will move accordingly. The equation of motion (14.96) is a model of the actual manipulator. However, there are many unmodeled phenomena that we cannot include them in our equation of motion or we cannot model them. Some are temperature, air pressure, exact gravitational acceleration, or even the physical parameters such as m and l that we assume to have good accuracy. So, applying a control command Q_c will move the manipulator and provide a real value for θ, $\dot{\theta}, \ddot{\theta}$, which are not necessarily equal to $\theta_d, \dot{\theta}_d, \ddot{\theta}_d$. Sensing is now important to measure the actual angle θ, angular velocity $\dot{\theta}$, and angular acceleration $\ddot{\theta}$ to compare with $\theta_d, \dot{\theta}_d, \ddot{\theta}_d$ and make sure that the manipulator is following the desired path. This is the reason why the feedback control systems and the error signal $e = \theta - \theta_d$ were introduced.

Robots can interact with the environment. Therefore, a robot needs two types of sensors: 1-sensing the robot's internal parameters, which are called *proprioceptors*, and 2-sensing the robot's environmental parameters, which are called *exteroceptors*. The most important interior parameters are position, velocity, acceleration, force, torque, and inertia.

14.4.1 Position Sensors

Rotary Encoders In robotics, almost all kinds of actuators provide a rotary motion encoder. Then we may provide a rotation motion for a revolute joint, or a translation motion for a prismatic joint by using gears. So, it is ideally possible to sense the relative position of connected links by joints based on the angular position of the actuators. A possible error in position sensing is due to non-rigidity and backlash. The most common position sensor is a *rotary encoder* that can be *optical, magnetic*, or *electrical*. As an example, when the encoder shaft rotates, a disk counting a pattern of fine lines interrupting a light beam. A photodetector converts the light pulses into a countable binary waveform. The shaft angle is then determined by counting the number of pulses.

Resolvers We may design an electronic device to provide a mathematical function of the joint variable. The mathematical function might be sine, cosine, exponential, or any combination of mathematical functions. The joint variable is then calculated indirectly by resolving the mathematical functions. Sine and cosine functions are more common.

Potentiometers Using an electrical bridge, the potentiometers can provide an electric voltage proportional to the joint position.

LVDT and RVDT LVDT/RVDT or a Linear/Rotary Variable Differential Transformer operates with two transformers sharing the same magnetic core. When the core moves, the output of one transformer increases while the other's output decreases. The difference of the current is a measure of the core position.

14.4.2 Speed Sensors

Tachometers A tachometer is a name for any velocity sensor. Tachometers usually provide an analog signal proportional to the angular velocity of a shaft. There are a vast amount of different designs for tachometers, using different physical characteristics such as magnetic field.

Rotary Encoders Any rotary sensor can be equipped with a time measuring system and become an angular velocity sensor. The encoder counts the light pulses of a rotating disk and the angular velocity is then determined by time between pulses.

Differentiating Devices Any kind of position sensor can be equipped with a digital differentiating device to become a speed sensor. The digital or numerical differentiating needs a simple processor. Numerical differentiating is generally an erroneous process.

Integrating Devices The output signal of an accelerometer can be numerically integrated to provide a velocity signal. The digital or numerical integrating also needs a processor. Numerical differentiating is generally a smooth and reliable process.

14.4.3 Acceleration Sensors

Acceleration sensors work based on Newton's second law of motion. They sense the force that causes an acceleration of a known mass. There are many types of accelerometers. Stress-strain gage, capacitive, inductive, piezoelectric, and micro-accelerometers are the most common. In any of these types, force causes a proportional displacement in an elastic material, such as deflection in a micro-cantilever beam, and the displacement is proportional to the acceleration.

Accelerometers are the most common sensors with great accuracy and many applications include measurement of acceleration, angular acceleration, velocity, position, angular velocity, frequency, impulse, force, tilt, and orientation.

Force and Torque Sensors Any concept and method that we use in sensing acceleration may also be used in force and torque sensing. We equip the wrists of a robot with at least three force sensors to measure the contact forces and moments with the environment. The wrist's force sensors are important especially when the robot's job is involved with touching unknown surfaces and objects.

Proximity Sensors Proximity sensors are utilized to detect the existence of an object, field, or special material before interacting with it. Inductive, capacitive, hall effect, sonic, ultrasonic, and optical are the most common proximity sensors.

The inductive sensors can sense the existence of a metallic object due to a change in inductance. The capacitive sensors can sense the existence of gas, liquid, or metals that cause a change in capacitance. Hall effective sensors work based on the interaction between the voltage in a semiconductor material and magnetic fields. These sensors can detect existence of magnetic fields and materials. Sonic, ultrasonic, and optical sensors work based on the reflection or modification in an emitted signal by objects.

14.5   Summary

In an open-loop control algorithm, we calculate the robot's required torque commands $\mathbf{Q}_c$ for a given joint path $\mathbf{q}_d = \mathbf{q}(t)$ based on the equations of motion

$$\mathbf{Q}_c = \mathbf{D}(\mathbf{q}_d)\,\ddot{\mathbf{q}}_d + \mathbf{H}(\mathbf{q}_d, \dot{\mathbf{q}}_d) + \mathbf{G}(\mathbf{q}_d). \tag{14.98}$$

However, there can be a difference between the actual joint variables and the desired values. The difference is called error $\mathbf{e}$

$$\mathbf{e} = \mathbf{q} - \mathbf{q}_d \tag{14.99}$$

$$\dot{e} = \dot{q} - \dot{q}_d. \tag{14.100}$$

By measuring the error command, we may define a control law and calculate a new control command vector

$$\mathbf{Q} = \mathbf{Q}_c + \mathbf{k}_D \dot{e} + \mathbf{k}_P \mathbf{e} \tag{14.101}$$

to compensate for the error. The parameters $\mathbf{k}_P$ and $\mathbf{k}_D$ are constant gain diagonal matrices.

The control law compares the actual joint variables $(\mathbf{q}, \dot{q})$ with the desired values $(\mathbf{q}_d, \dot{q}_d)$ and generates a command proportionally. Applying the new control command changes the dynamic equations of the robot to

$$\mathbf{Q}_c + \mathbf{k}_D \dot{e} + \mathbf{k}_P \mathbf{e} = \mathbf{D}(\mathbf{q})\,\ddot{q} + \mathbf{H}(\mathbf{q}, \dot{q}) + \mathbf{G}(\mathbf{q}). \tag{14.102}$$

This is a closed-loop control algorithm, in which the control commands are calculated based on the difference between actual and desired variables.

Computed torque control

$$\mathbf{Q} = \mathbf{D}(\mathbf{q})\left(\ddot{q}_d - \mathbf{k}_D \dot{e} - \mathbf{k}_P \mathbf{e}\right) + \mathbf{H}(\mathbf{q}, \dot{q}) + \mathbf{G}(\mathbf{q}) \tag{14.103}$$

is an applied closed-loop control law in robotics to make a robot follow a desired path.

14.6   Key Symbols

a, b	Real and imaginary parts of a complex number
A	Coefficient matrix
A, B	Real coefficients
B	Body coordinate frame
c	Damping
C_i	Complex coefficients
$\mathbf{e}$	Error, exponential function
$f_c, \mathbf{f}_c$	Actuator force control command
$f, \mathbf{f}$	Actual force command
g	Gravitational acceleration
G, B_0	Global coordinate frame, base coordinate frame
i	Imaginary unit number
$\mathbf{I} = [I]$	Identity matrix, moment of inertia
J	Jacobian
k	Stiffness
$\mathbf{k}_P$	Proportional constant control gain
$\mathbf{k}_D$	Derivative constant control gain
l	Length
m	Mass
$\mathbf{q}$	Actual vector of joint variables
$\mathbf{q}_d$	Desired path of joint
$\mathbf{Q}$	Actuators' actual command
$\mathbf{Q}_c$	Actuators' control command
$\mathbf{Q}_{fb}$	Feedback command
$\mathbf{Q}_{ff}$	Feedforward command
$\mathbf{r}$	Position vectors, homogenous position vector
r_i	The element i of $\mathbf{r}$
t	Time
x, y, z	Local Cartesian coordinates
X, Y, Z	Global Cartesian coordinates

Greek

δ	Small increment of a parameter
λ	Characteristic value, eigenvalue
θ	Rotary joint angle
ω_n	Natural frequency
ξ	Damping ratio

Symbol

DOF	Degree of freedom
$\mathbb{R}$	Real numbers set
Re	Real

Exercises

1. Response of second-order systems.

    Solve the characteristic equations and determine the response of the following second-order systems at $x(1)$, if they start from $x(0) = 1$, $\dot{x}(0) = 0$:

    (a)

$$\ddot{x} + 2\dot{x} + 5x = 0 \tag{14.104}$$

    (b)

$$\ddot{x} + 2\dot{x} + x = 0 \tag{14.105}$$

    (b)

$$\ddot{x} + 4\dot{x} + x = 0 \tag{14.106}$$

2. Modified PD control.

    Apply a modified PD control law

$$f = -k_P e - k_d \dot{x} \qquad e = x - x_d \tag{14.107}$$

    to a second-order linear system

$$m\ddot{x} + c\dot{x} + kx = f \tag{14.108}$$

    and reduce the system to a second-order equation in an error signal.

$$m\ddot{e} + (c + k_D)\dot{e} + (k + k_P)e = kx_D \tag{14.109}$$

    Then, calculate the steady state error for a step input.

$$x = x_d = const \tag{14.110}$$

3. Modified PID control.

    Apply a modified PD control law

$$f = -k_P e - k_d \dot{x} - k_I \int_0^t e\, dt \qquad e = x - x_d \tag{14.111}$$

    to a second-order linear system

$$m\ddot{x} + c\dot{x} + kx = f \tag{14.112}$$

    and reduce the system to a third-order equation in an error signal.

$$m\dddot{e} + (c + k_D)\ddot{e} + (k + k_P)\dot{e} + k_I e = 0 \tag{14.113}$$

    Then, find the PID gains such that the characteristic equation of the system simplifies to

$$\left(\lambda^2 + 2\xi\omega_n\lambda + \omega_n^2\right)(\lambda + \beta) = 0 \tag{14.114}$$

4. Linearization.

    Linearize the given equations and determine the stability of the linearized set of equations.

$$\dot{x}_1 = x_2^2 + x_1 \cos x_2 \tag{14.115}$$

$$\dot{x}_2 = x_2 + (1 + x_1 + x_2)x_1 + x_1 \sin x_2 \tag{14.116}$$

5. $2R$ manipulator control from the general robotic equations.

    Expand the control equations for a $2R$ planar manipulator using the following control law:

$$\mathbf{Q} = \mathbf{D(q)}\left(\ddot{\mathbf{q}}_d - \mathbf{k}_D\dot{e} - \mathbf{k}_P\mathbf{e}\right) + \mathbf{H(q, \dot{q})} + \mathbf{G(q)} \tag{14.117}$$

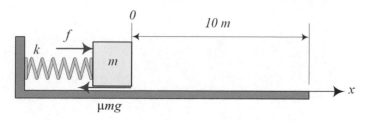

Fig. 14.7 A mass-spring oscillator on a rough surface and attached to a wall

6. One-link manipulator control.

   A one-link manipulator is shown in Fig. 14.4.
   (a) Derive the equation of motion.
   (b) Determine a rest-to-rest joint path between $\theta(0) = 45$ deg and $\theta(0) = -45$ deg.
   (c) Solve the time optimal control of the manipulator and determine the torque $Q_c(t)$ for the following date:

$$m = 1\,\text{kg} \qquad l = 1\,\text{m} \qquad |Q| \le 120\,\text{Nm} \tag{14.118}$$

   (d) Now assume the mass is $m = 1.01$ kg and solve the equation of motion numerically by feeding the calculated torques $Q_c(t)$. Determine the position and velocity errors at the end of the motion.
   (e) Design a computed torque control law to compensate the error during the motion.

7. ★ Mass-spring under friction control.

   Calculate the optimal control input f for the mass-spring system of Fig. 14.7. The optimal control command is limited $|f(t)| \le 100$ N, and moves the mass $m = 1$ kg rest-to-rest from $x(0) = 0$ to $x(t_f) = 10$ m. The mass is moving on a rough surface with coefficient $\mu = 0.5$ and is attached to a wall by a linear spring with stiffness $k = 5$ N/m.

   Increase the stiffness %10, and design a computed torque control law to eliminate error during the motion.

8. ★ $2R$ manipulator control.

   (a) Solve Exercise 12.19 and calculate the optimal control inputs.
   (b) Increase the masses by 10%, and solve the dynamic equations numerically.
   (c) Determine the position and velocity error in Cartesian and joint spaces by applying the calculated optimal inputs.
   (d) Design a computed torque control law to eliminate error during the motion.

9. ★ PR planar manipulator control.

   (a) Solve Exercise 13.13 and calculate the optimal control inputs.
   (b) Increase the gravitational acceleration by 10%, and solve the dynamic equations numerically.
   (c) Determine the position and velocity error in Cartesian and joint spaces by applying the calculated optimal inputs.
   (d) Design a computed torque control law to eliminate error during the motion.

10. Sensing and measurement.

    Consider the one DOF manipulator in Eq. (14.96). To control the manipulator, we need to sense the actual angle θ, angular velocity $\dot{\theta}$, and angular acceleration $\ddot{\theta}$ and compare them with the desired values of $\theta_d, \dot{\theta}_d, \ddot{\theta}_d$ to make sure that the manipulator is following the desired path. Can we measure the actual moment Q, that the actuator is providing, and compare with the predicted value Q_c instead? Does making Q equal to Q_c guarantee that the manipulator does what it is supposed to do?

In this appendix, the 12 combinations of triple rotation about global fixed axes are presented.

$$Q_{Z,\alpha} = \begin{bmatrix} \cos\alpha & -\sin\alpha & 0 \\ \sin\alpha & \cos\alpha & 0 \\ 0 & 0 & 1 \end{bmatrix} \tag{A.1}$$

$$Q_{Y,\beta} = \begin{bmatrix} \cos\beta & 0 & \sin\beta \\ 0 & 1 & 0 \\ -\sin\beta & 0 & \cos\beta \end{bmatrix} \tag{A.2}$$

$$Q_{X,\gamma} = \begin{bmatrix} 1 & 0 & 0 \\ 0 & \cos\gamma & -\sin\gamma \\ 0 & \sin\gamma & \cos\gamma \end{bmatrix} \tag{A.3}$$

1. $Q_{X,\gamma}\, Q_{Y,\beta}\, Q_{Z,\alpha}$

$$= \begin{bmatrix} c\alpha\,c\beta & -c\beta\,s\alpha & s\beta \\ c\gamma\,s\alpha + c\alpha\,s\beta\,s\gamma & c\alpha c\gamma - s\alpha\,s\beta\,s\gamma & -c\beta\,s\gamma \\ s\alpha\,s\gamma - c\alpha\,c\gamma\,s\beta & c\alpha s\gamma + c\gamma\,s\alpha\,s\beta & c\beta\,c\gamma \end{bmatrix} \tag{A.4}$$

2. $Q_{Y,\gamma}\, Q_{Z,\beta}\, Q_{X,\alpha}$

$$= \begin{bmatrix} c\beta\,c\gamma & s\alpha\,s\gamma - c\alpha\,c\gamma\,s\beta & c\alpha\,s\gamma + c\gamma\,s\alpha\,s\beta \\ s\beta & c\alpha\,c\beta & -c\beta\,s\alpha \\ -c\beta\,s\gamma & c\gamma\,s\alpha + c\alpha\,s\beta\,s\gamma & c\alpha c\gamma - s\alpha\,s\beta\,s\gamma \end{bmatrix} \tag{A.5}$$

3. $Q_{Z,\gamma}\, Q_{X,\beta}\, Q_{Y,\alpha}$

$$= \begin{bmatrix} c\alpha c\gamma - s\alpha\,s\beta\,s\gamma & -c\beta\,s\gamma & c\gamma\,s\alpha + c\alpha\,s\beta\,s\gamma \\ c\alpha\,s\gamma + c\gamma\,s\alpha\,s\beta & c\beta\,c\gamma & s\alpha\,s\gamma - c\alpha\,c\gamma\,s\beta \\ -c\beta\,s\alpha & s\beta & c\alpha\,c\beta \end{bmatrix} \tag{A.6}$$

4. $Q_{Z,\gamma}\, Q_{Y,\beta}\, Q_{X,\alpha}$

$$= \begin{bmatrix} c\beta\,c\gamma & -c\alpha\,s\gamma + c\gamma\,s\alpha\,s\beta & s\alpha\,s\gamma + c\alpha\,c\gamma\,s\beta \\ c\beta\,s\gamma & c\alpha\,c\gamma + s\alpha\,s\beta\,s\gamma & -c\gamma\,s\alpha + c\alpha\,s\beta\,s\gamma \\ -s\beta & c\beta\,s\alpha & c\alpha\,c\beta \end{bmatrix} \tag{A.7}$$

5. $Q_{Y,\gamma}\, Q_{X,\beta}\, Q_{Z,\alpha}$

$$= \begin{bmatrix} c\alpha\,c\gamma + s\alpha\,s\beta\,s\gamma & -c\gamma\,s\alpha + c\alpha\,s\beta\,s\gamma & c\beta\,s\gamma \\ c\beta\,s\alpha & c\alpha\,c\beta & -s\beta \\ -c\alpha\,s\gamma + c\gamma\,s\alpha\,s\beta & s\alpha\,s\gamma + c\alpha\,c\gamma\,s\beta & c\beta\,c\gamma \end{bmatrix} \tag{A.8}$$

6. $Q_{X,\gamma}\, Q_{Z,\beta}\, Q_{Y,\alpha}$

$$= \begin{bmatrix} c\alpha\,c\beta & -s\beta & c\beta\,s\alpha \\ s\alpha\,s\gamma + c\alpha\,c\gamma\,s\beta & c\beta\,c\gamma & -c\alpha\,s\gamma + c\gamma\,s\alpha\,s\beta \\ -c\gamma\,s\alpha + c\alpha\,s\beta\,s\gamma & c\beta\,s\gamma & c\alpha\,c\gamma + s\alpha\,s\beta\,s\gamma \end{bmatrix} \tag{A.9}$$

7. $Q_{X,\gamma}\, Q_{Y,\beta}\, Q_{X,\alpha}$

$$= \begin{bmatrix} c\beta & s\alpha\, s\beta & c\alpha\, s\beta \\ s\beta\, s\gamma & c\alpha\, c\gamma - c\beta\, s\alpha\, s\gamma & -c\gamma\, s\alpha - c\alpha\, c\beta\, s\gamma \\ -c\gamma\, s\beta & c\alpha\, s\gamma + c\beta\, c\gamma\, s\alpha & -s\alpha\, s\gamma + c\alpha\, c\beta\, c\gamma \end{bmatrix} \tag{A.10}$$

8. $Q_{Y,\gamma}\, Q_{Z,\beta}\, Q_{Y,\alpha}$

$$= \begin{bmatrix} -s\alpha\, s\gamma + c\alpha\, c\beta\, c\gamma & -c\gamma\, s\beta & c\alpha\, s\gamma + c\beta\, c\gamma\, s\alpha \\ c\alpha\, s\beta & c\beta & s\alpha\, s\beta \\ -c\gamma\, s\alpha - c\alpha\, c\beta\, s\gamma & s\beta\, s\gamma & c\alpha\, c\gamma - c\beta\, s\alpha\, s\gamma \end{bmatrix} \tag{A.11}$$

9. $Q_{Z,\gamma}\, Q_{X,\beta}\, Q_{Z,\alpha}$

$$= \begin{bmatrix} c\alpha\, c\gamma - c\beta\, s\alpha\, s\gamma & -c\gamma\, s\alpha - c\alpha\, c\beta\, s\gamma & s\beta\, s\gamma \\ c\alpha\, s\gamma + c\beta\, c\gamma\, s\alpha & -s\alpha\, s\gamma + c\alpha\, c\beta\, c\gamma & -c\gamma\, s\beta \\ s\alpha\, s\beta & c\alpha\, s\beta & c\beta \end{bmatrix} \tag{A.12}$$

10. $Q_{X,\gamma}\, Q_{Z,\beta}\, Q_{X,\alpha}$

$$= \begin{bmatrix} c\beta & -c\alpha\, s\beta & s\alpha\, s\beta \\ c\gamma\, s\beta & -s\alpha\, s\gamma + c\alpha\, c\beta\, c\gamma & -c\alpha\, s\gamma - c\beta\, c\gamma\, s\alpha \\ s\beta\, s\gamma & c\gamma\, s\alpha + c\alpha\, c\beta\, s\gamma & c\alpha\, c\gamma - c\beta\, s\alpha\, s\gamma \end{bmatrix} \tag{A.13}$$

11. $Q_{Y,\gamma}\, Q_{X,\beta}\, Q_{Y,\alpha}$

$$= \begin{bmatrix} c\alpha\, c\gamma - c\beta\, s\alpha\, s\gamma & s\beta\, s\gamma & c\gamma\, s\alpha + c\alpha\, c\beta\, s\gamma \\ s\alpha\, s\beta & c\beta & -c\alpha\, s\beta \\ -c\alpha\, s\gamma - c\beta\, c\gamma\, s\alpha & c\gamma\, s\beta & -s\alpha\, s\gamma + c\alpha\, c\beta\, c\gamma \end{bmatrix} \tag{A.14}$$

12. $Q_{Z,\gamma}\, Q_{Y,\beta}\, Q_{Z,\alpha}$

$$= \begin{bmatrix} -s\alpha\, s\gamma + c\alpha\, c\beta\, c\gamma & -c\alpha\, s\gamma - c\beta\, c\gamma\, s\alpha & c\gamma\, s\beta \\ c\gamma\, s\alpha + c\alpha\, c\beta\, s\gamma & c\alpha\, c\gamma - c\beta\, s\alpha\, s\gamma & s\beta\, s\gamma \\ -c\alpha\, s\beta & s\alpha\, s\beta & c\beta \end{bmatrix} \tag{A.15}$$

Local Frame Triple Rotation

B

In this appendix, the 12 combinations of triple rotation about local axes are presented.

$$A_{z,\varphi} = \begin{bmatrix} \cos\varphi & \sin\varphi & 0 \\ -\sin\varphi & \cos\varphi & 0 \\ 0 & 0 & 1 \end{bmatrix} \tag{B.1}$$

$$A_{y,\theta} = \begin{bmatrix} \cos\theta & 0 & -\sin\theta \\ 0 & 1 & 0 \\ \sin\theta & 0 & \cos\theta \end{bmatrix} \tag{B.2}$$

$$A_{x,\psi} = \begin{bmatrix} 1 & 0 & 0 \\ 0 & \cos\psi & \sin\psi \\ 0 & -\sin\psi & \cos\psi \end{bmatrix} \tag{B.3}$$

1. $A_{x,\psi} A_{y,\theta} A_{z,\varphi}$

$$= \begin{bmatrix} c\theta\,c\varphi & c\theta\,s\varphi & -s\theta \\ -c\psi\,s\varphi + c\varphi\,s\theta\,s\psi & c\varphi\,c\psi + s\theta\,s\varphi\,s\psi & c\theta\,s\psi \\ s\varphi\,s\psi + c\varphi\,s\theta\,c\psi & -c\varphi\,s\psi + s\theta\,c\psi\,s\varphi & c\theta\,c\psi \end{bmatrix} \tag{B.4}$$

2. $A_{y,\psi} A_{z,\theta} A_{x,\varphi}$

$$= \begin{bmatrix} c\theta\,c\psi & s\varphi\,s\psi + c\varphi\,s\theta\,c\psi & -c\varphi\,s\psi + s\theta\,c\psi\,s\varphi \\ -s\theta & c\theta\,c\varphi & c\theta\,s\varphi \\ c\theta\,s\psi & -c\psi\,s\varphi + c\varphi\,s\theta\,s\psi & c\varphi\,c\psi + s\theta\,s\varphi\,s\psi \end{bmatrix} \tag{B.5}$$

3. $A_{z,\psi} A_{x,\theta} A_{y,\varphi}$

$$= \begin{bmatrix} c\varphi\,c\psi + s\theta\,s\varphi\,s\psi & c\theta\,s\psi & -c\psi\,s\varphi + c\varphi\,s\theta\,s\psi \\ -c\varphi\,s\psi + s\theta\,c\psi\,s\varphi & c\theta\,c\psi & s\varphi\,s\psi + c\varphi\,s\theta\,c\psi \\ c\theta\,s\varphi & -s\theta & c\theta\,c\varphi \end{bmatrix} \tag{B.6}$$

4. $A_{z,\psi} A_{y,\theta} A_{x,\varphi}$

$$= \begin{bmatrix} c\theta\,c\psi & c\varphi\,s\psi + s\theta\,c\psi\,s\varphi & s\varphi\,s\psi - c\varphi\,s\theta\,c\psi \\ -c\theta\,s\psi & c\varphi\,c\psi - s\theta\,s\varphi\,s\psi & c\psi\,s\varphi + c\varphi\,s\theta\,s\psi \\ s\theta & -c\theta\,s\varphi & c\theta\,c\varphi \end{bmatrix} \tag{B.7}$$

5. $A_{y,\psi} A_{x,\theta} A_{z,\varphi}$

$$= \begin{bmatrix} c\varphi\,c\psi - s\theta\,s\varphi vs\psi & c\psi\,s\varphi + c\varphi\,s\theta\,s\psi & -c\theta\,s\psi \\ -c\theta\,s\varphi & c\theta\,c\varphi & s\theta \\ c\varphi\,s\psi + s\theta\,c\psi\,s\varphi & s\varphi\,s\psi - c\varphi\,s\theta\,c\psi & c\theta\,c\psi \end{bmatrix} \tag{B.8}$$

6. $A_{x,\psi} A_{z,\theta} A_{y,\varphi}$

$$= \begin{bmatrix} c\theta\,c\varphi & s\theta & -c\theta\,s\varphi \\ s\varphi\,s\psi - c\varphi\,s\theta\,c\psi & c\theta\,c\psi & c\varphi vs\psi + s\theta\,c\psi\,s\varphi \\ c\psi\,s\varphi + c\varphi\,s\theta\,s\psi & -c\theta\,s\psi & c\varphi\,c\psi - s\theta\,s\varphi\,s\psi \end{bmatrix} \tag{B.9}$$

© The Author(s), under exclusive license to Springer Nature Switzerland AG 2022
R. N. Jazar, *Theory of Applied Robotics*, https://doi.org/10.1007/978-3-030-93220-6

7. $A_{x,\psi} A_{y,\theta} A_{x,\varphi}$

$$
= \begin{bmatrix}
c\theta & s\theta\,s\varphi & -c\varphi\,s\theta \\
s\theta\,s\psi & c\varphi\,c\psi - c\theta\,s\varphi\,s\psi & c\psi\,s\varphi + c\theta\,c\varphi\,s\psi \\
s\theta\,c\psi & -c\varphi\,s\psi - c\theta\,c\psi\,s\varphi & -s\varphi\,s\psi + c\theta\,c\varphi\,c\psi
\end{bmatrix}
\tag{B.10}
$$

8. $A_{y,\psi} A_{z,\theta} A_{y,\varphi}$

$$
= \begin{bmatrix}
-s\varphi\,s\psi + c\theta\,c\varphi\,c\psi & s\theta\,c\psi & -c\varphi\,s\psi - c\theta\,c\psi\,s\varphi \\
-c\varphi\,s\theta & c\theta & s\theta\,s\varphi \\
c\psi\,s\varphi + c\theta\,c\varphi\,s\psi & s\theta\,s\psi & c\varphi\,c\psi - c\theta\,s\varphi\,s\psi
\end{bmatrix}
\tag{B.11}
$$

9. $A_{z,\psi} A_{x,\theta} A_{z,\varphi}$

$$
= \begin{bmatrix}
c\varphi\,c\psi - c\theta\,s\varphi\,s\psi & c\psi\,s\varphi + c\theta\,c\varphi\,s\psi & s\theta\,s\psi \\
-c\varphi\,s\psi - c\theta\,c\psi\,s\varphi & -s\varphi\,s\psi + c\theta\,c\varphi\,c\psi & s\theta\,c\psi \\
s\theta\,s\varphi & -c\varphi\,s\theta & c\theta
\end{bmatrix}
\tag{B.12}
$$

10. $A_{x,\psi} A_{z,\theta} A_{x,\varphi}$

$$
= \begin{bmatrix}
c\theta & c\varphi\,s\theta & s\theta\,s\varphi \\
-s\theta\,c\psi & -s\varphi\,s\psi + c\theta\,c\varphi\,c\psi & c\varphi\,s\psi + c\theta\,c\psi\,s\varphi \\
s\theta\,s\psi & -c\psi\,s\varphi - c\theta\,c\varphi\,s\psi & c\varphi\,c\psi - c\theta\,s\varphi\,s\psi
\end{bmatrix}
\tag{B.13}
$$

11. $A_{y,\psi} A_{x,\theta} A_{y,\varphi}$

$$
= \begin{bmatrix}
c\varphi\,c\psi - c\theta\,s\varphi\,s\psi & s\theta\,s\psi & -c\psi\,s\varphi - c\theta\,c\varphi\,s\psi \\
s\theta\,s\varphi & c\theta & c\varphi\,s\theta \\
c\varphi\,s\psi + c\theta\,c\psi\,s\varphi & -s\theta\,c\psi & -s\varphi\,s\psi + c\theta\,c\varphi\,c\psi
\end{bmatrix}
\tag{B.14}
$$

12. $A_{z,\psi} A_{y,\theta} A_{z,\varphi}$

$$
= \begin{bmatrix}
-s\varphi\,s\psi + c\theta\,c\varphi\,c\psi & c\varphi\,s\psi + c\theta\,c\psi\,s\varphi & -s\theta\,c\psi \\
-c\psi\,s\varphi - c\theta\,c\varphi\,s\psi & c\varphi\,c\psi - c\theta\,s\varphi\,s\psi & s\theta\,s\psi \\
c\varphi\,s\theta & s\theta\,s\varphi & c\theta
\end{bmatrix}
\tag{B.15}
$$

Principal Central Screws Triple Combination

In this appendix, the six combinations of triple principal central screws are presented.

1. $\check{s}(h_X, \gamma, \hat{I})\, \check{s}(h_Y, \beta, \hat{J})\, \check{s}(h_Z, \alpha, \hat{K})$

$$
= \begin{bmatrix}
c\alpha c\beta & -c\beta s\alpha & s\beta & \gamma p_X + \alpha p_Z s\beta \\
c\gamma s\alpha + c\alpha s\beta s\gamma & c\alpha c\gamma - s\alpha s\beta s\gamma & -c\beta s\gamma & \beta p_Y c\gamma - \alpha p_Z c\beta s\gamma \\
s\alpha s\gamma - c\alpha c\gamma s\beta & c\alpha s\gamma + c\gamma s\alpha s\beta & c\beta c\gamma & \beta p_Y s\gamma + \alpha p_Z c\beta c\gamma \\
0 & 0 & 0 & 1
\end{bmatrix}
\tag{C.1}
$$

2. $\check{s}(h_Y, \beta, \hat{J})\, \check{s}(h_Z, \alpha, \hat{K})\, \check{s}(h_X, \gamma, \hat{I})$

$$
= \begin{bmatrix}
c\alpha c\beta & s\beta s\gamma - c\beta c\gamma s\alpha & c\gamma s\beta + c\beta s\alpha s\gamma & \alpha p_Z s\beta + \gamma p_X c\alpha c\beta \\
s\alpha & c\alpha c\gamma & -c\alpha s\gamma & \beta p_Y + \gamma p_X s\alpha \\
-c\alpha s\beta & c\beta s\gamma + c\gamma s\alpha s\beta & c\beta c\gamma - s\alpha s\beta s\gamma & \alpha p_Z c\beta - \gamma p_X c\alpha s\beta \\
0 & 0 & 0 & 1
\end{bmatrix}
\tag{C.2}
$$

3. $\check{s}(h_Z, \alpha, \hat{K})\, \check{s}(h_X, \gamma, \hat{I})\, \check{s}(h_Y, \beta, \hat{J})$

$$
= \begin{bmatrix}
c\alpha c\beta - s\alpha s\beta s\gamma & -c\gamma s\alpha & c\alpha s\beta + c\beta s\alpha s\gamma & \gamma p_X c\alpha - \beta p_Y c\gamma s\alpha \\
c\beta s\alpha + c\alpha s\beta s\gamma & c\alpha c\gamma & s\alpha s\beta - c\alpha c\beta s\gamma & \gamma p_X s\alpha + \beta p_Y c\alpha c\gamma \\
-c\gamma s\beta & s\gamma & c\beta c\gamma & \alpha p_Z + \beta p_Y s\gamma \\
0 & 0 & 0 & 1
\end{bmatrix}
\tag{C.3}
$$

4. $\check{s}(h_Z, \alpha, \hat{K})\, \check{s}(h_Y, \beta, \hat{J})\, \check{s}(h_X, \gamma, \hat{I})$

$$
= \begin{bmatrix}
c\alpha c\beta & c\alpha s\beta s\gamma - c\gamma s\alpha & s\alpha s\gamma + c\alpha c\gamma s\beta & \gamma p_X c\alpha c\beta - \beta p_Y s\alpha \\
c\beta s\alpha & c\alpha c\gamma + s\alpha s\beta s\gamma & c\gamma s\alpha s\beta - c\alpha s\gamma & \beta p_Y c\alpha + \gamma p_X c\beta s\alpha \\
-s\beta & c\beta s\gamma & c\beta c\gamma & \alpha p_Z - \gamma p_X s\beta \\
0 & 0 & 0 & 1
\end{bmatrix}
\tag{C.4}
$$

5. $\check{s}(h_Y, \beta, \hat{J})\,\check{s}(h_X, \gamma, \hat{I})\,\check{s}(h_Z, \alpha, \hat{K})$

$$= \begin{bmatrix} c\alpha c\beta + s\alpha s\beta s\gamma & c\alpha s\beta s\gamma - c\beta s\alpha & c\gamma s\beta & \gamma p_X c\beta + \alpha p_Z c\gamma s\beta \\ c\gamma s\alpha & c\alpha c\gamma & -s\gamma & \beta p_Y - \alpha p_Z s\gamma \\ c\beta s\alpha s\gamma - c\alpha s\beta & s\alpha s\beta + c\alpha c\beta s\gamma & c\beta c\gamma & \alpha p_Z c\beta c\gamma - \gamma p_X s\beta \\ 0 & 0 & 0 & 1 \end{bmatrix} \tag{C.5}$$

6. $\check{s}(h_X, \gamma, \hat{I})\,\check{s}(h_Z, \alpha, \hat{K})\,\check{s}(h_Y, \beta, \hat{J})$

$$= \begin{bmatrix} c\alpha c\beta & -s\alpha & c\alpha s\beta & \gamma p_X - \beta p_Y s\alpha \\ s\beta s\gamma + c\beta c\gamma s\alpha & c\alpha c\gamma & c\gamma s\alpha s\beta - c\beta s\gamma & \beta p_Y c\alpha c\gamma - \alpha p_Z s\gamma \\ c\beta s\alpha s\gamma - c\gamma s\beta & c\alpha s\gamma & c\beta c\gamma + s\alpha s\beta s\gamma & \alpha p_Z c\gamma + \beta p_Y c\alpha s\gamma \\ 0 & 0 & 0 & 1 \end{bmatrix} \tag{C.6}$$

Most industrial robots are made by connected links that use one of the following joint configurations with their the associated Denavit–Hartenberg (DH) homogenous transformation matrices. The angle in parenthesis is α_i, the angle from z_{i-1} to z_i about x_i. The relative orientation of the axes z_{i-1} and z_i is indicated by a parallel sign ($\parallel$), if their joint axes are parallel; or by an orthogonal sign ($\vdash$), if their joint axes are intersecting at a right angle; or by a perpendicular sign ($\perp$), if their joint axes are at a right angle with respect to their common normal.

1	R $\parallel$ $R(0)$	or	R $\parallel$ $P(0)$
2	R $\parallel$ $R(180)$	or	R $\parallel$ $P(180)$
3	R $\perp$ $R(90)$	or	R $\perp$ $P(90)$
4	R $\perp$ $R(-90)$	or	R $\perp$ $P(-90)$
5	R $\vdash$ $R(90)$	or	R $\vdash$ $P(90)$
6	R $\vdash$ $R(-90)$	or	R $\vdash$ $P(-90)$
7	P $\parallel$ $R(0)$	or	P $\parallel$ $P(0)$
8	P $\parallel$ $R(180)$	or	P $\parallel$ $P(180)$
9	P $\perp$ $R(90)$	or	P $\perp$ $P(90)$
10	P $\perp$ $R(-90)$	or	P $\perp$ $P(-90)$
11	P $\vdash$ $R(90)$	or	P $\vdash$ $P(90)$
12	P $\vdash$ $R(-90)$	or	P $\vdash$ $P(-90)$

Denavit–Hartenberg homogenous transformation matrix $^{i-1}T_i$,

$$
^{i-1}T_i = \begin{bmatrix} \cos\theta_i & -\sin\theta_i\cos\alpha_i & \sin\theta_i\sin\alpha_i & a_i\cos\theta_i \\ \sin\theta_i & \cos\theta_i\cos\alpha_i & -\cos\theta_i\sin\alpha_i & a_i\sin\theta_i \\ 0 & \sin\alpha_i & \cos\alpha_i & d_i \\ 0 & 0 & 0 & 1 \end{bmatrix}
\tag{D.1}
$$

depends on four parameters according to the following table.

a_i	along x_i	from z_{i-1} to z_i
α_i	about x_i	from z_{i-1} to z_i
d_i	along z_{i-1}	from x_{i-1} to x_i
θ_i	about z_{i-1}	from x_{i-1} to x_i

Cases 1, 2. **Links with R$\parallel$R or R$\parallel$P**

$$a_i = const \quad \alpha_i = 0 \, or \, 180 \deg \quad d_i = 0 \quad \theta_i = variable$$

Figure D.1 illustrates a link R$\parallel$R(0), and Fig. D.2 illustrates a link R$\parallel$R(180). If the proximal joint of link (i) is revolute, the distal joint is either revolute or prismatic, and the joint axes at two ends are parallel, then $\alpha_i = 0$ or $\alpha_i = 180 \deg$, a_i is the

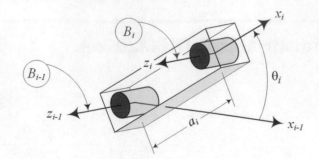

Fig. D.1   A link R∥R(0)

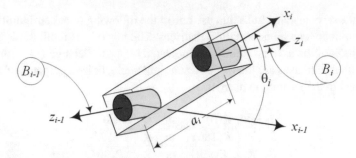

Fig. D.2   A link R∥R(180)

distance between the joint axes, and θ_i is the only variable parameter. The joint distance $d_i = const$ is the distance between the origin of B_i and B_{i-1} along z_{i-1}. We usually set (x_i, y_i) and (x_{i-1}, y_{i-1}) coplanar to get $d_i = 0$. The x_i and x_{i-1}-axes are parallel for a link R∥R at the rest position. Therefore, the transformation matrix $^{i-1}T_i$ for such a link with $\alpha_i = 0$ known as R∥R(0) or R∥P(0) is

$$R \parallel R(0) \text{ or } R \parallel P(0)$$

$$^{i-1}T_i = \begin{bmatrix} \cos\theta_i & -\sin\theta_i & 0 & a_i\cos\theta_i \\ \sin\theta_i & \cos\theta_i & 0 & a_i\sin\theta_i \\ 0 & 0 & 1 & d_i \\ 0 & 0 & 0 & 1 \end{bmatrix} \tag{D.2}$$

while for a link with $\alpha_i = 180\,\text{deg}$ and R∥R(180) or R∥P(180) is

$$R \parallel R(180) \text{ or } R \parallel P(180)$$

$$^{i-1}T_i = \begin{bmatrix} \cos\theta_i & \sin\theta_i & 0 & a_i\cos\theta_i \\ \sin\theta_i & -\cos\theta_i & 0 & a_i\sin\theta_i \\ 0 & 0 & -1 & d_i \\ 0 & 0 & 0 & 1 \end{bmatrix} \tag{D.3}$$

Cases 3, 4.   **Links with R⊥R or R⊥P**

$$a_i = const \qquad \alpha_i = 90\,\text{deg} \ or \ -90\,\text{deg}$$

$$d_i = 0 \qquad \theta_i = variable$$

Figure D.3 illustrates a link R⊥R(90), and Fig. D.4 illustrates a link R⊥R(−90). If the proximal joint of link (i) is revolute, the distal joint is either revolute or prismatic, and if the joint axes at two ends are perpendicular, then $\alpha_i = 90\,\text{deg}$ or $\alpha_i = −90\,\text{deg}$, a_i is the distance between the joint axes on x_i, and θ_i is the only variable parameter. The joint distance $d_i = const$ is the distance between the origin of B_i and B_{i-1} along z_{i-1}. We usually set (x_i, y_i) and (x_{i-1}, y_{i-1}) coplanar to get $d_i = 0$.

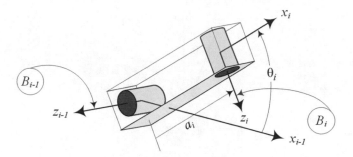

Fig. D.3 A R⊥R(90) link

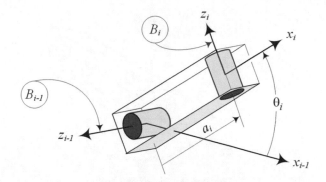

Fig. D.4 A R⊥R(−90) link

The R⊥R link is made by twisting the R∥R link 90 deg about its center line x_{i-1}-axis. The x_i and x_{i-1}-axes are parallel for a link R⊥R at the rest position. Therefore, the transformation matrix $^{i-1}T_i$ for such a link with $\alpha_i = 90$ deg known as R⊥R(90) or R⊥P(90) is

$$R \perp R(90) \text{ or } R \perp P(90)$$

$$^{i-1}T_i = \begin{bmatrix} \cos\theta_i & 0 & \sin\theta_i & a_i\cos\theta_i \\ \sin\theta_i & 0 & -\cos\theta_i & a_i\sin\theta_i \\ 0 & 1 & 0 & d_i \\ 0 & 0 & 0 & 1 \end{bmatrix} \tag{D.4}$$

while for a link with $\alpha_i = -90$ deg and R⊥R(−90) or R⊥P(−90) is

$$R \perp R(-90) \text{ or } R \perp P(-90)$$

$$^{i-1}T_i = \begin{bmatrix} \cos\theta_i & 0 & -\sin\theta_i & a_i\cos\theta_i \\ \sin\theta_i & 0 & \cos\theta_i & a_i\sin\theta_i \\ 0 & -1 & 0 & d_i \\ 0 & 0 & 0 & 1 \end{bmatrix} \tag{D.5}$$

Cases 5, 6. Links with R⊢R or R⊢P

$$a_i = 0 \qquad \alpha_i = 90 \deg \ or \ -90 \deg$$

$$d_i = 0 \qquad \theta_i = variable$$

Figure D.5 illustrates a link R⊢R(90), and Fig. D.6 illustrates a link R⊢R(−90). If the proximal joint of link (i) is revolute and the distal joint is either revolute or prismatic and the joint axes at two ends are intersecting orthogonal, then $\alpha_i = 90$ deg or $\alpha_i = -90$ deg, $a_i = 0$. $d_i = cte$ is the distance between the coordinate origin on z_{i-1}, and θ_i is the only variable parameter. It is possible to have or assume $d_i = 0$ at the rest position. When $d_i = 0$, the x_i and x_{i-1}-axes of a link R⊢R are coincident

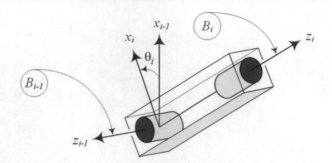

Fig. D.5 An R⊢R(90) link

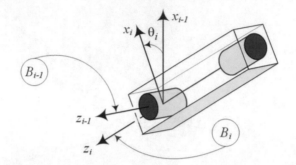

Fig. D.6 An R⊢R(−90) link

and when $d_i \neq 0$ they are parallel. Therefore, the transformation matrix $^{i-1}T_i$ for such a link with $\alpha_i = 90$ deg and R⊢R(90) or R⊢P(90) is

$$R \vdash R(90) \text{ or } R \vdash P(90)$$

$$^{i-1}T_i = \begin{bmatrix} \cos\theta_i & 0 & \sin\theta_i & 0 \\ \sin\theta_i & 0 & -\cos\theta_i & 0 \\ 0 & 1 & 0 & d_i \\ 0 & 0 & 0 & 1 \end{bmatrix} \tag{D.6}$$

while for a link with $\alpha_i = -90$ deg and R⊢R(−90) or R⊢P(−90) is

$$R \vdash R(-90) \text{ or } R \vdash P(-90)$$

$$^{i-1}T_i = \begin{bmatrix} \cos\theta_i & 0 & -\sin\theta_i & 0 \\ \sin\theta_i & 0 & \cos\theta_i & 0 \\ 0 & -1 & 0 & d_i \\ 0 & 0 & 0 & 1 \end{bmatrix} \tag{D.7}$$

Cases 7, 8. **Links with P∥R or P∥P**

$$a_i = const \qquad \alpha_i = 0 \text{ or } 180 \deg$$

$$d_i = variable \qquad \theta_i = 0$$

Figure D.7 illustrates a link P∥R(0), and Fig. D.8 illustrates a link P∥R(180). If the proximal joint of link (i) is prismatic, its distal joint is either revolute or prismatic, and the joint axes at two ends are parallel, then $\alpha_i = 0$ or $\alpha_i = 180$ deg, $\theta_i = 0$. $a_i = const$ is the distance between the joint axes on x_i, and d_i is the only variable parameter. It is possible to have $a_i = 0$. The transformation matrix $^{i-1}T_i$ for a link with $\alpha_i = 0$ and P∥R(0) or P∥P(0) is

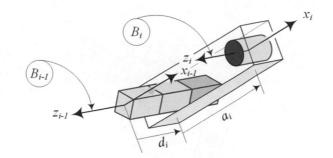

Fig. D.7 A P∥R(0) link

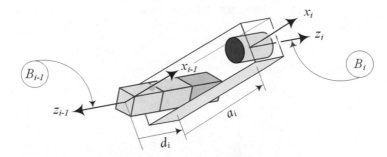

Fig. D.8 A P∥R(180) link

$$P \parallel R(0) \text{ or } P \parallel P(0)$$

$$^{i-1}T_i = \begin{bmatrix} 1 & 0 & 0 & a_i \\ 0 & 1 & 0 & 0 \\ 0 & 0 & 1 & d_i \\ 0 & 0 & 0 & 1 \end{bmatrix} \tag{D.8}$$

while for a link with $\alpha_i = 180\,\text{deg}$ and P∥R(180) or P∥P(180) is

$$P \parallel R(180) \text{ or } P \parallel P(180)$$

$$^{i-1}T_i = \begin{bmatrix} 1 & 0 & 0 & a_i \\ 0 & -1 & 0 & 0 \\ 0 & 0 & -1 & d_i \\ 0 & 0 & 0 & 1 \end{bmatrix} \tag{D.9}$$

The origin of the B_{i-1}-frame can arbitrarily be chosen at any point on the z_{i-1}-axis or parallel to the z_{i-1}-axis. A simple setup is to locate the origin o_i of a prismatic joint at the previous origin o_{i-1}. This sets $a_i = 0$ and furthermore sets the initial value of the joint variable $d_i = 0$, where d_i will vary when o_i slides up and down parallel to the z_{i-1}-axis.

Cases 9, 10. **Links with P⊥R or P⊥P**

$$a_i = const \qquad \alpha_i = 90 \text{ } or \text{ } -90\,\text{deg}$$

$$d_i = variable \qquad \theta_i = 0$$

Figure D.9 illustrates a link P⊥R(90), and Fig. D.10 illustrates a link P⊥R(−90). If the proximal joint of link (i) is prismatic and its distal joint is either revolute or prismatic with orthogonal, then $\alpha_i = 90\,\text{deg}$ or $\alpha_i = -90\,\text{deg}$, $\theta_i = 0$. $a_i = const$ is the distance between the joint axes on x_i, and d_i is the only variable parameter. The transformation matrix $^{i-1}T_i$ for a link with $\alpha_i = 90\,\text{deg}$ and P⊥R(90) or P⊥P(90) is

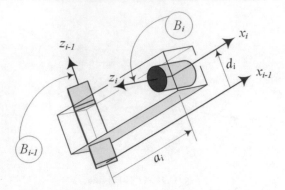

Fig. D.9 A P⊥R(90) link

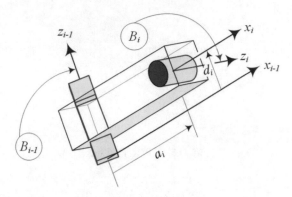

Fig. D.10 A P⊥R(−90) link

$$P \perp R(90) \text{ or } P \perp P(90)$$

$$^{i-1}T_i = \begin{bmatrix} 1 & 0 & 0 & a_i \\ 0 & 0 & -1 & 0 \\ 0 & 1 & 0 & d_i \\ 0 & 0 & 0 & 1 \end{bmatrix} \tag{D.10}$$

while for a link with $\alpha_i = -90\,\mathrm{deg}$ and P⊥R(−90) or P⊥P(−90) it is

$$P \perp R(-90) \text{ or } P \perp P(-90)$$

$$^{i-1}T_i = \begin{bmatrix} 1 & 0 & 0 & a_i \\ 0 & 0 & 1 & 0 \\ 0 & -1 & 0 & d_i \\ 0 & 0 & 0 & 1 \end{bmatrix} \tag{D.11}$$

Cases 11, 12. **Links with P⊢R or P⊢P.**

$$a_i = 0 \qquad \alpha_i = 90 \text{ or } -90 \deg$$
$$d_i = variable \qquad \theta_i = 0$$

Figure D.11 illustrates a link P⊢R(90), and Fig. D.12 illustrates a link P⊢R(−90). If the proximal joint of link (i) is prismatic and the distal joint is either revolute or prismatic and the joint axes at two ends are intersecting orthogonal, then $\alpha_i = 90\,\mathrm{deg}$ or $\alpha_i = -90\,\mathrm{deg}$, $\theta_i = 0$, and $a_i = 0$, and d_i is the only variable parameter. The x_i-axis must be perpendicular to the plane of the z_{i-1} and z_i-axes, and it is possible to have $a_i \neq 0$. Therefore, the transformation matrix $^{i-1}T_i$ for a link with $\alpha_i = 90\,\mathrm{deg}$ and P⊢R(90) or P⊢P(90) is

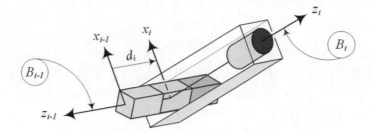

Fig. D.11 A P⊢R(90) link

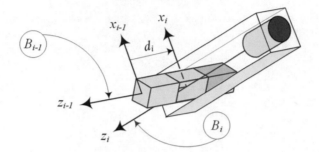

Fig. D.12 A link P⊢R(−90)

$$P \;\vdash\; R(90) \text{ or } P \vdash P(90)$$

$$^{i-1}T_i = \begin{bmatrix} 1 & 0 & 0 & 0 \\ 0 & 0 & -1 & 0 \\ 0 & 1 & 0 & d_i \\ 0 & 0 & 0 & 1 \end{bmatrix} \tag{D.12}$$

while for a link with $\alpha_i = -90\deg$ and P⊢R(−90) or P⊢P(−90) it is

$$P \;\vdash\; R(-90) \text{ or } P \vdash P(-90)$$

$$^{i-1}T_i = \begin{bmatrix} 1 & 0 & 0 & 0 \\ 0 & 0 & 1 & 0 \\ 0 & -1 & 0 & d_i \\ 0 & 0 & 0 & 1 \end{bmatrix} \tag{D.13}$$

An $m \times n$ matrix $\mathbf{A}$ consists of mn numbers arranged in m rows and n columns. The element in row i and column j of the matrix $\mathbf{A}$ is denoted by a_{ij}.

$$\mathbf{A} = \begin{bmatrix} a_{11} & a_{12} & \cdots & a_{1n} \\ a_{21} & a_{22} & \cdots & a_{2n} \\ \cdots & \cdots & \cdots & \cdots \\ a_{m1} & a_{m2} & \cdots & a_{mn} \end{bmatrix} = \begin{bmatrix} a_{ij} \end{bmatrix} \tag{E.1}$$

An $m \times 1$ matrix is a column vector $\mathbf{v}$ of dimension m, and a $1 \times n$ matrix is a row vector $\mathbf{w}$ of dimension n. An $m \times n$ matrix is called a square matrix if $m = n$.

$$\mathbf{v} = \begin{bmatrix} v_{11} \\ v_{21} \\ \cdots \\ v_{m1} \end{bmatrix} \tag{E.2}$$

$$\mathbf{w} = \begin{bmatrix} w_{m1} & w_{m2} & \cdots & w_{mn} \end{bmatrix} \tag{E.3}$$

Two matrices $\mathbf{A} = \begin{bmatrix} a_{ij} \end{bmatrix}$ and $\mathbf{B} = \begin{bmatrix} b_{ij} \end{bmatrix}$ are equal only if $a_{ij} = b_{ij}$.

Diagonal Matrix

A diagonal matrix is a square matrix $\boldsymbol{\Lambda} = \begin{bmatrix} \lambda_{ij} \end{bmatrix}$ such that $\lambda_{ij} = 0$ for $i \neq j$. The main diagonal of a square matrix are elements from top left to bottom right.

$$\boldsymbol{\Lambda} = \begin{bmatrix} \lambda_{11} & 0 & \cdots & 0 \\ 0 & \lambda_{22} & \cdots & 0 \\ \cdots & \cdots & \cdots & \cdots \\ 0 & 0 & \cdots & \lambda_{nn} \end{bmatrix} \tag{E.4}$$

A diagonal matrix may also be abbreviated as:

$$\boldsymbol{\Lambda} = \{\lambda_{11}, \lambda_{22}, \lambda_{22}, \cdots, \lambda_{nn}\} \tag{E.5}$$

A scalar matrix $\mathbf{S} = \begin{bmatrix} s_{ij} \end{bmatrix}$ is a diagonal matrix whose diagonal elements are equal $s_{ii} = s$.

$$\mathbf{S} = \begin{bmatrix} s & 0 & \cdots & 0 \\ 0 & s & \cdots & 0 \\ \cdots & \cdots & \cdots & \cdots \\ 0 & 0 & \cdots & s \end{bmatrix} \tag{E.6}$$

If all elements of the diagonal of $\mathbf{S}$ are unity, the matrix is the identity matrix, or unit matrix, and shown by $\mathbf{I}$.

$$\mathbf{I} = \begin{bmatrix} 1 & 0 & \cdots & 0 \\ 0 & 1 & \cdots & 0 \\ \cdots & \cdots & \cdots & \cdots \\ 0 & 0 & \cdots & 1 \end{bmatrix} \tag{E.7}$$

© The Author(s), under exclusive license to Springer Nature Switzerland AG 2022
R. N. Jazar, *Theory of Applied Robotics*, https://doi.org/10.1007/978-3-030-93220-6

A null matrix is a matrix of any shape in which every entry is zero.

$$\mathbf{0} = \begin{bmatrix} 0 & 0 & \cdots & 0 \\ 0 & 0 & \cdots & 0 \\ \cdots\cdots\cdots\cdots \\ 0 & 0 & \cdots & 0 \end{bmatrix} \tag{E.8}$$

Trace

Trace of a square matrix $\mathbf{A} = \begin{bmatrix} a_{ij} \end{bmatrix}$ is sum of its diagonal elements.

$$\operatorname{tr} \mathbf{A} = a_{11} + a_{22} + \cdots + a_{nn} \tag{E.9}$$

$$\operatorname{tr}[\mathbf{AB}] = \operatorname{tr}[\mathbf{BA}] \tag{E.10}$$

Matrix Arithmetic

$$\mathbf{A} + \mathbf{B} = \mathbf{B} + \mathbf{A} = \mathbf{C} \tag{E.11}$$

$$c_{ij} = a_{ij} + b_{ij} \tag{E.12}$$

$$\mathbf{AB} = \mathbf{C} \qquad c_{ij} = \sum_m a_{im} b_{mj} \tag{E.13}$$

$$\mathbf{AB} \neq \mathbf{BA} \tag{E.14}$$

$$[\mathbf{AB}]\,\mathbf{C} = \mathbf{A}\,[\mathbf{BC}] = \mathbf{D} \tag{E.15}$$

$$d_{ij} = \sum_{m,n} (a_{im} b_{mn})\, c_{jn} = \sum_{m,n} a_{im} \left(b_{mn} c_{jn} \right) \tag{E.16}$$

$$[\mathbf{A} + \mathbf{B}]\,\mathbf{C} = \mathbf{AC} + \mathbf{BC} = \mathbf{D} \tag{E.17}$$

$$d_{ij} = \sum_m (a_{im} + b_{im})\, c_{nj} = \sum_m \left(a_{im} c_{nj} + b_{im} c_{nj} \right) \tag{E.18}$$

Determinants

The determinant of a 2×2 matrix $\mathbf{A}$,

$$\mathbf{A} = \begin{bmatrix} a_{11} & a_{12} \\ a_{21} & a_{22} \end{bmatrix} = \begin{bmatrix} a_{ij} \end{bmatrix} \tag{E.19}$$

with elements a_{ij} comprising real or complex numbers, or functions, denoted by $|\mathbf{A}|$ or by $\det \mathbf{A}$, and is defined by:

$$\det \mathbf{A} = a_{11} a_{22} - a_{12} a_{21} \tag{E.20}$$

The determinant of a 3×3 matrix $\mathbf{A}$,

$$\mathbf{A} = \begin{bmatrix} a_{11} & a_{12} & a_{13} \\ a_{21} & a_{22} & a_{23} \\ a_{31} & a_{32} & a_{33} \end{bmatrix} = \begin{bmatrix} a_{ij} \end{bmatrix} \tag{E.21}$$

with elements a_{ij} comprising real or complex numbers, or functions, denoted by $|\mathbf{A}|$ or by $\det \mathbf{A}$, and is defined by:

$$\det \mathbf{A} = a_{11} a_{22} a_{33} + a_{12} a_{23} a_{31} + a_{21} a_{32} a_{13}$$

$$- a_{13} a_{22} a_{31} - a_{23} a_{32} a_{11} - a_{12} a_{21} a_{33} \tag{E.22}$$

To calculate the determinant of an $n \times n$ matrix $\mathbf{A}$,

$$\mathbf{A} = \begin{bmatrix} a_{11} & a_{12} & \cdots & a_{1n} \\ a_{21} & a_{22} & \cdots & a_{2n} \\ \cdots & \cdots & \cdots & \cdots \\ a_{n1} & a_{n2} & \cdots & a_{nn} \end{bmatrix} = \begin{bmatrix} a_{ij} \end{bmatrix} \tag{E.23}$$

we need first to define the minor M_{ij} associated with the element a_{ij} as determinant the order $(n-1)$ derived from by deletion of its ith row and jth column; and second, to define the cofactor C_{ij} associated with the element a_{ij} as:

$$C_{ij} = (-1)^{i+j} M_{ij} \tag{E.24}$$

The determinant of $\mathbf{A}$ may be based on expansion of elements of a row or of a column of $\mathbf{A}$. Expansion of $\det \mathbf{A}$ by elements of the ith row is

$$\det \mathbf{A} = \sum_{j=1}^{n} a_{ij} C_{ij} \qquad i = 1, 2, 3, \cdots, n \tag{E.25}$$

and expansion of $\det \mathbf{A}$ by elements of the jth column is

$$\det \mathbf{A} = \sum_{i=1}^{n} a_{ij} C_{ij} \qquad j = 1, 2, 3, \cdots, n \tag{E.26}$$

The expansion method to determine the determinant is called the Laplace expansion method.

As an example, let us calculate $\det \mathbf{A}$, where

$$\mathbf{A} = \begin{bmatrix} -2 & 1 & 3 \\ 1 & 2 & -1 \\ 4 & -1 & -2 \end{bmatrix} \tag{E.27}$$

Expanding $|\mathbf{A}|$ by elements of its second row gives

$$|\mathbf{A}| = C_{21} + 2C_{22} - C_{23} \tag{E.28}$$

$$C_{21} = (-1)^{2+1} \begin{vmatrix} 1 & 3 \\ -1 & -2 \end{vmatrix} = -1 \tag{E.29}$$

$$C_{22} = (-1)^{2+2} \begin{vmatrix} -2 & 3 \\ 4 & -2 \end{vmatrix} = -8 \tag{E.30}$$

$$C_{22} = (-1)^{2+3} \begin{vmatrix} -2 & 1 \\ 4 & -1 \end{vmatrix} = 2 \tag{E.31}$$

Therefore,

$$|\mathbf{A}| = -1 + 2(-8) - 2 = -19 \tag{E.32}$$

Alternatively, expanding $|\mathbf{A}|$ by elements of its third column gives

$$|\mathbf{A}| = 3C_{13} - C_{23} - 2C_{33} \tag{E.33}$$

$$C_{21} = (-1)^{1+3} \begin{vmatrix} 1 & 2 \\ 4 & -1 \end{vmatrix} = -9 \tag{E.34}$$

$$C_{22} = (-1)^{2+3} \begin{vmatrix} -2 & 1 \\ 4 & -1 \end{vmatrix} = 2 \tag{E.35}$$

$$C_{22} = (-1)^{3+3} \begin{vmatrix} -2 & 1 \\ 1 & 2 \end{vmatrix} = -5 \tag{E.36}$$

Therefore,

$$|\mathbf{A}| = 3(-9) - 2 - 2(-5) = -19 \tag{E.37}$$

If $\mathbf{A} = \left[a_{ij}\right]$, $\mathbf{B} = \left[b_{ij}\right]$ are two $n \times n$ matrices, then the following results are true for determinants of $\mathbf{A}$ and $\mathbf{B}$.

1. If any two adjacent rows (or columns) of $|\mathbf{A}|$ are interchanged, the sign of the resulting determinant is changed.
2. If any two rows (or columns) of $|\mathbf{A}|$ are identical, then $|\mathbf{A}| = 0$.
3. The value of a determinant is not changed if any multiple of a row (or column) is added to any other row (or column) of $|\mathbf{A}|$.
4. $|c\mathbf{A}| = c^n |\mathbf{A}|$, $c \in \mathbb{N}$.
5. $\left|\mathbf{A}^T\right| = |\mathbf{A}|$.
6. $|\mathbf{A}| \, |\mathbf{B}| = |\mathbf{AB}|$.
7. $\left|\mathbf{A}^{-1}\right| = 1/|\mathbf{A}|$.

Transpose, Symmetric, and Skew Matrices

The matrix $\mathbf{A}^T = \left[a_{ji}\right]$ obtained from $\mathbf{A} = \left[a_{ij}\right]$ by changing rows to columns is the transpose of $\mathbf{A}$.

$$\mathbf{A} = \left[a_{ij}\right] \qquad \mathbf{A}^T = \left[a_{ji}\right] \tag{E.38}$$

$$[\mathbf{A} + \mathbf{B} + \cdots + \mathbf{C}]^T = \mathbf{A}^T + \mathbf{B}^T + \cdots + \mathbf{C}^T \tag{E.39}$$

$$[\mathbf{AB} \cdots \mathbf{C}]^T = \mathbf{C}^T \cdots \mathbf{B}^T \, \mathbf{A}^T \tag{E.40}$$

$$[\mathbf{AB}]^T = \left[\sum_n a_{in} b_{nj}\right]^T = \left[\sum_n b_{ni} a_{jn}\right]^T = \mathbf{B}^T \, \mathbf{A}^T \tag{E.41}$$

A matrix $\mathbf{A}$ such that $\mathbf{A}^T = \mathbf{A}$ is symmetric matrix. A matrix $\mathbf{A}$ such that $\mathbf{A}^T = -\mathbf{A}$ is skew, skew symmetric, or alternating matrix.

Inverse

The inverse of a square matrix $\mathbf{A}$ is indicated by $\mathbf{A}^{-1}$,

$$\mathbf{A}^{-1} = \frac{\mathrm{adj} \, \mathbf{A}}{|\mathbf{A}|} \qquad \mathbf{A}^{-1} \, \mathbf{A} = \mathbf{A} \, \mathbf{A}^{-1} = \mathbf{I} \tag{E.42}$$

where $\mathrm{adj} \, \mathbf{A}$ is the adjoint of the matrix $\mathbf{A}$ and C_{ij} are cofactors associated to element a_{ij} of $\mathbf{A}$.

$$\mathrm{adj} \, \mathbf{A} = \left[C_{ij}\right]^T \tag{E.43}$$

$$[\mathbf{AB} \cdots \mathbf{C}]^{-1} = \mathbf{C}^{-1} \cdots \mathbf{B}^{-1} \, \mathbf{A}^{-1} \tag{E.44}$$

$$\left[\mathbf{A}^{-1}\right]^{-1} = \mathbf{A} \tag{E.45}$$

$$\left[\mathbf{A}^{-1}\right]^T = \left[\mathbf{A}^T\right]^{-1} \tag{E.46}$$

$$[k\mathbf{A}]^{-1} = k^{-1}\mathbf{A}^{-1} \tag{E.47}$$

Non-square matrices of order $m \times n$, $m \neq n$, do not have inverse. However, if $\mathbf{A}$ is $m \times n$ and the rank of $\mathbf{A}$ is equal to n, then $\mathbf{A}$ has a left inverse. An $n \times m$ matrix $\mathbf{B}$ such that $\mathbf{BA} = \mathbf{I}$. If $\mathbf{A}$ has rank m, then it has a right inverse. An $n \times m$ matrix $\mathbf{B}$ such that $\mathbf{AB} = \mathbf{I}$.

Sherman–Morrison–Woodbury formula.

$$\left[\mathbf{A} + \mathbf{v}^T \, \mathbf{w}\right]^{-1} = \mathbf{A}^{-1} - \mathbf{A}^{-1}\mathbf{v}\left[\mathbf{I} + \mathbf{v}^T\mathbf{A}^{-1}\mathbf{w}\right]^{-1} \mathbf{v}^T\mathbf{A}^{-1} \tag{E.48}$$

Differentiation and Integration

If $\mathbf{A}(t) = \left[a_{ij}(t)\right]$, and $\mathbf{B}(t) = \left[b_{ij}(t)\right]$ are two $n \times n$ matrices, then

$$\frac{d}{dt}\mathbf{A}(t) = \left[\frac{d}{dt}a_{ij}(t)\right] \tag{E.49}$$

$$\int_{t_0}^{t} \mathbf{A}(s)\ ds = \left[\int_{t_0}^{t} a_{ij}(s)\ ds\right] \tag{E.50}$$

$$\frac{d}{dt}[\mathbf{A}(t) \pm \mathbf{B}(t)] = \frac{d}{dt}\mathbf{A}(t) \pm \frac{d}{dt}\mathbf{B}(t) \tag{E.51}$$

$$\frac{d}{dt}[\mathbf{A}(t)\mathbf{B}(t)] = \left[\frac{d}{dt}\mathbf{A}(t)\right]\mathbf{B}(t) + \mathbf{A}(t)\left[\frac{d}{dt}\mathbf{B}(t)\right] \tag{E.52}$$

$$\frac{d}{dt}[\mathbf{A}(t)\mathbf{B}(t)]^T = \left[\frac{d}{dt}\mathbf{B}(t)\right]^T \mathbf{A}^T(t) + \mathbf{B}^T(t)\left[\frac{d}{dt}\mathbf{A}(t)\right]^T \tag{E.53}$$

If $|\mathbf{A}| \neq 0$, then

$$\frac{d}{dt}\mathbf{A}^{-1}(t) = -\mathbf{A}^{-1}(t)\left[\frac{d}{dt}\mathbf{A}(t)\right]\mathbf{A}^{-1}(t) \tag{E.54}$$

$$\frac{d}{dt}\mathbf{A}^T(t) = -\mathbf{A}^T(t)\left[\frac{d}{dt}\mathbf{A}(t)\right]\mathbf{A}^T(t) \tag{E.55}$$

If $\mathbf{A}(x) = \left[a_{ji}(x)\right]$, then

$$\frac{d|\mathbf{A}|}{dx} = \sum_{i,j=1}^{n} \frac{da_{ij}}{dx}C_{ij} \tag{E.56}$$

where C_{ij} are cofactors associated to element a_{ij} of $\mathbf{A}$.

The Matrix Exponential

If $\mathbf{A}$ is a square matrix, and z is any complex variable, then the matrix exponential $\exp(\mathbf{A}z)$ is defined as:

$$e^{\mathbf{A}z} = \mathbf{I} + \mathbf{A}z + \frac{\mathbf{A}^2 z^2}{2!} + \cdots + \frac{\mathbf{A}^n z^n}{n!} = \sum_{k=0}^{\infty} \frac{\mathbf{A}^k z^k}{k!} \tag{E.57}$$

$$e^0 = \mathbf{I} \qquad e^{\mathbf{I}z} = \mathbf{I}\,e^z \qquad e^{\mathbf{A}(z_1+z_2)} = e^{\mathbf{A}z_1}e^{\mathbf{A}z_2} \tag{E.58}$$

$$e^{-\mathbf{A}z} = \left[e^{\mathbf{A}z}\right]^{-1} \qquad e^{\mathbf{A}z}e^{\mathbf{B}z} = e^{(\mathbf{A}+\mathbf{B})z} \tag{E.59}$$

$$\frac{d^k}{dz^k}e^{\mathbf{A}z} = \mathbf{A}^k\,e^{\mathbf{A}z} = e^{\mathbf{A}z}\,\mathbf{A}^k \tag{E.60}$$

Numeric Values and Functions

$$\pi \simeq 3.14159265359\cdots \simeq \frac{355}{113} \simeq \frac{103993}{33102} \tag{F.1}$$

$$1\,\text{rad} \simeq 57.296\,\text{deg} \simeq 57°\ 17'\ 44.8'' \tag{F.2}$$

Trigonometric Functions

θ is an acute angle of a right-angled triangle, and the hypotenuse h is the long side that connects the two acute angles. The side b adjacent to θ is the side of the triangle that connects θ to the right angle. The third side a is opposite to θ.

$$h = hypotenuse \quad a = opposite \quad b = adjacent \tag{F.3}$$

$$\sin\theta = \frac{a}{h} \qquad \cos\theta = \frac{b}{h} \qquad \tan\theta = \frac{a}{b} = \frac{\sin\theta}{\cos\theta} \tag{F.4}$$

$$\csc\theta = \frac{1}{\sin\theta} = \frac{h}{a} \qquad \sec\theta = \frac{1}{\cos\theta} = \frac{h}{b} \qquad \cot\theta = \frac{b}{a} \tag{F.5}$$

Definitions in Terms of Exponentials

$$\cos z = \frac{e^{iz} + e^{-iz}}{2} = \cosh(iz) \tag{F.6}$$

$$\sin z = \frac{e^{iz} - e^{-iz}}{2i} = -i\sinh(iz) \tag{F.7}$$

$$\tan z = \frac{e^{iz} - e^{-iz}}{i\left(e^{iz} + e^{-iz}\right)} = \frac{1}{i}\tanh z \tag{F.8}$$

$$e^{iz} = \cos z + i\sin z \tag{F.9}$$

$$e^{-iz} = \cos z - i\sin z \tag{F.10}$$

$$(\cos\theta + i\sin\theta)^n = \cos n\theta + i\sin n\theta \tag{F.11}$$

Angle Sum and Difference

$$\sin(\alpha \pm \beta) = \sin\alpha\cos\beta \pm \cos\alpha\sin\beta \tag{F.12}$$

$$\cos(\alpha \pm \beta) = \cos\alpha\cos\beta \mp \sin\alpha\sin\beta \tag{F.13}$$

$$\sinh(\alpha \pm \beta) = \sinh \alpha \cosh \beta \pm \cosh \alpha \sinh \beta \qquad (F.14)$$

$$\cosh(\alpha \pm \beta) = \cosh \alpha \cosh \beta \mp \sinh \alpha \sinh \beta \qquad (F.15)$$

$$\tan(\alpha \pm \beta) = \frac{\tan \alpha \pm \tan \beta}{1 \mp \tan \alpha \tan \beta} \qquad (F.16)$$

$$\cot(\alpha \pm \beta) = \frac{\cot \alpha \cot \beta \mp 1}{\cot \beta \pm \cot \alpha} \qquad (F.17)$$

$$\tanh(\alpha \pm \beta) = \frac{\tanh \alpha \pm \tanh \beta}{1 \mp \tanh \alpha \tanh \beta} \qquad (F.18)$$

Periodicity

$$\sin(\alpha + 2n\pi) = \sin \alpha \qquad (F.19)$$

$$\cos(\alpha + 2n\pi) = \cos \alpha \qquad (F.20)$$

$$\tan(\alpha + 2n\pi) = \tan \alpha \qquad (F.21)$$

Symmetry

$$\sin(-\alpha) = -\sin \alpha \qquad (F.22)$$

$$\cos(-\alpha) = \cos \alpha \qquad (F.23)$$

$$\tan(-\alpha) = -\tan \alpha \qquad (F.24)$$

Displacement

$$\sin(\frac{\pi}{2} - \alpha) = \cos \alpha \qquad (F.25)$$

$$\cos(\frac{\pi}{2} - \alpha) = \sin \alpha \qquad (F.26)$$

$$\tan(\frac{\pi}{2} - \alpha) = \cot \alpha \qquad (F.27)$$

$$\sec(\frac{\pi}{2} - \alpha) = \csc \alpha \qquad (F.28)$$

$$\csc(\frac{\pi}{2} - \alpha) = \sec \alpha \qquad (F.29)$$

$$\cot(\frac{\pi}{2} - \alpha) = \tan \alpha \qquad (F.30)$$

Multiple Angles

$$\sin(2\alpha) = 2 \sin \alpha \cos \alpha = \frac{2 \tan \alpha}{1 + \tan^2 \alpha} \qquad (F.31)$$

$$\cos(2\alpha) = 2 \cos^2 \alpha - 1 = 1 - 2 \sin^2 \alpha = \cos^2 \alpha - \sin^2 \alpha \qquad (F.32)$$

$$\tan(2\alpha) = \frac{2 \tan \alpha}{1 - \tan^2 \alpha} \qquad (F.33)$$

$$\cot(2\alpha) = \frac{\cot^2 \alpha - 1}{2 \cot \alpha} \qquad (F.34)$$

$$\sin(3\alpha) = -4\sin^3\alpha + 3\sin\alpha \tag{F.35}$$

$$\cos(3\alpha) = 4\cos^3\alpha - 3\cos\alpha \tag{F.36}$$

$$\tan(3\alpha) = \frac{-\tan^3\alpha + 3\tan\alpha}{-3\tan^2\alpha + 1} \tag{F.37}$$

$$\sin(4\alpha) = -8\sin^3\alpha\cos\alpha + 4\sin\alpha\cos\alpha \tag{F.38}$$

$$\cos(4\alpha) = 8\cos^4\alpha - 8\cos^2\alpha + 1 \tag{F.39}$$

$$\tan(4\alpha) = \frac{-4\tan^3\alpha + 4\tan\alpha}{\tan^4\alpha - 6\tan^2\alpha + 1} \tag{F.40}$$

$$\sin(5\alpha) = 16\sin^5\alpha - 20\sin^3\alpha + 5\sin\alpha \tag{F.41}$$

$$\cos(5\alpha) = 16\cos^5\alpha - 20\cos^3\alpha + 5\cos\alpha \tag{F.42}$$

$$\sin(n\alpha) = 2\sin((n-1)\alpha)\cos\alpha - \sin((n-2)\alpha) \tag{F.43}$$

$$\cos(n\alpha) = 2\cos((n-1)\alpha)\cos\alpha - \cos((n-2)\alpha) \tag{F.44}$$

$$\tan(n\alpha) = \frac{\tan((n-1)\alpha) + \tan\alpha}{1 - \tan((n-1)\alpha)\tan\alpha} \tag{F.45}$$

Half Angle

$$\cos\frac{\alpha}{2} = \pm\sqrt{\frac{1 + \cos\alpha}{2}} \tag{F.46}$$

$$\sin\frac{\alpha}{2} = \pm\sqrt{\frac{1 - \cos\alpha}{2}} \tag{F.47}$$

$$\tan\frac{\alpha}{2} = \frac{1 - \cos\alpha}{\sin\alpha} = \frac{\sin\alpha}{1 + \cos\alpha} = \pm\sqrt{\frac{1 - \cos\alpha}{1 + \cos\alpha}} \tag{F.48}$$

$$\sin\alpha = \frac{2\tan\frac{\alpha}{2}}{1 + \tan^2\frac{\alpha}{2}} \tag{F.49}$$

$$\cos\alpha = \frac{1 - \tan^2\frac{\alpha}{2}}{1 + \tan^2\frac{\alpha}{2}} \tag{F.50}$$

Powers of Functions

$$\cos^2\alpha + \sin^2\alpha = 1 \tag{F.51}$$

$$\cosh^2\alpha - \sinh^2\alpha = 1 \tag{F.52}$$

$$\sin^2\alpha = \frac{1}{2}(1 - \cos(2\alpha))$$

$$\sin\alpha\cos\alpha - \frac{1}{2}\sin(2\alpha) \tag{F.53}$$

$$\cos^2\alpha = \frac{1}{2}(1 + \cos(2\alpha)) \tag{F.54}$$

$$\sin^3\alpha = \frac{1}{4}(3\sin(\alpha) - \sin(3\alpha)) \tag{F.55}$$

$$\sin^2\alpha\cos\alpha = \frac{1}{4}(\cos\alpha - 3\cos(3\alpha)) \tag{F.56}$$

$$\sin\alpha \cos^2\alpha = \frac{1}{4}\left(\sin\alpha + \sin(3\alpha)\right) \tag{F.57}$$

$$\cos^3\alpha = \frac{1}{4}\left(\cos(3\alpha) + 3\cos\alpha\right) \tag{F.58}$$

$$\sin^4\alpha = \frac{1}{8}\left(3 - 4\cos(2\alpha) + \cos(4\alpha)\right) \tag{F.59}$$

$$\sin^3\alpha \cos\alpha = \frac{1}{8}\left(2\sin(2\alpha) - \sin(4\alpha)\right) \tag{F.60}$$

$$\sin^2\alpha \cos^2\alpha = \frac{1}{8}\left(1 - \cos(4\alpha)\right) \tag{F.61}$$

$$\sin\alpha \cos^3\alpha = \frac{1}{8}\left(2\sin(2\alpha) + \sin(4\alpha)\right) \tag{F.62}$$

$$\cos^4\alpha = \frac{1}{8}\left(3 + 4\cos(2\alpha) + \cos(4\alpha)\right) \tag{F.63}$$

$$\sin^5\alpha = \frac{1}{16}\left(10\sin\alpha - 5\sin(3\alpha) + \sin(5\alpha)\right) \tag{F.64}$$

$$\sin^4\alpha \cos\alpha = \frac{1}{16}\left(2\cos\alpha - 3\cos(3\alpha) + \cos(5\alpha)\right) \tag{F.65}$$

$$\sin^3\alpha \cos^2\alpha = \frac{1}{16}\left(2\sin\alpha + \sin(3\alpha) - \sin(5\alpha)\right) \tag{F.66}$$

$$\sin^2\alpha \cos^3\alpha = \frac{1}{16}\left(2\cos\alpha - 3\cos(3\alpha) - 5\cos(5\alpha)\right) \tag{F.67}$$

$$\sin\alpha \cos^4\alpha = \frac{1}{16}\left(2\sin\alpha + 3\sin(3\alpha) + \sin(5\alpha)\right) \tag{F.68}$$

$$\cos^5\alpha = \frac{1}{16}\left(10\cos\alpha + 5\cos(3\alpha) + \cos(5\alpha)\right) \tag{F.69}$$

$$\tan^2\alpha = \frac{1 - \cos(2\alpha)}{1 + \cos(2\alpha)} \tag{F.70}$$

Products of sin **and** cos

$$\cos\alpha \cos\beta = \frac{1}{2}\cos(\alpha - \beta) + \frac{1}{2}\cos(\alpha + \beta) \tag{F.71}$$

$$\cos n\theta \cos\theta = \frac{1}{2}\cos(n + 1)\theta + \frac{1}{2}\cos(n - 1)\theta \tag{F.72}$$

$$\cos m\theta \cos n\theta = \frac{1}{2}\cos(m + n)\theta + \frac{1}{2}\cos(m - n)\theta \tag{F.73}$$

$$\sin\alpha \sin\beta = \frac{1}{2}\cos(\alpha - \beta) - \frac{1}{2}\cos(\alpha + \beta) \tag{F.74}$$

$$\sin\alpha \cos\beta = \frac{1}{2}\sin(\alpha - \beta) + \frac{1}{2}\sin(\alpha + \beta) \tag{F.75}$$

$$\cos\alpha \sin\beta = \frac{1}{2}\sin(\alpha + \beta) - \frac{1}{2}\sin(\alpha - \beta) \tag{F.76}$$

$$\sin(\alpha + \beta)\sin(\alpha - \beta) = \cos^2\beta - \cos^2\alpha = \sin^2\alpha - \sin^2\beta \tag{F.77}$$

$$\cos(\alpha + \beta)\cos(\alpha - \beta) = \cos^2\beta + \sin^2\alpha \tag{F.78}$$

Sum of Functions

$$\sin\alpha \pm \sin\beta = 2\sin\frac{\alpha\pm\beta}{2}\cos\frac{\alpha\pm\beta}{2} \tag{F.79}$$

$$\cos\alpha + \cos\beta = 2\cos\frac{\alpha+\beta}{2}\cos\frac{\alpha-\beta}{2} \tag{F.80}$$

$$\cos\alpha - \cos\beta = -2\sin\frac{\alpha+\beta}{2}\sin\frac{\alpha-\beta}{2} \tag{F.81}$$

$$\sinh\alpha \pm \sinh\beta = 2\sinh\frac{\alpha\pm\beta}{2}\cosh\frac{\alpha\pm\beta}{2} \tag{F.82}$$

$$\cosh\alpha + \cosh\beta = 2\cosh\frac{\alpha+\beta}{2}\cosh\frac{\alpha-\beta}{2} \tag{F.83}$$

$$\cosh\alpha - \cosh\beta = -2\sinh\frac{\alpha+\beta}{2}\sinh\frac{\alpha-\beta}{2} \tag{F.84}$$

$$\tan\alpha \pm \tan\beta = \frac{\sin(\alpha\pm\beta)}{\cos\alpha\cos\beta} \tag{F.85}$$

$$\cot\alpha \pm \cot\beta = \frac{\sin(\beta\pm\alpha)}{\sin\alpha\sin\beta} \tag{F.86}$$

$$\frac{\sin\alpha + \sin\beta}{\sin\alpha - \sin\beta} = \frac{\tan\frac{\alpha+\beta}{2}}{\tan\frac{\alpha-+\beta}{2}} \tag{F.87}$$

$$\frac{\sin\alpha + \sin\beta}{\cos\alpha - \cos\beta} = \cot\frac{-\alpha+\beta}{2} \tag{F.88}$$

$$\frac{\sin\alpha + \sin\beta}{\cos\alpha + \cos\beta} = \tan\frac{\alpha+\beta}{2} \tag{F.89}$$

$$\frac{\sin\alpha - \sin\beta}{\cos\alpha + \cos\beta} = \tan\frac{\alpha-\beta}{2} \tag{F.90}$$

Trigonometric Relations

$$\sin^2\alpha - \sin^2\beta = \sin(\alpha+\beta)\sin(\alpha-\beta) \tag{F.91}$$

$$= \cos^2\beta - \cos^2\alpha \tag{F.92}$$

$$\cos^2\alpha - \cos^2\beta = -\sin(\alpha+\beta)\sin(\alpha-\beta) \tag{F.93}$$

$$\cos^2\alpha - \sin^2\beta = \cos(\alpha+\beta)\cos(\alpha-\beta) \tag{F.94}$$

$$= \cos^2\beta - \sin^2\alpha \tag{F.95}$$

Inverse of Trigonometric Functions

$$\sin x = y \qquad \cos x = y \qquad \tan x = y \tag{F.96}$$

$$\arcsin y = (-1)^k x + k\pi \tag{F.97}$$

$$\arccos y = \pm x + 2k\pi \tag{F.98}$$

$$\arcsin y = x + k\pi \tag{F.99}$$

$$k = 0, 1, 2, 3, \cdots$$

$$-1 \le y \le 1 \quad -\frac{\pi}{2} \le \arcsin y \le \frac{\pi}{2}$$
$$-1 \le y \le 1 \quad 0 \le \arccos y \le \pi$$
$$-\infty \le y \le \infty \quad -\frac{\pi}{2} \le \arctan y \le \frac{\pi}{2} \tag{F.100}$$
$$-\infty \le y \le \infty \quad 0 \le \operatorname{arccot} y \le \pi$$

$$1 \le y \quad 0 \le \operatorname{arccsc} y \le \frac{\pi}{2}$$
$$y \le -1 \quad -\frac{\pi}{2} \le \operatorname{arccsc} y \le 0$$
$$1 \le y \quad 0 \le \operatorname{arcsec} y \le \frac{\pi}{2} \tag{F.101}$$
$$y \le -1 \quad \frac{\pi}{2} \le \operatorname{arcsec} y \le \pi$$

Derivatives of Trigonometric Functions

$$\frac{d}{d\theta} \sin \theta = \cos \theta \tag{F.102}$$

$$\frac{d}{d\theta} \cos \theta = -\sin \theta \tag{F.103}$$

$$\frac{d}{d\theta} \tan \theta = 1 + \tan^2 \theta = \sec^2 \theta \tag{F.104}$$

$$\frac{d}{d\theta} \csc \theta = -\csc \theta \cot \theta \tag{F.105}$$

$$\frac{d}{d\theta} \sec \theta = \sec \theta \tan \theta \tag{F.106}$$

$$\frac{d}{d\theta} \cot \theta = -\left(1 + \cot^2 \theta\right) = -\csc^2 \theta \tag{F.107}$$

Integrals of Trigonometric Functions

$$\int \sin \theta \, d\theta = -\cos \theta + C \tag{F.108}$$

$$\int \cos \theta \, d\theta = \sin \theta + C \tag{F.109}$$

$$\int \tan \theta \, d\theta = -\ln |\cos \theta| + C \tag{F.110}$$

$$\int \csc \theta \, d\theta = \frac{1}{2} \ln \left(-\frac{\cos \theta - 1}{\cos \theta + 1}\right) + C$$
$$= -\ln |\csc \theta + \cot \theta| + C \tag{F.111}$$

$$\int \sec \theta \, d\theta = \frac{1}{2} \ln \left(\frac{\sin \theta + 1}{\sin \theta - 1}\right) + C$$
$$= \ln |\sec \theta + \tan \theta| + C \tag{F.112}$$

$$\int \cot \theta \, d\theta = \frac{1}{2} \ln \left(2 - 2\cos 2\theta\right) + C = \ln |\sin \theta| + C \tag{F.113}$$

$$\int \cos^2 x \, dx = \frac{1}{2}x - \frac{1}{4}\sin 2x \tag{F.114}$$

$$\int \sin^2 x \, dx = \frac{1}{2}x + \frac{1}{4}\sin 2x \tag{F.115}$$

$$\int \cos mx \cos nx \, dx = \frac{\sin (m-n)x}{2(m-n)}$$

$$+ \frac{\sin (m+n)x}{2(m+n)} \quad m^2 \neq n^2 \tag{F.116}$$

$$\int \sin mx \sin nx \, dx = \frac{\sin (m-n)x}{2(m-n)}$$

$$- \frac{\sin (m+n)x}{2(m+n)} \quad m^2 \neq n^2 \tag{F.117}$$

$$\int \sin mx \cos nx \, dx = -\frac{\cos (m-n)x}{2(m-n)}$$

$$- \frac{\cos (m+n)x}{2(m+n)} \quad m^2 \neq n^2 \tag{F.118}$$

$$\int_{-\pi}^{\pi} \cos mx \cos nx \, dx = \pi \delta_{mn} m^2 \neq n^2 \quad m, n \in \mathbb{N} \tag{F.119}$$

$$\int_{-\pi}^{\pi} \sin mx \sin nx \, dx = \pi \delta_{mn} m^2 \neq n^2 \quad m, n \in \mathbb{N} \tag{F.120}$$

$$\int_{-\pi}^{\pi} \sin mx \cos nx \, dx = 0 \quad m^2 \neq n^2 \quad m, n \in \mathbb{N} \tag{F.121}$$

$$\int x \cos x \, dx = \cos x + x \sin x \tag{F.122}$$

$$\int x \sin x \, dx = \sin x - x \cos x \tag{F.123}$$

$$\int x^2 \cos x \, dx = 2x \cos x + \left(x^2 - 2\right) \sin x \tag{F.124}$$

$$\int x^2 \sin x \, dx = 2x \sin x - \left(x^2 - 2\right) \cos x \tag{F.125}$$

$$\int \cos mx \cos nx \, dx = \frac{\sin (m-n)x}{2(m-n)}$$

$$+ \frac{\sin (m+n)x}{2(m+n)} \quad m^2 \neq n^2 \tag{F.126}$$

$$\int \sin mx \sin nx \, dx = \frac{\sin (m-n)x}{2(m-n)}$$

$$- \frac{\sin (m+n)x}{2(m+n)} \quad m^2 \neq n^2 \tag{F.127}$$

$$\int \sin mx \cos nx \, dx = -\frac{\cos (m-n)x}{2(m-n)}$$

$$- \frac{\cos (m+n)x}{2(m+n)} \quad m^2 \neq n^2 \tag{F.128}$$

$$\int_{-\pi}^{\pi} \cos mx \cos nx \, dx = \pi \delta_{mn} \quad m^2 \neq n^2 \quad m, n \in \mathbb{N} \tag{F.129}$$

$$\int_{-\pi}^{\pi} \sin mx \sin nx \, dx = \pi \delta_{mn} \qquad m^2 \neq n^2 \qquad m, n \in \mathbb{N} \tag{F.130}$$

$$\int_{-\pi}^{\pi} \sin mx \cos nx \, dx = 0 \qquad m^2 \neq n^2 \qquad m, n \in \mathbb{N} \tag{F.131}$$

$$\int x \cos x \, dx = \cos x + x \sin x \tag{F.132}$$

$$\int x \sin x \, dx = \sin x - x \cos x \tag{F.133}$$

$$\int x^2 \cos x \, dx = 2x \cos x + \left(x^2 - 2\right) \sin x \tag{F.134}$$

$$\int x^2 \sin x \, dx = 2x \sin x - \left(x^2 - 2\right) \cos x \tag{F.135}$$

$$(a \pm b)^2 = a^2 \pm 2ab + b^2 \tag{G.1}$$

$$(a \pm b)^3 = a^3 \pm 3a^2b + 3ab^2 \pm b^2 \tag{G.2}$$

$$(a + b)^n = \sum_{k=0}^{n} \binom{n}{k} a^{n-k} b^k \tag{G.3}$$

$$\binom{n}{k} = \frac{n!}{k! \, (n-k)!} \tag{G.4}$$

$$(a + b + c)^2 = a^2 + b^2 + c^2 + 2(ab + bc + ca) \tag{G.5}$$

$$(a + b + c)^3 = a^3 + b^3 + c^3 + 3a^2(b + c)$$
$$+ 3b^2(c + a) + 3c^2(a + c) + 6abc \tag{G.6}$$

$$a^2 - b^2 = (a - b)(a + b) \tag{G.7}$$

$$a^2 + b^2 = (a - ib)(a + ib) \tag{G.8}$$

$$a^3 - b^3 = (a - b)(a^2 + ab + b^2) \tag{G.9}$$

$$a^2 + b^2 = (a + b)(a^2 - ab + b^2) \tag{G.10}$$

General Conversion Formulas

$$N^a m^b s^c \approx 4.448^a \times 0.3048^b \times lb^a ft^b s^c$$

$$\approx 4.448^a \times 0.0254^b \times lb^a in^b s^c$$

$$lb^a ft^b s^c \approx 0.2248^a \times 3.2808^b \times N^a m^b s^c$$

$$lb^a in^b s^c \approx 0.2248^a \times 39.37^b \times N^a m^b s^c$$

Conversion Factors

Acceleration

$$1 \, ft/s^2 \approx 0.3048 \, m/s^2 \quad 1 m/s^2 \approx 3.2808 \, ft/s^2$$

Angle

$$1 \, deg \approx 0.01745 \, rad \quad 1 \, rad \approx 57.307 \, deg$$

Area

$$1 \, in^2 \approx 6.4516 \, cm^2 \quad 1 \, cm^2 \approx 0.155 \, in^2$$
$$1 \, ft^2 \approx 0.09290304 \, m^2 \quad 1 \, m^2 \approx 10.764 \, ft^2$$
$$1 \, acre \approx 4046.86 \, m^2 \quad 1 \, m^2 \approx 2.471 \times 10^{-4} \, acre$$
$$1 \, acre \approx 0.4047 \, hectare \quad 1 \, hectare \approx 2.471 \, acre$$

Damping

$$1Ns/m \approx 6.85218 \times 10^{-2} \, lbs/ft \quad 1 \, lbs/ft \approx 14.594Ns/m$$
$$1Ns/m \approx 5.71015 \times 10^{-3} \, lbs/in \quad 1 \, lbs/in \approx 175.13Ns/m$$

Energy and Heat

$$1 \, Btu \approx 1055.056 \, J \quad 1 \, J \approx 9.4782 \times 10^{-4} \, Btu$$
$$1 \, cal \approx 4.1868 \, J \quad 1 \, J \approx 0.23885 \, cal$$
$$1 \, kWh \approx 3600 \, kJ \quad 1 \, MJ \approx 0.27778 \, kWh$$
$$1 \, ftlbf \approx 1.355818 \, J \quad 1 \, J \approx 0.737562 \, ftlbf$$

Force

$$1 \, lb \approx 4.448222N \quad 1N \approx 0.22481 \, lb$$

Fuel Consumption

$$1 \, l/100 \, km \approx 235.214583 \, mi/gal \quad 1 \, mi/gal \approx 235.214583 \, l/100 \, km$$
$$1 \, l/100 \, km = 100 \, km/l \quad 1 \, km/l = 100 \, l/100 \, km$$
$$1 \, mi/gal \approx 0.425144 \, km/l \quad 1 \, km/l \approx 2.352146 \, mi/gal$$

R. N. Jazar, *Theory of Applied Robotics*, https://doi.org/10.1007/978-3-030-93220-6

Length

$$1\,\text{in} \approx 25.4\,\text{mm} \qquad 1\,\text{cm} \approx 0.3937\,\text{in}$$
$$1\,\text{ft} \approx 30.48\,\text{cm} \qquad 1\,\text{m} \approx 3.28084\,\text{ft}$$
$$1\,\text{mi} \approx 1.609347\,\text{km} \qquad 1\,\text{km} \approx 0.62137\,\text{mi}$$

Mass

$$1\,\text{lb} \approx 0.45359\,\text{kg} \qquad 1\,\text{kg} \approx 2.204623\,\text{lb}$$
$$1\,\text{slug} \approx 14.5939\,\text{kg} \qquad 1\,\text{kg} \approx 0.068522\,\text{slug}$$
$$1\,\text{slug} \approx 32.174\,\text{lb} \qquad 1\,\text{lb} \approx 0.03.1081\,\text{slug}$$

Moment and Torque

$$1\,\text{lbft} \approx 1.35582\,\text{Nm} \qquad 1\,\text{Nm} \approx 0.73746\,\text{lbft}$$
$$1\,\text{lbin} \approx 8.85075\,\text{Nm} \qquad 1\,\text{Nm} \approx 0.11298\,\text{lbin}$$

Mass Moment

$$1\,\text{lbft}^2 \approx 0.04214\,\text{kg}\,\text{m}^2 \qquad 1\,\text{kg}\,\text{m}^2 \approx 23.73\,\text{lbft}^2$$

Power

$$1\,\text{Btu/h} \approx 0.2930711\,\text{W} \qquad 1\,\text{W} \approx 3.4121\,\text{Btu/h}$$
$$1\,\text{hp} \approx 745.6999\,\text{W} \qquad 1\,\text{kW} \approx 1.341\,\text{hp}$$
$$1\,\text{hp} \approx 550\,\text{lb}\,\text{ft/s} \qquad 1\,\text{lb}\,\text{ft/s} \approx 1.8182 \times 10^{-3}\,\text{hp}$$
$$1\,\text{lb}\,\text{ft/h} \approx 3.76616 \times 10^{-4}\,\text{W} \qquad 1\,\text{W} \approx 2655.2\,\text{lb}\,\text{ft/h}$$
$$1\,\text{lb}\,\text{ft/min} \approx 2.2597 \times 10^{-2}\,\text{W} \qquad 1\,\text{W} \approx 44.254\,\text{lb}\,\text{ft/min}$$

Pressure and Stress

$$1\,\text{lb/in}^2 \approx 6894.757\,\text{Pa} \qquad 1\,\text{MPa} \approx 145.04\,\text{lb/in}^2$$
$$1\,\text{lb/ft}^2 \approx 47.88\,\text{Pa} \qquad 1\,\text{Pa} \approx 2.0886 \times 10^{-2}\,\text{lb/ft}^2$$
$$1\,\text{Pa} \approx 0.00001\,\text{atm} \qquad 1\,\text{atm} \approx 101325\,\text{Pa}$$

Stiffness

$$1\,\text{N/m} \approx 6.85218 \times 10^{-2}\,\text{lb/ft} \qquad 1\,\text{lb/ft} \approx 14.594\,\text{N/m}$$
$$1\,\text{N/m} \approx 5.71015 \times 10^{-3}\,\text{lb/in} \qquad 1\,\text{lb/in} \approx 175.13\,\text{N/m}$$

Temperature

$$^\circ\text{C} = (^\circ\text{F} - 32)/1.8$$
$$^\circ\text{F} = 1.8\,^\circ\text{C} + 32$$

Velocity

$$1\,\text{mi/h} \approx 1.60934\,\text{km/h} \qquad 1\,\text{km/h} \approx 0.62137\,\text{mi/h}$$
$$1\,\text{mi/h} \approx 0.44704\,\text{m/s} \qquad 1\,\text{m/s} \approx 2.2369\,\text{mi/h}$$

$$1\,\text{ft/s} \approx 0.3048\,\text{m/s} \qquad 1\,\text{m/s} \approx 3.2808\,\text{ft/s}$$
$$1\,\text{ft/min} \approx 5.08 \times 10^{-3}\,\text{m/s} \qquad 1\,\text{m/s} \approx 196.85\,\text{ft/min}$$

Volume

$$1\,\text{in}^3 \approx 16.39\,\text{cm}^3 \qquad 1\,\text{cm}^3 \approx 0.0061013\,\text{in}^3$$
$$1\,\text{ft}^3 \approx 0.02831685\,\text{m}^3 \qquad 1\,\text{m}^3 \approx 35.315\,\text{ft}^3$$
$$1\,\text{gal} \approx 3.785\,\text{l} \qquad 1\,\text{l} \approx 0.2642\,\text{gal}$$
$$1\,\text{gal} \approx 3785.41\,\text{cm}^3 \qquad 1\,\text{l} \approx 1000\,\text{cm}^3$$

Bibliography

Front Matter

Jazar, R. N. (2011). *Advanced dynamics: Rigid body, multibody, and aero-space applications*. New York: Wiley.

Harithuddin, S., Trivailo M., & Jazar, R. N. (2015). On the Razi Acceleration. In: L. Dai, & R. Jazar (Eds.), *Nonlinear approaches in engineering applications*. New York: Springer.

Chapter 1: Introduction

Asimov, I. (1942). *Runaround, Astounding Science-Fiction*. New York: Street & Smith Publications, Inc.

Asimov, I. (1950a). *I, Robot*. New York: Gnome Press.

Asimov, I. (1950b). *Robots and Empire*. New York: Doubleday.

Aspragathos, N. A., & Dimitros, J. K. (1998). A comparative study of three methods for robot kinematics. *IEEE Transaction on Systems, Man and Cybernetic-PART B: CYBERNETICS, 28*(2), 115–145.

Čapek, K. (1994). *Tales from two pockets*, translated from the Czech and with an introduction by Norma Comrada. North Haven, CT: Catbird Press.

Chernousko, F. L., Bolotnik, N. N., & Gradetsky, V. G. (1994). *Manipulation Robots: Dynamics, Control, and Optimization*. Boca Raton, Florida: CRC Press.

Denavit, J., & Hartenberg, R. S. (1955). A kinematic notation for lower-pair mechanisms based on matrices. *Journal of Applied Mechanics, 22*(2), 215–221.

Dugas, R. (1995). *A History of Mechanics* (English translation), Switzerland, Editions du Griffon. New York: Central Book Co.

Erdman, A. G. (1993). *Modern Kinematics: Developed in the Last Forty Years*. New York: Wiley.

Fahimi, F. (2009). *Autonomous Robots: Modeling, Path Planning, and Control*. New York: Springer.

Fosdick, R. & Fried, E. (2016). *The Mechanics of Ribbons and Möbius Bands*. New York: Springer.

Harithuddin, S., Trivailo M., & Jazar, R. N. (2015). On the Razi Acceleration. In: L. Dai, & R. Jazar (Eds.), *Nonlinear Approaches in Engineering Applications*. New York: Springer.

Hunt, K. H. (1978). *Kinematic Geometry of Mechanisms*. London: Oxford University Press.

Husbands, P., Holland, O., & Wheeler M. (Eds.) (2008). *The Mechanical Mind in History*. Cambridge: The MIT Press.

Jazar, R. N. (2011). *Advanced Dynamics: Rigid Body, Multibody, and Aero-space Applications*. New York: Wiley.

Jazar, R. N. (2012). Derivative and coordinate frames. *Journal of Nonlinear Engineering, 1*(1), 25–34.

Jazar, R. N. (2017). *Vehicle Dynamics: Theory and Application* (3rd ed.). New York: Springer.

Jazar, R. N. (2019). *Advanced Vehicle Dynamics*. New York: Springer.

Jazar, R. N. (2020). *Approximation Methods in Science and Engineering*. New York: Springer.

Jazar, R. N. (2021). *Perturbation Methods in Science and Engineering*. New York: Springer.

Milne, E. A. (1948). *Vectorial Mechanics*. London: Methuen & Co. LTD.

Niku, S. B. (2020). *Introduction to Robotics: Analysis, Systems, Applications*. New York: Wiley.

Rosheim, M. E. (1994). *Robot Evolution: The Development of Anthrobotics*. New York: Wiley.

Shahinpoor, M. (1987). *A Robot Engineering Textbook*. New York: Harper and Row Publishers.

Tsai, L. W. (1999). *Robot Analysis*. New York: Wiley.

Veit, S. (1992). Whatever happened to … personal robots?. *The Computer Shopper, 12*(11), 794–795.

Chapter 2: Rotation Kinematics

Buss, S. R. (2003). *3-D Computer Graphics: A Mathematical Introduction with OpenGL*. New York: Cambridge University.

Cheng, H., & Gupta, K. C. (1989). A historical note on finite rotations. *Journal of Applied Mechanics, 56*, 139–145.

Coe, C. J. (1934). Displacement of a rigid body. *American Mathematical Monthly, 41*(4), 242–253.

Denavit, J., & Hartenberg, R. S. (1955). A kinematic notation for lower-pair mechanisms based on matrices. *Journal of Applied Mechanics, 22*(2), 215–221.

© The Author(s), under exclusive license to Springer Nature Switzerland AG 2022
R. N. Jazar, *Theory of Applied Robotics*, https://doi.org/10.1007/978-3-030-93220-6

Hunt, K. H. (1978). *Kinematic Geometry of Mechanisms*. London: Oxford University.
Jazar, R. N. (2011). *Advanced Dynamics: Rigid Body, Multibody, and Aerospace Applications*. New York: Wiley.
Jazar, R. N. (2019). *Advanced Vehicle Dynamics*. New York: Springer.
Jazar, R. N. (2020). *Approximation Methods in Science and Engineering*. New York: Springer.
Jazar, R. N. (2021). *Perturbation Methods in Science and Engineering*. New York: Springer.
Mason, M. T. (2001). *Mechanics of Robotic Manipulation*. Cambridge, MA: MIT Press.
Murray, R. M., Li, Z., & Sastry, S. S. S. (1994). *A Mathematical Introduction to Robotic Manipulation*. Boca Raton: CRC Press.
Nikravesh, P. (1988). *Computer-Aided Analysis of Mechanical Systems*. New Jersey: Prentice Hall.
Niku, S. B. (2020). *Introduction to Robotics: Analysis, Systems, Applications*. New York: Wiley.
Paul, B. (1963). On the composition of finite rotations. *American Mathematical Monthly, 70*(8), 859–862.
Paul, R. P. (1981). *Robot Manipulators: Mathematics, Programming, and Control*. Cambridge, MA: MIT Press.
Rimrott, F. P. J. (1989). *Introductory Attitude Dynamics*. New York: Springer.
Rosenberg, R. M. (1977). *Analytical Dynamics of Discrete Systems*. New York: Plenum Publishing Co.
Schaub, H., & Junkins, J. L. (2003). *Analytical Mechanics of Space Systems*. Reston, Virginia: AIAA Educational Series, American Institute of Aeronautics and Astronautics, Inc.
Suh, C. H., & Radcliff, C. W. (1978). *Kinematics and Mechanisms Design*. New York: Wiley.
Spong, M. W., Hutchinson, S., & Vidyasagar, M. (2020). *Robot Modeling and Control*. New York: Wiley.
Tsai, L. W. (1999). *Robot Analysis*. New York: Wiley.

Chapter 3: Orientation Kinematics

Buss, S. R. (2003). *3-D Computer Graphics: A Mathematical Introduction with OpenGL*. New York: Cambridge University Press.
Denavit, J., & Hartenberg, R. S. (1955). A kinematic notation for lower-pair mechanisms based on matrices. *Journal of Applied Mechanics, 22*(2), 215–221.
Hunt, K. H. (1978). *Kinematic Geometry of Mechanisms*. London U.K: Oxford University Press.
Jazar, R. N. (2011). *Advanced Dynamics: Rigid Body, Multibody, and Aerospace Applications*. New York: Wiley.
Jazar, R. N. (2019). *Advanced Vehicle Dynamics*. New York: Springer.
Jazar, R. N. (2020). *Approximation Methods in Science and Engineering*. New York: Springer.
Jazar, R. N. (2021). *Perturbation Methods in Science and Engineering*. New York: Springer.
Mason, M. T. (2001). *Mechanics of Robotic Manipulation*. Cambridge, MA: MIT Press.
Murray, R. M., Li, Z., & Sastry, S. S. S. (1994). *A Mathematical Introduction to Robotic Manipulation*. Boca Raton, Florida: CRC Press.
Nikravesh, P. (1988). *Computer-Aided Analysis of Mechanical Systems*. New Jersey: Prentice Hall.
Paul, B. (1963). On the composition of finite rotations. *American Mathematical Monthly, 70*(8), 859–862.
Paul, R. P. (1981). *Robot Manipulators: Mathematics, Programming, and Control*. Cambridge, MA: MIT Press.
Rimrott, F. P. J. (1989). *Introductory Attitude Dynamics*. New York: Springer.
Stanley, W. S. (1978). Quaternion from Rotation Matrix. *AIAA Journal of Guidance and Control, I*(3), 223–224.
Wittenburg, J. (2016). *Kinematics: Theory and Applications*. Berlin Heidelberg: Springer.
Rosenberg, R. M. (1977). *Analytical Dynamics of Discrete Systems*. New York: Plenum Publishing Co.
Schaub, H., & Junkins, J. L. (2003). *Analytical Mechanics of Space Systems*. Reston, Virginia: AIAA Educational Series, American Institute of Aeronautics and Astronautics, Inc.
Spong, M. W., Hutchinson, S., & Vidyasagar, M. (2020). *Robot Modeling and Control*. New York: Wiley.
Suh, C. H., & Radcliff, C. W. (1978). *Kinematics and Mechanisms Design*. New York: Wiley.
Tsai, L. W. (1999). *Robot Analysis*. New York: Wiley.
Wittenburg, J., & Lilov, L. (2003). Decomposition of a finite rotation into three rotations about given axes. *Multibody System Dynamics, 9*, 353–375.

Chapter 4: Motion Kinematics

Ball, R. S. (1900). *A Treatise on the Theory of Screws*. New York: Cambridge University Press.
Bottema, O., & Roth, B. (1979). *Theoretical Kinematics*. The Netherlands: North-Holland Publication, Amsterdam.
Chasles, M. (1830). Notes on the general properties of a system of 2 identical bodies randomly located in space; and on the finite or infinitesimal motion of a free solid body. *Bulletin des Sciences Mathematiques, Astronomiques, Physiques et Chimiques, 14*, 321–326.
Chernousko, F. L., Bolotnik, N. N., & Gradetsky, V. G. (1994). *Manipulation Robots: Dynamics, Control, and Optimization*. Boca Raton, Florida: CRC press.
Davidson, J. K., & Hunt, K. H. (2004). *Robots and Screw Theory: Applications of Kinematics and Statics to Robotics*. New York: Oxford University Press.
Denavit, J., & Hartenberg, R. S. (1955). A kinematic notation for lower-pair mechanisms based on matrices. *Journal of Applied Mechanics,22*(2), 215–221.
Hunt, K. H. (1978). *Kinematic Geometry of Mechanisms*. London: Oxford University Press.
Jazar, R. N. (2011). *Advanced Dynamics: Rigid Body, Multibody, and Aerospace Applications*. New York: Wiley.
Jazar, R. N. (2013). *Advanced Vibrations: A Modern Approach*. New York: Springer.
Jazar, R. N. (2017). *Vehicle Dynamics*. New York: Springer.
Jazar, R. N. (2019). *Advanced Vehicle Dynamics*. New York: Springer.

Jazar, R. N. (2020). *Approximation Methods in Science and Engineering*. New York: Springer.

Jazar, R. N. (2021). *Perturbation Methods in Science and Engineering*. New York: Springer.

Mason, M. T. (2001). *Mechanics of Robotic Manipulation*. Cambridge, Massachusetts: MIT Press.

Mozzi, G. (1763). Discorso matematico sopra il rotamento momentaneo dei corpi. In *Stamperia di Donato Campo, Napoli*.

Murray, R. M., Li, Z., & Sastry, S. S. S. (1994). *A Mathematical Introduction to Robotic Manipulation*. Boca Raton, Florida: CRC Press.

Niku, S. B. (2020). *Introduction to Robotics: Analysis, Systems, Applications*. New York: Wiley.

Plücker, J. (1866). Fundamental views regarding mechanics. *Philosophical Transactions, 156*, 361–380.

Schaub, H., & Junkins, J. L. (2003). *Analytical Mechanics of Space Systems*. Reston, Virginia: AIAA Educational Series, American Institute of Aeronautics and Astronautics, Inc.

Schilling, R. J. (1990). *Fundamentals of Robotics: Analysis and Control*. New Jersey: Prentice Hall.

Selig, J. M. (2005). *Geometric Fundamentals of Robotics* (2nd ed.). New York: Springer.

Sourin, A. (2021). *Making Images with Mathematics*. Cham, Switzerland: Springer.

Spong, M. W., Hutchinson, S., & Vidyasagar, M. (2020). *Robot Modeling and Control*. New York: Wiley.

Suh, C. H., & Radcliff, C. W. (1978). *Kinematics and Mechanisms Design*. New York: Wiley.

Chapter 5: Forward Kinematics

Asada, H., & Slotine, J. J. E. (1986). *Robot Analysis and Control*. New York: Wiley.

Ball, R. S. (1900). *A Treatise on the Theory of Screws*. USA: Cambridge University Press.

Bernhardt, R., & Albright, S. L. (2001). *Robot Calibration*. New York: Springer.

Bottema, O., & Roth, B. (1979). *Theoretical Kinematics*. The Netherlands: North-Holland Publication, Amsterdam.

Davidson, J. K., & Hunt, K. H. (2004). *Robots and Screw Theory: Applications of Kinematics and Statics to Robotics*. New York: Oxford University Press.

Denavit, J., & Hartenberg, R. S. (1955). A kinematic notation for lower-pair mechanisms based on matrices. *Journal of Applied Mechanics, 22*(2), 215–221.

Fahimi, F. (2009). *Autonomous Robots: Modeling, Path Planning, and Control*. New York: Springer.

Hunt, K. H. (1978). *Kinematic Geometry of Mechanisms*. London: Oxford University Press.

Jazar, R. N. (2011). *Advanced Dynamics: Rigid Body, Multibody, and Aerospace Applications*. New York: Wiley.

Jazar, R. N. (2013). *Advanced Vibrations: A Modern Approach*. New York: Springer.

Jazar, R. N. (2017). *Vehicle Dynamics*. New York: Springer.

Jazar, R. N. (2019). *Advanced Vehicle Dynamics*. New York: Springer.

Jazar, R. N. (2020). *Approximation Methods in Science and Engineering*. New York: Springer.

Jazar, R. N. (2021). *Perturbation Methods in Science and Engineering*. New York: Springer.

Mason, M. T. (2001). *Mechanics of Robotic Manipulation*. Cambridge, MA: MIT Press.

Paul, R. P. (1981). *Robot Manipulators: Mathematics, Programming, and Control*. Cambridge, MA: MIT Press.

Pieper, D. (1968). *The Kinematics of Manipulators under Computer Control*, Ph.D. thesis. Stanford: Stanford University.

Schilling, R. J. (1990). *Fundamentals of Robotics: Analysis and Control*. New Jersey: Prentice Hall.

Schroer, K., Albright, S. L., & Grethlein, M. (1997). Complete, minimal and model-continuous kinematic models for robot calibration. *Robotics and Computer-Integrated Manufacturing, 13*(1), 73–85.

Spong, M. W., Hutchinson, S., & Vidyasagar, M. (2020). *Robot Modeling and Control*. New York: Wiley.

Suh, C. H., & Radcliff, C. W. (1978). *Kinematics and Mechanisms Design*. New York: Wiley.

Tsai, L. W. (1999). *Robot Analysis*. New York: Wiley.

Wang, K., & Lien, T. (1988). Structure, design and kinematics of robot manipulators. *Robotica, 6*, 299–306.

Zhuang, H., Roth, Z. S., & Hamano, F. (1992). A complete, minimal and model-continuous kinematic model for robot manipulators. *IEEE Transactions on Robotics and Automation, 8*(4), 451–463.

Chapter 6: Inverse Kinematics

Asada, H., & Slotine, J. J. E. (1986). *Robot Analysis and Control*. New York: Wiley.

Fahimi, F. (2009). *Autonomous Robots: Modeling, Path Planning, and Control*. New York: Springer.

Jazar, R. N. (2011). *Advanced Dynamics: Rigid Body, Multibody, and Aerospace Applications*. New York: Wiley.

Jazar, R. N. (2013). *Advanced Vibrations: A Modern Approach*. New York: Springer.

Jazar, R. N. (2017). *Vehicle Dynamics*. New York: Springer.

Jazar, R. N. (2019). *Advanced Vehicle Dynamics*. New York: Springer.

Jazar, R. N. (2020). *Approximation Methods in Science and Engineering*. New York: Springer.

Jazar, R. N. (2021). *Perturbation Methods in Science and Engineering*. New York: Springer.

Paul, R. P. (1981). *Robot Manipulators: Mathematics, Programming, and Control*. Cambridge, Massachusetts: MIT Press.

Spong, M. W., Hutchinson, S., & Vidyasagar, M. (2020). *Robot Modeling and Control*. New York: Wiley.

Tsai, L. W. (1999). *Robot Analysis*. New York: Wiley.

Wang, K., & Lien, T. (1988). Structure, design and kinematics of robot manipulators. *Robotica, 6*, 299–306.

Chapter 7: Angular Velocity

Bottema, O., & Roth, B. (1979). *Theoretical Kinematics*. Amsterdam, The Netherlands: North-Holland Publications.
Geradin, M., & Cardonna, A. (1987). Kinematics and dynamics of rigid and flexible mechanisms using finite elements and quaternion algebra. *Computational Mechanics, 4*(2), 115–135.
Hunt, K. H. (1978). *Kinematic Geometry of Mechanisms*. London: Oxford University Press.
Jazar, R. N. (2011). *Advanced Dynamics: Rigid Body, Multibody, and Aero-space Applications*. New York: Wiley.
Jazar, R. N. (2019). *Advanced Vehicle Dynamics*. New York: Springer.
Jazar, R. N. (2020). *Approximation Methods in Science and Engineering*. New York: Springer.
Jazar, R. N. (2021). *Perturbation Methods in Science and Engineering*. New York: Springer.
Mason, M. T. (2001). *Mechanics of Robotic Manipulation*. Cambridge, MA: MIT Press.
Schaub, H., & Junkins, J. L. (2003). *Analytical Mechanics of Space Systems*. Reston, Virginia: AIAA Educational Series, American Institute of Aeronautics and Astronautics, Inc.
Spong, M. W., Hutchinson, S., & Vidyasagar, M. (2020). *Robot Modeling and Control*. New York: Wiley.
Suh, C. H., & Radcliff, C. W. (1978). *Kinematics and Mechanisms Design*. New York: Wiley.
Tsai, L. W. (1999). *Robot Analysis*. New York: Wiley.

Chapter 8: Velocity Kinematics

Carnahan, B., Luther, H. A., & Wilkes, J. O. (1969). *Applied numerical methods*. New York: Wiley.
Eich-Soellner, E., & Führer, C. (1998). *Numerical methods in multibody dynamics*. B.G. Teubner Stuttgart.
Gerald, C. F., & Wheatley, P. O. (1999). *Applied numerical analysis* (6th ed.). New York: Addison Wesley.
Hunt, K. H. (1978). *Kinematic geometry of mechanisms*. Oxford University Press.
Jazar, Reza N. (2011). *Advanced dynamics: Rigid body, multibody, and aero-space applications*. New York: Wiley.
Kane, T. R., Likins, P. W., & Levinson, D. A. (1983). *Spacecraft dynamics*. New York: McGraw-Hill.
Kane, T. R., & Levinson, D. A. (1980), *Dynamics: Theory and applications*. New York: McGraw-Hill.
Mason, M. T. (2001). *Mechanics of robotic manipulation*. Cambridge, MA: MIT Press.
Nikravesh, P. (1988). *Computer-aided analysis of mechanical systems*. New Jersey: Prentice Hall.
Rimrott, F. P. J. (1989). *Introductory attitude dynamics*. New York: Springer.
Schilling, R. J. (1990). *Fundamentals of robotics: Analysis and control*. New Jersey: Prentice-Hall.
Spong, M. W., Hutchinson, S., & Vidyasagar, M. (2020). *Robot modeling and control*. New York: Wiley.
Suh, C. H., & Radcliff, C. W. (1978), *Kinematics and mechanisms design*. New York: Wiley.
Talman, R. (2000). *Geometric mechanics*. New York: Wiley.
Tsai, L. W. (1999). *Robot analysis*. New York: Wiley.

Chapter 9: Acceleration Kinematics

Jazar, Reza N. (2011). *Advanced dynamics: Rigid body, multibody, and aero-space applications*. New York: Wiley.
Mason, M. T. (2001). *Mechanics of robotic manipulation*. Cambridge, MA: MIT Press.
Nikravesh, P. (1988). *Computer-aided analysis of mechanical systems*. New Jersey: Prentice Hall.
Rimrott, F. P. J. (1989). *Introductory attitude dynamics*. New York: Springer.
Spong, M. W., Hutchinson, S., & Vidyasagar, M. (2020). *Robot modeling and control*. New York: Wiley.
Suh, C. H., & Radcliff, C. W., (1978). *Kinematics and Mechanisms Design*, New York: John Wiley & Sons.
Tsai, L. W. (1999). *Robot analysis*. New York: Wiley.

Chapter 10: Applied Dynamics

Goldstein, H., Poole, C., & Safko, J. (2002). *Classical mechanics* (3rd ed.). New York: Addison Wesley.
Jazar, R. N. (2011). *Advanced dynamics: Rigid body, multibody, and aero-space applications*. New York: Wiley.
Jazar, R. N. (2013). *Advanced vibrations: A modern approach*. New York: Springer.
MacMillan, W. D. (1936). *Dynamics of rigid bodies*. New York: McGraw-Hill.
Meirovitch, L. (1970). *Methods of analytical dynamics*. New York: McGraw-Hill.
Nikravesh, P. (1988). *Computer-aided analysis of mechanical systems*. New Jersey: Prentice Hall.
Rimrott, F. P. J. (1989). *Introductory attitude dynamics*. New York: Springer.
Rosenberg, R. M. (1977). *Analytical dynamics of discrete systems*. New York: Plenum Publishing.
Schaub, H., & Junkins, J. L. (2003). *Analytical mechanics of space systems*. Reston, VA: AIAA Educational Series, American Institute of Aeronautics and Astronautics.
Thomson, W. T. (1961). *Introduction to space dynamics*. New York: Wiley.
Tsai, L. W. (1999). *Robot analysis*. New York: Wiley.
Wittacker, E. T. (1947). *A treatise on the analytical dynamics of particles and rigid bodies* (4th ed.). New York: Cambridge University Press.

Chapter 11: Robot Dynamics

Brady, M., Hollerbach, J. M., Johnson, T. L., Lozano-Prez, T., & Mason, M. T. (1983). *Robot motion: Planning and control*. Cambridge, MA: MIT Press.
Jazar, Reza N. (2011). *Advanced dynamics: Rigid body, multibody, and aero-space applications*. New York: Wiley.
Jazar, Reza N. (2013). *Advanced vibrations: A modern approach*. New York: Springer.
Kelly, R., Santibáñez, V., & Loría A., (2005). *Control of robot manipulators in joint space*. Leipzig, Germany: Springer.
Murray, R. M., Li, Z., & Sastry, S. S. S. (1994). *A mathematical introduction to robotic manipulation*. Boca Raton, FL: CRC Press.
Nikravesh, P. (1988). *Computer-aided analysis of mechanical systems*. New Jersey: Prentice Hall.
Niku, S. B. (2020). *Introduction to robotics: Analysis, systems, applications*. New York: Wiley.
Paul, R. P. (1981). *Robot manipulators: Mathematics, programming, and control*. Cambridge, MA: MIT Press.
Spong, M. W., Hutchinson, S., & Vidyasagar, M. (2020). *Robot modeling and control*. New York: Wiley.
Suh, C. H., & Radcliff, C. W. (1978). *Kinematics and mechanisms design*. New York: Wiley.
Tsai, L. W. (1999). *Robot analysis*. New York: Wiley.

Chapter 12: Path Planning

Asada, H., & Slotine, J. J. E. (1986). *Robot analysis and control*. New York: Wiley.
Biagiotti, L., & Melchiorri, C. (2008). *Trajectory planning for automatic machines and robots*. Berlin Heidelberg : Springer.
Fahimi, F. (2009), *Autonomous robots: Modeling, path planning, and control*. New York: Springer.
Murray, R. M., Li, Z., & Sastry, S. S. S. (1994). *A mathematical introduction to robotic manipulation*. Boca Raton, FL: CRC Press.
Niku, S. B. (2020). *Introduction to robotics: Analysis, systems, applications*. New York: Wiley.
Spong, M. W., Hutchinson, S., & Vidyasagar, M. (2006). *Robot modeling and control*. New York: Wiley.

Chapter 13: Time Optimal Control

Ailon, A., & Langholz, G. (1985). On the existence of time optimal control of mechanical manipulators. *Journal of Optimization Theory and Applications, 46*(1), 1–21.
Bahrami, M., & Jazar, R. N. (1991). Optimal control of robotic manipulators: Optimization algorithm. *Amirkabir Journal, 5*(18) (in Persian).
Bahrami M., & Jazar, R. N. (1992). Optimal control of robotic manipulators: Application. *Amirkabir Journal, 5* (19) (in Persian).
Bassein, R. (1989). An optimization problem. *American Mathematical Monthly, 96*(8), 721–725.
Bobrow, J. E., Dobowsky, S., & Gibson, J. S. (1985). Time optimal control of robotic manipulators along specified paths. *The International Journal of Robotics Research, 4*(3), 495–499.
Courant, R., & Robbins, H. (1941). *What is mathematics?* London: Oxford University Press.
Fahimi, F. (2009). *Autonomous robots: Modeling, path planning, and control*. New York: Springer.
Fotouhi, C. R., & Szyszkowski W. (1998). An algorithm for time optimal control problems. *Transaction of the ASME, Journal of Dynamic Systems, Measurements, and Control, 120*, 414–418.
Fu, K. S., Gonzales, R. C., & Lee, C. S. G. (1987). *Robotics, control, sensing, vision and intelligence*. New York: McGraw-Hill.
Gamkrelidze, R. V. (1958). *The theory of time optimal processes in linear systems* (Vol. 22, pp. 449–474). Izvestiya Akademii Nauk SSSR.
Garg, D. P. (1990). The new time optimal motion control of robotic manipulators. *The Franklin Institute, 327*(5), 785–804.
Hart P. E., Nilson N. J., & Raphael B. (1968). A formal basis for heuristic determination of minimum cost path. *IEEE Transaction Man, System & Cybernetics, 4*, 100–107.
Kahn, M. E., & Roth B. (1971). The near minimum time control of open loop articulated kinematic chain. *Transaction of the ASME, Journal of Dynamic Systems, Measurements, and Control, 93*, 164–172.
Kim B. K., & Shin K. G. (1985). Suboptimal control of industrial manipulators with weighted minimum time-fuel criterion. *IEEE Transactions on Automatic Control, 30*(1), 1–10.
Krotov, V. F. (1996). *Global methods in optimal control theory*. New York: Marcel Decker.
Golnaraghi, F., & Kuo, B. C. (2009). *Automatic control systems*. New York: Wiley.
Lee, E. B., & Markus, L. (1961). Optimal control for nonlinear processes. *Archive for Rational Mechanics and Analysis, 8*, 36–58.
Lee, H. W. J., Teo, K. L., Rehbock, V., & Jennings, L. S. (1997). Control parameterization enhancing technique for time optimal control problems. *Dynamic Systems and Applications, 6*, 243–262.
Lewis, F. L., & Syrmos, V. L. (1995). *Optimal control*. New York: Wiley.
Meier E. B., & Bryson A. E. (1990). Efficient algorithm for time optimal control of a two-link manipulator. *Journal of Guidance, Control and Dynamics, 13*(5), 859–866.
Mita T., Hyon, S. H., & Nam, T. K. (2001). Analytical time optimal control solution for a two link planar acrobat with initial angular momentum. *IEEE Transactions on Robotics and Automation, 17*(3),361–366.
Nakamura, Y. (1991). *Advanced robotics: Redundancy and optimization*. New York: Addison Wesley.
Nakhaie Jazar G., & Naghshinehpour A. 2005. Time optimal control algorithm for multi-body dynamical systems. *IMechE Part K: Journal of Multi-Body Dynamics, 219*(3), 225–236.
Pinch E. R. (1993). *Optimal control and the calculus of variations*. New York: Oxford University Press.
Pontryagin, L. S., Boltyanskii, V. G., Gamkrelidze, R. V., & Mishchenko, E. F. (1962). *The mathematical theory of optimal processes*. New York: Wiley.

Roxin, E. (1962). The existence of optimal controls. *Michigan Mathematical Journal, 9*, 109–119.

Shin K. G., & McKay N. D., (1985), Minimum time control of a robotic manipulator with geometric path constraints, *IEEE Transaction Automatic Control, 30*(6), 531–541.

Shin K. G., & McKay N. D. (1986). Selection of near minimum time geometric paths for robotic manipulators. *IEEE Transaction Automatic Control, 31*(6), 501–511.

Skowronski, J. M. (1986). *Control dynamics of robotic manipulator*. London, GB: Academic Press.

Slotine, J. J. E., & Yang, H. S. (1989). Improving the efficiency of time optimal path following algorithms. *IEEE Transactions on Robotics and Automation, 5*(1), 118–124.

Spong, M. W., Thorp, J. S., & Kleinwaks, J., M. (1986). The control of robot manipulators with bounded input. *IEEE Journal of Automatic Control, 31*(6), 483–490.

Sundar, S., & Shiller, Z. (1996). A generalized sufficient condition for time optimal control. *Transaction of the ASME, Journal of Dynamic Systems, Measurements, and Control, 118*(2), 393–396.

Takegaki, M., & Arimoto, S. (1981). A new feedback method for dynamic control of manipulators. *Transaction of the ASME, Journal of Dynamic Systems, Measurements, and Control, 102*, 119–125.

Vincent T. L., & Grantham W. J. (1997). *Nonlinear and Optimal Control Systems*. New York: Wiley.

Chapter 14: Control Techniques

Åström, K. J., & Hägglund, T. (1995). *PID controllers* (2nd ed.). ***Research Triangle Park, North Carolina: Instrument Society of America.

Fahimi, F. (2009). *Autonomous robots: Modeling, path planning, and control*. New York: Springer.

Fu, K. S., Gonzales, R. C., & Lee, C. S. G. (1987). *Robotics, control, sensing, vision and intelligence*. New York: McGraw-Hill.

Golnaraghi, F., & Kuo, B. C. (2009). *Automatic control systems*. New York: Wiley.

Jamshidi, M. (2008). *Systems of systems engineering: Principles and applications*. New York: Wiley.

Jeffrey, A., & Dai, H. H. (2008). *Handbook of mathematical formulas and integrals*. Burlington, MA: Academic Press.

Kelly, R., Santibáñez, V., & Loría A., (2005). *Control of robot manipulators in joint space*. Leipzig: Springer.

Lewis, F. L., & Syrmos, V. L., (1995), *Optimal control*. New York: Wiley.

Malek-Zavarei, M., & Jamshidi, M. (1986). *Linear control systems: A computer aided approach*. London: Pergamon Press.

Paul, R. P. (1981). *Robot manipulators: Mathematics, programming, and control*. Cambridge, MA: MIT Press.

Spong, M. W., Hutchinson, S., & Vidyasagar, M. (2020). *Robot modeling and control*. New York: Wiley.

Takegaki, M., & Arimoto, S. (1981). A new feedback method for dynamic control of manipulators. *Journal of Dynamic Systems, Measurements, and Control, 102*, 119–125.

Vincent T. L., & Grantham W. J. (1997). *Nonlinear and optimal control systems*. New York: Wiley.

Index

A

Acceleration
  angular, 489, 491, 501, 507, 508
  bias vector, 521
  body point, 384, 508, 510
  centripetal, 500, 504, 508, 510
  constant parabola, 705
  constant path, 694
  Coriolis, 510, 536, 563
  definition, 27
  discontinuous path, 700
  discrete equation, 740, 748
  end-effector, 499
  forward kinematics, 518, 519
  gravitational, 634, 660, 669
  inverse kinematics, 520
  jump, 689
  matrix, 490, 510, 516, 533
  mixed, 538, 540
  mixed Coriolis, 543
  mixed double, 543
  Razi, 543, 552
  recursive, 526, 528, 609
  rotational transformation, 490, 499
  sensors, 771
  simple, 534
  tangential, 500, 504, 508, 510, 536, 543
  transformation, 494, 541
  transformation matrix, 510
Accessory, 260
Active transformation, 79
Actuator, 7, 12
  force and torque, 610, 632, 673
  optimal torque, 750, 751
  torque equation, 617, 642, 747
Admissible path, 735
Algorithm
  floating-time, 739, 746
  inverse kinematics, 343
  LU factorization, 451
  LU solution, 451
  Newton-Raphson, 464
Angle
  Euler, 59
  nutation, 59
  precession, 59
  spin, 59
Angular acceleration, 489, 491, 508
  B-expression, 502
  end-effector, 499
  Euler parameters, 501, 507
  matrix, 490

  principal, 502
  quaternions, 507
  recursive, 533
  relative, 494, 500
  rotational transformation, 490, 492
  technical point, 504
  vector, 490
Angular jerk, 506
Angular momentum, 569–571
  2 link manipulator, 574
Angular velocity, 64, 66, 67, 97, 361, 364, 372, 491, 495, 502
  alternative definition, 384, 386
  combination, 364
  coordinate transformation, 372
  decomposition, 370
  elements of matrix, 376
  Euler frequencies, 370
  Euler parameters, 375
  instantaneous, 363
  instantaneous axis, 361, 364, 372, 495
  matrix, 362
  principal matrix, 369
  quaternions, 374
  rate, 361
  recursive, 418, 527
  rom B-frame, 363
  rom G-frame, 363
  rotation matrix, 371
  skew symmetric, 363
  vector, 361
Anthropomorphic hand, 7
Arm, 2
Articulated
  arm, 9, 252, 254
  manipulator, 9, 252, 254, 325, 433
  robot, 252
Articulated manipulator
  equations of motion, 647
  inverse kinematics, 319, 322, 323, 331
  inverse velocity, 447
  Jacobian matrix, 428, 472
  left shoulder configuration, 336
  right shoulder configuration, 336
Asimov, Isaac, 1
Associativity property, 76
Atan2 function, 317
Attitude angle, 58
Automorphism, 123
  property, 122
Auxiliary frame, 285
Avicenna, 278
Axis-angle

© The Author(s), under exclusive license to Springer Nature Switzerland AG 2022
R. N. Jazar, *Theory of Applied Robotics*, https://doi.org/10.1007/978-3-030-93220-6

matrix, 91
rotation, 91
transformation, 91
Axis-angle rotation, 91, 94–96, 110, 112, 114, 127

B
bac-cab rule, 19, 147
Bang-bang control, 731
Bank angle, 58
Bar, 2
Bernoulli, Jacob, 738
Bernoulli, Johann, 738
Binary link, 228
Block diagram, 760
Body frame, 14
Bong, 28
Brachistochrone, 737, 745
Bryant angles, 68

C
Camera
    inspection, 260, 263
    SSRMS, 260
    vision, 263
Capek, Karel, 1
Cardan
    angles, 68
    frequencies, 68
Cardano, Geronimo, 68
Cartesian
    angular velocity, 66
    coordinate space, 247
    coordinate system, 20
    end-effector position, 441
    end-effector velocity, 442
    manipulator, 9, 11
    orthogonality condition, 20
    path, 705
Cartesian manipulator
    equations of motion, 650
Catapults, 597
Central difference, 743
Centrifugal moments, 576
Centroid, 391
Chasles, Michel, 186
Chasles theorem, 187
China, 597
Christoffel
    Elwin Bruno, 658
Christoffel operator, 658
Christoffel symbol, 658
Classification
    industrial links, 236
Closure property, 75
Co-state variable, 732
Common normal, 226
Compound link, 2
Condition
    orthogonality, 20
Configuration path, 27
Control
    adaptive, 763
    admissible, 733
    bang-bang, 731, 733

characteristic equation, 762
closed-loop, 759
command, 759
computed force, 765
computed torque, 763, 764
derivative, 768
desired path, 759
directional control system, 263
error, 759
feedback, 760
feedback command, 765
feedback linearization, 763, 765
feedforward command, 765
gain, 760
gain-scheduling, 763
input, 764
integral, 768
linear, 763, 767
minimum time, 731
modified PD, 770
open-loop, 759, 764
path points, 707
PD, 770
proportional, 768
robots, 12
sensing, 770
stability of linear, 761
time-optimal, 738, 741, 746, 747, 751
time-optimal description, 747
time-optimal path, 745
Controller, 8
Control unit, 8
Coordinate
    cylindrical, 174
    non-Cartesian, 589
    non-orthogonal, 136
    parabolic, 589
    spherical, 175, 396
    system, 21
Coordinate frame, 14, 16, 21
    dummy, 278
    extra, 278
    global, 21
    intermediate, 278
    jump, 278
    local, 21
    main, 240, 278
    neshin, 277
    origin, 16
    orthogonal, 16, 20
    orthogonality condition, 20
    Sina, 240, 278
    spare, 278
    takht, 277
    temporary, 278
Coordinate space
    Cartesian, 247
    joint variable, 247
Coordinate system, 14, 21
Coriolis
    acceleration, 498, 510
    effect, 563
    force, 563
Coriolis, Gaspard, 563
Couple, 558
Crackle, 28

Critically-damped, 762
Cyclic interchanging, 14, 19
Cycloid, 738

D
Damping ratio, 762
Dead frame, 269
Decoupling technique, 313
    inverse kinematics, 313
    inverse orientation, 314
    inverse position, 314
Denavit-Hartenberg, 35
    construction steps, 231
    industrial links, 236
    linkage, 245
    mechanism, 245
    method, 225, 228, 293
    non-classical, 242
    nonstandard method, 245, 341
    notation, 225
    parameters, 225, 401, 405, 416, 528, 668
    proof of equation, 233, 234
    shortcomings, 232
    spherical robot, 244
    Stanford arm, 230
    3R PUMA robot, 229
    3R planar manipulator, 229
    transformation, 232, 235–239, 242, 289, 783–788
    2R planar manipulator, 235
Derivative
    coordinate frames, 376
    mixed, 377, 534
    mixed double, 543
    mixed second, 542
    simple, 376, 534
    transformation
        mixed, 387, 388
    transformation formula, 383, 541
Deviation moments, 576
Devol, George, 2
Differential
    transformation matrix, 402
Differential equation
    angular velocity, 383
    first-order vectorial kinematic, 383
    rotation matrix, 383
    vectorial kinematic, 383
Differential manifold, 76
Differentiating
    B-derivative, 376, 379, 381, 538
    coordinate frame, 376
    G-derivative, 376, 384
    second, 386
    second derivative, 534
    transformation formula, 383
Direct dynamics, 555
Direct kinematics, 247
Directed line, 202
Direction cosines, 53
Directional
    control system, 104, 263
    cosine, 18, 25, 72, 73, 127, 243
Directional cosine, 17
Directional line, 16
Directions principal, 578

Displacement, 23
Distal end, 225, 669
Dynamics, 13, 526, 555, 609
    actuator's force and torque, 632
    backward Newton-Euler, 626
    direct, 13, 555
    forward Newton-Euler, 627
    4 bar linkage, 624
    global Newton-Euler, 609
    inverse, 13, 555
    Lagrange, 632
    motion, 557
    Newton-Euler, 609
    Newtonian, 563
    one-link manipulator, 611
    recursive Newton-Euler, 609, 626
    robots, 609
    2R planar manipulator, 616–618, 628, 637

E
Earth
    effect of rotation, 563
    kinetic energy, 589
    moving vehicle, 498
    revolution, 589
    rotating, 498
    rotation, 589
    rotation effect, 498
Eigenvalue
    rotation matrix, 98, 142
Eigenvector
    rotation matrix, 98, 142
End-effector, 7
    acceleration, 518
    angular acceleration, 499
    angular velocity, 440
    articulated robot, 325
    configuration vector, 470, 518
    configuration velocity, 518
    force, 628
    frame, 231
    inverse kinematics, 313
    kinematics, 288
    link, 225
    orientation, 328, 441
    path, 708, 712
    position kinematics, 247
    position vector, 435
    rotation, 720
    SCARA position, 172
    SCARA robot, 256
    space station manipulator, 257
    speed vector, 419, 423
    spherical robot, 292
    time optimal control, 731
    velocity, 431, 442
Energy
    Earth kinetic, 589
    kinetic, 559, 572
    kinetic rotational, 568
    link's kinetic, 633, 660
    link's potential, 634
    mechanical, 589
    potential, 591
    robot kinetic, 633, 660

robot potential, 634, 660
Engelberger, Joseph, 2
Equation
    the most important, 19
Euclidean space, 126
Euler
    angles, 22, 62–64, 128
        integrability, 67
    coordinate frame, 66
    equation of motion, 567, 571, 611, 627
    frequencies, 64, 66, 370
    inverse matrix, 62
    -Lexell-Rodriguez formula, 93
    parameters, 109, 110, 112, 113, 116, 117, 119, 130, 374, 375
    rotation matrix, 62
    rotation theorem, 102
    theorem, 56, 102, 109
Euler angles, 59, 338
Euler-Chasles theorem, 187
Euler equation
    body frame, 571
Eulerian viewpoint, 391
Eulerian wrist, 105
Euler-Lagrange
    equation of motion, 735, 736
Euler, Leonhard, 56, 93, 130, 383
Euler theorem, 153
Exponent -1, 47
Exteroceptors sensor, 770

F
Feedback control, 760
Final rotation formula, 100, 101
First variation, 736
Fixed frame, 14, 21
Fleming, Sir John Ambrose, 15
Floating bar, 266
Floating time, 739
    analytic calculation, 744
    backward path, 741
    convergence, 743
    forward path, 741
    method, 738
    multi DOF algorithm, 746
    multiple switching, 751
    1 DOF algorithm, 739
    path planning, 745
    robot control, 746
Force, 557, 558
    action, 610
    actuator, 632
    body, 557
    conservative, 591
    contact, 557
    Coriolis, 563
    driven, 610
    driving, 610
    external, 557
    function, 563
    generalized, 586, 591, 635
    gravitational vector, 635
    internal, 557
    moment of, 558
    potential, 591
    potential field, 588

reaction, 610
resultant, 557
sensors, 771
shaking, 626
total, 557
Force system, 557, 558
    equivalent, 558
Formula
    derivative transformation, 383, 387, 388, 541–543
    derivative transport, 502
    geometric transformation, 359
    mixed-derivative, 388
    relative acceleration, 493, 494, 500
    relative angular velocity, 365
    Rodriguez, 132, 364
Forward kinematics, 36, 247
4 bar linkage dynamics, 624
4R planar manipulator statics, 669
Frame
    base, 231
    central, 565
    final, 231
    fixed, 21
    global, 21
    goal, 231
    moving, 21
    neshin, 277
    principal, 568, 571, 577, 578
    reference, 20
    special, 231
    station, 231
    takht, 277
    tool, 231
    transformation, 21
    world, 231
    wrist, 231
Function
    Heaviside, 322

G
Galilei, 738
Gauss, Johann Carl Friedrich, 130
Generalized
    coordinate, 584, 586, 588, 592
    force, 586, 588, 590, 591, 593, 595, 632
    inverse Jacobian, 467
Globalframe, 14
Grassmann, Hermann, 205
Grassmannian, 205
Gripper, 7
Gripper axis, 269, 270
Gripper frame, 269
Gripper wall, 270
Group properties, 75

H
Hamilton, Sir William Rowan, 120
Hamilton, William, 130
Hamiltonian, 731, 732
Hand, 7
Hayati-Roberts method, 298
Heading angle, 58
Heaviside, Oliver, 322
Heaviside function, 322

Helix, 28, 182
Hermann Grassmann, 205
Hodograph, 27
Home configuration, 228
Homogenous
    combined transformation, 167
    coordinate, 154, 159
    direction, 159
    general transformation, 159, 166
    inverse transformation, 162, 164, 165, 168
    position vector, 154
    reverse transformation, 162
    scale factor, 154
    transformation, 153–155, 157, 158, 162, 164

I

Identity matrix, 198
Identity property, 75
Industrial robot
    classification, 236
Inner automorphism, 122
Inspection camera, 260, 263
Integrability, 67
Inverse
    function symbol, 47
    trigonometric functions, 45, 46
Inverse dynamics, 555
Inverse kinematics, 36, 313
    articulated manipulator, 319, 322, 323, 331
    comparison of techniques, 347
    decoupling technique, 313
    Euler angles, 338
    Euler angles matrix, 340
    existence, 347
    general formulas, 328
    inverse transformation technique, 329
    iterative algorithm, 343
    iterative technique, 343
    multiple solutions, 315
    Newton-Raphson method, 343, 346
    nonstandard DH, 341
    numerical solution, 323
    Pieper technique, 331
    spherical robot, 334
    techniques, 347
    2R planar manipulator, 315
    uniqueness, 347
Inverse orientation, 314
Inverse position, 314
Inverse property, 76
Inverse symbol, 47
Inverse velocity, 442
Inverted pendulum, 765
Iteration technique, 346
Iterative technique, 343

J

Jacobi identity, 19
Jacobian, 343
    analytical, 441, 442
    angular, 441
    displacement matrix, 420, 423
    elements, 440
    generating rate vector, 533

generating vector, 429–431, 469, 533
    geometrical, 441
    inverse, 344, 467
    of link, 633
    matrix, 344, 346, 348, 419, 420, 423, 427, 429, 431, 433, 437, 438, 442, 446, 463, 466, 469, 471, 472, 518, 519, 523, 644
    polar manipulator, 423, 524
    rotational matrix, 420, 423
    spherical wrist, 446
    2R manipulator, 424
Jacobian matrix, 419, 420, 423
    displacement, 420
    rotational, 420–422
    systematic method, 422, 423
Jeeq, 28
Jerk, 506
    angular, 505
    body point, 506
    definition, 27
    global, 506
    matrix, 516
    rotational transformation, 505, 506
    transformation, 516, 517
    transformation matrix, 516
    zero path, 694
Johns Hopkins beast, 2
Joint, 3
    acceleration vector, 518
    active, 4
    angle, 227
    axis, 3
    coordinate, 3
    cylindrical, 296
    degree of freedom, 4
    distance, 227
    elbow, 229
    free, 4
    inactive, 4
    orthogonal, 9
    parallel, 9
    parameters, 228, 289
    passive, 4
    path, 708
    perpendicular, 9
    prismatic, 3
    revolute, 3
    rotary, 3
    shoulder, 229
    speed vector, 419, 431
    spherical, 269
    translatory, 3
    variable, 3
    waist, 229
Joint variable
    angle, 3
    coordinate space, 247
    distance, 3
Jolt, 28
Jounce, 28

K

Kinematic length, 227
Kinematic operation, 285
Kinematic pair, 3
Kinematic surgery, 285

Kinematics, 13, 35
    acceleration, 489, 491
    assembling, 276, 277
    direct, 247
    forward, 13, 36, 225, 247
    forward acceleration, 518
    forward velocity, 419
    inverse, 13, 36, 313, 329
    inverse acceleration, 520
    inverse velocity, 442
    motion, 149
    numerical methods, 448
    operation, 285
    orientation, 91
    rigid body, 149
    rotation, 37
    surgery, 285
    velocity, 415
Kinetic energy, 559
    Earth, 589
    link, 660
    parabolic coordinate, 590
    rigid body, 572
    robot, 633, 660
    rotational body, 568
Kronecker's delta, 73, 116, 568

L
Lagrange
    dynamics, 632
    equation, 659
    equation of motion, 584, 591
    mechanics, 591
    multiplier, 738
Lagrangean, 591, 661
    robot, 661
Lagrangean viewpoint, 391
Lagrange formula, 19
Lagrange multiplier, 738
Larz, 28
Law
    cosine, 33
    of motion, 558
    motion second, 565
    robotics, 1
    second of motion, 558, 559
    third of motion, 558, 559
Levi-Civita density, 116
Lexell, Anders Johan, 93
Lie group, 76
Linear space, 24
Line of action, 22
Link, 2
    angular velocity, 417
    binary, 228
    class 11 and 12, 238, 788
    class 1 and 2, 236, 783
    class 3 and 4, 237, 784
    class 5 and 6, 237, 785
    class 7 and 8, 238, 786
    class 9 and 10, 238, 787
    classification, 236
    compound, 2
    end-effector, 225
    Euler equation, 627

    kinetic energy, 633
    length, 227
    Newton-Euler dynamics, 609
    offset, 227
    parameters, 228, 289
    recursive acceleration, 526, 528
    recursive Newton-Euler dynamics, 626
    recursive velocity, 527, 528
    rotational acceleration, 526
    translational acceleration, 526
    translational velocity, 417
    twist, 227
    velocity, 415
Living frame, 269
Local frame, 14
Location vector, 184, 185
Louis Poinsot, 564
LU factorization method, 449, 452, 453, 456, 460
    matrix inversion, 458
    pivoting, 453
    uniqueness of solution, 456

M
Main coordinate frame, 240, 278
Mainframe, 7
Manganic, 597
Manipulator
    anthropomorphic, 9
    articulated, 9, 229
    Cartesian, 9
    cylindrical, 9
    definition, 5
    elbow, 9
    inertia matrix, 634
    one-link, 592
    one-link control, 768
    one-link dynamics, 611
    planar polar, 636
    PUMA, 229
    SCARA, 9
    space station, 257
    spherical, 9
    3R planar, 250
    transformation matrix, 325
    2R planar, 593, 642
Manjanic, 597
Manjaniq, 597
Mass center, 558, 560, 565
Mass moment
    matrix, 576
    principal, 578
Matrix
    displacement Jacobian, 420, 423
    Jacobian, 420
    orthogonality condition, 72
    rotational Jacobian, 420, 423
    skew symmetric, 92, 110, 137
Measure number, 16
Mechanics
    Newtonian, 563
Mechanism
    slider-crank, 245
    3D slider-crank, 266
Method
    Hayati-Roberts, 298

non Denavit-Hartenberg, 293
parametrically continuous convention, 298
Minimum time control, 731
Mixed derivative, 377, 388
Mobius strip, 15
Mobius surface, 15
Modular manipulator, 302, 304–307
Moment, 557, 558
  action, 610
  driven, 610
  driving, 610
  external, 571
  reaction, 610
  resultant, 557, 571
  total, 557
Moment of inertia, 576
  about a line, 583
  about a plane, 583
  about a point, 583
  characteristic equation, 582
  eigenvalues, 578
  elements, 576
  Huygens-Steiner theorem, 578
  matrix, 576
  parallel-axes theorem, 576–578
  polar, 576
  principal, 577
  principal axes, 568
  principal invariants, 582
  product, 576
  rigid body, 568, 570
  rotated-axes theorem, 576–578
Moment of momentum, 558
Momentum, 558
  angular, 558, 569–571
  linear, 558
  translational, 558
Motion, 13
Moving frame, 14
Mozzi, Giulio, 186
Multibody
  articulated manipulator, 180
  directional control system, 263
  order free rotation, 102
  order free transformation, 176
  spherical robot, 244
  spherical wrist, 280
  3D slider-crank mechanism, 266

N
Natural frequency, 762
Negative triad, 14
Neshin frame, 277, 279, 282
Newton
  equation of motion, 558, 559, 566, 584
Newton equation
  body frame, 566
  definition, 537
  global frame, 565
  Lagrange form, 586
  rotating frame, 563
Newton-Euler
  backward equations, 626
  equation of motion, 610, 627
  forward equations, 626, 627

global equations, 609
  recursive equations, 626
Newton-Raphson method, 343
Non-standard
  Denavit-Hartenberg method, 245
Non Denavit-Hartenberg
  methods, 293
Norm-infinity, 458
Norm of a vector, 24
Numerical methods, 448
  analytic inversion, 460
  Cayley–Hamilton inversion, 462
  condition number, 457
  consistent solution, 456
  ill-conditioned, 456
  inconsistent solution, 456
  Jacobian matrix, 469
  LU factorization, 449
  LU factorization with pivoting, 453
  matrix inversion, 458
  Newton-Raphson, 463, 466
  nonlinear equations, 463
  norm of a matrix, 458
  pivot element, 453
  undetermined solution, 456
  uniqueness of solution, 456
  well-conditioned, 456
Nutation, 59

O
Object manipulation, 173
Obstacle free path, 716
Onager, 597
Opposite triad, 14
Optimal control, 731
  a linear system, 733
  description, 747
  first variation, 736
  Hamiltonian, 732, 735
  Lagrange equation, 735
  objective function, 731, 735
  performance index, 735
  second variation, 736
  switching point, 733
Order-free theorem, 176
Origin, 16
Orthogonal exes, 226
Orthogonality condition, 20, 72
Orthogonal matrix, 71
Orthogonal transformation, 71
Orthogonal triad, 14
Over-damped, 762

P
Parallel exes, 226
Passive transformation, 79
Path
  Brachistochrone, 745
  Cartesian, 705
  configuration, 27
  constant acceleration, 694
  constant angular acceleration, 722
  control points, 707
  cubic, 687

cycloid, 704
harmonic, 703
higher polynomial, 692
jerk zero, 694
joint space, 708
non-polynomial, 702
planning, 687, 705
point sequence, 696
quintic, 692
rest-to-rest, 689, 690
rotational, 720
segment method, 697, 699, 700
spline, 701
splitting method, 697, 699–701
to-rest, 689
Pendulum
control, 765
inverted, 765, 770
linear control, 768
oscillating, 587
simple, 495, 586
spherical, 572, 574, 592
Permutation symbol, 116
Perpendicular exes, 226
Persia, 278, 543
Persian, 278, 597
Phase plane, 734
Physical
quantity
vectorial, 22
Physical quantity
scalaric, 24
vectorial, 23
Pieper technique, 331
Pitch angle, 58
Pivot element, 453
Pivoting method, 453
Plücker
angle, 208
axis coordinate, 205
classification coordinate, 206
coordinate, 202, 204, 206, 207
distance, 208
line, 202
line arrangement, 204
line coordinate, 202, 204, 205, 208, 211–213, 293
method, 202
moment, 208
ray coordinate, 204, 205
reciprocal product, 208
screw, 212
vector, 202
virtual product, 208
Plücker, Julius, 204
Plücker coordinate, 207
general case, 206
line at infinity, 206
line through origin, 206
Poinsot, Louis, 564
Poinsot theorem, 187
Point at infinity, 159
Polar manipulator
inverse acceleration, 524
Jacobian, 423
Pole, 192, 193, 391
Pontryagin principle, 732

Pop, 28
Positioning, 13
Position sensors, 771
Position vector, 16, 23
Positive triad, 14
Potential
energy, 559, 591
field, 559
force, 591
function, 559
Potential energy
robot, 634, 660
Precession, 59
Prince of physicians, 278
Principal
angular acceleration, 502, 503
angular velocity, 502
Principal directions, 578
Principle
conservation of energy, 559
superposition, 564
Prismatic
velocity coefficient matrix, 402
velocity transformation, 402
Problem
composition, 132
decomposition, 132
open, 57
Proprioceptors sensor, 770
Proximal end, 225, 669

Q
Quaternion
automorphism, 122
Quaternions, 117, 129
addition, 118
composition rotation, 122
flag form, 117, 123
inverse rotation, 120
matrix, 130
multiplication, 118
rotation, 119
unit, 130

R
Rabota, 1
Razi, Zakariya , 543
Razi acceleration, 543
Redundancy, 251
Reference frame, 14
Resolved rates, 442
Rest position, 228, 247
Revolute
velocity coefficient matrix, 402
velocity transformation, 402
Right hand rule, 14
Rigid body
acceleration, 508, 526
angular momentum, 569, 570
angular velocity, 97
Euler equation, 571
kinematics, 149
kinetic energy, 572
moment of inertia, 568, 570

motion, 149–151
motion classification, 195
motion composition, 152
non-central rotation, 169, 170, 195
off-center rotation, 169, 170, 195
principal rotation matrix, 581
rotational kinetics, 567
steady rotation, 574
translational kinetics, 565
velocity, 388, 389
Rigid body theorem, 153
Robot
    application, 13
    articulated, 9, 252, 254, 282, 325, 433, 438
    Cartesian, 11
    classification, 8
    closed-loop, 9
    control, 12, 13
    control algorithms, 763
    cylindrical, 9, 307
    dynamics, 13, 526, 609, 635, 642
    end-effector path, 712
    equation of motion, 662
    fixed sequence, 8
    forward kinematics, 247, 292
    gravitational vector, 635
    hybrid, 9
    inertia matrix, 634
    intelligent, 8
    kinematics, 13
    kinetic energy, 633, 660
    Lagrangean, 634, 658
    Lagrange dynamics, 632, 659
    Lagrange equation, 635
    link classification, 291
    manual handling, 8
    modified PD control, 770
    Newton-Euler dynamics, 609
    numerical control, 8
    open-loop, 9
    parallel, 9
    PD control, 770
    playback, 8
    potential energy, 634, 660
    recursive Newton-Euler dynamics, 626
    rest position, 226, 228, 252, 254, 283
    SCARA, 172, 255
    serial, 9
    spherical, 9, 230, 286, 292, 334, 432
    state equation, 734
    statics, 667
    time-optimal control, 734, 746
    variable sequence, 8
    velocity coupling vector, 635
Robotic
    geometry, 9
    history, 2
    laws, 1
Robotics Institute of America, 1
Rodriguez
    rotation formula, 93, 95, 111–115, 121, 127, 150, 185, 191, 196,
        201, 364, 372, 720
    vector, 115, 132, 133
Rodriguez rotation matrix, 116
Rodriguez, Benjamin, 130
Rodriguez, Olinde, 93

Roll angle, 58
Roll-pitch-yaw
    body rotation matrix, 58
    bovy angles, 58
    frequency, 58
    global angles, 50, 123
    global rotation matrix, 50, 123
Rotation, 35
    about global axes, 37, 43, 45, 46
    about local axes, 52, 55, 56
    acceleration transformation, 490, 492
    axis-angle, 91, 94–96, 110, 112, 114, 127, 372
    composition, 132
    decomposition, 132
    eigenvalue, 98, 142
    eigenvector, 98, 142
    exponential form, 114
    final formula, 100
    general, 71
    infinitesimal, 113
    instantaneous center, 391
    local versus global, 69
    matrix, 22, 127
    order free, 102
    pole, 391
    quaternion, 119
    representation, 91
    Rodriguez formula, 94, 372
    Rodriguez matrix, 116
    stanley method, 117
    Taylor expansion, 130
    triple global axes, 45, 46
    x-matrix, 38, 52
    y-matrix, 38, 52
    z-matrix, 38, 52
Rotational jerk, 506
Rotational path, 720
Rotation kinematics, 37
Rotations
    problems, 126
Rotation theorem, 102
Rotator, 94, 124, 126
Rule
    bac-cab, 19
    relative angular acceleration, 493
    relative angular velocity, 493
    right-hand, 14
Runaround, 1

S
Scalar, 24
    equal, 24
    equivalent, 24
Scale, 16
SCARA
    manipulator, 9
    robot, 172, 255
Screw, 182, 186, 195
    axis, 182, 392
    central, 183, 185, 188, 198, 202, 212, 228, 289, 291, 293
    combination, 200, 201
    coordinate, 182
    decomposition, 201, 202
    exponential, 200
    forward kinematics, 289

general, 199
instantaneous, 213
intersection, 293
inverse, 197–199, 201
left-handed, 182
link classification, 291
location vector, 182, 184
motion, 182, 188, 228, 289, 392
parameters, 182, 193
pitch, 182
Plücker coordinate, 212
principal, 186, 201, 202
reverse central, 183
right-handed, 14, 182
special case, 192
transformation, 185, 194
twist, 182
twist angle, 182
twist axis, 182
Screw motion, 182
Second derivative, 386
Second variation, 736
Sensor
acceleration, 771
position, 771
rotary, 771
velocity, 771
Sensors, 8
Shaking force, 626
Sheth method, 293
Sheth notation, 293
Shuttle arm, 2
Simple derivative, 376
Sina, Abu Ali, 278
Sina coordinate frame, 240, 278
Sina frame, 104, 240, 241, 263, 266, 267, 278, 285
Singular configuration, 347
Singularity, 298
Snap, 28
Sooz, 28
Space station
arm, 257
Spherical
pendulum, 572, 574
wrist, 280
Spherical coordinate, 175
Spherical joint, 269
Spherical manipulator, 283–285
Spherical robot, 244, 286
Spherical wrist, 105, 269, 280–282, 286, 446
dead frame, 269
forearm, 269
gripper frame, 269
hand, 269
Jacobian, 446
living frame, 269
tilt, 269
tool frame, 269
turn, 269
twist, 269
Spin, 59
Spin angle, 58
Spinor, 94, 124, 126
Spline, 701
SRMS, 257
SSRMS, 257, 260

Camera, 260
dimensions, 257
kinematic parameters, 257
Stanley method, 117
Stark effect, 590
Switching point, 733, 741
Symbols, xi
System of particles
motion equation, 560

T
Takht frame, 180, 277, 279, 283, 332
Tetrad, 15
Theorem
Euler, 102
Euler-Chasles, 187
Euler rigid body, 153
Euler rotation, 102
Huygens-Steiner, 578
order free rotations, 103
order free transformations, 176
parallel-axes, 576, 578
Poinsot, 187
rotated-axes, 576
3R planar manipulator
DH transformation matrix, 229
forward kinematics, 250
Tilt vector, 274
Time derivative, 376
Time optimal control, 731
Tool frame, 269
Top, 64
Torque, 558
Transformation, 35
active and passive, 79
general, 71
geometric, 504
homogenous, 153, 154
jerk, 516
kinematic, 504
order free, 176
orthogonal, 71
rotational acceleration, 504
rotational jerk, 505, 506
rotational velocity, 504
Transformation matrix
derivative, 400
differential, 402, 403
elements, 73
velocity, 392
Translation, 35
Trebuchet, 240, 595
Triad, 14
coordinate frame, 14
left-handed, 14
negative, 14
positive, 14
right-handed, 14
Trigonometric equation, 318, 319
first type, 318, 319
Trigonometric functions
inverse, 45, 46
Turn vector, 274
Twist vector, 274
2R planar manipulator

acceleration analysis, 512
assembling, 278
control, 766
DH transformation matrix, 235
dynamics, 593, 666
elbow down, 316
elbow up, 316
equations of motion, 595
forward acceleration, 519
ideal, 593
inverse acceleration, 522
inverse kinematics, 315, 316, 344, 464
inverse velocity, 443–445
Jacobian, 424
Jacobian matrix, 427
joint 2 acceleration, 509
joint forces, 623
joint path, 709
kinematic motion, 317
kinematics, 512
kinetic energy, 594
Lagrangean, 594
Lagrange dynamics, 642, 663
line path, 710
Newton-Euler dynamics, 616–618, 637
potential energy, 594
recursive dynamics, 628
singular configuration, 445
time-optimal control, 747
velocity analysis, 397
with massive joints, 617, 618, 637
with massive links, 663

U
Under-damped, 762
Unimates, 2
Unimation, 2
Unit system, xi
Unit vectors, 16

V
Variable
    scalar , 26
Variation
    first, 736
    second, 736
Vecface, 23
Vecfree, 23
Veclane, 23
Vecline, 23
Vecpoface, 23
Vecpoint, 23
Vecpolane, 23
Vecpoline, 23
Vecporee, 23
Vector
    absolute value, 16
    addition, 24
    associative property, 24
    axis, 22
    bounded, 23
    characteristics, 22
    commutative property, 24
    comparable, 23

components, 16
decomposed expression, 16, 26
decomposition, 16, 26, 136
definition, 22
derivative, 26
direction, 22, 23
end-effector speed, 419
end point, 22
free, 23
function, 26
gravitational force, 635, 659
inverse element property, 24
joint speed, 419
length, 16, 22
line, 23
line of action, 22
magnitude, 16
modulus, 16
natural expression, 16, 26
null element property, 24
physical quantity, 22
plane, 23
point, 23
point-free, 23
point-line, 23
point-plane, 23
position, 16
requirements, 22
sliding, 23
space, 24
start point, 22
surface, 23
tilt, 274
turn, 274
twist, 274
types, 22
variable, 25
variable direction, 25
variable length, 25
vecface, 23
vecfree, 23
vecline, 23
vecpoface, 23
vecpoint, 23
vecpolane, 23
vecpoline, 23
vecporee, 23
velocity coupling, 635, 659
Vector space, 25
Vector triple product, 19
Velocity
    coefficient matrix, 402
    definition, 27
    discrete equation, 740, 748
    inverse transformation, 394
    matrix, 516
    multiple frames, 390
    operator matrix, 400
    prismatic transformation, 402
    revolute angular matrix, 405
    revolute transformation, 402
    rigid body, 388
    sensors, 771
    simple, 534
    transformation matrix, 392–395, 400
Velocity kinematics, 415

forward, 415
    inverse, 415
Vision camera, 263

W
Walter
    William Grey, 2
Work, 559, 561
    virtual, 586
Work-energy principle, 559
Working space, 255
Workspace, 12
    dexterous, 12
    reachable, 12
World War II, 2
Wrench, 564
Wrist, 12, 13, 271
    classification, 270
    dead frame, 269
    decoupling kinematics, 314
    design, 276

    Eulerian, 272, 273
    forward kinematics, 269, 271
    frame, 231
    kinematics assembly, 282
    live frame, 269
    Pitch-Yaw-Roll, 272, 275
    point, 6, 269, 328
    position vector, 327
    possible types, 272
    Roll-Pitch-Roll, 270–273
    Roll-Pitch-Yaw, 272, 274
    spherical, 6, 230, 244, 269, 272, 274, 286, 438
    transformation matrix, 271, 282, 325
Wrist point, 263

Y
Yaw angle, 58

Z
Zero velocity point, 391

Printed in the United States
by Baker & Taylor Publisher Services